D1490456

The President John E. Murray, Jr.
Undergraduate Collection Enhancement Project
1996-1997

# BIOLOGY
## *Science and Life*

# ABOUT THE AUTHOR

MICHAEL R. CUMMINGS received his Ph.D. in Biological Sciences from Northwestern University in 1968. His doctoral work, conducted in the laboratory of Dr. R. C. King, centered on ovarian development in *Drosophila melanogaster.* After a year on the faculty at Northwestern, he moved to a teaching and research position at the University of Illinois at Chicago. Here, he established a research program on the developmental genetics of *Drosophila* and began teaching courses in genetics, developmental genetics, and evolution. Currently an associate professor in the Department of Biological Sciences and in the Department of Genetics, he has also taught at Florida State University.

About ten years ago, Dr. Cummings developed a strong interest in scientific literacy. In addition to teaching genetics to biology majors, he organized and currently teaches a course in human genetics for non-majors, and participates in teaching general biology. This text is an outgrowth of his interest in teaching biology to non-majors. He is now working to integrate the use of electronic resources such as the Internet and World Wide Web into the undergraduate teaching of genetics and general biology. His current research interests involve the role of the short arm/centromere region of human chromosome 21 in chromosomal aberrations. His laboratory is engaged in a collaborative effort to construct a physical map of this region of chromosome 21 to explore molecular mechanisms of chromosome interactions.

Dr. Cummings is the author and co-author of a number of widely used college textbooks, including *Human Heredity, Concepts of Genetics,* and *Essentials of Genetics.* He has also written sections on genetics for the *McGraw-Hill Encyclopedia of Science and Technology,* and has published a newsletter on advances in human genetics for instructors and students.

He and his wife, Lee Ann, are parents of two adult children, Brendan and Kerry, and have one grandchild, Colin. He is an avid sailor, enjoys reading and collecting books (biography, history), music (baroque, opera, and urban electric blues) and is a long-suffering Cubs fan.

# BIOLOGY

## *Science and Life*

MICHAEL R. CUMMINGS

UNIVERSITY OF ILLINOIS AT CHICAGO

DEPARTMENT OF BIOLOGICAL SCIENCES AND

DEPARTMENT OF GENETICS

WEST PUBLISHING COMPANY

MINNEAPOLIS/SAINT PAUL

NEW YORK

LOS ANGELES

SAN FRANCISCO

## ABOUT THE COVER

In the nineteenth century, coal miners took caged canaries with them into the mines. While they worked, the miners paid close attention to the canaries, which are sensitive to low concentrations of methane and other dangerous gases that accumulate in mines. If the canaries became unconscious, the miners left the mine before they too were overcome.

In the last decades of the twentieth century, there has been a worldwide decline in the number of amphibians, particularly frogs, toads, and salamanders. At various times in their life cycle, frogs live in water and on land, are herbivores, and then carnivores. As a result, they are in intimate contact with many parts of their ecosystem. Several factors are contributing to the accelerated extinction of frog species. These include depletion of the ozone layer and the associated increase in ultraviolet light reaching the earth, habitat destruction, air pollution in the form of acid rain, and pollution of water by herbicides and industrial chemicals.

The sensitivity of frogs to environmental degradation may make them the modern day equivalent of the miner's canary. Amphibians were able to survive the environmental changes that led to the extinction of the dinosaurs, but have not been able to adjust to the effects of human activity on the physical environment. Just as the canary did, perhaps frogs are warning us of danger to our own species.

PRODUCTION CREDITS

COPYEDITOR
Lorretta Palagi

INTERIOR DESIGN
Diane Beasley

ARTWORK
Precision Graphics, Carto-Graphics, Cyndie C. H.-Wooley, Wayne Clark, Darwen and Vally Hennings, Carlyn Iverson, Sandra McMahon, Elizabeth Morales-Denny, Publication Services, Rolin Graphics, Pat Rossi, John and Judy Waller, J/B Woolsey Associates

COMPOSITION
G & S Typesetters

INTERIOR ELECTRONIC PAGE LAYOUT
Diane Beasley

COVER DESIGN
Diane Beasley

INDEX
Schroeder Indexing Services

COVER IMAGE
James Carmichael/The Image Bank

PRODUCTION, PREPRESS, PRINTING, AND BINDING
West Publishing Company

BRITISH LIBRARY CATALOGING-IN-PUBLICATION DATA
A catalogue record for this book is available from the British Library.

COPYRIGHT © 1996       BY WEST PUBLISHING COMPANY
                       610 OPPERMAN DRIVE
                       P.O. BOX 64526
                       ST. PAUL, MN 55164-0526

All rights reserved

Printed in the United States of America

03  02  01  00  99  98  97  96        8  7  6  5  4  3  2  1  0

LIBRARY OF CONGRESS CATALOGING-IN-PUBLICATION DATA

Cummings, Michael R.
    Biology : science and life / Michael R. Cummings.
        p.   cm.
    Includes index.
    ISBN 0-314-07581-X (hard : alk. paper). — ISBN 0-314-06400-1 (soft : alk. paper)
    1. Biology.   I. Title.
QH308.2.C645   1996
574—dc20                                              95-45983
                                                     CIP

QH308.2
.C645
1996

PuR 3-10-97
DBCW ADA-9331

# WEST'S COMMITMENT TO THE ENVIRONMENT

In 1906, West Publishing Company began recycling materials left over from the production of books. This began a tradition of efficient and responsible use of resources. Today, 100% of our legal bound volumes are printed on acid-free, recycled paper consisting of 50% new paper pulp and 50% paper that has undergone a de-inking process. We also use vegetable-based inks to print all of our books. West recycles nearly 27,700,000 pounds of scrap paper annually—the equivalent of 229,300 trees. Since the 1960s, West has devised ways to capture and recycle waste inks, solvents, oils, and vapors created in the printing process. We also recycle plastics of all kinds, wood, glass, corrugated cardboard, and batteries, and have eliminated the use of polystyrene book packaging. We at West are proud of the longevity and the scope of our commitment to the environment.

West pocket parts and advance sheets are printed on recyclable paper and can be collected and recycled with newspapers. Staples do not have to be removed. Bound volumes can be recycled after removing the cover.

Printed with **Printwise**
Environmentally Advanced Water Washable Ink

TO THOSE WHO MEAN THE MOST,

LEE ANN, BRENDAN AND SHELLY,

KERRY AND TERRY, AND COLIN.

# BRIEF
# CONTENTS

# CONTENTS

# I  ORGANIZATION AND FUNCTION OF CELLS

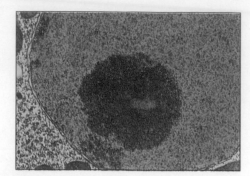

## 5
## CELL DIVISION 85

## 4
## CELLS AND ENERGY: PHOTOSYNTHESIS AND RESPIRATION 64

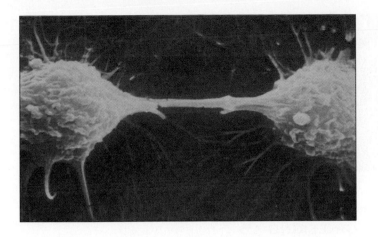

# II   CELLS AND HEREDITY

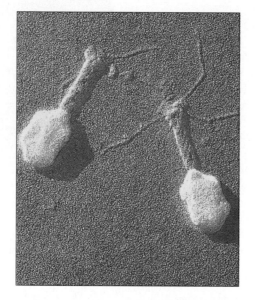

## III   EVOLUTION OF LIVING SYSTEMS

## 11
## EVOLUTIONARY PROCESSES: ABOVE THE SPECIES LEVEL 193

## 12
## VIRUSES, PROKARYOTES, AND PROTISTS 218

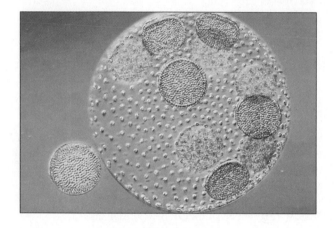

## 13
## DIVERSITY OF FUNGI AND PLANTS 238

## 14
## ANIMAL DIVERSITY 257

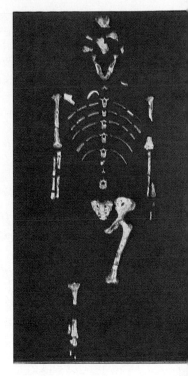

# IV   CONTROL SYSTEMS IN PLANTS

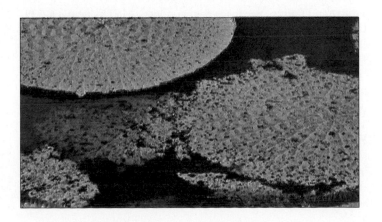

## 17
## PLANT NUTRITION, TRANSPORT, AND GAS EXCHANGE 319

# V    CONTROL SYSTEMS IN ANIMALS

## 18
## HOMEOSTASIS 334

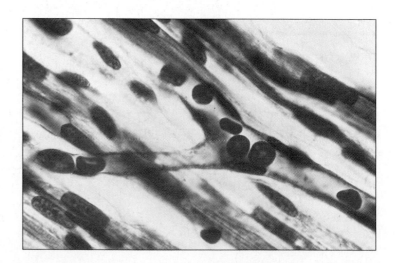

## 22
## THE IMMUNE SYSTEM 407

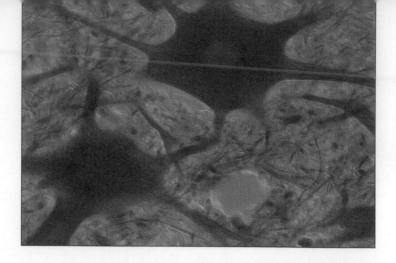

## 23
## DIGESTION AND NUTRITION 425

# VI  REPRODUCTION, DEVELOPMENT, AND BEHAVIOR

## 27
## PLANT REPRODUCTION AND DEVELOPMENT 514

## 28
## ANIMAL REPRODUCTION AND DEVELOPMENT 528

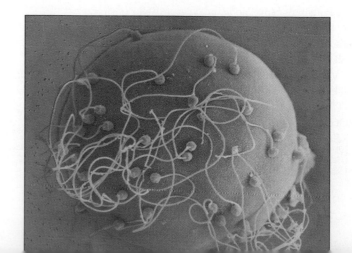

# VII  ECOLOGY AND THE ENVIRONMENT

# PREFACE

During the last few decades, biology has become the fastest growing area of science. Building on advances made in physics and chemistry early in this century, research in biology has generated massive amounts of information and created whole new areas of investigation, ranging from molecular genetics to biogeochemical cycles. In the process, biology has not only grown, but has essentially changed during the last fifty years from a science that describes and catalogs the natural world to one that provides a detailed explanation of the mechanics of the natural world. Even at the present level of understanding, biology is rapidly becoming a predictive science.

This transformation has altered the ways in which biology is studied and how it is taught. New findings are being generated at a rate that threatens to overwhelm those involved in teaching and research. For the student of biology, especially the nonmajor, the conceptual framework can easily be obscured by the overwhelming level of details.

## The Rationale for This Book

Mindful of the changes that have swept over biology, this text is written for a one-term introductory biology course for the nonmajor. Unlike other texts, it is not an abridged or diluted version of a textbook originally designed for majors. It has been written by an author who works with undergraduates on a daily basis, teaching basic courses in a biology curriculum to both majors and nonmajors.

In recognition of the need for nonscience students to acquire knowledge of basic biological concepts, rather than an array of facts, each chapter is organized around one or two central ideas. For example, the organizing concept in Chapter 5 is the cell cycle. Once the cell cycle has been explained, the process of mitosis, meiosis, cancer, and aging are interpreted in light of this concept. Without question, new discoveries will continue to refine the concept of the cell cycle and the related processes. If students have a firm grounding in the cell cycle, they will be equipped with a way of thinking about and interpreting newly discovered details, and be able to integrate them into a conceptual framework.

The speed at which technology is being developed from scientific knowledge and the pressure of human population growth make it clear that difficult, informed decisions need to be made at all levels, from the personal to the global. A great many of these decisions, ranging from genetic screening to global warming, will require knowledge of biological principles. Elected leaders and citizens outside the community of biologists need to have a working knowledge of biology in order to shape and support the decisions regarding biological developments that will need to be made in the coming decades. A course and a text based on a hierarchical approach, which links a basic set of biological concepts, can provide the basis for transmitting these principles without unnecessary detail or jargon.

This text has been written with several premises in mind:

1. The use of a limited number of clearly presented, interlinked concepts is the best approach to learning a complex, detailed subject such as biology.
2. A text for nonmajors should be written in a straightforward, clear fashion with relevant examples that students can apply to themselves.
3. The figures and photographs should teach rather than merely illustrate the ideas under discussion.
4. The concepts must be linked by a small number of basic, organizing themes.

## The Interwoven Themes

Areas of biology that were once islands of knowledge are rapidly being integrated into a larger picture. As this process continues, a small number of organizing ideas are proving useful in interpreting biological processes. Two of the most powerful of these themes are genetics and evolution. During the last four decades, genetics has illuminated with great clarity the physical and mechanistic bases for the key characteristics of living organisms: how they reproduce, develop, and function.

Evolution is the result of natural selection of genetic variation. All organisms have an evolutionary history that can be extended backward in time to the origin of life on Earth. As forms diverged under natural selection, they carried genetic evidence of their relationship. The recent discovery that the gene controlling eye formation is an ancient genetic innovation links all organisms with eyes to a common ancestor.

The themes of genetics and evolution are used throughout the texts to organize and link the concepts presented in each chapter. In each chapter, the concepts are presented in an evolutionary context. For example, excretion in multicellular organisms is associated with a tubular structure lined with epithelial cells. This basic unit functionally links the flame cells of the rotifer with the human kidney. A discussion of the genetics of kidney disorders provides an insight into the nature and function of genes involved in excretion.

## The Organization

The text is divided into seven parts. Part 1 presents background to understanding the origin, organization, and basic functions of all organisms. The chemical and biological bases of life and the basic properties of all living systems including structure of cells, energy transfer, and cellular reproduction are discussed.

Part II deals with the basic principles of genetics, from Mendelian patterns of inheritance and gene action to applications of genetic engineering. These principles are related to organisms ranging from bacteria to humans to everyday events such as having your cat vaccinated and buying milk.

In Part III, the development of the theory of evolution is discussed and then related to the origin of species, and the events above the species level, emphasizing the relationship between the physical environment and adaptations that allow fitter individuals to survive and leave more offspring.

Many biology texts, especially those for nonmajors, give only cursory treatment to the biology of plants. The emphasis of Part IV is on terrestrial plants and how they developed different adaptations to the same environmental challenges as animals, namely, seasonal variation, protection, desiccation, and gravity.

Part V deals with control systems in animals, briefly describing the evolutionary history of adaptations for establishing and maintaining homeostasis, with a focus on the structure and function of body systems in humans. This serves two purposes: It provides students with an evolutionary perspective and motivates them to learn about concepts to which they can personally relate.

In Part VI, the chapters cover reproduction in plants and animals to contrast again how plants and animals have solved some basic reproductive problems: fertilization in the absence of water, protection of the developing embryo, and the developmental strategies employed in producing the structures of the adult organism.

Part VII deals with organisms and their interactions with the physical environment. This section begins with a description of the structure and life cycle of populations, and how populations form communities. It stresses the important point that natural communities enhance and reinforce the lives of the species that make up the community, and that disruption of any species or of the physical environment can have devastating consequences on the community. The last chapter emphasizes the fact that human populations have grown to the point where the limited resources of the planet are being stressed, and that we need to plan now to manage our physical and biological environment in order to sustain human society.

Although the text is organized in a format that begins with the chemical level and ends with cycles and the biosphere, the text can be used in a variety of formats, including those that place ecology and the environment first. Most chapters have been written to stand alone, so that the parts and chapters can be used in an interchangeable format, providing the instructor with maximum flexibility.

## The Pedagogy

Each chapter opens with a short vignette that relates to the concept or concepts to be discussed in the chapter. These episodes have been selected to relate the concepts discussed in the chapter to a real situation, either historical or contemporary, and to kindle student interest in the material to be covered.

As described earlier, each chapter is centered around a limited number of concepts, woven together with a discussion of their evolutionary and genetic aspects. The connections to other concepts and applications of the concepts are also outlined.

Within the chapter, headings are written as descriptive, summary statements that preview the material and often summarize the point of the section. This provides the student with an overview and a convenient way of reviewing the material in the chapter.

The chapters end with a summary that restates the major ideas covered in the chapters. Beginning each chapter with an example of the concept and then ending with a restatement of the concepts and their applications will help focus the students' attention on the conceptual framework and minimize the chance that they will attempt to learn by rote memorization of facts.

At the end of each chapter, questions test the students' knowledge of the facts and their ability to reason from the facts to conclusions. Some questions relate biology to our society and our culture.

Several elements of the text are designed to serve as a built-in study guide for the student. Chapter 1 provides an overview of biology and the organization of the book. This chapter is followed by a section on developing study skills. It has been positioned here to make it visible, to provide a list of study skills and how to acquire them, and to assist the student in preparing for examinations. This is followed by the opening vignette of Chapter 2 as a lead-in to the material on the chemistry and origin of life. All chapters use opening vignettes to set the stage for the topics to be discussed. Other elements in chapters serve as part of a built-in study guide. These include the use of statements as primary and secondary headings, chapter summaries in list form, a list of key terms in the chapter, all of which are defined in the text and in the glossary, and the end-of-chapter questions that test recall of facts and applications of the chapter's concepts to social issues.

Throughout the book, sidebars are used to highlight applications of concepts, discuss oddities of nature, present controversial ideas, and communicate the latest results in a field, without interrupting the flow of the text.

## The Art Program

The art program is based on more than 25 years of classroom experience teaching undergraduates. To be effective, classroom illustrations must condense concepts and communicate major points effectively. This experience has been reinforced by more than a decade of preparing illustrations for textbooks of genetics, a field that underpins and informs much of modern biology. For this text, the result is uncluttered and effective illustrations that parallel the presentation of material in the text. To emphasize the parallel importance of text and art as pedagogical devices, the text and its relevant illustrations are presented on the same page or same two-page spread. Students will not have to turn pages to find figures or photos that relate to the point being discussed. In addition, colored icons in the text highlight the figure callouts, allowing the student to return quickly to the appropriate place in the text after studying the figure.

Like the text, the art program is concept driven, and the resulting illustrations are clear summaries of a concept under discussion, without unnecessary text, labels, or detail. This is accomplished by having the illustrations supported by a detailed explanation of the concept in the text, allowing the art to be an unencumbered summary, often in the form of a flow diagram of a process or concept.

## The Human Connection

The book covers many aspects of biology, and presents examples from organisms ranging from prokaryotes to primates. Because it is easiest for students to relate biological concepts to their own lives, several features of the text have been constructed to connect the concepts and the process of science to the everyday experience of the students. Some of these features are as follows:

1. **Science and Society.** At the end of each chapter, one or more open-ended questions ask the students to apply what they have learned to common situations. These questions are wide ranging, and include considering how the diet of spiders might be beneficial to humans, whether it is ethical to use zoos to preserve endangered species of primates, or how to balance the risks of using radioactive isotopes with diagnosis and treatment of disease.
2. **Beyond the Basics.** Boxed material in each chapter relates concepts in the chapter to topics that either elaborate ideas presented in the chapter or are interesting, but tangential, examples that should be of interest to the students. These include heat-generating plants, speculation that early hominids used tools as killer frisbees, and the use of insects found on corpses to establish the time of death.
3. **Guest Essays.** To emphasize that science is a human endeavor, and is always in the process of refinement, the book contains essays written by scientists, describing how they became interested in science, what they study, and how their research relates to the larger context of human society. These essays are not just about the scientists, but are written *by* the scientists, providing students with insights into the lives, thoughts, and motives of biologists.

The text and coordinated ancillaries also incorporate other features, which are described in the following section.

## ANCILLARIES

To assist you in teaching this course and supplying your students with the best in teaching aids, West Publishing Company has prepared a complete supplemental package available to all adopters.

## For the Instructor

The comprehensive instructor's manual and test bank, prepared by Judith Lanum Mohan of Case Western University, includes teaching ideas, chapter overviews, learning objectives, discussions of common student misconceptions, audiovisual and multimedia sources, Internet sources, and the test bank containing approximately 3000 multiple choice, true/false, fill-in-the-blank, matching, short answer, analogy, and quantitative questions.

The entire test bank is provided on diskette along with WESTEST, a computerized testing package. Using WESTEST 3.1, instructors can generate examinations containing questions they select or have questions randomly generated by the computer. Instructors can also use the WESTEST 3.1 edit function to modify theses questions, add new questions, or delete existing questions. Additionally, West's Classroom Management software allows student data to be recorded, stored, and used for various reports.

West provides an on-line update service through its HomePage containing articles by the text author. Dr. Cummings reports monthly on new and interesting developments in the field of biology since the publication of the text, West's Instructor's Reference CD-ROM, which contains all of the instructors' printed supplements, WESTEST 3.1, the Classroom Management Software, the electronic presentation package, and all of the art from the text, provides instructors with a single source for the supplements.

A slide set and full-color transparency acetate set provide clear, effective illustrations of 150 of the most important artwork and maps from the text.

An electronic slide presentation package using Persuasion, prepared by Michael Farabee of Estrella Mountain Community College, provides text outlines, guiding questions, all artwork, with 3–4 available as animations, and photos for each chapter of *Biology: Science and life.* The presentations are available for Macintosh and for Windows. Instructors with Persuasion software can customize the outlines, art, and order of presentation. The text's art has been converted to a visual format suitable for computer projection in the classroom.

West also has available its Biology Videodisc which includes animations of intricate processes, illustrations, and video footage. Used in conjunction with Lecture Builder software you can create, edit, store and build upon your own personalized lectures.

Understanding Human Evolution software, by Ronald Wetherington, contains 13 interactive evolution exercises with graphics. For example, in one interactive simulation, students compare dominant/recessive selection and balanced polymorphisms. It's available for IBM PCs or compatibles with at least 256k.

Other multimedia is available to qualified adopters. Please contact the local West sales representative for more details.

## For the Student

*Current Perspectives in Biology,* edited by Shelly Cummings of the University of Chicago, is a collection of approximately 50 very current articles chosen from general interest and science magazines to supplement material that students will encounter in their course work. West can make this supplement available with the text as a set, or it can purchased separately.

*Laboratory Manual for Introductory Biology,* Second Edition by Jay M. Templin is a concise, easy to follow laboratory manual that covers all of the major themes of contemporary biology. The labs emphasize the discovery process by challenging students to predict what will happen or explain what has happened during an experiment. The second edition provides a more even balance of animal and plant exercises. In addition, human fluids are no longer used in any experiment. When needed, artificial fluids are substituted. The lab manual includes full color photos of pig dissections. An Instructor's Manual is also available.

A study guide, written by Richard Friedman Drexel University, is closely tied to the main text. The guide provides students with an overview of each chapter, learning objectives, exercises to test the students on key terms, chapter concept questions, experiments and activities, diagram labeling exercises, and practice exams consisting of multiple-choice, true/false, fill-in-the-blank, and matching questions.

*Internet Activity Booklet,* tied to the West Publishing home page for address updates, provides projects for students to get the latest information and data on topics such as recent research in genetics.

*Exploring Critical Issues in Biology* by Andrea Huvard (California Lutheran University) contains 34 biological topics and issues (genetic engineering, AIDS, evolution and human intelligence) presented to encourage students to understand practical applications of their studies in biology. Questions at the end of each discussion provide a springboard for critical analysis.

*Study Skills for Science Students* by Daniel Chiras is an 86-page booklet which offers practical tips on notetaking, test-taking, reading efficiency, concept mapping, and using computer software tutorials.

A Student Note Taking Guide contains printed copies of all of the art that is used in the electronic slide presentation package. Bound with perforated, 3-hole punched pages, it allows students to take notes as the slides are shown in lecture. The complete text of Study Skills for science students is also included.

# ✵ ACKNOWLEDGMENTS

A book of this type represents a major undertaking for both the author and the publisher. At West Publishing, there are a myriad of people to thank for their vision, energy, and dedication to this book. Rex Jeschke, manager of the College and School Division, helped put the full resources of the publisher behind this project. I am grateful to him for his efforts and for his work with the American Publishers Association on behalf of all textbook authors. My editor, Jerry Westby, is a man of great patience and forbearance. This book had a long gestation period, and throughout, his enthusiasm and encouragement never flagged. His creative contributions, ability to look at things from the side of the student, and commitment to undergraduate education helped to shape the text and many of its features. As before, his friendship and advice have been among the most valuable benefits of this project. Dean DeChambeau, the developmental editor, was, as usual, unflappable in the face of crises. His work provided a clear overview of the strengths and weaknesses of the manuscript. Dean was also instrumental in developing much of the ancillary material that expertly serves the needs of both students and instructors. Special thanks to Halee Dinsey, the developmental editor who worked closely with Michael Farabee of Estrella Mountain Community College to develop the state-of-the-art electronic lecture presentation software and Internet activities supplement described earlier. Betsy Friedman was a miracle worker in assembling the list of guest essayists, and nudging them when necessary to get the essays in on time. She made it look easy, even though it was, I am sure, an often frustrating task. The copyeditor, Loretta Palagi, worked diligently to keep my grammar and punctuation within the limits of acceptable use.

The transformation of a manuscript into a book requires the efforts and strong support of a production staff. In this case, I stood by in awe as the production team poured their creative energies and time into this book. The production editors, Sandy Gangelhoff and Matt Thurber, believed in this project and were relentless in their insistence on excellence in all aspects of manuscript preparation, the art program, and the photo program. The design of the book, the cover, and page layout were created by Diane Beasley. Their enthusiasm, persistent efforts and attention to the smallest detail are responsible for the remarkable book you are now reading. I am grateful to have had the opportunity to work with these talented individuals.

Ann Hillstrom and Ellen Stanton spearheaded the development of the marketing program that brings the innovations, pedagogy, and themes of this book to the attention of potential adopters.

I would also like to thank my colleagues who contributed to this book. Judy Verbeke provided valuable insight into undergraduate teaching and did the first drafts of Chapters: 16, 17, and 27. Mike McKinney lent his expertise in reviewing most of the chapters in Part II, and prepared the first drafts of Chapters 30, 31, and 32. Their efforts have greatly enhanced the book. Special thanks are due to Shelly Cummings, who wrote the Questions and Problems and the Science and Society questions at the end of each chapter. She also researched and wrote the sidebars for all of the chapters. In the face of tight schedules and looming deadlines it was always reassuring to know that her optimism and diligence would carry the day. Patricia Lewis took on the daunting task of preparing the glossary, an important study aid for students struggling with the vocabulary of biology.

This book went through a number of meticulous reviews for both content and style. The comments and suggestions of the following reviewers has enhanced the focus and presentation of the material.

Dawn Adrian Adams
*Baylor University*

Thomas G. Balgooyen
*San Jose State University*

Stephen W. Banks
*Louisiana State University-Shreveport*

Rolf W. Benseler
*California State University, Hayward*

Robert D. Bergad
*University of St. Thomas*

Clyde E. Bottrell
*Tarrant County Junior College*

David M. Brumagen
*Morehead State University*

Harold W. Burton
*SUNY–Buffalo*

Kathleen Burt-Utley
*University of New Orleans*

Roy Coomans
*North Carolina Agricultural and Technical State University*

Lee Couch
*Albuquerque Technical-Vocational Institute*

John W. Crane
*Washington State University*

Loren L. Denney
*Southwest Missouri State University*

Jean DeSaix
*University of North Carolina at Chapel Hill*

William C. Dickison
*University of North Carolina at Chapel Hill*

Jamin Eisenbach
*Eastern Michigan University*

James H. Eley
*University of Colorado*

Thomas C. Emmel
*University of Florida*

Larry Frolich
*University of St. Thomas*

Sally Frost-Mason
*University of Kansas*

Larry Fulton
*American River College*

Geoffrey W. Gearner
*Morehead State University*

Douglas Gelinas
*University of Maine*

Edward J. Greding, Jr.
*Del Mar College*

Robert D. Griffin
*City College of San Francisco*

William H. N. Gutke
*Memphis State University*

John P. Harley
*Eastern Kentucky University*

Terry P. Harrison
*University of Central Oklahoma*

Jean Helgeson
*Collin County Community College*

George A. Hudock
*Indiana University*

Andrea L. Huvard
*California Lutheran University*

Aaron E. James
*Gateway Community College*

Miles F. Johnson
*Virginia Commonwealth University at Indianapolis*

Florence Juillerat
*Indiana University-Purdue University at Indianapolis*

Judy Kaufman
*Monroe Community College*

Timothy R. Kelley
*Valdosta State University*

Eugene N. Kozloff
*University of Washington*

Dana M. Krempels
*University of Miami*

Bruce F. Lindgren
*Normandale Community College*

Doug McElroy
*Western Kentucky University*

Michael L. McKinney
*University of Tennessee–Knoxville*

Jon R. Maki
*Eastern Kentucky University*

Kathy Martin
*Central State University*

James E. Mickle
*North Carolina State University*

Glendon R. Miller
*Wichita State University*

Neil A. Miller
*University of Memphis*

Herbert Monoson
*Bradley University*

Joseph Moore
*California State University, Northridge*

Donald B. Pribor
*University of Toledo*

David J. Prior
*Northern Arizona University*

Sylvia Jane Rigby
*University of Guelph*

Sonia J. Ringstrom
*Loyola University, Chicago*

Franklin L. Roberts
*University of Maine*

O. Tacheeni Scott
*California State University, Northridge*

Prem P. Sehgal
*East Carolina University*

Marilyn Shopper
*Johnson County Community College*

Bruce N. Smith
*Brigham Young University*

H. Eldon Sutton
*University of Texas at Austin*

Sarah H. Swain
*Middle Tennessee State University*

Haven C. Sweet
*University of Central Florida*

Darrell J. Weber
*Brigham Young University*

Daniel E. Wivagg
*Baylor University*


I extend my appreciation to my colleagues, Suzanne McCutcheon and Jeffrey Doering, for their patience and for the support they provided to my research program during the preparation of this manuscript, and to Michael Westfall for his encouragement and good cheer. Thanks to you all.

# LEARNING GUIDE

Before you become absorbed in this textbook, take a moment to look over the next few pages. We've provided an overview of the built-in learning devices you'll find throughout the book. Becoming familiar with these unique features can make it easier to navigate through the material.

## GETTING STARTED

**Chapter outlines will get you started with a quick overview of what topics you can expect to cover.**

**Prologues open each chapter. These engaging short stories provide a fascinating glimpse into a number of diverse subjects. Interested in the big bang theory? Turn to the prologue to Chapter 2. Ever wonder what your body has in common with a car? Check out the prologue to Chapter 4. All of the prologues were written to make your journey into biology as interesting and rewarding as possible.**

At the Marine Biological Laboratories at Woods Hole, Massachusetts, scientists gather each summer as they have for more than 100 years to teach, to learn, and to conduct experiments using the marine plants and animals that can be collected there. In the summer of 1982, a group of young scientists led by Tim Hunt began experiments designed to study biochemical changes that take place after fertilization of the sea urchin egg. They decided to examine new gene products (gene products are proteins) made in the hours immediately after fertilization, a time when a series of cell divisions take place. They fertilized a batch of sea urchin eggs and, at 10-minute intervals, analyzed the newly made proteins during the first 2 to 3 hours of development. The fertilized egg first divides at about 60 minutes, and again about 2 hours after fertilization, resulting in a four-cell embryo.

Several new proteins appeared almost immediately after fertilization, including one that exhibited unusual behavior. This protein was continuously synthesized, but was unexpectedly destroyed just before each round of cell division. Because of its cyclic behavior, this protein was called **cyclin.** Experiments using another species of sea urchin revealed two cyclins that periodically disappear during the cell divisions associated with early development. Investigating other organisms available at Woods Hole, this group discovered that the developing eggs of clams also contain cyclins. Because these proteins disappear...
(continued)

**BEAUTIFUL AND INSTRUCTIONAL COLOR**
As you look through the text, you'll notice the consistent use of vibrant colors. Not only are the colors beautiful to look at, but they also make for clearer and more informative illustrations.

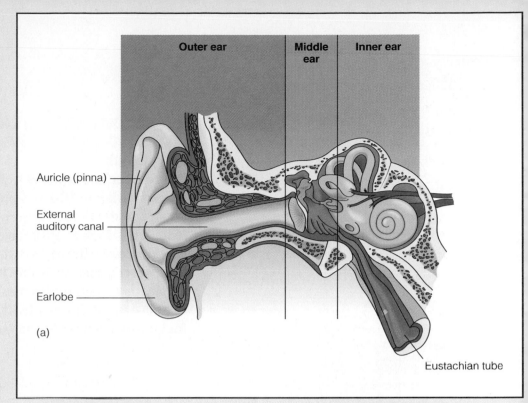

(a)

**EVERYTHING IS
RIGHT WHERE YOU NEED IT**
If you've ever found yourself flipping back and forth between pages, trying to find the illustration that the text refers to, you'll appreciate the effort that went into the page layout of this book. Whenever possible, illustrations and photographs appear on the same page or spread as the text they accompany. This makes it much easier to go back and forth from text to visuals.

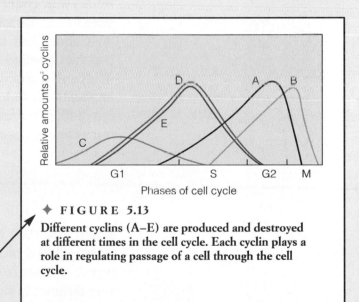

✦ **FIGURE 5.13**
Different cyclins (A–E) are produced and destroyed at different times in the cell cycle. Each cyclin plays a role in regulating passage of a cell through the cell cycle.

**NOW, WHERE WAS I?**
You just spent a few minutes studying an illustration and now you can't remember where you left off reading. This little red diamond appears next to figure references within the text. It's a little feature, but a real time-saver.

The signals that are part of this switch point in the cell cycle were discovered as a result of the work started at Woods Hole in 1982. These signals are generated by the synthesis and action of cyclins (✦ Figure 5.13). At the G1 control point, a cyclin combines with another protein, generating a signal that moves the cell from G1 into S. What signals cause the production and destruction of cyclins are not yet known.

Cancer cells have disabled this and other control points, and divide continuously. Mutations in genes that control the synthesis or action of cyclins are important in generating the transformation of a normal cell into a cancer cell.

## Sidebar

### CELL DIVISION AND SPINAL CORD INJURIES

Many highly differentiated cells, like those of the nervous system, do not divide. They are sidetracked from the cell cycle into an inactive state called G0. As a result, injuries to nervous tissue such as the spinal cord cause permanent loss of cell function and paralysis. For years, scientists have worked to learn how to stimulate growth of spinal cord cells so that injuries can be repaired. Recent work suggests that it may soon be possible to get nerves in the spinal cord to reconnect to their proper targets and to restore function in nerve cells that are damaged, but not cut. Researchers have shown that the severed spinal cords of young rats can be reconstructed by transplantation of the corresponding section of spinal cord from rat embryos. When the rats reached adulthood, most of the sensory and movement was restored.

Other researchers have isolated a growth factor found only in the central nervous system that causes cell growth from the ends of severed spinal cords. Whether such growth can result in reconnection of nerves to their proper muscle targets, and whether function can be restored are unresolved questions.

## WHAT DO DEEP SEA VENTS HAVE TO DO WITH CONVICTING A SUSPECTED MURDERER?

The *Sidebar* in Chapter 4 explains how bacteria living in 100°C water at the vents provide proteins used in recombinant DNA techniques such as DNA fingerprinting. Other *Sidebars* throughout the text provide brief descriptions of human health applications, new discoveries, and quirks of nature—and, at the very least, they're great for impressing friends and family.

## GUEST ESSAYS

Here's your chance to read what scientists have to say about the roots of their interest in science, their careers, and basic research that has resulted in important, but often unintended, discoveries with beneficial applications. *Guest Essays* are fun and interesting to read—you might be tempted to read all of them first.

## WHEN YOU JUST HAVE TO KNOW MORE

If we included in-depth coverage of every interesting topic in biology, this text would be very difficult to finish in one term, much less to carry. But some topics are *so* compelling, that you want to know more. *Beyond the Basics* boxes feature topics of special interest and develop them a little further and in more detail.

## A Life of Questions: The Questions of Life

NANCY KEDERSHA

Most children go through a "What?" stage, progress on through "Why?" and "How?" and eventually grow into adults concerned with increasingly specific questions, such as "Will you marry me?" and "How much does this cost?" I never gave up asking the "What? How?" questions, and they have led me into science. How do seeds grow into plants? How do robins find worms? How do living things grow and move? These questions that I asked as a child are still meaningful to me as an adult. Living things are both so beautiful and so interesting that I could never decide whether I preferred to pursue art or science full time.

My college years were spent shuttling between the biology labs and the university theater, so I was usually either looking through a microscope at diatoms I had cultured or focusing lights on scenery I had built and painted. After college, I was still reluctant to commit myself exclusively to science. I deferred graduate school and I took a job as a lab technician in a biomedical research lab. Working on cells "hands on" I gradually came to realize that being a scientist was much more of an art than I had realized. Glassblowing, photography, assembling posters, and redesigning equipment were as much a part of my job as growing cells or purifying proteins. The lab was a place where skill and "good hands" were needed as well as knowledge, and where someone who asked a lot of questions was just doing her job proficiently rather than being a pain in the neck.

In science, as in life, my curiosity has led me to a series of serendipitous discoveries. In graduate school, I saved one of the "contaminant" proteins I was purifying away from my main "thesis-related" protein, and found that the contaminant was a new form of the well-known protein actin, differing in its location within cells as well as its biochemistry. While learning to use the electron microscope, I found some strange and beautiful structures in the "garbage" fractions that normally would have gone down the drain. For months afterwards, any hapless scientist visiting the lab was handed photographs of these structures and ruthlessly grilled: "Have you ever seen anything like THESE before??" No electron microscopist for miles (we asked a lot of them!) could identify these particles. We named them "vaults" because of their beautiful and intricate morphology, reminiscent of the multiple arches that form vaulted cathedral ceilings.

One question that people ask me is "If there are so many of these in all cells, why didn't anyone ever see them before you?" One answer is that vaults can't be seen in samples prepared for electron microscopy using normal stains, and are in essence invisible in conventionally prepared sections of cells. The technique I was using enabled me to see them, and another new technique allowed me to purify them. Vaults were in fact actually photographed and appeared in publications as contaminants of coated vesicle preparations. No one realized they were intact new structures instead of fragments of coated vesicles until I purified them. So perhaps the most complete reason why no one ever saw them was because nobody looked using the right tools, and when they did use the right tools, they didn't ask the right questions. The "right" question I asked was "What are these things?" And now that we know more about what they are, the question is "What do they do?"

Powerful new techniques, instrumentation, and computers have resulted in an explosion of data that is coming in faster than we can digest or integrate. Science is a process to make sense out of data and understand the principles behind events. Like a microscope or a telescope, science is also a way of viewing the world, and it evolves like a living organism as does our understanding. The more I look, the more I see; the more questions I answer, the more new ones I perceive. I find such beauty in the answers that for me, art and science are one.

*Nancy Kedersha is a research scientist at Immunogen, Inc., in Cambridge, Massachusetts. She has published dozens of articles on topics in cell biology, and her photos of cancer cells have won numerous prestigious international awards. Her work has been the subject of feature articles in several national and international magazines.*

---

## Beyond the Basics

## GENES, CELL DIVISION, AND CANCER

Cancer is a disorder of cellular reproduction. This condition arises from within the cell, causing cell division to be out of control; cells continually move through the cell cycle and divide, giving rise to daughter cells that also undergo continuous division. Cells may also break away from the original tumor mass and establish new tumor sites elsewhere in the body (a process known as **metastasis**). Because the ability to divide in an uncontrolled fashion is passed on to the progeny of cancer cells, it is logical to assume that this property is under genetic control. The question is, if genes are involved in cancer, which ones are they and what do we know about them?

Until recently, these questions were difficult if not impossible to answer. Now however, scientists are uncovering two types of genes that are associated with the development of cancer, **tumor suppressor genes** and **oncogenes.** The first group contains genes that normally keep cell division in check, preventing the cell from responding to external and internal commands to divide. If one or more of these genes are damaged or absent, restraints on cell division are removed, leading to the formation of tumors.

The second group of genes, the oncogenes, are able to activate cell division in cells that normally do not divide or do so only slowly. In other words, when these genes are damaged, or make gene products at inappropriate times or in the wrong amounts, resting cells begin to divide in an uncontrolled fashion, causing tumor formation.

At the present time, fewer than 10 tumor suppressor genes have been discovered, but more than 75 oncogenes have been described. The importance of oncogenes in relation to cancer can be shown by the fact that scientists working with oncogenes were awarded Nobel Prizes for their efforts in 1966 and 1989. New research on cancer genes is aimed not only at discovering more of these genes, but at understanding how the action of these genes at the molecular level fits into the network of external and internal signals that bombard the cell and result in a decision to move through the cell cycle and divide.

### Some Cells Divide by Binary Fission

In the cells of prokaryotic organisms, including blue-green algae and other bacteria, cell division follows a period of growth during which the cell enlarges by synthesizing new sections of the cell wall and by making more cytoplasm. During this period of growth, the circular DNA molecule that serves as a chromosome is copied. Each copy of the chromo[...] ner surface of [...] sides of the cel[...] gates by the ad[...] the attachmen[...] move further a[...] the cell wall an[...] ward. Eventual[...]

### Meiosis I Reduces the Chromosome Number

In the first stage of prophase I, the chromosomes coil, thicken, and become visible under the microscope. As the chromosomes condense, the nucleoli and nuclear membrane disappear and the spindle becomes organized. Each chromosome physically associates with its homologue, and the two chromosomes line up side-by-side, a process known as **synapsis** (Figure 5.15). During synapsis, each chromosome is composed of two sister chromatids held together at the centromere, and the four chromatids in the cluster are often referred to as a **tetrad**. During this stage, evidence for a physical exchange of chromosome material between homologous... (continued)

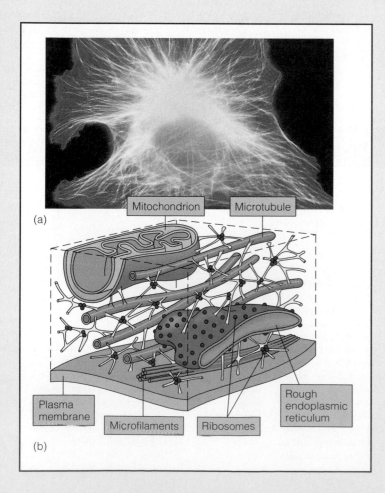

(a)

Mitochondrion   Microtubule

Plasma membrane
Microfilaments   Ribosomes   Rough endoplasmic reticulum

(b)

## SUBSECTION HEADINGS ARE FULL SENTENCES
Instead of traditional topic titles for each subsection (for example, Cell Division or Meiosis I), full statements like the ones shown at the left foreshadow the main points to come. It may not seem like a big deal, but it can really improve your comprehension.

## BUILT-IN STUDY AIDS
Need a quick memory refresher for an upcoming exam? Chapter summaries reinforce the main topics covered in each chapter.

Cytokinesis? Diploid? Gametes? Important key terms are listed at the end of each chapter. For complete definitions just flip to the glossary in the back of the book. It's easy to find—there's a colored band on the edge of each glossary page so you can open right to it.

*Questions and Problems* at the end of each chapter are another great way to test your knowledge of the subject matter.

Brief objective questions (with answers in the back of the book) along with the *Developing Your Abilities To Think and Study* section between Chapters 1 and 2, provide all the basics for a short-form study guide. *Science and Society* questions require the use of analytical skills to discuss scientific and societal issues with which we may be confronted in coming years.

## THE WHOLE PICTURE
In many cases photographs and illustrations are used together (as in the example to the left) to visually explain concepts.

# SUMMARY

1. Cells are the units of structure and reproduction in all organisms. The period from one nuclear division to another encompasses one cell cycle. In eukaryotes, during the interphase portion of the cycle, a duplicate set of chromosomes is made. Once the chromosomes have been replicated, the cell proceeds through the cell cycle to mitosis. Mitosis is a process of nuclear division that distributes a complete set of genetic information to two daughter cells.

2. The process of mitosis has been divided into four stages: prophase, metaphase, anaphase, and telophase. In prophase, the replicated chromosomes coil and become visible and the nuclear membrane disappears. In metaphase, the chromosomes migrate to the equator of the cell and become attached to the spindle fibers at the kinetochores of the centromere. In anaphase, the centromeres split longitudinally, allowing the sister chromatids (now called chromosomes) to migrate toward opposite poles of the spindle. In telophase, the chromosomes at each pole are incorporated into a nucleus and uncoil into a dispersed condition.

3. Cytokinesis divides the cytoplasm, producing two cells, each of which contains a complete set of chromosomes. In animal cells, cytokinesis involves the formation of a cleavage furrow, whereas in plant cells, a partition, known as the cell plate, is involved in distributing the cytoplasm.

4. Meiosis is a form of nuclear division that produces haploid cells containing only one member of each chromosome pair. In prophase I, homologous chromosomes synapse. At metaphase I, the paired homologous chromosomes become aligned at the equator of the cell and attach to the spindle fibers. In anaphase I, homologous chromosomes are separated from each other and migrate to opposite poles of the cell. In meiosis II, there is no chromosome replication, and the unpaired chromosomes align at the metaphase plate, the centromeres split longitudinally, and the sister chromatids (now called chromosomes) move toward opposite poles of the cell. In telophase II, nuclei reform and cytokinesis divides the cell. The result is four cells, each containing the haploid number of chromosomes.

# KEY TERMS

alternation of generations
anaphase
assortment
cell cycle
cell plate
centromere
chiasma, chiasmata
chromatids
cleavage furrow
cyclin
cytokinesis
diploid

gametes
gametophytes
haploid
homologues
interphase
kinetochores
meiosis
metaphase
metastasis
mitosis
mitotic spindle
oncogenes

prophase
recombination
reductional division
sister chromatids
somatic
sporophyte
synapsis
telophase
tetrad
tumor suppressor genes
unicellular
zygote

# QUESTIONS AND PROBLEMS

## TRUE/FALSE

1. Somatic cells divide during each part of the cell cycle.
2. A cell that contains a single basic complement of chromosomes is said to be diploid or $2n$.
3. In the fern plant, the diploid stage of the life cycle is known as the sporophyte, whereas the multicellular haploid plants are known as gametophytes.
4. In the cell cycle, all cells pass through all stages leading to division.
5. Mitosis is duplication division, whereas meiosis is reduction division.

## MULTIPLE CHOICE

1. Four chromatids are the same as a pair of _____.
   a. chromomeres
   b. centromeres
   c. centrioles
   d. homologous chromosomes
2. The cells of prokaryotic organisms undergo a unique form of cell division. This splitting in half of the cell to form two daughter cells is called _____.
   a. meiosis
   b. mitosis
   c. binary fission
   d. cytokinesis

## SCIENCE AND SOCIETY

1. Because of recombination in meiosis I, DNA sequences on the X and Y chromosomes normally exchange in a specific region. In rare instances, however, genetic recombination occurs between the X and Y chromosomes outside this region, and this aberrant exchange can produce two very rare abnormalities: XY females and XX males. How would you tell the parents of a XX male child that their "son" is genotypically female? Is this something that would affect the physical development of the child?

# BIOLOGY
*Science and Life*

# 1

# MODERN BIOLOGY AND ITS UNIFYING THEMES

**OPENING IMAGE**

*Biological systems depend on the*

*physical environment for air,*

*water, and light.*

Welcome to the study of biology. What is biology? Simply put, it is the study of living organisms. Biology is an exciting, vibrant field, using its own methods, as well as those of the physical sciences, to develop and test ideas about the living world. Almost each day newspapers contain one or more articles about new findings and discoveries in biology, related to the environment, medicine, endangered species, genetics, or DNA fingerprinting in criminal trials. These advances are quickly incorporated into the base of knowledge that serves as the foundation for even more discoveries. Because of the pace of discoveries, and the importance of what we are learning, this is an exhilarating time to be a biologist or to study biology.

Beginning with this chapter, you are going to explore a subject that will teach you things you will remember all your life. Some of these things may help you make informed personal and professional decisions that affect you, your family, and the public. Others may help you appreciate the complexity and fragility of life.

What do biologists do? They describe the diversity of living organisms, trace the ancestry of these organisms through the fossil record, learn about their structure and function, how they interact with each other and with the physical environment. Biologists also study human population growth and how the development of technology impacts the biological and physical world. As the century draws to a close, it is clear that human activity is beginning to have global consequences. Knowledge about the large scale and long term effects of population growth, and the use of natural resources will be important in making decisions that, hopefully, will keep the planet habitable.

*To do all this, biologists use their sense of curiosity to observe, pose questions, and collect facts about living organisms. The information they gather is organized into ideas that can be tested by controlled experiments. Over time, these ideas are coalesced into sets of generally accepted principles. In this chapter, we will present some of the basic principles biologists have developed about the living world, inquire about the nature of science and how scientists ask questions and obtain answers. Finally, we will explore three unifying themes of biology that underlie the approach of this book.*

 ## ORGANISMS SHARE BASIC CHARACTERISTICS

### Organisms Are Made of Cells

Some organisms are composed of one cell, while others are made up of aggregations of cells. In all cases, cells are the basic unit of life. Cells are composed of highly ordered sets of molecules. The structure and function of cells has a chemical basis.

### Energy Transfer Is Essential for Life

Organisms obtain energy from their environment and expend this energy to maintain the organization, growth, and reproduction of their bodies. Energy in almost all living systems originates in the sun. Green plants capture energy from sunlight to maintain themselves. All other organisms acquire energy

from their diet, whether they eat plants, animals, or both.

### Organisms Maintain a Relatively Stable Internal Environment

Cells obtain nutrients and oxygen from their environment and these materials must be replenished for life to continue. Waste materials that are the by-products of cellular processes must be removed and other components of the cell must be maintained within certain limits. In multicellular organisms, the chemical and physical state of the internal environment must be maintained within narrow limits. The ability to maintain a relatively constant internal environment is called **homeostasis.**

### Response to Stimuli Is Essential for Survival

Both plants and animals respond to environmental stimuli. Plants bend in response to the direction and intensity of light. Roots grow downward and shoots grow upward in response to gravity. Animals are able to sense changes in the environment and respond by behavior that can include movement, aggression, or flight.

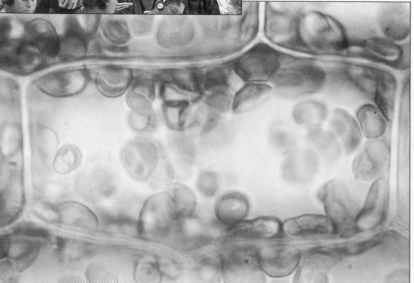

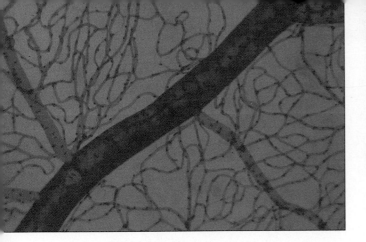

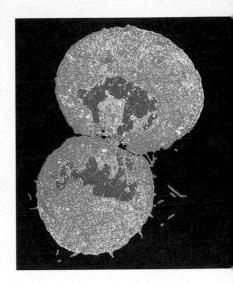

## All Organisms Are Capable of Growth and Reproduction

New cells arise only from preexisting cells, and new multicellular organisms arise only from the reproduction of one or more parents. Asexual reproduction can occur by simple cell division or by budding. Sexual reproduction results from the fusion of sex cells of different mating types. Most organisms use sexual reproduction.

## Adaptation and Evolution Have Changed Organisms

Over time, a combination of naturally occurring differences in organisms and environmental factors has acted to enhance survival and reproduction of some members of a population more than others. In this way, favorable variations spread through the population over generations, and organisms become better adapted to the environment. As the environment changes, organisms change (or evolve), often resulting in the appearance of new species from common ancestors.

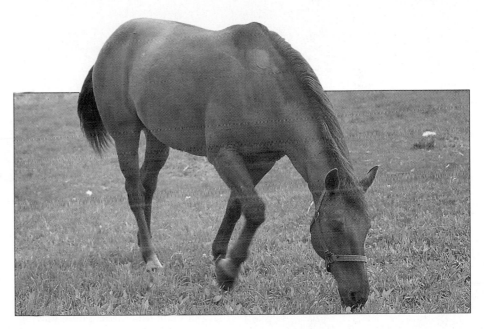

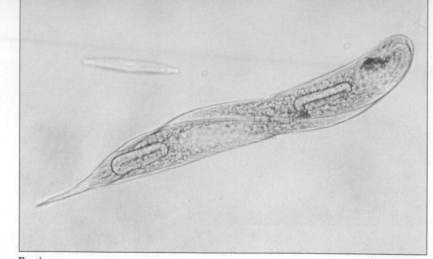
Protista

Monera

Fungi

Plantae

Animalae

## ORGANISMS CAN BE GROUPED INTO FIVE KINGDOMS

The overwhelming diversity of life is represented by the millions of species that populate the Earth. There may be 30 to 50 million species now alive, and millions more, now extinct, have preceded them. Based on common characteristics, organisms are grouped into five broad categories called **kingdoms.** The five kingdoms are Monera, Protista, Plantae, Fungi, and Animalae. Different characteristics are used to define each kingdom. For example, photosynthetic organisms with cell walls made of cellulose are grouped into the Plantae, whereas organisms in the Animalae lack cell walls, eat other organisms, and are capable of moving about.

Within each kingdom, subcategories are based on similarities in anatomy, physiology, development, and behavior. Within a kingdom, the broadest category is a **phylum** (botanists often use the term division instead of phylum). The phyla are divided into **classes.** Classes in turn are divided into **orders,** and orders into **families.** Within each family, are **genera** (sing., genus), and each genus is made up of one or more **species.**

The scientific name of a species is made up of two parts. The first part designates the genus, and the second designates the species within that genus. In this system, the housefly is called *Musca domestica.* Note that the genus is always capitalized, and the name of the species always begins with a lower-

Groups of organisms form populations and communities.

case letter. Scientific names are always underlined or italicized because they are Latin words or derivatives, and foreign words are underlined or italicized when they appear in English.

##  LIFE HAS DISTINCT LEVELS OF ORGANIZATION

The physical and natural worlds are characterized by hierarchical levels of organization. An **organism** like the housefly occupies one level in this hierarchy. Below the level of the organism are **organ systems** (nervous system, reproductive system, etc.), each made up of one or more **organs.** The reproductive system in female flies is made up of several organs: ovaries, spermathecae, several glands, ducts, and the vagina. Organs are composed of **tissues,** groups of cells organized for one or more specific functions. Below the level of **cells,** are the **organelles,** components of cells that carry out individual functions. Organelles are composed of **molecules,** which are clusters of **atoms** held together by chemical bonds.

In the hierarchy above the organism is the **population,** groups of individuals of the same species living in the same area at the same time. All species living in the same area make up a **community,** and the community and its physical environment make up an **ecosystem.** Taken together, all ecosystems compose the **biosphere,** which includes our planet,

its crust, waters, and atmosphere on and in which organisms exist.

## THE NATURE OF SCIENCE AND SCIENTIFIC INQUIRY

To many people, science is an esoteric, abstract and complex activity that seems to have no bearing on everyday life. When asked to define science, students often speak vaguely of some arcane methods that scientists use to pursue their goals, but cannot pin down what science is. As for scientists, popular images have ranged from presenting scientists as high priests or priestesses of technology to the demented individuals portrayed in movies such as *Frankenstein* and *Jurassic Park.*

In studying science (biology included), it is reasonable to ask about its nature. Simply put, science is knowledge. The word science is derived from the Latin word *scientia,* which means "to know." The knowledge gained by scientists in studying the physical world can be distinguished from other forms

**Two views of science.**
**(a) Chemistry as a goddess supporting technology.**
**(b) A mad scientist creating the Frankentomato.**

Scientific inquiry requires that ideas have an internal logic and that they can be tested by experimentation. As they are constantly challenged by new observations, scientific ideas are modified and sharpened. If they fail to meet the test of experimental verification, they are discarded and replaced by new ideas. The power of science lies not in the knowledge it generates, but in the methods by which it generates knowledge of the universe around us.

The structured process of scientific inquiry is undertaken to answer a question or to test an idea so it can be supported or rejected. This testing often takes the form of planned, controlled experiments. The knowledge gained from these tests is organized into conceptual schemes with varying degrees of certainty. These are known as hypotheses, theories, principles, and laws. A **hypothesis** is an idea, one that can be experimentally tested. The most useful hypotheses can be tested directly. One of the keys to science is being able to deduce what other things should be true if a hypothesis is correct. Such deductions form the basis for the design of experiments. The ability to make these deductions and to test them experimentally in a way that provides a clear answer is often the hallmark of a distinguished scientist.

The results of efforts to test hypotheses are reported by scientists at meetings and in papers published in journals or books. The methods used to gather their data are also reported in detail so that the experiments can be repeated or extended by others. A hypothesis that has withstood extensive testing by a variety of methods is referred to as a **theory.** In everyday speech, the word theory is often used to refer to something based on speculation or a hunch. In science, the term has a more restricted meaning. Theories are formed only after a long history of experimentation, and are widely accepted ideas. Scientific theories are useful because they are usually simple, they organize a large number of previously unrelated ideas, and they make definite predictions about future observations. The development and acceptance of theories often mark

of knowledge by the way it is acquired. In science, knowledge is acquired by demonstrating what causes things to happen or what causes them to be the way they are. Science, therefore, is also a way of knowing. As a result, information gained through this structured process has a degree of certainty that sets it apart from other forms of knowledge.

The degree of certainty and the objective nature of the knowledge provided by science are often downplayed and reduced to the status of a narrative produced by cultural, economic, and social factors that has no more validity than other forms of knowledge. Consider for a moment Devil's Rock, a landform rising out of what is now Wyoming. This rock rises some 260 m above the landscape, and is marked by a series of vertical striations. One story about the origin of the rock and its markings is that a group of six brothers and a woman were being pursued by a giant grizzly bear. When the youngest brother sang a song, a rock he carried grew, lifting the party to the top, away from the bear. The marks on the side of the rock were left by the bear unsuccessfully attempting to claw his way to the top. Another version of how the rock originated is that some 45 to 50 million years ago, one or more masses of molten rock were intruded into the Earth's crust, and cooled, forming the tower. The vertical striations are the intersections of columnar joints formed during cooling and contraction of the molten rock. Both versions account for the formation of Devil's Rock, but differ in the degree of certainty that can be ascribed to them.

(a)

(b)

(a) Charles Darwin and (b) Alfred Russel Wallace co-discovered evolution by natural selection. Their landmark discovery helped transform biology.

significant turning points in a science. In biology, the theory of evolution, the cell theory, and the theory of inheritance are landmarks that transformed both the ways of thinking and methods of natural science.

## Science and Social Values

During the last three centuries, biology has changed. It began with observation, description, and classification and has progressed to mechanism, causality, and prediction. Biology, and all science, has also become enmeshed with technology. Advances such as nuclear power, antibiotics, and pesticides represent technological applications of scientific knowledge. Whether technology is used wisely is outside the scope of science and involves decisions of public policy. Technology and the growing predictive power of science advance rapidly, whereas discussion, policy formulation, and legislation often lag far behind. This gap often poses ethical and social dilemmas.

Recombinant DNA technology allows prenatal testing to be used to determine the genetic status of a fetus as early as 8 to 16 weeks of pregnancy. Such testing can determine whether a fetus has a genetic disorder such as Huntington disease, which will become apparent only after age 40. This disorder is associated with progressive neurodegeneration and premature death. There is no treatment for this disease. Science alone is unable to resolve whether such testing should be used, whether insurance companies or prospective employers should have access to the results of such testing, or even whether the affected individual should be informed of his or her condition, and if so, at what age this information should be given.

Science can provide two types of guidance in making decisions about social, ethical, and political issues. First, the facts as currently known can be used as the basis for informed decisions. Second, science can provide a way of thinking about new information as it becomes available. Subjecting new information to careful analysis can separate beliefs from knowledge, relate cause and effect, and provide the basis for making fair and effective public policy decisions. It is the responsibility of all citizens to become informed about scientific knowledge and to participate in decisions using this knowledge to achieve the greatest good for the greatest number.

## ✿ UNIFYING THEMES

We began this chapter with the statement that biology is the study of living organisms. Throughout the following chapters, three fundamental themes will be emphasized. These themes have been selected for several reasons. Taken together, evolution and genetics constitute the organizing theme of biology over the last century.

1. *The process of science.* As outlined earlier, the ideas and methods of science are often perceived as mysterious and incomprehensible. To illustrate that the methods of science are basic tools of understanding, examples woven through the text, the Concepts and Controversies features, and the sidebars show how deductive reasoning is used in biology. For example, if an experiment presents several possible explanations, begin by examining the one that is simplest and will require the fewest number of steps to test.

Theories provide organizing themes in biology. The cell theory states that all organisms are composed of at least one cell. Acceptance of this idea allows work to proceed on understanding how cells grow, reproduce, and interact with other cells and with their environment. The methods of science allow the knowledge gained to have a high degree of certainty and to be predictive.

2. *Genetics and evolution.* The transmission of genes from parent to offspring is the thread that connects life as it passes from generation to generation. The properties of cells and organisms, including their structure, shape, function, and even their life span, are encoded in their genes. Throughout the text, examples of genetic analysis or the discovery of mutations will be used to emphasize that the function of molecules, organelles, cells, tissues, and the organism can be studied using genetics.

The division of organisms into the five kingdoms is based on the evolutionary history of life on our planet. Evolution is one of the organizing themes of biology at all levels. Throughout the book, examples will document how structures are adapted and modified over time to serve new functions, and how natural selection shapes species to adapt to their environments. Evolution works by selection of genetically controlled variations. The close relationship between genetics and evolution is illustrated by the discovery of a master gene for eye formation.

Based on the diversity of eyes found in animals, biologists had concluded that eyes evolved independently as many as 40 different times. Early in 1995, geneticists discovered a master regulator for eye formation in the fruit fly, *Drosophila.* This gene may control the expression of up to 2500 other genes involved in eye formation. The mouse version of this gene can control the formation of eyes in fruit flies. This suggests that the gene's role in controlling eye formation is ancient, and dates to a time when flies and mice shared a common ancestor— half a billion years ago. Although eyes evolved under natural selection into the diverse forms seen today, in all cases, their development remains under control of a single master gene. Evolution through natural selection is the engine that drives the survival of life through changing environmental conditions.

3. *Scientists and science as a human endeavor.* Scientists are frequently portrayed as antisocial loners. Often, their work is thought of as being concerned with esoteric topics unworthy of attention. A U.S. senator used to hold press conferences to award symbolic "golden fleece" prizes to scientists whose research grant titles were held up to ridicule and scorn. Yet research on the seemingly obscure question of how soil bacteria resist viral infections led directly to the development of the multibillion dollar biotechnology industry. Rather than fleecing the government, the research dollars invested in this work have benefitted the economy, provided new methods

**Fluid is withdrawn from the amniotic sac surrounding the fetus and used for prenatal testing.**

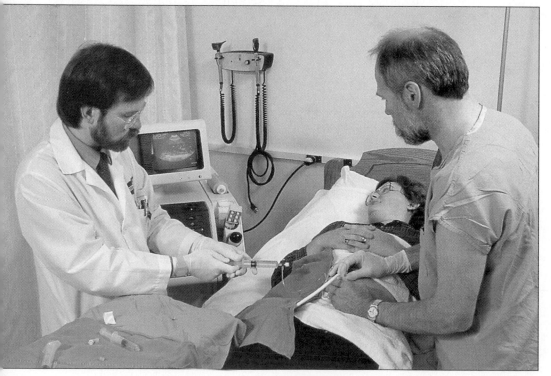

## Frogs and Human Health

MICHAEL ZASLOFF

For the past ten years, I have been involved in a very exciting mission—developing antibiotics. I founded a pharmaceutical company and I still work at the bench with my hands.

Ten years ago, I was working in my lab studying tRNA gene expression in the oocytes of the African clawed frog, very large single cells, and I noticed that the frogs I had sutured were all healing. They had an immune system, but no redness or inflammation. How could this be? I was trained as a pediatrician and knew that a wound on our skin would look angry after such treatment. My best guess to explain this property was that the frog had to be making a powerful antimicrobial agent in its skin, displacing a need for its immune system. To identify such a system, I simply asked whether an antibiotic could be found in the skin.

One day, after many experiments on which I failed to see any antibiotic, it struck me that perhaps there was an antibiotic in the skin, but it was being destroyed by enzymes liberated when I extracted it to do tests. To address this problem, I added a few simple chemicals to the extraction which inhibited such degrading enzymes. Following this attempt, a dramatic amount of antibiotic activity could be demonstrated! This discovery created considerable excitement that continues to this day. We have come to realize that the frog is really not a terribly special creature in this way. Antibiotics, or simple substances that kill microbes, are to be found in abundance in all animals, from insects to man.

The discovery of the frog antibiotics has taken me through a journey to other animals, including man. Recently, I began a search for antibiotics in the shark because this is a very primitive animal with respect to immunity. Yet no shark viral diseases are known. It rarely gets cancer. Why? This led to a search and eventual discovery of a new class of antibiotics present in all tissues of its body. These chemicals, which are steroids, are called squalamines. They exhibit powerful antiviral and anticancer activity, and we believe that they will become very important drugs.

Several principles guide me as a professional scientist, and it's a bit different from what is expected from a student. I work hard at trying to understand what it is we really don't know. I continue to work at a laboratory bench with my colleagues, and very much need to see experiments to understand them. I have a great memory for certain types of data (experiments I've done, down to the finest detail) and I read scientific texts avidly, and generally reread constantly.

Failure is a critical key to scientific pursuits. A scientist must fail, and must regard that failure as did Edison—as simply a body of experience, as information where not to travel. One must develop a strong appreciation that failure is part of the discovery process. Also important is the skill of observation, especially for a biologist. Many of my insights derive from my own direct observations, not through the secondhand reports of another.

For me, the experience of scientific discovery is very much like that of an explorer. I happen upon things that are part of nature, and have the good fortune to see some of these wonders before anyone else does. What makes this experience especially exciting is that I am now in a position to take some of these discoveries and develop them into substances that can be used for the treatment of disease and the maintenance of health.

*MICHAEL ZASLOFF earned his M.D. and Ph.D. at New York University School of Medicine. He studied pediatrics at the Boston Children's Hospital, and joined the National Institutes of Health for postdoctoral studies in molecular genetics, eventually becoming chief of genetics of the National Institutes of Child Health and Human Development. He remained at NIH for 13 years, leaving for the position of professor of pediatrics and genetics at the University of Pennsylvania School of Medicine. He also founded Magainin Pharmaceuticals. At the present time he devotes full time to the discovery and development of antibiotics and anticancer agents as president of the Magainin Research Institute of Magainin Pharmaceuticals.*

of medical diagnosis and treatment, and created a new industry.

Essays placed throughout the book serve to introduce scientists and their work. Biologists are people with a highly developed sense of curiosity, who pursue that curiosity, often in a single-minded fashion. While dispassionate analysis and objectivity are the hallmarks of science, they are not always the hallmarks of scientists. Most scientists are passionate in their approach to science, and this passion—coupled with reason—is what drives science. On the other hand, scientists should not be taken too seriously. They are subject to all the foibles and habits of human nature, and competition, ego, and personalities also play roles in science. The essays allow individual scientists to present their work, their motivations, and the applications of their findings to other areas of our society, including agriculture, medicine, and the environment.

# KEY TERMS

atoms

biosphere

cells

classes

community

ecosystem

families

homeostasis

hypothesis

kingdoms

molecules

orders

organ systems

organelles

organism

organs

phylum

population

species

theory

tissues

# DEVELOPING YOUR ABILITIES TO THINK AND STUDY

## Introduction

The overview presented in Chapter 1 provided a glimpse of what biology and this book are all about. To succeed in college and in this course, you may need to make adjustments in the way you approach class work, papers, and other assignments and studying.

College is a time when you will be challenged by exposure to new ideas and philosophies presented in lectures, in readings, and in conversations with faculty and fellow students. These ideas need to be analyzed, evaluated, and critiqued before you accept all or parts of them. Reading assignments, tests, and term papers, along with work and recreation, place demands on your schedule. Learning to manage your time is one of the most important lessons you can learn in college. Acquiring time management and study skills early in your college career will help lighten the load of assignments, improve your knowledge and understanding, and increase your chances of getting good grades.

This section offers suggestions for managing your time, improving your study skills, and learning to think analytically. Like most things, acquiring new skills is often slow and frustrating at first. Once mastered, these skills will pay dividends not just in college, but throughout your life. Begin by reading the following sections. Some of the suggestions may be appropriate, and yet others may already be part of your daily life. Pick some of the ones that apply to you and put them into practice. The best students are usually busy with course work, extracurricular activities, and outside jobs. Part of their success is based on their ability to manage their time and to study efficiently. If you take the time to learn and apply these skills, you will have more time for yourself, get better grades, and find college a more rewarding experience.

One of the first keys to success is to set clear, specific goals and work toward them on a regular basis. In many classes, some of the brightest students do not get the highest grades because they are disorganized, they procrastinate, and they substitute last-minute "all nighters" to try to learn everything they have let go all semester. It is better to work efficiently for a short period of time on a single subject, and then switch to another subject.

## General Study Skills

At the beginning of a course, especially a general course, some of the material may seem familiar, and it is tempting to think of it as a boring rehash of elementary facts. Science courses usually build on the material presented in the first lectures, and it is important to attend lectures and keep up with the readings.

- As trite as it may seem, develop the habit of studying on a daily basis.
- Set aside a regular time each day for study and stick to this schedule. Some people are more alert in the mornings, others at night. Figure out what part of the day is best for you and use that time of day for work that requires concentration.
- Let others know your study schedule and ask them to respect that time.
- Study in a quiet, well-lit space. Use a desk or table, not a bed or couch. Have all necessary supplies on hand. Don't get up to leave your study area to track down a pencil. Nothing breaks up a study session faster than simple distractions.
- Study for short periods. Take a break about once every hour. Exercise or move around during your break; it will help you stay alert.
- Study each subject each day, or on the day of the class. Review your lecture notes from a class as soon as you can, and fill in any gaps.
- Take the time to look up new terms and words. Become familiar with the vocabulary of the course, using textbook glossaries or a dictionary. Learning the language of the course is essential to learning the material. It has been said that students in a full-year biology course learn as many new words as students in a foreign language course.

## Taking Notes

Getting the most from a course means learning to take good notes. This does not mean you should be a stenographer, copying down every word spoken by the instructor. One of the secrets to note taking is learning what to write down and what to leave out.

From the beginning, try to determine whether the lecture closely follows the textbook. Reading the textbook on a regular basis is one way to do this. If much of the lecture material is in the textbook, then your notes do not have to be as detailed or extensive as when the lecture will be the only source for the subject. The following suggestions will help you be a better note taker:

- Look over the reading assignment to be covered in the lecture beforehand. This will give you some idea of how much material is going to be covered, and will make the concepts somewhat familiar to you before lecture.
- Review your notes before lecture to refresh your memory about the past material and to establish a context for your new notes.
- Develop a style of note taking that incorporates your own set of abbreviations and symbols: w/o (without), w/ (with), = (equals), ↑ (above or increases), ↓ (below or decreases), < (less than), > (greater than), & (and), u (you), etc.
- Biology terms lend themselves to abbreviations that will help speed up your note taking. In sections of the course dealing with human physiology, human genetics, etc., abbreviate human to H. Muscles can be abbreviated to ms. or msl. Instead of writing "million years ago," write "mya."
- Be alert for verbal signals from your instructor. If your professor says: "This is an important point, . . ." take it seriously and write it down. Some professors have a habit of repeating important points twice.
- Learn how the instructor paces the lecture, and be aware when a new subject is being discussed. Some lecturers put an outline of the lecture on the board or screen. Copy it down. Most instructors *are* using an outline. The challenge is to transfer the outline from their notes at the lectern to your notes even if you never get to see the outline.
- Attend all classes and sit near the front to avoid distractions and to make it easier to see and hear.
- Check on any unclear points in your notes with a classmate or look them up in the textbook, and write them into the notes. Jot down the page numbers in the text that cover this material, or that give good examples you want to consult at a later time.
- Some people find it helpful to tape record the lectures, but often, this becomes a substitute for paying attention in class. If you want to tape record, it is a matter of courtesy (and, in some cases, a matter of law) to ask permission.
- If available, buy class lecture notes. In many cases, these are taken by a graduate student who is familiar with the subject. Use these notes as a supplement, not as a substitute for your own notes.
- Ask questions. Too many students are reluctant to ask the instructor to clarify a point, especially in a large lecture hall. If you are unsure of something, chances are good that other people in the lecture hall are also confused. If questions are not permitted during lecture, talk to the professor after the lecture or during office hours. If your course has a discussion section with a teaching assistant, attend the discussion, and ask questions of the assistant. Don't wait until the week before the exam to visit your teaching assistant or professor during office hours to ask questions or clarify points.

## Improving Your Reading Comprehension

To get the most out of a course, read and study the textbook. In combination with the notes, it will greatly increase your understanding of the subject and reinforce what you have learned in lecture.

- Before you begin a chapter, read the chapter outline to get an idea of what material is covered and how it is organized. If time is available, skim through the chapter to get a taste of the topics.
- Pay attention to the headings. They encapsulate what the section is all about. Give attention to the figures, tables, and graphs. Many concepts in biology are visual, and the figures will help you visualize what is being discussed. Read the examples and try to apply the concepts to other examples of your own.
- Make notes in the margin or on a separate sheet of paper. If you highlight key concepts or terms, do so sparingly. Too many students highlight almost every line on a page, making it an ineffective study tool. Don't skip terms that you don't understand. Take the time to look them up in the glossary or dictionary. Work to improve your vocabulary in the subject.
- Use whatever study aids are in the text. Read the chapter summary and be sure you understand the concepts and the vocabulary. If there are end-of-chapter questions, use them to test your understanding of the material. Don't say "Yeah, yeah, I know all that." Take the time to write out the answers as if you were taking an examination. Look up answers to problems that confuse you.

## Analytical Thinking and Healthy Skepticism

One of the most important skills you can acquire in college is the ability to think critically and not accept everything you hear or read at face value. This is particularly important when learning new material and relating it to what you already know. You must learn to question effectively and arrive at conclusions that are consistent with the facts.

- Have a clear understanding of all terms. In discussions, insist on having all the terms defined. If someone is unable to provide a clear, concise definition of terms, the idea they are advocating is probably not clear either.
- Determine how the facts were derived. Were they derived from rigorous, thorough experimentation using a number of different methods, free from bias? Have the experiments been repeated by others? The recent controversy over cold fusion is an example. Two scientists, using simple laboratory equipment, claimed to have achieved atomic fusion on a benchtop. No one else has been able to repeat these experiments with the same results.
- Do not accept statements at face value. What or who is the source of the information? How reliable is the source? Beware of self-proclaimed experts who have a hidden agenda or a vested interest in their statements. Scientists working for tobacco companies may not be the best source for information on the medical effects of smoking.
- Carefully consider whether the conclusions are supported by the facts at hand. If the facts do not appear to support the conclusions, start asking questions. Is the argument flawed, or are there other facts available that can resolve the discrepancy?
- Learn to distinguish between correlation and causation. Just because two things appear to be related does not mean that one causes the other. There is almost a perfect correlation between which National Football League division wins the Super Bowl and the course of the stock market in that same year. If a team from the old National Football League (before merger with the American Football League) wins the Super Bowl, the stock market is almost always higher at year end (about 90% of the time). If a team from the old American Football League wins, the stock market is lower. Does this mean that one causes the other? Obviously not, but in other cases, it can be more difficult to distinguish between causality and correlation.
- Be open to new ideas. Not everything is black or white. Ambiguity exists and you must be able to examine an issue from all sides and still come to a logical conclusion. Some scientists think that the atmospheric temperature is warming as a result of the release of carbon dioxide and other gases. Others argue that the atmosphere warms and cools slightly anyway. Still others are not sure, and do not want to draw conclusions at this time.
- Look at the big picture. Look for multiple causes and effects, hidden effects, or previously unknown relationships. For example, if global warming does occur, what are the consequences for ocean levels, species distributions, crop yields, population growth, and economic development? With massive changes like global warming, it is easier to look at large-scale consequences, but even small events can have multiple outcomes or cause cascades of other events.

## Preparing for Examinations

Examinations are a critical part of any course. Preparation is the key to success of all examinations. Getting ready for exams should start on the first day of the course. Several guidelines will help you focus on preparation.

- Don't fall behind. Review your lecture notes as often as possible. Keep up with the reading assignments and study regularly. Cramming at the last minute rarely works.
- Review the chapter summaries in the textbook. If you have time, outline your notes and/or the assigned readings. Outlines can help you understand how things fit together, help identify the major concepts, and understand how the supporting evidence builds the case for an idea. Draw diagrams or figures illustrating the key points.
- Form study groups. This is one of the most effective ways of learning material. Make sure the group is serious about the subject at hand and not interested in a social hour. If you can explain a concept to someone else in a clear and concise way, you probably understand it. Quiz each other, go over possible questions, and review the material.
- If review sessions are offered, be sure to attend. Study before the review session, and be an active participant. Ask questions and go over the major concepts and issues.
- If available, look over old tests to see how the test is organized, what is emphasized, and what types of questions are asked.

## Taking Examinations

The first rule of test taking is simple: don't panic. Test anxiety is a universal phenomenon, and almost all students are affected by it to some extent. The following suggestions can help reduce stress and allow you to concentrate on the matter at hand.

- Eat well and get plenty of rest before the examination. Arrive at the examination early or on time.
- When you get the test, look it over to determine the format, length, and point distribution. This will allow you to budget your time in the most efficient manner.
- Begin by answering questions you are sure you know. Don't spend too much time on any one question or on questions that are worth only a few points.
- If the exam is a combination of objective questions such as multiple choice and essay questions, answer the multiple-choice questions first. If the question requires a written answer, read the question and make sure you understand it. Organize your ideas either mentally or by notes on the back of the exam. Answer only what is asked—get right to the point. Save time by not repeating the question in the first line of your answer.
- If you don't understand the question, ask the examiner. Above all, do not make assumptions about what is wanted. It is your grade that will suffer if you misinterpret the question.

## In Conclusion

Take a few minutes now to go back over the lists given here and mark off the items that you think are most important to making improvements in your study habits. Work on incorporating them into your daily routine. Old habits may have to be broken in order to form new ones, but persistence pays. A regular schedule with new habits can quickly become a familiar routine. Keep track of your progress, and take corrective action when necessary.

Following these suggestions will make you a better student, and will help you later in life. Learning does not stop at the graduation ceremony when you are awarded your diploma. The skills you acquire in college will help you to become an informed citizen.

# I

# ORGANIZATION AND
# FUNCTION OF CELLS

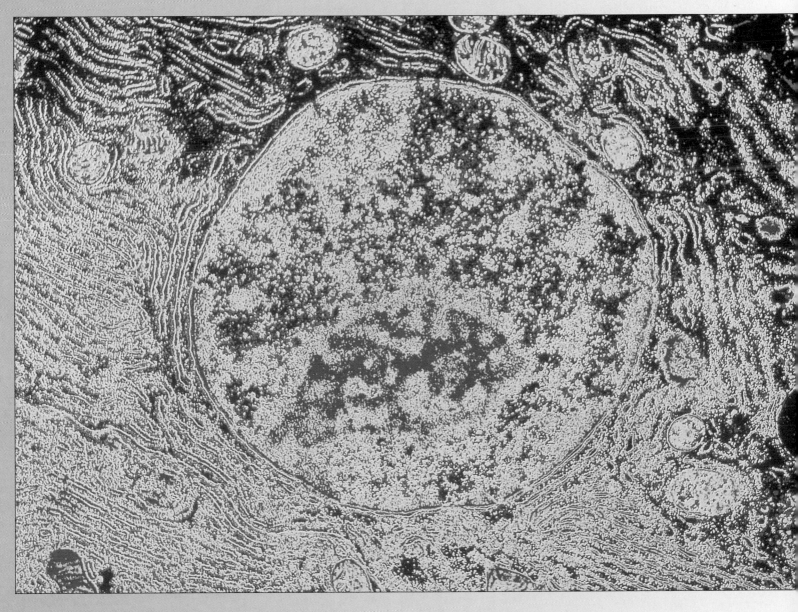

A COLORIZED MICROGRAPH OF A MAMMALIAN CELL, TAKEN
WITH A TRANSMISSION ELECTRON MICROSCOPE, SHOWING
DETAILS OF THE INTERNAL CELLULAR STRUCTURE.

# 2

# THE ORIGIN AND CHEMISTRY OF LIFE

O P E N I N G   I M A G E

*Photosynthetic cyanobacteria form mats known as stromatolites. These stromatolites, photographed in Western Australia, were a common part of the landscape some 3 billion years ago.*

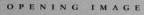

*he most widely held theory on the origin of the universe is called the **big bang theory.** According to this idea, at one time, about 10 to 20 billion years ago, all the matter and energy of the universe was concentrated in an infinitely dense, infinitely hot point somewhere in space. This point of matter and energy suddenly exploded in a big bang. In the events that followed, time is measured in fractions of a second. In the first fraction of a second (about $10^{-43}$ seconds, a decimal point followed by 42 zeros and a 1) the universe consisted only of energy.*

*As the universe expanded in the next fraction of a second, its temperature decreased enough to allow the formation of particles from energy. These particles were the precursors of atoms and their components. As the universe cooled during the next half million years, atoms of hydrogen and helium formed, and over time, heavier elements like carbon and oxygen and iron formed. These elements were not distributed randomly;*

◆

**✦ FIGURE 2.1**

After the Big Bang, matter condensed in the universe to form the stars seen in this galaxy.

**✦ FIGURE 2.2**

For the first billion or so years after Earth's formation, widespread volcanic activity and massive lightning storms occurred. The atmosphere was composed of poisonous gases and little surface water.

*instead they formed clusters that condensed into large galaxies (✦ Figure 2.1). The galaxy we inhabit, the Milky Way, was created by condensation of elements a few billion years after the big bang.*

*Our solar system slowly condensed in an arm of the Milky Way some 7 to 10 billion years ago. The sun formed first, from a condensed mass of hydrogen, helium, and other elements. As its mass increased, its gravitational field compressed the elements, generating heat and light. The gravity field of the sun attracted matter, which formed a disk of material, slowly rotating around the sun. Gradually, matter in the disk condensed into the planets. Planet Earth formed about 5 billion years ago as a cold rock with no atmosphere. About 4 or 4.5 billion years ago, volcanic activity within the earth produced the first atmosphere, which included water vapor, carbon dioxide, nitrogen, hydrogen sulfide, and hydrogen. Little or no oxygen was present in this early atmosphere.*

*As the earth cooled, the water vapor in the atmosphere condensed, and rains began. These rains lasted tens of thousands of years and gradually filled the oceans (✦ Figure 2.2). The rains also dissolved carbonates and other salts present in the rocks, making the ocean salty and slightly acidic. Sometime about 4 billion years ago, life arose on earth.*

*All that you see around you, all that is living and nonliving, from other humans, to plants, animals, buildings, rocks, all this and the air, is composed of atoms and elements created during the formation of the universe. In this chapter we explore atoms, elements, molecules, and some basic chemistry related to living systems. We also emphasize the importance of water, acids, and bases to life, and describe the chemistry of the major organic molecules found in living systems.*

*Finally, we consider some of the ideas about how life originated on earth, and the origins of prokaryotic and eukaryotic cells.*

---

## MATTER, ELEMENTS, AND ATOMS

As the energy of the big bang dispersed, what formed in its wake is generally called **matter.** Matter is anything that occupies space and has mass (➤ Table 2.1). Because everything we can observe including rocks, water, plants, and animals is composed of matter, understanding the nature of matter has long been a fundamental question in science and philosophy. An ancient Greek philosopher, Democritus, proposed that the basic building block of all matter was the **atom,** the smallest indivisible particle that can have a separate existence. Others, including Aristotle, proposed that matter in the physical and biological world was composed of four substances: earth, fire, water, and air, mixed in different combinations. The views of Aristotle prevailed among European philosophers and scientists for more than a thousand years. However, as investigators began to observe, measure, and experiment with matter, it became clear that matter was com-

## TABLE 2.1 STATES OF MATTER

| PHASE | CHARACTERISTICS | EXAMPLES |
|---|---|---|
| Solid | Rigid substance that retains its shape unless distorted by a force | Minerals, rocks, iron, wood |
| Liquid | Flows easily and conforms to the shape of the containing vessel; has a well-defined upper surface and greater density than a gas | Water, lava, wine, blood, gasoline |
| Gas | Flows easily and expands to fill all parts of a containing vessel; lacks a well-defined upper surface; is compressible | Helium, nitrogen, air, water vapor |

## TABLE 2.2 PROPORTIONS OF ELEMENTS IN THE EARTH'S CRUST AND IN THE HUMAN BODY

| EARTH'S CRUST | % | HUMAN BODY | % |
|---|---|---|---|
| Oxygen | 46.6 | Oxygen | 65 |
| Silicon | 27.7 | Carbon | 18 |
| Aluminum | 8.1 | Hydrogen | 10 |
| Iron | 5.0 | Nitrogen | 3 |
| Calcium | 3.6 | Calcium | 2 |
| Sodium | 2.8 | Phosphorus | 1.1 |
| Potassium | 2.6 | Potassium | 0.35 |
| Magnesium | 2.1 | Sulfur | 0.25 |
| | | Sodium | 0.15 |
| | | Magnesium | 0.05 |
| | | Iron | 0.004 |

posed of **elements,** each of which has a unique set of properties, and that the elements, in turn, were composed of smaller components, or atoms, as originally proposed by Democritus.

## ✦ FIGURE 2.3

**Atoms are composed of a nucleus, containing protons and neutrons, with a surrounding cloud of electrons.**

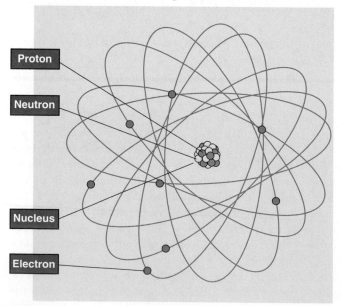

Proton

Neutron

Nucleus

Electron

## Chemical Elements Have Unique Properties

More than 100 different elements are recognized by chemists; each element is a substance that cannot be broken down in ordinary chemical reactions. Many elements such as oxygen, hydrogen, carbon, and aluminum are rather familiar to most of us. On Earth, eight elements account for about 98% of all matter (Table 2.2). The elemental composition of the human body is somewhat different than that of the Earth, illustrating the chemical requirements that are essential for living systems ( ➠ Table 2.2).

Each element is composed of atoms, the smallest component of matter. The organization of atoms confers a set of chemical and physical properties on each element. Atoms consist of three kinds of particles, called **subatomic particles: protons, neutrons,** and **electrons** ( ✦ Figure 2.3). Each atom contains a core or **nucleus,** which, in turn, contains protons. Nuclei (except for hydrogen) also contain one or more uncharged neutrons. Negatively charged electrons orbit the nucleus at a speed close to the speed of light. Because all atoms are electrically neutral, the number of electrons is equal to the number of protons in the nucleus.

The chemical and physical properties of an element depend on the protons, neutrons, and electrons in its atoms. The protons and neutrons remain closely packed in the nucleus. The less energetic electrons orbit close to the nucleus, whereas the more energetic electrons circle the nucleus at greater distances. The higher energy electrons in the outer orbits are most able to interact with other atoms ( ✦ Figure 2.4). Elements can be characterized by their proton and electron content and then arranged in a chart or table called the **periodic table of the elements.**

All atoms of an element contain the same number of protons. This number is called the **atomic number.** The periodic table arranges elements in order of increasing atomic number, using an abbreviation to symbolize each element ( ➠ Table 2.3).

| I | II | III | IV | V | VI | VII | VIII |
|---|---|---|---|---|---|---|---|
| 1<br>H<br>Hydrogen<br>1.0 | Atomic number<br>Atomic symbol<br>Mass number<br>(Atomic weight) | | | | | | 2<br>He<br>Helium<br>4.0 |
| 3<br>Li<br>Lithium<br>7.0 | 4<br>Be<br>Beryllium<br>9.0 | 5<br>B<br>Boron<br>11.0 | 6<br>C<br>Carbon<br>12.0 | 7<br>N<br>Nitrogen<br>14.0 | 8<br>O<br>Oxygen<br>16.0 | 9<br>F<br>Fluorine<br>19.0 | 10<br>Ne<br>Neon<br>20.2 |
| 11<br>Na<br>Sodium<br>23.0 | 12<br>Mg<br>Magnesium<br>24.3 | 13<br>Al<br>Aluminum<br>27.0 | 14<br>Si<br>Silicon<br>28.1 | 15<br>P<br>Phosphorus<br>31.0 | 16<br>S<br>Sulfur<br>32.1 | 17<br>Cl<br>Chlorine<br>35.5 | 18<br>Ar<br>Argon<br>40.0 |
| 19<br>K<br>Potassium<br>39.1 | 30<br>Ca<br>Calcium<br>40.1 | | | | | | |

*Note that each element is represented by a one- or two-letter symbol. Elements are listed according to their atomic number (the number of protons). Their atomic mass is also shown.*

The table also lists the **atomic weight** of each element, which is approximately equal to the sum of the weights of its protons and neutrons. Thus, hydrogen, which has only one proton, has an atomic weight of 1; helium, with two protons and two neutrons, has an atomic weight of 4. Atomic weight is expressed in units of atomic mass. One unit of atomic mass is equal to $\frac{1}{12}$ of the weight of a carbon atom, and protons and neutrons each have an atomic mass of 1.

## Isotopes Are Different Physical Forms of Elements

Although the number of protons remains constant for all the atoms of a given element, the number of neutrons can change. The nuclei of all carbon atoms contain six protons, and most contain six neutrons. Some carbon atoms contain seven neutrons, and still others can have eight neutrons in the nucleus. Atoms with the same atomic number (that is, the same number of protons; six in the case of carbon) but different numbers of neutrons are called **isotopes.** To indicate an isotope, the mass number is added to the name of the element, for example, carbon-12 or oxygen-18, which can also be written as $^{12}C$ or $^{18}O$. ✦ Figure 2.5 shows the nuclear structure and electron shell for hydrogen. The isotope $^1H$ has one proton and no neutrons in its nucleus, $^2H$ has one proton and one neutron, and $^3H$ has one proton and two neutrons.

Usually one isotope is more plentiful than any of the others. For hydrogen, 99.985% of the element is the $^1H$ isotope, with the remaining amount di-

✦ **FIGURE 2.4**

**When the outer shell of an atom can accommodate more electrons (shown by the dotted circles), the atom is reactive, and can donate, receive, or share electrons in chemical reactions.**

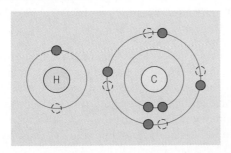

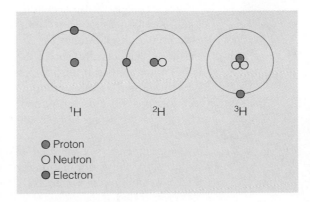

● Proton
○ Neutron
● Electron

✦ **FIGURE 2.5**
**Isotopes are different forms of the same element. The isotopes of hydrogen differ in having zero, one, or two neutrons in the atomic nucleus.**

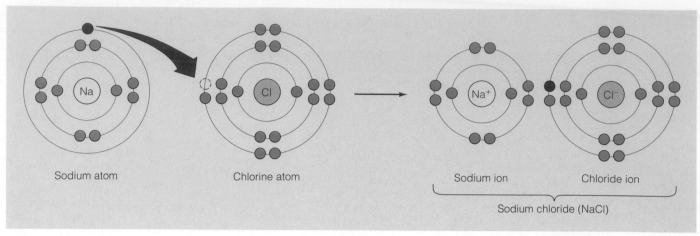

Sodium atom          Chlorine atom                    Sodium ion          Chloride ion

Sodium chloride (NaCl)

◆ FIGURE 2.6

**Sodium tends to give up an electron to atoms like chlorine, making sodium positively charged and chlorine negatively charged. The attraction between opposite charges creates an ionic bond, forming sodium chloride.**

vided between $^2$H and $^3$H. The addition of neutrons makes some nuclei unstable, and these isotopes emit energy from the nucleus in the form of **radiation** as they convert to a more stable condition. Radioactive isotopes have many uses in medicine and science, and we will discuss some of them in later chapters.

Occasionally, electrons in the outer shell of atoms are so energetic they escape from orbit, and are lost from the atom. This process of electron loss is known as **oxidation** (because the lost electron is often captured by oxygen). As we will learn in Chapter 4, electron loss is often accompanied by the transfer of a proton. Since a proton and an electron constitute a hydrogen atom, oxidation often means losing a hydrogen atom. Correspondingly, the gain of an electron or a hydrogen atom is called **reduction.** The processes of oxidation and reduction are essential parts of many chemical reactions in living systems.

◆ FIGURE 2.7

**In a crystal of sodium chloride, each sodium ion is surrounded by six chloride ions and each chloride ion is surrounded by six sodium ions.**

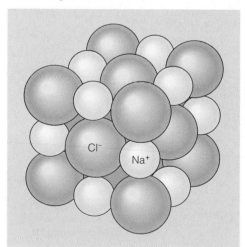

## ELEMENTS FORM COMPOUNDS AND MOLECULES

In nature, atoms are combined with other atoms, sometimes of the same kind, sometimes with atoms of other elements. Natural materials such as bone, wood, or stone are combinations of different kinds of atoms. Elements are pure substances composed of atoms that all have the same atomic number. Combinations of different atoms are either molecules or compounds.

When elements are combined in a fixed ratio, the combination is called a **compound.** Water is a

compound composed of 88.8% oxygen and 11.2% hydrogen. Other compounds include sugar (sucrose) and salt (sodium chloride). Many compounds are composed of **molecules,** which are separate units of two or more atoms chemically bonded together. Molecules are represented by formulas that indicate how many of each type of atom is present. A water molecule is composed of two hydrogen atoms and one oxygen atom, and the formula is written as $H_2O$:

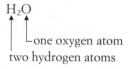

$H_2O$

one oxygen atom
two hydrogen atoms

Symbols written without subscripts represent a single atom. Many molecules found in living systems are large and complex, and have more complex formulas. A molecule of glucose contains 20 atoms and is written as:

$$C_6H_{12}O_2$$

Other compounds are nonmolecular, and contain atoms or **ions** arranged in a network lattice or pattern that extends throughout the material. Ions are formed when atoms gain or lose electrons from their outer shell (◆ Figure 2.6). Positively charged ions have more protons in their nucleus than electrons in orbit, whereas negatively charged ions contain more electrons than protons. Ions are symbolized by superscript positive or negative signs:

$Na^+$ indicates a sodium ion
$Cl^-$ indicates a chloride ion

Sodium chloride or table salt consists of a lattice of sodium ions alternating with chloride ions, arranged so that each sodium ion is surrounded by six chloride ions, and each chloride ion is surrounded by six sodium ions (◆ Figure 2.7). A salt crystal has equal numbers of sodium and chloride ions, so the compound itself is electrically neutral. The formula for sodium chloride is written as NaCl, but there are no molecules of sodium chloride in the salt crystal. Instead, the crystal is held together by a type of chemical bond called an **ionic bond.**

## Ionic Bonds Are Formed by Charged Atoms

In the formation of the lattice in sodium chloride, electrons are transferred from one atom to another. Sodium atoms give up an electron, and acquire a positive charge, and the chlorine atoms take up an electron and become negatively charged. These atoms of opposite charge attract each other, and the force of this attraction, called **electrostatic attraction,** is known as an ionic bond. Within the salt crystal, the charged atoms are arranged so that attractive (opposite) charges are maximized, and repulsive (identical) charges are minimized. It is the sum of these ionic bonds that holds the salt crystal together. When the salt crystal is dropped into water, the ionic forces holding the sodium and chlorine atoms are destroyed (✦ Figure 2.8). Sodium and chloride ions at the surface of the crystal break away and become surrounded by water molecules. This happens because the attractive forces between water molecules and the sodium and chlorine atoms are stronger than the forces between the sodium and chlorine atoms. Thus, in solution, the sodium and chlorine atoms exist as separate atoms designated as $Na^+$ and $Cl^-$.

## Covalent Bonds Involve Electron Sharing

In molecular compounds, the atomic components are held together by an interaction known as a **covalent bond.** In its simplest form, a covalent bond consists of a pair of electrons shared between two atoms (✦ Figure 2.9). In a hydrogen atom, a single electron orbits the atomic nucleus. When two hydrogen atoms combine to form hydrogen gas ($H_2$), they each share their electron with the other atom (Figure 2.9a).

It is possible for two atoms to form more complex covalent bonds by sharing two or more electrons. For example, oxygen atoms can share two electrons, forming a double covalent bond. In the example shown in Figure 2.9b, two electrons from one oxygen atom are shared with a second oxygen atom. In combining hydrogen and oxygen atoms to form water, each of the hydrogen atoms shares two electrons with oxygen, forming the covalent bonds that hold the molecule together (Figure 2.9c).

Sometimes electrons in a bond are shared in an unequal fashion, forming a **polar covalent bond.** In a water molecule, oxygen forms a covalent bond with each of two hydrogen atoms. The electrons in these bonds spend more time orbiting the oxygen atom than the hydrogen atom. The result is a slight negative charge on the oxygen atom, and a slight positive charge on the hydrogen atom. This electron distribution makes the water molecule a polar molecule, with a more negatively charged region and a more positively charged region (✦ Figure 2.10). Polar molecules have important chemical properties, as discussed in the next section.

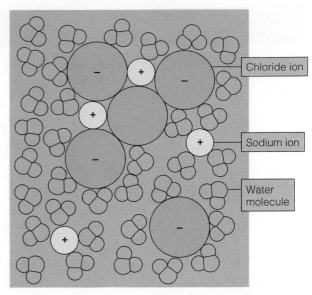

Chloride ion

Sodium ion

Water molecule

✦ **FIGURE 2.8**
**When sodium chloride is placed in water, molecules of water attract ions from the salt crystal, and the crystal dissolves.**

## Hydrogen Bonds Are Weak Interactions

A third type of atomic interaction involves a weak attraction known as a **hydrogen bond.** In living systems, hydrogen bonds make a significant contribution to the three-dimensional structure and, therefore, the functional capacity of biological molecules, so it is important to know something about this form of chemical bonding. These bonds are weak interactions between two atoms (one of which is hydrogen) carrying partial but opposite electrical charges. They are usually represented in chemical formulas by dotted or dashed lines connecting two atoms (✦ Figure 2.11).

Although individual hydrogen bonds are weak and easily broken, they function to hold molecules together by force of sheer numbers (✦ Figure 2.12). As we will see in Chapter 8, hydrogen bonds hold the two strands of DNA together, and are responsible for the three-dimensional shape of protein molecules.

## CHEMICAL REACTIONS TRANSFER ENERGY AND REARRANGE ATOMS

Living systems require energy to sustain life, and this energy is obtained directly or indirectly from cosmic energy present in the universe. Energy can be acquired from atoms and molecules, or from solar radiation through photosynthesis. Energy trans-

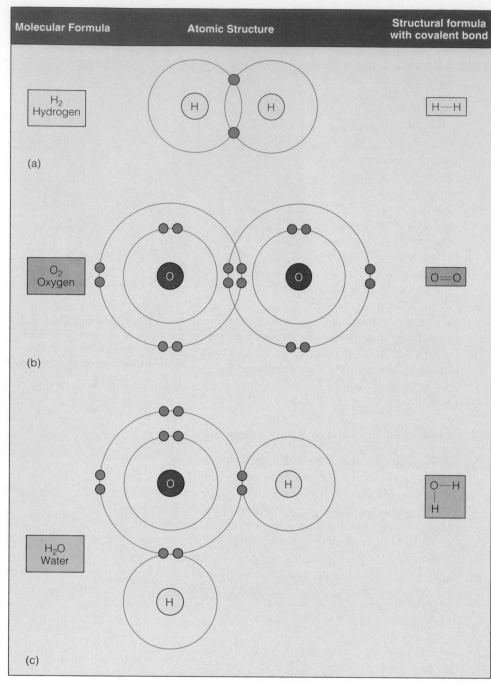

| Molecular Formula | Atomic Structure | Structural formula with covalent bond |
|---|---|---|
| $H_2$ Hydrogen | | H—H |
| (a) | | |
| $O_2$ Oxygen | | O=O |
| (b) | | |
| $H_2O$ Water | | O—H $\;\;\;\;\;$ H |
| (c) | | |

◆ FIGURE 2.9

Covalent bonds share electrons more or less equally, resulting in a strong chemical bond.

fers in living systems involve chemical reactions that transfer energy in controlled steps so that it can be effectively utilized to drive cellular functions.

## Chemical Bonds Store Energy

All atoms and molecules are in constant motion. When an atom is subjected to an increase in energy, the electrons in the atom absorb some of this energy and move to a higher energy level. The electron can then transfer this energy and move to a lower energy level. Occasionally, so much energy is transferred that the electron escapes from the atom and is itself transferred. As we will see in Chapter 4, many cellular functions such as photosynthesis and cellular respiration depend on such energy transfers.

The three types of chemical bonds we just discussed represent a form of energy. The amount of energy in each type of bond is different and can be measured by calculating how much energy it takes to break the bonds. Energy in chemical bonds is measured in **kilocalories.** A kilocalorie is the energy needed to heat 1000 grams of water from 14.5 to 15.5°C. Covalent bonds typically contain 80 to 100 kilocalories,

◆ FIGURE 2.11

**The attraction between differently charged regions of polar molecules creates a force called hydrogen bonds.**

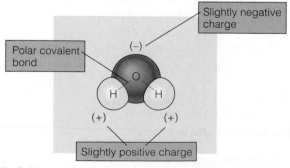

◆ FIGURE 2.10

**In a water molecule, oxygen atoms tend to pull electrons from the attached hydrogen atoms, making one end of the molecule slightly negatively charged, and the other end slightly positively charged. This creates a polar molecule.**

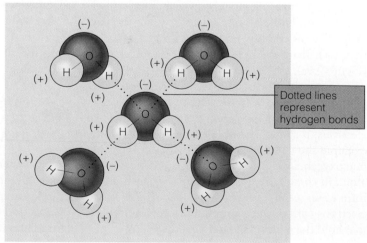

ionic bonds about 5 to 6 kilocalories, and hydrogen bonds less than 5 kilocalories.

Cells can store energy by forming covalent bonds, and can transfer energy by breaking covalent bonds. In cells, the sum of all chemical reactions is called **metabolism.** The reactions in which new chemical bonds are formed and new molecules are made are called **anabolic reactions.** Reactions in which existing chemical bonds are broken and molecules are broken down are called **catabolic reactions.**

## Chemical Reactions Require Several Conditions

Consider two molecules moving toward each other on a collision course. When they collide, will they undergo a chemical reaction or simply bounce off each other like two billiard balls? The answer to this question depends on several factors. If certain conditions are met, a reaction will take place. These conditions are as follows:

1. The reacting molecules must collide or interact with one another.
2. A minimum amount of energy must be present in order for the interaction to result in a chemical reaction.
3. Often the reactants must be oriented in a certain way for the reaction to occur.

The first requirement seems obvious enough. To react, the two molecules must be physically brought together. If enough energy is present, the two molecules will react. The minimum amount of energy that must be present for a reaction to occur is called the **energy of activation** (✦ Figure 2.13). This energy is necessary to disrupt the existing chemical bonds, and because of differences in bonds strengths, this energy varies from reaction to reaction. In other words, some chemical reactions occur

✦ **FIGURE 2.12**

**Some insects are able to take advantage of the hydrogen bonds that hold water molecules together and walk across the surface.**

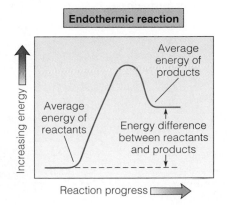

easily; others require large inputs of energy. The energy of activation is necessary in order to increase the kinetic energy of the reactants to a level that allows existing chemical bonds to break and new ones to form.

In a chamber filled with hydrogen gas and oxygen, this activation energy can be supplied by a simple spark. The resulting explosion breaks the chemical bonds between hydrogen atoms and between oxygen atoms and allows the formation of new bonds to produce water:

$$2H_2 + O_2 \rightarrow 2H_2O$$

Chemical reactions are usually written in this form, called a chemical equation, with reactants given on the left and products on the right. In this equation, two molecules of hydrogen react with one molecule of oxygen to form two molecules of water. Note that the total number of hydrogen and oxygen atoms on both sides of the arrow is the same.

An energy diagram for this reaction indicates that the product, water, is at a lower energy state than the reactants (✦ Figure 2.14a). In other words, the reaction gives off energy to its surroundings; such reactions are known as **exothermic** reactions. In other reactions (Figure 2.14b), the products have a higher energy level than the reactants; this type of reaction is called an **endothermic** reaction.

## Reactions Can Reach an Equilibrium

Cells release carbon dioxide as a by-product of metabolism. In organisms with circulatory systems,

✦ **FIGURE 2.13**

**An energy diagram for a typical chemical reaction.**

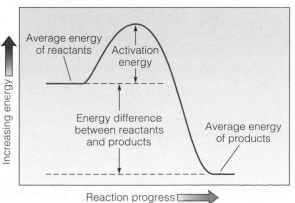

✦ **FIGURE 2.14**

**Energy diagrams for exothermic and endothermic reactions.**

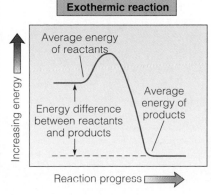

# CHICKEN COOP CHEMISTRY OR PASS THE PERRIER

In hot summer weather, chickens tend to lay eggs with thinner shells. This poses a problem for chicken farmers who sell or transport eggs to market or who want to hatch new chicks to replenish their stock.

As background, two facts about the biology of chickens are important. First, chickens release carbon dioxide into their blood as a product of metabolism. Second, chickens do not sweat. Instead, they cool themselves by panting, much as dogs do. In hot weather, chickens pant to prevent their body temperature from causing metabolic and systemic distress.

Let us first consider the reactions and equilibria established as a result of the release of carbon dioxide into the blood:

$$CO_2 + H_2O \rightleftharpoons H_2CO_3 \rightleftharpoons 2H^+ + CO_3^{2-}$$

The carbon dioxide released into the blood forms carbonic acid, which can reversibly dissociate into hydrogen ions and carbonate ions. Two equilibria are established: one (to the left) between carbonic acid and carbon dioxide and water, and a second (to the right) between carbonic acid and the two ions shown at the right side of the equation.

Next, consider how chickens make eggshells. A major component of chicken shells is calcium carbonate ($CaCO_3$). This is formed by a reaction between the carbonate ion and a calcium ion:

$$CO_3^{2+} + Ca^{2+} \rightarrow CaCO_3$$

In hot weather, when a chicken pants, more carbon dioxide is removed from the body. When this happens, the equilibrium in the reaction with carbonic acid and the carbonate ion shifts to the left, lowering the concentration of carbonate ion in the blood. With less carbonate ion available, the eggshells become thinner.

To correct this problem, at least two remedies are available. First, you could install an air-conditioning system in the chicken coop to ensure that the chickens are comfortable so they will not pant as much. Alternatively, you could provide them with carbonated drinking water. Such water contains dissolved carbon dioxide. This increases the carbon dioxide concentration in the blood, shifting the equilibrium to the right, raises the concentration of the carbonate ion, and makes the eggshells thicker.

---

the carbon dioxide reacts with water to form carbonic acid:

$$CO_2 + H_2O \rightarrow H_2CO_3$$

Suppose we have a situation in which cells begin to release carbon dioxide. Initially, no carbonic acid is present, but after a short time, some of the carbon dioxide present in the blood reacts with water to produce carbonic acid. The reverse reaction can also occur, with carbonic acid breaking down to form carbon dioxide and water. Eventually, the forward reaction and reverse reaction will reach a point where the two reaction rates are equal. From that point on, the concentrations of carbon dioxide, water, and carbonic acid remain constant (not necessarily equal, but unchanging). This situation is known as **chemical equilibrium,** and is represented by double arrows in a chemical equation:

$$CO_2 + H_2O \rightleftharpoons H_2CO_3$$

A number of factors can change the equilibrium condition. For example, if the concentration of molecules on either side of the equation changes, there will be a change in reaction rates until a new equilibrium is established. If, for example, cells release more carbon dioxide, the equilibrium will shift to the right, producing more carbonic acid (see Beyond the Basics: Chicken Coop Chemistry or Pass the Perrier, above). Eventually, the altered reaction rates will produce a new equilibrium. Chemical equilibria play an important role in many biological processes.

## Energy-Producing and Energy-Requiring Reactions Are Linked

Chemical reactions in living systems are usually organized into a series of reactions known as a **metabolic pathway** (✦ Figure 2.15). Individual reactions

✦ **FIGURE 2.15**
Individual chemical reactions in the cell are coupled together to form metabolic pathways. Each reaction is catalyzed by an enzyme. The enzymes in this pathway are hexokinase, phosphoglucose isomerase, and phosphofructokinase.

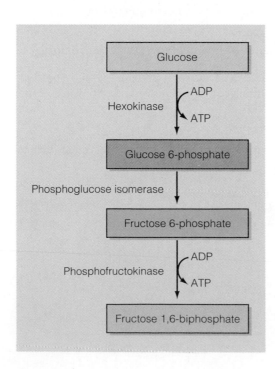

in a pathway may serve no immediate function, but the collection of reactions that make up the pathway has one or more important functions. A series of reactions that involve the breakdown of a compound such as glucose constitutes a catabolic pathway. These pathways usually produce energy, involve oxidation (loss of electrons), and lead to a decrease in order.

Reaction pathways that lead to the synthesis of new compounds are anabolic pathways. They usually require energy, involve reduction (acceptance of electrons), and lead to an increase in atomic order.

Catabolic reactions produce energy that is stored in molecules of **adenosine triphosphate** or **ATP.** Anabolic reactions use the energy stored in ATP to carry out chemical conversions. This linking or coupling of energy-producing and energy-requiring metabolic reactions is a fundamental property of chemical reactions in living systems, and is an important aspect of energy generation and energy flow in all cells.

## WATER, pH, AND LIFE

Water is a molecule made up of two hydrogen atoms joined to an oxygen atom by covalent chemical bonds. As we have seen, the polar nature of water molecules allows hydrogen bonds to form between adjacent water molecules (Figure 2.11). This combination of polar covalent bonds and the network of hydrogen bonds extending between water molecules gives water several chemical properties that are important to living systems.

### The Properties of Water Are Critical to Living Systems

Water makes up between 60 to 90% percent of all cells. The chemical reactions necessary for life take place in water. Without water, cells either die or become inactive (as in the case of seeds and spores). Although water has many chemical properties that are advantageous to living systems, several of these properties are critical and include the following:

1. *Water is a solvent.* As a polar molecule, water serves as a fluid in which polar and ionic substances can be dissolved. Such fluids are known as **solvents.** When sodium chloride is placed in water, the surface atoms of sodium and chlorine lift off from the crystal as ions and interact with water molecules. The positive sodium ions become surrounded by the partly positive (hydrogen) ends of water molecules, and the negative chloride ions become surrounded by the partly negative (oxygen) ends of other water molecules. Nonpolar molecules do not dissolve in

water, but form interfaces, such as when oil is mixed with water. These boundaries are important in determining the structure and organization of cellular components (discussed in Chapter 3).

2. *Water is cohesive.* **Cohesion** is the force that holds molecules of a substance together. In water, this force is generated by hydrogen bonds between water molecules. Although each hydrogen bond is a weak force, the sum of such forces is large, making water a highly cohesive liquid. The cohesion of water and its **adhesion** (ability to stick to different substances) are responsible for the behavior of water in capillary action. This is how water moves up the small transport tubules in plants, or penetrates through fine particles of soil.

3. *Water resists temperature changes.* The hydrogen bonds that form networks among water molecules contribute to the ability of water to moderate temperature changes. To change the temperature of water, a large amount of heat must be used. Before the water molecules can acquire more kinetic energy and reach a higher temperature, a large amount of energy must be invested to break the hydrogen bonds that hold water molecules together. Because cells are largely water, they gain or lose heat slowly and have a built-in resistance to changes in the environmental temperature.

## pH, ACIDS, BASES, AND BUFFERS

Water molecules can dissociate, forming hydrogen ions and hydroxide ions:

$$H_2O \rightarrow H^+ + OH^-$$
$$\text{hydrogen} \quad \text{hydroxide}$$
$$\text{ion} \quad \text{ion}$$

In a sample of pure water, the number of hydrogen ions is the same as the number of hydroxide ions, and in this situation, the sample is chemically neutral (✦ Figure 2.16). This state of neutrality can be changed by adding hydrogen ions or hydroxide ions in the form of acids or bases. **Acids** are substances that add to the number of hydrogen ions in a solution. If hydrochloric acid (HCl) is added to a sample of pure water, it dissociates into hydrogen ions ($H^+$) and chloride ions ($Cl^-$), increasing the hydrogen ion concentration and making the water

*Sidebar*

**PET SCANS**

Radioactive isotopes are often used in biological research and in medical diagnosis and treatment. Isotopes that emit positively charged electrons called positrons are used to study brain function. In use, positron-emitting isotopes are detected by computer-controlled imaging instruments, and the method is called positron emission tomography or PET scanning. To study brain activity, a positron-emitting isotope is coupled to glucose, a molecule used by the brain as an energy source. Active areas of the brain use more glucose and emit more positrons. A computer coupled to the isotope detector produces colored images, with darker colors signaling more activity. PET scans are used to identify areas of the brain that are affected by seizure disorders, strokes, tumors, and to study brain activity in mental disorders such as manic depression and schizophrenia. A PET scan of a normal brain is shown here.

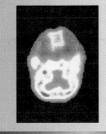

Neutral solution    Acidic solution    Basic solution

✦ **FIGURE 2.16**

**In a neutral solution, the number of positively charged ions is balanced by the number of negatively charged ions. In acids, there is an excess of hydrogen ions. Bases have a shortage of hydrogen ions.**

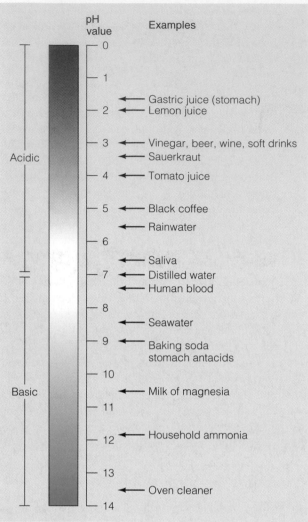

ion concentration of solutions are known as **bases.**

The hydrogen ion concentration of a solution can be measured by using the pH scale. On this scale, neutral substances have a pH of 7. Acids have an excess of hydrogen oxygen ions and, therefore, a pH of less than 7. Bases have fewer hydrogen ions than neutral substances (and have an excess of hydroxide ions) and, therefore, have a pH higher than 7. The pH scale ranges from 0 to 14, and is an exponential scale. When the pH number changes by one unit, the hydrogen ion concentration undergoes a tenfold change. This means that a solution of pH 4 has 10 times more hydrogen ions than a solution at pH 5. ♦ Figure 2.17 shows the pH scale and the pH values of some common substances.

In cells, most chemical reactions take place at pH values between 6 and 8. In humans, the pH value of blood and other body fluids (with the exception of stomach secretions) is about 7.4. Many chemical reactions in cells generate hydrogen ions and hydroxide ions. The pH of cells does not fluctuate greatly, however, because chemicals in cells and blood act as **buffers,** absorbing or releasing hydrogen ions to maintain the pH value within narrow limits. In blood, carbonic acid dissociates into hydrogen ions and bicarbonate ions:

$$H_2CO_3 \quad \rightleftharpoons \quad H^+ \quad + \quad HCO_3^-$$

carbonic acid    hydrogen ion    bicarbonate ion

more acidic. If sodium hydroxide (NaOH) is added to a sample of pure water, it dissociates into sodium ions (Na$^+$) and hydroxide ions (OH$^-$). The hydroxide ions react with hydrogen ions to form water, lowering the concentration of hydrogen ions (Figure 2.16). Substances that lower the hydrogen

This relationship is sensitive to changes in hydrogen ion concentration, and it shifts accordingly. If hydrogen ions are added (an action that tends to lower pH values), the reaction shifts to the left, raising the concentration of carbonic acid and lowering the hydrogen ion concentration. If hydrogen ions are lost (an action that tends to raise pH values), the reaction shifts to the right, raising the hydrogen ion concentration and lowering the pH.

Although many chemical reactions generate hydrogen or hydroxide ions, buffers in the cellular and intercellular fluids prevent drastic changes in pH. This is important because most biochemical reactions can take place within a narrow range of pH values, usually around pH 7. A few enzymes can operate at pH values near either end of the pH scale (♦ Table 2.4).

On a larger scale, the necessity of pH control in living systems is dramatically visible by observing the effects of acid rain on lakes, streams, and forests in the eastern United States and Canada. In this case, acid produced by industrial processes, power plants, and automobile exhaust is discharged into

### TABLE 2.4 pH RANGES FOR BIOCHEMICAL REACTIONS

| ENZYME | SOURCE | OPTIMUM pH |
|---|---|---|
| Pepsin | Gastric mucosa | 1.5 |
| β-glucosidase | Almond | 4.5 |
| Sucrase | Intestine | 6.2 |
| Urease | Soybean | 6.8 |
| Catalase | Liver | 7.3 |
| Succinate dehydrogenase | Beef heart | 7.6 |
| Arginase | Beef liver | 9.0 |
| Alkaline phosphatase | Bone | 9.5 |

the atmosphere and is washed from the air by rain and snow. Some lakes, streams, and woodlands have little or no buffering capacity and are unable to resist pH changes that kill or damage organisms (✦ Figure 2.18).

## THE CHEMISTRY OF BIOLOGICAL SYSTEMS

Although eight elements account for just over 98% of all matter on earth, the distribution of elements in the human body and most other living systems is somewhat different (Table 2.1). Carbon, hydrogen, oxygen, and nitrogen account for about 96% of all elements in our bodies. Other elements commonly found in the body include phosphorus and sulfur. This section explores the role of these elements in the structure and function of biologically important molecules.

### Organic Molecules Contain Carbon

Life on earth is based on the chemistry of carbon. This element is highly versatile and can form up to four covalent bonds with other atoms, including other carbon atoms. Carbon compounds can form long polymeric chains or ring structures (✦ Figure 2.19). Carbon-containing molecules assembled by cells are called **organic** compounds. Many small organic molecules (containing up to about 15 or 20 carbon atoms) are used as subunits of larger molecules (often called **macromolecules**). Smaller subunit molecules include simple sugars, fatty acids, amino acids, and nucleo-

## Sidebar

### LIPOSOMES

In water, phospholipids form water-filled microspheres known as liposomes. As shown, the polar, hydrophilic regions of the phospholipid molecules are in contact with water, while the nonpolar, hydrophobic regions are oriented inward. Liposomes can be made from natural products that contain lipids, such as egg yolks or soybeans.

Liposomes are used to carry compounds into the body for a number of uses, such as cancer chemotherapy. Anticancer drugs such as doxorubicin kill cancer cells but cause damage to heart muscle. Encapsulating the drug in liposomes allows the drug to act against cancer cells, but significantly reduces heart damage. Liposomes are also used in daily life in cosmetics where they provide controlled time release of wrinkle creams and skin softeners.

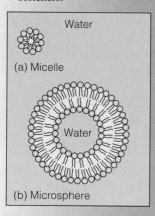

## TABLE 2.5 SUMMARY OF THE FOUR CLASSES OF ORGANIC MOLECULES FOUND IN ORGANISMS

| CLASS | SUBCLASSES | EXAMPLES | FUNCTIONS |
|---|---|---|---|
| **Carbohydrates** | Monosaccharides (simple sugars) | Glucose | Energy source, and as polymers, structural materials |
| | Oligosaccharides (simple sugars) | Sucrose | Sugar in plants |
| | | Lactose | Sugar in milk |
| | Polysaccharides (complex carbohydrates) | Starch | Energy storage in plants |
| | | Cellulose | Structural component in plants |
| | | Glycogen | Energy storage in animals |
| **Lipids** | *Fatty Acids and Derivatives:* Glycerides: one, two or three fatty acids attached to glycerol | Fats, oils | Energy storage |
| | Waxes: long-chain fatty acids attached to alcohol | | Waterproof coatings on leaves, hair, feathers, chitin |
| | Phospholipids: fatty acids, glycerol, and phosphate group | Phosphatidylcholine | Plasma membrane component |
| | *Lipids Without Fatty Acids:* Steroids: Four carbon rings | Cholesterol | Plasma membrane component |
| | | Testosterone | Sex hormone |
| | | Estrogen | Sex hormone |
| **Proteins** | Fibrous: water-insoluble proteins that are structural components | Keratin | Structural component of hair, nails |
| | Globular: multifunctional molecules | Enzymes | Catalyze chemical reactions |
| | | Hemoglobin | Transports oxygen |
| | | Growth hormone | Regulates growth, development |
| | | Antibodies | Immunological defense against infection |
| **Nucleotides and nucleic acids** | Adenosine, phosphates | ATP | Energy storage molecule |
| | Nucleic acids | RNA | Storage, transfer, and transmission of genetic information |
| | | DNA | |

tides. The polymers or macromolecules formed from these subunits include carbohydrates, lipids, proteins, and nucleic acids. These macromolecules constitute the four main classes of organic molecules found in cells of all organisms (◆ Table 2.5 provides a description of each class).

## Carbohydrates Are Energy Sources and Structural Building Blocks

In the biological world, **carbohydrates** have two important functions: energy sources and structural materials. As sources of energy, carbohydrates are found in two forms, simple sugars such as glucose and fructose, and storage forms such as glycogen and starch. As structural building blocks, carbohydrates in the form of cellulose are important in forming the cell walls of plants.

The simplest carbohydrates are **monosaccharides.** Most monosaccharides have a five- or six-carbon skeleton (◆ Figure 2.20). Five-carbon sugars are important components of nucleic acids, and glucose and fructose are among the most common six-carbon sugars. As we will see in Chapter 4, glucose is important as an energy source in almost all cell types. Note in Figure 2.20 that glucose and fructose have different molecular structures, and thus belong to different families of monosaccharides, but both contain the same number of carbon, hydrogen, and oxygen atoms.

# GUEST ESSAY: RESEARCH AND APPLICATIONS

## Life and Its Origin

CLIFFORD MATTHEWS

My interest in the phenomenon of life and its origin began isoon after the end of World War II. Arriving from the Far East where I had been a prisoner of war in my hometown of Hong Kong, I enrolled at the University of London. One professor, John Desmond Bernal, was much admired for the extraordinarily wide range of his scientific activities. Most inspiring were his lectures on how life could have started on Earth, probably the first public discussions in the English language of that fascinating topic. I saw that all branches of science would be required to piece together the story of life's origins.

After completing my studies, I moved to the United States with my wife in 1950. After a stimulating year working for a small chemical firm, I entered Yale University as a graduate student. Here I learned to read the scientific literature critically, and gained experience in carrying out research. In those days, grad students did their experiments in a large laboratory, where we had the chance to interact with each other day and night, learning a lot in the course of a year or two together. Hard work, but fun! It seemed that important discoveries in biochemistry were occurring everywhere, including the experimental results of Stanley Miller and Harold Urey. In 1953 they showed that amino acids and other important biomolecules were readily formed by the action of high energy sources on mixtures of methane, ammonia, and water, simulating the atmosphere of the primitive Earth.

After earning my Ph.D. (in 1953), I began a journey of research in industry, and most exciting for me, developed a controversial idea for the origin of proteins on the early Earth, which I continue to investigate to this day. The experiments of Miller and Urey showed how amino acids might have formed in the primitive oceans, but how did these assemble into the long chains we call proteins? I proposed that the action of sunlight on ammonia and methane in the Earth's early atmosphere produced clouds of hydrogen cyanide (HCN) which polymorized to form chains of amino acids upon exposure to water. To test the unorthodox view that proteins came before amino acids, Bob Moser and I sparked a mixture of methane and ammonia (without water), and got a brownblack sticky mass. Treatment with water yielded a mixture of at least six amino acids.

The results of these and many other experiments are best explained by the cyanide model rather than the Miller-Urey approach. In 1969, I joined the chemistry faculty at the University of Illinois at Chicago. My students, colleagues, and I have conducted numerous tests on this hypothesis, and whether it will stand up to criticism and further tests remains to be seen. In December of 1995, the Galileo spacecraft will parachute instruments into the atmosphere of Jupiter, which has a reducing atmosphere (similar to the primitive Earth) to investigate the yellow-orange-brown-red streaks seen there. I suggest these are the results of HCN polymerization, since these are the colors seen in our reaction flasks.

Which leads to the profound realization that life is an inherent property of matter, most probably existing on Earth-like planets throughout the universe. As my late friend and fellow explorer Cyri Ponnamperuma said, "The universe is in the business of making life."

*CLIFFORD N. MATTHEWS was born in Hong Kong in 1921. His early education was interrupted by his experiences as a prisoner of war there and in Japan from 1941 to 1945. After the war he completed his undergraduate studies in England at the University of London and then moved to the United States for graduate work in chemistry at Yale University. After several years in industry, mostly at Monsanto carrying out fundamental chemical research, he became in 1969 a professor of chemistry at the University of Illinois at Chicago, where his research on cosmochemistry and the origin of life led him to use the unifying theme of cosmic evolution in all his teaching activities. As emeritus professor since 1992, he continues to work on his controversial cyanide model for the simultaneous origin of proteins and nucleic acids.*

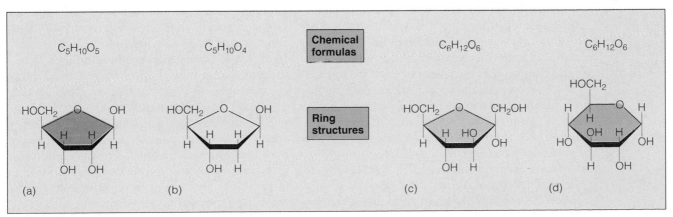

**✦ FIGURE 2.20**

The formulas and structures for a number of common monosaccharides. (a) Ribose. (b) Deoxyribose. (c) Fructose. (d) Glucose.

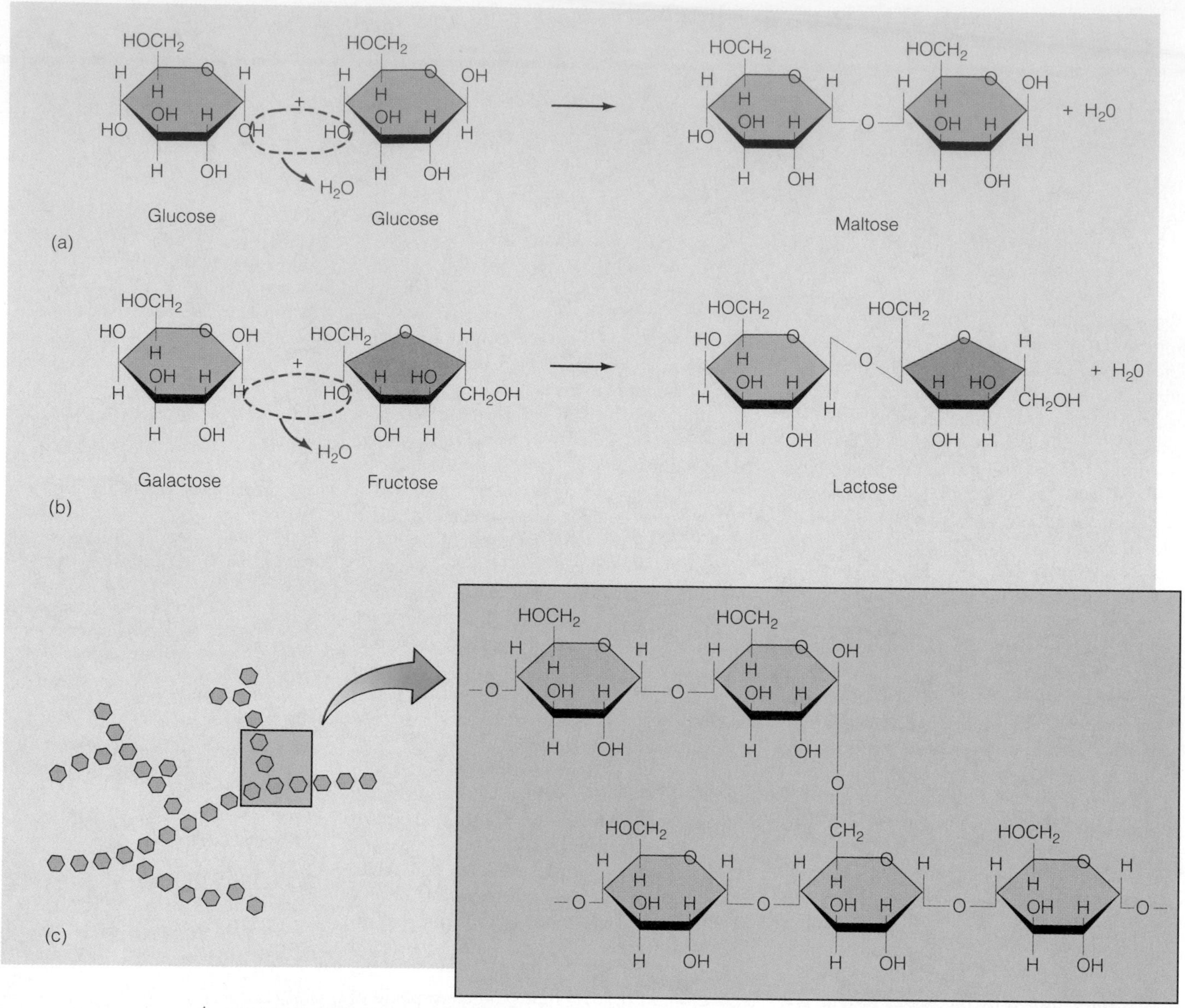

**✦ FIGURE 2.21**

**Disaccharides and polysaccharides. Monosaccharides can combine to form disaccharides. (a) Maltose. (b) Lactose. (c) Polysaccharides are long chains, made up of monosaccharide units.**

**Disaccharides** are combinations of two monosaccharides held together by a covalent bond (✦ Figure 2.21). Common disaccharides include sucrose and lactose. Table sugar is sucrose, and is extracted from sugarcane plants or from the roots of the sugar beet. In the spring, when plants mobilize sucrose from storage to supply energy for new growth, plant sap is rich in sucrose. Maple sugar is traditionally made in the spring by collecting and boiling down the sap from maple trees.

Lactose is found only in milk. It is formed by covalently linking glucose and galactose, and it is digested by first splitting the two monosaccharides apart.

In nature, most carbohydrates exist as long chains of monosaccharide units known as **polysac-**

**charides.** The most important polysaccharides are glycogen, starch, and cellulose (Figure 2.21). Each consists of glucose units, linked to each other in different ways. Glycogen is produced in animal cells as a storage form of glucose. In the human body, glucose is removed from the blood by the liver (and to a lesser extent by muscle cells) and converted into highly branched molecules of glycogen. When energy is required, glucose subunits are removed from glycogen and transferred to the blood to be transported and used as an energy source. Starch is assembled in the cells of land plants from glucose molecules made during photosynthesis.

Cellulose is the major structural carbohydrate of plants, and close to 50% of all the carbon atoms in plants are found in cellulose molecules. In cellu-

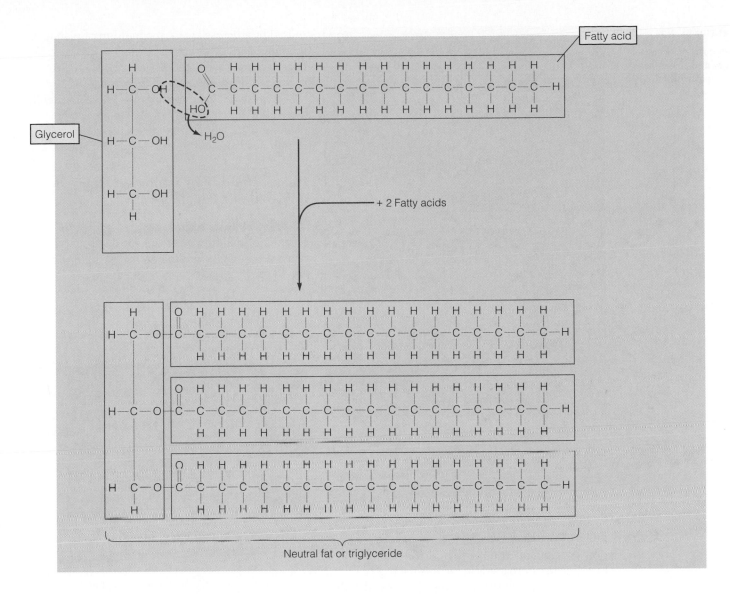

losc, glucose units are joined into long, unbranched chains. Cellulose is insoluble in water, indigestible in the human intestine, and is widely known as the dietary component called fiber. High-fiber diets are those that contain a great deal of cellulose.

Polysaccharides formed by derivatives of sugars, or sugars linked to proteins (called **glycoproteins**) or lipids (**glycolipids**), are also found in many organisms. Chitin is made up of long, cross-linked chains of a glucose derivative that form rigid skeletal structures in crustaceans (crabs), insects, and some fungi.

The outer surface of the cell membrane in higher organisms contains glycoproteins and glycolipids. These polysaccharides serve to anchor and orient membrane components involved in interactions with the extracellular environment. Bacterial cells have an outer cell wall that is composed of a meshwork of polysaccharides combined with proteins to provide protection for the cell.

## Lipids Are Structurally and Functionally Diverse

The lipid class of biological molecules is so diverse that it is difficult to define in structural terms. Lipids are usually defined in terms of their insolubility in polar solvents such as water and of their solubility in nonpolar organic solvents such as ether and benzene. The lipids of biological significance can be classified into several categories: fatty acids and glycerides, waxes, phospholipids, and steroids. Lipids have a variety of biological functions: They are a major structural element of membranes. Certain lipids serve as energy reserves, and others function as vitamins and hormones. In addition, lipid derivatives in the form of bile acids assist in the digestion and absorption of lipids.

The simplest lipids are fatty acids; they consist of long chains of carbon atoms with attached hydrogen atoms (✦ Figure 2.22). Fatty acids with one or

✦ **FIGURE 2.22**
**Triglycerides are formed from glycerol and fatty acids.**

## FIGURE 2.23

Cross sections of (a) a normal artery and (b) an artery almost closed with atherosclerotic plaque.

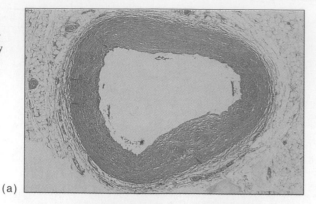

(a)

(b)

caused by atherosclerosis, the buildup of plaque on the inner surface of arterial walls (◆ Figure 2.23). Diets rich in unsaturated fats can reduce the risk of such heart disease.

Waxes are long-chain fatty acids covalently linked to long-chain alcohols. They are secreted by plants and animals to provide a waterproof coating on hair, feathers, chitin, and leaves (◆ Figure 2.24).

## Phospholipids Play Important Roles in Membrane Function

This class of lipids is similar to fats, except that one or more of the fatty acids are replaced by a phosphate group and a choline group (◆ Figure 2.25).

## ◆ FIGURE 2.25

**Phospholipids are composed of a glycerol backbone, fatty acids, phosphate, and choline.**

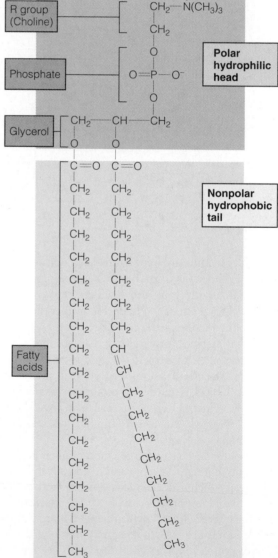

more double bonds between carbon atoms are called **unsaturated** fatty acids. Those molecules without any double bonds between carbon atoms are called **saturated** fatty acids.

Fatty acids are most often combined with other molecules such as glycerol to form monoglycerides, diglycerides, or triglycerides. Most animals and many plants store lipids in the form of triglycerides, which serve as energy reserves. Triglycerides that are solid at room temperature are **fats,** whereas those that are liquid at room temperature are **oils.** Human diets rich in saturated fats (usually animal fats) are one of the risk factors for heart disease

## ◆ FIGURE 2.24

**Many plants have protective waxes on their surfaces to prevent water loss.**

Steroids consist of a skeleton of four rings with different side groups attached. (a) Cholesterol. (b) Testosterone. (c) Structural formula of testosterone.

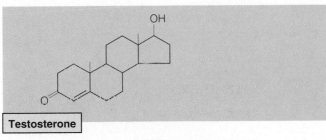

**Cholesterol**

(a)

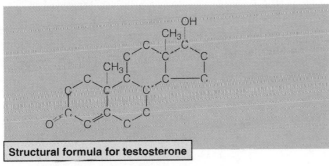

**Testosterone**

(b)

**Structural formula for testosterone**

(c)

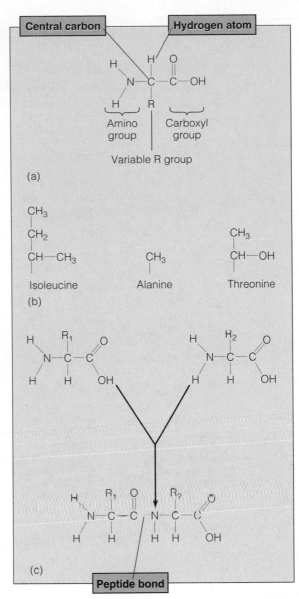

✦ FIGURE 2.27

Amino acids are small molecules with three characteristic groups attached to a central carbon atom. One of these groups (the R group) is variable. In forming a covalent bond, the carboxyl group of one amino acid reacts with the amino group of another amino acid. The resulting bond is called a peptide bond. (a) Amino acid. (b) R groups. (c) Formation of peptide bond.

Phospholipids are important structural components of biological membranes and play important roles in membrane function. Although they are referred to as a single group, they are in fact a complex class of molecules.

## Cholesterol and Sex Hormones Are Steroids

**Steroids** are based on a four-ring carbon backbone (✦ Figure 2.26) and differ from each other in the type, number, and placement of chemical groups attached to this ring structure. One of the most common steroids in the human body is cholesterol, an important component of cell membranes. Cholesterol is a precursor for the synthesis of many different hormones, including the sex hormones estrogen and testosterone (Figure 2.26). Included in this class are anabolic steroids, hormones that promote synthetic metabolic pathways, which leads to a buildup of muscle mass. Athletes and weight lifters often use anabolic steroids to increase their strength and endurance, but metabolic side effects can produce liver damage and adverse psychological effects.

## Amino Acids Are Linked Together to Form Proteins

**Proteins** represent the most diverse class of biological molecules. Proteins have a wide array of cellular functions, some of which are summarized in ➡ Table 2.6 (see page 34). Proteins are polymers, assembled from subunits known as **amino acids** (✦ Figure 2.27). Twenty different amino acids are used in assembling proteins.

## TABLE 2.6  BIOLOGICAL FUNCTIONS OF PROTEINS

| PROTEIN FUNCTION | EXAMPLES | OCCURRENCE OR ROLE |
| --- | --- | --- |
| Catalysis | Lactate dehydrogenase | Oxidizes lactic acid |
| | Cytochrome c | Transfers electrons |
| | DNA polymerase | Replicates and repairs DNA |
| Structural | Viral-coat proteins | Sheath around nucleic acid of viruses |
| | Glycoproteins | Cell coats and walls |
| | $\alpha$-keratin | Skin, hair, feathers, nails, and hoofs |
| | $\beta$-keratin | Silk of cocoons and spider webs |
| | Collagen | Fibrous connective tissue |
| | Elastin | Elastic connective tissue |
| Storage | Ovalbumin | Egg-white protein |
| | Casein | A milk protein |
| | Ferritin | Stores iron in the spleen |
| | Gliadin | Stores amino acids in wheat |
| | Zein | Stores amino acids in corn |
| Protection | Antibodies | Form complexes with foreign proteins |
| | Complement | Complexes with some antigen–antibody systems |
| | Fibrinogen | Involved in blood clotting |
| | Thrombin | Involved in blood clotting |
| Regulatory | Insulin | Regulates glucose metabolism |
| | Growth hormone | Stimulates growth of bone |
| Nerve impulse transmission | Rhodopsin | Involved in vision |
| | Acetylcholine receptor protein | Impulse transmission in nerve cells |
| Motion | Myosin | Thick filaments in muscle fiber |
| | Actin | Thin filaments in muscle fiber |
| | Dynein | Movement of cilia and flagella |
| Transport | Hemoglobin | Transports $O_2$ in blood |
| | Myoglobin | Transports $O_2$ in muscle cells |
| | Serum albumin | Transports fatty acids in blood |
| | Transferrin | Transports iron in blood |
| | Ceruloplasmin | Transports copper in blood |

The amino acid sequence of proteins determines the three-dimensional shape and functional capacity of proteins. The sequence of amino acids in a protein is known as its **primary structure** (✦ Figure 2.28). The formation of hydrogen bonds between different amino acids results in the formation of a pleated or coiled **secondary structure.** Interactions between side groups in the amino acid chain can fold the helix or sheet into its **tertiary structure.** A few proteins exhibit another level of structure, **quaternary structure,** through interactions with other polypeptide chains (Figure 2.28).

## Enzymes Act as Catalysts in Biological Reactions

One of the most important classes of proteins is **enzymes,** molecules that act as catalysts in bio-chemical reactions. Enzymes accelerate the rate of a chemical reaction, but themselves remain unchanged by the reaction. Molecules that bind to the enzyme and undergo a chemical reaction are known as **substrates.** Enzymes are usually named for their substrate, with the suffix "ase" added. The enzyme that catalyzes the breakdown of the sugar lactose is named lactase, and the enzyme that catalyzes the conversion of the amino acid phenylalanine to the amino acid tyrosine is called phenylalanine hydroxylase.

As discussed earlier, a certain amount of energy must be present for chemical reactions to take place. Enzymes function to lower the amount of this energy, thus allowing reactions to occur at relatively lower energy levels. This is accomplished partly because the enzyme acts to bind physically to its substrate, forming an enzyme-substrate complex. The

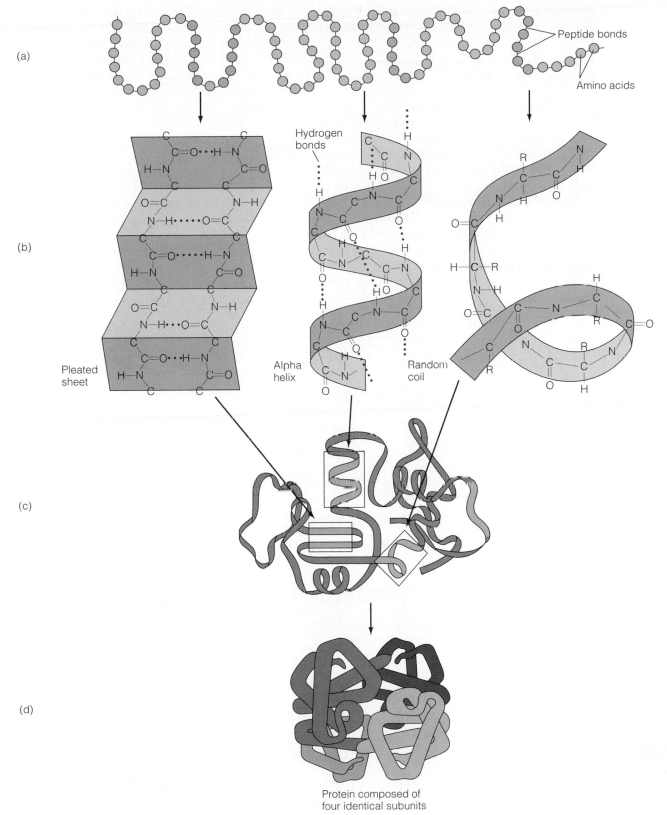

(a)

Peptide bonds

Amino acids

(b)

Hydrogen bonds

Pleated sheet

Alpha helix

Random coil

(c)

(d)

Protein composed of four identical subunits

✦ FIGURE 2.28
Proteins can have four levels of structure. (a) The primary structure is the amino acid sequence and, to a large extent, it determines other levels of structure. (b) At the secondary level, the structure formed can be a pleated sheet, alpha helix, or a random coil. (c) Tertiary structure is the folding of the secondary structure into a functional three-dimensional configuration. (d) Some proteins can interact with other proteins to form a fourth level of structure.

role of enzymes in metabolism and energy production will be studied in greater detail in Chapter 4.

## Nucleic Acids Store and Transfer Genetic Information

The biological molecules with the most unique set of functions are the **nucleic acids.** These polymers are composed of subunits called **nucleotides** (✦ Figure 2.29). Nucleotides themselves serve important functions in energy transfer reactions within cells, including ATP.

**Deoxyribonucleic acid (DNA)** is a nucleic acid composed of two polynucleotide strands wound around a central axis to form a double helix (✦ Figure 2.30). DNA has an important function as the repository of genetic information. The information is encoded in the sequence of nucleotides within the molecule.

RNA or ribonucleic acid is a single strand of polynucleotides (Figure 2.31). RNA transfers genetic information within the cell. In some viruses, RNA is the nucleic acid that acts as the repository of genetic information. The details of nucleic acid structure and function will be discussed in Chapters 7 and 8.

## ORIGIN OF LIFE: HOW DID IT ALL BEGIN?

We have described in some detail the chemical components of life without touching on one of the basic questions in all biology: How did life arise on this planet?

The question of how life originated cannot be easily answered, because we do not know precisely the conditions or circumstances that existed billions of years ago. This uncertainty makes it impossible

✦ **FIGURE 2.29**
(a) **Nucleotides are composed of three components: (b) a phosphate, (c) a sugar, and (d) a base. Nucleotides are the building blocks of DNA and RNA.**

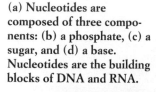

(a)

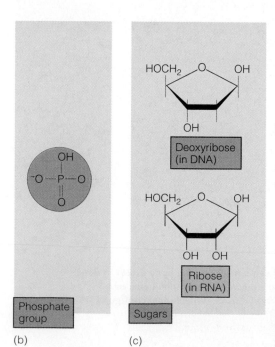

(b)  (c)

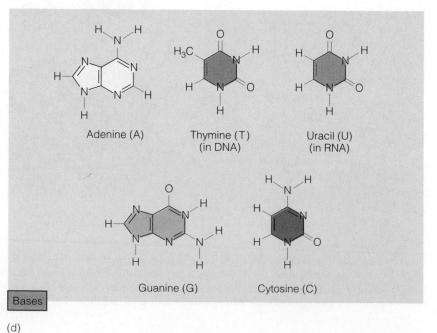

(d)

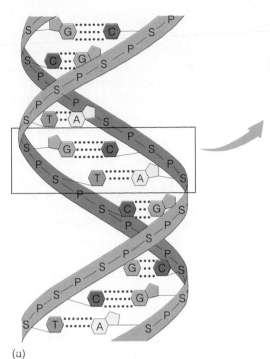

(a)

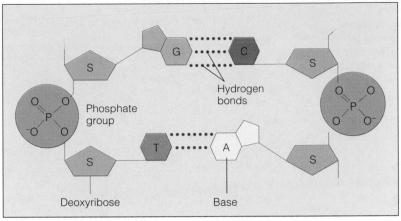

Hydrogen bonds

Phosphate group

Deoxyribose

Base

✦ FIGURE 2.30

(a) DNA is a double-stranded molecule of two polynucleotide chains coiled around a central axis.
(b) RNA is a single-stranded polynucleotide chain.

Key to the bases:
A = Adenine     G = Guanine
T = Thymine     U = Uracil (RNA)
C = Cytosine

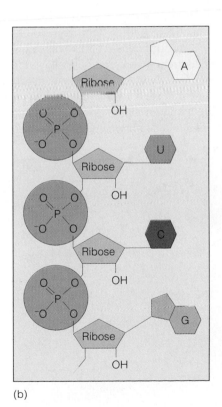

(b)

to design and carry out experiments that duplicate those conditions. Nevertheless, enough is known generally about the history of the Earth to generate hypotheses about the origin of life on this planet.

Evidence indicates that life originated sometime during the first billion years after the formation of the planet. Almost all hypotheses agree that simple chemicals present on earth formed more complex organic compounds, polymers and aggregates. Somewhere along the line, metabolism, biological information systems, and autotrophy (the ability to make food from inorganic sources) developed. The debate is about the order in which these events occurred, and the mechanisms by which they occurred. Rather than debate these ideas, we summarize the landmarks associated with the origin of life.

The first step in the evolution of life, the transition to complex organic molecules from simple inorganic components, has been replicated in experiments. In 1953, Stanley Miller built an apparatus (✦ Figure 2.31) containing ammonia, methane, hydrogen, and water vapor. The gases were exposed to electrical sparks (simulating lightning), heated to simulate volcanic action, and condensed from vapors to liquid. After several weeks, the mixture was analyzed and found to contain four different amino acids. In a variation of this experiment, Clifford Matthews found that ammonia and methane can form polymers of hydrogen cyanide, which when reacted with water, form six different amino acids. Other variations resulted in the production of nucleotides as well as amino acids.

Once formed, these organic molecules may have attached themselves to the charged surfaces of clay

◆ **FIGURE 2.31**
Schematic of the apparatus
used by Miller. Amino acids
associated with organisms
were formed during his experiments.

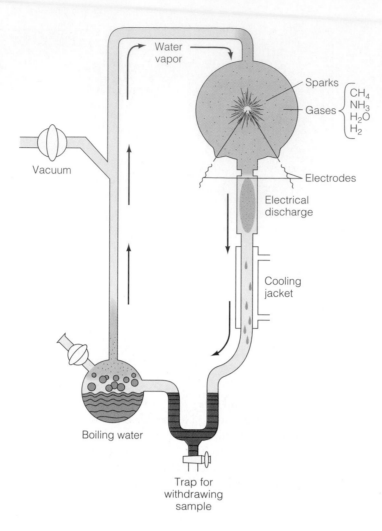

or other minerals in tide pools or
aqueous environments. Minerals such
as clay can promote the formation of
polymers, especially under conditions
in which evaporation concentrates the
organic molecules (such as in tide
pools under a hot sun). Polymers
of amino acids, called **proteinoids,**
formed in this manner have enzyme-
like properties, and can catalyze
chemical reactions (◆ Figure 2.32).
RNA molecules formed by the polymerization of nucleotides can also act
as catalysts to splice other RNA molecules and to polymerize short RNA
chains into longer ones.

Perhaps aggregates of self-replicating RNA molecules and proteinoids
were able to catalyze a wider range
of reactions than either component
alone, leading to the proliferation of
aggregates composed of RNA and
primitive proteins.

At some point, chemical selection
favored aggregate systems that were
able to trap and use other molecules
for energy transfer and were also able
to use DNA rather than RNA to encode information. At this transition,
reproduction and metabolism defined
a self-replicating chemical system, setting the stage for the evolution of cellular forms of life (◆ Figure 2.33). How a chemical
aggregate gave rise to a primitive cell, probably an

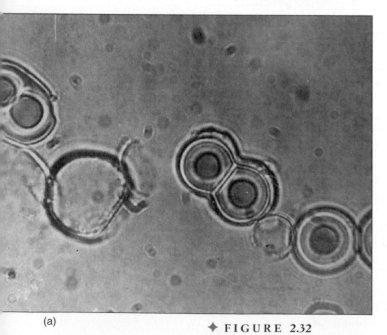

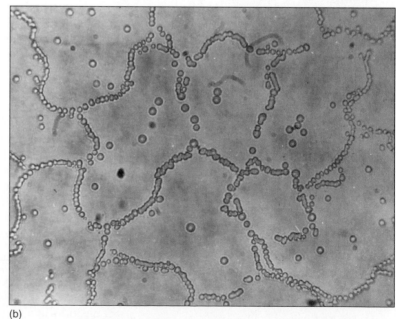

(a)

(b)

◆ **FIGURE 2.32**

**Synthetic proteins spontaneously form cell-like structures called proteinoids. (a) Proteinoids at high magnification.
(b) Proteinoid microspheres aggregate in chains.**

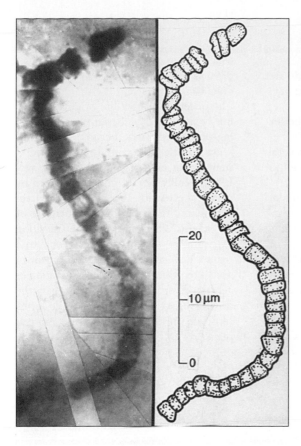

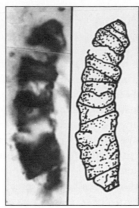

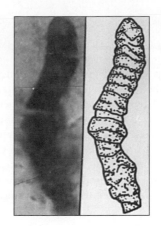

◆ FIGURE 2.33
Fossil microbes recovered from rocks in Australia. The rocks are at least 3.4 billion years old, suggesting that life evolved soon after the earth was formed some 4.5 billion years ago.

RNA-containing bacteria now extinct, cannot be determined.

What is increasingly clear is that eukaryotic cells originated as symbiotic associations among free-living prokaryotes, giving rise to the mitochondria, plastids, and cilia present as organelles in current eukaryotic cells.

Once formed, eukaryotic cells developed sexual reproduction, producing offspring with genetic variability. This variability was acted on by natural selection, driving the diversification of organisms and giving rise to the myriad of species that have existed during the last 700 million years (◆ Figure 2.34).

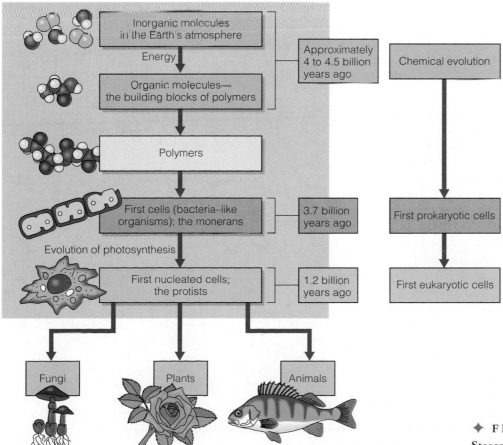

◆ FIGURE 2.34
Stages in the evolution of life on Earth.

# SUMMARY

1. Elements that formed as the universe evolved are the basis for the physical and chemical properties of all living organisms. The basic chemical unit is the atom; different types of atoms, characterized by the number of protons and electrons, make up elements.

2. Combinations of atoms form compounds and molecules. Molecules are held together by atomic forces called bonds. Depending on their nature, bonds can be ionic, covalent, or hydrogen bonds. The making and breaking of chemical bonds is the basis of chemical reactions.

3. Energy transfers in living systems involve chemical reactions that transfer energy in controlled steps so that it can be effectively utilized to drive cellular functions.

Chemical reactions in living systems are organized into a series of reactions known as a metabolic pathway.

4. A combination of polar covalent bonds and a network of hydrogen bonds between water molecules gives water several chemical properties that are important to living systems.

5. Life on earth is based to a great extent on the chemistry of carbon, oxygen, hydrogen, and nitrogen. Many small organic molecules (containing up to about 15 or 20 carbon atoms) are used as subunits of larger macromolecules. The polymers or macromolecules formed from these subunits include carbohydrates, lipids, proteins, and nucleic acids. These macromolecules constitute the four main classes of

organic molecules found in cells of all organisms.

6. The question of how life originated cannot be answered by designing and carrying out experiments, because we do not know precisely the conditions or circumstances that existed billions of years ago. It is generally agreed that life evolved by chemical evolution, but there is disagreement about the order and mechanisms of the stages. When and how the transition to primitive prokaryotic cells occurred remain unanswered. The development of sexual reproduction and the generation of genetic diversity among offspring is the driving force that produced the life-forms recorded in the history of life on this planet.

# KEY TERMS

acids
adenosine triphosphate (ATP)
adhesion
amino acids
anabolic reactions
atom
atomic number
atomic weight
bases
buffers
carbohydrates
catabolic reactions
chemical equilibrium
cohesion
compound
covalent bond
deoxyribonucleic acid (DNA)
disaccharides
electrostatic attraction
elements
endothermic

energy of activation
enzymes
exothermic
fats
glycolipids
glycoproteins
hydrogen bond
ionic bond
ions
isotopes
kilocalories
macromolecules
matter
metabolic pathway
metabolism
molecules
monosaccharides
nucleic acids
nucleotides
nucleus
oils

organic
oxidation
periodic table of the elements
polar covalent bond
polysaccharides
primary structure
proteinoids
proteins
quaternary structure
radiation
reduction
saturated
secondary structure
solvents
subatomic particles: protons, neutron, electrons
substrates
tertiary structure
unsaturated

## COMPARE AND CONTRAST

Briefly define and explain the similarities/differences between the following:

1. Protons, neutrons, and electrons
2. Atomic number and atomic weight
3. Oxidation and reduction
4. Atoms, elements, compounds, and molecules
5. Ionic, covalent, polar covalent, and hydrogen bonds
6. Acids, bases, and buffers
7. Monosaccharides, disaccharides, and polysaccharides
8. Primary, secondary, tertiary, and quaternary structure
9. DNA and RNA

## MULTIPLE CHOICE

Select the best response for each statement.

1. Matter is composed of _____.
   a. earth, fire, water, and air in various combinations
   b. elements
   c. carbohydrates
   d. phospholipids
2. The smallest particle of an element is an _____
   a. atom
   b. ion
   c. isotope
   d. enzyme
3. Atoms of each element are composed of a particular number of _____ while the number of _____ may vary.
   a. electrons . . . neutrons
   b. protons . . . electrons
   c. neutrons . . . protons
   d. protons . . . neutrons
4. In the formation of a double covalent bond between carbon and oxygen atoms, _____.
   a. an unequal number of electrons are shared between the carbon and oxygen atoms

   b. two atoms carrying opposite electrical charges are shared
   c. two electrons from carbon are shared with an oxygen atom and two electrons from oxygen atom are shared with carbon atom
   d. neutrons are shared
5. _____ are substances that add to the number of hydrogen ions in a solution, and _____ lower the hydrogen ion concentration of solutions.
   a. bases . . . acids
   b. neutrons . . . electrons
   c. lipids . . . nucleic acids
   d. acids . . . bases

## TRUE/FALSE

1. All atoms of an element contain the same number of protons.
2. Weight of protons + weight of neutrons = atomic weight.
3. New chemical bonds are formed and new molecules are made during catabolic reactions.
4. The energy of activation is the maximum amount of energy necessary to disrupt chemical bonds in a reaction.
5. Once a chemical reaction has reached a chemical equilibrium, it can never be altered.
6. All proteins are enzymes.

## MATCHING

Match each type of molecule/solution with the appropriate description.

1. ATP
2. buffers
3. fatty acids
4. DNA
5. monosaccharides

_____ simplest lipid
_____ genetic information storage molecule
_____ pH regulators
_____ simplest carbohydrate
_____ energy storage molecule

## SCIENCE AND SOCIETY

1. Radioactively labeled molecules are biological molecules that contain radioactive molecules. The labeled molecule emits radioactive decay products that allow scientists to track its pathway through an organism. Over the last 50 years, radiolabeling has provided science and medicine with important information that has resulted in a better understanding of human physiology. What are the risks and benefits to the use of radioactive tracers in the diagnosis and treatment of disease?
2. Currently, national attention is focused on ways of preserving the environment, by recycling goods, using non-aerosol sprays, and the reduced use and production of harmful products such as styrofoam. Organic wastes such as grass clippings and kitchen wastes dumped in landfills and covered by dirt are slowly decomposed by bacteria. These landfills in turn, are converted into parks and other recreational areas. However, some research indicates that garbage is not decomposing in these landfills as fast as originally thought. Do you think decomposition in landfills is the same in the Midwestern United States as it is in the Southwestern United States? What physical, chemical, and geological factors contribute to the rate of garbage decomposition? Is converting landfills into public parks a wise decision if evidence suggests that the decomposition process is not complete? Why or why not? What effects could this have on the environment, human health and safety, and industry?

# THE STRUCTURE OF CELLS

OPENING IMAGE

*A color-enhanced transmission*

*electron micrograph showing a*

*eukaryotic cell's internal components.*

A long the shores of Lake Superior, the waters of the lake wash up on rock
formations in Western Ontario that were laid down billions of years ago.
These rocks, known as the Gunflint Iron formation, contain belts of a
dense, black sediment known as Gunflint chert. Geologists, paleontologists, and biologists
have long studied the chert, looking for unmistakable evidence that life existed almost
two billion years ago. In the laboratory, sections of the rock are polished into thin slivers,
or etched to show surface details that are examined and photographed using light and electron microscopes.

What kind of evidence would be universally accepted as confirmation of the presence
of living organisms on Earth two billion years ago? Clearly, the best kind of evidence is
the presence of fossil cells. Why? Because scientists today recognize that cells are such a
unique property of living systems that their presence is accepted as undisputed evidence for
the existence of life. The Gunflint chert was apparently deposited underwater as fine mud
or silt, and has proven to be a rich source of preserved cells, including bacteria, cyanobacteria, algae, and fungi. More than a dozen species of plants have been identified from their
remains laid down in the chert.

Although the fossil cells found along the shores of Lake Superior are a rich source of
early cells, much older fossils have been found elsewhere. Remains of bacterial cells
almost 3.5 billion years old have been recovered in other formations. In Gunflint chert

*and elsewhere, the search for early forms of life is almost always directed at the identification of cells as evidence for the existence of life on the primitive Earth.*

*The relationship between cells and life may seem obvious today, but the notion that all living things are composed of one or more cells emerged just over 150 years ago. The development of the cell theory as a fundamental principle of the biological sciences represents an example in which a new theory organizes and changes the direction of thought.*

*In this chapter, we consider the emergence of the cell theory and review its meaning and impact on biology. Next, we outline the important features of cell structure and show why an understanding of cell structure and function is essential if we hope to explain the workings of plants and animals.*

##  THE CELL THEORY STARTED A REVOLUTION IN BIOLOGY

The **cell theory** states that all living things are composed of at least one cell and that the cell is the fundamental unit of function in all organisms. Related to this are two corollaries: First, that all cells are fundamentally alike in their chemical composition and second, that all cells arise from preexisting cells through a process of cell division.

### How the Cell Theory Developed

The cell theory is often credited to Matthias Schleiden and Theodor Schwann, based on their work in the 1830s. Many other individuals made contributions to this concept, which evolved gradually during a period of about 200 years. In the 1600s, Robert Hooke used a microscope to describe the structure of cork (a form of tree bark), and first used the term **cells** to describe what he saw (✦ Figure 3.1). In the 1820s, Robert Brown described the existence of the nucleus in a variety of cell types. Shortly after Brown's work, Schleiden (a botanist) and Schwann (an anatomist) made a series of observations about cells and organisms. They suggested that fundamental properties of organisms such as growth, metabolism, and reproduction are actually properties of the individual cells that make up an organism. As a result, the study of cells would provide information about the organization and function of the intact organism. The cell theory emerged as a conclusion from these proposals.

### The Cell Theory and the Origin of Life

Like many important scientific ideas, the cell theory is simple, it organizes a large number of observations (namely, that all organisms examined to date contain cells), and makes definite predictions about future observations (all organisms examined in the future will be composed of cells). A previous theory, **abiogenesis** (*a* = without, *bio* = life, *genesis* = coming into being), first articulated by the ancient Greeks, held that some organisms originated from nonliving material. As a theory, abiogenesis can explain how some organisms reproduce (they are generated anew from nonbiological material), but it cannot make reliable predictions about which organisms will arise by abiogenesis and which ones will reproduce biologically, nor can it predict what organisms will arise from nonliving material.

One of the corollaries of the cell theory is that all cells arise from preexisting cells. According to this idea, cells cannot arise by abiogenesis from nonliving material; instead, reproduction is a cellular function. Louis Pasteur provided evidence for this idea when he showed that in microorganisms, reproduction is a cellular process. These and similar findings allowed the cell theory to replace ideas of abiogenesis, and provided the foundation for modern biology.

##  CELL SIZE IS PARTLY A PROBLEM OF PHYSICS AND MATHEMATICS

The development and acceptance of the cell theory paralleled the development of technology in the form of lenses and microscopes, which allowed investigators to observe and describe cells and their internal components (see Beyond the Basics: Microscopy, on pages 50–51). In multicellular organisms such as a fly or a human, most cells are too small to be seen with the unaided eye, and their dimensions are measured in micrometers (✦ Table 3.1). In multicellular organ-

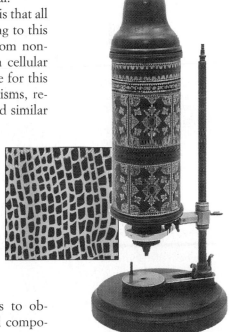

✦ **FIGURE 3.1**
**Cells were first described by Robert Hooke in 1665.**

 **TABLE 3.1 MEASUREMENT EQUIVALENTS**

1 meter = $10^2$ cm = $10^3$ mm = $10^6$ $\mu$m = $10^9$ nm
1 centimeter (cm) = 1/100 meter = 10 mm
1 millimeter (mm) = 1/1,000 meter = 1/10 cm
1 micrometer ($\mu$m) = 1/1,000,000 meter = 1/10,000 cm
1 nanometer (nm) = 1/1,000,000,000 meter = 1/10,000,000 cm

**Range of cell sizes. The largest are bird eggs, and red blood cells are among the smallest. The size range for most cell types is between 10 to 100 micrometers.**

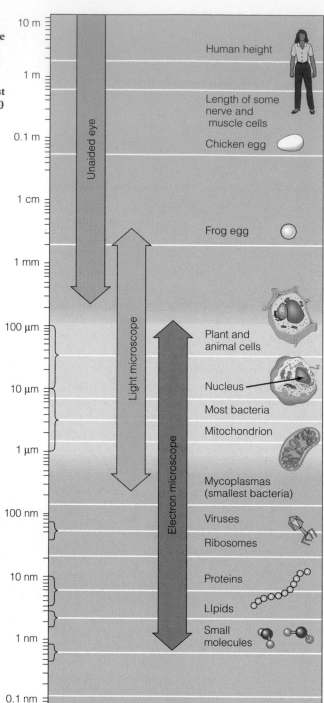

♦ FIGURE 3.3

**Following fertilization in amphibians, cell division distributes a constant amount of cytoplasm among daughter cells so that the sixteen-cell embryo is the same size as the fertilized egg. At a later stage, cell volume will increase between divisions.**

isms, cells range between 0.5 and 30 micrometers in diameter. Some cells are, of course, exceptions to this rule. Eggs are often the largest cell produced by an organism, and many are visible without the use of lenses: a human egg is about 80 to 100 micrometers in diameter, and is just visible to the unaided eye. A frog's egg is about 1500 micrometers, and an ostrich egg is almost 152,000 micrometers (about 6 inches) in length. Other types, such as nerve cells, are only a few micrometers in width, but may be several feet in length (♦ Figure 3.2).

The limited range of cell sizes means that large multicellular organisms must be composed of many cells. Robert Hooke measured cell sizes in cork and made some calculations as follows:

> ". . . there were usually about threescore [60] of these small cells placed endways in the eighteenth part of an inch in length, whence I concluded there must be near eleven hundred of them, or more than a thousand in the length of an inch and therefore in a square inch above a million, or 1,166,400, and in a cubic inch, above twelve hundred million, or 1,259,712,000, a thing almost incredible, did not our microscope assure us of it by ocular demonstration . . ."

Hooke's observations are essentially correct. Many multicellular organisms contain literally billions of cells. For example, a newborn human has more than ten trillion cells ($2 \times 10^{12}$), produced from a single fertilized egg over a 38-week period.

Why is it that cells such as the egg are so large, but the cells that arise from it during development are so small (♦ Figure 3.3)? The functional capacity of a cell depends on the ability to absorb nutrients and to transport waste products into the surrounding environment. This exchange of material takes place at the cell surface. As the size of a cell decreases, the surface area increases more rapidly than the volume (♦ Figure 3.4), making it easier to provide for an adequate rate of exchange of material between the cell and its environment. Cells with high metabolic rates tend to be smaller than

(a) The fertilized egg

(b) Four-cell stage

(c) Eight-cell stage

(d) Sixteen-cell stage

cells with lower metabolic rates (such as unfertilized eggs).

Before considering some aspects of the organization and function of cell, remember that just as the cell theory predicts that the properties of organisms can be explained in terms of cells, as their components, it also predicts that the structure and function of cells can be understood in terms of their molecular components.

In the following discussion, we refer to both the structural and molecular organization of cellular structures, which will serve as a foundation for an exploration of cellular functions in later chapters. The emphasis is on answering the questions "What is a cell?" and "How are cells related to life?"

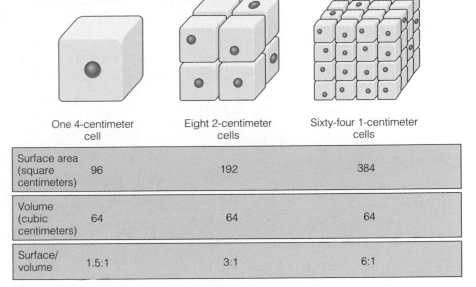

| | One 4-centimeter cell | Eight 2-centimeter cells | Sixty-four 1-centimeter cells |
|---|---|---|---|
| Surface area (square centimeters) | 96 | 192 | 384 |
| Volume (cubic centimeters) | 64 | 64 | 64 |
| Surface/volume | 1.5:1 | 3:1 | 6:1 |

✦ **FIGURE 3.4**

The relationship between cell volume and surface area.

 ## THERE ARE TWO TYPES OF CELLS: PROKARYOTIC AND EUKARYOTIC

There are two basic types of cells: **prokaryotic** and **eukaryotic.** These cell types have different evolutionary histories. Prokaryotic cells include those found in bacteria and blue-green algae. These organisms belong to the kingdom Monera (Chapter 12). Eukaryotes include single-celled organisms called **protists** (Chapter 12) and multicellular fungi, plants, and animals.

### Prokaryotic Cells Have a Simple Organization

Prokaryotic cells are enclosed by a plasma membrane, but do not contain a membrane-bounded nucleus (✦ Figure 3.5). Instead, the genetic information (in the form of a DNA molecule) is in direct contact with the cytoplasm. In addition, prokaryotic cells do not contain any of the membranous organelles found in eukaryotic cells, as described in the following sections.

### Eukaryotic Cells Are Larger and Have a Complex Internal Structure

Eukaryotic cell types include single-celled organisms called protists (described in Chapter 12) and multicellular fungi, plants, and animals (✦ Figure 3.6). Whatever the source, eukaryotic cells have a common set of structural features that distinguish them from prokaryotic cells. Eukaryotic cells contain a nucleus and are divided into a number of compartments by internal membrane systems

Cell membrane

Cell wall

Ribosomes

DNA

DNA

✦ **FIGURE 3.5**

**Prokaryotic cells, like those of this bacterium, are small and have a less complex structure than eukaryotic cells.**

✦ **FIGURE 3.6**

**In the single-celled protists, such as these ciliates, the cell is equivalent to the organism.**

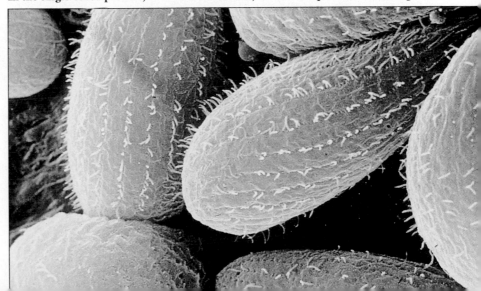

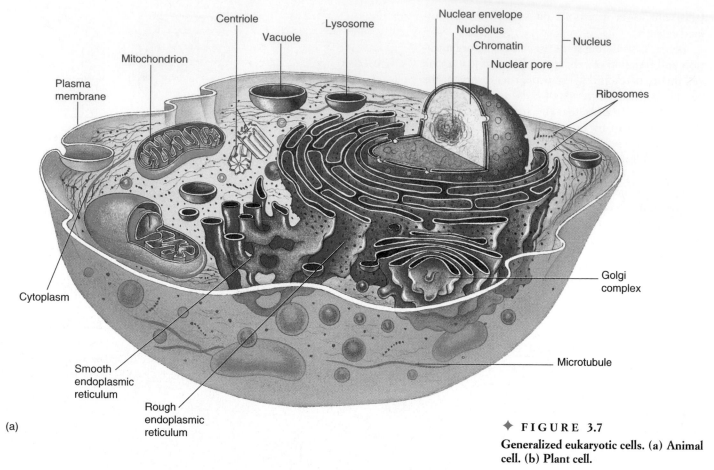

Plasma
membrane

Mitochondrion

Centriole

Vacuole

Lysosome

Nuclear envelope

Nucleolus

Chromatin

Nuclear pore

Nucleus

Ribosomes

Cytoplasm

Golgi
complex

Microtubule

Smooth
endoplasmic
reticulum

Rough
endoplasmic
reticulum

(a)

✦ F I G U R E  3.7

**Generalized eukaryotic cells. (a) Animal
cell. (b) Plant cell.**

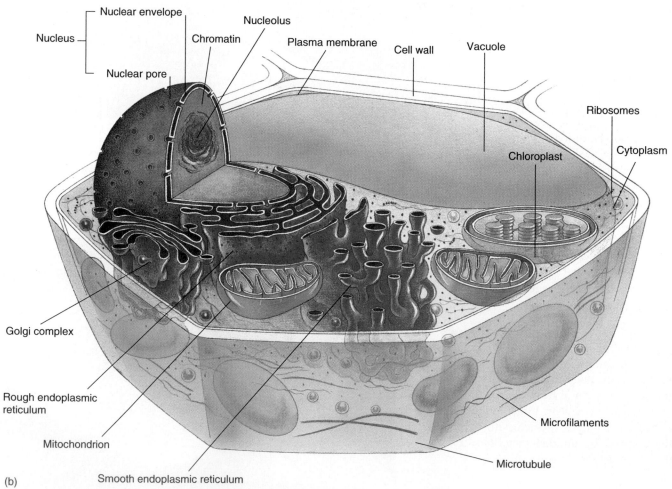

Nuclear envelope

Nucleolus

Nucleus

Chromatin

Plasma membrane

Cell wall

Vacuole

Nuclear pore

Ribosomes

Chloroplast

Cytoplasm

Golgi complex

Rough endoplasmic
reticulum

Microfilaments

Mitochondrion

Microtubule

Smooth endoplasmic reticulum

(b)

## TABLE 3.2 COMPARISON OF PLANT AND ANIMAL CELLS

| FEATURE | PLANT CELL | ANIMAL CELL |
| --- | --- | --- |
| Plasma membrane | Yes | Yes |
| Cell wall | Yes | No |
| Nucleus | Yes | Yes |
| Mitochondria | Yes (generally few) | Yes |
| Chloroplast | Yes | No |
| Endoplasmic reticulum | Yes | Yes |
| Golgi complex | Yes | Yes |
| Ribosomes | Yes | Yes |
| Lysosomes | Yes (rare) | Yes |
| Central vacuole | Yes | No |
| Cytoskeleton | Yes | Yes |
| Centrioles | No | Yes |

## TABLE 3.3 OVERVIEW OF PLASMA MEMBRANE FUNCTIONS

Ensures the cell's structural integrity.

Regulates the flow of molecules and ions into and out of the cell.

Maintains the chemical composition of cytoplasm and extra cellular fluid.

Participates in cellular communication.

Forms a cellular identification system.

(◆ Figure 3.7). Major differences between plant and animal cells are summarized in ◆ Table 3.2.

In the remainder of this chapter, we discuss the structural components of eukaryotic cells and their organization into tissues and organs, beginning with the outer boundary, the plasma membrane.

## THE PLASMA MEMBRANE: LIFE'S BOUNDARY

Bordering all cells is a double-layered membrane that is a dynamic and active component of cell function (◆ Figure 3.8). The plasma membrane has two functions: to act as a barrier between the living material inside the cell and the nonliving environment outside the cell, and to regulate the exchange of materials at this border. Plasma membranes are similar in structure to the membranes inside the cell, although they may perform vastly different functions (◆ Table 3.3).

### The Fluid Mosaic Model of Membranes

Membranes in biological systems are composed of lipids and proteins. The most widely accepted model for the organization of membranes is called the **fluid-mosaic** model (◆ Figure 3.9). In a fluid-

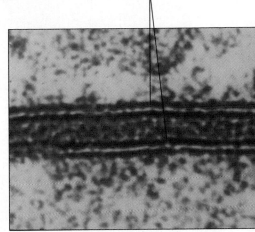

Plasma membranes

◆ **FIGURE 3.8**
The plasma membrane is a double-layered structure at the cell boundary.

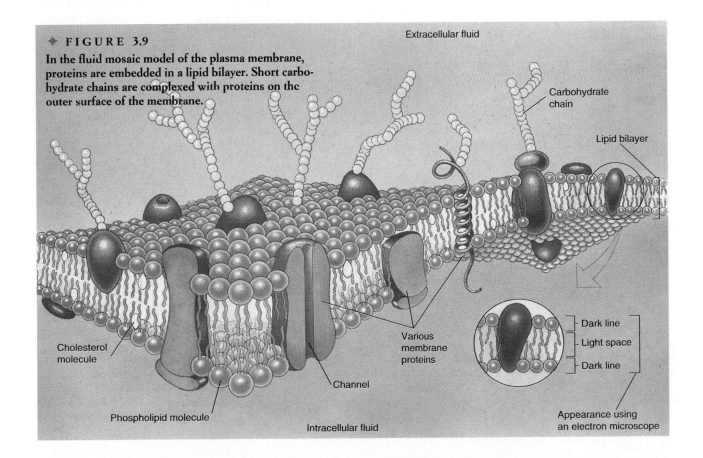

◆ **FIGURE 3.9**
In the fluid mosaic model of the plasma membrane, proteins are embedded in a lipid bilayer. Short carbohydrate chains are complexed with proteins on the outer surface of the membrane.

Extracellular fluid

Carbohydrate chain

Lipid bilayer

Cholesterol molecule

Various membrane proteins

Channel

Phospholipid molecule

Intracellular fluid

Dark line
Light space
Dark line

Appearance using an electron microscope

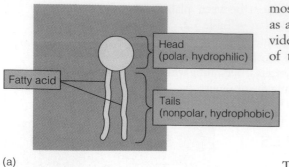

(a)

Head (polar, hydrophilic)

Fatty acid

Tails (nonpolar, hydrophobic)

(b)

Polar heads (hydrophilic)

Nonpolar tails (hydrophobic)

Lipid bilayer

Polar heads (hydrophilic)

✦ **FIGURE 3.10**

**Phospholipid molecules (a) are composed of a hydrophilic end and a hydrophobic end. (b) When in contact with water, phospholipids organize themselves into a lipid bilayer, with the hydrophobic ends away from the surface in contact with water.**

mosaic membrane, lipids act as a fluid component and provide the structural framework of the membrane. Distributed throughout the lipid components, a patchwork (a mosaic) of different proteins gives the membrane many of its functional characteristics.

The majority of lipids in membranes are **phospholipids,** asymmetrical molecules with a **hydrophilic** (water-loving) head, and a **hydrophobic** (water-fearing) tail (✦ Figure 3.10). If phospholipids are surrounded by water, they aggregate together so their hydrophilic heads interact with water, and their hydrophobic tails interact with each other. In an aqueous environment, exposure of hydrophobic groups to the surrounding water molecules is limited when phospholipids form lipid bilayers. Phospholipid bilayers are the basic structural component of all biological membranes, although other lipids, including cholesterol, are also present.

The molecular components of membranes are not fixed, but can move laterally or even flip from one side of the membrane to the other, making the membrane behave somewhat like a fluid, allowing it to flex as conditions change. The protein components of membranes are present on the external surface, the internal surface, or extend through both surfaces of the membrane (Figure 3.9).

## Membrane Components Have Many Functions

In spite of the fact that both surfaces of a membrane are composed of lipids and proteins, they are very different from one another. On the outside of the membrane, carbohydrate molecules are linked to proteins and lipids. On the inner surface, cytoplasmic proteins bind to the membrane, providing contact between the cytoplasm and the plasma membrane. These differences in chemical organization are related to the functional roles of each side of the membrane. In general, membrane proteins function as transport channels, pumps, enzymes, and receptors. The carbohydrate/protein complexes (called **glycoproteins**) on the outer surface act as receptors that bind molecular signals originating outside the cell. These receptors also serve as sites of recognition in blood types and in other aspects of the immune system.

## Passive Transport and Osmosis Are Types of Diffusion

Membranes regulate the exchange of materials between cells and the environment (➤ Table 3.4). Substances cross a plasma membrane in several ways. **Diffusion** is the movement of particles from an area of higher concentration to an area of lower concentration. It is powered by the random, kinetic movement of molecules (✦ Figure 3.11). When diffusion occurs across a plasma membrane, it is called **passive transport,** because the cell expends no energy in this process. Passive transport is an important process in all cells. In the lungs, oxygen moves into red blood cells by diffusion, and carbon

**TABLE 3.4 OVERVIEW OF PLASMA MEMBRANE TRANSPORT**

| PROCESS | DESCRIPTION |
|---|---|
| Simple diffusion | Flow of ions and molecules from high concentrations to low. Water-soluble ions and molecules probably pass through pores; water-insoluble molecules pass directly through the lipid layer. |
| Facilitated diffusion | Flow of ions and molecules form high concentrations to low concentrations with the aid of protein carrier molecules in the membrane. |
| Active transport | Transport of molecules from regions of low concentration to regions of high concentration with the aid of transport proteins in the cell membrane and ATP. |
| Endocytosis | Active incorporation of liquid and solid materials outside the cell by the plasma membrane. Materials are engulfed by the cell and become surrounded in a membrane. |
| Exocytosis | Release of materials packaged in secretory vesicles. |
| Osmosis | Diffusion of water molecules from regions of high water (low solute) concentration to regions of low water (high solute) concentrations. |

dioxide, a metabolic by-product, diffuses out of cells into the blood.

**Osmosis** is a type of diffusion in which water moves across a membrane that is permeable to water, but not to molecules (solutes) dissolved in the water (Figure 3.11). If there are different concentrations of a solute on either side of a membrane, water will move across the membrane until the solutes are at equal concentrations. Solutions with a higher concentration of solutes are **hypertonic,** and those with a lower solute concentration are **hypotonic.** Solutions of equal solute concentration are **isotonic.**

To prevent excess gain or loss of water, cells must be able to regulate movement of water by osmosis. This control is called **osmoregulation** and will be discussed in Chapter 24.

## Active Transport Can Move Molecules Against a Concentration Gradient

Cells can transport specific molecules across the plasma membrane *against* a concentration gradient by the process of **active transport.** This process requires energy and uses proteins in the membrane as molecular pumps to move solutes (✦ Figure 3.12). Many cells, including those of the nervous system, function by using active transport.

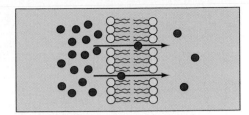
(a)

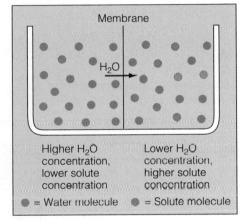
(b)

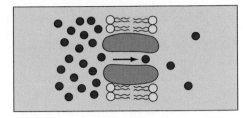

Membrane

$H_2O$

Higher $H_2O$ concentration, lower solute concentration

Lower $H_2O$ concentration, higher solute concentration

● = Water molecule    ● = Solute molecule

(c)

### ✦ FIGURE 3.11

**Substances can move across plasma membranes in several ways. (a) Simple diffusion; lipid-soluble materials move directly through the membrane in response to a concentration gradient. (b) Water-soluble materials passively diffuse through pores formed by proteins. (c) Osmosis is the movement of water across a membrane in response to differences in solute concentration.**

### ✦ FIGURE 3.12

**Membrane transport can be accomplished by processes that expend energy. (a) Active transport uses energy from ATP to move materials across a membrane against a concentration gradient. (b) Cells can engulf and internalize particles by endocytosis. (c) Exocytosis is used to secrete materials from the cell.**

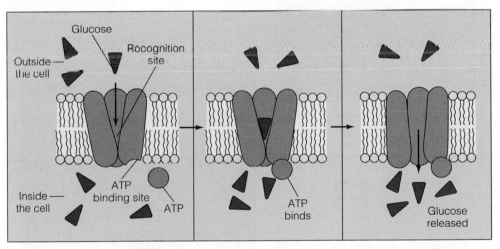

Glucose

Recognition site

Outside the cell

Inside the cell

ATP binding site    ATP

ATP binds

Glucose released

(a)

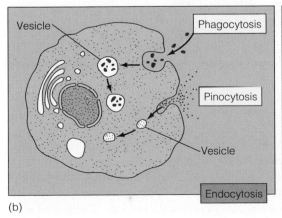

Vesicle

Phagocytosis

Pinocytosis

Vesicle

Endocytosis

(b)

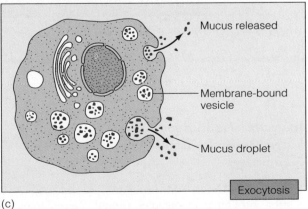

Mucus released

Membrane-bound vesicle

Mucus droplet

Exocytosis

(c)

# MICROSCOPY

The cell theory came into existence partly because advances in the design and construction of microscopes allowed investigators to learn more about the structure and organization of cells. Advances in the design of microscopes in the twentieth century have provided opportunities to study details of the subcellular and even macromolecular organization of cells. Because most of what we know about cells has been accumulated through the use of microscopes, it is important to understand something about how they work.

In addition to the obvious property of magnification, microscopes have two other important properties that determine their usefulness: resolution and contrast. Magnification is the ability to enlarge the image of what is being viewed, **resolution** is the ability to view adjacent objects as distinct structures, and **contrast** gives different densities to different structures. Of these, the most critical factor in microscopy is resolution. The unaided eye has a resolving power of 100 micrometers; that is, if two objects are closer together than 100 micrometers, they will be perceived as a single object. On the other hand, if two objects are further apart than 100 micrometers, they will be seen as two distinct objects.

Three different types of microscopes are widely used in the study of cell structure: the light microscope, the transmission electron microscope, and the scanning electron microscope. The light microscope has a resolving power of about 0.2 micrometers (about 500 times better than the unaided eye). Because most eukaryotic cells are between 10 and 50 micrometers, they are easily seen with the light microscope. Contrast in light microscopes is provided by the use of stains and dyes that differentially stain structures within cells. The most important factor in determining resolution is the wavelength of light that is used to illuminate the image. In some light microscopes, ultraviolet light (which has shorter wavelengths than white light) is used to provide higher resolution.

The transmission electron microscope (TEM) depends on a beam of electrons instead of light to provide illumination. The electron beam has a wavelength much shorter than light, and provides much higher resolution. The electron beam is focused with electromagnetic lenses and passes through a specimen that has been sliced very thin. Because the human eye cannot see objects illuminated with electrons, the image is transferred to a fluorescent screen or a piece of photographic film. The contrast of the image in the transmission electron microscope can be enhanced by staining the specimen with heavy metals that differentially bind to cellular components. The transmission electron microscope has a resolving power of 0.2 nanometers, about 1,000 times better than the light microscope and about 10,000 times better than the human eye. Much of what we know about cell structure has been learned with the transmission electron microscope.

The scanning electron microscope (SEM) uses electrons to provide images of the surfaces of objects such as cells. A fine beam of electrons is used to scan back and forth across the entire object, and electrons scattered by the object are collected and displayed on a monitor.

**Biologist using a light microscope.**

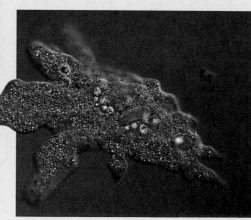

**Amoeba proteus.**

## Exocytosis and Endocytosis Move Large Molecules Across Plasma Membranes

Small molecules can diffuse across plasma membranes, but larger molecules cannot. Very large molecules, including large proteins, move out of and into cells by **exocytosis** and **endocytosis**. In exocytosis, membrane-enclosed vesicles move through the cytoplasm to the plasma membrane. The vesicles fuse with the membrane, depositing their con-

The instrument has a resolving power of about 10 nanometers, about 50 times less than the transmission electron microscope. Contrast is provided by coating the specimen with evaporated atoms of a heavy metal such as gold or platinum. The SEM is used for images of the three-dimensional structure of cells and subcellular components.

Operator using a scanning electron microscope (SEM).

Scientist using a transmission electron microscope (TEM).

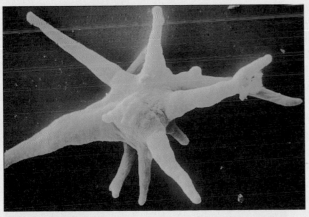

Amoeba (SEM).

Amoeba Axopodia (TEM).

tents outside the cell (Figure 3.12). Many glands in the body use exocytosis to secrete their products. When you salivate at the smell of food, the saliva has been exported from cells in the salivary glands by exocytosis.

Endocytosis is the reverse process. Cells take in large molecules or particles by forming vesicles at the plasma membrane surface (Figure 3.12). There are several forms of endocytosis, including **phagocytosis,** the process by which white blood cells

*Sidebar*

### A FATAL MEMBRANE FLAW

Cystic fibrosis is a genetic disorder that leads to an early death. Affected individuals have thick, sticky secretions of the pancreas and lungs. Diagnosis is often made by finding elevated levels of chloride ions in sweat. According to folklore, midwives would lick the forehead of newborns. If the sweat was salty, they would predict that the infant would die in childhood. Despite intensive therapy and drug treatments, the average survival is only about 25 years.

Cystic fibrosis is caused by a functional defect in a membrane protein that controls the movement of chloride ions into and out of the cell. In normal cells, this protein functions as a pore or channel controlling the flow of chloride, but in cystic fibrosis, the channel is unable to open. This causes chloride ions to accumulate inside the cell. To balance the chloride ions, the cells absorb excess sodium. In secretory glands, this leads to decreases in fluid production, resulting in blockage of flow from the pancreas and the accumulation of thick mucus in the lungs. The symptoms and premature death associated with this disorder point out the important role of membranes in controlling cell function.

surround and engulf invading bacteria or viruses (✦ Figure 3.13).

## CELL COATS AND CELL WALLS OFFER PROTECTION AND SUPPORT

In many cells, extensions of the plasma membrane or materials deposited outside the plasma membrane form structures known variously as coats, shells, walls, and capsules. Animal cells do not produce cell walls; they are present only in bacteria, fungi, and higher plants.

Many animal cells such as insect eggs are surrounded by coats or shells which are secreted by cells that surround the maturing egg. These shells provide a covering that protects the egg from water loss and provides a means for gas exchange during development. Other animal cells are surrounded by a halo of complex molecules extending from the cell surface (✦ Figure 3.14). The molecules in this halo act as recognition sites to receive molecular signals, such as hormones, and as sites for interaction with other cells.

Outside the plasma membrane, plant cells are surrounded by a cell wall (✦ Figure 3.15). The **primary cell wall** is composed of a polysaccharide called **cellulose,** which is synthesized and secreted by the plant cell. Where two plant cells meet, a layer called the **middle lamella,** containing a substance known as **pectin,** cements them together. In woody plants, another wall, a **secondary cell wall,** forms inside the primary wall. This wall, also secreted by the cell, contains alternating layers of cellulose and **lignin,** which strengthens and stiffens the wall.

In response to unfavorable conditions, or as part of the reproductive cycle, the cells of some bacteria, plants, and animals can form impervious capsules known as **spores** (✦ Figure 3.16). Spores isolate and protect the cell from the environment until conditions become more favorable for growth. Encapsulated cells cannot exchange material with the environment and are functionally inactive until they emerge from the capsule.

## THE NUCLEUS: A GENETIC INFORMATION CENTER

Inside the plasma membrane, the **cytoplasm** is a viscous semiliquid that contains many types of macromolecules in solution and suspension, as well as a collection of structures known as **organelles** (✦ Table 3.5). The largest and most prominent organelle in most eukaryotic cells is the **nucleus** (✦ Figure 3.17). The nucleus is separated from the

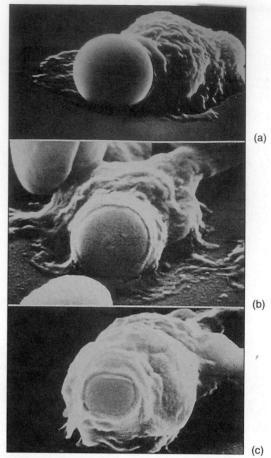

(a)

(b)

(c)

✦ **FIGURE 3.13**

Phagocytosis is a special form of endocytosis. (a–c) A white blood cell surrounds and engulfs a bacterium.

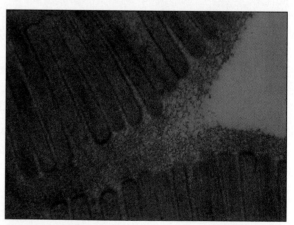

✦ **FIGURE 3.14**

Many animal cells are surrounded by a halo of carbohydrate/protein complexes.

cytoplasm by a nuclear membrane or nuclear envelope. As mentioned earlier, prokaryotic cells contain a region, called a **nuclear area,** not surrounded by a membrane.

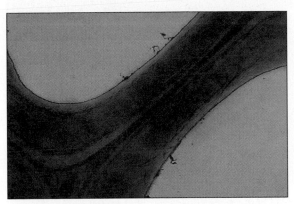

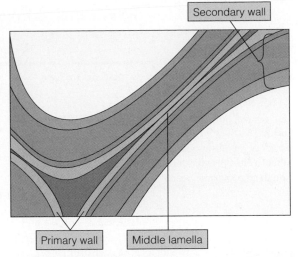

Secondary wall

Primary wall    Middle lamella

✦ FIGURE 3.15

Two adjacent plant cell walls, made up of the primary wall, the middle lamella, and the secondary wall. Plant cells grow by elongating in a single direction.

## The Nuclear Envelope Encloses and Separates the Nucleus from the Cytoplasm

The nuclear envelope is made up of an inner and outer membrane, each composed of a phospholipid bilayer. This double membrane is studded with **nuclear pores,** which allow exchange of molecules between the nucleus and cytoplasm (✦ Figure 3.18). The nuclear envelope is a dynamic structure that enlarges or contracts as the nucleus changes size, and it is broken down during cell division (Chapter 5).

## Chromosomes Are Carriers of Genetic Information

One of the properties that distinguishes living from nonliving systems is the ability to reproduce. In eukaryotic cells, this property resides in nuclear structures known as **chromosomes.** In nondividing cells, chromosomes are uncoiled and consist of thin strands of DNA complexed with proteins to form **chromatin** (✦ Figure 3.19). The nucleus regulates the cell's internal functions and its responses to stimuli from the external environment. The nucleus contains a set of genetic instructions encoded in the chromosomes that determine the organization and shape of the cell and also the range of functions exhibited by the cell.

The number of chromosomes present in the nucleus is characteristic for a given species: For example, the fruit fly *Drosophila melanogaster* has 8, corn plants have 20, human cells have 46 chromosomes, and dogs carry 78. ◆ Table 3.6 lists the chromosome number characteristic for several species of plants and animals.

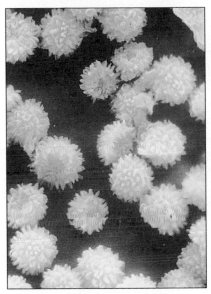

✦ FIGURE 3.16

Some cells can form impervious structures called spores.

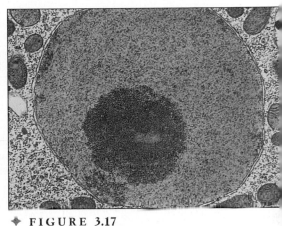

✦ FIGURE 3.17

The nucleus of a eukaryotic cell. The nucleus contains a prominent nucleolus, a structure that is the site of synthesis of ribosomes.

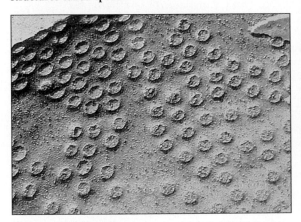

✦ FIGURE 3.18

Nuclear pores allow exchange of materials between the nucleus and the cytoplasm.

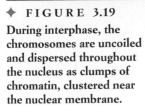

✦ FIGURE 3.19

During interphase, the chromosomes are uncoiled and dispersed throughout the nucleus as clumps of chromatin, clustered near the nuclear membrane.

| ORGANELLE | STRUCTURE | FUNCTION |
|---|---|---|
| Nucleus | Round or oval body; surrounded by nuclear envelope. | Contains the genetic information necessary for control of cell structure and function. DNA contains hereditary information. |
| Nucleolus | Round or oval body in the nucleus consisting of DNA and RNA | Produces ribosomal RNA. |
| Endoplasmic reticulum | Network of membranous tubules in the cytoplasm of the cell. Smooth endoplasmic reticulum contains no ribosomes. Rough endoplasmic reticulum is studded with ribosomes. | Smooth endoplasmic reticulum (SER) is involved in the production of phospholipids and has many different functions in different cells; rough endoplasmic reticulum (RER) is the site of the synthesis of lysosomal enzymes and proteins for extracellular use. |
| Ribosomes | Small particles found in the cytoplasm; made of RNA and protein. | Aid in the production of proteins on the RER and ribosome complexes (polysomes). |
| Golgi complex | Series of flattened sacs usually located near the nucleus. | Sorts, chemically modifies, and packages proteins produced on the RER. |
| Secretory vesicles | Membrane-bound vesicles containing proteins produced by the RER and repackaged by the Golgi complex; contain protein hormones or enzymes. | Store protein hormones or enzymes in the cytoplasm awaiting a signal for release. |
| Vaults | Octagonal-shaped membrane-enclosed vesicle. | Shuttles material between nucleus and cytoplasm. |
| Food vacuole | Membrane-bound vesicle containing material engulfed by the cell. | Stores ingested material and combines with lysosome. |
| Lysosome | Round, membrane-bound structure containing digestive enzymes. | Combines with food vacuoles and digests materials engulfed by cells. |
| Mitochondria | Round, oval, or elongated structures with a double membrane. The inner membrane is thrown into folds. | Complete the breakdown of glucose, producing NADH and ATP. |
| Cytoskeleton | Network of microtubules and microfilaments in the cell. | Gives the cell internal support, helps transport molecules and some organelles inside the cell, and binds to enzymes of metabolic pathways. |
| Cilia | Small projections of the cell membrane containing microtubules. | Propel materials along the surface of a cell. |
| Flagella | Large projections of the cell membrane containing microtubules. | Provide motive force for cells. |
| Cell wall | Layer of cellulose fibers lying outside the plasma membrane of plant cells. | Provides support and protection. |
| Chloroplast | Ovoid or disk-shaped organelle in plant cells: delimited by two membranes. Its inner membrane is not infolded like the inner membrane of mitochondrion. It contains numerous stacks of thylakoid disks, called grana. | Captures solar energy and produces ATP and carbohydrates. |
| Central vacuole | Large membrane-bound cavity in plant cells. | Stores water, ions, toxic materials, pigments, protein, and starch. |

## The Nucleolus Is the Site of Ribosome Synthesis

The nucleus of eukaryotic cells also contains one or more dense regions known as **nucleoli** (sing., **nucleolus**) (Figure 3.19). Under the electron microscope these regions appear as a collection of fibers and granules, and they are associated with the production of cytoplasmic organelles known as **ribosomes.** Ribosomes are the most abundant organelle in the cytoplasm and are found in all prokaryotic and eukaryotic cells.

## THE CYTOPLASM IS DIVIDED INTO COMPARTMENTS BY MEMBRANES

The organization and structural appearance of the cytoplasm in eukaryotes is closely related to the functional state of the cell, and it differs from cell to cell. Cytoplasmic organelles are formed from membranes and divide the cell into a number of functional compartments. Organelles can be classified into four categories, based on functional similarities.

## The Endoplasmic Reticulum Synthesizes Proteins for Secretion

The **endoplasmic reticulum (ER)** (*endo* = within, *reticulum* = network) is a network of membranous channels and vesicles located within the cytoplasm (✦ Figure 3.20). Although in electron micrographs it appears to be a series of unconnected sacs and channels, it is thought to be a single, highly folded, continuous membrane system. The inside of the ER, called the **lumen,** is a separate compartment within the cell, where proteins are modified and prepared for transport.

When the outer surface of the ER is studded with ribosomes it is known as rough ER. The rough ER is predominant in cells that make proteins, especially proteins that will be secreted. Segments of ER lacking ribosomes are known as smooth ER (Figure 3.20). In some cell types, the smooth ER is involved in the synthesis of lipids, including hormones, and is especially abundant in hormone-producing cells, including those of the testis and ovary. However, in most cell types, the smooth ER acts to package proteins.

Proteins synthesized by ribosomes on the rough ER move into the lumen and are transported to a region of smooth ER where sacs of membrane containing the protein bud off to form vesicles. The vesicles move through the cytoplasm to fuse with the plasma membrane and release proteins into the extracellular environment, or they fuse with the membranes of other organelle systems such as the Golgi complex.

Much of the new membrane in cells is produced in the ER and can be incorporated into the nuclear membrane (which is connected to the ER), the plasma membrane, or other membranous organelles such as the Golgi complex. The ER functions in the synthesis and transport of proteins, in the replacement and maintenance of its internal and external membrane system, and in the separation of materials from each other in the cytoplasm.

**TABLE 3.6  CHROMOSOME NUMBER IN SELECTED ORGANISMS**

| ORGANISM | DIPLOID NUMBER (2N) | HAPLOID NUMBER (N) |
|---|---|---|
| Human (*Homo sapiens*) | 46 | 23 |
| Chimpanzee (*Pan troglodytes*) | 48 | 24 |
| Gorilla (*Gorilla gorilla*) | 48 | 24 |
| Dog (*Canis familiaris*) | 78 | 39 |
| Chicken (*Gallus domesticus*) | 78 | 39 |
| Frog (*Rana pipiens*) | 26 | 13 |
| Housefly (*Musca domestica*) | 12 | 6 |
| Onion (*Allium cepa*) | 16 | 8 |
| Corn (*Zea mays*) | 20 | 10 |
| Tobacco (*Nicotiana tabacum*) | 48 | 24 |
| House mouse (*Mus musculus*) | 40 | 20 |
| Fruit fly (*Drosophila melanogaster*) | 8 | 4 |
| Nematode (*Caenorhabditis elegans*) | 12 | 6 |

✦ **FIGURE 3.20**

**(a) Three-dimensional representation of the endoplasmic reticulum showing the relationship between the rough and the smooth ER. (b) Electron micrograph of ribosome-studded rough endoplasmic reticulum.**

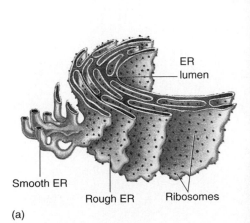

(a)

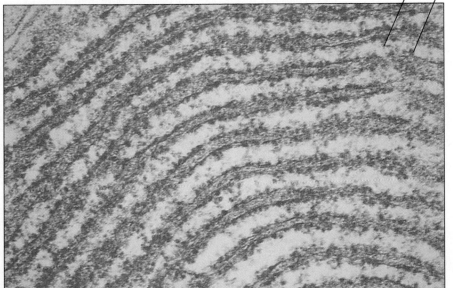

(b)

## Molecular Sorting Occurs in the Golgi Complex

Animal and plant cells contain organelles composed of four to eight flattened membrane sacs arranged as a cluster of stacked disks (✦ Figure 3.21). In animal cells, they are called **Golgi complexes,** and in plant cells they are known as **dictyosomes.** Golgi complexes function as centers for the modification and distribution of material synthesized by the endoplasmic reticulum. Vesicles from the ER fuse with the receiving face of the Golgi complex, emptying their protein contents. Within the Golgi, the proteins are modified by enzyme action as they are transported through each of the Golgi membrane sacs. At the last stage, the proteins are repackaged into vesicles that pinch off the opposite side of the Golgi complex for delivery to other organelles or to the plasma membrane for export from the cell. These vesicles and their protein contents are targeted for their destinations by molecular markers attached to the proteins and by specific molecules on the membranes of the vesicles (Figure 3.21).

The Golgi complexes also serve as a source of membrane for the plasma membrane and other organelles. Secretion vesicles pinch off the outer face of the Golgi and move to the plasma membrane where they fuse with the cell membrane, discharging their contents into the extracellular environment. As a result, membranes of vesicles become part of the cell membrane.

## Lysosomes and Peroxisomes Are Cytoplasmic Disposal Sites

**Lysosomes** are membrane-enclosed vesicles that store a collection of digestive enzymes (✦ Figure 3.22). The hydrolytic enzymes stored here originate in the ER, are transported to the Golgi, packaged, and bud off in vesicles that form lysosomes. The membrane of the lysosome is modified so that it becomes resistant to the action of the enzymes it contains. Lysosomal enzymes degrade a wide range of materials, including proteins, fats and carbohydrates, and invading viruses that might enter the cell.

The importance of lysosomes in cellular maintenance is underscored by several genetic disorders that disrupt or halt lysosomal function. These disorders, including Tay-Sachs disease and Pompe's disease, act at the level of organelles within the cell, but their outcomes can include severe mental retardation, blindness, and death by the age of 3 or 4 years. These disorders illustrate the point made earlier that the functioning of the organism can be explained by events within cells, as outlined in the cell theory.

**Peroxisomes** are membrane-bound vesicles similar to lysosomes that contain potent oxidative enzymes. In cells of green plants, peroxisomes are called **glyoxysomes.** In both plants and animals, these enzymes catalyze oxidative (electron-removing) chemical reactions that produce hydrogen peroxide ($H_2O_2$), a strong oxidant that could be disruptive to the cell if present in the cytoplasm. Hydrogen peroxide is destroyed within the peroxisome by an enzyme that converts it to oxygen and water.

✦ **FIGURE 3.21**

**(a) Three-dimensional representation of a Golgi complex. Transport vesicles carrying proteins move from the endoplasmic reticulum to the Golgi complex, and fuse with the membrane. After modification, the proteins are packaged for distribution. (b) Electron micrograph of Golgi complex.**

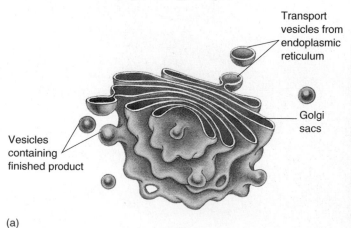

Transport vesicles from endoplasmic reticulum

Golgi sacs

Vesicles containing finished product

(a)

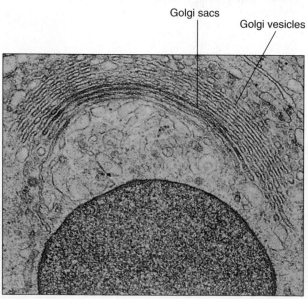

Golgi sacs

Golgi vesicles

(b)

## Vesicles and Vacuoles Are Sites of Storage and Transport

Almost all animal and plant cells contain a series of small (usually less than 100 nanometers in diameter) membrane-bound spaces known as **vesicles.** Vesicles transport macromolecules into and out of the cell and shuttle materials between the various organelles within the cell. One type of vesicle, called a **vault,** has recently been discovered (✦ Figure 3.23). Vaults are octagonal-shaped vesicular barrels (the interior arches of the vesicle walls resemble vaulted ceilings, hence the name) of the same size and shape as nuclear pores. Based on their shape and size, it has been postulated that vaults may be positioned at nuclear pores, fill with materials to be transported to the cytoplasm, then deliver molecules made in the nucleus to specific cytoplasmic destinations.

**Vacuoles** are membrane-bound, fluid-filled spaces found in both plant and animal cells, although they are more frequently found in plant cells. In the cells of some protists such as *Tetrahymena,* vacuoles function to remove water and waste products from the cell, and to store food that is ingested from the environment (✦ Figure 3.24). In plant cells, vacuoles are part of a storage system, allowing the cell to accumulate food products or collect waste products into separate compartments. Fluid-filled plant vacuoles also contribute to plant structure by exerting outward pressure on the cell walls.

✦ **FIGURE 3.22**

**Lysosomes contain digestive enzymes secreted by the endoplasmic reticulum and packaged by the Golgi. Lysosomes fuse with and digest the contents of vesicles pinched off from the cell membrane.**

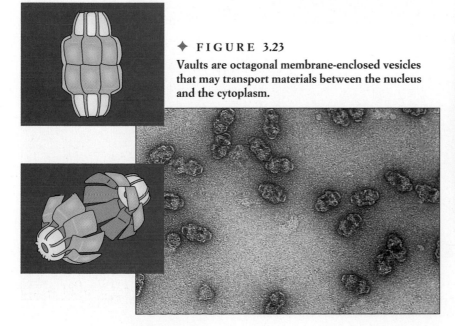

✦ **FIGURE 3.23**

**Vaults are octagonal membrane-enclosed vesicles that may transport materials between the nucleus and the cytoplasm.**

✦ **FIGURE 3.24**

**Vacuoles in ciliated protozoa help maintain water balance in the cell.**

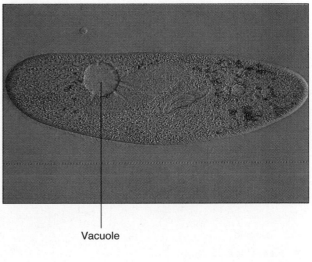

Vacuole

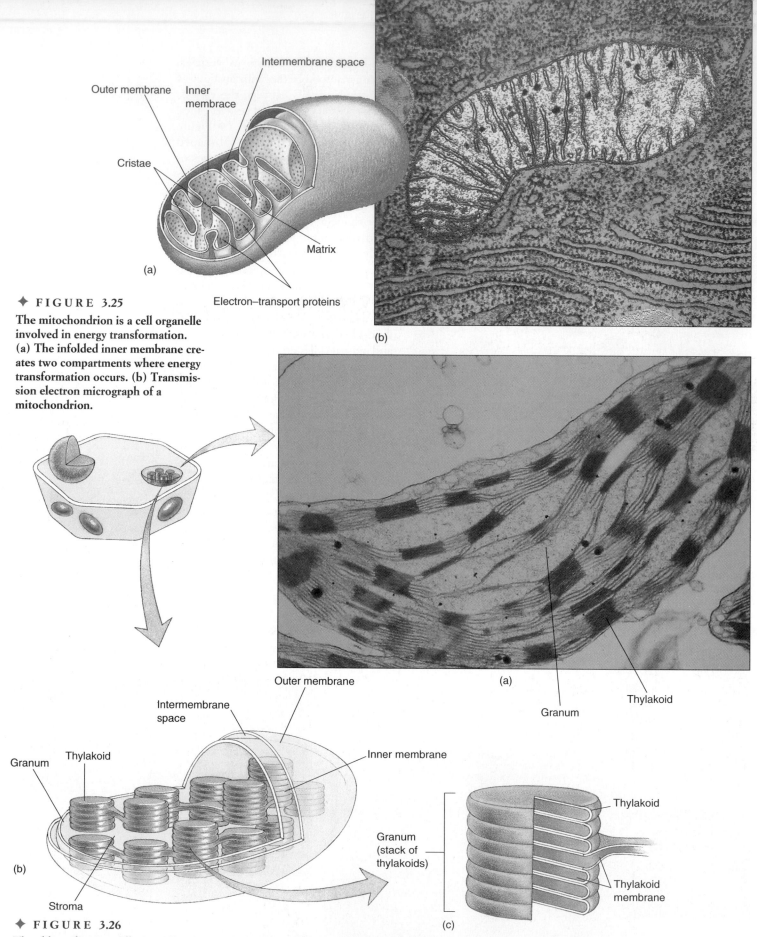

**◆ FIGURE 3.25**

The mitochondrion is a cell organelle involved in energy transformation. (a) The infolded inner membrane creates two compartments where energy transformation occurs. (b) Transmission electron micrograph of a mitochondrion.

**◆ FIGURE 3.26**

The chloroplast is a cell organelle involved in converting light energy into food. (a) Electron micrograph of the chloroplast. (b) Three-dimensional representation of the chloroplast. Stacks of thylakoids fill the interior of the chloroplast.

## ENERGY CONVERSION OCCURS IN MITOCHONDRIA

**Mitochondria** (sing., **mitochondrion**) are membrane-bound cytoplasmic organelles found in both plant and animal cells (✦ Figure 3.25). The outer membrane of the mitochondrion is in contact with the cytoplasm; the inner membrane is usually folded inward to form structures called **cristae** (sing., **crista**). Mitochondria contain genetic information in the form of DNA molecules, and they are self-replicating organelles. During cell growth, or under changing cellular conditions, mitochondria divide to reproduce themselves. As we will see in a later chapter, this fact is useful in studying some evolutionary problems including the origin of eukaryotes and the origin of the human species.

Within mitochondria, energy is transferred from a variety of molecules to produce adenosine triphosphate (ATP), the molecule that serves as the energy source for many biochemical reactions in the cell. This energy transfer and ATP formation is called **cellular respiration.** In this process, oxygen is consumed and carbon dioxide is generated:

glucose + oxygen → carbon dioxide + water + ATP

Cells with higher energy requirements, such as muscle cells, have more mitochondria than cells with lower energy requirements, such as skin cells. Within cells, mitochondria are located at sites where energy is most needed. In gland cells, for example, they are found near the plasma membrane where material is being secreted.

## CHLOROPLASTS CONVERT SOLAR ENERGY INTO FOOD

Cells of many plants contain membrane-bound organelles known as **plastids,** which function in storage or food production. Plastids are classed into a number of categories, one of which is the **chloroplast.** Chloroplasts are disk-like organelles surrounded by a double membrane (✦ Figure 3.26). The inner membrane is usually arranged as a series of stacked disks called **grana.** The matrix surrounding the grana is called the **stroma.** A green pigment, chlorophyll, is found in the grana (Figure 3.26). Chloroplasts are the site of **photosynthesis,** the process of using solar energy to form ATP molecules that, in turn, are used to synthesize food molecules such as glucose. This overall process can be summarized as follows:

radiant energy (sunlight) + carbon dioxide + water → carbohydrate + oxygen

Like mitochondria, chloroplasts also contain genetic information and have the ability to make their own gene products. The structure and function of chloroplasts will be considered in more detail in Chapter 4.

## CILIA AND FLAGELLA PROVIDE LOCOMOTION

The cell surface of many eukaryotic cells contains hair-like extensions known as **cilia** (sing., **cilium**) (✦ Figure 3.27). These organelles function in locomotion, either moving the cell, or, if the cell is anchored, moving objects in the environment past the cell. Some single-celled organisms such as *Tetrahy-*

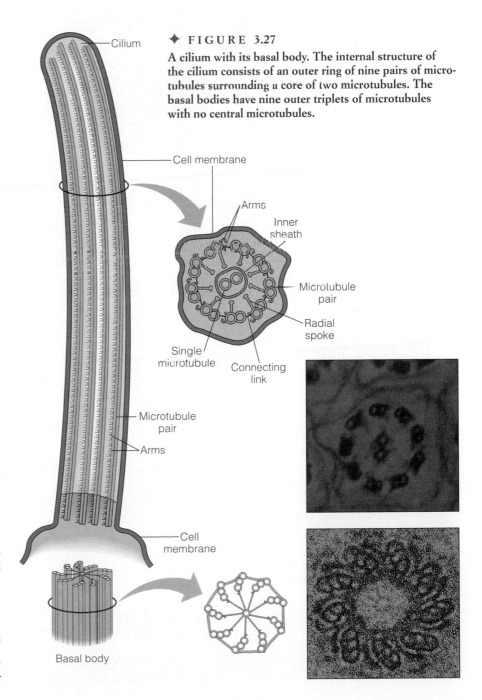

✦ **FIGURE 3.27**

**A cilium with its basal body. The internal structure of the cilium consists of an outer ring of nine pairs of microtubules surrounding a core of two microtubules. The basal bodies have nine outer triplets of microtubules with no central microtubules.**

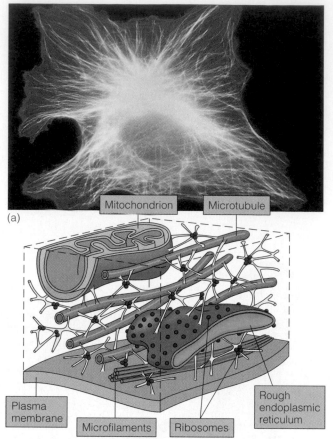

Mitochondrion · Microtubule

(a)

Rough
endoplasmic
reticulum

Plasma
membrane

Microfilaments · Ribosomes

(b)

✦ **FIGURE 3.28**
(a) Photomicrograph of the cytoskeleton of a fibroblast. Microtubules are stained in yellow, and microfilaments are stained in red. (b) Three-dimensional representation of the cytoskeleton showing its major components.

ments known as the **cytoskeleton,** which establishes and maintains cell shape, anchors internal structures including the organelles, and functions in cell movement and cell division (✦ Figure 3.28). The cytoskeleton is composed of three types of fibers. The largest (about 25 nanometers in diameter) are **microtubules** (Figure 3.28). They are assembled from protein subunits known as **tubulins.** Microtubules play several important roles in the cell: Those located beneath the plasma membrane help maintain cell shape; microtubules assembled during cell division are involved in movement of chromosomes. As described earlier, microtubules are also structural components of cilia and flagella.

The other components of the cytoskeleton are **intermediate filaments** and **microfilaments.** Intermediate filaments can be constructed from several types of protein subunits. They play a role in anchoring organelles, including the nucleus. Microfilaments are rods composed of a protein called **actin.** They play an important role in cell division and cell movement. During cell division in animals, actin microfilaments organize to help pinch the cytoplasm in two. In both plant and animal cells, microfilaments are associated with cytoplasmic streaming or cell movement such as the extension of cytoplasmic fingers.

## CELL-CELL INTERACTIONS COORDINATE CELL FUNCTION

In multicellular organisms, the ability of a group of cells to act in a specific manner depends not only on the orderly arrangement of cells into tissues, but also on the coordinated action of cells. Various structures facilitate cell–cell interactions, maintain tissue organization, and promote the combined action of cells. In animal cells, **tight junctions** join the plasma membranes of adjacent cells to form a barrier that prevents passage of material between cells (✦ Figure 3.29). Such junctions between cells of the skin prevent materials from entering the body by passing between cells. **Desmosomes** are circular regions of membrane cemented to an adjacent membrane by a molecular glue made of polysaccharides. Each joint is backed with a plate composed of proteins. Desmosomes are found in tissues that undergo stretching such as the skin and lining of the stomach.

Cell–cell interactions can also be used to transfer molecules and ions between adjacent cells. **Gap junctions** in the membranes of animal cells provide a direct means of communication between the cytoplasm of adjacent cells.

In higher plants, cells are separated from each other by cell walls as well as plasma membranes. Junctions known as **plasmodesmata** (sing., **plasmo-**

*mena* or *Paramecium* are covered with rows of cilia that are used to propel the cell through an aqueous environment. Cells of the human oviduct have cilia on their surface that move the egg from the ovary through the oviduct to the uterus.

Other cells such as *Chlamydomonas* or the sperm of multicellular organisms use a whip-like flagella for locomotion.

Both cilia and flagella have a similar pattern of organization, but flagella are usually much longer than cilia. Each cilium and flagellum consists of a membrane continuous with the plasma membrane that surrounds a space called the matrix. The matrix contains a circular arrangement of microtubules. A ring of nine doublet microtubules is arranged around a pair of central microtubules. At the base of the cilium or flagellum is a **basal body,** consisting of nine triplet microtubules arranged in a circle, with no central microtubules. The basal body is thought to play a role in organizing the structure of the cilia and flagella.

## THE CYTOSKELETON GIVES CELLS SHAPE AND SUPPORT

The cytoplasm of eukaryotic cells contains a three-dimensional network of protein tubules and fila-

**desma**) penetrate the cell walls and plasma membranes, allowing direct communication between the cytoplasm of adjacent cells. Cell–cell interactions are important in the action of tissues and organs, and we will discuss the role of these structures in functions such as muscle contraction later in the text (Chapter 19).

## CELLS ARE ORGANIZED INTO TISSUES AND ORGANS

Many eukaryotic organisms such as ciliated protozoa and some algae are composed of single cells that must carry out all biological functions such as digestion, respiration, and reproduction. Multicellular eukaryotes, on the other hand, are more than just a collection of cells. The cells of these organisms are organized into groups that carry out only one or a small number of specialized functions. **Tissues** are groups of similar cells organized to carry out a specific function. Tissues of different types are organized into units of function known as **organs.** Finally, at the highest level of organization, a number of organs that perform related functions constitute an **organ system,** such as the digestive system or nervous system.

No matter how complex its structure or function, remember that each organ system and tissue is a structure that is a variation on a theme: All are composed of cells, each with a plasma membrane, nucleus, and cytoplasm, and that all of life's processes can be described by events that take place within and through the action of cells. The structure and function of organ systems will be covered in Section V.

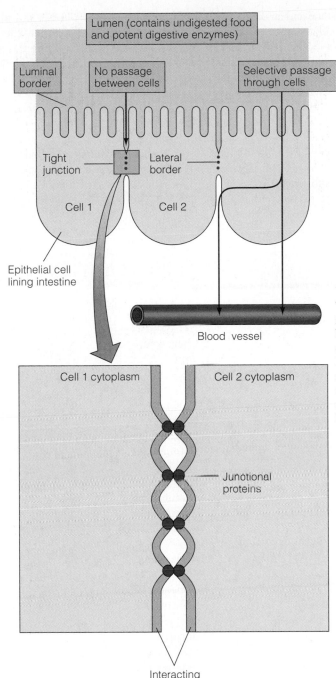

✦ **FIGURE** 3.29
**Tight junctions form impermeable barriers between epithelial cells of the small intestine. Materials are able to move selectively from the lumen through the epithelial cell into a blood vessel.**

# SUMMARY

1. The cell theory is one of the fundamental organizing principles of biology. It states that all living organisms are composed of at least one cell. Related to this are the ideas that all cells are alike in their chemical composition and are derived from preexisting cells. Properties of organisms such as growth, metabolism, and repro-

duction are also properties of the cells that make up the organism. The study of structure and function in cells provides information about the organization and function of the intact organism.

2. All cells are bordered by a double-layered membrane that separates the cell from the nonliving environment and regulates the

exchange of material with the environment. Each layer is composed of lipids and proteins. Many cells are also surrounded by an outer coat or capsule. Cells of plants and some bacteria and fungi are surrounded by cell walls.

3. Within many cells, the most prominent structure is the nucleus. Organisms hav-

ing a nuclear membrane and an organized nucleus are called eukaryotes; those having no nuclear membrane are prokaryotes. In eukaryotes, the nucleus contains chromatin that condenses into chromosomes during cell division. Genetic information carried in the chromosomes determines the structural and functional characteristics of the cell.

4. The cytoplasm is a viscous semiliquid that contains many types of macromolecules as well as cellular substructures known as organelles. The cytoplasm is organized by a framework of tubules and filaments known as the cytoskeleton. The organelles of the cytoplasm divide the cell into a series of compartments, including vesicles and vacuoles, the endoplasmic reticulum, and the Golgi complex. The cytoplasm also contains ribosomes, which represent the sites of protein synthesis within the cell.

5. Mitochondria are membranous organelles found in both plant and animal cells that are the centers of cellular respiration. The cells of many plants contain chloroplasts, which function in photosynthesis.

6. Although some organisms are composed of single cells, others are composed of many cells organized into tissues, which are groups of structurally similar cells that carry out one or a small number of functions. Tissues of different types are combined to form organs, and organs are often organized into a system such as the digestive system or circulatory system that performs a closely related set of functions.

 KEY TERMS

abiogenesis
actin
active transport
basal body
cells
cell theory
cellular respiration
cellulose
chloroplast
chromatim
chromosomes
cilia, cilium
cristae, crista
cytoplasm
cytoskeleton
desmosomes
dictyosomes
diffusion
endocytosis
endoplasmic reticulum (ER)
eukaryotic
exocytosis
fluid-mosaic
gap junctions

glycoproteins
glyoxysomes
Golgi complexes
grana
hydrophilic
hydrophobic
hypertonic
hypotonic
intermediate filaments
isotonic
lignin
lysosomes
microfilaments
microtubules
middle lamella
mitochondria, mitochondrion
nuclear area
nuclear pores
nucleoli, nucleolus
nucleus
organelles
organs
organ system
osmoregulation

osmosis
passive transport
pectin
peroxisomes
phagocytosis
phospholipids
photosynthesis
plasmodesmata, plasmodesma
plastids
primary cell wall
prokaryotic
protists
ribosomes
secondary cell wall
spores
stroma
tight junctions
tissues
tubulins
vacuoles
vault
vesicles

## TRUE/FALSE

1. Biological cells were first discovered by Hooke in 1665.
2. The Greeks' belief in abiogenesis provided the foundation for modern-day biology.
3. There are three basic types of cells: prokaryotic, eukaryotic, and heteroplasmic.
4. Phospholipid membranes consist of a hydrophilic head and a hydrophobic tail.
5. Diffusion accounts for the majority of substances passing across the cell membrane.
6. All organisms have the same number of chromosomes because we evolved from the same ancestor.
7. Mitochondria and chloroplasts have the ability to self-replicate.

## MATCHING

Match each organelle with its function.

1. nucleolus        ____ intracellular digestion
2. ribosome         ____ organization of DNA
3. endoplasmic      ____ protein synthesis
   reticulum        ____ movement
4. lysosome         ____ protein modification
5. mitochondria          and lipid synthesis
6. cilia            ____ ribosomal subunit
7. nucleus               assembly
                    ____ "powerhouse" of
                         the cell

## SHORT ANSWER/MULTIPLE CHOICE

The following questions pinpoint important material in this chapter. Supply an answer for each.

1. The differences between animal and plant cells are distinct. animal cells contain mitochondria. Plant cells contain chloroplasts. Do animal cells contain chloroplast? Why or why not? What function do chloroplasts serve?
2. The most widely accepted model for the organization of membranes is the .
   a. Big Bang model
   b. fluid-mosaic model
   c. cell movement model
   d. cellular compartmentalization model
3. State the three tenets of the cell theory.
4. On average, the size of the cells of a human and a house fly are the same small size; humans just have many more of them. What is the benefit of cells being so small?

a. Small cells can replicate easier than larger cells.
b. Small cells have more surface area.
c. Small cells are more rigid structurally.
d. Small cells are able to provide for an adequate rate of exchange with the environment.
e. Small cells have lower metabolic rates.
5. What distinguishes a prokaryote from a eukaryote?
6. The polypeptide chains of proteins are assembled in the cytoplasm of the cell. What is the most likely pathway of the proteins from where they are made to the plasma membrane?
   a. smooth ER . . . rough ER . . . plasma membrane
   b. Golgi complex . . . nucleus . . . plastid . . . plasma membrane
   c. chloroplasts . . . chromatid . . . microtubule . . . plasma membrane
   d. vesicle . . . lysosome . . . plasma membrane
   e. none of the above

## SCIENCE AND SOCIETY

1. The cell theory states that all living organisms consist of at least one cell. Once this seemingly simple theory was supported by observations on a variety of organisms, it was extrapolated to all creatures on Earth. The living world is divided between two forms—prokaryotic and eukaryotic.

   The study of evolution would be more intricate if there were more than these two subdivisions of life. For example, it is almost indisputable that life only exists on Earth in our solar system. However, it is conceivable that conditions exist on other planets that have resulted in the evolution of life. Organisms on other planets may be vastly different from our own due to different atmospheric conditions, gravitational forces, and various other contributing factors.

   Speculation about life on other planets helps provide an overview about cellular life on Earth. If we learned that extraterrestrial life does exist, what could we theorize about it from our knowledge of prokaryotes and eukaryotes?

2. The Greek philosopher and scientist Aristotle offered spontaneous generation as an explanation for the origins of living things. This theory asserted that organisms arise spontaneously from inert matter. According to advocates of the theory, it was possible for people to emerge from a worm that developed from the dredges of a mud puddle. It was not until 1861 when Louis Pasteur published the results of an experiment that this theory was laid to rest. Pasteur placed sterilized broth in sterilized swan-necked flasks and allowed them to incubate. Only when the broth was open directly to the air did bacteria appear in the broth. Pasteur clearly showed that bacteria did not arise spontaneously. Why did bacteria appear when the flask was exposed directly to the air? What was the purpose of using a sterilized broth solution and sterilized flasks? Why do you think scientists believed the theory of spontaneous generation for so long? Many of the ideas that led to Pasteur's discovery were built on a foundation of scientific research by many other scientists. Should Pasteur take all the credit for his discovery? Why or why not?

3. One of the central tenets of biology is that all organisms consist of cells and that all cells arise from preexisting cells—except the very first cells that formed on Earth about 3.5 billion years ago. Most organisms living today resemble one another fundamentally, having the same kinds of membranes and hereditary systems. Not all living organisms are the same in these respects, however. Sheltered from evolutionary alterations in unchanging environments resembling those of ancient times, these organisms are living relics, the surviving representatives of the first stages of life on earth. Most organisms now living are descendants of a few lines of bacteria. Why do you suppose certain forms of bacteria became extinct? What do you think contributed to their extinction? What do you think the early bacteria were like? Fungi possess cell walls that are as rigid as those of bacteria. No bacterium is multicellular, however, whereas fungi are. Why do you suppose that this is so?

# CELLS AND ENERGY: RESPIRATION AND PHOTOSYNTHESIS

OPENING IMAGE

*Plants convert solar energy into food molecules.*

As you slide into the driver's seat, turn on the ignition, and slip the car's gearshift into drive, you are taking advantage of energy conversion to provide yourself with a means of transportation. In the car's battery, chemical energy is converted into electrical energy to power the starter. Inside the engine, energy stored in chemical bonds between atoms of gasoline molecules is released by a combination of pressure generated by the piston and electrical energy provided by a sparkplug. As the energy is released in an explosion known as combustion, it is converted into mechanical energy that rotates the driveshaft and makes the wheels turn. This energy conversion is not complete; some bonds in the fuel molecules remain unbroken, and the remaining chemical products of combustion are released as exhaust fumes. Other energy is present in the form of heat generated by the explosion; this energy is absorbed by the fluid in the cooling system and dispersed by the radiator.

As you sit at the wheel of the car, a process of energy conversion similar to combustion is taking place in the trillions of cells that make up your body. This energy conversion, known as metabolism, enables a cell to maintain its existence, grow, function and reproduce. In cells, energy is provided by breaking the chemical bonds in the atoms of organic food molecules. Unlike combustion, metabolism releases energy in stages, basically one chemical bond at a time. This enables the cell to capture more of the energy stored in the chemical bonds in a controlled way, whereas in combustion, the energy is released all at once, and much is wasted.

*In this chapter we examine how cells are able to store energy from the sun in food molecules, and how they extract energy from food molecules and use this energy for cellular functions. The sum of all these chemical reactions in the cell is called metabolism, and it involves thousands of possible operations. The steps in the conversion of light energy into food molecules are parts of* **photosynthesis.** *The three steps involved in the production of energy from food molecules are known as* **cellular respiration.** *After reviewing the basic process of metabolism, we will consider each of the steps in photosynthesis and cellular respiration, evaluate the efficiency of this energy conversion, and discuss how the process is controlled. In later chapters, we will see how the energy extracted from food is converted by cells into other forms of energy including electricity, mechanical energy (motion), heat, and even light.*

## METABOLISM IS A METHOD OF ENERGY CONVERSION

In the physical and biological worlds, energy exists in a variety of forms, including chemical energy and electrical energy, as well as heat and light. **Energy** can be defined as the ability to bring about change or, as physicists say, to do work. The study of energy is called thermodynamics, and the behavior of energy in both the physical and biological worlds is governed by basic principles called the laws of thermodynamics. The first of these laws is that energy is always conserved; it cannot be created or destroyed, only changed from one form to another (✦ Figure 4.1). When chemical bonds in food molecules are broken, the energy stored in them can be used to make other chemical bonds, or released as heat, but no new energy is created and no energy is destroyed.

Second, in the transfer of energy from one form to another, the process is inefficient, and some of the energy will be dissipated as heat. As the amount of energy in a system decreases, the disorder in the system increases. The term **entropy** is used to describe the measurement of this disorder; as the amount of disorder increases, the amount of entropy increases.

Because cells and organisms are highly organized systems that convert energy from one form to another, they have low entropy and need a constant supply of energy to maintain order and life itself. This flow of energy into living systems ultimately originates in the sun (✦ Figure 4.2). Solar energy in the form of light is generated in thermonuclear reactions in the interior of the sun and then radiates through space. Although much of this energy strikes the earth and is converted to heat, a very small portion of light is captured by photosynthesis as the first step in a series of energy exchanges that drives most living systems on this planet. As we noted earlier, the sum of these energy exchanges is known as **metabolism.**

### Biochemical Reactions and Metabolism

At a given moment inside a cell, thousands of chemical reactions are occurring simultaneously. At first glance, this may seem like a chaotic situation. A closer look at metabolism, however, reveals a defi-

✦ FIGURE 4.1

**Combustion efficiently releases the stored energy in the wood and converts it to heat and light, illustrating the first law of thermodynamics.**

✦ FIGURE 4.2

**In biological systems, energy flows from sunlight through photosynthesis to power all cellular functions via ATP.**

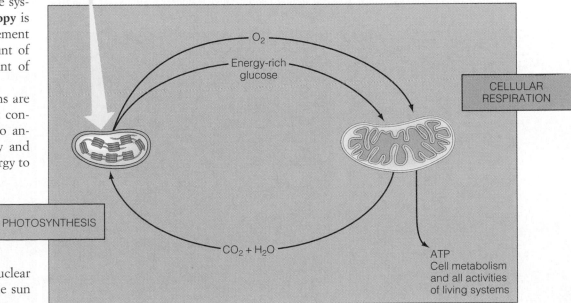

$O_2$

Energy-rich glucose

CELLULAR RESPIRATION

PHOTOSYNTHESIS

$CO_2 + H_2O$

ATP
Cell metabolism and all activities of living systems

nite order to the operation of the cell. First, each metabolic reaction is mediated by an enzyme that controls the specificity of the reaction and, in some cases, its speed. Second, enzymatic reactions are organized into clusters known as metabolic pathways, which result in an end product. Third, the cell is subdivided into a series of compartments (mitochondria, nuclei, endoplasmic reticulum, lysosomes, etc.) where enzymes in a metabolic pathway are segregated to carry out their functions.

The overall energy equations for the metabolic pathways of photosynthesis and respiration are as follows:

photosynthesis:   $CO_2 + H_2O + energy \rightarrow sugar + O_2$

respiration:   $sugar + O_2 \rightarrow CO_2 + H_2O + energy$

Photosynthesis uses energy from sunlight to combine carbon dioxide and water into food molecules in the form of carbohydrates, releasing oxygen in the process. Respiration uses oxygen in the breakdown of carbohydrates into carbon dioxide and water, releasing energy for cellular functions. These equations show that sunlight is the driving force in energy transfer, first in the conversion to food molecules and then as the ultimate source of energy released during respiration. In effect, your body is solar powered.

✦ **FIGURE 4.3**
ATP molecules have three components. Energy is stored in the phosphate bonds.

Phosphate groups

Adenine

Ribose

## Metabolic Pathways Are Organized Sets of Reactions

As discussed previously, enzymatic reactions within a cell do not occur at random: Most are organized into one or more series of reactions known as biochemical pathways or metabolic pathways. In such a pathway, the product of one reaction serves as the substrate for the next reaction. This results in greater efficiency in energy transformation, because there is little or no accumulation of intermediate products in the pathway. This overall efficiency is increased when the enzymes are grouped into clusters attached to the inside or outside of organelles or when enzymes are secured to the cytoskeleton. Photosynthesis takes place in molecular complexes organized within chloroplasts, and major parts of cellular respiration take place within the membrane system of mitochondria.

## ATP Transfers Energy Within Cells

Within the cell, energy transfer almost always involves a molecule called **adenosine triphosphate** or **ATP.** Although glucose has energy stored in its chemical bonds, this energy cannot be used directly to power chemical reactions in the cell. This energy must first be transferred into the chemical bonds of an ATP molecule. In turn, the ATP can transfer the energy to drive energy-absorbing (endergonic) reactions of metabolism.

ATP consists of a nucleotide with three phosphate groups (✦ Figure 4.3), and it is formed in an endergonic reaction as follows:

✦ **FIGURE 4.4**
**Photosynthetic organisms provide the link between sunlight and biological energy systems. (a) tulip tree. (b) pond lilies. (c) field of lupine.**

(a)

(b)

(c)

$$\text{ADP} \quad + \quad \text{P}_i \quad + \quad \text{energy} \quad = \quad \text{ATP}$$

adenosine    inorganic                adenosine
diphosphate   phosphate            triphosphate

Some energy is transferred from ATP when the chemical bond holding the third phosphate group to the molecule is broken:

$$\text{ATP} \rightarrow \text{ADP} + \text{energy} + \text{P}_i$$

The energy transferred in breaking this chemical bond is what drives many of the metabolic reactions in the cell. To prevent the depletion of ATP, it is constantly synthesized by the breakdown of food molecules (usually from glucose).

## PHOTOSYNTHESIS CONVERTS SOLAR ENERGY INTO FOOD

Photosynthesis is the process whereby sunlight, oxygen, and water are combined to harvest solar energy and store it in the chemical bonds of sugar molecules. Organisms that can photosynthesize include some bacteria, algae, and green plants (✦ Figure 4.4). These organisms are key factors in the evolution and preservation of life on this planet. All the oxygen in the atmosphere has been produced as a by-product of photosynthesis. This oxygen is utilized by organisms that obtain energy via cellular respiration. Oxygen has made possible the evolution of active, multicellular organisms with a high demand for ATP. All food molecules utilized by photosynthetic and nonphotosynthetic organisms ultimately originate from photosynthesis (exceptions to this are described later).

There are two main processes in photosynthesis:

1. *Conversion of light into chemical energy.* This is accomplished by harvesting energy from sunlight and transferring this energy into the chemical bonds of ATP. The reactions in this process are called the **light reactions,** because they take place only in light.
2. *Conversion of chemical energy into food molecules.* This is accomplished by a series of chemical reactions that uses ATP from the light reaction to form organic molecules from carbon dioxide (from the atmosphere). The reactions that form the food (sugar) molecules are called the **dark reactions** because they can occur in the dark as long as ATP is present.

### Light Is Converted into Chemical Energy

Photosynthesis takes place in specialized membrane structures called **thylakoids,** flattened membrane sacs or vesicles (✦ Figure 4.5). In prokaryotes,

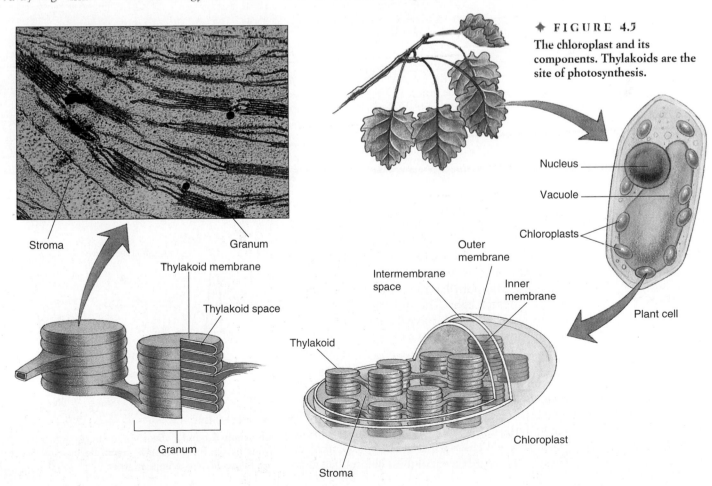

✦ **FIGURE 4.5**
The chloroplast and its components. Thylakoids are the site of photosynthesis.

Stroma

Granum

Thylakoid membrane

Thylakoid space

Granum

Nucleus

Vacuole

Chloroplasts

Plant cell

Outer membrane

Intermembrane space

Inner membrane

Thylakoid

Chloroplast

Stroma

# GUEST ESSAY: RESEARCH AND APPLICATIONS

## Exploring Membranes

ANNE WALTER

While growing up it seemed to me that a biologist could be only a physician or a quintessential naturalist with sensible shoes, binoculars, and field notebook. Yet even though I loved biology, neither of these futures felt right. Fortunately, two wonderful teachers, Miss Hill and Miss Strosneider, taught biology in the Washington, D.C., public schools. With a paper chromatography experiment to extract plant pigments and a lab on enzymes that I remember to this day, these two remarkable teachers were the first to show me that the basis for much of biology was the precise and intricate interactions of specialized molecules. This was amazing! My interest must have showed, because I was encouraged to compete for an American Heart Association research opportunity that resulted in a summer at George Washington University working on lipid metabolism. Little did I expect lipids to be in my future.

My experiences in college helped me decide that I was interested in physiology and I started graduate studies with Maryanne Hughes, who wanted to understand how seagulls are able to drink seawater. These birds have several special adaptations including a gland that secretes an incredibly salty solution after they've had a salty drink. I had learned that because water permeability across cell membranes is high, cells equilibrate rapidly with their external medium. So why didn't the seagull salt glands rapidly lose water and shrink? I concluded that

there must be something unusual about their membrane lipids and went on to study at Duke University in a department that specialized in membranes. My research problem was to define the permeability properties of the lipid bilayer in the absence of protein. My results confirmed that hydrophobic solutes like $CO_2$, ethanol, and aspirin all penetrate the membrane rapidly, whereas hydrophilic molecules such as glucose or amino acids penetrate very slowly. Part of my research was to test our methods thoroughly to ensure our values reflected the true permeabilities, allowing us to be very confident in our conclusion that the membrane behaves like a hydrocarbon that is very constrained . . . like an oil, but not like an oil. One potential application of this research may be a phospholipid "sponge" for cleaning up hydrocarbon pollutants. If lipid bilayers can be a sponge to soak up other molecules they can also release molecules. This idea is being used by the cosmetics and pharmaceutical industries to develop phospholipid dispersions as safe, slow-release systems.

My current research asks whether the behavior of transmembrane proteins is affected by their environment. With Neal Rote's research group at Wright State University, I have helped explore the possibility that the autoimmune disease called "antiphospholipid antibody syndrome" might be due to antibodies against phosphatidylserine that react when this

lipid is exposed on the outside of the cell during platelet activation and possibly during placenta formation. I never would have guessed that expertise with membranes and lipids would be important in trying to figure out a disease process that has as one of its main symptoms blood clotting disorders and poor placental development.

Lipid bilayers are essential to all living cells. In fact, it has been suggested that the "primordial soup" contained lipids that spontaneously formed closed vesicles and bilayer surfaces that both protected and concentrated the protoenzymes as one of the first steps in the origin of living cells. Discovering the molecular basis for these properties is a puzzle that is turning out to be quite exciting to put together.

*Anne Walter is an associate professor in the Biology Department of St. Olaf College in Northfield, Minnesota. She received a B.A. in biology in 1973 from Grinnell College in Iowa and a Ph.D. in physiology and pharmacology in 1981 from Duke University, North Carolina.*

these structures can be part of an extensive internal membrane system (as in *Anabena*) or be embedded in the cell membrane. In eukaryotes, thylakoids are contained in **chloroplasts** (Figure 4.5). Within thylakoids, light is absorbed by pigments. Green plants use **chlorophyll** as the pigment for capturing energy from sunlight. The two chemical forms of chlorophyll (called chlorophyll *a* and *b*) are used in photosynthesis (✦ Figure 4.6).

Pigments in a thylakoid are organized into clusters of several hundred molecules called **photosystems.** There are two different types of photosystems in eukaryotes, photosystem I and II. When light is absorbed by a photosystem II cluster (✦ Figure 4.7), the absorbed energy is transferred to specially arranged chlorophyll *a* molecules within the cluster. The arrangement of these chlorophyll mole-

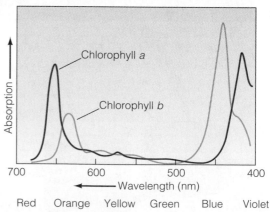

✦ **FIGURE 4.6**

**Chlorophyll *a* and *b* absorb light energy across the spectrum, except for yellow and green. Because these colors are reflected, leaves appear green.**

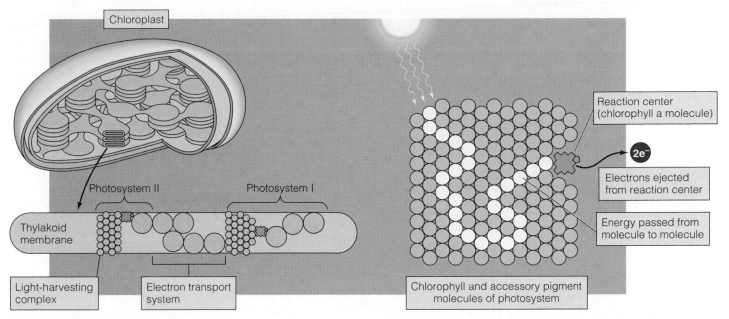

Chloroplast

Photosystem II    Photosystem I

Thylakoid membrane

Light-harvesting complex

Electron transport system

Reaction center (chlorophyll a molecule)

2e⁻

Electrons ejected from reaction center

Energy passed from molecule to molecule

Chlorophyll and accessory pigment molecules of photosystem

✦ **FIGURE 4.7**

**Photosynthesis takes place in two photosystems in the thylakoids. In both photosystems, energy absorbed from light ejects electrons from chlorophyll molecules.**

cules acts much like a satellite dish or antenna to collect, amplify, and focus the energy in sunlight. As the collected energy reacts with the chlorophyll molecules, electrons from the chlorophyll are transferred to a molecule known as an **electron acceptor** (✦ Figure 4.8).

The electron acceptor is part of an energy transfer system in photosystem II through which the energy-carrying electron is moved. Because each energy transfer is inefficient, some energy is lost at each transfer. As electrons are transferred from molecule to molecule in the transfer system, some of its energy is lost. The remaining energy is used to pump protons (H⁺ ions) into the interior of the thylakoid. This step is essential for the formation of ATP, as we discuss later.

## Energy Transfer Is Mediated by NADPH

At the end of the transfer system in photosystem II, the electron is depleted of much of its energy and is transferred to photosystem I. The electron entering photosystem I is moved from a pigment molecule to an electron acceptor (Figure 4.8). This transport system transfers electrons to **nicotine adenine dinucleotide phosphate (NADP⁺),** which combines with a proton to form NADPH:

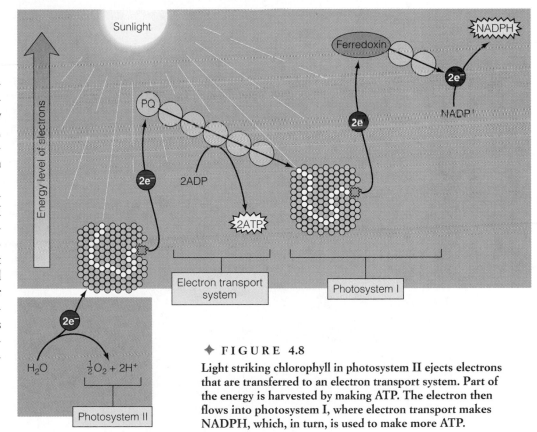

Sunlight

Energy level of electrons

NADPH

Ferredoxin

2e⁻

NADP⁺

2e

PQ

2e⁻

2ADP

2ATP

Electron transport system

Photosystem I

2e⁻

$H_2O$    $\frac{1}{2}O_2 + 2H^+$

Photosystem II

✦ **FIGURE 4.8**

**Light striking chlorophyll in photosystem II ejects electrons that are transferred to an electron transport system. Part of the energy is harvested by making ATP. The electron then flows into photosystem I, where electron transport makes NADPH, which, in turn, is used to make more ATP.**

$$NADP^+ + 2e + H^+ \rightarrow NADPH$$

Like ATP, NADPH serves as a storage form of energy, though it is less versatile than ATP. Sunlight causes the flow of electrons (an electrical current) from photosystem II to photosystem I to NADP⁺. The electrons lost from photosystem II are replaced by splitting water molecules into electrons, oxygen, and protons:

$$H_2O \rightarrow 2 \text{ electrons} + \tfrac{1}{2}O_2 + 2H^+$$

Electrons released from the water are transferred to photosystem II, and the oxygen gas is released into the environment (Figure 4.8).

The electron flow generated by sunlight through the photosystems to NADPH is accompanied by the transfer of protons into the inner lumen of the thylakoid and the formation of ATP. The entire process is shown in ◆ Figure 4.9.

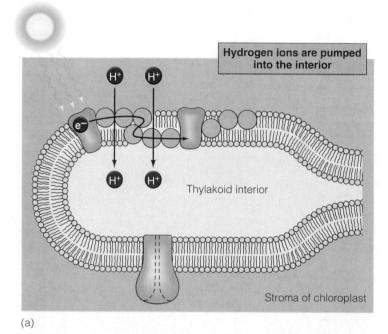

(a)

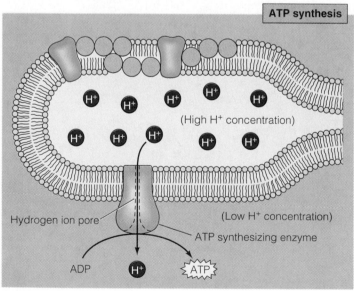

(b)

◆ **FIGURE 4.9**

**ATP synthesis in thylakoids. (a) Electron flow through the photosystems generates transfer of protons into the thylakoid. (b) As these ions leak outward, they drive the synthesis of ATP.**

## What the Light Reaction Accomplishes

Sunlight striking a thylakoid causes protons to be pumped into the inner space of the thylakoid using energy from the transport of electrons along the transfer system. More protons are produced in the inner space when water molecules are split to release protons and contribute electrons to photosystem II.

As protons accumulate inside the thylakoid, they begin to flow outward through channels in the thylakoid membrane in response to a concentration gradient. As they do so, they pass through an array of proteins in the channel wall that use the energy of proton movement to convert ADP into ATP. This step represents the primary source of ATP generated by photosynthesis.

At this stage, it is fair to ask why any other steps in photosynthesis are necessary. After all, if molecules like ATP and NADPH are able to drive most of the metabolic reactions of the cell, and the first stage of photosynthesis results in the synthesis of ATP and NADPH, why is it necessary to convert this energy into carbohydrate molecules? The answers are that conversion to food molecules is a more convenient and more stable way to store the energy trapped from sunlight, and that molecular synthesis adds directly to the mass of the plant, resulting in growth. We examine this process of conversion to food molecules next.

## The Dark Reaction Converts Chemical Energy into Food

The second stage of the conversion of sunlight to sugar, called the **dark reactions,** utilizes the energy transferred to ATP and NADPH in the light reactions to make sugar molecules using carbon dioxide. Plants such as algae use carbon dioxide present in the surrounding water, whereas land plants use carbon dioxide obtained directly from the air. The dark reactions take place on the *outer* surface of the thylakoid membrane.

These reactions require the following components: an energy source in the form of ATP and NADPH, a five-carbon sugar molecule, ribulose biphosphate (RuBP) to which carbon dioxide can be attached, and the enzymes to carry out the reactions (◆ Figure 4.10). The first reaction attaches carbon dioxide to RuBP to form an unstable six-carbon molecule that splits into two molecules of **PGA (phosphoglycerate).** Each three-carbon PGA molecule is then converted to **PGAL (phosphoglyceraldehyde),** using ATP and NADPH made in the light reactions.

The dark reactions form a circular metabolic pathway, in which some PGAL molecules are converted to six-carbon sugars, including glucose, and others are used to regenerate RuPB. The glucose produced in this pathway can be used to synthesize starch, cellulose, pectin, and other plant products. PGAL can also be used to synthesize amino acids and lipids. More than half of the PGAL is used to form more RuBP, using ATP made in the light reactions (Figure 4.10).

In the molecular bookkeeping of energy conversion, for every three molecules of RuBP that have a carbon added from carbon dioxide, the yield is one molecule of PGAL. The cost of this molecule of PGAL is three molecules of carbon dioxide, nine molecules of ATP, and six molecules of NADPH. The overall equation for the production of food molecules from the ATP and NADPH produced by solar energy in the cycle is:

$$3 \text{ RuBP} + 3 \text{ CO}_2 + 9 \text{ ATP} + 6 \text{ NADPH} \rightarrow$$
$$3 \text{ RuBP} + 9 \text{ ADP} + 9 \text{ P}_i + 6 \text{ NADP}^+ + 1 \text{ PGAL}$$

## CHEMOSYNTHESIS: AN ALTERNATIVE TO PHOTOSYNTHESIS

In 1977, scientists exploring the sea floor to study seams in the earth's tectonic plates discovered oases of life-forms living at depths below 3000 meters (9800 feet) (✦ Figure 4.11). Because sunlight cannot penetrate to these depths, organisms living in these communities are completely independent of photosynthesis. These ecosystems are clustered around vents in the sea floor, where superheated water (at up to 350°C) rich in dissolved minerals emerges, cools, and disperses. Along with water, large amounts of hydrogen sulfide are released through these vents. Bacteria present in and near these vents use the sulfur in hydrogen sulfide and combine it with oxygen to generate energy. These

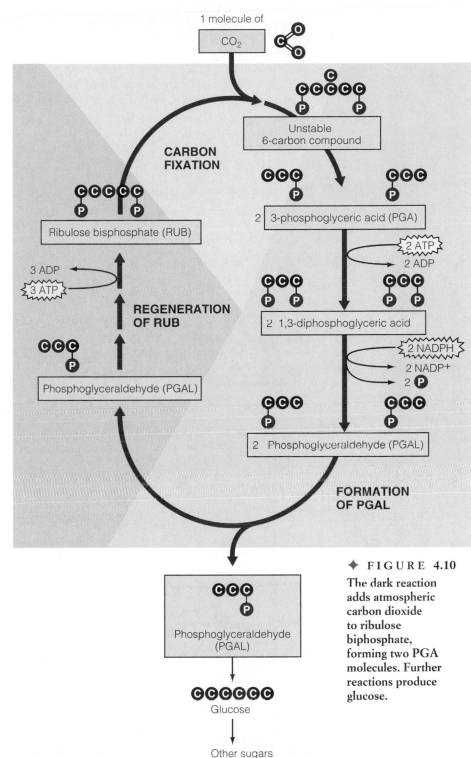

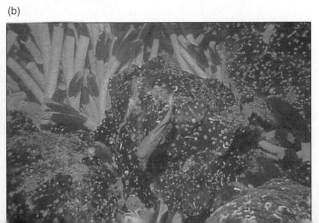

✦ FIGURE 4.10

The dark reaction adds atmospheric carbon dioxide to ribulose biphosphate, forming two PGA molecules. Further reactions produce glucose.

✦ FIGURE 4.11

Organisms around thermal vents in the ocean floor rely on the transfer of electrons from sulfur to oxygen to generate energy. These organisms are not dependent on light as an energy source. (a) tube worms. (b) tube worms and fish.

(a)

(b)

bacteria are able to live at temperatures where the water is so hot that only the extreme pressure keeps it liquid.

In addition to bacteria, the vents support a rich array of organisms including beardworms, mussels, clams, fish, and crabs. Bacteria live within the tissues of the beardworms and produce energy and organic compounds used by the worms. Other organisms feed on free-living bacteria, and still others are scavengers and predators. Many of the species found at these depths are unrelated to those living at shallower levels, and represent previously unknown life-forms. These large and diverse commu-

nities exist without dependence on photosynthesis or direct input of solar energy.

## CONVERTING FOOD INTO ENERGY IS A THREE-STAGE PROCESS

All cellular functions require energy. To carry out muscle contraction, nerve impulse conduction, endocytosis, active transport, and other operations, animal cells need a constant supply of energy. Aside from photosynthesis, plant cells require energy to carry out cellular functions. The usual source of this

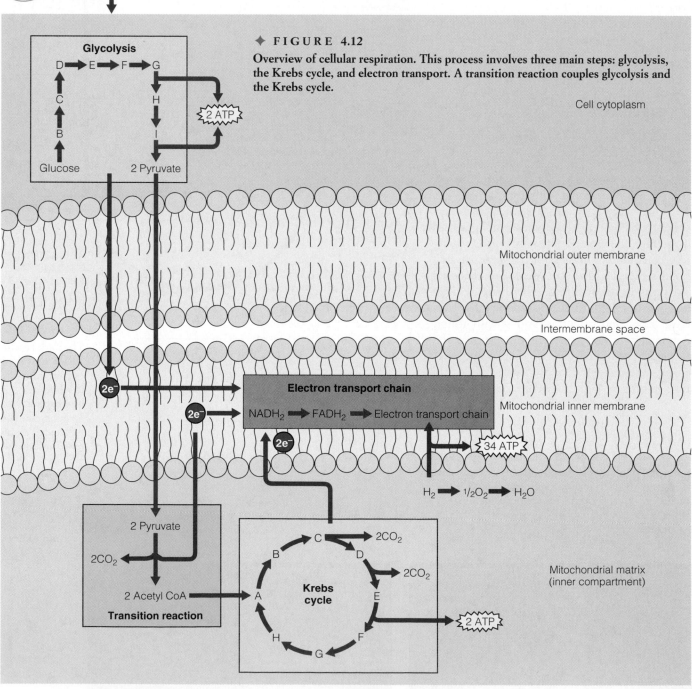

✦ **FIGURE 4.12**

**Overview of cellular respiration. This process involves three main steps: glycolysis, the Krebs cycle, and electron transport. A transition reaction couples glycolysis and the Krebs cycle.**

energy is ATP. ATP is produced by breaking the chemical bonds in food molecules, especially the bonds in molecules of glucose and related carbohydrates. This breakdown is a complex process that begins in the cytoplasm and is completed in the mitochondria (◆ Table 4.1). The overall reaction for the breakdown of glucose is:

glucose + oxygen → carbon dioxide + water + energy

$$C_6H_{12}O_6 + 6O_2 \rightarrow 6CO_2 + 6H_2O + \text{ATP and heat}$$

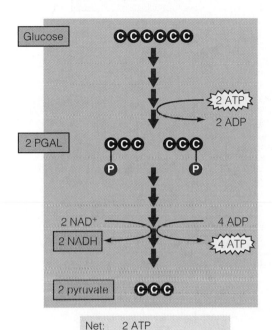

◆ **FIGURE 4.13**

**In glycolysis, a glucose molecule is broken down into two molecules of pyruvate, with a net gain of two ATP and two NADH molecules.**

The process is called cellular respiration because "to respire" means to breathe, that is, to exchange oxygen and carbon dioxide with the atmosphere. During the breakdown of glucose, cells take in oxygen and give off carbon dioxide. This process can be divided into three stages:

1. Glycolysis, the breakdown of glucose (in the cytoplasm)
2. The Krebs cycle (in the mitochondria)
3. Electron transport and ATP formation (in the mitochondria).

These steps are shown in ◆ Figure 4.12.

## Glycolysis Converts Glucose to Pyruvate

The first step in the breakdown of glucose is a set of metabolic reactions that takes place in the cytoplasm of the cell. In these reactions, a six-carbon glucose molecule is broken down into two three-carbon molecules of pyruvate (also called pyruvic acid). These steps require the use of some ATP, but there is a net gain of two ATP molecules for every molecule of glucose that is split (◆ Figure 4.13). Electron and proton transfer during pyruvate formation results in the synthesis of NADH from NAD⁺. These protons and electrons will play a role later in the production of ATP. The products of glycolysis include the following:

1. *Pyruvate.* Two molecules of pyruvate are formed from each molecule of glucose. Most of the chemical energy contained in a glucose molecule is still present in the pyruvate molecules. (Each pyruvate molecule will be broken down in the Krebs cycle, during which more ATP is produced.)
2. *ATP.* Each glucose molecule that is broken down results in a net gain of two ATP molecules.

*Sidebar*

**DEEP-SEA VENTS AND BIOTECHNOLOGY**

Communities of plants and animals are found clustered around deep-sea vents, far below levels where light penetrates. The vents are formed by movement of tectonic plates, which allows seawater to penetrate into the underlying hot magma. The water is superheated, and pushed back to the seafloor. Ecosystems around these vents have a food chain that starts with bacteria that use hydrogen sulfide dissolved in the superheated water for energy transfer. Small planktonic organisms eat the bacteria, and the plankton are, in turn, eaten by larger animals such as limpets and snails.

The bacteria growing in and around deep-sea vents are hyperthermophilic, that is, they can grow at temperatures beyond 100°C (the boiling point of water). These bacteria contain enzymes that function at high temperatures and are useful in a number of applications in biotechnology. Samples of bacteria from deep-sea vents have been recovered using robot submarines and then grown in the laboratory. Proteins extracted from these bacteria are commercially available for use in recombinant DNA techniques such as DNA fingerprinting.

| TABLE 4.1 | OVERVIEW OF CELLULAR ENERGY PRODUCTION | |
|---|---|---|
| REACTION | LOCATION | DESCRIPTION AND PRODUCTS |
| Glycolysis | Cytoplasm | Breaks glucose into two three-carbon compounds, pyruvate; nets two ATPs; nets two NADH molecules. |
| Transition reaction | Mitochondrion | Removes one carbon dioxide from each pyruvate, producing two acetyl CoA molecules; produces two NADH molecules. |
| Krebs cycle | Inner compartment of mitochondrion | Completes the breakdown of acetyl cycle CoA; produces two ATPs per glucose; produces numerous NADH and FADH molecules. |
| Electron transport | Inner membrane of mitochondrion | Accepts electrons from NADH and FADH, generated by previous system reactions; produces 34 ATPs. |

✦ **FIGURE 4.14**

**The chemical reactions involved in glycolysis.**

This metabolic pathway has two parts. In the first phase (steps 1–3), glucose is activated by the addition of ATP in the first and third reactions. In the process, the glucose molecule becomes rearranged, and transformed into an unstable, highly reactive molecule of fructose-1,6 diphosphate. This activation process makes it easier to harvest energy from the bonds holding the glucose molecule together.

In the second phase, the fructose-1,6 diphosphate is split into two three-carbon intermediates (step 4). The DHAP is converted into PGAL in step 5. The two PGAL molecules enter steps 6–9, so these reactions occur twice for each glucose molecule entering the pathway. In step 6, phosphates are added to each PGAL molecule (without using ATP), resulting in the formation of NADH.

In reactions 7–9, the phosphates are stripped off, producing four molecules of ATP and two molecules of pyruvate for each glucose molecule entering the pathway. Since two molecules of ATP were used in the first phase, the net gain to the cell is two molecules of ATP for each molecule of glucose that enters glycolysis. The NADH produced in step 6 is used to produce more ATP in later reactions.

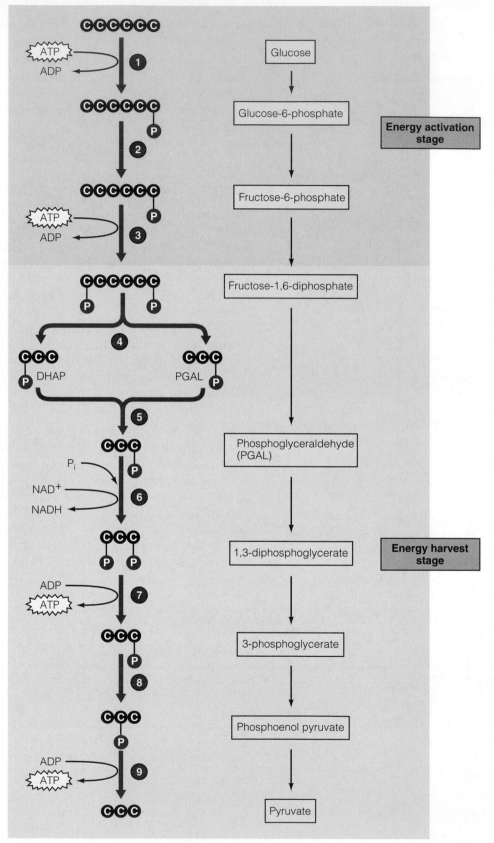

3. *NADH.* Two NADH molecules are produced when glucose is broken down. These molecules pass into the electron transport chain, where energy is released to make ATP.

Details of the chemical reactions in glycolysis are shown in ◆ Figure 4.14.

## The Krebs Cycle Completes the Breakdown of Glucose

In the second stage of glucose breakdown, the three-carbon pyruvate molecules are broken down (chemically speaking, they are oxidized: at this point you may want to review the chemistry of oxidation/ reduction reactions in Chapter 2) into carbon dioxide and water. Breaking the chemical bonds in pyruvate results in the production of more ATP molecules. This second stage of breakdown, called the Krebs cycle, takes place in the inner compartment of the mitochondria. This compartment is formed by the highly folded inner membrane of the mitochondrion (◆ Figure 4.15). Enzymes that carry out the reactions of the Krebs cycle are found in this compartment.

Before entering the Krebs cycle, pyruvate undergoes a transition reaction in which a carbon atom is removed and released as a molecule of carbon dioxide. The remaining two-carbon fragment attaches to a carrier molecule (called CoA or Coenzyme A) to form **acetyl CoA** (◆ Figure 4.16). During these reactions, a molecule of NADH is formed, which will be used later to make ATP. Acetyl CoA is an important intermediate in converting food into energy.

The Krebs cycle is a cyclic series of metabolic reactions in which the last step produces a molecule that represents the starting point for the cycle (Figure 4.16). In glycolysis each glucose molecule gives rise to two molecules of pyruvate, each of which is broken down into an acetyl group for entry into the Krebs cycle. This means that it takes two turns of the cycle to break down the chemical bonds in one molecule of glucose and release six carbon molecules as carbon dioxide. These carbon dioxide molecules leave the cell and, in the case of humans, enter the bloodstream to be transported to the lungs, where it is exhaled into the atmosphere. All that remains of the glucose molecule at this stage is the energy recovered from its chemical bonds, stored in a few ATP molecules and a number of electron carriers such as NADH.

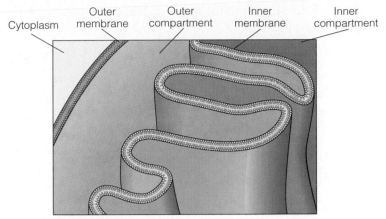

◆ FIGURE 4.15

**The outer and inner membrane of the mitochondrion divides the organelle into two compartments. The Krebs cycle takes place in the inner compartment. Electron transport pumps protons from the inner compartment to the outer compartment, forming ATP.**

The energy dividend produced in the Krebs cycle is greater than that of glycolysis and includes ATP and electron acceptors.

1. *ATP.* Two molecules are produced from each glucose molecule put through the cycle.
2. *Electron acceptors.* In the Krebs cycle, energy stored in the chemical bonds of glucose is transferred to NADH and FADH₂. Each glucose molecule put through the Krebs cycle generates six molecules of NADH and two molecules of FADH₂.

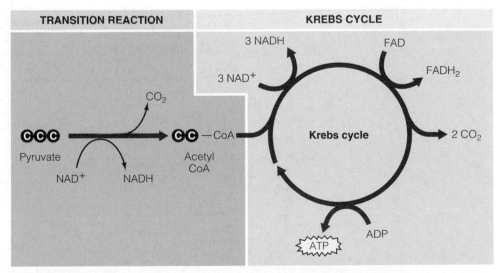

◆ FIGURE 4.16

**The Krebs cycle generates two ATP molecules for each glucose molecule entering glycolysis, but its main products are NADH and FADH, which carry high-energy electrons to the electron transport system.**

Details of the Krebs cycle are presented in ◆ Figure 4.17.

Details of the Krebs cycle are presented in ◆ Figure 4.17.

## ATP Synthesis Is the Third Stage in Converting Food into Energy

After the Krebs cycle, only a fraction of the energy recovered from the breakage of chemical bonds in glucose has been used to form ATP. Most of the energy is present as high-energy electrons in NADH and $FADH_2$ that will enter the electron transport chain to generate ATP. In this third stage in the breakdown of glucose, energy is released in a series of steps as electrons are passed from one electron carrier to another. Some of this energy is captured and used to convert ADP into ATP; some is lost as heat.

Electron transport begins in the inner mitochondrial compartment where NADH and $FADH_2$ transfer their high-energy electrons to an electron-accepting molecule that is part of an electron transfer chain. The molecules of the electron transfer chain are embedded in the inner membrane of

◆ **FIGURE 4.17**
**The reactions of the Krebs cycle.**

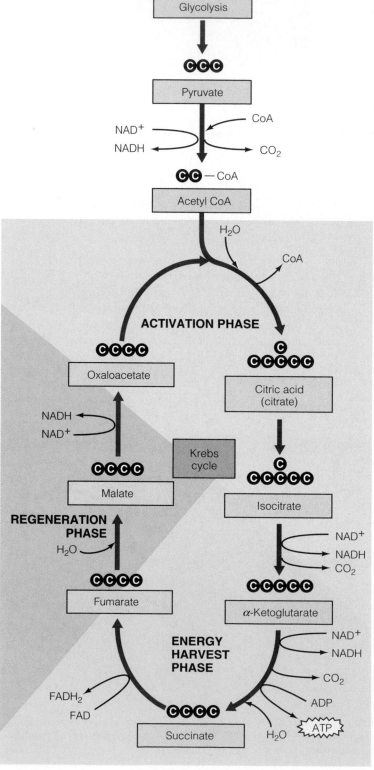

The Krebs cycle can be divided into three stages, an activation stage, an energy-harvesting stage, and a regeneration stage. In the activation stage, acetyl-CoA reacts with the four-carbon compound oxalacetate to form citric acid. CoA is regenerated for use in the transition reaction. Citric acid is rearranged into isocitrate, a reactive molecule.

In the second stage, energy is harvested by successively stripping carbon dioxide and protons from the acetyl group. Energy is also harvested in the form of ATP in the conversion of alpha ketoglutarate into succinate. In the last reaction of this stage, energy is harvested by electron transfer to form FADH. Because two pyruvates are produced for each molecule of glucose entering glycolysis, the yield in the Krebs cycle is four carbon dioxide molecules and two ATPs per molecule of glucose.

In the regeneration reactions, fumarate is rearranged into oxalocetic acid, completing the cycle. The last step in this stage generates NADH, which is used to make ATP.

the mitochondrion (✦ Figure 4.18). Because energy transfer is always inefficient (one of the laws of thermodynamics), some energy is lost as heat in each transfer. The transfer process produces protons (H⁺ ions), and some of the energy is used to pump these protons through the inner membrane into the outer compartment. At the end of the chain, the electrons are passed to oxygen, which combines with protons present in solution to produce water.

## Chemiosmosis Is Used to Form ATP

As in photosynthesis, the proteins of the mitochondrial electron transfer chain have two functions: to move electrons from molecule to molecule, and to pump protons from the inner compartment into the outer compartment. This last step, called **chemiosmosis,** is essential for the formation of ATP.

As protons accumulate in the outer compartment of the mitochondrion, they begin to flow back into the inner compartment through channels in the inner membrane. As they do so, they pass clusters of enzymes that use the energy of proton movement to add phosphate to ADP to generate ATP (Figure 4.18).

It is useful to recall that photosynthesis works in a similar way. As the concentration of protons increases in the thylakoid, they begin to flow outward into the chloroplast and, as they do so, this proton movement is used to convert ADP into ATP.

## Bacteria Use Phosphorylation to Generate ATP

As prokaryotes, bacteria do not have mitochondria to serve as energy centers for the generation of ATP. Instead, electron transport systems are located in the plasma membrane of the bacterial cell. Electron flow through the transport system pumps protons out of the bacterial cell into the surrounding environment. As a result, there are a relatively large number of protons outside the cell, and relatively few protons inside the cell. The cell membrane contains complexes of proteins that act as channels for protons to flow back into the cell. This proton flow drives the synthesis of ATP in the bacterial cell. In effect, the entire bacterial cell acts as a mitochondrion.

## Calculating the Energy Yield in Converting Glucose to ATP

When a single glucose molecule has been broken down into carbon dioxide and water, about 36 molecules of ATP are generated. To calculate how this energy is generated from breaking the bonds in a

✦ **FIGURE 4.18**

**The electron transport system in the inner membrane of the mitochondrion produces ATP via chemiosmosis.**

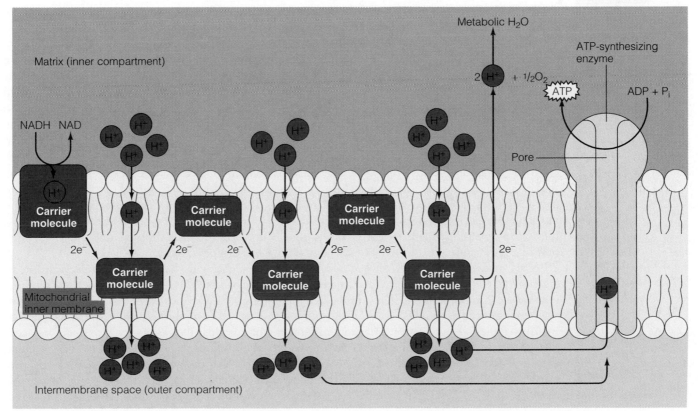

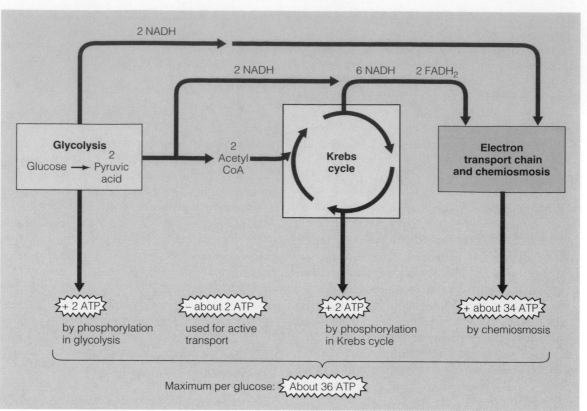

**✦ FIGURE 4.19**

**Summary of energy production in cellular respiration.**

*Sidebar*

**BACTERIAL BATTERIES**

The breakdown of glucose in glycolysis and the Krebs cycle generates electrons. Normally, these electrons flow to complexes of molecules that produce ATP, and end up combining with oxygen or pyruvate. Several groups are attempting to divert this electron flow to create electricity, making a bacterial culture into a battery. In prototypes, sheets of cloth containing a wire screen are placed into a bacterial culture. When sugar is added, the electrons generated by glucose metabolism are collected and carried through wires attached to the screen into a second chamber, where the electrons are transferred to oxygen. In the 1930s, batteries powered by microbes were able to deliver 35 volts (AA batteries deliver 1.5 volts), but were room-sized arrays of cultures. The efficiency of bacterial batteries has been greatly enhanced since then, and small biobatteries about an eighth of an inch square are able to power digital clocks for a short period of time. Further refinements are aimed at capturing more of the energy stored in the sugar that powers the battery. If this work is successful, someday you might recharge batteries by adding sugar.

molecule of glucose to produce ATP, the events of cellular respiration are summarized in ✦ Figure 4.19. During glycolysis, some ATP is directly generated, and NADH is produced. Each molecule of NADH produces 3 ATP molecules in the mitochondria. Since 1 ATP molecule must be used to get the NADH into the mitochondria, the net gain from NADH is reduced. The net yield in the conversion of 1 molecule of glucose into 2 molecules of pyruvate during glycolysis is 6 ATP molecules.

At the beginning of the Krebs cycle, the conversion of 2 pyruvate molecules into acetyl CoA generates 2 NADH molecules, which produce 6 ATP molecules by electron transport. In the Krebs cycle, the two pyruvate fragments result in the production of 2 ATP molecules, 6 NADH molecules, and 2 FADH molecules. The 6 NADH molecules generate a total of 18 molecules of ATP. Electron transport by the 2 molecules of $FADH_2$ results in a total of 4 more molecules of ATP.

All together, then, electron transport generates 32 molecules of ATP. Four are formed in glycolysis and the Krebs cycle, giving a total of 36 ATPs per molecule of glucose (Figure 4.19). This calculation assumes that the system is working at peak effi-

ciency. The yield varies somewhat from cell to cell, depending on a variety of conditions.

How much energy does 36 ATP molecules represent? Can we determine the energy content of food in terms of ATP molecules and total energy needs for the body? First, as we have discussed, remember that energy transfer is somewhat inefficient. The 36 ATP molecules represent about 263 kilocalories. The energy content in the chemical bonds in glucose is about 686 kilocalories, and only about 40% of the energy released when glucose is broken down is conserved in the high-energy bonds of ATP. This may seem inefficient, but by comparison only about 20 to 25% of the energy in gasoline molecules is used in powering a car's engine. The rest escapes as heat.

Figuring in the amount of energy captured from glucose as ATP and the energy used by an active adult (kilocalories/day), this means that the equivalent of 3 to 5 pounds of ATP are used in the course of a normal day. However, the body only contains about one-fifth of an ounce of ATP, meaning that each ATP molecule must be recycled and regenerated many times each day in order to provide the energy necessary to sustain life.

## Converting Fats and Proteins into Energy

Normally, the body uses glucose as an energy source to provide the ATP for cellular functions. Under conditions where glucose is not available, fats and proteins can be used as molecular sources of energy. Dietary fats are digested in the intestine and transported for storage in specialized cells found in fat tissue.

When the supply of food energy available in the diet is too low, fat is released from the storage cells and transported through the body to other tissues. Inside cells, fats are metabolized through a series of biochemical reactions, ending in the formation of acetyl CoA (✦ Figure 4.20). The acetyl CoA enters the Krebs cycle and electron transport chain and results in ATP production. The major fat used as a source of energy is palmitic acid. Although cells of the heart and skeletal muscle are able to use fat as a source of energy, not all cells can use this source. For example, brain cells *must* have glucose in order to function. Thus, lack of blood glucose can affect brain function, even if adequate supplies of fat are available.

Proteins can also serve as an energy source for the production of ATP. The initial step in the use of proteins as energy sources occurs in the cytoplasm with the removal of amino groups from the backbone of the protein molecule. These amino groups are the major source of the nitrogenous wastes excreted by animals in urine (see Chapter 24). Most amino acids are metabolized into acetyl CoA, which enters the Krebs cycle and electron transport chain for ATP synthesis. Other amino acids are degraded to metabolic compounds that are part of the Krebs cycle, so they enter the cycle at places other than the starting point (Figure 4.20), and still others are converted to oxaloacetate, and participate in the initial reaction of the Krebs cycle.

### ❀ FERMENTATION: ENERGY CONVERSION IN THE ABSENCE OF OXYGEN

We require oxygen to survive because we use oxygen as the ultimate electron acceptor in cellular respiration to generate the protons necessary to make ATP. If oxygen levels in the cell are depleted, the electron transport chain stops because there is no oxygen to act as an acceptor. This stops ATP synthesis, and backs up the Krebs cycle, which comes to a stop. In humans, cells of the skeletal muscles can carry out respiration for short periods under conditions of

oxygen deprivation (anaerobic respiration). Other cells, such as brain cells, begin to die if deprived of oxygen for more than a minute or two.

Organisms such as yeast can carry out ATP synthesis in the absence of oxygen, and certain bacteria use no oxygen at all in respiration. In these cases, ATP is synthesized by **fermentation** (✦ Figure 4.21). In fermentation, glycolysis is the sole source of ATP synthesis. For this process to work, the NADH generated in glycolysis must be converted to NAD+ by donating electrons to an electron acceptor. In fermen-

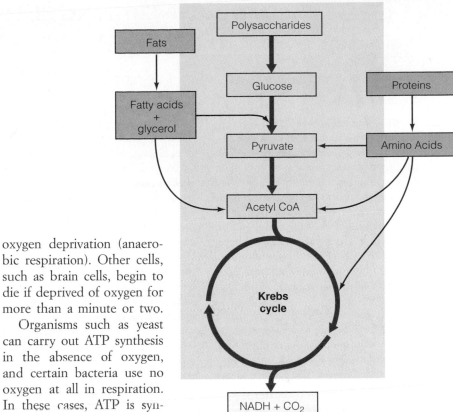

✦ **FIGURE 4.20**

**Energy for cellular respiration can be obtained from proteins and fats as well as carbohydrates.**

✦ **FIGURE 4.21**

(a) Fermentation in yeast cells can produce pyruvate and ethanol as a by-product. (b) When oxygen levels are low in animals cells, lactic acid is produced.

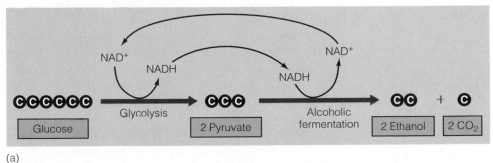

## Beyond the Basics

# HEAT-GENERATING PLANTS

In most plants, heat generated through metabolism is lost to the environment. However, in plants belonging to a number of different families, heat generated by altering metabolism can raise the internal temperature of the plant 15–20°C above ambient levels. Such plants are called **thermogenic** plants. Some of these heat-producing species are related to familiar house plants, including relatives of the philodendron and dieffenbachia. The Eastern skunk cabbage, a plant native to the U. S., is also thermogenic. In the voodoo lily, native to Southeast Asia, temperatures in flower parts can reach as high as 44°C or 110°F.

The timing and metabolic sources of heat production in thermogenic plants has been the subject of intensive investigation. Heat production is not constant, but is present only at certain times in the life cycle. The Eastern skunk cabbage is one of the first plants to sprout in the spring, and its thermogenic properties assist it in melting through frosty soil layers and help protect the plant from freezing. In the voodoo lily, heat production is used at the time of flower production to enhance the attraction of pollinating insects.

Normally, in cellular metabolism, glucose is broken down and some of the energy released by breaking chemical bonds is transferred to ATP, and the rest is lost to the environment. In thermogenic plants, such as the voodoo lily, an alternate metabolic pathway is used, one which by-passes ATP production, and directly releases the energy from breaking chemical bonds as heat. Heat production is regulated by the release of a hormone-like substance that switches metabolic pathways. In 1987, this substance was identified as salicylic acid (aspirin is a related compound, acetylsalicylic acid). Experiments on the voodoo lily show that salicylic acid levels increase by about 100 fold on the day before flowering. The following day, there is a 7–8 hour burst of heat production, which boosts the temperature in the pollination chamber by more than 12°C. A second round of salicylic acid production causes a 14 hour round of heat production in another part of the flower.

The alternate metabolic pathway by which thermogenic plants produce heat is not well understood. It is not yet clear whether salicylic acid works by switching off normal respiration, switching on the alternate pathway, or both. In addition, the steps in the alternate pathway and the electron acceptors are as yet unknown. Research into thermogenic plants is likely to produce new insights into how metabolism and energy production are linked. If the genes for the thermogenic pathway can be identified and isolated, it may be possible to transfer them to crop plants and prevent freeze damage, extending the growing season and increasing food production.

---

tation, other molecules act to accept electrons from NADH, producing a supply of NAD⁺.

Yeast is able to carry out alcoholic fermentation, the conversion of glucose to carbon dioxide and ethanol (◆ Table 4.2). In wine or beer making, yeasts break down sugar (in the grape juice to be made into wine, or in the sugars in the grain used to make beer) by glycolysis, converting the sugar to pyruvate, and making two ATP molecules in the process (Figure 4.21). Carbon dioxide is split off from each pyruvate molecule, converting it to acetaldehyde. The acetaldehyde acts as an electron acceptor from NADH, converting acetaldehyde to ethanol, a two-carbon alcohol (the alcohol in beer and wine is ethanol). This regenerates NAD⁺, which can then participate again in glycolysis, making more ATP in the process.

A second form of fermentation, called lactate fermentation, occurs in skeletal muscle during periods of peak exertion. During periods of intense physical activity, the oxygen supply needed for ATP synthesis (which powers muscle contraction) cannot keep up with demand. As the supply of oxygen falls, the pyruvate produced by glycolysis acts as an acceptor, takes up an electron and a proton, and becomes converted into lactic acid (Figure 4.21). When the period of physical activity is over, the lactate is moved to the liver where it is converted back into pyruvate and metabolized in the Krebs cycle.

## HOMEOSTASIS AND METABOLISM

In your car, the rate of energy conversion depends on how much fuel is made available to the engine. As you press the accelerator, more fuel is delivered to the combustion chambers, the rate of energy production increases, and the car's speed increases. In your body, this would mean that energy conversion in the form of metabolism and ATP production would increase after a meal and then decline greatly between meals. This does not happen because the rate of metabolism is controlled at a number of levels.

As we calculated earlier, only about 40% of the energy contained in the chemical bonds of the glucose molecule are captured in the high-energy bonds of ATP. Some of the rest is used to pump protons across membranes, but most escapes as heat. The majority of plants and animals rapidly lose this heat to the surrounding environment, and are unable to regulate their body temperature. Through the use of insulating materials such as fat, fur, and feathers, some animals, including mammals, are able to use this heat to regulate their body temperature and to maintain an internal temperature that varies over a narrow range (see Beyond the Basics: Heat-Generating Plants).

The control of metabolic rate and body temperature is an example of **homeostasis,** the capacity for maintaining a regulated internal environment. We will discuss the mechanisms of homeostasis in Chapter 18, but will touch on the regulation of metabolism and homeostasis here in this section.

## Metabolism Is Regulated in Several Ways

There are several ways in which the regulation of metabolism can be considered. First, in single-celled organisms and most plants, when nutrients are plentiful, metabolism will direct the synthesis of glucose and its storage forms, glycogen (mostly in animals) or starch (mostly in plants). When nutrients are in short supply, the process is reversed. Glucose is split off from starch or glycogen, and glycolysis serves as the starting point for ATP synthesis.

In animals, the liver plays an important role in regulating metabolism. After a meal, blood glucose levels are high. Glucose is removed and converted into glycogen. Later, when blood glucose levels fall, glycogen is broken down in the liver and supplied via the blood to the cells of the body for energy

conversion. This interplay is regulated by a series of factors, including hormones and other molecules that transmit signals.

Within cells, the levels of ATP regulate metabolism. When a cell does not require much energy, the internal level of ATP rises. When ATP levels rise above a certain level, glycolysis slows or stops (✦ Figure 4.22). If the energy requirements of the cell increase, ATP levels fall, and glycolysis increases. By controlling the activity of key enzymes in many different metabolic pathways, cells are able to balance the need for energy with the ability to store energy.

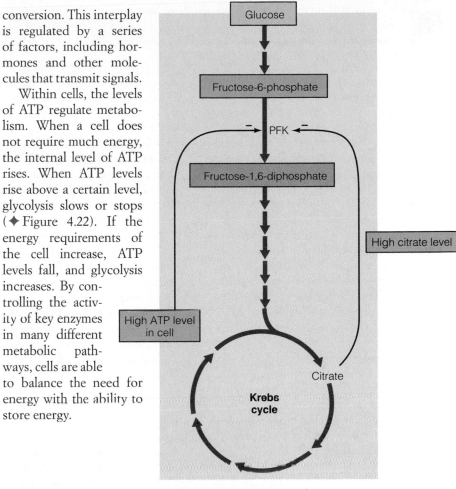

✦ **FIGURE 4.22**
Levels of citric acid and ATP control the rate of cellular respiration by interacting with an enzyme in glycolysis (phosphofructokinase).

## TABLE 4.2   FOODS PRODUCED BY FERMENTATION

| FOOD OR PRODUCT | STARTING MATERIAL | FERMENTING AND FLAVOR-CONTRIBUTING MICROORGANISMS | |
| --- | --- | --- | --- |
| | | Bacteria | Fungi |
| **Breads** | | | |
| Cakes, rolls | Wheat flours | | Saccharomyces cerevisiae |
| Sourdough bread | Wheat flours | Lactobacillus sanfrancisco | Saccharomyces exiguu |
| **Dairy Products** | | | |
| Cheeses (ripened) | Milk curd | Brevibacterium spp. Lactic acid bacteria Micrococcus caseolyticus Propionibacteria Streptococcus cremoris Streptococcus lactis Streptococcus thermophilus | |
| Cultured buttermilk | Milk | Leuconostoc cremoris Streptococcus cremoris Streptococcus lactis | Geotrichum spp., Penicillium camemberti, Penicillium roqueforti |
| Yogurt | Milk and milk solids | Lactobacillus bulgaricus Streptococcus thermophilus | |
| **Meat Products** | | | |
| Country cured hams | Pork hams | | Aspergillus spp. Penicillium spp. |
| Dry sausages | Pork, beef | Pedicoccus cerevisiae | |

1. Energy transformations in the physical and biological worlds are governed by the laws of thermodynamics. Energy cannot be created or destroyed, only changed from one form to another. In living systems, these transformations are somewhat inefficient, and energy is dissipated at each step, increasing the amount of entropy in the system. As a result, biological systems require a constant input of energy in order to maintain life.

2. Inside cells, energy transformation takes place through biochemical reactions mediated by enzymes. Enzymes act as catalysts to lower the energy required to carry out a biochemical reaction, and also control the specificity of the reaction. Reactions are classified as endergonic if they require energy input, and exergonic if they release energy.

3. In biological systems, two main types of reactions that involve energy transfer are photosynthesis (energy in) and cellular respiration (energy out). In photosynthesis, solar energy is captured and converted into chemical energy in the form of chemical bonds in glucose molecules. This series of chemical reactions takes place in thylakoid membranes arranged as stacked disks inside chloroplasts. Solar energy activates electrons in pigment systems, and the electrons are transferred to a transport system that pumps protons into the inner matrix of the thylakoid. As these protons flow out of this matrix, ATP is synthesized, capturing the energy in the chemical bonds of the ATP molecule.

4. The glucose molecules produced in photosynthesis can be converted into other storage forms such as starch or used to synthesize other components for the growth of the plant.

5. In cellular respiration, energy stored in the chemical bonds of glucose molecules is released. Glucose is broken down into pyruvate. Pyruvate enters the mitochondria and is converted to a two-carbon fragment called acetyl CoA (with the liberation of carbon dioxide). It then enters the Krebs cycle. The Krebs cycle is a series of biochemical reactions during which protons and electrons stripped from pyruvate are transferred to $NAD^+$ and FAD.

6. The energy yield from cellular respiration is incorporated into the chemical bonds of ATP molecules. Glycolysis and the Krebs cycle generate a net yield of 36 ATP molecules per molecule of glucose metabolized.

7. Although the breakdown of glucose is the main source of energy to make ATP, fats and proteins can also be used as energy sources. Fats are metabolized through a series of pathways to produce fragments that are converted to acetyl CoA, which enters the Krebs cycle, and produces electrons and protons that are cycled through the electron transport chain to make ATP. Proteins are broken down into amino acids, and these, in turn, are metabolized into a variety of compounds, some of which end up as acetyl CoA. Others are converted into components of the Krebs cycle.

8. Some cells use other pathways to make ATP in the absence of oxygen as an electron acceptor. In each case, the only ATP that is made is generated in glycolysis, yielding only two ATP molecules per molecule of glucose. In alcoholic fermentation, pyruvate is converted to acetaldehyde, which accepts electrons from NADH to form ethanol, regenerating $NAD^+$. In lactate fermentation, pyruvate itself acts as an electron acceptor and is converted to lactic acid, again regenerating $NAD^+$.

9. Control of metabolic reactions is exerted at several levels in multicellular organisms. The liver acts to absorb excess glucose from the blood and to convert amino acids from dietary proteins into glycogen. In times of energy need, glycogen is broken down in the liver into glucose, which is released into the blood, and transported to cells for conversion to ATP. Within cells, the process of ATP formation can be regulated by the levels of ATP present in the cytoplasm. Excess ATP accumulation inhibits part of glycolysis, shutting down energy conversion. When the level of ATP falls, glycolysis restarts, and ATP production is resumed.

10. The conversion of glucose into ATP is inefficient, and most of the energy is dissipated as heat. Some animals, especially birds and mammals, have the ability to use this metabolic heat to regulate body temperature.

# KEY TERMS

acetyl CoA
adenosine triphosphate
  (ATP)
cellular respiration
chemiosmosis
chlorophyll
chloroplasts

dark reactions
energy
electron acceptor
entropy
fermentation
light reactions
metabolism

nicotine adenine dinucleotide phosphate
  (NADP+)
PGA (phosphoglycerate)
PGAL (phosphoglyceraldehyde)
photosynthesis
photosystems
thylakoids

# QUESTIONS AND PROBLEMS

## SHORT ANSWER

1. Describe the laws of thermodynamics.
2. Are the pigments used in the conversion of light to chemical energy inside, outside, or on the thylakoids?
3. Describe the two main processes in photosynthesis.
4. Complete the diagram outlining the way a cell obtains energy from food.

```
_____ → amino acids _____
carbohydrates → _____ →
energy + acetyl CoA _____→
                                   ↓
fats → glycerol + _____ ──────────→ Energy
```

5. Describe how enzymes function in catabolic and anabolic reactions.
6. An exergonic reaction is accompanied by energy loss, whereas an endergonic reaction requires additional energy from outside sources. Is photosynthesis an endergonic or exergonic reaction?
7. Describe how organisms can survive without dependence on photosynthesis or direct sunlight.
8. The breakdown of food into energy begins in the cytoplasm and is completed in the mitochondria. Name the three stages that result in energy production from food.
9. At the end of the electron transport chain, electrons are passed to oxygen, which combines with protons present in solution. What is the end product of this reaction?
10. Describe how chemiosmosis and photosynthesis are alike.

## TRUE/FALSE

1. The primary source of ATP generated by photosynthesis occurs through the out-ward flow of protons from thylakoids resulting in the conversion of ADP into ATP.
2. The oxidation of glusose results in a total of 36 ATP molecules.
3. Prolonged glycolysis cannot occur without fermentation in an anaerobic environment.
4. As the amount of disorder increases, the amount of entropy decreases.
5. Glycolysis results in greater energy production than the TCA cycle.
6. Prokaryotes generate ATP in their mitochondria.

## MATCHING

Identify each reaction with the correct name.

_____ glycolysis
_____ photosynthesis
_____ TCA cycle
_____ respiration

1. a. sugar + $O_2$ → $CO_2$ + $H_2O$ + energy

b. glucose → 6 carbon molecule → 6 carbon molecule

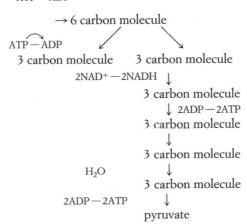

c.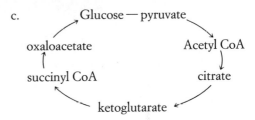

d.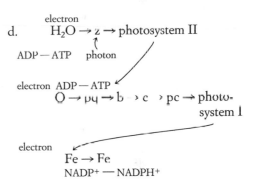

2. Match the reactions with the location of their occurrence. Answers may be used more than once.

a. glycolysis
b. TCA cycle
c. electron transport and ATP formation
d. fat digestion
e. lactate fermentation
f. enzyme

_____ intestines
_____ mitochondria
_____ substrate
_____ skeletal muscle
_____ cytoplasm
_____ nucleus

## SCIENCE AND SOCIETY

1. Solar energy is used to run solar-powered calculators and heating systems. Plants have been using solar energy and converting it to chemical energy for millions of years. Plants are believed to use less than 1% of the visible light that reaches the earth. There is obviously plenty of solar

energy left over, and slowly we are learning to convert it into electricity. What advantages and disadvantages are there to using solar energy instead of fossil fuels? Would constructing solar cells to convert solar energy to electricity destroy more land than power stations? What safety issues need to be explored before we convert to solar energy as a future energy source? Can you think of any examples where solar energy would be more convenient then electrical energy?

2. Leaves of plants contain chloroplasts which trap sunlight and use solar energy to produce the organic molecules that sustain life. These molecules come in a variety of shapes and sizes, illustrating the adaptations that have emerged by natural selection. Photosynthetic organisms emerged more than 3 billion years ago, and they turned out to have a profound effect on evolution. How have photosynthetic organisms contributed to evolution? How would life exist on Earth without photosynthesis? Why do you think there is such diversity in plants? What contributes to this diversity? What roles do plants play in the ecosystem?

# 5

# CELL DIVISION

**OPENING IMAGE**

*During prometaphase, chromosomes,*

*or more precisely, the kinetochores,*

*migrate to the equator of the cell.*

A t the Marine Biological Laboratories at Woods Hole, Massachusetts, scientists gather each summer as they have for more than 100 years to teach, to learn, and to conduct experiments using the marine plants and animals that can be collected there. In the summer of 1982, a group of young scientists led by Tim Hunt began experiments designed to study biochemical changes that take place after fertilization of the sea urchin egg. They decided to examine new gene products (gene products are proteins) made in the hours immediately after fertilization, a time when a series of cell divisions take place. They fertilized a batch of sea urchin eggs and, at 10-minute intervals, analyzed the newly made proteins during the first 2 to 3 hours of development. The fertilized egg first divides at about 60 minutes, and again about 2 hours after fertilization, resulting in a four-cell embryo.

Several new proteins appeared almost immediately after fertilization, including one that exhibited unusual behavior. This protein was continuously synthesized, but was unexpectedly destroyed just before each round of cell division. Because of its cyclic behavior, this protein was called **cyclin.** Experiments using another species of sea urchin revealed two cyclins that periodically disappear during the cell divisions associated with early development. Investigating other organisms available at Woods Hole, this group discovered that the developing eggs of clams also contain cyclins. Because these proteins disappear *just before mitosis, the investigators concluded that cyclins might be involved in the control of cell division.*

*Subsequent investigations showed that cyclins are found in the dividing cells of many organisms, and act as important control switches in cell division. Sea urchins have only one cyclin, but humans and other mammals may have as many as eight different cyclins, each controlling one or more steps in cell division. Some forms of cancer involve loss of control over the action of cyclins.*

*The lesson here is that sometimes experiments designed to study one event, such as development in the sea urchin, end up being of major importance to all of biology. Also, sometimes what you do on your summer vacation is worth writing about.*

---

## CELL DIVISION IS A PROPERTY OF LIVING SYSTEMS

As described in Chapter 3, cells are the basic structural unit of all living systems. Here we examine how new cells are produced. Individual cells grow in size by incorporating materials from the environment to produce new components. Cells grow in number by undergoing a process of division, resulting in the production of new cells (✦ Figure 5.1). Single-celled (**unicellular**) organisms produce new individuals at each cell division (✦ Figure 5.2). Multicellular organisms begin life as a single cell, and by division, produce cells that form the tissues and organs of the body. These cells, called **somatic** cells, divide by **mitosis.** Within the reproductive organs of multicellular organisms, groups of **germ** cells divide by **meiosis** to form **gametes,** cells that participate in the formation of the next generation.

Cell division is essential for reproduction in all organisms. To reproduce effec-tively, a cell must ensure that all offspring receive a complete set of the genetic information carried by the parent cell, and a supply of cytoplasm. A basic feature of cell division, therefore, is a process that ensures that each daughter cell carries a complete set of genetic information, and a process that provides each daughter cell with enough cytoplasm to survive. Because the ability to reproduce is an essential characteristic of all living organisms, the process of cell division represents one of the fundamental attributes of life.

✦ **FIGURE 5.2**

**Many organisms divide by fission, producing offspring identical to the parents.**

✦ **FIGURE 5.1**

**During mitosis, each daughter cell receives about half the cytoplasm of the dividing cell.**

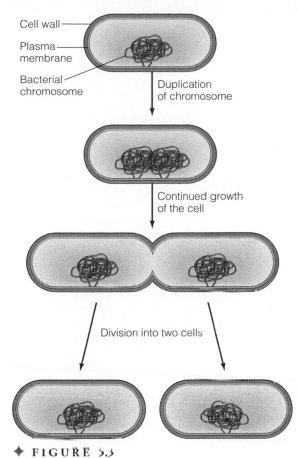

Cell wall

Plasma membrane

Bacterial chromosome

Duplication of chromosome

Continued growth of the cell

Division into two cells

✦ **FIGURE 5.3**

**In bacteria, the chromosome duplicates when attached to the cell wall. Growth of the wall separates the copies, which are incorporated into the daughter cells.**

## Some Cells Divide by Binary Fission

In the cells of prokaryotic organisms, including blue-green algae and other bacteria, cell division follows a period of growth during which the cell enlarges by synthesizing new sections of the cell wall and by making more cytoplasm. During this period of growth, the circular DNA molecule that serves as a chromosome is copied. Each copy of the chromosome becomes attached to the inner surface of the plasma membrane at opposite sides of the cell (✦ Figure 5.3). As the cell elongates by the addition of new membrane between the attachment sites, the two chromosomes move further apart. In the first stage of division, the cell wall and the plasma membrane pinch inward. Eventually, the cell becomes pinched in two, with each daughter cell receiving one copy of the chromosome, and about half of the cytoplasmic components. In this way, each cell produced by binary fission receives a complete set of genetic information *and* the cytoplasmic components necessary for survival and growth.

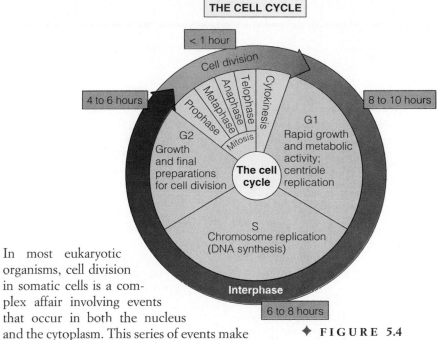

< 1 hour

Cell division

4 to 6 hours

8 to 10 hours

Prophase
Metaphase
Anaphase
Telophase
Cytokinesis
Mitosis

G2
Growth and final preparations for cell division

**The cell cycle**

G1
Rapid growth and metabolic activity; centriole replication

S
Chromosome replication (DNA synthesis)

**Interphase**

6 to 8 hours

✦ **FIGURE 5.4**

**The cell cycle has two parts: interphase and mitosis. Interphase is divided into three parts. Times shown are representative for cells grown in the laboratory.**

In most eukaryotic organisms, cell division in somatic cells is a complex affair involving events that occur in both the nucleus and the cytoplasm. This series of events make up the cell cycle.

## THE CELL CYCLE DESCRIBES THE LIFE HISTORY OF A CELL

Cells of eukaryotic organisms have a characteristic life history in which periods of division alternate with periods of nondivision. The interval between divisions can vary from minutes in embryonic cells to months or even years in some cells of adults. The sequence of events from one division to the next is called the **cell cycle** (✦ Figure 5.4). The period of division itself consists of two processes, **mitosis** and **cytokinesis**. Mitosis is the division of the nucleus and nuclear material; cytokinesis is the division or partitioning of the cytoplasm. The period between divisions is known as **interphase** (✦ Table 5.1).

### Interphase Has Three Stages

A good place to begin a discussion of the cell cycle is with a cell that has just been formed by division, and is entering interphase. After a cell divides, the resulting daughter cells are about one-half the size of the parental cell. Before these cells can divide again, they must undergo a period of growth and preparation for division. These events occur during the three stages of interphase: G1, S, and G2 (Figure 5.4). The G1 stage (known as gap 1) begins immediately after division and is a period of growth during which many cytoplasmic components are replaced. Cellular structures such as the plasma membrane, Golgi apparatus, lysosomes, and vesicles made during G1 are derived from precursors in the endoplasmic reticulum (ER). Cytoplasmic

## A Life of Questions: The Questions of Life

NANCY KEDERSHA

Most children go through a "What?" stage, progress on through "Why?" and "How?" and eventually grow into adults concerned with increasingly specific questions, such as "Will you marry me?" and "How much does this cost?" I never gave up asking the "What? How?" questions, and they have led me into science. How do seeds grow into plants? How do robins find worms? How do living things grow and move? These questions that I asked as a child are still meaningful to me as an adult. Living things are both so beautiful and so interesting that I could never decide whether I preferred to pursue art or science full time.

My college years were spent shuttling between the biology labs and the university theater, so I was usually either looking through a microscope at diatoms I had cultured or focusing lights on scenery I had built and painted. After college, I was still reluctant to commit myself exclusively to science. I deferred graduate school and I took a job as a lab technician in a biomedical research lab. Working on cells "hands on" I gradually came to realize that being a scientist was much more of an art than I had realized. Glassblowing, photography, assembling posters, and redesigning equipment were as much a part of my job as growing cells or purifying proteins. The lab was a place where skill and "good hands" were needed as well as knowledge, and where someone who asked a lot of questions was just doing her job proficiently rather than being a pain in the neck.

In science, as in life, my curiosity has led me to a series of serendipitous discoveries. In graduate school, I saved one of the "contaminant" proteins I was purifying away from my main "thesis-related" protein, and found that the contaminant was a new form of the well-known protein actin, differing in its location within cells as well as its biochemistry. While learning to use the electron microscope, I found some strange and beautiful structures in the "garbage" fractions that normally would have gone down the drain. For months afterwards, any hapless scientist visiting the lab was handed photographs of these structures and ruthlessly grilled: "Have you ever seen anything like THESE before??" No electron microscopist for miles (we asked a lot of them!) could identify these particles. We named them "vaults" because of their beautiful and intricate morphology, reminiscent of the multiple arches that form vaulted cathedral ceilings.

One question that people ask me is "If there are so many of these in all cells, why didn't anyone ever see them before you?" One answer is that vaults can't be seen in samples prepared for electron microscopy using normal stains, and are in essence invisible in conventionally prepared sections of cells. The technique I was using enabled me to see them, and another new technique allowed me to purify them. Vaults were in fact actually photographed and appeared in publications as contaminants of coated vesicle preparations. No one realized they were intact new structures instead of fragments of coated vesicles until I purified them.

So perhaps the most complete reason why no one ever saw them was because nobody looked using the right tools, and when they did use the right tools, they didn't ask the right questions. The "right" question I asked was "What are these things?" And now that we know more about what they are, the question is "What do they do?"

Powerful new techniques, instrumentation, and computers have resulted in an explosion of data that is coming in faster than we can digest or integrate. Science is a process to make sense out of data and understand the principles behind events. Like a microscope or a telescope, science is also a way of viewing the world, and it evolves like a living organism as does our understanding. The more I look, the more I see; the more questions I answer, the more new ones I perceive. I find such beauty in the answers that for me, art and science are one.

*Nancy Kedersha is a research scientist at Immunogen, Inc., in Cambridge, Massachusetts. She has published dozens of articles on topics in cell biology, and her photos of cancer cells have won numerous prestigious international awards. Her work has been the subject of feature articles in several national and international magazines.*

components including enzymes and ribosomes are produced at this time. Mitochondria and chloroplasts reproduce using the genetic information carried in the cell's nucleus and the genetic information contained within the organelle. The synthetic activity and growth that take place in G1 results in a doubling of cell size and the replacement of components lost in the previous division.

G1 is followed by the S (synthesis) phase, during which a duplicate copy of each chromosome is made via DNA replication. The S phase is followed by G2 (gap 2), a stage in which the cell manufactures the molecular components needed for mitosis. By the end of G2, the cell has grown even larger, and is ready to begin a new round of division (Figure 5.4). The time spent in the interphase portion of the cell cycle (remember that G1, S, and G2 together make up the interphase) varies from 18 to 24 hours in cultured cells of animals and from 10 to 30 hours in plants. The events of mitosis usually takes less than an hour and, therefore, somatic cells spend most of their time in interphase.

The life history of cells and their relationship to the cell cycle varies for different cell types. Most unicellular organisms and some cell types in plants and animals pass through the cell cycle continuously, and divide on a regular basis. At the other extreme, some cell types become permanently arrested in G1, and never pass through the rest of the cycle to undergo division. In between are cell types that are arrested in G1 or G2 but *can* divide under certain circumstances.

## CHROMOSOMES HAVE A CHARACTERISTIC STRUCTURE

Although interphase is a period of intense synthetic activity in both the cytoplasm and nucleus, aside from an increase in size there is little change in the appearance of the cell. Inside the nuclear membrane, nucleoli are present as distinct structures, but the chromosomes are dispersed and evident only as threads and clumps of chromatin. Recall from Chapter 3 that the number of chromosomes present in the nucleus is species specific (Table 3.6).

The chromosomes of most eukaryotic organisms occur in pairs. One member of each chromosome pair is derived from the female parent, and the other from the male parent. Members of a chromosome pair are known as **homologues.** Cells that contain homologous pairs of chromosomes are known as **diploid** cells, and the number of chromosomes carried in such cells is known as the diploid or 2*n* number. In humans, there are 23 different types of chromosomes, with two of each type present in a cell, resulting in a diploid (2*n*) number of 46.

Each chromosome contains a specialized region known as the **centromere** (✦ Figure 5.5). The position of the centromere divides the chromosome into two arms, and its location is characteristic for a given chromosome.

### TABLE 5.1  PHASES OF THE CELL CYCLE

| PHASE | CHARACTERISTICS |
|---|---|
| Interphase | |
| G1 (gap 1) | Stage begins immediately after mitosis. |
| | RNA, protein, and other molecules are synthesized. |
| S (synthesis) | DNA is replicated. |
| | Chromosomes become double stranded. |
| G2 (gap 2) | Mitochondria divide; precursors of spindle fibers are synthesized. |
| Mitosis | |
| Prophase | Chromosomes condense. |
| | Nuclear envelope disappears. |
| | Centrioles divide and migrate to opposite poles of the dividing cell. |
| | Spindle fibers form and attach to chromosomes. |
| Metaphase | Chromosomes line up on equatorial plate of the dividing cell. |
| Anaphase | Chromosomes begin to separate. |
| Telophase | Chromosomes migrate or are pulled to opposite poles. |
| | New nuclear envelope forms. |
| | Chromosomes uncoil. |
| Cytokinesis | Cleavage furrow forms and deepens. |
| | Cytoplasm divides. |

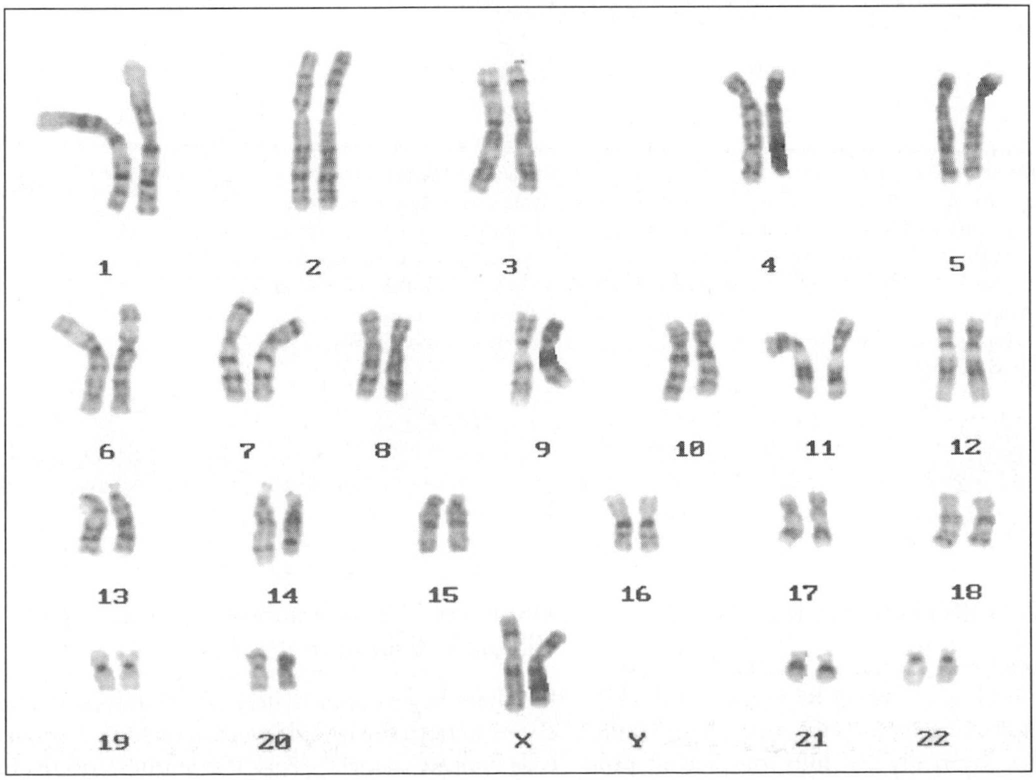

✦ **FIGURE 5.5**

**The human karyotype, showing replicated chromosomes isolated from a cell in metaphase. Humans have 46 chromosomes.**

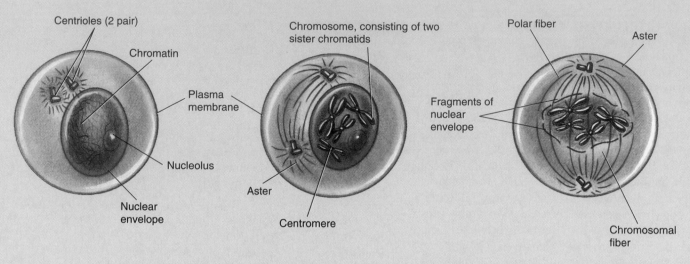

Centrioles (2 pair)

Chromatin

Plasma membrane

Nucleolus

Nuclear envelope

(a) Interphase

Chromosome, consisting of two sister chromatids

Aster

Centromere

(b) Prophase

Polar fiber

Aster

Fragments of nuclear envelope

Chromosomal fiber

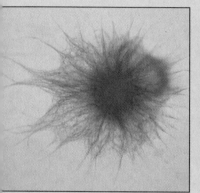

Interphase.

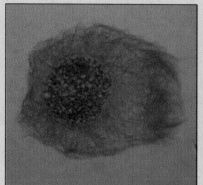

Early prophase.

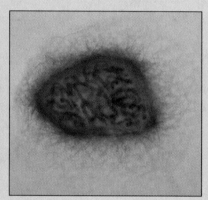

Midprophase.

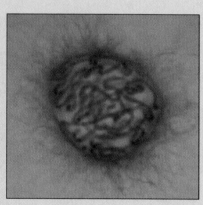

Late prophase.

✦ **FIGURE 5.6**

**Stages of mitosis. During interphase (a), replication of chromosomes takes place. In prophase (b), the chromosomes coil and become visible as thread-like structures. Originally single, they become visible as double structures, with sister chromatids joined at a single centromere. At the end of prophase, the nuclear membrane breaks down. In metaphase (c), chromosomes become aligned at the equator of the cell. In anaphase (d), the centromeres divide, converting sister chromatids into chromosomes, which move toward opposite sides of the cell. At telophase (e), the chromosomes uncoil, the nuclear membrane reforms, and the cytoplasm divides.**

## MITOSIS HAS FOUR STAGES

When the cell reaches the end of the G2 stage, it is ready to undergo division by mitosis. Although mitosis is a continuous process, this phase of the cell cycle has been divided into four stages: **prophase, metaphase, anaphase,** and **telophase** (✦ Figure 5.6).

## Duplicated Chromosomes Become Visible in Prophase

Prophase begins immediately after interphase, and events in both the cytoplasm and nucleus of animal cells indicate that cell division is about to occur. In the cytoplasm of animal cells, the centrioles, which replicated during interphase, separate and migrate

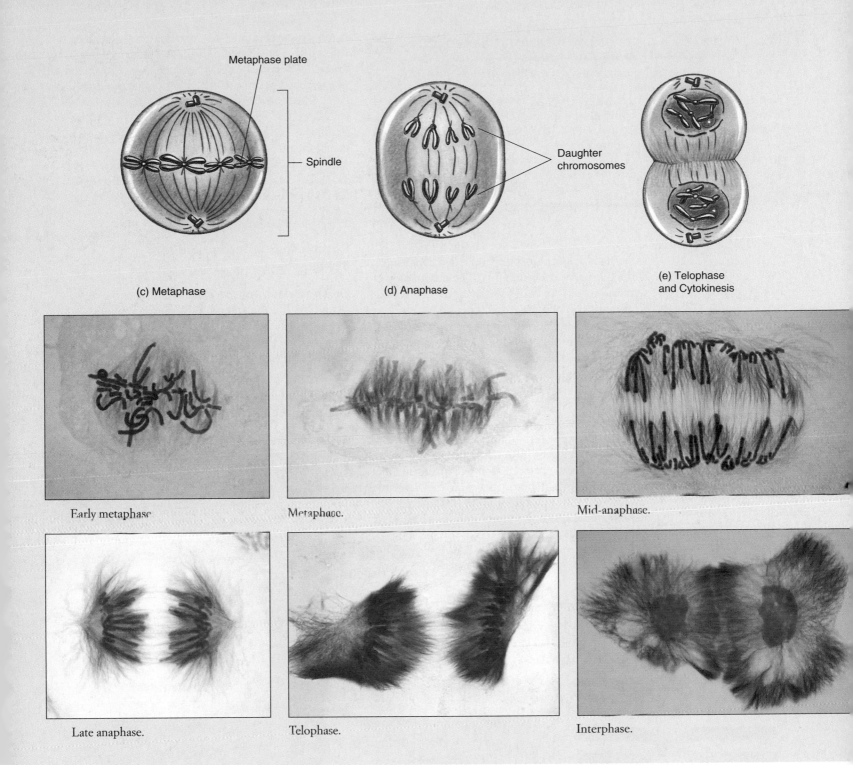

Metaphase plate

Spindle

Daughter chromosomes

(c) Metaphase

(d) Anaphase

(e) Telophase and Cytokinesis

Early metaphase

Metaphase.

Mid-anaphase.

Late anaphase.

Telophase.

Interphase.

to opposite sides of the nucleus where they will act as organizing centers for spindle fibers. In the nucleus, the long fibers of chromatin begin to coil up on themselves to form the shorter and thicker structures recognizable as chromosomes. When they can first be seen under a light microscope, the chromosomes appear as long, intertwined threads; their

appearance marks the beginning of prophase (Figure 5.6).

The condensation of chromosomes at the beginning of prophase serves an important function. In a shortened and contracted form, the chromosomes untangle from each other and move freely during mitosis. Near the end of prophase, each chromo-

| | |
|---|---|
| Chromatin | General term referring to a strand of DNA and associated histone protein |
| Chromatin fiber | Strand of chromatin |
| Chromosome | Structure consisting of one or two chromatin fibers |
| Chromatid | Generally used to refer to one of the chromatin fibers of a replicated chromosome |
| Sister chromatid | Identical chromatids joined by a common centromere |
| Centromere | Region of each chromatid to which a sister chromatid attaches |

reach from centriole to centriole, while other, shorter fibers reach from the centriole to structures known as **kinetochores** located at the centromere of each chromosome. Other fibers radiate outward from the centrioles and may serve to attach the spindle to the plasma membrane.

At the end of prophase, the nucleolus disappears, the nuclear membrane breaks down, and the chromosomes migrate to the center or equator of the cell and become distributed within the spindle. Each chromosome is attached to a spindle fiber and ready to begin the next stage of mitosis.

## Chromosomes Become Aligned in Metaphase

During metaphase, the chromosomes are moved around within the spindle until they are aligned at the equator of the cell (Figure 5.8). In this position, they are roughly at right angles to the spindle fibers, and the sister chromatids face opposite centrioles.

## Duplicated Chromosomes Separate in Anaphase

At the beginning of anaphase the centromeres joining sister chromatids divide, converting each of the sister chromatids into a separate chromosome (Figure 5.6). Because each chromatid was formed by DNA replication, the two chromosomes derived from sister chromatids will be genetically and struc-

some can be seen to consist of two longitudinal strands known as **chromatids**. The chromatids are separate structures held together at the centromere. Chromatids joined by a common centromere are known as **sister chromatids** (✦ Figure 5.7). The chromatids visible at this stage of prophase were actually formed during the S phase of the cell cycle. During S phase, DNA replication makes a copy of the genetic information carried in the DNA of each chromosome. Sister chromatids represent identical sets of genetic information. ✦ Table 5.2 summarizes chromosome terminology.

Near the end of prophase, microtubules and other cytoplasmic proteins condense to form a network of fibers extending between the centrioles known as the **mitotic spindle** (✦ Figure 5.8). Within the elliptically-shaped spindle, some fibers

### ✦ FIGURE 5.7

**Chromosomes replicate during the S phase. While attached at the centromere, the replicated chromosomes are called sister chromatids.**

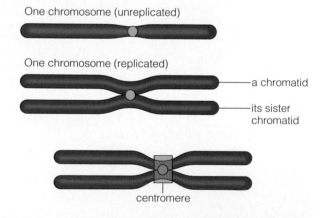

One chromosome (unreplicated)

One chromosome (replicated)

— a chromatid

— its sister chromatid

centromere

### ✦ FIGURE 5.8

**Microtubules form the mitotic spindle during prophase. Some of these fibers connect to the centromeres.**

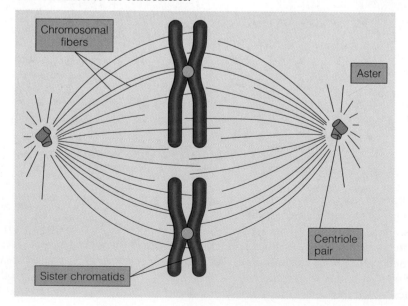

Chromosomal fibers

Aster

Centriole pair

Sister chromatids

turally identical. The essence of mitosis is the separation and correct distribution of two identical sets of chromosomes now present in the cell. In anaphase, the chromosomes derived from sister chromatids migrate toward opposite centrioles (Figure 5.6). Movement of chromosomes depends on the correct attachment of spindle fibers to the centromere. Spindle fibers attached to kinetochores shorten, moving the centromere first. The chromosome arms move more slowly, making the chromosomes look J-shaped or V-shaped. At the end of anaphase, a complete set of chromosomes is present at each pole of the cell.

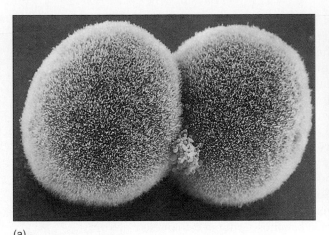
(a)

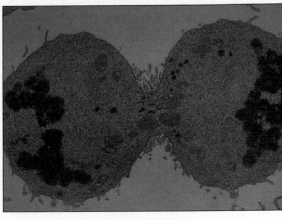
(b)

✦ **FIGURE 5.9**

Cytokinesis in an animal cell. (a) Scanning electron micrograph of the cleavage furrow as seen from outside the cells. (b) Transmission electron micrograph of cytokinesis in a cross-section of a dividing cell.

## Cytoplasmic Cleavage Occurs in Telophase

The final stage of mitosis is telophase (✦ Figure 5.9). As the chromosomes reach opposite poles of the cell, the spindle fibers begin to break down into subunits, which are stored in the cytoplasm, or become incorporated into the cytoskeleton. Membranous buds from the endoplasmic reticulum fuse to form a new nuclear membrane. The centriole migrates to take up a position outside the newly formed nucleus, and begins to replicate. By the time the cell has passed through the next interphase, two pairs of centrioles will be present.

Inside the newly formed nucleus, the chromosomes begin to uncoil and become dispersed as chromatin. As the chromosomes become more diffuse and indistinct, the nucleolus reappears. The uncoiling of chromosomes represents another functionally specialized state. It is probable that genetic information in an uncoiled chromosome is more accessible, and able to direct the synthesis of gene products during interphase.

The division of the cytoplasm, by cytokinesis, usually accompanies the division of the nucleus in mitosis and occurs during telophase. In animal cells, cytokinesis begins with formation of the **cleavage furrow,** a constriction of the cell membrane at the equator of the cell (Figure 5.9). This furrow gradually deepens and divides the cell into two new cells. Rings of microfilaments made of actin are present in the cytoplasm at the site of the furrow. Contraction of these filaments draws the constriction furrow tighter and tighter until the cytoplasm is split into two cells.

## MITOSIS IN ORGANISMS WITH CELL WALLS

While the general features of mitosis described for animal cells also apply to organisms with cell walls (some algae, fungi, and more advanced plants), some important differences should be noted. First, in plants, although spindle fibers are formed from microtubules, no centrioles are present (✦ Figure 5.10). This indicates that the presence of centrioles is not a prerequisite for the formation of the spindle. Second, because plant cells are surrounded by cell walls outside their plasma membrane, the process of cytokinesis is different from that in animal cells. In plants, the division of cytoplasm begins with the

✦ **FIGURE 5.10**

In plants, cytokinesis is accompanied by the formation of a new primary cell wall made from vesicles secreted by the Golgi apparatus.

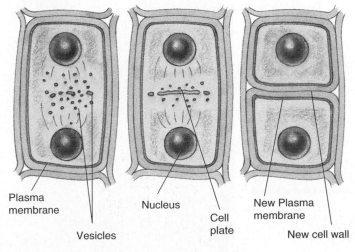

Plasma membrane

Vesicles

Nucleus

Cell plate

New Plasma membrane

New cell wall

synthesis of vesicles in the Golgi apparatus that migrate to the region of the cell equator. These aligned vesicles fuse to produce a membrane-bound space known as the **cell plate.** This plate grows outward to fuse with the plasma membrane of the cell, dividing the cell into two compartments (Figure 5.10). A new middle lamella forms in the space between the two cells, and new cell wall material is laid down by each daughter cell.

As is the case with certain insect embryos, cytokinesis in plants does not always follow mitosis. Inside the developing coconut, for example, the tissue called coconut "milk" is a liquid containing many nuclei formed by mitosis without cytokinesis. As the coconut matures, cell walls begin to form around individual nuclei, forming the solid "meat" of the coconut.

 ## MITOSIS IS ESSENTIAL FOR GROWTH AND CELL REPLACEMENT

Mitosis is an essential process in both unicellular and multicellular organisms. In multicellular organisms, growth and development from a fertilized egg to the adult is mediated by mitosis. In humans, a total of 40 to 44 rounds of mitosis accompany the transition of the fertilized egg or **zygote** into a newborn. Throughout adult life, mitosis is involved in cell replacement.

### Sidebar

**CELL DIVISION AND SPINAL CORD INJURIES**

Many highly differentiated cells, like those of the nervous system, do not divide. They are sidetracked from the cell cycle into an inactive state called G0. As a result, injuries to nervous tissue such as the spinal cord cause permanent loss of cell function and paralysis. For years, scientists have worked to learn how to stimulate growth of spinal cord cells so that injuries can be repaired. Recent work suggests that it may soon be possible to get nerves in the spinal cord to reconnect to their proper targets and to restore function in nerve cells that are damaged, but not cut. Researchers have shown that the severed spinal cords of young rats can be reconstructed by transplantation of the corresponding section of spinal cord from rat embryos. When the rats reached adulthood, most of the sensory and movement was restored.

Other researchers have isolated a growth factor found only in the central nervous system that causes cell growth from the ends of severed spinal cords. Whether such growth can result in reconnection of nerves to their proper muscle targets, and whether function can be restored are unresolved questions.

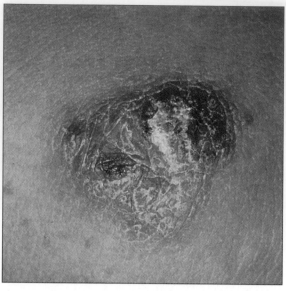

✦ **FIGURE 5.12**

**Cell division at the edges will slowly close the wound, repairing the damage to the tissues.**

✦ **FIGURE 5.11**

**More than 32 million new red blood cells are produced each day.**

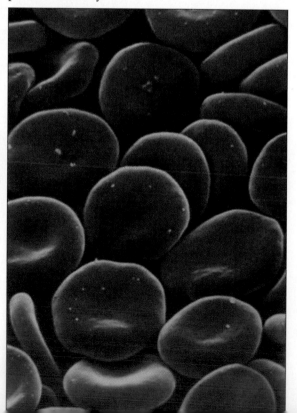

Cells in the bone marrow rapidly divide to replace worn out red blood cells. On average, between 2 and 2.5 million cell divisions are required each *second* to replace red blood cells removed from the circulation (✦ Figure 5.11). Overall, each day, hundreds of billions of cell divisions take place in the body. By contrast, most cells in the nervous system do not undergo division. As a result, when nerves are damaged or destroyed, they are not replaced. For this reason, many injuries to the spinal cord result in permanent paralysis.

Wound healing and the replacement of dead and dying cells also depend on cell division (✦ Figure 5.12). Most cells are unable to continue division indefinitely, and the aging process is associated with a slowdown or halt in the replacement of worn-out cells. On the other hand, cells that are able to divide indefinitely are often associated with malignant conditions such as cancer.

### Cancer and Control of the Cell Cycle

Some nondividing cells are arrested in G1. The mechanism that determines whether cells move through the cycle operates in G1. A critical switch point commits a cell to enter the S phase and proceed to mitosis, or causes the cell to leave the cycle and become quiescent. The nature of this switch point, unquestionably one of the central regulatory mechanisms in all of biology, is slowly being revealed by research in genetics and cell biology.

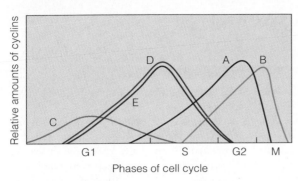

**◆ FIGURE 5.13**

**Different cyclins (A–E) are produced and destroyed at different times in the cell cycle. Each cyclin plays a role in regulating passage of a cell through the cell cycle.**

The signals that are part of this switch point in the cell cycle were discovered as a result of the work started at Woods Hole in 1982. These signals are generated by the synthesis and action of cyclins (◆ Figure 5.13). At the G1 control point, a cyclin combines with another protein, generating a signal that moves the cell from G1 into S. What signals cause the production and destruction of cyclins are not yet known.

Cancer cells have disabled this and other control points, and divide continuously. Mutations in genes that control the synthesis or action of cyclins are important in generating the transformation of a normal cell into a cancer cell.

## CELL DIVISION BY MEIOSIS: THE BASIS OF SEX

Cells usually contain a species-specific chromosome number: humans have 46, corn plants have 20, dogs have 78, and so forth. Mitosis distributes two copies of each chromosome into each daughter cell. The result is two cells, each genetically identical to the parental cell, with a diploid chromosome number that remains constant from division to division. How is it, then, that the chromosome number remains constant from parent to offspring when sperm and egg fuse to form the fertilized egg known as a **zygote?** If the sperm and egg were produced by mitosis, the chromosome number would double from generation to generation, and quickly become astronomical. However, the chromosome number in a given species remains constant from generation to generation. The chromosome number remains constant because a form of cell division known as **meiosis** reduces the number of chromosomes in sperm and eggs.

In meiosis, members of a chromosome pair separate from each other, and the resulting cells contain only one copy of each chromosome. Such cells are **haploid** cells, and contain the haploid or $n$ number of chromosomes. At fertilization, the fusion of two haploid gametes ($n + n$) to form a single cell restores the diploid ($2n$) chromosome number (◆ Figure 5.14). Meiosis also produces a reshuffling of genetic information, so that the offspring are not genetically identical to the parents. All multicellular organisms that exhibit sexual reproduction undergo meiosis as a precursor to gamete formation.

In most animals, meiosis occurs at the time of gamete formation. In organisms with cell walls, for example, ferns, meiosis can be separated from gamete formation by time (weeks or months) and by several mitotic divisions. Nonetheless, meiosis has the same function in these organisms as it does in animals: to reduce the chromosome number from diploid to haploid, and generate genetic diversity by reshuffling genetic information. Meiosis in plants and other organisms with cell walls is discussed in a later section.

**◆ FIGURE 5.14**

**Diploid germ cells undergo meiosis to produce haploid gametes. Fusion of gametes at fertilization restores the diploid number to the offspring.**

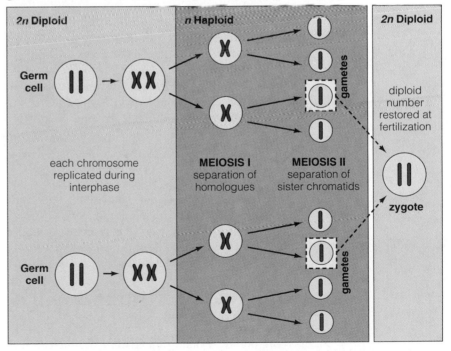

*Beyond the Basics*

# GENES, CELL DIVISION, AND CANCER

Cancer is a disorder of cellular reproduction. This condition arises from within the cell, causing cell division to be out of control; cells continually move through the cell cycle and divide, giving rise to daughter cells that also undergo continuous division. Cells may also break away from the original tumor mass and establish new tumor sites elsewhere in the body (a process known as **metastasis**). Because the ability to divide in an uncontrolled fashion is passed on to the progeny of cancer cells, it is logical to assume that this property is under genetic control. The question is, if genes are involved in cancer, which ones are they and what do we know about them?

Until recently, these questions were difficult if not impossible to answer. Now however, scientists are uncovering two types of genes that are associated with the development of cancer, **tumor suppressor genes** and **oncogenes.** The first group contains genes that normally keep cell division in check, preventing the cell from responding to external and internal commands to divide. If one or more of these genes are damaged or absent, restraints on cell division are removed, leading to the formation of tumors.

The second group of genes, the oncogenes, are able to activate cell division in cells that normally do not divide or do so only slowly. In other words, when these genes are damaged, or make gene products at inappropriate times or in the wrong amounts, resting cells begin to divide in an uncontrolled fashion, causing tumor formation.

At the present time, fewer than 10 tumor suppressor genes have been discovered, but more than 75 oncogenes have been described. The importance of oncogenes in relation to cancer can be shown by the fact that scientists working with oncogenes were awarded Nobel Prizes for their efforts in 1966 and 1989. New research on cancer genes is aimed not only at discovering more of these genes, but at understanding how the action of these genes at the molecular level fits into the network of external and internal signals that bombard the cell and result in a decision to move through the cell cycle and divide.

## AN OVERVIEW OF MEIOSIS: CONVERTING DIPLOID CELLS INTO HAPLOID CELLS

The process of meiosis involves two rounds of cell division in succession, producing four haploid cells (✦ Figure 5.15). The first division results in a reduced number of chromosomes in the daughter cells (and is known as a **reductional division**). In the second division, the centromeres divide *for the first time,* and sister chromatids are converted to individual chromosomes, which migrate to opposite poles of the cell. Each round of meiosis has four phases: prophase, metaphase, anaphase, and telophase.

The distribution of chromosomes in meiosis is an exact process. Each haploid cell contains not just a random selection of chromosomes amounting to half the diploid number, but rather, one member of each chromosome pair. The precise sorting of chromosomes is made possible by events that occur in the prophase of the first meiotic division.

## THERE ARE TWO DIVISIONS IN MEIOSIS

Our bodies contain diploid cells destined to give rise to gametes. These diploid germ cells undergo meiosis. DNA replication occurs in the S phase before meiosis. Cells begin the two meiotic divisions with a replicated set of chromosomes that will be distributed to four haploid nuclei. Because cells in meiosis pass through one round of chromosomal replication followed by two rounds of division, they are not in the cell cycle at this time.

### Meiosis I Reduces the Chromosome Number

In the first stage of prophase I, the chromosomes coil, thicken, and become visible under the microscope. As the chromosomes condense, the nucleoli and nuclear membrane disappear and the spindle becomes organized. Each chromosome physically associates with its homologue, and the two chromosomes line up side-by-side, a process known as **synapsis** (Figure 5.15). During synapsis, each chromosome is composed of two sister chromatids held together at the centromere, and the four chromatids in the cluster are often referred to as a **tetrad.** During this stage, evidence for a physical exchange of chromosome material between homologous chromosomes may be visible (Figure 5.15). The genetic significance of these events is discussed in the next section. At the end of meiotic prophase I, the synapsed pairs of chromosomes move to the equator of the cell.

In mitosis, each member of a chromosome pair behaves independently in moving to the equator of the cell. In meiosis, however, members of a chromosome pair are physically associated or synapsed, and they move to the cell equator as a pair during metaphase I. When the chromosome pairs are aligned at the equator of the cell, one member of each pair (consisting of two sister chromatids and a

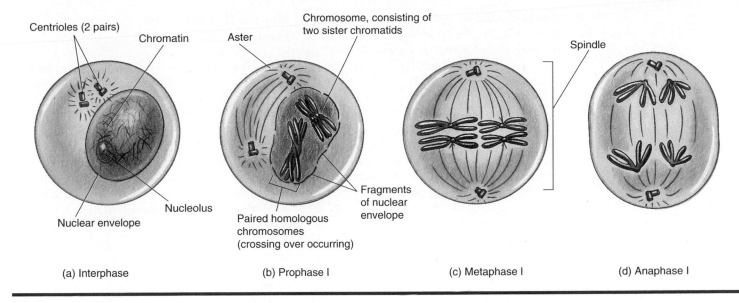

(a) Interphase  (b) Prophase I  (c) Metaphase I  (d) Anaphase I

Centrioles (2 pairs)
Chromatin
Aster
Chromosome, consisting of two sister chromatids
Spindle
Nucleolus
Nuclear envelope
Fragments of nuclear envelope
Paired homologous chromosomes (crossing over occurring)

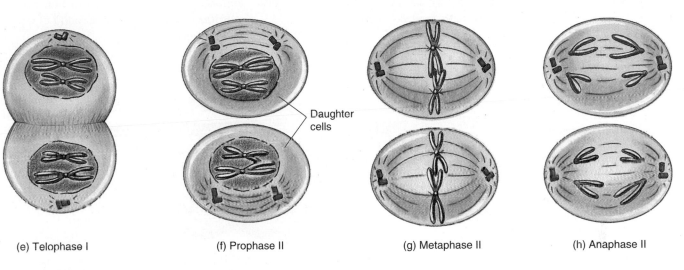

Daughter cells

(e) Telophase I  (f) Prophase II  (g) Metaphase II  (h) Anaphase II

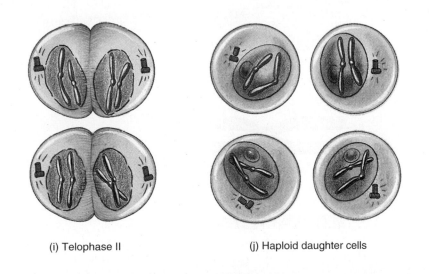

(i) Telophase II  (j) Haploid daughter cells

◆ FIGURE 5.15
The stages of meiosis.

centromere) is adjacent to the other member of that pair (Figure 5.15). The order of this alignment of homologous chromosomes is purely random. Its significance is discussed later. By the time the homologous pairs are at the cell's equator, the centromere of each chromosome has already attached to a separate set of spindle fibers.

In anaphase I, the centromeres remain intact, and one member of the chromosome pair (consisting of two sister chromatids joined at the centromere) moves toward one pole of the spindle, while the other member of the pair moves toward the opposite pole (Figure 5.15). In this way, each of the daughter cells receives one member of each chromosome pair, in a **reductional division.** The double-stranded chromosomes uncoil in telophase I and cytokinesis occurs, producing two haploid daughter cells (Figure 5.15). In some species, the nuclear membrane reforms; in others, the daughter cells immediately enter the second division.

## Meiosis II Begins with Haploid Cells

In prophase II, the chromosomes coil and become visible, a spindle forms, and the nuclear membrane (if present) dissolves. The chromosomes cannot synapse because there are no homologous chromosomes with which to pair. Each chromosome moves to the equator of the cell and becomes attached to spindle fibers in metaphase II. At the beginning of anaphase II, centromeres divide and the chromosomes formed from the sister chromatids move toward opposite poles of the cell. In this process, one copy of each chromosome is distributed to each daughter cell. In telophase II, cytokinesis occurs, and the chromosomes become organized into haploid nuclei (Figure 5.15).

The process of meiosis is now complete. One diploid cell containing pairs of homologous chromosome has undergone one round of chromosome

✦ **FIGURE 5.16**

**Summary of chromosome movements in meiosis. Homologous chromosomes appear and pair in prophase I. At metaphase I, members of a homologous pair separate from each other. In meiosis II, centromeres split, and sister chromatids are converted into chromosomes. The resulting haploid cells have one set of chromosomes.**

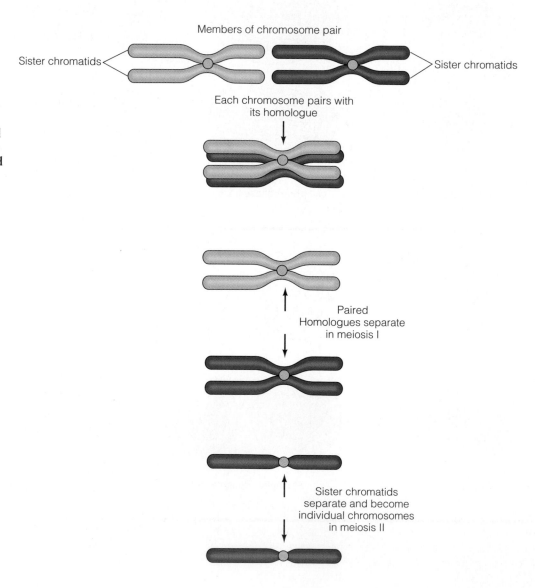

Members of chromosome pair

Sister chromatids — — Sister chromatids

Each chromosome pairs with its homologue

Paired Homologues separate in meiosis I

Sister chromatids separate and become individual chromosomes in meiosis II

replication and two rounds of cell division to produce four haploid daughter cells, each containing one copy of each chromosome (✦ Figure 5.16).

## ✳ MEIOSIS AND GENETICS

Meiosis is linked to sexual reproduction in both plants and animals. It serves to keep the chromosome number constant from generation to generation, and serves to produce new combinations of genes and chromosomes in the offspring. As a result, the offspring produced by sexual reproduction are never genetically identical to either parent. These new genetic combinations are important in generating genetic diversity—the material for evolution by natural selection.

### Meiosis Produces New Combinations of Genes in Two Ways

Meiosis produces new combinations of genetic information in two ways: by the **assortment** of chromosomes in meiosis I and by **recombination,** the physical exchange of homologous chromosome parts that also occurs in meiosis I. When a diploid individual is formed by fusion of haploid gametes, one member of each chromosome pair is received from the mother, and the other member from the father. When synapsed chromosome pairs line up on the cell's equator in metaphase I of meiosis, there is no fixed pattern of arrangement of the maternal and paternal chromosomes (✦ Figure 5.17). In other words, whether the maternal member of a chromosome pair lines up on the left or the right is at random. As a result, each daughter cell is more likely to receive an assortment of maternal and paternal chromosomes than a complete set of maternal or paternal chromosomes.

The number of maternal and paternal chromosomal combinations produced by meiosis is equal to $2^n$, where 2 represents the chromosomes in each pair, and $n$ represents the number of different chromosomes present in cells of the species (the haploid number). For example, humans have 46 chromosomes, or 23 different types. In the cells humans produce by meiosis, $2^{23}$ or 8,388,608 different combinations of maternal and paternal chromosomes are possible in haploid cells. On top of this diversity, any combination from one parent can be united with any combination from the other parent during fertilization, meaning that more than $7 \times 10^{13}$ different combinations of chromosomes are possible in the offspring from one pair of parents.

This impressive number does not take into account the combinations of genes produced by the second mechanism responsible for increasing genetic diversity during meiosis via **recombination.**

In prophase I homologous chromosomes synapse and become paired. At this stage, the chromosomes are each composed of two sister chromatids joined by a common centromere. Frequently, a chromatid from one chromosome and a chromatid from its homologue will break at similar points and the two broken parts will then be joined to the opposite chromatid, so that there is an actual physical

✦ **FIGURE 5.17**

**Orientation of members of chromosome pairs at meiosis is random. Here three chromosomes (1, 2, and 3) have four possible alignments (maternal chromosomes are light blue, paternal chromosomes are dark blue). There are eight possible combinations of maternal and paternal chromosomes in the resulting meiotic cells.**

**Combinations Possible:**

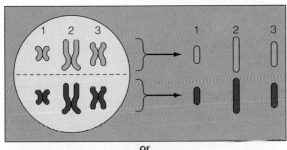

or

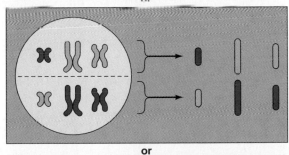

or

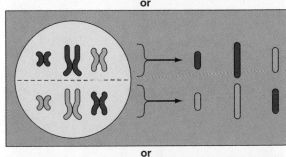

or

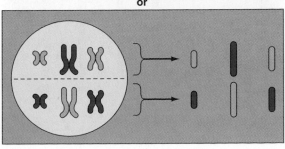

Crossing over increases genetic variation by combining genes from both parents on the same chromatid. The result is an increase in genetic variability in the gametes produced.

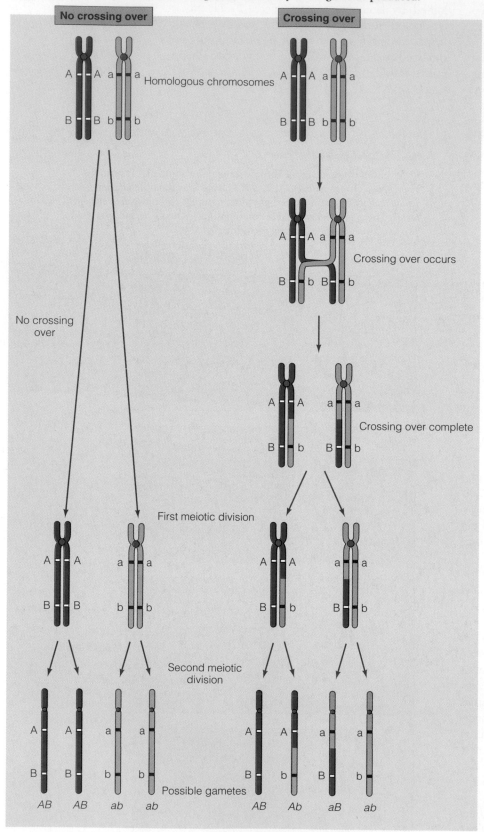

exchange of chromosome parts (♦ Figure 5.18). The result is that genes from both parents end up being combined on the same chromatid. If the homologous chromosomes carry different forms of one or more genes, then new combinations of parental genes are produced. At a later time in prophase, the chromatids involved in crossing over can be seen to be joined at a point called a **chiasma** (pl., **chiasmata**) (♦ Figure 5.19). Because chiasmata are frequently observed in human meiosis and in meiosis in most other eukaryotes, crossing over must be relatively common.

## Sexual and Asexual Reproduction Are Different Evolutionary Strategies

Obviously, meiosis and sexual reproduction generate an enormous amount of genetic diversity. If environmental conditions change, the presence of genetic diversity helps ensure that at least some members of the population will have combinations of genetic traits that will enable them to survive under the new conditions and reproduce successfully. Sexual reproduction also entails some cost, because in an asexual population each individual can produce offspring, making the reproductive capacity twice as efficient.

Asexual reproduction is useful when the environment is stable, and allows a species to produce offspring rapidly that are well adapted to exploiting that environment. Some organisms retain the best of both worlds by having the capacity to reproduce asexually or sexually, as the environmental conditions dictate.

## MEIOSIS AND THE LIFE CYCLE

Although meiosis is characteristic of the majority of eukaryotic species, it can occur at different times in the life cycle. Further, the cells produced by meiosis do not always function only as gametes. In some algae and fungi, the diploid stage of the life cycle is represented only by the zygote. Here, meiosis occurs immediately after fertilization and produces haploid spores. These spores divide by mitosis and the resulting cells act as individual organisms (as in the alga *Chlamydomonas*) (♦ Figure 5.20), or form a multicellular plant as in *Spirogyra* and *Chara*. In such species, most of the life cycle is spent in the haploid stage, and gametes are not produced until long after the original spore was generated by mitosis. In *Chlamydomonas,* the haploid cells are of two mating types, designated as plus (+) or minus (−). Because spores of opposite mating types look and act exactly alike, they are called plus and minus, rather than male and female. Gametes produced by the mitotic descendants of these spores can unite when they are of opposite mating types to produce a diploid zygote that undergoes meiosis to produce spores.

Many algae, fungi, and all land plants have a life cycle in which a multicellular haploid stage is followed by a multicellular diploid stage, which is again followed by a haploid stage. This type of life cycle is known as **alternation of generations.** The life cycle of a fern illustrates alternation of generations (✦ Figure 5.21). The

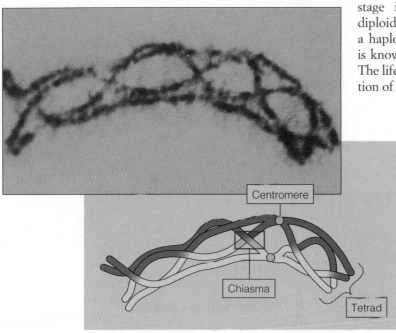

✦ **FIGURE 5.19**

**Crossing over is the exchange of chromosome segments between homologous chromosomes. The X-shaped regions are the site of crossing over.**

✦ **FIGURE 5.20**

**Well-adapted asexually reproducing organisms can rapidly produce a large number of offspring to exploit a suitable environment. Two such multicellular organisms are (a)** *Hydra* **(b)** *Chara***.**

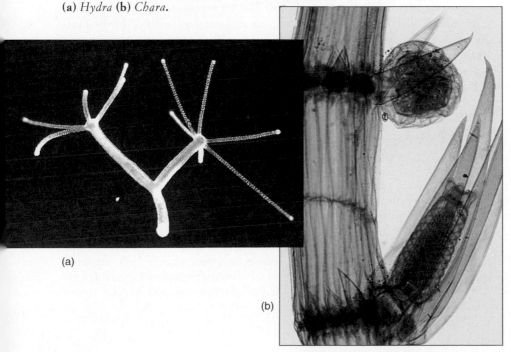

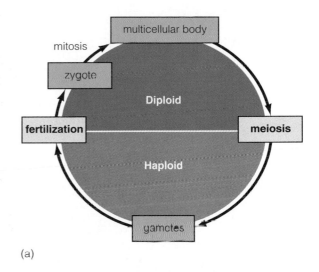

(a)

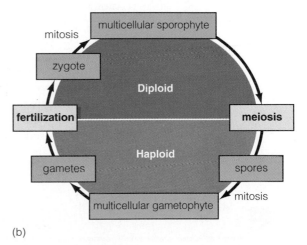

(b)

✦ **FIGURE 5.21**

**The timing of meiosis in the life cycle is different for animals (a) and multicellular plants (b). Plants often have both haploid and diploid phases of the life cycle.**

In the fern, meiosis produces haploid spores that develop into thumbnail-sized haploid (*n*) gametophytes. Haploid gametes develop on the underside of the gametophyte and fuse to form a zygote. From the zygote, a diploid (*2n*) sporophyte develops. Meiosis occurs in the mature sporophyte, beginning the cycle again.

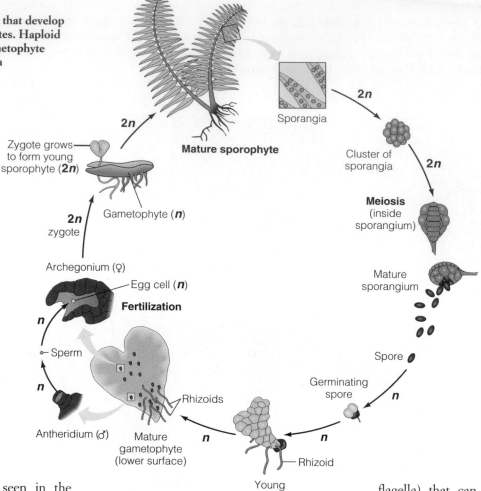

fern plant seen in the forest or greenhouse is the diploid stage of the life cycle known as the **sporophyte.** On the underside of the fern leaves, certain cells undergo meiosis to form haploid spores. The spores drop to the ground and germinate to form small, multicellular haploid plants known as **gametophytes.** The gametophyte produces haploid gametes by mitosis. These gametes can be distinguished as eggs (large and stationary, enclosed in specialized structures in the gametophyte), and sperm (small, and equipped with

flagella) that can swim to fertilize the eggs when water is available (✦ Figure 5.22).

Gametes fuse and give rise to the diploid sporophyte, beginning the cycle once again.

In humans, as in most animals and a few algae and fungi, haploid gametes fuse to form a diploid zygote that divides by mitosis to form a multicellular organism. Specialized germ cells divide by mitosis, and then enter meiosis to produce the gametes. In these organisms, the haploid phase is represented only by the gametes.

## SUMMARY

1. Cells are the units of structure and reproduction in all organisms. The period from one nuclear division to another encompasses one cell cycle. In eukaryotes, during the interphase portion of the cycle, a duplicate set of chromosomes is made. Once the chromosomes have been replicated, the cell proceeds through the cell cycle to mitosis. Mitosis is a process of nuclear division that distributes a complete set of genetic information to two daughter cells.

2. The process of mitosis has been divided into four stages: prophase, metaphase, anaphase, and telophase. In prophase, the replicated chromosomes coil and become visible and the nuclear membrane disappears. In metaphase, the chromosomes migrate to the equator of the cell and become attached to the spindle fibers at the kinetochores of the centromere. In anaphase, the centromeres split longitudinally, allowing the sister chromatids (now called chromosomes) to migrate toward

opposite poles of the spindle. In telophase, the chromosomes at each pole are incorporated into a nucleus and uncoil into a dispersed condition.

3. Cytokinesis divides the cytoplasm, producing two cells, each of which contains a complete set of chromosomes. In animal cells, cytokinesis involves the formation of a cleavage furrow, whereas in plant cells, a partition, known as the cell plate, is involved in distributing the cytoplasm.

4. Meiosis is a form of nuclear division that produces haploid cells containing only one member of each chromosome pair. In prophase I, homologous chromosomes synapse. At metaphase I, the paired homologous chromosomes become aligned at the equator of the cell and attach to the spindle fibers. In anaphase I, homologous chromosomes are separated from each other and migrate to opposite poles of the cell. In meiosis II, there is no chromosome replication, and the unpaired chromosomes align at the metaphase plate, the centromeres split longitudinally, and the sister chromatids (now called chromosomes) move toward opposite poles of the cell. In telophase II, nuclei reform and cytokinesis divides the cell. The result is four cells, each containing the haploid number of chromosomes.

5. Genetic diversity is produced in meiosis by crossing over and by the alignment of members of chromosome pairs at the cell equator during metaphase I. This genetic diversity explains why offspring produced by sexual reproduction are not identical copies of their parents.

6. Meiosis occurs at different times in the life cycle of different organisms. In some organisms, the diploid stage is represented only by the zygote, and one or more haploid cells constitute the body of the organism. Many land plants exhibit a life cycle with a more conspicuous alternation of generations, involving a diploid sporophyte and a haploid gametophyte. In most animals, including humans, the zygote develops into a diploid organism, and the haploid phase is represented only by the gametes.

7. Meiosis is an important process that provides a mechanism for maintaining a constant chromosome number from generation to generation and creating genetic diversity that provides the raw material for natural selection and evolution.

# KEY TERMS

| | | |
|---|---|---|
| alternation of generations | gametes | prophase |
| anaphase | gametophytes | recombination |
| assortment | haploid | reductional division |
| cell cycle | homologues | sister chromatids |
| cell plate | interphase | somatic |
| centromere | kinetochores | sporophyte |
| chiasma, chiasmata | meiosis | synapsis |
| chromatids | metaphase | telophase |
| cleavage furrow | metastasis | tetrad |
| cyclin | mitosis | tumor suppressor genes |
| cytokinesis | mitotic spindle | unicellular |
| diploid | oncogenes | zygote |

# QUESTIONS AND PROBLEMS

## TRUE/FALSE

1. Somatic cells divide during each part of the cell cycle.
2. A cell that contains a single basic complement of chromosomes is said to be diploid or 2n.
3. In the fern plant, the diploid stage of the life cycle is known as the sporophyte, whereas the multicellular haploid plants are known as gametophytes.
4. In the cell cycle, all cells pass through all stages leading to division.
5. Mitosis is duplication division, whereas meiosis is reduction division.

## MULTIPLE CHOICE

1. Four chromatids are the same as a pair of _____.
   a. chromomeres
   b. centromeres
   c. centrioles
   d. homologous chromosomes
2. The cells of prokaryotic organisms undergo a unique form of cell division. This splitting in half of the cell to form two daughter cells is called _____.
   a. meiosis
   b. mitosis
   c. binary fission
   d. cytokinesis

3. At the completion of mitosis, the daughter cells will consist of genetic information that is _____ and has _____ chromosome number as the parent cell.
   a. identical to the parent cell . . . a quarter
   b. altered . . . half
   c. rearranged . . . same
   d. identical to the parent cell . . . with the same
4. The diploid number of chromosomes for humans is _____.
   a. 46
   b. 92
   c. 22
   d. 23

5. The production of new combinations of genes by the assortment and recombination of chromosomes occurs in _____.
   a. mitosis
   b. asexual reproduction
   c. meiosis
   d. binary fission

## SHORT ANSWER

1. What is the distinguishing difference between asexual reproduction and sexual reproduction? What are the benefits of each?
2. Explain the differences between chromatin, chromosome, chromatid, and centromere.
3. Explain what happens during each part of the cell cycle.
4. What are the differences in mitosis between organisms with cell walls and animal cells?
5. The X-shaped chromosomal configuration that results from crossing over is called a _____.
6. In meiosis, the physical exchange of chromosome parts in _____ results in genes from both parents being combined on the same chromatid.

## SCIENCE AND SOCIETY

1. Because of recombination in meiosis 1, DNA sequences on the X and Y chromosomes normally exchange in a specific region. In rare instances, however, genetic recombination occurs between the X and Y chromosomes outside this region, and this aberrant exchange can produce two very rare abnormalities: XY females and XX males. Each of these sex-reversal disorders occurs with a frequency of approximately 1 in 20,000 births. How do you think this abnormality occurs during cell division? How would you tell the parents of a XX male child that their "son" is genotypically female? Is this something that would affect the physical development of the child? Why or why not?

2. Electric or magnetic fields from power lines and household appliances have been proposed as a source of childhood cancers, especially leukemia and brain cancer. Studies have shown a slight increase in the incidence of these cancers in children living in homes where there is an elevated magnetic field caused by the neighborhood power line. However, these data are confusing because the relationship is stronger with indirect rather than direct measures of the magnetic field. Opponents of the idea that magnetic fields cause cancer point out the inconsistencies seen in epidemiological studies. Some of these studies used data from the state of Connecticut, which has the oldest cancer registry in the United States, and reported that leukemia rates over the last half-century remained constant despite the increased electric power consumption. What scientific evidence do you need to have before forming an opinion about the effects of magnetic fields? What would you propose as an accurate means for measuring the effects of magnetic fields? Can you think of other health problems that may result from this form of exposure? Are there other variables that could contribute to an individual's susceptibility to magnetic field problems? Do you think this should be a prime concern for society? Why or why not?

# II

# CELLS AND HEREDITY

THE AFRICAN BLOOD LILY (*HAEMANTHUS KATHERINAE* BAK.) HAS LARGE METAPHASE CHROMOSOMES AND IS AN IDEAL ORGANISM IN WHICH TO STUDY MITOSIS.

# 6

# PRINCIPLES OF HEREDITY

O P E N I N G   I M A G E

*Pea plants were carefully selected*

*by Mendel for his experiments*

*on heredity.*

C*harles Darwin is perhaps best known for his monumental book on the subject of evolution,* The Origin of Species. *Less well known is the fact that other members of his family also made significant contributions to our understanding of genetics and evolution in the latter part of the nineteenth century. One of these was Darwin's cousin, Francis Galton. Galton was concerned that Darwin's account of evolution did not explain how small variations in appearance were generated.*

*Darwin himself subscribed to the accepted idea of the time that parental traits are blended in the offspring. According to the blending idea, if a plant with red flowers is crossed to a plant with white flowers, the offspring should have pink flowers (a blend of red and white). However, if traits were blended generation after generation, variation would be reduced instead of enhanced. To get around this troubling consequence, Darwin turned to another idea about how traits were inherited. According to this idea, instructions to form body tissues were contained in particles called "gemmules" that moved through the blood to the reproductive organs and were transmitted from there to the offspring. The instructions in the gemmules were called pangenes, and this theory of inheritance was called pangenesis. According to pangenesis, gemmules from each parent make a contribution to the growth and form of the organs and tissues of the offspring.*

*Galton decided to put the idea of gemmules to a test. He took rabbits with different coat colors, and transfused blood between them. He thought that a transfusion should mix the gemmules from the two rabbits, altering the ability of the rabbits to transmit coat*

color to their offspring. He then bred the transfused rabbits. If blood from black-coated rabbits was transfused into rabbits with white coats, then crosses between the transfused white rabbits should produce at least some offspring with gray coats (a blend of black and white).

Galton's experiments did not support the idea of pangenesis. He presented his results to the Royal Society, the highest scientific body in England, on March 30, 1871, and said: "The conclusion from this large series of experiments is not to be avoided, that the doctrine of Pangenesis, pure and simple, as I have interpreted it, is incorrect." The report showed that traits were not transmitted by pangenesis, but it left the question of how traits were inherited unanswered.

Galton made other significant contributions to the study of inheritance. He established the mathematical foundation for the study of traits controlled by several genes, emphasizing the importance of twin studies in human genetics.

Unfortunately, it appears that neither Galton nor Darwin read the work of Gregor Mendel on the inheritance of traits in the garden pea published in 1866. This work, titled "Experiments in Plant Hybrids" is one of the most important scientific papers ever published. Here Mendel departs from the usual reporting of observations and experimental results and takes the additional step of fitting his findings into a conceptual framework that could be used to explain the mechanism of heredity in any organism, not just the garden pea. In this chapter, we reconstruct the experiments of Mendel and show how he moved from recording his results to drawing conclusions about the principles of heredity.

## HEREDITY: HOW DOES IT WORK?

Up to the end of the nineteenth century, philosophers and scientists often considered several questions about heredity. One question asked "How are specific traits handed down from parent to offspring?" Speculations and observations on this topic covered a wide range of ideas and generated ideas about heredity that influenced biology until just over 100 years ago. Pangenesis was one of these ideas. First proposed by the ancient Greeks, it was thought that both males and females formed pangenes. According to this idea, pangenes were formed in every organ, moved through the blood to the genitals, where they were transmitted from parents to children.

A relic of this idea persists in the use of terms such as "blood relative" or "full-blooded" and descriptions of persons like Prince Charles as being of "royal blood." Although such notions may seem far-fetched today, the idea lasted thousands of years, and was experimentally tested by Galton as recently as the 1870s.

By observation alone, it can be seen that children resemble their parents. Another question about heredity asked "Does each parent makes an equal contribution to the traits of the offspring?" There was much disagreement on this point. By the middle of the last century, cytologists using a new generation of microscopes discovered that, in many organisms, the female gamete is much larger than the male gamete. On the basis of these findings, some biologists argued that the female's genetic contribution to the offspring must be greater than the male's. The opposite viewpoint was taken by some of the ancient Greeks, who believed that only the male parent contributed to the characteristics of the offspring. This idea was expressed by Aeschylus in 458 B.C.:

> The mother of what is called her child is no parent of it, but nurse only of the young life sown in her. The parent is the male, and she but a stranger, a friend, who, if fate spares his plant, preserves it till it puts forth.

A third question about heredity asked "Can physical characteristics acquired through experience or accident be transmitted to the offspring?" From ancient times, this idea was often invoked when parents produced children with physical defects. Such birth defects were attributed to emotional shocks suffered by the mother during pregnancy.

Questions about heredity that had been unresolved for thousands of years were dramatically changed by the work of Gregor Mendel. Working with pea plants in the late part of the nineteenth century, Mendel showed that traits are passed from parent to offspring through the inheritance of what we now call **genes.** He reasoned that each parent contributes one factor to each trait shown in the offspring. He also concluded that the two members of each pair of factors separate from each other during the formation of sperm and eggs. His work restructured our ideas about heredity by (1) discounting blending inheritance, (2) showing that males and females contribute equally to the traits in the offspring, and (3) showing that acquired traits are not inherited. How he came to these conclusions by crossing varieties of the garden pea and analyzing the results of these crosses is the subject of this chapter.

## MENDEL'S EXPERIMENTAL APPROACH RESOLVED MANY UNANSWERED QUESTIONS

Mendel's success in uncovering the mechanisms of inheritance was not the result of blind luck or accident, but the product of carefully planned experiments. Having determined what he wanted to investigate, Mendel set about choosing an organism for these experiments. Near the beginning of his landmark paper on inheritance, Mendel wrote:

> The value and validity of any experiment are determined by the suitability of the means as well as by the way they are applied. In the present case as well, it cannot be unimportant which plant species were chosen for the experiments and how these were carried out.
>
> Selection of the plant group for experiments of this kind must be made with the greatest possible care if one does not want to jeopardize all possibility of success from the very outset.

He then listed the properties that an experimental organism should have. First, it should have a number of differing traits that can be studied; second, the plant should be self-fertilizing and have a flower structure that minimizes accidental contamination with foreign pollen; and last, the offspring of self-fertilized plants should be fully fertile so that further crosses can be made.

He paid particular attention to a plant group known as the legumes because their flower structure allows self-pollination or cross-pollination with a minimum chance of accidental pollination by other plants. Among the legumes, he noted that 34 varieties of pea plants with different traits were available to him from seed dealers. Peas have a relatively short growth period, can be grown in the ground or in pots in the greenhouse, and can be self-fertilized or artificially fertilized by hand when necessary (✦ Figure 6.1).

He tested all 34 varieties of pea plants for two years to ensure that the characteristics they carried were **true-breeding;** that is, self-fertilization gave rise to the same traits in all offspring, generation after generation. From these, 22 varieties were planted annually for the next eight years to provide plants for his experiments (✦ Figure 6.2). For his work, he selected seven characters that affected the seeds, pods, flowers, and stems of the plant (✦ Table 6.1). Each character he studied was represented by two distinct forms or traits: plant height by tall and short, seed shape by wrinkled and smooth, and so forth.

To avoid errors caused by small sample sizes, he planned experiments on a large scale. During the next eight years, Mendel's experiments used some 28,000 pea plants. In all experiments, he kept track of each character separately for several generations. He began by studying one pair of traits at a time and repeated his experiments for each of the traits to confirm his results. Using his training in physics and mathematics, Mendel analyzed his data according to the principles of probability and statistics. His methodical and thorough approach to his work and his lack of preconceived notions were the secrets of his success.

(a)

✦ **FIGURE 6.1**

**Traits in pea plants provided the material for Mendel's work on heredity. (a) Smooth peas in pod. (b) Constricted pods with purple flowers. (c) White flower.**

(b)

(c)

| TABLE 6.1 | TRAITS SELECTED FOR STUDY BY MENDEL | |
|---|---|---|
| STRUCTURE STUDIED | DOMINANT | RECESSIVE |
| SEEDS | | |
| Shape | Smooth | Wrinkled |
| Color | Yellow | Green |
| Seed Coat Color | Gray | White |
| PODS | | |
| Shape | Full | Constricted |
| Color | Green | Yellow |
| FLOWERS | | |
| Placement | Axial (along stems) | Terminal (top of stems) |
| STEMS | | |
| Length | Long | Short |

✦ **FIGURE 6.2**
**The monastery garden where Mendel carried out his experiments in plant genetics.**

## CROSSING PEA PLANTS: THE PRINCIPLE OF SEGREGATION

To illustrate how Mendel's ideas about inheritance developed, we first describe some of his experiments and outline the results he obtained. Then, we follow the reasoning Mendel used in reaching his conclusions and outline some of the further experiments that confirmed his ideas.

In the first set of experiments, Mendel followed the inheritance of seed shape. Since this cross in-volved only one character (seed shape) he called it a **monohybrid cross.** He took plants with smooth seeds and crossed them to a variety with wrinkled seeds (✦ Figure 6.3). In making this cross, flowers from one variety were fertilized by hand using pollen from the other variety. In this first experiment, Mendel performed 60 fertilizations on 15 plants. The seeds that formed as a result of these fertilizations were all smooth. This result was true whether the pollen was contributed by a plant with smooth or wrinkled peas. The next year, Mendel planted the smooth seeds from this cross. When the plants ma-

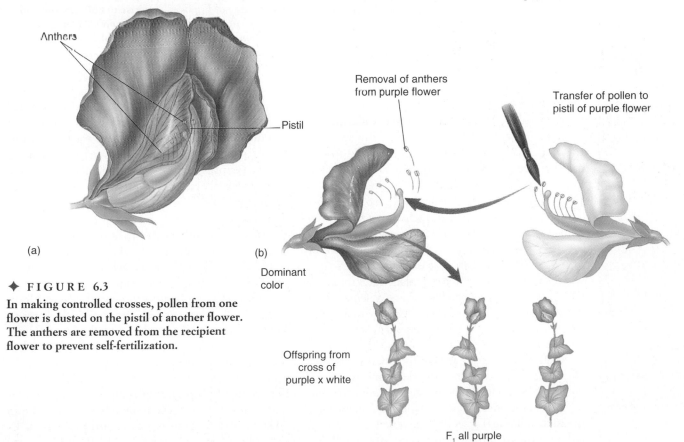

Anthers

Pistil

(a)

✦ **FIGURE 6.3**

**In making controlled crosses, pollen from one flower is dusted on the pistil of another flower. The anthers are removed from the recipient flower to prevent self-fertilization.**

Removal of anthers from purple flower

Transfer of pollen to pistil of purple flower

(b)

Dominant color

Offspring from cross of purple x white

F₁ all purple

plants, which produced a total of 7324 seeds. Of these, 5474 were smooth and 1850 were wrinkled.

He conducted similar experiments on the inheritance of the other six characters. In his experiments, Mendel designated the parental generation as the $P_1$ generation and the offspring as the $F_1$ (first filial) generation. The second generation, produced by self-fertilizing the $F_1$ plants, was called the $F_2$ (or second filial) generation. The experiments with seed shape are summarized in ✦ Figure 6.4.

$P_1$ : Smooth × wrinkled
$F_1$ : All smooth
$F_2$ : 5474 smooth and 1850 wrinkled

## Results and Conclusions from Mendel's First Series of Crosses

The results from experiments with all seven characters were similar to those seen in the cross with smooth and wrinkled seeds, and are summarized in ✦ Figure 6.5. In all crosses, the following results were obtained:

1. The $F_1$ offspring showed only one of the two parental traits, and always the same trait.
2. In all crosses, it did not matter which variety served as the male parent (that is, served to donate the pollen). The results were always the same.
3. The trait not shown in the $F_1$ offspring, reappeared in about 25% of the $F_2$ offspring.

The results of these crosses were the basis for Mendel's first discoveries. The traits remained unchanged as they passed from parent to offspring: Traits did *not* blend together in any of the offspring. On the contrary, they were transmitted in a discrete fashion, and although they might be unexpressed, they remained unchanged from generation to generation. This convinced him that inheritance did not work by blending the traits of the parents in the offspring; rather, traits behaved as separate units.

In all of his experiments, reciprocal crosses were made so that the variety used as a male plant in one set of crosses was used as the female plant in another set of crosses. In all cases, it did not matter whether the male or female plant had round or wrinkled seeds; the results were the same. From these results he concluded that each parent makes an equal contribution to the genetic makeup of the offspring.

Based on the results of his experiments with each of the seven characters, Mendel came to several conclusions. First, the evidence indicated that factors that determine traits can be hidden or unexpressed. All the $F_1$ seeds resembled the smooth parent, but when these seeds were grown and self-

*Sidebar*

**MENDEL AND TEST ANXIETY**

Mendel entered the Augustinian monastery in 1843 and took the name Gregor. While studying at the monastery, he served as a teacher at the local technical high school. In the summer of 1850 he decided to take the examinations that would allow him to have a permanent appointment as a teacher. The exam was in three parts. Mendel passed the first two parts, but failed one of the sections in the third part. In the fall of 1851, he enrolled at the University of Vienna to study natural science (the section of the exam he flunked). He finished his studies in the fall of 1853, returned to the monastery and, again, taught at a local high school.

In 1855 he applied to take the teacher's examination again. The test was held in May of 1856, and Mendel became ill while answering the first question on the first essay examination. He left, and never took another examination. As a schoolboy and again as a student at the monastery, Mendel suffered bouts of illness, all associated with times of stress. In an analysis of his illnesses made in the early 1960s, a physician concluded that Mendel had a psychological condition that today would probably be called "test anxiety." If you are feeling stressed at exam time, take some small measure of comfort in the fact that for Mendel, it was probably worse.

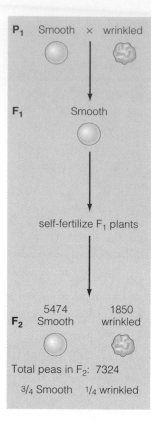

✦ **FIGURE 6.4**

**One of Mendel's crosses. Pure-breeding varieties (smooth and wrinkled) were used as the $P_1$ generation. The offspring in the $F_1$ had all smooth seeds. Self-fertilization of $F_1$ plants gave rise to both smooth and wrinkled progeny in the $F_2$ generation. About $\frac{3}{4}$ of the offspring were smooth and about $\frac{1}{4}$ were wrinkled.**

fertilized, they produced some plants with wrinkled seeds. This means that the $F_1$ seeds contained a hereditary factor for *wrinkled* that was present but not expressed. The trait that is not expressed in the $F_1$ but *is* expressed in the $F_2$ he called the **recessive** trait. The trait expressed in the $F_1$ he called the **dominant** trait. Mendel called this phenomenon **dominance.**

Second, a comparison of the $P_1$ smooth plants and the $F_1$ smooth plants showed that, despite identical appearances, their genetic makeup must be different. When $P_1$ plants are self-fertilized, they give rise only to plants with smooth seeds. However, when $F_1$ plants are self-fertilized, they give rise to plants with smooth and wrinkled seeds. Mendel realized that it was important to make a distinction between the appearance of an organism and its genetic constitution. The term **phenotype** refers to the observed properties or outward appearance of a trait, and the term **genotype** refers to the genetic makeup of an organism with regard to the trait. In our example, it is apparent that the $P_1$ and $F_1$ plants

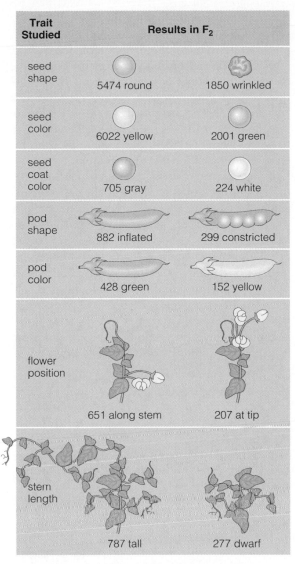

| Trait Studied | Results in F₂ | |
|---|---|---|
| seed shape | 5474 round | 1850 wrinkled |
| seed color | 6022 yellow | 2001 green |
| seed coat color | 705 gray | 224 white |
| pod shape | 882 inflated | 299 constricted |
| pod color | 428 green | 152 yellow |
| flower position | 651 along stem | 207 at tip |
| stem length | 787 tall | 277 dwarf |

**✦ FIGURE 6.5**

**Results of Mendel's monohybrid crosses in peas. The numbers represent the F₂ plants showing a given trait. On average, ¾ of the offspring showed one trait, and ¼ showed the other (a 3:1 ratio).**

with smooth seeds have identical phenotypes, but must have different genotypes.

The results of these self-fertilization experiments indicate that the F₁ plants must have contained factors for smooth and wrinkled traits, since both types of seeds are present in the F₂ generation. The question is, how *many* factors for seed shape are carried in the F₁ plants? From the results of his crosses, Mendel reasoned that both the male and female parent contributed equally to the traits of the offspring, since it did not matter which parent was smooth or wrinkled. In view of this, the *simplest* interpretation is that each F₁ plant received two hereditary factors, one for smooth that was expressed, and one for wrinkled that remained unex-

pressed. By extension of this reasoning, each P₁ and F₂ plant must also contain two factors that determine seed shape.

Traditionally, uppercase letters are used to represent the dominant factor, and lowercase letters are used to represent the recessive factor. For example, $S$ = smooth and $s$ = wrinkled. Using this shorthand, we can reconstruct the genotypes and phenotypes of the P₁ and F₁ as shown in ✦ Figure 6.6.

## Inheritance of a Single Trait: The Principle of Segregation

If factors that determine traits exist in pairs, then some mechanism must exist to prevent these factors from being doubled in each succeeding generation. That is, if each parent has two factors for a given trait, why doesn't the offspring have four? Mendel reasoned that members of a pair of factors must separate or segregate from each other during gamete formation. In doing so, each gamete receives

**✦ FIGURE 6.6**

**The phenotypes and genotypes of the parents and offspring in Mendel's cross involving seed shape.**

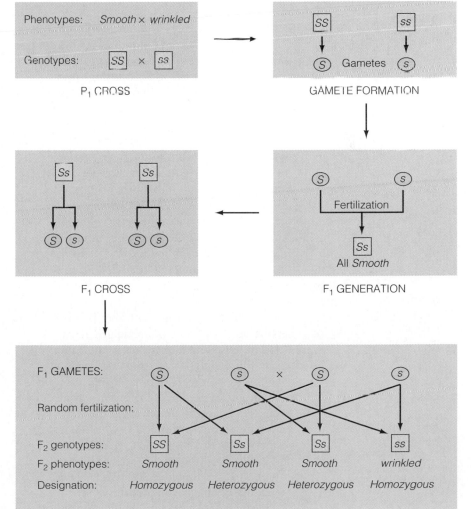

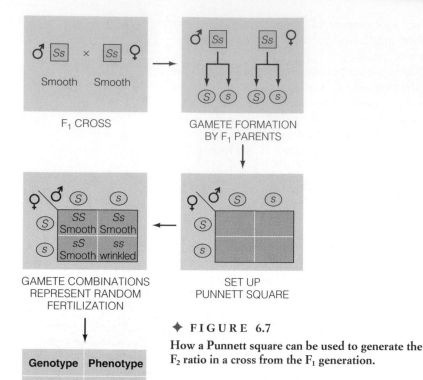

F₁ CROSS

GAMETE FORMATION
BY F₁ PARENTS

GAMETE COMBINATIONS
REPRESENT RANDOM
FERTILIZATION

SET UP
PUNNETT SQUARE

| Genotype | Phenotype |
|----------|-----------|
| 1 SS<br>2 Ss | } 3/4 Smooth |
| 1 ss | } 1/4 wrinkled |

F₂ RATIO

✦ **FIGURE 6.7**

**How a Punnett square can be used to generate the F₂ ratio in a cross from the F₁ generation.**

self-fertilized. In fact, Mendel fertilized a number of plants from the F₂ generation and five succeeding generations to confirm these predictions.

Mendel carried out his experiments before the discovery of mitosis and meiosis, and before the discovery of chromosomes. As we describe in a later section, his deductions about the way traits are inherited are, in fact, descriptions of the way chromosomes behave in meiosis. Seen in this light, his discoveries are all the more remarkable.

Today we call Mendel's factors **genes** and refer to the alternate forms of a gene as **alleles.** In the example we have been discussing, the gene for seed shape has two alleles, smooth and wrinkled. Individuals carrying identical alleles of a given gene (*SS* or *ss*) are said to be **homozygous** for the gene in question. Similarly, when two different alleles are present in a gene pair (*Ss*), the individual is said to have a **heterozygous** genotype. The *SS* homozygotes and the *Ss* heterozygotes will show dominant phenotypes (because *S* is dominant to *s*), and *ss* homozygotes will show recessive phenotypes.

only one of the factors for a given trait (✦ Figure 6.7). The separation of paired factors during gamete formation results in each gamete receiving one member of a pair, and is called the **principle of segregation,** or Mendel's first law.

As shown in Figure 6.7, members of a gene pair separate (or segregate) from each other so that only one or the other is included in each gamete. In the F₁ generation, the heterozygous parents each make two kinds of gametes in equal proportions. At fertilization, the random combination of these gametes produces the genotypic combinations shown in the Punnett square. The F₂ genotypic ratio of 1 *SS* : 2 *Ss* : 1 *ss* is expressed as a phenotypic ratio that is $\frac{3}{4}$ dominant and $\frac{1}{4}$ recessive. This is usually abbreviated as a 3 : 1 ratio.

Mendel's experiments with the six other sets of traits can also be explained in this way. His reasoning also makes a prediction about the genotypes of the F₂ generation. One-fourth of the F₂ plants should carry only smooth factors (*SS*) and give rise only to smooth plants when self-fertilized. One-half of the F₂ plants should carry hereditary factors for both smooth and wrinkled (*Ss*) and give rise to smooth and wrinkled progeny in a $\frac{3}{4}$ to $\frac{1}{4}$ ratio when self-fertilized (✦ Figure 6.8). Finally, one-fourth of the F₂ plants should be homozygous for wrinkled (*ss*) and give rise to all wrinkled progeny if

✦ **FIGURE 6.8**

**Self-crossing the F₂ plants demonstrates that there are two different genotypes among the plants with smooth peas in the F₂ generation.**

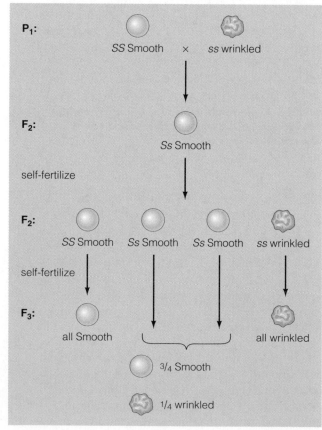

## MORE CROSSES WITH PEA PLANTS: THE PRINCIPLE OF INDEPENDENT ASSORTMENT

Mendel realized the need to extend his studies on the inheritance from monohybrid crosses to more complex situations. He wrote:

> In the experiments discussed above, plants were used which differed in only one essential trait. The next task consisted in investigating whether the law of development thus found would also apply to a pair of differing traits. . . .

Two sets of experiments were used to investigate the inheritance of two or more characters simultaneously. For this work, he selected seed shape and seed color, because as he put it: "Experiments with seed traits lead most easily and assuredly to success." A cross that involves two sets of characters is called a **dihybrid cross**.

### Crosses with Two Traits

As in the first set of crosses, we will analyze the actual experiments of Mendel, outline the results, and summarize the conclusions he drew from them. From previous crosses, it is known that in seeds, smooth is dominant to wrinkled and yellow is dominant to green. In our reconstruction of these experiments, we will represent smooth by an uppercase *S*, wrinkled by a lowercase *s*, yellow by an uppercase *Y* and green by a lowercase *y*.

### Methods, Results, and Conclusions

Mendel selected true-breeding plants with smooth, yellow seeds and crossed them to true-breeding plants with wrinkled, green seeds (✦ Figure 6.9). The seeds of the F₁ plants were all smooth and yellow, confirming that smooth and yellow are dominant traits. He then self-fertilized the F₁ and produced an F₂ generation. These F₂ plants produced seeds of four types, with all four often found together in a single pod. From 15 plants, he counted a total of 556 seeds with the following distribution:

| | |
|---|---|
| 315 | Smooth and Yellow |
| 108 | Smooth and green |
| 101 | wrinkled and Yellow |
| 32 | wrinkled and green |

The F₂ phenotypes include not only the parental phenotypes, but two new combinations (smooth green and wrinkled yellow).

To determine the mode of inheritance of the two genes in a dihybrid cross, Mendel first analyzed the results in the F₂ for each trait separately, as if the

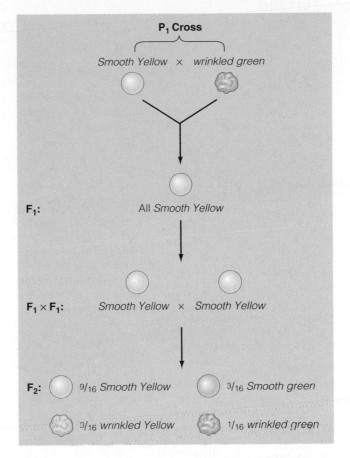

✦ FIGURE 6.9

The phenotypic distribution in a dihybrid cross. The F₂ generation contains the parental phenotypes and two new phenotypic combinations.

other trait were not present (✦ Figure 6.10). If we consider only seed shape (smooth or wrinkled) and ignore color, we expect to obtain ¾ smooth and ¼ wrinkled offspring in the F₂. Analyzing the actual results, we find that the total number of smooth offspring is $315 + 108 = 423$. The total number of wrinkled seeds is $101 + 32 = 133$. The proportion

✦ FIGURE 6.10

**Analysis of a dihybrid cross for the separate inheritance of each trait.**

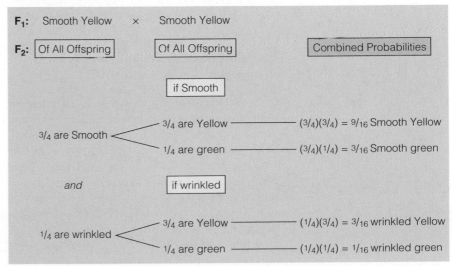

In solving genetics problems, several steps must be followed to ensure success. The process of analyzing and solving these problems depends on several steps: (1) Analyze each problem carefully to determine what information is provided and what information is asked for, (2) translate the terms and words of the problems into symbols, and (3) solve the problem using logic.

The most basic problems involving Mendelian inheritance usually provide some information about the parental generation ($P_1$), and ask you to employ your knowledge of Mendelian principles to come to conclusions about the genotypes or phenotypes of the $F_1$ or $F_2$ generation. The solution utilizes several steps:

1. Carefully read the problem and establish the genotype of each parent; assign letter symbols if necessary.
2. Based on their genotypes, determine what types of gametes can be formed by each parent.
3. Unite the gametes from the parents in all combinations. Use a Punnett square if necessary. This will automatically give you all possible genotypes and their ratios for the $F_1$ generation.
4. If necessary, use all combinations of $F_1$ individuals as parents for the $F_2$, and repeat steps 2 and 3 to derive the genotypes and phenotypes of the F generation.

As an example, consider the following problem. The recessive allele *wrinkled* (s) causes peas to appear wrinkled when homozygous. The dominant allele *smooth* (S) causes peas to appear smooth when homozygous or heterozygous. In the following cross, what phenotypic ratio would you expect in the offspring?: One parent is a plant that bears wrinkled seeds. The other is a plant that bears smooth seeds and is the offspring of a cross between true-breeding smooth and wrinkled parents.

The solution to this problem depends on an understanding of the principle of segregation and the relationship between dominance and recessiveness. To derive the genotypes of the parental plants, the following is relevant:

1. Since one parental plant bears wrinkled seeds, this plant is homozygous for the recessive allele (ss). The other parental plant bears smooth seeds and carries at least one dominant allele (S). Since this parent is the offspring from a cross between true-breeding smooth and true-breeding wrinkled plants, it must have received a wrinkled allele, and therefore must be heterozygous (Ss).
2. The cross is therefore Ss × ss. The gametes each parent can make and their combinations in fertilizations are shown here:

| Genotype of $P_1$ | Ss × ss |
| Gametes of $P_1$ | ⑤ ⑤ ⑤ |
| Genotypes of $F_1$ | Ss ss |
| Phenotypes of $F_1$ | smooth wrinkled |

3. In this cross, half of the $F_1$ offspring will be heterozygous smooth individuals (Ss), and half will be homozygous wrinkled (ss). The phenotypic ratio will be 1 smooth : 1 wrinkled.

---

of smooth to wrinkled seeds (423:133) is close to a ratio of 3:1. Similarly, if we consider only seed color (yellow or green), there are 416 yellow seeds (315 + 101) and 140 green seeds (108 + 32) in the $F_2$ generation. These results are also close to a 3:1 distribution.

Once he established a pattern of 3:1 inheritance for each trait separately (consistent with the principle of segregation), he then considered the inheritance of both traits simultaneously.

## The Principle of Independent Assortment

Before we discuss what is meant by independent assortment, let us consider how the phenotypes and genotypes of the $F_1$ and $F_2$ plants were generated. The $F_1$ plants with smooth yellow seeds were heterozygous for both seed shape and seed color. Therefore, the genotype of the $F_1$ plant must have been SsYy, with the S and Y alleles dominant to s and y. Mendel had earlier postulated that members of a gene pair separate or segregate from each other during gamete formation. In this case, involving two pairs of genes, the segregation of the S and s al-

leles must have occurred independently from the segregation of the Y and y alleles (✦ Figure 6.11).

Because of independent assortment, the gametes formed by the $F_1$ plants contained all combinations of these alleles in equal proportions: SY, Sy, sY, and sy. If fertilizations involving the four types of male and female gametes occurred at random (as expected), 16 possible combinations would result (Figure 6.11). An inspection of the 16 combinations in the Punnett square shows the following:

- Nine have at least one copy of each dominant allele, S and Y.
- Three have at least one copy of the dominant allele S and are homozygous yy.
- Three have at least one copy of the dominant allele Y and are homozygous ss.
- One combination is homozygous for ss and yy.

In other words, the 16 combinations of fertilization events (genotypes) fall into four phenotypic classes:

$\frac{9}{16}$ Smooth and Yellow

$\frac{3}{16}$ Smooth and green

$\frac{3}{16}$ wrinkled and Yellow

$\frac{1}{16}$ wrinkled and green

These phenotypic combinations correspond to the number of phenotypic classes seen in the $F_2$ and to the proportions of progeny seen in each class (Figure 6.11). For example, 315 of 556 seeds were smooth and yellow, corresponding to about $\frac{9}{16}$ of the total number of offspring, 108 of 556 seeds were smooth and green, corresponding to about $\frac{3}{16}$ of the offspring, and so forth. This distribution of offspring in the $F_2$ corresponds to a phenotypic ratio of 9 : 3 : 3 : 1.

The results of this cross can be explained by assuming (as Mendel did) that during gamete formation, alleles in one gene pair segregate into gametes independently of the alleles belonging to other gene pairs. This results in the production of gametes containing all combinations of alleles. This second fundamental principle of genetics outlined by Mendel is called the **principle of independent assortment,** or Mendel's second law.

After 10 years of experimentation involving thousands of pea plants, Mendel presented his results at the February and March 1865 meetings of the Natural Science Society, in what is now Brno, the Czech Republic. The text of these lectures was published in the following year in the *Proceedings* of the society. Although his work was cited in several bibliographies, and copies of the journal were widely read, the significance of Mendel's findings was unappreciated. Even Charles Darwin, continuing his search for a mechanism to explain heredity and its role in natural selection, failed to realize the significance of Mendel's work. Darwin's own copy of the *Proceedings* has many notes scribbled in the margin of the paper adjacent to Mendel's, but not one pencil mark anywhere in Mendel's paper.

Despite its importance, Mendel's work was overlooked for 40 years. Finally, in 1900, three workers independently investigating the mechanism of heredity confirmed Mendel's findings, stimulating a great interest in the study of what is now called **genetics.** Unfortunately, Mendel died in 1884 unaware that he had founded an entire scientific discipline.

## MENDELIAN INHERITANCE IN HUMANS

After 1900, the principles of segregation and independent assortment discovered by Mendel were studied in a wide range of organisms. Although it was believed by some that inheritance of traits in humans might be an exception to these principles, the first Mendelian trait (a hand deformity called *brachydactyly*) was identified in 1905; since then, more than 3500 such traits have been described. To illustrate how segregation and independent assortment apply to the inheritance of human traits, we

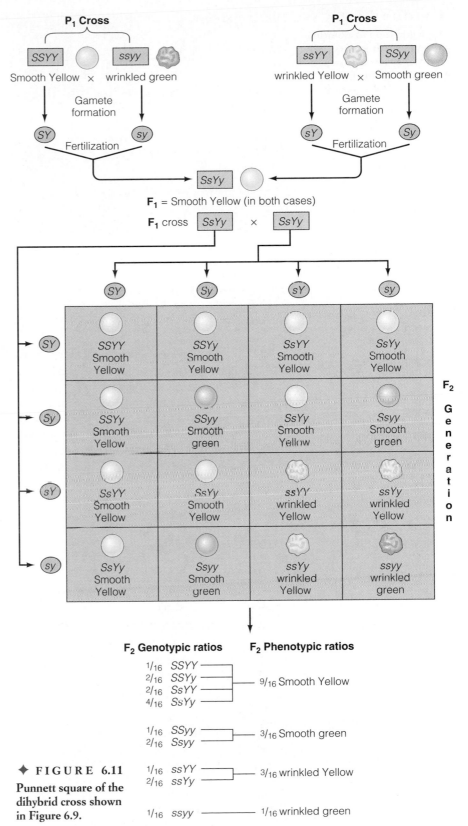

✦ **FIGURE 6.11** **Punnett square of the dihybrid cross shown in Figure 6.9.**

will follow the inheritance of a recessive trait called albinism (*a*). Individuals who are homozygous (*aa*) for this recessive trait have no pigment in skin, hair, and eyes, and thus have very pale white skin, white

The segregation of albinism, a recessive trait in humans. As in pea plants, alleles of a gene pair separate from each other during gamete formation.

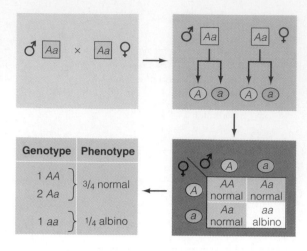

| Genotype | Phenotype |
|----------|-----------|
| 1 *AA*<br>2 *Aa* | ¾ normal |
| 1 *aa* | ¼ albino |

hair, and colorless eyes (actually, the eyes may appear pink because of the blood vessels in the iris). The dominant allele (*A*) controls normal pigmentation.

In the example we consider, both parents have normal pigmentation, but are heterozygous for the recessive allele causing albinism (♦ Figure 6.12). In each parent, the dominant and recessive alleles separate or segregate from each other at the time of gamete formation. Because each parent can produce two different types of gametes (one containing the dominant allele *A* and another type carrying the re-

cessive allele *a*), there are four possible combinations of these gametes at fertilization. These four possible types of fertilization events result in a predicted phenotypic ratio of 3 pigmented : 1 albino offspring, and a genotypic ratio of 1 *AA* : 2*Aa* : 1*aa* (Figure 6.13). In other words, segregation of alleles during gamete formation produces the same outcome in both pea plants and humans. This does not mean than in every such family with four children there will be one albino child and three normally pigmented children. It does mean that in a mating between heterozygotes, there is a 25% chance that a child will be albino and a 75% chance that it will have normal pigmentation.

The simultaneous inheritance of two traits in humans follows the Mendelian principle of independent assortment (♦ Figure 6.13). To illustrate, consider a situation in which each parent is heterozygous for albinism (*Aa*) and for another recessive trait, hereditary deafness (*Dd*). As in albinism, the normal allele (*D*) is dominant and will be expressed in the homozygous dominant (*DD*) or heterozygous condition (*Dd*). At the time of gamete formation, members of each gene pair will assort into gametes independently of all other gene pairs. As a result, each parent will produce equal proportions of four different types of gametes. If during fertilization, four types of gametes combine in all possible ways, there can be a total of 16 different combinations, with a possibility of four different phenotypic classes (Figure 6.14).

An examination of the possible genotypes shows that there is a 1 in 16 chance that a child will be both deaf *and* an albino.

## Pedigree Analysis in Human Genetics

In pea plants and other organisms such as *Drosophila*, genetic analysis can be performed by experimental crosses. In the case of humans, geneticists must base their work on crosses that have already taken place, and cannot design crosses to directly test a hypothesis. One of the basic methods in human genetics is to follow a trait for several generations in a family to determine how the trait is inherited. This method is called **pedigree analysis.** A pedigree is the orderly presentation of family information in the form of an easily readable chart. The symbols used to construct a pedigree are shown in ♦ Figure 6.14. In pedigrees, males are represented by squares, females by circles. The generations are indicated by Roman numerals, and the individuals within a generation are indicated by Arabic numbers. Pedigrees consistent with the inheritance of dominant and recessive traits are shown in ♦ Figure 6.15.

♦ FIGURE 6.13

Independent assortment for two traits in humans.

|  | ♀ **AD** | **Ad** | **aD** | **ad** |
|---|---|---|---|---|
| ♂ **AD** | **AADD**<br>Pigment<br>Hearing | **AADd**<br>Pigment<br>Hearing | **AaDD**<br>Pigment<br>Hearing | **AaDd**<br>Pigment<br>Hearing |
| **Ad** | **AADd**<br>Pigment<br>Hearing | **AAdd**<br>Pigment<br>Deaf | **AaDd**<br>Pigment<br>Hearing | **Aadd**<br>Pigment<br>Deaf |
| **aD** | **AaDD**<br>Pigment<br>Hearing | **AaDd**<br>Pigment<br>Hearing | **aaDD**<br>Albino<br>Hearing | **aaDd**<br>Albino<br>Hearing |
| **ad** | **AaDd**<br>Pigment<br>Hearing | **Aadd**<br>Pigment<br>Deaf | **aaDd**<br>Albino<br>Hearing | **aadd**<br>Albino<br>Deaf |

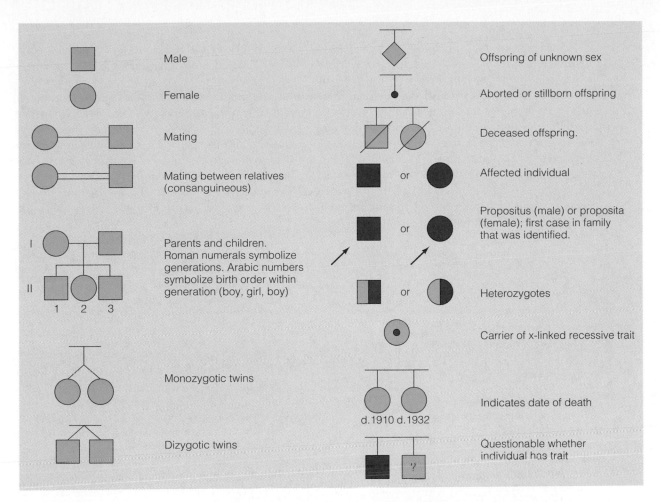

Symbols used in pedigree analysis.
Affected individuals are shown
in red.

✦ FIGURE 6.15

Pedigrees showing inheritance for
(a) dominant and
(b) recessive traits.
Affected individuals are shown
in red.

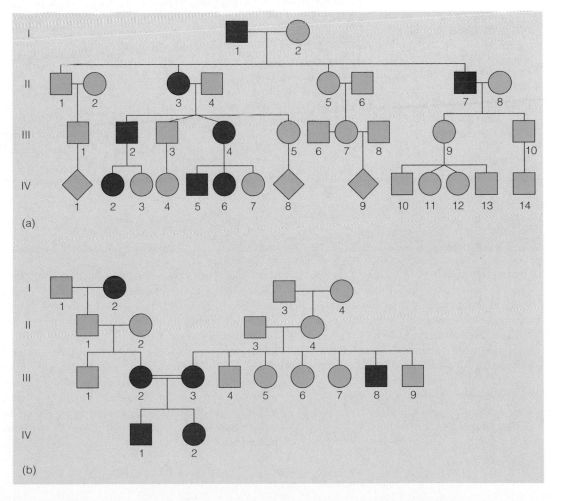

Knowledge of the principles of Mendelian inheritance is used in human genetics to determine which conditions are inherited, and to predict the chances of having offspring affected by genetic disorders (◆ Table 6.2). In the example given, the genotypes of the parents were known beforehand, but in most actual situations, it can be determined or inferred only after the birth of affected children. Establishing genotypes of parents and predicting the chances of having affected children is part of **genetic counseling**, a topic that is discussed in Chapter 9.

Shortly after the rediscovery of Mendel's work, genetic investigations on humans and other organisms turned up instances in which phenotypic patterns did not fit cleanly into the categories of dominant or recessive phenotypes, and cases where the phenotypic ratios seemed distorted. In the next chapter, we will discuss some of these variations in phenotypic expression, and show that even these apparent exceptions follow the principles of inheritance outlined by Mendel. Following that, we will discuss how genes act to produce phenotypes.

**TABLE 6.2  SOME GENETIC TRAITS IN HUMANS**

| RECESSIVE | | DOMINANT | |
|---|---|---|---|
| Albinism | Absence of pigment in skin, eyes, hair | Achondroplasia | Dwarfism associated with defects in growth regions of long bones |
| Ataxia telangiectasia | Progressive degeneration of nervous system | Brachydactyly | Malformed hands with shortened fingers |
| Bloom syndrome | Dwarfism, skin, rash, increased cancer rate | Campodactyly | Stiff, permanently bent little fingers |
| Cystic fibrosis | Mucous production that blocks ducts of certain glands, lung passages; often fatal by early adulthood | Crouzon syndrome | Defective development of midface region, protruding eyes, hook nose |
| Fanconi anemia | Slow growth, heart defects, high rate of leukemia | Ehler-Danlos syndrome | Connective tissue disorder, elastic skin, loose joints |
| Galactosemia | Accumulation of galactose in liver; mental retardation | Familial hypercholesterolemia | Elevated levels of cholesterol; predisposes to plaque formation, cardiac diseases; may be most prevalent genetic disease |
| Phenylketonuria | Excess accumulation of phenylalanine in blood, mental retardation | Familial polycystic kidney disease | Formation of cysts in kidneys; leads to hypertension, kidney failure |
| Sickle cell anemia | Abnormal hemoglobin, blood vessel blockage, early death | Huntington disease | Progressive degeneration of nervous system, dementia, early death |
| Thalassemia | Improper hemoglobin production; symptoms range from mild to fatal | Hypercalcemia | Elevated levels of calcium in blood serum |
| Xeroderma pigmentosum | Lack of DNA repair enzymes, sensitivity to UV light, skin cancer, early death | Marfan syndrome | Connective tissue defect; death by aortic rupture |
| Tay-Sachs disease | Improper metabolism of gangliosides in nerve cells, early death | Nail-patella syndrome | Absence of nails, kneecaps |
| | | Porphyria | Inability to metabolize porphyrins, episodes of mental derangement |

# SUMMARY

1. In the centuries before Gregor Mendel experimented with the inheritance of traits in the garden pea, several competing theories attempted to explain how traits were passed from generation to generation. In his decade-long series of experiments, Mendel established the foundation for the science of genetics.

2. Mendel studied crosses in the garden pea that involved one pair of alleles, and demonstrated that the phenotypes associated with these traits are controlled by pairs of factors, now known as genes. These factors separate or segregate from each other during gamete formation and exhibit dominant/recessive relationships.

3. In later experiments, Mendel discovered that members of one gene pair separate or segregate independently of other gene pairs. This principle of independent assortment leads to the formation of all possible combinations of gametes with equal probability in a cross between two individuals.

4. The principles of segregation and independent assortment proposed by Mendel apply to all sexually reproducing organisms, including humans.

5. Because genes for human genetic disorders exhibit segregation and independent assortment, the inheritance of certain human traits in predictable, making it possible to provide genetic counseling to those at risk of having children affected with genetic disorders.

# KEY TERMS

alleles
dihybrid cross
dominance
dominant
genes

genetic counseling
genotype
heterozygous
homozygous
monohybrid cross

pedigree analysis
phenotype
principle of independent assortment
principle of segregation
recessive

# QUESTIONS AND PROBLEMS

## TRUE/FALSE

1. The basic principles of inheritance were developed by the Austrian monk Gregor Mendel.
2. The outward appearance of an organism is its genotype and is determined by its phenotype.
3. Alleles are alternate forms of a chromosome.
4. Darwin was the first to describe the phenomena of independent assortment and segregation.
5. Variation and heredity occur mainly within boundaries of a particular species.

## MATCHING

1. Darwin          _____ pea plant
2. $P_1$           _____ gemmules
3. Mendel          _____ heterozygous state
4. ss              _____ parental generation
5. $F_1$           _____ monohybrid cross
6. AaBb × AaBb     _____ first filial generation
7. Ss              _____ homozygous state
8. AA × aa         _____ dihybrid cross

## SHORT ANSWER

1. According to Mendel what are three properties that an experimental organism should exhibit?
2. Describe Mendel's first law of heredity (law of segregation).
3. Freckles are inherited in a dominant fashion. A woman with freckles (*Ff*) has a baby with a man who is without freckles (*ff*). What are the chances that this child will have freckles?

4. Two fair-skinned adults marry and have a child with albinism. What are the genotypes of the parents?
5. Attached earlobes (*A*) are a dominant trait. The (*A*) allele is dominant to the (*a*) allele, which results in unattached earlobes in homozygous recessive individuals. A woman with freckles and attached earlobes (*FfAa*) marries a man who has freckles and attached earlobes (*FfAa*). Draw a Punnett square showing the gametes as well as the phenotype.
6. The ability to curl up the sides of your tongue into a U-shape is a dominant trait. Having freckles is also a dominant trait. Suppose a woman who has freckles and can roll her tongue marries a man who does not have freckles and cannot roll his tongue. Their first child does not have freckles and cannot roll the tongue.
   a. What are the genotypes of the mother, father, and child?
   b. What is the chance that their next child will have freckles and be unable to roll the tongue?

## SCIENCE AND SOCIETY

1. Researchers are currently attempting to transfer sperm-making cells from male mice into infertile male mice in the hopes of learning more about reproductive abnormalities. These donor spermatogonia cells have developed into mature spermatozoa in 70% of the cases and some recipients have gone on to father pups. This new advance opens the way for a host of experimental genetic manipulations. It also offers enormous potential for correcting human genetic disease. Another human application might be the use of this procedure to treat infertile males who wish to be fathers. Do you foresee any ethical or legal problems with the implementation of this technique? If so, elaborate on them. Could this procedure have the potential for misuse? If so, explain how. Should donors be screened? If so, what kind of screening measures should be used for donors prior to their participation? What is your opinion on this issue?

2. A pedigree illustrates a family history and shows the inheritance pattern of a genetic trait or disease. In many cases this pattern has features that suggest a dominant or a recessive basis for the characteristic. Most pedigrees include a relatively small number of individuals. The number of offspring produced in a generation is therefore usually too small for the observed phenotypic ratio to be a reliable indicator of expected results. How much do you know about the genetic history of your family? Should you have the right to inspect the genetic history of your future spouse? Should a genetic history be included in your medical records? Do you think one's genetic history should be available to an employer or insurance agency? Why or why not? What are the advantages and disadvantages of knowing more about your family's genetic history?

# 7

# GENES AND CHROMOSOMES

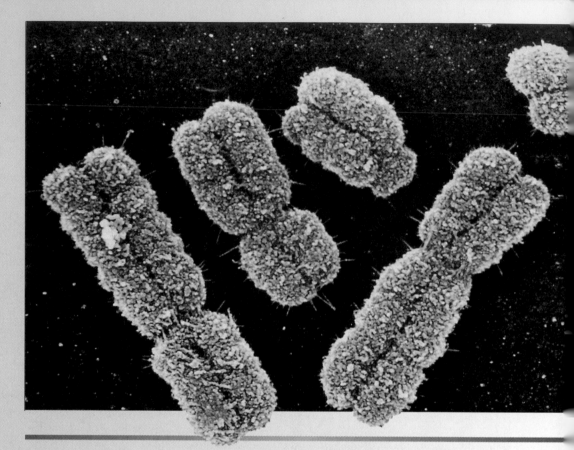

OPENING IMAGE

*Scanning electron micrograph of human metaphase chromosomes.*

Early in the spring of 1902, Walter Sutton, a graduate student in the laboratory of Professor E. B. Wilson at Columbia University, announced that he had discovered why yellow dogs are yellow. This statement was rather confusing to Professor Wilson because he had assigned Sutton to study the behavior of chromosomes during meiosis in the grasshopper, and he later admitted that he was a little puzzled about Sutton's announcement. In the summer of that year, both men worked on research projects at the seashore, first in South Carolina and later in Maine. During their conversations during the course of that summer, Wilson came to understand not only what Sutton was trying to tell him, but realized that the idea represented a major breakthrough in biology. By the fall, Sutton had fully developed his idea and his evidence, and published two brief papers, one in 1902 and the other in 1903.

What was this revolutionary idea? Simply put, Sutton proposed that the Mendelian processes of segregation and independent assortment could be explained by proposing that genes are located on chromosomes. At a single stroke, this simple yet powerful idea provided a mechanism to explain many of the events associated with the transmission of traits from generation to generation.

Sutton came to this idea through a careful study of meiosis during spermatogenesis in the grasshopper. He observed that the chromosomes were grouped into pairs, and that each of these chromosome pairs consisted of a maternally derived and a paternally derived member. From his study of the arrangement of chromosome pairs at metaphase during each of the meiotic divisions, and their separation into different gametes, he concluded that the behavior of chromosomes during division and the behavior of Mendelian factors (genes) in heredity have the same essential features. He further proposed that each chromosome must obviously carry a number of genes, each of which can be present in the form of a dominant or a recessive allele.

His idea, now known as the chromosome theory of inheritance, is one of the central ideas in modern biology. This theory led directly to the efforts of T. H. Morgan and his colleagues to map the location of specific genes on the chromosomes of the fruit fly, Drosophila.

In another form, this effort continues today in the Human Genome Project, started in October 1990, with the objective of identifying and determining the chromosomal location of the 50,000 to 100,000 genes carried on the human chromosome set.

In this chapter, we review the variations on Mendelian inheritance discovered in the early years after 1900, the chromosome theory of inheritance, and gene mapping. We also consider the phenotypic consequences of changes in the number and structure of the chromosomes.

 ## VARIATIONS ON A THEME BY MENDEL

In the pairs of traits selected by Mendel for analysis, heterozygotes and dominant homozygotes both had the same phenotype. This type of inheritance is called complete dominance. After Mendel's conclusions were confirmed and brought to wider attention, geneticists in the early years of this century turned up cases in which the phenotypes of the F1 offspring did not resemble one or the other of the parents. In some cases, the offspring had a phenotype intermediate between that of the parents. These findings led to a debate about whether these cases of phenotypic blending in the offspring could be reconciled with Mendelian principles of inheritance, or whether there might be another, separate mechanism of inheritance that did not follow the laws of segregation and independent assortment.

This debate was not clearly resolved for almost 30 years, as scientists struggled to understand the relationship between phenotypes and genotypes in cases that appeared to defy Mendelian principles. Eventually, experimental work conclusively demonstrated that these cases were not exceptions to Mendelian inheritance and could be explained by the way in which genes act to produce phenotypes. In this section, we discuss some of these phenotypic variations, and show that although the phenotypes may not follow the predicted Mendelian ratios for complete phenotypic dominance, the outcome of crosses involving these traits can be predicted according to the Mendelian distribution of genotypes.

✦ FIGURE 7.1
Incomplete dominance in snapdragon flower color. Red-flowered snapdragons crossed with white-flowered snapdragons produce pink offspring in the $F_1$. In heterozygotes, the allele for red flowers is incompletely dominant over the allele for white.

### Incomplete Dominance Has a Distinctive Phenotype

In the case of **incomplete dominance,** the heterozygote has a phenotype intermediate to those of the homozygous parents. An example of this type of inheritance is flower color in snapdragons (✦ Figure 7.1). If a true-breeding variety bearing red flowers is crossed to a variety that produces white flowers, the $F_1$ offspring will all have pink flowers. The phenotype of the $F_1$ is different from that of either parent, and is intermediate to the phenotypes of the parents. When the $F_1$ plants are self-fertilized, they produce plants with red, pink, and white flowers in a 1:2:1 ratio (Figure 7.1). This would not be expected if true blending occurred, since crossing plants with pink flowers should produce only offspring with pink flowers.

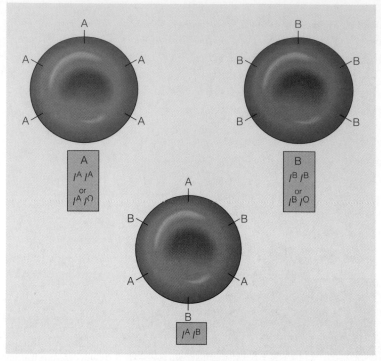

**◆ FIGURE 7.2**

**Codominant genes are fully expressed in the heterozygous condition. Type A blood has A antigens on the cell surface, and type B has B antigens on the surface. In type AB, both the A and B antigen are present on the cell surface. In type O blood ($I^O I^O$), neither antigen is expressed.**

The results of this cross can, however, be explained by Mendelian principles. Note that each genotype has a distinct phenotype, and therefore in this case, the phenotypic ratio of 1 red : 2 pink : 1 white is the same as the genotypic ratio of 1 *RR* : 2 *Rr* : 1 *rr*. It takes two doses of the *R* allele to produce red flowers. One dose will result in pink flowers, and the *R* allele is said to be incompletely dominant over the *r* allele. The *r* allele is unable to produce color; the absence of *R* alleles results in white flowers.

## Many Genes Have More Than Two Alleles

So far our discussion of genes has been confined to genes with only two alleles. But since alleles represent different forms of a gene, there is no reason why a given gene cannot have more than two alleles. In fact, many genes have multiple alleles. In humans, the ABO blood groups are an example of multiple alleles. Blood types are determined by the presence of different glycoproteins (proteins with polysaccharides attached) on the surface of human red blood cells. These glycoproteins serve

to provide the cell with an identity tag recognized by the body's immune system.

The *A* and *B* alleles (of the gene *I*) each encode a slightly different form of a glycoprotein (called the A antigen and B antigen, respectively). Individuals who are homozygous for the *A* allele (*AA*) carry only the A antigen on cells, and have blood type A. Those who are homozygous for the *B* allele (*BB*) carry only the B antigen and are type B. Individuals homozygous for the *O* allele (*OO*) carry neither the A nor the B antigen. The *O* allele is recessive to the *A* and *B* alleles. Because there are three alleles, there are six possible genotypes including $I^O I^O$ (◆ Figure 7.2).

Blood type can be determined by a simple test, and it is important to match blood types in transfusions. The ABO blood groups are also used as evidence in paternity cases. Blood typing can provide evidence that rules out a man as the father of a given child.

## Codominant Alleles Are Both Expressed

In **codominance,** heterozygotes fully express *both* alleles. In the ABO blood type, *AB* heterozygotes have both the *A* and *B* antigens on their cell membranes and are blood type AB. In *AB* heterozygotes, neither allele is dominant over the other, and since each allele is fully expressed, they are said to be codominant (Figure 7.2).

## Some Traits Are Controlled by More Than One Gene

When Mendel crossed tall pea plants with short pea plants, all the offspring had a single phenotype: tall. When these $F_1$ heterozygotes were intercrossed, $\frac{3}{4}$ of the offspring were tall, and $\frac{1}{4}$ were short. The tall and short phenotypes in the $F_2$ pea plants are easily classified into two distinct categories, tall (about 84 in.) and short (about 18 in.). These two distinct phenotypes are an example of **discontinuous variation** (◆ Figure 7.3).

In measuring height in humans, it is difficult to set up only two phenotypes. Instead, height is an example of **continuous variation.**

Unlike Mendel's pea plants, people are not either 84 in. or 18 in. tall; they fall into a series of overlapping phenotypic classes. We now know that a trait showing continuous variation is usually controlled by the additive effects of two or more separate gene pairs. A trait controlled by several gene pairs is an example of **polygenic inheritance** (◆ Figure 7.4).

Experimental work with plants early in this century demonstrated that traits determined by a number of different genes exhibit a continuous distribution of phenotypes in the $F_2$ generation. This is true

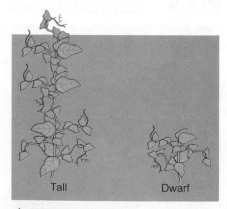

Tall      Dwarf

**◆ FIGURE 7.3**

**In crosses with pea plants, the $F_2$ phenotypes of traits controlled by a single gene can be sorted into two distinct phenotypic classes, an example of discontinuous variation.**

even though the inheritance of *each* gene follows rules of Mendelian inheritance. This distribution follows a bell-shaped curve, and contains some individuals with extreme phenotypes (very short or very tall, for example), but most individuals show a range of phenotypes between these extremes. The continuous variation produced by polygenic inheritance has several distinguishing characteristics:

- Traits are usually quantified by measurement rather than counting.
- Two or more gene pairs contribute to the phenotype.
- Phenotypic expression of polygenic traits varies within a wide range. This variation is best studied in populations rather than individuals.

The number of phenotypic classes increases with the number of genes involved, and as the number of classes increases, there is less and less phenotypic difference between classes (✦ Figure 7.5). As you can see, in polygenic systems with two genes, there are five phenotypic classes. With four genes, the number of phenotypic classes has increased to nine.

✦ **FIGURE 7.5**

**The number of F₂ phenotypic classes increases as the number of genes controlling the trait increases. This relationship allows geneticists to estimate the number of genes controlling a polygenic trait.**

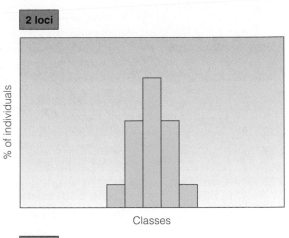

2 loci

% of individuals

Classes

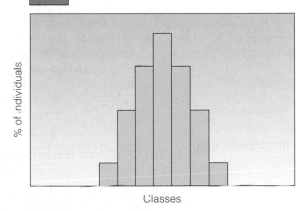

3 loci

% of individuals

Classes

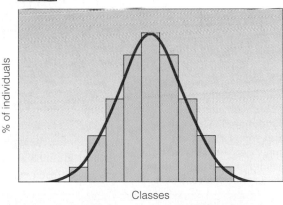

4 loci

% of individuals

Classes

✦ **FIGURE 7.4**

(a) **The continuous variation in polygenic phenotypes results in a bell-shaped curve. (b) Skin color is probably controlled by three to four genes, producing a wide range of phenotypes.**

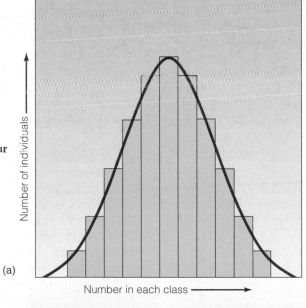

(a)

Number of individuals

Number in each class

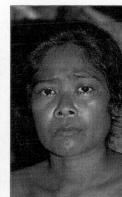

(b)

Polygenic inheritance plays an important role in human genetics. Traits such as height, weight, intelligence, skin color, and many forms of behavior appear to be under the control of two or more genes. In addition, many polygenic traits show a strong degree of interaction with environmental factors that affect the ultimate expression of the genes controlling these traits. Genetic variation within populations is a prerequisite for evolution. The role of variation in evolution is discussed in Chapter 10.

## Gene Expression Can Be Affected by External Factors

Many genes, such as the one controlling albinism, exhibit a regular and consistent pattern of expression. In other cases where a trait is controlled by a single gene, the phenotype can be highly variable. In some cases, a mutant genotype may be present but remain unexpressed, resulting in a normal phenotype. This variation in phenotypic expression is caused by interaction with other genes in the genotype and by interactions between genes and the environment.

In humans, some individuals carry a dominant allele for **campodactyly,** a trait that causes the improper attachment of muscle to bones in the little finger. The result is a permanently bent little finger. In some individuals, both fingers are bent, in others, only one little finger is affected, and in a small percentage of cases, neither finger is bent (✦ Figure 7.6). In the pedigree shown in Figure 7.6, one individual (III-4) does not express the trait, but must have carried the gene, since he passed it on to his children. In this case, the environmental factors that affect the expression of the gene are still unknown.

Some genes depend on well-known environmental factors for their phenotypic expression. Siamese cats and Himalayan rabbits are light colored, with dark fur on their paws, nose, ears, and tail (✦ Figure 7.7). In these cases, a gene that controls pigment production is able to function at the lower temperatures found in the extremities, but not at the slightly higher temperatures throughout the rest of the body. All cells of these animals carry the genes

✦ **FIGURE 7.7**

Animals such as Siamese cats and Himalayan rabbits have dark fur at the extremities of the nose, ears, and paws. These colors are the result of expression of an allele that is active only at the slightly lower body temperatures found in the extremities.

for pigment production, but the environment (in the form of the temperature) determines the phenotypic pattern of expression.

While many genes are active throughout life, the expression of other genes is age-dependent. An example in humans is the gene that causes **Huntington disease** (HD). This disorder, which is controlled by a single dominant allele, has its onset between the ages of 30 and 50 years. Affected individuals undergo a progressive degeneration of the nervous system, causing uncontrolled jerky movements of the head and limbs as well as mental deterioration. Death usually ensues some 5 to 15 years after onset. This disorder is particularly insidious because expression usually occurs after the affected person has started a family and because each child has a 50% chance of developing HD. Research in molecular genetics has now made it possible to identify those who will develop the disorder. The tests used to diagnose this disorder are discussed in Chapter 9.

The variations in gene **expression** that we have discussed are all the result of the relationship between a gene and the mechanisms that produce the gene's phenotype. The inheritance of these genes follows the predictable pattern worked out by Mendel for traits in the pea plant, but the expression of these genes is complicated by temperature, age, and other, unknown, factors. In the following section, we consider *how* genes are passed from generation to generation and see that the patterns of inheritance described by Mendel are the result of the precise choreographed behavior of chromosomes.

## THE IDEA THAT GENES ARE LOCATED ON CHROMOSOMES

When Mendel performed his experiments, he concluded that factors for specific traits were present

✦ **FIGURE 7.6**
Pedigree of campodactyly, a dominant trait. In this pedigree, those with two affected hands are indicated by fully shaded red symbols. Those with affected left hands have the left half of the symbol shaded in red, and those with affected right hands have the right side of the symbol shaded in red.

in cells and were passed into gametes for transmission to the progeny, but he did not speculate as to *where* these factors were located in the cell, nor *how* they were transmitted to produce the predictable ratios in the $F_2$ generation. While Mendel was working on the analysis of genetic traits, other scientists were studying **cytology** or cell structure using a new generation of microscopes, instruments that provided an unparalleled increase in magnification and clarity. In 1875, cytologists discovered chromosomes, and observed the fusion of sperm and egg nuclei within the egg at fertilization.

In the years that followed, the longitudinal splitting of chromosomes during cell division and the reduction of chromosome number during germ cell formation were observed. Between 1884 (the year Mendel died) and 1888, the details of mitosis and meiosis were reported, and the cell nucleus was identified as the repository of genetic information. Based on the behavior of chromosomes in cell division, the developmental biologist Wilhelm Roux even proposed that linearly arranged "qualities" on chromosomes were equally transmitted to daughter cells at mitosis.

## Behavior of Genes and Chromosomes: Generation to Generation

At the turn of the century, when Mendel's work became widely known, and the fundamental aspects of Mendelian inheritance were found to operate in many organisms, it became apparent that chromosomes and genes had much in common. In 1903 Walter Sutton and Theodore Boveri formally proposed that chromosomes are the cellular components that physically contain genes. The resulting **chromosome theory of inheritance** merged aspects of cytology with genetics and has been confirmed by many experiments over the following decades. This theory is one of the foundations of genetics and explains the principles of Mendelian inheritance (◆ Figure 7.8).

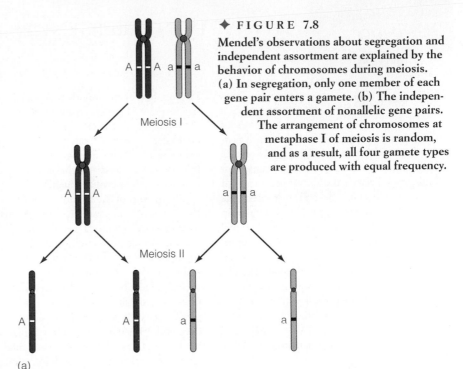

◆ **FIGURE 7.8**
Mendel's observations about segregation and independent assortment are explained by the behavior of chromosomes during meiosis. **(a)** In segregation, only one member of each gene pair enters a gamete. **(b)** The independent assortment of nonallelic gene pairs. The arrangement of chromosomes at metaphase I of meiosis is random, and as a result, all four gamete types are produced with equal frequency.

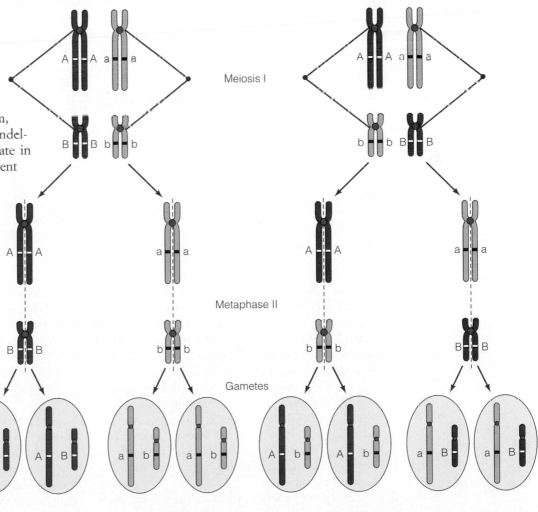

## Beyond the Basics

# WHAT IS INTELLIGENCE?

The idea that intelligence is a single entity that can be measured began in the nineteenth century with the idea that intelligence was related to brain size. *Craniometry*, the measurement of brain size, became the dominant means of assessing intelligence. Several measurements were used to assess intelligence, including brain weight, cranial capacity, and cranial index (ratio of the maximum width to maximum length of the skull). Brain weight was measured by removing the brains of the dead and weighing them. Cranial capacity was measured by filling the cranium of a skull with small lead shot (about the size of a BB). The shot was then emptied into a graduated cylinder and the cranial capacity was calculated in cubic centimeters.

The results of such misguided attempts to measure intelligence were often used to justify cultural biases. For example, in 1879, Gustave Le Bon, one of the founders of social psychology, wrote: "All psychologists who have studied the intelligence of women, as well as poets and novelists, recognize today that they represent the most inferior forms of human evolution and that they are closer to children and savages than to an adult, civilized man."

At the turn of the twentieth century, Alfred Binet abandoned craniometry and turned to psychological methods rather than physical methods to measure intelligence. He began by trying to identify children whose poor performance in the classroom indicated a need for special education. Binet developed a series of tasks he thought were related to basic mental processes, such as comprehension, classification, and error correction. He assigned an age level to each task, and the age assigned to the last task that the child was able to complete was assigned as the child's mental age. Later, the mental age was divided by the chronological age, and the result became known as the *intelligence quotient*, or *IQ*.

The use of IQ tests became associated with the idea that intelligence is a single, genetically determined property of the brain that can be quantitatively measured. Now, however, it is recognized that intelligence has many components, including verbal reasoning, quantitative reasoning, and abstract reasoning. Others reject the idea that the concept of intelligence should be limited to mental competence. They argue that intelligence should include linguistic, mathematical, visual–spatial, musical, body–kinesthetic (the ability to control body movements), and personal (the ability to understand ourselves and others) intelligence. Standard intelligence tests measure only the first three. Broadening the definition of intelligence may lead to the development of new ways of assessing intelligence and of increasing the predictive powers of such tests.

---

In the following sections, we explore how the chromosome theory has been confirmed through observation and experimentation, and how this information led to the discovery of the relationship between abnormal chromosome numbers and genetic disorders.

## Chromosomes Can Help Determine Sex

In studying chromosome structure, cytologists discovered that the males and females of some animal species have slightly different chromosome sets. For example, cells from females of a certain species of grasshopper have two copies of a chromosome called the X chromosome, while males have only one copy. In other species (as is the case in most mammals, including humans) females have two X chromosomes and males have one X chromosome and one copy of another chromosome called the Y chromosome. These chromosomes are called the **sex chromosomes.**

The chromosomes other than the sex chromosomes are called **autosomes.** The discovery of sex chromosomes provided an explanation for how the

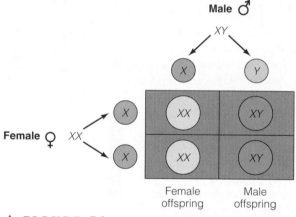

**Male ♂**

**Female ♀**

Female offspring — Male offspring

### ✦ FIGURE 7.9

**The segregation of sex chromosomes into gametes and the random combination of X-bearing or Y-bearing sperm with an X-bearing egg produces on average, a 1:1 ratio of males:females.**

sex of the offspring is determined. In humans, all gametes produced by females carry an X chromosome (✦ Figure 7.9). Gametes produced by the male contain either an X chromosome or a Y chromosome. An egg fertilized by an X-bearing sperm results in an XX fertilized egg or **zygote** that will develop as a female. Fertilization by a Y-bearing sperm will produce an XY (male) zygote. In humans, the Y chromosome is present only in males, and is passed from father to son. Consequently, the sex of the offspring depends on whether the fertilizing sperm carries an X or a Y chromosome.

Although XX/XY sex determination is widespread among species, it is not the only mechanism for determining the sex of an individual. In the grasshopper, females are XX and males are XO, with the O symbolizing the absence of another chromosome. In birds, females have a pair of unmatched chromosomes, and males have a pair of matched chromosomes. The system of sex determination in birds is a ZW-female and ZZ-male system, where the sex chromosomes are represented by Z and W instead of X and Y. In certain fish and reptiles, environmental factors play a key role in sex determination. In these animals, sex is determined by the incubation temperature at which the egg develops (✦ Figure 7.10).

In organisms with the XX/XY mechanism of sex determination, the question still remains as to how the chromosome constitution determines maleness or femaleness. Is a male a male because he has a Y chromosome, or because he does not have two

✦ FIGURE 7.10

(a–c) Animals have several mechanisms of sex determination involving chromosomes. (d) In some reptiles, the temperature at which the egg is incubated determines the sex of the offspring.

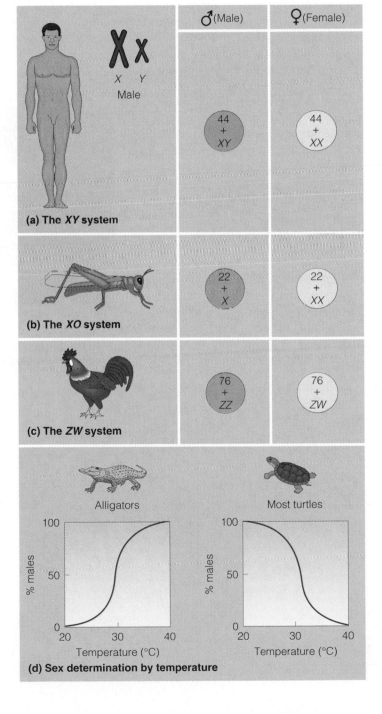

(a) The *XY* system

(b) The *XO* system

(c) The *ZW* system

(d) Sex determination by temperature

**Sidebar**

**SEX DETERMINATION**

The Y chromosome in humans carries very few genes, but among them is the SRY gene, which initiates male sexual development. The presence of this gene's product (testis-determining factor, TDF) dictates that an individual will develop testes; its absence produces ovaries. The hormones secreted from the testes and ovaries stimulate the development of male or female secondary sexual characteristics, i.e., pubic, facial, and underarm hair, breast development, etc.

Scientists studying sexual development in individuals with mutations in SRY are attempting to understand the developmental switch that triggers the differentiation of an indifferent gonad into a testis. The components identified in this pathway can help define mechanisms underlying the molecular events that lead to sexual differentiation. The SRY gene, located on the short arm of the Y chromosome, acts as a regulatory gene. This gene encodes a DNA-binding protein that binds to specific sites in DNA and bends the DNA. Abnormal DNA bends induced by SRY binding can create a mutation associated with clinical sex reversal (males into females). Delineating the mechanisms that underlie sexual differentiation may prove valuable in understanding clinical cases of sex reversal and possibly one's sexual orientation.

X chromosomes? Humans normally carry 46 chromosomes, including an XX or XY pair. Rarely, individuals are born with 45 chromosomes. They carry a single X as the only sex chromosome, resulting in a condition known as **Turner syndrome** (✦ Figure 7.11). Turner syndrome results from **nondisjunction,** the failure of chromosomes to separate properly during cell division. Affected individuals have short stature, and are generally infertile, *but are always female.* Males carrying two X chromosomes and a Y chromosome (47,XXY), have a condition known as **Klinefelter syndrome** (Figure 7.11). Affected individuals are male, and typically tall and infertile. From the study of individuals with sex chromosome abnormalities it became clear that some females may have only one X chromosome, and that some males may have more than one X chromosome. In humans, the male phenotype is associated with the presence of a Y chromosome, and the absence of a Y chromosome results in the female phenotype. However, two X chromosomes are required for normal female development, and a single X chromosome is required for normal male development. In addition to providing insights into how sex is determined, the distinction between sex chromosomes and autosomes is crucial in providing evidence that genes are on chromosomes.

## Sex Linked Genes Have a Unique Pattern of Inheritance

In organisms with the XX/XY form of sex determination, a distinct pattern of inheritance is associated with the sex chromosomes. The Y chromosome is present only in males and is passed directly from father to son. The X chromosome of males is always passed to daughters, who, in turn, can pass this on to their sons and daughters. Recognition of this distinctive pattern of inheritance provided some of the first evidence that genes are located on chromosomes, and that specific genes are located on the X chromosome.

The evidence came from work on the fruit fly, *Drosophila,* started in the first decade of this century by Thomas Morgan and his colleagues at Columbia University (✦ Figure 7.12). The small flies are easy to grow in the laboratory, pass through a complete generation in about 12 days, and a single female can produce up to a thousand eggs in her lifetime. In *Drosophila,* females are XX and males are XY.

The eyes of *Drosophila* are typically a brick-red color, but in one culture, a white-eyed male was discovered. To investigate how this trait was inherited,

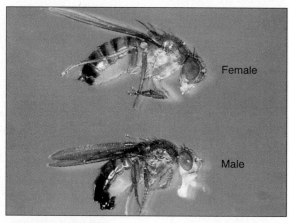

✦ **FIGURE 7.12**

Adult female and male *Drosophila melanogaster,* **an organism widely used in genetic research.**

✦ **FIGURE 7.11**

(a) **Karyotype of Turner syndrome, showing single X chromosome. (b) Karyotype of Klinefelter syndrome, with XXY chromosome constitution.**

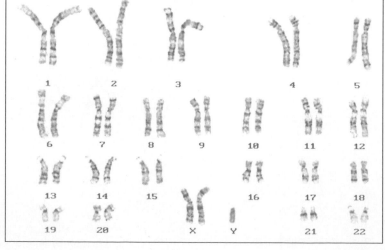

(a)

(b)

a true-breeding strain of white-eyed flies was established. When a white-eyed male was crossed to a red-eyed female, the offspring were all red-eyed, indicating that the allele for white is recessive to the dominant allele for red. However, when a homozygous white-eyed female was crossed to a red-eyed male, the results were somewhat unexpected (✦ Figure 7.13). All the $F_1$ male offspring were white-eyed (like their mother), and all the $F_1$ female offspring were red-eyed (like their father).

The results of this second cross can be explained by assuming that the gene for eye color is located on the X chromosome (Figure 7.13). The white-eyed *Drosophila* males carry only one copy of the X chromosome, and inherit it from their mother. Along with the X chromosome, these males inherit the white-eye allele, and have white eyes. The red-eyed $F_1$ females also receive an X chromosome carrying a white allele, but have red eyes because they receive an X chromosome from their father carrying the dominant allele for red eyes. In this cross, the inheritance of the eye color gene is coupled to the inheritance of a specific chromosome as it is passed from mother to son. When the inheritance of a chromosome is coupled to that of a given gene, the condition is called **linkage**. In this case, because the chromosome involved is one of the sex chromosomes, this is a case of **sex linkage**.

As a final note, it is necessary to explain that the $F_1$ males have white eyes because genes on the X chromosome have no alleles on the Y chromosome. The X and Y chromosomes pair at meiosis, but have few regions in common. As a result, genes on the X chromosome are expressed in males, whether they are dominant or recessive. *Drosophila* males (and males of all species with XX/XY sex determination, including humans) cannot be homozygous or heterozygous for any genes on the X chromosome, since they only carry one copy of all such genes. Instead, males are said to be **hemizygous** for these genes.

In humans, conditions such as red-green color blindness and hemophilia show X-linked inheritance (✦ Figure 7.14). Red-green color blindness is caused by defective cells in the eyes. Affected individuals are unable to distinguish red or green. They see these and some other colors as gray.

Hemophilia is a disorder of blood clotting. Hemophiliacs cannot form clots to stop bleeding from cuts and scrapes; they lack a biochemical factor necessary for clot formation. Seriously affected hemophiliacs are in danger of bleeding to death from even minor cuts.

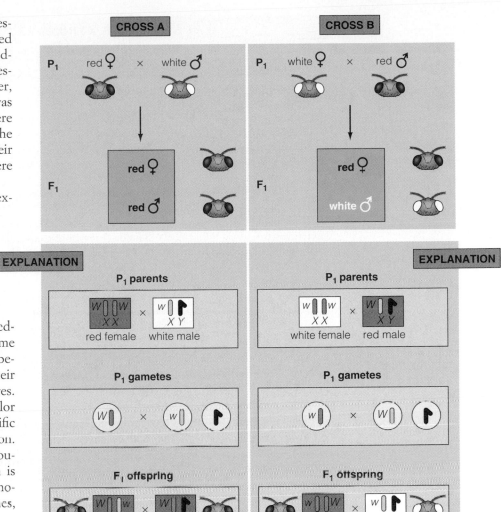

✦ **FIGURE 7.13**

The results of reciprocal crosses involving the X-linked *white* mutation in *Drosophila*. In cross A, homozygous red-eyed females are crossed to white-eyed males. The $F_1$ offspring all have red eyes. In cross B, homozygous white-eyed females are crossed to red-eyed males. In the $F_1$, all the females have red eyes and all the males have white eyes. The results of the crosses can be explained by assuming that the gene for *white* eyes is on the X chromosome.

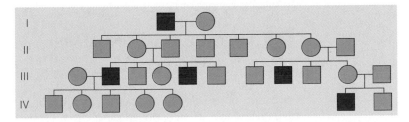

✦ **FIGURE 7.14**

Pedigree showing a pattern of inheritance typical for an X-linked recessive trait. Affected individuals are shown in red.

TABLE 7.1  SOME X-LINKED RECESSIVE GENETIC TRAITS

| TRAITS | PHENOTYPE |
|---|---|
| Adrenoleukodystrophy | Atrophy of adrenal glands, mental deterioration; death 1 to 5 years after onset |
| Color blindness | |
|    Deuteranopia | Insensitivity to green light; 60% to 75% of color blindness |
|    Protanopia | Insensitivity to red light; 25% to 40% of color blindness |
| Fabry disease | Metabolic defect caused by lack of enzyme alpha-galactosidase A; progressive cardiac, renal problems, early death |
| Glucose-6-phosphate dehydrogenase deficiency | Benign condition that can produce severe, even fatal, anemia in the presence of certain foods, drugs |
| Hemophilia A | Inability to form blood clots; caused by lack of clotting factor VIII |
| Hemophilia B | "Christmas disease"; clotting defect caused by lack of factor IX |
| Icthyosis | Skin disorder causing large, dark scales on extremities, trunk |
| Lesch-Nyhan syndrome | Metabolic defect caused by lack of enzyme hypoxanthine-guanine phosphoribosyl transferase (HGPRT); causes mental retardation, self-mutilation, early death |
| Muscular dystrophy | Duchenne-type, progressive; fatal condition accompanied by muscle wasting |

These and other X-linked traits (➤ Table 7.1) have a distinctive pattern of inheritance. The characteristics of X-linked traits include the following: (1) Phenotypic expression is much more common in males than females (males have only one X chromosome, and a recessive allele will be expressed, while females can be heterozygous for such alleles). (2) Sons cannot inherit an X-linked recessive allele from their fathers, but daughters can. Sons of heterozygous females have a 50% chance of inheriting the recessive allele.

Genes on the Y chromosome are inherited directly from father to son. Such genes show Y linkage. In humans, only a small number of genes have been mapped to the Y chromosome. Among these is the testis-determining factor (TDF) that promotes the development of the male phenotype during embryonic development. The action of the Y chromosome in testis development is discussed in Chapter 28.

The discovery of linkage led to the development of chromosome maps, allowing genes to be assigned to specific regions of chromosomes.

## 🌸 LINKAGE AND GENE MAPS

In our discussion of Mendel's work, emphasis was placed on the segregation and independent assortment of genes into gametes. The discovery that genes are located on specific chromosomes modified the concept of independent assortment. If two genes are located on the same chromosome, they will tend to move into the same gamete. In all but one case, the traits Mendel selected for analysis are controlled by genes located on different chromosomes that show independent assortment. In the case where Mendel unknowingly selected traits whose genes are located on the same chromosome, the consequences of another genetic phenomenon (discussed later) produces results that are indistinguishable from independent assortment.

## Linked Genes

As mentioned earlier, *Drosophila* has four pairs of chromosomes. In the first few years of work on *Drosophila,* more than 80 different genes, controlling traits such as eye color, wing shape, body color, and bristle patterns, were discovered. Because the number of genes was much greater than the number of chromosomes, it follows that each chromosome must carry a number of different genes. When two or more genes occur on the same chromosome, they are said to be **linked.** Linked genes are arranged in a linear series along the length of a chromosome (✦ Figure 7.15). The concept of linkage applies to genes on the sex chromosomes and to genes on any autosome. The general question is "How do we know when two genes are linked, and do genes on the same chromosome always remain linked?"

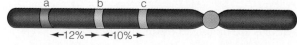

✦ **FIGURE 7.15**

**Linked genes are carried on the same chromosome. Recombination frequencies can be used to construct genetic maps, giving the distance between genes on a chromosome.**

Because X-linked genes are carried on the X chromosome, what results might be expected in a cross involving two X-linked genes? A cross involving a female *Drosophila* homozygous for white eyes and yellow body color and a wild-type male ("wild type" refers to the dominant alleles) is shown in ◆ Figure 7.16. The F₁ females are phenotypically wild type, and heterozygous for both traits; the F₁ males have yellow bodies and white eyes.

Although linked genes tend to be inherited together, this coupling is not always retained. Linked genes often separate from one another, as evidenced by the phenotypes of some of the offspring. In the cross involving *yellow body* and *white eyes,* two new phenotypic combinations are produced, with some flies showing only one or the other of the recessive traits (either yellow bodies *or* white eyes). These combinations arose through the physical exchange of segments of paired X chromosomes during meiosis in the female parent

◆ FIGURE 7.16

(a) A cross between a female *Drosophila* homozygous for white eyes and yellow body and a male with wild-type eyes (red) and normal body color (gray). The offspring in the F₁ follow the expected pattern of inheritance for recessive, X-linked traits. In the F₂, two new phenotypic classes appear.
(b) These classes arise because crossing over between the locus for body color and eye color during meiosis in the F₁ produces new combinations of alleles. The wild-type alleles are represented by upper-case letters. The mutants (yellow body color, white eye color) are represented by lower case letters.

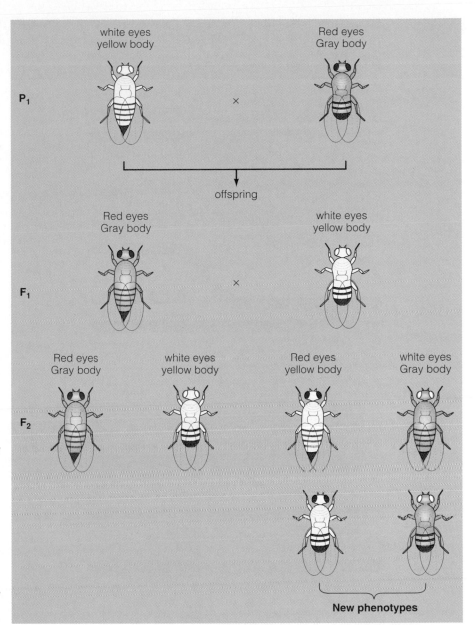

(a)

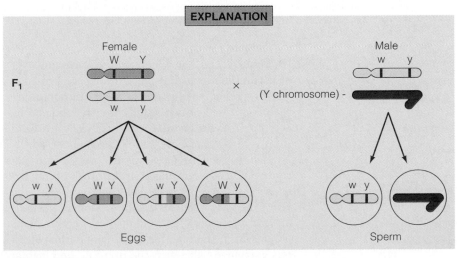

(b)

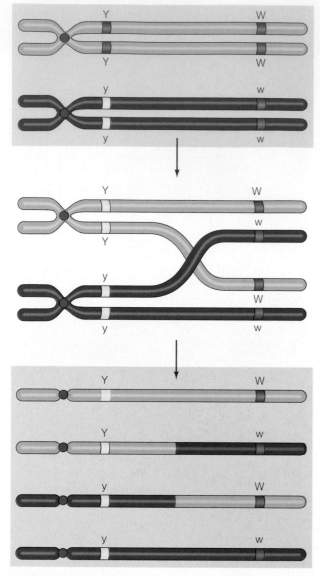

**✦ FIGURE 7.17**

**Crossing over (recombination) involves the physical exchange of chromosome parts. In this case, crossing over between the locus of the gene for body color (y) and the gene for eye color (w) produces two new allele combinations. The frequency of crossing over is proportional to the distance between genes, allowing genetic maps to be constructed.**

(✦ Figure 7.17). The phenomenon of **crossing over** or **recombination** takes place during the first meiotic prophase. At this stage, the chromosomes are each composed of two sister chromatids joined by a common centromere. Occasionally, a chromatid from one chromosome and a chromatid from its homologue will break at similar points and become physically exchanged. If the break occurs between the genes for body color and eye color, the *yellow* and *white* alleles will be separated, generating chromosomes that contain new combinations of alleles. The chromatids that do not participate in this exchange will retain the original combination of alleles.

## Genetic Maps Can Be Constructed Using Crossover Information

Alfred Sturtevant, working in Morgan's laboratory, soon realized that the closer together two genes are on a chromosome, the less likely it is that they will be separated by a crossing-over event. He concluded that the amount of crossing over between two genes can be used to determine the distance between them. If crossing over is very infrequent, the two genes must be close together; conversely, if crossing over occurs frequently, the two genes must be relatively far apart. He used this information from genetic crosses to construct **genetic maps** for the chromosomes of *Drosophila,* giving the order and distance between genes on each of the four chromosomes (✦ Figure 7.18). This basic principle of map construction has been used to construct genetic maps for a wide range of diploid organisms including humans.

What about Mendel's traits controlled by genes located on the same chromosome? Why didn't he detect linkage instead of independent assortment? In this case, he unknowingly selected two traits controlled by genes at opposite ends of a chromosome. Such genes are almost always separated by one or more crossover events, and the outcome is indistinguishable from that seen in independent assortment.

## CHROMOSOME ABNORMALITIES

During the first meiotic division, members of a chromosome pair segregate from each other and enter separate cells. In the second meiotic division, chromosomes composed of sister chromatids align at the metaphase plate. In anaphase, the centromeres split, and the sister chromatids are converted into chromosomes (single chromatids) and distributed to daughter cells. In a small percentage of cases, nondisjunction occurs in the process of chromosome separation, resulting in some gametes that contain two copies of a given chromosome and others that contain no copies of that same chromosome. Turner syndrome and Klinefelter syndrome are the result of nondisjunction of the sex chromosomes.

In other cases, chromosomes undergo changes brought about by breakage and realignment of chromosome fragments, or the attachment of chromosomes to each other. These changes lead to alterations in chromosome structure.

### Changes in Chromosome Number

Variations in chromosome number involving one or a small number of chromosomes are called **aneuploidy.** The most common cases are the result of

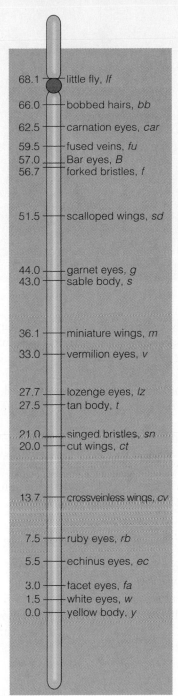

**✦ FIGURE 7.18**

A partial genetic map of the X chromosome in *Drosophila*, showing some of the genes on this chromosome.

Since the discovery that Down syndrome is caused by a chromosomal imbalance, studies have shown that alterations in chromosome number caused by meiotic errors are fairly common in humans and are a major cause of reproductive failure. Approximately 30 to 50% of all conceptions are aneuploid, and about 50% of all spontaneous abortions have chromosome abnormalities. About 1 in every 200 live births involves aneuploidy, and from 5 to 7% of all early childhood deaths in the United States are related to aneuploidy.

**Polyploidy** involves abnormal variations in the number of chromosome sets. In a triploid individual, three copies of each chromosome are present; in a tetraploid, four copies are present; and so forth. The most common form of polyploidy in humans is triploidy, which is observed in 12 to 15% of all spontaneous abortions. Such errors can arise during meiosis, mitosis, or fertilization. Most cases of human triploidy arise as a result of the fertilization of a haploid egg by two haploid sperm. Although biochemical changes that accompany fertilization normally prevent such events, the system is not fail-safe.

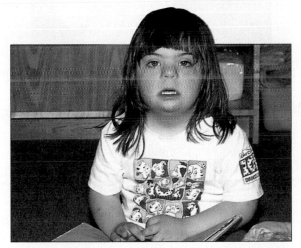

(a)

**✦ FIGURE 7.19**
**(a) Child with Down syndrome. (b) This condition is most often caused by the presence of three copies of chromosome 21.**

the gain or loss of a single chromosome. The first example of aneuploidy in humans was described in 1959 when it was discovered that Down syndrome is associated with the presence of three copies of chromosome 21, a condition also known as trisomy 21 (✦ Figure 7.19). In almost all cases, Down syndrome is the result of nondisjunction during meiosis. Affected individuals have distinctive characteristics: They are shorter than normal, mentally retarded to a greater or lesser degree, and can have a large, fissured tongue, and abnormal creases in the palm.

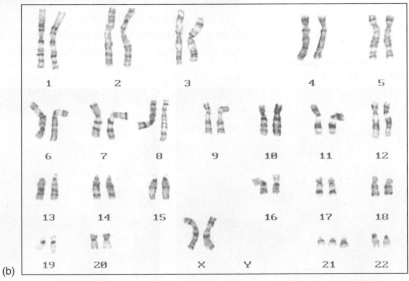

(b)

(a)

(c)

✦ FIGURE 7.20
About 75% of all flowering plants are polyploid. (a) Chrysanthamums. (b) Pansies. (c) Day lilies.

(b)

✦ FIGURE 7.21
Many common vegetables and fruits are polyploid, such as potatoes, peanuts, and strawberries.

While polyploidy is not a viable condition in most animals, it is fairly common among higher plants. There are several reasons for this, including the fact that plants can often be self-fertilized and can be propagated by asexual means. It has been estimated that the majority of all flowering plants have evolved through polyploidy (✦ Figure 7.20).

Polyploidy can be induced in plants by the use of chemical compounds, such as colchicine, that inhibit chromosome separation during cell division.

Polyploidy can also be produced in plants by genetic crosses between strains of related plants with differing chromosome numbers. A cross between a tetraploid variety of watermelon and a diploid variety will produce a sterile triploid seedless strain. Many of the plants people cultivate for food are polyploid derivatives of wild strains that have been manipulated by human intervention; polyploid strains produce larger fruits, vegetables, or grain kernels. Potatoes, wheat, peanuts, bananas, grapes, and watermelons are all examples of polyploid derivatives of ancestral plants (✦ Figure 7.21).

## Changes in Chromosome Structure

Changes in chromosome structure result from the breakage and reunion of chromosomal segments. These changes can produce abnormal chromosomes without altering the number of chromosomes present in the cell. Breaks can occur spontaneously or be produced by environmental agents such as viruses, chemicals, and radiation. The resultant changes in chromosome structure include **deletions** (✦ Figure 7.22), or loss of a chromosome segment; **duplications**, or extra copies of a chromosome segment; **translocations**, which move a segment from

# GUEST ESSAY: RESEARCH AND APPLICATIONS

## *Neurogenetics: From Mutants to Molecules*

BARRY S. GANETZKY

From the time I was young, I was fascinated by living things and enjoyed reading books on natural history. But I really had no idea how biologists earned a living. I just knew that I was about 100 years too late to become a naturalist. So I planned on studying chemistry, although it did not give me the same sheer pleasure as biology. In introductory biology I became interested in genetics and molecular biology, and realized that biology was what I wanted to immerse myself in. I have never regretted that choice.

In my junior year, I signed up to do an honor's research project. My mentor was a new young professor (a certain Michael R. Cummings) who at the time was studying egg development in *Drosophila* (fruit flies). This was my first exposure to research. The challenge of using one's creativity, reason, and imagination to discover answers to some of nature's secrets was the most exciting thing I had ever done. Although my project was to last only 10 weeks, I remained for the next two years.

I pursued a Ph.D. in genetics at the University of Washington with the late Larry Sandler, who was so intellectually gifted that I knew in-

stinctively no one could provide me with better graduate training.

As a postdoctoral fellow with Seymour Benzer, I was interested in the molecular basis of the signaling mechanisms in neurons. Ion channel proteins were known to play key roles in nerve impulses, but little was known about their molecular structure or how they worked. I isolated mutations that were defective in neuronal signaling to identify the genes encoding ion channels.

The trick was to find the right mutations. I began by screening for mutants that became paralyzed when exposed to elevated temperatures. In a stroke of luck, one of the first paralytic mutations I found caused a complete block of action potentials. This mutation led us to identify other mutants with neuronal defects. After taking a faculty position at the University of Wisconsin, my colleagues and I succeeded in cloning these genes. We now have the largest collection of mutations affecting ion channels in any organism, and they are providing us with new insights into the molecular basis of neuronal activity. One of the human genes we identified, because of its similarity to

a *Drosophila* gene, turns out to be defective in a heritable form of cardiac arrhythmia. Identification of the affected gene opens the way to identifying individuals at risk. It is gratifying to know that work pursued primarily because it was interesting and fun is also important and useful.

*Barry S. Ganetzky received a B.S. in biology from the University of Illinois–Chicago and a Ph.D. in genetics from the University of Washington. He has been a faculty member in the Laboratory of Genetics at the University of Wisconsin, Madison, since 1979 where he is now the Steenbock Professor of Biological Sciences. After spending more than half his life working with fruit flies he still derives great pleasure from discovering new mutations with interesting and unusual phenotypes.*

✦ **FIGURE 7.22**

**Karyotype of child with Cri du Chat syndrome, caused by a partial deletion of chromosome 5. As infants, affected individuals have a distinctive cry that sounds like a cat.**

one chromosome to another; and **inversions,** the reversal in the gene order of a chromosome segment.

The consequences of these alterations in chromosome structure are variable, but usually deleterious. For instance, almost all forms of human leukemia are associated with translocations that involve specific chromosomes. While the cause-and-effect relationship in this situation is not yet clear, it appears that the translocation event is responsible for the changes in growth control that trigger the onset of these cancers. In the fruit fly *Drosophila,* on the other hand, large inversions of one or more chromosomes reduce fertility, but have no major impact on viability.

# SUMMARY

1. Work by Mendel established that genes exhibit dominant/recessive relationships, undergo segregation during gamete formation, and that members of a gene pair separate or segregate independently of other gene pairs. Once Mendel's work was confirmed and became widely known, work on other organisms uncovered cases where phenotypes of the parents apparently blended together in the offspring, or phenotypes in which the traits of both were expressed. These results caused a widespread debate as to the universal application of Mendel's findings, and led to speculation that there might be other, separate mechanisms of inheritance not covered by the laws of segregation and independent assortment.

2. One by one, cases that appeared to violate Mendelian principles were shown to be situations where the outcome of crosses could be predicted according to the Mendelian distribution of genotypes, but where the phenotypic distribution might not follow the predicted Mendelian ratios. These cases include incomplete dominance and codominance.

3. More difficult to resolve was polygenic in-
heritance, which required understanding that a single phenotype can be brought about by the action of two or more genes. Once these problems were resolved, it became apparent that whether the phenotype exhibits incomplete dominance, codominance, or polygenic inheritance, the genes in question are inherited according to Mendelian principles.

4. In addition to variations in phenotype, genes interact with environmental factors to determine phenotypic patterns. These factors can include the age of the individual, nutrition, exposure to chemicals, and even temperature.

5. The chromosome theory of inheritance merged mainstream components of genetics and cytology and served as a guidepost in the discovery of linkage, the mechanism of sex determination, and the development of genetic mapping. Perhaps the greatest modification of Mendelian principles was the recognition that independent assortment holds true only if the genes in question are on separate pairs of chromosomes.

6. Linked genes were shown to exhibit a distinctive pattern of inheritance in the $F_2$
generation. Exceptions to this pattern produced new phenotypic combinations of the parental traits. These combinations arose through the physical exchange of chromosome segments (crossing over or recombination). The degree of crossing over between two gene loci can be used to determine the relative distance between genes along the chromosome. This information can be used to construct genetic maps, giving an overview of the organization and anatomy of the genome.

7. Phenotypic variations produced by abnormalities of chromosome number and structure represent situations where the abnormal phenotype is caused by changes in the amount or arrangement of genetic information, not the presence of abnormal alleles. In humans, trisomy 21 or Down syndrome is caused by the presence of an extra copy of chromosome 21.

8. Changes in chromosome number also play an important role in the evolution of new species, especially in plants.

9. Alterations in chromosome structure change the arrangement but not the amount of genetic information that is present in a cell.

# KEY TERMS

aneuploidy
autosomes
chromosome theory of inheritance
complete dominance
compodactyly
continuous variation
crossing over
cytology
deletions
discontinuous variation
duplications

expression
genetic maps
hemizygous
Huntington disease
incomplete dominance
inversions
karyotype
Klinefelter syndrome
Leber's optic atrophy
linkage
linked

maternal inheritance
MERRF
nondisjunction
polygenic inheritance
polyploidy
recombination
sex chromosomes
sex linkage
translocations
Turner syndrome

## MATCHING

Match the terms appropriately.

1. dominant
2. incomplete dominance
3. codominance
4. discontinuous variation
5. cytology
6. zygote
7. linkage
8. autosomes
9. continuous variation
10. nondisjunction

____ failure of chromosomes to separate during division

____ study of cell structure

____ chromosomes 1 through 22

____ allele always expressed in heterozygote

____ small degree of phenotypic variation that occurs over a range

____ each allele in heterozygote is separately expressed

____ two alleles, neither one dominant over other

____ phenotypic variation that fits into distinct categories

____ fertilized egg

____ situation of genes located on same chromosome stay together during meiosis and end up together in same gamete.

Match the condition with the inheritance pattern.

1. Klinefelter syndrome
2. intelligence
3. Down syndrome
4. Leber's optic atrophy
5. Turner syndrome
6. Hemophilia A
7. Huntington disease

____ mitochondrial disorder

____ XO

____ X-linked

____ 47, XXY

____ polygenic inheritance

____ autosomal dominant

____ aneuploidy

## TRUE/FALSE

1. A change in a chromosome number that results in less than a complete set of chromosomes is called ploidy.
2. Different glycoproteins on the human red blood cell surface are used by the body's immune system as a means for identifying foreign cells.
3. The number of phenotypic classes increases with the number of genes involved, and as the number of classes increases, there is more phenotypic difference between classes.
4. Genetic disorders can be caused by chromosome duplications, inversions, translocations, and gene mutations.
5. In humans the male phenotype will only be expressed if there is one Y chromosome and one X chromosome.

## SHORT ANSWER

1. Outline the distinguishing characteristics of polygenic inheritance.
2. How do you explain the high level of variability in phenotypic expression in certain genetic traits?
3. An individual born with only one sex chromosome (X) can survive. However, an individual born with only one Y and no X chromosome cannot survive. Why do you suppose there is this difference?
4. The bleeding disorder, hemophilia A, is an X-linked recessive condition. Why are only females known to be carriers for this recessive allele?

## SCIENCE AND SOCIETY

1. James Watson and Francis Crick published a paper in 1953 in the journal *Nature* that proposed a hypothesis for the structure of DNA. They suggested that DNA is a double helix and explained its nucleotide pairing. The work and efforts of many scientists over many years led to this discovery. A vital piece of information came from Rosalind Franklin, who produced X-ray photographs of purified DNA. Franklin's work was performed in the laboratory of Maurice Wilkins in Britain and suggested that the DNA molecule was helical. Another vital piece of information came from the work of Erwin Chargaff, who showed that the ratio of purine to pyrimidine was always 1:1. In 1962, Watson, Crick, and Wilkins were awarded the Nobel Prize for their discovery. Franklin was not included in the celebration because she had died of cancer in 1958, and Nobel Prizes are awarded only to living scientists. In your opinion should Nobel Prizes be awarded to scientists even after their death? Why or why not? Do you think the contributions of Chargaff were significant enough for him to receive a Nobel Prize? What criteria do you feel an investigator needs to meet before being awarded a Nobel Prize? What differences would we see in the Watson-Crick model without Franklin's photographs or Chargaff's data?

2. Studies have shown that male creative writers exhibit high rates of mood disorders and alcholism. Psychiatrists have now refined the picture with a study of women writers. The results suggest that a range of psychopathologies—as well as a history of childhood sexual abuse—seem to go hand in hand with being creative. Results from this study showed that twice as many writers as nonwriters had some form of mental disorder. These data suggest that in women writers, a state of general unease and tension is conducive to creative activity. What about this conclusion? Are there other factors that should be considered before reaching this conclusion? If so, what are they? Do you think the methods used to determine creativity and psychiatric problems are important to this study's results? Do genetic factors play a role in psychopathologies? What is your opinion on this issue?

# MOLECULAR BASIS OF INHERITANCE

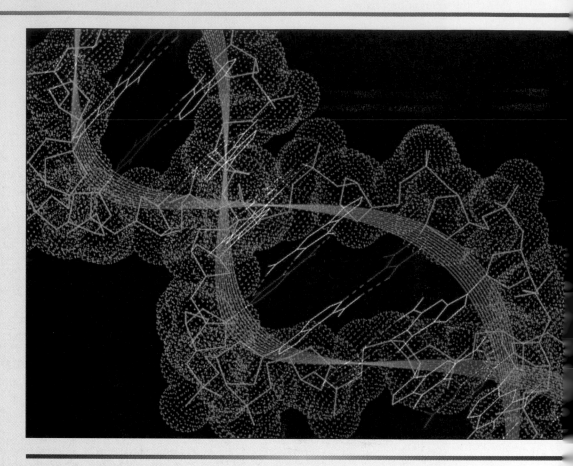

*Just after World War II, Charles Dent began using the technique of paper chromatography to separate the amino acid components of human urine. He applied a sample spot near one side of a square of filter paper and dried it thoroughly. The paper was placed in a chamber containing a small amount of solvent, with the side carrying the sample nearest the solvent. The top of the container was sealed and, gradually, the solvent moved up the filter paper by capillary action. As the solvent moved upward, the chemical components in the urine sample also migrated upward, but at different rates, thus separating from each other. When the solvent reached the top of the paper, the filter paper was removed, and sprayed with a reagent to make the amino acids visible as a series of spots.*

*In conjunction with Harry Harris, Dent used this technique to investigate the amino acids excreted in the urine of patients with cystinuria. In this disorder, a defect in kidney function causes large quantities of the amino acid cystine to be excreted in the urine. In some case, the concentration is so high that cystine crystals or even stones are excreted. Forty years earlier, Archibald Garrod had postulated that this disorder was the result of abnormal metabolism, but until Dent and Harris began their work, the biochemistry was confusing and there was no good evidence that the disorder was heritable.*

*Harris and Dent showed that there are two forms of cystinuria, one characterized by the excretion of cystine, lysine, and three other amino acids, and a second form marked by the excretion of just cystine and lysine. Through a combination of genetic and biochemical*

OPENING IMAGE

*Computer-generated image of DNA.*

studies, Harris established that the first type was caused by a homozygous recessive genotype, and the second by a heterozygous condition. This work had three important consequences: (1) It demonstrated that some genetic disorders have biochemical phenotypes, (2) it helped establish human biochemical genetics as part of genetics, and (3) it helped forge the link between genes and phenotypes through biochemical processes.

In this chapter, we consider the relationship between genes and proteins and the role of DNA as a carrier of genetic information. We explore the structure of DNA, its mechanism of replication, and its role in storing genetic information. We also discuss the transfer of genetic information from the sequence of nucleotides in DNA into the sequence of amino acids in protein. Finally, we will see how mutations in DNA lead to alterations in gene products and how these altered gene products generate alterations in phenotype.

## GENES AND PROTEINS: UNRESOLVED QUESTIONS

By the 1930s, genetics had surged forward on the strength of the chromosome theory of inheritance, and genetic maps were being prepared for several experimental organisms, including corn, *Drosophila,* and mice. Several important questions remained unanswered, often frustrating biologists who attempted to investigate these unresolved issues. What exactly are genes? How do they work? What produces the unique phenotype associated with a specific allele? In spite of the advances made using the newly developed techniques of genetics and cytology, these questions could not be answered with these methods. The answers came from other fields, including chemistry and physics and from the study of infectious disease.

Just as the merging of cytology and genetics produced the chromosome theory of inheritance, the fusion of these other disciplines with genetics gave rise to molecular genetics, an area that currently dominates research in genetics. In this chapter, we examine some of the explanations provided for the questions posed above, beginning with some background information about what genes do.

### Genes Control Metabolism

As outlined in Chapter 2, biochemical reactions are organized into chains of reactions known as **metabolic pathways.** The product of one reaction in a pathway serves as the substrate for the following reaction (✦ Figure 8.1). **Enzymes** are protein molecules that serve as biological catalysts to carry out specific biochemical reactions (see Chapter 2 to review the functions of proteins). Loss of activity in a single enzyme can disrupt an entire biochemical pathway, causing an alteration in metabolism.

The relationship between metabolism and human genetic disease was first proposed by Archibald Garrod in 1902, through his study of a condition called alkaptonuria. Affected individuals can be

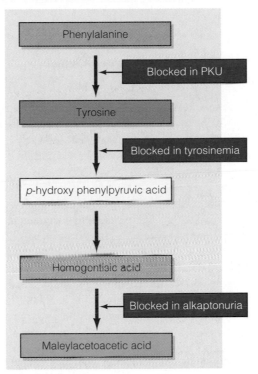

✦ **FIGURE 8.1**

**A metabolic pathway beginning with the amino acid phenylalanine. In humans, metabolic blocks caused by mutations lead to genetic disorders such as phenylketonuria (PKU), tyrosinemia, and alkaptonuria.**

identified soon after birth because their urine turns black. Garrod found that alkaptonuria is associated with the excretion of large quantities of a compound he called alkapton (now called homogentisic acid). He reasoned that in unaffected individuals, homogentisic acid is metabolized and does not build up in the urine. In alkaptonuria however, the metabolic pathway is blocked, causing the buildup of homogentisic acid, which is excreted in the urine. He called this condition an "inborn error of metabolism." Garrod also discovered that alkaptonuria is inherited as a recessive Mendelian trait.

✦ FIGURE 8.2

A culture of the mold *Neurospora*, an organism used by Beadle and Tatum in their work on biochemical genetics.

## The Relationship between Genes and Enzymes

The connection between inborn errors of metabolism and enzymes was established by George Beadle and Edward Tatum in the late 1930s and early 1940s through their work on *Neurospora*, a common bread mold (✦ Figure 8.2).

Using X rays to cause genetic mutations, Beadle and Tatum induced the formation of mutant strains of *Neurospora* that were unable to carry out a specific biochemical reaction. The mutation was inherited in Mendelian fashion as a single gene, and was caused by the loss of activity in a single enzyme.

Beadle and Tatum concluded that since each reaction in a metabolic pathway is regulated by a single enzyme, the production of that enzyme must be controlled by a single gene. Mutation in a gene whose product is involved in a metabolic pathway changes the ability of a cell to carry out a particular biochemical reaction, and disrupts the whole pathway. This idea was expressed as the **"one gene, one enzyme hypothesis"** for which Beadle and Tatum received the Nobel Prize in 1958.

What is it, then, that genes do? A gene directs the synthesis of a gene product in the form of a protein. The action of this protein, in turn, results in a particular phenotype. In some cases, the protein acts as an enzyme to catalyze a biochemical reaction. The blocking of a biochemical reaction can disrupt a metabolic pathway, leading to an altered phenotype, as in alkaptonuria. In other cases, proteins function as transport molecules, hormones, structural components of cells, the contractile fibers of muscles, and in numerous other ways. Because some proteins are composed of different polypeptide chains encoded by separate genes, the "one gene, one enzyme hypothesis" is now expressed as the **"one gene, one polypeptide hypothesis."** A mutation in a gene encoding a specific polypeptide can alter the ability of the encoded protein to function and, in turn, produce an altered phenotype.

## DNA AS A CARRIER OF GENETIC INFORMATION

Research in the first few decades of this century established that genes are carried on chromosomes. But what is a gene? As is often the case in science, the answer to this question came from an unexpected direction: the study of an infectious disease.

At the beginning of this century, pneumonia was a serious public health threat and the leading cause of death in the United States. Medical research was directed at understanding the nature of this infectious disease as a step toward developing an effective treatment, perhaps even a vaccine. The unexpected outgrowth of this research was the discovery of the chemical nature of the gene.

### Transfer of Genetic Traits in Bacteria

By the 1920s, it was known that pneumonia is caused by a bacterial infection, and that one form of pneumonia is caused by *Streptococcus pneumoniae.* Frederick Griffith studied the difference between two strains of this bacterium. In one strain (the S strain), the cells were surrounded by a capsule. This strain was infective and caused pneumonia (that is, it was a virulent strain). The other strain (the R strain) did not have a capsule and was not infective. The results of Griffith's experiment are straightforward and easily interpreted.

Griffith showed that mice injected with live cells of strain R did not develop pneumonia, but mice injected with live cells from strain S developed pneumonia and soon died. Mice injected with heat-killed S cells survived and did not develop pneumonia (✦ Figure 8.3). However, if mice were injected with a mixture of heat-killed S cells and live cells from strain R, they developed pneumonia and died. From the bodies of the dead mice Griffith recovered live, S strain bacteria with capsules. When grown in the laboratory, the progeny of these transformed cells were always strain S. After further experiments, Griffith concluded that the living cells from strain R were transformed into S cells in the bodies of the mice. Griffith explained his results by proposing that hereditary information had passed from the dead S cells into the R cells, allowing them to make a capsule and become virulent. He called this process **transformation,** and the unknown material the **transforming factor.**

In 1944, after almost a decade of work, a team at the Rockefeller Institute in New York that included

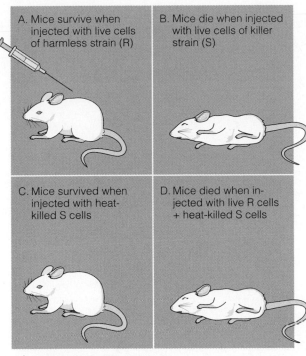

A. Mice survive when injected with live cells of harmless strain (R)

B. Mice die when injected with live cells of killer strain (S)

C. Mice survived when injected with heat-killed S cells

D. Mice died when injected with live R cells + heat-killed S cells

**✦ FIGURE 8.3**

**Griffith discovered that the ability to cause pneumonia (a genetic trait) could be passed from one strain of bacteria to another.**

Oswald Avery, Colin MacLeod, and Maclyn Mc-Carty discovered that the transforming factor is DNA. The story of this discovery is recounted in a book by Maclyn McCarty entitled *The Transforming Principle: Discovering that Genes are Made of DNA* (New York: Norton, 1985).

Although the evidence was strong, many in the scientific community were not persuaded, and remained convinced that proteins were the only molecule complex enough to be carriers of genetic information. A few years later, conclusive evidence for the idea that DNA carries genetic information came from the study of viruses.

## Reproduction in Bacterial Viruses Involves DNA

In the late 1940s and early 1950s scientists began working on a group of viruses that attack and kill bacterial cells (✦ Figure 8.4). These viruses, known as **bacteriophages** (or phages for short), include one that infects and reproduces within *Escherichia coli,* the bacterium that inhabits the human intestinal tract. After gaining entry to the bacterial cell, a phage can reproduce rapidly, and in 20 to 25 minutes, about a hundred new phages burst from the ruptured bacterium, ready to invade other bacterial cells.

Phages are composed only of DNA and protein, making them ideal candidates to help resolve the question of which molecule carries genetic information. In 1952, Alfred Hershey and Martha Chase labeled phage DNA with radioactive phosphorus (there is little or no phosphorus in proteins) and

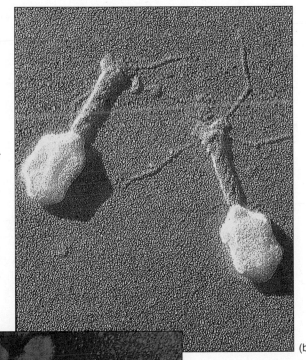

(b)

(a)

**✦ FIGURE 8.4**

**(a) Bacteriophage attack and kill bacteria. (b) A phage as it appears in the electron microscope.**

141

Phage contain only DNA and protein. Hershey and Chase used radioactively labeled phage protein or labeled phage DNA. They concluded that only the phage DNA entered the bacterial cell and directed the synthesis of new phages. This strengthened the conclusion that DNA, not protein, encoded genetic information.

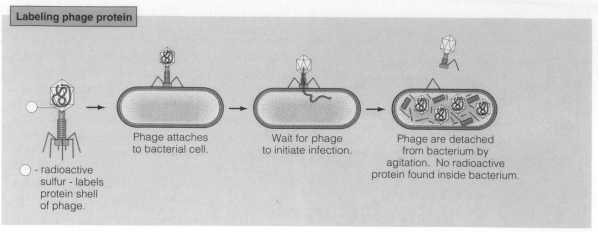

**Labeling phage protein**

◯ - radioactive sulfur - labels protein shell of phage.

Phage attaches to bacterial cell.

Wait for phage to initiate infection.

Phage are detached from bacterium by agitation. No radioactive protein found inside bacterium.

(a)

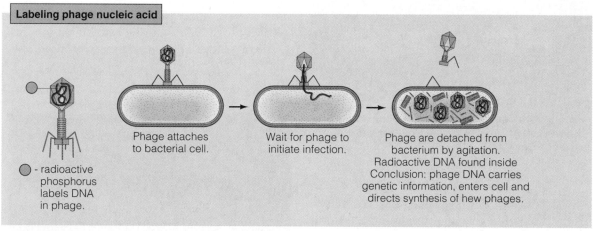

**Labeling phage nucleic acid**

● - radioactive phosphorus labels DNA in phage.

Phage attaches to bacterial cell.

Wait for phage to initiate infection.

Phage are detached from bacterium by agitation. Radioactive DNA found inside Conclusion: phage DNA carries genetic information, enters cell and directs synthesis of hew phages.

(b)

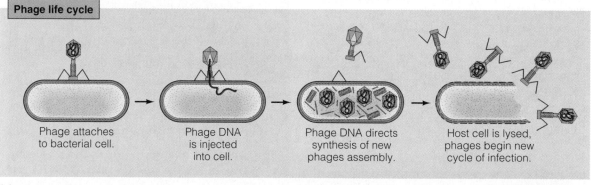

**Phage life cycle**

Phage attaches to bacterial cell.

Phage DNA is injected into cell.

Phage DNA directs synthesis of new phages assembly.

Host cell is lysed, phages begin new cycle of infection.

(c)

labeled phage protein with radioactive sulfur (there is no sulfur in DNA). Using phage with labeled protein or labeled DNA, they demonstrated that only radioactive phosphorus (contained in DNA) enters the bacterial cell and directs the production of new phage (◆ Figure 8.5). This work confirmed and extended the evidence from bacterial transformation that DNA, not protein, is the carrier of genetic information.

## WATSON, CRICK, AND THE STRUCTURE OF DNA

In 1953, a short paper written by James Watson and Francis Crick appeared in the journal *Nature*. It began with the statement "We wish to suggest a structure for the salt of deoxyribose nucleic acid (D.N.A.). This structure has novel features which are of considerable biological interest." This understated introduction to their paper on the structure of DNA resulted in a Nobel Prize for Watson, Crick, and Maurice Wilkins. Present-day applications of this model, including genetic engineering, gene mapping, and gene therapy, can be traced directly to this paper.

To build their model, Watson and Crick sifted through and organized the information already available about DNA. They used the accumulated data to construct a model of DNA structure.

When they began building models, DNA was known to be a linear molecule, composed of nucleotide subunits (◆ Figure 8.6). The sugars and phosphates in nucleotides are chemically linked to form long chains called **polynucleotides.** Two types of bases are found in the nucleotides of DNA: purines and pyrimidines. The double-ringed purines are adenine (A) and guanine (G). The pyrimidines, composed of single rings, are cytosine (C) and thymine (T).

### DNA Is a Double Helix

The Watson-Crick model is based on information about the chemistry of DNA and information from X-ray analysis about the physical arrangement of the nucleotides in DNA. Chemical analysis by Erwin Chargaff and his colleagues using DNA from many different organisms established that the amount of A always equals the amount of T, and that the amount of C always equals the amount of G. X-ray diffraction studies done by Maurice Wilkins and Rosalind Franklin showed that DNA has a helical shape of constant diameter.

The model of DNA structure devised by Watson and Crick consists of two polynucleotide chains running in opposite directions (◆ Figure 8.7). In each chain, the alternating sugar and phosphate groups are chemically linked to form the backbone of the chain. The bases face inward, where they are paired by hydrogen bonds to bases in the opposite chain (Figure 8.7). In the model, base pairing between the strands is highly specific: A pairs only with T, C pairs only with G. This base-pairing arrangement is based on the results from Chargaff's analysis of the chemical composition of DNA. The two polynucleotide chains are coiled around a central axis to form a **double helix.** This part of the model fits the X-ray results obtained by Wilkins and Franklin.

### DNA IS COILED TOGETHER WITH PROTEINS TO FORM CHROMOSOMES

Understanding the structure of DNA was an important advance in genetics, but provided little evidence about how a chromosome is organized or

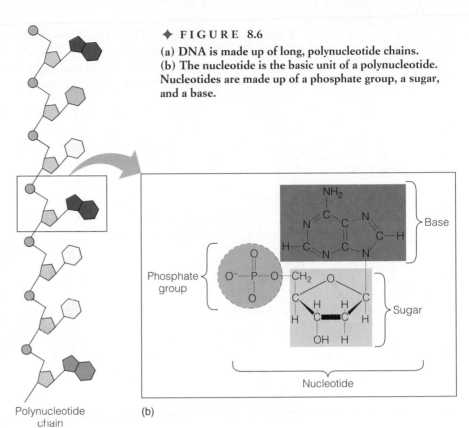

◆ **FIGURE 8.6**
(a) DNA is made up of long, polynucleotide chains.
(b) The nucleotide is the basic unit of a polynucleotide. Nucleotides are made up of a phosphate group, a sugar, and a base.

Polynucleotide chain
(a)

(b)

about what regulates the cycles of condensation and decondensation that accompany cell division and subsequent cell growth. This problem is significant because the spatial arrangement of DNA may play an important role in regulating the expression of genetic information. Models of human chromosome organization have to explain how more than

◆ **FIGURE 8.7**
**Models of DNA structure.**

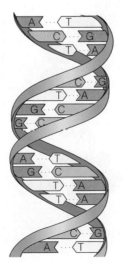

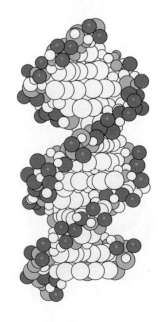

2 meters (2,000,000 μm) of DNA can be coiled into 46 chromosomes inside a nucleus about 5 μm in diameter. The model must also explain how chromosomes uncoil for replication and gene expression during interphase.

The current model for chromosome organization in eukaryotes envisions a complex of DNA and protein organized to form **chromatin** (✦ Figure 8.8). Chromatin consists of DNA molecules wound around clusters of histone proteins to form spherical bodies known as **nucleosomes.** The nucleosomes are, in turn, coiled into larger and larger structures, which form the loops and fibers seen in electron micrographs of chromosomes (✦ Figure 8.9).

## DNA REPLICATION RELIES ON BASE PAIRING

Between cycles of division, all cells replicate their DNA so that each daughter cell receives a complete set of genetic information. In their paper on the structure of DNA, Watson and Crick note that "It has not escaped our notice that the specific pairing we have postulated immediately suggests a possible copying mechanism for the genetic material." In a subsequent paper, they proposed a mechanism for DNA replication that depends on the complementary base pairing in the polynucleotide chains of DNA. If the DNA helix is unwound, each strand can serve as a template or pattern for the synthesis of a new, complementary strand (✦ Figure 8.10). This process is known as **semiconservative replication,** because one old strand is conserved in each new molecule.

DNA replication in all cells, from bacteria to humans is a complex process, requiring the action of more than a dozen different enzymes. The main role is filled by **DNA polymerase,** an enzyme that links the complementary nucleotides together to form the newly synthesized strand.

Each chromosome contains one double-stranded DNA helix running from end to end. When replication is finished, the chromosome consists of two sister chromatids, joined at a common centromere. Each chromatid contains a DNA molecule consisting of one old strand and one new strand. When the centromeres divide at the beginning of anaphase, each chromatid becomes a separate chromosome containing an accurate copy of the genetic information present in the parental chromosome.

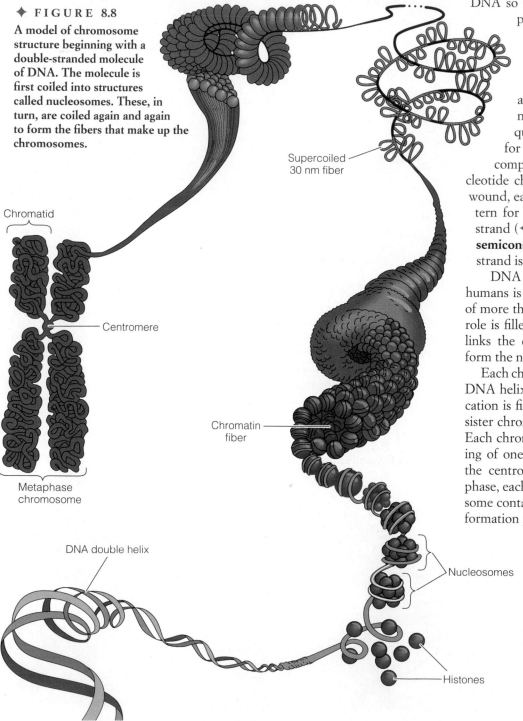

✦ **FIGURE 8.8**
**A model of chromosome structure beginning with a double-stranded molecule of DNA. The molecule is first coiled into structures called nucleosomes. These, in turn, are coiled again and again to form the fibers that make up the chromosomes.**

Chromatid

Centromere

Metaphase chromosome

Supercoiled 30 nm fiber

Chromatin fiber

DNA double helix

Nucleosomes

Histones

## HOW IS GENETIC INFORMATION STORED IN DNA?

One of the intriguing features of the Watson-Crick model of DNA is that it offers an explanation for how genetic information is encoded in the structure of the molecule. Watson and Crick proposed that genetic information could be represented as the sequence of nucleotide bases in DNA.

The amount of information stored in any cell is related to the number of base pairs in the DNA carried within the cell. This number ranges from a few thousand base pairs in some viruses, to more than 3 billion base pairs in humans, and more than twice that amount in some amphibians and plants.

A gene typically consists of hundreds or thousands of nucleotides. Each gene has a beginning and end, marked by specific nucleotide sequences, and a molecule of DNA can contain thousands of genes.

In an earlier section we reviewed the evidence that genes produce phenotypes by controlling the production of specific polypeptides. The next question therefore is "How do genes (in the form of DNA) control the production of proteins?" Recall from Chapter 2 that proteins are composed of amino acid subunits. Twenty different types of amino acids can be used to assemble proteins. The diversity of proteins found in nature results from the number of possible combinations of these 20 different amino acids. Because each amino acid position in a protein can be occupied by any of 20 amino acids, the number of different combinations is $20^n$, where $n$ is the number of amino acids in the protein. In a protein composed of only 5 amino acids, $20^5$, or 3,200,000 combinations are possible, each with a different amino acid sequence, and a potentially different function. Most proteins are actually composed of several hundred amino acids, so literally billions and billions of combinations are possible.

Given that genes are linear sequences of nucleotides and proteins are linear sequences of amino acids, the question posed earlier can be rephrased as "How is the linear sequence of nucleotides in a gene converted into the linear sequence of amino acids in a protein?" In eukaryotes, the bulk of the cell's DNA is found in the nucleus, and almost all proteins are found in the cytoplasm. This means that the process of information transfer from gene to gene product must be indirect.

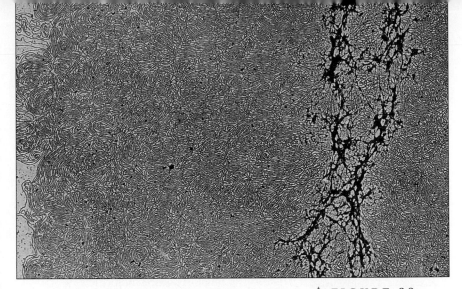

✦ **FIGURE 8.9**

Electron micrograph of a human chromosome, showing the lateral loops. The X-shaped structure at the right is the core of the chromosome. Tightly coiled loops of chromatin extend from the core.

✦ **FIGURE 8.10**

(a) In DNA replication, the two strands uncoil, and each serves as a template for the synthesis of a new strand. (b) As the strands uncoil, complementary bases pair with the template strand, and are linked together by enzymes to form a new polynucleotide strand.

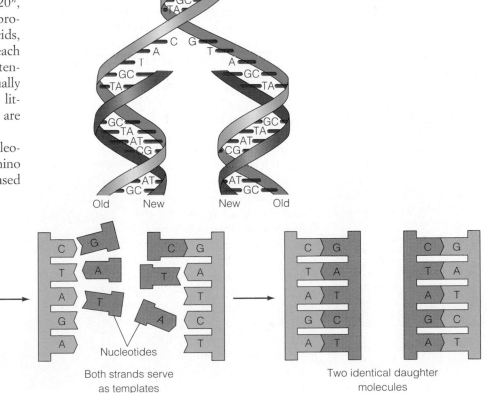

(a)

Old   New   New   Old

(b)

DNA to be replicated → Both strands serve as templates → Two identical daughter molecules

Nucleotides

## THE FLOW OF GENETIC INFORMATION IS A MULTISTEP PROCESS

The transfer of genetic information from DNA nucleotides into the amino acids of a protein has two main steps: **transcription** and **translation** (✦ Figure 8.11). In transcription, a single-stranded molecule of RNA is synthesized at an unwound section of DNA, with one of the DNA strands serving as a template for the assembly of the RNA. The product is called an RNA transcript or mRNA molecule. RNA differs from DNA in three ways:

1. RNA is a single-stranded molecule.
2. Nucleotides in RNA contain the sugar ribose rather than deoxyribose.
3. RNA contains uracil in place of thymine.

The differences between DNA and RNA are reviewed in ✦ Table 8.1. To summarize, genetic information encoded in DNA is transferred to RNA during transcription. RNA, in turn, carries this information to the cytoplasm and is directly involved with the synthesis of proteins.

In translation, mRNA moves to the cytoplasm and interacts with ribosomes to synthesize a protein containing a linear series of amino acids specified by the sequence of nucleotides in the RNA. In Figure 8.11, the brackets below the RNA indicate that genetic information is encoded in a sequence of three nucleotides, called **codons.** The four nucleotides of mRNA can be arranged to form 64 codons, each containing three letters ($4^3$ combinations), more than enough to specify the 20 amino acids used in proteins. Codons and the genetic code are examined in a later section.

### Transcription Produces Genetic Messages

In the nucleus, transcription begins when a section of a DNA double helix unwinds, and one strand acts as a template for the formation of an mRNA molecule. An enzyme called **RNA polymerase** binds to a specific nucleotide sequence in the DNA, which marks the beginning of a gene. This sequence is called a **promoter** region. Nucleotides that will make up the mRNA molecule form hydrogen bonds with complementary nucleotides in the template DNA strand. The

enzyme RNA polymerase joins the RNA nucleotides together to form a polynucleotide chain (✦ Figure 8.12). In humans, 30 to 50 nucleotides per second are added to an mRNA molecule. In bacteria, up to 500 nucleotides per second are incorporated into mRNA.

The end of the gene is marked by a nucleotide sequence called a **terminator** region. When the RNA polymerase reaches this point, it detaches from the DNA template strand, the mRNA molecule is released, and the DNA strands reform a double helix (Figure 8.12).

### Translation Requires the Interaction of Several Components

Conversion of the information in the codons of an mRNA molecule into the amino acids of a polypeptide chain is accomplished by interaction of the mRNA with two other components of the cell: **ribosomes** and **transfer RNAs** (**tRNAs**).

**TABLE 8.1 CHEMICAL DIFFERENCES BETWEEN DNA AND RNA**

|  | DNA | RNA |
|---|---|---|
| Bases | Adenine | Adenine |
|  | Thymine | Uracil |
|  | Guanine | Guanine |
|  | Cytosine | Cytosine |
| Sugar | Deoxyribose | Ribose |

✦ **FIGURE 8.11**

**The transfer of genetic information. One strand of DNA is transcribed into a strand of mRNA. The mRNA moves to the cytoplasm, where its nucleotide sequence is converted into the amino acid sequence of a polypeptide.**

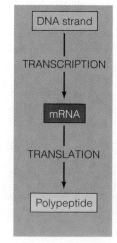

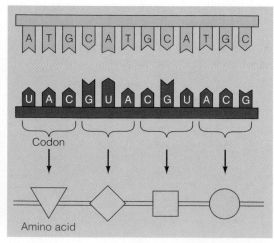

### Sidebar

### DNA ORGANIZATION AND DISEASE

Huntington disease (HD) is an autosomal dominant neurodegenerative disease characterized by chorea (involuntary movements), psychiatric and mood disorders, and dementia. The mutation responsible for HD is an expansion of a trinucleotide (CAG) repeat. This mutation tends to expand further when passed on by the father.

Those at risk for HD can now be identified by measuring the length of the CAG repeat sequence. In the HD gene, normal individuals have 10–29 CAG repeats, and those with HD have 40 or more repeats. Many people with 36–39 repeats develop the disease. Individuals with 30–35 repeats do not appear to get HD, but males with 30–35 repeats may pass on an expanded gene to their offspring, who may become affected.

Individuals from families with HD can have presymptomatic testing that follows recommendations established by the Huntington Disease Society and involves pretest and post-test visits with a neurologist, geneticist, and psychiatrist or psychologist.

Ribosomes are cellular organelles (described in Chapter 3) composed of two subunits, each containing RNA combined with proteins. Ribosomes, either free in the cytoplasm or bound to the membranes of the endoplasmic reticulum, are the site of protein synthesis.

Transfer RNA molecules help convert the mRNA codons into the amino acid sequence of a polypeptide. A tRNA molecule is a small (about 80 nucleotides) single-stranded molecule folded back on itself to form several looped regions (✦ Figure 8.13). Molecules of tRNA act as adaptors to match the codons in mRNA with the proper amino acids for incorporation into a polypeptide chain. As adaptors, tRNA molecules have two tasks: (1) Bind to the appropriate amino acid and (2) recognize the proper codon in mRNA. The structure of tRNA molecules allows them to perform both tasks. A loop at one end of the molecule contains a triplet nucleotide sequence called an **anticodon.** The anticodon recognizes and pairs with a specific codon in an mRNA molecule. The other end of the tRNA contains a site that can bind to the appropriate amino acid (Figure 8.13).

There are 20 different amino acids, and at least 20 different types of tRNA. Each tRNA carries an anticodon with a specific nucleotide sequence. If, for example, the anticodon in a tRNA molecule is CCC, the amino acid glycine

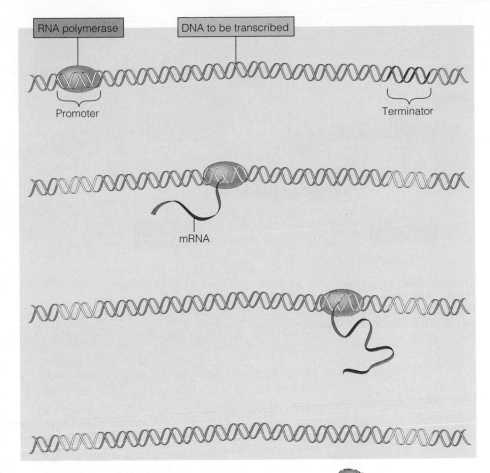

Completed mRNA

RNA polymerase

✦ **FIGURE 8.12**

**Transcription begins when an enzyme, RNA polymerase, attaches to the promoter sequence that marks the beginning of a gene. One strand is transcribed into a complementary mRNA molecule. Transcription ends when the RNA polymerase reaches a terminator sequence that marks the end of the gene.**

✦ **FIGURE 8.13**

**(a) Transfer RNA acts as a molecular adaptor and can recognize mRNA codons (at the anticodon loop) and also bind the appropriate amino acid (at the amino acid binding site). (b) The three-dimensional structure of a tRNA molecule.**

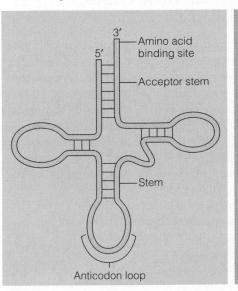

(a)

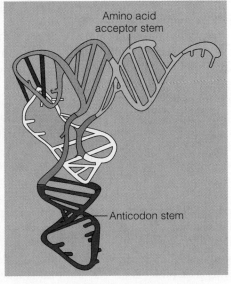

(b)

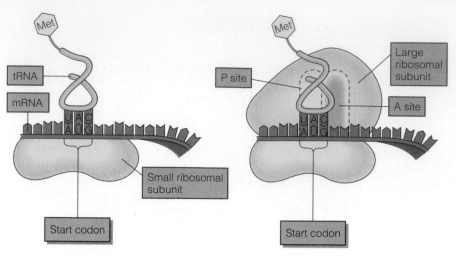

**✦ FIGURE 8.14**

Translation begins with initiation. In the cytoplasm, an mRNA molecule binds to a small ribosomal subunit. A tRNA carrying an amino acid (usually methionine) binds to the start codon. In a second step, a large ribosomal subunit binds to the small subunit.

should bind at the other end of the molecule. But by themselves, tRNA molecules cannot recognize amino acids. This task is carried out by an enzyme that recognizes a specific amino acid and its proper tRNA. Because there are 20 amino acids, there is a family of 20 such enzymes to bind amino acids to the proper tRNAs. In turn, because of base-pairing rules, tRNAs with the anticodon CCC can only bind to the codon GGG on an mRNA molecule. In this way, the three nucleotides in each mRNA codon are matched with the proper amino acid to form the new polypeptide.

Translation takes place in a series of steps: **initiation, elongation,** and **termination.** In the first step, mRNA, the small ribosomal subunit, and a tRNA carrying the first amino acid join to form a complex (✦ Figure 8.14). Each mRNA molecule has codons marking the beginning of the message (**start codon**) and the end of the message (**stop codon**). The small ribosomal subunit binds at the start codon (AUG), and the anticodon (UAC) of a tRNA carrying methionine binds to the start codon. To complete initiation, a large ribosomal subunit binds to the small subunit.

Once initiation is complete, the polypeptide elongates by the addition of amino acids. Ribosomes have two binding sites for tRNA, one called the P site and the other, the A site. A tRNA carrying methionine binds to the P site during initiation. Elongation begins when a tRNA molecule, carrying the second amino acid, pairs with the mRNA codon in the A site (✦ Figure 8.15). When the second amino acid is in position, an enzyme associated with the ribosome forms a **peptide bond** to join the two amino acids. When the peptide bond is

formed, the tRNA in the P site is released and moves away from the ribosome.

The tRNA in the A site (with its attached amino acids) is moved to the P site (Figure 8.15). This movement brings the third mRNA codon into the A site, where it is recognized by the anticodon of a tRNA carrying the third amino acid. A peptide bond is formed between the second and third amino acid, and the process repeats itself, adding amino acids to the growing polypeptide chain. Elongation continues until the ribosome reaches the stop codon. Stop codons do not code for amino acids, and no tRNA anticodon binds to these codons. At this point, no more amino acids are added to the polypeptide, and the polypeptide, mRNA, and tRNA are released from the ribosome.

The process of protein synthesis occurs rapidly in prokaryotes, where the polypeptide grows at the rate of 15 to 20 amino acids per second, and more slowly in eukaryotes, where 2 to 10 amino acids per second are added.

Once formed, the polypeptide chain can have several fates. In prokaryotes, the polypeptide folds into a three-dimensional shape determined by its amino acid sequence, and becomes functional. In eukaryotes, if the polypeptide is synthesized on ribosomes associated with endoplasmic reticulum (ER), it is released into the cisterna (the inside of the ER) where it can be chemically modified and transported to the Golgi complex for packaging and secretion from the cell. Alternately, if the polypeptide is synthesized on ribosomes free in the cytoplasm it usually assumes a three-dimensional shape and becomes functional when released from the ribosome.

## THE GENETIC CODE: THE KEY TO LIFE

The information specifying the order of amino acids in a given polypeptide is encoded in the nucleotide sequence of a gene. Because DNA is composed of only four different nucleotides, at first glance it may seem difficult to envision how the information for literally billions of different combinations of 20 different amino acids can be carried in DNA. If each nucleotide encoded the information for 1 amino acid, only 4 different amino acids could be inserted into proteins (four nucleotides, taken one at a time, or $4^1$). If a sequence of two nucleotides encoded 1 amino acid, only 16 combinations would be possible (four nucleotides, taken two at a time or $4^2$). On the other hand, a sequence of three nucleotides allows 64 combinations (four nucleotides, taken three at a time or $4^3$), 44 more than the 20 required.

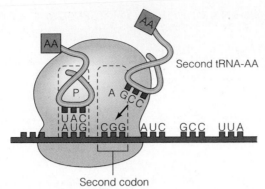

Second tRNA-AA

P A GCC

UAC
AUG CGG AUC GCC UUA

Second codon

(a) As the first step in elongation, a second tRNA-AA complex binds to the codon in the A site.

◆ **FIGURE 8.15**

After initiation, protein synthesis continues in stages. (a) Codon recognition. The anticodon of the second tRNA recognizes and binds to the mRNA codon in the A site of the ribosome. (b) Peptide bond formation. The two amino acids are covalently linked by a peptide bond. The tRNA in the P site is released. (c)–(d) Translocation. The tRNA carrying the growing polypeptide chain moves to the P site, and the ribosome moves down the mRNA by one codon. The A site is filled by another tRNA, which binds to the mRNA codon. This process continues until a stop codon is reached. At the stop codon, the ribosome comes apart, and the mRNA and polypeptide are released.

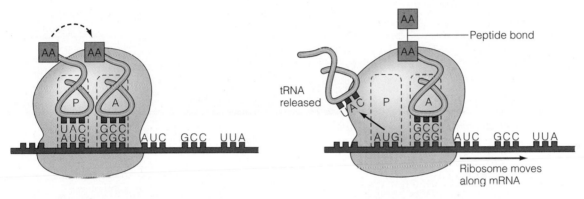

Peptide bond

tRNA released

P A

UAC
AUG GCC
CGG AUC GCC UUA

UAC
AUG GCC
CGG AUC GCC UUA

Ribosome moves along mRNA

(b) An enzyme catalyzes the formation of a peptide bond between the two amino acids. The dipeptide that forms is attached to the second tRNA. This frees up the first tRNA, which vacates the P site.

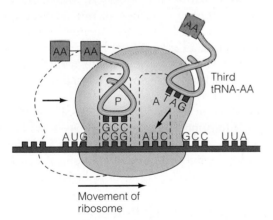

Third tRNA-AA

P A TAG

GCC
AUG CGG AUC GCC UUA

Movement of ribosome

(c) The ribosome moves down the mRNA, transferring the tRNA-dipeptide to the P site and opening up the A site to a third amino acid.

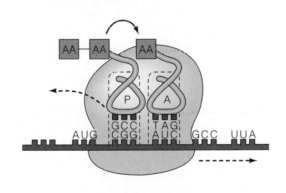

P A

GCC
AUG CGG
TAG
AUC GCC UUA

(d) The dipeptide is linked by a peptide bond to the third amino acid, forming a tripeptide. This frees the second tRNA. The ribosome moves down one more codon, exposing the A site and freeing it up for the addition of another tRNA-AA. This process repeats itself until the terminator codon is reached.

The triplet nature of the **genetic code** was confirmed by Francis Crick, Sidney Brenner, and colleagues in a series of experiments using mutants of a gene in a bacteriophage called T4. They also proposed that some amino acids could be specified by more than one combination of three nucleotides, using most of the remaining 44 combinations. This work established that the genetic code consists of a linear series of nucleotides, read three at a time, and that each triplet specifies an amino acid.

The code itself was soon deciphered, and the coding nature of all 64 triplets was established (◆ Table 8.2). By convention, the genetic code is written in codons. Of these, 61 actually code for amino acids, and the other 3 (UAA, UAG and UGA) serve as **stop** codons or **terminator** codons. The

TABLE 8.2   CODONS ON MESSENGER RNA AND THEIR CORRESPONDING AMINO ACIDS

| CODON | AMINO ACID | CODON | AMINO ACID | CODON | AMINO ACID | CODON | AMINO ACID |
|---|---|---|---|---|---|---|---|
| AAU<br>AAC | Asparagine | CAU<br>CAC | Histidine | GAU<br>GAC | Aspartic acid | UAU<br>UAC | Tyrosine |
| AAA<br>AAG | Lysine | CAA<br>CAG | Glutamine | GAA<br>GAG | Glutamic acid | UAA<br>UAG | Terminator* |
| ACU<br>ACC<br>ACA<br>ACG | Threonine | CCU<br>CCC<br>CCA<br>CCG | Proline | GCU<br>GCC<br>GCA<br>GCG | Alanine | UCU<br>UCC<br>UCA<br>UCG | Serine |
| AGU<br>AGC | Serine | CGU<br>CGC | | GGU<br>GGC | | UGU<br>UGC | Cystine |
| AGA<br>AGG | Argenine | CGA<br>CGG | Arginine | GGA<br>GGG | Glycine | UGA<br>UGG | Terminator*<br>Tryptophan |
| AUU<br>AUC<br>AUA | Isoleucine | CUU<br>CUC<br>CUA | Leucine | GUU<br>GUC<br>GUA | Valine | UUU<br>UUC | Phenylalanine |
| AUG | Methionine** | CUG | | GUG | | UUA<br>UUG | Leucine |

*Terminator codons signal the end of the formation of a polypeptide chain.
**Codon has two functions: specifies the amino acid methionine and serves as the start codon, marking the beginning of a polypeptide chain.

AUG codon serves two functions: It specifies the amino acid methionine and serves as the **start codon,** marking the beginning of a polypeptide chain.

An interesting feature of the genetic code is that the same codons are used for the same amino acids in all life-forms, from viruses, bacteria, algae, fungi, and ciliates to multicellular plants and animals. The universal nature of the genetic code means that the genetic code was established early in the evolution of life on this planet. The existence of a universal code is regarded as strong evidence that all living things are closely related and may have evolved from a small number of common ancestors.

There are some rare exceptions to the universal nature of the genetic code, involving stop codons. In some species of ciliates including *Tetrahymena* and *Paramecium,* UAG and UAA encode glutamine. Other more substantial alterations in the genetic code have taken place in mitochondria, and point to a divergent evolutionary pathway for this organelle.

## MUTATIONS ARE CHANGES IN DNA

The nucleotide sequence of DNA determines the amino acid sequence of proteins. If protein function is to be maintained from cell to cell and generation to generation, the nucleotide sequence of a gene must remain unchanged. At the molecular level, any change in the nucleotide sequence of DNA is a **mutation.** Mutations can involve substitution, insertion, or deletion of one or more nucleotides in a DNA molecule. Mutations can occur as the result of an error during DNA replication or as a result of the action of environmental agents such as radiation or chemicals (Beyond the Basics: The Ozone Layer and Mutation).

Mutations in the nucleotide sequence are not the only way to bring about changes in DNA. We have previously considered other alterations including genetic recombination and chromosomal aberrations that are detectable at the cytological level.

## Mutant Proteins Can Produce Mutant Phenotypes

Alterations in the nucleotide sequence of DNA produce mutations, which in turn produce mutant gene products. Using a human genetic disorder called sickle cell anemia as an example, it is possible to trace the disorder from the level of DNA nucleotides to an altered gene product and to the clinical phenotype.

Sickle cell anemia is an autosomal recessive disorder. Affected individuals have an abnormal type of hemoglobin. Hemoglobin, a protein found in red blood cells, transports oxygen from the lungs to the cells and tissues of the body. The abnormal hemoglobin in the red blood cells of those with sickle cell

# THE OZONE LAYER AND MUTATION

More than two billion years ago, the process of photosynthesis caused organisms in the ocean to begin releasing oxygen. This oxygen drifted into the upper reaches of the atmosphere where it reacted to form ozone ($O_3$). Eventually a layer of ozone developed that shields the earth from almost all the harmful ultraviolet radiation coming from the sun. It is thought that development of the ozone layer was an essential step in the evolution of life on land. In the mid–1970s, it was discovered that *chlorofluorocarbons* (CFCs), a class of chemicals used in air conditioning, aerosol sprays, insulation, and an array of other products, were destroying the ozone layer. As CFCs are released at ground level, they move into the upper atmosphere and are broken down by sunlight. The chlorine released in this process reacts with and destroys ozone molecules, depleting the ozone layer. Because of atmospheric conditions, the ozone layer over Antarctica has been reduced by about 10% and in certain seasons by almost 50%. Over North America, the average decline is about 3%.

Why should we be concerned about the depletion of the ozone layer? Among other effects, ultraviolet radiation acts as a mutagen, causing base changes in DNA. Thinning of the ozone layer allows more ultraviolet radiation to reach the surface of the earth, causing an increase in mutations leading to an increase in the number of cases of skin cancer. It has been estimated that for every 1% decline in the ozone layer, there will be a 2 to 4% increase in skin cancer cases. In the United States, this translates into an additional 50,000 to 100,000 cases of skin cancer and up to 9000 additional cancer deaths per year.

In response to this threat, the Montreal Conference of 1987 produced an agreement signed by 56 nations to phase out gradually the use and production of CFCs, with a 50% reduction by 1999. Since then, the gap in the ozone layer has expanded much faster than predicted, and other chemicals including carbon tetrachloride (used in dry cleaning) and methyl chloroform have been recognized as threats to the ozone layer. In 1990, the United Nations convened the London Conference to speed the elimination of CFCs and related compounds. The representatives of almost 100 nations who attended the conference agreed to phase out all CFCs completely by the year 2000, and to eliminate the production and use of other depleting chemicals by 2010. Even so, it is expected that it will take a century for the ozone layer to repair itself.

---

anemia aggregates into long fibers (✦ Figure 8.16) and causes red blood cells to become crescent or sickle-shaped (✦ Figure 8.17). The sickled blood cells are fragile and break easily, causing a reduction in the oxygen-carrying capacity of the blood (anemia). The deformed blood cells also clog small blood vessels, causing intense pain in affected areas. Blockage of blood vessels in the brain can lead to strokes and partial paralysis.

✦ **FIGURE 8.16**

A sickled red blood cell is broken open, revealing the aggregated fibers of hemoglobin within.

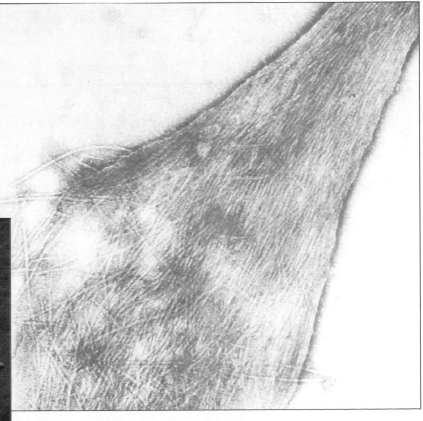

✦ **FIGURE 8.17**

Scanning electron micrographs of normal (left) and sickled (right) red blood cells.

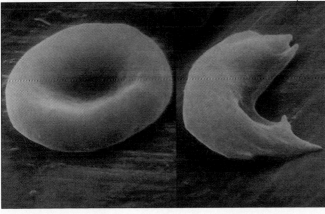

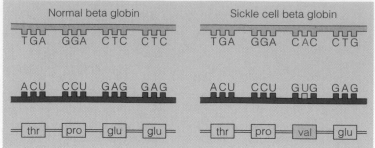

## FIGURE 8.18

A single base change in the beta globin gene is responsible for an amino acid substitution in the mutant form of beta globin associated with sickle cell anemia.

Figure 8.18 shows the DNA, mRNA, and amino acid sequences for positions 4 through 7 of beta globin, the hemoglobin polypeptide that is altered in sickle cell anemia. The normal and abnormal forms of beta globin differ only in the amino acid at position 6: Normal beta globin has glutamic acid (glu), and sickle cell globin has valine (val). This change in 1 out of 144 amino acids is caused by a single nucleotide substitution in DNA. The DNA sequence that encodes glutamine is CTC; in sickle cell anemia this has been mutated to CAC. The change from T to A causes the production of

an altered gene product that changes the shape of red blood cells.

All the symptoms of this disease, including the formation of hemoglobin fibers, the fragile and deformed red blood cells, the clogging of blood vessels, and associated pain and organ damage, result from a single nucleotide substitution in DNA (◆ Figure 8.19). In this case, there is a clear link between a gene, its protein product, and the phenotype. Because hemoglobin is composed of two different polypeptide chains (alpha and beta chains), the mutation in sickle cell anemia illustrates the one gene, one polypeptide hypothesis.

Although the mutation in sickle cell anemia is deleterious, not all mutations are harmful. Occasionally a mutation produces a new and improved version of a protein, or results in a new function for a protein. These alterations can confer an advantage on the organism carrying the mutation, allowing it to be more successful in survival and reproduction. Such mutations are the raw material for natural selection and evolution, and are responsible for the myriad life-forms around us. The role of mutation and natural selection in evolution will be discussed in Chapter 10.

## FIGURE 8.19

The cascade of consequences of the sickle cell mutation. Affected homozygotes have effects at the molecular, cellular, and organ levels.

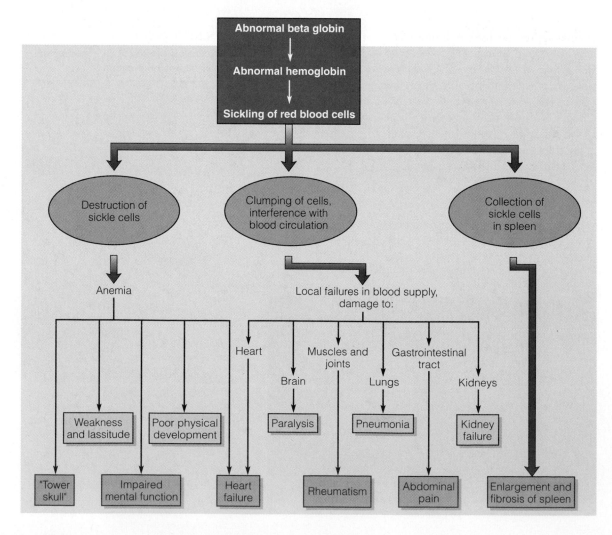

1. Early in this century, the studies of Archibald Garrod on the human genetic disorder alkaptonuria provided the first hints that genes can control biochemical reactions. In the 1930s, Beadle and Tatum showed that the generation of single mutations led to the loss of activity in a single enzyme. Their work established that genes act to produce phenotypes by specifying a gene product in the form of a protein.

2. DNA was first identified as the carrier of genetic information through experiments that demonstrated that DNA is able to transfer a genetic trait between two strains of bacteria.

3. Watson and Crick proposed that DNA consists of two polynucleotide strands wound around a central axis. Each nucleotide consists of a sugar (deoxyribose), a base, and a phosphate group.

4. In higher organisms, DNA and proteins are complexed to form chromatin. In chromatin, DNA is wound around clusters of histone proteins to form nucleosomes. Nucleosomes are coiled and recoiled to form the loops and whorls seen in the electron microscope as lateral elements of chromosomes.

5. In replication, the two strands of DNA unwind, and each strand serves as a template for the replication of a new strand. After chromosomal replication, each chromosome consists of two chromatids joined by a common centromere. Each chromatid contains a DNA molecule composed of one old strand and one newly synthesized strand of DNA.

6. Genetic information is stored as a sequence of nucleotides in DNA, and is converted into a linear sequence of amino acids in a polypeptide.

7. The sequence of nucleotides in a gene is transcribed into a complementary series of nucleotides in mRNA. The genetic information in mRNA is contained in groups of three nucleotides called codons. Messenger RNA moves to the cytoplasm and serves as a template for protein synthesis.

8. Transfer RNA molecules contain an anticodon region that recognizes mRNA codons, and they deliver the appropriate amino acid to the ribosome for insertion into the growing polypeptide chain.

9. At the molecular level, any change in the nucleotide sequence of DNA constitutes a mutation. Mutations can occur during DNA replication or through the action of environmental agents such as ultraviolet light, radiation, or chemicals.

10. Studies of disorders such as sickle cell anemia provide evidence for a direct link between a gene, a gene product, and a specific phenotype.

# KEY TERMS

anticodon
bacteriophages
chromatin
codons
DNA polymerase
double helix
elongation
enzymes
genetic code
initiation
metabolic pathways

mutation
nucleosomes
"one gene, one enzyme hypothesis"
"one gene, one polypeptide
  hypothesis"
peptide bond
polynucleotides
promoter
ribosomes
RNA polymerase
semiconservative replication

start codon
stop codon
termination
terminator codon
transcription
transfer RNAs (tRNAs)
transformation
transforming factor
translation

## FILL IN THE BLANK

1. Loss of activity in a single _____ can disrupt an entire biological pathway.

2. Beadle and Tatum won the Nobel Prize for the "_____" hypothesis.

3. DNA is a double helix, consisting of two _____ joined by _____ bonds between _____ and pyrimidine bases on opposite strands.

4. Each nucleotide in the DNA molecule is joined by _____ bonds between _____ groups and _____ molecules.

5. The cell contains three types of RNA: _____, _____, _____. All play a role in _____ synthesis.

6. Base pairing in RNA is similar to base pairing in DNA except _____ pairs with adenine instead of _____ in RNA.

7. The first step in protein synthesis is _____ followed by _____.

8. tRNA contains a three-base region called the _____ carrying a specific nucleotide sequence and binds to the _____ on a mRNA molecule.

9. In transcription an enzyme called _____ binds to a specific base sequence called a _____ region that marks the beginning of a gene.

10. Chromatin consists of DNA molecules wound around clusters of proteins known as _____ to form spherical bodies known as _____.

11. _____ _____ is the formation of a complementary strand of a double helix.

12. Alterations in the nucleotide sequence of DNA produce _____, which in turn produce altered _____ _____.

## SHORT ANSWER

1. Outline the synthesis of DNA. Explain how the cell ensures the accurate replication of DNA.

2. List the three types of RNA and describe their role in protein synthesis.

3. Using the genetic code in Table 8.2, translate the mRNA sequence UAUCGCAC-CUCAGGAUGAGAU.

4. Compare and contrast RNA and DNA.

5. Pair the appropriate RNA bases to the following DNA bases: TACGACCTGATG.

6. How many amino acids are coded for in this mRNA chain?: GCAACCAUGC-GGAUCGCC.

7. Transcription is to nucleus as translation is to _____.

8. If a protein is composed of 15 amino acids, how many different combinations of these amino acids are possible?

## SCIENCE AND SOCIETY

1. Genetics and lifestyle choices are frequently associated with some diseases. One example is a genetic disorder called familial hypercholesterolemia. In this autosomal dominant condition, the liver fails to take up cholesterol from the bloodstream. Heterozygous individuals have half the normal number of liver cell cholesterol receptors; no receptors are present in individuals who are homozygous for the mutant gene. Heterozygous individuals usually die of heart attacks in their early to late twenties while homozygotes die during childhood. Heart attacks also occur in people who do not have hypercholesterolemia condition, often due to poor diet, lack of exercise, smoking, elevated stress levels, and high salt, fat, and cholesterol intake. When physicians treat heart attack patients, should their approach be dependent upon the cause of the heart attack? Would it be ethical to have different care modalities for someone who has hypercholesterolemia and is likely to die from this disorder than someone who is not? Should employers and insurance agencies have access to information about those who have hypercholesterolemia? How would you tell someone that they carry the gene for this condition?

2. The Human Genome Project is rapidly accomplishing its goal of determining the nucleotide sequences of all human genes. This knowledge might be used to treat genetic diseases or produce life-saving technology. The National Institutes of Health (NIH) and private companies have applied for patents on their discoveries. The most common type of gene therapy—the transfer of genes by a retrovirus into cells that have been temporarily removed from a patient's body (ex vivo gene therapy)—has been granted as an exclusive patent to Gene Therapy, Inc. of Gaithersburg, Maryland, through a patent awarded to the NIH. Should individuals and companies be able to patent genes, gene products, and experimental therapies? Who is to gain from patenting this technique? Do you think a patent should be awarded before a therapeutic benefit is seen for the technique? What problems do you think could arise from awarding patents to researchers? How do you propose that scientists solve the problems caused by patenting widely used techniques?

# RECOMBINANT DNA AND GENETIC ENGINEERING

OPENING IMAGE

*Bacterial plasmids, shown on the*

*screen behind a group of scientists,*

*are one of the foundations of*

*recombinant DNA technology and*

*the biotechnology industry.*

*S*ome 17 to 20 million years ago, a magnolia leaf fluttered into the cold water of
a lake in what is now Idaho. Magnolias and other plants found in warm, humid,
temperate climates flourished along the shores of the lake that formed when a
volcanic lava flow blocked the stream bed of a valley. The leaf sank into the cold, still
water down to the oxygen-poor bottom, and quickly became covered with mud. Because
of the lack of oxygen, the leaf did not deteriorate, but instead over a period of millions of
years, formed what is known as a compression fossil as the mud turned to shale.

As researchers from the twentieth century split the shale layers, the exposed fossil was
still green, although its color faded rapidly in the oxygen-containing atmosphere. They
scraped the fossilized leaf into a mortar, and ground it with dry ice to form a fine powder.
From this powder, the DNA of the fossil leaf was extracted and purified. Using a recombinant DNA technique, the researchers isolated a chloroplast gene from the fossil leaf. This
gene encodes the information for the production of an enzyme, which, in turn, controls
the production of a sugar that is an intermediate in photosynthesis (Chapter 3).

The nucleotide sequence of this fossil gene was determined and then compared with
the nucleotide sequence of similar genes from present-day species of magnolias. This
analysis indicated that although there have been some nucleotide changes between the
fossil gene and the contemporary gene, there has been little overall change in this gene
as it has traveled through the intervening generations. The differences in gene sequence
between the fossil magnolia and present-day species were used to establish its relationship

*to the species we know today. The results indicate that the fossil species of Magnolia is closely related to a species of Magnolia that grows in the eastern Unites States.*

*The ability to isolate and characterize a specific gene from a plant that lived some 17 to 20 million years ago and compare it to the gene from species alive today demonstrates the power of recombinant DNA technology. The data gathered in this experiment allow a direct measurement of the rate of mutation in an individual gene over evolutionary time, and assist in establishing the evolutionary relationship between present species and their fossil ancestors.*

*Recombinant DNA technology is having a dramatic impact on many other areas, including molecular biology, pharmaceutical manufacturing, agriculture, diagnosis and treatment of genetic disorders, and criminal investigations. This revolution is just beginning, and the impact of recombinant DNA technology and the accompanying biotechnology industry will continue to spread and profoundly change the lives of many individuals.*

*In this chapter, we discuss recombinant DNA technology and gene cloning. Following this, we discuss some of the changes this technology has brought about in medicine, genetics, and agriculture.*

## GENETIC MANIPULATION BY SELECTIVE BREEDING

Domesticated plants and animals have been manipulated genetically for thousands of years. This has been accomplished by selecting those individuals with desirable traits for use in breeding. In each succeeding generation, individuals of the desired type are selected as parents for the next generation. Over many generations, this practice, known as **selective breeding,** leads to the development of stocks or strains with the desired characteristics.

### Selective Breeding Alters Plants and Animals

Most of the plants and animals used in agriculture, gardening, horse racing, dog and cat breeding, etc., have been developed by selective breeding. All recognized breeds of dogs belong to a single species, *Canis familiaris,* but selective breeding has generated a wide range of phenotypes and abilities to perform tasks (✦ Figure 9.1).

The origin of selective breeding is lost in antiquity. Some strains or breeds of domesticated animals (sheep and cattle) and plants (corn and wheat) originated thousands of years ago, whereas others, such as thoroughbred horses, minks, and certain strains of apples and roses were developed only in the last century or two.

The strategy of selective breeding also can be used on microorganisms. Scientists screen natural populations of bacteria gathered from soil samples, swamps, and sludge for organisms that produce useful chemical compounds, or show an ability to degrade toxic industrial waste products. These organisms are then selectively bred into commercially useful strains. Even today, microbiologists roam the world searching for strains of bacteria and fungi that produce compounds that can be used in medicine.

### What Is Recombinant DNA Technology?

As useful as selective breeding has been for thousands of years, it is rapidly being overshadowed by a new technology that began developing in the mid-1970s. This collection of techniques, known as **recombinant DNA technology,** has moved from the research laboratory to the pharmaceutical, agricultural, and plant and animal breeding industries with dazzling speed. We are now in the beginning stages of the biotechnology revolution. Recombinant DNA techniques are used to generate knowledge about basic biological processes such as gene regulation, to develop and produce commercial products such as human growth hormone, to create disease-resistant plants, and to diagnose and treat genetic disorders.

A small group of techniques forms the basis of recombinant DNA technology. These methods are used to:

1. Produce DNA fragments using enzymes that cut DNA at specific base sequences.
2. Link these DNA fragments to self-replicating forms of DNA (called **vectors**) to create recombinant DNA molecules.
3. Replicate the recombinant DNA molecule in a host organism to create hundreds or thousands of exact copies (**clones**) of the inserted DNA segment.
4. Retrieve the cloned DNA segment in quantities large enough for further study or modification.
5. Produce and purify the gene product encoded by the cloned DNA segment.

**✦ FIGURE 9.1**
Selective breeding has produced a wide range of phenotypes in dogs.

## CLONING GENES IS A MULTISTEP PROCESS

### Restriction Enzymes Cut DNA at Specific Sites

In the early 1970s, scientists discovered a series of enzymes that attach to DNA molecules at specific nucleotide sequences and cut both strands of DNA at that site. These enzymes, known as **restriction enzymes,** are a key component in recombinant DNA technology. The recognition and cutting site for an enzyme isolated from *Escherichia coli,* a bacterium that lives in the human intestine, is shown in (✦ Figure 9.2. This enzyme, *Eco*RI, has a recognition and cutting sequence that reads the same on either strand of the DNA. Such sequences are said to be **palindromes,** because they read the same in either direction. Words like "mom" "pop" and "radar" are palindromes as are phrases such as "live not on evil" or "was it Ararat I saw?"

Restriction enzymes such as *Eco*RI create single-stranded tails when they cut DNA. These "sticky ends" can form hydrogen bonds with other DNA fragments with complementary tails. The fragments

**✦ FIGURE 9.2**
The recognition and cutting sites for the enzyme *Eco*RI.

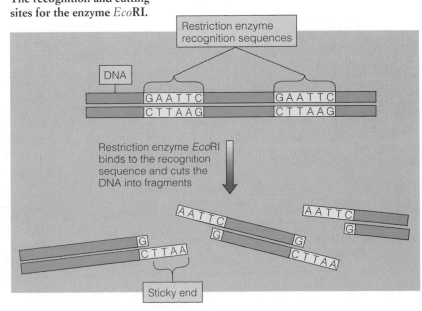

held together by hydrogen-bonding are sealed together by an enzyme called **DNA ligase.** The result is the formation of new combinations of DNA fragments, called **recombinant DNA molecules** (✦ Figure 9.3).

molecules found within bacterial cells. One such plasmid is shown in ✦ Figure 9.4. The diagram shows the position of sites recognized and cut by restriction enzymes. These sites can be used for the insertion of DNA segments to be cloned.

## Vectors Serve as Carriers of Genes for Cloning

**Vectors** are DNA molecules that can be joined with DNA fragments that are to be cloned. Many of the vectors used in recombinant DNA research are derived from **plasmids,** self-replicating, circular DNA

## Steps in the Cloning Process

When DNA molecules from a plasmid vector and from human cells are cut with a restriction enzyme and placed together in solution, their sticky ends can reassociate and be sealed together by treatment with DNA ligase. This creates a recombinant DNA

*Sidebar*

### SELECTIVE BREEDING GONE BAD

Purebred dogs are the result of selective breeding over many generations, and worldwide, there are now over 300 recognized breeds and varieties of dogs. Selective breeding to produce dogs with desired traits, such as long noses and closely set eyes in collies, or the low, sloping hindlegs of German shepherds, can have unintended side effects. About 70% of all collies suffer from hereditary eye problems and more than 60% of German shepherds are at risk for hip dysplasia. It is estimated that 25% of all purebred dogs have a serious genetic disorder.

The high level of genetic disorders in purebred dogs is a direct result of selective breeding. Over time, selective breeding has caused a decrease in genetic variability and an increase in animals homozygous for recessive genetic disorders.

Outbreeding is a simple genetic solution to this problem. In the world of dogs, this means mixed-breed dogs, or mutts. Crosses between collies and German shepherds or between Labrador retrievers and German shepherds combine the looks and temperaments of both breeds, but reduce the risk of offspring with genetic disorders.

✦ **FIGURE 9.3**
**Recombinant DNA molecules can be created using DNA from two sources, a restriction enzyme and DNA ligase. DNA from two sources is cut with a restriction enzyme, and the fragments mixed. In the mixture, the ends of the two types of DNA will associate by forming hydrogen bonds. The fragments can be covalently joined together by DNA ligase, creating a recombinant DNA molecule.**

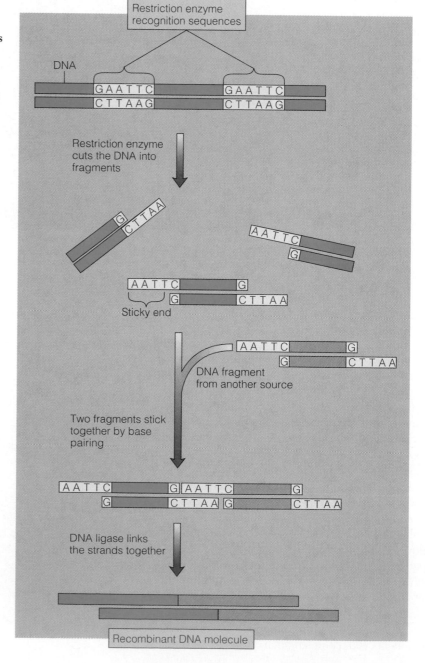

molecule composed of plasmid DNA and human DNA (✦ Figure 9.5). The recombinant plasmid can be inserted into bacterial cells, and each time the bacterial cell divides, the recombinant plasmid is copied. Because bacterial cells can divide every 20 minutes, large numbers of recombinant plasmids can be produced in only a few hours. Each copy of the recombinant plasmid is a **clone,** and the process is known as **cloning.**

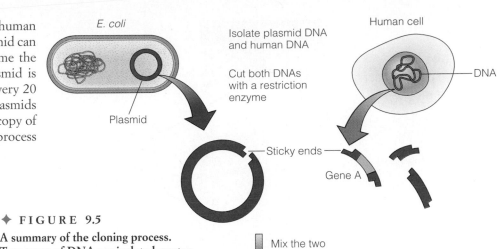

Isolate plasmid DNA and human DNA

Cut both DNAs with a restriction enzyme

E. coli

Plasmid

Human cell

DNA

Sticky ends

Gene A

✦ **FIGURE 9.5**

A summary of the cloning process. Two types of DNA are isolated: vector DNA in the form of a plasmid and human DNA carrying the A gene. Both DNAs are cut with a restriction enzyme. The cut DNAs are mixed and, after pairing, are linked by DNA ligase. The recombinant plasmid DNA is inserted into a bacterial host cell. At each bacterial cell division, the plasmid is replicated, producing many copies, or clones, of the DNA segment carrying the A gene.

✦ **FIGURE 9.4**

Bacterial plasmids, such as pBR322, contain restriction sites that can be used to insert DNA for cloning.

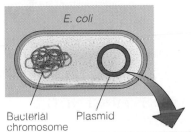

E. coli

Bacterial chromosome

Plasmid

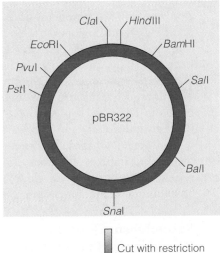

ClaI    HindIII

EcoRI    BamHI

PvuI

PstI    SalI

pBR322

BalI

SnaI

Cut with restriction enzyme BamHI

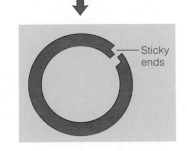

Sticky ends

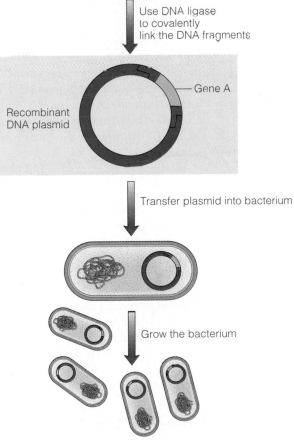

Mix the two DNAs, allow them to join by their sticky ends

Use DNA ligase to covalently link the DNA fragments

Recombinant DNA plasmid

Gene A

Transfer plasmid into bacterium

Grow the bacterium

Bacteria carrying cloned copies of human gene A

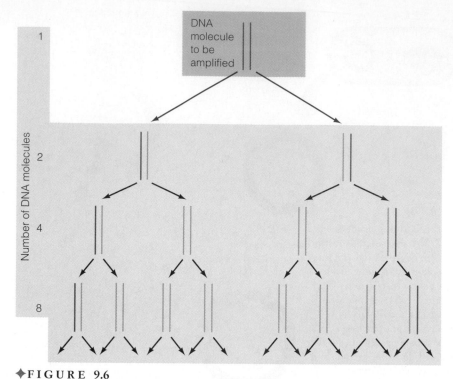

**◆FIGURE 9.6**

In the PCR reaction, DNA is replicated in a series of cycles, doubling the number of molecules in each cycle.

method for producing copies of DNA molecules was developed and adopted in a number of fields including molecular biology, genetics, medicine, animal conservation, evolution, and forensics. This method is called the **polymerase chain reaction** (**PCR**). It is a relatively simple, fast way of amplifying or copying specific DNA fragments, and can be used on infinitesimally small amounts of DNA (◆ Figure 9.6). In fact, the DNA contained in the cells at the base of a single hair or in an individual cell can be used as the starting material for PCR amplification.

In the PCR technique, the DNA to be amplified is mixed with DNA polymerase, nucleotides, and other essential components. The DNA is copied in this mixture, producing two daughter molecules. The two daughter molecules are copied in a second round of replication, and the process continues, with the amount of DNA doubling in each round of replication. Beginning with a single DNA molecule, 25 cycles of amplification will produce more than a million copies of the original DNA in a few hours. This process has been automated, and microchip-programmed machines, loaded with a DNA sample (hair, blood cells, mummified remains, and even fossils) can produce millions or billions of DNA copies with little effort.

## APPLICATIONS OF GENETIC ENGINEERING

Applications of recombinant DNA technology have had a significant impact on medicine, law, agriculture, and industry. In part, this impact is derived from the ability to identify and isolate genes that control heritable disorders, to produce commercial quantities of human proteins for therapeutic purposes. In addition, it is possible to transfer genes between individuals and species in order to treat human genetic disorders, create disease-resistant plants, and improve livestock. In this section, we review some of these developments to illustrate how this new technology is being put to use.

### Recombinant DNA Techniques Have Revolutionized Genetic Mapping

Humans carry between 50,000 to 100,000 genes. Of these, about 4000 are known to be associated with genetic disorders. To understand how the symptoms of a genetic disorder are produced, it is essential to identify the nature and action of the protein encoded by the gene. Once this is known, it is often possible to design therapeutic strategies to treat the disorder. Knowing where the gene is located is the first goal in investigating a genetic disorder. Using recombinant DNA techniques, genes

After growth, the bacterial cells can be broken open, and the recombinant plasmids extracted. The inserted human DNA can be released from the plasmid with the same restriction enzyme used in cloning. The cloned human DNA can be used in further experiments, or transferred via specialized vectors called expression vectors to another bacterial cell for synthesis of the protein encoded by the human DNA.

The steps in DNA cloning can be summarized as follows:

1. The DNA to be cloned is isolated and treated with a restriction enzyme to produce segments ending in specific sequences.
2. These segments are linked to vector DNA molecules, producing a recombinant DNA molecule.
3. Vectors carrying DNA to be cloned are transferred into bacterial host cells where they replicate, producing many copies or clones of the inserted DNA.
4. This cloned DNA can be recovered from the bacterial host cells and used in further experiments or in the production of gene products.

### PCR Can Be Used to Amplify DNA

One of the major uses of recombinant DNA technology is the production of many copies of genes or specific DNA segments. Cloning involves a large number of steps, is labor intensive, and can take weeks or months. About ten years ago, a new

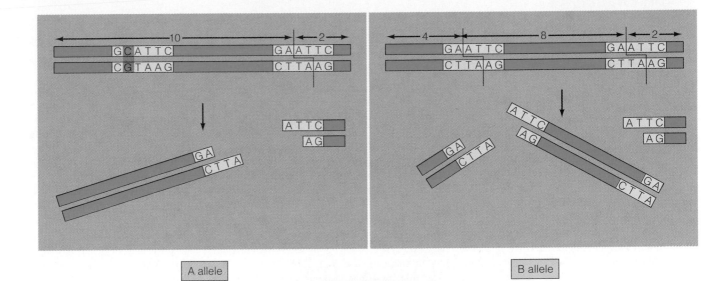

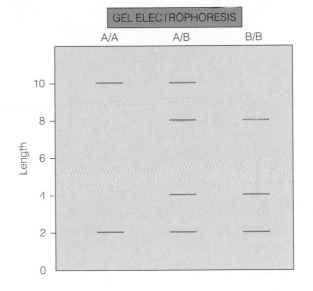

**✦ FIGURE 9.7**

The A and B alleles represent regions of DNA from homologous chromosomes. Arrows indicate recognition and cutting sites for restriction enzymes. Because of variation in nucleotide sequence (highlighted), a cutting site present in B is missing in A. This variation produces differences in the number and length of DNA fragments produced in A and B by cutting with a restriction enzyme. The fragments can be separated and detected by a technique called gel electrophoresis (right). Since the variations in fragment length are inherited as codominant traits, they produce three distinct patterns that serve as phenotypes and also identify genotypes. These variations are called restriction fragment length polymorphisms or RFLPs.

can be *mapped* even when the nature of the gene product is unknown.

Mapping genes using recombinant DNA depends on two things: (1) the fact that restriction enzymes cut DNA at specific base sequences and (2) the availability of large, multigenerational families in which a genetic disorder is expressed.

Nucleotide sequences in long stretches of human DNA are subject to variation. This variation is not associated with any deleterious phenotype. In some cases, this variation in nucleotide sequence creates or destroys a cutting site for a restriction enzyme. This can create a heritable difference in the length and number of DNA fragments produced by cutting with a restriction enzyme. This difference is known as a **restriction fragment length polymorphism** or **RFLP** (✦ Figure 9.7). RFLPs are passed from generation to generation in a codominant fashion (review codominant inheritance in Chapter 7), and serve as genetic markers for specific chro-

mosomes. Because of linkage, any genes near these markers are usually inherited along with the marker. Thousands of RFLP variations have been mapped in humans, and many such markers have been assigned to each chromosome (see Beyond the Basics: DNA Fingerprints).

The use of RFLPs in mapping human genes can be illustrated by considering the case of the gene for cystic fibrosis (CF). Cystic fibrosis is a recessive condition with an incidence of 1 in 2000 births among those of Caucasian ancestry. CF causes the production of a thick mucus that blocks the airways in the lungs, and clogs the ducts that carry secretions from the pancreas, reducing the effectiveness of digestion. CF homozygotes develop obstructive lung disease and infections that lead to premature death. The mapping of the gene for CF was accomplished in several steps.

Large families with a history of CF were studied to find the link between CF and RFLP markers rep-

# DNA FINGERPRINTS AND THE LAW

In RFLP analysis, interest is often focused on nucleotide sequence variations in a single gene. A version of RFLP analysis can detect nucleotide variations in short, repeated DNA sequences scattered at 8 to 15 sites on different chromosomes. The resulting pattern of 25 to 60 bands is unique to each individual (except for identical twins), and the term "DNA fingerprint" is used to describe this characteristic pattern of DNA fragments. In the last few years, DNA fingerprinting has been used in criminal cases and immigration disputes. It has also been used to solve problems in archaeology, ecology, preservation of endangered species, and animal breeding.

For use in criminal cases, DNA is extracted from biological material left at the crime scene. This can include blood, tissue, hair, skin fragments, and semen. The DNA is cut with restriction enzymes and the resulting pattern of fragments is analyzed and compared with the DNA fingerprints of the victim and any suspects in the case. In England, DNA fingerprinting was used to solve a case involving the rape and murder of two teenage girls. More than 4000 men were typed during the investigation, and the results freed an innocent man who had been jailed for the crime and led to a confession from the killer. This case was described in the best selling book, *The Blooding,* by Joseph Wambaugh.

In the United States, DNA fingerprinting was first used in a criminal case in 1987, and since then has been used in more than 1000 similar cases. DNA fingerprinting came to widespread public attention in the murder trial of O. J. Simpson. In that trial, 400 alleles were analyzed from the evidence, and all were found to match either the defendant or the victims.

The use of DNA fingerprints in legal cases requires care in the collection and handling of samples, accurate identification of bands on the gel, methods for determining when bands match, and reliable calculations for the probability of matching bands in the evidence with those of the defendant.

DNA fingerprints have also been used in other applications, including analysis of DNA recovered from an Egyptian mummy, and in the settlement of disputes over the bloodlines of purebred dogs. It has also been used to determine the degree of relatedness among members of endangered species, and on the frozen remains of a wooly mammoth recovered in Siberia to determine the evolutionary relationships between mammoths and modern-day elephants.

---

resenting specific chromosomes. The result of these initial studies produced an exclusion map, or a list of chromosomes where the gene was *not* linked to a marker. Finally, it was discovered that the CF allele was inherited along with RFLP markers on chromosome 7. Further work established that the CF gene is located in a region near the tip of the long arm (✦ Figure 9.8).

In the last stage, researchers cloned, sequenced, and scanned more than 500,000 base pairs of DNA from the region, identifying the CF gene from a number of candidates by finding a base sequence that encoded information for an unknown protein. From the base sequence, the amino acid sequence of the protein was predicted. Normal and mutant alleles of the gene were cloned, and the normal and mutant proteins were isolated. The protein, known as CFTCR, is inserted in the cell membrane and functions in the transport of chloride ions into and out of the cell (✦ Figure 9.9). In CF, the protein is defective in function.

The mapping, identification, and cloning of the CF gene was accomplished in about 5 years, beginning with no knowledge of the nature of the gene product, the location of the gene, or the kind of muta-

tional event that results in cystic fibrosis. The successful cloning of the CF gene has several consequences. Studies on the structure and function of the CFTCR protein can lead to new and effective therapies for CF. Secondly, heterozygotes can be detected in the population (4% of the Caucasian population), and those couples at risk for having a child with CF can be identified and counseled. Lastly, it confirms the value of using recombinant DNA methods in gene mapping.

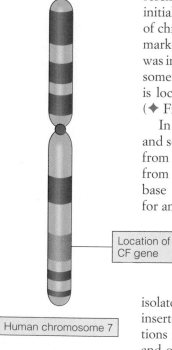

Location of CF gene

Human chromosome 7

✦ **FIGURE 9.8**

**The gene for cystic fibrosis was mapped to the long arm of chromosome 7 using RFLP analysis.**

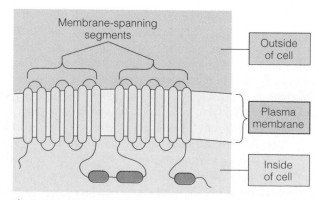

Membrane-spanning segments

Outside of cell

Plasma membrane

Inside of cell

✦ **FIGURE 9.9**

**Cystic fibrosis is caused by a defective membrane protein called CFTCR, which controls passage of chloride ions into and out of the cell.**

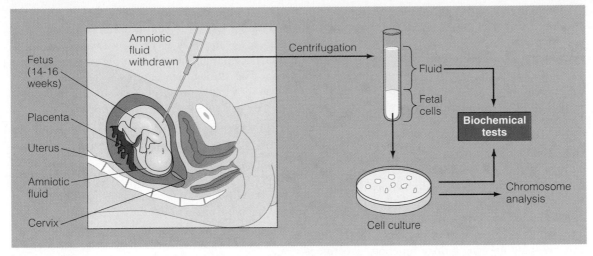

(a)

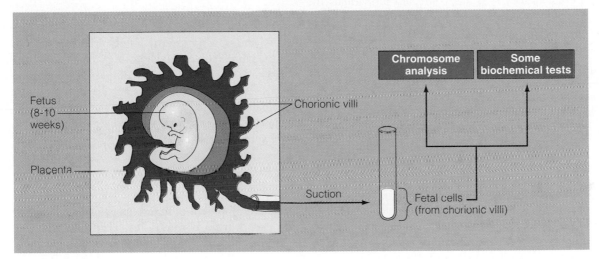

(b)

✦ FIGURE 9.10

(a) Amniocentesis removes a sample of the fetal cells and fluid surrounding the fetus. Many fetal genetic disorders can be detected by analysis of the fluid and/or the cells. This test can be performed beginning at the sixteenth week of pregnancy. (b) Chorionic villus sampling obtains fetal cells at an earlier stage of pregnancy (8 weeks).

## Prenatal and Presymptomatic Diagnosis of Genetic Disorders

Using recombinant DNA techniques, it is possible to do both prenatal and presymptomatic genetic testing. **Prenatal testing** can detect the presence of a genetic disorder in an embryo or fetus. The most common method of prenatal testing (**amniocentesis**) involves the insertion of a needle into the uterus to withdraw amniotic fluid (✦ Figure 9.10). The fluid and the fetal cells contained in the fluid can be analyzed to detect biochemical or chromosomal disorders (such as Down syndrome, Chapter 7) or for testing by recombinant DNA methods. An alternative procedure for collecting samples for prenatal diagnosis is called **chorionic villi sampling** (**CVS**) (Figure 9.10). For CVS, a thin tube is inserted into the uterus through the cervix and is used to retrieve tissue from the fetal side of the placenta (the chorionic villi) (Figure 9.10). The cells recovered

can be analyzed for chromosomal, biochemical, or molecular defects.

Prenatal screening for sickle cell anemia can be used to show how a combination of amniocentesis and recombinant DNA methods can be used in prenatal diagnosis.

As described in Chapter 8, sickle cell anemia is a recessive trait found in families with ancestral origins in areas of West Africa and the lowland regions around the Mediterranean Sea. Sickled red blood cells obstruct blood flow in small vessels and capillaries, causing episodes of intense pain and organ damage. It also increases the risk of skin ulcers, strokes, and heart attacks. The defective gene product (beta globin, a polypeptide component of hemoglobin), is not produced before birth, and the condition cannot be prenatally diagnosed using conventional techniques.

The mutation that causes sickle cell anemia destroys a restriction enzyme cutting site, changing the length and number of DNA fragments generated by the enzyme (✦ Figure 9.11). By analyzing the fragments in DNA from fetal cells collected by amniocentesis, the genotypes of normal homozygous, heterozygous, and affected homozygous individuals can be determined prenatally.

**Presymptomatic screening** involves the detection of genetic disorders that only become apparent later in life, for example, individuals with a progressive genetic disorder leading to premature death known as polycystic kidney disease (PKD). The symptoms of PKD develop when affected individuals are between 35 and 50 years of age. This rare condition, inherited as a dominant trait, is characterized by the development of cysts in the kidney that gradually destroy kidney function. Treatment includes kidney dialysis and transplantation of a normal kidney, but many affected individuals die prematurely. Because PKD is a dominant trait, anyone heterozygous or homozygous for the gene is affected. The gene for PKD is located on chromosome 16, and RFLP markers can be used to determine which family members are most likely to develop PKD. This testing can be done prenatally or at any age before (or after) the condition appears.

The availability of presymptomatic testing for disorders such as PKD raises social and ethical issues that are being addressed by the medical profession, patients and their families, and the public at large. Should parents in families with such disorders have their children tested? Should children be informed of their status with respect to these and other conditions? Questions and issues such as

these serve to illustrate that the effects of the recombinant DNA revolution have far-reaching consequences, and may affect decisions you make in your personal life or as a member of your community. The ethical consequences of recombinant DNA technology are discussed later in this chapter.

## THE HUMAN GENOME PROJECT IS AN INTERNATIONAL EFFORT

The development of recombinant DNA technology and its application to gene mapping led to the establishment of the Human Genome Project, a large-scale coordinated effort to determine the location of the 50,000 to 100,000 genes in the human genome (a **genome** is the set of genes carried by an individual), and to analyze the nucleotide sequence of these genes. Originally established by the U.S. Congress, and now an international effort coordinated by the Human Genome Organization (HUGO), the project is scheduled to be completed by 2005 at a cost of about $3 billion dollars (about $1 per nucleotide).

The major steps in the project are shown in ✦ Figure 9.12. Although the work is shown as a series of separate projects, work is progressing more or less simultaneously on all stages. Briefly, the objectives are as follows:

1. Development of high-resolution genetic maps for each human chromosome. Using the methods described earlier, geneticists are working to pro-

✦ **FIGURE 9.11**

Prenatal diagnosis of sickle cell anemia. In sickle cell anemia, the mutation destroys a cutting site for a restriction enzyme. The resulting alteration in the number and size of DNA fragments can be used to read directly the genotypes of family members. The first child (II–1) is a heterozygous carrier, the second (II–2) is affected with sickle cell anemia, and the unborn fetus (II–3) is homozygous for the normal alleles, and will be unaffected by sickle cell anemia.

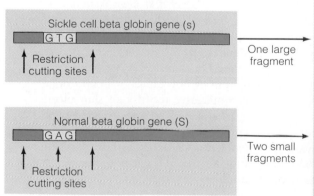

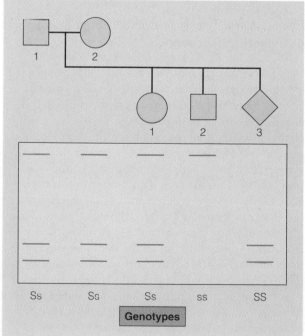

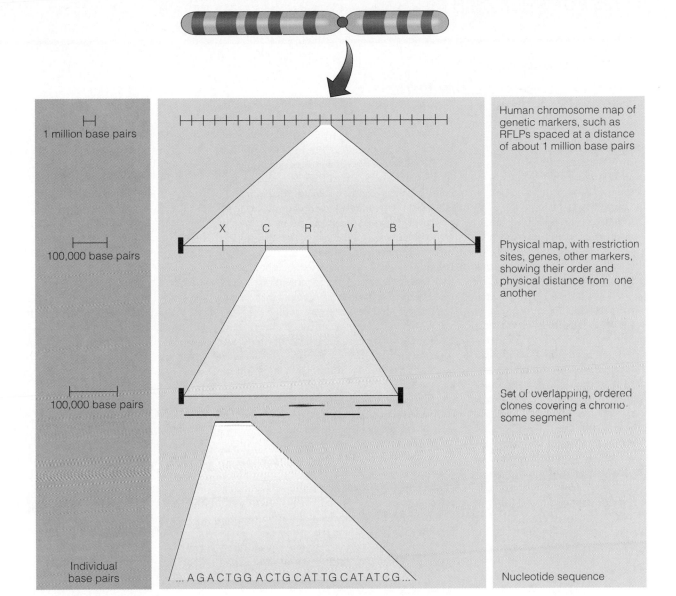

| 1 million base pairs | | Human chromosome map of genetic markers, such as RFLPs spaced at a distance of about 1 million base pairs |

X    C    R    V    B    L

| 100,000 base pairs | | Physical map, with restriction sites, genes, other markers, showing their order and physical distance from one another |

| 100,000 base pairs | | Set of overlapping, ordered clones covering a chromosome segment |

| Individual base pairs | ...A G A C T G G A C T G C A T T G C A T A T C G... | Nucleotide sequence |

♦ **FIGURE 9.12**

**The Human Genome Project is proceeding in four major steps. The first is the production of a genetic map of each chromosome, with markers spaced at a distance of about one million base pairs. The second is the construction of a high-resolution physical map, with markers assigned to chromosomal sites about every 100,000 base pairs along the chromosome. Recovering a series of overlapping clones covering regions between markers is the third step in the project. Each clone will be sequenced, and the nucleotide sequence of the genome will be stored in a computer data bank.**

duce maps (using about 3000 genes and other markers) with RFLP and other markers spaced evenly along each chromosome. Having high-resolution maps will make it easier to map new genes by testing for linkage to the markers spaced on the chromosomes. Genetic maps for many chromosomes are already complete, and this phase of the project is drawing to a close.

2. Physical mapping of each chromosome. Physical maps show the order and distance between genes and markers, expressed in base pairs of DNA.

The goal of this phase of the project is the production of physical maps for each chromosome, with markers spaced every two million base pairs of DNA. This phase of the project is ahead of schedule, and may be completed soon.

3. Cloning each chromosome. In the third stage, overlapping clones covering the length of each chromosome will be developed. The problem is not generating such clones, but in identifying and putting in order a collection of clones that overlap with each other and cover the entire chromosome.

# GUEST ESSAY: RESEARCH AND APPLICATIONS

## The Human Genome Project: Reading Our Own Genetic Blueprint

FRANCIS SELLERS COLLINS

I developed an early love for mathematics and science and, after learning about the elegant features of the structure of the atom and the chemical bond in a high school chemistry class, I became convinced that I wanted to be a chemist. Accordingly, I majored in chemistry at the University of Virginia, and immediately after graduation went into a Ph.D. program in physical chemistry at Yale, never glancing around to notice what other exciting areas of science I might be missing. Almost by accident, however, I was introduced to molecular biology. To my surprise, I learned that all of my biases about biology being chaotic, intellectually unsatisfying, and devoid of the principles that I prized so much in the physical sciences were really quite unjustified. In fact, it was clear that biology was poised on the threshold of a remarkable revolution, as it became possible to analyze the information content of DNA and the nature of the genetic code.

After a somewhat prolonged and torturous route, including medical school, I resolved to pursue a career that combined clinical practice and research in medical genetics. I eventually got a "real job" as a junior faculty member at the University of Michigan, splitting my time between organizing a research program in human genetics and seeing patients with genetic diseases. Often frustrated by the lack of information available on the fundamental basis of many genetic diseases, I was particularly drawn to those disorders where the action of a single mutant gene caused a great deal of sickness and distress, but where the cell biology had not been worked out sufficiently to allow a direct identification of the responsible gene. Cystic fibrosis (CF) was a particularly puzzling and heart-breaking example of this situation. In collaboration with another researcher in Toronto, my research group began an intense effort to try to identify the CF gene by mapping it to the proper chromosome, and then narrowing down the region where the gene must be until finally the correct candidate was identified. This process, known as positional cloning, had never really succeeded for a problem of this complexity, so it was both gratifying and sobering to participate in such an effort. Many times we thought the problem was just too difficult and were tempted to give up. But finally the strategy worked, and the successful cloning of the cystic fibrosis gene in 1989 made better diagnosis available and opened new doors to the design of therapies, including gene therapy.

After this, similar efforts in which my research group participated yielded up the genes for other puzzling human genetic disorders, including neurofibromatosis (often erroneously referred to as the Elephant Man disease) and Huntington disease. At about the same time an international effort to map and sequence all of the human DNA, known as the Human Genome Project, was getting under way. This would make it possible to find *all* of the genes responsible for human disease, and so I was happy to have the chance to participate by setting up a Human Genome Center at the University of Michigan. When the first U.S. director of the Human Genome Project, Dr. James Watson (the same Watson who with Francis Crick discovered the structure of DNA in 1953) resigned, I was asked to take on this role. Since 1993 I have been director of the National Center for Human Genome Research of the National Institutes of Health, which is the lead agency in the United States responsible for this ambitious and historic effort to map and sequence all the human DNA by the year 2005. So far the project is doing very well—actually running ahead of schedule and under budget.

While my own career path was far from focused, I can now look back on all the experiences that I had in science and medicine and see how they have helped prepare me for overseeing this remarkable project.

*Francis Sellers Collins has been the director of the National Center for Human Genome Research of the National Institutes of Health since 1993. In addition to numerous honors and awards, certifications, and society memberships, he is an associate editor of a number of scientific journals in the area of genetics. He received a Ph.D. from Yale University and an M.D. with honors from the University of North Carolina School of Medicine, Chapel Hill.*

This will be done by dividing a chromosome into a number of segments, and identifying clones that cover one segment at a time.

4. Sequencing the genome. The final stage of the Genome Project will be determining the exact order of each nucleotide in each chromosome. The overlapping clones from stage 3 will be used as a starting point. The result will be a catalog of the more than three billion base pairs in the human genome. This part of the project will probably be time consuming, possibly requiring 5 to 7 years to complete.

The implications of the project are far reaching, and will have an impact on clinical medicine, genetic counseling, and treatment for the more than 4000 genetic disorders that affect humans. Information from the project is also expected to have an impact on the identification of genes for polygenic traits such as adult-onset diabetes, some forms of cancer, mental illnesses such as manic depression and schizophrenia, high blood pressure, and for conditions where familial tendencies interact with one or more environmental factors.

The Human Genome Project has raised a number of related legal, ethical, and moral issues, including the use of and access to information about an individual's genetic status or predisposition. Within the project, a program has been established to identify and discuss such issues. This program, called ELSI (Ethical, Legal and Social Implications), draws on the expertise of scientists, lawyers, ethicists, philosophers, and others to consider how the information generated by the Human Genome Project impacts individuals and society. It also examines the possible uses of information and techniques generated by the project, and develops public policy options to ensure that the information is used for the benefit of individuals and society.

## GENE TRANSFER TECHNOLOGY HAS MANY APPLICATIONS

Originally, recombinant DNA technology was used to transfer foreign genes into bacterial cells as a way of cloning large amounts of a specific gene or DNA sequence for further study. Gene transfer is not limited to the use of bacteria as host cells; it is also possible to transfer genes between higher organisms. In this section, we review some of the current applications of gene transfer technology.

### Proteins Can Be Manufactured by Recombinant DNA Technology

One of the first commercial applications of recombinant DNA technology was the production of gene products (in the form of proteins) for use in the therapeutic treatment of human disease. Previously, such proteins were collected from animals, pooled blood samples, or even human cadavers. Most of these methods produce only a limited amount of therapeutic proteins, and specific risks are associated with the use of such products. For example, individuals afflicted with hemophilia, an X-linked recessive disorder, are unable to manufacture a clotting factor and, hence, suffer episodes of uncontrollable bleeding. Before blood testing became routine, many hemophiliacs treated with clotting factor prepared in this way were infected with the human immunodeficiency virus (HIV).

In humans, an enzyme deficiency is associated with heritable forms of emphysema, a progressive and fatal respiratory disorder. The defective enzyme is alpha-1-antitrypsin, and it can be produced for use in treating emphysema by recombinant DNA. The human alpha-1-antitrypsin gene has been cloned into a vector next to a DNA sequence that regulates expression of milk proteins. Vectors carrying this gene were microinjected into fertilized sheep eggs, and the eggs were implanted into foster mothers (✦ Figure 9.13). The resulting transgenic sheep developed normally, and the females produce milk that contains up to one-third of a pound of alpha-1-antitrypsin per gallon of milk. This method is still under development, but it is not hard to imagine that a small herd of sheep could supply the world's need for this protein.

The use of recombinant DNA technology to manufacture specific proteins such as insulin, clotting factors, and alpha-1-antitrypsin ensures a supply of a product free from contamination by disease-causing agents. Human insulin derived from a cloned gene inserted into bacteria was first marketed in 1982 and is used by many diabetics. When your cat is inoculated against feline leukemia virus, it receives a vaccine produced by recombinant DNA technology; farmers routinely use recombinant DNA derived vaccines to prevent diseases in hogs. ✦ Table 9.1 lists some genetically engineered products now available or in testing.

The use of such proteins is not without controversy, however. There is considerable disagreement about the use of recombinant DNA-derived bovine growth hormone to boost milk production in dairy cows. The use of this hormone was approved by the U.S. Food and Drug Administration in February

✦ **FIGURE 9.13**

**Transgenic sheep carrying the human gene for alpha-1-antitrypsin secrete the gene product into their milk.**

| TABLE 9.1 SOME PRODUCTS MADE BY RECOMBINANT DNA TECHNOLOGY | |
|---|---|
| **PRODUCT** | **USE** |
| Atrial natriuetic factor | Treatment for hypertension, heart failure |
| Bovine growth hormone | Improve milk production in dairy cows |
| Cellulase | Break down cellulose in animal feed |
| Colony stimulating factor | Treatment for leukemia |
| Epidermal growth factor | Treatment of burns, improve survival of skin grafts |
| Erythropoieten | Treatment for anemia |
| Hepatitis B vaccine | Prevent infection by hepatitis B virus |
| Human insulin | Treatment for diabetes |
| Human growth hormone | Treatment for some forms of dwarfism, other growth defects |
| Interferons (Alpha, gamma) | Treatment for cancer, viral infections |
| Interleukin-2 | Treatment for cancer |
| Superoxide dismutase | Improve survival of tissue transplants |
| Tissue plasminogen activator | Treatment of heart attacks |

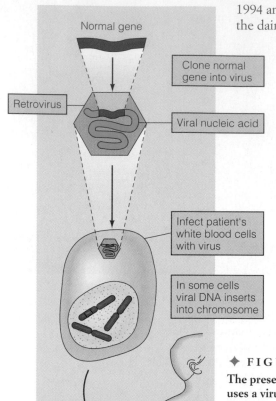

Normal gene

Clone normal gene into virus

Retrovirus

Viral nucleic acid

Infect patient's white blood cells with virus

In some cells viral DNA inserts into chromosome

Inject cells into patient

✦ **FIGURE 9.14**

**The present method of gene therapy uses a virus to insert a normal gene into the white blood cells of a patient. The normal gene becomes active, and the cells are reinserted into the affected individual, curing a genetic disorder. Future work will transfer genes into bone marrow cells for a one-time treatment.**

1994 and is now used on about 15% of the dairy cows in the United States. On the one hand, use of the hormone, known as BST, benefits farmers by increasing milk production by about 10%, resulting in lowered fixed costs. On the other hand, opponents charge that the technology is unnecessary, that cows given this hormone are at greater risk of disease, and that the risks of long-term use of this protein are unknown. Nevertheless, it is estimated that by the year 2000, 70% of the nation's milk supply will be produced with BST.

## Defective Genes Can Be Replaced Via Gene Therapy

The use of genetically engineered proteins (such as insulin) for therapy is an indirect method of treating genetic disorders. The development of gene transfer techniques has made it possible to use recombinant DNA techniques to treat human genetic disorders by inserting normal copies of genes in place of defective ones. A number of methods for transferring cloned genes into human cells are under investigation, including the use of viruses as vectors, chemical methods that aid transfer of DNA across the cell membrane, and physical methods such as microinjection or fusion of cells with vesicles carrying cloned DNA sequences.

Viral vectors, especially viruses known as retroviruses, are being currently used for human gene-transfer experiments. In these clinical trials, some non-essential genes are removed from the viral genome, allowing a human gene to be inserted (✦ Figure 9.14). Gene therapy began in 1990, when a human gene was inserted into a retrovirus, and then transferred into the white blood cells of a young girl suffering from a genetic disorder called **severe combined immunodeficiency** or **SCID.** Affected individuals have no functional immune system and are prone to infections, many of which can be fatal. The normal gene, inserted into her blood cells, encodes an enzyme that allows cells of the immune system to mature properly. As a result, she has a functional immune system and is leading a normal life.

Gene replacement therapy eventually will become a standard method for treating certain genetic disorders. One form of diabetes is currently treated by daily injection of a gene product, insulin. In the future, treatment by transfer of the insulin gene itself would represent a one-time treatment to correct this condition.

## New Plants and Animals Can Be Created by Gene Transfer

Gene transfer has been used for some time in agriculture as a way of improving crop plants and farm animals. Corn is a major cereal crops in the United States, with annual market value of about $22 billion dollars. Corn is used as animal feed, as a source of sweeteners for foods and beverages, and for the production of ethanol, which may become an alternative to gasoline.

Gunpowder

Gun

Plastic bullet

DNA-coated pellets

Plant cells

(b)

✦ **FIGURE 9.15**

**(a) Gene gun for transferring DNA into plant cells. (b) In the gene gun, plastic bullets are used to drive DNA-coated pellets into cells.**

(a)

To transfer genes into corn, cloned genes for a desired trait are blasted into cultured cells with a "gene gun" that forces the genes into the cell (✦ Figure 9.15). The cells are cultured to form a tissue mass that will grow into a corn plant. Field testing is under way on corn plants that have been transformed with a gene for herbicide resistance. If successful, these plants will be used in selective breeding experiments to transfer the gene to other strains of corn.

Before the end of this decade, strains of corn that have been genetically modified using recombinant DNA techniques will be planted, harvested, and sold commercially. This first generation of genetically engineered corn will probably be resistant to herbicides, insect pests, and plant diseases. The second generation of genetically engineered corn strains may have altered levels of protein, carbohydrate, and oils. Field testing of genetically modified strains of other crop plants including potatoes, tomatoes, and cotton are under way, and these will reach the market in a few years.

Genetically altered tomatoes, called Flavr-Savr tomatoes (✦ Figure 9.16), are currently on the market in many areas of the United States. They have been modified using recombinant DNA techniques to slow the softening that accompanies the ripening process. This allows tomatoes to be left on the vine longer, and makes refrigeration in shipping unnecessary. The result is a tomato with much more flavor.

Through genetic engineering, it is possible to customize the genetic makeup of crop plants (like the tomato) for specific purposes, such as delayed ripening, or the development of high protein strains for food uses, high oil content for sweeteners, etc. Manipulating organisms by genetic engineering achieves the same goals as selective breeding, but is more efficient and faster than traditional methods.

Genetic engineering of crop plants is accompanied by parallel developments in genetic alterations of domestic animals. Gene transfer technology has been used to transfer human and cow growth hormone genes into pigs in an attempt to develop leaner, faster growing hogs (✦ Figure 9.17). The genes were transferred by injection into newly fertilized eggs that were implanted into a foster mother. Although the transgenic (carrying a transferred gene) pigs grow faster on a high protein diet, they have a number of problems including ulcers, arthritis, sterility, and premature death resulting from excessive production of growth hormone in their tissues. Further work is needed to control the expression of the transferred gene in order to eliminate these problems and pave the way for the appearance of genetically engineered pork in the meat case at the supermarket.

✦ FIGURE 9.16
Transgenic tomatoes are now on the market. The tomato has been genetically altered to slow fruit softening and improve flavor.

✦ FIGURE 9.17
This transgenic pig has received human growth hormone genes.

## GENE TRANSFER TECHNOLOGY RAISES ETHICAL QUESTIONS

The ability to transfer genes into human cells paves the way for a new method of treating genetic disorders, and also opens the possibility that we can direct the evolution of our species by modifying the genetic makeup of individuals and their offspring. The gene therapy now being used involves somatic cells as target for genes. Genes can also be transferred to germ cells, the parent cells of gametes. Transfer of genes into these cells will alter the genetic makeup of future generations. There is serious disagreement about whether germ line therapy is ethical. A related question is whether it is ethical to create new forms of plants and animals by gene transfer.

There is also the possibility that gene therapy will be used as a form of self-improvement rather than a treatment for a genetic disorder. Should children who are very short receive growth hormones produced by recombinant DNA, or even a growth hormone gene to allow them to be taller adults? If it is acceptable to treat abnormally short children, how about treating children of normal height to enhance their chances of becoming professional basketball or volleyball players?

The current guidelines of the U.S. Food and Drug Administration do not require identifying labels on food produced by recombinant DNA technology. Should such food be labeled and, if so, why? On one hand, it can be argued that food products have been genetically manipulated for thousands of years and that gene transfer is simply an extension of past practices. On the other hand, it can be claimed that consumers have a need and a right to know that food products have been altered by gene transfer.

The legal, moral, and ethical implications of gene transfer technology need to be considered carefully. As outlined earlier, some of these issues are being addressed by the Human Genome Project, but in other areas, discussion, education, and policy formation lag far behind the technology.

## SUMMARY

1. Genetic manipulation through selective breeding has been used to produce strains of plants and animals that carry a desired set of genetic traits. Although this method is effective, it is slow and the results are not always predictable. Transferring a gene from one strain to another by selective breeding can require several generations and many genetic crosses.

2. Selective breeding is being replaced by recombinant DNA technology.

3. Recombinant DNA technology developed from the discovery of a series of enzymes called restriction endonucleases that recognize and cleave DNA at specific base sequences. Use of these enzymes coupled with vector molecules such as plasmids or viruses permits DNA molecules to be transferred to host cells and cloned into thousands of copies per cell. The cloned DNA segments can be recovered and used in gene mapping, chromosome mapping, disease diagnosis, and the production of drugs, vaccines, and agricultural and industrial chemicals.

4. Although the commercial applications of genetic technology are predicted to grow into an important industry, currently the most significant use of genetic engineering is the diagnosis and treatment of human genetic disorders and the commercial production of human gene products for therapeutic purposes.

5. An ever-growing list of genetic disorders can be diagnosed by RFLP analysis. Mutant genes and heterozygous carrier status can be detected in adults, children, fetuses, embryos, and even gametes. The first clinical trials involving the insertion of genes into humans have already started, and although some technical barriers remain, gene therapy will probably become a common form of medical treatment within the next decade.

6. Genetic engineering is also being used to transfer genes into crop plants, conferring resistance to herbicides, insect pests, and plant diseases. Gene transfers to alter the protein, carbohydrate, and oil content of cereal crops will dramatically alter farming practices in the next century. Experiments to produce genetically altered farm animals used in meat and milk production are under way, and the products from these experiments should reach consumers within the next decade.

amniocentesis
chorionic villi sampling (CVS)
clones
DNA ligase
genome
palindromes
plasmids

polymerase chain reaction (PCR)
prenatal testing
presymptomatic screening
recombinant DNA molecules
recombinant DNA technology
restriction enzymes

restriction fragment length polymorphism (RFLP)
selective breeding
severe combined immunodeficiency (SCID)
vectors

# QUESTIONS AND PROBLEMS

## SHORT ANSWER

1. What is recombinant DNA technology? What are some of its advantages and potential disadvantages?
2. What is the difference between selective breeding and genetic engineering?
3. Outline the techniques used in recombinant DNA technology.
4. Describe the method of producing DNA clones by means of the polymerase chain reaction (PCR).
5. The mapping of genes using recombinant DNA depends on what two things?
6. What are the four major objectives of the international Human Genome Project?
7. Using examples cited in this chapter, describe the potential advantages of recombinant DNA technology in combating or diagnosing genetic disorders.

## FILL IN THE BLANK

1. Double stranded DNA molecules that occur naturally and replicate within bacterial cells are called _____.
2. RFLPs are inherited in a _____ fashion and can serve as _____ _____ for specific chromosomes.
3. The technique of using recombinant DNA to directly treat human genetic diseases by replacing defective genes with copies of normal genes is called _____.
4. A _____ is a special type of RNA virus that carries an enzyme enabling it to produce complementary strands of _____ on the RNA template.
5. Gene transfer techniques have been successfully used in the treatment of _____,

which wipes out the immune system of affected individuals.

6. The cloning of animals has been possible for many years, and involves the transplantation of entire gene sets, in the form of _____, rather than single _____.

## MATCHING

1. restriction enzymes
2. cloning vehicles
3. Huntington disease
4. amplification of DNA fragments
5. BST
6. heritable differences in DNA fragments
7. missing/defective protein in cystic fibrosis

_____ synthetic growth hormone
_____ DNA cutting enzymes
_____ RFLP
_____ vectors
_____ gene located on chromosome #4
_____ PCR
_____ CFTCR

## SCIENCE AND SOCIETY

1. Fetal cell transplantation is a process where healthy cells of a fetal organ or organ system are transplanted into an adult whose own organ does not function properly. One example of this is diabetes mellitus type I. Insulin injections are able to treat this disorder, although they offer no cure. Transplanting healthy fetal pancreatic cells into diabetics who can not produce insulin on their own may now serve as a cure for this condition. Fetal cell transplantation offers promise in other areas of medicine as well. In 1992, Congress passed a law authorizing funding of fetal

cell transplant research, but it was vetoed by President Bush. After taking office, President Clinton signed an order allowing federal funds for fetal transplantation research. Despite Federal approval, this technique continues to raise considerable controversy in the United States. Where do the tissues come from for fetal cell transplantation? Do you feel this technique is ethical? Why or why not? What other medical conditions can you think of that might benefit from fetal cell transplantation?

2. Biological research is frequently conducted and strongly dependent upon animal studies. These studies have provided knowledge that allows health care providers to improve the quality of life for humans and animals, investigate the effects of new treatments and drugs, and ease pain and suffering due to diseases. Some groups, such as In Defense of Animals, question the usefulness of animal research for studying human disease and oppose certain practices performed on research animals. The Foundation for Biomedical Research focuses on informing the public of the importance of humane animal research and the benefits of this research. What ethical issues do you feel are relevant for animal research studies? What are the advantages and disadvantages of using animals for research? Why do you think opponents to the use of animal research feel that animals are an inadequate model for studying human disease?

# III

# EVOLUTION OF LIVING SYSTEMS

THESE ANIMALS AT AN AFRICAN WATERHOLE ARE THE PRODUCT OF MANY GENERATIONS OF NATURAL SELECTION.

# 10

# EVOLUTION: NATURAL SELECTION AND GENETICS

OPENING IMAGE

The Komodo dragon is one of many
endangered species.

On a small island in the middle of Indonesia, an Englishman lay in a hut
shivering with a malaria fever. It was February, and even though the temperature was near 88°F, he was wrapped in blankets. While shivering, he began to think about how species evolved. Something led him to think about an essay he read several years before on human populations, and the effects of disease, accidents, and famine on the growth of populations. It occurred to him that in the case of animals, those individuals lost from the population in the so-called "struggle for existence," must, on average, have a lower capacity for survival. Suddenly, the idea of survival of the fittest flashed into his mind. Over the long run, this process would necessarily improve the population, by allowing more of the best adapted individuals to survive and reproduce. He later wrote: "The more I thought it over the more I became convinced that I had at long last found the long-sought-for law of nature that solved the problem of the origin of species."

In a burst of effort during the next two hours, he worked out the entire theory of the origin of species by natural selection. Later that evening, after his shivering fit had passed, he quickly sketched out a draft of his idea, and during the next two nights, wrote it out in full and sent it off to England by the next packet ship.

The man whose burst of intellectual effort provided this insight was Alfred Russel Wallace. The year was 1858. The paper he dispatched to England was sent to Charles Darwin. Darwin and his colleagues arranged for Wallace's paper to be read at a meeting

*of the Linnean Society of London along with some notes and an abstract of a letter by Darwin on the same subject.*

*Immediately thereafter, Darwin set about writing what he termed an abstract of a much longer work. This so-called abstract, published the following year, was the monumental book* The Origin of Species, *published in 1859. Within a span of less than two years, the life's work of these two men created an intellectual flux that proposed a mechanism by which evolutionary changes take place. This idea is one of the greatest achievements in the history of the natural sciences and forms one of the cornerstones of modern biology.*

*This chapter explores the background to the theory of evolution, the travels and thoughts of these two men, and the role of natural selection in bringing about evolutionary change.*

##  EVOLUTION AS AN IDEA

Although Greek philosophers, among others in the ancient world, speculated on the idea that different kinds of organisms might have arisen from ancestral forms, serious inquiry into the progressive development of animal and plant species from preexisting forms began only in the eighteenth century. The intellectual movement called the Enlightenment was sweeping across Europe, encouraging rational analysis of natural and social phenomena, and emphasizing the role of learning in understanding the world. This new attitude provided an atmosphere in which many unique and radical ideas were advanced, debated, and either adopted or discarded. Several of these ideas provide the background for the nineteenth century synthesis of the concept of organic evolution.

### The Natural Sciences and the Idea of Evolution

One of the important eighteenth-century ideas that helped the natural sciences emerge as a scientific discipline was the use of **taxonomy** as a way of cataloging the diversity of life-forms. Carl von Linné, who called himself by the Latin version of his name, Carolus Linnaeus, developed a method of classification by which all plants and animals could be named in an orderly fashion. This system, developed in the early 1700s, is called the **binomial system of nomenclature** (✦ Figure 10.1). It is useful in naming all

✦ **FIGURE 10.1**

**In the binomial system of nomenclature, related plants and animals are grouped according to their evolutionary relationships. The tiger (*Panthera tigris*) and lion (*Panthera leo*) are related species, and the housecat (*Felis domestica*) and the jaguar (*Felis panthera*) are related species.**

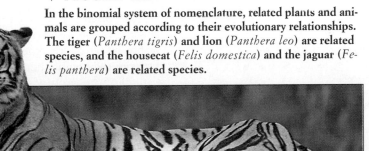

✦ **FIGURE 10.2**

**In the eighteenth century, naturalists cataloged and named thousands of species. Their descriptions were often accompanied by elaborate engravings. These three engravings are taken from a work by Buffon.**

known organisms in a simple and unambiguous way, and can also be used to name newly discovered organisms.

His contemporary, Comte Georges-Louis Leclerc de Buffon, devoted decades to the writing of a natural history book (it eventually expanded to 44 volumes) that described all known plants and ani-

mals in detail, and included meticulous engravings to accompany the descriptions (✦ Figure 10.2). Although Linneaus did not speculate on the origin of the species he described, Buffon sometimes argued that species evolved from other, preexisting species, but at other times he advocated the idea that each species was the separate creation of a Supreme Being.

The origin of species through evolutionary change was an idea advanced several times in the eighteenth and early nineteenth centuries. Erasmus Darwin, the grandfather of Charles Darwin, was a naturalist who wrote widely on zoology and botany. Using his observations of vestigial organs (structures that are reduced and nonfunctional) and the structural changes that occur during development before birth, Erasmus Darwin proposed that present-day species arose through changes in preexisting species.

In 1809, Jean-Baptiste Lamarck concluded that organisms of higher complexity had evolved from preexisting, less complex organisms. Lamarck advanced the idea that individuals can acquire traits during their lifetime that adapt them to their environment. Once acquired, these traits can be passed on to offspring, resulting in changes from generation to generation that eventually produce new species. In support of his idea, Lamarck suggested that ancestral giraffes had short necks, but to compete successfully for food, individuals in each generation stretched their necks to reach leaves higher up in trees (✦ Figure 10.3). The elongated neck

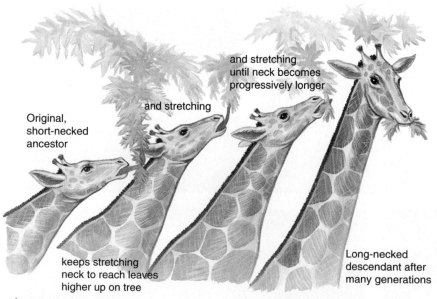

and stretching until neck becomes progressively longer

and stretching

Original, short-necked ancestor

keeps stretching neck to reach leaves higher up on tree

Long-necked descendant after many generations

✦ **FIGURE 10.3**

**In Lamarck's theory of species formation, ancestral, short-necked giraffes stretched their necks to reach leaves higher up in trees. Their longer necks were an acquired characteristic which was passed on to descendants. Gradually, giraffes became long necked.**

represented an acquired trait passed on to the next generation, resulting over time in giraffes with long necks.

Although it is easy to ridicule this idea today, remember that in Lamarck's time, little was known about the mechanism of heredity. It was the work of Gregor Mendel in the 1860s and the discovery of meiosis in the 1880s that enables us to discount Lamarck's idea of how species evolve. Lamarck did propose the important idea that species of plants and animals arise from preexisting species. Although the later ideas of Charles Darwin and Alfred Russel Wallace, coupled with knowledge about genetics, displaced Lamarckian notions about the mechanism of evolution, sporadic efforts to substantiate Lamarck's proposals continued well into this century, most notably in the Soviet Union under the guidance of the Soviet plant breeder Troffim Lysenko.

## Geology, the Age of the Earth, and Evolution

**Geological time** is the span of time that has passed since the formation of the Earth and its physical structures. In the mid-seventeenth century, James Ussher calculated the age of the Earth by adding the life spans of individuals listed in the genealogies contained in the book of Genesis in the Bible. By doing so, he concluded that the Earth was created on October 4, 4004 B.C.

In the late eighteenth and early nineteenth centuries, the age of the Earth was calculated in various ways using approaches fostered by the Enlightenment. Geologists attempted to measure the rate at which sedimentary rocks were formed, and others calculated the age of the oceans based on the rate at which dissolved salts were carried into the oceans by rivers (✦ Figure 10.4).

Errors in assumptions or in measurements caused these calculations to be imprecise (estimates ranged from several million years to more than one billion years). Using a variety of methods, we now calculate the Earth to be about 4.6 billion years old. These early attempts to measure the age of the Earth changed the way naturalists and scientists thought about such problems. The effect was that scientific methods (part of the heritage of the Enlightenment) were developed to discover the underlying principles of geology, and recognition that the Earth is much older than 6000 years.

The second important advance in geology that paved the way for the idea of species evolution is the concept of **uniformitarianism.** This idea, proposed by James Hutton in 1795, holds that geological processes observable at the present time, such as erosion, wave action, and sediment movement,

✦ **FIGURE 10.4**

Nineteenth-century geologists reasoned that if the oceans originated shortly after the Earth was formed, they could be used to date the age of the Earth. If the ocean water was originally fresh, and the salt washed into the oceans from rivers, then by measuring the amount of salt in river water and comparing it to the amount of salt in ocean water, the age of the Earth could be calculated. This calculation placed the age of the Earth at 90 million years, considerably younger than the 4.6 billion years now accepted, but the earlier calculation did serve to expand the concept of geological time.

are responsible for the geological changes of the past. In other words, the physical processes that shape and change the geological landscape have remained uniform over time. This idea was taken up by Charles Lyell, a geologist, mentor, and close friend of Charles Darwin. Lyell refined and restated the idea of uniformitarianism to include the idea that slight, almost undetectable changes taking place over long periods of time can have large-scale consequences. This linked the idea of geological time with uniformitarianism. He wrote a book, titled *Principles of Geology,* that had a major influence on both Darwin and Wallace.

## NATURAL SELECTION IS THE MECHANISM OF EVOLUTION

Acceptance of the idea that species are not fixed but can change or evolve was limited by the fact that no one could explain how such changes took place or what forces of nature help bring about the evolution of species. Before this problem was solved, both Darwin and Wallace traveled extensively, observing plants and animals in their natural settings. They developed ideas based on these observations that set the stage for the concept of natural selection.

**✦ FIGURE 10.5**

**An engraving of the HMS** *Beagle* **on its voyage.**

than 25,000 miles (✦ Figure 10.6). As naturalist aboard the ship, Darwin spent time ashore at each landing, collecting plants, animals, and fossils and carefully noting their geographic distribution. He read Lyell's *Principles of Geology* while on the voyage, and made extensive geological observations at the many locations he visited.

While on the voyage, Darwin was impressed with the relationship between the distribution of living forms and fossils of similar species in various locations in South America (✦ Figure 10.7). He also visited many island groups, including the Cape Verdes, the Falklands, the Galapagos, Keeling Island, and the Maldives.

Darwin returned home in 1836 and the next year moved to a country house to begin work on the material he had collected. As he says in the introduction to *The Origin of Species,*

> When on board HMS 'Beagle' as naturalist, I was much struck with certain facts in the distribution of the organic beings inhabiting South America and in the geological relations of the present to the past inhabitants of the continent. These facts, as will be seen in the latter chapters of this volume, seemed to throw some light on the origin of species—that mystery of mysteries, as it has been called by one of our greatest philosophers. On my return home, it occurred to me, in 1837 that something could be made out on this question by patiently accumulating and reflecting on all sort of facts which could possibly have any bearing on it.

### Darwin's Voyage on the *Beagle*

At the age of 22, Charles Darwin was a graduate of Christ's College, Cambridge, with a degree in theology. While a student, Darwin was an avid naturalist and collector. At Cambridge, he attended the lectures of John Henslow, a professor of botany. Henslow befriended Darwin and recommended him for the position of naturalist aboard the *Beagle,* a British ship that was to undertake a surveying expedition in the Southern Hemisphere (✦ Figure 10.5). The *Beagle* sailed on December 27, 1831, on a voyage that lasted five years and covered more

**✦ FIGURE 10.6**

**The route of the** *Beagle* **during its 1831–1836 voyage.**

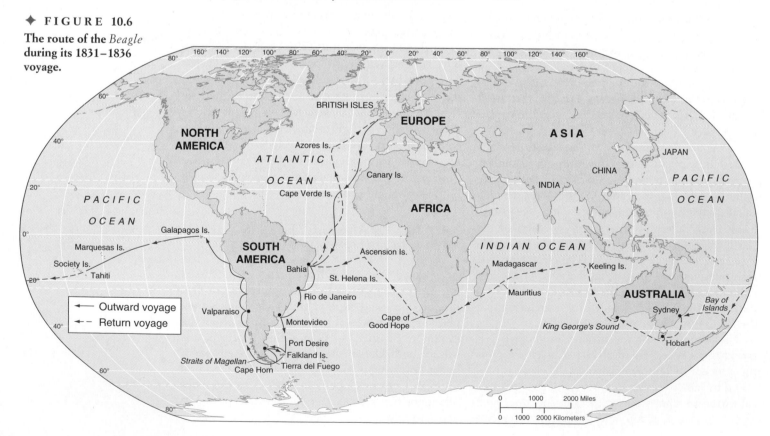

Darwin labored over this problem for the next 22 years before publishing his book *The Origin of Species*. In fact, as we will see later, he was not ready to publish even then, but was spurred to do so by the work of Alfred Russel Wallace.

## Wallace's Explorations in South America and Indonesia

Alfred Russel Wallace received his formal education in a one-room schoolhouse in a small English village. At the age of 13, he was apprenticed to his brother, a surveyor. In his spare time, he studied botany and collected plants. At age 21 he took a job as a teacher and continued his studies at the local library, reading books by naturalists and explorers. He became friends with another amateur naturalist, Henry Bates, and they decided to undertake a trip to the Amazon River basin, making their living collecting plants and animals and selling them through an agent in London.

During the next four years, Wallace explored the region along the Amazon River and its tributary, the Rio Negro, collecting and classifying animal and plant materials. His travels took him thousands of miles across the continent from the Atlantic Ocean to the foothills of the Andes Mountains. Although some of his collections were shipped back to England periodically, he took the bulk of his material with him aboard ship, which sailed for England in July 1852. Three weeks into the voyage, the ship caught fire and the crew and passengers were forced to abandon ship. After drifting for 10 days in lifeboats, Wallace and the others were rescued by another vessel. Unfortunately, Wallace's collection, the result of four years of painstaking work, was lost in the fire and sinking of the ship. He salvaged part of his journal and some notes, from which he wrote an account of his adventures,

*Travels on the Amazon and Rio Negro,* first published in 1853.

Undaunted by circumstances, Wallace left England in 1854 to study the natural history of Indonesia. He spent seven years among the islands of Malaysia and Indonesia, collecting animal and plant specimens, and recording his thoughts on the mechanism of species formation (✦ Figure 10.8). It was

✦ **FIGURE 10.7**

An armadillo (top) and a reconstruction of a fossil armadillo. Darwin was struck by the similarities between the two species and their similar geographic distribution.

✦ **FIGURE 10.8**

The route of Wallace's trip in Indonesia, which lasted from 1854 to 1862.

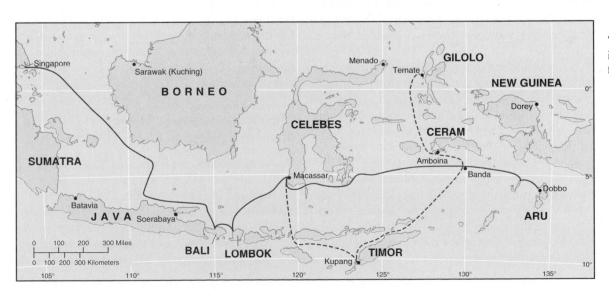

## *Nothing in Biology Makes Sense without Evolution*  MICHAEL ROSE

The best course I took in high school, maybe the best course I've ever taken, was comparative vertebrate anatomy. I loved taking apart the bodies, learning the names of each part, carefully removing material to see little holes and tiny nerves. But before the gore came the scientific theory: fossils, Lamarck, Darwin, Mendel. The theory animated the corpses, made them meaningful, gave them some sense. At the time, I was a big science fiction fan, a genre in which every story had to have a portentous meaning. Evolution is the deep meaning behind life, and without it biology becomes a lot of details.

When I started my doctoral studies, I was given the task of showing experimentally that evolution could make sense of biological aging. At first I was dismayed. Aging, throughout human history, has been a mystery. The people who talk or write most about it are charlatans, quacks, and hustlers. I feared that my career would be aborted by the combination of my failure to accomplish the task and my spattering with the mud of aging quackery. So far, my fears have proven erroneous.

Consider two genetic diseases, progeria and Huntington disease. It is now thought that each of these is caused by a specific mutation at a single copy of a normal human gene, one gene for each disease. Progeria strikes children between 5 and 10 years of age and is an extremely rare disorder. Huntington disease, on the other hand, strikes almost entirely after the age of 30 and affects thousands of individuals around the world.

Progeria strikes children, prevents their reproduction, and also completely precludes transmission of progeria into the next generation. With 100% success, natural selection screens out new mutations for progeria. With Huntington disease, the patients may already have children before they show any symptoms. Natural selection fails to screen the Huntington mutation out of the population because the gene has effects primarily at later ages, not earlier ages. The key is that *the force of natural selection falls with adult age.*

Genes that have effects on the health or development of young animals are sharply scrutinized by natural selection. This fundamental idea was first proposed by evolutionary biologists in the 1930s and 1940s. My role has been to test this idea. When this kind of reproductive pattern is imposed on laboratory populations of the fruit fly *Drosophila,* they evolve an increased life span over dozens of generations. This happens because we are artificially strengthening natural selection at later ages. Fruit flies that can't survive to reproduce in their middle age, in this experiment, are selected against. Thus natural selection alters the genetic basis of aging and increases the life span.

The flies that live longer are interesting beasts. They have a lot of physiological differences that can be related to their ability to live longer. They reproduce less when young, even when they have opportunity to do so. They resist stresses better, including starvation and desiccation. They move around more when they are older, whether walking or flying. They can reproduce more at later ages. Longer lived flies appear, for now, to be superior organisms—"superflies!"

When I present these superflies to audiences, people want to know if I will ever be able to do the same things for them. Literally, the answer is no. But there is the possibility of learning more about the genetics of postponed aging in fruit flies in the hope that we can apply our findings to humans. If this is ever done, it will transform the human life cycle, as aging becomes something that can be controlled—instead of merely endured.

*Michael Robertson Rose is a professor of evolutionary biology at the University of California, Irvine. He received B.S. and M.S. degrees from Queen's University, Ontario, Canada, and a Ph.D. from University of Sussex in England. He is a member of the editorial board of the* Journal of Evolutionary Biology *and the* Journal of Theoretical Biology. *In 1992 he won the President's Prize from the American Society of Naturalists.*

from his camp on the island of Ternate that Wallace dispatched a manuscript to Charles Darwin describing the role of natural selection in species formation.

Darwin and his colleagues arranged for the Wallace paper to be read at the July 1, 1858, meeting of the Linnean Society along with extracts from an unpublished manuscript and a letter on the same subject, written by Darwin. Wallace's paper, published in the *Journal of the Proceedings of the Linnean Society,* was the first to define the role of natural selection in species formation, and the third major paper published by Wallace on the topic of species formation. Galvanized to action by Wallace's work, Darwin plunged into writing *The Origin of Species,* which remains one of the most important and influential books ever written.

Wallace's trip to Indonesia remains a turning point in the development of evolutionary biology. On this trip he developed the idea of natural selection, started the science of biogeography, collected more than 125,000 specimens, and dabbled in social anthropology.

### The Wallace–Darwin Theory of Natural Selection

The journeys of both Darwin and Wallace were important in the development of their ideas about the mechanism of evolution. Darwin's ideas about the evolution of species were the *result* of his voyage, whereas pursuit of natural history specimens and a search for evolutionary mechanisms were the *reasons* for Wallace's travels. The idea that natural selection is the driving force behind evolution was the result of careful field observations on organisms and their environments made by both Wallace and Darwin, coupled with their readings in geology and the essay on populations by Malthus.

The Reverend Thomas Malthus' work, *The Essay on the Principle of Population as It Affects the Future Improvement of Society,* influenced both Darwin and Wallace in formulating the principle of natural selection. In this essay, first published in 1798, Malthus noted that populations grow in an exponential fashion, with the human population capable of doubling every 25 years. Resources to accommodate the population, such as living space and food are more limited and expand very slowly, if at all. Eventually, population growth outstrips the ability of the environment to support a growing population. If human population size is not controlled by measures such as delayed marriage and birth control, then factors such as disease, war, and pestilence would inevitably increase the death rate and lower the population size.

In adapting the ideas of Malthus about how limited resources affect populations, Wallace and Darwin concentrated their attention on those that lived and reproduced rather than those that perished. This provided a key part to their proposal about the role of selection in evolution.

The Wallace–Darwin theory on the role of natural selection in evolution can be summarized in a series of statements:

1. Among individuals of a species, many variations exist (✦ Figure 10.9). These might include differences in agility, size, and ability to obtain food. Such variations can be very small and might even seem insignificant to a casual observer. Some of this variation is heritable and is passed on to offspring.

2. If left unchecked, organisms tend to reproduce in an exponential fashion. More offspring are produced than can survive. This leads individuals of a species to compete for scarce resources, engaging in a struggle for existence.

3. In the struggle for existence, some individuals will be more successful than others, allowing them to survive to a reproductive age. Often these individuals have some small variation in phenotype that gives them an advantage.

4. Those individuals that survive to reproduce because of some phenotypic variation leave behind more offspring than those with less advantageous variations, eventually eliminating some variants.

5. Over time there can be heritable changes in the phenotype (and genotype) of a species, trans-

✦ **FIGURE 10.9**
**Members of a single species often show wide variations in phenotype.**

forming it into a new species, similar to, but distinct from, the original species it has replaced.

Both Wallace and Darwin proposed that formation of new species requires populations of organisms and long periods of time. The work of Lyell and others on geological time and the antiquity of life as seen in the fossil record provided a timescale over which these changes could occur. In spite of the fact that natural selection provided an explanation of *how* evolution occurs, neither Wallace nor Darwin could explain the origin of the variations that provide the raw material for selection, nor how such variations were maintained in a population. The answer to that question was not forthcoming until the principles of Mendelian genetics were applied to populations, and evolution was understood to involve changes in the genetic structure of a population over time. In the twentieth century, the concept of natural selection was united with genetics in a new view of the evolutionary process, called *neo-Darwinism.*

## LINKING NATURAL SELECTION TO GENETICS

The key to understanding the mechanism of evolution lies in knowing the factors that generate genetic variation among members of a species, and how those variations are acted on by natural selection. Without variation, natural selection has nothing on which to act. In this section, we explore how variation is produced, and in the next section, we show how natural selection brings about changes in the distribution of genetic variation within and between populations.

A **population** is a group of individuals living in the same geographical area and sharing a common

◆ FIGURE 10.10
Species live in distinct populations over a geographic range.

fined as any heritable change in DNA. Mutational changes can be as small as a single nucleotide substitution or can involve large-scale changes such as chromosome fusion or rearrangement or exchange of chromosome segments. Whether a given mutation is helpful, neutral, or harmful depends on circumstances, namely, how the mutation affects the survival and reproductive success of a given individual in a given environment. Because most organisms are generally well suited to their environments, many mutations are detrimental and make the individual less adapted to the environment. However, if the environment changes, mutations that were previously neutral or harmful can become beneficial, giving individuals carrying such mutations an advantage in the struggle for existence and reproduction. In this situation, the new environment allows a previously detrimental allele and its phenotype to spread through the population.

The rate of mutation varies among species, and even among genes in an individual. Many mutations are caused by errors in DNA replication and by environmental factors including radiation and chemicals (natural and synthetic).

Although mutation is the ultimate source of all genetic variation, mutation rates alone have only a minimal impact on the genetic variability found within a population. It is only when mutation is combined with factors that reshuffle the gene pool, including sex (meiosis and fertilization), selection, genetic drift, and migration, that large-scale effects occur.

gene pool (◆ Figure 10.10). The **gene pool** is the sum of all the genetic information carried by members of a population. Because of genetic differences, individuals within a population differ to a greater or lesser extent in the distribution of phenotypic traits. These traits might include such things as height, beak size, leaf arrangement, or any of the features that are part of an individual's phenotype. Several factors are involved in the generation and maintenance of phenotypic variation in a population, some of which are described in the following sections.

## New Alleles Are Generated by Mutation

All genetic variation within a population is ultimately generated by mutation. **Mutation** can be de-

## Sex Reshuffles the Gene Pool

Selection does not act on individual genes, but rather on individuals, each of which contains a

◆ FIGURE 10.11
Independent assortment and recombination shuffle the parental gene pool so that family members resemble each other, but are not identical.

unique combination of alleles in its genotype. Sexual reproduction increases variation by reshuffling the genetic information from parents into new combinations in the offspring (this topic can be reviewed in Chapter 5).

As a result of the reshuffling that takes place during meiosis and sexual reproduction, genetic diversity is generated. The result is that even though all offspring receive 50% of their genes from each parent, no two offspring (aside from identical twins) are genetically identical (✦ Figure 10.11). Other factors also alter the amount of variation in a population's gene pool. Among these are small gene pools and migration.

## Genetic Variation Can Be Introduced by Drift and Migration

If a gene pool is very small (as in small populations), the frequency of alleles can change from generation to generation as a result of chance alone. This phenomenon is called **genetic drift.** As an example, ✦ Figure 10.12 shows a population of two heterozygous individuals who produce only two offspring. Because three genotypes are possible (*AA, Aa,* and *aa*), either the *A* or the *a* allele can be eliminated by chance in one generation.

In the biological world, genetic drift is often accentuated by **founder effects.** When a population is started by one (a pregnant female) or a few individuals who randomly separate from a larger population, chance may dictate that allele frequencies in the new population will become very different from those in the original population. This difference in gene pools is called a founder effect. Many of the species of plants, insects, and birds found on islands or island groups (including those visited by Darwin) originated from one or a few individuals who were carried there by wind or ocean currents (✦ Figure 10.13).

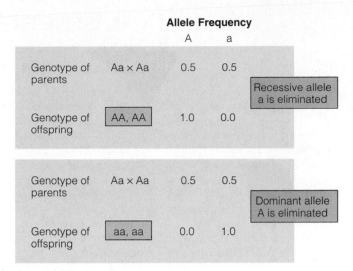

| | Allele Frequency | | |
| --- | --- | --- | --- |
| | | A | a |
| Genotype of parents | Aa × Aa | 0.5 | 0.5 |
| Genotype of offspring | AA, AA | 1.0 | 0.0 |

Recessive allele a is eliminated

| | | A | a |
| --- | --- | --- | --- |
| Genotype of parents | Aa × Aa | 0.5 | 0.5 |
| Genotype of offspring | aa, aa | 0.0 | 1.0 |

Dominant allele A is eliminated

✦ **FIGURE 10.12**

**When a breeding population is small, alleles can be eliminated by chance.**

Another way in which chance can affect a gene pool is through drastic short-term reductions in population size caused by natural disasters, disease, or predators. These reductions are called **bottlenecks.** By chance, the genetic information carried by the survivors may represent only a small fraction of the gene pool present in the original population (as in a founder effect). Even when the population increases to its previous size, a portion of its genetic variation remains lost (see Beyond the Basics: Cheetahs, Genetic Diversity, and Bottlenecks, p. 184).

The genetic structure of a population can also change when individuals migrate into or out of the population. Migration acts to break down genetic differences between neighboring populations. If individuals migrate between two neighboring populations, mutations that originate in either population will spread to the adjoining population, reducing genetic differences between the populations. In a sense, migration is similar to mutation in that it can serve to introduce new alleles into a population.

✦ **FIGURE 10.13**

**The finches on the Galapagos are thought to be descended from a small founding population.**

## Beyond the Basics

# CHEETAHS, GENETIC DIVERSITY, AND BOTTLENECKS

Selection acts on individuals in a population, and those with variations that give an advantage in survival and reproduction are said to be fitter than others in the population. Populations with small amounts of variation in their gene pool are at risk for long-term survival, because they are unable to respond to changes in selection, making them at greater risk of extinction.

A genetic survey started in 1983 of cheetah populations in Southern Africa revealed a complete lack of genetic variation at 47 loci tested. Further tests confirmed a general lack of heterozygosity in cheetah populations in the wild. The genetic condition of the cheetah raises several concerns about the potential for extinction of the species, conservation practices, and survival of populations during disease epidemics. One of the first questions raised is how did the cheetah population become so inbred?

Work using both nuclear and mitochondrial genetic markers supports the idea that a bottleneck is responsible for the lack of genetic diversity in this species. Calculations on the amount of diversity found in present-day populations and the rate of genetic change in the markers used is consistent with the idea that the species underwent a large reduction in population near the end of the last ice age. This event occurred in the late Pleistocene, about 10,000 years ago. At the same time, numerous species of large vertebrates became extinct on several continents. In this case, it appears that the cheetah came close to the brink of extinction, but that a small number of individuals survived. Present-day cheetahs are the descendants of these survivors.

The lack of genetic variation in the cheetah has been used to explain why there are so few cheetahs in wild populations. According to this idea, increased homozygosity would expose deleterious genes, which would raise mortality and make populations more susceptible to infectious diseases. Although the reasons for the decline in cheetah populations may be genetic, recent field research indicates there may be other reasons as well. In a study of breeding in wild cheetah populations, genetic defects, neonatal mortality, and defective reproductive physiology were ruled out as the reason for high mortality rates among juvenile cheetahs. Predation by lions and spotted hyenas on cheetah cubs was found to be a significant factor in the low density of the cheetah population.

These findings emphasize that genetic and environmental factors interact to determine population size and that selective factors need to be carefully identified.

---

Together, genetic drift and migration can have a dramatic impact on allele frequency in a population. Around 1800, a French nobleman and his wife, fleeing the aftermath of the French Revolution, settled on the island of Mauritius in the Indian Ocean. They brought with them a mutant allele for Huntington disease (HD), a progressive and fatal neuropsychiatric disorder inherited as a dominant trait. As of the early 1990s, at least 6 descendants of this couple were afflicted with HD and another 25 family members may be at risk for the disorder. Because of the small size of the European community on this island, the frequency of Huntington disease in this population is among the highest in the world.

### What Is Natural Selection?

In formulating their thoughts on the mechanism of evolution, Wallace and Darwin recognized that not all members of a given population have equal chances for survival or reproduction, given the competition for food and mates and the effects of predation and disease. By virtue of the small phenotypic variations present in the population, some individuals will be better adapted to the environment than others. These better adapted individuals have an increased chance of surviving to a reproductive age and of leaving more offspring than

those who are less well adapted. **Fitness** is a measure of an individual's ability to survive and reproduce. Those with the highest fitness are those most able to survive and reproduce. As a result, they make a larger contribution to the gene pool of the next generation than other, less fit individuals.

In time, these different rates of reproduction lead to changes in allele frequencies within the population. The process of differential reproduction of fitter genotypes is known as **natural selection.** Both Darwin and Wallace argued that natural selection is the primary mechanism leading to evolutionary divergence and the formation of new species. Note that the basis of selection is the differential survival and reproduction of the most adapted individuals, not the popularized notion of survival of the fittest. In the final analysis, it is the fact that the better adapted individuals survive and leave more offspring that is important.

### NATURAL SELECTION ACTS ON VARIATION PRESENT IN THE POPULATION

Because natural selection acts on individuals within a population, it can exert a range of effects on the genotypes and their related phenotypes present in the population. Selection can be classified as stabilizing, directional, or disruptive (✦ Figure 10.14).

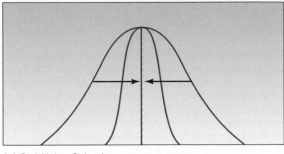

(a) Stabilizing Selection

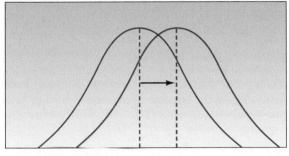

(b) Directional Selection

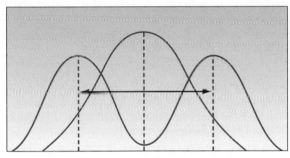

(c) Disruptive Selection

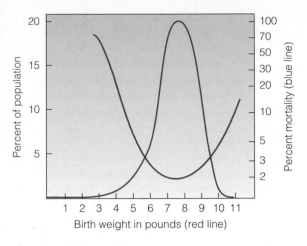

Birth weight in pounds (red line)

✦ **FIGURE 10.15**

**Distribution of birth weight (red) and survival rate (blue) for 13,730 infants. Stabilizing selection favors a birth weight between 7 and 8 pounds, and selects against phenotypes at either extreme.**

✦ **FIGURE 10.14**

**(a) Stabilizing selection narrows the phenotypic range and favors an intermediate phenotype. (b) Directional selection shifts the phenotype toward one extreme. (c) Disruptive selection favors phenotypes at both ends of a phenotypic range.**

## Stabilizing Selection Narrows the Range of Phenotypes

**Stabilizing selection** tends to favor genotypic combinations that produce an intermediate phenotype. An example is the relationship between birth weight and survival in humans. ✦ Figure 10.15 shows the distribution of birth weights for some 13,000 children and the rate of infant mortality at four weeks of age. The survival curve shows that infants weighing significantly less or significantly more than 7 to 7.5 pounds have higher rates of infant mortality. In this case, selection works against phenotypes at both

extremes, and promotes survival of genotypes that produce birth weights between 7 and 8 pounds.

## Directional Selection Favors an Extreme Phenotype

**Directional selection** tends to favor phenotypes at one extreme of the phenotypic range. Insecticide resistance is an example of directional selection. DDT was developed as an insecticide between 1939–1945. It was effective on a large number of insect species, and its use was so widespread that it was sprayed over cities from airplanes, applied to clothing, and even incorporated into paint. After several years of use, DDT lost its effectiveness on houseflies. Subsequent investigations determined that DDT resistance is a genetic trait in houseflies. The presence of DDT in the environment acted as a selective agent, leading to the differential survival and reproduction of DDT-resistant flies, eventually rendering the insecticide useless.

Another well-documented case of directional selection involves the peppered moth (*Biston betularia*). Before the Industrial Revolution, a light-gray speckled form of the moth was common in the English countryside, its wings peppered with black spots and lines. During the daytime, the moths rest on tree trunks, where they are preyed on by birds. The light-colored lichen and moss-covered trunks provide camouflage for the light-gray forms of the moth. A rare form, with a dark gray body and black wings was first reported in 1848, near the city of Manchester. The black moths were much more visible on tree trunks (◆ Figure 10.16) and were frequently eaten by birds.

As England industrialized, soot and deposits from factories killed the lichens and mosses on tree trunks, making the white moths more visible to predators. But this same darkening of the landscape offered more protection for the dark forms (Figure 10.16), making them better adapted to the new environment. By 1893 (a 50-year period), the black form of the moth, which originally accounted for less than 1% of the population, became the predominant type, often accounting for more than 98% of all specimens captured. In rural areas where there was no industrialization and no soot deposited on the landscape, the black moths remained rare.

Body and wing color in the moth are genetically controlled phenotypic traits, and predation by birds was the agent of selection. Before the Industrial Revolution, the black form was selected against, and remained rare. In regions where factories polluted the vegetation, the selection was reversed. The black form was better adapted, and the white form was selected against. It has been calculated that in such an environment, the selective advantage of black moths was at least 50%. That is, black moths had a 50% higher chance than white moths of passing their genes on to the next generation.

Changes in the genetic structure of a population brought about by selection are not necessarily permanent. Following the enactment of strict pollution laws beginning in the early 1950s, tree trunks near industrial cities have been repopulated with lichens, providing a selective advantage for the lighter, speckled moths. As a result, the frequency of the light moths is increasing and the frequency of the dark moths is decreasing. In other words, as environmental conditions change, selection acts to favor those genotype/phenotype combinations that are fittest for the new conditions.

## Disruptive Selection Favors Individuals at Both Extremes of a Phenotypic Range

**Disruptive selection** favors individuals at *both* extremes of a phenotypic range. This results in a discontinuity in the pattern of variation present in a population, which is often represented as two distinct phenotypes or **morphs.** One example involves certain African butterflies. Individuals of several species of these butterflies are protected from predation by birds because they secrete a chemical that makes them extremely distasteful. After repeated encounters with these butterflies, birds learn to leave them alone. These butterflies are distinguished from other species by bright coloration and elongated wings, bodies, and antennae. They do not blend in with the environment; their bright coloration, signaling their unsuitability as food, is their best protection.

In the African swallowtail butterfly, *Papilio dardanus,* disruptive selection has produced two very different phenotypes (◆ Figure 10.17). Each phenotype resembles or mimics a different species of distasteful butterfly. In this way, each morph or phenotypic variant gains protection from predators, although it remains quite edible. Although it may seem odd to mimic two different species in-

*Sidebar*

**POPULATION GENETICS OF SICKLE CELL GENES**

Sickle cell anemia is a genetic disorder caused by an alteration in the coded message for hemoglobin which changes the normal allele (HbA) to a mutant form (HbS). Individuals homozygous for sickle cell often die at younger ages (median age approximately 45.6 years).

The high frequency of heterozygotes in Africa indicates that they have a competitive edge over homozygotes in certain environments. West Africa is an area where malaria has served as an agent of selection. Resistance to malaria is about 25% greater in heterozygotes than in homozygous individuals.

In the United States, the gene for hemoglobin S is decreasing due to early screening and testing of those at risk and because of the lack of malaria as a selective agent.

◆ **FIGURE 10.16**

**Directional selection in moths. (a) On light-colored backgrounds, the speckled form of the peppered moth is less visible to predators than the dark-colored variant. (b) On trees covered with soot and industrial pollution, the dark-colored form is less visible.**

(a)

(b)

stead of one, there is a selective advantage in this scheme. To be successful, the number of edible mimics must be far fewer in number than the inedible butterflies. If the number of mimics is large, birds will not learn to avoid the distasteful species or have their avoidance reinforced. As a result, if *P. dardanus* mimics two different species, more members of the mimic species are protected from predation. In this case, disruptive selection has caused phenotypic extremes to develop as mimics of two different model species. The genetics of this mimicry are somewhat complex, but nevertheless, selection is acting to produce two different genotype/phenotype combinations.

The cumulative action of natural selection over time causes a population of organisms to become better adapted to its environment. As environmental conditions change, previously insignificant variations and/or genotypes can become essential to the survival of the population. As populations diverge from one another under the force of natural selection, they may form similar but related species. The question is, when do divergent populations of related organisms reach the stage that enables us to call them separate species? The answer to that question depends in part on how species are defined. In general, when populations are no longer able to interbreed (exchange genetic information), they are said to be separate species. In the following section, we discuss the stages of divergence that accompany the formation of species.

(a)

(b)

(c)

(d)

✦ **FIGURE 10.17**

**Disruptive selection in the butterfly** *Papilio dardanus.* **(a) A noxious butterfly,** *Danaus chrysippus.* **(b) A female** *P. dardanus,* **which mimics** *D. chrysippus.* **(c) A noxius butterfly,** *Amauris niavius dominicanus.* **(d) A female** *P. dardanus* **that mimics** *A. niavius.* **The two butterflies on the right (b and d) are females of the same species, with very different phenotypes produced by disruptive selection.**

## NATURAL SELECTION AND THE FORMATION OF SPECIES

As outlined in earlier sections, forces such as mutation, genetic drift, gene flow, and selection operate to bring about changes in the gene pool of a population. For gene pools to diverge, different populations of a species must encounter different environmental conditions and be subjected to selection. When populations occupy subenvironments, they will diverge as natural selection acts to adapt them to these subenvironments. Eventually, these differentiated populations can form different **races, subspecies,** and eventually different species. To determine when this event has occurred, we need to define the term species.

### What Is a Species?

A **species** can be defined as one or more populations of interbreeding or potentially interbreeding organisms that are reproductively isolated in nature from all other organisms. Note that the definition depends on the separation of gene pools. As long as two populations share a common gene pool, they belong to a single species; only when they are no longer able to interbreed and exchange genetic information are they defined as two separate species. This division of a single gene pool into two or more separate gene pools can be accompanied by obvious changes in morphology or physiology, but these are not necessary components of speciation.

As mutation, genetic drift, and selection operate on the gene pool of a population, separation from the gene pools of other populations will be reflected in the appearance of new alleles, changes in allele frequency, or larger scale events such as new chromosome arrangements. This process, called **genetic divergence,** has several possible outcomes. If the barriers to gene flow between the populations are eliminated, the populations can merge into a single gene pool. On the other hand, the gene pools of separated populations may have diverged to the point where individuals from these populations cannot interbreed when brought back together, indicating that speciation has occurred. To accomplish this degree of divergence, populations must usually be isolated in some fashion.

## Geographical Isolation and Species Formation

As mentioned populations begin to diverge when gene flow between them is restricted. The first step in this process is often geographic isolation. Isolation can occur rapidly, as when earthquakes cause rivers to change courses or volcanic eruptions divide a habitat, or more slowly, as when islands form and disappear or as continental drift separates landmasses. As isolated populations diverge through changes in their gene pools, they acquire a set of new biological or behavioral characteristics. Some of these characteristics can reduce or prevent potential interbreeding if secondary contact is made with other populations. When these characteristics, called **reproductive isolating mechanisms,** make their appearance, they reduce or prevent fertilization or give rise to sterile or nonviable hybrids in matings with other populations.

Several types of isolating mechanisms are possible, all of which prevent gene flow between populations. Populations may become isolated by changes in the complex rituals of courtship that precede matings in many animals. In two closely related species of wolf spiders (✦ Figure 10.18), males from one species use a series of foot taps in the mating ritual. If he is courting a female of his own species, he often is successful in mating, but if he is courting a female of a related species, courtship is broken off, and mating does not take place.

(a)

(b)

**✦ FIGURE 10.19**

**(a)** *Bufo americanus* **and (b)** *Bufo fowleri,* **two species who live in overlapping geographic areas, but remain isolated because they mate at different times of the year.**

**✦ FIGURE 10.18**

**Wolf spider. Mating rituals in this species isolate it from closely related species.**

Seasonal or temporal differences in breeding patterns can also serve as isolating mechanisms. Two species of toads, *Bufo americanus* and *Bufo fowleri* (✦ Figure 10.19), will interbreed in the laboratory, but in nature, *B. americanus* breeds early in the spring, and *B. fowleri* breeds later in the season, preventing interbreeding. In cases where breeding might occur, the production of sterile offspring also serves as an isolating mechanism. Breeding between zebras and horses produces a hybrid animal called a **zebroid** (✦ Figure 10.20). These sterile hybrids are unable to reproduce. Consequently, the gene pools of these two closely related species are maintained as separate reservoirs of genetic information.

## The Pace of Evolutionary Change Is Not Always Slow

Although the theory of geographic speciation is widely accepted, the slow pace of evolutionary change has prevented observation of all the steps in this process in a sequential fashion. An alternative approach has been to look for cases where all of the stages are still represented and can be studied in detail. Studies on natural populations of several closely related species of the fruit fly, *Drosophila,* provide evidence for the stages in the formation of new species. One species, *D. pseudoobscura,* exhibits the initial stages of species formation through the formation of distinct races or **ecotypes.** In addition, there is evidence that the critical step, reproductive isolation, has taken place, forming a closely related sibling species, *D. persimilis* (✦ Figure 10.21). The two species are almost identical in appearance, and coexist over much of the U.S.

✦ **FIGURE 10.20**

A zebroid is a sterile hybrid offspring of a horse and a zebra. Sterility in hybrids helps keep species from breaking down.

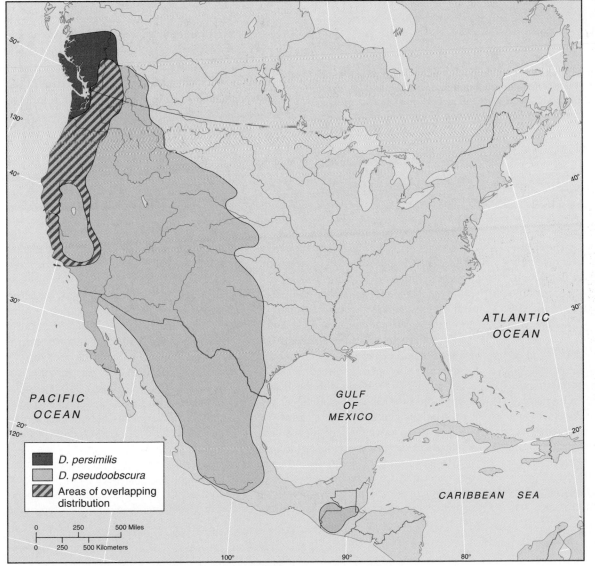

D. persimilis
D. pseudoobscura
Areas of overlapping distribution

PACIFIC OCEAN

PACIFIC OCEAN

GULF OF MEXICO

ATLANTIC OCEAN

CARIBBEAN SEA

0   250   500 Miles
0   250   500 Kilometers

✦ **FIGURE 10.21**

*Drosophila persimilis* and *Drosophila pseudoobscura* have overlapping ranges covering an area from British Columbia to the mountains of California. Although almost identical in appearance, no hybrids between these species have been recovered. Courtship and behavioral mechanisms keep them from mating.

West Coast region. However, no hybrids between the two species have been found among the thousands of individuals examined.

The classical theory of evolution espoused by Darwin and fully developed in the early part of this century is based on the notion that species formation is a gradual event, resulting from the accumulation of small genetic differences over long time periods under the influence of natural selection. More recently, scientists have questioned the idea that only slow changes can result in species formation, and several alternative schemes have been put forward, differing not in the type of events that occur, but only in the timescale required to accomplish these changes. These are collectively called **quantum models of speciation.**

The Hawaiian Islands, stretching from northwest to southeast across the central Pacific, are the products of volcanic action, with the southeastern island of Hawaii the youngest at about 700,000 years of age. Studies on the evolution of *Drosophila* species in these islands indicate that two species on Hawaii originated from a common ancestral species on the neighboring older island of Maui (✦ Figure 10.22), and that these speciation events occurred over a relatively short time period. Laboratory experiments using Hawaiian *Drosophila* confirm that stages of speciation including the development of isolating mechanisms can occur over the short time span of 15 generations (less than a year).

According to another model based on the fossil record, called **punctuated equilibrium,** the evolutionary process is characterized by long periods with little or no change, interspersed or punctuated by short periods of rapid speciation. In terms of the fossil record, "short" means thousands or hundreds of thousands of years. This model is discussed in more detail in the next chapter.

✦ FIGURE 10.22

**Pathway of colonization and speciation in Hawaiian** *Drosophila.* **The open circle represents a population ancestral to the three species shown. It is postulated that one or a few flies crossed to Hawaii from Maui and diverged rapidly into the two species found there today. Laboratory tests indicate that only about 15 generations are needed to show significant divergence.**

# SUMMARY

1. Developments in the eighteenth century in the physical and natural sciences provided the background for the nineteenth-century idea that species were not immutable creations but can and do undergo change. These developments include the establishment of geology as a science, and the concept of geological time, providing a timescale over which evolution of organisms could take place. In the natural sciences, the system of classification proposed by Linneaus provided a framework for establishing the relationships between organisms, grouping them together by shared or similar characteristics.

2. Against this background, the idea of the evolution of species was proposed several times, once by Charles Darwin's grandfather, among others. Its acceptance was limited by the lack of an explanation of how such changes could occur.

3. Both Darwin and Wallace embarked on long expeditions to explore, collect, and observe plants and animals in their natural settings, and to record their findings about the geology and geographical distributions of species. Darwin returned from his voyage and began to contemplate the problem of how species formed. Wallace, who embarked on his trip with the idea of solving the problem of the origin of species, continued his field investigations.

4. Wallace and Darwin independently formulated the idea that species arise through the struggle for existence. In this struggle, organisms that are better adapted to their environment survive and leave more offspring. As conditions change, organisms respond by changing and adapting. Over time, this chain of changes and adaptations leads to the development of new species.

5. All populations contain genetic variation in their gene pools. These variations ultimately arise through mutation, and a number of factors including genetic drift, migration, and natural selection act to change the frequency of this variation.

6. Selection can act in several ways on a population, and these are classified as disruptive, directional, and stabilizing, depending on the phenotype/genotype combination selected for.

7. The transition from a population to a variety to a race to a species can be gradual and results from the gradual accumulation of genetic changes over a long time

period. The process is complete when a new species has been formed. The ultimate criterion for speciation is the lack of interbreeding between two populations that formerly exchanged genes via matings. Physical isolation plays a key role in this process, allowing the gene pools to diverge.

8. In the classical idea of speciation, the pace of evolutionary change is gradual, and occurs over long time periods. More recently, examinations of the fossil record and observations of living species suggest that speciation can also occur rapidly, perhaps in a few generations. These forms of speciation are called quantum speciation.

# KEY TERMS

binomial system of nomenclature
bottlenecks
directional selection
disruptive selection
ecotypes
fitness
founder effects
gene pool

genetic divergence
genetic drift
geological time
morphs
mutation
natural selection
population
punctuated equilibrium

quantum models of speciation
races
reproductive isolating mechanisms
species
stabilizing selection
subspecies
taxonomy
uniformitarianism

# QUESTIONS AND PROBLEMS

## MATCHING

Match the concepts of evolution appropriately.

1. mutation
2. genetic drift
3. gene flow
4. natural selection
5. recombination
6. directional selection

_____ random fluctuations in allele frequency due to chance
_____ shifts in allele frequency in response to the environment
_____ exchange of genetic material
_____ heritable change in DNA
_____ change or stabilization of allele frequency due to differential production of variant members of a population
_____ change in allele frequency due to immigration and/or emigration

Match the scientists with their work.

7. Carolus Linnaeus
8. Comte de Buffon
9. James Hutton
10. Alfred Russel Wallace
11. Greek philosophers
12. Erasmus Darwin
13. Jean-Baptiste Lamarck
14. Charles Darwin
15. Charles Lyell

_____ speculated about evolution
_____ taxonomy
_____ *The Origin of Species*
_____ linked geological time with uniformitarianism
_____ advocated that each species was individually created by a Supreme Being
_____ uniformitarianism
_____ Enlightenment
_____ first to publish the role of natural selection in species formation
_____ simple organisms give rise to complex organisms

## FILL IN THE BLANK

1. An example of stabilizing selection in which genotype combinations produce an intermediate phenotype is _____.
2. The compound _____ is an example of an environmental agent that creates _____ selection.
3. The discontinuity in the pattern of variation present in a population, which is frequently represented by two distinct morphs, is created by _____ selection.
4. The process of long periods with little or no change, interspersed with short periods of rapid speciation, is the evolutionary process called _____ _____.
5. It was not until the work of _____ and the discovery of _____ that scientists were able to discount Lamarck's work on the evolution of species.

## TRUE/FALSE

1. The calculated approximate age of the earth is about 1 billion years old.
2. Darwin and Wallace successfully explained the origin of variations in species and how these variations were maintained in a population.
3. Mutational rates vary among species.
4. The gene pool is the total complement of genes in a population.
5. The foundation of selection is based on the notion of only the strong will survive.

## SCIENCE AND SOCIETY

1. A cherry tree grower discovered that his trees were infested with an insect that was rapidly reducing his crop yield. He sprayed the trees with an insecticide, which killed almost all of the pests. As he continued spraying over a period of years, he noticed that fewer insects died while the remainder seemed unaffected by the insecticide. Can you explain why the insects were almost all killed the first time? After a few years, why were the insects resistant? Could this same phenomenon occur in plants? Would it be fair to say that this insect's environment influenced its genetic constitution? Why or why not? If the farmer had used a larger amount of in-

secticide would he have killed more insects? Why or why not?

2. The early twentieth century saw the development of a new approach to human heredity—eugenics. Eugenics is selective human breeding for the purpose of improving the human species. Eugenics includes positive eugenics (encouraging the breeding of people with "desirable" traits) and negative eugenics (discouraging the breeding of people with "undesirable" traits). Eugenics was sparked by Darwin's theory of evolution through natural selection, elaborated on by his cousin, Francis Galton. The use of negative eugenics was made most famous in the 1940s when Nazi Germany tried to create "the perfect race." Since World War II eugenics has fallen into disfavor, but recent genetic technologies have triggered new interest in eugenics. What practical applications could eugenics serve? Who benefits from eugenic discrimination? Although Nazi Germany was an extreme example of negative eugenics, do you think this form could be advantageous in some cases of human disease? What guidelines should be set to ensure that genetic technologies are not being used for eugenic purposes?

# EVOLUTIONARY PROCESSES: ABOVE THE SPECIES LEVEL

I n the Middle Ages, one of the hazards of life was thought to include small triangu-
lar "stones" that fell from the sky and killed people. These structures were known
as tonguestones or in Latin as Glossopterae, because they resembled small
pointed tongues. Although they could be found embedded in rock and along beach areas,
the only logical explanation for their appearance seemed to be that they fell from the sky.
Since they were pointed and sharp, it was obvious that they could inflict injury or death if
they struck someone. In 1667, however, Nicholas Steno, an anatomist, found a dead shark
washed ashore on a beach in Italy and identified tonguestones as fossilized shark's teeth.

Physicians in the Middle Ages used highly prized toadstones as cures for several kinds
of ailments. These were believed to come from the heads of certain types of toads, and to
have healing powers. In our time, we would recognize toadstones as fossilized teeth of
another type of fish, the ray.

This inability to identify fossils as the remains of once living organisms and to place
fossils into the wider context of evolution was not unusual in the Middle Ages. Then as
now, fossils were found in rock formations, and it was difficult to understand what fossils
were and how they originated because there was little understanding of how rocks were
formed.

OPENING IMAGE

*Dinosaurs were the dominant land*

*vertebrates for over 140 million years.*

*Nicholas Steno not only identified the origin of tonguestones, he was responsible for the development of some of the basic principles of geology in the last part of the seventeenth century. These principles, along with the work of later geologists such as Lyell, provided the context for the interpretation of fossils as the remains of ancient organisms.*

*In Darwin's time, the fossil record became an important form of evidence for the continuity of relationships among species during the process of evolution. In this chapter, we consider the fossil record and other forms of evidence, ranging from geography to molecular biology, that provide clues to the evolutionary events that have shaped the living world we observe.*

## THE FOSSIL RECORD AS IT APPLIES TO EVOLUTION

To interpret the fossil record as it relates to evolution, we need first to consider what constitutes a fossil and how fossils are formed, dated, and interpreted. With this background, we can reconstruct major events through the fossil record as they have occurred over time.

### Types of Fossils and How Fossils Are Formed

Fossils can be defined as the remains, traces, or impressions of prehistoric organisms preserved in the earth's crust (✦ Figure 11.1). These can include **body fossils,** the skeletal remains of organisms such as dinosaurs that can be reconstructed to form skeletons, or **trace fossils,** such as tracks, burrows, or nests that provide evidence about the behavior, gait, and other activities of extinct organisms. Even fossilized feces, called **coprolites,** provide valuable information about the diet, size, and distribution of extinct animals. Although fossils are abundant, not all forms of life are equally represented as fossils. The soft body parts of plants and animals are rarely preserved, and even hard body parts such as shells, bones, and teeth must be protected from weathering by being buried in a protective rock layer.

Whether or not a given type of rock contains fossils depends on its geological history. Rocks formed from sand or mud in riverbeds or sea bottoms are often rich in fossil remains. These are **sedimentary rocks,** and form from solid particles and dissolved minerals (✦ Figure 11.2). Sedimentary rocks are deposited in a variety of marine and terrestrial environments, and form layers (or strata) that reflect the rate of sediment deposition. These layers provide a physical history of the earth, with the oldest layers at the bottom, and the youngest at the top. The position of fossils in layers of sedimentary rocks is a guide to their relative age. Each stratum usually contains a unique collection of fossils, and with each successive layer of younger and younger strata, the fossilized organisms become more complex.

(a)

(b)

### ✦ FIGURE 11.1

Fossils can form in various ways. (a) Shells of marine invertebrates can be chemically altered or replaced with other minerals during fossilization. (b) Insects trapped in amber represent unaltered remains.

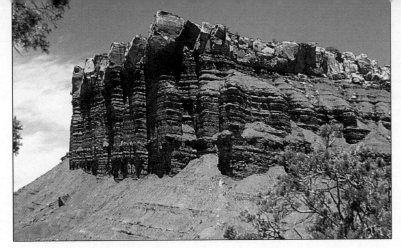

**✦ FIGURE 11.2**

Stratified layers in sedimentary rocks from Capitol Reef National Park, Utah.

Rocks such as lava or rock that has been heated or formed by crystallization of material from the earth's core are unlikely to contain fossils. As a result, the fossil record of organisms with hard body parts and those that lived in or near shallow lakes, oceans, or riverbeds are preserved where sediments were deposited. Altogether, the fossil remains of some 250,000 different species ranging over a time span of over three billion years have been recovered to date, with more waiting to be discovered. It is estimated that one to three billion species existed during that time span, and approximately 99% of them are extinct. Compared to the billions of species that have existed, fossils provide only a glimpse of past life-forms. On the other hand, it is the only record we have of the vast diversity of past organisms.

## Relationship of Fossils to Biological Time

Fossils are clearly the remains of ancient life-forms. However, Darwin and other scientists of his time had no way of establishing the age of any fossils or the rocks in which they were encased. It was accepted that in an area where many rock layers were exposed, the fossils in the lower rock layers were older, and those in higher layers were younger. In Great Britain, the number and thickness of the known strata indicated that evolutionary events took place on a vast time scale. Although this evidence provided a relative timescale, it was impossible to date fossils or rocks with any degree of accuracy. The discovery that some elements undergo radioactive decay provided geologists and paleontologists (scientists who study fossils) with a tool for measuring the age of fossils and rock layers.

In radioactive decay, an unstable atom of one element is transformed into an atom of another, more stable element. The rate of decay is expressed as a **half-life,** the time it takes for one-half of the original unstable element to be converted into a more stable daughter element. The half-lives of elements are constant under all conditions and range from a few billionths of a second to billions of years. By measuring the amount of an unstable element and its daughter element present in a sample of rock or fossil, and knowing the half-life of the element, it is possible to date the specimen with a high degree of accuracy. To reduce error, the decay levels of two different elements are routinely measured in the same sample. ✦ Table 11.1 lists some pairs of elements that are commonly used in dating rocks and fossils and lists the dating ranges over which they are used.

Other techniques are used to date recent, non-fossilized specimens. Carbon dating (measuring the decay of carbon-14 to carbon-12), can be used only on the biological remains of organisms, and has an accuracy that is limited to specimens up to 60,000 years old. As a result, this method is useful only for dating fossils from the recent past and in ar-

| **TABLE 11.1  FIVE OF THE PRINCIPAL LONG-LIVED RADIOACTIVE ISOTOPE PAIRS USED IN DATING ROCKS** | | | |
|---|---|---|---|
| ISOTOPES | | HALF-LIFE OF PARENT (YEARS) | EFFECTIVE DATING RANGE (YEARS) |
| *PARENT* | *DAUGHTER* | | |
| Uranium 238 | Lead 206 | 4.5 billion | 10 million to 4.6 billion |
| Uranium 235 | Lead 207 | 704 million | |
| Thorium 232 | Lead 208 | 14 billion | |
| Rubidium 87 | Strontium 87 | 48.8 billion | 10 million to 4.6 billion |
| Potassium 40 | Argon 40 | 1.3 billion | 100,000 to 4.6 billion |

chaeological studies. Dating during the last 14,000 or 15,000 years can be done by using tree-ring studies. This method, known as **dendrochronology,** is discussed in Chapter 16.

By combining radioisotope dating and the fossil record, the history of the Earth and its organisms has been divided into four main eras (◆ Figure 11.3). Each division is characterized by abrupt changes in the type, number, and distribution of fossils. Radioisotope dating has been used to determine the age and duration of each of these eras. These eras are the **Archean/Proterozoic, Paleozoic, Mesozoic,** and **Cenozoic.** The time span covered by these eras ranges from the origin of the earth, some 4.6 billion years ago, to the present. The major geological and biological features of each of these eras are discussed

in a subsequent section, after we examine several forms of evidence for the evolutionary process.

## EVOLUTION: WHAT IS THE EVIDENCE?

The formation of a species from a preexisting one, or the divergence of separated populations into new species is a small-scale evolutionary event, called **microevolution.** On a larger scale, the combination of events associated with the origin, diversification, extinction, and interactions of organisms that produced the species that presently inhabit the planet is collectively known as **macroevolution.** If, as Darwin proposed, living species arose by descent with modification from ancestral forms, there should be

◆ **FIGURE 11.3**
**The geological and biological timescale. Some major geological and biological events are shown on the right side of the scale.**

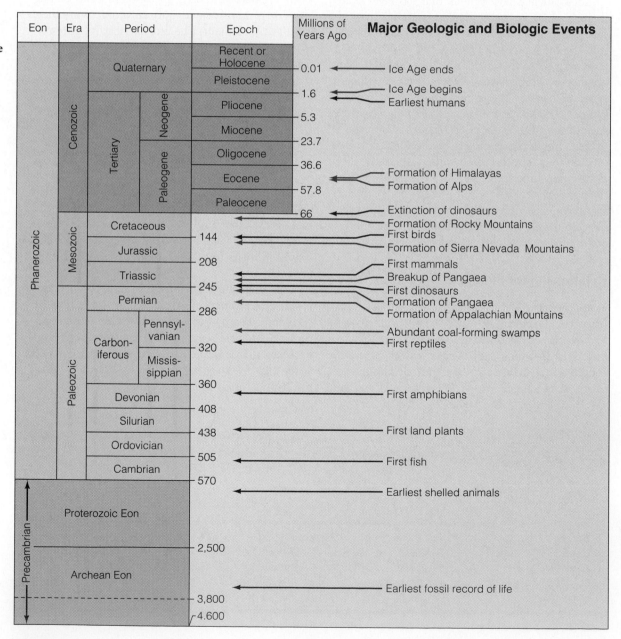

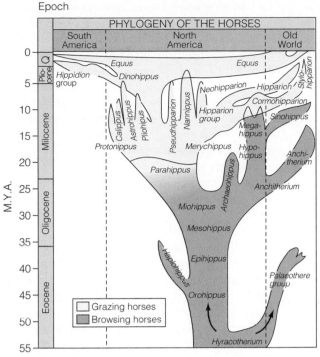

## ✦ FIGURE 11.4

Evolution of the horse. Above is a chart summarizing the major events. Early horses lived in forests and were browsers. During the Oligocene, two major lines of horses formed, one leading to three-toed browsing horses and the other to one-toed grazing horses. Major trends in the evolution of the horse are shown to the right. These include an increase in size, loss of toes, and an increase in size and complexity of teeth as an adaptation to grazing.

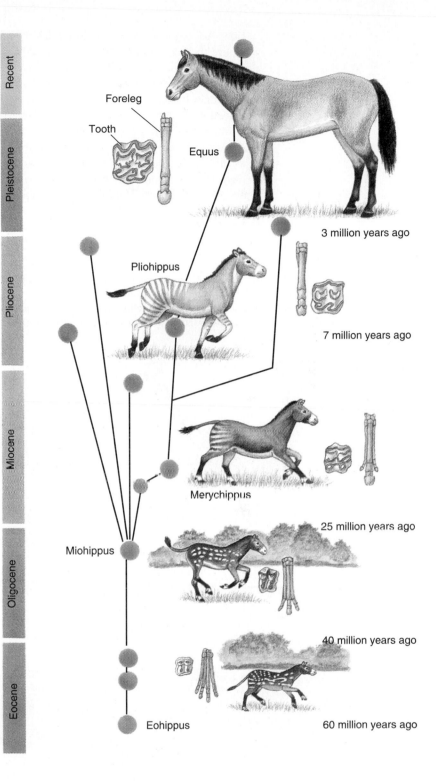

underlying similarities, with closely related species having more similarities than distantly related ones. It is confirmation of these underlying similarities that we shall seek by examining Darwin's evidence for evolution, as well as the evidence amassed by modern biochemical and molecular techniques.

Darwin arrived at the idea of evolution by natural selection through a combination of inductive and deductive reasoning. The evidence he presented for the process of evolution included geographic distribution of organisms, the fossil record, comparative embryology and anatomy, and behavior. He selected these areas through deduction, concluding that if evolution operated by natural selection, then the similarities among organisms in each of these areas can best be explained as evidence for the process of evolution. Since Darwin's time, facts from genetics, biochemistry, immunology, and molecular biology have added to the evidence for evolution.

## Comparative Anatomy Can Be Used to Trace Evolutionary Change

When the structures of related organisms, both living and extinct, are compared, many similarities

become apparent. For example, the evolution of the modern horse can be traced through an exceptionally well-documented series of fossils (✦ Figure 11.4). In comparing the anatomy of the now extinct ancestral forms, the legs show an overall increase in length and diameter, a gradual increase in the size of the central toe, and a reduction and eventual loss of the side toes. These changes occurred in response to the pressure of natural selection. Ancestral forms of the horse lived in forests, where they could hide from predators. In the face of climatic changes, the forests slowly gave way to grasslands. In the grassland habitat, horses could no longer hide, and had to outrun their predators. Selection therefore favored horses with longer and stronger legs suitable for running.

At first glance, the forelimbs of vertebrates such as birds, humans, bats, whales, and cats appear to have little in common (✦ Figure 11.5). In reality, however, they are **homologous structures** com-posed of the same bones, and have similar ar-rangements of muscles, blood vessels, and nerves. The vertebrate forelimb is an example of morpho-logical divergence from a common ancestral struc-ture found in an early reptile. Over time, the basic structure of the limb was modified through selection into wings in extinct (pterosaurs) and living forms (birds and bats), into the the flippers of whales and porpoises, into the forelimbs of horses and cats, and into the arms of humans.

Anatomical similarities cannot always be used as evidence of evolutionary descent from a common ancestor. Insects, birds, and bats all have wings, but insects are not closely related to birds and bats. The wings of bats and birds are homologous structures having a common ancestral origin, but the wings of insects have no bones. Instead they are composed of a thin membrane stiffened by wing veins composed of an armor-like material, chitin (Fig-ure 11.5). Thus, insect wings rep-resent a case of **morphological convergence,** the evolution of basically dissimilar structures to serve a common function, flight. Structures that have dif-ferent origins but serve similar functions are called **analogous structures.**

**Vestigial structures** are non-functional remnants of organs that were functional in ancestral species, but are often non-functional in present-day species. They rep-resent structures that are in the pro-cess of being lost. In humans, the ap-pendix is a small finger-like projection of a functionally unimportant sac called the caecum. In mammals that live mainly on plants (herbivores) the caecum is a large, important structure filled with microorganisms that assist in the diges-tion of plant material such as cellulose. Similarly, dogs have a small, vestigial structure called a dewclaw on each foot. During the evolution of dogs, ancestral species walked on their toes, and the fifth digit (equivalent to the big toe or thumb) did not contact the ground. Over time, this structure lost its function, became re-duced in size, and is a vestigial structure in present-day dogs and related species.

Both whales and snakes evolved from four-footed ancestors. In the whale, the forelimbs have been retained as power-ful flippers, but the hindlimbs have been lost. However, in their skeleton, whales retain a remnant of the pelvic structures

## ✦ FIGURE 11.5

**(a) The forelimbs of vertebrates are homologous organs that serve different functions but are composed of the same elements and undergo similar embryological development. (b) An insect wing serves the same function as the wings of bats and birds, but has a different embryological origin, and is an analogous structure.**

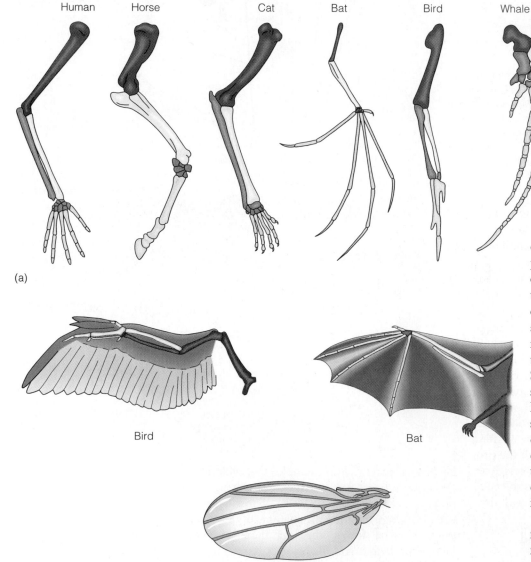

Human  Horse      Cat      Bat      Bird      Whale

(a)

Bird        Bat

Fly

(b)

### Beyond the Basics

# HEN'S TEETH: GONE BUT NOT FORGOTTEN

Sometime in the late Triassic or early Jurassic age (180–200 million years ago), reptiles began to exploit a new habitat: the air. The pterosaurs, representing a now extinct group, developed modifications of the forelimbs and elongated "fingers" that served as a framework for thin membranes of skin. Another group of reptiles developed elongated, modified scales that, over time, became transformed into feathers. From this group, modern birds evolved. During the transition from reptiles to birds, the intermediate species retained many reptilian characteristics. For example, fossil remains of primitive birds such as *Archeopteryx* have teeth. Later, as birds became a distinct class, these characters were lost.

The evolution of new adaptations such as wings involves genetic change, as does the loss of old adaptations such as teeth. The nature of such changes are often difficult to identify, but the question of what happened to bird's teeth may be answerable. Teeth are formed by the interaction of two embryonic tissue layers. The underlying layer (mesenchyme) forms the core of the tooth, and the overlying layer (epithelium) forms the enamel and outer tooth structure.

When the overlying layer from chick embryos is placed on the underlying layer from mouse embryos, the two layers can successfully interact and form an enameled tooth. This means that the chicken tissue still retains the ability to stimulate underlying layers to begin tooth formation. The reason birds do not form teeth is that the underlying tissue does not respond to genetically programmed signals from the overlying layer during embryonic development. Hence, genetic information for the formation of teeth remains intact in the chicken, as does the ability to initiate tooth formation. What has been lost during the transition from reptiles to birds is the genetic information that allows the mesenchyme to respond to the initial signals from the epithelium. Alteration or loss of a single developmental switch that controls tissue interaction is responsible for a major phenotypic change in evolution. The lesson of hen's teeth for students of evolution is that not all evolutionary change is of necessity a gradual process. In this case, what may be a single mutational event has produced a dramatic change.

---

to which hind limbs attach. In fact, some whales with small rear limbs have been captured. Snakes have lost both the forelimbs and hindlimbs, but some, like the boa constrictor, retain small pelvic and hind limb bones. The presence of these vestigial structures and their occasional development indicates that the genetic information to form such structures may be retained, but these genes are not normally expressed during development.

## Comparative Embryology Can Establish Common Ancestors

Evidence for evolution can also be found by comparing mechanisms of embryonic development in animals. Fish, amphibians, reptiles, birds, and mammals are all vertebrates, with a common ancestral background. The most primitive of these vertebrates are fish. During embryonic development in fish, the circulatory system forms a series of arched blood vessels that will exchange oxygen in the gills. Similarly, gill pouches and gill slits develop and form the gills of adult fish.

Early embryos of reptiles, birds, and mammals all develop fish-like structures including the arched blood vessels and gill slits, even though these gill slits never become functional (✦ Figure 11.6). The presence of these structures during development demonstrates that certain developmental processes remained constant during the evolution of vertebrates. These similar patterns in embryonic

✦ **FIGURE 11.6**

Early embryos of vertebrates, including humans, have gill arches at an early stage of development.

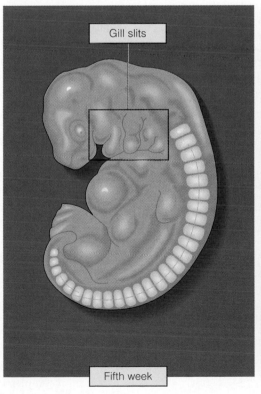

Gill slits

Fifth week

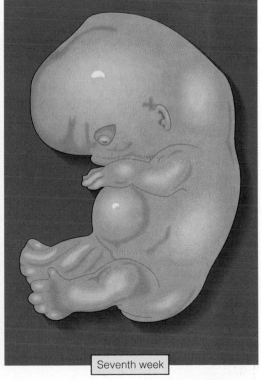

Seventh week

development mean that evolution in the vertebrates has come about by adapting and remodeling pre-existing structures to serve new functions, not by changing the basic mechanisms of development. In fish, embryos form arched structures that support the adult gills. In vertebrates that evolved later, the arches have been modified to form parts of the lower jaw, parts of the inner ear, and other structures in the head and neck.

## Comparative Biochemistry Provides Molecular Insight into Evolution

Just as anatomical structures and embryonic processes reveal evolutionary relationships, biochemistry can be used to reconstruct evolutionary history. Cytochrome *c* is an important protein in mitochondrial respiration. In most higher organisms, it is composed of 104 amino acids that have changed very little over the course of evolution. The amino acid sequence of cytochrome *c* in humans and chimpanzees is identical; between humans and rhesus monkeys, there is only one amino acid difference. Differences in amino acid sequence can be used to construct **evolutionary trees** to indicate the origin of species from common ancestors. In this tree, organisms with fewer amino acid differences are placed closer together, and those with more differences are placed further apart. An evolutionary tree constructed from the amino acid sequences in cytochrome *c* is shown in ✦ Figure 11.7a.

Cytochrome *c* has changed very little over the course of evolution (about a 1% change every 20 million years), and is not useful for examining relationships among recently evolved species. To examine closely related species, it is necessary to study changes in rapidly evolving proteins. One such protein is the enzyme carbonic anhydrase, which is 115 amino acids in length. Differences in the amino acid sequence of this protein have been used to study evolutionary relationships among humans and other higher mammals (✦ Figure 11.7b).

Remarkably, evolutionary trees constructed from differences in amino acid sequence agree very well with similar diagrams constructed using the fossil record, comparative morphology, and embryology. The advantage of using protein phylogenics to study evolution is that an ever-increasing number of proteins is being sequenced and information on them is available through computerized data banks.

## Comparative DNA Studies Link Organisms by Genetics

Evolutionary patterns can also be revealed by directly comparing the nucleotide sequence of genes. This is accomplished through a technique known as **DNA hybridization.** DNA is heated until it dissociates into single strands. When allowed to cool, complementary sequences rejoin to form double-stranded DNA molecules (✦ Figure 11.8). In mo-

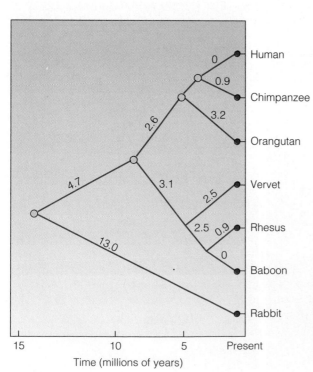

(a)

(b)

✦ **FIGURE 11.7**

**(a) Phylogenetic relationships established by the amino acid sequence of cytochrome *c* span 30 million years of descent from a common ancestor. On this scale, humans are most closely related to monkeys and least related to yeast. (b) Evolutionary relationships among mammals are established by using changes in the amino acid sequence of carbonic anhydrase, which mutates at a faster rate than cytochrome *c*. The numbers along the line are the estimated number of nucleotide changes that have occurred during evolution.**

lecular evolution studies, DNA from one species is prepared and dissociated into single strands. These single strands are mixed with dissociated DNA from another species, and the two DNAs are allowed to cool and form double-stranded hybrid DNA molecules. These are called hybrid molecules because each of the two strands comes from a different species. More closely related species will have more complementary sequences in common, and therefore form a larger number of stable double-stranded DNA molecules. The degree of relatedness can then be calculated by measuring the amount of double-stranded molecules formed and their stability.

The gorilla, chimpanzee and humans are so closely related that studies using anatomical, embryological, and even amino acid sequences have failed to resolve the evolutionary relationship among these species. Using DNA hybridization, and incorporating information from other studies including the fossil record, the results indicate that humans are more closely related to chimpanzees than to gorillas. The evolutionary relationships among humans and other primates are discussed in Chapter 15.

DNA has been preserved in many types of fossils; recovery and use of this genetic material in hybridization studies has permitted insight into the relationships between living and extinct species. Woolly mammoths once roamed over North America and much of Northern Asia, but became extinct some 10,000 to 20,000 years ago. Their frozen remains have been retrieved from the permafrost in Siberia. DNA recovered from their tissues has been used to determine that they were closely related to elephant species alive today.

The quagga, a horse-like animal, became extinct just over a century ago. DNA extracted from quagga hides in museum collections has established that quaggas are more closely related to the zebra than to the horse. DNA cloning and nucleotide sequencing has been used to establish the evolutionary standing of fossilized material as outlined in the opening vignette of Chapter 8. As in other areas of biology, further use of recombinant DNA technology promises to be a powerful tool in the study of evolution.

## Biogeography Divides the World into Biological Zones

One of the lines of evidence that Darwin used to support the idea of evolution by natural selection was the distribution of similar species in neighbor-

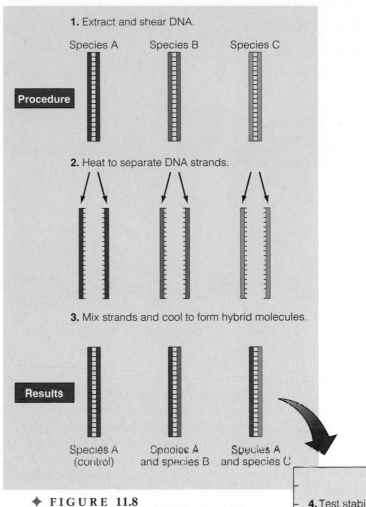

**◆ FIGURE 11.8**

**In DNA hybridization (above), DNA from different species is heated to separate it into single strands. Single strands from two or more species are mixed, and allowed to form double-stranded hybrid molecules. The hybrids are tested by heating again (right), to measure how much energy is required to break the bonds in the hybrid molecules and convert them into single strands. More closely related species require more energy to break the bonds. In this case, species A is more closely related to species C than to species B.**

ing regions. **Biogeography** is the study of the distribution of plants and animals across the earth. As Darwin journeyed down the coast of the South American continent, he repeatedly found examples of closely related species that replaced each other in adjacent geographic zones. He speculated that this distribution was not a random one, but might be the result of evolution and adaptation. For example, he saw a bird called the greater rhea in the northern regions of the continent. As he moved south, he saw the greater rhea replaced by a related species, the lesser rhea. He proposed that these two species evolved from a common ancestor and had become adapted to their respective environments through

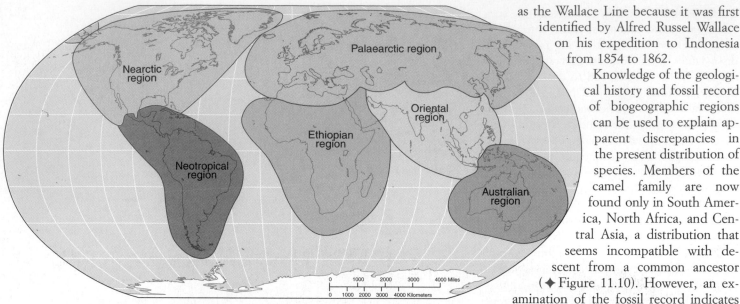

◆ FIGURE 11.9

The six major biogeographic zones are defined by the distribution of plants and animals within them.

◆ FIGURE 11.10

Camels originally evolved in North America, and spread from there to Europe, Asia, and South America. After their extinction in North America, camels and their relatives (llama, alpaca) are widely separated in their distribution.

as the Wallace Line because it was first identified by Alfred Russel Wallace on his expedition to Indonesia from 1854 to 1862.

Knowledge of the geological history and fossil record of biogeographic regions can be used to explain apparent discrepancies in the present distribution of species. Members of the camel family are now found only in South America, North Africa, and Central Asia, a distribution that seems incompatible with descent from a common ancestor (◆ Figure 11.10). However, an examination of the fossil record indicates that camels originally evolved in North America and later migrated into South America and across a land bridge through Siberia into North Africa and Central Asia. Later, camels became extinct in North America, but descendants remain in South America (vicunas, llamas, and alpacas) and in Africa and Asia (camels and dromedaries).

natural selection, thus explaining their geographic distribution.

By grouping similar organisms into geographic regions, the landmasses of the world can be divided into six geographic zones (◆ Figure 11.9). These divisions reflect both the geological and biological histories of the zones and the organisms that evolved, migrated into, and underwent adaptation in these zones. Most zones are separated from each other by oceans, but others are delineated by geological and geographical features. For example, the Ethiopian region is separated from the Palaearctic and Oriental regions by the Sahara and Arabian Deserts. The Oriental region is separated from the Palaeartic by the Himalayas and other mountain ranges that separate climate regions. The division between the Oriental and Australian regions that threads through the islands of Indonesia is known

## Artificial Selection Is Analogous to Natural Selection

Although the process of natural selection is often difficult to observe under natural conditions, Darwin realized that the domestication of animals represented a case of artificial selection, with humans as the selecting agent. In **artificial selection,** the breeder chooses the variants to be used in producing succeeding generations, resulting in an acceleration of the results of selection. As Darwin wrote:

No doubt the strawberry had always varied since it was cultivated, but the slight varieties had been neglected. As soon, however, as gardeners picked out individual plants with slightly larger, earlier, or better fruit, and raised seedlings from them, and again picked out the best seedlings and bred from them, then (with some aid by crossing distinct species) those many admirable varieties of the strawberry were raised which have appeared during the last half-century.

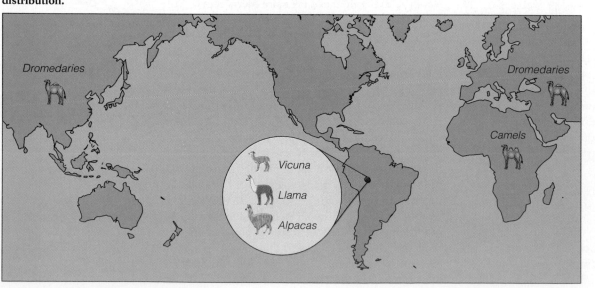

Artificial selection serves as a model (although somewhat imperfect) of the events that take place in nature under the direction of natural selection. As in natural selection, certain individuals will survive and leave a greater number of offspring, but it differs from natural selection in that the "fittest" individuals have been selected by human intervention. Artificial selection does, however, provide some insight into the amount of genetic variation that is present in natural populations. Artificial selection is important as a form of evidence for evolution because it is a parallel process that documents the degree of change that can take place under the influence of selection.

## THERE ARE PATTERNS IN THE EVOLUTIONARY PROCESS

For a population to survive, it must be adapted to its present environment, but must also contain enough genetic variation so it can respond to changes in environmental conditions. The level of adaptation is maintained by constant shuffling of genotypes in response to natural selection. Scanning across a wide range of species reveals certain patterns in the way in which organisms have evolved in response to changing environmental conditions. In this section, we briefly survey some of these patterns and describe how they have shaped the number and distribution of species now present on the planet.

### Adaptive Radiation Fills Empty Habitats by Rapid Speciation

The development of a variety of species from a single ancestral form is called **adaptive radiation.** A burst of speciation takes place when a new habitat or geographical region is made available to populations of an organism. Darwin recorded 13 species of finches in the Galapagos Islands, which lie some 600 miles west of South America. Some species are restricted to a single island; others share one or more islands. Darwin speculated that the ancestral species of finch originated on the mainland and, on reaching the islands, was presented with an array of new, unoccupied habitats. Descendants of these original colonizers gradually became adapted to the new environments and subenvironments through natural selection, and diverged to the point where new species were formed (✦ Figure 11.11). Today, there are several species of ground-dwelling finches

✦ **FIGURE 11.11**

**Adaptive radiation in finches on the Galapagos Islands. All but one species (the Cocos Island finch) are thought to have arisen from a small number of colonizing birds. As the birds adapted to the environments on the islands, the group on the right became ground dwellers, and the group on the left adapted to life in the trees.**

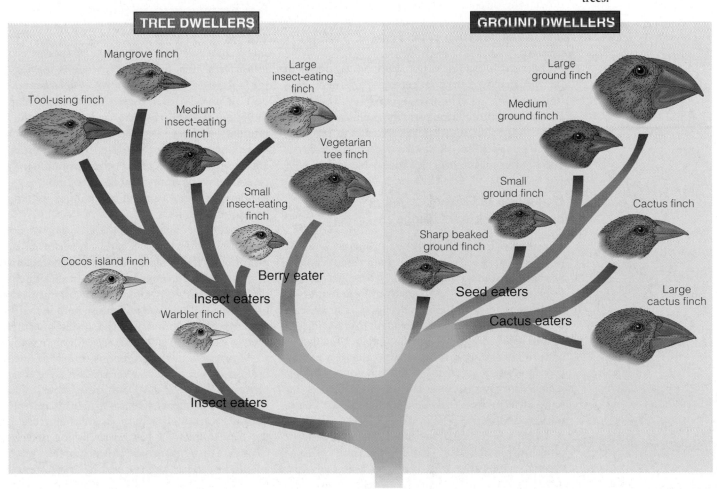

## TABLE 11.2  FIVE MASS EXTINCTIONS

| EXTINCTION | TIME (MILLIONS OF YEARS BEFORE PRESENT) | POSSIBLE CAUSES |
|---|---|---|
| Ordovician/Silurian | 438 | Glacier formation in Gondwanda |
| Late Devonian | 370 | Global cooling |
| Permian/Triassic | 245 | Continental movements, low sea level |
| Triassic/Jurassic | 208 | Continental movements, low sea level |
| Cretaceous/Tertiary | 65 | Climate changes, meteorite impact |

whose beaks differ in size and shape depending on the seeds they employ as a food source. Another species has evolved a long, sharp beak for boring holes into tree bark to search for insects. These birds use cactus thorns or small twigs to probe the holes, and as insects emerge, they are seized and eaten.

Similarly, Darwin discovered that the species and varieties of tortoises, iguanas, insects, and plants he collected in the Galapagos chain were often confined to a single island. This profusion of species contrasted with their isolated distribution and puzzled him, because all the islands were within a 50- to 60-mile range, often in sight of each other. The discovery of the phenomenon of adaptive radiation was instrumental in Darwin's idea that natural selection is the agent of evolutionary change.

Adaptive radiation is not confined to species in the Galapagos, but is a general pattern of evolutionary change. The marsupials of Australia, the honeycreepers of Hawaii, and the evolution of birds, bats, and pollinating insects are all the result of adaptive radiation. Adaptive radiations also occurred after mass extinction (discussed later).

### Extinction Is Caused by Environmental Change

The opposite side of adaptive radiation and the proliferation of species is the rapid disappearance of species groups by extinction. There is a pattern to extinctions that is evident from examination of the fossil record. Extinction occurs on a small scale when individual species fail to adapt to changing environmental conditions and disappear. On a larger scale, mass extinctions cause large numbers of species to disappear (such as the dinosaurs). Small-scale extinctions occur constantly and, as an essentially local process, can be considered part of the background to the evolutionary process.

Large-scale extinctions affect major groups of often unrelated species and are the result of cataclysmic events that take place on a global scale.

These extinctions are caused by environmental changes on a scale that exceed the ability of any population or members of a species to adapt to the new conditions. When environmental changes overwhelm the ability of a species to respond, the result is extinction.

Five major extinctions have occurred during evolutionary history ( ◆ Table 11.2). The causes of these extinctions are debated, but a meteorite impact may account for one of these events. The underlying causes of the great extinctions are discussed in a later section.

In the course of mass extinctions and background extinctions, the randomness of change and the role played by pure chance in evolution are now becoming appreciated as major factors in evolution. The role of luck in evolutionary history is explored in Steven Jay Gould's book *A Wonderful Life,* an account of the rich diversity of life found in the fossils of the Burgess Shale, a 500-million-year-old Canadian rock formation. In the book, he speculates on what present life-forms would be like if chance favored the survival of other species instead of those that survived and became ancestral to existing species.

### Convergent and Parallel Evolution Are Responses to Selection

The pattern of evolution that takes place when a single interbreeding population or species evolves into several new types of descendent species, like the Galapagos finches, is known as **divergent** evolution. As noted earlier, divergence through adaptive radiations is a well-characterized pattern in the evolutionary process. Divergent evolution leads to a variety of organisms that differ markedly from the ancestral population. Two other evolutionary processes, **convergent** and **parallel** evolution, result in the development of similar adaptations in different groups of organisms.

Convergent evolution is the evolution of similar structures in distantly related organisms. Parallel evolution is the evolution of similar characteristics in closely related organisms. It is often difficult to distinguish between closely and distantly related organisms, so what we will stress here is that both patterns of evolution are the result of different species adapting to similar environments and/or strategies of life. Brachiopods are marine invertebrates adapted to life in a variety of marine environments ( ◆ Figure 11.12a). Some live in burrows, and others attach themselves to structures on the sea floor. As a group, brachiopods are characterized by the development of hinged shells for protection. Clams are distantly related to brachiopods, and have adapted to life in marine and freshwater environ-

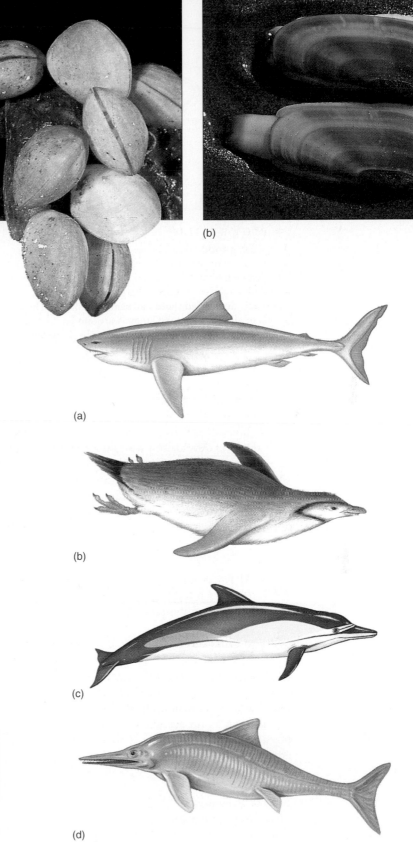

✦ **FIGURE 11.12**

**(a) Brachiopods. (b) Clams. Although distantly related, both brachiopods and clams developed shells as a protective adaptation. The development of shells as an adaptation to similar environments is an example of convergent evolution.**

(a)

(b)

ments. They have also evolved hinged shells as a means of protection (✦ Figure 11.12b). Because of the distant evolutionary relationship between brachiopods and clams, the development of shells as an adaptation to similar environments is an example of convergent evolution.

Convergent evolution can also be seen in the evolution of body shape in marine organisms. Sharks, penguins, porpoises, and dolphins have body shapes adapted for swimming rapidly in pursuit of food. Here, the convergence of body form involves distantly related organisms such as fish (shark), birds (penguins), and mammals (porpoises) (✦ Figure 11.13). Convergence can also operate across widely separated gaps in the evolutionary record. Ichthyosaurs were marine reptiles with a lifestyle like that of the porpoise. Their body shape was remarkably similar to that of present-day dolphins and porpoises that occupy the environmental habitat that was vacant for some 30 million years after the extinction of the ichthyosaurs.

Convergent evolution is also evident when comparing placental and marsupial animals. These animals occupy similar niches in their respective environments, but are distinct in their geographic distribution and reproductive strategies (✦ Table 11.3).

(a)

(b)

(c)

(d)

✦ **FIGURE 11.13**

**Convergent evolution in body shape. (a) Shark. (b) Penguin. (c) Dolphin. (d) Ichthyosaur. Although distantly related, these organisms adapted to the marine environment by developing similarly shaped, streamlined bodies. Ichthyosaurs were fish-eating reptiles that became extinct at the end of the Mesozoic Era. Their niche remained empty until dolphins and porpoises evolved some 30 million years later.**

**TABLE 11.3   CONVERGENT EVOLUTION AMONG MARSUPIAL MAMMALS AND PLACENTAL MAMMALS**

| MARSUPIALS | PLACENTAL MAMMALS |
|---|---|
| Kowari | House mouse |
| Marsupial mole | Coast mole |
| Greater glider | Flying squirrel |
| Tasmanian Devil | Wolf/Fox |
| Wallaby | Rabbit |
| Koala bear | Sloth |
| Wombat | Badger/porcupine/hedgehog |

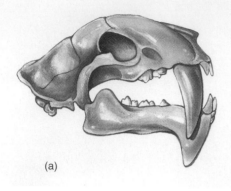
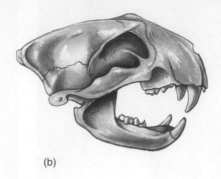

(a)                                    (b)

### ✦ FIGURE 11.14

**Large saber-toothed carnivores evolved several times
and in different places. In North America, saber-
toothed members of the cat family evolved twice.
In South America, a cat-like marsupial (a primitive
mammal) developed saber-like canine teeth. (a) The
skulls of all saber-toothed animals are similar, and differ from those of modern cats, such as the mountain
lion (b).**

An example of parallel evolution is the independent appearance of saber-toothed animal species.
Saber-toothed carnivores (✦ Figure 11.14) and
their relatives developed long canine teeth that may
have been used to inflict slashing wounds to the
throat of prey. The evolution of these teeth included modification of the skull and jaw structures to allow such enlarged canines to be carried
and used.

## WE CAN RECONSTRUCT THE LARGE-SCALE EVENTS IN EVOLUTION

With this background information about evolution, we can turn to a discussion of the geological
and biological events that define the four great
time-based divisions of evolutionary history on our
planet. In each division, the evolution of life-forms
was preceded and accompanied by changes in the
physical environment. The changes in the physical
environment include widespread and large-scale
changes in the geology, geography, and climate of
the Earth.

This coevolution of the physical and biological domains is an important force in shaping the
world as we now see it. Even though our emphasis
will be on past events, these geological forces are
still at work shifting the organization of oceans and
continents and the diversity of life over the entire
Earth's surface. In the following discussion, three
themes will be apparent. First, the surface of the
Earth is composed of a series of plates, and the motion of these plates (**plate tectonics**) is the backdrop

for the evolution of life (✦ Figure 11.15). Second,
life on Earth evolved in a complex series of events
starting from organic precursors and, third, the geological and biological events involved in the evolution of the Earth have taken place over a vast
timescale.

### Archean/Proterozoic Era Began with the Formation of the Earth

The Archean/Proterozoic Era began with the formation of the Earth some 4.6 billion years ago. By
about 3.8 billion years ago, a crust had formed
on the Earth's surface, and the earliest rocks had
formed. The Archean covers some 45% of all geological time. As the Earth cooled, the primitive
atmosphere of helium and hydrogen was supplemented with volcanically derived gases including
water vapor, sulfur, nitrogen, carbon dioxide, sulfur dioxide, methane, and ammonia. There was little or no oxygen in the early atmosphere, and
therefore no ozone layer to protect the surface from
the intense bombardment of ultraviolet light by the
sun. The water vapor in the early atmosphere condensed as the earth cooled, and the oceans began
to form.

In this inhospitable environment of a hot earth,
with turbulent volcanic activity and prolonged
rainfall, life evolved. The theories and mechanisms by which this event occurred were considered in Chapter 2. Recall that living systems probably evolved from a prebiotic environment rich in
organic material. The earliest organisms evolved in
an atmosphere devoid of oxygen, and were **anaerobic.** In addition, it is likely that these organisms derived nutrition from the surrounding environment,
and were **heterotrophic,** as opposed to the later,
**autotrophic** organisms, which could synthesize all
their nutrients. The early unicellular prokaryotic
organisms probably utilized ATP for energy transfer, and fermentation for ATP synthesis may have
been the first metabolic pathway to evolve.

A second key event in the Archean Era was the
evolution of photosynthesis. The first photosynthetic cells were probably anaerobic bacteria. By
evolving a method of synthesizing nutrients, they
became autotrophic and independent of the nutrient content of the environment. These organisms,
which resembled the blue-green algae or cyanobacteria of the present, became one of the predominant organisms of the early oceans.

As the Proterozoic Era began, some 2.5 billion
years ago, life had evolved into complex and diverse prokaryotic communities. As envisioned by
Lynn Margulis, the landscape and the oceans were
covered:

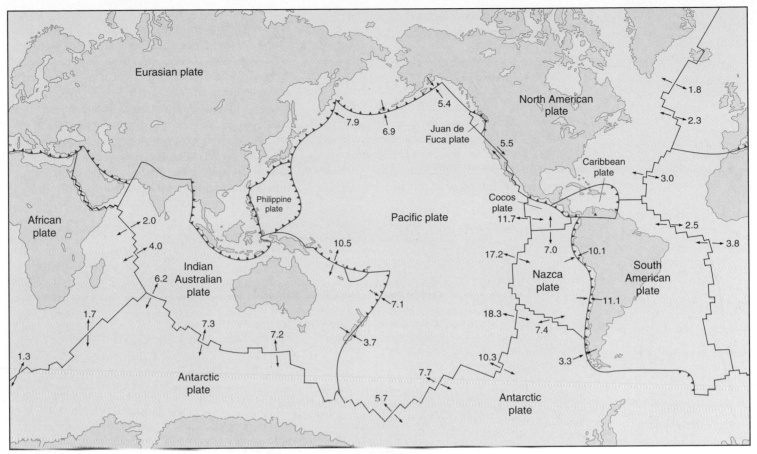

**✦ FIGURE 11.15**

**The Earth's surface is composed of a series of tectonic plates. Movement of the plates shapes the continents. The map shows the plates, the average rate of movement in centimeters per year, and the direction of movement. If you were born in Southern California, the region will move almost 15 feet to the northwest during your lifetime.**

by the blue-green and purple scum of photosynthetic bacteria underlain by yellow, brown and black layers of nonphotosynthetic and anaerobic bacteria.

Several important evolutionary events occurred during The Archean Era, which covers 42% of all geological time. These include the accumulation of oxygen in the atmosphere, the appearance of eukaryotic cells, and the origin of multicellular organisms.

The evolution of the oxygen-containing atmosphere provided a new environment to be exploited by the development of aerobic forms of respiration, and also led to the formation of the ozone layer to protect evolving life-forms from the damaging effects of ultraviolet light.

The origin of eukaryotic cells is one of the most important events in evolutionary history, because it allowed life to develop levels of organizational and cellular complexity lacking in prokaryotic systems. The earliest fossil eukaryotic cells are algae, found in 1.2- to 1.4-billion-year-old rocks from Southern California. Fossils showing evidence of mitosis and meiosis are found in Australian rocks more than 1 billion years old; other rocks of similar age reveal

fossils from dozens of species of algae and fungi, indicating that eukaryotes underwent a burst of speciation involving several major groups. The evidence for the eukaryotic nature of these fossils is based on the size, relative complexity and multicellular nature of the preserved cells.

Eukaryotic cells may have evolved from a **symbiotic** relationship between two or more prokaryotic organisms. Symbiosis is a long-term interactive association between two or more species. The origin of eukaryotes by symbiosis is discussed in Chapter 12.

Multicellular organisms are found in rock formations from the late Proterozoic Era, but there is little evidence in the fossil record about the transition from unicellular eukaryotes to multicellular forms. Colonial organisms such as sponges represent intermediates between unicellular and multicellular forms. By the end of the Proterozoic Era, multicellular animals had evolved, as evidenced by fossilized burrows and skeletonized remains. In addition, impressions from soft-bodied forms such as jellyfish are present in fossils from about 700 million years ago.

## Paleozoic Era Was a Period of Shifting Landmasses

The Paleozoic Era spanned the period from 570 million years ago to 245 million years ago and covers some 7.1% of geologic time. It is subdivided into the Cambrian, Ordovician, Silurian, Devonian, Carboniferous, and Permian Periods (◆ Table 11.4).

During the Proterozoic Era, geological changes increased the size of the continental landmasses, and tectonic movements rearranged the landmasses to some degree, but these events had little effect on the evolved life-forms of the time. At the beginning of the Paleozoic Era, there were six major continental masses, and the movement of these continental masses resulted in dramatic shifts in environments and changes in the climate (◆ Figure 11.16). These changes included mountain building, continental collisions, glacier and ice sheet formation, and major shifts in ocean levels. These changes, coupled with the evolution of the major animal and plant groups, produced alternating radiations and extinctions of life-forms in the major animal groups.

**TABLE 11.4   MAJOR EVOLUTIONARY EVENTS OF THE PALEOZOIC ERA**

| AGE (MILLIONS OF YEARS) | GEOLOGIC PERIOD | | INVERTEBRATES | VERTEBRATES | PLANTS |
|---|---|---|---|---|---|
| 245 | Permian | | Largest mass extinction event to affect the invertebrates. | Acanthodians, placoderms, and pelycosaurs become extinct. Therapsids and pelycosaurs the most abundant reptiles. | Gymnosperms diverse and abundant. |
| 286 | Carboniferous | Pennsylvanian | Fusulinids (a type of foraminifera) diversify. | Reptiles evolve. Amphibians abundant and diverse. | Coal swamps with flora of seedless vascular plants and gymnosperms. |
| 320 | Carboniferous | Mississippian | Crinoids, lacy bryozans, become abundant. Renewed adaptive radiation following extinctions of many reef-builders. | | Gymnosperms appear (may have evolved during Late Devonian). |
| 360 | Devonian | | Extinctions of many reef-building invertebrates near end of Devonian. Reef building continues. | Amphibians evolve. All major groups of fish present—Age of Fish. | First seeds evolve. Seedless vascular plants diversify. |
| 408 | Silurian | | Major reef building. Diversity of invertebrates remains high. | Ostracoderms common. Acanthodians, the first jawed fish, evolve. | Early land plants—seedless vascular plants. |
| 438 | Ordovician | | Extinctions of a variety of marine invertebrates near end of Ordovician. Major adaptive radiation of all invertebrate groups. Suspension feeders dominant. | Ostracoderms diversify. | Plants move to land? |
| 505 | Cambrian | | Many trilobites become extinct near end of Cambrian. Trilobites and brachiopods are most abundant. | Earliest vertebrates—jawless fish called ostracoderms. | |
| 570 | | | | | |

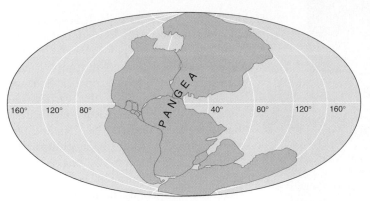

(a) 225 million years ago: Triassic

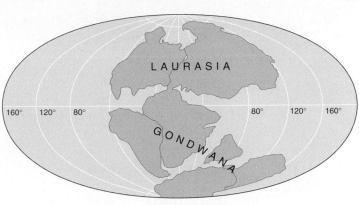

(b) 135 million years ago: Cretaceous

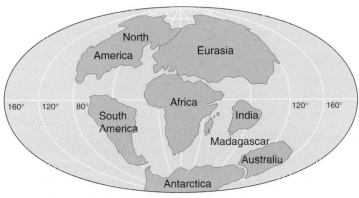

(c) 65 million years ago: Paleocene

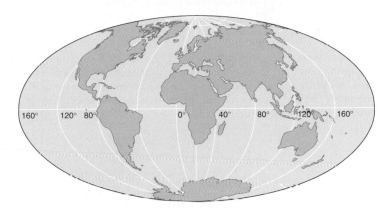

(d) Present

✦ FIGURE 11.16

The shapes and locations of major landmasses change over time. The accompanying changes in climate altered the distribution of plants and animals. (a) About 225 million years ago, the continents formed a supercontinental landmass called Pangaea. (b) By 135 million years ago, in the Cretaceous, two major landmasses, Laurasia and Gondwana, had formed. (c) In the Paleocene, 65 million years ago, the continents began to assume familiar forms. North America was connected to Eurasia, and separated from South America. (d) The present distribution of land masses.

The Cambrian Period was marked by the development of almost all the major animal groups, although animal life was dominated by just three groups: trilobites, brachiopods, and archaeocyathids. Trilobites (✦ Figure 11.17) were a dominant group, and may have composed almost half of all the total animal life during the Early and Middle Cambrian. They suffered mass extinction near the end of the Cambrian. It has been suggested that their habitat in warm, shallow seas was destroyed by a temporary cooling of the oceans, perhaps caused by a mixing with waters from the colder, deep oceans. Archaeocyathids were organisms that built conical or tubular skeletal structures, and probably fed in the same way as sponges. Like trilobites, they lived in warm shallow water, and became extinct near the end of the Cambrian.

✦ FIGURE 11.17

Trilobites were a common life-form in the oceans some 500 to 600 million years ago.

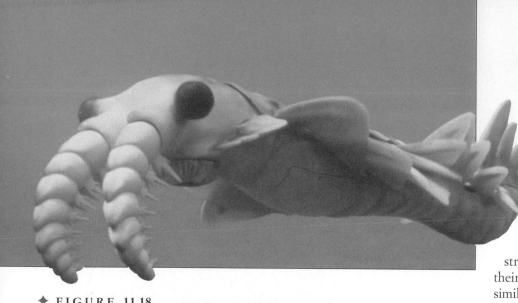

**✦ FIGURE 11.18**

Large predators like this reconstructed *Anomalocarus* cruised the oceans in the Cambrian, some 500 million years ago. In the Cambrian, organisms with complex body plans suddenly appeared. Species with exoskeletons, legs, joints, and antennae became common. The first large predators appeared, with some species of *Anomalocarus* reaching 2 meters (6–7 feet) in length.

The Ordovician Period was marked by significant geological changes and corresponding changes in the number and distribution of animal life-forms (✦ Figure 11.18). There was a dramatic increase in the number and diversity of animal species, especially shelled marine invertebrates and reef building organisms such as corals. The organisms in the marine community began to establish complex ecological relationships (✦ Figure 11.19). The end of the Ordovician Period was marked by a period of mass

extinctions of marine animals. It is speculated that glacier and ice sheet formation may have lowered ocean levels, causing mass extinctions. More than 100 major groups of marine organisms became extinct at this time.

The extinctions of the Ordovician were followed by another major radiation that took place during the Silurian and Devonian Periods. Many of the groups that had been decimated by the mass extinctions of the Ordovician such as brachiopods and corals flourished and diversified in the following periods. Reef structures were a major feature of the oceans, and their organization and structure were remarkably similar to those in modern oceans.

The evolution of fish, which began at the Cambrian/Ordovician junction, underwent a major radiation during the Devonian Period (✦ Figure 11.20). These fish were the first **vertebrates,** and the rapid speciation of fish signaled the beginning of the diversification of vertebrates. During the Devonian radiation, cartilaginous fish, represented today by sharks and rays, first appeared, as did the bony fish. The bony fish are the most numerous and varied of all fishes, and gave rise to the amphibians. The bony fish are divided into two groups: ray-finned fish and lobe-finned fish. The earliest amphibians are known from fossils in the Devonian and had many structural similarities to the lobe-finned fish from which they had evolved.

Amphibians were not the first organisms to invade the land. They were preceded by plants that

**✦ FIGURE 11.19**

In the Ordovician, adaptive radiations of bryozoans and corals created diverse and complex communities of organisms.

**✦ FIGURE 11.20**

Fish underwent a major adaptive radiation during the Devonian, and their rapid speciation signaled the beginning of the diversification of vertebrates.

## Palynology—A Link between Geology and Biology

REED WICANDER

I became a geologist with a specialty in palynology by a circuitous route that involved a certain amount of luck and the good fortune to have had classes from some very inspirational professors. When I began college, I intended to go into forestry because I had always had an interest in biology and the outdoor life. Along with an introductory biology course, I also enrolled in an introductiory geology course. One of the things about geology that appealed to me was that it applied all of the sciences (including biology) to the study of the Earth. It wasn't until I took a paleontology course that I realized I could combine my interest in geology and biology into a profession that offered, for me, the best of both worlds. I also realized that I had better take as much biology as I could so that I could apply my knowledge of modern ecosystems to the study of ancient ones.

Palynology is the study of organic microfossils called *palynomorphs*. These include such familiar things as spores and pollen (both of which cause allergies for many people), but also acritarchs, dinoflagellates (marine and freshwater, single-celled phytoplankton, some species of which are responsible for red tides), chitinozoans (marine, vase-shaped microfossils of unknown origin), scolecodonts (jaws of marine annelid worms), and microscopic colonial algae. Fossil palynomorphs are extremely resistant to decay and are extracted from sedimentary rocks by dissolving the rock in a variety of acids.

My particular area of palynologic research involves acritarchs, which lived in the world's ancient oceans approximately 1.8 billion years ago to the present, but were most abundant about 570 to 360 million years ago, and during the time they flourished, they formed the base of the marine food chain. Fluctuations in their diversity and abundance during that time thus influenced the evolution of the marine ecosystem. Therefore, research into their origins, diversification, and extinctions is an integral part of any study of the world's ancient marine ecosystem and the interrelationship between the earth's various systems.

The occurrence and distribution of acritarchs allows geologists to plot the location and movement of continents and ocean basins during the past. Furthermore, because of the short geologic range of many acritarch species, their occurrence in rocks can be used in age-dating locally, regionally, and globally. Acritarchs are also important to the petroleum industry because they are the primary constituent in petroleum formed during the time interval when they flourished. Their concentration and subsequent conversion into petroleum has, in part, helped fuel the industrialization that occurred during the past century.

During the past 20 years, I have focused primarily on the usefulness of acritarchs in age-dating, correlation of contemporaneous rock units throughout the world, delineation of ocean basins, determination of ancient marine environments, and on attempting to determine their biological affinities, evolutionary history, and impact on ecosystem stability through time. My research has allowed me to travel all over the world collecting samples and meeting other researchers.

Another area of palynology that attracts many biologists and geologists is the study of spores and pollen, which can tell us when plants colonized the Earth's surface, which, in turn, influenced weathering and erosion rates, soil formation, and changes in the composition of atmospheric gases. Furthermore, because plants are not particularly common as fossils, the study of spores and pollen can frequently reveal the origin and extinction of various plant groups.

Because of the importance of plants in the world's ecosystem, geology and biology are linked in the field of palynology. Because of this linkage, I have found the study of palynology personally and professionally satisfying.

*Reed Wicander is a professor in the Geology Department at Central Michigan University. He has coauthored five books and numerous papers on various aspects of Paleozoic palynology. He was president of the American Association of Stratigraphic Palynologists in 1994–1995.*

colonized the land at the Ordovician/Silurian transition. In addition, insects, spiders, and snails were land dwellers before the amphibians. Amphibians underwent adaptive radiations during the Carboniferous and Permian periods, but were limited in distribution by their need to return to water for reproduction. The evolution of an egg with compartmented sacs (one a liquid-filled sac surrounding the embryo, one a food sac, and one for waste accumulation) allowed land vertebrates to reproduce without fertilization and development in water. The evolution of this structure, called the **amniote egg,** marks the emergence of the reptiles (✦ Figure 11.21). Reptiles were successful at displacing amphibians from many habitats, and diversified into plant-eating and meat-eating groups.

✦ **FIGURE 11.21**
**The adaptive radiation of reptiles was associated with the development of an egg that could survive in a dry environment. This reptile egg, recovered from the Gobi Desert in Mongolia, shows skeletal components similar to those of birds.**

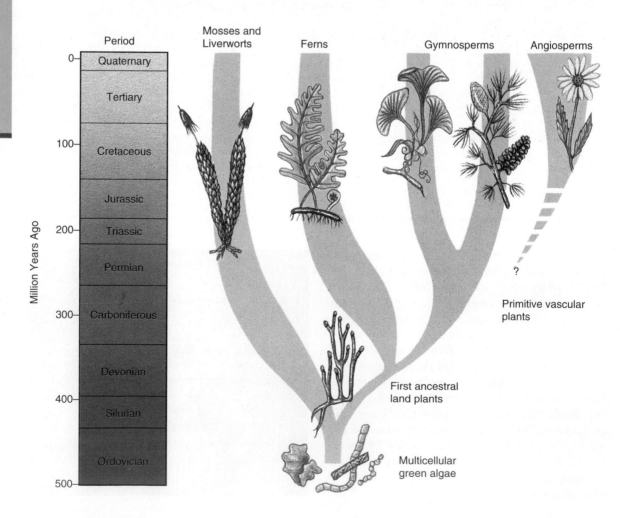

As mentioned earlier, plants colonized the land before amphibians. The transition to land depended on solving the same problems faced later by animals: skeletal support, protection from desiccation, and the increased effects of gravity. Plants with specialized cells for the movement of water and nutrients, root systems, and leaves were most successful in the terrestrial environment. These vascular plants underwent a large radiation during the Silurian Period (◆ Figure 11.22). Seedless vascular plants are similar to amphibians in that they require water for fertilization. The development of the pollen tube and seeds allowed plants to escape the need for water during reproduction. Although the seedless vascular plants were a dominant part of the landscape during the Carboniferous Period, the first seed plants, the **gymnosperms,** were also present.

The Paleozoic Era came to an end with a mass extinction of animal groups, the largest of the five extinctions by far. About 50% of marine invertebrate species became extinct, as did 75% of the amphibians and 80% of the reptiles. Some plant species became extinct at this time, but most survived. Continental collisions caused the formation of a giant landmass, Pangaea, and the consequent changes in climate, ocean levels, and mountains played a major role in these extinctions.

## Mesozoic Era Was the Age of the Dinosaurs

The Mesozoic Era comprised some 3.9% of geological time, beginning some 245 million years ago, and ending some 65 million years ago. It is subdivided into the Triassic, Jurassic, and Cretaceous Periods. During the Mesozoic, complex geological changes caused the breakup of Pangaea. North America separated from Africa, South America separated from North America, and the recognizable continental outlines of the present world began to form. These events had a profound effect on the climate and atmospheric circulation, resulting in a gradual warming of the earth.

In the oceans, survivors of the Paleozoic extinctions repopulated the invertebrate habitats, with mollusks becoming diversified and much more dominant than in the Paleozoic. The cartilaginous fishes including the sharks became abundant during the Mesozoic, but the lobe-finned bony fish de-

*Sidebar*

### THE EVOLUTION OF BIRDS

Near the end of the Paleozoic Era the major groups of amniote vertebrates appeared over a relatively short span of time. This rapid appearance of such diverse groups of vertebrates makes it difficult to estimate their relationships to one another. However, the fossil bird, *Archaeopteryx,* provides strong evidence for a link between birds and dinosaurs.

Until recently support for a bird–crocodilian relationship rested primarily in fossil data. Now studies have shown that DNA sequences from slowly evolving genes (mitochondrial 12S and 16S rRNA, tRNA$^{val}$, and nuclear alpha-enolase) support a close relationship between birds and reptiles. Recombinant DNA techniques in conjunction with the fossil record conclude a bird–reptile relationship is likely.

◆ **FIGURE 11.22**
**The adaptive radiation of plants, beginning with the first ancestral land plants.**

✦ FIGURE 11.23

Dinosaurs became abundant during the Jurassic and Cretaceous Periods, and were successful life forms for almost 140 million years. In contrast, our species has been in existence for about 200,000 years.

clined and became almost extinct during this era. Although frogs and salamanders appeared during the Mesozoic, as a group, amphibians continued to decline during the Mesozoic.

In the minds of most people, the Mesozoic is primarily the age of the reptiles, particularly the dinosaur. There was a rapid diversification of reptiles during the Mesozoic (✦ Figure 11.23) giving rise not only to the dinosaurs, but also to the ancestors of the birds and mammals as well. Four major groups of reptiles are represented in the fauna of today: the turtles, snakes and lizards, tuataras, and crocodiles. Dinosaurs appeared during the Late Triassic Period but became most abundant during the Jurassic and Cretaceous Periods, and were the dominant animal life-form for almost 140 million years.

One of the debates about dinosaurs involves the question of whether they had ability to regulate body temperature. Living reptiles are all **ectotherms,** whose body temperature is determined by the environment. **Endotherms** are warm-blooded animals and have the ability to maintain a constant body temperature over a wide range of environ-

mental conditions. Were dinosaurs ectotherms or endotherms? This controversy is more than an academic argument, and relates to questions of metabolic rates, food consumption by predators, brain size, and physical activities such as running and flying. There is no direct way to answer this question, but using indirect evidence, it seems likely that the small dinosaurs ancestral to birds were endothermic or partly endothermic. Larger herbivores may have been ectothermic, or somewhat intermediate between the two states, with some ability to regulate body temperature.

Plant species continued to diversify during the Mesozoic, with the development of flowering plants, the **angiosperms,** taking place in the Cretaceous. This now dominant plant group rapidly spread into almost every available land habitat.

The close of the Mesozoic was marked by another round of mass extinctions. Dinosaurs, marine reptiles, and marine invertebrates are among the groups that became extinct at this time. Although still debated, it appears that this extinction occurred following the crash of a meteorite near the

(a)

**✦ FIGURE 11.24**

(a) The end of the Cretaceous was marked by major extinctions. The extinctions may have been associated with the impact of a meteorite, near the Yucutan Peninsula in Mexico. (b) The Cretaceous-Tertiary boundary is visible in geological formations as a layer of thin, white clay.

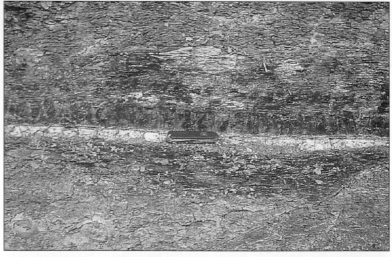

(b)

Yucutan region of Mexico (✦ Figure 11.24), which triggered climatic changes and the collapse of food chains.

None of the other four major extinctions shows any connection to meteorite impacts, and all appear to be associated with changes in climate or sea levels.

## Cenozoic Era Includes the Present Day

The Cenozoic Era, which began some 65 million years ago and encompasses the present, makes up 1.4% of geological time. During the Cenozoic, tectonic movements shaped the continents into the dispersed forms we observe today. India moved northward, collided with Asia, and formed the Himalayas. Other mountain ranges were formed at this time, including the Andes and the Alps. Mountain building caused significant shifts in the climate, creating new habitats in the form of vast stretches of dry grasslands.

During the early Cenozoic, birds rapidly evolved into the species we observe today, and placental

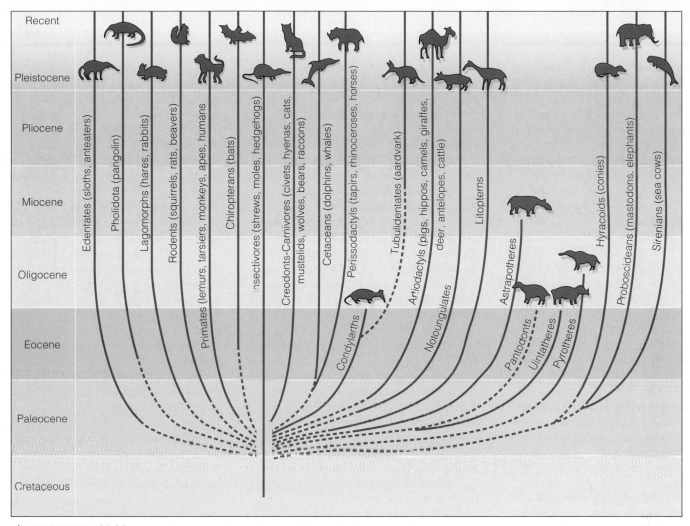

**✦ FIGURE 11.25**

**During the Paleocene the placental mammals underwent an adaptive radiation that lead to the major groups that populate the world today.**

mammals underwent a great adaptive radiation (✦ Figure 11.25), and the first rodents and primates appeared. The primates include lemurs, monkeys, apes, and humans. During this time, primates developed three major adaptations: opposable thumbs, stereoscopic vision, and enlarged brains.

The flowering plants continued their diversification during the Cenozoic, and formed complex communities in rain forests and jungles. Major climatic changes caused these communities to retreat to the more tropical regions of Central and South America.

## THE EVOLUTIONARY PROCESS AND THE FIVE KINGDOMS

In the preceding sections, observations from several disciplines including comparative anatomy, embryology, biochemistry, and DNA sequences provide evidence for the evolutionary origins and relatedness of living and extinct organisms.

The reconstruction of evolutionary relatedness among living organisms is part of **systematics.** Systematics uses information provided by observational and experimental methods to classify organisms. Currently, a system that divides organisms into five kingdoms is widely used. These five kingdoms, Monera, Protista, Plantae, Fungi, and Animalae, are discussed in the remaining chapters in this part.

# SUMMARY

1. An understanding of the formation and types of fossils is important in interpreting the past events in the evolution of life on Earth. Fossils are the remains, traces, or impressions of prehistoric organisms preserved in the Earth's crust.

2. Dating of fossils employs various techniques including measuring the rate of radioactive decay of isotopes. The history of the Earth has been divided into four main eras, each characterized by abrupt changes in the fossil record: Archean/Proterozoic, Paleozoic, Mesozoic, and Cenozoic. They span the time from the origin of the Earth some 4.6 billion years ago to the present.

3. The large-scale events in the origin, diversification, and extinction of species of plants and animals make up macroevolution. Evidence at this level is based on large-scale events and trends in the pattern of evolution.

4. Comparative anatomy compares the structural features of living and extinct organisms to document the changes in function that a basic structure has undergone during evolution. Such anatomical similarities are evidence for descent from a common ancestor.

5. Comparative embryology examines the preadult stages of development to record the nature and number of developmental processes and the modifications and adaptations of these processes during the course of evolution. Such relationships are used to ascertain evolutionary relationships and the mechanisms by which modifications in structure arose.

6. Comparative biochemistry and DNA studies use molecular techniques to establish evolutionary relationships and can provide quantitative information about the evolutionary distance between individuals species or higher taxonomic groups.

7. Biogeography and artificial selection are two of the classic lines of evidence used to support and document the events in evolution. Biogeography takes note of the geographic distribution of species, and uses this information to reconstruct the evolutionary history of related and neighboring species. Artificial selection mimics the course of natural selection, demonstrating that the genotypes of organisms contain a large amount of genetic variability.

8. Large-scale events such as adaptive radiations, extinctions, and convergent and parallel evolution are patterns of evolutionary change that many groups of organisms have undergone in response to natural selection, again demonstrating the genetic variability present in the gene pools of many species.

9. The large-scale events in evolution include the coevolution of the geological and biological worlds. In each of the major eras, dramatic changes occurred in both the geology and biology of the earth. In the Archean Era, which covers 45% of all geological time, life evolved on earth, and photosynthesis began oxygenating the atmosphere. In the Protozoic, which covers some 42% of geological time, eukaryotic cells arose and multicellular organisms appeared, and the land was colonized.

10. In the Paleozoic Era, spanning some 7% of geological time, life-forms diversified, and large-scale geological changes rearranged the shapes and sizes of the continents. Major radiations and extinctions occurred during this era. In the Mesozoic Era, geological changes produced the major continents of today, causing drastic changes in the world climate. Dinosaurs and other reptiles diversified and became dominant life-forms, but a mass extinction at the end of the Mesozoic greatly reduced the reptiles, and the dinosaurs became extinct.

11. In the Cenozoic Era, which began some 65 million years ago and composes some 1.4% of geological time, birds, mammals, and flowering plants underwent a burst of speciation. Mountain building and changes to a cooler and drier climate eliminated many species of gymnosperms, but allowed the flowering plants to become the dominant form of plant life.

# KEY TERMS

adaptive radiation
amniote egg
anaerobic
analogous structures
angiosperms
Archean/Proterozoic Era
artificial selection
autotrophic
biogeography
body fossils
Cenozoic Era
convergent evolution

coprolites
dendrochronology
divergent evolution
DNA hybridization
ectotherms
endotherms
evolutionary trees
gymnosperms
half-life
heterotrophic
homologous structures
macroevolution

Mesozoic Era
microevolution
morphological convergence
Paleozoic Era
parallel evolution
plate tectonics
sedimentary rocks
symbiotic
systematics
trace fossils
vertebrates
vestigial structures

## SHORT ANSWER

1. What are the distinguishing differences between macroevolution and microevolution?
2. Explain how Darwin and other scientists during his time used fossils to describe evolutionary events. Detail how fossil dating is currently done.
3. Why is carbon dating only used on the biological remains of organisms? Speculate why this method can be used only on specimens up to 60,000 years old.
4. Describe the DNA hybridization technique as it applies to evolutionary patterns.
5. What fundamental problems do plants and animals need to overcome before successfully colonizing land?

## MATCHING

Match the evolution terms with the appropriate example.

1. homologous structures
2. analogous structures
3. morphological convergence
4. vestigial structures
5. adaptive radiation
6. convergent evolution

___ brachiopods and clams
___ appendix and dewclaw
___ Galapagos Island finches
___ limbs of gophers and limbs of elephants
___ bird wings and insect wings
___ wings of birds and fins of penguins

Match the evolutionary period appropriately.

7. Archean
8. Proterozoic
9. Paleozoic
10. Cambrian
11. Mesozoic
12. Cenozoic
13. Devonian

___ age of the dinosaur
___ rapid speciation of fish
___ no ozone layer, evolution of photosynthesis
___ encompasses the present
___ accumulation of oxygen, appearance of eukaryotic cells
___ shifting continental masses
___ development of major animal groups

## FILL IN THE BLANK

1. Similarities in embryonic development in vertebrates has come about by _____ bodily structures for new functions and not by _____ basic developmental processes.
2. Evolutionary trees arrange organisms closer together who share _____ amino acid differences and places those with _____ differences further apart.
3. Humans are more closely related to _____ than to _____.
4. The study of plant and animal distribution across the earth is called _____.
5. The result of a species' inability to adapt to environmental changes is _____.
6. The theory that large segments of the outer part of the earth move relative to one another is called _____.
7. Body temperature is regulated by the environment in _____, whereas _____ have the ability to maintain a constant body temperature independent of environmental conditions.

## SCIENCE AND SOCIETY

1. Livestock growers have added antibiotics to cattle and pig feed for over 30 years to protect animals confined in close quarters from infectious diseases. Antibiotics were first given to fight diseases, which could spread rapidly in confined areas. Farmers soon learned that antibiotics also accelerated the rate of growth in these animals. Microbiologists fear that widespread use of antibiotics in livestock feed could promote the emergence of new strains of bacteria that are resistant to antibiotics. Do microbiologists have a legitimate argument? Would reducing the amount of antibiotics added to livestock feed reduce the risk of the evolution of superstrains of bacteria? Could the bacteria in livestock be transferred to humans? Do you know of any antibiotic-resistant strains of bacteria? Do the benefits of using antibiotics in livestock outweigh the potential health effects? Would reducing or banning antibiotic use have an impact on the economy?
2. The debate between proponents of evolution and creationism has been long-standing. The theory of evolution states that organisms arise by descent and modification from previously existing organisms. It states that the Earth is very old (about 4.5 billion years old) and that over time species have changed through natural selection. Some species went extinct, and others gave rise to entirely new species. Creationism is based on interpretations of the Bible. Some proponents of creationism say the Earth is very young (about 6,000 years old). Furthermore, creationists believe that God designed different species using similar blueprints and that the presence of common anatomical features among diverse organisms and the presence of analogous structures lend support to their ideas. They believe all creatures arose independently and therefore are unrelated to any other organism. Should creationism be taught in the classroom along with the theories of evolution? Why or why not? Who should decide whether creationism is taught in the classroom? Do you think the interpretation of the Bible is a scientific way to study nature? Do you think there are enough gaps in our knowledge of evolution that the current theories should be questioned? Would the beliefs of creationism fill these gaps?

# VIRUSES, PROKARYOTES, AND PROTISTS

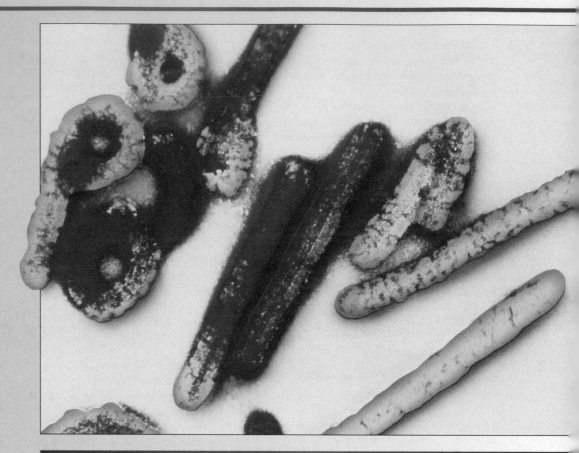

## OPENING IMAGE

*Ebola virus is one of a number of
viruses adapting to the human species
as a new host. These emerging
viruses are responsible for new
diseases.*

I n the fall of 1989, a nondescript building in an office park in Reston, Virginia, a
suburb of Washington, D.C., became the center of intense activity. The building
housed monkeys imported into the United States for medical research. That fall, a
shipment of crab-eating monkeys arrived from the Philippine Islands. The monkeys were
quarantined in the building for a month to prevent the spread of infectious diseases they
may have carried. Shortly after their arrival, the monkeys began to die. The staff veterinar-
ian turned to the scientists at the Army Medical Institute at Fort Detrick in nearby Frederick, Maryland, for assistance. A preliminary diagnosis indicated that the monkeys were
ill with a viral infection called simian fever. This virus infects monkeys but not humans.

After further tests, the Army Medical Institute determined that the virus was not
simian hemorrhagic virus (SHV), but instead was a virus called the Ebola virus, which
can infect humans. Ebola virus is one of the deadliest viruses ever discovered. Outbreaks
in Sudan and Zaire in 1976 killed almost 600 people, and an outbreak in Zaire in 1995
killed another 200. The Zaire strain is the deadlier of the two; it kills 75% of those it infects. It was feared that an outbreak in the heavily populated area around Reston could
kill thousands or tens of thousands of people.

Staff from the Centers for Disease Control and the Army Medical Institute worked to
contain a possible outbreak. The building was closed, the monkeys destroyed, and the
facility was decontaminated by workers in special suits. Although some of the workers at

*the monkey facility became ill, none developed Ebola fever, and no outbreak occurred in the human population. This particular strain of Ebola was named Reston, and became one of the four known strains of this filovirus. Why the Reston strain does not infect humans while others kill almost everyone they infect is not yet known.*

*Viruses such as the filoviruses belong to a group known as emerging viruses. These are either previously unknown viruses coming to our attention in outbreaks, or known viruses that are expanding their host range by jumping into new hosts, including the human population.*

*In this and the next two chapters, we will explore the diversity of known life-forms, and those on the edge of life, such as viruses. Although classification systems are in place to describe organisms and help define their evolutionary origins and history, many thousands—perhaps hundreds of thousands—of species and strains remain to be discovered. Some come to our attention in dramatic fashion as agents of disease like Ebola viruses, whereas others grow quietly in locations as familiar as the sand bottoms near our coastal beaches, awaiting discovery.*

## SPECIES AND CLASSIFICATION: THE FIVE KINGDOMS

For many years, organisms were classified into two kingdoms, Animalae and Plantae. Organisms that moved from place to place and obtained nutrition by eating other organisms (heterotrophs) were animals, and those that did not move and synthesized their own nutrients (autotrophs) were plants. In this scheme, protozoa (which move and eat other organisms) were animals, and fungi were grouped with plants.

Around 1970, a new, five-kingdom classification scheme was proposed by Robert Whittaker. This system is based on grouping organisms by their cellular organization and how they obtain nutrients. The five kingdoms in this system are the Monera, Protista, Fungi, Plantae, and Animalia. Since then, other schemes employing additional kingdoms have been proposed, but for the present, the five-kingdom system is the most widely accepted, and is the one we use in this book.

The Monera are the prokaryotes, and are classified on the basis of their cellular structure (see Chapter 3) and metabolism (see Chapter 4). One group of ancient bacteria, the Archaebacteria, are so different from other prokaryotes that they may eventually be placed in a separate kingdom. The Protists are eukaryotic organisms that are difficult to classify, and include a diverse array of organisms that are not fungi, plants, or animals. This kingdom contains some organisms that are autotrophs and others that are heterotrophs. Most protists are single celled, but some scientists consider multicellular algae to be protists, because they are similar to single celled forms of algae.

The other kingdoms contain multicellular eukaryotes distinguished primarily by how they obtain nutrients. Fungi decompose dead and dying organisms and absorb nutrients from them. Plants produce nutrients by photosynthesis, and animals eat other organisms. ◆ Table 12.1 lists some of the characteristics of the five kingdoms. In this and the next two chapters, we will focus on members of each kingdom. Before discussing the Monera, we will consider viruses, viroids, and prions. Viruses do not have all the properties of organisms; instead, they infect and use living cells for reproduction.

| | MONERA | PROTISTA | FUNGI | PLANTAE | ANIMALIA |
|---|---|---|---|---|---|
| Cell type | Prokaryotic | Eukaryotic | Eukaryotic | Eukaryotic | Eukaryotic |
| Cell wall | Present | Present in some forms | Present | Present | Absent |
| Mitochondria | Absent | Present | Present | Present | Present |
| Chloroplasts | Absent | Present in some forms | Absent | Present | Absent |
| Nutrition | Autotrophic Heterotrophic | Photosynthetic Heterotrophic | Heterotrophic | Photosynthetic | Heterotrophic |
| Multicellular forms | Absent | Present in some forms | Present in most forms | Present in all forms | Present in all forms |

**TABLE 12.1 CHARACTERISTICS OF THE FIVE KINGDOMS**

## VIRUSES ARE INFECTIOUS PARTICLES, BUT ARE NOT ALIVE

In a system of classification that depends on cell structure and mode of nutrition, viruses are excluded. A **virus** is a submicroscopic, infectious particle composed of a protein coat and a nucleic acid core (✦ Figure 12.1). Like cells, however, viruses carry genetic information, can undergo genetic events such as mutation, and can reproduce. Unlike cells, virus particles are unable to carry out functions such as metabolism or energy transfer. As a result, viruses are **intracellular parasites** that enter a host cell and take over the host's cellular machinery to produce new viral particles. These new viruses are released, often killing the host cell.

Two features are used in classifying viruses: the type of nucleic acid they contain and the shape of the protein capsule that surrounds the nucleic acid (Figure 12.1). Although these characteristics are used in classifying viruses, they may or may not be based on evolutionary relationships.

## Viruses Are Intracellular Parasites

Several types of replication cycles are known in viruses. The simplest form is illustrated by viruses that invade and kill bacterial cells. These **bacteriophages** attach to the outer surface of a bacterial cell and inject their nucleic acid. Once inside, the nucleic acid is transcribed, and the viral gene products make dozens of copies of the viral nucleic acid and the protein capsule. These components are assembled into new viral particles that are released by destruction of the host cell. Other bacteriophage can infect a bacterial cell and insert their DNA into the host cell's DNA (✦ Figure 12.2). When the host bacterial cell replicates its DNA, the hitchhiking viral DNA is replicated and passed on to all descendants. Under certain conditions, the viral DNA can detach from the host chromosome and direct the synthesis of multiple virus particles that destroy the bacterial cell as they are released.

Some viruses that invade animal cells replicate without immediately killing the host cell. New virus particles are released by budding off from the host cell's plasma membrane (✦ Figure 12.3), so that the host cell continues to survive for some time as a vi-

✦ **FIGURE 12.1**

**Viruses consist of a nucleic acid core surrounded by a protein coat. (a) Globular influenza virus. (b) Polyhedral adenovirus. (c) Cylindrical tobacco mosaic virus. (d) Odd-shaped T4 bacteriophage.**

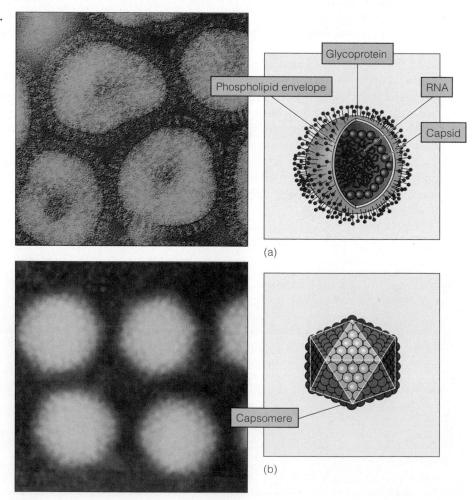

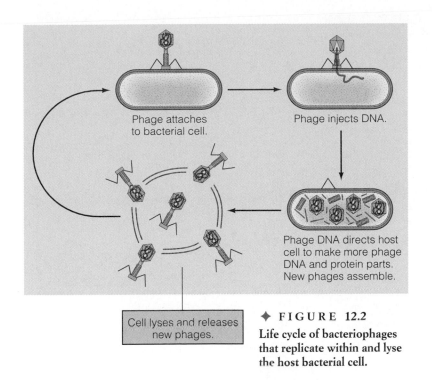

Phage attaches to bacterial cell.

Phage injects DNA.

Phage DNA directs host cell to make more phage DNA and protein parts. New phages assemble.

Cell lyses and releases new phages.

◆ **FIGURE 12.2**
**Life cycle of bacteriophages that replicate within and lyse the host bacterial cell.**

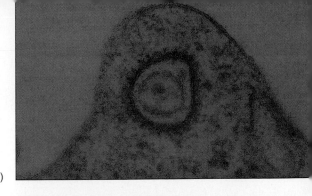

(a)

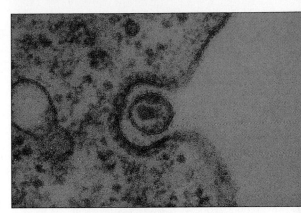

(b)

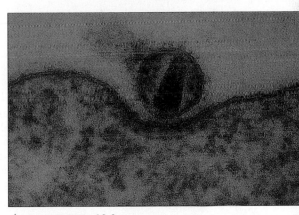

(c)

◆ **FIGURE 12.3**
**Copies of the HIV virus bud off from the host cell. (a) A virus particle near the cell surface. (b) The virus is released from the cell. (c) The virus freed from the cell surface can infect another cell.**

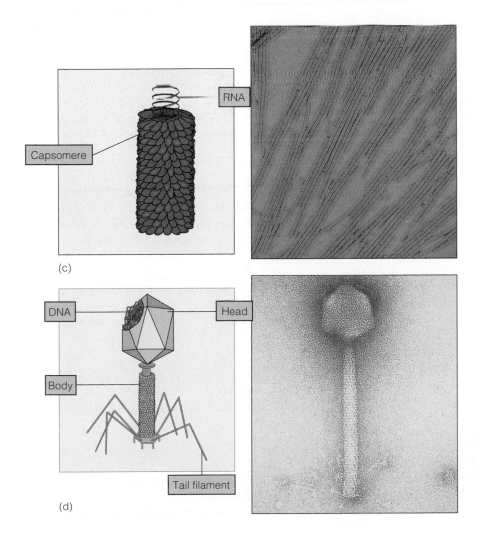

RNA

Capsomere

(c)

DNA

Head

Body

Tail filament

(d)

## Sidebar

### VIRAL DISEASES IN PLANTS

Mention viral diseases, and most everyone thinks of viral infections in humans or animals. But at least a thousand plant diseases are caused by viral infections. Many of these diseases are transmitted by insects such as aphids and leafhoppers, which suck fluids from plants. In plants, viral diseases usually cause areas of dead tissue called necroses, or mosaic diseases, in which parts of the plant have a mottled light green or yellow color. In leaves and other tissues containing chloroplasts, this is often caused by reducing or eliminating these organelles. Plant viral diseases are a serious agricultural problem in many parts of the world because they reduce crop yields.

Not all viral diseases in plants are destructive, however. The white or yellow borders on some ornamental plants are caused by viral infections, and the streaked flower color of tulips is caused by a viral infection.

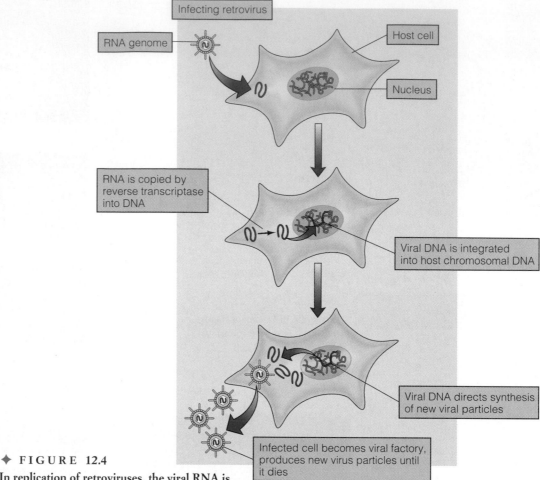

✦ **FIGURE 12.4**

**In replication of retroviruses, the viral RNA is copied into a double-stranded DNA, which can insert into a chromosome of a host cell. From there, the viral DNA can direct the synthesis of RNA and viral proteins that are assembled into new viral particles.**

ral production line. The human immunodeficiency virus (HIV), a **retrovirus** associated with acquired immunodeficiency syndrome (AIDS), replicates in this way.

Retroviruses contain single-stranded RNA as their genetic material and employ an unusual method of reproduction. Once inside the host cell, the RNA is copied into a complementary DNA strand using a unique enzyme called **reverse transcriptase.** (The discovery of this enzyme and work on its importance for viral replication resulted in a Nobel Prize being awarded to Howard Temin and David Baltimore.) The single strand of DNA is copied into a complementary strand that combines with its template to form a double-stranded DNA. This double-stranded DNA is inserted into a chromosome of the host cell, where it is transcribed into new viral RNA genomes, which are incorporated into newly-synthesized protein capsules (✦ Figure 12.4).

## Viruses Cause a Wide Range of Diseases

Viruses are important agents of disease among many types of organisms, including invertebrates, plants, animals, and humans (✦ Table 12.2). Viral diseases include influenza, the common cold, herpes, measles, chicken pox, and encephalitis. Antibiotics are not effective against viral diseases, but protection against some viral diseases is offered by vaccination (Chapter 22). Disease-causing viruses are usually highly specific, invading only the cells of one or a small number of species, or only a few cell types. For example, the polio virus can only infect

| TABLE 12.2   SOME INFECTIOUS DISEASES CAUSED BY VIRUSES | |
|---|---|
| Measles | Rabies |
| Chicken pox | Encephalitis |
| Shingles | Hepatitis (A, B, and C) |
| Colds | AIDS |
| Influenza | Herpes |
| Viral pneumonia | |

cells of higher primates, such as apes and humans, and the adenovirus that causes colds infects only the cells of the respiratory tract.

During the last decade or so, a number of retroviruses have moved from their normal (mostly unknown) hosts into the human population. The best known of these viruses is HIV, associated with AIDS. HIV is closely related to a virus that infects monkeys, and apparently jumped into the human population somewhere in East Africa and spread from there to other parts of the world. Other potentially lethal viruses include the Marburg virus, the Ebola viruses, and the hantaviruses, which infect and kill in a matter of days. In some cases, such as the Marburg and Ebola viruses, outbreaks are sporadic and after an infected population dies, the virus is no longer detectable. In other cases, such as HIV, mutations in the virus have allowed it to adapt to humans as a new host. Given the mutation rate in viruses and the speed of world travel, outbreaks of new viral diseases remain a possibility.

## Viroids and Prions Are Infective Molecules

Although the union of a protein and a nucleic acid to form a virus appears to be the simplest form of reproductive system, it is not. Infective forms of nucleic acid without a protective coat of protein, called **viroids,** are known to be responsible for some plant diseases. Infective forms of proteins, known as **prions,** are responsible for fatal diseases of the nervous system in humans and other mammals.

## MONERA: THE KINGDOM OF BACTERIA

The kingdom Monera contains all species of bacteria, and these are among the most abundant and diversified life-forms on Earth. They have exploited

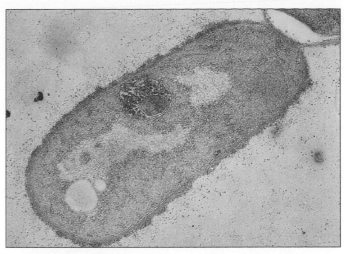

✦ FIGURE 12.5

Bacteria are characterized by the absence of a nuclear membrane and membranous organelles including mitochondria. The DNA is confined to the nucleoid region.

almost every available habitat on Earth, including volcanic vents at the bottom of the ocean, geysers, arctic snows, the water cooling jackets of nuclear reactors, the fuel tanks of jet aircraft, the folds, nooks, and crannies inside and outside the human body, and almost every other imaginable place.

Bacterial cells are characterized by the absence of a nuclear membrane and membrane-bound organelles (✦ Figure 12.5). They possess a type of flagella different than those of eukaryotes. Bacterial cells generally assume one of three shapes: rods (bacilli), spirals (spirilla), and spheres (cocci) (✦ Figure 12.6). Often, cells stick together to form filaments or clusters. Almost all bacteria have a cell

✦ FIGURE 12.6

The three most common shapes in bacteria are (a) rod-shaped, (b) spiral, and (c) spherical.

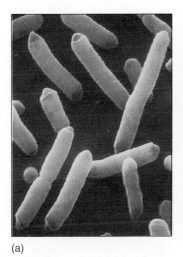

(a)

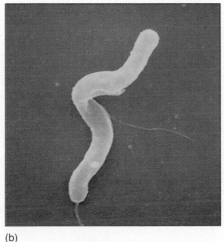

(b)

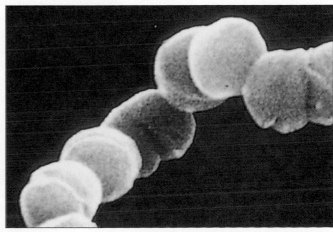

(c)

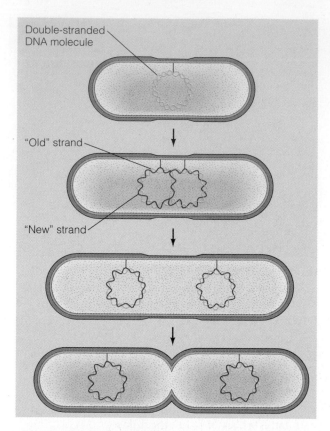

Double-stranded DNA molecule

"Old" strand

"New" strand

✦ FIGURE 12.7

**Bacteria reproduce by binary fission. The circular DNA molecule is replicated prior to cell division. Following replication of the DNA, the cell splits into two new bacteria.**

wall that is a combination of proteins and polysaccharides. Many have a sticky outer capsule that attaches the cell to a surface. Bacteria reproduce by a process called **binary fission.** Under proper growth conditions, bacterial cells enlarge, replicate their genetic information, and split into two identical cells (✦ Figure 12.7). In the laboratory, this event can occur every 20 minutes or so.

Nutritionally, bacteria are versatile organisms, and are able to obtain and transfer energy in a variety of ways. Some bacteria are able to synthesize organic molecules from simpler inorganic compounds, and are classified as **autotrophs.** Most bacteria require organic molecules as a source of nutrition and energy, and are classified as **heterotrophs.**

Several methods have been used to classify bacteria, including cell size, shape, the presence or absence of filaments, energy sources, reaction to stains, and biochemical tests.

## Archaebacteria Are Among the Most Ancient Organisms on Earth

This group of Monerans contains the oldest and most primitive organisms known. They are so different from other bacteria in the composition of their cell walls and biochemical features that they may belong in a separate, sixth kingdom, the Archaea. There are three types of archaebacteria: methanogens, halophiles, and thermoacidophiles. They live in extreme habitats that have probably changed little since early in the history of the planet.

Methanogens are anaerobic organisms that live in mud, stagnant swamps, sewage treatment plants, and the intestines of animals (including humans). These bacteria metabolize organic molecules and produce methane gas as a metabolic by-product. They are an important link in the recycling of carbon (discussed in Chapter 31). The halophiles and thermoacidophiles are aerobic organisms that also inhabit habitats with extreme conditions. Halophiles are found in salty environments such as the Great Salt Lake, the Dead Sea, and brackish ponds, where the concentration of salt can reach 25%. As their name implies, thermoacidophiles (*thermo* = hot, *acido* = acidic, *philes* = loving) inhabit hot, acidic environments such as hot springs, where the temperatures can reach 120 to 165°C (remember that water boils at 100°C), and the pH can be as acidic as 2 (with pH 1 being the lowest possible).

## Most Bacteria Are Classified as Eubacteria

The majority of bacteria are true bacteria (*eu-* is a prefix meaning true). They are subdivided by the means used to obtain energy: chemosynthetic, photosynthetic, and heterotrophic. The mechanism of energy transfer is the primary characteristic used in their classification. Chemosynthetic bacteria are autotrophic and are able to synthesize organic nutrients from inorganic compounds without the use of sunlight. They obtain energy from the oxidation of inorganic compounds such as ammonia:

$$2NH_4^+ + 3O_2 \rightarrow 2NO_2^- + 2H_2O + 4H^+ + \text{energy}$$

Other chemosynthetic bacteria are able to utilize energy released in the conversion of nitrite ($NO_2$) to nitrate ($NO_3$), or the oxidation of sulfur to sulfate:

$$2S + 3O_2 + 2H_2O \rightarrow 2SO_4^{--} + 4H^+ + \text{energy}$$

These organisms capture energy released by the oxidation of inorganic materials and use this energy in the synthesis of carbohydrates. Chemosynthetic bacteria that oxidize nitrogen compounds play an important role in the recycling of nitrogen in the biosphere.

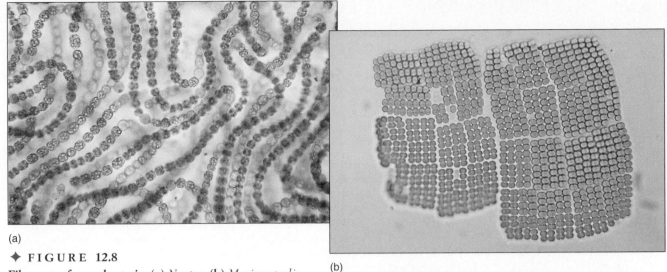

(a)

(b)

**✦ FIGURE 12.8**

**Filaments of cyanobacteria. (a)** *Nostoc.* **(b)** *Merismopedia.*

Photosynthetic bacteria utilize the energy of sunlight to synthesize ATP and carbohydrates. **Cyanobacteria,** or blue-green bacteria, are unicellular or filamentous chains of cells, often embedded in a gelatinous matrix (✦ Figure 12.8). Photosynthesis is carried out in thylakoids, but these are not contained in chloroplasts (✦ Figure 12.9). The chlorophyll carried by cyanobacteria is chlorophyll *a,* as in higher plants, and like higher plants, some cyanobacteria generate molecular oxygen as a by-product of photosynthesis. Ancient forms of cyanobacteria were probably responsible for oxygenating the early atmosphere. Other pigments including phycocyanin (blue) and phycoerythrin (red) are present in these cells. The Red Sea is periodically covered with mats of cyanobacteria containing the red pigment, giving the water its characteristic color.

Some filamentous species of cyanobacteria can directly fix atmospheric nitrogen (✦ Figure 12.10) under anaerobic conditions. To carry out this function, a specialized cell, the **heterocyst,** is formed. The heterocyst loses its nuclear region and acquires a thick cell wall that excludes oxygen. These cells also lose their photosystem II because they generate their own oxygen. Heterocysts are unable to couple ATP formation to carbohydrate synthesis, and must be supplied with nutrients from adjacent cells. The ability to function under low oxygen conditions may be a relic from several billion years of evolution under low oxygen conditions.

The heterotrophic eubacteria constitute a large and diverse group of organisms. Heterotrophs cannot manufacture their own high-energy compounds from inorganic molecules and must derive energy from complex molecules obtained in feeding. There are two main types of bacterial heterotrophs: **sapro-**

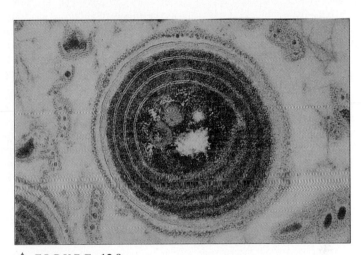

**✦ FIGURE 12.9**

**Thylakoids in cyanobacteria carry out photosynthesis, but are not organized into chloroplasts.**

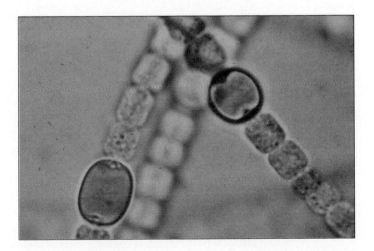

**✦ FIGURE 12.10**

**Filaments of** *Anabaena.* **The large, spherical cells are heterocysts, which can fix nitrogen directly from the atmosphere.**

# GUEST ESSAY: RESEARCH AND APPLICATIONS

## Microbiology: A Learn By Doing Science

RAÚL J. CANO

While in school I was not certain of my career goals, but that changed rather suddenly when I took a genetics course from Dr. Norman Vigfusson, a professor with a zest for learning and the ability to pass it on. During those early days of my training as a biologist, life was exciting and hectic. We started a joint project investigating the sexual life cycle of a fungus known as *Neurospora crassa* (the white mouse of the fungal world) and found out some very exciting things about the chemical regulation of sexuality in this fungus.

I was awed by the impact microorganisms have on our everyday life. Never mind that microorganisms are known as the killers of humankind—that's just to sell books. Granted, there are a few microorganisms that take their job of recycling nutrients a bit too close to heart and can't wait until we die to carry out decomposition, but the vast majority of them carry out *essential* functions on our planet, without which life on earth would not be possible—not to mention the beer, wine, cheese, etc., that they help make.

After I finished my master's work, I went on to the University of Montana to do my Ph.D. in microbiology with a remarkable teacher by the name of John J. Taylor, a mycologist (a rare breed of humans who like to study fungi). During my two years there I became the microbiologist that I am today. I was interested in the evolution of host-parasite relationships, al-though I did not verbalize it until much later at Cal Poly, San Luis Obispo.

In 1985 I took my first sabbatical leave and went to Spain to work in the area of molecular biology along with a group of microbiologists at the University of Sevilla Medical School. Upon my return, I had found a new love, molecular microbiology. It still was a new concept and I wanted to pioneer it at Cal Poly.

My students and I began by asking this question: Can DNA be preserved in amber and if so for how long? We applied the methods for extracting DNA that I learned in Spain and made up a few ones of our own until we were successful in retrieving ancient DNA. I was amazed at the tenacity of life and the ability of amber to preserve complex biomolecules. So the thought occurred to me that if DNA can be preserved in amber in such good condition, how about microorganisms? I knew from experience that microorganisms are inveterate survivors. They were the first life-form to colonize this planet more than 3500 million years ago and are still the dominant life-form on this planet. They produce all kinds of spores, which they use for dispersal and which they use for survival under adverse conditions. It has been shown that microorganisms can survive thousands of years in their spore form, so why not millions of years if their spores are further protected by the amber? In December 1991, my students and I discovered that we could re-cover viable bacterial spores from within the amber and study them in detail in the laboratory. We proceeded to perform experiments that would disprove the hypothesis that these organisms are of ancient origin and not contaminants. The results pointed to the ancient origin of the microorganisms and finally we became convinced ourselves. After three years of work, we published our findings in *Science*.

My zest for knowledge and discovery was passed on to me by great teachers and scientists with whom I share the philosophy of learning by doing. They taught me microbiology through experimentation. The laboratory was their classroom, as it is mine.

*Raúl J. Cano earned his Ph.D. in microbiology from the University of Montana in 1974. He is a Fellow in the American Academy of Microbiology and has been a member of the microbiology faculty at California Polytech-nic State University, San Luis Obispo, where he is now a professor and chairman of the Microbiology Division. He specializes in studies of ancient DNA and the evolution of host-parasite relationships.*

---

**phytic** and **symbiotic.** Saprophytes feed on dead or decaying material and are important in recycling organic material in the biosphere. The decaying matter on the forest floor contains a complex community of saprophytic bacteria. These bacteria speed the process of decomposition and make nutrients available to neighboring fungi and plants (✦ Figure 12.11).

A symbiotic relationship with bacteria may sound unhealthy, but many nondisease-causing bacteria live in harmony with their hosts and can contribute to the host's nutrition. Cows and other grazing animals have bacterial colonies in their digestive systems that convert the cellulose in food into glucose. The glucose is absorbed by the cow and utilized in metabolism. Note, however, that many disease-causing bacteria are heterotrophs (➤ Table 12.3), so that exposure to the wrong heterotroph can induce an infection and a disease. Symbiosis among different types of bacteria may have been beneficial to the organisms involved, and more importantly set the stage for the evolution of eukaryotes.

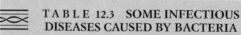

| TABLE 12.3 SOME INFECTIOUS DISEASES CAUSED BY BACTERIA | |
|---|---|
| Strep throat | Cholera |
| Rheumatic fever | Toxic shock syndrome |
| Bacterial pneumonia | Gonorrhea |
| Diphtheria | Syphilis |
| Tuberculosis | Chlamydia |
| Meningitis | |

## Bacterial Diseases Are a Serious Health Threat

Along with viruses, bacterial infections are capable of causing infectious disease. Bacteria can cause disease by destroying cells, by producing toxins, by contaminating food, or by the reaction of the body to the infecting bacteria (see Beyond the Basics, p. 228). Rheumatic fever, caused by a strain of *Streptococcus,* can cause arthritis and permanently damage heart valves. Diphtheria is caused by a bacterium that is inhaled and which produces a toxin that damages cells in the respiratory system. The toxin can travel to other parts of the body, producing destruction of cells in the nervous system and kidneys.

Bacterial infections can be controlled by vaccination and by drug treatment, especially by **antibiotics.** Antibiotics are substances produced by microorganisms that kill or inhibit the growth of bacteria. Antibiotics are effective because they inhibit some process in the infecting bacterial cells, without affecting the host's cells. Penicillin, for example, inhibits the synthesis of cell walls in certain types of bacteria.

For the last 50 years antibiotics have been an effective treatment for bacterial infections and have saved millions of lives. But the use of antibiotics, especially the indiscriminate and overuse of these drugs, has served as a selective agent. In response to this selection, antibiotic resistant strains of bacteria are becoming common. If this trend continues, bacterial diseases may again become a leading cause of death.

 ## THE TRANSITION TO EUKARYOTES OCCURRED ABOUT A BILLION YEARS AGO

In the previous chapter we considered the origin of eukaryotes as an evolutionary event that took place in the Middle Protozoic, perhaps some 1.2 billion years ago. The fossil record is incomplete, and it can be difficult to tell from fossil material whether the cell being examined is eukaryotic rather than prokaryotic. It seems certain that eukaryotes evolved from prokaryotes, but how many times eukaryotes independently arose from prokaryotic ancestors, the number of steps in the process, and the order in which these events occurred is not clear (✦ Figure 12.12).

Eukaryotic cells are characterized by

✦ **FIGURE 12.11**

Saprophytic bacteria are important in recycling organic material in the biosphere. The forest floor contains saprophytic bacteria and fungi that break down dead and decaying material.

a nuclear membrane and membrane-bound organelles (mitochondria and chloroplasts). On average, eukaryotic cells are larger than prokaryotic cells, and eukaryotes have evolved into both unicellular and multicellular forms. Clues to the origin of eukaryotes can be obtained by examination of membrane-bound organelles and by studying some living eukaryotes.

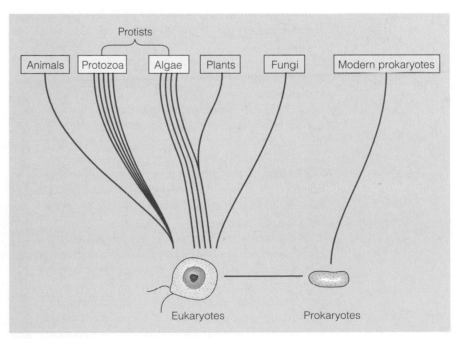

✦ **FIGURE 12.12**

In the five kingdoms, fungi, plants, protists, and animals evolved from early eukaryotic forms. Members of the algae and the protozoa may have had multiple origins from single-celled eukaryotic ancestors.

# ARE THEY REALLY FLESH-EATING BACTERIA?

Heterotrophic bacteria are found in many habitats, including the soil, freshwater, and on the skin and mucus membranes of warm-blooded vertebrates, including humans. One group of heterotrophs, called streptococci, includes beneficial species used in manufacturing food products, such as Cheddar cheese. Other species in this group are responsible for serious infectious diseases. One such variety, called streptococcus A, causes strep throat infections and rheumatic fever.

Recently, infections associated with the destruction of muscle and flesh have caused one streptococcus A strain to be called "flesh-eating" bacteria. This deadly strain is not new; it was well known to be a problem during World War II. After the war, infections caused by this strain subsided, but a new cycle of infection began in the 1980s. During that decade, small, isolated outbreaks were reported in Australia, Scandinavia, and some western states in the United States. An upsurge in infections began about 1990, and there are now about 15,000 cases reported each year in this country (Jim Henson, creator of the Muppets, was one victim). The present cycle of infection is only sporadic, and infections are likely to remain rare.

One unresolved question for scientists is whether the current wave of infection is caused by the reemergence of an old strain, or whether a new invasive strain has arisen. Researchers discovered that a common streptococcal strain called M1 was infected by a virus sometime in the late 1970s. Toxin-producing genes were apparently transferred from the virus to some M1 cells, making them highly infective.

The presence of newly acquired toxin genes has turned the M1 strain into a flesh-destroying variety of streptococcus. The toxin produced by the bacteria breaks down and destroys cells of the skin and connective tissue that would ordinarily protect against infection. Once on the skin, the toxin-secreting bacteria destroy surface cells. Inside the body, the bacteria secrete more toxin and kill cells by dissolving them from the outside. Bacteria then swarm into and reproduce in the liquified waste from the destroyed cells and tissue. Because the strep cells in this deadly soup double in number every 20 minutes or so, and each can produce more toxin, the infection spreads rapidly. In that sense, yes, the bacteria are flesh eating. The only effective treatment in some cases is surgical removal of the dead tissue, often by amputation of an infected limb.

From an evolutionary perspective, the transfer of toxin genes to streptococcus A represents an adaptation that confers a selective advantage. This advantage allows toxin-producing cells to penetrate the body faster (gain access to a productive habitat), and promotes reproduction without the need to invade cells of the body. Eventually, technology may allow us to counter this development by providing us with a new adaptation of our own in the form of a vaccine. As elsewhere in the history of infectious disease, the advantage will probably be temporary.

## Some Eukaryotic Organelles May Have Been Prokaryotic Symbionts

One of the theories about the origin of mitochondria and chloroplasts is that they originated as prokaryotes living in a symbiotic relationship within larger host cells. **Symbiosis** is an interactive association between two or more species living together. This association may be parasitic, where one species (the parasite) benefits at the expense of the host, or one species may benefit and the other is neither helped nor harmed (commensalism), or the association could be of mutual benefit to both species (mutualism).

According to the symbiotic model, eukaryotes evolved in several stages. The first stage was the formation of a mutualistic relationship between two or more prokaryotic organisms. The second stage was the development of a progressive interdependence over many generations until the symbionts could not survive without their partners. This transition may have involved the loss of duplicate genes in one partner, and/or the transfer of genes between partners. These symbiotic relationships may have evolved more than once, and mitochondria and chloroplasts may have independent origins. The end result of this process is a eukaryotic cell (✦ Figure 12.13).

Evidence to support the symbiotic theory of eukaryotic origins comes from the observation that

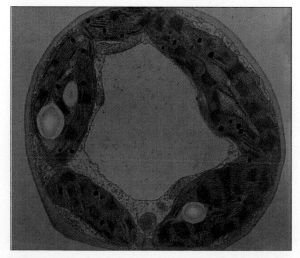

✦ **FIGURE 12.13**

**Eukaryotic cells probably arose from communities of symbiotic prokaryotic cells. The chloroplasts and mitochondria of eukaryotic cells are the remnants of prokaryotic symbionts.**

mitochondria and chloroplasts are each about the size of bacterial cells, they contain circular DNA molecules as do bacteria, their ribosomes are similar to those of prokaryotes, and they reproduce by binary fission like bacteria. In addition, their genetic coding system is more prokaryotic than eukaryotic.

Other evidence comes from observations of living eukaryotic cells. The amoeba *Pelomyxa* (✦ Figure 12.14) is a single-celled eukaryote living in the mud of freshwater ponds. This organism lacks mitochondria, and has formed a symbiotic relationship with two types of bacteria that live within the cell and perform the respiratory functions of mitochondria. About two billion years ago, this same type of relationship may have resulted in the emergence of eukaryotic cells.

The origin of multicellular eukaryotes is a separate event in eukaryotic evolution, and it followed or paralleled the development of the eukaryotic cell. The organization and diversity of unicellular eukaryotes will be considered before we discuss the foundations of multicellularity.

 ## PROTISTS ARE A DIVERSE GROUP OF EUKARYOTES

The kingdom Protista is a collection of diverse eukaryotes that includes phototrophs, heterotrophs, and organisms that can be either heterotrophs or phototrophs depending on environmental conditions. Protists have adapted to many different habitats in soil, freshwater, saltwater, and as symbionts within the bodies of multicellular eukaryotes.

For the first several hundred million years after their appearance, eukaryotes were unicellular. Multicellular forms appeared about 600 million years ago. For about half of the 1.2 billion years that eukaryotes have been around, protists were the only form that existed.

The classification of protists is the subject of much debate. They have often been divided into the animal-like and plant-like protists. Another method of classification divides the protists into three broad categories: unicellular heterotrophs (protozoa), unicellular and multicellular phototrophs (algae), and slime molds. Slime molds are organisms that are both unicellular and multicellular, and resemble protists in some ways, but in other ways resemble fungi and higher plants. Because protists have such a diverse evolutionary history and may not fit neatly into a single kingdom, we will not attempt to resolve their phylogenetic relationships but simply describe their characteristics.

Because of their complex evolutionary history, it is difficult to generalize about the cellular organization and physiology of the Protista. Some, like the diatoms, are encased in shells, others (algae) have plant-like cell walls containing cellulose, and still others such as amoeba have only plasma membranes. The mode of reproduction in protists can involve a mitotic division of the nuclear material followed by division of the cytoplasm, or meiosis and sexual reproduction.

### Protozoa Are Single-Celled, Motile Organisms

As a group, protozoans are motile, unicellular heterotrophs. Most digest their food in vacuoles, and many use vacuoles to control their internal osmotic environment. Reproduction is usually by binary fission after mitotic division of the nucleus, but many protozoans can also reproduce sexually following meiosis. The 60,000 to 70,000 species are distinguished from one another by the mechanisms employed for locomotion.

### Amoeboid Protozoa Use Pseudopods for Movement

The amoeboid protozoa group includes the familiar freshwater *Amoeba* and other amoeba-like organisms that move by means of cytoplasmic extensions called **pseudopodia** (✦ Figure 12.15). The pseudo-

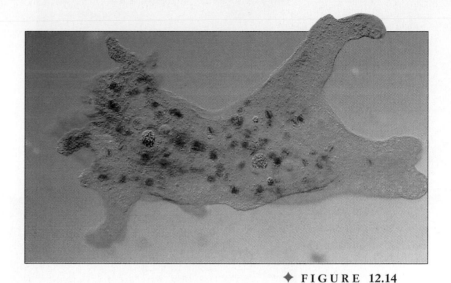

✦ FIGURE 12.14

The amoeba *Pelomyxa* lacks mitochondria, and has a symbiotic relationship with bacteria, which perform respiratory functions. Similar relationships may have given rise to eukaryotic cells.

✦ FIGURE 12.15

*Amoeba* move by extending pseudopodia.

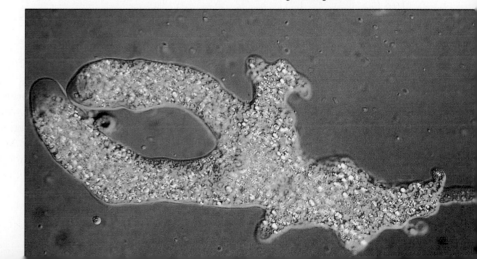

(c)

(a)

(b)

✦ **FIGURE 12.16**

(a,b) Foraminifera are single-celled protists with a shell. The shells of dead foraminifera accumulate on the ocean floors and form chalk deposits. (c) The white cliffs of Dover, England, are composed of foraminifera chalk.

form of amoeba, *Entamoeba histolytica,* is parasitic in humans, and when ingested (usually in contaminated water) causes amoebic dysentery.

## Many Flagellated Protozoa Cause Disease

Another group of protozoa use one or more whip-like flagella for locomotion. The flagella of these organisms have the 9 + 2 arrangement of microtubules discussed in Chapter 3. This group includes free-living heterotrophs as well as symbionts, including some human parasites. They typically lack a cell wall, do not contain chloroplasts, and are found in freshwater. Parasitic trypanosomes are flagellates that require two hosts, an insect and a mammal (including humans) in order to complete their life cycle (✦ Figure 12.17). As a group, trypanosomes are responsible for African sleeping sickness, Chagas disease, and leishmaniasis, diseases that afflict millions of people in Africa, Asia, and South America. Trypanosomes have evolved a system of shifting cell surface proteins to elude the immune system of their hosts.

Symbiotic flagellates include trichonymphs (✦ Figure 12.18), which live in the intestinal tracts of termites and digest the cellulose eaten by the termites. On its own, the termite is unable to break down cellulose and the other structural components of wood, so it is completely dependent on its symbiont for nutrition.

## Sporozoans Are Parasitic Protists

Sporozoans are parasitic protists with a complex life cycle that includes an infectious spore-like stage. The sporozoans lack cilia for locomotion and are

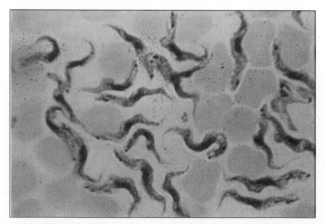

✦ **FIGURE 12.17**

Trypanosomes are human parasites associated with diseases such as sleeping sickness, Chagas disease, and leishmaniasis.

pods are also used to surround and engulf food, including bacteria, other protozoans, or algae.

Most members of this group are ocean-dwelling and are encased in hard shells (✦ Figure 12.16a) pierced with holes, through which long thin, pseudopods are extended. Shells of these **foraminifera** sink to the ocean floor when they die, producing thick layers of chalk. The cliffs of Dover, on the southeast coast of England (✦ Figure 12.16b) are composed of foraminiferan shells. Although most amoebas and foraminifera are free living, one

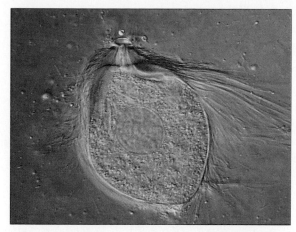

(a)

(b)

**✦ FIGURE 12.18**

(a) Trichonymphs live in the intestines of termites and convert cellulose into glucose. (b) Wood-eating termites obtain nutrients from symbiotic protists.

nonmotile. Among these is an intracellular parasite that causes a disease called toxoplasmosis. This sporozoan can infect any warm-blooded animal. Most individuals infected with this parasite develop only mild, flu-like symptoms, but the disease can be progressive and fatal in infected fetuses or infants. The sexual stage of the life cycle occurs only in the intestinal cells of cats; transmission to humans occurs by exposure to soil contaminated with cat feces or by eating undercooked, infected meat. Surveys of local human populations worldwide indicates that from 7 to 72% of individuals are infected with this parasite, depending on the geographic area.

Sporozoan parasites are also responsible for malaria, a disease characterized by recurring episodes of chills, fever, and sweating with chronic anemia, lethargy, and enlargement of the liver. The sporozoans responsible for malaria require certain mosquitoes to complete their life cycle (✦ Figure 12.19). Many regions of the world are affected by malaria, and it is estimated that more than 300 million humans are infected with this parasite. In addition to mosquitoes, the disease can be spread by blood transfusions from infected people and by drug addicts sharing syringes and needles.

## Ciliated Protozoans Are Complex Protists

Ciliates are heterotrophic organisms without a cell wall that use cilia for locomotion and food ingestion (✦ Figure 12.20). They are among the most structurally complex Protists. Most of the 8000 or more species of ciliates are found in freshwater, although some are marine. In addition to cilia, many have barbed, thread-like organelles called **trichocysts** that are discharged for defense or for captur-

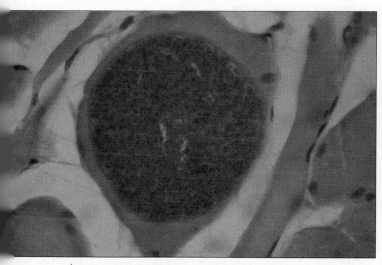

**✦ FIGURE 12.19**

Intracellular sporozoan parasites are responsible for malaria.

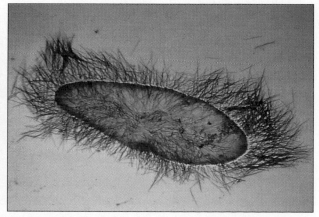

**✦ FIGURE 12.20**

Some ciliated protozoans have trichocysts, which can be discharged to capture prey or are used for defense against predators.

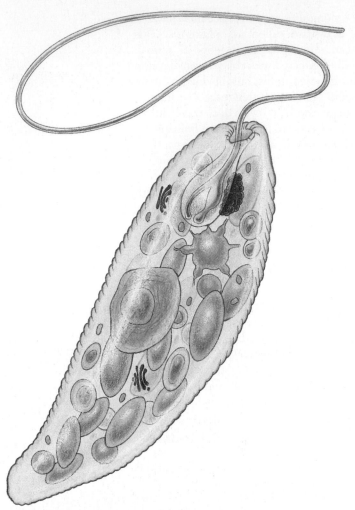

**✦ FIGURE 12.21**

*Euglena* **are protists with some characteristics of plants (chloroplasts) and animals (lack of cell wall, flagellum for locomotion).**

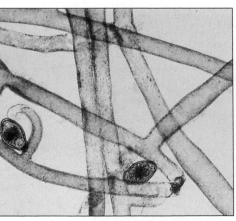

(a)

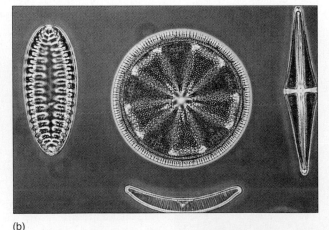

(b)

**✦ FIGURE 12.22**

**(a,b) Chrysophytes are photosynthetic protists that have yellow carotenoid accessory pigments that give them their characteristic color.**

ing prey (Figure 12.20). Ciliates have two nuclei. One is the macronucleus, which carries up to several hundred copies of the genome and controls cell metabolism and asexual reproduction. The other nucleus, the micronucleus, contains a single copy of the genome and controls sexual reproduction.

## Algae Evolved Several Times from Protist Ancestors

Several unrelated groups of photosynthetic eukaryotes were present in the early Paleozoic (more than 450 million years ago). Algae utilize photosynthesis for nutrition at least part of the time. They inhabit freshwater and marine habitats, and are a major part of the **phytoplankton,** a floating layer of photosynthetic organisms that are an important source of oxygen in the earth's atmosphere. Phytoplankton also serve as a source of food for aquatic organisms and are at the base of aquatic food chains. Most are unicellular, although some are colonial. Based on the presence or absence of a cell wall, the type of cell wall, photosynthetic pigments, and method of food storage, the unicellular phototrophs have been divided into several divisions, which are the equivalent of phyla.

The *Euglenoids* are a unique form of phototrophs. They are characterized by their lack of a cell wall and the presence of two flagella emerging from one end of the cell, and they possess chloroplasts, containing chlorophyll *a* and *b.* Euglenoids can live as heterotrophs or phototrophs, and as a result, their classification is often a matter of dispute. In addition to light-sensitive molecules in the chloroplasts, Euglenoids also possess an **eyespot,** a pigmented photoreceptor that senses light and orients the cell for maximum rates of photosynthesis. When switched into the dark, one phototrophic species, *Euglena,* becomes heterotrophic and loses its chloroplasts (✦ Figure 12.21). When light becomes available again, the chloroplasts are regenerated and photosynthesis begins to supply carbohydrates.

Chrysophytes are photosynthetic organisms found in freshwater, the marine environment, and occasionally in terrestrial habitats. They live as unicellular organisms or simple colonies (✦ Figure 12.22a). Included in this group are the golden-brown algae and diatoms. These organisms store food reserves as oils and use chlorophyll *a* and *c* and yellow carotenoid pigments in photosynthesis. Diatoms are enclosed in two-part shells made mostly of silica (✦ Figure 12.22b). The shells are pierced with holes that allow exchange of material with the environment. Fossilized deposits of diatoms are known as **diatomaceous earth,** a material that is used for abrasives, polishes, and as a filtering agent.

A third group of phototropic protists, the dinoflagellates (✦ Figure 12.23), are characterized by flagella, cell walls, and starch as a food reserve. Most are free living, although some are symbionts in the marine environment. These organisms employ chlorophyll and other pigments including carotenoids in photosynthesis, and their colors, ranging from red to yellow-green to brown depends on which combination of pigments are present.

Some members of this group are **bioluminescent.** The phosphorescent wake that trails behind boats on the ocean at night is due to light emitted by dinoflagellates. During bursts of growth, dinoflagellates can lower oxygen levels to the point that fish die. The red color of some dinoflagellates has given the name "red tide" to population explosions of this organism (Figure 12.23). Members of one genus secrete a neurotoxin that is dangerous to vertebrates. These toxins do not harm shellfish, but accumulated toxins can cause illness and death when contaminated shellfish are eaten by humans.

## Red and Brown Algae May Represent Ancient Life-Forms

Red algae are so called because of the presence of large amounts of the photosynthetic pigment phycoerythrin. Because of differences in pigment distribution, many red algae actually appear green or almost black. This group includes around 4000 species, encompassing unicellular forms and multicellular species of various shapes, some of which attain lengths of more than one meter. It is thought that red algae are an ancient life-form that evolved from cyanobacteria. This conclusion is based on similarities in photosynthetic pigments and the structure of the photosynthetic apparatus in the two groups. Present-day species of red algae are mostly marine, and are common in deep tropical waters, an environment that has been stable for millions of years.

Red algae attach to a substrate, and some species form calcium carbonate skeletons for this purpose.

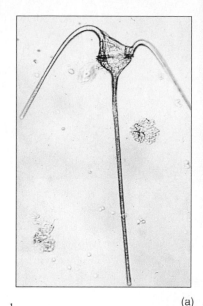

(a)

(b)

✦ **F I G U R E  12.23**

(a) **Dinoflagellate with its characteristic flagella.**
(b) **A red tide caused by a population explosion of dinoflagellates.**

By depositing continuous layers of this material over generations, red algae are now recognized as important contributors to the building of coral reefs in tropical oceans (✦ Figure 12.24). Although you may never have seen red algae, chances are that you have probably eaten parts of this plant, because red algae are harvested from the ocean and used in food preparation. Carrageenan, a stabilizer used in puddings and ice cream, is extracted from red algae (read the labels). If you have ever eaten *nori* at home or in a Japanese restaurant, you have consumed dried sheets of red algae.

Brown algae are multicellular organisms characterized by the presence of the pigment *fucoxanthin,* which gives the plants a golden-brown color. The 1500 or so species of brown algae include giant kelp, an ocean plant that can reach more than 100 meters in length. In cold temperate regions along the ocean, *Fucus,* a species of brown algae, is commonplace (✦ Figure 12.25). *Fucus* is an interesting organism in that examination of its body plan shows the be-

✦ **F I G U R E  12.24**
The red algae *Porphyra,* which live in warm tropical waters.

✦ **F I G U R E  12.25**
*Fucus* **is a brown algae that lives along the shoreline in cooler climates.**

(a)

(b)

**✦ FIGURE 12.26**

**Some varieties of green algae.**
(a) *Spirogyra.* (b) *Acetabularia.* (c) *Ulothrix.*

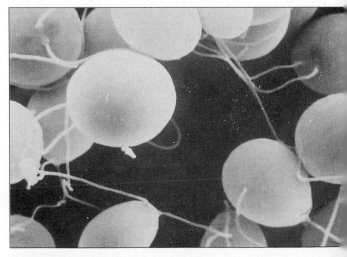

(c)

cell wall component called alginate. Alginate is used as a thickening or emulsifying agent in commercially prepared foods such as ice cream, salad dressing, and candies.

## Green Algae Have Characteristics of Higher Plants

The green algae are a diverse group of organisms with about 7000 known species. Their body forms are highly variable, including species with unicellular, colonial, and multicellular shapes (✦ Figure 12.26). Green algae share several characteristics with plants: Cellulose is the main component of the cell walls; chlorophyll *a* and *b* are used in photosynthesis; and food reserves are stored as carbohy-

ginnings of tissue differentiation in plants. The base of the plant consists of a *holdfast,* specialized for attachment of the plant to a surface. From the holdfast, one or more stem-like structures radiate, carrying leaf-like blades. The blades are supported by gas-filled bladders that keep the plant close to the surface where more light is available for photosynthesis. Another brown alga, *Sargassum,* survives on the ocean surface by floating on air bladders. The mid-Atlantic region known as the Sargasso Sea encompasses several million square miles, and is named for the masses of *Sargassum* found floating there.

The reproduction of brown algae involves an alternation of generations. In some species, the gametophyte and sporophyte are of similar size, whereas in others such as the giant kelp, the dominant sporophyte is more than 100 meters in length, and the gametophyte is of microscopic size.

Like the red algae, brown algae are harvested as a food source and as a source of ingredients used in prepared foods. Kelp are harvested for use as animal feeds and fertilizers and for the extraction of a

**✦ FIGURE 12.27**

The gametes of *Chlamydomonas* are of two different mating types although they are of equal size and appearance.

drates in the form of starch. For these reasons, the green algae are regarded as the organisms from which higher plants have evolved.

Sexual reproduction in green algae shows a number of variations in the form of gametes. These include species in which the two gametes are of equal size and have flagella (✦ Figure 12.27), species in which the gametes are nonflagellated and of equal size, and species possessing large, nonmotile female gametes fertilized by small, flagellated male gametes.

## Slime Molds Represent a Transition between Protists and Multicellular Organisms

To complete our survey of the protists, we consider one final group of organisms, the slime molds (✦ Figure 12.28). Like the euglenoids, the classification of these organisms is often a matter of dispute. They are most often regarded as protists, because they spend most of their life cycle as single-celled, amoeba-like heterotrophic organisms. However, in some species, cells aggregate to form a multicellular, differentiated structure that produces spores with cellulose walls. On this basis, they could be classified as fungi. **Spores** are haploid cells that can survive unfavorable conditions and germinate into new haploid individuals, or act as gametes in fertilization.

✦ **FIGURE 12.28**
The fruiting body of a slime mold.

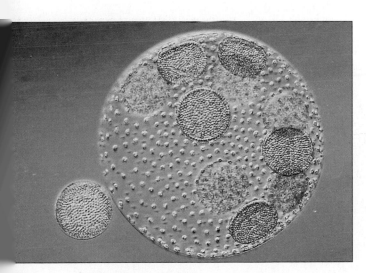

✦ **FIGURE 12.29**
*Volvox,* a colonial algae, may represent a step in the development of multicellular organisms.

The slime molds occupy a position between single-celled protists and multicellular plants and animals belonging to other kingdoms. They are not closely related to other groups of organisms. After the origin of eukaryotes, the transition to multicellularity was a second key step in the evolutionary process. Multicellular organisms are distinct from colonial organisms. Many protists, like the *Volvox,* live in colonies (✦ Figure 12.29), which are aggregates of independent cells physically joined into a larger structure. Each cell in a colony is usually in contact with the external environment and carries on all functions of metabolism, growth, reproduction, etc. In a multicellular organism, specialized groups of cells carry out a subset of functions. This division of labor has enabled multicellular organisms to achieve increases in size and complexity far beyond that of the protists. In the next chapters we will see how fungi, plants, and animals have solved the problem of division of labor by evolving ever more specialized groups of cells that carry on a smaller and smaller number of specialized functions.

1. The five-kingdom system of classification attempts to classify organisms by their evolutionary relationships and, while imperfect, is the most comprehensive system currently available.

2. Viruses, which fall outside of the five-kingdom classification system, are infective particles composed of a protein coat and a nucleic acid core. Viruses are classified according to the type of nucleic acid they contain and by the shape of their protein capsule. They are important agents of disease, and infect a wide range of organisms, including humans.

3. The kingdom Monera contains all bacterial species. Bacteria are characterized by the lack of a nuclear membrane and membranous organelles. Nutritionally, bacteria are heterotrophic or autotrophic.

4. Archaebacteria are the oldest and most primitive bacteria. They are divided into three groups, and all live in habitats characterized by extremes of heat, cold, or salt concentration.

5. Eubacteria is the division of Monera that contains the majority of bacterial species. They are subdivided by the means they employ to obtain energy: photosynthetic, heterotrophic, and chemosynthetic. Chemosynthetic bacteria obtain energy from the oxidation of inorganic compounds such as ammonia and sulfur and play an important role in the recycling of nitrogen. Photosynthetic bacteria utilize the energy of sunlight to synthesize ATP and carbohydrates. Heterotrophs obtain energy and nutrients from dead or decaying matter, or they live as symbionts.

6. The transition from prokaryotic organisms to eukaryotic organisms was a key event in evolution because it allowed organisms to become more complex and more efficient. Eukaryotes are characterized by a nuclear membrane and the presence of membrane-enclosed organelles. The compartmentalization of the cell allowed subcellular regions to carry out specialized functions. The origin of eukaryotic cells is unknown, but may have evolved from symbiotic relationships between two or more prokaryotes.

7. The kingdom Protista is a collection of single-celled eukaryotes that have only tenuous evolutionary relationships with each other. These organisms probably evolved separately from the prokaryotes.

8. Protozoans are unicellular heterotrophs that are classified on the basis of the mechanisms they use for locomotion. Amoeboids use cytoplasmic extensions called pseudopods for locomotion, and engulf food by means of phagocytosis. Most are encased in hard shells and are free living. Flagellated protozoans use one or more whip-like flagella for locomotion, and this group includes significant human parasites. The ciliated protozoans use cilia for locomotion and feeding. Sporozoans are nonmotile parasitic protists whose life cycles include a spore-like stage. This group includes the organisms that cause malaria.

9. Many unicellular prototrophs utilize photosynthesis for energy transfer and constitute the phytoplankton found on the surface and upper layers of the oceans and many freshwater sources. Euglena, golden-brown algae, diatoms, and dino-flagellates are all unicellular prototrophs. The algae also include multicellular forms, with separate evolutionary origins.

10. Slime molds are classified as protists, but share may characteristics of other organisms including fungi, algae, plants and animals. They occupy a unique evolutionary position between single-celled protists and the multicellular plants and animals in other kingdoms.

# KEY TERMS

antibiotics
autotrophs
bacteriophages
binary fission
bioluminescent
cyanobacteria
diatomaceous earth
eyespot

foraminifera
heterocyst
heterotrophs
intracellular parasites
phytoplankton
prions
pseudopodia
retrovirus

reverse transcriptase
saprophytic
spores
symbiosis
symbiotic
trichocysts
viroids

## FILL IN THE BLANK

1. A _____ is a submicroscopic, infectious particle composed of a protein coat and a nucleic acid core.
2. _____ are infective forms of proteins that are responsible for fatal diseases in humans and _____ are infective forms of nucleic acids that are responsible for plant diseases.
3. The three types of archeabacteria are _____, _____, and _____.
4. A type of bacterial heterotroph that feeds on dead or decaying material is called a _____.
5. Protozoa are distinguished from one another by their means of _____.
6. _____ is a floating layer of photosynthetic organisms that are an important source of oxygen in the earth's atmosphere.
7. Slime molds represent a transition between single cell _____ and multicellular _____ belonging to other kingdoms.

## TRUE/FALSE

1. Ancient forms of cyanobacteria were probably responsible for oxygenating the early atmosphere.
2. According to the symbiotic model, eukaryotes evolved in a single step.
3. Protozoa, like other single-celled organisms, only reproduce by binary fission.
4. Sporozoan parasites use trichocysts for defense or for capturing prey.
5. Spores germinate into new haploid individuals or act as gametes in fertilization.

## MATCHING

1. viruses
2. autotrophs
3. heterotrophs
4. methanogens
5. halophiles
6. thermo-acidophiles
7. cyanobacteria
8. symbiosis
9. slime molds

____ resemble protists and fungi
____ involved in recycling carbon
____ chlorophyll *a*
____ intracellular parasites
____ interactive relationship
____ require organic molecules
____ found in salty environments
____ synthesize organic molecules
____ found in hot, acidic environments.

## SHORT ANSWER

1. Explain how bacteria reproduce.
2. How are bacteria currently classified?
3. Outline the evolution of eukaryotes according to the symbiotic model.

## SCIENCE AND SOCIETY

1. A prion is an infectious particle that consists only of protein, and exists in extracts prepared from the brains of diseased animals. The name is derived from the first letters of *proteinaceous* (pr) and *infectious* (i). Prions cause a neurological disorder in sheep and goats called scrapie, and Creutzfeldt-Jacob disease, a rare neurological disorder in humans. They also are suspected as the causative agent in such human diseases as multiple sclerosis, Parkinson's disease, Lou Gehrig's disease and Alzheimer's disease. So far, it has not been possible to infect animals with any of these last-mentioned diseases. Prions are able to retain their infectiveness when exposed to heat, chemicals, or radiation. What makes prions resistant to agents that would normally destroy infectious agents? How do you think a prion replicates since it contains no DNA or RNA? How do you think the immune system is affected by prions? Is the immune system's response to prions a typical property seen in viruses?

2. Toxic shock syndrome (TSS) is a rare and sometimes fatal disease that is caused by certain strains of *Staphylococcus aureus*. TSS is characterized by fever, skin rashes, and serious drops in blood pressure. Certain strains of *Staphylococcus aureus* produce a bacterial toxin that causes the disease. Superabsorbant tampons containing polyester foam and polyacrylate rayon have been found to absorb high concentrations of magnesium from surrounding blood and tissue. This elevated magnesium concentration enhances production of the bacterial toxin. Tampons containing these fibers have been taken off the market and the incidence of TSS has been reduced dramatically. What other measures could women take to reduce their risk of TSS? Do you think the toxin this bacterium produces has a benefit to the bacterium itself? Why or why not? TSS responds well to antibiotics once the disease has been identified. Is it possible for this bacterium to become antibiotic-resistant?

# 13

# DIVERSITY OF FUNGI AND PLANTS

OPENING IMAGE

*A temperate rain forest is a rich source of diverse plant species.*

Across New England in the late summer and early fall of 1741, hundreds, perhaps thousands of people experienced trances, visions, and seizures. Symptoms included muscular spasms and contractions, loss of consciousness, and stupors during which individuals experienced hallucinations including out-of-body sensations. While in these trances, people reported they saw visions of Heaven and Hell, and sensations of great heat alternating with terrible cold. Others who did not suffer trances or visions had numbness, difficulty in speaking, painful urination, wide swings in mood, and demented behavior.

Diaries indicate that the sickness reached epidemic proportions in some towns. The trances and visions were associated with a religious revival called the Great Awakening that swept across New England and reached a peak in late 1741. Records show that the churches of Connecticut enrolled many new members in 1741 and 1742. The cause of the trances and other symptoms divided the clergy and even the physicians of these towns. Some regarded the symptoms as inspired by God, whereas others felt they had been confronted with an illness that affected the nervous system. No satisfactory explanation of what might have caused the symptoms has ever been put forward.

Recently, these abnormal symptoms have been reinterpreted in light of what is now known about the physical effects of toxins produced by fungi growing on harvested grain. In particular, attention has been focused on a plant disease called ergot, found in rye and rye flour made from grain contaminated with the fungus, Claviceps purpurea. Ergot

*contains several chemical compounds, including lysergic acid amide (LSD), that affect the central nervous system. Symptoms include tremors, spasms, loss of consciousness, confusion, and hallucinations. The description, severity, and duration of the symptoms of ergotism are remarkably similar to those reported across New England in 1741.*

*Searches of agricultural records indicate that rye was a cereal crop grown in New England, and accounted for about two-thirds of the grain listed on inventory sheets. From other records and tree ring analysis, it is known that the weather in the late summer and fall of 1741 was damp, providing perfect conditions for the growth of* Claviceps. *There were also food shortages that year, caused by two previous severe winters. Export of grain was forbidden and people were forced to use rotten grain for food. Some perceptive individuals even recorded in their diaries that the illnesses began just following the rye harvest.*

*This bit of historical detective work is outlined in the book* Poisons of the Past, *by Mary Kilbourne Matossian. It explores the relationship between fungal toxins and events in human history. This analysis of historical events illustrates the close relationship that has evolved between human society, crop plants, and plant diseases. The evolution of our culture, and to some extent, our history has been shaped by fungi and plants, and this chapter is devoted to understanding the evolutionary origins and diversity of these organisms.*

## FUNGI ARE EUKARYOTES UNRELATED TO PLANTS

The evolution of multicellular eukaryotes brought about changes in the size and complexity of organisms. This adaptation also allowed the exploitation of the terrestrial habitat, removing the need to live in an aqueous environment in order to obtain nutrients and reproduce. Although fungi probably evolved in water, they successfully made the transition to land, often in the company of other organisms, both prokaryotic and eukaryotic. The transition to land in both the fungi and plants was accompanied by the development of specialized structures to prevent desiccation and to allow for reproduction and dispersal to new habitats.

Originally classified as plants, fungi are so unlike other organisms that they are now assigned to their own kingdom.

### Fungi Have Diverse Body Plans and Nutritional Requirements

Some taxonomists regard fungi as a transitional group between unicellular and multicellular organisms because they contain unicellular, multinucleate, and multicellular species. Fungi have two characteristic structures that aid in their classification: reproductive **spores** that can be dispersed to new habitats by air or water, and multinucleate or multicellular filaments known as **hyphae** (✦ Figure 13.1). The cell walls of fungi contain an armor-like polysaccharide called **chitin,** which also forms the external skeleton of insects.

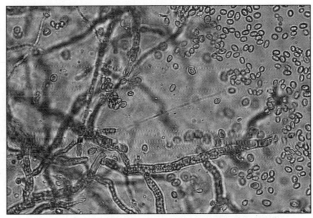

(a)

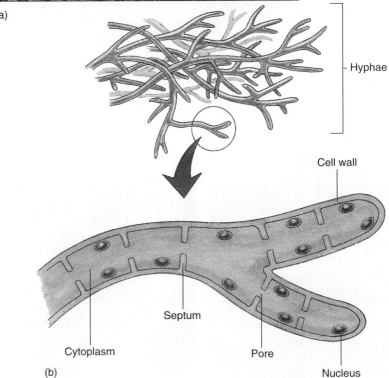

(b)

✦ FIGURE 13.1
(a) Hyphal filaments form interwoven mats called mycelia.
(b) Within hyphae, multinuclear cells are connected by pores.

Hyphae

Cell wall

Septum

Cytoplasm

Pore

Nucleus

# FUNGI AS PREDATORS

Most fungi are heterotrophs that live on or in something, and obtain nutrients by secreting digestive enzymes that break down large complex molecules into smaller, simpler molecules that are absorbed by the hyphae. Some years ago it was discovered that fungi can employ more aggressive means of obtaining nutrients. A few species of fungi trap, kill, and digest small animals such as nematode worms. To do this, one species forms small, spherical knobs on the hyphae. If nematodes or other animals such as insect larvae come in contact with these

knobs, they become stuck to them. Hyphal filaments respond by attaching to the surface of the trapped animal, secreting digestive enzymes that slowly digest the prey from the outside and release nutrient molecules that are absorbed by the fungus.

Other species hunt by forming small loops in their hyphae and awaiting passing prey. If a worm or other animal enters the loop, the cells in the loop quickly swell, contracting the ring and trapping the prey. The hyphal loops can contract in less than one-tenth of a second. The trapped animal is then slowly digested by enzyme secretion after hyphae penetrate its body.

The nutrients released by digestion are absorbed by the fungal hyphae.

The complex nature of the predator/prey relationship in this case can be illustrated by growing the loop-forming fungus in a culture medium in the laboratory. Under these growth conditions, loops are formed only rarely. If nematodes or extracts of nematodes are added to the culture medium, the fungus responds by forming loops. Somehow the presence of prey is sensed by the fungus, stimulating the predator to develop traps.

---

The body plan of fungi is directly related to their mode of nutrition. Fungi are absorptive heterotrophs. They secrete digestive enzymes to break down large molecules in their environment so they can absorb the resulting small molecules as food. The structure of hyphae as long, narrow filaments with small volumes and large surface areas is well suited to the absorption of nutrients. Hyphae grow together in an interwoven mass called a **mycelium** (pl., mycelia). Some parasitic fungi have specialized hyphae that penetrate plant and animal cells to obtain nutrients. Most fungi are **saprophytes,** obtaining nutrients from the decaying bodies of plants and animals; others are **parasites.** Interestingly, some fungi are predators and have evolved mechanisms for capturing protists or animals as food sources (see Beyond the Basics, above).

Fungi are important in both the production of human food and as a source of food. Edible fungi include morels, truffles, and mushrooms (✦ Figure 13.2). The blue flecks in Gorgonzola, Roquefort, Stilton, and blue cheese are fungal colonies that give the cheese its distinctive flavor. Beer and wine are produced through the action of yeast, another fungus. Fungi are also the source of medical products, including some antibiotics. On the other hand, fungi also attack crop plants, are a major cause of spoilage of food, and cause infectious disease.

## Reproduction and Classification of Fungi

More than 60,000 species of fungi have been identified. Their classification is based partly on their method of reproduction. Most fungi can reproduce both asexually and sexually. Asexual methods include the growth and fragmentation of hyphae, or the formation of asexual spores. Sexual reproduction in fungi involves fusion of hyphae from different mating types. In the following sections, the characteristics of the major fungal divisions are outlined.

### Zygomycetes Are Multinucleate Species

Zygomycetes have hyphae that are not divided into separate cells, but consist of long, filamentous strands that contain multiple haploid nuclei distributed throughout the cytoplasm. Members of this group can reproduce asexually by haploid spores formed in structures called **sporangia** (✦ Figure 13.3). Hyphae can belong to one of two different mating types. In sexual reproduction, haploid hyphae of two different mating types fuse to form a diploid zygote that develops to form a **zygospore.**

✦ **FIGURE 13.2**

**Some fungi are poisonous, such as (a) these mushrooms, and some are edible, such as (b) morels.**

(a)

(b)

**Life cycle of a black bread mold (zygomycete).** Asexual reproduction (center) takes place through the production of haploid spores that germinate to form new hyphae. In sexual reproduction, hyphae from two mating types (+ and −) fuse and form haploid gametes. These gametes fuse to form a diploid zygospore that can remain dormant for months. Meiosis occurs in the zygospore, and haploid hyphae germinate from the spore.

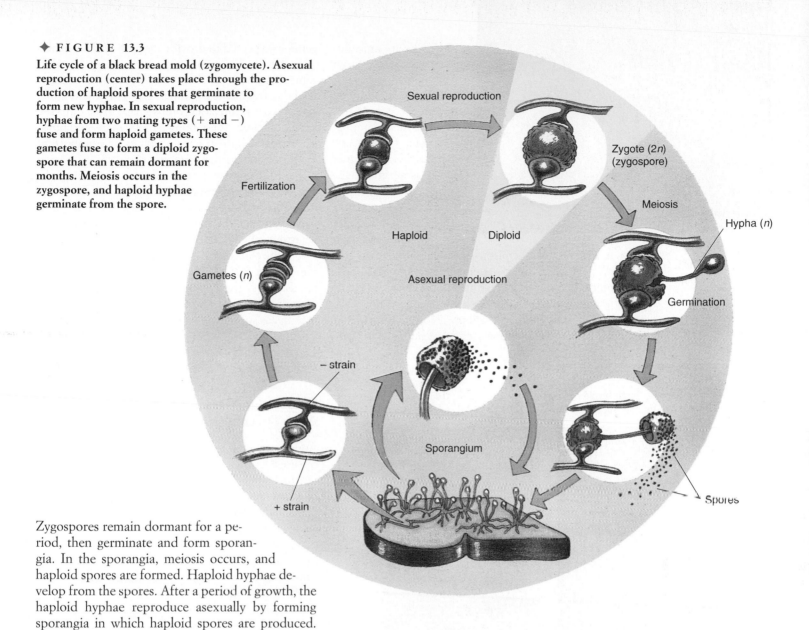

Sexual reproduction

Zygote (2n)
(zygospore)

Meiosis

Hypha (n)

Fertilization

Haploid    Diploid

Gametes (n)

Asexual reproduction

Germination

− strain

+ strain

Sporangium

Spores

Zygospores remain dormant for a period, then germinate and form sporangia. In the sporangia, meiosis occurs, and haploid spores are formed. Haploid hyphae develop from the spores. After a period of growth, the haploid hyphae reproduce asexually by forming sporangia in which haploid spores are produced. The spores are dispersed by air currents and form new hyphae in a suitable environment.

There are less than 1000 species in this group, although they are widely dispersed. Spores from this group have been recovered from far-flung regions of the world, including the Arctic. Most zygomycetes feed on dead or decaying plant and animal material. You may have seen species belonging to this group growing on moldy bread. A few species are parasites of plants and animals.

## The Sac Fungi (Ascomycetes) Include Yeasts

This diverse group of fungi contains more than 30,000 species including unicellular and multicellular forms (✦ Figure 13.4). This group is not well characterized: Additional species have yet to be described and classified, some of which may be of economic importance.

✦ FIGURE 13.4

**A representative sac fungus.**

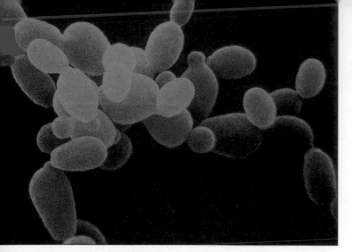

**✦ FIGURE 13.5**
Yeast can reproduce by a process of budding.

Yeasts are unicellular forms of sac fungi that reproduce asexually by budding (✦ Figure 13.5) and form asci for sexual reproduction. One yeast, *Saccharomyces cerevisiae* is an important organism in genetic research, where it has been used as a model for studying the genetics and biochemistry of eukaryotes. Yeasts are part of the Human Genome Project in two important ways. Because yeast has been extensively studied genetically, sequencing the yeast genome is a part of the Human Genome Project. Comparing the organization and sequences of human and yeast genes is expected to provide insights into the evolution of gene function. In addition, yeast chromosomes have been modified to serve as vectors, carrying large fragments of human DNA. These vectors, called yeast artificial chromosomes (YACs) are the vector of choice for making genetic maps and physical maps of human chromosomes, two important phases of the Human Genome Project.

Sac fungi are important in the decomposition and recycling of organic molecules from dead and decaying organisms. The group also includes pathogenic species that cause Dutch Elm disease and Chestnut blight. These plant diseases have decimated the elm tree and chestnut tree populations in the United States, altering the look of cities and towns across the country. Sac fungi are also used in industrial processes including wine and beer making, as sources of antibiotics (penicillin), and in gourmet cooking (truffles and morels). The fungus *Claviceps,* described at the beginning of the chapter, is a member of this group of fungi.

## Club Fungi (Basidiomycetes) Are a Diverse Group

Among the club fungi, or Basidiomycetes, perhaps the most familiar forms are mushrooms, toadstools, puffballs, and shelf fungi (✦ Figure 13.6). The body plans of species in this division include an inconspicuous mycelial mass that forms a large, visible reproductive structure called a fruiting body.

Sexual reproduction results in the formation of **basidiospores** on specialized, club-shaped cells called **basidia.** Basidia form along the slits or gills under the cap of the edible mushroom. When mature, spores fall to the ground where they generate new mycelia (✦ Figure 13.7).

Club fungi are of great economic importance both as a commercial crop for human consumption and as disease-causing agents that destroy significant portions of the world's grain harvest. The most popular commercial mushrooms include *Agaricus campestris,* the field mushroom, and *Lentinus edodes,* the shiitake mushroom. The value of these crops is estimated to be more than $14 billion per year. Not all basidiomycetes are edible, however. The most poisonous of all mushrooms, *Aminita phalloides,* is a basidiomycete. Certain mushrooms with hallucinogenic properties are important in the religous rituals of some groups of Indians in Central and South America. These mushrooms contain psilocybin, a drug chemically related to those found in ergot. Two groups of basidiomycetes, the rusts and smuts, are important causes of diseases in plants. A fungal infection (corn smut) almost destroyed the entire U.S. corn crop in the 1970s.

**✦ FIGURE 13.6**
Club fungi. (a) Shelf fungus growing on a tree trunk. (b) Puffballs. (c) Mushrooms on the forest floor.

(a)

(b)

(c)

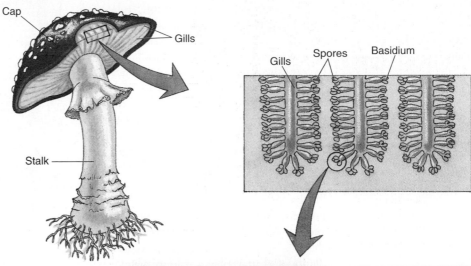

## FIGURE 13.7

In the club fungi, spores are produced in gills on the underside of the fruiting body. The spores are formed on club-like structures called basidia.

## Lichens Are Symbiotic Organisms

Some fungi form symbiotic, mutualistic relationships with protists and with plants. From an examination of the fossil record, it appears that this type of relationship is an ancient one. Two significant forms of these symbiotic relationships have survived, and because of their unique evolutionary origins, do not fit neatly into many classification systems. These groups include the **lichens** and the **mycorrhizae.**

Lichens are autotrophic organisms composed of a fungus (sac or club fungus) and a photosynthetic unicellular organism (which can be a cyanobacterium or algae). The fungus anchors the organism to a substrate (such as rocks or tree trunks) and may be able to obtain minerals and some nutrients from the substrate, while the photosynthetic cyanobacteria or algae synthesize carbohydrates (✦ Figure 13.8). Lichens grow where few other organisms survive and are resistant to extremes of cold and drought. For example, large regions of the desolate Arctic tundra are covered with lichens. Because the lichens here serve as an important food source for reindeer, caribou, and other animals, they are called reindeer moss (✦ Figure 13.9).

## ✦ FIGURE 13.8

Lichens are symbiotic organisms composed of a fungus and a protist such as cyanobacteria or unicellular algae.

## ✦ FIGURE 13.9

In the arctic tundra, lichens serve as a food for large animals such as the caribou and the reindeer.

## Sidebar

### PHYTOREMEDIATION

Certain plants have the ability to remove toxic compounds such as nickel, copper, mercury, lead and cadmium from the soil and concentrate them in their own tissues. This strategy of soil cleansing, known as phytoremediation, can be used at abandoned mines laced with zinc and lead, military sites loaded with lead, municiple waste heaps and dump sites for sewage.

Metal-scavenging plants are common in the subtropics and tropics where metal accumulation serves as a defense mechanism against plant-eating insects and microorganisms. Botanists have known for several years that certain plants, like the mustard plant, have the ability to concentrate toxic metals at levels that would kill most plant species. Ecologists and botanists deduced that the plant's ability to accumulate metal is an evolutionary response to soils of a specific geological region.

Although these hyperaccumulating plants show great promise for restoring damaged soil, some plants have characteristics that prevent their widespread application. Some plants are small and grow very slowly, making them inefficient tools for removing the large quantities of metals found in most contaminated sites. Recent genetic engineering techniques have been applied to these plants in hopes that the genes that code for the plant's ability to accumulate metals can be transferred to faster-growing plants. Phytoremediation is just one of the many benefits that plant biodiversity can offer the ecosystem.

The second form of symbiotic relationship involving a fungus is called a mycorrhiza, or "fungus root." In this case, a fungus, usually a zygomycete or basidiomycete, becomes woven around and sometimes into the roots of a plant (✦ Figure 13.10). Fungal hyphae grow out into the soil from the plant roots and absorb nutrients and minerals, especially phosphorus, that are passed to the plant roots. In turn, the fungus obtains carbohydrate nutrients from the plant roots. Plants with mycorrhizal roots grow more rapidly, and are often able to grow in nutrient-depleted soils where plants without such fungal colonies fail to survive. Mycorrhizae are found in most groups of vascular plants, and may play a role in maintaining soil fertility.

## PLANTS ARE MULTICELLULAR, PHOTOSYNTHETIC ORGANISMS

The kingdom Plantae contains photosynthetic eukaryotic organisms with multicellular body plans adapted for life on land. They use sunlight, water, and inorganic nutrients to synthesize complex carbohydrates and, as a by-product, release oxygen into the atmosphere. The body plan of plants is highly diversified, but all plant cells contain organelles called plastids and are surrounded by cell walls containing cellulose.

The life cycle of plants has evolved so that a haploid stage alternates with a diploid stage. Alternation of generations and other evolutionary adaptations are collectively responsible for the success of the more than 300,000 species of plants now in existence.

## THE EVOLUTION AND ADAPTATION OF PLANTS BEGAN WITH ALGAL ANCESTORS

Evidence from the fossil record indicates that present-day plants are descended from multicellular algae. In the Precambrian era some 700 million years ago, algae flourished in the mild climate and vast oceans, perhaps forming large, thick mats covering miles of the water's surface. Sometime in the following 200 million years, plants made the transition to land, and began spreading through terrestrial habitats. Around 350 million years ago, the ancestors to modern, stemmed (vascular) plants appeared, and by the late Paleozoic (roughly 300 million years ago), vast forests spread over the earth.

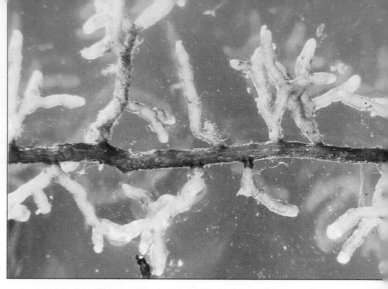

✦ **FIGURE 13.10**

**Mycorrhizal fungi live in the soil and attach to the roots of plants. The fungi aid the plant in the absorption of water and minerals from the soil and, in turn, obtain nutrients from the plant.**

Following this stage, plants with seeds evolved and became dominant, culminating with the development of flowering plants about 100 million years ago.

The evolution of plants occurred in distinct stages, with each stage representing the acquisition of a new and important adaptation that enabled plants to continue their spread over the land. These adaptations included protection from the drying effects of the atmosphere, development of leaves for increasing the efficiency of photosynthesis, roots for retrieving moisture and minerals from the earth, and stems to elevate the photosynthetic parts above other plants competing for sunlight. The evolution of large, multicellular bodies also required transport systems to move water and nutrients through the plant. Perhaps most importantly, the transition to land was aided by changes in the life cycle and methods of reproduction.

In life cycles of organisms such as humans, the body is composed of diploid somatic cells and haploid cells present as gametes (sperm and eggs) produced through meiosis. In most plants, a somewhat different life cycle has evolved. In the plant life cycle, a haploid phase, called the **gametophyte** alternates with a diploid phase, the **sporophyte** (✦ Figure 13.11a). This succession of phases in the life cycle is called **alternation of generations.** The haploid gametophyte generation produces gametes (without meiosis) that fuse to form the diploid sporophyte plant. Sporophytes grow and undergo meiosis to produce haploid spores. Germination and mitosis of these spores gives rise to a generation of gametophyte plants. As we will see in the

*Acetabularia* life cycle

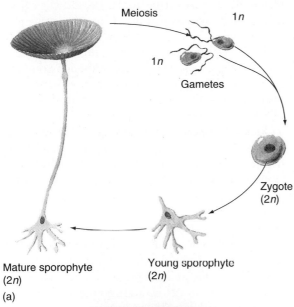

(a)

*Ulva* life cycle

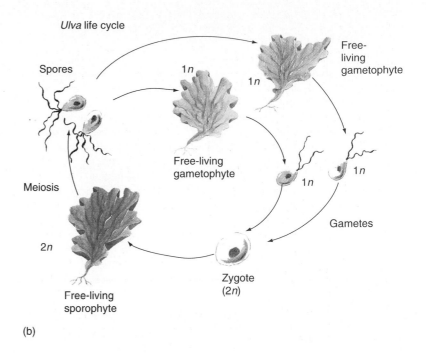

(b)

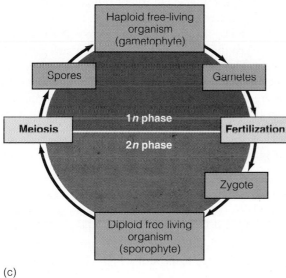

(c)

✦ **FIGURE 13.11**

**(a) The life cycle of** *Acetabularia,* **a green alga. In this species, the gametes are the only haploid cells in the life cycle. (b) The life cycle of** *Ulva,* **a multicellular green alga, illustrating the alternation of generations. (c) A generalized plant life cycle, showing alternation of generations. The life cycles of green algae were adapted by plants invading land.**

following sections, adaptations of this basic life cycle, which evolved in aquatic plants (green algae), permitted plants to become adapted to the terrestrial environment (Figure 13.11).

## THE TRANSITION TO LAND REQUIRED SEVERAL ADAPTATIONS

The transition from an aqueous environment to land required several major adaptations (✦ Table 13.1). First, on land, water is lost by evaporation and protective layers are required to conserve water. The development of waxy coverings or **cuticles** on stems and other aboveground parts and pores or **stomata** on leaves that can be opened or closed are specializations that prevent water loss (✦ Figure 13.12). Second, in a terrestrial environment, water is not readily available as a medium for movement of gametes, for fertilization, or for nourishment of the developing embryo. The reduction of the gametophyte generation and the development of **pollen** as a gametophyte dispersed by insects or air and the development of multicellular reproductive organs to shield the early stages of embryo development helped to solve this problem.

Third, air does not offer the buoyancy of water to help support the body of the plant, and the development of supportive tissues with rigid cell walls represented an important adaptation. Fourth, moisture and nutrients are available only to those plant parts in contact with the soil. The development of **roots** to absorb nutrients and water, and a **vascular system** to transport materials within the plant are adaptations to overcome this restriction.

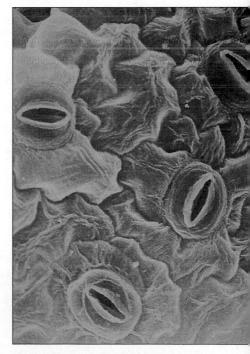

✦ **FIGURE 13.12**

**Stomata on the underside of leaves can be opened or closed to control gas exchange and water loss.**

| TABLE 13.1 AQUATIC ENVIRONMENT VERSUS TERRESTRIAL ENVIRONMENT | |
|---|---|
| **WATER AS AN ENVIRONMENT** | **LAND AS AN ENVIRONMENT** |
| Water surrounds the plant, prevents drying out. | Water must be obtained, conserved. |
| Water is available as a medium for movement of gametes, fertilization, and nourishment of embryo. | Water must be provided for reproduction, or gametes must be adapted for transport by other means. |
| Water provides buoyancy, reduces effect of gravity on plant. | Air provides no buoyancy, rigid support tissues required for upright growth. |
| Water provides medium for dissolved nutrients to reach the plant and be absorbed. | Terrestrial environment provides water and nutrients via soil. Plants need roots for absorption, vascular system for transport. |

## BRYOPHYTES ARE A TRANSITIONAL GROUP OF PLANTS

Many of the features of land plants described in the last section are previewed in the bryophytes. Bryophytes are intermediates in the transition from an aqueous to a terrestrial environment, and include the mosses, liverworts, and hornworts. Bryophytes evolved about 500 million years ago, and present-day species represent a small remnant of what was a large, ancient plant group.

Most bryophytes are small, with body plans that have some of the adaptations necessary for the transition to land. The sporophyte generation has a cuticle and pore-like stomata to prevent water loss. Bryophytes have multicellular organs for sexual reproduction, and the embryo begins development within the protective environment of the female gametophyte. However, bryophytes have motile sperm and require water for fertilization. Bryophytes do not have true roots; instead they have filamentous **rhizoids** that attach to a substrate and absorb moisture. Most bryophytes are flat, leaf-like, photosynthetic organisms (✦ Figure 13.13). They do not have a vascular system, although some do have cells for conducting water and nutrients. As a result of these limitations, they grow most successfully in moist or damp environments.

Mosses are perhaps the most familiar and one of the most common bryophytes. The moss plants you usually see represent the haploid gametophyte generation (✦ Figure 13.14). At the tips of the gametophytes, reproductive structures produce flagellated sperm (in an antheridium) or a nonmotile egg (in an archegonium). Water is required for the sperm to swim to the archegonium and fuse with the egg. A diploid zygote formed following fertilization is the first stage of the sporophyte generation. The embryo develops inside the archegonium to form the mature sporophyte. This plant remains attached to and dependent on the gametophyte. Haploid spores produced by the diploid sporophyte via meiosis fall to the ground and germinate to form a new gametophyte. The structural features of the bryophytes and their dependence on water for fertilization demonstrate that they bridge the water and terrestrial environment, but require a bit of both for survival.

## VASCULAR PLANTS: CONQUERING THE LAND

The most complex forms of plant life are the vascular plants. Here we will classify them as follows: lower vascular plants and seed plants. The seed plants are divided into two groups, the gymnosperms and the angiosperms.

✦ **FIGURE 13.13**
**Representative bryophytes. (a) Moss. (b) Liverwort. (c) Hornwort.**

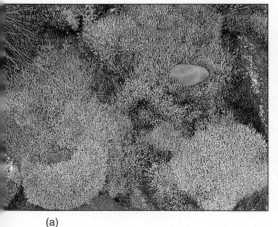

(a)

(b)

(c)

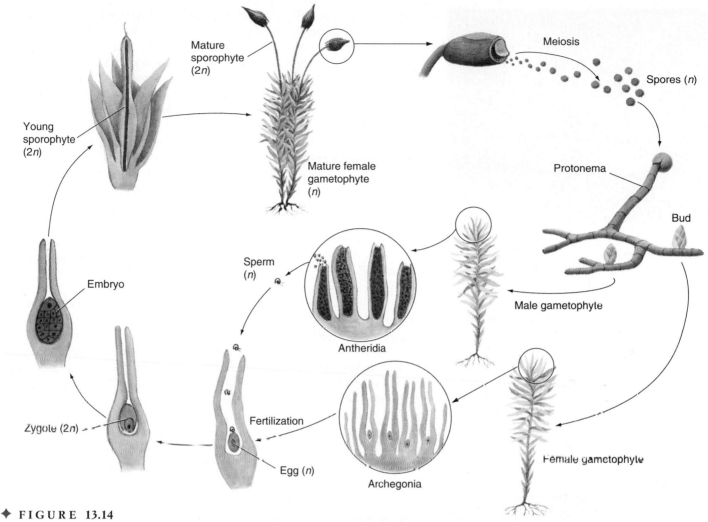

Labels in figure:
Mature sporophyte (2n)
Meiosis
Spores (n)
Young sporophyte (2n)
Mature female gametophyte (n)
Protonema
Embryo
Sperm (n)
Bud
Antheridia
Male gametophyte
Zygote (2n)
Fertilization
Egg (n)
Archegonia
Female gametophyte

♦ **FIGURE 13.14**

**Life cycle of moss. Haploid spores (upper right) germinate to form protonema. Haploid male and female gametophytes form on the protonema. In the presence of water, sperm formed in the male gametophyte (antheridia) swim to the female gametophyte (archegonia) for fertilization. The zygote develops into a diploid gametophyte, from which a sporophyte forms. Meiosis in the sporangia form haploid spores, which germinate to form protonema.**

## Vascular Plants Resulted from Several Major Evolutionary Advances

Several major evolutionary advances are present in vascular plants: the development of the **root-leaf-vascular system axis**, a **reduction in the size of the gametophyte generation,** and the **development of seeds** in some members of this group. Vascular plants evolved in the Silurian period, more than 400 million years ago. Fossil plants from this period, such as *Cooksonia,* show that early vascular plants had a central cluster of tissue specialized for the transport of water and other nutrients up the plant stem, and for the movement of photosynthetic products back down the stem.

As plants increased in size, a more efficient system evolved for the absorbtion and distribution of water and mineral nutrients throughout the plant. By selective advantage, roots replaced rhizoids as structures for anchoring the plant and for absorbing water and nutrients. At the other end of the plant, leaves evolved to increase the surfaces available for photosynthesis. Connecting these two systems was an increasingly complex transport system composed of **xylem,** which moved water and dissolved nutrients from the roots to the leaves, and the **phloem,** which moved dissolved sugars and other products of photosynthesis from the leaves to other regions of the plant. The success of this axis in plant evolution is evident by considering that giant redwoods, reaching a height of more than 300 feet, are among the largest organisms in the world, and that the tallest reaches of the tree are supplied with water from the roots via the xylem.

The second major feature of the vascular plants has been the development of a dominant sporophyte generation and an associated reduction of the gametophyte generation to microscopic size. In less

**✦ FIGURE 13.15**

*Psilotum,* **which resembles** *Cooksonia* **and other early vascular plants, has a well-developed stem, but no leaves.**

**✦ FIGURE 13.16**

**The fern sporophyte has a well-developed leaf system and a vascular system.**

complex vascular plants such as the fern, the gametophyte is an independent plant. In the gymnosperms and angiosperms, the gametophyte is not only reduced to microscopic size, but is nutritionally dependent on the sporophyte. The gametophyte grows and reproduces entirely within the sporophyte (reproduction in higher plants is discussed in Chapter 27). A dominant diploid sporophyte equipped with a root-leaf-vascular system axis offers both genetic and functional advantages for success in a terrestrial environment.

The third evolutionary advance present in the vascular plants is the development of seeds. Seeds are structures that contain the embryonic sporophyte and a food reserve encased in a protective coat. Seed plants were in existence some 345 million years ago, near the end of the Paleozoic era, although the dominant land plants of the time were the nonseed vascular plants such as the ferns. The formation of mountain ranges in the late Paleozoic brought about climatic changes leading to glacier formation and changes in sea levels. The earth grew cooler and drier, and mass extinctions occurred among the nonseed vascular plants. In addition to climate changes, it appears that other selective advantages favored the seed plants and allowed them to displace the nonseed plants. Just as amphibians of the time were less adapted to the new environment and gave way to reptiles, plants without the protection offered by seeds gave way to the seed plants, which became the dominant form of plant life, a situation that has continued to the present.

## Ferns Are Seedless Vascular Plants with a Dominant Sporophyte

The seedless vascular plants are classified into four groups based on their evolutionary origins and relationships. Three of these divisions are represented by only a handful of species, because most members of these groups are extinct. One group, represented by the species *Psilotum* (✦ Figure 13.15), is notable for its resemblance to the fossil *Cooksonia.*

The most familiar lower vascular plants are probably the ferns, represented by some 12,000 species (✦ Figure 13.16). The body plan of a typical fern includes a horizontal stem (**rhizome**) with upright,

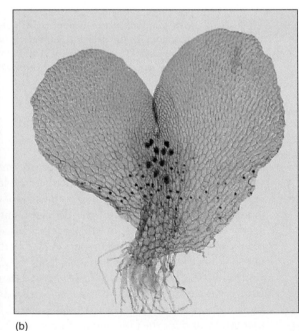

(a)                (b)

**✦ FIGURE 13.17**

**(a) On the underside of fern leaves, sporangia are clustered in circular arrays called** *sori.* **(b) The fern gametophyte, called a prothallus, is a free-living plant.**

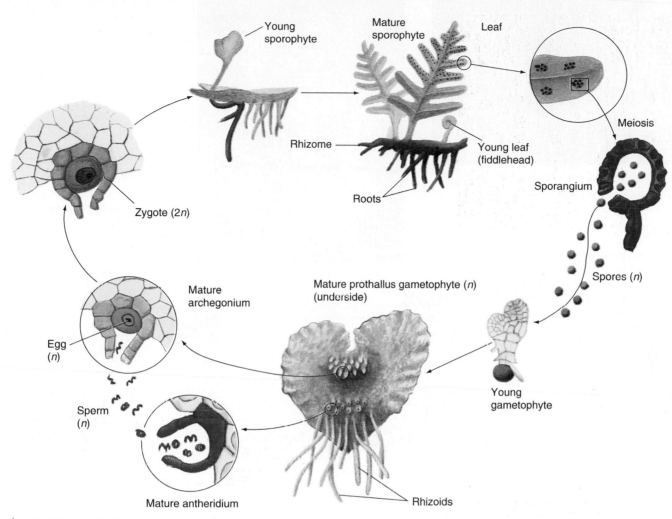

Young
sporophyte

Mature
sporophyte

Leaf

Meiosis

Rhizome

Young leaf
(fiddlehead)

Sporangium

Roots

Zygote (2*n*)

Spores (*n*)

Mature
archegonium

Mature prothallus gametophyte (*n*)
(underside)

Egg
(*n*)

Young
gametophyte

Sperm
(*n*)

Mature antheridium

Rhizoids

✦ **FIGURE 13.18**

**The fern life cycle. On the fronds of the mature sporophyte (upper right), haploid spores are produced by meiosis in
the sporangia. The haploid spores fall to the ground and germinate to form the mature gametophyte (prothallus).
On the underside of the gametophyte, male gametes are produced in the antheridia, and female gametes are produced
in the archegonia. In the presence of water, a male gamete swims to and fertilizes a female gamete. The resulting
diploid zygote grows into the sporophyte.**

true leaves containing vascular tissue. This plant is
the dominant sporophyte generation. On the underside of certain leaves, reproductive structures
known as **sporangia** develop (✦ Figure 13.17a).
Meiosis within these sporangia gives rise to spores,
which are shed and fall to the ground. Each spore
can germinate into a small, heart-shaped gametophyte, called the **prothallus** (✦ Figure 13.17b).
Male and female reproductive structures develop,
and water is needed for the motile, flagellated sperm
to fertilize the egg (✦ Figure 13.18). The zygote develops in the archegonium and grows out of the
prothallus by developing a root and shoot.

## Gymnosperms Are Seed Plants

The term gymnosperm means naked seed (*gymno*
= naked). The gymnosperms were a dominant

plant form in the Mesozoic era, but have undergone a series of extinctions since then and have
now been reduced to around 700 living species.
The gymnosperms are classified into four divisions
(✦ Table 13.2), three of which are represented by

| TABLE 13.2 CLASSIFICATION OF THE GYMNOSPERMS | | |
|---|---|---|
| | COMMON NAME | APPROXIMATE NUMBER OF SPECIES |
| Coniferophyta | Conifers | 550 |
| Cycadophyta | Cycads | 100 |
| Ginkophyta | Ginkos | 1 |
| Gnetophyta | Gnetae | 70 |

**◆ FIGURE 13.19**

Cycads are tropical gymnosperms that resemble palm trees.

**◆ FIGURE 13.20**

Bristlecone pines are gymnosperms that may be among the oldest plants on earth.

**◆ FIGURE 13.21**

Life cycle of the pine. The mature sporophyte (a pine tree) produces pollen (male) and ovulate (female) cones. Meiosis produces male gametophytes in the form of pollen grains. Female gametophytes are formed after meiosis in the ovule. The female gametophyte produces eggs within archegonia. During fertilization, the male gametophyte produces a pollen tube to transport the sperm. An embryo develops from the zygote and is dispersed as a seed. The seed germinates to form a new sporophyte generation.

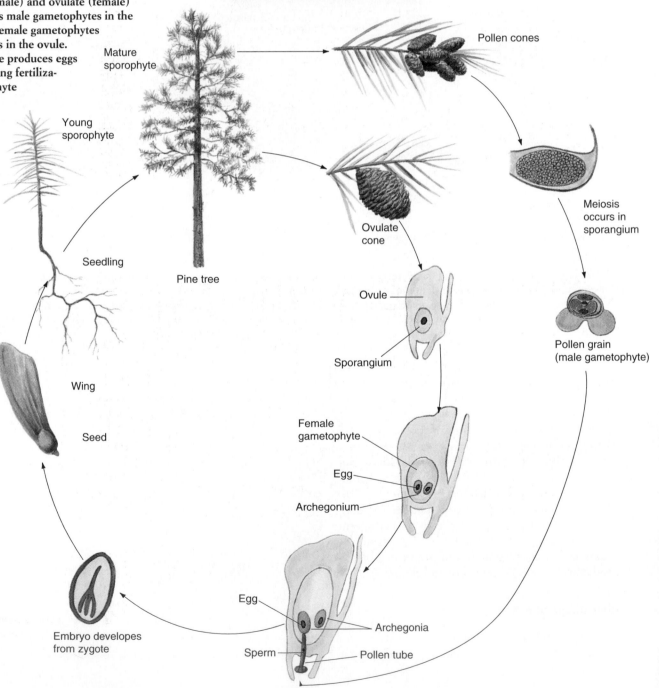

Mature sporophyte

Young sporophyte

Seedling

Pine tree

Wing

Seed

Embryo developes from zygote

Pollen cones

Ovulate cone

Ovule

Sporangium

Meiosis occurs in sporangium

Pollen grain (male gametophyte)

Female gametophyte

Egg

Archegonium

Egg

Archegonia

Sperm

Pollen tube

Fertilization

only a few surviving species (✦ Figure 13.19). The fourth division is the conifers, which includes the pines, firs, the giant redwoods, and the giant sequoias, as well as the bristlecone pines, some of which are among the oldest living organisms on Earth (✦ Figure 13.20). One tree in Nevada, for example, is about 5000 years old.

A conifer tree represents the dominant sporophyte generation; the gametophyte is reduced to microscopic size and develops in the cones (✦ Figure 13.21). In male cones, a pollen grain represents the male gametophyte (✦ Figure 13.22) and, after pollination, releases a nonflagellated sperm as a gamete. The female gametophyte develops in female cones, in a protective structure called the **ovule** (Figure 13.22). Meiosis takes place in the ovule, and produces a haploid spore. The spore develops into the female gametophyte, which produces eggs.

The pollen grain is released into the air, and carried by air currents to the female **ovule,** where fertilization occurs. This separation from dependence on water for reproduction is one of the most recent stages in the evolution of plants, and is marked by the appearance of nonflagellated male gametes. Along with the seeds that develop from the ovule following fertilization, these adaptations provide a reproductive advantage. The seed nourishes and protects the developing embryo, which grows into a sporophyte (Figure 13.21).

## Angiosperms Are Flowering Plants

All flowering plants are by definition, **angiosperms.** This division is distinguished from all other plants by the presence of two features: flowers and fruit. Angiosperms are currently the dominant form of plant life, with more than 235,000 known species. They range from the plants just under one millimeter such as *Wolffia* and *Lemna* (✦ Figure 13.23) to giant trees such as the eucalyptus. In angiosperms,

(a)

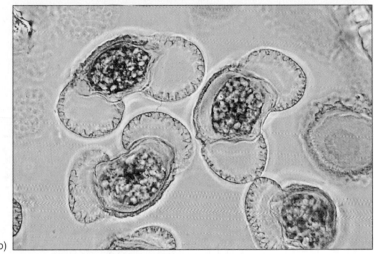
(b)

✦ **FIGURE 13.22**

(a) An intact ovulate cone is shown on the left, and a cross section of an ovulate cone, showing the ovules, is at the right. (b) Pollen grains of pine.

✦ **FIGURE 13.23**

*Lemna,* seen floating on this pond, is among the smallest angiosperms, at about one millimeter in diameter.

# GUEST ESSAY: RESEARCH AND APPLICATIONS

## Biology and the Pursuit of the Aesthetic

PAUL ALAN COX

As a biology professor, I believe there are four levels of learning. The first level is the mere transmission of information that this and other textbooks do well. A higher level is developing curiosity. Nobel prizes are given to women and men, not because they master standardized tests, but because they have diligently pursued interesting areas of research. An even higher level of learning than curiosity, however, is to approach the aesthetic. I fear that sometimes we immerse our students so deeply in the details of biology that we fail to communicate the aesthetic of biology, the sheer grandeur of life on this planet. Yet for most biologists it is precisely this search for this aesthetic that originally led them to enter the field. It certainly led me to study evolutionary ecology and ethnobotany.

Some of my earliest memories are of plants. I remember as an infant feeling the prickly leaves of juniper trees and smelling the intoxicating fragrance of lilacs in front of our home. Late, as a boy, I began collecting and drawing flowers. At a very early age I learned that plants are beautiful but vulnerable. When I was twelve, I wrote and obtained permission from the U.S. Forest Service to collect several living specimens of *Darlingtonia californica,* the insectivorous Cobra lily, from California. As my parents drove me across the Great Basin toward the coastal Redwood forest, I remember the excitement I felt. Soon we drove down Gasquet Creek, where I saw for the first time the Cobra lilies, rising like strange green snakes from the wetland. However, my delight turned to alarm when I looked across the population and saw heavy machinery being used to build a road

right through it. At that precise moment, I decided to devote a major portion of my life to conservation.

My scientific training has neither dulled my thrill with the beauty of plants, nor reduced my ardor for conserving them. I now know a lot more about the biochemistry, genetics, and physiology of *Darlingtonia californica* than I did as a twelve-year-old, but I still feel a boyish awe in the presence of such plants. Scientific details should not deaden our sensitivity to the beauty of life, rather they should increase it.

For the last decade I have studied how seagrasses, monocotyledons that grow in intertidal areas, are pollinated. Although I was led to this study by theoretical considerations as diverse as diffusion physics and the mathematics of search theory, it is the beauty of the plants that has carried me through the long weeks of field observations. Many of the plants have extraordinary pollen that is long and noodle-like. Their pollen is so long that it can be seen floating on the water without magnification. When many flowers release their pollen simultaneously, it is like someone has cast white confetti on the surface of the sea. The process is breathtaking. But it is the careful study of the details themselves that has given me the most delight. As a biologist you learn to find hidden beauty where no one has discovered it before.

My ethnobotanical work has been similarly delightful. I have spent my time in tropical rain forests studying how indigenous people use plants for medicine. I have been deeply impressed with the biological sophistication of indigenous people. Unlike so many of us in the West, they have never lost sight of the sacred-

ness of the world. To walk beneath the rain forest canopy and see light filtering through the iridescent leaves is for them to see the very face of God. Conservation for them is but recognition of the beauty and sacredness of the entire planet.

They believe it is wrong to destroy a rain forest for the same reason that we believe it is wrong to vandalize a mosque, synagogue, temple, or other sacred edifice. They believe that the act itself is a sacrilege. I believe they are right. On both an intellectual and spiritual level as humans we are clearly part of the fabric of this planet. Despite considerable speculation, we have yet to find convincing evidence for the existence of life anywhere else in the entire universe. We clearly live on a very lonely and extraordinarily fragile planet. It behooves us to take care of it. If your biology class does nothing more than open your eyes to the beauty of Earth, this book, your professor, and all who have labored so hard to help you in your education will have succeeded.

*Paul Alan Cox is dean of General Education and Honors and professor of botany at Brigham Young University. He received his Ph.D. in biology from Harvard University in 1981. His most recent ethnobotanical discoveries include the isolation of Prostratin, a compound from a Samoan plant, that is now being developed as a treatment for AIDS.*

the vascular system is refined and enlarged, increasing the efficiency of transport.

The first angiosperms known from the fossil record are from the Cretaceous, some 120 million years ago. Angiosperms underwent a burst of evolution beginning in the late Mesozoic, perhaps 100 million years ago, spreading to many different habitats.

The evolutionary advance that allowed angiosperms to spread and achieve their dominance is related to the development of flowers as reproductive structures. Gymnosperms transfer male pollen to female ovules by dispersal through the air. Pollen grains are captured by drops of sap (which is why some pine cones are sticky) exuded

by the female ovules. In the gymnosperm ancestors to angiosperms, insects were presumably attracted to this agglomeration of sap and pollen grains as a food source and ended up assisting in the transfer of pollen as they moved from plant to plant. This relationship increased the efficiency of plant reproduction as a by-product of the insect's search for food.

The selective value of this relationship led to the evolution of structures called **nectaries,** nectar-secreting organs that served as insect feeding stations. A series of evolutionary changes presumably under the influence of selection culminated in the development of **flowers** as reproductive structures (✦ Figure 13.24).

Angiosperm plants are sporophytes; the gametophyte generation is contained in the flower. The evolution of this reproductive adaptation in plants and the evolution of flower-feeding insects was and remains closely intertwined. The adaptive radiation of the angiosperms, beginning about 65 million years ago was coupled with a similar adaptive radiation in the insects. Flowering plants appeared to have coevolved with certain groups of insects.

Today, more than 235,000 species of angiosperms are known, with more to be described. They so dominate the landscape that they cover almost 90% of the earth's surface that can support plant life. Trees, shrubs, and flowers dominate most of the landscape. The grains, cereals, vegetables, and fruits in the human diet are all products of angiosperms. The cultural, economic, and political consequences of angiosperms cannot be underestimated. The domestication of grain and the conversion from a hunter-gatherer to an agriculture-based society has been documented many times by anthropologists. However, the role of other plants in human history is just beginning to be appreciated (see Beyond the Basics, p. 254). Combined with the historical view related in the story of ergot poisoning in the opening part of this chapter, it is apparent that the coevolution of human culture and angiosperms is an active area of investigation.

##  TRENDS IN PLANT EVOLUTION

Terrestrial plants are thought to have evolved from multicellular green algae with a life cycle showing alternation of generations, with a sporophyte and a gametophyte phase to the life cycle. One of the major evolutionary trends in the transition to land is the gradual reduction of the gametophyte generation and the dominance of the sporophyte generation (✦ Figure 13.25).

A second trend in plant evolution is the development of the root-shoot-leaf axis and specialized tissues for the transport of water and nutrients. These adaptations are found in the now dominant vascular plants.

A third trend is the development of the seed, and the associated mechanisms of sexual reproduction. Flowers allow the gametophyte generation, the gametes and the embryo to be protected within structures and coverings produced by the sporophyte generation. Fruits provide a means for dispersing offspring to new environments.

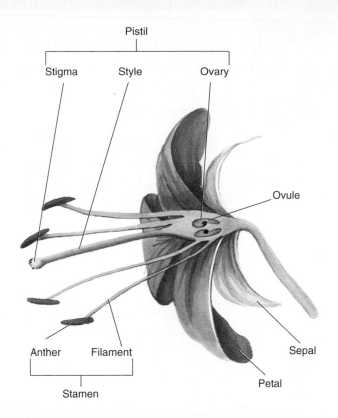

✦ FIGURE 13.24

A typical flower. The pistil, made up of the ovary, style, and stigma, contains female reproductive structures. The stamen, composed of the anther and the filament contains the male reproductive structures.

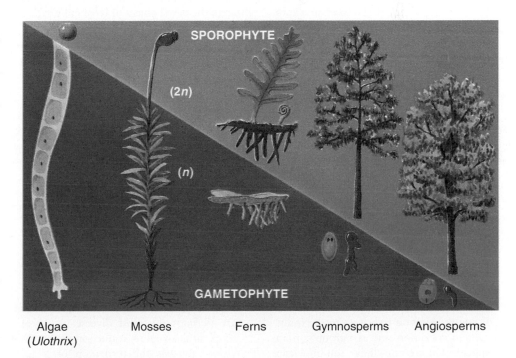

✦ FIGURE 13.25

Evolutionary trends in plants. Adaptation to a terrestrial environment is marked by a reduction in size of the gametophyte generation, and a dominant sporophyte.

# SEEDS OF CHANGE

The role of angiosperms in the development of human culture has been researched and described by anthropologists, historians and biologists in many books and periodicals. Most of the emphasis has been placed on the domestication of grain and the development of agricultural societies in the ancient world. However, global explorations of the seventeenth and eighteenth century brought to light new plants and new opportunities for commercial and economic use. In his book *Seeds of Change,* Henry Hobhouse details how five new plants, discovered in the Western Hemisphere and Eastern Asia brought about significant changes in world affairs. The plants and a brief synopsis of their effects follow.

## Quinine

Malaria is a serious disease, affecting hundreds of millions of people worldwide. The discovery by the Spanish that natives used the bark from cinchona trees growing in the Andes Mountains to treat this infective disease allowed Europeans and others to settle in vast areas of Africa, Asia, and the Middle East, which had until then been only sparsely populated. The colonial empires in these regions were built partially on quinine extracted from tree bark.

## Sugar

Sugar cane was transplanted from India and Eastern Asia to the Caribbean Islands and formed an important part of the Triangle Trade: exportation of trade goods from Britain to Africa, the exportation of slaves from Africa to the Caribbean, and the exportation of rum and sugar from the Caribbean to England, leading to a complex relationship between British naval power and mercantile policy.

## Tea

Developed in China, competed for in trade by the British, Dutch, and Portuguese, tea transformed the cultures and economies of several countries in Asia, including China, India, and Sri Lanka. At a more subtle level, tea influenced the design of merchant ships, and the development of the porcelain industry in China and Europe.

## Cotton

Two species of cotton are native to the Western Hemisphere. Planted in the Southern states, cotton became an economic mainstay for the United States and England in the nineteenth century. In England, the Industrial Revolution brought the development of large mills, which precipitated the transition from spinning and weaving as cottage industries to the development of large factories in urban industrial settings. Similarly, the mill towns of New England became dependent on cotton as the base of the economy.

## Potato

Introduced into Ireland indirectly from South America, the potato became part of the Irish diet by 1620. During the next 200 years the potato became the main crop and often the sole item in the Irish diet. When a fungus killed off the potato crop in the 1840s, starvation ravaged Ireland. More than one million Irish starved to death in 1845–46, and another 1.5 million emigrated, many to the United States.

This short discussion excludes other plants with profound effects on human culture, including pepper, which stimulated much of the explorations of the fifteenth and sixteenth centuries (and was a reason for Columbus's first voyage); tobacco, the profits from which helped finance the American Revolution; and in the twentieth century, the development of corn as a food source and mainstay of the U.S. economy. Although we tend not to think about it, angiosperms have had a dramatic impact on the development and direction of human culture and society.

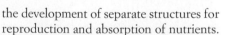

# SUMMARY

1. One of the key steps in evolution was the development of multicellular organisms. This allowed increases in the size and complexity of organisms and opened new habitats for exploitation, in particular, the land. One of the groups of organisms that coupled an increase in complexity with the invasion of the land was the eukaryotic fungi.

2. Fungi represent a transition to the multicellular condition, and as a group contain species that are unicellular, multinucleate (partially multicellular), and multicellular. Fungi show specialization of function in the development of separate structures for reproduction and absorption of nutrients.

3. More than 60,000 species of fungi have been identified, and they are classified by their method of reproduction. Major groups include the zygospore fungi, such as black bread mold. These fungi reproduce sexually to form a dormant zygospore. The sac fungi or ascomycetes are perhaps the most diverse group of fungi, and sexual reproduction in this division involves the formation of a sac called an ascus, within which ascospores develop. Yeast belong to this group. A third major group of fungi are the club fungi or Basidiomycetes. These fungi reproduce sexually through the formation of basidiospores.

4. Fungi form symbiotic relationships with plants. Examples of this interrelationship to form a hybrid organism include the lichens, which are composed of a fungus and a unicellular photoautotroph. Another form of symbiosis involving fungi are mycorrhizae, interactions between a fungus and the roots of higher plants. In this relationship, the fungus supplies nutrients in the form of water and minerals,

and the plant roots provide carbohydrate nutrients to the fungus, an example of mutualism.

5. A second group of organisms that made the transition from water to land are the plants. These eukaryotes are characterized by their ability to convert the energy from sunlight into food molecules and by the presence of cell walls containing cellulose and organelles called plastids. The life cycle of these organisms shows an alternation of generations in which a haploid phase alternates with a diploid phase.

6. Present-day land plants probably evolved from multicellular green algae in distinct stages. Each stage was marked by the acquisition of new traits that assisted the transition to land and increased the adaptation to the terrestrial environment. These adaptations include the development of a protective cuticle, leaves, roots, stems, vascular systems, seeds, and flowers.

7. Bryophytes represent plants that have been successful on land, but have not completely escaped the need for an aqueous environment. This group includes the mosses and liverworts, which have evolved protective cuticles and rhizoids, but still require water for fertilization and have not evolved vascular systems. Based on this combination of characteristics, they are viewed as a transitional group.

8. Vascular plants are the most complex and successful forms of plant life. The lower vascular plants, including the ferns, show development of vascular systems, true leaves, and stems, but still require water for fertilization and sexual reproduction.

9. The higher vascular plants, including the gymnosperms and angiosperms, have successfully completed the transition to land and escaped the need for water as a medium for reproduction. Innovations in this group include the development of the seed as protection for the embryo, the reduction of the gametophyte generation, and the evolution of flowers.

## KEY TERMS

alternation of generations
angiosperms
basidia
basidiospores
chitin
cuticles
development of seeds
flowers
gametophyte
hyphae
lichens

mycelium
mycorrhizae
nectaries
ovule
parasites
phloem
pollen
prothallus
reduction in the size of the gametophyte
    generation
rhizoids

rhizome
root-leaf-vascular system axis
roots
saprophytes
sporangia
spores
sporophyte
stomata
vascular system
xylem
zygospore

## QUESTIONS AND PROBLEMS

### TRUE/FALSE

1. Fungi absorb their food after digesting it by secreting enzymes.
2. Zygomycetes comprise the smallest division of fungi.
3. Club fungi play an enormous role as decomposers and recyclers of organic molecules in the biosphere.
4. The diversity of the plant kingdom is demonstrated by the fact that some plants contain plastids and others do not.
5. Species of algae are placed into three divisions.
6. Mosses are the most familiar green algae.
7. In plant life, the gametophyte is the haploid phase, and the sporophyte is the diploid phase.

### MATCHING

Match the groups appropriately.

1. fungi
2. mycorrhiza
3. brown algae
4. red algae
5. green algae
6. bryophytes
7. gymnosperm
8. angiosperm

___ flowering plants
___ "fungus root"
___ absorptive heterotrophs
___ fucoxanthin
___ seed plants
___ rhizoids
___ phycoerythrin
___ chlorophyll *a* and *b*

Match each structure with its function.

9. nectaries
10. xylem
11. phloem
12. stomata
13. holdfast
14. basidiospores

___ reproductive structures
___ anchoring system for *Fucus*
___ pores on leaves that prevent water loss
___ vascular transport system moving nutrients from roots to leaves
___ secreting organs that serve as insect feeding stations
___ vascular transport system moving food through plant

### SHORT ANSWER

1. If you had samples of fungal hyphae, without characteristic reproductive structures, how would you determine to which major group the organism belonged?
2. Describe the evolution of plants that led

2. Describe the evolution of plants that led to their transition to land.
3. What characteristics are used to classify the three divisions of algae?
4. Name the three major evolutionary advances present in vascular plants.

## SCIENCE AND SOCIETY

1. Ferns and mosses can be found in a variety of ecosystems. They are usually the pioneering colonizers in regions where the normal plant life has been destroyed by flooding or fires. How do you think mosses and ferns travel to areas where environmental disasters have occurred? What makes mosses and ferns such good travelers? Do you think their evolutionary advances or their reproductive life cycles play a role in their effectiveness to inhabit broad geographical areas? Why or why not?

2. Traveler's diarrhea is one of the most common diseases of tourists who come from areas where infectious diarrhea is rare into areas where it is common. Traveler's diarrhea is also accompanied by fever and chills, nausea, and vomiting. *Escherichia coli* is a colon bacterium that is the most frequent cause of traveler's diarrhea. The strains of *E. coli* which are normally present in the gut do not cause diarrhea even though about 25% of the mass of normal feces contains *E. coli.* Why would one strain of *E. coli* cause problems when another strain does not? What contributes to the differences between strains? Would taking prophylactic antibiotics prior to traveling prevent traveler's diarrhea? Why or why not? How do you think the disease could be avoided (besides not traveling)?

# 14 ANIMAL DIVERSITY

OPENING IMAGE

*Penguins are well adapted for life in*

*the harsh conditions of Antarctica.*

Late in the summer of 1909, a party of paleontologists on a fossil-collecting trip was in the Canadian Rockies on the eastern edge of British Columbia, close to the railroad town of Field. Led by Charles Walcott, the group was near a ridge connecting two mountains when, as the story goes, Mrs. Walcott's horse stumbled on a slab of shale. Examination of the shale revealed a number of fossil impressions of soft-bodied organisms. A search revealed that the block fell from a seven-foot-wide band higher up on the mountain slope. Walcott returned to this site the following summer and for years afterward to collect fossils. Because the fossil bed was reached by traveling through Burgess Pass, he named the deposit the Burgess Shale. In subsequent years, he quarried large numbers of fossil specimens from the Burgess Shale, and shipped them to the National Museum of Natural History in Washington, D.C.

The remarkably preserved specimens found in the Burgess Shale were formed when animals living on soft mud bottoms in shallow water near the base of a sheer vertical cliff were buried by mud slides. Preserved from degradation in cold, deep water deprived of oxygen, the bodies from these Cambrian marine communities became progressively compressed by subsequent mud slides, and slowly turned into fossil impressions in shale.

Aside from their state of preservation, why are these fossils so remarkable? The scientific papers describing and cataloging these finds were dry and rather unremarkable. However, by the late 1960s, it had become clear that the Burgess Shale fossils are among the most remarkable collections of fossils ever discovered. The impressions left by these *soft-bodied organisms are a window into what an animal community on the ocean floor looked like some 530 million years ago.*

*Many of the fossils from the Burgess Shale do not belong to groups known from other fossil sites, or to any groups of animals alive today. Up to 15 of the species from this collection do not fit into any known phylum and each may represent separate phyla. The shale also contains fossils that can be classified into known groups, such as the arthropods (a group that includes present-day shrimp, spiders, and insects), but some 20 types of arthropod fossils from the shale cannot be placed in any known arthropod taxonomic group. The story of these fossils has been told by Stephen Jay Gould in his book* Wonderful Life.

*The Burgess Shale fossils remind us that the animals we know from the world around us represent only a tiny fraction of the groups that evolved, lived, and died on earth. The fossils from a few hundred square feet of shale in the mountains of British Columbia may contain fossils with as much diversity as all the animals in existence today. This diversity is the product of an evolutionary process that through chance and selection produced the organisms we know as animals. In the following pages, we review the history of present-day animal life, but keep the Burgess Shale in mind.*

## ANIMAL BODY PLANS, ORGANIZATION, AND DEVELOPMENT

Animals are characteristically multicellular heterotrophs. As discussed in Chapter 11, the development of multicellularity was an important step in evolution. Over time, this adaptation allowed the exploitation of the terrestrial environment and freed organisms from the need for water as a medium for support and reproduction. Animals have an evolutionary history that parallels that of plants in two ways: the development of multicellularity and the exploitation of the terrestrial environment. In the transition to land, both groups faced similar problems: preventing desiccation, developing skeletal elements for support, and providing internal transport systems. Solutions arose in both plants and animals, although in different ways. In plants, cell walls, cuticle, or bark provides protection, whereas in animals, chitin, skin, fur, claws, and feathers developed to serve this function. Vascular plants have woody xylem and lignified cells for structural support, invertebrates use exoskeletons, and vertebrate animals use bone and connective tissues for support. Plants use xylem and phloem as a transport system; most animals use a circulatory system.

Unlike plants, which are autotrophs, animals are heterotrophs. Animals obtain nutrition by ingestion or absorption of organic nutrients. As autotrophs, plants can synthesize nutrients via photosynthesis.

✦ FIGURE 14.1

**Animals are heterotrophs that must find their food.**

Animals must find food, and hence movement of all or parts of the body is closely tied to nutrition and survival (✦ Figure 14.1). In fact, one of the characteristic features of animals is the ability to move during all or some part of the life cycle.

The ability to move in search of food is associated with the development of sensory systems to identify food, muscle and skeletal elements to move toward food, and some form of integration between sensory and muscle elements, usually in the form of a nervous system. The distinction between autotrophs and heterotrophs is important, and this difference is responsible for many of the structural and functional dissimilarities between plants and animals. It is also responsible for a number of evolutionary trends within and between animal groups.

Besides being multicellular, heterotrophic, and motile at some stage of the life cycle, animals share several other characteristics with each other and with plants. These include a life cycle with a preadult stage that (usually) includes embryonic development and a diploid genome. (See Chapter 7 for a review of chromosomes and the haploid/diploid state.)

### Evolutionary Origins and Classification of Animals

Like plants, animal life probably arose from marine protists. Evidence about the origins of animals from the fossil record is sketchy at best, and no group of protists has been identified as ancestral to the animals. Recall from Chapter 11 that the Proterozoic Era extended from some 2.5 billion years ago to about 570 million years ago. Unicellular eukaryotes appeared in the fossil record about 1.2 to 1.4 billion years ago, and colonial and multicellular eukaryotic organisms appeared in the late Protero-

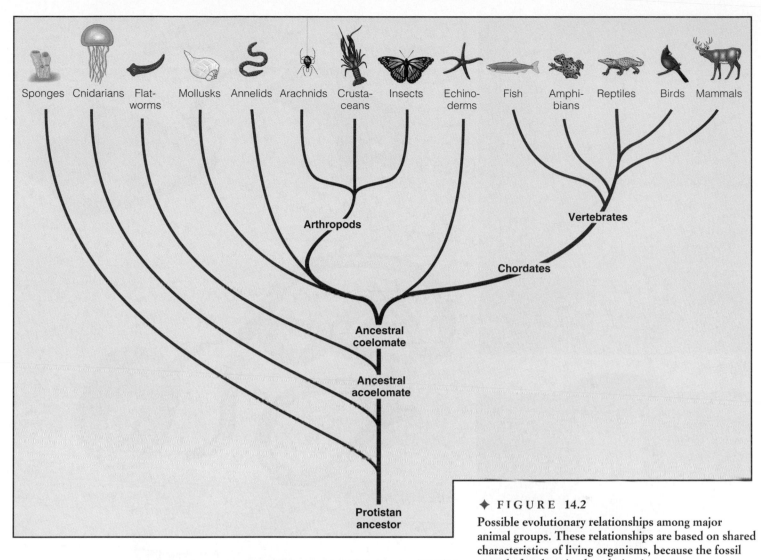

✦ FIGURE 14.2

Possible evolutionary relationships among major animal groups. These relationships are based on shared characteristics of living organisms, because the fossil record of early animal evolution is scanty.

zoic. Multicellular animal fossils in the form of impressions and burrows first appeared some 700 million years ago. These burrows were presumably made by organisms that could tunnel into the floor of the seabed, and are taken as signs of animal life. All known animal fossils from the Proterozoic were soft-bodied; animals with external skeletons in the form of shells and other skeletal materials appeared about 570 million years ago, at the end of the Proterozoic.

Present-day animals are classified into 30 to 35 phyla (depending on which classification scheme is used), and all modern phyla are represented in the fossil record by the beginning of the Paleozoic Era, some 570 million years ago. Most of the present-day phyla live in marine environments.

The presumed evolutionary relationships among some of the major animal groups are shown in ✦ Figure 14.2. These relationships are based mainly on characteristics shared by living forms, because animals arose over a short period at the Precambrian/Cambrian boundary and because evidence in the fossil record for this period is scarce. As our

knowledge of animal evolution improves, major divisions of animals may turn out to have multiple origins among ancestral protists.

### TRENDS IN ANIMAL EVOLUTION

Several major trends in animal evolution are apparent. First, animals use one of two basic body plans. These can be described as either **sac-like** or a **tube within a tube**. Most animals have bodies that follow the tube-within-a-tube plan. This body plan uses two body openings, one for food, and the other for the elimination of digestive waste (✦ Figure 14.3). Organisms with this body plan have a specialized

✦
**FIGURE 14.3**

This worm exhibits the tube-within-a-tube body plan, with a digestive system inside the cylindrical body.

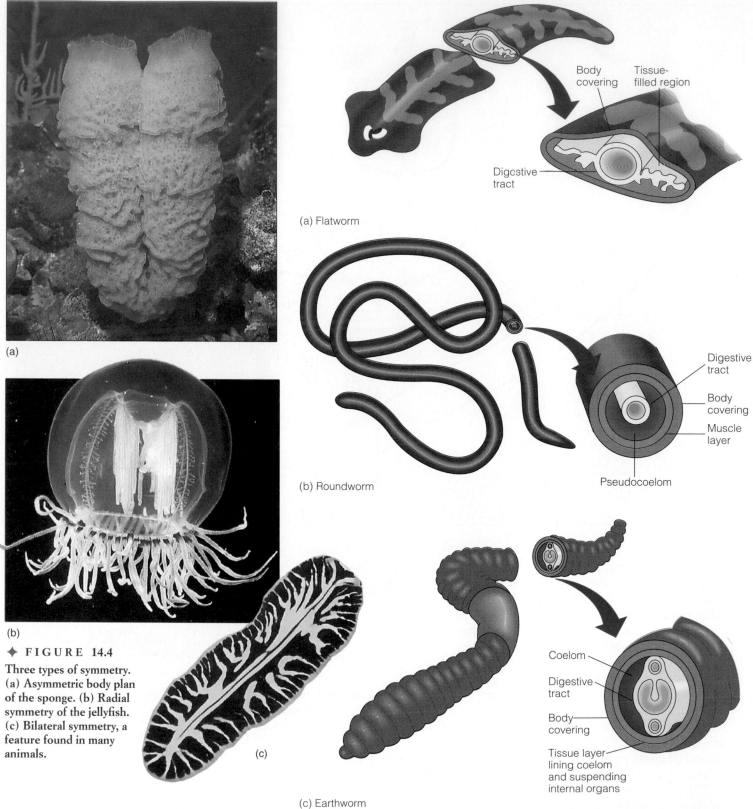

Body
covering

Tissue-
filled region

Digestive
tract

(a) Flatworm

Digestive
tract

Body
covering

Muscle
layer

Pseudocoelom

(b) Roundworm

Coelom

Digestive
tract

Body
covering

Tissue layer
lining coelom
and suspending
internal organs

(c) Earthworm

(a)

(b)

✦ **FIGURE 14.4**
**Three types of symmetry.
(a) Asymmetric body plan
of the sponge. (b) Radial
symmetry of the jellyfish.
(c) Bilateral symmetry, a
feature found in many
animals.**

(c)

✦ **FIGURE 14.5**
**The development of a body cavity represents a significant advance in animal evolution.
The first three groups we will consider, the sponges, cnidarians, and flatworms, lack a
body cavity as shown in (a). They are acoelomate animals. (b) Roundworms represent
animals with a body cavity in direct contact with the outer muscular layer of the body.
This cavity is called a pseudocoelom. (c) An earthworm has a body cavity called a
coelom, with the inner and outer surfaces lined with a tissue layer.**

digestive system and are more efficient (by about 10%) at recovering nutrients from food than animals with the sac-like plan.

In the sac-like body plan, there is only one body opening, which is used for both intake of food and discharge of wastes. Animals with this body plan lack a highly specialized digestive system, and are less efficient at food processing.

The second general trend is the progressive organization of the body into a symmetrical pattern (✦ Figure 14.4). **Asymmetrical** animals have no general body plan nor an axis of symmetry that divides the body into mirror-image halves. **Radially symmetrical** animals have body parts organized around a central axis, and tend to be circular or cylindrical in shape. Such animals have multiple axes of symmetry and can be divided into mirror-image halves by any number of longitudinal cuts that pass through the center of the body. **Bilaterally symmetrical** animals have a single axis of symmetry, and only one longitudinal slice through the body will produce mirror-image halves (✦ Figure 14.4). By definition, animals with bilateral symmetry have a head (anterior), a tail (posterior), a top (dorsal side) and a bottom (ventral side), as well as left and right halves.

The presence of a body cavity or **coelom** that forms during preadult development was an important step in animal evolution (✦ Figure 14.5). **Acoelomate** organisms do not have such a body cavity, and **pseudocoelomate** organisms have a body cavity, but it does not arise by splitting of the mesoderm. **Coelomate** animals have a body cavity lined with mesoderm.

Some animals have body parts composed of repeating units or **segments** (✦ Figure 14.6). The presence of segments offers an increase in the efficiency of the body plan because adjacent segments can take on different functions.

We discuss a selected set of the animal phyla, chosen to illustrate features such as body plans, their relationship to function, complexity, evolutionary relationships, and the origins of adaptations that were important in the evolution of animals.

## SPONGES: ASYMMETRY AND CELL SPECIALIZATION

Overall, sponges (members of the phylum *Porifera*) have to be rated as evolutionary survivors. Fossil sponges from the Cambrian period some 500 million years ago closely resemble living species. The evolutionary origin of sponges is uncertain. They may have arisen from a colonial form of a ciliated or flagellated protist independent of other animal groups. Sponges are aquatic organisms, living mostly in the ocean, although some freshwater

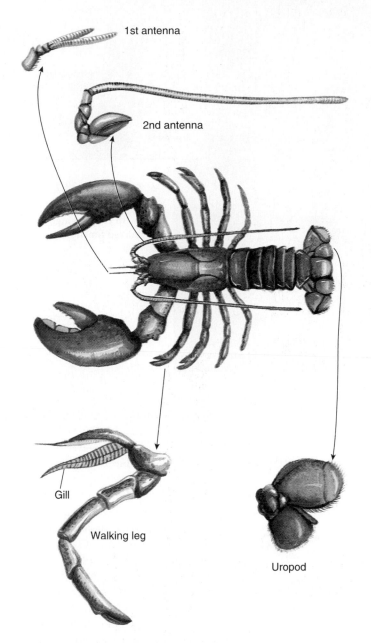

1st antenna

2nd antenna

Gill

Walking leg

Uropod

✦ **FIGURE 14.6**

Lobsters have appendages on body segments that have become specialized for a variety of tasks, including sensation, respiration, walking, and swimming.

forms are known. As adults, they are **sessile** (attached to something) organisms with an asymmetrical body plan and a sac-like organization. Sponges range from small organisms a centimeter across, to large structures a meter or more in length, and are often brightly colored.

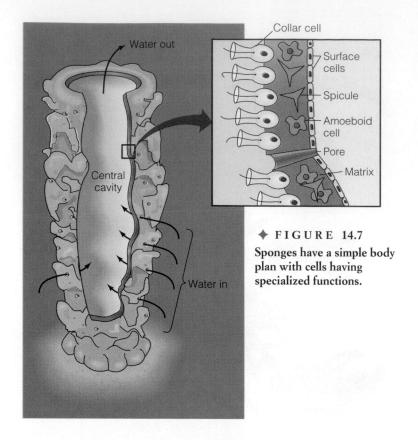

**✦ FIGURE 14.7**

**Sponges have a simple body plan with cells having specialized functions.**

Sponge bodies contain specialized cells, but there are no true tissues (✦Figure 14.7). The sponge body is made up of a thin layer of outer **epidermal** cells and an inner layer of **collar** cells. The collar cells are equipped with flagella to help move water through the body. Between these layers is a gel-like matrix containing needle-shaped **spicules** that serve as skeletal elements, and motile amoeboid cells called **amoebocytes** that transport nutrients.

Sponges feed by remaining in place and moving large volumes of water through pores in their body wall. Currents created by the flagella on the collar cells move water into and out of the body. Bacteria and other food particles are engulfed by the collar cells, and passed to the ameobocytes for digestion and nutrient distribution. Water exits through the large opening at one end of the body, carrying away wastes. Because sponges feed by removing food from water that filters through their body, they are called **filter feeders.**

Sponges reproduce asexually by budding off fragments that grow into new sponges. In areas of the world where sponges are commercially harvested, divers often scatter fragments over the seafloor to replace the sponges they have collected. Sponges can also reproduce sexually, producing a motile, ciliated preadult **larval** stage. Using cilia, the larvae can swim some distance away, attach to a rock or other substrate, and grow into a new sponge.

## JELLYFISH, CORALS, AND SEA ANEMONES: CELLS ORGANIZE INTO TISSUES

Members of the phylum *Cnidaria* are a group of approximately 10,000 species characterized by adult bodies with radial symmetry (✦Figure 14.8) and a sac-like organization. This group is aquatic, and most are ocean dwellers. There are two variations of the body plan in cnidarians: an attached sessile form called a **polyp,** and a motile, bell-shaped form called a **medusa** (Figure 14.8). In both, the opening of the sac is surrounded by a ring of tentacles, with the mouth directed upward in polyps and downward in medusa forms.

Animals of this phylum exhibit several important adaptations. The first is the development of cells organized into tissues. In this phylum, organisms have an outer epithelial layer for protection, an inner epithelial layer that functions in digestion, contractile cells for movement and feeding, and nerve nets that are connected to sensory cells and to contractile cells. In addition, the spaces between the outer and inner epithelial layers are filled with a gel-like matrix called **mesoglea** that contains **interstitial** cells.

In most species, the adult is radially symmetrical, with organs arranged around a central axis of symmetry, like spokes radiating from the hub of a wheel. Radial symmetry allows the animal to capture food as it approaches from any direction. Cap-

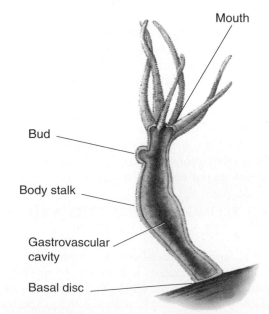

**✦ FIGURE 14.8**

**Cnidarians have a radial symmetry and two variations of a body plan. The freshwater hydra represents the polyp body plan.**

tured prey is forced into the gastrovascular cavity
(✦ Figure 14.9) and digested. Nutrients are distributed to the inner and outer cells of the
body wall.

The typical life cycle of cnidarians
involves both asexual and sexual
forms of reproduction Figure 14.9).
Zygotes produced by sexual
reproduction typically develop
into an immature, motile
form called a **planula,** a free-
swimming stage that serves
to disperse members of
the species. Planulae select
a suitable habitat, settle
down, and develop into
polyps. Mature polyps can
reproduce asexually to
form medusas. In turn, the
medusas form gametes and
reproduce sexually to give
rise to planula larvae, starting the life cycle all over
again.

Corals belong to this group
of animals, but have lost the
medusa stage of the life cycle. Inside their calcium-hardened external skeletons, coral live as a series
of interconnected polyps. Coral polyps
form the basis for the coral reef, a complex
oceanic community found mostly in tropical
waters (✦ Figure 14.10).

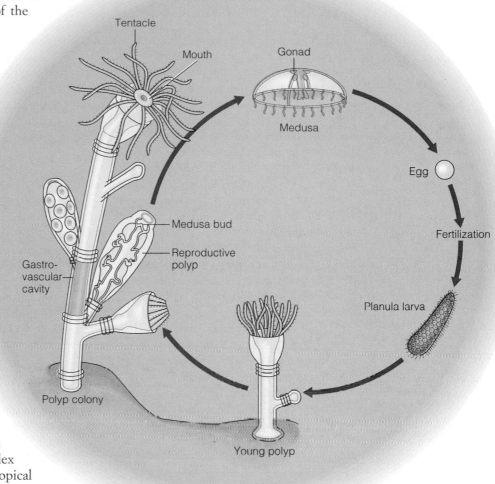

✦ FIGURE 14.9

Life cycle of *Obelia,* a cnidarian. Buds on the stationary polyp form a medusa, which reproduces
sexually to form a planula larva. The free-swimming planula settles down and forms a polyp,
beginning the cycle again.

✦ FIGURE 14.10

Corals are cnidarians that have
lost the medusa stage of the life
cycle, and remain stationary.

## FLATWORMS HAVE BILATERAL SYMMETRY AND CEPHALIZATION

Animals commonly known as flatworms (the phylum *Platyhelminthes*) are grouped into three classes: two that live mainly as parasites, and free-living forms such as **planarians** (◆ Figure 14.11). The adult worms are bilaterally symmetrical. This body plan is associated with movement and **cephalization,** the concentration of sensory tissues in the anterior part of the body. Planaria have two light-sensitive spots at the anterior end of the body and two small elongated projections on either side of the head that are sensitive to chemical stimuli.

Free-living forms have no respiratory or circulatory systems, but do have muscles, a nerve cord that terminates in the head, and organs for digestion, excretion, and reproduction. Flatworms have three distinct tissue layers: an outer **ectoderm,** an inner **endoderm,** and a middle **mesoderm.** Coupled with these advances in tissue organization and specialization, flatworms retain the sac-like body plan with a single opening. Flatworms represent an intermediate level of body organization, with a combination of advanced and more primitive features.

Freshwater **planaria,** often used as experimental organisms or in classroom demonstrations are an example of a free-living flatworm. They are small (a few centimeters in length), with body colors ranging from colorless to mottled brown or black. They feed by means of a retractable tube called the pharynx through which they capture small organisms or feed on dead or decaying tissues.

The other two classes in this phylum are parasites: tapeworms and flukes. Both are thought to have originated from free-living forms that adapted to a parasitic life cycle. During adaptation, certain tissues and organs were lost or reduced. Because parasites absorb nutrients from the bodies of their hosts, a digestive system and mouth are unnecessary. Since there is no need to search for food, cephalization has been reduced, and the head of the animal serves mainly to attach the parasite to the host with hooks and/or suckers.

Tapeworms attach to the wall of the intestine and bud off segments that pass out of the host's body with the feces and infect new hosts. Flukes are parasitic flatworms that have an elongated body with a sucker at one end for attachment to the host. They infect a variety of organs, including the lungs, liver, and vessels of the circulatory systems.

## NEMATODES: A TUBE-WITHIN-A-TUBE BODY PLAN

Nematodes (roundworms) have cylindrical bodies with smooth outer walls, and they are tapered at each end (◆ Figure 14.12). The phylum *Nematoda* consists of perhaps hundreds of thousands of species, most of which have not been studied or classified. Most nematodes are free living, but some species are parasitic, with humans providing habitats for between 40 to 80 species. Pinworms are one of the most common nematode parasites in the United States, and it is estimated that they infect 30% of all children.

There are literally millions of nematodes in a spadeful of soil. Because of their density, it has been said that if everything on Earth, including the planet, were to disappear, leaving only the nematodes, the outlines of the continents and the location of major cities (thanks to the parasitic nematodes) would be visible as concentrated masses of nematodes.

Nematodes have adult bodies derived from three layers of embryonic cells. Nematodes have a cavity located between the ectoderm and mesoderm, forming a body that has a tube-within-a-tube plan. This cavity, called a **pseudocoelom,** is a closed, fluid-containing space that acts as a hydrostatic skeleton to maintain body shape, circulate nutrients, and hold the major body organs. Nematodes lack an organized circulatory system, but most have a highly specialized digestive system.

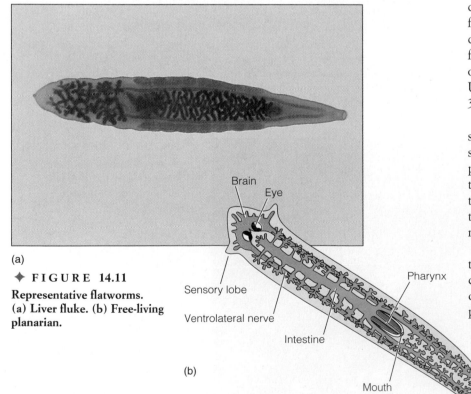

(a)

(b)

Brain
Eye
Sensory lobe
Ventrolateral nerve
Intestine
Mouth
Pharynx

◆ **FIGURE 14.11**
**Representative flatworms. (a) Liver fluke. (b) Free-living planarian.**

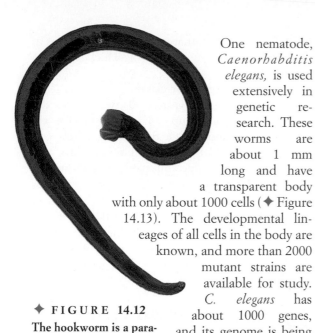

✦ **FIGURE 14.12**

**The hookworm is a parasitic nematode.**

One nematode, *Caenorhabditis elegans,* is used extensively in genetic research. These worms are about 1 mm long and have a transparent body with only about 1000 cells (✦ Figure 14.13). The developmental lineages of all cells in the body are known, and more than 2000 mutant strains are available for study. *C. elegans* has about 1000 genes, and its genome is being investigated as a model eukaryote system. As part of this effort, the *C. elegans* genome is being mapped and sequenced as part of the Human Genome Project.

4. The skeleton served to anchor muscle attachments, allowing refined patterns of locomotion.

The second major advance that took place in the Cambrian Period was the development of the **coelom,** an internal body cavity between the body wall and the digestive system (Figure 14.5). The presence of a coelom allowed the development of more efficient organ systems. In animals with coeloms, two different strategies evolved for the formation of body parts from embryonic tissues. One strategy is used by mollusks, annelids, and arthropods. The other strategy is used by the echinoderms and chordates.

## MOLLUSKS: ADVANCED ORGAN SYSTEMS APPEAR

The mollusks (phylum *Mollusca*) are a large group, with more than 100,000 species (✦ Figure 14.14).

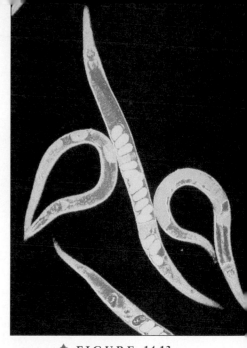

✦ **FIGURE 14.13**

**The nematode *C. elegans* is used to study genetics, behavior, and development in a simple eukaryotic organism.**

## AN EVOLUTIONARY TRANSITION: ANIMALS WITH INTERNAL BODY CAVITIES

Most major groups of animals evolved during the Cambrian Period (550 to 505 million years ago) and were confined to the oceans. During this time, two important evolutionary advances took place. First, instead of hydroskeletons, fossils with new types of skeletons appeared over a period extending from the Cambrian into the Ordovician Period. External skeletons were the first to develop, and conferred several advantages:

1. Secretion of a mineral-containing shell allowed animals to maintain a mineral balance in the body by putting minerals into the skeleton when they were plentiful, and withdrawing them from the skeleton when necessary.

2. External skeletons provided protection from dessication in tidal zones, allowing exploitation of the shallow water environment.

3. Hard body parts offered protection against predators.

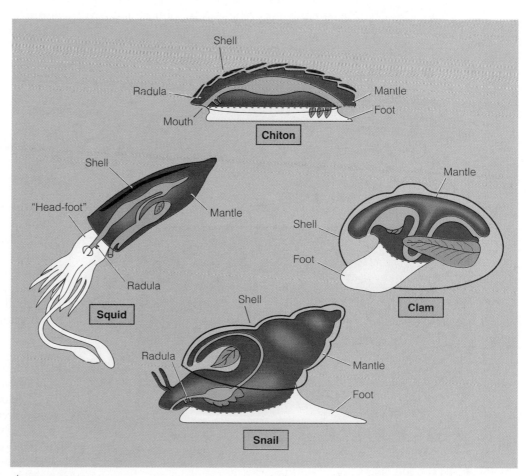

✦ **FIGURE 14.14**

**Mollusks have a variety of shapes, but common elements to their body plan, including a mantle, shell, and foot.**

**✦ FIGURE 14.15**
The giant Pacific octopus.

Fossil mollusks first appeared in the Cambrian Period. The body plans of mollusks are diverse, but they all have three distinct body parts:

1.   A **visceral mass** that contains the organs for digestion, excretion, and reproduction.

2. A **mantle** that surrounds but does not completely cover the visceral mass. This membranous or muscular structure secretes the shell if one is present.

3. A **head/foot** region that contains sensory organs, and a muscular structure used for movement.

Mollusks exhibit well-developed organ systems, which consist of two or more organs that work in coordination to accomplish a single function. For example, in the octopus, the nervous system is highly developed. It includes multiple components: eyes with lenses and photoreceptor cells, nerve fibers that connect the eyes to a brain, and nerve fibers from the brain that connect to the muscles in the mantle and the modified foot (which forms the tentacles).

There are three major classes of mollusks. The largest of these is the gastropods, which includes the snails and slugs (Figure 14.14). The bivalves are a second class, and include the clams, mussels, and oysters (Figure 14.14). These mollusks have two shells (thus their name, bivalves), hinged on one side. The third class is the cephalopods, perhaps the most highly evolved of all mollusks. This class includes the squid, octopus, and cuttlefish (✦ Figure 14.15). In this class, the shell is reduced and internalized as a skeletal element. There is a highly developed circulatory system and an advanced nervous system.

## ANNELIDS: SEGMENTED BODY PLANS

The annelids (phylum *Annelida*) are a phylum of segmented worms, with the segments visible as rings circling the body (✦ Figure 14.16). Bodies composed of repeating units in the form of segments are an important evolutionary advance. The presence of repeating segments allows the development of specialized functions in different segments.

Annelids also have an enlarged coelom, which allows larger and more complex organ systems to develop. There are about 10,000 species of annelids, living in marine, freshwater, and terrestrial environments. There are similarities in the preadult stages of development in both annelids and mollusks, suggesting that annelids may have evolved from marine mollusks. Annelids are divided into three classes, one of which is represented by the earthworm. Although a terrestrial species, the earthworm is poorly adapted to land, and desiccates easily. It lives in moist earth and feeds by ingesting dead or decaying plant and animal material. The digestive system is a specialized, tube-within-a-tube system.

Movement in the earthworm is aided by sets of bristles that extend from each segment. Circular and longitudinal muscles allow the worm to control segments independently, pushing bristles on one side of the body forward, locking them into the earth, and then contracting muscles to pull the body forward, segment by segment.

Annelids are characterized by the presence of a closed circulatory system (✦ Figure 14.17), which is more efficient in the distribution of nutrients and the removal of wastes than an open system (one without blood vessels). The nervous system is composed of a brain, a collection of nerve cells, and a ventral nerve cord, with nerves entering and leaving the cord in each segment.

(a)

(b)

**✦ FIGURE 14.16**
Representative annelids. (a) Leech. (b) Tubeworm.

The closed circulatory system of the earthworm. Closed systems consist of one or more hearts to pump fluids, and a system of blood vessels to move blood to all parts of the body.

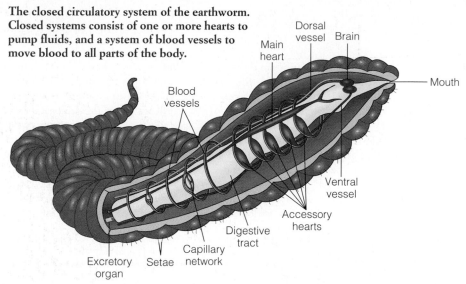

## ARTHROPODS: JOINTED APPENDAGES ON SEGMENTS

The term *arthropod* means "joint-footed" and applies to animals with jointed appendages on the body (✦ Figure 14.18). In many respects, arthropods can be regarded as the most successful animal life-form to have evolved on the planet. They occupy every possible habitat: the ocean, freshwater, the land, and the air. They live in every climatic zone; the animal life of Antarctica is composed almost entirely of arthropods. There are well over 1,000,000 species of arthropods, but because so many species remain to be discovered, this is regarded as a conservative estimate. The biologist E. O. Wilson (who studies ants) estimates that there are about 10,000,000 species of organisms on earth and, of these, 9,000,000 are arthropods.

Not only are arthropods successful in their diversity, (numbers of species), they are also successful in terms of population size. At any given time, there are probably about 1 trillion arthropods alive on, over, and under the earth's surface. The success of this group in general and one subgroup, the insects, in particular is due to several factors, including:

1. A hard exoskeleton
2. The presence of jointed appendages
3. A complex nervous system
4. A unique respiratory system
5. A life cycle that is complex and adaptable to varied environments.

**Sidebar**

**KING OF THE HILL**

The red fire ant, *Solenopsis invicta,* has become a major pest in most of the southern United States. Since its accidental introduction into the United States in the 1930s nothing has stopped the ant's migration; in fact, insects used against it only killed off native ant species, making it easier for this newcomer to succeed. The invading insect wipes out native ant species and limits food sources for some birds. Regions where this ant has been introduced suffer losses in biodiversity. Because ants play an important role in the food chain, biologists are waiting for the serious consequences of this invasion to become evident.

The success of the fire ants has been largely due to the fact that all of its natural enemies were left behind in Brazil. One of those enemies may serve as a possible solution to the problem. In Brazil, a species of fly co-evolved with *S. invicta* and became its parasite. The female lays eggs on the thorax of the ants. As the larvae hatch, they eat through the ant's body to its head, where they continue to devour all tissue. Meanwhile, the infected ants go about their normal routine until they stop moving and their head falls off. The larvae use the empty head case of the ant as a cocoon for three weeks before an adult fly emerges. Entomologists studying this relationship are hopeful that this parasite may be the means to extinguish the fire ants.

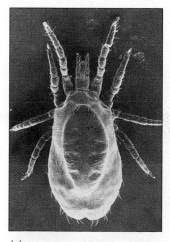

(a)

(b)

✦ FIGURE 14.18

Arthropods occupy almost all habitats on earth, including the polar regions. (a) Mite. (b) Scorpion.

# CREEPING CRUSTACEANS: THE GIANT KING CRAB

The attention-grabbing dinosaurs and their large size usually leave people with the impression that most of the giant forms of animal life existed in the distant past. Although, in the case of whales, this is demonstrably not true, still, the feeling lingers that among lower forms, the age of giants is well past. Discoveries such as those of fossil crocodiles more than thirty feet long help reinforce this idea.

Recent studies on the evolution of the king crab have identified its evolutionary origins.

Using analysis of the similarities in the molecular organization of ribosomal RNA, the time of origin and ancestral group of the king crab has been established. Previously, the lack of fossil king crabs prevented a determination of the time of origin of this group. The results of these studies indicate that king crabs evolved from hermit crabs, an ancient group that appeared in the fossil record some 150 million years ago. Hermit crabs are not covered by hard exoskeletons, but instead have flexible abdomens that allow them to curl up inside the empty shells from snails and other gastropods. This lifestyle

has prevented hermit crabs from evolving to be larger than the largest available shells. Sometime between 13 and 25 million years ago (only a short time on the geological and evolutionary scale), a group of hermit crabs abandoned shell-living and, over many generations, increased in size and became covered with a hard exoskeleton. Present-day king crabs reach leg spans of four to five feet, representing the largest form of the taxon that has ever lived and placing the king crab among the largest known arthropods.

---

The exoskeleton of the arthropod body serves as protection from predators, prevents desiccation, and allows for the attachment of muscles used in locomotion. The segmented body of many arthropods has fused to form three regions: the **head, thorax,** and **abdomen.** The jointed appendages attached to the body segments are highly specialized for a range of functions, including sensory reception, walking, swimming, flying, feeding, and reproduction.

## Two Arthropod Systems Signal Evolutionary Advances

The arthropod nervous system is equipped with highly evolved sensory organs for sight and smell, a brain composed of fused nerve cells (ganglia), a ventral nerve cord, and a system of hormones that regulates behavior and reproduction. This nervous system allows complex behavioral responses to environmental stimuli and also allows social behavior.

Some arthropods retain a constant body form throughout life, growing and shedding their exoskeleton several times before adulthood. Others undergo a more complex life cycle that includes a series of changes in body form. These include a feeding larval stage, an intermediate pupal stage, and a reproductive adult stage (✦ Figure 14.19). The process of changing from one form to another is known as **metamorphosis.** In insects, some larvae or pupae cease development in unfavorable conditions, and resume when conditions become more favorable. Metamorphosis is an important adaptation because it reduces competition between individuals at different stages of the life cycle.

✦ **FIGURE 14.19**

**Insects are arthropods that have a complex life cycle that can include several stages: (a) eggs, (b) larva, (c) pupa, and (d) adult.**

(a)         (b)         (c)         (d)

**✦ FIGURE 14.20**

Crustaceans inhabit marine, freshwater, and terrestrial environments. (a) Pillbug. (b) Bristly cave crayfish.

**Crustaceans** include shrimps, lobsters, and crabs (see Beyond the Basics: Creeping Crustaceans: The Giant King Crab). Although most crustaceans live in aquatic environments that include both saltwater and freshwater, some are terrestrial, like the pill bug (✦ Figure 14.20).

The most diverse group of arthropods is, without question, the insects. There are probably well over 1,000,000 species of insects. Appendages on the head are specialized for sensory reception (antennae), or biting, chewing, or puncturing (mouth parts). Thoracic appendages include three pairs of legs, and may include one or two pairs of wings (✦ Figure 14.21).

The evolution of insects accompanied the development of flowering plants. As angiosperms evolved, new habitats opened for insects, including host–parasite relationships, predator–prey relationships, and mutualistic relationships involving pollination. The success of insects in using plants for food and reproduction also makes them major pests of agricultural crops. In many areas, insects devour around 30% of a crop before harvesting, and can destroy a significant fraction of food while it is in storage or transport. Other insects feed on decaying animals (see Beyond the Basics: Bugs on Bodies).

## ECHINODERMS: THE DEUTEROSTOME BODY PLAN

Echinoderms are the first phylum of animals we have encountered that are **deuterostomes.** In deuterostomes, the opening that first appears in the embryo becomes the anus, and the mouth appears at the other end of the developing digestive tract. In the **protostomes** (mollusks, annelids, and arthropods) the first opening in the embryo develops into the mouth. These differences in embryonic development represent two major trends in animal evolution.

**✦ FIGURE 14.21**

Insects are among the most diverse group of animals. Their bodies have a head, thorax, and abdomen, with appendages for specialized functions. (a) Ladybird beetle. (b) Cecropia moth. (c) Leaf-cutting ant.

## BUGS ON BODIES

Most species of insects have one of two characteristic life cycles: Eggs hatch into miniature versions of the adult, grow, and shed their exoskeleton several times before reaching adult size. Cockroaches and other lower insects employ this life cycle. The second life cycle is employed by insects such as flies. In this case, the egg hatches into a white, worm-like organism, the larva or maggot, that begins feeding immediately. After a day or so, the larva molts, shedding its skin, and continues feeding. This process is repeated a second time after a few days, producing a mature larva. After a 2- or 3-day period, the mature larva crawl to a drier location and begin the transformation into an adult. First, the larva contracts, and the outer skin hardens and darkens, with color changes occurring at predictable intervals, resulting in the formation of the pupa. Inside the pupa, most larval tissues are broken down, and the larva is remodeled into the adult. After 3 to 5 days, adults emerge from the pupal case, pump up their wings, and fly away. The time for this life cycle to be completed is dependent on the species of fly and on the temperature. For example, in the fruit fly *Drosophila,* the life cycle from egg to adult is 10 to 12 days at 25°C; at room temperature (about 21°C), the process takes about two weeks.

The biological clock that controls the fly life cycle is so exquisitely sensitive to temperature that it has been used in criminal cases to determine the time of death. Entomologists are scientists who study insects, and forensic entomologists are those who apply entomology to criminal cases. Suppose that a homicide occurs. Shortly after death, flies are attracted to the body by the first signs of decomposition. Eggs are deposited, and the larva hatch. The time required to complete the life cycle depends on the environmental temperature. If the body is discovered during this time, a forensic entomologist will often be called in to assist in determining the time of death. First the larvae, pupae, and adults on the body are examined to determine their species. Next, the entomologist catalogs the species in the surrounding area, and consults weather records to establish the temperature ranges over the preceding time periods, measured in hours or days. From these three types of information, the approximate time of death can be established with some degree of accuracy. It isn't pretty, but it works, and has been used in dozens of criminal cases from coast to coast and, in some cases, has served as a defense in establishing an alibi for those accused of murder.

Echinoderms are marine animals. There are about 6000 species, belonging to five different classes. This phylum includes the sea urchins, starfish, sea stars, brittle stars, and sea cucumbers (✦ Figure 14.22). The body plan of most adult echinoderms involves radial symmetry. Because preadult larvae are bilaterally symmetrical, the radial body plan of adults appears to be a secondary adaptation. This adaptation may be related to the fact that echinoderms are all marine species, mostly bottom dwellers, and the radial body plan offers advantages in feeding from any direction. A result of radial symmetry is that cephalization is reduced. As a consequence, the nervous system in adult echinoderms is also reduced, and no brain is present.

✦ **FIGURE 14.22**

**Echinoderms, including the starfish, have radial symmetry, and a water vascular system connected to their tube feet for locomotion.**

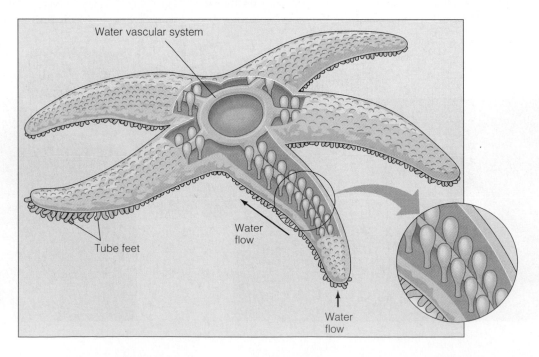

Echinoderms have a vascular system that pumps water into tube feet, allowing the animal to move, to attach to substrates, and to grip and open prey such as clams and oysters.

## CHORDATES: ORIGIN OF THE VERTEBRATES

Chordates are animals that at some point in their life cycle have three characteristics (✦ Figure 14.23):

1. A **notochord,** or dorsal support in the form of a cellular rod that runs the length of the body
2. A dorsally placed, hollow **nerve cord**
3. **Gill slits** or pouches that may be present only in early developmental stages (as in humans) or retained in the adult (as in fish).

The origins of chordates is largely unknown because early species did not have hard skeletons or body parts; as a result, there are few fossils. Evidence from comparative embryology and biochemistry indicates that there is a close relationship between the echinoderms and the chordates, and the two groups may have shared a common ancestor. Presently, there are some 40,000 species of chordates.

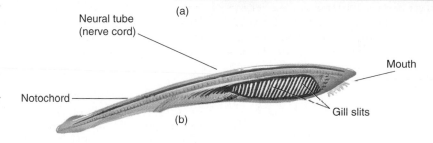

✦ **FIGURE 14.23**
**Invertebrate chordates. (a) Sea squirt. (b) Drawing of a lancelet, showing major features of the body plan of chordates.**

## VERTEBRATES ARE ANIMALS WITH BACKBONES

The largest and most recognizable subphylum of the chordates is the vertebrates, characterized by a backbone and an internal skeleton. The backbone is composed of individual vertebrae. Bony processes extending from the vertebrae enclose and protect the dorsal nerve cord or spinal column. The seven classes of vertebrates include fish (three classes), amphibians, reptiles, birds, and mammals.

### Fish: Vertebrates with Jaws

The most primitive vertebrates are the fish. They first appeared in the fossil record during the Cambrian Period, but whether they evolved in a marine or a freshwater environment is still unresolved.

The most ancient class of fish are the jawless fish (Agnatha). During the Silurian and Devonian Periods these fish were a large and important group, but most became extinct at the end of the Devonian, some 360 million years ago. Today, the jawless fish are represented only by hagfishes and lampreys. They have long, cylindrical bodies and skeletons made of cartilage. They lack jaws and paired fins. Most species of lamprey feed by attaching their mouths to the body of other fish and sucking out

body tissues (✦ Figure 14.24). Hagfish are scavengers, feeding on dead and decaying organisms.

One of the major advances in vertebrate evolution took place in fish: the development of jaws. The evolution of jaws is an example of existing structures being modified for new functions. Jaws evolved from the first three gill arches, structures that support the thin, soft-tissue gills of jawless fish.

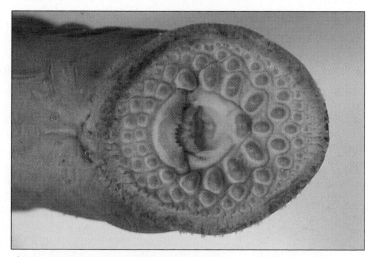

✦ **FIGURE 14.24**
**The jawless fish are represented by the Atlantic hagfish and the sea lamprey. These animals feed by attaching their mouths to their food.**

## Saving Endangered Species

OLIVER A. RYDER

I can remember clearly the day that my high school biology teacher told us that DNA was somehow responsible for specifying all genetic information. This idea was presented as such an astounding and novel finding by my teacher that I resolved to learn more about cellular biology and its governing molecule. Pursuing this interest in college, I became involved in the genetics and biochemistry of DNA replication in the bacterium *Escherichia coli.*

While I was in graduate school I was living in an undeveloped area on one of the coastal mesas in the San Diego area. I identified and began learning about some of the birds, discovering that my backyard was habitat for threatened and endangered species. While I was experimenting with *E. coli,* I was also learning about blue whales, giant pandas, pocket mice, and rufous-sided towhees.

In graduate school, my mentor was Professor Kurt Benirschke, a pathologist and medical geneticist who had delved enthusiastically into comparative studies and thereby became established as an expert on reproduction of endangered species. I had thought of a career in veterinary medicine, but was not accepted into the program at UC Davis. Instead, I started my postdoctoral work at the newly founded Center for Reproduction of Endangered Species at the San Diego Zoo under his direction.

I was given a chance to see what a molecular biologist could do in support of conservation of endangered species. I learned a lot in those days: a new literature, new disciplines and professional organizations, and a new environment, literally a zoo. I was surprised at how relatively easily basic research findings became relevant to genetic characterization and genetic management of endangered species.

I never really expected that my work would literally take me around the world, but now I know many airports and how to say "thank you" in many languages. Recognizing that in the developing countries facing imminent loss of species such as rhinos, snow leopards, and gazelles, that it would be the local scientists who could provide the most effective counsel to wildlife protection agencies, I began to carry out programs of training and technology transfer. It was because I had the chance to travel to east Africa and sleep under the starry African sky that I received the inspiration and resolve to attempt such a new program. Although it might decrease my scientific productivity in traditional terms, I was convinced it was a useful and necessary task.

Lately, I have been directing efforts toward China, and its quintessential charismatic megavertebrate, the giant panda. Giant pandas have been endangered because of poaching, but also because of habitat fragmentation. Necessary for their survival in the wild is an interlinked system of reserves that will ensure viable populations. I find that I have returned to interests I discovered as a graduate student in San Diego living in the midst of a habitat that was becoming fragmented because of the burgeoning human presence; these are the same problems that face the giant panda in China. Neither can be ignored. San Diego County has more threatened and endangered species of plants and animals than any county in the continental United States. My latest challenges include linking concern for wildlife around the world—the kind of regard that zoos can foster—with concern for habitat preservation to ensure species persistence, once again in my own backyard.

*Oliver A. Ryder received his doctorate from the University of California, San Diego, in 1975. He began his scientific career with the Zoological Society as a research fellow at the Center for Reproduction of Endangered Species. In 1979 he accepted a permanent staff position as geneticist. He is an associate editor of the* Journal of Heredity, *serves on the editorial board of* Conservation Biology, *and is a member of the Council of the American Genetics Association.*

The arches were modified into the supports for jaws. The development of jaws allowed the exploitation of new habitats, enabling fish to become active predators ranging through all depths, feeding on other organisms. There are two types of jawed fish, the cartilaginous and the bony fish.

Cartilaginous fish include the sharks, skates, and rays (✦ Figure 14.25). These fish have jaws, cartilaginous (rather than bony) skeletons, paired fins, and a number of gill slits on either side of the head. This class arose during the Devonian Period and underwent a major expansion in the Carboniferous and Permian Periods. Many species became extinct at the end of the Permian, but a large group of species survives today and they are important members of the marine fauna.

The bony fishes (the third class of fish), of which there are at least 15,000 species, are found in both saltwater and freshwater. The bony fish are divided into two groups: the **ray-finned** fish and the **lobe-finned** fish (✦ Figure 14.26). The ray-finned fish include most of the familiar species such as trout, salmon, tuna, bass, and perch. These first appeared in the Devonian and, despite some minor extinctions, have continued to diversify since then. They are presently the dominant vertebrates in marine and freshwater environments.

Ray-finned fish have an internal swim bladder, a gas-filled sac that can be filled or emptied to change the overall buoyancy of the body, allowing the fish to change depths. The term *ray-finned* refers to the thin, bony supports that hold the fins away from the body.

The lobe-finned fish are distinguished by the presence of muscular fins (Figure 14.26b). These fins do not have thin radiating bones like the ray-

(a)

(b)

✦ **FIGURE 14.25**

The cartilaginous fish include (a) the manta ray and (b) spoonbill paddlefish.

finned fish, but contain large, jointed bones that attach to the body. This combination of bones and muscles in the lobe fin is another structure that underwent modification for a new function, this time as legs in the amphibians. The lobe-finned fish are divided into two groups: the **lungfish** and the **crossopterygians.** Only a small number of lungfish species exist at present, and these are found in freshwater in Africa, Australia, and South America.

The crossopterygians are important because they are regarded as ancestral to the amphibians. One group of crossopterygians is represented by the coelacanth, a fish once thought to be extinct (✦ Figure 14.27). The other group of crossopterygians were freshwater predators in the Devonian Period. These fish had a streamlined body with a powerful tail, strong, lobed fins that could be used to walk short distances on land, and lungs that allowed air breathing.

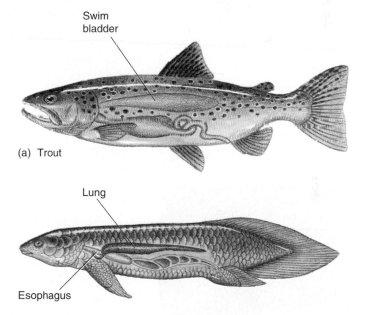

(a) Trout

(b) Australian lung fish

✦ **FIGURE 14.26**

(a) Ray-finned bony fish have an internal swim bladder that changes the buoyancy of the body. (b) The lungfish are able to breathe air and move short distances across land using their lobe fins for support.

✦ **FIGURE 14.27**

The coelacanth represents a group of lobe-finned fish that are thought to be ancestral to the amphibians.

(a)

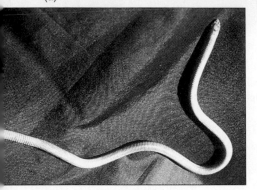

(b)

✦ **FIGURE 14.28**
**Amphibians have a moist skin that is used for gas exchange. (a) Salamander. (b) Caecilian.**

## Amphibians: Vertebrates Move Ashore

This class includes some 4000 species of animals that are partially terrestrial and partly aquatic (*amphi-* is a term that means both; when coupled with *bios,* it means both lives, or two lives). As outlined at the beginning of this chapter, the transition to land required solving several problems. These included desiccation, dependence on water for reproduction, a skeleton to maintain body shape out of the aquatic environment, and a mechanism for obtaining oxygen from the air rather than the water.

Amphibian species evolved mechanisms to solve some of these problems, but not all. Early amphibians closely resembled the crossopterygian fish, and had a skeleton and lungs. In addition to retaining these features, early amphib-ians developed legs. Later, they evolved into the animals we are familiar with, including newts, salamanders, mud puppies, frogs, and toads (✦ Figure 14.28).

Most amphibians have a bony internal skeleton, and moist skin that is used to supplement the gas exchange that occurs in the small, inefficient lungs. Almost all species of amphibians return to water to deposit gametes for fertilization, and the preadult stages are spent in water (✦ Figure 14.29). For these reasons, amphibians are regarded as a transitional group in the evolution of terrestrial vertebrates.

## Reptiles Reproduce without Water

The reptiles are a class of about 6000 species that includes snakes, lizards, turtles, alligators, and crocodiles. Reptiles do not need water for reproduction. Two adaptations account for this independence. First, although some reptiles bear live young, many reproduce by means of a fluid-filled egg, which has a shell and internal membranes to protect the embryo as it develops (✦ Figure 14.30). Second, because reptiles have internal fertilization, the gametes are independent of water as a medium for fertilization. The adaptations that allowed rep-

✦ **FIGURE 14.29**
**Amphibians are dependent on water for fertilization, and the preadult stages (tadpoles) develop in water.**

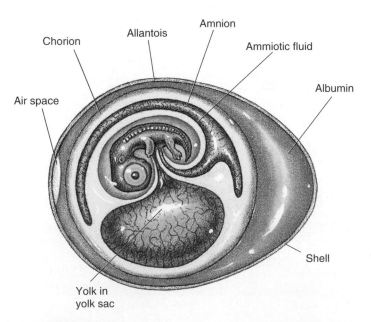

✦ **FIGURE 14.30**
**The reptilian egg provides a protected, aqueous environment for development. This adaptation was partly responsible for the success of reptiles in the terrestrial environment.**

tiles to become independent of water for reproduction parallels the development of seeds, which allows vascular plants to reproduce without water as a medium.

Reptiles evolved from amphibians sometime in the late Carboniferous Period, some 330 million years ago (✦ Figure 14.31). The reptiles developed rapidly into a large and diverse group of animals, partially displacing amphibians from many habitats. The success of the reptiles was due in part to their advantage in reproduction, but also to the development of more efficient jaws and teeth for seizing and biting prey, and a dry, rough skin that re-

sists water loss. In addition, changes in the limb bones allow reptiles to pursue prey over long distances. Some reptiles can move short distances on the two hind legs, freeing the front legs for grasping. A major extinction at the end of the Cretaceous ended the domination of the reptiles.

Only three major groups of reptiles remain today: the turtles, the snakes and lizards, and the crocodiles and their relatives. The turtles, tortoises, and terrapins are an ancient group of reptiles, characterized by a shell that protects the body. They are found in freshwater, marine, and terrestrial environments. Crocodiles are living remnants from the

**Sidebar**

**REFINING IDEAS ABOUT TURTLE EVOLUTION**

Turtles are one of the three major groups of reptiles that survived the mass extinction at the end of the Cretaceous. The fossil record of turtle groups is fragmentary, but it indicates that turtles, along with frogs, crocodiles, dinosaurs, and mammals, made their appearance in the Early Triassic. A recent discovery in the northwest region of Argentina extends the record of South American turtle fossils by 60 million years, to 210 million years ago. These turtles, recovered from what was probably a muddy side channel of a riverbed, may have been partly terrestrial. This finding, along with fossils from other sites, confirms that by the end of the Triassic (about 208 million years ago), turtles were a diverse group of animals with a widespread geographic distribution.

These findings strengthen the idea that turtles, along with the dinosaurs, probably first appeared early in the Triassic (which began 245 million years ago), between the formation and the breakup of Pangea.

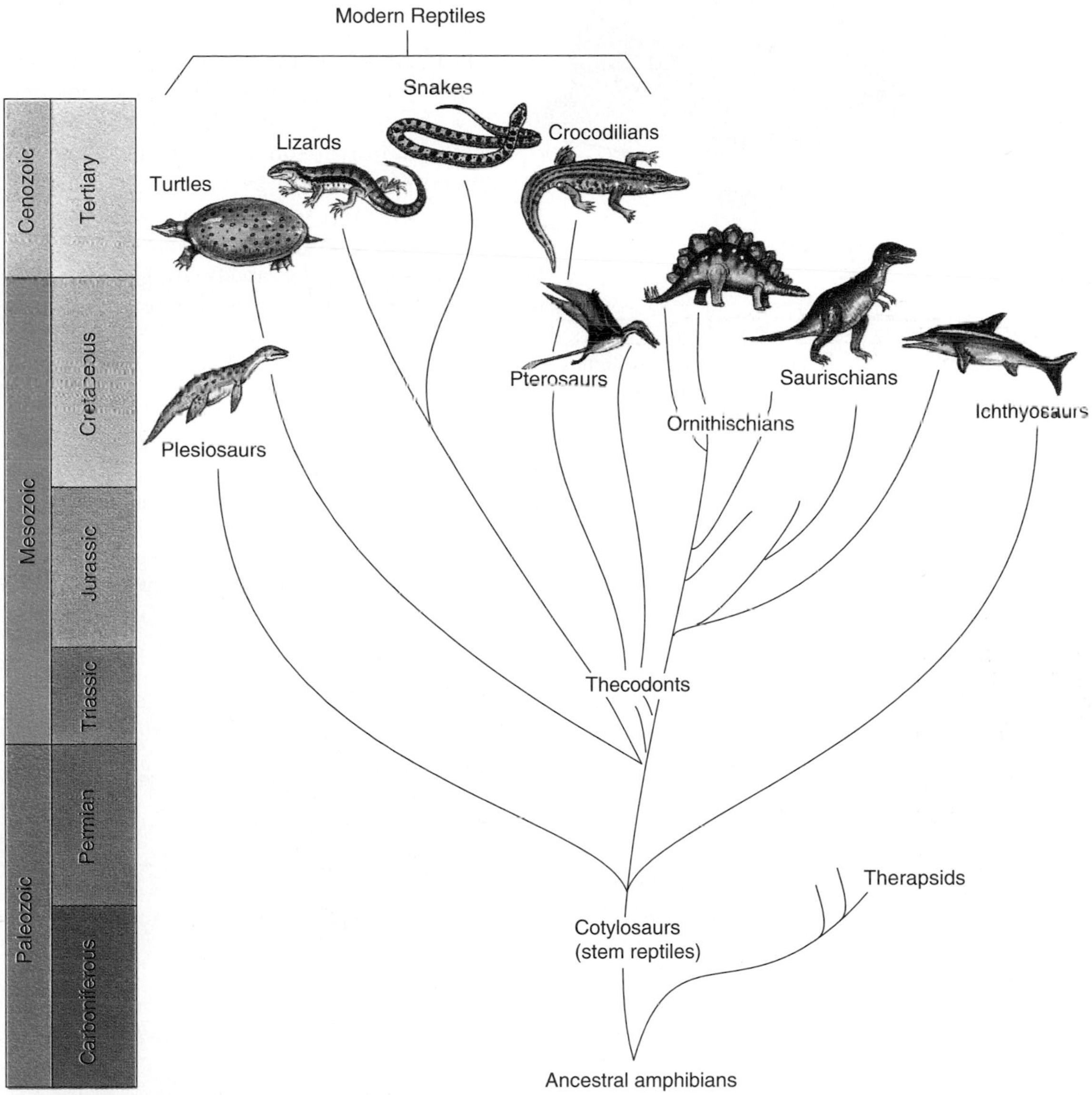

✦ **FIGURE 14.31**

A family tree of the reptiles, showing their probable evolutionary relationships. Modern reptiles are represented by three major groups: turtles, lizards and snakes, and crocodiles.

(a)

(b)

(c)

✦ **FIGURE 14.32**

**Representative reptiles. (a) Collared lizard. (b) Boa constrictor. (c) Alligator.**

age of the dinosaurs, and are well adapted as carnivores (✦ Figure 14.32). Recent findings indicate that crocodiles prospered as dinosaurs became extinct, indicating that they were able to adapt to changing conditions at the end of the Cretaceous.

## Birds Control Internal Body Temperature

Birds evolved from reptilian ancestors sometime in the Jurassic Period (150 to 200 million years ago). This class contains about 9000 species. Reconstructions from the fossil record indicate that birds originated from small reptiles that became specialized for flight. The earliest bird fossils, such as *Archeopteryx*, closely resemble reptiles (✦ Figure 14.33). In fact, some specimens were originally classified as dinosaurs until impressions of feathers were noticed.

Early bird-like reptiles such as *Archeopteryx* or an earlier fossil called *Protoavis* have a combination of reptilian and avian characteristics. This retention of ancestral characteristics in combination with new ones illustrates the concept of **mosaic evolution.** *Archeopteryx,* for example, has some dinosaur-like characteristics including teeth, tail, structure of the hind limbs, and brain size. It also has characteristics of birds, including feathers and a wishbone. Later, the long reptilian tail became reduced, and other reptilian features such as teeth became lost. The distinguishing feature of birds is the presence of feathers, which provide a means of flight, as well as insulation.

Feathers are an important evolutionary advance because they allow the exploitation of a new habitat, the air. They permit a new method of locomotion (flight), without restricting other methods such as running or walking. Feathers also provide insulation, and contribute to the trend toward **endothermy,** or internal control of body temperature. The ability to generate and maintain internal body heat distinguishes birds

✦ **FIGURE 14.33**

**The *Archeopteryx* retains many features of reptiles, such as teeth, and has some features of birds, such as feathers.**

from most reptiles, and has allowed birds to occupy colder regions of the world (✦ Figure 14.34).

Modern birds appeared in the Early Tertiary Period, some 65 million years ago, and the basic body plan has not changed much since then. The adaptations necessary for flight, such as hollow bones to reduce weight and internal air sacs to reduce density, have kept the body plan of birds conservative and, as a result, birds show less variability in body plan than other vertebrates.

Birds show complex social behavior, including elaborate mating rituals and extended care for the young. Associated with this behavior as well as the interaction of sensory receptors and muscles required for flight, the brains of birds are highly developed, large, and complex.

## Mammals Nourish Their Young with Milk

The evolutionary history of mammals is well known, thanks to a rich fossil record. Mammals arose from reptiles in the early Mesozoic (✦ Figure 14.35). After the Mesozoic extinctions that eliminated the dinosaurs, mammals underwent a large radiation beginning some 65 million years ago, and occupied

✦ **FIGURE 14.34**

**Feathers played a role in allowing birds to occupy new habitats, including Antarctica.**

✦ **FIGURE 14.35**

**Mammals have diverse body plans. (a) Sea otter. (b) Humpback whale. (c) Giraffe.**

(a)

(b)

(c)

277

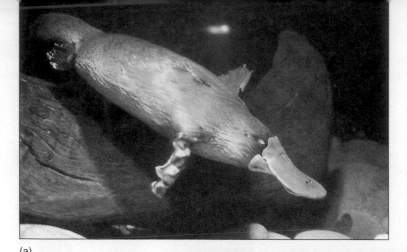

(a)

✦ **FIGURE 14.36**

**(a) Monotremes, such as the duck-billed platypus, are egg-laying mammals.
(b) Marsupials nurture their young in pouches. The koala is a marsupial.**

(b)

*Sidebar*

**A NEW SPECIES OF WHALE**

Given the size of whales and the fact that they have been intensively hunted for several hundred years, it would seem likely that all species of whales have been identified. Yet scientists have recently described a new species, found in the Pacific Ocean off Peru. The first clues about this species came from the discovery of a partial skull and some vertebrae on the coast of Peru in 1976. Between 1985 and 1989, the partial remains of nine additional whales were recovered from the central and southern coast of Peru. After years of studying the evidence, scientists recently announced the discovery of a previously unknown whale.

The new species, *Mesoplodon peruvianus,* belongs to a group of beaked whales that have a dolphin-like snout. The animals are gray on top, with a gradual shading to a light gray on the abdomen. Adults are about 12 feet long and have elongated jaws and few teeth. This species is distinguished from other members of the genus by its small size and tiny brain case.

many of the habitats vacated by dinosaurs. In this process, mammals became larger and more complex. During their evolution, mammals developed several adaptations that contributed to their success. These include rapid forms of locomotion, larger brains, more efficient control over internal body temperatures, and specialized teeth: canines for biting and tearing, incisors for cutting, and molars for chewing.

There are about 4500 species of living mammals. The characteristics that distinguish mammals from other vertebrates include the presence of hair, and mammary glands for nourishing young. The links to reptiles have not been completely broken, however; some mammals such as the platypus retain the reptilian trait of egg laying, and reptilian-like scales are found on the tails of some rodents. Mammals are classified by their reproductive methods. Among present-day mammals, only the **monotremes** lay eggs (✦ Figure 14.36). Representative monotremes include the echidna or spiny anteater and the duck-billed platypus of Australia. This transitional group retains the reptilian characteristic of laying eggs

with soft shells, but shares with higher mammals the characteristic of nourishing the young with milk. The second group of mammals is the **marsupials,** represented by the kangaroo, opossum, and the koala (Figure 14.36). Marsupials develop internally, but are born while still in an embryonic stage, and crawl to a pouch on the mother's abdomen where they complete development. They are nourished by milk secreted by the mother.

The largest and most diverse group is the placental mammals. The placenta is a structure formed by the interaction of embryonic tissue with the wall of the uterus and serves as the point of exchange for nutrients and waste during embryonic and fetal development. After birth, the young are nourished by milk from the mammary glands. The placental mammals include 12 major orders (✦ Figure 14.37). Although somewhat less specialized than other mammals, the primates (✦ Figure 14.38) have several adaptations including opposable thumbs, enlarged brains, and stereoscopic vision that have contributed to their success. The evolution of primates will be described in the next chapter.

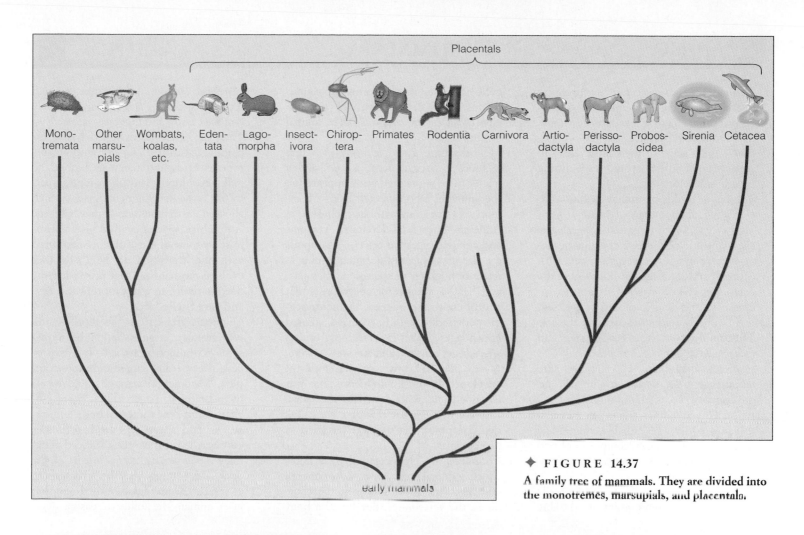

Placentals

Mono-
tremata | Other marsu-pials | Wombats, koalas, etc. | Eden-tata | Lago-morpha | Insect-ivora | Chirop-tera | Primates | Rodentia | Carnivora | Artio-dactyla | Perisso-dactyla | Probos-cidea | Sirenia | Cetacea

early mammals

✦ **FIGURE 14.37**

A family tree of mammals. They are divided into the monotremes, marsupials, and placentals.

(a)

(b)

✦ **FIGURE 14.38**

Primates have several adaptations including stereoscopic vision. (a) Lemur. (b) Bush-baby.

# SUMMARY

1. Animals are heterotrophic multicellular organisms that first appeared in the fossil record some 700 to 900 million years ago. Most animals reproduce sexually, are motile at some stage of their life cycle, and pass through an embryonic stage during preadult development.

2. The major trends in animal evolution include the development of bilateral symmetry, cephalization, paired appendages, the tissue and organ level of cellular organization, a one-way digestive system with a mouth and anus, the development of the coelom to allow internal organs to increase in size and complexity, the segmented body, and an increase in the size and complexity of the nervous system and its associated organ, the brain.

3. Porifera, including the sponges, are among the most primitive animals, and are composed of a limited number of cell types. The cnidarians, including the jellyfish and corals, are radially symmetrical organisms with a body composed of two cell layers and an internal sac-like digestive cavity.

4. Animals in all other phyla are bilaterally symmetrical at some point in the life cycle, including the flatworms and roundworms. The development of the coelom was an important step in animal evolution.

5. Mollusks have a simple body plan that includes a visceral mass, a foot, and a mantle. The segmented body appeared in the annelids, who diverged from a common ancestor along with the mollusks.

6. Annelids have well-developed coeloms and tubular digestive tracts. Arthropods are the most successful animal group in terms of numbers of species and habitats, and have a number of adaptations that contribute to that success, including segmented bodies with specialized, jointed appendages, hard exoskeletons, and a specialized sensory and nervous system.

7. A new plan for embryonic development developed in the Echinoderms that was extended and refined by the chordates. Animals that employ this plan are called deuterostomes; the rest are classified as protostomes.

8. Chordates are characterized by a notochord, a dorsal nerve cord, and a pharynx with gill slits. A subgroup of the chordates, the vertebrates, developed a bony vertebral column, jaws, limbs, refined sensory systems, and an increasingly complex brain and nervous system.

9. There are seven classes of vertebrates, including fish (three classes), amphibians, reptiles, birds, and mammals. One of the major evolutionary advances represented by fish is the development of jaws, which allowed the exploitation of new habitats. Amphibian species evolved mechanisms to solve some of the problems associated with the transition to land. Reptiles evolved mechanisms that precluded the need to return to water for reproduction and gave rise to birds.

10. Mammals arose from the reptiles, and underwent a large radiation beginning some 65 million years ago. In the process, they became larger and more complex. Mammals developed rapid forms of locomotion, larger brains, more efficient control over internal body temperatures, and specialized teeth. Primates are regarded as the most advanced mammals based on the development of the opposable thumb and the enlargement of the cerebrum region of the brain that gives humans the ability to reason.

# KEY TERMS

abdomen
aceolomate
amoebocytes
asymmetrical
bilaterally symmetrical
cephalization
coelom
coelomate
collar
crossopterygians
crustaceans
deuterostomes
ectoderm
endoderm
endothermy
epidermal
filter feeders

gill slits
head
head/foot
interstitial
larval
lobe-finned
lungfish
mantle
marsupials
medusa
mesoderm
mesoglea
metamorphosis
monotremes
mosaic evolution
nerve cord
notochord

planaria
planarians
planula
polyp
protostomes
pseudocoelom
pseudocoelomate
radially symmetrical
ray-finned
sac-like
segments
sessile
spicules
thorax
tube within a tube
visceral mass

## FILL IN THE BLANK

1. Animals are _____, which means they cannot manufacture their own food while plants are _____ and can synthesize their needed nutrients.
2. Animals that are _____ have no general body plan. Organisms organized like a wheel exhibit _____ symmetry. Humans are an example of a body plan that is _____ symmetrical.
3. The thin outer layer of the sponge body is made up of _____ cells and the inner layer is composed of _____ cells.
4. An attached sessile body plan called a _____ and a motile, bell-shaped form called a _____ are the two variations of body plans seen in cnidarians.
5. The _____ are a phylum of segmented worms with visibly segmented rings circling the body.
6. In _____ the opening that first appears in the embryo becomes the anus, and the mouth appears at the other end of the developing digestive tract.
7. Sharks have a skeleton of _____ but they evolved from bony ancestors.
8. The hallmark of reptiles is the _____ egg.
9. The feathers of birds provide an important evolutionary advance and contribute to _____ or internal control of body temperature.

10. The _____ is an organ in the uterus composed of embryonic and maternal tissue that allows nutrients to reach the embryo and wastes to be carried away.

## SHORT ANSWER

1. Describe the tube-within-a-tube body plan.
2. What are the differences between acoelomate, pseudocoelomate, and coelomate organisms?
3. Name the five factors that contributed to the success of the arthropods.
4. What are the characteristics that distinguish chordates from other phylums?
5. Give two examples of how a body structure underwent modification for a new function.
6. What factors contributed to the success of the reptiles?
7. What distinguishes mammals from other vertebrates?
8. Since primates are less specialized than other mammals, why are they regarded as the most advanced mammals?

## MATCHING

Match the groups accordingly.

1. sponges
2. arthropods
3. echinoderms
4. Agnatha
5. crossopterygians
6. crocodiles
7. monotremes

_____ jawless fish
_____ ancestral to amphibians
_____ mammals that lay eggs
_____ related to birds
_____ filter feeders
_____ joint-footed
_____ deuterostomes

## SCIENCE AND SOCIETY

1. Parasites that require two or more hosts to complete their life cycle often produce very large numbers of offspring. Why do you think this is so? What advantages and disadvantages would this present to them? Do you know of any parasites that cause human diseases?
2. Spider webs can be found everywhere—in woods, gardens, and even in the corners of your home. Spiders make these webs by producing different kinds of silk from a set of paired glands in their abdomen. One gland produces the strong silk that holds the web to surfaces; another produces the silk that wraps the prey that become trapped in the web. Different species of spiders produce different web designs. What causes the diversity among spider web designs? Why do spiders produce different web styles? Most spiders have very poor vision. How do you think spiders know when prey have become trapped in their web? Do spiders provide any benefits to humans?

# 15 HUMAN EVOLUTION

OPENING IMAGE

*The development of art was an*

*important stage in human evolution.*

One day, about three and a half million years ago, two human-like creatures, one larger, one smaller, walked across a muddy field of volcanic ash in a region of what is now Tanzania. Then, as now, the area contained small lakes, patches of vegetation, and a volcano. The pair passed through the field during or just after a rain, and shortly after an eruption covered the ground with a layer of volcanic ash. Along the way, the smaller one stopped, turned away to look at something, then resumed walking. Soon after these creatures passed, their footprints were covered by new layers of volcanic ash. This unremarkable event was later to provide important clues about several aspects of human evolution.

After about three and a half million years, erosion by wind and rain partly exposed the fragile footprints, and they were discovered in 1978 by a team of fossil hunters led by Mary Leakey. Altogether, more than 50 prints covering a distance of about a hundred feet were recovered and preserved as casts.

These prints demonstrate that ancestral primates, known as hominids, had feet with well-developed arches, heels, and toes. More importantly, these prints reveal that one of the key events in human evolution, upright posture and walking, evolved more than three million years ago. These findings along with other evidence from fossil remains suggest that these small, primitive hominids had ape-like faces with receding foreheads and small

*brains. However, their pelvis and leg anatomy allowed them to stand upright and walk much like modern humans, freeing their hands for other tasks such as carrying food or young.*

*In this chapter, we consider the evidence for how and where our species,* Homo sapiens, *arose from primate ancestors through a series of evolutionary adaptations over a period spanning millions of years. The broad outlines of this evolutionary process are now well established. New discoveries and interpretations of the evidence are constantly refining our knowledge of the nature and sequence of the events in the history of our own species. Although much of the evidence consists of bits of fossilized bones dug out of rocks, there is a direct and thrilling connection between us and our ancestors when viewing human-like footprints that have survived for more than three million years.*

## DEFINING PRIMATES AND THEIR EVOLUTIONARY ADAPTATIONS

As a species, humans are members of the class Mammalia, a group of vertebrates that arose from primitive reptiles during the age of the dinosaurs (in the Triassic, some 200 to 245 million years ago). From the fossils they left behind, these early mammals were probably small, meat-eating, and active at night. Among species living today, the shrew probably most closely resembles these first mammals (✦ Figure 15.1).

After the dinosaurs and large reptiles became extinct in the Mesozoic (about 65 million years ago), mammals evolved rapidly and occupied the habitats vacated by the reptiles. The major orders of placental mammals arose at this time, including the carnivores, herbivores, rodents, and primates. Humans are classified as placental mammals belonging to the primate order, along with lemurs, tarsiers, monkeys, gibbons, and apes (✦ Figure 15.2). Although humans have characteristics that separate us from other primates, including upright walking, large brain capacity, and complex social and cultural practices, our history as a species is intertwined with that of the primates.

## EVOLUTION IN PRIMATES IS LINKED TO CHANGES IN THE ENVIRONMENT

The traits used to distinguish humans from other primates developed as a result of adaptive changes that began early in the process of primate evolution, and which accumulated and became more specialized at evolutionary branch points leading to our species. Based on single adaptations, it can be argued that primates are less specialized than other mammals; however, the combination of adaptations found in primates also place them among the most advanced group of mammals. These adaptations include highly developed forelimbs and hands, stereoscopic color vision, versatile teeth, upright posture, large brains, and, in one species, language and culture.

This chapter focuses on adaptations that led to the formation of our species. As with speciation in other groups, the driving force behind these adaptations is environmental change. About 20 million years ago, the central region of Africa from east to west was densely forested. About 15 million years ago, a combination of tectonic plate movements and episodes of global cooling drastically altered the landscape, and the forest was gradually replaced with regions of savannah mixed with clumps of open woodland. These geological and climatic changes set the stage for hominid evolution in East Africa that culminated in the emergence of our species, *Homo sapi-*

(a)

(b)

✦ **FIGURE 15.2**

**Representative primates. (a) Lemur. (b) Baboon.**

✦ **FIGURE 15.1**

**Organisms resembling this shrew are thought to be ancestral to primates.**

(a)

(b)

**✦ FIGURE 15.3**

(a) About 15 million years ago, a mosaic of savannah mixed with clumps of open woodland replaced the forest (b). These changes set the stage for hominid evolution in East Africa.

*ens.* During hominid evolution, cyclic changes in the climate caused fragmentation of the habitat (✦ Figure 15.3), leading in some cases to bursts of speciation, and in others, to bursts of extinction. In the following sections, important adaptations and the evolutionary trends that resulted from these climatic changes are described.

## Limbs and Hands Are Highly Developed in Primates

Early primates were adapted to life in the trees. In central Africa 20 million years ago, this was the major habitat. Over time, modifications in limb structure allowed primates to become highly specialized for success in this new habitat. The forelimb in reptiles, birds, and mammals has two long bones, the radius and the ulna (✦ Figure 15.4). Unlike reptiles, primates can turn the radius over the ulna so that the hand can rotate without moving the elbow or upper arm. In addition, some primates, including lemurs, apes, and humans, can rotate the arm in the shoulder socket. These two skeletal adaptations offer obvious advantages to tree-living animals that swing from limb to limb, but later, also permitted the development and use of tools.

The mammalian forelimb is derived from an ancestral limb with five digits. By adaptive modifications of the hand, primates evolved the ability to grasp objects, in what is known as **prehensile movement** (✦ Figure 15.5) (some primates also have pre-

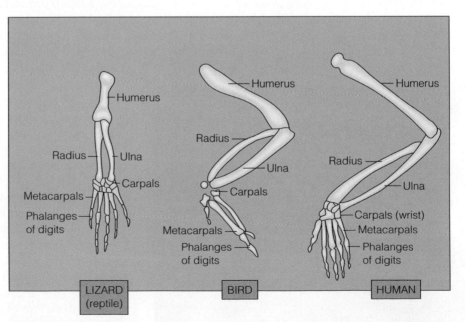

**✦ FIGURE 15.4**

The forelimb in reptiles, birds, and primates has two long bones, the radius and the ulna. Primates can turn the radius over the ulna so that the hand can rotate without moving the elbow or upper arm.

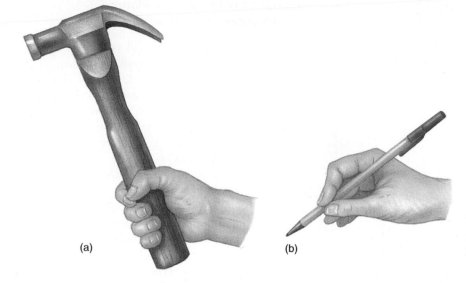

The development of opposable thumbs
gives primates the ability to have (a) a
power grip and (b) a precision grip.

(a)　　　　　　　　　　　(b)

hensile toes on their hind limbs). A second modification makes one of the digits **opposable,** allowing
the tips of the fingers and the thumb to touch
(Figure 15.5). In addition, most primates have
nails instead of claws. This leaves one surface of
the digit free and increases the sense of touch.
These adaptations provide the ability to grip and
manipulate a wide range of objects, allowing the
secondary development of tools as a basis for human technology.

## Stereoscopic and Color Vision Are Adaptations to Primate Habitats

Jumping or swinging from branch to branch in
trees places a premium on depth perception and
visual acuity. Early primates had eyes that were situated on each side of the head, providing little in
the way of depth perception. As primates became
more adapted to their environment, selection favored eyes located on the front of the head, providing more depth perception and stereoscopic vision
(♦ Figure 15.6). This evolutionary trend was accompanied by the changes in the location and number of light receptors in the eyes. Primates have two
types of light receptors, **cones** and **rods.** Cones provide color vision and visual acuity. In the primates,
cones are concentrated in areas called **foveas,** producing high-resolution color images. High densities
of rods at the periphery of the primate eye provide
increased vision in dim light (which is why sometimes you can see objects better in dim light if you
don't look at them directly). These evolutionary
changes in the visual system helped adapt primates
to their habitat, and kept them off the ground,
where predators awaited.

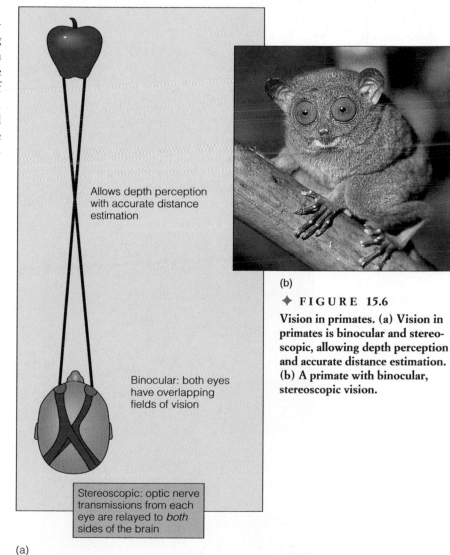

Allows depth perception
with accurate distance
estimation

Binocular: both eyes
have overlapping
fields of vision

Stereoscopic: optic nerve
transmissions from each
eye are relayed to *both*
sides of the brain

(a)

(b)

♦ FIGURE 15.6

Vision in primates. (a) Vision in
primates is binocular and stereoscopic, allowing depth perception
and accurate distance estimation.
(b) A primate with binocular,
stereoscopic vision.

(a)

## Upright Posture Was a Major Development in Primate Evolution

Even casual observation will confirm that primates such as monkeys are well adapted to life in the trees. They use all four limbs for climbing, walking, and running. However, monkeys can also sit upright. This posture allows monkeys to view their surroundings from a vertical position, scanning for food, companions, mates, and predators. Upright posture also frees the hands for other tasks. In the apes, adaptations in the skeletal system allow them to sit upright *and* maintain a semiupright posture when standing or walking. In the lineage leading to humans, skeletal modifications to the backbone, pelvis, femur, and knees (✦ Figure 15.7) allowed the body to be held upright when standing or sitting, and allows walking on two legs (bipedal locomotion). The gradual trend to an upright body posture in the course of primate evolution allowed the development of bipedal locomotion; it also permitted the arms and hands to become more independent, and freed them to take on other tasks including the development and use of tools.

To summarize this important adaptation, it is important to point out that the change from four-legged walking in monkeys and apes to the upright posture and two-legged walking in humans occurred in stages. Almost all primates can sit upright; some can stand upright; and one present-day primate (our own species) can walk and run upright (✦ Figure 15.8).

(b)

✦ **FIGURE 15.7**

**Comparison between quadrepedal and bipedal locomotion. (a) In the gorilla pelvis, the ischium is long, and the pelvis tilts horizontally, keeping the upper body forward. (b) In humans, the ischium is shorter, and the pelvis is vertical, keeping the upper body upright.**

✦ **FIGURE 15.8**
**Humans can not only walk upright, but can run long distances.**

## Primate Jaws and Teeth Are Adaptations to Changing Diets

The size, shape, and pattern of teeth in a jawbone and even the jaw itself can provide important clues about the food eaten by a species, as well as the behavior associated with food gathering. In primate evolution, several trends in jaw and tooth development are evident. First, the jaw became shorter from front to back, and deeper from top to bottom. This progressive change had the effect of altering the face from the elongated snout of primitive primates (look at the face of the ground shrew in Figure 15.1) to the flat face of modern humans (✦ Figure 15.9). A parallel series of changes in tooth shape and size led to a dentition pattern that is efficient in grinding food.

The tooth surface is the first site of contact with food, and patterns of wear on teeth provide clues about diet and behavior. Examination using

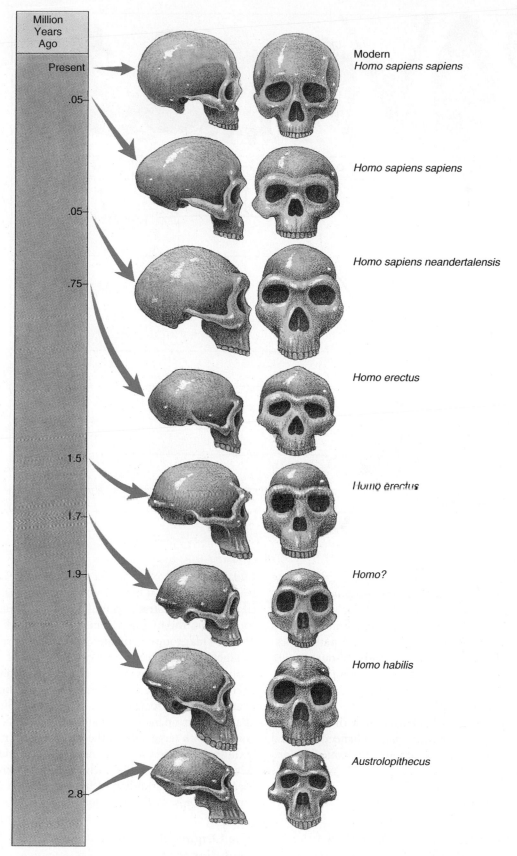

♦ **FIGURE 15.9**

**Skull changes during hominid evolution.** Beginning with *Australopithecus,* the hominid skull underwent many changes. The timescale refers to the ages of the skulls, not dates of divergence.

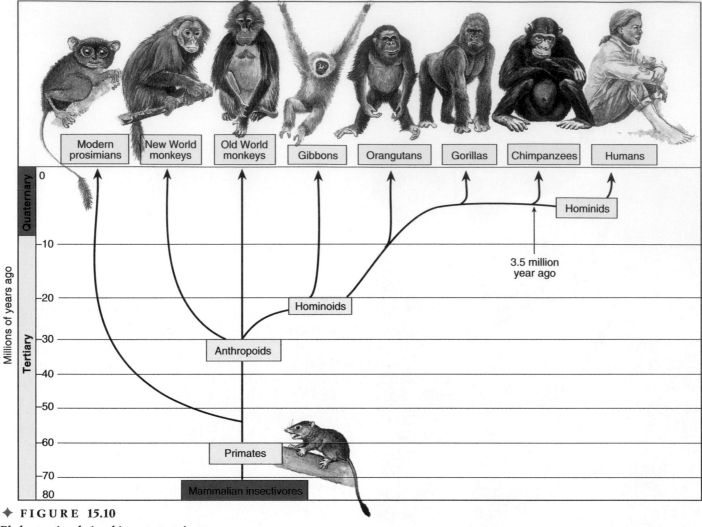

Modern prosimians · New World monkeys · Old World monkeys · Gibbons · Orangutans · Gorillas · Chimpanzees · Humans

Hominids

3.5 million year ago

Hominoids

Anthropoids

Primates

Mammalian insectivores

Millions of years ago

Quaternary · Tertiary

✦ FIGURE 15.10

**Phylogenetic relationships among primates.**

scanning electron microscopy on fossil primate teeth has shown that the diet was mainly fruit, consistent with life in trees. Sometime after the development of bipedal locomotion, primate fossils show a dramatic change in tooth wear patterns, especially in the immediate ancestors of our species. These changes occurred about 1.5 million years ago, and point to a diet composed of vegetables and fruit and the use of meat as a food source. This abrupt change in diet probably correlates with a change in food-gathering behavior that enhanced survival and population growth among our immediate primate ancestors.

### STAGES IN PRIMATE EVOLUTION: ORIGIN OF APES AND HOMINIDS

Based on the fossil record, it appears that a divergence in primate evolution took place about 30 million years ago in Africa (✦ Figure 15.10). One branch leads to primate species now classified as

Old World and New World monkeys. The other leads to groups that include apes and humans. Fossil primates ancestral to both humans and apes are found from the middle to late Miocene (✦ Figure 15.11). This group of primates, known as *hominoids,* originated in Africa and underwent a series of rapid evolutionary changes, giving rise to an interconnected array of species. These rapid evolutionary changes were associated with a series of climatic changes that opened new habitats. This diversity, coupled with large gaps in the fossil record, makes it difficult to draw conclusions about evolutionary relationships among the early hominoids and to identify the ancestral line or lines leading to modern hominoids, including humans.

### The Origins of Hominids Is an Evolutionary Puzzle

Up until a few years ago, it was thought that a group of hominoids known as ramapiths gave rise to the *hominids,* a line of primates leading to hu-

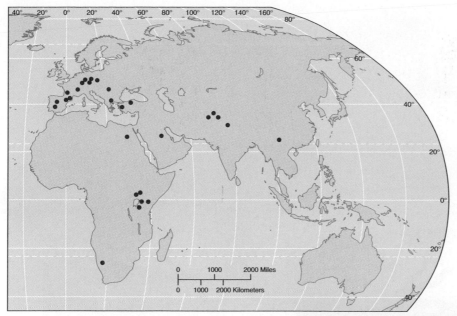

(a)

✦ **FIGURE 15.11**

**(a) The distribution of Miocene (22.5 to 5 million years ago) hominoid fossil sites discovered thus far. (b) An early Miocene hominoid skull from Africa.**

(b)

mans. The ramapiths lived partly in the trees and partly on the ground in a habitat that was a combination of woods and open grasslands. Hominids can be distinguished from ramapiths by several characteristics: differences in jaw structure and teeth probably accompanied by dietary changes; larger brain size; and upright walking. More recently, however, several lines of evidence ranging from molecular biology to paleontology indicate that ramapiths are ancestral to the orangutans, one of the great apes. The line leading to humans did not branch off from the ramapiths, but instead arose from an as-yet unknown ancestor. Although there is no direct evidence from the fossil record, the genetic distances between modern apes, chimpanzees, and humans indicates that the line leading to gorillas on the one hand, and the hominids, leading to our species, on the other, probably split from each other about six to eight million years ago.

## Human-like Hominids Appeared about Four Million Years Ago

The earliest human fossils (hominids) are from a species that first appeared 3.6 to 4.0 million years

ago, called *Australopithecus afarensis*. This species, which lived in East Africa, had a combination of ape-like and human-like characteristics. Members of this species walked upright (they were the creatures described in the opening vignette of the chapter), but the body proportions were ape-like, with

**✦ FIGURE 15.12**

**A Pliocene landscape showing a band of** *Australopithecus afarensis* **gathering and eating fruits and seeds.**

short legs and relatively long arms (✦ Figure 15.12). The arm bones of australopithecines were long and curved like those of chimpanzees, but have human-like elbows that could not support body weight while knuckle-walking (which is how chimpanzees and gorillas move around). This combination of features is well suited for a species that spends time climbing in trees, but walks on two legs across the ground. The skulls of australopithecines (✦ Figure 15.13) are ape-like and have a small brain, receding face, and large, canine teeth. Perhaps the most famous specimen of *Australopithecus* is a 40% complete skeleton recovered in the 1970s and known as Lucy (named after the Beatles song, *Lucy in the Sky with Diamonds*) (✦ Figure 15.14). A new set of fos-

sils, older and anatomically more ape-like than Lucy, has recently been discovered in East African rock layers dating to 4.4 million years ago. This species has been named *Australopithecus ramidus,* and its relationship to *A. afarensis* remains to be established.

There is general, though not universal, agreement that *Australopithecus afarensis* is the ancestral stock from which all other known hominid species are derived (✦ Figure 15.15). For more than a million years after they appeared in the fossil record, this species was remarkably stable. About two million years ago however, there was a burst of evolutionary change, probably related to climatic changes that accompanied the advancing glaciers in

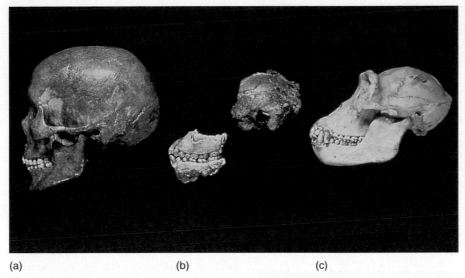

(a)                    (b)                    (c)

✦ FIGURE 15.13

Comparison of the skulls of (a) modern humans, (b) *Australopithecus* fragments, and (c) chimpanzees.

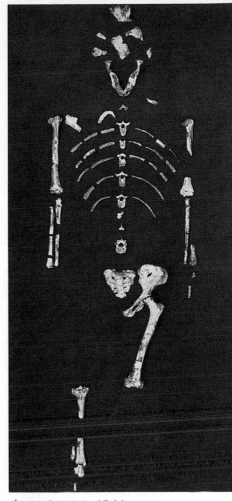

✦ FIGURE 15.14

The skeleton of "Lucy," a 3.5 million-year-old female *Australopithecus afarensis*.

Europe. The fossil record indicates that at least three and perhaps as many as six or more species of hominids occupied different or overlapping habitats at this time. These hominid species can be placed into two groups. One was composed of several species of more graceful (gracile), small-brained forms including *Australopithecus africanus*, as well as heavier, more robust species including *Australopithecus robustus*. This group gradually died out, with the last species surviving until about one million years ago.

The second group was characterized by relatively large brains, a smaller facial structure, and reduced teeth. These relatively large-brained, less robust species are grouped in the genus *Homo*, the genus to which our species belongs.

While the fossil record is incomplete, it is commonly recognized that the australopithecines are ancestral to the later hominids. By examining the fossils of the australopithecines as well as earlier hominoids, it is clear that the evolutionary adaptations in primates developed at different rates, showing a pattern described earlier as **mosaic evolution.** Skeletal changes leading to walking and dental changes leading to diet diversification were early adaptations. Later changes include an increase in brain size accompanied by the development of tools, the use of fire, and the origins of language.

## The Genus *Homo* Appeared about Two Million Years Ago

The earliest humans classified as members of the genus *Homo* appeared as part of a cluster of hominid species that developed about 2 to 2.5 mil-

lion years ago. This group contains several species, including one classified as *Homo habilis* (some scientists now think that this classification should actually be split into two or more species). Members of this species or species cluster had a brain size about 20% larger than that of the australopithecines. They also had a differently shaped skull, and teeth distinct from those of australopithecines. At least one and probably several species of the genus

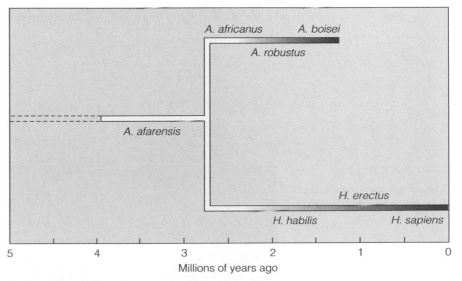

✦ FIGURE 15.15

A little less than three million years ago, the *A. afarensis* line split, with one line leading to several species of australopithecines, and the other line leading to *Homo sapiens*.

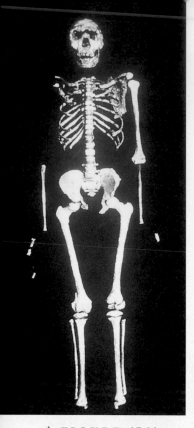

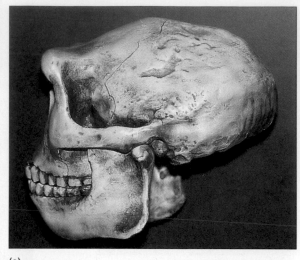

(a)

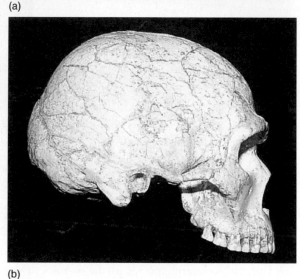

(b)

✦ FIGURE 15.17

Skulls of (a) *Homo erectus* and (b) *Homo sapiens*.
The skull of *H. erectus* is low, and pointed in rear,
while the skull of *H. sapiens* is high and dome-shaped.

✦ FIGURE 15.16

A skeleton of *Homo erectus*.
This is the most complete
skeleton of *H. erectus* found
to date, and is of a boy who
lived some 1.6 million years
ago.

*Homo* lived alongside the robust australopithecines in East Africa for about a million
years, while other members of the genus *Homo*
lived in South Africa at the same time. After this
period, the australopithecines became extinct.

Sometime during this period, a species that
became our immediate ancestor arose within the
*Homo* line. This hominid, *Homo erectus* (✦ Figure 15.16) represents an important turning point
in human evolution (see Beyond the Basics, p.
293).

## *Homo erectus* Originated in Africa and Migrated to Europe and Asia

African fossils of *Homo erectus* date to about 1.8
million years ago, although as discussed later,
the species may actually be much older. The
"Turkana boy," discovered in 1985, is one of the
most complete skeletons of *H. erectus* ever recovered (Figure 15.16). Although *H. erectus*
originated in Africa, soon after its origins, members of this species migrated into Asia and Europe. Fossils of *H. erectus* have been recovered
from Indonesia (Java Man) and China (Peking
Man) as well as sites in North Africa. The recent dating of Indonesian fossils of *H. erectus*
to about 1.8 million years ago indicates that
as a species, *H. erectus* is probably older than
2 million years.

Physically, *Homo erectus* is different from
*Homo habilis* in several respects; increased brain
size, a flatter face, and the presence of prominent brow ridges. There are physical similarities
between *Homo erectus* and modern humans, including height and walking patterns, but there
are also some significant differences. In some
populations of *Homo erectus*, the skull is low
and pointed in the back, not dome-shaped as in
*Homo sapiens* (✦ Figure 15.17). In addition, *H.
erectus* had a receding chin, a prominent brow
ridge, and some differences in teeth.

✦ FIGURE 15.18

A landscape showing members
of *H. erectus* using fire and
stone tools.

# TOOL TIME: DID *HOMO ERECTUS* USE KILLER FRISBEES?

The appearance of *Homo erectus* about two million years ago represents a turning point in human evolution. Studies of tooth wear patterns indicate that meat became an important component in the diet of *Homo erectus*. This species was also the first hominid to move out of Africa into Europe and Asia, exploiting new habitats as resources. Evidence indicates that these and other changes were accompanied by technological innovation in the form of new tools. The use of tools predates *Homo erectus* by about a million years, but these early tools (called Oldowan tools) were small, were used mainly for chopping, and remained unchanged over a span covering a million years.

The tools of *Homo erectus* (called Acheulian tools) were large, with two cutting faces, and included hand axes, cleavers, and picks (see Figure). The typical toolbox also included scrapers, as well as trimmers for producing and sharpening new tools. Over time, Acheulian tools were gradually refined, but changed little in the million years they were used, and few new tools were added to the basic toolbox. These tools disappeared about 200,000 years ago, at a time when tool making entered a period of technological innovation and re-

finement that marks the Middle Paleolithic (Middle Stone Age).

Of all the tools in the kit of *Homo erectus,* the possible uses of the hand axe have remained a subject of speculation. Modern-day anthropologists have learned how to make such tools and have used them as small axes for chopping, or as heavy-duty knives for slicing animal hides or skinning carcasses. Examination of fossil hand axes by electron microscopy indicates that these tools may have had a range of uses on many materials including hide, meat, bone, and even wood. One form of the handaxe is ovoid, with a pointed end. Their size and shape has led to the suggestion that they may have been thrown like a frisbee into a herd of small animals, stunning or wounding one, which could be overtaken and killed. Flight tests of fossils and replicas support the idea of killer frisbees, but this use is not widely accepted by anthropologists

and remains among the most speculative proposals for the function of hand axes.

Whatever their uses, the tools of *Homo erectus* were associated with dramatic changes in behavior, including diet, migration, systematic hunting, the use of fire, and the establishment of home bases or camps. The role of new technology in promoting or supporting these new behaviors remains an area of intense investigation in paleoanthropology.

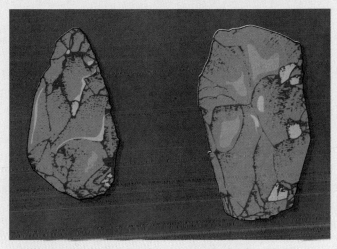

**Acheulian hand-axe.**　　**Acheulian cleaver.**

## *Homo erectus* Developed Social and Cultural Systems

*H. erectus* is the first hominid species to provide evidence for social and cultural aspects of human evolution: the first use of fire, the appearance of organized food gathering in hunts, the first use of permanent settlements, and the first evidence for a prolonged period of growth and maturation after birth (✦ Figure 15.18).

A census taken about a million years ago would have found populations of *H. erectus* living in Africa, Europe, and Asia. Sometime later, perhaps 100,000 to 500,000 years ago, populations of *H. erectus* disappeared, and were replaced by our species, *Homo sapiens*. Although there is general agreement that this new species lived on all three continents, disagreement exists about where, when, and how the transition to *Homo sapiens* took place.

## THE APPEARANCE AND SPREAD OF *HOMO SAPIENS*

Tracing the origins of our species has become a multidisciplinary task, using the tools and methods of anthropology, paleontology, and archaeology, and more recently, satellite mapping from space, and the techniques of genetics and recombinant DNA technology. These methods are being used to reconstruct the origins and ancestry of populations of *Homo sapiens,* and to answer questions about how and when our species originated and became dispersed across the globe.

### Two Theories Differ on How and Where *Homo sapiens* Originated

From the evidence provided by fossils and artifacts, there is consensus that groups of *H. erectus* moved out of Africa about two million years ago and spread through parts of Europe and Asia. What currently divides paleoanthropologists is the question of how and where members of our species, *Homo sapiens,* originated. In a general sense, there are two opposing views about the origin of modern humans. One idea (often called the **out of Africa hypothesis**) argues that after *H. erectus* moved out of Africa, populations that remained behind continued to evolve, and gave rise to *H. sapiens* about 100,000 to 200,000 years ago. From this single source, somewhere in Africa, modern humans migrated to all parts of the world, displacing populations of *H. erectus,* which then became extinct. According to this model, modern human populations are all derived from a single speciation event that took place in a restricted region within Africa, and should show a relatively high degree of genetic relatedness.

This model of human evolution is based on evidence that African populations show the greatest amount of genetic diversity as measured by different mitochondrial types (based on genetic differences in their DNA) from thousands of individuals tested worldwide. The underlying assumption in this study is that mutational changes in mitochondrial DNA accumulate at a constant rate, and provide a "molecular clock" that can be calibrated by studying the fossil record. According to this reasoning, since African populations show the greatest amount of diversity in mitochondrial types, modern humans (*H. sapiens*) have lived longer in Africa than in other regions of the world. Phylogenetic trees constructed from nucleotide differences in mitochondrial DNA lead back to a single ancestral mitochondrial lineage for our species, originating in Africa. Calculations using the rate at which new mutational changes occur indicates that our species began to branch about 200,000 years ago from an African population that could have consisted of as few as 10,000 individuals.

The second idea about the origin of *H. sapiens* postulates that after *H. erectus* spread from Africa over most of Europe and Asia, modern humans arose at multiple sites as part of an interbreeding network of lineages descended from the original colonizing populations of *H. erectus.* The evidence to support this model (often called the **regional continuity hypothesis**) is derived from a combination of genetic and fossil evidence that shows a gradual transition from archaic to modern humans. Based on the continuity of anatomical characteristics seen across geographic regions, this model suggests that the transition took place at multiple sites outside of Africa, and that *H. erectus* became gradually transformed into *H. sapiens.* The *H. erectus* populations in Asia became the Asian form of *H.* sapiens, and the European forms of *H. erectus* became the European form of *H. sapiens.*

The two opposing ideas have been hotly debated and have received a great deal of attention in the press and other media. These alternate explanations for the appearance of modern humans show that scientists can sometimes reach different conclusions about the same problem. Often this is caused by the limitations of the methods being used or by having only a small amount of information available. In this case, part of the problem revolves around which information (mitochondrial DNA or the fossil record) is regarded as most important.

The debate over the origins of *Homo sapiens* has not been resolved. Further work has called into question the accuracy of the molecular clock and the method used to construct mitochondrial phylogenetic trees, the main evidence for the out of Africa hypothesis. If the clock is inaccurate, perhaps genetic divergence in mitochondrial DNA took place after the migration of *H. erectus* from Africa, and before the appearance of *H. sapiens.* In addition, a phylogenetic tree leading back to a single ancestral mitochondrial lineage is not the only possibility. The same data used to construct the phylogenetic tree that supports the out of Africa hypothesis can be used to construct other phylogenetic trees that are shorter and do not lead back to a single origin. These findings weaken the argument for the out of Africa hypothesis.

On the other hand, recent work on genetic divergence in a region of the Y chromosome weaken the argument for the regional continuity hypothesis. Mitochondrial DNA is passed from mother to all children. As a result, the out of Africa hypothesis is based on maternal inheritance. The Y chromosome is passed only from father to sons. Information about genetic diversity on the Y chromosome can provide evidence from paternal inheritance about the origin of our species.

Analysis of nucleotide differences in Y chromosome DNA from a world-wide sample of men indicates a single point of origin for modern humans. The time of origin of our species based on the Y chromosome results places the date at about 270,000 years ago. While the results of this work cannot pinpoint the geographical origin of the Y chromosome lineage, the single point of origin and the time of origin are consistent with the out of Africa hypothesis.

Much work remains to be done before the questions of when, where and how our species originated can be answered. Although we have considered only two models, others are possible. Each new model and each new idea will have to be scrutinized using the information available, and re-evaluated in the light of new information.

## STONE AGE FABRICS

When did humans begin weaving fabrics? This is a difficult question to answer because materials made of plant fibers decay and leave little in the way for archaeologists to discover. Until recently, twisted fibers found in Israel that date to about 19,000 years ago were the earliest indications that humans used cord or fibers. Recently, pottery excavated from a site in the Czech Republic between 1952 and 1972 has provided evidence that textile production in Central Europe was well developed some 27,000 years ago.

Examining pottery fragments from the site, researchers spotted imprints of fabric, pressed into the clay of the vessel while it was still wet. These impressions were made by loosely woven cloth or flexible basketry. The texture of the material suggests that it might have been made using a loom. Because of the sophisticated weave of the material, archaeologists suggest that weaving and fabric use had a beginning much earlier than 27,000 years ago, and must have been used in cultures before the development of agriculture.

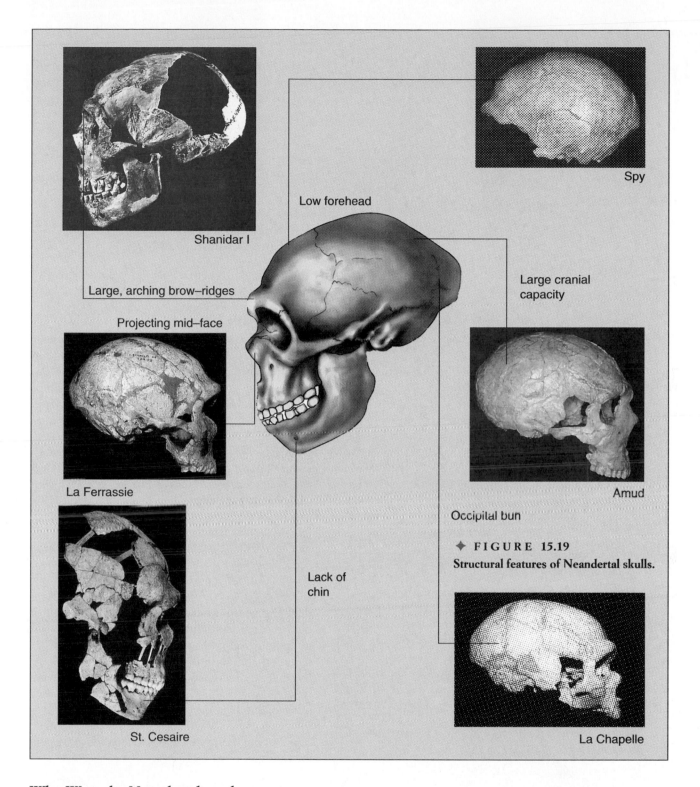

Shanidar I

Spy

Low forehead

Large, arching brow-ridges

Large cranial capacity

Projecting mid-face

La Ferrassie

Amud

Occipital bun

✦ FIGURE 15.19

Structural features of Neandertal skulls.

Lack of chin

St. Cesaire

La Chapelle

## Who Were the Neandertals and What Happened to Them?

Populations of *H. sapiens* that lived from about 500,000 to 30,000 years ago are often referred to as archaic forms of our species, because they retained some physical characteristics of *H. erectus* mixed in with features of modern humans. One archaic form of *H. sapiens* is the Neandertals, a group that lived in Europe and Western Asia for about 100,000 years and became extinct between 30,000 and 50,000 years ago. This subspecies lived during the last interglacial period, beginning about 125,000 years ago.

Several skeletal features of the Neandertals distinguish them from modern *H. sapiens*. Neandertals were large-brained (larger on average than modern humans), with low forehead, prominent brow ridges, and a receding chin (✦ Figure 15.19).

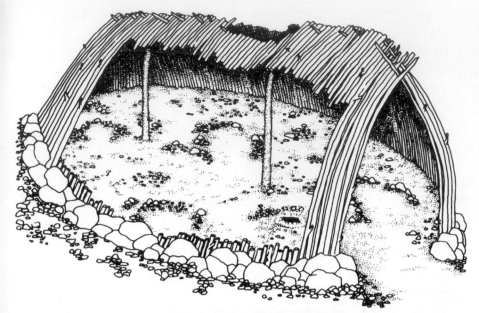

✦ FIGURE 15.20

**Cutaway view of reconstructed Neandertal hut. This shelter included workspaces for making tools, a hearth, ridgepoles, and stones shoring up the sides.**

vances in culture and technology were developed by these archaic humans. Neandertals were the first humans to bury their dead, and burials often included grave goods such as flowers, beads, and animal bones. This practice indicates the ability for abstract thought and perhaps a belief in an afterlife. Neandertals lived in settlements as well as caves and rock shelters, and built huts of bone and skin (✦ Figure 15.20). They developed the technology of tools far beyond that of *H. erectus,* including the use of handles on stone tools.

## Are Neandertals Our Ancestors or a Separate Evolutionary Line?

The question of whether adaptations over time produced anatomically modern *H. sapiens* from Neandertals, or whether modern humans replaced this archaic form, is unresolved. Part of the problem is that there is a gradation of features between archaic and modern forms; as a result, it is impossible to draw a line that clearly separates the forms of our species. The regional continuity hypothesis holds that modern humans arose from archaic populations through a graded series of intermediates, with Neandertal genes represented in modern humans. The out of Africa hypothesis holds that modern *H. sapiens* arose separately from the Neandertals, replaced them across Europe and Western Asia, eventually driving them into extinction. In the multiregional view, Neandertals are a subspecies (*Homo sapiens neandertalensis*), whereas in the out of Africa view, they are a separate species (*Homo neandertalensis*) that went extinct, making little or no genetic contribution to our species.

Perhaps the most distinctive feature of these people is that the face projected from the skull as if it were pulled forward, ending with a prominent nose. The remainder of the Neandertal body was very robust and well muscled. Adult males probably averaged 5 feet 6 inches and weighed about 150 pounds, females were about 5 feet 2 inches and 120 pounds.

In spite of the popular image of Neandertals as cave-dwelling, dim-witted brutes, important ad-

Whatever the outcome of this debate, it is clear that migrations have been a constant feature of human evolution for more than a million years, and in conjunction with other forces, such as genetic mutation and natural selection, have helped shape the course of human evolution. In the following sections we discuss how genetics in tandem with other techniques has been used to study patterns of human migrations, and how present-day gene distributions represent evolutionary relics of past migrations.

## The Transition to Agriculture Is a Cultural Adaptation

Since the appearance of *H. sapiens,* the most recent transitions in human evolution involve cultural adaptations rather than physical changes, such as methods of locomotion or brain size. One of the most important, starting about 10,000 years ago, is known as the Neolithic transition. It involved a change from hunter-gatherer cultures to those

✦ FIGURE 15.21

**The Fertile Crescent in the Middle East, where agriculture is thought to have started and then spread to other regions.**

based on agriculture and animal breeding. Evidence suggests that this transition began in several regions of the world, but probably arose first in the Middle East (✦ Figure 15.21). How this cultural transition spread to other parts of the world is not yet understood. Was the spread of farming only cultural (the technique spread from region to region, but people essentially stayed in place), or was the technique spread by groups of migrating people who moved from the site of origin through existing populations?

Although these events took place before recorded history, the genetic structure of present-day populations provides a clue to how the transition to farming spread from its center of origin. If farming spread by the movement of farmers through existing populations, such migrations would tend to blur the genetic differences between populations, and might provide some clues about the direction of migrations. Genetic maps that plot the degree of genetic variability have been constructed for much of the world's population (✦ Figure 15.22). These maps correlate well with maps of the spread of agriculture, constructed from archaeological data, indicating that agriculture probably moved from its origin by the migration of farmers through hunter-gatherer cultures.

## Human Migration to the Americas Is a Recent Event

The origins, migrations, and interrelationships of the original inhabitants of the Americas have been the subject of much controversy. It is generally

✦ **FIGURE 15.22**

**Studies of allele frequencies have established regions of genetic discontinuities. In this case, there is a gradient of the B blood type from east to west, consistent with known patterns of migration.**

agreed that North America and, in turn, Central and South America were populated by migrations from Asia during the last period of glaciation, but there has been a great deal of disagreement about the number of migrations and their timing. The dispute centers around whether the earliest migrations into North America occurred around 30,000 to 35,000 years ago or whether humans first arrived on this continent about 15,000 years ago. Techniques of archaeology, anthropology, linguistics, and genetics have been used to study the prehistory of the Americas, but the origins of the first Americans are still obscure.

A land bridge, called Berengia, connected Asia and North America between 15,000 to 25,000 years ago, and allowed passage between the continents along the shoreline or overland (◆ Figure 15.23). This land bridge existed for thousands of years, allowing time for the passage of different groups of Asians at different times into North America.

Using linguistics and anthropology, scientists have identified at least three groups of American Indians: (1) the Amerinds, representing tribes spread across North and South America, (2) the Na-Denes, a cluster of tribes found in the northwestern region of North America, and (3) the Eskaleuts, composed of the Eskimos and Aleuts, who live across the far northern regions of North America.

Analysis of genetic differences using mitochondrial DNA (mtDNA) and nuclear genes indicate that all American Indian groups share a common lineage with populations in Siberia, the presumed origin of migrants who populated the Americas. The mtDNA studies reveal the presence of four mtDNA lineages among American Indians. Results indicate that the present-day Amerinds are derived from two or more population groups founded at different times; the oldest of these may have crossed from Siberia 21,000 to 42,000 years ago. The Na-Denes and Eskaleuts are of more recent origin, having arrived in North America some 8500 to 12,000 years ago.

Genetic studies point to several waves of migration beginning more than 20,000 years ago. This work provides strong support for previously controversial classifications of American Indians based on linguistics and other anthropological standards.

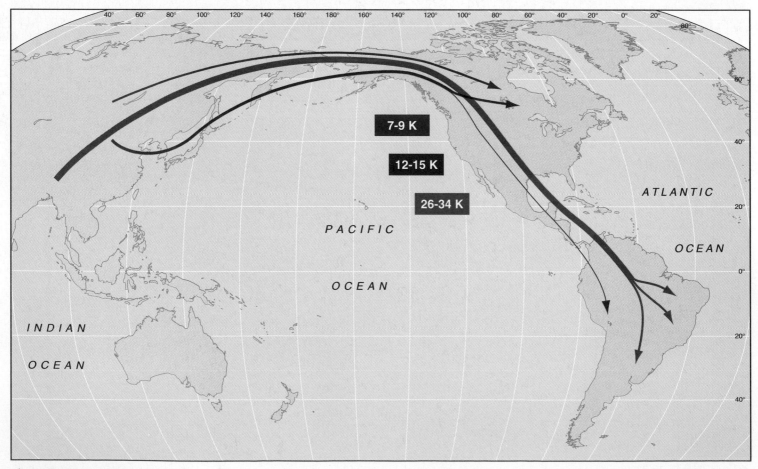

◆ FIGURE 15.23

A land bridge that temporarily connected North America to Asia allowed waves of migration.

The archaeological evidence, on the other hand, still generates disputes. There is agreement that sites based on skeletal remains and artifacts are able to trace human habitation in the Americas to 13,000 to 15,000 years ago. Other sites, indicating much earlier settlement, are the most controversial, but if confirmed, would indicate that humans were present in the Americas some 20,000 years ago.

At the present time, the genetic evidence, although incomplete in some regards, can be used to reconstruct the flow of *Homo sapiens* into the Americas. A genetic model, based on several assumptions, approximates how the genetic variation, population density, and geographic distribution of Native Americans could have been generated. The model assumes that several waves of migration crossed from Siberia, and that the migrants were in bands of no more than 500 individuals. If the bands reproduced and split off into new bands, doubling in size every 25 years, and migrated on average five miles each year, North, Central, and South America could have been occupied in a few thousand years. Once established, neighboring populations underwent numerous fusions and splits, allowing gene flow to spread new alleles across geographic regions.

Although this model remains speculative, it reinforces the earlier point that genetic analysis of present-day populations can provide evidence for the origin, migration, and interrelationships of human populations. Joined with the techniques of anthropology, archaeology, and linguistics, genetics can be a powerful tool in reconstructing the history of our species.

  # SUMMARY

1. As a species, humans are classified as mammals, belonging to the order of primates, and the family *Hominidae,* which it shares with gorillas and chimpanzees. The history of our species is tied to the evolution of the primates, a group of placental mammals.
2. The driving forces that helped shape the evolution of primates include the forces of geology and climate. Environmental changes that took place about 20 million years ago in Africa brought about ecological changes that helped drive primate evolution. Subsequent cycles of glacial advances and retreats in the northern latitudes caused climatic fluctuations in Africa associated with bursts of speciation and extinction.
3. Trends in primate evolution in the line leading to our species include the development of highly specialized limbs and hands, stereoscopic color vision, upright posture, bipedal locomotion, increased brain size, social organization, culture, and language.
4. Fossil primates that are ancestral to both humans and apes are found in the middle to late Miocene between 5 and 10 million years ago. These hominoids diverged rapidly into a number of species. Because of the gap in the fossil record and the diverse structure of species origins during this time period, it is difficult to identify the ancestral line or lines leading to modern hominoids, including humans.
5. A group of human-like hominids appeared in the fossil record about four million years ago. One species, *Australopithecus afarensis,* had a combination of ape-like and human-like characteristics. Members of this species walked upright like humans, but had ape-like bodies and arms. This species is generally regarded as the ancestral line that gave rise to humans.
6. The earliest humans classified as members of our genus arose about 2 to 2.5 million years ago. From this line, the species *Homo erectus* represents an important turning point in human evolution. This species, which arose about 2 million years ago, had some physical similarities to modern humans, developed social and cultural systems, used meat in the diet, and had a prolonged period of growth and maturation of the young, implying some social organization. Members of this species also migrated over vast distances, from Africa into regions of Europe and Asia.
7. The origin of our species, *Homo sapiens,* is somewhat controversial. Two theories differ on the origins of *H. sapiens;* one proposes that the species developed in Africa from populations of *H. erectus,* another holds that groups of *H. erectus* were gradually transformed into *H. sapiens* at multiple sites in Europe and Asia.
8. The position of one subspecies, *H. neandertalensis,* in the appearance of modern humans remains unresolved.
9. The appearance of humans in the Americas is a relatively recent event, resulting from a series of migrations from Siberia across a land bridge to the North American continent.

  # KEY TERMS

cones
define
foveas

mosaic evolution
opposable

prehensile movement
rods

## SHORT ANSWER

1. Humans belong to what genus? species? order? family? class?
2. What combination of adaptations found in primates makes them the most advanced group of mammals?
3. What adaptive trends in teeth and jaws are seen in primate evolution?
4. Explain why it is difficult to trace definitively the evolutionary relationships among early hominoids and modern hominoids.
5. Compare and contrast *Homo erectus* and *Homo sapiens.*
6. Describe the opposing views about the origin of modern humans.
7. What skeletal features distinguish Neanderthals from modern *Homo sapiens?*
8. Outline the genetic model that has been used to reconstruct the flow of *Homo sapiens* into the Americas.

## MULTIPLE CHOICE

1. The single most important factor influencing the evolutionary divergence of primates was the adaptation to living in _____.
   a. grasslands
   b. trees
   c. water
   d. forests
2. The earliest appearance of known human fossils dates from the _____.
   a. Triassic, 200–245 million years ago
   b. Mesozoic, 65 million years ago
   c. Miocene, 25–30 million years ago
   d. Pliocene, 3.6–4.0 million years ago
3. Skeletal remains and artifacts date human habitation in the Americas to about _____ years ago.
   a. 13,000–15,000
   b. 21,000–42,000
   c. 8500–12,000
   d. 3000–10,000
4. ____ are examples of hominids and ____ are examples of hominoids.
   a. Lemurs; gorillas and their ancestors
   b. Chimpanzees; humans and their recent ancestors
   c. Ramapiths; gorillas
   d. Humans and their recent ancestors; apes and humans
5. The first known hominids were _____.
   a. bipedal with larger brains than their predecessors
   b. tall with long arms
   c. solely carnivores
   d. without teeth

## FILL IN THE BLANK

1. Primates have the ability to grasp objects in a fashion known as _____. _____ digits allow the tips of the fingers and the thumb to touch.
2. Skeletal adaptations allowed primates to sit _____ and maintain a _____ posture when walking or standing.
3. Changes in the _____ may have been part of the selection pressure that lead to bipedal locomotion.
4. There is general agreement, although not universal, that _____ _____ is the ancestral stock from which all known hominid species are derived.
5. _____ _____ is the term that describes the pattern of evolutionary adaptations in primates.
6. *Homo erectus* is different from *Homo habilis* by having an _____ brain size, a _____ face, and the presence of _____ brow ridges.
7. _____ were the first humans to bury their dead.
8. In the multiregional view, Neanderthals are a subspecies _____, whereas in the out of Africa view, they are a separate species _____ that went extinct.
9. Since the appearance of *H. sapiens,* the most recent transition in human evolution involves _____ adaptations rather than _____ changes.
10. _____ was a land bridge that connected Asia and North America and allowed passage between the continents.

## SCIENCE AND SOCIETY

1. Studying other primates, our closest relatives, can teach us about many aspects of the evolution of our own behavior patterns. There are about 190 species of living primates other than human beings. Approximately 170 of them live in tropical rain forests. Human population growth and industrialization in these forests have caused these areas to be cut, cleared, and destroyed. What effects will clear-cutting of the rain forests have on primates? How will this affect the study of human evolution and behavior? Can you think of other areas of science that would be affected by a massive loss in the number of species of primates? Do you think the species of primates in zoos will prevent mass extinction of these animals? Why or why not? Should measures be taken to protect these animals before they become endangered species?
2. Neandertals are an archaic form of *Homo sapiens* that lived in Europe and Western Asia for about 100,000 years and became extinct between 30,000 and 50,000 years ago. They developed important advances in culture and technology that were far beyond those employed by *Homo erectus.* There are many skeletal features of Neandertals that distinguish them from modern *H. sapiens.* Still the question remains as to whether Neandertals are our ancestors or a separate evolutionary line. How did Neandertal people differ from modern humans? Can you hypothesize why Neandertals became extinct? Give reasons supporting your hypothesis. Do you think Neandertals are the early ancestors of the modern human? Why or why not? If Neandertals are our ancestors, how did they get to North America?

SCANNING ELECTRON MICROGRAPH (SEM) OF CORN ROOT.

# 16

# PLANT STRUCTURE AND FUNCTION

**OPENING IMAGE**

The design of the Crystal Palace was

based on the structure of a water

lily leaf.

L ate in 1849, Prince Albert of England appointed a Royal Commission to oversee the design and construction of a building to house the Great Exhibition of 1851 in London. One of his aims was to have a building that was more spectacular than the one which housed the Paris International Exhibition of 1849. The twenty-four member Commission produced a design for a brick building that was denounced by the newspapers and the public as an architectural monstrosity that would have destroyed many trees in Hyde Park, the site of the Exhibition.

At the point where there was only ten months left before the Exhibition, one of the Commission members, Joseph Paxton, stepped forward with a radical design for a building to be built of glass with metal framing, like a greenhouse. The Commission opposed the design, so Paxton took his plans to a newspaper, which published them, and helped rally public support for the design. With some protests, the design was accepted, and construction began in the last week of September, 1850. The building was built with a modular metal framework surrounding large glass panes. When completed, its arches soared over 100 feet into the air. In its final form, the building contained more than a million square feet of glass, covered nineteen acres, and was built around trees in the park. The building and the Exhibition were a huge success.

*The designer, Joseph Paxton, was a man with no formal training in architecture, and was a gardener for the Duke of Devonshire. How was he able to produce this architectural masterpiece? The Duke was an avid collector of plants, and Paxton was originally hired to supervise his greenhouses and gardens. In this capacity, he undertook a project to grow a giant water lily collected in the Amazon River in 1837. In nature, the leaves of this plant (named* Victoria regina *in honor of Queen Victoria) are five to ten feet across, with flower buds the size of large watermelons. Paxton was successful in getting the lily to grow, and in 1849, the plant bloomed. The event was a sensation in the world of horticulture, and a leaf and flower were sent to the Queen. His success got him appointed to the Royal Commission.*

*In reflecting on the design of a building to house the Exhibition, Paxton studied the leaf anatomy of the giant water lily. The plant had an arching, cantilevered arrangement of veins that enabled the leaf to spread out to an enormous size and still float on the surface of the water. For the framework of his building, Paxton used the vein arrangement of the water lily leaf. Thus, the giant water lily, with its structure refined over time by natural selection, was the inspiration for one of the greatest buildings of the nineteenth century.*

*In this chapter, we will consider the structure of flowering plants, and how form and function are interrelated in these organisms. The lessons learned have been applied not only to understanding how plants have adapted to the terrestrial environment, but to other areas, including architecture.*

##  FORM AND FUNCTION IN PLANTS

Can you tell the difference between a giant saguaro cactus and a fern? What distinguishes a redwood tree from green algae? Although these plants differ vastly in size and are separated by perhaps hundreds of millions of years of evolution, all of these organisms make their own food through photosynthesis and obtain chemical energy through respiration. They convert carbon dioxide into the food used by the five billion humans and many other creatures sharing this planet; they generate the oxygen needed to preserve our ecology; and, despite tremendous differences in their appearance, they perform these important physiological processes in essentially the same way and even share the same basic architectural plan.

When ancient plants made the transition from water to land, they faced the same problems animals did: desiccation, support, and dependence on water for reproduction. Aquatic plants were not adapted to cope with the demands of life on dry land; they were vulnerable to water loss and had no means of upright growth. Plants, however, have a unique advantage over other multicellular organisms—they are capable of both sexual and asexual reproduction. A new plant can form simply by budding off from an existing plant. In adapting to life on land, plants could use asexual reproduction when conditions were unfavorable for sexual reproduction.

Green algae, from which land plants evolved, exchange gases and absorb minerals and nutrients from the water that surrounds their body surface. Sexual reproduction involves the release of gametes into the water. These gametes meet and fuse in the drifting currents and are carried away to anchor somewhere. These same functions must be performed by land plants, but with adaptations to the terrestrial environment.

Evolution in early land plants involved selection for adaptations that reduced water loss and provided structures for support. A consequence of selection for these adaptations was the need for a vascular system to absorb moisture and nutrients and transport them to all parts of the plant. Most land plants are vascular plants, with a system of specialized cells for movement of water and nutrients. The nonvascular plants, such as the mosses and liverworts, lack such a system, and are typically small and restricted to moist environments.

The dominant vascular plants are those that have seeds, either in cones (the gymnosperms) or in fruits (the angiosperms). Because the angiosperms are the most familiar and the most common land plants, they will serve as the focus of this chapter.

## DIVIDING THE ANGIOSPERMS: MONOCOTS AND DICOTS

Flowering plants can be divided into two classes, the Monocotyledones (monocots) with about 65,000 species and the Dicotyledones (dicots), with about 170,000 species. The monocots include plants such as grass, cattails, lilies, and palm trees. The dicots include trees (except the conifers) and most ornamental and crop plants. The similarities between these groups are as great or greater than their differences, and some botanists feel it is unnecessary to distinguish between the two groups. There are, however, some important differences between these groups that bear on the discussions in this chapter. The major differences between monocots and dicots are summarized in ✦ Figure 16.1. The names **monocot** and **dicot** refer to a leaf-like structure, the **cotyledon,** which is present in the seed and appears during seed germination. Monocots have one cotyledon, and dicots have two. There are also differences in the leaves, stems, roots, and flowers (✦ Figure 16.2).

## THE BASIC ARCHITECTURAL PLAN OF PLANTS

Because each plant species differs in how it harnesses available resources—such as water, light, and pollinators—each does better or worse in certain conditions. These differences are a result of variations on a common body plan (✦ Figure 16.3). Above ground, plants have shoots, consisting of the stem, branches, leaves, and flowers. Below ground, plants have roots that penetrate the soil, absorb water and nutrients, and store food. The shoots and roots are composed of three major cell types organized into three main tissue systems.

### Plants Have Three Major Cell Types

The three major cell types found in plants are parenchyma, collenchyma, and sclerenchyma. **Parenchyma** cells have relatively few distinctive charac-

teristics, but do have thin walls that are usually multisided. Parenchyma cells are unspecialized, but carry on photosynthesis and respiration and can store food. They are regarded as the basic cell type, from which all other, more specialized cell types have evolved. In practice, any cell type that cannot be assigned to another functional classification is considered to be parenchyma. Parenchyma cells form the bulk of the plant body. They are found in the fleshy tissues of fruits and seeds, the photosynthetic cells of leaves, and the vascular system.

**Collenchyma** cells are elongated and have thicker walls than parenchyma cells. Their main function is to provide support, and they are usually arranged in strands. They are most often found in the region of the plant that is growing.

**Sclerenchyma** cells have thickened, rigid, secondary walls that are hardened with lignin (the main component of wood). As the cells grow, they elongate and form thick cell walls. When mature, these cells are dead, and contain no cytoplasm. Sclerenchyma cells give support to the plant. There are two basic types of sclerenchyma cells. One is a long, slender cell called a **fiber.** The primary function of fibers is support, but they have also been widely used by humans. Fiber from plant bodies is used to make linen, rope, and burlap. **Sclerids** are the other type of sclerenchyma cell. They are shorter than fibers and have an irregular, hard, secondary wall. The shells of many nuts contain sclerids.

✦ **FIGURE 16.1**

**The major differences between monocots and dicots.**

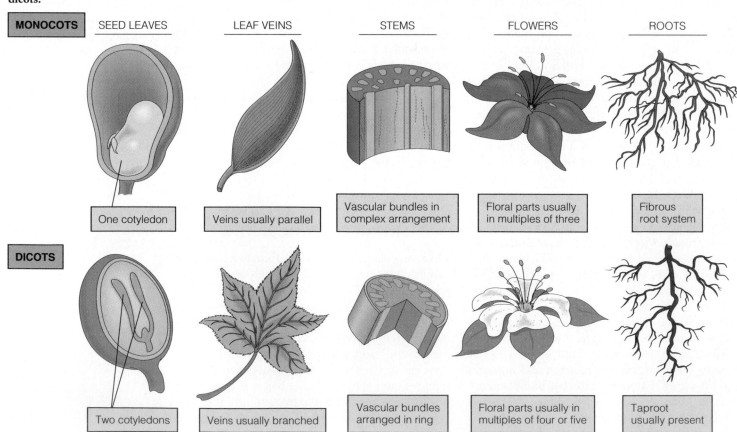

| MONOCOTS | SEED LEAVES | LEAF VEINS | STEMS | FLOWERS | ROOTS |
|---|---|---|---|---|---|
| | One cotyledon | Veins usually parallel | Vascular bundles in complex arrangement | Floral parts usually in multiples of three | Fibrous root system |
| DICOTS | Two cotyledons | Veins usually branched | Vascular bundles arranged in ring | Floral parts usually in multiples of four or five | Taproot usually present |

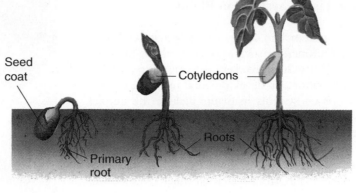

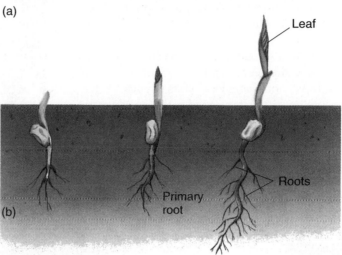

Seed coat

Cotyledons

Roots

Primary root

(a)

**✦ FIGURE 16.2**

(a) During germination in beans (a dicot), the entire seed is brought above ground. (b) In corn (a monocot), the seed remains underground during germination.

Leaf

Roots

Primary root

(b)

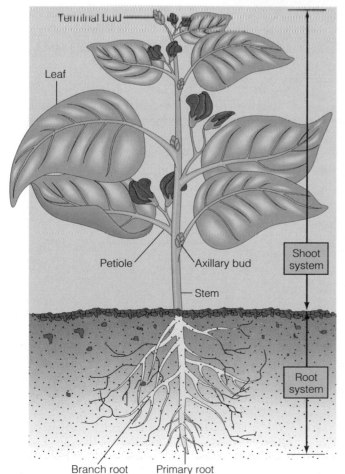

Terminal bud

Leaf

Petiole

Axillary bud

Stem

Shoot system

Root system

Branch root    Primary root

**✦ FIGURE 16.3**

General body plan of a flowering plant. The two main systems are the root and the shoot, with the shoot subdivided into the stem and the leaves. Flowers are modified leaves.

*Sidebar*

**POPULAR POTATOES**

The potato has been culti-vated for over 6,000 years and has sustained millions of people from the Incas to the Irish. However, it is only within the last decade that the potato has achieved worldwide appreciation and become an important dietary component in over 130 countries.

The potato is the fourth most important food crop worldwide after wheat, rice, and corn. If scientific ma-nipulations succeed in con-quering diseases and ex-panding this tuber's natural ability to grow almost any-where, the potato is ex-pected to supply an even larger proportion of the world's calories and nutri-tional requirements.

This ancient crop has benefitted from one of the most modern techniques of genetic engineering, the in-troduction of genes that provide resistance to insects and diseases while increas-ing the nutritional value of the plant. So far, scientists have manufactured 80 genes, many of which greatly in-crease the potato plant's resistance to invading organ-isms. These genetically engi-neered plants are being tested in the greenhouse be-fore the cream of the crop are moved into field tests.

Europeans snubbed this lowly food source when Christopher Columbus brought it from the New World. The potato now is gaining international notori-ety thanks to the spread of American fast food. Nearly every fast-food meal has a side order of "American fries," as they are known in the Far East. Whether "American fries" are better than "French fries" is a question that remains to be answered.

## Four Tissue Systems in Plants

Tissue systems are groups of cells that perform a common function, such as photosynthesis or transport. Plant tissues are classified according to structure, function, or developmental origin. The four main tissue systems in plants include the dermal system, which forms the outer covering of organs; the vascular system, which conducts nutrients and water; the fundamental or ground tissue, which provides support and storage of food; and the meristematic tissue, from which all other tissues arise. These tissues are continuous throughout the body of the plant (Figure 16.3) and are found in all vascular plants.

The **dermal system** forms a protective epidermis that covers the leaves and the young roots and stems. It prevents water loss, offers physical protection, and blocks the invasion of pathogens. In many older plants, the epidermis gives rise to **cork,** which replaces the epidermis.

The **vascular system** transports water and nutrients and plays a role in supporting the plant. The vascular system is composed of xylem and phloem. **Xylem** conducts water and dissolved minerals and nutrients from the roots to the rest of the plant body. **Phloem** consists of cells called sieve tubes that transport sugars made in photosynthesis to other parts of the plant for storage, and carries sugars from storage tissues to actively growing regions.

In a young plant, the **ground system** makes up most of the plant. It occupies the space between the epidermis and the vascular system. The ground system is made up of parenchyma cells, but also contains some collenchyma and sclerenchyma cells. The ground system has several functions, including photosynthesis, water and food storage, and physical support of the plant body.

The **meristematic tissue** is located at the tips of stems and roots, although in some plants, it is located along the entire length of the stem and root. Meristematic tissue is an embryonic tissue, and can divide to give rise to new cells. The meristems at the tip of roots and stems give rise to the parenchyma and sclerenchyma cells.

✦ **FIGURE 16.4**

In monocots, such as corn, the base of the leaf wraps around the stem to form a sheath.

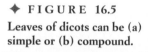

✦ **FIGURE 16.5**

Leaves of dicots can be (a) simple or (b) compound.

(a)

(b)

## PLANTS HAVE THREE MAIN ORGAN SYSTEMS

The major organs of plants are the leaves, shoots, and roots. The **leaves** are the main sites of photosynthesis in the plant, and have evolved many adaptations to the environment. The **shoot** or stem of the plant provides support for the leaves and flowers. The **roots** anchor the plant in the soil. They also absorb water and nutrients from the soil, and act as storage sites for water and food. Each organ system is described in the following sections.

### Structure, Variability, and Function in Leaves

Leaves show great variety in shape. The leaves of monocots are usually flat, with an expanded base forming a **sheath** that encircles the stem, as in grass or corn plants (✦ Figure 16.4). Dicots usually have an expanded **blade** and a **stalk** (the petiole) that attaches to the stem. Leaves may be simple or compound, with a whole range of intermediates (✦ Figure 16.5). A simple leaf consists of the blade (the photosynthetic region), the stalk and the base, which often contains outgrowths, or scale-like structures.

Variations in leaf structure are related to the habitat of the plant; the availability of water and the need to prevent water loss are important factors affecting leaf shape. Based on water needs, leaves are classified as **mesophytic** (abundant soil and water and a relatively humid atmosphere), **xerophytic** (low levels of soil and water and a dry atmosphere) or **hydrophytic** (abundant moisture, or grow in water) (✦ Figure 16.6).

Leaves can be greatly modified, and perform unusual functions. Leaf size can be greatly reduced, as in cacti. In these plants, the leaves are hard, dry, nonphotosynthetic spines that serve to prevent water loss and help protect the plant from predators. The usual leaf function of photosynthesis is carried out by the stem. In certain plants, such as the Venus flytrap, which grows in the nitrogen-poor soils of swamps of North Carolina, the leaves are modified for trapping insects. The insects are digested by enzymes secreted by the leaves, and the nutrients, including nitrogen, are absorbed by the plant (✦ Figure 16.7).

(b)

(a)

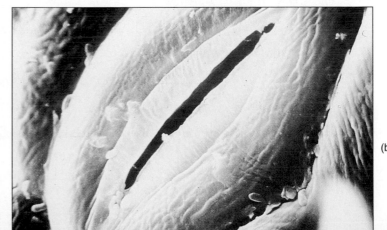
(c)

✦ **FIGURE 16.6**

Based on water requirements, plants and their leaves show a series of adaptations. (a) Leaf of a mesophyte (maple tree). (b) Leaf of a xerophyte (jade plant). (c) Leaf of a hydrophyte (water lily).

## Internal Structure of Leaves

Leaf tissue contains an epidermis, a vascular system, and ground tissue, called mesophyll. The epidermis covers both sides of the leaf, and is made up of compact cells, arranged in a single layer. The outside of epidermal cells is covered with a **cuticle** that prevents water loss. It contains wax and cutin, a fatty acid polymer similar to varnish. Openings in the epidermis called pores or **stomata** (sing., stoma) are surrounded by guard cells (✦ Figure 16.8). The guard cells control the movement of gases, including water vapor, into and out of the leaves.

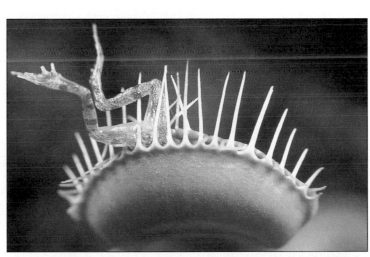

✦ **FIGURE 16.7**

Some leaves are adapted to catching and digesting insects to provide a source of nitrogen to the plant. Here a leaf from a Venus flytrap has captured a small frog.

✦ **FIGURE 16.8**

Stomata are openings in leaves that control transpiration and gas exchange. (a) Open stomata. (b) Closed stomata.

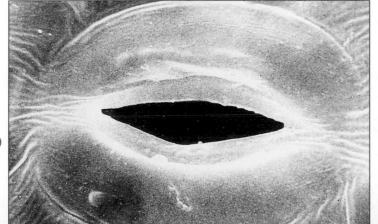
(a)

(b)

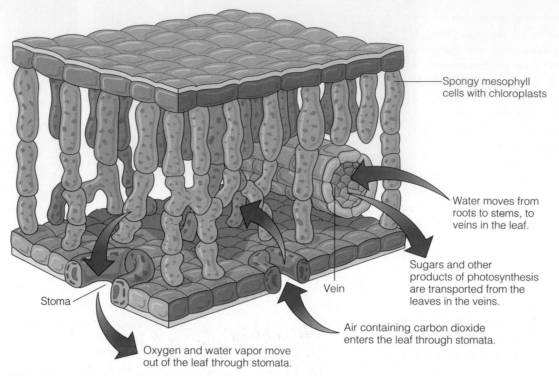

Spongy mesophyll cells with chloroplasts

Water moves from roots to stems, to veins in the leaf.

Sugars and other products of photosynthesis are transported from the leaves in the veins.

Stoma

Vein

Air containing carbon dioxide enters the leaf through stomata.

Oxygen and water vapor move out of the leaf through stomata.

✦ **FIGURE 16.9**

**Diagram showing internal structure of a leaf. Most of the tissue inside the leaf is mesophyll. Note that the stomata are on the underside of the leaf.**

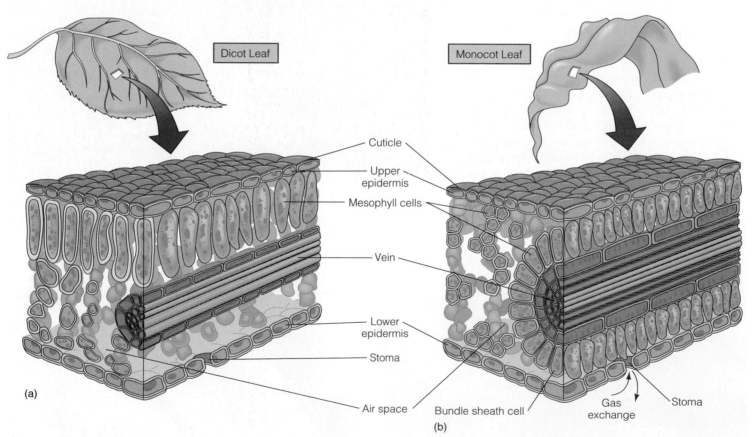

Dicot Leaf

Monocot Leaf

Cuticle

Upper epidermis

Mesophyll cells

Vein

Lower epidermis

Stoma

(a)

Air space

Bundle sheath cell

Gas exchange

Stoma

(b)

✦ **FIGURE 16.10**

**Cross section of the vascular system in (a) a dicot leaf and (b) a monocot leaf. Note that the vascular bundle in the monocot leaf is surrounded by a layer of parenchyma that is missing in the dicot leaf.**

The mesophyll of the leaf is specialized for photosynthesis. Some leaves (especially mesophytic) have two types of mesophyll: (1) a closely packed mesophyll with rectangular cells called palisade mesophyll, and (2) the loosely packed and more irregularly arranged spongy mesophyll (✦ Figure 16.9). As a rule, leaves with two types of mesophyll are oriented more or less horizontally on the plant, whereas leaves with a uniform mesophyll are oriented more or less vertically.

The vascular system of a leaf consists of bundles of vascular tissue in the form of veins, that form a network throughout the leaf blade (✦ Figure 16.10). As mentioned earlier, veins in monocot leaves are arranged mostly in parallel, whereas in dicots, the veins form a net-like, branching arrangement. The vascular system of the leaf is continuous with the vascular system of the stem. It transports water and nutrients to the leaf cells and carries away sugars and other materials for storage.

## Structure and Function of Plant Stems

The main functions of the stem are leaf display, support, and conduction. Leaves are attached to stems at regions called **nodes;** the stem regions between nodes are called **internodes** (✦ Figure 16.11). At the end of all stems are terminal buds that contain meristematic tissue. Mitosis in the meristem forms the leaves and the tissue systems of the stem. Axillary buds at nodes form branches and new leaves; some axillary buds form flowers.

Internally, stems consist of several tissues. The epidermis is present in all stems as a protective outer layer. The vascular tissues form bundles that run through the surrounding ground tissue.

There are significant differences in the arrangement of vascular bundles in monocots and dicots. In most dicots, the vascular bundles are arranged

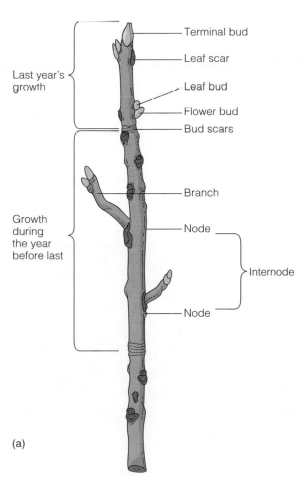

(a)

(b)

(c)

## ✦ FIGURE 16.11

(a) The external features of a woody stem. Buds occur (b) at the tips and (c) along the stem. The terminal bud scars show the position of previous terminal buds. The distance between terminal bud scars is a measure of growth in previous years. A node is the stem region where one or more leaves attach.

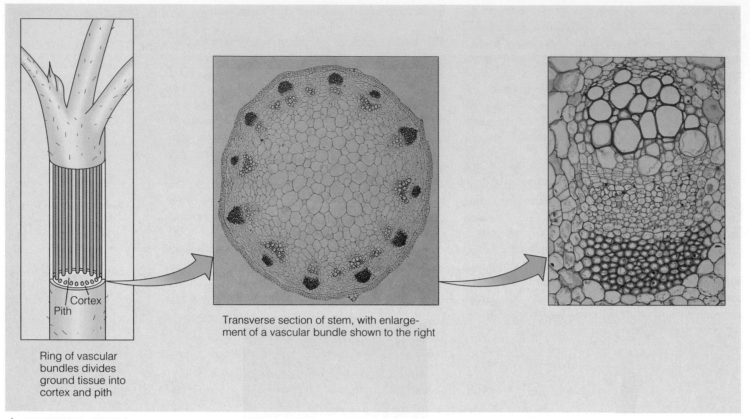

Ring of vascular
bundles divides
ground tissue into
cortex and pith

Cortex
Pith

Transverse section of stem, with enlargement of a vascular bundle shown to the right

✦ FIGURE 16.12

**Organization of vascular bundles in a dicot stem.**

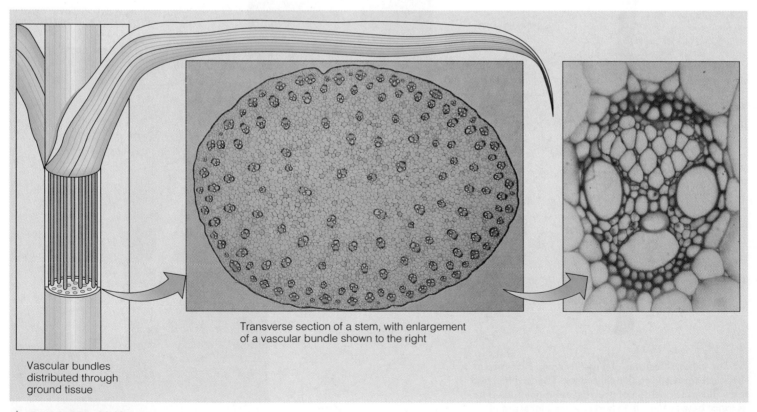

Vascular bundles
distributed through
ground tissue

Transverse section of a stem, with enlargement
of a vascular bundle shown to the right

✦ FIGURE 16.13

**Organization of vascular bundles in a monocot stem.**

(a)

✦ FIGURE 16.14

**Modified stems. (a) Rhizomes are underground stems, growing horizontally. (b) Potatoes are tubers, which are the terminal portion of rhizomes. (c) Bulbs, such as the onion, are modified stems that store food made in the leaves.**

(b)

(c)

in an outer ring (✦ Figure 16.12). This divides the ground tissue into two zones: an outer cortex and an inner pith (Figure 16.12). The cortex is composed of parenchyma, which contains some chloroplasts; it functions to carry out photosynthesis and also stores starch. The inner pith is composed of parenchyma and often plays a role in food storage. In most monocots, the vascular bundles are scattered throughout the ground tissue (✦ Figure 16.13).

In both monocots and dicots, each vascular bundle is composed of xylem and phloem. Within a bundle, the phloem is positioned nearest the stem surface, and the xylem closest to the center of the stem.

In some plants, part or even all of the stem is located underground (✦ Figure 16.14). These modified stems function in food storage. The potato and the onion are examples of such modified stems.

## Roots Provide Support and Storage

Roots typically are the underground organ of a plant. They function in absorption, storage, and anchorage. The first root formed by a plant is called the **primary root** (✦ Figure 16.15a). In dicots, this becomes a **tap root,** which grows downward. As it grows, small lateral roots branch off the tap root. The root of a dandelion is an example of a tap root system. In monocots, the primary root dies, and

(a)

(b)

✦ FIGURE 16.15

**Root systems in flowering plants. (a) Tap root in a dandelion. (b) Fibrous roots in grass.**

the root system develops from the stem in the form of **adventitious roots.** Branches from the adventitious root form a **fibrous root** system, in which one root is about the same size and length as another (Figure 16.15b). Tap root systems can penetrate the soil to great depths, while fibrous root systems are shallow.

**✦ FIGURE 16.16**
**Prop roots on this banyan tree hold up the horizontal branches.**

**✦ FIGURE 16.17**
**Prop roots on this corn plant help support the stem.**

**✦ FIGURE 16.18**
**The knees of a cypress plant are roots that grow out of the water to help support the plant and obtain oxygen.**

Modified roots include aerial roots, which support the plant (✦ Figure 16.16). Banyan trees form large, pillar-like roots that grow downward from horizontal branches to prop up the branches. Corn plants form prop roots at the base of the stem to support the plant (✦ Figure 16.17) If these roots enter the soil, they function in the absorption of water and minerals. The prop roots of cypress trees, known as "knees," are specialized roots that help support the plant in the soft ground of the swamp and also serve to aerate the plant (✦ Figure 16.18).

The tip of the root is covered by a root cap (✦ Figure 16.19). This thimble-like structure consists of living cells that protect the tip as it grows through the soil. Most root tips secrete substances to lubricate the tip, helping it move through the soil.

The root epidermis is highly modified to permit absorption of water and minerals. Some of these modifications include (1) lack of cuticle; (2) the presence of root hairs, which increase the surface area for absorption; (3) fibrils at the cell wall, which increase the absorbing surface area; and (4) mutualistic associations with microorganisms. Just inside the epidermis is a layer known as the cortex, which is made up of parenchymal cells. These may contain plastids specialized for storage of starch. At the inner area of the cortex is the endodermis, a single layer of cells that controls the movement of materials between the cortex and the interior of the root.

In the area between the tip and the root hair zone is the meristematic region. Elongation of roots results from division and enlargement of cells in the meristem (Figure 16.19). The vascular tissue in the root forms a central column known as a **vascular cylinder** (Figure 16.19), which is surrounded by parenchymal ground tissue. A ring called the endoderm, made up of a mixture of dead and living cells, surrounds the cylinder (Figure 16.19).

The bulk of the root is a thick layer of ground tissue. Parenchymal cells in this layer may contain plastids for food storage.

## MERISTEMS AND PLANT GROWTH

Plant growth is the result of cell divisions in meristematic tissue. Mitosis in primary meristems makes plants grow taller; division in secondary or lateral meristems makes plants grow thicker.

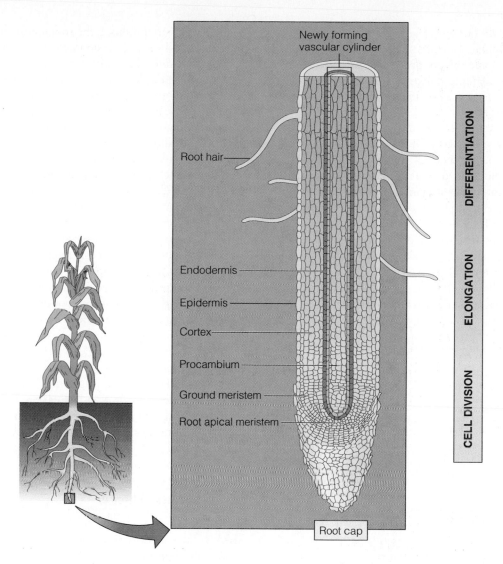

Newly forming
vascular cylinder

Root hair

DIFFERENTIATION

Endodermis

Epidermis

ELONGATION

Cortex

Procambium

Ground meristem

Root apical meristem

CELL DIVISION

Root cap

✦ **FIGURE 16.19**

**The root tip of a plant contains zones of division, elongation, and differentiation.**

## Primary Growth Elongates the Shoots and Roots

When cell division takes place in primary meristems, the result is primary growth: the elongation of shoots and roots and the production of new roots, shoots, and leaves (✦ Figure 16.20). The **apical meristem** is located at the tip of the shoot. Parts of the plant body that are derived directly from divisions in the apical meristem are called the primary body of the plant. Most monocots are nonwoody, and consist only of primary tissue, whereas dicots (woody plants) have both primary and secondary tissue.

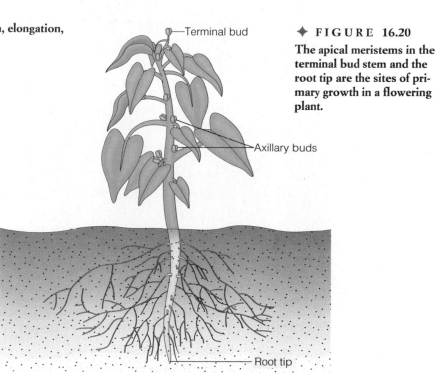

Terminal bud

Axillary buds

Root tip

✦ **FIGURE 16.20**

**The apical meristems in the terminal bud stem and the root tip are the sites of primary growth in a flowering plant.**

The shoot apical meristem is a dome of cells in the terminal buds of stems (✦ Figure 16.21). Cells in this dome divide, producing cells that form the stem and lateral structures that become leaves. Just below the apical meristem is a region of tissue differentiation.

The stem differentiates into nodes, which are regions where leaves attach, and into internodes, the regions between nodes. Later, cells in the internode region elongate, increasing stem length. At the base of the internode region, xylem and phloem differentiate, and connect with the more differentiated regions of the stem below.

Leaves arise from clusters of cells called leaf primordia, which form as lateral outgrowths of the apical dome. The first visible changes in the leaf primordia are the formation of (1) the leaf base, where the leaf attaches to the stem, and (2) the leaf blade, which is the flat part of the leaf.

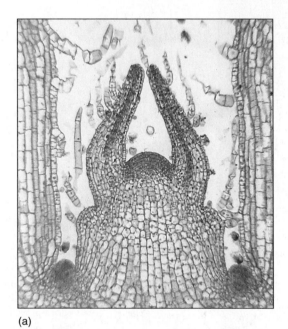

(a)

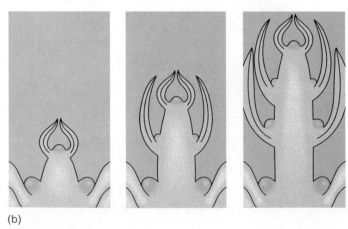

(b)

✦ **FIGURE 16.21**

**(a) The apical meristem at the tip of a stem. (b) Cell divisions in the apical meristem are responsible for elongation of the stem.**

In contrast to the shoot, the root apical meristem is not dome shaped. It is composed of a cluster of small, many-sided cells with large nuclei and dense cytoplasm (✦ Figure 16.22). Division of the cells in the meristem give rise to the cells of the root tip on one side of the meristem and to the procambium on the other (Figure 16.22). This area is called the **region of division.** The procambium differentiates into the vascular tissues. Behind the region of cell division is the **region of elongation.** Cells in this area grow by elongating, which increases the length of the root. As a result, only the tip of the root is pushed through the soil. Behind the region of elongation is the **region of maturation.** Most cells of the primary tissues develop in this region. This is also a region of root hair development. Root hairs are outgrowths of epidermal cells that serve to increase greatly the surface area of the root tip.

The division of the root tip into three regions is a generalization, and all three processes can be found to some extent in each region, largely due to the fact that different tissues develop at different rates (Figure 16.21). As a result, some cells begin to elongate while in the region of division; others differentiate in the region of elongation. Nonetheless, division, elongation, and maturation occur in a stepwise fashion, even if the zones overlap.

## Secondary Growth Makes Woody Plants Thicker

In the life cycle of most monocots and some dicots, there is little secondary growth; these are the nonwoody (herbaceous) plants. Many dicots and gymnosperms are woody plants and show secondary growth. Flowering plants can be divided into three groups based on their growth characteristics:

*Annuals:* Plants that complete their life cycle in one year or one growing season, with little or no secondary growth. Crops such as corn and wheat and flowers such as impatiens and marigolds are annuals.

*Biennials:* plants that complete their life cycle in two growing seasons. Stems, roots, and leaves form the first year, and the plant flowers in the second year. Stems and roots may undergo some secondary growth. Sugar beets and carrots are biennials.

*Perennials:* plants in which the vegetative structures live year after year, and flowering structures are formed only at maturity. Some perennials have secondary growth, and others do not. Most trees and shrubs are perennials.

## Beyond the Basics

# MESSAGES WRITTEN IN THE RINGS

In the early 1930s, Dr. Andrew Ellicot Douglas used tree rings to determine the age of artifacts recovered from an ancient Indian site. The study of tree rings, called *dendrochronology,* has now expanded from archeology to the study of ancient climates, volcanic eruptions, and fires.

Tree rings are the results of secondary growth in the vascular cambium. Xylem forms on the inner surface of the vascular cambium. Most woody plants grow seasonally rather than continuously, so divisions in the vascular cambium exhibit seasonal patterns. The xylem cells formed in the spring are larger, have thinner walls, and are lighter in color than the xylem cells formed during the summer, which are smaller and have denser, darker appearing walls. These alternating layers of xylem form the annual rings or growth rings.

To use tree rings effectively over long time periods, scientists have constructed unbroken series of tree rings. The chronology of a tree ring series begins by counting rings in a freshly cut tree. The date of the last ring (the outermost ring) is the year the tree was cut. By counting back, the age of the tree can be determined. The chronology can be extended by adding in the tree rings of older, overlapping specimens collected in the same area. In the southwestern United States, trees such as the bristlecone pine, which can live up to 4500 years, are used to construct such chronologies. A chronology constructed in Ireland extends from the present to 7272 years ago. This Irish project took 14 years and more than 6000 overlapping specimens to construct. A similar chronology constructed in the United States from bristlecone pines extends over a period of 8500 years.

By comparing rings in chronologies from different parts of the world, local and global events can be distinguished. From a cluster of stunted rings found in European and American samples that date to 1628 B.C., scientists have established the year of the volcanic eruption that destroyed the Greek island of Santorini and the flourishing Minoan civilization. This has aided archeologists in dating other events in this and other ancient civilizations.

Tree rings also provide a year-by-year record of temperature and precipitation. This record substantiates the observations of medieval writers that the period from 1100 to 1375 A.D. was extraordinarily warm in Europe, even allowing widespread farming in the Viking settlements in Greenland. This was followed by a period called the Little Ice Age, with abnormally cold weather lasting from 1450 to 1850.

The rings also indicate that compared to past centuries, the western United States, especially California, has been much wetter than normal in this century. If the climate swings back to a drier pattern, the recent droughts in California will seem wet by comparison. These findings have serious implications for population growth and development in the entire state, and for the agriculture industry in the Central Valley, where much of the nation's winter fruits and vegetables are grown.

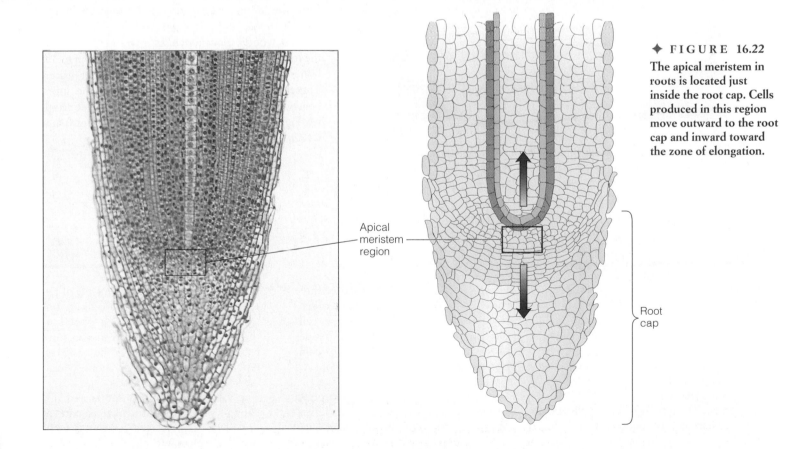

**◆ FIGURE 16.22**

**The apical meristem in roots is located just inside the root cap. Cells produced in this region move outward to the root cap and inward toward the zone of elongation.**

Apical meristem region

Root cap

✦ FIGURE 16.23

**Annual rings are the result of xylem production during a growing season.**

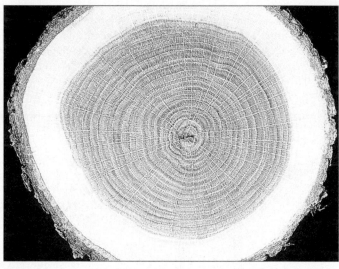

✦ FIGURE 16.24

**Cross section of a tree trunk showing the inner, darker heartwood, which is nonconducting xylem. The outer, lighter sapwood is made up of layers of conducting xylem.**

The stems of woody plants consists of an outer layer of **bark** and the inner **wood.** The bark is made up of an outer layer of dead cells called cork, and an inner layer of phloem. The wood consists of the xylem. Secondary growth results from division in two lateral meristematic tissues: the cork cambium and the vascular cambium. The **cork cambium** is a thin layer of cells that separates the cork and the phloem, and the **vascular cambium** separates the xylem and the phloem.

The cork cambium gives rise to new cork, making the bark thicker. The corks used in wine bottles are cut from the bark of a species of oak that grows in Portugal and nearby Mediterranean countries. These trees have a very active cork cambium, which produces a thick cork layer that can be stripped off without harming the tree.

Secondary growth occurs on both sides of the vascular cambium. Xylem forms on the inner surface of the vascular cambium, and phloem forms on the outer surface. Most woody plants grow seasonally rather than continuously, so divisions in the vascular cambium exhibit seasonal patterns.

The xylem that is formed each growing season remains, and the layers of xylem grow thicker each year. Xylem cells formed at the beginning of the year are larger and have thinner walls. The xylem formed later in the growing season are smaller, and have denser walls. These alternating layers of xylem form the annual rings or growth rings (✦ Figure 16.23).

As a woody plant, such as a tree, grows older, the inner rings of xylem become clogged with metabolic byproducts and no longer transport water. These clogged cells darken, and form the inner **heartwood** visible in a cross section of a tree trunk (✦ Figure 16.24). The outer, lighter colored **sapwood** is xylem that is still functional.

## SUMMARY

1. Ancient plants made the transition from water to land and developed adaptations to solve the problems of terrestrial life: desiccation, support, and dependence on water for reproduction.

2. The dominant vascular plants are seed plants. Their seeds are carried in cones (the gymnosperms) or in fruits (the angiosperms). The angiosperms are divided into two classes, the monocots and the dicots.

3. Flowering plants are composed of three basic cell types: parenchyma, collenchyma, and sclerenchyma. These cells can be organized into simple tissues (made up of only one cell type) or complex tissues (made up of two or more cell types).

4. The angiosperms have three main organ systems: the root, the stem, and the leaves. They are composed mainly of ground, vascular, and dermal tissues.

5. Ground tissue is composed mainly of parenchyma cells. Parenchyma cells are unspecialized, but can carry on photosynthesis and respiration and can also store food. They are regarded as the basic cell type, from which all other, more specialized cell types have evolved.

6. The vascular system transports and distributes water and nutrients to all parts of the plant. The vascular system is composed of xylem and phloem. Xylem con-

ducts water and dissolved minerals and nutrients from the roots to the rest of the plant body. The phloem transports sugars made in photosynthesis to other parts of the plant for storage, and carries sugars from storage tissues to actively growing regions.

7. The dermal system forms a protective cellular layer, the epidermis, that covers leaves, young roots, and stems. It prevents water loss, offers physical protection, and blocks the invasion of pathogens. In older plants, the epidermis produces cork, which replaces the epidermis.

8. Plant growth results from mitotic cell divisions in meristematic tissue. Division in primary meristems makes plants grow taller, and mitosis in secondary or lateral meristems makes plants grow thicker.

# KEY TERMS

adventitious roots
apical meristem
bark
collenchyma
cork
cork cambium
cotyledon
cuticle
dermal system
dicot
fiber
fibrous root
ground system
heartwood

hydrophytic leaves
internodes
leaves
meristematic tissue
mesophytic leaves
monocot
nodes
parenchyma
phloem
primary root
region of division
region of elongation
region of maturation
roots

sapwood
sclerenchyma
sclerids
sheath blade
shoot
stalk
stomata
tap root
vascular cambium
vascular cylinder
vascular system
wood
xerophytic leaves
xylem

# QUESTIONS AND PROBLEMS

## TRUE/FALSE

1. Flowering plants can be divided into two distinct classes, the monocots and the angiosperms.
2. Most woody plants grow seasonally rather than continuously.
3. Phloem is a vascular tissue that conducts food.
4. Herbaceous plants are mostly composed of secondary tissues.
5. In general, leaves with two types of mesophyll are orientated horizontally on the plant.
6. The stomata is specialized for photosynthesis.
7. A fibrous root system is more efficient at food storage than a taproot system.
8. Mitosis in primary meristems make plants grow taller, division in secondary meristems make plants grow thicker.
9. Root hairs grow in the region of division.

## MATCHING

Match the description with the appropriate term.

____ cork cambium
____ annuals
____ parenchyma
____ endodermis
____ roots
____ sclerids
____ ground tissue
____ xylem
____ collenchyma
____ xerophytic

1. Cell type found in tissues of fruits and seeds and the vascular system.
2. Cell type that provides support for plant.
3. Type of sclerenchyma cell that has a hard, irregular secondary wall.
4. Transports water and dissolved minerals from the roots to the rest of the plant body.
5. Completes life cycle in one growing season.
6. Major functions include water and nutrient absorption, food storage, and anchoring the plant.
7. Contains parenchyma, collenchyma and sclerenchyma.
8. Leaf structure that requires little soil water and a dry atmosphere.
9. A single layer of cells that controls the movement of materials between the cortex and the interior of the root.
10. Separates the cork and the phloem.

## COMPARE AND CONTRAST

For each characteristic listed below, tell whether it is seen in dicotyledons, monocotyledons, or both:

1. Vascular bundles arranged in an outer ring
2. The xylem lies in a ring surrounding the parenchyma
3. Stomata on both leaf surfaces
4. Roots with root caps
5. Vascular cambium
6. Leaves are usually flat with parallel venation

## SCIENCE AND SOCIETY

1. In 1980, Mount St. Helens in southwestern Washington state exploded and spread its lava over the landscape. Within seconds, millions of old growth trees were destroyed. Layers of pumice and ash turned the previously lush forest into a barren strip of land. This natural disaster provided a scary glimpse of what the world would be like without plants. What do

you imagine would be the effects if plant life were dramatically reduced? Who would suffer more, humans or animals? Would the land feel the effects of life on earth without plants? If so, how and why? Is it reasonable to protect "endangered plants" the same way endangered animals are protected? Do you feel it is too late to protect plant life?

2. The world's plants either directly or indirectly nourish other organisms and make the soil habitable. The leaves of plants are by far the most extensively used structures by man. Besides providing the atmosphere with oxygen and decorating the landscape, what other useful purposes do plants serve? Would the medical field be affected if leaves were suddenly unavailable? Can you think of examples of how leaves are used for medicinal purposes? How would the economy be affected by a significant loss in plant life?

# PLANT NUTRITION, TRANSPORT, AND GAS EXCHANGE

**OPENING IMAGE**

*An Amazon hunter uses darts tipped*

*with curare, a plant-derived poison.*

Late in the sixteenth century, Sir Walter Raleigh, returning to England from a voyage to South America, brought back a small wooden vial. It contained a black, thick substance which was applied to the tips of hunting arrows by the natives of the Orinoco and Amazon regions. This substance, known as woorali, urari, and other names, was a deadly poison extracted from plants. Charles Waterton, who traveled in South America at the beginning of the nineteenth century, described its use by a hunter using blowguns and poison-tipped arrows to hunt birds: ". . . taking a poisoned arrow from his quiver, he puts it in the blow-pipe and collects his breath for the fatal puff. . . . Silent and swift the arrow flies, and seldom fails to pierce the object at which it is sent. Sometimes the wounded bird remains in the same tree where it was shot, and in three minutes falls down at the Indian's feet."*

Hunters prepare the drug from a jungle vine. Boiling water is poured repeatedly over strips of bark from this vine, and the liquid is concentrated by boiling. As it cools, the liquid turns into a dark, thick syrup or paste. The paste is used to coat the last few inches of blowgun darts, and when dried, the darts are ready for use.*

Over 150 years were to pass after Waterton's account before the medical properties of this substance, now called curare, were widely recognized. Today, curare is used in conjunction with an anesthetic in abdominal surgery. Curare acts at nerve endings, and blocks signals for muscle contraction. As a result, it serves as a powerful muscle relaxant, and only a small dose is required. The same degree of relaxation can be obtained by spinal* *anesthesia, or by large doses of anesthetics, but both of these can have complications or potentially serious side effects.*

*Today, scientists search for new drugs and medicines by studying the use of plants in traditional societies. Many drugs, including digitalis, aspirin, and anticancer agents such as vincristine were first identified by their use as folk remedies. Over 100 drugs are derived from flowering plants, including taxol, discovered in 1992 and used in the treatment of ovarian cancer and metastatic breast cancer. Of the more than 250,000 species of angiosperms, less than one half of one percent have been evaluated as sources of medical compounds.*

*In this chapter, we consider how plants obtain nutrition, transport and store nutrients, and how they survive predators. Many of the compounds synthesized by plants for protection are also used by humans as medicines. Other plant products in the form of coffee, tea and tobacco are used in everyday life, often without a second thought as to their origin.*

## PLANTS HAVE DIFFERENT NUTRITIONAL NEEDS THAN ANIMALS

Plants need more than just sunlight and water to grow. Animals derive nutrition from their diet. Plants must obtain their nutritional requirements from the atmosphere and the soil. Unlike animals, plants do not require sugars or other compounds as an energy source, nor do they need certain amino acids or vitamins. Using sunlight as an energy source, most green plants can transform water and carbon dioxide into all the organic compounds they need to survive. Plant nutrition involves uptake of raw materials, the distribution of these materials within the plant body, and the use of these materials in metabolism.

## A Balanced Diet for Plants

Plants need a number of elements in order to survive (Table 17.1). Carbon, hydrogen, and oxygen are considered to be **essential elements,** acquired by plants from the atmosphere during photosynthesis. The remaining elements are obtained from the soil, and occasionally by parasitism or predation. Nitrogen, phosphorus, and potassium are known as **primary macronutrients** because they are needed in relatively large quantities. Calcium, magnesium, and sulfur are **secondary macronutrients** because they are needed in smaller quantities. Iron, manganese, molybdenum, copper, boron, zinc, and chlorine are known as **micronutrients** because they are needed in very small quantities, and are toxic in large quantities.

Natural sources of these mineral elements are the atmosphere, irrigation water, rainfall, and the breakdown of the soil itself. Humans supply these elements to crop plants by applying natural or synthetic fertilizers. A complete fertilizer contains the three primary macronutrients and some of the secondary or micronutrients. When you go to the store to buy fertilizer for your garden or your houseplants, look at the label. On it, some numbers will be prominently displayed (such as 5–10–5). These numbers list the percentage by weight of each of the primary macronutrients: 5% nitrogen, 10% phosphorus, and 5% potassium.

## SOILS AND PLANT NUTRITION

Soil is the growth medium for plants and the source of most nutrients. Soil is really the weathered superficial layer of the Earth's crust. Its components are decomposed and partly decomposed rock and associated organic material (✦ Figure 17.1). Fertile soil contains all the chemical elements needed for plant growth in readily available form.

| TABLE 17.1 | MINERAL NUTRIENTS IN PLANTS |
|---|---|
| **MACRONUTRIENTS** | **FUNCTION/COMPONENT OF** |
| Potassium | Amino acids, enzymes, activation of enzymes. Opening, closing of stomata. |
| Calcium | Cell walls, permeability, enzyme cofactor. As component of calmodulin, regulates enzyme activity. |
| Phosphorus | ATP, nucleic acids, phospholipids, coenzymes. |
| Magnesium | Chlorophyll, enzyme activator. |
| Sulfur | Coenzyme A, some amino acids. |
| **MICRONUTRIENTS** | **FUNCTION/COMPONENT OF** |
| Iron | Chlorophyll synthesis, cytochromes for electron transfer. |
| Chlorine | Osmotic and ionic balance. |
| Copper | Activates some enzymes. |
| Manganese | Activates some enzymes. |
| Zinc | Activates some enzymes. |
| Molybdenum | Nitrogen fixation. |
| Boron | Unknown. |

## Plants Mine the Soil to Obtain Minerals

Soil is made up of solid matter and the space around soil particles. Water is present in soil mostly as a film on the surface of soil particles. The roots of plants act as "miners" of the Earth's crust; they supply inorganic nutrients essential for growth and reproduction (✦ Figure 17.2). Mineral nutrients enter the biosphere at the plant's root system, and then move to various parts of the plant for use in biological functions.

Plants utilize mineral elements essential for nutrition in four basic ways: (1) as structural components, such as the carbon found in cellulose and the nitrogen found in protein; (2) for organic molecules important in metabolism, such as the magnesium found in chloroplasts or the phosphorus found in ATP; (3) for enzyme activators, like potassium, which is thought to affect more than 50 enzymes; and (4) to help maintain osmotic balance. The movement of water into and within the plant is dependent on solute concentrations within cells. Uptake or release of ions can result in movement of water into or out of cells. The availability of essential elements and the energy from sunlight allows plants to synthesize all the compounds they need for normal growth and reproduction.

## Symbiosis and Mineral Uptake

Plants and animals depend on nitrogen for the synthesis of proteins, nucleic acids, and other biological compounds, including certain hormones. Many species of bacteria have the ability to convert nitrogen from the atmosphere into either nitrate or ammonia. Some plants, such as the legumes (peas, beans, alfalfa, and others), have formed symbiotic relationships with nitrogen-fixing bacteria. The roots of these plants have swellings, called root nodules, in which symbiotic bacteria convert atmospheric nitrogen to ammonia (✦ Figure 17.3). Other bacteria in the soil, through some intermediate reactions, convert the ammonia into nitrate, which is the form absorbed by plants. This adaptation allows these species of plants to grow in nitrogen-poor soil. All the nitrogen in living organisms, including the nitrogen atoms in your body, were at one time removed from the atmosphere by bacteria and absorbed by plants.

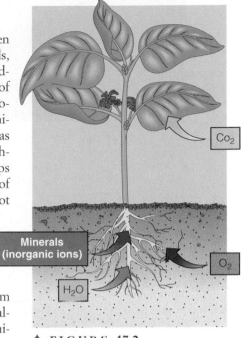

✦ **FIGURE 17.2**

Nutrients enter the plant in several ways. Minerals, water, and oxygen enter through the roots; carbon dioxide enters through the leaves. All minerals enter the biosphere through the roots of plants.

✦ **FIGURE 17.1**

A cross section of soil showing three layers or horizons. The upper layer is the topsoil, containing a mixture of inorganic and organic material. The middle layer consists mainly of inorganic material including iron oxide and clay, and a small amount of organic material that has washed down from the A layer. The lower layer consists mainly of broken-down weathered rock and minerals.

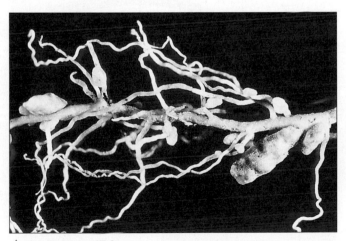

✦ **FIGURE 17.3**

The roots of legumes (peas, peanuts, beans, and other plants with seed pods) have nodules that contain nitrogen-fixing bacteria.

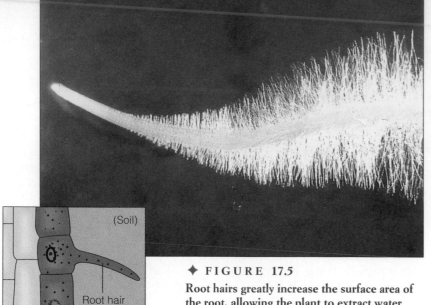

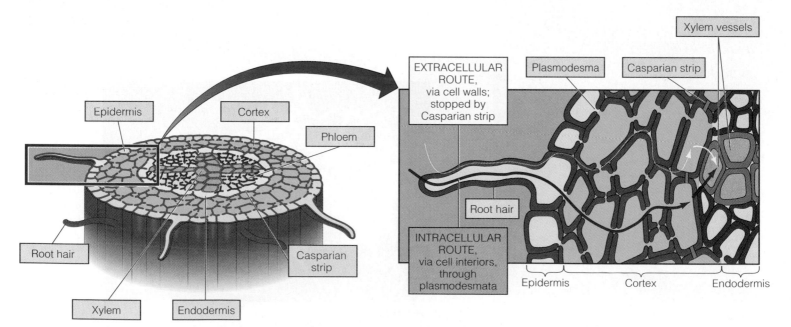

**✦ FIGURE 17.5**

Root hairs greatly increase the surface area of the root, allowing the plant to extract water and minerals from the soil.

**✦ FIGURE 17.4**

Root hairs are cytoplasmic extensions of epidermal cells.

Other plants use root hairs to effectively increase the surface area of the root. These hairs are cytoplasmic extensions of epidermal cells (✦ Figure 17.4).

## HOW PLANTS TAKE UP WATER AND NUTRIENTS

Like animals, plants transport water and nutrients within the body. Animals use a circulatory system equipped with a pump (the heart), but the vascular system of plants operates without a pumping system. In plants, transport is primarily dependent on concentration gradients, some of which require the expenditure of energy. Ions that can move freely across plasma membranes in response to a concentration gradient are transported without energy expenditure. Ions or molecules that move against a concentration gradient require energy expenditure, in the form of ATP.

### Water and Minerals Enter Roots by Two Pathways

The root hair zone on plant roots contains thousands of root hairs (✦ Figure 17.5). The root hairs provide a large and efficient surface for absorbing materials. Water and dissolved minerals enter the plant in two ways: an *intracellular route* and an *extracellular route* (✦ Figure 17.6). In the intracellular route, water and selected solutes cross the plasma membrane of a root hair cell, and pass from the epidermal cell, through other cells, to the xylem

**✦ FIGURE 17.6**

Water has two paths of entry into the plant. In the intracellular route, water moves through a root hair and the inside of other cells into the xylem. Water that enters by the extracellular route moves between cells until it reaches the Casparian strip. To enter the xylem, water must pass through a cell in the endoderm layer.

# GUEST ESSAY: RESEARCH AND APPLICATIONS

## A Journey through Science

BRUCE AMES

As a youngster, I was always interested in biology and chemistry, and read the books left lying around the house by my father, who was chairman of a high school chemistry department, and later supervisor of science for all the New York City public schools. During the summers, my sisters and I explored the natural world and collected animals at our family's summer cabin on a lake in the Adirondack Mountains. Throughout my childhood, I read voraciously, and would come back from the library with a whole stack of books.

I attended the Bronx High School of Science, where I conducted my first scientific experiments, studying the effect of plant hormones on the growth of tomato root tips. Motivated by my experiences with research, I enrolled at Cornell University to study chemistry and biology. After graduating, I headed west to study at the California Institute of Technology. In the early 1950s, there was an exceptional group of faculty and students at Cal Tech, many of whom were learning about genes by studying biochemical genetics. I joined the laboratory of Herschel K. Mitchell, and began using mutant strains to work out the steps used by the bread mold Neurospora to make the amino acid histidine.

After completing my studies at Cal Tech, I moved to the National Institutes of Health at Bethesda, Maryland, to do research on gene regulation, using mutant strains of the bacterium Salmonella. By 1964, I was married and had two children, a daughter, Sofia, and a son, Matteo. Sometime in that same year, I happened to read the list of ingredients on a box of potato chips, and began to think about all the new synthetic chemicals being used, and wondered if they might cause genetic damage to human cells.

To test the ability of chemicals to cause mutation, I devised a simple test, using strains of Salmonella. Chemicals that cause mutations in bacteria may cause mutation in human genes. Using this test, over 80 percent of cancer-causing substances were shown to cause mutations. The Salmonella test, widely known as the Ames test, is now used by more than 3,000 laboratories as a first step in identifying chemicals that might cause cancer.

In 1968, I joined the faculty at the University of California at Berkeley, where my students and I developed genetically engineered bacterial strains that can be used to identify what types of DNA changes are caused by mutagens. My colleague Lois Gold and I have assembled a database on the results of animal cancer tests. I have also studied the mechanisms of aging and cancer.

The accumulated evidence indicates that there is no epidemic of cancer caused by synthetic chemicals. In fact, pollution appears to account for less than 1 percent of human cancer. Several factors, including tobacco and diet, have been identified as the major contributors to cancer in the U.S. The use of tobacco contributes to about one third of all cases of cancer, and the quarter of the population with the lowest dietary intake of fruits and vegetables compared to the highest quarter, have twice the cancer rate for most types of cancer. Hormonal factors contribute to most breast cancer cases. Decreases in physical activity and recreational exposure to the sun have also contributed importantly to increases in some cancers. These results strongly suggest that a large portion of cancer deaths can be avoided by using knowledge at hand to modify lifestyles.

*Bruce Ames is a professor of biochemistry and molecular biology at the University of California, Berkeley. He has been the international leader in the field of mutagenesis and genetic toxicology for over 20 years. His work has had a major impact on, and changed the direction of, basic and applied research on mutation, cancer, and aging. He earned his B.A. from Cornell University and his Ph.D. from California Institute of Technology.*

---

vessels in the root's vascular cylinder. In the intracellular route, materials cross only the plasma membrane of the root hair epidermal cell. All other cells are interconnected by channels called plasmodesmata that are present in the walls of adjacent cells (Figure 17.6).

In the extracellular route, water and its solutes enter the cell wall of a root epidermal cell and move between cells by passing through the cell wall or intracellular spaces. Around the outside of the vascular cylinder is a layer of cells called the **endodermis** (Figure 17.6). Between the cells of the endodermis is a waxy, impermeable layer called the **Casparian strip,** which stops water and solutes from entering the xylem. To enter the xylem, water and solutes must first cross the plasma membrane of an endodermal cell. The plasma membrane of the endo-dermal cell selects ions for transport, controlling the type and amount of solutes that will enter the plant and be distributed throughout the plant body. In the intracellular route, this selective function is carried out by the plasma membrane of the epidermal cell.

## Xylem, Water Transport, and Transpiration

**Xylem** is the water and mineral transporting system in plants. It is a complex tissue, composed of several cell types, each of which plays a role in movement of water and solutes. For this discussion, two cell types are important: **tracheids** and **vessel elements.** Tracheids are long, tapered cells with rigid, wood walls. The walls of tracheids are studded with

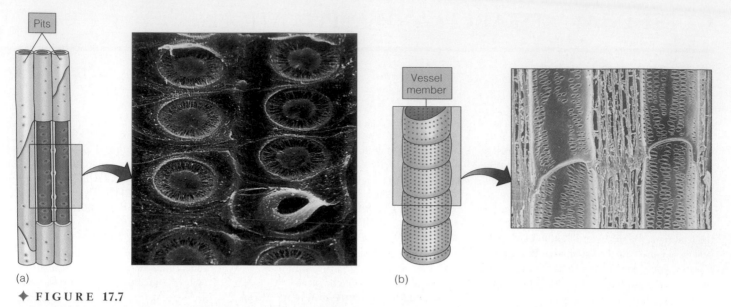

(a)                                                                        (b)

✦ **FIGURE 17.7**

**Xylem tracheids and vessels. (a) Tracheids have tapered ends and pits along their side walls. (b) Vessels are shorter and thicker than tracheids, and are arranged end to end, forming water-conducting tubes.**

pits (✦ Figure 17.7a). Vessel elements are wider and shorter than tracheids (Figure 17.7b). Both tracheids and vessel elements are arranged end to end and form a system of tubes that move water and solutes from the roots to all parts of the plant. When mature, both cell types are dead and hollow, and only their walls remain.

Water is not pumped through the xylem, instead it is pulled by the force of **transpiration** (✦ Figure 17.8). More than 90% of the water that enters a plant through the roots passes through the xylem, into the leaves and stems, and evaporates into the atmosphere from the stomata, or pores present in the epidermis. In fact, a mature corn plant can transpire four gallons of water a week. As water evaporates from the leaves and passes out through the open stomata, solute concentration in the mesophyll cells of the leaf increases. Water in the xylem cells near the mesophyll moves from the xylem into the mesophyll in response to an osmotic gradient. This creates a "pull" or tension on the column of water in the xylem, drawing water from one xylem cell to another, all the way through an entire column of xylem cells. Because of transpiration, the pressure in the leaves is decreased relative to the pressure in the roots, and water (and dissolved nutrients) move upward in response to this pressure gradient.

The chemical and physical properties of water molecules (which are discussed in Chapter 2) help move water through the plant, and form the basis of the **cohesion–adhesion theory** of water movement (Figure 17.8). Cohesion is the ability of molecules of the same kind to stick together. Water molecules are asymmetrical in shape, with a slight

positive charge on one end and a slight negative charge on the other end. As a result, water is a polar molecule, and water molecules form hydrogen bonds with each other, causing them to be cohesive, or stick together. Cohesion of water molecules makes them stick together in a chain, extending from the leaf all the way to the roots.

Adhesion is the ability of molecules of different kinds to stick together. This property of water causes it to stick to the cellulose walls of the tracheid and vessel element cells. This counteracts the force of gravity, and helps lift the column of water from the roots to the leaves.

The movement of water in transpiration according to the cohesion–adhesion theory can be summarized as follows: (1) Transpiration exerts a pulling force on water molecules in the xylem, forming a water column by cohesion that extends from the leaves to the roots. (2) As water molecules escape from the leaves, they are replaced by water molecules that move from the water column into the mesophyll cells. (3) As water is drawn up the xylem, adhesion to the cell walls facilitates the upward flow of the water molecules.

## Guard Cells Regulate Transpiration

Plants need carbon dioxide for photosynthesis. As this gas enters a leaf through epidermal pores, water is lost by transpiration. Plants are faced with the problems of maintaining an adequate supply of carbon dioxide and regulating water loss by transpiration. If the rate of water loss exceeds the rate of water uptake, metabolic activity is disrupted and the plant wilts. To control water loss, plants secrete

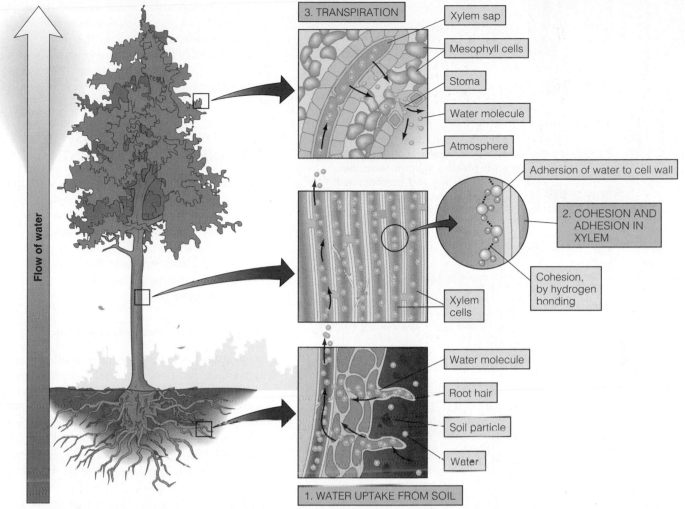

**✦ FIGURE 17.8**

Movement of water from roots to leaves in transpiration. Water enters the xylem in the roots and is carried upward in the tracheids and vessel elements. Water is pulled upward in the xylem by two forces: cohesion of water molecules to each other by hydrogen bonds, and by adhesion, the sticking of water molecules to the cellulose in the cell walls. Water moves from the sap, through the leaf, and into the atmosphere.

cuticle, a waxy covering described in Chapter 16. Most of the water loss in plants occurs through stomata, openings in the epidermis surrounded by two guard cells (✦ Figure 17.9). They are also the site where most of the carbon dioxide enters the plant.

Guard cells can change shape to open and close the stomata. The walls of guard cells are thicker along the inner walls surrounding the stoma (Figure 17.9). When water moves into the guard cell by osmosis in response to a solute gradient, the cells bulge outward, opening the stoma. When transpiration rates are high, water moves out of the cells, the guard cells become relaxed, and the stoma closes. Stomata occupy only about 1% of the leaf surface, but about 90% of the water lost in transpiration passes through them.

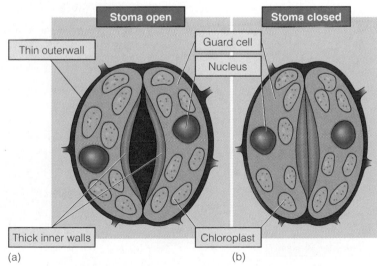

**✦ FIGURE 17.9**

Opening and closing of a stoma. (a) When a guard cell takes up potassium ions, water moves into the cell in response to a concentration gradient. This causes the guard cell to swell and become kidney-shaped, opening the stoma. (b) When the guard cells lose potassium, water flows out of the cell by osmosis, the cell sags, becomes more elliptically shaped, and the stoma closes.

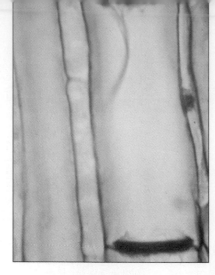

✦ **FIGURE 17.10**

A transmission electron micrograph of a sieve tube member, one of the cell types in phloem. When mature, these cells remain alive, and are connected to one another by perforated plates (sieve plates) between cells.

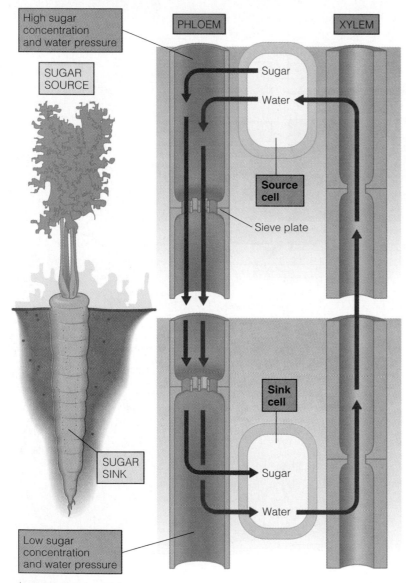

✦ **FIGURE 17.11**

The pressure-flow mechanism for the movement of nutrients in phloem. Sugar formed in the leaf by photosynthesis moves from a source cell in the leaf into the phloem, along with water. The water and sugar move through the phloem to a sink cell located in the root. The sugar is retained and converted into starch for storage. The water moves from the sink cell and enters the xylem, where it is carried upward, back to the leaf.

Transpiration makes water available to all parts of the plant, but also helps distribute minerals and dissolved nutrients throughout the plant. Transpiration also cools the plant, and plays an important part in the cycle of water movement from the soil to the atmosphere, where it returns to Earth as rain or snow.

## HOW PLANTS TRANSPORT AND STORE NUTRIENTS

Plants make sugars by photosynthesis, and some of these are used to provide energy for metabolic reactions. Others are used to make proteins, fats, steroids, and other biological molecules. The rest are converted to storage forms, and stored in other areas of the plant, including the roots and fruits. How do the nutrients produced by photosynthesis move through the plant?

### Phloem, Sugar, and Translocation

To move nutrients made by photosynthesis, plants use another transport system, the phloem. Phloem consists of several cell types, including **sieve elements, companion cells,** and the **vascular parenchyma.** Sieve elements are tubular cells, with thin walls, and are among the most unusual cells in the plant (✦ Figure 17.10). Unlike the cells of the xylem, sieve elements remain alive at maturity. Most sieve elements lose their nuclei and most organelles at maturity, leaving an empty cell, surrounded by a functional plasma membrane. The plasma membrane is needed to maintain osmotic pressure, believed to be the underlying force that moves solutions through the phloem.

Sieve elements are connected to one another by pores in their end walls, called **sieve plates** (Figure 17.10). Sieve elements form a system of tubes that extends from leaves to the roots. Sieve elements are usually connected to one or more highly specialized companion cells by plasmodesmata. The companion cells load sugars into the sieve elements and help maintain a functional plasma membrane in the sieve elements.

Unlike xylem, where solutions move only upward from the roots, fluids can move up or down in phloem, and are **translocated** from one place in the plant to another. What determines the direction of movement is whether a given location in the plant is producing or consuming sugars. Locations where sugar is being produced, either by photosynthesis in leaves, or the breakdown of starch in a root, are **sources.** Places in the plant where sugar is being consumed, either in metabolism, or by conversion into starch are **sinks.** Food, in the form of sugars, always moves from a source to a sink.

**✦ FIGURE 17.12**

In the spring, sugar moves upward from storage in the roots to other parts of the plant. Phloem sap from sugar maple trees can be boiled down and concentrated to produce maple syrup.

## Food Moves through the Phloem by a Pressure-Flow Mechanism

Sugar accumulates in the leaves as it is produced by photosynthesis (✦ Figure 17.11). The sugar is moved from the mesophyll cells into the sieve elements of the smallest leaf veins, in an energy-requiring step. This causes an increase in the solute concentration in the phloem. To counteract this, water flows into the sieve elements by osmosis. As more water enters the phloem, the pressure increases at the source end of the phloem, forcing the fluid, now called phloem sap, through the sieve elements toward the lower pressure at a sink.

At the sink, the sugar is removed from the phloem in another energy-requiring process (Figure 17.11). As sugar is removed, the solute concentration in the phloem falls, and water leaves by osmosis. This lowers the pressure at the sink end of the phloem. The water leaving the phloem diffuses into the xylem, where it can be transported back to the sink. Because phloem is basically a closed system, the combination of higher pressure at the source end and lower pressure at the sink end causes water and dissolved sugar to flow from source to sink.

During the growing season, the bulk of flow in the phloem is from the leaves to storage sinks, such as the roots, where sugars are converted to starch and stored. In the spring, when growth resumes, starch is converted to sugar, and sugar flows up-

ward to provide energy for leaf formation. To take advantage of this phenomenon, taps are inserted into the phloem of sugar maple trees to collect sap, which is concentrated by boiling to make maple syrup (✦ Figure 17.12).

## PLANTS RESPOND TO ENVIRONMENTAL STIMULI

Plants have evolved a number of adaptations to detect and respond to alterations in the environment. A response that involves plant parts moving toward or away from a stimulus is a **tropism.** If the response is toward the stimulus, the tropism is positive; if the response is away from the stimulus, the response is a negative tropism. Other responses to stimuli have a direction that is independent of the direction of the stimulus. These are called **nastic movements.**

### Alterations in Growth Patterns Generate Tropisms

Charles Darwin and his son Francis studied the familiar reaction of young stems of plants bending toward light, a reaction known as **phototropism** (✦ Figure 17.13). The Darwins discovered that the tops of the shoots bent toward the light first, and that the bending extended gradually toward the base. To test whether the tops of the plants controlled the response to light, they covered the tips of one group of plants with a metal foil, and another group with transparent caps. Those with foil caps did not bend toward the light, while those with transparent caps did. Charles and Francis Darwin concluded that some factor was transmitted from the tip of the plant to lower regions, causing the plant to bend toward the light.

Phototropism is generated by a plant hormone, **auxin.** Normally, auxin is present in equal concentrations in all regions of the growing tip. If the plant is exposed to light from only one direction, auxin is transported to the darker side of the stem. This causes cells on the shaded side to elongate more than the cells on the lighted side, and the plant bends toward the light.

Plants also exhibit **geotropism,** a response to gravity (✦ Figure 17.14). The roots of a plant show positive geotropism and grow downward, while the shoots of a plant have a negative response to gravity, and grow upward.

**✦ FIGURE 17.13**

In a phototropic response, bean seedlings bend toward light from a single direction.

**✦ FIGURE 17.14**

The stem of this pine tree bends upward in a geotropic response.

✦ FIGURE 17.15
✦ FIGURE 17.15
Tendrils of a vine will wrap around objects in a touch tropism known as thigmotropism.

Geotropisms were originally thought to be controlled by auxin, but this idea is now in doubt. In monocots such as corn, auxins are distributed asymmetrically in response to gravity, but it is not clear that this is true in dicots. Another idea proposes that the settling and distribution of starch-containing plastids in specialized cells of the root and shoot control geotropism. How the distribution of plastids is converted into a hormonal reaction controlling elongation and growth is not yet known.

Some plants also exhibit **thigmotropism,** a response to contact with a solid object (✦ Figure 17.15). Tendrils of vines wrap around objects they come in contact with, allowing the vine to attach and grow upward.

## Nastic Movements Are Nondirectional Responses to Stimuli

Nastic movements result from several types of stimuli, including light and touch. In legumes, which include the pea and the bean family, the leaves of the plant respond to the daily rhythms of light and dark by moving their leaves up and down. During daylight, the leaves are horizontal, and at night, the leaves fold up to a vertical position (✦ Figure 17.16). This response, also known as **sleep movement,** is brought about by changes in the size of ground parenchyma cells at the base of the leaf stem (the petiole). Water moves in and out of the cells in response to the movement of potassium ions, and as the cells change size, the leaf moves up and down.

Movement in response to touch is especially dramatic in the *Mimosa* plant (✦ Figure 17.17). When touched, individual leaflets or entire leaves fold and droop. The leaves are held upright by water pressure in the ground parenchyma cells at the base of the leaf. When touched, potassium ions flow out of the cells, changing the osmotic gradient. In response, water flows out of the cells, and the leaves fold.

The adaptive value of this response of the *Mimosa* plant is not clear. The plant grows in dry environments, and exposure to wind can have a drying effect on the plant. Action of wind on the leaves may cause them to fold, conserving water. A second idea is that folding of the leaves prevents attacks by leaf-eating insects.

## Plants Respond to Light and Cold

Some plants are able to sense the relative amounts of light and dark in a 24-hour period. This response

(a)                                                                                      (b)

✦ FIGURE 17.16
Sleep movement in leaves. (a) During the day, leaves are lowered to a horizontal position to maximize capture of sunlight for photosynthesis. (b) During the night, some plants fold their leaves vertically to conserve water.

(a)

(b)

## ✦ FIGURE 17.17

The leaves of the *Mimosa* plant are sensitive to touch. (a) Leaves in an open position. (b) After being touched, the leaves fold to a closed position.

is known as **photoperiodism,** and it controls the on set of flowering in many plants. Based on their response to light–dark cycles, plants fall into one of three groups: long-day, short-day, and day-neutral (✦ Figure 17.18). The **short-day plants,** such as strawberry, poinsettia, dandelions, and some chrysanthemums, flower during early spring or fall, when nights are relatively long and the days are short. **Long-day plants** flower mostly in the summer, when nights are short and the days are long. Spinach, onions, wheat, and some potatoes are long-day plants. Day-neutral plants flower without respect to day length. Plants of this type include sunflowers, rice, corn, the garden pea, and the snapdragon.

Photoperiodism is partly controlled by a plant pigment called **phytochrome.** Phytochrome in the leaves of plants detects day length and generates a

(b)

(c)

(a)

## ✦ FIGURE 17.18

(a) Short-day plants include poinsettia. (b) Wheat is an example of a long-day plant. (c) Corn is a day-neutral plant.

### Sidebar

#### PLANT BIOENGINEERING

Recombinant DNA techniques are changing the concept of plant-based raw materials. Plants make a tremendous variety of chemicals that could be used in industry. Genetic engineers have their sights set on plant-based bio-degradable plastics, soaps, detergents, industrial lubricants, drugs, pharmaceutical and vaccines.

A genetically engineered rapeseed plant is one of the first genetically transformed plants with great potential to achieve large-scale commercial success. A single gene from the California bay tree was introduced into the rapeseed plant, making it able to produce lauric acid, a 12-carbon fatty acid used to make detergents and soaps. The transplanted gene stopped fatty acid synthesis after 12 carbons rather than allowing the acid to continue to grow to the typical 18-carbon length, while having little effect on productivity. While obtaining laurate oils from plant products (they are now acquired from coconuts and palm kernel oils) is not a new procedure, it is a novel approach and a new source for this needed product.

Rapeseed is also being investigated for its other potential industrial benefits. Eurcic acid, which is used as a lubricant, is produced by gene transfer into rapeseed. Rapeseed may also be used to produce petroselinic acid, which can be used to make margarine and shortening.

# DO PLANTS HAVE AN IMMUNE SYSTEM?

Most flowering plants don't move. They are unable to forage for food, seek shelter in bad weather, or escape predators or diseases. To deter predators and to resist diseases, plants use metabolic by-products as chemical defenses. Until recently, it was thought that plants had no way of fighting infections caused by viruses, bacteria, fungi, or insects. During the last few years, however, it has become clear that plants have a sophisticated and complex defense system against infectious disease that has some similarities to the immune system of vertebrates.

Like the immune system, the chemical defense system of plants has several levels, ranging from a local to a systemic response. The plant defense systems also include a feature similar to immunological memory, that is, they develop resistance to reinfection by a disease agent. In vertebrates, the immune system is composed of specialized white blood cells. Plants have no such cells, and so do not have an immune system as such.

How does the defense system of a plant work? The soybean plant has been used as a model system to study the response to an invading microbe. If the soybean plant is infected by a fungus, hyphae from the fungus grow into the plant, penetrating cells and sucking nutrients from them. In response, within minutes after infection, the soybean begins cross-linking cell wall proteins in the vicinity of the infection, to make the cells resistant to the fungus. In the next stage of resistance, infected cells and healthy cells in the vicinity of the infection die and break open. This suicide reaction helps prevent spread of the infection.

Leaf damage causes nearby cells to produce antibiotics called **phytoalexins.** These antifungal agents attack and kill hyphae of the fungus. Cells at an increasing distance are also mobilized to produce and secrete phytoalexins, in case the fungus spreads.

In the next stage, cells begin the production and release of enzymes called chitinases and glucanases that break down the walls of the fungal hyphae. Gradually, areas of the plant far from the site of infection begin to cross-link cell walls and produce antibiotics and antifungal enzymes.

Scientists who study resistance to infection in plants have also discovered that some of the microbes that infect plants are related to pathogens that infect humans. This suggests that infection of plant and animal cells takes place by similar mechanisms, and that the study of how infections occur in plants may have some application to the study of human diseases.

---

response (◆ Figure 17.19). Phytochrome exists in two forms: phytochrome red (Pr) and phytochrome far-red (Pfr). During daylight, Pr is converted into Pfr. At the end of the day, the Pfr is converted over a period of several hours into Pr. Pfr is the active form of phytochrome, and generates a biological response. The response depends on how much Pfr is produced, and how long it is available.

Because different species of plants can have different photoperiod responses, flowering in these species can be precisely controlled by manipulating the light cycle. In the fall, flowering chrysanthemum plants are widely available. Commercial growers time the flowering process in this short-day plant by switching on the light in the middle of the night for a short period, getting all the plants to flower at the same time. Lilies and poinsettias are also brought into flower for Easter and Christmas by manipulating the light–dark cycle in greenhouses.

Cold is another environmental factor that can affect flowering. Apples, pears, winter wheat, Christmas cactus, and some perennial grasses need to be exposed to temperatures near freezing in order to flower. Winter wheat is usually planted in the fall, and flowers in the spring, about seven weeks after growth begins. If the wheat is planted in the spring, it may not flower until late in the year. Winter wheat can be planted in the spring if the seeds are first exposed to cold. This treatment, called **vernalization,** allows the wheat to germinate and produce seeds in one growing season.

## PLANT SECONDARY COMPOUNDS: DEFENSE MECHANISMS

We have already learned that plants produce compounds such as sugars and proteins that are used in metabolism, and also form the base of our own food web. Plants also produce a large array of **secondary compounds** that are not important in meta-

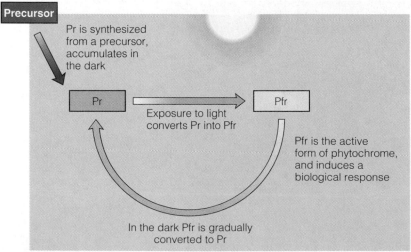

Precursor

Pr is synthesized from a precursor, accumulates in the dark

Pr

Exposure to light converts Pr into Pfr

Pfr

Pfr is the active form of phytochrome, and induces a biological response

In the dark Pfr is gradually converted to Pr

◆ **FIGURE 17.19**

**The Pr form of phytochrome is synthesized from a precursor, and accumulates in the plant during the night. Exposure to light converts Pr into Pfr, the active form of phytochrome. The Pfr form induces a biological response. At the end of the day, the level of Pfr declines as it is converted to Pr in the dark.**

bolism, but serve to attract animals for pollination, repel predators, kill parasites and prevent infectious diseases (see Beyond the Basics: Do Plants Have an Immune System?). In fact, it has been estimated that 99.5% of all pesticides in our diet are naturally produced by plants as a defense against predation.

Pea plants (the same species used by Gregor Mendel) produce a substance called pisatin that protects them from most strains of parasitic fungi. As an example of the constant battle between predator and prey fueled by natural selection, some strains of one parasitic fungus (*Fusarium*) contain enzymes that inactivate pisatin, allowing these strains to infect pea plants.

Other plants produce chemicals that are distasteful to insects, chemicals that interfere with the life cycle of predatory insects, or act as insect repellents. Pyrethrum is an insecticide produced by a species of chrysanthemum and is commercially available to gardeners.

Some plants produce secondary compounds only when under attack. When the leaf of a tomato plant is chewed by an insect, the leaf damage causes the production of a wound hormone. The hormone circulates through the plant and stimulates the production of secondary chemicals called **antinutrients.** These compounds inhibit the action of enzymes in the insect's digestive system. This makes it difficult for the insect to digest the tomato tissue. Antinutrient production continues and even intensifies while the tomato plant is under attack.

More than 10,000 defensive chemicals have been identified to date, including caffeine, phenol, tannin, nicotine, and morphine. By identifying such secondary defensive chemicals and the genes that encode them, it may be possible to transfer the ability to produce defensive chemicals to crop plants, cutting down on the use of synthetic pesticides.

## Plant Secondary Compounds Are Used by Humans

Some plant secondary compounds such as pyrethrum or rotenone are useful to humans as insecticides. Others are important as medicines. Salicylic acid, the main ingredient in aspirin, was first discovered in the bark of the white willow and, in synthetic form, is widely used for relief from pain and inflammation. It is estimated that about 25% of all prescriptions issued in the United States contain at least one plant compound.

The active compounds of many stimulating beverages we drink also come from plants. Coffee, tea, and hot chocolate all contain caffeine, which causes a range of physiological effects in humans. It stimulates the central nervous system and mimics the effects of adrenaline. Caffeine is the most widely used psychoactive drug in the world, not only because of the large numbers of coffee and tea drinkers, but because it is added to soft drinks and medications.

Even chewing gum has its origins in plant compounds. Its use in the United States can be traced to Antonio Lopez de Santa Anna, a political leader, who served 11 terms as president of Mexico. On a visit to the United States, Santa Anna brought with him chicle, a compound from the sapodilla tree that he had chewed when he was young. Gum chewing was fairly common in the United States at that time, but most gum was made of sweetened paraffin, which is not very chewy. Thomas Adams, who was acquainted with Santa Anna, added sugar to chicle, and formed it into small balls of chewing gum. His product was an instant success. Although a small amount of chewing gum is still made with chicle, synthetic vinyl resins and sugar substitutes are used to make almost all the chewing gum sold today.

 # SUMMARY

1. Plants obtain their nutritional requirements from the atmosphere and the soil. Most green plants transform water and carbon dioxide into all the organic compounds they need to survive, including vitamins and amino acids. Plant nutrition involves uptake of raw materials, and the distribution of these materials within the plant body.

2. Soil is really the weathered superficial layer of the earth's crust. Fertile soil contains all the chemical elements needed for plant growth in readily available form. Plant roots mine the soil to obtain nutrients essential for growth and reproduction. Mineral nutrients enter the biosphere at the plant's root system.

3. Water and dissolved minerals enter the plant by an intracellular route and an extracellular route. Xylem transports water and minerals in plants, and is a complex tissue, composed of several cell types, each of which plays a role in movement of water and solutes. Water is not pumped through the xylem, instead it is pulled by the force of transpiration. The cohesion and adhesion of water molecules helps move water through the xylem.

4. Nutrients produced by photosynthesis are moved through the plant by the phloem. Unlike xylem, where solutions can move only upward, fluids can move up or down in the phloem, and are translocated from one place in the plant to another. Nutrients move through the phloem from a source to a sink by a pressure-flow mechanism.

5. Plants have a number of adaptations that allow them to detect and respond to environmental changes. Tropisms are responses that involve plant parts moving toward or away from a stimulus. Plant

tropisms include responses to the direction of light, gravity, and touch. Nastic movements are responses that are independent of the direction of the stimulus. The folding of a *Mimosa* leaf in response to touch is a nastic response.

6. To protect themselves from predators and diseases, plants use a class of chemicals called secondary compounds. These chemicals defend the plant against infection by viruses, bacteria, or fungi. These compounds include antibiotics that kill the invading pathogen, or that trigger destruction of infected plant cells. Plants also produce compounds that defend the plant against predation by insects or other animals.

7. Some of the secondary compounds produced by plants are used by humans. These compounds include salicylic acid (the active ingredient in aspirin), caffeine, and chicle, used in chewing gum.

# KEY TERMS

antinutrients
auxin
Casparian strip
cohesion–adhesion theory
companion cells
endodermis
essential elements
geotropism
long-day plants
micronutrients
nastic movement

photoperiodism
phototropism
phytoalexins
phytochrome
primary macronutrients
secondary compounds
secondary macronutrients
short-day plants
sieve elements
sieve plates
sink

sleep movement
source
thigmotropism
tracheids
translocate
transpiration
tropism
vascular parenchyma
vernalization
vessel elements
xylem

# QUESTIONS AND PROBLEMS

## MATCHING

Match the description with the appropriate term.

1. Nitrogen, phosphorus, potassium
2. Materials cross only the plasma membrane of the root hair epidermal cell
3. The loss of water molecules from leaves
4. The property that counteracts gravity and allows water to be lifted from the roots to the leaves
5. Location where sugar is produced
6. Reaction of plants to light
7. Reaction of plants to solid objects
8. Plant products not important in metabolism
9. Depends on pressure gradients between source and sink regions
10. Artificial exposure of seeds or seedlings to cold to overcome the cold requirements for flowering

 a. translocation
 b. transpiration
 c. vernalization
 d. phototropism
 e. primary nutrients
 f. secondary compounds
 g. source
 h. intracellular route
 i. adhesion
 j. thigmotropism

## TRUE/FALSE

1. The atmosphere is the primary source of nutrients for plants.
2. The transportation of water and nutrients within the plant body never requires the expenditure of energy.
3. Fluids can move up or down in the phloem.
4. During the spring, starch is converted to sugar, and sugar flows upward to provide energy for leaf formation.
5. Long-day plants flower mostly in the summer.

## SHORT ANSWER

1. List the four ways in which plants utilize minerals and nutrients essential for nutrition.
2. What prevents water and solutes from entering the xylem?
3. What three cell types make up the phloem?
4. What would be the effect on the rate of transpiration if the humidity were high?
5. How is flower production regulated in a short-day plant?

## SCIENCE AND SOCIETY

1. Most plants do not store much food in the leaves, where it is made. Instead, they transport food through the phloem to other parts of the plant. The food storage areas of plants include roots, underground stems, and sometimes underground leaves. Can you think of examples of plants with each of these different food storage areas? What would be the advantage of storing food underground? Would food storage sites change during plant reproduction? If yes, what purpose would this change serve?

2. New plants are being developed by means of genetic engineering. Researchers are developing new strains of plants that have increased nutritional value and at the same time are more resistant to pests and pathogens. These hybrid plants could help combat malnutrition in countries like Mexico, Central America, and Africa, where hunger and starvation are common. Although these new plants present an optimistic solution to world hunger, what other issues should be considered before they are commercially available? Will Third World countries be able to afford these new plants? What effects will these plants have on the soil?

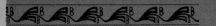

# V CONTROL SYSTEMS IN ANIMALS

HIGHLY COORDINATED MOVEMENTS DEPEND ON THE ACTION OF MANY CONTROL SYSTEMS IN THE BODY.

# 18

# HOMEOSTASIS

O P E N I N G   I M A G E

*Homeostasis allows each system*

*in Hakeem Olajuwon's body to*

*function at maximum capacity.*

About 200 years ago, an experiment by Dr. Charles Blagden, an English physician, showed that organisms can survive extreme conditions by using mechanisms that maintain a relatively constant internal environment. Dr. Blagden had a room constructed with the walls, ceiling and floor made of iron panels. When the room was completed, he had it heated to a temperature of 121°C (250°F). He then entered the room with several associates, a dog and a raw beefsteak. The group remained in the room for 45 minutes. The dog, Dr. Blagden and his rather dedicated associates emerged unharmed. Although the beefsteak was similar in chemical composition to the dog and the humans in the room, it was cooked through in 45 minutes. This demonstration dramatically illustrates the fact that mammals (and other organisms) have mechanisms that respond to and compensate for environmental change, and maintain a steady state. This steady state is called homeostasis.

Humans usually maintain an internal core temperature of about 37.7°C (100°F). If the internal body temperature rises a few degrees, nerve malfunctions occur. With an internal temperature of about 41°C (106°F), convulsions begin, and at 43.3°C (110°F), death occurs. How did Dr. Blagden, the dog and others survive for 45 minute at a temperature almost three times the lethal level? To maintain body temperature at a level below 41°C, many mechanisms that control heat production and heat loss were adjusted. These

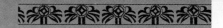

*processes are regulated through a region of the brain, the hypothalamus, that receives sensory input from the environment and the body and responds by lowering internal heat production (reducing muscle tone) and by increasing heat loss (dilation of blood vessels in the skin, and sweating).*

*Homeostasis is one of the basic principles of physiology. This section of the book describes the major body systems and how they function. One function shared by all systems is a contribution to homeostasis, which is essential for the survival of the cells that make up the body. This chapter discusses the concept of homeostasis, and explores in detail how one system,the skin, contributes to the maintenance of a constant internal environment.*

## HOMEOSTASIS MAINTAINS THE INTERNAL ENVIRONMENT SHARED BY ALL CELLS

The term **homeostasis** can be defined as the maintenance of a stable internal environment (✦ Figure 18.1). A ciliated protozoan such as *Tetrahymena* is a single-celled organism that can take up nutrients and oxygen directly from the environment, and transfer metabolic wastes and carbon dioxide to the external environment (✦ Figure 18.2). In multicellular organisms, most cells are isolated from the external environment. These cells are surrounded by an aqueous **internal environment,** which is outside the cells, but inside the body (Figure 18.2). The internal environment must be maintained in a state that allows the cells of the body to function at maximum efficiency. This means that the temperature and chemical composition of the internal environment can fluctuate only within narrow limits.

For cells to survive, nutrients and oxygen must be constantly supplied to cells, and waste products

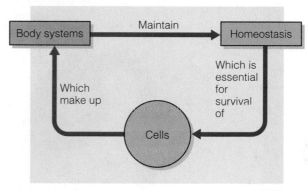

✦ **FIGURE 18.1**

**The relationships of cells, body systems, and homeostasis.**

and carbon dioxide must be constantly removed. Maintenance of the internal environment within the narrow limits compatible with life is homeostasis. Many organ systems contribute to exchanges between the external and internal environments in order to maintain an internal environment that permits cells to function properly.

✦ **FIGURE 18.2**

(a) Single-celled organisms such as *Tetrahymena* are able to interact directly with the environment to obtain nutrients and dispose of wastes. (b) In multicellular organisms, cells are in contact with an internal environment through the blood and other body fluids.

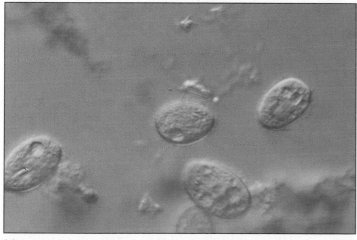

(a)

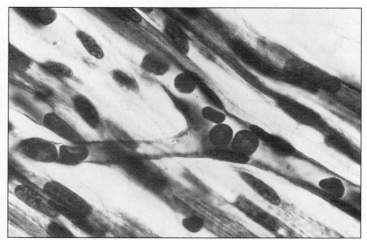

(b)

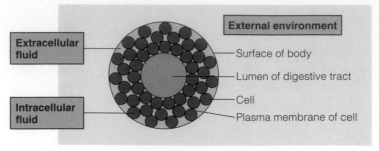

(a)

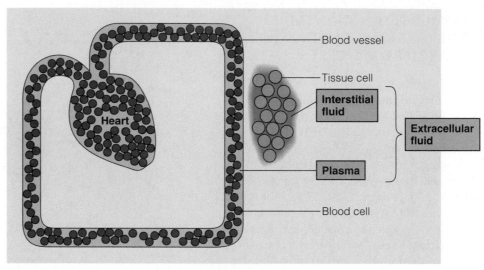

(b)

✦ FIGURE 18.3

(a) The fluid compartments of the body. Extracellular fluid surrounds cells, and intracellular fluid is inside cells. (b) The extracellular fluid has several components. Within the heart and blood vessels, it is called plasma, but outside the circulatory system it is called interstitial fluid.

*Sidebar*

### HUMAN OBESITY AND GENETICS

Obesity increases the risk of life-threatening conditions such as high blood pressure, high blood cholesterol, heart attack, and diabetes. Human obesity is thought to be about 60% determined by genetic factors, yet genes controlling obesity have not yet been localized.

Mutational studies in mice demonstrated that animals carrying the obesity gene or *ob* become grossly obese—almost three times the weight of a normal mouse—and develop a form of diabetes similar to type II (non-insulin-dependent) diabetes that is seen in humans late in life. The gene was mapped to mouse chromosome 6. Screening in normal mice proved that this gene's protein product was expressed only in adipose tissue.

The overproduction of this defective protein in mutant mice suggests that the obesity-regulating factor is a hormone. It is not yet clear where or how this hormone might act, although the hypothalamus is a likely candidate. The hypothalamus controls food intake and energy expenditure, which are major factors regulating fat deposition.

Researchers found an 84% homology between the protein encoded by the comparable human gene to the mouse protein. This finding raises the possibility that human obesity may be treatable by administering the normal gene product, much like diabetes is treated with insulin.

External forces and internal events constantly exert changes on the internal environment and threaten to disrupt homeostasis. For example, if the external temperature falls, the internal body temperature will begin to drop. As the body cools, it moves the internal environment away from its optimal temperature. The change in internal body temperature is monitored by a sensory system, integrated by the nervous system, and an appropriate response is generated. In cold-blooded animals this may involve movement to a sunny location where the absorption of solar energy can passively raise the internal temperature. In warm-blooded animals, this may trigger the thyroid gland to release hormones that raise the level of metabolism, producing more heat as a by-product.

In homeostasis, small, incremental adjustments are made to keep the internal environment within the optimal limits. For cells to survive and function, homeostasis is essential. Remember that tissues, organs, organ systems and, in fact, the organism are ultimately composed of cells, and if these systems are to survive and function, homeostasis must allow the cells to function.

## The Internal Environment Has Two Components

Within the body of multicellular animals, cells exchange materials with the aqueous internal fluid. The internal environment with which cells are in contact is the extracellular fluid. There are two principal types of extracellular fluids in the bodies of multicellular animals: the **extracellular** fluid that surrounds and bathes the cells, and **plasma,** the liquid portion of the blood or circulatory fluid (✦ Figure 18.3).

In humans, water is the main component of body fluids, and makes up from 50 to 60% of body weight. About two-thirds of the body fluid is in the cytoplasm within cells, with the remaining one-third in the extracellular compartment.

Several important elements of the internal environment must be maintained to achieve homeostasis, including the following:

1. The concentration of oxygen and carbon dioxide. Most cells require oxygen to carry out the chemical reactions of energy conversion. The process of energy conversion generates carbon dioxide, which must be removed.
2. The pH of the internal environment. Carbon dioxide production causes an increase in acidity of the intracellular and extracellular fluids. This must be balanced by removal of carbon dioxide and by the buffering capacity of the plasma.
3. The concentration of nutrients and waste products. Even though an animal does not eat continuously, cells need a constant supply of nutrient molecules for energy conversion. The molecular waste products from biochemical degradation and energy conversion must be removed from the cellular environment and the body.
4. The concentration of salts and other electrolytes. Within the body, the concentration of NaCl and other salts must be maintained to ensure that cells are not swollen or shrunken by osmotic pressure. In addition, sodium and other electrolytes such as potassium are essential in conduction of nerve impulses and muscular contraction.
5. The volume and pressure of the extracellular fluid. The amount of circulating plasma and the regulation of pressure in the circulatory system are vital to supporting the exchange of nutrients and oxygen and the disposal of waste products and carbon dioxide.

## Cybernetics and Control Systems Used in Animals

Cybernetics is the science that deals with control systems in mechanical devices and living organisms. There are two main types of control systems: open systems and closed systems. Open systems are linear (✦ Figure 18.4) and have no feedback. When you turn on a light switch, a signal in the form of electricity flows to the light. The output of the light bulb in the form of light energy does not affect the switch; there is no feedback of information to confirm that the light is on.

In a closed system, there are at least two components: a **sensor** and an **effector.** The sensor detects change and signals the effector, which initiates a response (Figure 18.4). In a house, the thermostat acts as a sensor, sending a signal to an effector, the furnace, to initiate a response in the form of heat. As the temperature in the house rises, it acts as a feedback system to the thermostat, which then shuts off its signal to the furnace.

Most physiological systems in the body employ feedback to maintain the internal environment within the optimal range. The properties of homeostatic mechanisms and their feedback systems are discussed in the following section.

## HOMEOSTATIC SYSTEMS HAVE SEVERAL PROPERTIES

In homeostasis, regulation is for the most part extrinsic. In this form of regulation, organs and body systems are controlled from outside. In higher animals, the nervous system and the endocrine system are the major extrinsic control systems. These extrinsic control systems use reflexes to exert homeostatic control, with some important differences between them.

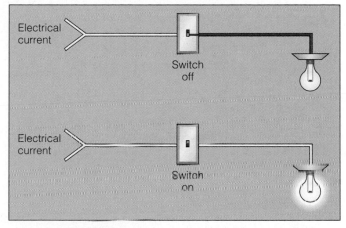

(a) Open control system

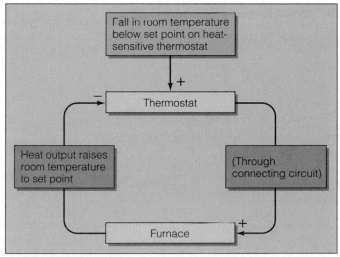

(b) Closed control system

✦ **FIGURE 18.4**

**Open and closed systems. (a) A switch and light bulb represent an open system. There is no feedback to confirm light is on. (b) A thermostat in a home heating system is a closed system, with the thermostat acting as a sensor. This closed system uses negative feedback. Closed systems can also use positive feedback.**

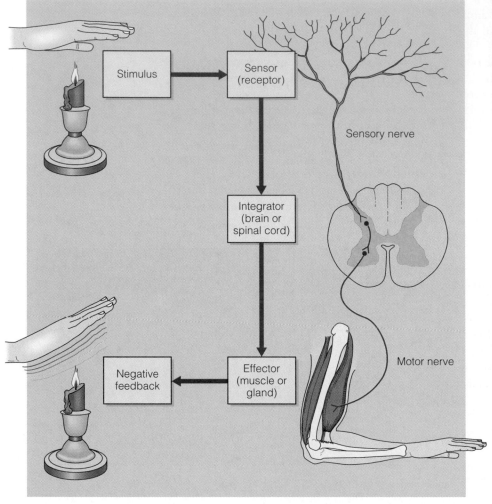

✦ **FIGURE 18.5**

Nervous system reflex. A reflex involves a stimulus, a sensor, an integrator, and an effector. The sensor receives the stimulus and sends a signal to the integrator, which causes the effector to respond. Negative feedback eliminates the stimulus.

## Reflexes Are Mediated by the Nervous System

The nervous reflex depends on sensors, such as pain or touch receptors in the skin, or sensory organs such as the eyes or ears to receive stimuli and transmit a signal to the brain or spinal cord through the nervous system. In the brain and spinal cord, sensory input from many sources is processed and integrated, and a response is generated in the form of a signal to an effector. For example, when you put your hand over a candle, pain and temperature receptors in the hand generate a signal that is sent over nerves to the spinal cord (✦ Figure 18.5). The input is integrated, processed, and a signal is sent to an effector system, the muscles in your arm. As a re-

sponse, you move the hand away from the flame. In this case, the movement of the hand is extrinsically regulated, and occurs within the span of less than a second.

## The Endocrine System Mediates Extrinsic Control by Hormone Secretion

The second type of extrinsic control, through the endocrine system, involves a chemical component in the reflex. In this reflex, sensors detect a change within the body and transmit a chemical signal to the nervous system or directly to cells that are effectors (✦ Figure 18.6). The parathyroid glands synthesize and release a parathyroid hormone (PTH) when blood levels of calcium are low. PTH is carried by the blood to bone, where it stimulates the activity of osteoclasts, bone-destroying cells. The action of osteoclasts raises blood levels of calcium. As the blood level of calcium rises, it shuts down the secretion of PTH. In this case, the cells of the parathyroid cells are the sensors, and the osteoclasts are the effectors.

In other cases, signals are routed through both the nervous system and the endocrine system. The thyroid gland synthesizes and releases a hormone, thyroxin, into the blood (✦ Figure 18.7). Thyroxin acts to control the rate of metabolism. When blood levels of thyroxin fall, receptors in the brain are stimulated. The brain receptors signal a region of the brain called the hypothalamus to release a hormone that travels to the pituitary gland (part of the endocrine system). The pituitary gland, in turn, releases a hormone (called thyroid-stimulating hormone) that travels through the blood and stimulates the thyroid to release more thyroxin. While somewhat complicated, the reflex in this case employs the two basic components of homeostatic control: a sensor and an effector.

## Homeostasis Also Uses Intrinsic Control Systems

Although most homeostatic systems rely on extrinsic control, local or intrinsic control is also employed. These controls are local, and involve only one tissue or organ. Consider what happens when a skeletal muscle contracts rapidly. Oxygen is consumed and carbon dioxide is produced. The lowered oxygen concentration and the increase in carbon dioxide concentration exert an effect on the diameter of blood vessels within the muscle, causing them to dilate, allowing more blood flow to supply additional oxygen and remove carbon dioxide. This local or intrinsic control affects only the blood vessels in the contracting muscle, not the entire circulatory system.

## Negative and Positive Feedback Systems Operate in Homeostasis

Most of the body's homeostatic systems operate on the principle of **negative feedback control.** It is called negative feedback because the information produced by the feedback reverses the direction of the response. The regulation of hormonal secretions provides a good example of negative feedback in homeostasis. As described earlier, the pituitary gland secretes a hormone called thyroid-stimulating hormone (TSH), which causes the thyroid to secrete thyroid hormone. As the blood levels of thyroid hormone rise, further secretion of TSH is inhibited. In this example, blood levels of thyroid hormone serve as the feedback information, reversing the direction of TSH secretion.

Although negative feedback control is widely used in maintaining homeostasis, **positive feed-**

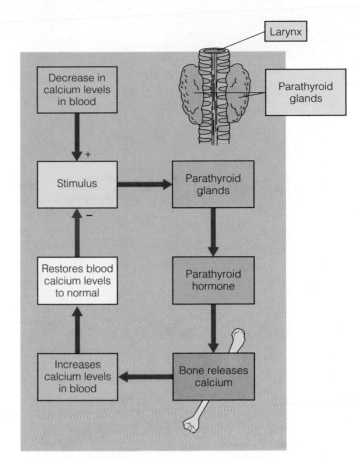

✦ **FIGURE 18.6**

**Endocrine system reflex. In this system, a decrease in blood calcium level is the stimulus. Cells in the parathyroid gland are the sensors and integrators. Secretion of a hormone causes action by an effector (osteoclasts), increasing calcium levels in the blood. A plus sign indicates a stimulus that increases calcium levels, and a minus sign indicates a stimulus that decreases calcium levels.**

✦ **FIGURE 18.7**

**A neuroendocrine reflex. This type of reflex involves both the nervous system and the endocrine system. The sensory system in this reflex is the cells in the hypothalamus of the brain.**

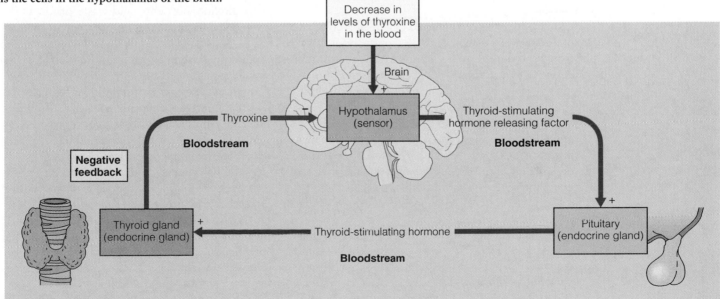

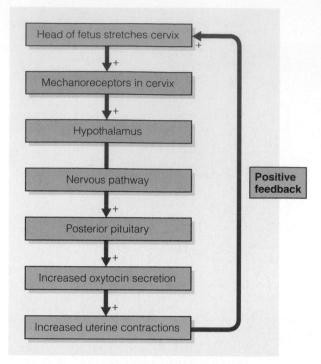

```
┌─────────────────────────────┐
│ Head of fetus stretches cervix │◄──────────┐
└─────────────────────────────┘          +  │
              │ +                            │
              ▼                              │
┌─────────────────────────────┐             │
│ Mechanoreceptors in cervix   │             │
└─────────────────────────────┘             │
              │ +                            │
              ▼                              │
┌─────────────────────────────┐             │
│ Hypothalamus                 │             │
└─────────────────────────────┘             │
              │                    ┌──────────────┐
              ▼                    │  Positive    │
┌─────────────────────────────┐   │  feedback    │
│ Nervous pathway              │   └──────────────┘
└─────────────────────────────┘             │
              │ +                            │
              ▼                              │
┌─────────────────────────────┐             │
│ Posterior pituitary          │             │
└─────────────────────────────┘             │
              │ +                            │
              ▼                              │
┌─────────────────────────────┐             │
│ Increased oxytocin secretion │             │
└─────────────────────────────┘             │
              │ +                            │
              ▼                              │
┌─────────────────────────────┐             │
│ Increased uterine contractions │──────────┘
└─────────────────────────────┘
```

✦ **FIGURE 18.8**

**A positive feedback system during childbirth. Pressure on the uterus from the head of the fetus sends signals to the hypothalamus, which, in turn, stimulates the posterior pituitary gland. This leads to secretion of a hormone, oxytocin, that increases uterine contractions, increasing the pressure of the head on the uterus.**

back control is used in some circumstances (✦ Figure 18.8). In positive feedback, input increases and accelerates the response. In general, reflexes that empty body cavities, such as urination, defecation, ejaculation, and childbirth, are under positive control.

In childbirth, the uterus begins a series of rhythmic contractions. As these contractions grow stronger, the head of the fetus is forced against the uterine wall at the cervix. Receptors in the uterine wall transmit a nerve impulse to the hypothalamic region of the brain, causing the release of the hormone **oxytocin.** Oxytocin causes an increase in the frequency and strength of uterine contractions, causing the fetus to press harder on the uterine wall, further stimulating the receptors. This, in turn, causes the release of more oxytocin, and so on, until the fetus is expelled from the uterus, and the cycle stops.

The role of feedback systems in maintaining homeostasis means that conditions in the internal environment are not maintained to an absolute level, but oscillate within a range that is optimal (✦ Figure 18.9). The feedback necessary to maintain homeostasis balances input from sensory systems with output from effector systems. Homeo-

stasis depends on the action and interaction of a number of body systems, and these are considered in the next section.

## FUNCTIONAL ASPECTS OF HOMEOSTASIS

### All Major Body Systems Are Involved in Homeostasis

Eleven major body systems are present in animals. Not all animal groups have all these systems, and there are variations in systems between different animal groups. Whatever the number and form of body system present in an animal, each has a specific role in maintaining internal homeostasis (See ✦ Figure 18.10 on pages 342-343). The major systems and their role in homeostasis are as follows:

1. The **muscular system** allows movement and locomotion. It moves animals toward food or mates and away from predators. Muscles also power the circulatory system, digestive system, and respiratory system. Muscle contraction also plays a role in temperature regulation.

2. The **skeletal system** provides support for the body and protection for internal organs. The skeleton also serves as attachment points for the muscular system, and with the muscular system allows movement and locomotion.

3. The **respiratory system** is an adaptation of multicellular animals that moves oxygen from the external environment into the internal environment, and removes carbon dioxide from the body.

4. The **circulatory system** transports oxygen, carbon dioxide, nutrients, and waste products between cells and the respiratory system. In addition, it carries chemical signals from the endocrine system, and is important in maintaining internal homeostasis.

5. The **immune system,** present in vertebrates, defends the internal environment against invading microorganisms and viruses. It also provides defense against the growth of cancerous cells.

6. The **digestive system** receives food from the external environment. It converts food into nutrient molecules by a combination of mechanical and chemical processes. These nutrient molecules along with water and ions are absorbed into the internal environment and transported by the circulatory system for storage and energy conversion. The digestive system also eliminates solid wastes into the environment.

7. The **excretory system** regulates the volume of the internal body fluids, and the molecular and ionic constitution of these fluids. It eliminates

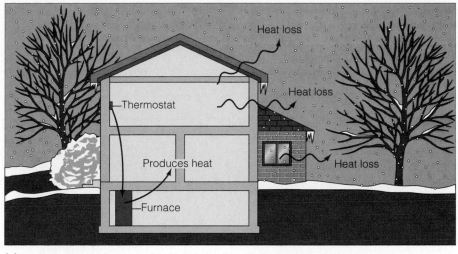

(a)

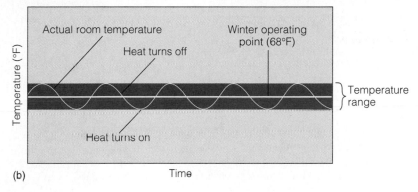

(b)

✦ FIGURE 18.9

**Homeostatic systems maintain systems within an optimal range, rather than at a single point. (a) In the house shown, heat is lost in various ways, lowering the temperature. The thermostat switches the furnace on to raise the temperature, and then switches the furnace off when heat is provided. (b) The action of the thermostat maintains the temperature within an optimal range.**

metabolic waste products from the internal environment.

8. The **nervous system** coordinates and controls actions of internal organs and body systems. It also receives and processes sensory information from the external environment, and coordinates short-term reactions to these stimuli.

9. The **endocrine system** works with the nervous system in controlling the activity of internal organs, especially the kidney, and in coordinating the long-range response to external stimuli.

10. The **reproductive system,** controlled mostly by the endocrine system, is responsible for survival of the species.

11. The **skin** or **integument** is the outermost protective layer protecting multicellular animals from the loss or exchange of internal fluids and from invasion of foreign microorganisms into the body.

To demonstrate how one organ system plays multiple roles in homeostasis, we will examine the organization and role of the skin in these facets of homeostasis. Other organ systems will be considered in the next several chapters.

## THE SKIN AND ITS DERIVATIVES MAKE UP THE INTEGUMENTARY SYSTEM

The skin and its derivatives—hair, nails, feathers, horns, antlers, and glands—make up the **integumentary system.** In humans, the skin is the largest organ in the body, and amounts to 12 to 15% of body weight, with a surface area of between 1 to 2 square meters. The skin is continuous with, but structurally distinct from, the mucous membranes that line the mouth, anus, urethra, and vagina.

**The role of body systems in homeostasis.**

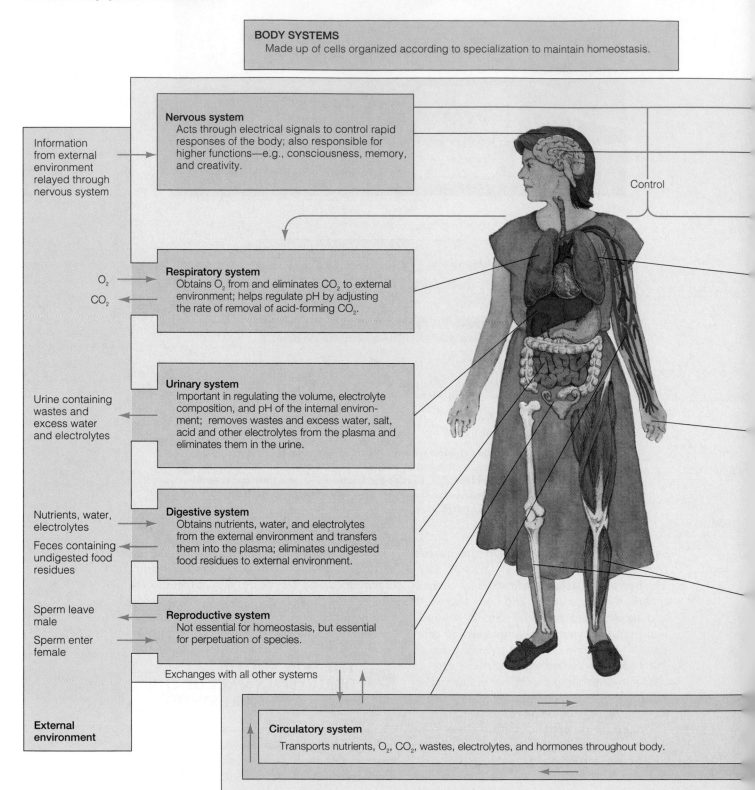

**BODY SYSTEMS**
Made up of cells organized according to specialization to maintain homeostasis.

**Nervous system**
Acts through electrical signals to control rapid responses of the body; also responsible for higher functions—e.g., consciousness, memory, and creativity.

Information from external environment relayed through nervous system

Control

**Respiratory system**
Obtains $O_2$ from and eliminates $CO_2$ to external environment; helps regulate pH by adjusting the rate of removal of acid-forming $CO_2$.

$O_2$
$CO_2$

**Urinary system**
Important in regulating the volume, electrolyte composition, and pH of the internal environment; removes wastes and excess water, salt, acid and other electrolytes from the plasma and eliminates them in the urine.

Urine containing wastes and excess water and electrolytes

**Digestive system**
Obtains nutrients, water, and electrolytes from the external environment and transfers them into the plasma; eliminates undigested food residues to external environment.

Nutrients, water, electrolytes

Feces containing undigested food residues

**Reproductive system**
Not essential for homeostasis, but essential for perpetuation of species.

Sperm leave male

Sperm enter female

Exchanges with all other systems

**External environment**

**Circulatory system**
Transports nutrients, $O_2$, $CO_2$, wastes, electrolytes, and hormones throughout body.

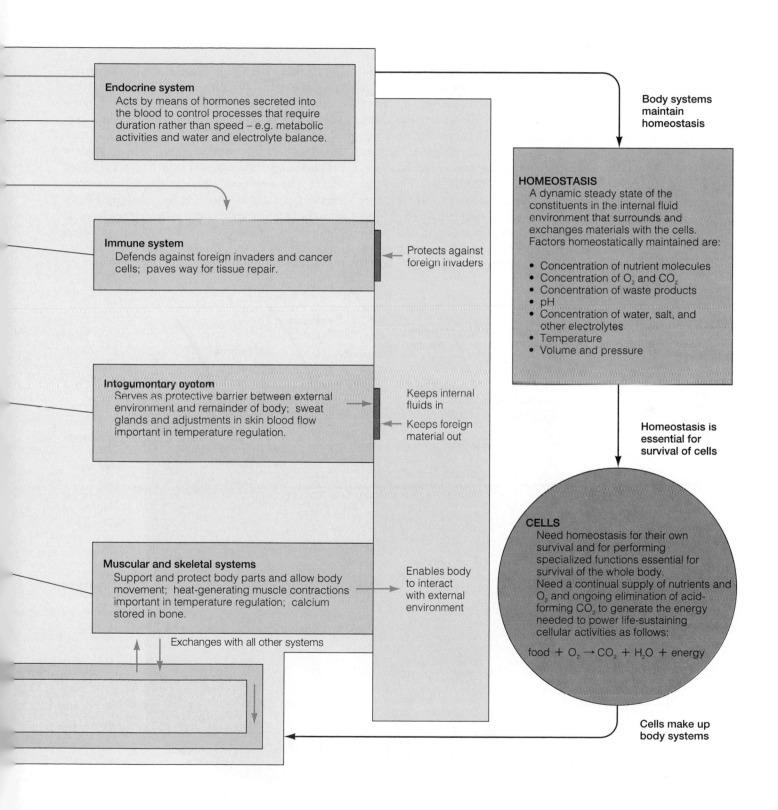

**Endocrine system**
Acts by means of hormones secreted into the blood to control processes that require duration rather than speed – e.g. metabolic activities and water and electrolyte balance.

**Immune system**
Defends against foreign invaders and cancer cells; paves way for tissue repair.

Protects against foreign invaders

**Integumentary system**
Serves as protective barrier between external environment and remainder of body; sweat glands and adjustments in skin blood flow important in temperature regulation.

Keeps internal fluids in

Keeps foreign material out

**Muscular and skeletal systems**
Support and protect body parts and allow body movement; heat-generating muscle contractions important in temperature regulation; calcium stored in bone.

Enables body to interact with external environment

Exchanges with all other systems

Body systems maintain homeostasis

**HOMEOSTASIS**
A dynamic steady state of the constituents in the internal fluid environment that surrounds and exchanges materials with the cells. Factors homeostatically maintained are:

- Concentration of nutrient molecules
- Concentration of $O_2$ and $CO_2$
- Concentration of waste products
- pH
- Concentration of water, salt, and other electrolytes
- Temperature
- Volume and pressure

Homeostasis is essential for survival of cells

**CELLS**
Need homeostasis for their own survival and for performing specialized functions essential for survival of the whole body.
Need a continual supply of nutrients and $O_2$ and ongoing elimination of acid-forming $CO_2$ to generate the energy needed to power life-sustaining cellular activities as follows:

$$food + O_2 \rightarrow CO_2 + H_2O + energy$$

Cells make up body systems

## The Skin Is Composed of Two Layers

The skin is composed of two distinct layers: the **epidermis** and the **dermis** (✦ Figure 18.11). At birth, the epidermis consists of several layers of epithelial cells. The basic cell type in the epidermis is the **keratinocyte.** The keratinocytes are produced by division of the innermost cells of the epidermis, adjacent to the dermis, the **basal cells.** The inner layer of the epidermis also contains **melanocytes,** which produce a pigment called **melanin.** Melanin gives the skin color and protects the underlying layers against damage from ultraviolet light. Keratinocytes are locked to each other by desmosomes. As they mature and move toward the skin surface, they become filled with **keratin,** a fibrous protein. When they reach the surface, the outer layer of keratinocytes becomes elongated and flattened to form a protective layer. Eventually these cells die, and form a thin, outer layer of dead cells that are continually sloughed off the body surface.

The **dermis** is a connective tissue layer under the epidermis that contains elastic fibers, collagen fibers, capillary networks, and nerve endings. The blood vessels supply the dermis and, by diffusion, exchange nutrients and wastes with the epidermis.

## Infoldings of the Epidermis Form Follicles and Glands

Pouches and infoldings of the epidermis penetrate into the dermis and form the hair follicles, sebaceous glands, and sweat glands of the skin (Figure 18.11). Hair follicles are lined with cells that synthesize proteins which form hair. Associated with each hair follicle is a sebaceous gland, a capillary network, nerve endings, and a small muscle (Figure 18.11). The sebaceous glands of the follicle secrete an oily coating that coats the hair shaft and flows onto the skin surface to form a waterproof layer. If the duct of the sebaceous gland becomes plugged with excessive secretions and then becomes infected, the result is a skin blemish, or pimple.

✦ **FIGURE 18.11**

**The skin consists of two layers, the epidermis and the dermis. Infoldings of the outer layer form glands and hair follicles. The skin is anchored to underlying muscle by a fat-containing layer of connective tissue, called the hypodermis.**

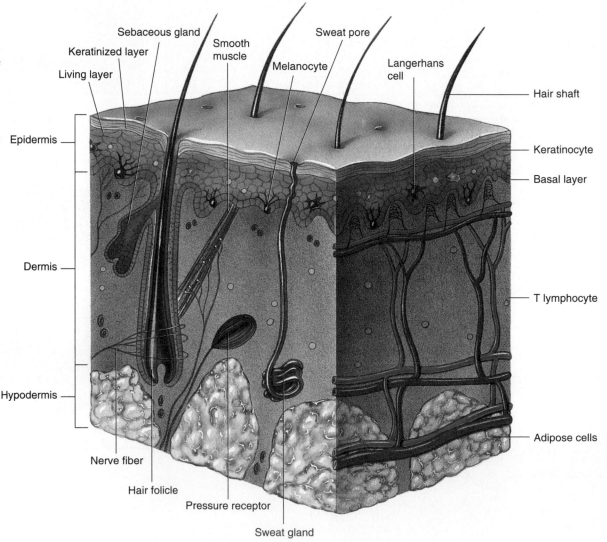

Sebaceous gland
Smooth muscle
Keratinized layer
Sweat pore
Melanocyte
Langerhans cell
Living layer
Epidermis
Hair shaft
Keratinocyte
Basal layer
Dermis
T lymphocyte
Hypodermis
Adipose cells
Nerve fiber
Hair folicle
Pressure receptor
Sweat gland

The sweat glands open to the surface through a skin pore (Figure 18.11). There are two types of sweat glands. One type (**eccrine glands**) is linked to the sympathetic nervous system. These glands are widely distributed over the body surface and produce a watery, salty solution. The second type of sweat gland (**apocrine glands**) is larger and located primarily in the armpits and groin areas. These glands produce a complex solution that, when acted on by surface bacteria, cause what is commonly known as body odor.

## Hair and Nails Are Skin Derivatives

Animals have several types of structures derived from skin, including hair, nails, scales, feathers, and horns. Antlers are derived from bone, and are not related to horns or other skin structures. In this section, we discuss only two derivatives, hair and nails.

The human body is covered with hair except for the eyelids, lips, nipples, parts of the external genitals, palms of the hand, and soles of the feet. The **hair shaft** extends above the skin surface (✦ Figure 18.12), and a **hair root** extends from the surface to the base or **hair bulb** (✦ Figure 18.13). Cells in the bulb are supplied with a capillary network and divide mitotically to produce columns of hair cells that become keratinized. Several aspects of hair are genetically controlled, including pattern baldness, hair color, and whether the hair is curly or straight.

Each hair follicle has a muscle, called the **arrector pili,** running from the follicle to the dermis (Figure 18.13). When this muscle contracts, it pulls the hair follicle upright, causing the hair to stand up perpendicular to the skin surface, forming what is commonly called "goose pimples." In animals with fur coats, this reaction is used to conserve body heat, and is also used as a form of communication in confrontations.

Nails consist of highly keratinized, modified epidermal cells. The nail rests on an epidermal layer known as the nail bed, which is thickened at the base of the nail to form the **lunula** or little moon (✦ Figure 18.14). Cells in this thickened base, called the nail matrix, constantly divide to form cells that are linked together and become keratinized to form the nail.

✦ **FIGURE 18.12**

**Hair projects from follicles, which are infoldings of the skin.**

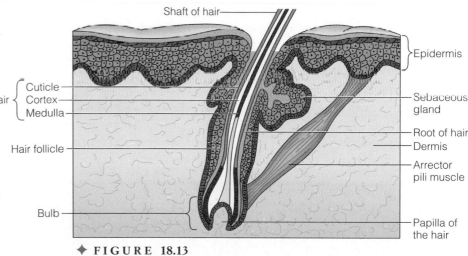

✦ **FIGURE 18.13**
**A hair follicle and its components.**

✦ **FIGURE 18.14**
**Nails. (a) External view of fingernail. (b) Section through nail showing the fingernail and the nail bed.**

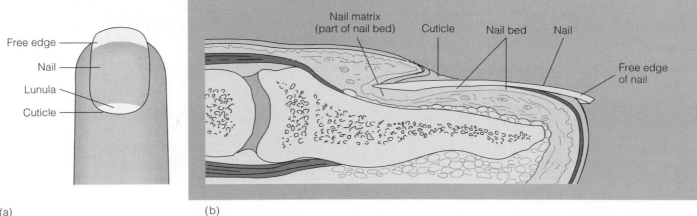

(a)

(b)

## THE SKIN PLAYS AN IMPORTANT ROLE IN HOMEOSTASIS

The roles of the skin in homeostasis are related to its structure, organization, and derivatives. The skin functions in many homeostatic roles, including protection, regulation of body temperature (**thermoregulation**), sensory reception, water balance (**osmoregulation**), synthesis of vitamins and hormones, and absorption of materials.

### The Skin Is a Protective Barrier between the Internal and External Environments

One of the primary functions of the skin is to serve as a barrier to prevent microorganisms from entering the body, and to prevent loss or exchange of extracellular fluids to the environment. The multiple layers of epithelial cells of the epidermis form a physical barrier that prevents the multitudes of surface bacteria and fungi from gaining entry to the body. In addition, the acidic nature of secretions by skin glands retards the growth of surface bacteria and fungi.

A second type of barrier is formed by the melanocytes and their production of melanin (✦ Figure 18.15). This protects the mitotically active cells of the epidermis and the underlying dermis from the damaging effects of ultraviolet radiation present in sunlight (See Beyond the Basics, p. 348).

If the skin is penetrated by an infectious microorganism, cells in the epidermis and dermis participate in what is known as an **inflammatory response.** This response (described in Chapter 22) activates the body's immune system to fight the infection. Wandering white blood cells called **macrophages** and epidermal cells called **Langerhans' cells** engulf invading microorganisms. They also release chemicals that produce both local and long-range mobilization of cells of the immune system to fight the infection.

### Temperature Regulation Is Controlled by Blood Flow in the Skin

The temperature of the body is regulated, in part, by heat and cold receptors in the skin. These sensory receptors feed a region of the brain known as the **hypothalamus.** When the body temperature rises, the hypothalamus, through the nervous system, signals the sweat-producing glands in the skin, resulting in the release of sweat in quantities of up to 1 to 2 liters per hour, cooling the body by the evaporation of sweat.

As the body temperature rises, conduction and radiation are accomplished by the hypothalamic-induced dilation of blood vessels in the dermis (✦ Figure 18.16). This allows more blood to be carried to the skin, and the heat from the blood is conducted to the body surface and radiates into the atmosphere. Convection is a constant process of heat transfer between the skin and the environment.

If the body temperature falls, skin receptors signal the hypothalamus, resulting in the shut down of sweat glands, and the constriction of dermal blood vessels, conserving heat. Heat loss from the skin occurs by a series of mechanisms, including evaporation, conduction, radiation, and convection. If the body's core temperature falls even after blood flow through the skin has been restricted, the hypothalamus can engage in **thermiogenesis,** or heat generation by raising the body's metabolic rate, and by shivering.

### The Skin Helps Regulate Water Balance

Cells of the body do not function properly when they shrink or swell because of imbalances in the distribution of water inside cells and in the extracellular fluids. Water balance is maintained by controlling input and output. The skin is fairly impermeable to water in both directions. Water loss occurs through the skin in two ways, by

✦ **FIGURE 18.15**
Melanin production is higher in (a) black skin than in (b) white skin.

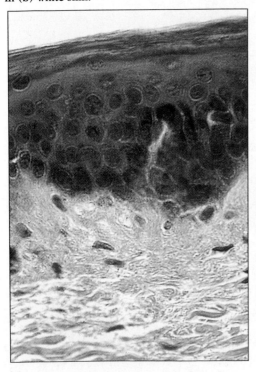
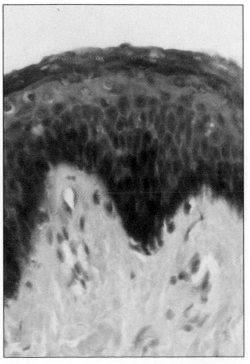

(a)  (b)

evaporation and by sweating. Normally, only a small amount of water (about 100 ml) is lost by these mechanisms. In hot weather, up to 4 liters per hour can be lost, to help cool the body by evaporation.

When the protective barrier of the skin is lost by burns, or accidents, fluid loss can be greatly increased. Often, fluid loss through damaged skin can cause serious fluid imbalances, and can be life-threatening if not treated.

## The Skin Is a Primary Organ for Sensory Reception

The skin contains several types of receptors, including those for temperature, touch or pressure, and pain. Free endings of nerve cells are found associated with hairs and in between epidermal cells. When activated, they generate signals that the nervous system interprets as touch or pain. Deeper in the epidermis are touch receptors called **Meissner's**

**✦ FIGURE 18.16**

**Heat transfer. (a) Radiation is the transfer of energy from a warmer object to a cooler one. (b) Conduction is the transfer of heat by direct contact. (c) Convection is the transfer of heat energy by air currents. (d) Evaporation is the cooling of a surface by converting a liquid into a vapor.**

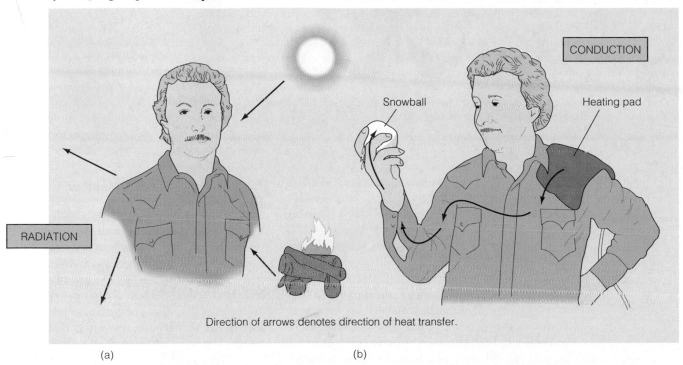

(a)                                        (b)

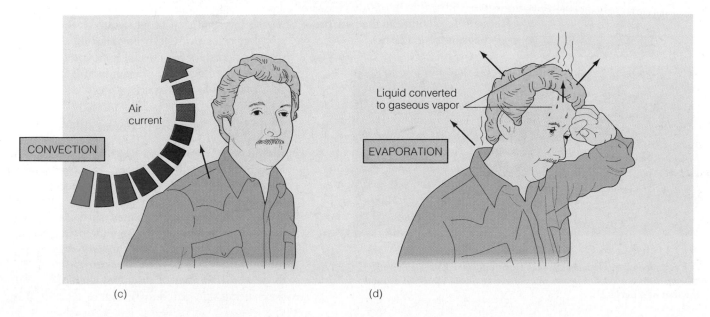

(c)                                        (d)

# SUNBURN AND CELL SUICIDE

Most people are familiar with the effects of overexposure to the sun. The result: sunburn. The sun emits a wide range of ultraviolet (UV) radiation, but it is the UVB rays (from 280 to 320 nm in wavelength) that produce sunburn. Usually, signs of sunburn appear within 24 hours and include skin reddening. More severe overexposure can result in pain, tenderness, swelling, and blisters. Often, within a few days, the skin begins to peel, and the skin underneath is tender and sensitive to sunlight for a few days. Although blondes and redheads are most susceptible to sunburn, blacks can become sunburned with prolonged exposure.

The dead skin cells that result from sunburn may actually be a defensive reaction of the skin to UV exposure. In fact, skin cells are not killed by the sun; they may, instead, commit suicide.

This process may be important as a safety mechanism, in which skin cells die rather than become cancerous.

Ultraviolet light is absorbed strongly by the DNA in skin cells. This absorbed energy causes damage to DNA in a number of ways, including strand breaks and mispairing of bases. Normally, a cell pauses before beginning mitosis, to repair any damage to DNA. The repair process prevents damaged DNA from being passed on to a new generation of cells. A gene called p53 directs the pause and the repair of damaged DNA. In cases where the DNA is severely damaged (as in sunburn), the p53 gene can initiate a process leading to cell death rather than repair the damaged DNA. This programmed cell death is termed **apoptosis.**

In cases where UV exposure damages the p53 gene in a skin cell, the cell loses its ability to undergo apoptosis and to repair its DNA.

While other cells die because of damage to their DNA, the p53-deficient cell survives, divides, and passes on its damaged DNA. These mutated cells continue to grow, forming a precancerous lesion. Further exposure to sunlight will not kill these cells, and may result in the accumulation of additional mutations leading to skin cancer.

Mutation of the p53 gene is associated with many forms of cancer, and may be the most common genetic change observed in human cancers. More than half the cases of breast, colon, lung, and bladder cancer examined carry a p53 mutation. This relationship emphasizes the link between genes and cancer. It also suggests that it is worthwhile to protect the p53 genes in your skin cells, because more than 400,000 new cases of skin cancer are reported each year.

---

**corpuscles.** These receptors are concentrated in the tips of fingers and lips, and make these areas very sensitive to touch. Other receptors called **Pacinian corpuscles** respond to pressure (✦ Figure 18.17).

Temperature receptors are also located in the skin, and can detect heat and cold. Cold receptors are more numerous than heat receptors, and are more numerous in fingertips and lips than elsewhere on the skin. Signals from these receptors travel via nerves to the spinal cord, and from there to the brain. In the brain, the signals are interpreted as heat or cold, and their surface origins recorded. The sensations we call an itch or a tickle are thought to originate in the bare nerve endings in the epidermis.

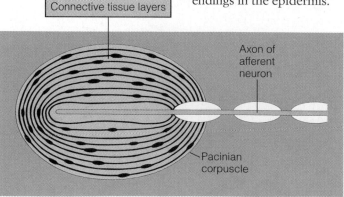

Connective tissue layers

Axon of afferent neuron

Pacinian corpuscle

✦ **FIGURE 18.17**

**A Pacinian corpuscle is a skin receptor that detects pressure and vibration. It consists of layers of connective tissue wrapped around the end of a nerve fiber.**

## The Skin Synthesizes a Number of Important Compounds

Cells of the skin synthesize a number of compounds, including melanin, and other pigments such as carotenes, that give skin its characteristic color. The skin assists in the synthesis of vitamin D from a dietary sterol, called dehydrocholesterol. In the presence of sunlight, this compound is converted to **cholecalciferol,** which is converted to vitamin D. Vitamin D acts as a hormone, and plays an important role in maintaining homeostatic levels of calcium and phosphorus in the body. Vitamin D acts on the intestine as a target organ to promote absorption of calcium and phosphorus, compounds that are necessary for normal bone growth and repair. Children deficient in vitamin D develop bone abnormalities known as rickets.

## The Skin Is Selectively Permeable

Although the skin forms a protective layer, it is not impermeable to all substances. The skin is selectively permeable, especially to substances that are fat soluble, such as certain vitamins (A, D, E, and K), as well as steroid hormones (such as estrogen). These materials cross the skin and enter the circulatory system through the capillary networks in the dermis. This property of skin has been used to deliver a number of drugs for therapeutic purposes using skin patches (✦ Figure 18.18). Drugs delivered in this manner include estrogen, scopolamine (for motion sickness), nitroglycerin (for heart patients), and nicotine (for those trying to quit smoking).

## The Integument Has Multiple Roles in Homeostasis

The integumentary system has multiple roles in homeostasis, including protection, temperature regulation, sensory reception, and biochemical synthesis. The other major body systems discussed in this chapter also play important and interlocking roles in maintaining the internal environment of the body within the limits necessary to support life. All body systems work in an interconnected manner to establish and maintain a homeostatic environment, just as the skin works in conjunction with the nervous system, the circulatory system, and the immune system, among others, to fulfill its homeostatic role.

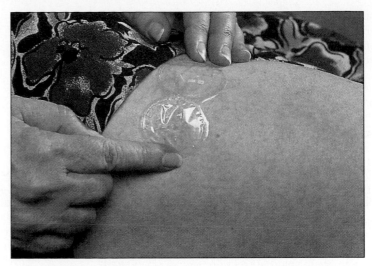

**◆ FIGURE 18.18**

**The ability of the skin to absorb certain chemicals is used to deliver controlled doses of drugs to the body.**

 SUMMARY

1. Homeostasis maintains the composition of the internal environment (body fluids) in a steady state. Maintaining homeostasis is essential for the survival and function of cells. Each cell of the body, as part of an organ system, contributes to homeostasis.

2. Several important elements of the internal environment must be maintained to achieve homeostasis. These include the concentration of oxygen and carbon dioxide, the pH, the concentration of nutrients and waste products, the concentration of salts and other electrolytes, and the volume and pressure of the extracellular fluid.

3. Two main types of control systems are used in homeostasis: open systems and closed systems. Open systems are linear and have no feedback. Closed systems have two components: a **sensor** and an **effector.** The sensor detects change and signals the effector, which initiates a response.

4. Homeostatic regulation is for the most part, extrinsic. In higher animals, the nervous system and the endocrine system are the major extrinsic control systems. Local or intrinsic control is also employed in maintaining homeostasis. These controls are local, and involve only one tissue or organ.

5. Most homeostatic systems operate on the principle of negative feedback control. In negative feedback, information produced by the feedback reverses the direction of the response. Positive feedback control is also used in some circumstances. In positive feedback, input increases and accelerates the response.

6. The skin and its derivatives—hair, nails, feathers, horns, and glands—make up the integumentary system. In humans, the skin is the largest organ in the body, and amounts to 12 to 15% of body weight.

7. The roles of the integument in homeostasis are related to the structure, organization, and derivatives of the skin. The skin functions in several homeostatic roles, including protection, regulation of body temperature (**thermoregulation**), sensory reception, water balance (**osmoregulation**), synthesis of vitamins and hormones, and absorption of materials.

8. All body systems work in an interconnected manner to establish and maintain a homeostatic environment, just as the skin works in conjunction with the nervous system, the circulatory system, and the immune system, among others, to fulfill its homeostatic role.

 KEY TERMS

apocrine glands
apoptosis
arrector pili
basal cells
circulatory system
cholecalciferol
dermatoglyphics

dermis
digestive system
eccrine glands
effector
endocrine system
epidermis
excretory system

extracellular
hair bulb
hair root
hair shaft
homeostasis
hypothalamus
immune system

inflammatory response
integument
integumentary system
internal environment
keratin
keratinocyte
Langerhans' cells
lunula
macrophages

Meissner's corpuscles
melanin
melanocytes
muscular system
negative feedback control
nervous system
osmoregulation
oxytocin
Pacinian corpuscles

plasma
positive feedback control
reproductive system
respiratory system
sensor
skeletal system
skin
thermiogenesis
thermoregulation

# QUESTIONS AND PROBLEMS

## TRUE/FALSE

1. Multicellular organisms are capable of maintaining their internal environment within broad limits that are compatible with proper cell functioning.
2. The integumentary system includes feathers, nails, glands, antlers, and skin.
3. Increases in the carbon dioxide level in blood raises pH levels.
4. Positive feedback control increases and accelerates a response.
5. Skin cells are capable of synthesizing bodily compounds.

## FILL IN THE BLANK

1. The skin aids in synthesis of _____ from _____. This compound is converted to _____, and then to vitamin D.
2. Skin contains touch receptors called _____ and pressure receptors called _____.
3. In _____ systems the information produced reverses the direction of the response.
4. Glands linked to the sympathetic nervous system are _____ and are widely distributed over the body. _____ are primarily located in the armpits and groin regions and contribute to body odor.
5. The epidermal layer of skin arises from the _____ of the embryo and the dermal layer originates from the _____ .

## MATCHING

Match the system with its appropriate function.

1. circulatory
2. digestive
3. endocrine
4. excretory
5. integument
6. immune
7. muscular
8. nervous
9. reproductive
10. respiratory
11. skeletal

____ body support
____ movement
____ species survival
____ controls activity of internal environment
____ transports gases, nutrients, and wastes
____ moves oxygen and removes carbon dioxide
____ receives food
____ coordinates actions of body
____ regulates internal fluid volumes
____ outermost protection
____ defends internal environment against intruders

## SHORT ANSWER

1. Outline the structural components and the development of the skin cell layers.
2. When cats feel threatened, their fur stands on end along their spine and tail. Explain how this defense mechanism occurs.
3. Name the homeostatic roles of skin.
4. Describe the involvement of the hypothalamus in regulating body temperature.

## SCIENCE AND SOCIETY

1. Albinism is one of the most widely recognized and striking genetic conditions. There are many forms of albinism and all but one involve generalized depigmentation of the skin. Albinism is the result of abnormal melanin metabolism. Most forms of this condition are inherited as autosomal recessive traits. What homeostatic problems do you think people with albinism experience? Are there ways for these individuals to regulate these problems? Melanin is responsible for skin coloration in humans. Can people with albinism stimulate melanin production in their skin by taking drugs? Why or why not? If both parents are unaffected but carry the same recessive gene for albinism, what is the risk for each pregnancy for an affected child?

2. Cancer is not a single disease. All cancers have in common uncontrollable cell growth and, when untreated, have a lethal outcome. Early diagnosis and treatment are vital to prevent death of the affected individuals. Most cancers occcur later in life. However, those that occur in childhood tend to have a greater degree of malignancy. Essentially, cancer results from mutations in somatic cells and its progression is often due to mutations in genes usually produced by environmental factors. Although cancer at any age has a genetic component, cancer in childhood is more likely to be associated with a hereditary predisposition. Why do you suppose uncontrollable cell growth is bad for an organism? What homeostatic mechanisms have gone awry when cancer develops? Why do you think childhood cancers are more likely to be associated with a genetic predisposition than late-onset forms? Why do you suppose there is a higher degree of malignancy in childhood cancers than in adult forms? It has been estimated that nearly 1 in 3 adults will be affected with some form of cancer during their life. How does this number appear to you?

# 19

# MUSCULAR/SKELETAL SYSTEMS

OPENING IMAGE

*The Venus de Milo is an artistic*

*expression of human beauty.*

Anyone who visits the Louvre Museum in Paris and sees the statue of the Venus de Milo is struck by this example of human beauty. The archaeologist who uncovered the statue in 1820 was similarly moved, and wrote about her beauty and perfection. Measurements of the statue revealed that it is not a model of perfection, because there is a lack of symmetry between the left and right sides of the body. This finding inspired many European physicians, who measured their patients and found that slight bilateral asymmetry is common. Does this mean that ideal beauty in men and women is related to symmetry, or is a slight asymmetry considered part of beauty? The relationship between beauty and symmetry is not easy to resolve, but scientists including ecologists, evolutionary biologists, and geneticists have turned to the study of symmetry as a trait that may play a role in performance, mate selection, and evolutionary change.

Humans are bilaterally symmetrical. This means that if a line is drawn vertically from the top of the head down through the body, the right side should be an exact mirror image of the left side. In reality, however, almost all humans have a degree of fluctuating asymmetry, so called because there is no trend in whether the right or left side varies.

Is fluctuating asymmetry important, or is it just an example of variation? That question is difficult to answer, but research is providing some interesting clues. Two English scientists compared asymmetry in young racehorses with their performance. Using 10 structural characteristics they measured differences between the left and right sides of the body. Horses with the largest differences performed poorly, according to an examination *of racing records. Why there is a correlation between symmetry and performance has not yet been determined. Perhaps symmetry in sensory systems such as eyes and ears is essential in judging distance and the position of other horses. Symmetrical bodies may be more efficient in athletic performance, or symmetry may be an indicator of overall fitness.*

*Studies in human newborns indicate that large differences between the right and left sides of the body are associated with defects of the heart and circulatory system. Currently, investigators are studying asymmetry in the bones of adult hands, to see whether asymmetry is associated with diseases such as osteoporosis.*

*In this chapter we consider the structure and function of the skeleton and muscles of the body. In animals with bilateral symmetry, these are the systems that make a large contribution to symmetry. Even though we are concentrating on the components of the skeleton and muscle system, on a larger scale, these systems may make subtle contributions to overall health and fitness.*

## SKELETAL SYSTEMS PROVIDE SUPPORT AND PROTECTION

The single-celled ancestors of animals were organisms whose weight was entirely supported by ocean water, and they required little more than cilia to provide a means of active movement. The evolution of multicellular animals was accompanied by the development of systems to support and maintain body shape and to assist in movement.

In particular, animals that live on land require structural support to maintain body shape and to allow movement from place to place. In the terrestrial environment, higher plants achieved large size only when a rigid support system in the form of xylem developed.

In addition to support, skeletons provide protection. Vertebrate skulls protect the brain, and ribs protect the heart and lungs. The shells of mollusks offer the organism protection from predators.

## THERE ARE THREE MAIN TYPES OF SKELETAL SYSTEMS

Animals are characterized by movement. Movement depends on the contraction of muscles attached to a skeleton. Muscle contractions are transmitted to fins, legs, wings, arms, and other body parts through the skeleton. This skeleton can consist of a fluid-filled body cavity, an external arrangement of hard plates in the form of a shell, an outer skeleton, or an internal set of bones, made up of connective tissue hardened by calcium deposition.

### Hydrostatic Skeletons Use Fluid to Provide Support

At first glance, animals such as jellyfish and earthworms do not seem to have recognizable skeletons. However, these organisms do have a skeletal system; they employ internal **hydrostatic skeletons** to give support and shape to the body. Hydrostatic skeletons consist of fluid-filled closed chambers. Internal pressure generated by a combination of fluid in a compartment and muscular contraction can be used for maintaining body shape and for movement.

In cnidarians such as the sea anemone, a network of internal fluid-filled chambers provides support and maintains body shape (✦ Figure 19.1). Movement of the tentacles is achieved by action of contractile fibers embedded in the walls of the body and tentacles. These fibers run longitudinally in the outer layer of body cells, and in a circular pattern in the inner layer of cells. The anemone can elongate or contract its body by contracting its circular or longitudinal muscles (✦ Figure 19.2).

✦ **FIGURE 19.1**
**Sea anemones have a hydrostatic skeleton.**

In annelids such as the earthworm, each segment contains a fluid-filled compartment and a set of circular and longitudinal muscles. This combination of fluid-filled compartments and muscles forms a hydrostatic skeleton. In the earthworm (✦ Figure 19.3), contractions of the longitudinal muscles shorten and thicken the animal, whereas contractions of the circular muscles elongate the animal, and reduce the body diameter. Because of the segmented body plan, it is possible for one group of segments to be elongating while an adjacent group contracts. Rhythmic waves of muscle contraction and relaxation along the length of the body allow the earthworm to move forward. Movement is aided by groups of bristles called **setae** that extend from the body. The setae grip the earth and pull the body forward during muscle contraction (Figure 19.3).

## Exoskeletons Provide Support and Protection

Many arthropods (the group of animals that includes the insects) have an external skeleton or **exoskeleton** (✦ Figure 19.4). These exoskeletons encase the body in a hard, jointed covering, and all muscles and organs are enclosed by the skeleton. The skeleton is thinner and more flexible at joints, allowing for a range of body movements. Muscles for movement attach to the inner surface of the exoskeleton. In terrestrial animals the outer surface of the exoskeleton is often covered by a thin layer of waxy material to prevent water loss. Some mollusks such as clams have exoskeletons in the form of shells made from calcium carbonate (✦ Figure 19.5).

One of the limitations of exoskeletons is that they restrict the growth of the organism. As a result, the exoskeleton must be shed periodically, in a process called **molting,** and a new, larger exoskeleton must be formed. A new, soft exoskeleton develops under the old one. As the old exoskeleton is shed, the new one expands and then hardens, providing room for growth. Individuals of some species molt only a few times before reaching adult size, while others, including lobsters and crabs, molt at intervals throughout their life. In clams, the exoskeleton is not shed, but grows by depositing calcium carbonate at the edge of the shell.

The bulk and weight of exoskeletons and the associated mechanical problems of moving these skeletons limits the size of organisms. Although arthropods are among the

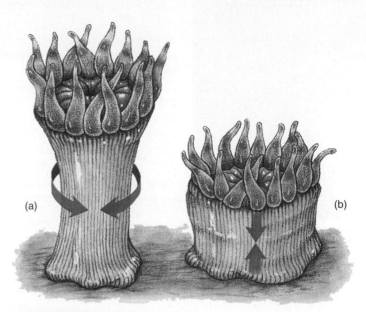

✦ **FIGURE 19.2**

Muscular contractions in sea anenome. (a) Contraction of circular muscles elongates the body. (b) Contraction of the longitudinal muscles shortens the body.

✦ **FIGURE 19.3**

Earthworms move by alternate contractions of circular and longitudinal muscles.

✦ **FIGURE 19.4**

The exoskeleton of this beetle offers protection from predators and prevents water loss. Muscles attached to the inside of the skeleton allow movement.

✦ **FIGURE 19.5**

Some mollusks have a shell that serves as an exoskeleton.

**✦ FIGURE 19.6**
Spiders walk by a combination of muscle contraction and hydrostatic pumping.

most successful organisms on Earth, they are usually smaller than organisms with internal, hard skeletons.

In some cases, animals use a combination of skeletal types for support and movement. Spiders have a hardened exoskeleton to provide protection and body shape, but can also use hydrostatic pressure for movement (✦ Figure 19.6). Their leg motion is accomplished partly by muscles attached to the exoskeleton and partly by hydrostatic fluid pumping.

Spiders flex (bend) their legs by the action of muscles, but extend (straighten) their legs by pumping blood into the leg to generate hydrostatic pressure. As a spider walks, muscle action alternates with fluid movement, extending and retracting each leg.

## Endoskeletons Are a Form of Connective Tissue

Vertebrates have an internal supporting **endoskeleton,** with muscles on the outside (✦ Figure 19.7). Some vital organs are enclosed and protected by the skull and the rib cage. The endoskeleton of vertebrates is composed of cartilage or a combination of cartilage and bone. Cartilage and bone are composed of connective tissue, a composite of cells embedded in a matrix. The spaces between the unspecialized cells are filled with a matrix that may contain gelatinous materials, elastic fibers, and inorganic materials. The matrix is secreted by cells of the connective tissue.

Skeletons of species such as sharks, rays, and paddlefish are composed entirely of cartilage; most adult vertebrates have skeletons that contain cartilage and bone. Bone is the main skeletal component in humans, and cartilage is found only in joints and flexible structures including the nose, external ear, and walls of the larynx, trachea, and bronchi.

## The Musculoskeletal System Is Multifunctional

Taken together, the skeleton and muscles form the musculoskeletal system. In the remaining sections of the chapter, we examine the components of the human skeleton, the structure and function of muscle, and the interaction of these components in producing motion.

The musculoskeletal system, as is the case with other organ systems, plays an important role in homeostasis. This system allows the organism to move to more favorable conditions. In addition, certain bones produce cells which become the immune system. Bones also regulate blood levels of calcium (important in muscular contraction), and rapid muscular contractions in the form of shivering help maintain body temperature.

## THE HUMAN SKELETON HAS TWO MAIN COMPONENTS

### The Axial and Appendicular Skeletons

The bones of the human skeleton (and those of all vertebrates) are grouped into two categories: (1) the **axial skeleton,** composed of the skull, vertebral column, and rib cage, and (2) the **appendicular skele-**

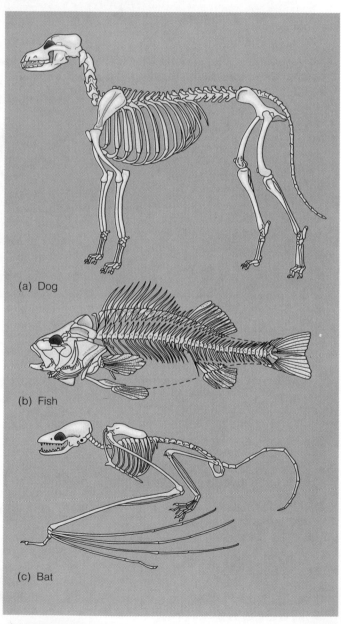

(a) Dog

(b) Fish

(c) Bat

**✦ FIGURE 19.7**
Endoskeletons of vertebrates consist of a skull, spinal column, ribs, and appendages.

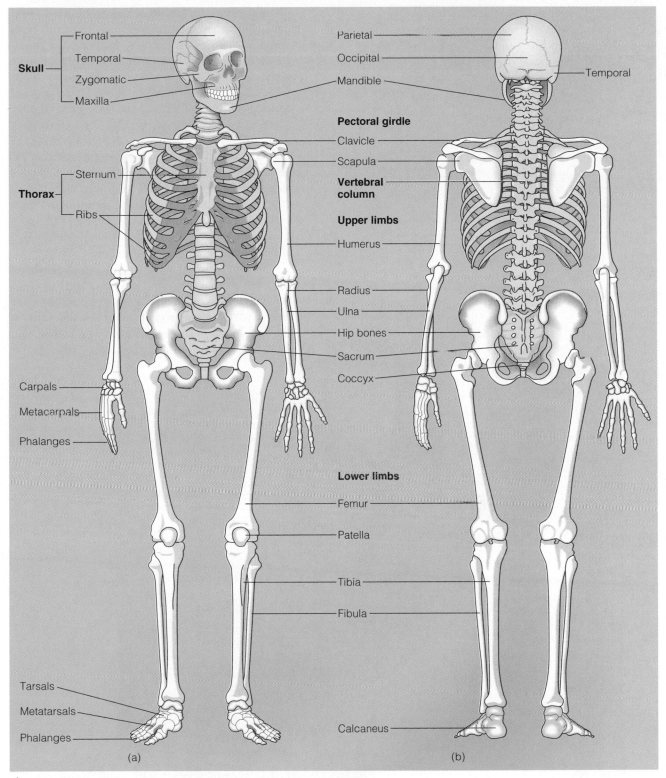

Skull
— Frontal
— Temporal
— Zygomatic
— Maxilla

Thorax
— Sternum
— Ribs

Carpals

Metacarpals

Phalanges

Tarsals

Metatarsals

Phalanges

(a)

Parietal

Occipital

Mandible

Temporal

**Pectoral girdle**
Clavicle
Scapula

**Vertebral column**

**Upper limbs**
Humerus

Radius
Ulna
Hip bones
Sacrum
Coccyx

**Lower limbs**
Femur
Patella

Tibia

Fibula

Calcaneus

(b)

✦ FIGURE 19.8

**The human skeleton. The region shaded in gold is the axial skeleton, and the unshaded region is the appendicular skeleton. Cartilage is shaded blue. (a) Anterior view. (b) Posterior view.**

ton, made up of the bones of the appendages (wings, legs, or fins) and the bones of the pectoral and pelvic girdles that join the appendages to the rest of the skeleton (✦ Figure 19.8).

In humans, the skull, and in particular the braincase or **cranium**, is composed of a large number of bones fitted tightly together at immovable joints. Some of the human cranial bones are not com-

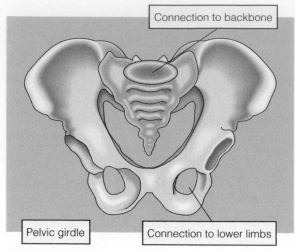

**FIGURE 19.9**

**The pelvis transfers weight from the upper skeleton to the legs.**

pletely formed at birth, and membranous areas called **fontanels** or "soft spots" do not form bony structures until the age of 14 to 18 months. The vertebral column is composed of 33 **vertebrae** that are separated from one another by connective tissue pads called disks. The disks allow for movement, making the intervertebral joint somewhat flexible. With age, disks weaken and can slip out of position, or rupture, producing lower back pain as the disk presses on nerves in or near the spinal cord. Attached to the vertebral column are the 12 pair of ribs (Figure 19.8). All but the lowest two pair of ribs are joined at the front to the **sternum** or breastbone. At this junction, cartilage connects the ends of the ribs to the sternum, providing flexibility that allows movement during respiration.

The appendicular skeleton contains the bones of the arms and legs, and the bones that form the pectoral and pelvic girdles. The upper skeleton of the arms and legs consists of single bones (humerus in the arm, femur in the leg) attached at the elbow and knee to the two bones of the lower limbs (radius

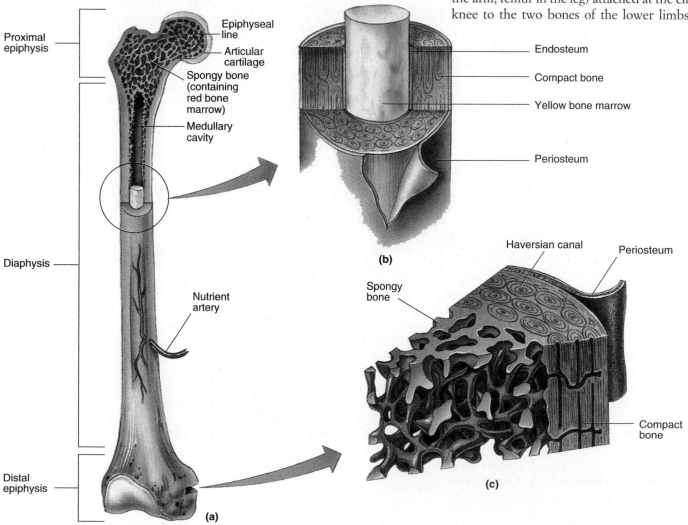

**FIGURE 19.10**

**Structure of a long bone. (a) Frontal section showing the epiphyses and diaphysis. (b) Enlarged view of the diaphysis showing the medullary cavity filled with yellow marrow. (c) The spongy and compact bone in the epiphysis.**

Bone fractures are usually repaired by resetting the bone pieces, putting the limb in a cast, and waiting for the body to repair the damage by knitting together the broken bones. Anyone who has had to wear a cast for six to eight weeks or more can tell you this is a slow process. More complicated breaks and some hip fractures are repaired by surgical procedures. One option for repairing these fractures uses stainless steel plates, screws, and pins. Another method uses bone grafts and a chemical substance called hydroxyapatite as a cement. Hydroxyapatite is chemically similar to bone, but is not widely used because it is difficult to

work with and offers little in the way of mechanical strength.

A new treatment for many types of fractures is now being used. This method is based on an understanding of how coral reefs are formed. In humans and other vertebrates, bone formation occurs in a matrix of connective tissue protein, which becomes embedded in the crystalline structure of the newly deposited bone. Coral organisms (cnidarians) form their mineralized skeleton without proteins. A new material, called Norian SRS, mixed into a paste, quickly hardens in place and enhances bone replacement at fracture sites.

In use, the fractured bones are realigned, and the paste is injected. After a short wait, the

paste hardens and forms a mineral called dahlite, which is naturally found in bones. The paste hardens to the same strength as bone in about 12 hours, rather than the six to eight weeks required for conventional healing. In addition, the body treats the injected material as if it were a natural part of the bone. Blood vessels grow into the dahlite, and the bone is remodeled in the same way as normal bones.

The SRS system is being tested in 12 hospitals across the country, and is expected to be of use in repairing fractures of the hip, shin, wrist, and some joints. If this trial proves successful in the near future, many bone fractures will be repaired in about 12 hours by injection of bone glue.

---

and ulna in the arm, and tibia and fibula in the leg). The bones of the lower limbs are connected to the joints of the wrists and ankles. Each of these joints is composed of a number of bones called the **carpals** in the wrists and the **tarsals** in the ankle (Figure 19.8). The bones in these joints are arranged to allow a complex series of motions for the hands and feet. Each hand and foot ends in a set of five digits (fingers and toes) composed of the metacarpals (hands) or the metatarsals (feet) and phalanges (fingers and toes).

The limbs are attached to the rest of the skeleton by a collection of bones known as a girdle. The **pectoral girdle** is composed of the **clavicle** (collar bone) and the **scapula** (shoulder blade). These are held together by ligaments and muscles. The top end of the humerus fits into a socket in the scapula, and is held in place by the clavicle and muscles (Figure 19.8). A dislocated shoulder results when the end of the humerus slips out of the socket in the scapula, stretching the ligaments and attached muscles; the bones are usually restored to their normal positions by manipulation.

The **pelvic girdle** is composed of two **hipbones** attached to each other to form a hollow cavity, the **pelvis** (✦ Figure 19.9). At its top, the pelvis is anchored to a lower part of the backbone called the sacrum, and at its lower end, the pelvis is connected to the **femur,** the upper leg bone. The body's weight is borne by the pelvis, and transmitted through the femur, lower legs, ankles, and feet to the ground. Vertebrates such as fish that are not exposed to the full force of gravity have only primitive pectoral and pelvic girdles. The transition to a terrestrial habit was associated with the enlargement and strengthening of these girdles to support the full weight of land animals.

## Bones Are a Living Tissue

Bones are a living tissue. In humans, bones range in size from those of the inner ear to the femur, the longest bone in the body. Although they vary widely in size and shape, bones share structural similarities. Bones are complex structures composed of cells embedded in a matrix of collagen fibers. The fibers allow bones to resist twisting and stretching and give bone its tensile strength. The hardness of bone and its ability to resist compression is due to the presence of inorganic salts, mainly in the form of calcium phosphate, in the matrix.

A typical bone (✦ Figure 19.10), has an outer dense layer called **compact bone,** and an inner layer of **spongy bone.** Compact bone forms the shafts of the long bones. It is made up of packed concentric layers of mineral deposits surrounding a central opening (**Haversian canal**). The concentric layers contain bone cells (**osteocytes**) that lay down new bone. The Haversian canals form a network of interconnected channels that contain nerves and blood vessels that nourish the osteocytes.

Spongy bone, found at the ends of long bones, is much less dense than compact bone, and contains an irregular, interlocking lattice of plates. The spongy bone in the femur, humerus, and sternum contains red marrow, in which stem cells divide to form the cellular components of blood and the immune system. The central cavity of long bones con-

## ◆ FIGURE 19.11
**Stages in bone formation.**

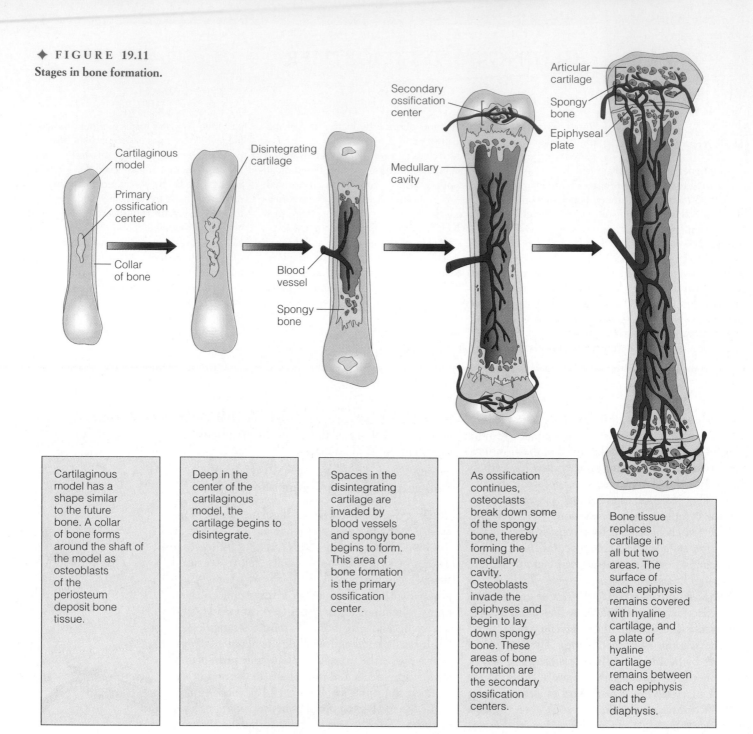

Cartilaginous model
Primary ossification center
Collar of bone

Disintegrating cartilage
Blood vessel
Spongy bone

Secondary ossification center
Medullary cavity

Articular cartilage
Spongy bone
Epiphyseal plate

| Cartilaginous model has a shape similar to the future bone. A collar of bone forms around the shaft of the model as osteoblasts of the periosteum deposit bone tissue. | Deep in the center of the cartilaginous model, the cartilage begins to disintegrate. | Spaces in the disintegrating cartilage are invaded by blood vessels and spongy bone begins to form. This area of bone formation is the primary ossification center. | As ossification continues, osteoclasts break down some of the spongy bone, thereby forming the medullary cavity. Osteoblasts invade the epiphyses and begin to lay down spongy bone. These areas of bone formation are the secondary ossification centers. | Bone tissue replaces cartilage in all but two areas. The surface of each epiphysis remains covered with hyaline cartilage, and a plate of hyaline cartilage remains between each epiphysis and the diaphysis. |

tains yellow marrow, which is used for the storage of fat (Figure 19.10b).

Bones are covered with an outer fibrous membrane called the **periosteum,** which serves as the site of attachment for skeletal muscles (Figure 19.10c). The inner layer of the periosteum contains bone-forming cells that lay down new bone and alter existing bone to meet changing conditions. The periosteum contains a rich supply of nerves, blood vessels, and lymphatic vessels that enter and leave

the bone at many locations. When fractured or bruised, pain impulses from bones are carried by the nerves that pass through the periosteum.

## Bones Grow Until Adolescence

Most of the bones in the human body are formed from cartilage in a process called **endochondral ossification.** Early in development, cartilage is laid down in a shape that resembles the bone to be formed. Near the center of the cartilage, cells enlarge and begin to deposit the mineral components of bone around the fibers in the extracellular matrix (✦ Figure 19.11). The matrix is compressed into a series of bars, plates, and partitions, and eventually will form the spongy bone. Bone-forming cells called **osteoblasts** attach to these structures and lay down the mineral portions of spongy bones. As this center of bone formation enlarges, cells called **osteoclasts** remove material to form the central cavity of the long bones. In the long bones such as the humerus and femur, secondary centers of bone deposition form at the ends of the bone, growing in toward the center.

While these changes are taking place, a connective tissue layer (the **perichondrium**) forms around the cartilage, and cells within the perichondrium begin to lay down a peripheral layer that will become compact bone (Figure 19.11). Blood vessels grow into the perichondrium and penetrate the cartilage, carrying stem cells into the interior and supplying nutrients to the cells embedded in the developing bone.

As bone growth occurs from each end toward the center, two bands or plates remain as cartilage (✦ Figure 19.12). During childhood, new cartilage forms in these plates, allowing for bone elongation during body growth. In late adolescence, the rate of cartilage formation in these plates slows and then stops. However, bone formation continues from each side, gradually eliminating the cartilage, and halting any further increase in bone length.

## Adult Bones Are Constantly Being Remodeled

During adulthood bone undergoes constant remodeling to adapt to physical stresses generated by physical activity. Exercise can increase the diameter and strength of bone; conversely, inactivity can decrease the size and strength of bones. Age is also a factor in bone loss; millions of older individuals, primarily women, suffer from **osteoporosis,** a disorder in which bone density decreases (✦ Figure 19.13). This occurs most often in postmenopausal women, indicating a role for the hormone estrogen in preventing bone loss. Increasing calcium intake, reducing protein intake, exercise, and low doses of estrogen are all effective in treating this condition.

## Joints Connect Bones

Where individual bones meet, a joint is formed. Joints can be classified functionally into three types. The bones of the cranium are examples of **immovable joints;** the bone edges tightly interlock, and are held together by fibers or bony processes that cross

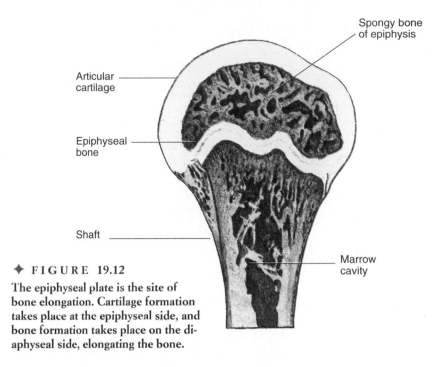

✦ **FIGURE 19.12**
The epiphyseal plate is the site of bone elongation. Cartilage formation takes place at the epiphyseal side, and bone formation takes place on the diaphyseal side, elongating the bone.

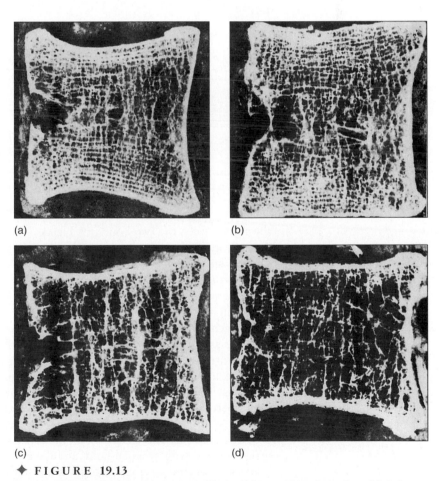

✦ **FIGURE 19.13**
In osteoporosis, the mineral portion of bone is lost, making it weak and brittle. (a) Section of vertebra from a 29-year-old woman. (b) A vertebra from a 40-year-old woman shows some thinning. (c) Severe bone loss in a specimen from an 84-year-old woman. (d) Bone loss is most severe in a 92-year-old woman.

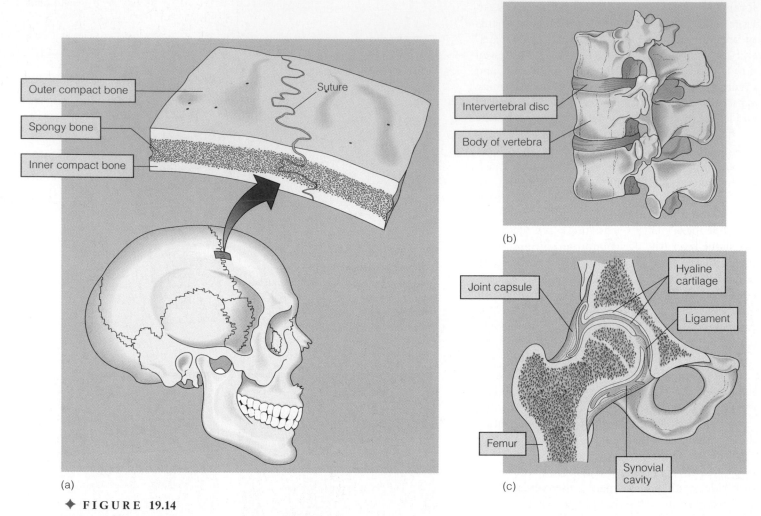

**Outer compact bone**
**Spongy bone**
**Inner compact bone**
Suture

(a)

**Intervertebral disc**
**Body of vertebra**

(b)

**Joint capsule**
Hyaline cartilage
Ligament
**Femur**
Synovial cavity

(c)

✦ **FIGURE 19.14**

Joints. **(a) Immovable joints in the skull are called sutures. (b) The disks between vertebrae allow for slight movement. (c) Synovial joints are flexible. A cross section through the hip joint.**

the gap between the two bones (✦ Figure 19.14a). A **partly movable joint** allows some degree of flexibility (Figure 19.14b). Joints of this type usually have cartilage between the bones, allowing for a slight degree of movement. The joints between vertebrae in the backbone are partly movable. Between each pair of vertebrae is an intervertebral disk that acts as a cushion and permits some flexibility. Similarly, most ribs are joined to the sternum by cartilage, allowing the ribs to flex during breathing. The most movable type of joint, called a **synovial joint,** permits the greatest degree of movement (Figure 19.14c). In these joints, the bones are covered by a sheath of connective tissue, the interior of which is filled with synovial fluid, and the ends of bones are covered with cartilage pads that reduce friction.

The outer surface of the sheath in synovial joints contains **ligaments,** dense parallel bundles of connective tissue fibers that strengthen joints and hold the bones in position. The inner surface of the

sheath, called the synovial membrane, contains cells that produce synovial fluid that lubricates the joint and prevents the two cartilage caps on the bones from rubbing together.

Some joints also contain **tendons,** which are bundles of connective tissue that link muscle to bone. **Bursae** are small sacs lined with synovial membrane and filled with synovial fluid. They act as small cushions to reduce friction between tendons and bones. The knee joint, for example, contains 13 bursae. The most common forms of synovial joints are hinge joints, such as the knee, and the ball-and-socket joint, such as the hip joint and shoulder joint.

## Skeletal Disorders Are Common

Joints are subject to a variety of problems, including injury, degenerative wear and tear, and inflammatory disorders (✦ Table 19.1). The most common injuries to joints are sprains, which result when mechanical stresses cause ligaments to rip or become separated from a bone. Sprains heal only slowly because ligaments do not have a rich supply of blood vessels; often, torn ligaments must be repaired surgically. Tendinitis and bursitis are inflammations of tendon sheaths and bursae and can result from heavy use, injury, or infection. The condition known as "tennis elbow" is tendinitis in the elbow joint resulting from heavy exercise of the joint.

One of the most common joint problems is **osteoarthritis,** a degenerative condition associated with the wearing away of the protective cap of cartilage at the ends of bones. As the cartilage erodes, bony growths or spurs develop, restricting movement and causing pain. The cause of osteoarthritis is not known, but may result from normal age-related changes in connective tissues, or simply the result of "wear and tear" that joints are subject to in everyday life. **Rheumatoid arthritis,** on the other hand, is a severely damaging and crippling form of arthritis that begins with inflammation and thickening of the synovial membrane, followed by bone degeneration and disfigurement (✦ Figure 19.15). This condition affects women more frequently than men, and tends to affect joints in the hands, feet, and knees. Some individuals may have a genetic predisposition to this disorder. In severe cases, joints may need to be replaced by artificial prosthetic devices (✦ Figure 19.16), restoring function to a failed joint.

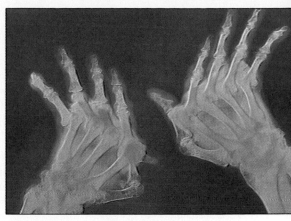

✦ **FIGURE 19.15**
Color-enhanced X ray of hands disfigured by rheumatoid arthritis.

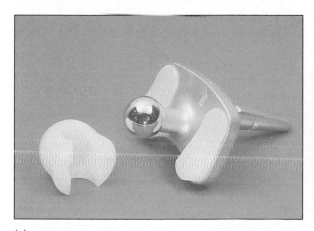

(a)

✦ **FIGURE 19.16**
Artificial joints. (a) Knee joint. (b) Hip joint.

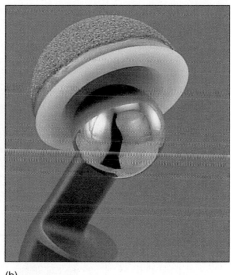

(b)

| TABLE 19.1 COMMON INJURIES OF THE JOINTS | | |
|---|---|---|
| INJURY | DESCRIPTION | COMMON SITE |
| Sprain | Partially or completely torn ligament; heals slowly; must be repaired surgically if the ligament is completely torn. | Ankle, knee, lower back, and finger joints |
| Dislocation | Occurs when bones are forced out of a joint; often accompanied by sprains, inflammation, and joint immobilization. Bones must be returned to normal positions. | Shoulder, knee, and finger joints |
| Cartilage tears | Cartilage may tear when joints are twisted or when pressure is applied to them. Torn cartilage does not repair well because of poor blood supply. It is generally removed surgically; this operation makes the joint less stable. | Cartilage in the knee |

No discussion of joints would be complete without considering what it means to be "double-jointed." Some people, especially when younger, have the ability to flex joints much further than others, and are often said to be double jointed. These individuals do not really have two joints where other have only one, but do have longer and/or more flexible ligaments surrounding joints, allowing the joint to be much more flexible than normal. For example, some people can hyperextend their thumb (hitchhikers thumb), or bend their wrists and fingers to allow their fingers to touch the back of the hand.

 **SKELETAL MUSCLE SYSTEMS WORK IN PAIRS**

In vertebrates, movement is the result of skeletal muscles acting on bones (✦ Figure 19.17). Tendons anchor many skeletal muscles across joints, so that when muscles contract, one bone in the joint moves, while the other bone remains stationary. Muscles generally work in pairs to produce opposite effects. These pairs or groups of opposing muscles are called **antagonistic muscles;** when one member of an antagonistic pair contracts, the other relaxes. For example, upper arm movement is controlled by an antagonistic pair of muscles, the biceps and the triceps (Figure 19.17). When the biceps contracts, the triceps relaxes, and the upper arm bends (flexes). To straighten the upper arm (extension), the biceps relaxes, and the triceps contracts.

Muscles are a distinctive tissue in that they possess two basic characteristics: electrical activity and mechanical activity. Muscle cells have an electrical gradient across the plasma membrane, with the outside of the cell being more positive than the inside. Under stimulation from the nervous system, this gradient can undergo an instantaneous reversal, initiating muscular contraction. As the muscle shortens, it produces a mechanical effect, such as a twitch or a movement.

### Humans Have Three Types of Muscle

There are three types of muscles in vertebrates: **smooth muscle, cardiac muscle,** and **skeletal muscle** (✦ Figure 19.18). Smooth muscle forms part of the walls of the digestive tract, bladder, and some blood vessels. In the digestive system, contraction of smooth muscles moves food through the stomach and intestines; contraction of smooth muscles empties the bladder; and in the blood vessels, contraction and relaxation of smooth muscle helps regulate blood pressure. Most smooth muscles are stimulated to contract by impulses from the nervous system or by hormones.

Individual smooth muscle cells are long and thin, and are joined together to form sheets of muscle tissue. In some organs, smooth muscle cells are connected to each other by gap junctions, allowing the entire muscle to contract as a single unit. Smooth muscle contracts in response to changes in the body's internal environment. Contractions of smooth muscle are slower than those of skeletal muscle. Smooth muscle can contract gradually and remain contracted longer than skeletal muscle.

Cardiac muscle, as the name implies, is found in the vertebrate heart. The cells of cardiac muscle are interconnected to form branching networks. End-to-end connections of cardiac muscle cells form structures called intercalated disks. Cardiac muscles contract in response to electrical impulses that origi-

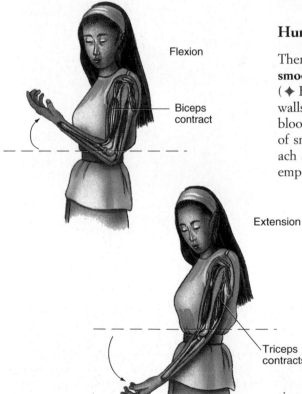

✦ **FIGURE 19.17**

**The biceps and triceps move the same body part, but in opposite directions. The biceps flexes the forearm and the triceps extends the forearm. These muscles work in pairs.**

Flexion

Biceps contract

Extension

Triceps contracts

*Sidebar*

**BONE INJURY DUE TO EXERCISE**

A combination of aerobic and weight-lifting exercise is the best approach to ensuring strong bones while at the same time decreasing the risk of injury during exercise. A proper diet, adequate warmup, proper shoes, and appropriate exercise surfaces also contribute to an effective and safe exercise program. When these or a combination of these elements is not followed, bone injuries may occur.

Shin splints are one of the most common exercise-related bone injuries. The intensity, duration, and frequency of exercise are directly proportional to the risk of injury in most athletic muscular and skeletal injuries. Shin splints are either inflamed tendons that attach muscle to the tibia (shin bone) or an inflammation of the periosteum (dense connective tissue that covers bone) of the tibia at the region where muscle attaches to it. Shin splints result from exercise on hard or uneven surfaces, repeated overuse, poor warmup, and wearing improper shoes. They typically require about three weeks to heal and rarely reoccur if proper conditioning takes place after healing.

nate in the sinoatrial node, the heart's pacemaker, located in the wall of the right atrium. The action of the pacemaker and contractions of the heart are discussed in Chapter 21.

The focus in this chapter is on skeletal muscle and its role in working with the skeleton to produce changes in body position and locomotion. Skeletal muscle consists of long, multinucleated cells or **muscle fibers.** Each fiber arises by the fusion of several embryonic muscle cells. Skeletal muscle fibers appear striated when viewed with a light microscope (✦ Figure 19.19). An intact muscle is surrounded by a connective tissue sheath that is elongated at the ends of the muscle to form a tendon that, in turn, attaches the muscle to a bone. Under this sheath is another layer of connective tissue that divides the muscle into a series of compartments (✦ Figure 19.20), each of which contains bundles of muscle fibers. Within each bundle, individual muscle fibers are surrounded by a thin connective tissue sheath. Blood vessels and nerves are embedded in these connective tissue sheaths, eventually reaching the individual muscle fibers.

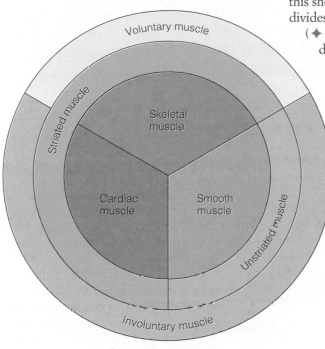

✦ **FIGURE 19.18**
**Muscle types.**

✦ **FIGURE 19.19**
**Light micrograph of muscle showing striations.**

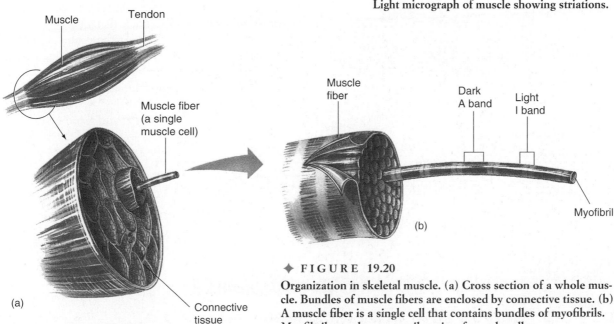

✦ **FIGURE 19.20**
**Organization in skeletal muscle. (a) Cross section of a whole muscle. Bundles of muscle fibers are enclosed by connective tissue. (b) A muscle fiber is a single cell that contains bundles of myofibrils. Myofibrils are the contractile units of muscle cells.**

(a)                                                    (b)

✦ FIGURE 19.21

(a) Light micrograph of muscle showing striations and nuclei. (b) Electron micrograph of muscle fiber.

✦ FIGURE 19.22

Enlarged view of myofibril, showing the molecular components, including the thick myosin filaments and the thin actin filaments.

Muscle fiber

Dark A band    Light I band

Myofibril

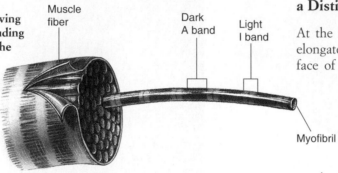

Z line    A band    I band

Portion of myofibril

M line    ←Sarcomere→    H zone

Thick filament
Thin fillament

A band    I band

Cross bridges

M line

H zone    Z line

Myosin

Thick filament

Actin

Thin filament

## Skeletal Muscle Fibers Have a Distinct Structure

At the microscopic level, each muscle fiber is an elongated cell with several nuclei just under the surface of the plasma membrane (✦ Figure 19.21). Most of the cell is filled with hundreds or thousands of striated, thread-like **myofibrils.** Within each myofibril, dense **Z lines** are a distinctive feature (Figure 19.21b). Beginning at the Z lines, each muscle can be divided into a series of repeating segments known as **sarcomeres,** which are the functional units of the skeletal muscle cell. To understand how muscles contract, it is necessary to explore the organization of the sarcomere.

Each sarcomere contains thick and thin filaments (✦ Figure 19.22). The thick filaments, made of a protein called **myosin,** occupy the center section of each sarcomere. The thin filaments, made of the protein **actin,** are anchored in the Z lines and extend toward the center of the sarcomere, but do not meet in the middle.

## Muscle Contraction Results from Filament Sliding

When a body part such as an arm moves, it does so because of muscular contraction. When muscles contract, they shorten. Since the functional unit of muscles is the sarcomere, a muscle can shorten only if its sarcomeres shorten. During contraction, actin filaments at each side of the sarcomere slide past the myosin filaments toward the center of the sarcomere, until they meet in the middle. This shortens the sarcomere, putting the Z lines closer together. In this model, called the **sliding filament model** of muscle contraction, the filaments remain the same length, but the ends of the actin fibrils move closer together until they touch (✦ Figure 19.23).

The question, of course, is how do actin filaments slide past myosin filaments and move toward the center of the sarcomere? Myosin molecules have club-shaped heads that project toward actin filaments (Figure 19.23a). The myosin heads attach to binding sites on the actin filaments, forming a bridge between the two filaments. The myosin heads then swivel toward the center of the sarcomere. This moves the actin filaments inward toward the center of the sarcomere. The myosin heads detach and reattach to the next available actin binding site, and swivel again, moving the actin filaments still closer to the center of the sarcomere. Each cycle of attachment, swiveling, and detachment shortens the sarcomere by about 1%, meaning that many cycles of myosin–actin interaction are needed in a single muscle contraction. Each cycle occurs rapidly, allowing hundreds of such cycles to take place each second.

Energy for muscle contraction is provided by ATP (see Chapter 4 for a discussion of ATP as an energy source). ATP binds to the cross bridges between myosin heads and actin binding sites. ATP is then converted into ADP and phosphate, resulting in a transfer of energy. This energy is captured with a high degree of efficiency to power the swiveling of the cross bridges. Muscles contain only enough ATP to support contraction for a few seconds, so ATP must be recycled quickly. Muscles store creatine phosphate, which reacts with ADP to regenerate ATP. Ultimately, of course, ATP for muscular contraction, as in all other cellular activities, is provided by the metabolism of glucose.

In addition, calcium ions ($Ca^{2+}$) are required for each cycle of myosin–actin interaction. When a muscle is stimulated to contract, calcium is released into the sarcomere, where it exposes the actin binding sites, allowing myosin to form cross bridges to the actin. When the muscle is no longer stimulated to contract, calcium ions are pumped from the sar-

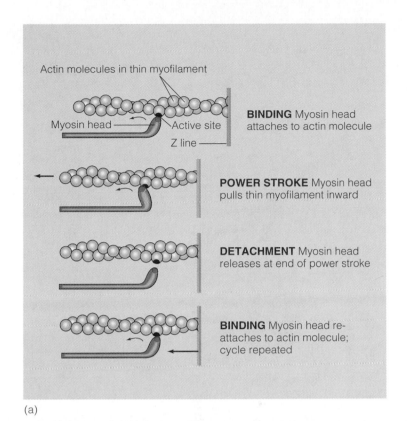

Actin molecules in thin myofilament

Myosin head — Active site

Z line

**BINDING** Myosin head attaches to actin molecule

**POWER STROKE** Myosin head pulls thin myofilament inward

**DETACHMENT** Myosin head releases at end of power stroke

**BINDING** Myosin head re-attaches to actin molecule; cycle repeated

(a)

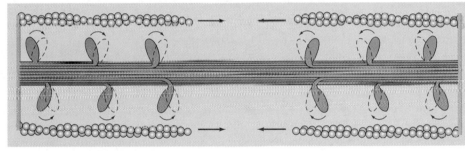

(b)

Thin myofilament          Thick myofilament

(c)

### ✦ FIGURE 19.23

**Muscle contraction. (a) In the presence of calcium ions, myosin filaments bind to a site on actin, pull, and then move to an adjacent site to repeat contraction. This series of contractions moves the actin myofilament toward the center of the sarcomere. (b) The movement of all myosin filaments is directed toward the center of the thick myofilament. (c) The action of the myofilaments pulls the thin actin myofilaments inward, causing muscle contraction.**

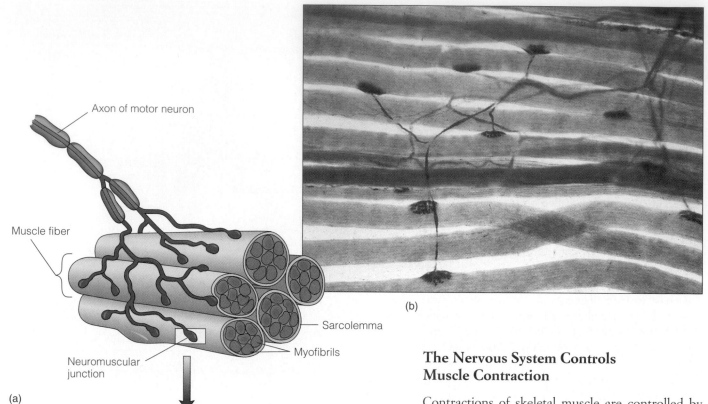

Axon of motor neuron

Muscle fiber

Sarcolemma

Myofibrils

Neuromuscular junction

(a)

(b)

**✦ FIGURE 19.24**

**(a) Muscles receive the axons of motor neurons.**
**(b) Light micrograph showing neuromuscular junctions in muscle fibers.**

comere back into storage. As the calcium is withdrawn, binding of myosin to the actin sites can no longer occur, and the muscle relaxes. The components of muscle fibers and their roles in contraction are summarized in ◆ Table 19.2.

## The Nervous System Controls Muscle Contraction

Contractions of skeletal muscle are controlled by the nervous system. **Neuromuscular junctions** (✦ Figure 19.24) are the point where a motor neuron attaches to a muscle cell. When a nerve impulse reaches this junction, it triggers the release of a chemical called **acetylcholine** from the axon ending. Acetylcholine binds to receptors on the surface of the muscle cell plasma membrane, causing a wave of electrical activity to travel along the plasma membrane. This electrical impulse releases calcium ions from the endoplasmic reticulum (ER) of the muscle cell (in muscle cells the ER is called the

| TABLE 19.2 COMPONENTS OF MUSCLE CONTRACTION | |
|---|---|
| MUSCLE FIBER COMPONENT | FUNCTIONAL ROLE |
| Plasma membrane | Conducts impulse from motor neuron |
| Sarcoplasmic reticulum | Releases stored calcium, which stimulates contraction; absorbs calcium to end contraction |
| Tropomyosin | Blocks binding sites on actin filament, preventing contraction |
| Troponin | Holds tropomyosin in place on actin filament, blocking contraction; binds to calcium, releasing tropomyosin to permit contraction |
| Actin filaments | Slide toward center of sarcomere during contraction |
| Myosin filaments | Pull the actin filaments toward center of sarcomere during contraction |
| Heads of myosin molecules (cross bridges) | Bind to actin and pull actin filaments; contain binding site for ATP; contain myosin ATPase, which catalyzes the breakdown of ATP |
| Calcium ions | Released from the sarcoplasmic reticulum; bind to troponin, causing it to release tropomyosin from binding sites on the actin filaments |
| ATP | Binds to cross bridges of myosin filaments; broken down by ATPase in cross bridges, providing energy for muscle contraction |

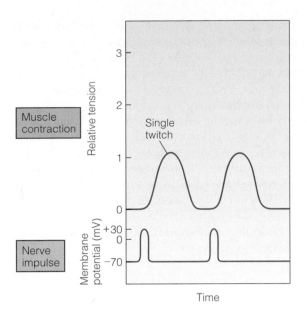

Muscle contraction

Nerve impulse

**✦ FIGURE 19.25**

**An impulse from a nerve cell causes calcium release and brings about a single short, muscle contraction called a twitch.**

sarcoplasmic reticulum). Calcium unmasks the binding sites on actin, initiating the events leading to muscle contraction. An impulse from a motor neuron causes a swift muscle contraction called a twitch; removal of calcium ions causes a subsequent relaxation of the muscle fiber (✦ Figure 19.25).

Obviously a different amount of force is required for a biceps muscle to pick up a piano than to pick up a pencil. If there is an all-or-nothing contraction response to a stimulus from the nervous system, how are such graded responses possible?

The answer is that skeletal muscles are organized into hundreds of **motor units,** each of which consists of a motor neuron and a group of muscle fibers (✦ Figure 19.26). Depending on the circumstances, a graded response to the amount of muscle force required can be achieved by controlling the number of motor units and the frequency of nerve impulses to these motor units. So, while individual muscle fibers contract on an all-or-nothing basis, whole muscles can contract on a graded basis because of their organization into motor units.

## Nonmuscular Cells Can Contract

In skeletal muscles, contraction is the result of interaction between microfilaments of actin and myosin. Surprisingly, actin and myosin are found in the cytoplasm of a wide range of eukaryotic nonmuscle cells, often comprising a significant fraction of the cell's total protein (✦ Figure 19.27). In these cells, instead of being anchored in Z lines, actin is attached to the inner surface of the plasma membrane. Movement in nonmuscle cells can result from the interaction of cytoplasmic actin and myosin. In cells of the intestine, one surface of the cell is folded into a series of projections known

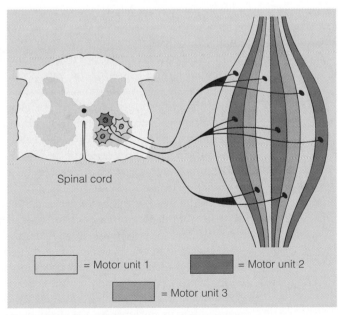

Spinal cord

☐ = Motor unit 1

■ = Motor unit 2

▨ = Motor unit 3

**✦ FIGURE 19.26**

**Motor units in a skeletal muscle.**

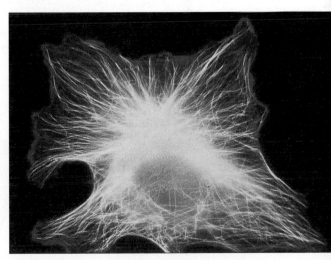

**✦ FIGURE 19.27**

**Actin is found in many cell types.**

**✦ FIGURE 19.28**

In some fish, muscles are arranged to create electricity that is used to stun prey and for defense.

as microvilli. Ordered arrangements of actin and myosin interact to produce contractions that cause the microvilli to move back and forth in a waving motion, allowing their surface areas to be exposed for the absorption of nutrients.

### Electric Organs Are Modified Muscles

In some fish (✦ Figure 19.28), modified muscles are specialized for the discharge of electricity. The electric organs of these fish consist of muscle fibers organized into structures called **electroplates.** These electric organs have evolved independently in a number of species of fish in both marine and freshwater environments. Electroplates develop from the same embryonic tissue that gives rise to skeletal muscle. As they develop, the cells become multinucleated. They elongate and enlarge laterally to form arrays of electroplates stacked in columns and organized to produce bioelectricity. The South American electric eel has more than 6000 plates organized into about 70 columns.

The plates are innervated by nerves that branch repeatedly to make numerous synaptic contacts with many modified muscle fibers. These plates are synchronously activated by a single nerve impulse. The amount of current generated (amperes) and its rate of flow (voltage) depends on the size and number of plates. Each plate produces only a

**✦ FIGURE 19.29**

Muscles and bones work together to act as levers, allowing Frank Thomas to generate enough force to hit a home run.

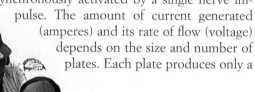

small amount of electricity, but when added together, the result can be spectacular. In the electric eel, more than 600 volts can be generated, but the amount of current is relatively weak. However, at maximum effort, more than 100 watts (amperes × volts) of power can be produced.

Other species of fish produce low-level electric fields, not from specialized electric organs, but from skeletal muscles. These include most species of freshwater fish. Fish that produce powerful electrical discharges use them for self-defense or for stunning prey; low-power discharges are used for a variety of functions such as locating objects in the surrounding environment, or for establishing and maintaining a territory.

### INTERACTION OF MUSCULAR AND SKELETAL SYSTEMS

Movements as delicate as those of a surgeon replacing a heart valve or as powerful as a sumo wrestler pushing an opponent out of the ring require an interaction between muscles and the skeleton. Muscles act by exerting force on a skeleton. When the skeletal muscles of an earthworm contract, they exert force on the fluid-filled interior of a body segment, which serves as a skeleton. Similarly, when a spider moves its legs, the muscles attached to the inner surface of the exoskeleton are exerting force on the hardened outer cuticle, causing the leg to move. In vertebrates, the bones and joints of the body act as levers when moved by contraction of attached muscles.

Two principles related to levers are important for producing motion. The first is that levers can amplify or increase the force and/or the velocity of motion. Second, the amount of this amplification is determined by the length of the lever. Swinging a baseball bat shows how the muscle and skeletal systems interact to produce the effects associated with levers. When you swing a bat, muscle contractions move the bones, generating force and velocity. Because of the length of the arm and the bat, the tip of the bat moves with a greater velocity, force, and distance than does the elbow. In the hands of a skilled batsman, the head of the bat achieves a velocity of 100 ft/sec (about 68 mph) as it crosses the plate (✦ Figure 19.29). The increase in the length of the lever (by holding the bat) amplifies the effect of muscular contraction. These principles are seen in all interactions between muscles and skeletons, and represent an adaptive mechanism that has evolved several times in response to the need for support and motion in animal bodies. Although many materials are used as skeletons, muscles derived from the mesodermal tissue layer in embryos remain a constant in the animal world.

# SUMMARY

1. Body support systems in the form of skeletons are important to multicellular animals, especially for life in a terrestrial environment. Several types of support systems in the form of skeletons have evolved independently, including hydrostatic skeletons, exoskeletons, and endoskeletons. In conjunction with muscular systems, skeletons also provide a means of locomotion, a point that distinguishes plants from animals.

2. Hydrostatic skeletons are found in marine invertebrates, such as jellyfish, and in terrestrial invertebrates, such as flatworms and annelids. Arthropods use an external skeleton for support, with the muscles arranged within the skeleton. This provides great strength, but limits the ultimate size of such organisms.

3. In vertebrates, two types of endoskeletons are found: those composed mainly of cartilage and those composed mainly of bone. The skeletons of vertebrates are connective tissue, which is characterized by widely spaced cells embedded in a matrix that may be gel, fibrous, or mineral. In lower vertebrates, the skeleton in the adult is made of cartilage, while in the higher vertebrates, including humans, most of the skeleton is laid down first in the form of cartilage, which is replaced by bone. In addition to providing support, the skeleton of higher vertebrates is important in homeostasis, helps regulate the ionic composition of blood, and serves as a repository for fat and the stem cells that give rise to the cells of the blood and immune system.

4. The vertebrate skeleton has two components; the axial and appendicular skeleton. The appendicular skeleton is joined to the body by sets of bones that form the pectoral girdle and the pelvic girdle. Lower vertebrates such as fish have only primitive girdles, but large, terrestrial animals have highly specialized girdles to offer support and strength to combat the pull of gravity.

5. The outer layer of bone contains cells that build or remodel bone. Centrally, bones contain a cavity that is filled with marrow, housing fat or blood-forming cells. In humans, the skeleton is laid down in the form of cartilage, which is replaced by bone during development, up to late adolescence. Joints are classified by the amount of movement they allow, and serve as articulation sites for the attachment of muscles.

6. The vertebrate body contains three types of muscle: smooth, cardiac, and skeletal. All muscles have two properties in common: electrical activity and mechanical activity.

7. The cells of skeletal muscles contain bundles of fibrils organized longitudinally into sarcomeres, the functional unit of muscles, which contain filaments of actin and myosin.

8. When stimulated to contract, calcium flows into the sarcomere, allowing myosin to form cross bridges and bind to actin. In the presence of ATP, energy is provided to allow the actin filaments to slide past the myosin filaments, shortening the sarcomere and contracting the muscle.

9. Nerves innervate muscle fibers and divide the muscle into a number of contractile units called motor units. A graded muscular contraction is possible by stimulating different numbers of motor units in a muscle.

10. The electrical nature of muscle contraction is emphasized by the fact that modified muscles are able to produce electrical fields that can be emitted periodically or continuously by lower vertebrates. These fields are used for protection, the capture of prey, and electrolocation of objects.

11. Muscles act on the skeleton to cause the skeleton to act as a lever, amplifying the effect of the muscle.

# KEY TERMS

acetylcholine
actin
antagonistic muscles
appendicular skeleton
axial skeleton
bursae
cardiac muscle
carpals
clavicle
compact bone
cranium
electroplates
endochondral ossification
endoskeleton
exoskeleton
femur
fontanels
Haversian canal

hipbones
hydrostatic skeleton
immovable joint
ligament
molting
motor units
muscle fibers
myofibrils
myosin
neuromuscular junctions
osteoarthritis
osteoblasts
osteoclasts
osteocytes
osteoporosis
partly movable joint
pectoral girdle
pelvic girdle

pelvis
perichondrium
periosteum
rheumatoid arthritis
sarcomeres
scapula
setae
skeletal muscle
sliding filament model
smooth muscle
spongy bone
sternum
synovial joint
tarsals
tendon
vertebrae
Z lines

## SHORT ANSWER

1. Voluntary muscles are those that are under the conscious control of the individual. Classify the three basic muscle types based on their form of control.
2. Describe the three types of muscle.
3. Outline and give examples of the three classifications of joints.
4. Compare and contrast the three main types of skeletal systems. Give examples of organisms with each.
5. Describe in detail the contraction of skeletal muscles.
6. How does the skeletal system participate in homeostasis?
7. What is the difference between osteoarthritis and rheumatoid arthritis?

## MULTIPLE CHOICE

1. Which substance provides skeletal muscle fibers with a means of rapidly forming ATP immediately following the initiation of muscular activity?
   a. fatty acids
   b. creatinine phosphate
   c. calcium
   d. acetylcholine
   e. none of the above
2. During skeletal muscle contraction, ATP occupies _____ .
   a. actin
   b. ADP
   c. myosin
   d. perichondrium
   e. a and c
3. The bones of the human skeleton are grouped in two categories:
   a. appendicular and axial
   b. axial and vertebral
   c. hydrostatic and cartilaginous
   d. exoskeletal and appendicular
   e. none of the above
4. The scapula is the same as the _____ .
   a. collar bone
   b. metacarpal
   c. shoulder blade
   d. sacrum
   e. a and c
5. Endochondral ossification involves _____ .
   a. bone growth from each end toward the center
   b. interaction between osteoblasts and osteoclasts
   c. growth of blood vessels into perichondrium
   d. laying down of cartilage
   e. all of the above

## FILL IN THE BLANK

1. The vertebral column is composed of _____ vertebrae. Attached to the vertebral column are _____ pair of ribs.
2. Carpals are located in the _____ and the ankle is composed of _____ .
3. Compact bone is made up of layers of mineral deposits surrounding the central opening called the _____ .
4. The femur, humerus, and sternum contain _____ , which form the components of blood and the immune system. The central cavity of long bones contain _____ , which stores fat.
5. Tendons link _____ to bone while _____ link bone to _____ .

## SCIENCE AND SOCIETY

1. The aging process results in the loss of calcium and collagen deposition in bone tissue, so that more bone is composed of inorganic material. Predict the effects that this process has on the bone's ability to resist breakage.
2. There are several structural differences between male and female pelvises. For example, females have a pelvic girdle that is more brittle, broader, and tilted forward. Explain the functional value of the anatomical differences between male and female pelvises.
3. Muscles that are not routinely used gradually experience atrophy or diminish in size and strength. The biological mechanisms that induce atrophy are unknown but it is believed to be due to the lack of customary forcefulness of contraction. Predict the effect seen in muscle mass in astronauts who spend extended periods of time in an environment with near-zero gravity.

# 20

# RESPIRATION

**OPENING IMAGE**

*Respiratory system of a mammal*

*showing larynx, trachea, bronchi,*

*and lungs.*

I n most states, it is illegal to drive if your blood alcohol level is over 0.1%. Law-enforcement officials do not ask drivers to donate a blood sample to measure alcohol levels; instead, they use a breath test. In use, hand-held breath testers sample about one cubic centimeter of breath exhaled by the donor. The exhaled gases are drawn into a cell where they are oxidized. A microchip calculates the alcohol concentration and displays the results as a digital readout.

In a similar way, the lingering odor on your breath after you eat garlic is not coming from your mouth, but from your lungs. The characteristic odor of garlic is caused by a compound called methyl mercaptan. After a meal rich in garlic, this chemical is absorbed into the bloodstream. As blood circulates to the lungs, some of the methyl mercaptan moves from the blood into the lungs in response to a concentration gradient. As you exhale, others may be aware of what you had for lunch.

Analysis of the breath can provide valuable information about metabolism and the state of an individual's health. A thin membrane is all that separates the air in the lungs from the blood. Compounds dissolved in the blood move into the lungs across a gradient and are exhaled. In the late nineteenth century, uncontrolled forms of diabetes were diagnosed by having patients breathe into a device that bubbled the exhaled air through a chemical solution. Uncontrolled diabetes results in acetone production as a by-product of glucose metabolism. When present in the exhaled breath, the acetone caused a color change in the chemical solution.

*Currently, breath tests are used to diagnose conditions that affect the gastrointestinal system, including the stomach, small intestine and the pancreas. The inability to absorb lactose (the sugar in milk) as an adult is a genetic trait. When affected individuals drink milk, they develop gas, cramping, and diarrhea. To diagnose this condition, physicians administer a measured dose of lactose, which passes into the colon, where bacteria absorb and metabolize the sugar. In the process, the bacteria release hydrogen. The hydrogen is absorbed into the blood and exhaled in the lungs. Analysis of hydrogen levels in the exhaled air can identify those with this disorder.*

*More than 25 years ago, Linus Pauling (a two-time Nobel Prize winner) developed a simple method for microanalyzing the chemical components in exhaled breath. He was able to observe more than 250 different compounds, and he proposed that sophisticated medical testing might be done by breath analysis. Applications of his work have been slow to develop, but his findings emphasize the importance of the respiratory system in homeostasis.*

## RESPIRATION INVOLVES THE EXCHANGE OF GASES

In cellular respiration (see Chapter 4), carbohydrates, fats, and proteins are broken down to release energy for cellular functions. During this process, carbon dioxide is released, oxygen is consumed, and water is formed. For energy transfer to occur, cells require a continuous supply of oxygen. When oxygen levels are not sufficient, metabolism slows down or stops. In addition, carbon dioxide generated during metabolism must be removed. To survive, cells must have a system for the **exchange of gases.** In large animals with a complex body plan, this is accomplished by breathing, gas transport, and exchange of gases at the cellular level. These processes encompass what is called **external respiration,** in contrast to the events of cellular respiration discussed earlier.

## Body Plans and Size Are Related to Respiration

In single-celled marine organisms, the entire body surface is in contact with the environment and is available for gas exchange. In these organisms oxygen enters the cell by diffusion from the aqueous medium, and carbon dioxide is removed by diffusion into the environment. In water (and therefore in cytoplasm), oxygen diffuses some 25 times more slowly than carbon dioxide, and the slow diffusion rate of oxygen is a limiting factor in the size of single-celled organisms. Diffusion of oxygen is so slow that with a normal rate of metabolism, oxygen cannot penetrate more than 0.5 mm. This limits the size of single-celled organisms to about 1 mm at largest. The body plans of multicelluar organisms without highly developed respiratory and/or transport systems are also adapted to promote gas exchange. Flattened, tubular, and thin shapes are the most efficient for the inward diffusion of oxygen and the release of carbon dioxide (✦ Figure 20.1).

✦ **FIGURE 20.1**
**Organisms without organized respiratory systems have body shapes that promote gas exchange with the environment.**
**(a) Planaria. (b) Hydra. (c) Jellyfish.**

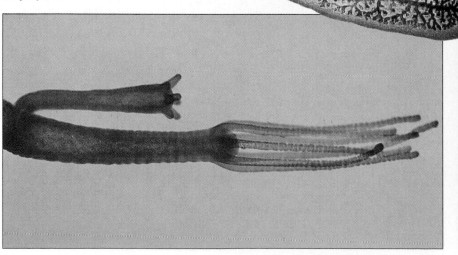

(a)

(b)

(c)

## Gas Exchange Occurs across a Respiratory Surface

In animals with large, compact body plans, gas exchange cannot be maintained by diffusion, and an organized system for gas exchange is necessary. This adaptation permits an increase in body size and an accompanying increase in mobility. In all animals, gas exchange occurs across a **respiratory surface,** an interface that allows oxygen to cross a thin, moist, epithelial surface into the body, and carbon dioxide to move out of the body (✦ Figure 20.2). Oxygen and carbon dioxide can cross cell membranes only when they are dissolved in water or an aqueous fluid, so the respiratory surface must be kept moist. In terrestrial animals, cellular secretions prevent respiratory surfaces from drying out. The respiratory surface must be large enough to process the gas exchange needs of the organism, and to deal with environmental limitations. Several types of animal respiratory systems have evolved. In spite of their apparent diversity, they all employ a thin, moist epithelium as a respiratory surface.

##  SEVERAL METHODS OF RESPIRATION HAVE EVOLVED

As mentioned earlier, some groups of animals obtain oxygen directly from the surrounding medium, without the aid of a specialized respiratory system. These groups are aquatic, and include sponges and jellyfish. Although these animals do not have a well-defined respiratory system, water is conducted through channels in the body. Movement of water past cells and cytoplasmic streaming within cells facilitates gas transport and exchange. Oxygen dissolved in the water diffuses into the cells and carbon dioxide diffuses out in response to differences in concentration.

### The Body Surface Is a Respiratory Organ in Many Aquatic and Terrestrial Animals

In animals such as flatworms and annelids, the entire outer surface of the body acts as the respiratory system. In the earthworm, a single thin layer of mucus-coated epidermal cells covers a network of small blood vessels called **capillaries.** In the capillaries under the respiratory surface, oxygen diffuses into the blood and carbon dioxide is released by diffusion into the environment. As blood travels through the body, oxygen diffuses from the blood into the oxygen-poor cells, and carbon dioxide moves from a higher concentration in the cells into the blood (✦ Figure 20.3).

✦ **FIGURE 20.2**
Respiratory surfaces such as the external gills on this axolotl facilitate gas exchange.

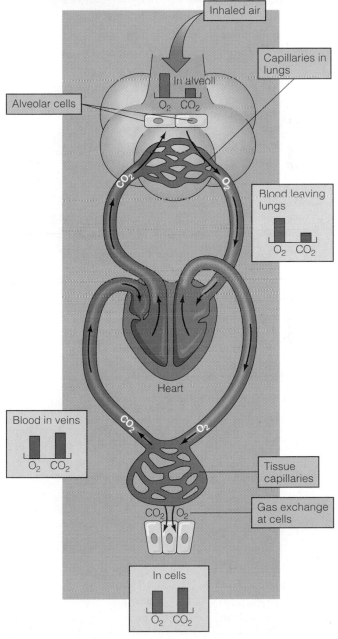

✦ **FIGURE 20.3**
Gas exchange in the lungs and tissues. Blood leaving the lungs has high levels of oxygen and low levels of carbon dioxide. After exchange in the tissues, blood has high levels of carbon dioxide and low levels of oxygen.

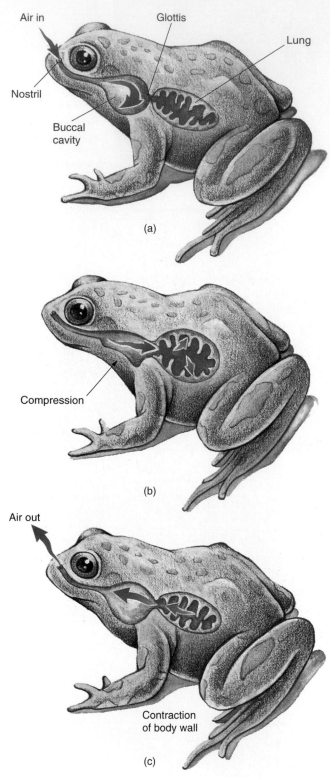

Air in

Glottis

Lung

Nostril

Buccal
cavity

(a)

Compression

(b)

Air out

Contraction
of body wall

(c)

**✦ FIGURE 20.4**

**Frogs use both skin and lungs for gas exchange. (a,b) Air is
drawn into the mouth and then pushed back into the lungs.
(c) Contraction of muscles in the body wall forces air out of the
lungs.**

Even in animals with more specialized respiratory systems, the body surface can play an important role in respiration. In the frog, elimination of carbon dioxide through the skin is almost 2.5 times greater than that through the lungs (✦ Figure 20.4). Eels obtain up to 60% of their oxygen supply through the skin, even though they possess a primary respiratory system in the form of gills. In humans, only about 1% of the carbon dioxide exchanged in respiration is eliminated through the skin.

## Gills Increase the Surface Area Available for Respiration

Gills are specialized structures that greatly increase the surface area available for respiration. Most people associate gills with fish, but they are also used as respiratory structures in other animal groups (annelids, aquatic and terrestrial crustaceans, aquatic insects, and amphibians).

A typical gill (✦ Figure 20.5) is a convoluted outgrowth containing blood vessels covered by a thin layer of cells. Organized into a series of plates, gills provide a large surface area for the uptake of oxygen and the disposal of carbon dioxide. Gills may be external (as they are in some amphibians) or internal (as in crabs and fish).

Oxygen is more difficult to obtain for organisms in an aqueous environment than it is for terrestrial organisms. Water contains less than one-twentieth the amount of oxygen that is present in air. As a result, aquatic organisms must process 20 times more water to obtain the same amount of oxygen as air-breathers. Gills are remarkably efficient at extracting oxygen from water.

The respiratory surface of a gill is a moist, single cell layer that is in contact with water on one side and with blood vessels on the other side (Figure 20.5). In fish, oxygen is obtained from water by **countercurrent flow,** an arrangement in which water moves over the gill in one direction, and blood flows through capillaries in the gill in the opposite direction. The countercurrent flow system is an adaptation that maximizes the transfer of oxygen from the water to the blood.

## Tracheal Systems Deliver Oxygen Directly to Cells

In many terrestrial animals, the respiratory system is folded into the body and connected to the outside by a series of tubes (✦ Figure 20.6). This adaptation helps keep the internal respiratory surfaces moist, promoting gas exchange. **Tracheae** (sing., **trachea**) consist of a series of tubes that carry air directly to cells for gas exchange. Tracheae are

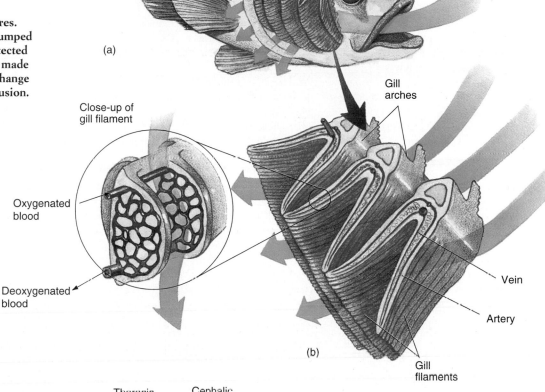

(a)

**✦ FIGURE 20.5**

Fish have gills as respiratory structures. (a) Water enters the mouth and is pumped across the gill arches, which are protected by gill plates. (b) The gill arches are made up of feather-like filaments. Gas exchange occurs across these filaments by diffusion.

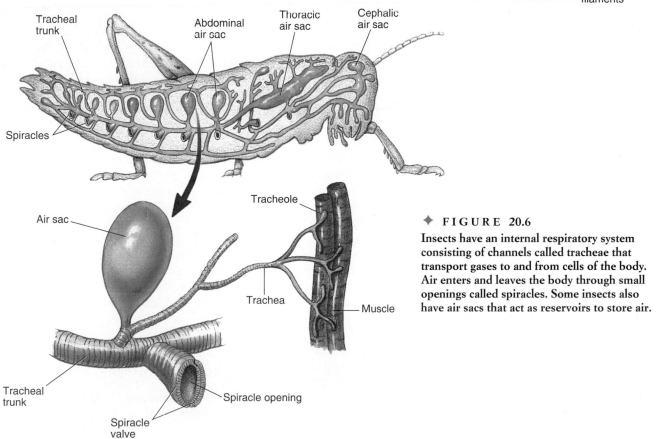

**✦ FIGURE 20.6**

Insects have an internal respiratory system consisting of channels called tracheae that transport gases to and from cells of the body. Air enters and leaves the body through small openings called spiracles. Some insects also have air sacs that act as reservoirs to store air.

# GUEST ESSAY: RESEARCH AND APPLICATIONS

## Medicine—A Scientific Safari

M. MICHAEL GLOVSKY

What is it that motivates a person to become a physician or a scientist? Chance and the desire for adventure and the opportunity to discover the unknown are important ingredients. Let me explain.

Growing up in a small town in central Massachusetts, I attended Tufts University. Chemistry was a favorite subject. I learned that it was possible to question why certain chemical reactions occurred and that several different approaches could be used to solve problems. The desire to work in a rewarding and stimulating profession was also important. I then entered Tufts Medical School. Most medical students are overwhelmed by the amount of knowledge that is available. I was not an exception. Yet, after the initial shock, coupled with long hours of study, I adapted to the need to absorb the most important information in anatomy, physiology, and biochemistry.

During my second year I became intrigued with the problems in immunology. How is the fetus able to tolerate the foreign environment of the mother's uterus? Why do habitual abortions occur? These were two questions I had an opportunity to explore in the laboratory of Dr. Wadi Bardawell, an obstetrical pathologist. In the final two years of medical school, patient contact and the reality of the consequences of disease were opportunities to integrate knowledge gained in the basic sciences.

After graduation from medical school, I spent almost four years learning about complement and immunoglobulins. I was fortunate to have as mentors Dr. Elmer Becker at the Walter Reed Army Institute of Research and Hugh Fudenberg, M.D., at the University of California, San Francisco. My initial research addressed what chemicals could block the inflammatory reactions in the skin when complement and immunoglobulins interact.

Because of my interest in immunology, I became an allergist. Can chemicals derived from the structure of IgE, the allergy antibody, interfere in the allergic response? We were fortunate to have as collaborators Drs. Bergitt Helm and Hannah Gould from King's College, London. They provided recombinant proteins synthesized from the known structure of the heavy chain regions of the IgE myeloma proteins (N.D.). Together, we were able to show that the IgE recombinant fragment (301–376) was able to block ragweed allergens and grass allergen reactions in human skin and in blood basophils. Further studies were performed to pinpoint the binding site of the allergy antibody, IgE, binding to human basophils.

More recently we have studied substances in air pollution that are important in causing the nose to run and bronchial tubes to constrict. We are exploring whether latex, an ingredient in natural rubber, produces allergenic proteins that are important in asthma. Together with Dr. Ann Miguel and Dr. Glenn Cass at Caltech, we have demonstrated latex allergen in air samples and roadside dust samples. Latex allergen, present in natural rubber gloves, balloons, and tire dust, is one of the most potent allergens of the last 10 years. We are studying whether latex allergy is important in the increasing symptoms of asthma in the last decade.

Medical science has evolved to provide sophisticated tools, especially in molecular biology, to address important questions and seek relevant answers. Biologic science is a safari through the wilderness. Every path leads to adventure and exploration with both frustration and rewards. But the trail is always interesting and provocative.

*M. Michael Glovsky is director of the Asthma and Allergy Center at Huntington Memorial Hospital in Pasadena, California.*
*He received a B.S. degree from Tufts University in 1957 and an M.D. degree from Tufts Medical School in 1962.*

found in insects and spiders, and consist of an opening at the body surface (a spiracle) and a series of extensively branching tracheae inside the body. As they branch, the tracheae become smaller and eventually form small tubes called tracheoles that penetrate into tissues. This system is so extensive that body cells are only one or two cells away from a tracheole.

Tracheal respiration depends on gas diffusion to move oxygen and carbon dioxide into and out of the body. Alternatively, the process of gas exchange can be sped up by compressing the abdomen in a rhythmic fashion. In addition, abdominal movement during walking or flight helps move air through the tracheal system. Even though oxygen diffuses in air some 10,000 times faster than it does in water, tracheae are efficient for respiration only when they do not exceed a certain length. This restriction imposes a limit on the body size of insects, to about 5 cm in length and width. Because an increase in environmental temperature accelerates the rate of gas diffusion, insects with body sizes of 5 cm are found only in tropical climates, where the increased diffusion rate allows tracheal tubes to be longer than in insects from the temperate zones. This is why cockroaches and other insects of the tropics are often much larger than those of the temperate zones.

## Lungs Provide an Internal Body Surface for Gas Exchange

Several types of respiratory structures are grouped together and classified as **lungs.** Lungs are ingrowths of the body wall and connect to the outside through a series of tubes and a small opening. The fossil record indicates that fish ancestral to amphibians had lungs, so lung breathing probably evolved around 400 million years ago. Lungs are still found in some modern species of fish, including lungfish of the Amazon and Nile Rivers (✦ Figure 20.7). Lungs are not limited to vertebrates; respiratory structures similar to those of frogs are found in some terrestrial snails.

In vertebrates, the structure of the lung increases in complexity from a simple sac in salamanders to the partitioned air sacs of birds and the alveolar lungs of mammals (✦ Figure 20.8). Each stage of evolutionary development represents an adaptation that increases the internal surface area available for gas exchange, permitting an increase in body size.

## HUMAN RESPIRATION USES PRINCIPLES FOUND IN OTHER SYSTEMS

Respiration provides oxygen for cellular energy conversion and the elimination of carbon dioxide produced as a by-product of metabolism. As outlined earlier, several types of respiratory systems have evolved in large multicellular animals. Most of them work using a few common principles:

1. Movement of an oxygen-containing medium (water or air) so that it contacts a moist membrane overlying blood vessels
2. Diffusion of oxygen from the medium into the blood

✦ FIGURE 20.7
Lungfish are able to breathe air. This adaptation allows them to live in water with low levels of oxygen.

3. Transport of the oxygen to the tissues and cells of the body
4. Diffusion of the oxygen from the blood into cells.

Carbon dioxide produced as a by-product of metabolism follows the reverse route from the cells to the external environment.

In humans and other terrestrial vertebrates, the respiratory system moves air to the respiratory surface by breathing, where the exchange of oxygen and carbon dioxide takes place. On the other side of the respiratory surface, the circulatory system carries gases for transport into and out of cells.

### The Human Respiratory System

The human respiratory system includes the lungs, the airways that connect the lungs to the environment, and the structures of the chest involved

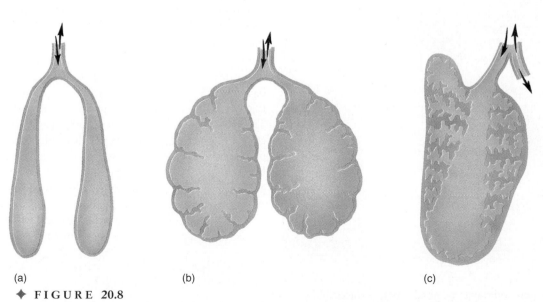

(a)          (b)                    (c)

✦ FIGURE 20.8
A trend in vertebrate evolution is an increase in the surface area of the lungs, making gas exchange more efficient, and allowing larger body sizes. (a) Salamander. (b) Frog. (c) Reptile.

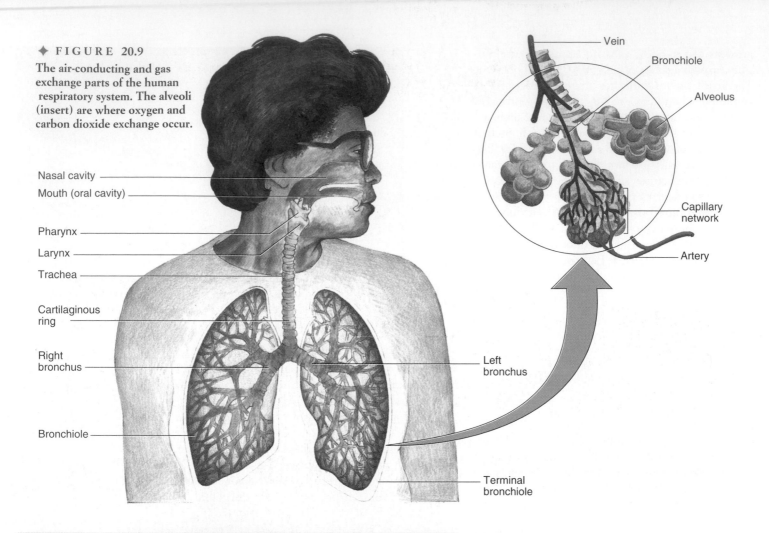

✦ **FIGURE 20.9**

The air-conducting and gas exchange parts of the human respiratory system. The alveoli (insert) are where oxygen and carbon dioxide exchange occur.

Vein

Bronchiole

Alveolus

Nasal cavity

Mouth (oral cavity)

Pharynx

Larynx

Trachea

Cartilaginous ring

Right bronchus

Bronchiole

Capillary network

Artery

Left bronchus

Terminal bronchiole

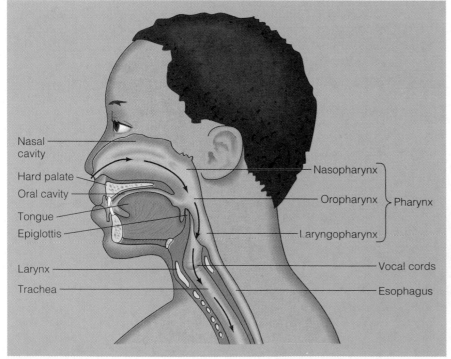

Nasal cavity

Hard palate

Oral cavity

Tongue

Epiglottis

Larynx

Trachea

Nasopharynx

Oropharynx

Laryngopharynx

Pharynx

Vocal cords

Esophagus

✦ **FIGURE 20.10**

The upper respiratory system. As air passes through the nasal cavity, particles in the inhaled air settle on the mucus coating lining the passages. Air passes through the pharynx and larynx. The epiglottis prevents food from entering the respiratory system.

in moving air into and out of the lungs (✦ Figure 20.9).

Air enters the body through the nose (the openings in the nose are called the **nares** or nostrils). The nose is divided into paired nasal cavities (✦ Figure 20.10). Air is warmed and filtered as it passes through the chambers of the nasal cavity. The respiratory system has several adaptations that screen out inhaled particles and infectious agents such as bacteria and fungi. Nasal passages contain hairs that serve to screen out large airborne particles. The nasal cavity is lined with mucus-producing cells and ciliated epithelium that also filter and remove dirt and microorganisms from the inhaled air. Further down the respiratory system, the tonsils and adenoids provide immunological defenses against air-borne pathogens. Finally, cilia in the respiratory airways beat upward in a wave-like motion, carrying inhaled particles out of the respiratory system.

Inhaled air passes into the **pharynx,** which serves as a common passage for food and air. Air can also reach the pharynx through the mouth. Two tubes open from the pharynx: the esophagus, which carries food to the stomach (Figure 20.10), and an air-conducting tube, the trachea. A small flap of tissue, the **epiglottis,** opens and closes in a reflex reaction during swallowing to prevent food from entering the trachea.

The beginning of the trachea contains the **larynx,** which protrudes from the front of the neck to form the "Adam's apple." Two bands of folded tissue, the vocal cords, extend across the opening of the larynx (✦ Figure 20.11). These bands of tissue are elastic. As air passes over them, they may vibrate to produce speech or sounds. In the chest, the trachea divides into two large tubes, the **bronchi** (sing., **bronchus**), that carry air in and out of the lungs (✦ Figure 20.12). The bronchi are rein-

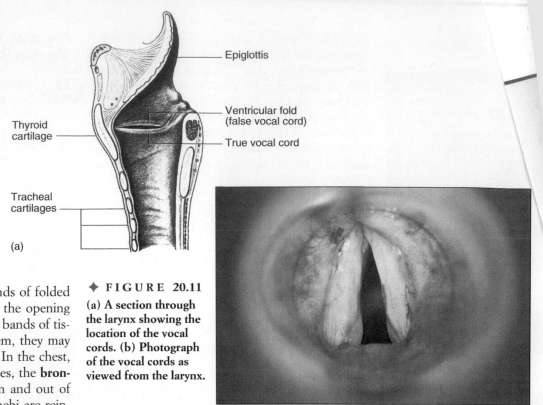

(a)

✦ **FIGURE 20.11**

**(a) A section through the larynx showing the location of the vocal cords. (b) Photograph of the vocal cords as viewed from the larynx.**

(b)

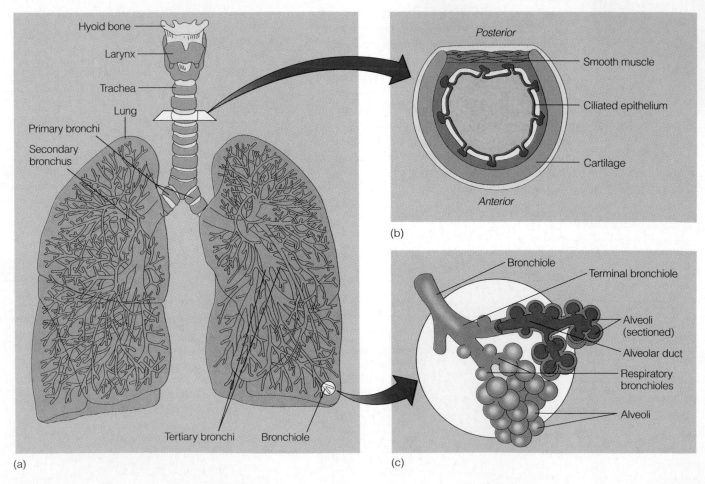

(a)

(b)

(c)

✦ **FIGURE 20.12**

**(a) The trachea and bronchial system. (b) A cross section of the trachea. (c) An enlarged view of the end of a bronchiole, showing the alveoli.**

*Sidebar*

**THE AIR WE BREATHE**

The polluted urban air that many of us breathe is benign compared to the noxious "air" that 65 million Americans voluntarily inhale from cigarettes. It is estimated that 1000 people die from the adverse effects of smoking each day. These effects include heart attacks, lung cancer, and emphysema. Nonsmokers are also affected by the smoke of others. Research shows that passive smokers are more likely to develop lung cancer than individuals who manage to stay away from smokers. Studies on Japanese women whose husbands were smokers showed that they were as likely to develop lung cancer as someone who smokes half a pack of cigarettes a day!

Smoking has been shown to decrease fertility in women. Women smoking more than a pack of cigarettes a day (greater than 20 cigarettes) are half as fertile as nonsmokers. The surgeon general warns that women who smoke several packs a day during pregnancy have a higher chance of miscarriage and may give birth to smaller children. Mental aptitude tests of children from women who smoked during pregnancy were generally lower than children whose mothers did not smoke during pregnancy. Children from families where both parents are smokers suffer twice as many respiratory infections than children from nonsmoking families.

Despite the numerous adverse effects of smoking, an estimated 65 million Americans continue this habit. Cigarette smoking is on the decline in the United States, but is increasing in Third World countries.

forced with rings of cartilage to prevent their collapse and are lined with a ciliated epithelium and mucus-producing cells. Particles and microorganisms are trapped in the mucus, which is swept upward and out of the respiratory system. The bronchi branch into smaller and smaller tubes called **bronchioles.** These terminate in grape-like clusters of thin-walled, inflatable sacs, the **alveoli** (sing., **alveolus**). The components of the respiratory system and their roles in respiration are summarized in Table 20.1.

The alveolar walls are made up of a single layer of thin, flattened epithelial cells. The network of capillaries surrounding each alveolus is composed of a single, thin cell layer (◆ Figure 20.13). Because the cells in the alveoli and their surrounding capillaries are so thin, only about 0.2 $\mu$m separates air in the alveoli and the blood (about 50 times thinner than a sheet of tracing paper).

The lungs are large paired organs divided into several lobes, each of which is connected to the bronchi. The lungs consist of the internal airways, the alveoli, the pulmonary circulatory vessels, and elastic connective tissue. The lungs are situated in the chest or **thoracic cavity** (◆ Figure 20.14). The inside of the thoracic cavity and the outer surface of the lungs are covered with a thin sheet of epithelium, the **pleura.** The space between the two sheets of pleura is called the **pleural cavity.** A thin layer of fluid covers the pleural surfaces and reduces friction during breathing. The bottom of the thoracic cavity is formed by the **diaphragm,** a dome-shaped sheet of muscle that separates the thoracic cavity from the abdominal cavity.

## Breathing and External Respiration Ventilate the Lungs

The mechanics of breathing in and out is called **ventilation** and involves the diaphragm and muscles in the wall of the chest cavity. When you breathe in (inspiration), muscles in the chest wall contract, lifting the ribs and pulling them outward (◆ Figure 20.15). While this is happening, the diaphragm contracts, and moves downward, toward the stom-

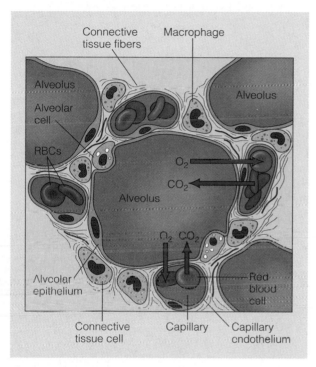

**◆ FIGURE 20.13**

**A section through an alveolus. Oxygen moves by diffusion from the alveolus into the capillary. Carbon dioxide diffuses in the opposite direction, and is expelled from the alveolus during exhalation.**

| TABLE 20.1 SUMMARY OF THE RESPIRATORY SYSTEM | |
|---|---|
| ORGAN | FUNCTION |
| **AIR CONDUCTING** | |
| Nasal cavity | Filters, warms, and moistens air; also transports air to pharynx |
| Oral cavity | Transports air to pharynx; warms and moistens air; helps produce sounds |
| Pharynx | Transports air to larynx |
| Epiglottis | Covers the opening to the trachea during swallowing |
| Larynx | Produces sounds; transports air to trachea; helps filter incoming air; warms and moistens incoming air |
| Trachea and bronchi | Warm and moisten air; transport air to lungs; filter incoming air |
| Bronchioles | Control air flow in the lungs; transport air to alveoli |
| **GAS EXCHANGE** | |
| Alveoli | Provide area for exchange of oxygen and carbon dioxide |

**A model of the thoracic cavity and the lungs. (a)** Pushing a fist into a large balloon is a model for how a pleural membrane surrounds a lung. **(b)** In the chest, the pleural membrane is a continuous, closed, fluid-filled sac. Downward movement of the diaphragm during breathing reduces pressure in the pleural cavity, allowing the lungs to fill. The upward movement of the diaphragm during exhalation causes the pleural membrane to squeeze air out of the lungs.

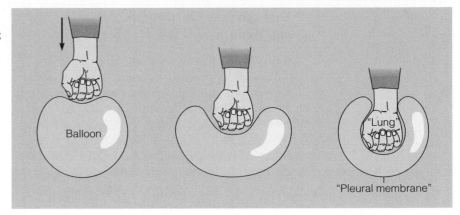

ach. These actions cause the chest cavity to enlarge, reducing the pressure in the pleural cavity. This reduced pressure causes the lungs to expand, lowering the air pressure within the alveoli. As a result, air flows from the atmosphere into the lungs.

When you exhale (expiration), the steps in this process are reversed. The muscles of the chest wall and the diaphragm relax, causing the pleural cavity to become smaller. Increased pressure in the pleural cavity raises the air pressure in the alveoli. The lungs contract, and air is forced from the lungs into the atmosphere. Normally, young adults move about 500 ml of air into and out of the lungs with each breath. At the end of a breath, there is still more than 2 liters of air still left in the lungs. Even when you try to empty all the air out of your lungs there is still a residual volume of about 1 liter, allowing gas exchange to occur continuously, even during maximum exhalation.

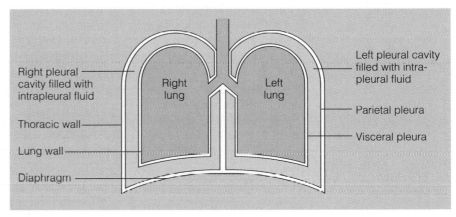

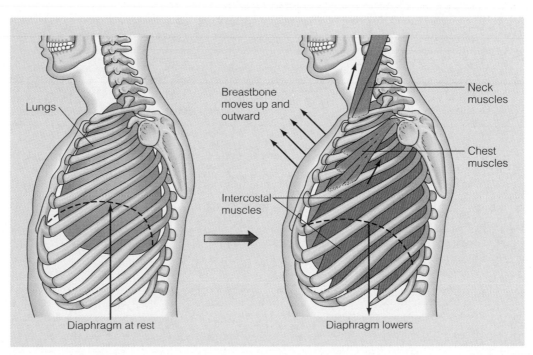

✦ FIGURE 20.15

**Muscles and bones work together during breathing. During inhalation, the diaphragm moves downward. Intercostal muscles in the chest and muscles in the neck contract, moving the breastbone and ribs upward and outward. This enlarges the thoracic cavity, and the lungs fill. During exhalation, the process is reversed, causing the lungs to contract and expel air.**

## Diseases Can Obstruct the Airways and Reduce Air Flow

Although the flow of air is determined primarily by differences in pressure between the lungs and the atmosphere, another important factor is the condition of the airways. In some diseases, airways become obstructed, causing a reduction in airflow (◆ Table 20.2). Trying to breath when you have a cold, with swollen nasal airways plugged with mucus is a familiar example of how alterations in the condition of an airway produce a reduced airflow. Other diseases can cause chronic, often life-threatening, reductions in airflow. **Asthma** narrows the smaller airways in the lungs by causing allergy-induced spasms in the surrounding muscles, swelling of the airway walls, or by clogging the airways with mucus. Asthma is a serious health problem, and affects about 10% of all children and 5% of all adults. The number of asthma cases in the United States is increasing, and may be related to a general degradation in air quality.

**Bronchitis** is an inflammatory response caused by long-term exposure to irritants such as cigarette smoke, air pollutants, or allergens. It results in excess mucus production and swelling of the bronchiole walls, which in turn impairs the movement of air, causing a reduction of gas exchange in the lungs.

**Cystic fibrosis** (CF) is a genetic disorder that causes production of a thick, viscous mucus that clogs the airways of the lungs and the ducts of the pancreas and other secretory glands. Affected individuals have difficulty breathing and are susceptible to respiratory infections. The gene for CF was identified and cloned by recombinant DNA techniques in 1989. A strategy for treating this disease by inserting a normal CF gene into the cells of the respiratory system is being developed.

## The Alveoli Are Centers of Gas Exchange and Transport

In the alveoli, gas exchange between air and blood takes place by diffusion. Recall from Chapter 4 that **diffusion** is the spontaneous movement of a substance from a region of higher concentration to a region of lower concentration. In respiration, the differences in oxygen and carbon dioxide concentrations are measured by partial pressures. The greater the difference in partial pressure, the greater the rate of diffusion. At sea level, the atmosphere supports a column of mercury 760 mm high. Because about 21% of the atmosphere is oxygen, oxygen's contribution to that pressure is about 160 mm ($760 \times 0.21 = 160$) (◆ Table 20.3). The partial pressure of carbon dioxide in the atmos-

---

**T A B L E  20.2  COMMON RESPIRATORY DISEASES**

| DISEASE | SYMPTOMS | CAUSE | TREATMENT |
|---|---|---|---|
| Emphysema | Breakdown of alveoli; shortness of breath | Smoking and air pollution | Administer oxygen to relieve symptoms; quit smoking; avoid polluted air. No known cure. |
| Chronic bronchitis | Coughing, shortness of breath | Smoking and air pollution | Quit smoking; move out of polluted area; if possible, move to warmer, drier climate. |
| Acute bronchitis | Inflammation of the bronchi; shortness of breath | Many viruses and bacteria | If bacterial, take antibiotics, cough |
| Sinusitis | Inflammation of the sinuses; mucus discharge; blockage of nasal passageways; headache | Many viruses and bacteria | If bacterial, take antibiotics and decongestant tablets; use vaporizer. |
| Laryngitis | Inflammation of larynx and vocal cords; sore throat; hoarseness; mucus buildup and cough | Many viruses and bacteria | If bacterial, take antibiotics, cough medicine; avoid irritants, like smoke; avoid talking. |
| Pneumonia | Inflammation of the lungs ranging from mild to severe; cough and fever; shortness of breath at rest; chills; sweating; chest pains; blood in mucus | Bacteria, viruses, or inhalation of irritating gases | Consult physician immediately; go to bed; take antibiotics, cough medicine; stay warm. |
| Asthma | Constriction of bronchioles; mucus buildup in bronchioles; periodic wheezing; difficulty breathing | Allergy to pollen, some foods, food additives; dandruff from dogs and cats; exercise | Use inhalants to open passageways; avoid irritants. |

| TABLE 20.3 COMPOSITION OF AIR | |
|---|---|
| GAS | PERCENTAGE COMPOSITION |
| Nitrogen ($N_2$) | 78 |
| Oxygen ($O_2$) | 21 |
| Argon (Ar) | 0.9 |
| Carbon dioxide ($CO_2$) | 0.03 |
| Water vapor ($H_2O$) | Variable (0–4) |
| Pollutants | Variable |

phere is about 0.3 mm. In the alveoli, oxygen moves from the air into the blood, and carbon dioxide moves from the blood into the air in response to differences in concentration gradients.

In addition to gas exchange, other interactions between the respiratory and circulatory systems occur in the lungs. As blood passes through the lungs, small blood clots present in the blood are trapped and dissolved in the lungs, removing them from circulation. The blood-borne hormone angiotensin II regulates the concentration of sodium ions in the blood and lymph. A precursor of this hormone is synthesized in the liver and secreted into the blood. As the blood passes through the lungs, the precursor is converted into the active hormone.

## Hemoglobin Carries Oxygen to the Tissues

Because blood is mostly water, only a small amount of dissolved oxygen can be carried in the blood. Respiratory pigments increase the oxygen-carrying capacity of the blood. In humans, this red-colored pigment is **hemoglobin,** present inside red blood cells. In response to a concentration gradient, oxygen moves from the air in the alveolus into the blood plasma across the membrane of the red blood cell, where it combines with hemoglobin. The presence of hemoglobin increases the oxygen-carrying capacity of blood some 65 to 70 times. Each red blood cell carries about 250 million molecules of hemoglobin (about one-third of the volume of the cell), and each milliliter of blood contains some 5 million red blood cells. This means that each milliliter of blood in the body contains some $1.25 \times 10^{15}$ molecules of hemoglobin. As blood leaves the lungs, it is about 97% saturated with oxygen. In the tissues, the oxygen concentration in cells is low (✦ Figure 20.16), and oxygen diffuses from the hemoglobin, through the blood plasma, and into the cells, where it is consumed during energy conversion.

## Carbon Dioxide Is Transported in Red Blood Cells

Carbon dioxide is present at a higher concentration in cells that are metabolically active than in the surrounding blood plasma. In response to this concentration gradient, carbon dioxide diffuses from the cells into the blood plasma, where it is combined with water to form bicarbonate (✦ Figure 20.17).

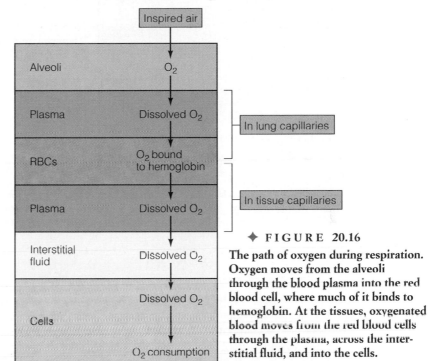

✦ **FIGURE 20.16**

The path of oxygen during respiration. Oxygen moves from the alveoli through the blood plasma into the red blood cell, where much of it binds to hemoglobin. At the tissues, oxygenated blood moves from the red blood cells through the plasma, across the interstitial fluid, and into the cells.

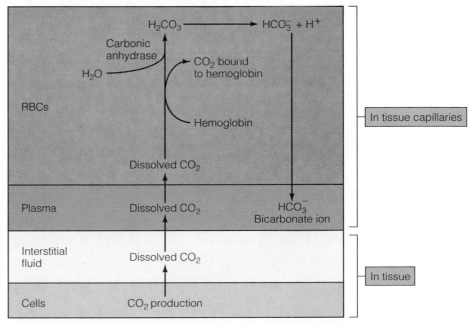

✦ **FIGURE 20.17**

The path of carbon dioxide during respiration. Carbon dioxide diffuses out of cells, through the interstitial fluid and into the plasma. About 25% of the carbon dioxide binds to hemoglobin, and some dissolves into the plasma. Most is converted to carbonic acid in the red blood cells. Once formed, carbonic acid dissociates and forms carbonate ions and hydrogen ions, which move to the plasma for transport.

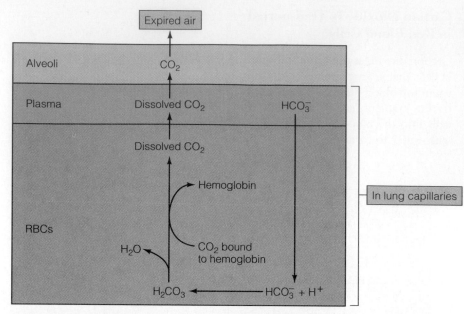

**✦ FIGURE 20.18**

**As the carbon-dioxide-containing venous blood reaches the lungs, the bicarbonate ion combines with hydrogen to form carbonic acid. The carbonic acid dissociates, forming dissolved carbon dioxide, which diffuses out of the red blood cell, through the plasma into the alveoli.**

Conversion to the bicarbonate ion ($HCO_3^-$) increases the efficiency of transport; it removes carbon dioxide from the plasma, allowing more carbon dioxide to diffuse from the tissues into the blood. In the alveoli, the chemical reaction is reversed. The bicarbonate ion combines with a proton to form carbonic acid; the carbonic acid breaks down into water and carbon dioxide, which diffuses from the blood into the alveolus (✦ Figure 20.18).

## RESPIRATION IS CONTROLLED BY THE BRAIN

The muscles of the chest and the diaphragm involved in inhalation are skeletal muscles; these contract only when stimulated by nerves (review this in Chapter 19). When these muscles receive a nerve impulse, they contract, expanding the pleural cavity, and causing the lungs to fill with air. When the nerve impulses stop, the chest muscles and diaphragm relax, causing the pleural cavity to become reduced in size, and the respiratory gases are expelled in the process called exhalation.

The normal cycle of breathing in and out depends on periodic stimulation of the muscles controlling inhalation. These muscles can be stimulated from two respiratory centers in the brain: one for automatic control of breathing, and one for conscious control (✦ Figure 20.19). The center for conscious control allows you to alter the rate and depth of breathing, and to hold your breath. When you hold your breath for a period of time, the concentration of carbon dioxide in the blood rises, and the level of oxygen falls. These changes cause the automatic system to override the conscious control center and restore breathing. Changes in blood levels of carbon dioxide and oxygen are monitored by receptors in the brain and in the walls of blood vessels. These receptors are stimulated when carbon dioxide levels rise, sending nerve impulses to the respiratory centers. In turn, motor neuron signals transmitted from the respiratory centers to the chest muscles and the diaphragm increase the rate of breathing.

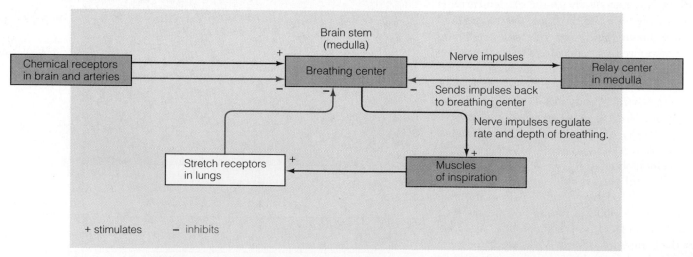

**✦ FIGURE 20.19**

**The breathing center in the brain controls respiration. Impulses from this center cause muscle contractions that result in inhalation. Impulses to and from the relay center control the rate and depth of inhalation and the relaxation of muscles.**

## Beyond the Basics

# SUDDEN INFANT DEATH SYNDROME

The unexpected death of an infant during sleep is called sudden infant death syndrome (SIDS). In the United States, almost 7000 infants, usually between the ages of 2 weeks and 1 year, die of SIDS each year. SIDS is the leading cause of death in infants, and 1 in 500 infants will die from this disorder each year. The causes of SIDS are not yet fully understood, but more than likely they involve failure of automatic respiratory control. Many victims show changes in the circulatory system that are evidence of an oxygen deficit. These changes include thickening of the arteries in the lungs, elevated erythrocyte production, and slow growth.

Anatomical changes indicate that before death, SIDS victims have a history of low blood oxygen levels and elevated carbon dioxide levels. It appears that the respiratory center in the brain of SIDS infants does not respond properly when blood gas levels fall outside the normal range. What remains unknown is *why* this occurs. Currently, much attention is centered on the structure and developmental state of the brain respiratory centers in SIDS patients. There is some evidence that the maturation of this brain region may be delayed in cases of SIDS, and that certain cell types may proliferate excessively. To date, however, these changes have been found only in a small number of cases.

Other approaches to understanding SIDS have searched for genetic and/or environmental factors. In general, pedigree studies have failed to identify families with a higher incidence of SIDS than the general population. However, there is a higher incidence of SIDS in infants with the B blood type, indicating that some genetic factors may be involved. Environmental factors that have been identified as increasing risk for SIDS include smoking or use of barbiturates during pregnancy, bacterial infections of the amniotic fluid, and premature birth. Again, these conditions are present only in a fraction of SIDS cases, indicating that more research is needed to understand this disorder.

## The Respiratory Control Centers Can Malfunction

Although the automatic system of regulation allows you to breathe while sleeping, it is subject to malfunctions. In **apnea,** breathing stops during sleep for periods longer than 10 seconds. In affected individuals, these episodes may occur more than 300 times each night. Each one of these episodes is a serious and potentially lethal disorder of the respiratory system. Apnea can be caused by failure of the automatic respiratory center to respond to elevated blood levels of carbon dioxide. Neural signals for inhalation become shorter and shorter, often stopping altogether. This failure can be caused by viral infections of the brain, by tumors, or it may develop spontaneously. A malfunction in the respiratory centers of newborns can result in **sudden infant death syndrome** or **SIDS**.

## Respiration Control Can Adjust to Changing Conditions

The respiratory control centers in the brain can adapt to changes in environmental conditions. As altitude increases, atmospheric pressure decreases, so that at an elevation of 18,000 feet, atmospheric pressure is 380 mm Hg, about half of what it is at sea level (✦ Figure 20.20). As a result, the pressure of oxygen at this altitude is only half what it is at sea level. Above 10,000 feet, the decreased oxygen pressure causes loading of oxygen into hemoglobin to fall off rapidly, resulting in lowered oxygen levels in blood. To compensate, the respiratory centers increase the breathing rate to move more air into the lungs. This can have side effects (called mountain sickness) including nausea and loss of appetite. These side effects are not caused by oxygen deprivation, but by the increased loss of carbon dioxide that accompanies the increased breathing rate. As the body increases the rate of breathing in order to obtain more oxygen, carbon dioxide is removed faster, causing the symptoms of mountain sickness. After a few days at higher elevations, balance is restored by increasing production of red blood cells, by unloading oxygen more easily at the tissue level, and by other compensatory mechanisms.

✦ **FIGURE 20.20**

At higher altitudes, oxygen levels are low, making physical activity difficult. Mountain climbers often live at high altitudes before climbing in order to adjust to the differences in oxygen concentration.

# SUMMARY

1. Cellular respiration involves the breakdown of nutrient molecules that consume oxygen and generate carbon dioxide in the transfer of energy for cellular function. Multicellular organisms use a respiratory system to exchange gases to support cellular metabolism.

2. In all animals, gas exchange occurs across a respiratory surface, an interface that allows oxygen to cross a thin, moist, epithelial surface into the body, and carbon dioxide to move out of the body.

3. Several types of respiratory systems have evolved. Some animals such as earthworms use their body surface to exchange gases, while others use specialized respiratory structures. These include gills, tracheae, and lungs.

4. Humans and other terrestrial vertebrates use lungs for respiration. In humans, the respiratory system consists of the nasal cavities, the pharynx and larynx, tracheae, bronchi, bronchioles, and alveoli. The alveoli form the respiratory surface of the lungs.

5. Breathing moves oxygen-rich air into the lungs and removes carbon-dioxide-rich air from the lungs. Inhalation expands the chest cavity, reducing air pressure in the alveoli, which causes air to enter the lungs. During exhalation, the rib muscles relax and the diaphragm moves upward, forcing air out of the lungs.

6. In the alveoli, oxygen diffuses into the lung capillaries and combines with hemoglobin in red blood cells. In the tissues, oxygen diffuses out of the blood into metabolically active cells.

7. Carbon dioxide diffuses from cells into the bloodstream where it is converted into bicarbonate. In the lungs, the bicarbonate is converted into carbon dioxide, and the gas diffuses across the respiratory surface into the lungs, from which it is exhaled.

8. Control centers in the brain maintain breathing rates according to body needs. Changes in blood levels of carbon dioxide and oxygen are monitored by receptors in the brain and in the walls of blood vessels. These receptors are stimulated when carbon dioxide levels rise, sending nerve impulses to the respiratory centers. The respiratory centers adjust the breathing rate to maintain proper levels of carbon dioxide in the blood.

# KEY TERMS

alveoli
apnea
asthma
bronchi
bronchiole
bronchitis
capillaries
countercurrent flow
cystic fibrosis

diaphragm
diffusion
epiglottis
exchange of gases
external respiration
hemoglobin
larynx
lungs
nares

pharynx
pleura
pleural cavity
respiratory surface
sudden infant death syndrome (SIDS)
thoracic cavity
tracheae
ventilation

# QUESTIONS AND PROBLEMS

## SHORT ANSWER

1. What is the difference between cellular respiration and external respiration?

2. Arrange the following structures in the proper order that traces the pathway of air flow through the respiratory system.
   a. alveoli
   b. bronchi
   c. bronchioles
   d. external nares
   e. larynx
   f. nasal cavity
   g. pharynx
   h. trachea

3. Describe the mechanism employed by fish to obtain oxygen from water.

4. Summarize the processes of inhalation and exhalation.

5. Emphysema is a disease that results from airway obstruction and causes abnormal exhalation. In this condition the alveolar wall breaks down, which causes several normal alveoli to appear as fewer larger ones. Describe how going from several small alveoli to fewer but larger ones would affect breathing.

6. What is the effect on blood from holding your breath?

7. What are the major principles of respiratory systems that are shared by large multicellular organisms?

8. Describe how the respiratory system screens out substances.

9. What effect does altitude have on respiration?

## TRUE/FALSE

1. Increases in environmental temperature accelerate the rate of gas diffusion.

2. The lungs are located in the pleural cavity and are surrounded by the diaphragm.

3. Forced expiration eliminates all air from lungs.

4. In respiration, the greater the difference in partial pressure, the greater the rate of diffusion.

5. Oxygen moves from the air in the alveolus into the blood plasma, to the red cell membrane, where it combines with he-

moglobin in response to a concentration gradient.

## MATCHING

Match each respiratory disease appropriately.

1. apnea
2. cystic fibrosis
3. asthma
4. pneumonia
5. SIDS
6. bronchitis

____ infection of lung tissue

____ newborn respiratory abnormality

____ inflammatory response from irritants

____ periods of breathing cessation during sleep

____ thick, viscous mucus clogging airways

____ narrowing of smaller lung airways

## SCIENCE AND SOCIETY

1. Tobacco smoke contains numerous hazardous substances that damage the delicate lining of the respiratory system. One component that contributes to this destruction is carbon monoxide. Carbon monoxide in cigarette smoke is responsible for angina attacks. Carbon monoxide in cigarette smoke takes the place of oxygen in hemoglobin molecules. Describe how the respiratory system responds to this. Would this replacement of oxygen with carbon monoxide effect other parts of the body? What effects would smoking have at elevated altitudes? How would someone with asthma handle this overload of carbon monoxide? Describe the effects seen in blood pressure and blood pH.

2. Asbestos refers to a diverse group of naturally occurring fibers that are ubiquitous in the environment and mined for their use in insulation and floor and ceiling tile. Asbestos exposure has been associated with lung cancer, mesothelioma (tumor associated with occupational asbestos exposure), gastrointestinal cancers, and cancers in other areas. Active politicians, union activists, mining companies, scientists, and lawyers have been instrumental in the air monitoring of asbestos fibers in buildings, homes, and schools. In 1986 Congress passed the Asbestos Hazard Emergency Act (AHERA) which required inspections of asbestos in public and private schools. The Environmental Protection Agency in 1989 initiated a ban on asbestos, which was reversed by the Fifth U.S. Circuit Court, that would eliminate industrial use of asbestos over a 10-year period.

   Passive tobacco smoking has been estimated to have almost 400 times the cancer-risk of low-level asbestos exposure. No legal or large-scale efforts to protect the public from tobacco smoke has ever taken place. Why do you think the attention to asbestos is so much greater than that to tobacco smoke? Given the above information, do you think asbestos removal in schools and the ban on its industrial use is warranted? Why or why not? Do you think controlling urban air pollution would be an area that is more worthy of attention than controlling asbestos? Why or why not? Do you agree or disagree that the single most effective way of reducing deaths in the United States would be to ban cigarette smoking? Explain your reasoning.

# THE CIRCULATORY SYSTEM

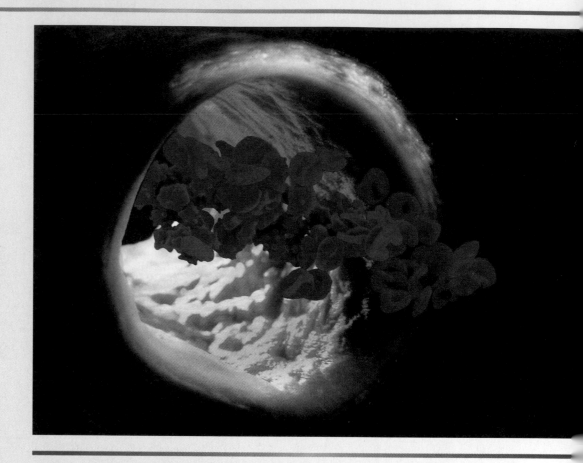

I In the 1620s, William Harvey carried out a series of observations about the circulatory system that revolutionized our understanding of how blood is pumped through the body. Harvey's work identified the heart as a pump that moves a fixed volume of blood in a single direction through a closed system of vessels, and he provided experimental results to support his conclusions.

To demonstrate that blood flows in a single direction through the veins, Harvey applied a tourniquet to the arm of a subject, causing the veins to swell with blood. The valves in the veins appeared as nodules along the swollen veins. Pressing on a vein just below a nodule with one finger, and using a second finger to squeeze the blood upward toward the heart caused the squeezed part of the vein to remain empty. No backflow occurred through the valve, but releasing the finger just below the valve allowed the valve and the vein to fill.

Harvey also calculated that blood was not consumed by the tissues, since the volume of blood pumped by the heart in an hour was greater than the entire volume of blood in the body. He also proposed that blood traveled to the lungs where it was mixed with air, and that the arteries were connected to the veins.

O P E N I N G   I M A G E

*The flow of red blood cells in the*

*circulatory system.*

*By the standards of the seventeenth century, Harvey's work was unusual. At the time, anatomists followed the practices of the ancient Greeks and limited themselves to descriptions of organisms or organ systems. They only speculated about the functions of the systems they studied. Harvey used an experimental approach to reach conclusions about the functions of the circulatory system. In doing so, he challenged long-standing theories about the heart, the blood vessels, and blood. Another century was to pass before Harvey's ideas were incorporated into a new theory of the circulatory system, and experimental methods were accepted as a way of explaining how living systems operate.*

*In this chapter, we explore the structure and function of the circulatory system in animals.*

 ## CIRCULATORY SYSTEMS TRANSPORT NUTRIENTS, WASTES, AND GASES

The ability to transport nutrients, wastes, and gases is essential in order to maintain cell function and homeostasis. When an organism consists of a single cell, the entire body surface is in contact with the external environment, allowing direct exchange of materials with the environment. In multicellular organisms, transport systems circulate nutrients and oxygen to all cells. Circulatory systems also transport metabolic by-products and carbon dioxide away from cells to the kidney and lungs for elimination.

### Several Types of Transport Systems Have Evolved in Animals

The first multicellular animals had a body made up of a double layer of cells, similar to the sponges and jellyfish of today. Even these relatively simple organisms have transport systems. In sponges, seawater is the medium of transport; it is propelled into and out of the body through ciliary action (✦ Figure 21.1).

The cells of larger and more complex animals are not in direct contact with the external environment, so they require a mechanism to obtain nutrients, dispose of metabolic wastes, and exchange gases. These animals have a **circulatory system.** Components of this system include:

*Blood:* a connective tissue composed of liquid (plasma) and cells
*Heart:* a muscular pump that moves the blood throughout the body by its contractions
*Blood vessels:* a system of vessels (including arteries, capillaries, and veins) that delivers blood to all the tissues and cells of the body, and returns the blood to the heart for repumping.

Two basic types of circulatory systems have evolved. Insects and other invertebrates such as crayfish and some mollusks have an **open circulatory system** (✦ Figure 21.2). In these animals, the

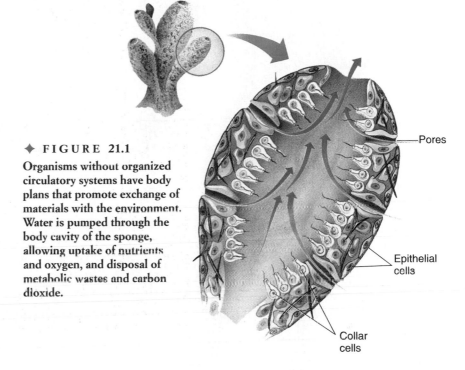

**✦ FIGURE 21.1**

**Organisms without organized circulatory systems have body plans that promote exchange of materials with the environment. Water is pumped through the body cavity of the sponge, allowing uptake of nutrients and oxygen, and disposal of metabolic wastes and carbon dioxide.**

Pores

Epithelial cells

Collar cells

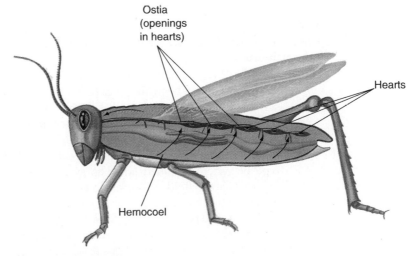

Ostia (openings in hearts)

Hearts

Hemocoel

**✦ FIGURE 21.2**

**Insects have an open circulatory system. The insect heart pumps blood into vessels that empty into a body cavity (hemocoel). Organs are bathed in the blood in a hemocoel.**

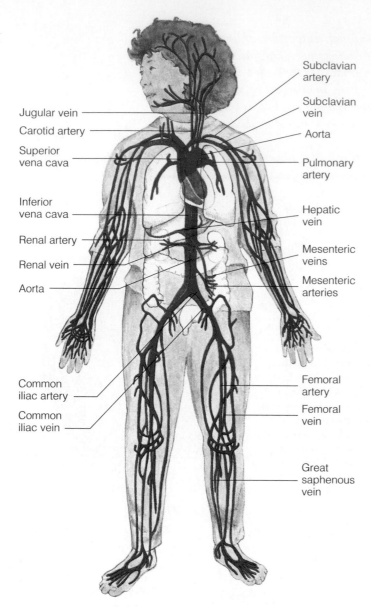

**◆ FIGURE 21.3**

**The human circulatory system consists of the heart and the vessels that transport blood to and from the heart.**

Labels on figure:
Jugular vein
Carotid artery
Superior vena cava
Inferior vena cava
Renal artery
Renal vein
Aorta
Common iliac artery
Common iliac vein
Subclavian artery
Subclavian vein
Aorta
Pulmonary artery
Hepatic vein
Mesenteric veins
Mesenteric arteries
Femoral artery
Femoral vein
Great saphenous vein

move into the spaces between cells. A secondary circulatory system, the **lymphatic circulation,** collects some of this fluid and returns it to the main circulatory system. Specialized cells of the circulatory system also function to defend the body against infectious diseases. These cells and associated components of the circulatory and lymphatic system form the **immune system,** which will be discussed in the next chapter.

## The Circulatory System Has Undergone Evolutionary Adaptations

The transition from an aquatic to a terrestrial environment was associated with a major evolutionary adaptation in the circulatory system as gill breathing changed to lung breathing. The circulatory system of a fish consists of a single circuit (◆ Figure 21.4a), with two networks of thin-walled capillary vessels.

In mammals, the cardiovascular system has two loops (Figure 21.4b), one associated with the lungs (the pulmonary loop), and the other with the rest of the body (a systemic loop).

A cardiovascular system with two circuits separates oxygen-rich from oxygen-depleted blood, and allows the systemic loop to operate at much higher pressure. The evolutionary adaptations associated with the transition from an aquatic to a terrestrial environment also involved the development of lungs for the exchange of gases. These adaptations were considered in the last chapter. The evolution and function of the circulatory system are explored in the following sections.

## ORGANIZATION OF THE HEART AND ITS EVOLUTIONARY ADAPTATIONS

Blood is pumped through the circulatory system by a specialized, muscular structure known as a **heart,** a multicellular chambered structure that alternately contracts and relaxes in a rhythmic pattern. Structurally, there are several types of hearts: the chambered hearts found in mollusks and vertebrates, the tubular hearts of arthropods, the pulsating vessels of annelids, and accessory hearts, which are used as boosters to supplement a main or systemic heart. Insects have accessory hearts at the base of the legs and wings. Fish, amphibians, and reptiles all have **lymph hearts,** contractile enlargements of vessels that pump lymph back into the veins. The accessory hearts of cephalopods (squid and octopus) gather blood from the body via veins and pump it through the gills and into the systemic heart.

The heart is the part of the circulatory system that has undergone the greatest amount of struc-

system includes a heart that pumps a fluid into vessels from which it is released into the spaces between tissues. Open-ended vessels collect the fluid from the tissues and direct it to the gills for gas exchange, and then to the heart for recirculation. This type of system is called an open system because the circulating fluid is not enclosed in vessels at all times.

The second type of system, found in echinoderms and vertebrates, is a **closed circulatory system.** A closed system uses a continuous series of vessels of different sizes to deliver blood to cells of the body and return it to the heart. In humans, this is often called the **cardiovascular system** (◆ Figure 21.3). Although the closed system has a continuous series of vessels, the fluids and cells of the blood can also

tural modification as a result of evolutionary adaptation. In vertebrates, the chambered heart of most fish consists of two chambers, an **auricle,** where blood is received from the body via the veins, and a **ventricle,** which pumps the blood into the vessels, which carry blood away from the heart (✦ Figure 21.5a). In fish, blood is pumped from the ventricle through a capillary bed in the gills where the blood is oxygenated. From the gills, blood is pumped through vessels to a second capillary bed in the tissues. A more complex, two-sided heart, characteristic of amphibians and higher vertebrates first appeared in the air-breathing fish of the Devonian.

In amphibians, the two-sided heart is represented by the presence of two atria (Figure 21.5b). One atrium receives blood from the veins of the body, the other receives newly oxygenated blood from the lungs. Both atria share a single ventricle. In some species, folds of ventricular tissue reduce mixing of deoxygenated and oxygenated blood.

Two evolutionary adaptations should be noted in the transition from fish to amphibians. First, a two-sided or two-channeled heart permits higher blood pressure in the systemic circulation, allowing the blood to overcome the resistance that develops as blood passes through vessels of decreasing size. Secondly, the respiratory loop pumps blood through the capillary bed in the lungs at lower pressure, then returns the blood to the heart for pumping at higher pressure through the body.

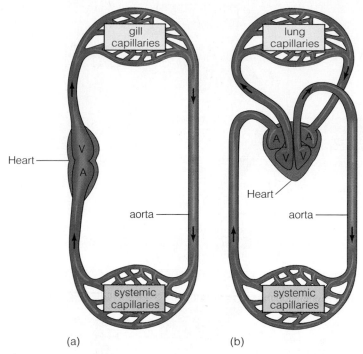

(a)                                      (b)

✦ **FIGURE 21.4**

(a) In fish, the circulatory system consists of a single circuit. The heart has two chambers, the ventricle (V) and the auricle (A). Blood is pumped from the heart to the gills, and from here, moves directly to capillaries in the tissues without returning to the heart. (b) In mammals, the heart consists of four chambers and two circuits. One circuit pumps blood to and from the lungs, and the other pumps blood to and from the tissues of the body.

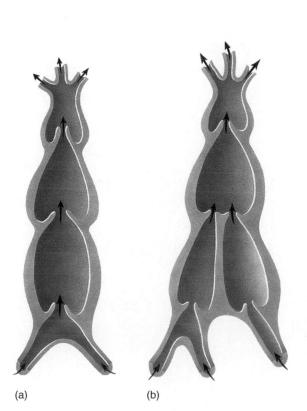
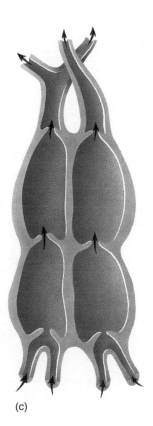

✦ **FIGURE 21.5**

Stages in the evolution of vertebrate hearts. (a) In fish, there is one atrium and one ventricle. The atrium receives blood from the body. The ventricle pumps blood through two capillary beds, one in the gills and the other in the tissues. (b) In amphibians, there are two atria. One receives oxygen-rich blood from the lungs, and the other receives oxygen-depleted blood from the body. These are mixed in the ventricle and pumped through the capillary beds of the lungs and the body. (c) In birds and mammals, there are two atria and two ventricles. This adaptation increases the efficiency of the circulatory system by separating the circuit to and from the lungs from the circuit pumping blood to and from the body.

(a)                    (b)                    (c)

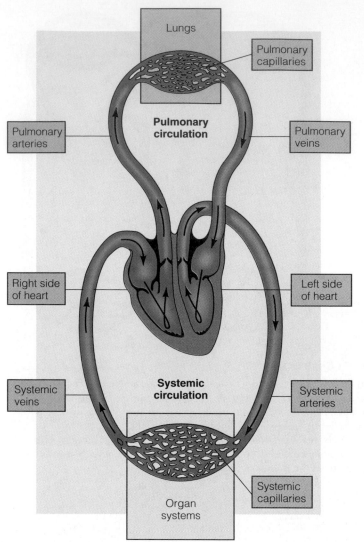

**FIGURE 21.6**

The pulmonary and systemic circuits. In humans, as in other mammals and in birds, the circulatory system consists of two circuits, one that carries blood to and from the lungs (pulmonary circuit) and one to and from the body (systemic circuit).

These dual circuits, the **pulmonary circuit** and the **systemic circuit** completely separate oxygenated and deoxygenated blood and allow blood to be pumped through each circuit at different pressures (✦ Figure 21.6). The high metabolic rates necessary to maintain a constant body temperature in all birds and mammals is possible with such a circulatory system.

The hearts of birds and mammals have four chambers, two atria and two ventricles, representing a further development of the two-sided heart (Figure 21.5c). These hearts are equipped with valves separating the chambers to prevent backflow of blood.

## The Human Heart Has Two Sides and Four Chambers

The human heart (✦ Figure 21.7) is a two-sided, four chambered structure with muscular walls, lined with a specialized epithelial layer. Between each auricle and ventricle is a flap of tissue known as an **atrioventricular valve** (or AV valve). Between each ventricle and its connected artery is the arterial valve or **semilunar valve.** During your lifetime, your heart will undergo about 3 billion cycles of contraction and relaxation, with the valves opening and closing with each cycle. The muscular walls of the heart are supplied with nutrients by a capillary bed that branches from two small arteries.

**FIGURE 21.7**

Blood flow into and out of the human heart. Oxygen-depleted blood (which is also rich in carbon dioxide) flows into the right atrium (blue) from the body. From here, it is pumped to the right ventricle and then to the lungs, where carbon dioxide is removed, and oxygen added to the blood. The oxygen-rich (red) blood returns from the lungs to the left atrium. From here it is pumped to the left ventricle and then to the systemic circulation through the aorta.

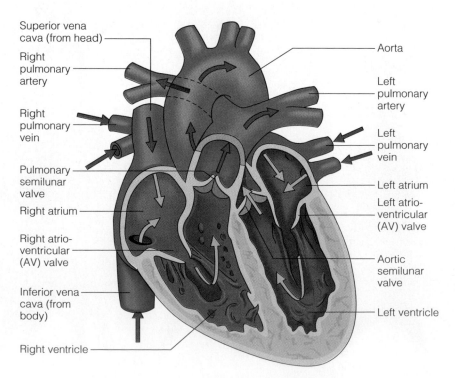

## Beyond the Basics

# HOW DOES EXERCISE AFFECT THE HEART?

The benefits of regular exercise for building muscles and increasing endurance have been widely publicized. Muscles respond to exercise by becoming larger and stronger. Exercise involving repetitive muscle contractions, such as weight lifting or use of an exercise machine will increase muscle mass. Aerobic exercises such as swimming do not increase muscle mass, but do increase endurance, the ability to sustain muscular effort.

What are the effects of exercise on the heart? Since the heart is a muscle, it responds to exercise in the same way that skeletal muscle does. Exercise increases oxygen demand to meet the requirements for increased metabolism, especially in muscle cells. To meet this demand, the heart increases its output by two mechanisms. The first is based on the Frank-Starling principle. Exercise causes muscles to contract veins, increasing blood flow to the heart. Increased return of blood to the heart leads to increased ventricular filling, and stronger ventricular contractions. This mechanism can increase cardiac output threefold and is adequate to handle oxygen demands during moderate exercise. During more strenuous exercise, the heart receives signals from the nervous system to increase the heart rate and strength of contractions, which leads to a further increase in cardiac output. From a resting rate of about 72 beats per minute, the hearts of healthy young adults can reach a maximum rate of about 200 beats per minute, achieving a cardiac output that is five to six times the resting output.

As in the skeletal muscles, over time, the heart responds to exercise by increasing in size, as the ventricular walls thicken. This allows the heart to pump more efficiently during exercise and lowers the resting heart rate. Well-conditioned athletes often have resting heart rates around 50 beats per minute. Regular exercise benefits the cardiovascular system in several ways. Moderate aerobic exercise two or three times a week is beneficial in most cases of hypertension, and offers protection against the development of hypertension. In addition, a regular exercise program can raise blood levels of high density lipoproteins (HDL). HDL is partly responsible for controlling the flow of cholesterol between the liver and cells of the body. Malfunctions in this control system lead to cholesterol accumulation in blood vessels, potentially blocking blood flow. Elevated levels of HDL are associated with a low risk of atherosclerotic heart disease.

## THERE ARE THREE GENERAL TYPES OF BLOOD VESSELS

In the closed cardiovascular system of mammals, blood is conducted through a series of blood vessels that are structurally specialized to carry out their functions. In humans, there are three types of blood vessels:

*Arteries:* thick-walled vessels that carry blood away from the heart
*Capillaries:* thin-walled vessels that exchange materials with cells
*Veins:* thin-walled vessels that carry blood back to the heart.

### Arteries Have Thick, Flexible Walls

As blood leaves the heart, it is pumped through **arteries,** which branch repeatedly to form smaller and smaller vessels, the smallest of which are called **arterioles.** Arteries have thick flexible walls, made up of three layers (✦ Figure 21.8). The inner layer is composed of thin-walled cells. The middle layer contains smooth muscle fibers and elastic connective tissue that allows the vessel to expand and contract as blood pressure changes. The outer layer contains connective tissue with elastic fibers. This outer layer is relatively thin in arteries, but is the thickest layer in veins.

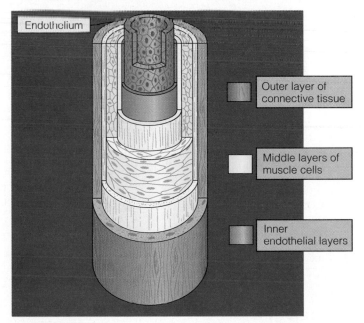

Endothelium

Outer layer of connective tissue

Middle layers of muscle cells

Inner endothelial layers

✦ **FIGURE 21.8**

**General structure of arteries and veins. Both arteries and veins are made up of three layers. The inner layer is a thin layer of endothelial cells. The middle layer is primarily smooth muscle, and the outer layer is mostly connective tissue. Veins tend to be thinner than arteries and have thinner middle layers.**

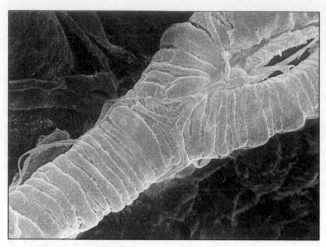

**FIGURE 21.9**

A scanning electron micrograph of arterioles with a surrounding layer of smooth muscle cells.

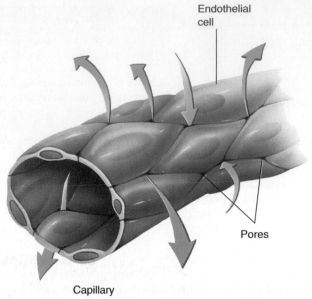

Endothelial cell

Pores

Capillary

**FIGURE 21.10**

Structure of capillaries. Materials move into and out of capillaries via pores between adjacent endothelial cells.

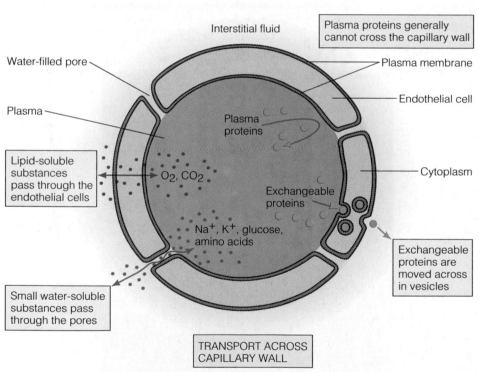

Interstitial fluid

Plasma proteins generally cannot cross the capillary wall

Water-filled pore

Plasma membrane

Plasma proteins

Plasma

Endothelial cell

Lipid-soluble substances pass through the endothelial cells

$O_2$, $CO_2$

Cytoplasm

Exchangeable proteins

Na⁺, K⁺, glucose, amino acids

Exchangeable proteins are moved across in vesicles

Small water-soluble substances pass through the pores

TRANSPORT ACROSS CAPILLARY WALL

**FIGURE 21.11**

Exchanges across capillaries. A cross section through a capillary. Small, water-soluble molecules move from the capillary through the pores into the interstitial fluid, which surrounds the cells of the body. Lipid-soluble substances are transported across the capillary walls by moving through the capillary cells. Larger molecules, including many proteins, move across the capillary by transport in vesicles. In general, plasma proteins cannot cross the capillary wall.

As the arteries branch and grow smaller, the middle layer becomes thinner and has fewer elastic fibers but more muscle fibers. These muscle fibers are circular and surround the arterioles (◆ Figure 21.9). Arterioles play an important role in controlling blood pressure. Contraction of the muscle fibers in arterioles reduces the diameter of the vessel, shutting down the flow of blood; relaxation of the fibers opens the vessels, increasing the flow of blood. The regulation of blood pressure as a homeostatic mechanism is brought about by controlling the relaxation and contraction of the muscles in these smaller arteries.

## Capillaries Are One Cell Thick

The smallest arterioles branch into a network of vessels called the **capillary bed.** The walls of capillaries consist of a single layer of thin, flattened endothelial cells (◆ Figure 21.10). In some tissues, there are small pores between these cells to allow materials to flow in and out of the capillaries with greater ease, and to allow white blood cells to squeeze out into the intercellular spaces.

The main functions of the circulatory system are carried out in the capillary bed. Oxygen and nutrients leave the capillaries as the blood flows into the tissues, and carbon dioxide, ammonia, and other wastes move from the tissues into the capillaries. Materials cross into and out of the capillaries by passing through or between the cells of the capillary walls (✦ Figure 21.11). Most materials pass in and out by diffusion (passing from an area of higher concentration to one of lower concentration), but fluids are forced out by the internal pressure of the capillaries. Some fluids or components of the fluids reenter by osmosis.

In order to receive nutrients and remove wastes, all cells of the body are close to a capillary. The extensive system of capillaries in the human body has a total length of somewhere between 50,000 to 60,000 *miles*. The capillary network has a series of shortcuts or **thoroughfare channels** to allow blood

to bypass a capillary bed (✦ Figure 21.12). These channels have rings of muscle fibers that can open and close, controlling blood flow. Thus, after a meal, capillary beds in the digestive system open and divert more blood to the gastrointestinal tract, while other parts of the body receive less blood flow.

## Veins Operate at Lower Pressure and Are Thinner Than Arteries

As blood leaves the capillary bed, it flows into **venules,** or small veins. Venule walls have three layers, but the middle layer is much thinner than in the arterial system. Venules join to form **veins,** which carry blood back to the heart (✦ Figure 21.13). Because pressure in the venous system is low, veins depend on the contraction of nearby major muscles to move blood along. In the larger veins, valves prevent the backflow of blood. No valves are present in arteries.

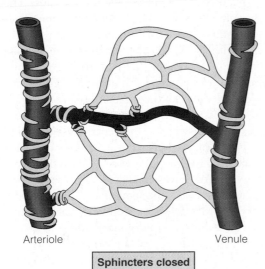

✦ FIGURE 21.12
**Thoroughfare channels allow blood flow to bypass the capillary beds. Circular muscles surrounding small arterioles called metarterioles act as sphincters to control blood flow to capillary beds.**

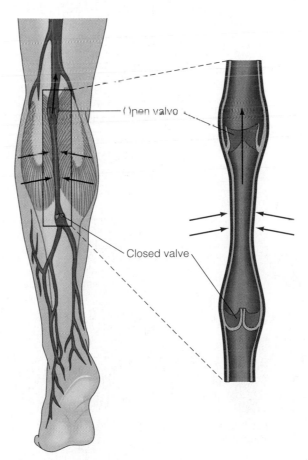

✦ FIGURE 21.13
**The combination of low pressure in the veins and the contraction of skeletal muscles moves blood in the veins back toward the heart. One-way valves in the veins prevent blood from flowing backward in the veins.**

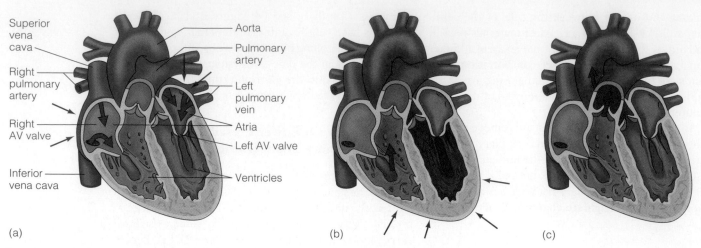

Superior vena cava
Aorta
Pulmonary artery
Right pulmonary artery
Left pulmonary vein
Right AV valve
Atria
Left AV valve
Inferior vena cava
Ventricles

(a)                    (b)                    (c)

✦ FIGURE 21.14

**The dynamics of blood pumping by the heart. (a) Both atria fill at the same time, from the systemic and pulmonary circuits. When the atria are full, they contract and pump blood into the ventricles. (b) Both ventricles fill simultaneously, then contract simultaneously, (c) moving blood into the pulmonary and systemic circuits.**

## BLOOD FLOWS THROUGH THE BODY IN A SINGLE DIRECTION

In lower vertebrates such as fish, blood is pumped from the heart to the gills, and from there directly to the tissues of the body. In mammals, blood is pumped through both the pulmonary circuit and the systemic circuit simultaneously. The role of the

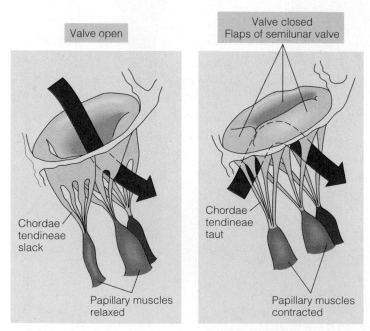

Valve open

Valve closed
Flaps of semilunar valve

Chordae tendineae slack

Chordae tendineae taut

Papillary muscles relaxed

Papillary muscles contracted

✦ FIGURE 21.15

**The valves in the heart make blood flow in one direction. In a semilunar valve, muscles (papillary muscles) attached to connective tissue (chordae tendineae) control action of the valve. When the valve is open, the muscles are relaxed. When the valve is closed, the muscles are contracted, holding the valve closed against the back pressure of blood in the ventricle.**

pulmonary circuit in respiration was discussed in Chapter 20. This chapter considers the function of the systemic circuit.

### Heart Muscle Pumps Blood through the Body

Blood flows through the heart in a single direction, from veins, to atria, to ventricles, to arteries. Blood flow is limited to a single direction by the four heart valves that act as one-way doors. One heartbeat, called a **cardiac cycle,** includes atrial contraction and relaxation, ventricular contraction and relaxation, and a short pause. At a resting state of 72 beats per minute, the cardiac cycle is about 0.8 seconds.

At the heart, blood from the body flows into two veins, the superior and inferior vena cava. These veins empty into the right atrium (✦ Figure 21.14). The atrium is relatively thin walled, and expands as it fills with blood. At the same time, oxygenated blood returning from the lungs via the **pulmonary vein** fills the expanding left atrium (Figure 21.14). The muscles in both the right and left atria contract at the same time, forcing blood through the AV valves into the right and left ventricles, respectively. The filling of the ventricles with blood is called **diastole.** As the pressure in the ventricles rises, it forces the AV valves to close. Ventricular **systole** (contraction) opens the semilunar valves and forces blood into the arteries. From the right ventricle, blood flows into the **pulmonary artery** toward the lungs. On the left side of the heart, the ventricular contraction opens the semilunar valves to pump oxygenated blood from the left ventricle into the **aorta,** for distribution to the tissues of the body (Figure 21.14). After the ventricles empty, the semilunar valves snap shut, preventing blood from flowing back into the heart.

With each cardiac cycle, a characteristic "lub-dub" heart sound can be heard with a stethoscope.

When heart valves close, the sudden change in pressure causes the flaps of the valve to vibrate. This vibration is a heart sound. Although there are four valves in the heart, there are only two major heart sounds because the two AV valves close at the same time and, later, the two semilunar valves close together. The first heart sound ("lub") is associated with closure of the AV valves. This sound marks the beginning of systole. The second sound ("dub") is associated with the closing of the semilunar valves and is the beginning of diastole (✦ Figure 21.15).

At rest, the human heart beats about 70 times per minute, but heart rates vary widely across the animal kingdom. Generally, heart rates in smaller animals are faster than in larger animals. For example, elephants have a heart rate of 25 to 40 beats/minute, and mice have rates of 300 to 500 beats/minute. In animals that do not regulate body temperature, heart rate is sensitive to changes in temperature, generally increasing by a factor of two to three times for every 10°C rise.

## A Pacemaker Regulates the Heartbeat

Unlike skeletal muscle, contractions of cardiac muscle originate from within the heart itself. In humans, the heartbeat originates in an area of the right atrium known as the **sinoatrial (SA) node** or **pacemaker** (✦ Figure 21.16). This region contains modified muscle cells that undergo spontaneous contractions and generate a signal that resembles a nerve impulse. This signal passes to other muscle cells, causing them to contract. Contractions in the sinoatrial node spread to both atria, which contract simultaneously. The signal spreads across the border between atria and ventricles to the **atrioventricular (AV) node** (Figure 21.16). Impulses from the atrioventricular node are carried by bundles of specialized cells (the bundle of His) to each ventricle. At the end of the bundles, small fibers branch through the ventricles (Purkinje fibers). Signals from the AV node travel through the bundles and fibers and cause the ventricles to contract simultaneously. Signals from the AV node are slightly delayed (by about 0.1 second), allowing the atria to contract and empty before the ventricles contract.

## Electrocardiograms Record the Electrical Activity of the Heart

The waves of contraction impulses that pass over the surface of the heart are monitored on an electrocardiogram (ECG) by measuring changes in

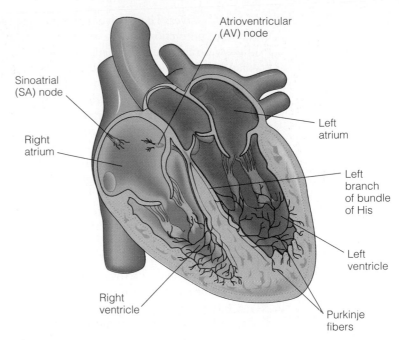

✦ **FIGURE 21.16**

The impulse-conducting system of the heart. This system includes the sinoatrial (SA) node, the atrioventricular (AV) node, the bundles of His, and the Purkinje fibers.

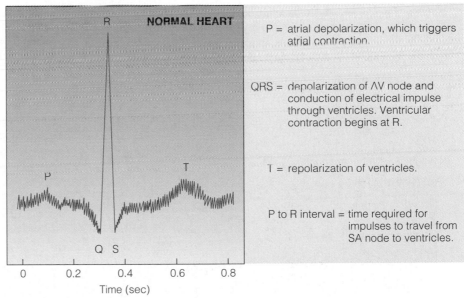

P = atrial depolarization, which triggers atrial contraction.

QRS = depolarization of AV node and conduction of electrical impulse through ventricles. Ventricular contraction begins at R.

T = repolarization of ventricles.

P to R interval = time required for impulses to travel from SA node to ventricles.

✦ **FIGURE 21.17**

A normal ECG contains three components in each cardiac cycle (heartbeat). The P wave is a signal that initiates contraction of the atria. The QRS complex is caused by signal transmission to the ventricles. Ventricular contraction begins at R. The T wave is caused by ionic exchange in ventricular muscles, bringing about relaxation of the ventricular muscles.

electrical potential across the heart. In a typical ECG (✦ Figure 21.17) there are a series of slow, upward (negative) changes called the P, R and T waves, and downward (positive) deflections called the Q and S waves. The P wave represents conduction of the contraction impulse across the atria; the PQ interval represents the pause at the atrioventricular junction, allowing the atria to contract before any ventricular contractions begin. The QRS complex is generated by contraction in the ventricles, with ventricular contraction beginning at R. The T wave represents electrical changes that take place as the ventricles relax.

ECGs are useful in diagnosing abnormal heart rates and identifying the location and extent of heart muscle damage caused by heart attacks. Abnormal rhythms or heart damage is reflected in the pattern of an ECG (✦ Figure 21.18).

If the pacemaker region of the heart is damaged by heart disease, heart rhythms become abnormal. To restore a normal heartbeat, an artificial pacemaker can be implanted near the AV node. This device sends a timed electrical signal to the cardiac muscle, restoring a normal pattern of cardiac muscle contractions.

## Heart Attacks Result from the Death of Cardiac Muscle

Muscle fibers of the heart are supplied with nutrients and oxygen by a system of **coronary arteries.** Blood flow is adjusted according to the heart's activity. During exercise, blood flow can be increased by up to five times over that of the resting rate.

Blockage of blood flow through the coronary arteries can cause death of cardiac muscle, resulting in a heart attack (✦ Figure 21.19). Blockage of the coronary circulation is most often caused by the gradual buildup of lipids and cholesterol on the inner wall of a coronary artery (atherosclerosis) or a blood clot that suddenly blocks a blood vessel.

Gradual blockage of coronary arteries can result in occasional chest pain or **angina pectoralis,** especially during periods of physical exertion or emotional stress. This pain is usually temporary, and can be reversed by use of drugs such as nitroglycerin, which enlarge (dilate) blood vessels. Angina is an indication that cardiac oxygen demands are greater than blood flow, and that a heart attack might occur in the future.

Sudden blockage of a coronary artery by a blood clot can result in symptoms that include a crushing pressure under the sternum, pain down the left shoulder and arm, and shortness of breath, sweating, and nausea. Blockage of a coronary artery results in oxygen deprivation and death of the heart muscle served by that vessel. If cardiac muscle fibers die, they are not replaced because heart muscle cells do not divide.

Heart disease and coronary artery disease are the leading causes of deaths in the United States. The outcome of a heart attack depends on the location and amount of cardiac muscle that is destroyed (Figure 21.19).

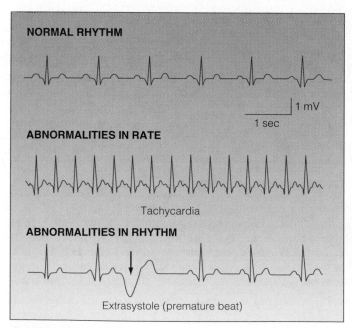

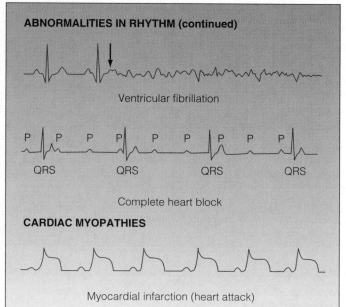

✦ FIGURE 21.18

A variety of cardiac problems can be detected by ECGs.

Area of cardiac muscle deprived of blood supply if coronary vessel blocked at point A

Area of cardiac muscle deprived of blood supply if coronary vessel blocked at point B

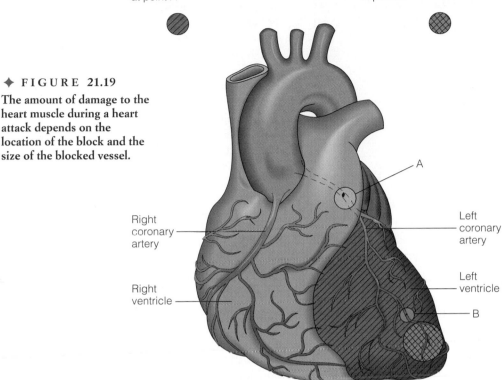

**✦ FIGURE 21.19**

**The amount of damage to the heart muscle during a heart attack depends on the location of the block and the size of the blocked vessel.**

Right coronary artery

Right ventricle

A

Left coronary artery

Left ventricle

B

## Blood Flow Exerts Pressure on Blood Vessels

The circulatory system in vertebrates, including humans, is a closed system consisting of vessels, a pump, and fluid (blood). Ventricular contraction propels blood into the arteries with force, generating a pressure. The amount of pressure generated depends on the strength of the contraction, friction in the blood vessels, the volume of blood being pumped, and several other factors. By convention, this pressure is measured by the height of a column of mercury. In healthy young adults, the pressure generated by ventricular contractions (ventricular systole) is around 120 mm, and 80 mm when the ventricles are relaxed (ventricular diastole) (✦ Figure 21.20). This is written as 120/80, a familiar way of expressing blood pressures.

Blood pressures in other animals can be much different. Blood pressure is higher in larger mammals, ranging from 190 in the horse to 25 to 30 in the cat. Pressures in cold-blooded animals are generally lower, about 30 mm in the frog, for example. In animals with an open circulatory system, blood pressure is even lower. In the lobster, pressure is so low it

is measured with a column of water rather than mercury. Even with water, a normal blood pressure reading for a lobster would be 12/1. Low pressure means a correspondingly low rate of blood circulation. One volume of blood takes about 8 minutes to circulate through the body of a lobster, but only about 20 seconds in humans.

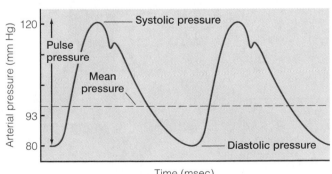

**✦ FIGURE 21.20**

**Arterial blood pressure. Systolic pressure is the force of blood being pumped into arteries during ventricular contraction (systole). The diastolic pressure is the lowest pressure remaining in the arteries when blood is draining from the arterial system between ventricular contractions.**

*Sidebar*

### GENETICALLY ENGINEERED BLOOD

The long search for an effective man-made substitute for blood that is both easy to use and free from contamination may be coming to a close. Companies are working to develop a blood substitute that will satisfy the need for the estimated 13 million units of blood transfused in the United States each year. That translates into a domestic market of approximately $2 billion and a global market as large as $8 billion a year.

Commercial research is aimed at producing blood substitutes. One approach to producing blood substitutes involves genetically engineered hemoglobin. Genes that control hemoglobin production are inserted into a bacterium and become incorporated into the bacterium's genetic material. This genetic material containing the hemoglobin gene is inserted into *E. coli*, a commonly used bacterium. During fermentation, *E. coli* produces hemoglobin, which is purified and packaged for commercial sale. However, this approach has yet to be perfected in laboratory experiments.

The obvious advantage to genetically engineered hemoglobin is that it could have the efficiency of natural blood, be manufactured in unlimited quantities through genetically altered bacteria, and it would not contain contaminants, especially viruses such as HIV and hepatitis.

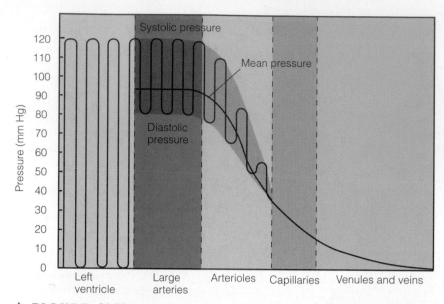

**✦ FIGURE 21.21**

**Systemic blood pressure. Arterial blood pressure is the same throughout the larger arteries. Because of the resistance offered by the decreased diameter of the walls in arterioles, the pressure is reduced dramatically, and the fluctuations in systolic and diastolic pressure are reduced and gradually disappear. Pressure continues to decline as blood flows through the capillaries and veins.**

Blood pressure is not the same throughout the circulatory system. As blood gets further from the heart, pressure is reduced, and so as blood returns to the heart, pressure in the right atrium is very low (✦ Figure 21.21). In the lungs, capillary walls flex easily, keeping pressure low in the pulmonary circuit. Ventricular contractions send blood coursing through the systemic circuit at higher pressure. Blood pressure measurements record this systemic pressure.

Systemic blood pressure is sensed by receptors (baroreceptors) embedded in the walls of the atria and arteries. Nerve impulses from these pressure receptors feed the **medulla,** a region of the brain (✦ Figure 21.22). Signals from this center regulate blood pressure by controlling the expansion or contraction of the circular muscles in the arterioles, and by controlling the thoroughfare circuits in the capillary beds.

**Hypertension** or high blood pressure occurs when the blood pressure is consistently above 140/90. At this range, the risk for cardiovascular disease is increased. In 90% of cases, the cause or causes are unknown. There is a genetic predisposition to hypertension that can be intensified by factors such as obesity, high salt intake, stress, and smoking, but this predisposition accounts for only a small number of cases.

**✦ FIGURE 21.22**

**Pressure in the circulatory system is monitored by receptors in the walls of blood vessels in the arterial system.**

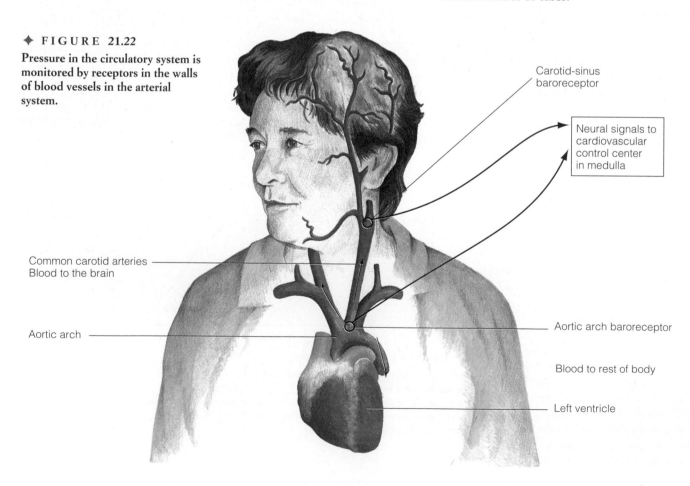

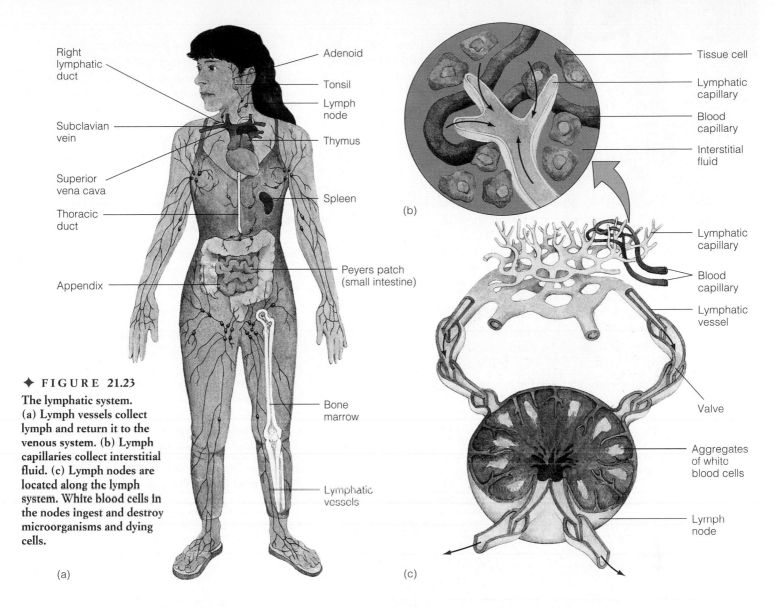

**✦ FIGURE 21.23**

**The lymphatic system. (a) Lymph vessels collect lymph and return it to the venous system. (b) Lymph capillaries collect interstitial fluid. (c) Lymph nodes are located along the lymph system. White blood cells in the nodes ingest and destroy microorganisms and dying cells.**

Labels in figure (a): Right lymphatic duct, Subclavian vein, Superior vena cava, Thoracic duct, Appendix, Adenoid, Tonsil, Lymph node, Thymus, Spleen, Peyers patch (small intestine), Bone marrow, Lymphatic vessels

Labels in figure (b): Tissue cell, Lymphatic capillary, Blood capillary, Interstitial fluid

Labels in figure (c): Lymphatic capillary, Blood capillary, Lymphatic vessel, Valve, Aggregates of white blood cells, Lymph node

Hypertension is a silent killer, because there are no obvious symptoms. The increased workload eventually results in damage to the heart and blood vessels, often leading to heart failure. Under sustained high pressure, blood vessels can rupture, leading to strokes and hemorrhages. Hypertension can also result in kidney failure and progressive blindness caused by damage to blood vessels. It is estimated that 10 to 20% of the adult population of the United States suffers from hypertension, making this a serious health problem.

## ✦ THE LYMPHATIC SYSTEM IS A SECONDARY CIRCULATORY SYSTEM

The **lymphatic system,** an extensive network of vessels and glands, begins in the tissues of the body with lymph capillaries, vessels about the same size as blood capillaries (✦ Figure 21.23). Under pressure from the circulatory system, water and plasma

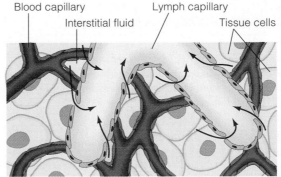

Labels: Blood capillary, Interstitial fluid, Lymph capillary, Tissue cells

**✦ FIGURE 21.24**

**The lymphatic capillaries are thin-walled vessels that collect excess interstitial fluid. Cells in the walls of the lymph capillaries push inward, allowing fluid to enter the capillary.**

are forced from the capillaries into the spaces between cells. This **interstitial fluid** is used to transport gases, nutrients, and wastes between cells and the capillaries. Some of the interstitial fluid enters blood capillaries after circulating around cells of the body (✦ Figure 21.24). The rest collects in lymph capillaries (once in the system, the interstitial fluid is called **lymph**) and is passed to larger lymph vessels. Somewhat like veins, lymph vessels have walls

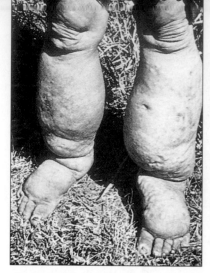

**✦ FIGURE 21.25**

A parasite that invades the lymph system blocks the flow of lymph, causing a buildup of tissue fluids. The result is a condition known as elephantiasis.

with smooth muscle fibers and internal valves to prevent backflow. Lymph vessels merge into larger and larger vessels throughout the body, emptying into the thoracic duct and the right lymphatic duct, which connect with the venous system near the base of the neck (Figure 21.23).

Lymph organs include the lymph nodes, tonsils, and the spleen. The lymph nodes are small, irregularly shaped masses of spongy tissue through which lymph vessels flow (Figure 21.23). Clusters of lymph nodes are found in the groin, armpits, and neck. Lymph nodes consist of several chambers intermixed with fibers and channels. Cells of the immune system line the channels and ingest and destroy any invading viruses and bacteria traveling in the lymph. When you are sick and have swollen lymph nodes, it is because the nodes are filled with cells responding to infection.

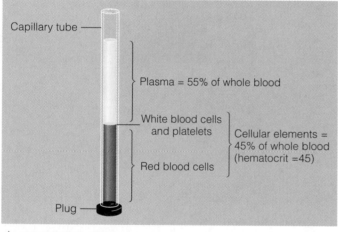

**✦ FIGURE 21.26**

Whole blood placed in a tube and centrifuged separates into two major components: red blood cells and plasma. The red blood cells pack into a lower layer, and white blood cells and platelets form a thin layer at the boundary of the red blood cells and the plasma.

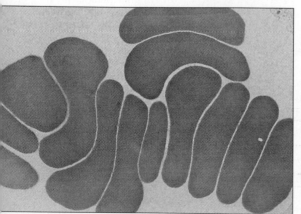

(a)

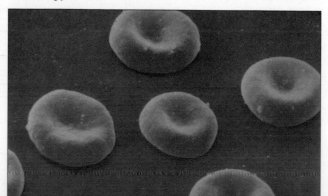

(b)

**✦ FIGURE 21.27**

Red blood cells as seen by (a) transmission electron microscopy and (b) scanning electron microscopy.

Lymph is pumped through the lymphatic circulatory system as various muscles of the body contract during stretching and movement, and by the contraction of muscles in the walls of the lymph vessels. Some forms of parasitic infection block lymph nodes, impeding the flow of lymph and causing it to accumulate in tissues. The result can be a disfiguring swelling of tissues known as **elephantiasis** (✦ Figure 21.25).

## BLOOD IS A TISSUE COMPOSED OF FLUID AND CELLS

In mammals, including humans, blood is composed of two components: cells and plasma. If a blood sample is taken, placed into a test tube, and allowed to settle, or spun briefly in a centrifuge, two main layers will form: a top layer made up of plasma (about 50 to 60% of the total) and a layer containing cells (about 40 to 50% of the volume) (✦ Figure 21.26).

The plasma component of blood is a straw-colored fluid composed of about 90% water. The other 10% is made up of dissolved materials including proteins, nutrients (such as glucose and amino acids), ions, hormones, and gases.

### Red Blood Cells Transport Oxygen

Red blood cells (**erythrocytes**) are somewhat flattened, doubly concave cells about 7 micrometers in diameter (✦ Figure 21.27). They are formed from stem cells in bone marrow, and during the process of maturation, lose their nucleus, mitochondria, and other organelles, yet remain as living cells. The mature red cell has a life span of about 120 days, and during that time, it functions to transport oxygen to the cells and tissues of the body. Humans have about 5 million red blood cells in every cubic milliliter of blood, and around 25 trillion such cells (about one-third of all cells) in the body. Each *second*, about 2 million red blood cells are produced, replacing 2 million that are lost by turnover.

The red color of erythrocytes is due to the presence of hemoglobin, a protein complexed with iron. In the lungs, oxygen moves into the red blood cells in response to a pressure gradient, where it complexes with the iron in hemoglobin for transport to the cells and tissues. As a result, oxygenated blood is bright red, and blood that is depleted of oxygen is darker in color, and is blue when seen inside veins.

Recycling of erythrocytes takes place in the spleen and liver. Here, hemoglobin is broken down and its amino acids are used to make other proteins. The iron released from dying erythrocytes is transported to the bone marrow where it is used in making new cells.

A genetic disorder caused by a mutation in hemoglobin is responsible for the symptoms of sickle cell anemia. In affected individuals, the altered hemoglobin changes the shape of the red blood cell (✦ Figure 21.28). Sickled red cells block capillaries and small blood vessels, causing pain and other symptoms including ulcers and strokes. This debilitating disorder is prevalent in Americans whose ancestors lived in West Africa, the lowlands around the Mediterranean Sea, or parts of the Middle East and India.

**✦ FIGURE 21.28**

**In the recessive disorder sickle cell anemia, mutant hemoglobin molecules distort the shape of red blood cells, causing them to take on a sickled appearance.**

## White Blood Cells Are Part of the Body's Defense System

**Leukocytes** or white blood cells make up less than 1% of the blood volume (there is one white blood cell for every 1000 erythrocytes). Leukocytes are produced from stem cells in bone marrow. There are five types of leukocytes, based on their size, staining characteristics, and the shape of their nuclei (✦ Table 21.1). Among the leukocytes, neutrophils and monocytes are capable of engulfing and

**TABLE 21.1  SUMMARY OF BLOOD CELLS**

| NAME | LIGHT MICROGRAPH | DESCRIPTION | CONCENTRATION (NUMBER CELLS/MM³) | LIFE SPAN | FUNCTION |
|---|---|---|---|---|---|
| **Red blood cells** (RBCs) | | Biconcave disk; no nucleus | 4 to 6 million | 120 days | Transports oxygen and carbon dioxide |
| **White blood cells** | | | | | |
| Neutrophil | | Approximately twice the size of RBCs; multi-lobed nucleus; clear-staining cytoplasm | 3000 to 7000 | 6 hours to a few days | Phagocytizes bacteria |
| Eosinophil | | Approximately same size as neutrophil; large pink-staining granules; bilobed nucleus | 100 to 400 | 8 to 12 days | Phagocytizes antigen–antibody complex; attacks parasites |
| Basophil | | Slightly smaller than neutrophil; contains large, purple cytoplasmic granules; bilobed nucleus | 20 to 50 | Few hours to a few days | Releases histamine during inflammation |
| Monocyte | | Larger than neutrophil; cytoplasm grayish-blue; no cytoplasmic granules; U- or kidney-shaped nucleus | 100 to 700 | Lasts many months | Phagocytizes bacteria, dead cells, and cellular debris |
| Lymphocyte | | Slightly smaller than neutrophil; large, relatively round nucleus that fills the cell | 1500 to 3000 | Can persist many years | Involved in immune protection, either attacking cells directly or producing antibodies |
| **Platelets** | | Fragments of megakaryocytes; appear as small dark-staining granules | 250,000 | 5 to 10 days | Play several key roles in blood clotting |

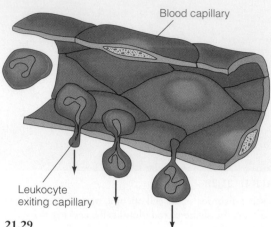

Blood capillary

Leukocyte
exiting capillary

✦ **FIGURE 21.29**

**White blood cells (leukocytes) can squeeze out of the capillaries to move to the sites of infection.**

✦ **FIGURE 21.30**

**Scanning electron micrograph of a blood clot. The fibrin of the clot has trapped red blood cells and platelets. Clots help seal off damaged blood vessels.**

destroying microorganisms, including viruses and bacteria. Such cells are classified as **phagocytes** (literally, "cell eaters"). Lymphocytes are important components of the immune system and will be described in more detail in the next chapter.

Leukocytes squeeze out of the circulatory system into the intercellular spaces where they function in fighting infectious diseases (✦ Figure 21.29). At the site of infections, white blood cells become distended with engulfed microorganisms and die.

### Platelets Help Form Blood Clots

**Platelets** are cell fragments that bud off from cells in bone marrow known as **megakaryocytes.** These fragments do not contain nuclei or organelles, but they do carry a number of chemicals needed for the formation of blood clots. Platelets survive for about ten days, and are removed by the liver and the spleen. There are about 150,000 to 300,000 platelets in each cubic millimeter of blood. Platelets prevent blood loss from damaged vessels in two ways.

The surface of platelets is sticky, and they adhere to rough spots, irregularities or tears in blood vessels, forming a plug. In addition, clotting factors released by the platelets activate proteins in the plasma to form a clot (✦ Figure 21.30).

**Hemophilia** is a genetic disorder of the clotting mechanism. In this disorder, even small cuts and bruises can cause serious bleeding and even death. One form, caused by a mutation carried on the X chromosome, affects about 1 in every 10,000 males.

---

 SUMMARY

---

1. The cells of larger and more complex animals are not in direct contact with the external environment; hence, they require a circulatory system to obtain nutrients, dispose of metabolic wastes, and exchange gases.

2. Two basic types of circulatory systems have evolved. Insects and other invertebrates such as crayfish and some mollusks have an open circulatory system. The sec-

ond type of system, found in echinoderms and vertebrates, is a closed circulatory system.

3. A cardiovascular system with two circuits, such as those in humans, separates oxygen-rich from oxygen-depleted blood, and allows the systemic system to operate at much higher pressure.

4. Blood is pumped through the human circulatory system by a two-sided, four-

chambered structure with muscular walls, lined with a specialized epithelial layer.

5. In the closed cardiovascular system of mammals, blood is conducted through a series of blood vessels that are structurally specialized to carry out their functions. In humans, there are three types of blood vessels: arteries, veins, and capillaries.

6. In humans, the heartbeat originates in an area of the right atrium known as the

sinoatrial (SA) node or pacemaker. Muscle contractions in the sinoatrial node spread to both atria, and after a short delay, to the ventricles.

7. Muscle fibers of the heart are supplied with nutrients and oxygen by a system of coronary arteries. Blockage of blood flow through the coronary arteries can cause cardiac muscle to die, resulting in a heart attack.

8. The lymphatic system is a secondary circulatory system composed of lymph capillaries, vessels, and organs including lymph nodes and the spleen. The lymph system empties into the circulatory system through a duct system near the base of the neck.

9. In mammals, including humans, blood is a tissue composed of two components:

cells and plasma. The red blood cells transport oxygen from the lungs to the cells in the tissues. White blood cells are components of the immune system, and they protect the body from infection. Blood platelets assist in the formation of blood clots.

# KEY TERMS

angina pectoralis
aorta
arteries
arterioles
atrioventricular (AV) node
atrioventricular (AV) valve
auricle
capillary bed
cardiac cycle
cardiovascular system
circulatory system, closed
circulatory system, open
coronary arteries
diastole

elephantiasis
erythrocytes
heart
hemophilia
hypertension
immune system
interstitial fluid
leukocytes
lymph hearts
lymph
lymphatic circulation
lymphatic system
medulla
megakaryocytes

pacemaker
phagocytes
platelets
pulmonary artery
pulmonary circuit
pulmonary vein
semilunar valve
sinoatrial node
systemic circuit
systole
thoroughfare channels
veins
ventricle
venules

# QUESTIONS AND PROBLEMS

## FILL IN THE BLANK

1. The _____ circuit supplies blood to the lungs. The vessels of the _____ circuit supply blood to the rest of the body.
2. _____ have a two-sided heart with two atria sharing a single ventricle.
3. A flap of tissue known as the _____ _____ separates each auricle and ventricle in the human heart.
4. Vessels that carry blood away from the heart are called _____. The vessels responsible for blood flow back to the heart are called _____.
5. The exchange of nutrients and wastes occurs in the _____, thin-walled vessels that form extensive networks in the body tissue. These vessels empty into _____, the smallest veins.
6. The heartbeat in humans originates in an area of the right _____ known as the _____ node.
7. The _____ valves in the base of the aorta and pulmonary artery prevent

blood from flowing back into the _____.

8. _____ is the gradual buildup of lipids on the inner wall of a coronary artery.
9. _____ is the liquid portion of the blood and is about _____ water.
10. _____ recycling of erythrocytes takes place in the _____ and _____.

## MATCHING

Match each component of the circulatory system appropriately.

1. red blood cells
2. white blood cells
3. phagocytes
4. platelets
5. lymph nodes

_____ small masses of spongy tissue
_____ cell eaters
_____ erythrocytes
_____ clot formers
_____ leukocytes

Match each condition appropriately.

6. angina pectoralis
7. hypertension

_____ clotting disorder

8. elephantiasis
9. sickle cell anemia
10. hemophilia

_____ chest pain
_____ hemoglobin mutation
_____ blood pressure consistently above 140/90
_____ swelling of tissues

## SHORT ANSWER

1. Describe the general structure of the arteries and veins. How are they similar? How are they different?
2. Explain the role of the lymphatic system in circulation.
3. Explain how a blood clot forms.
4. Describe the ways in which the circulatory system functions in homeostasis.
5. Describe the two basic types of circulatory systems.

## SCIENCE AND SOCIETY

1. Cadmium is a heavy metal once used by physicians to treat syphilis and malaria, a remedy abandoned when toxic effects

were seen. Today, cadmium is used for commercial purposes. The most common use is as a chemical stabilizer used to make children's toys. It is also combined with gold and other metals to make jewelry.

Cadmium is released into the air and water during the manufacturing of products. Incinerators that burn garbage release cadmium from rubber tires, plastic bottles, and other items. It has been estimated that over 4 million pounds of cadmium are released into the air each year in the United States.

Cadmium is toxic to all body systems of humans and animals. It is easily absorbed into the body, and its levels increase as we age. Hypertension is one probable result from cadmium exposure. Humans with hypertension secrete 40 times more cadmium in their urine than people with normal blood pressure. Given the rise of cadmium use and the increase in recycling and incineration of trash, what do you project will happen to the number of individuals suffering from hypertension? Can you propose a safer means of disposing of garbage? Do the advantages gained from recycling outweigh the disadvantages? Based on the information given, do you agree or disagree with the statement that most cases of hypertension are due to cadmium exposure? Why or why not? What additional information do you need before you can form an opinion on the relationship between hypertension and cadmium exposure?

2. The stem cell in bone marrow is the focus of intense research by scientists who believe that by manipulating the cell they can discover better ways to treat such conditions as leukemia, sickle cell anemia, osteopetrosis (bone disorder resulting in overly dense bones), and even AIDS. Basic research is investigating what causes the cell to replicate and differentiate. Experiments performed in the laboratory have proven very promising. Although a great deal of basic research remains to be done, many scientists believe they are on the path to a new form of treatment for many untreatable diseases. Can you think of any problems with these experiments? What other influences might contribute to stem cell differentiation? Can you think of possible problems with injecting someone's stem cells into the bone marrow of another person?

# THE IMMUNE SYSTEM

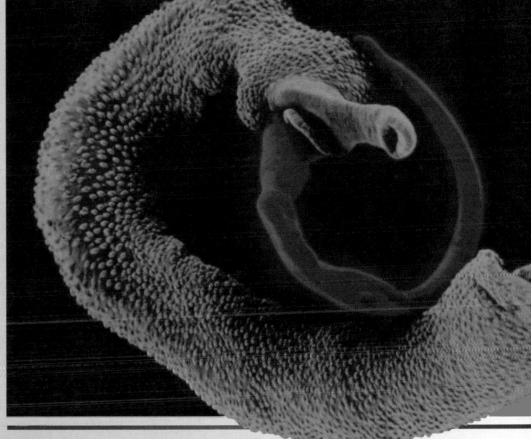

**OPENING IMAGE**

*Schistosomes are blood flukes that*

*can bypass the immune system and*

*cause infection in humans.*

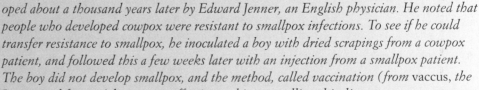

Approximately 2500 years ago, a mysterious plague swept through Athens, Greece, killing thousands of residents. The historian Thucydides wrote an account of the epidemic, and noted that those who had recovered from the disease could care for the sick without becoming ill a second time. In ancient China, physicians observed that people who contracted smallpox became resistant to the disease. Records from eighth-century China indicate that partially successful attempts were made to transfer resistance to uninfected individuals by injecting them with fluid obtained from smallpox victims.

The first safe and successful method of transferring resistance to a disease was developed about a thousand years later by Edward Jenner, an English physician. He noted that people who developed cowpox were resistant to smallpox infections. To see if he could transfer resistance to smallpox, he inoculated a boy with dried scrapings from a cowpox patient, and followed this a few weeks later with an injection from a smallpox patient. The boy did not develop smallpox, and the method, called vaccination (from vaccus, the Latin word for cow) became an effective tool in controlling this disease.

The World Health Organization led a worldwide vaccination campaign against smallpox and eradicated the disease by 1980. Because the virus that causes smallpox cannot reproduce outside the body of an infected individual, the disease has disappeared. The only remaining samples of the virus are stored in research facilities in the United States and Russia.

*Vaccines are used to prevent a number of diseases, including diphtheria, whooping cough, measles, and hepatitis B. However, the study of the body's resistance to disease is a young and developing science, and new diseases still arise to remind us of our limited knowledge. In the early 1970s, women began to develop a condition known as toxic shock syndrome. By 1980, several thousand cases and more than 25 deaths had been reported. Careful investigation determined that certain brands of highly absorbent tampons led to vaginal infections with certain strains of* Staphylococcus aureus. *The bacteria secrete a toxin that produce the symptoms of fever, rash, low blood pressure, and, in some cases, death.*

*In 1981, an infectious disease known as acquired immunodeficiency syndrome (AIDS) appeared in the United States. Affected individuals exhibit rare forms of cancer, pneumonia, and other infections. In all cases, there is a complete and irreversible breakdown of the immune system, associated with infection by the human immunodeficiency virus (HIV). This breakdown allows the development of infections, one or more of which proves fatal.*

## THE IMMUNE SYSTEM DEFENDS THE BODY AGAINST INFECTION

Vertebrates have evolved a series of mechanisms to defend against the entry of foreign organisms and infectious agents into their internal environment. Infections caused by viruses, bacteria, and fungi remain a health problem in industrialized nations, and are a major cause of death in Third World nations. The responses of the body to infection include a nonspecific response and a highly specific set of responses known as the immune reaction.

Nonspecific responses are designed to (1) block the entry of disease-causing agents into the body, and (2) block the spread of infectious agents if they are successful in gaining entry to the body. The immune system has two types of responses to invading organisms: antibody-mediated immunity and cell-mediated immunity. Each of these is highly specific and consists of two stages, a primary response and a later-developing secondary response. The secondary response is characterized by a quick reaction after a second exposure to the same invading organism. In addition, the immune system is associated with the success or failure of blood transfusions and organ transplants. In this chapter, we examine the cells of the immune system and how they are mobilized to mount an immune response.

We also consider the role of the immune system in determining blood groups, mother–fetus incompatibility, how cell-surface markers are matched in organ transplants, and how these markers can be used in a predictive way to determine risk factors for a wide range of diseases. Finally, we will describe a number of disorders of the immune system, including how acquired immunodeficiency syndrome (AIDS) acts to cripple the immune response of infected individuals.

## THE FIRST RESPONSES TO INFECTION ARE NONSPECIFIC

The skin represents a passive barrier to infectious agents such as viruses and bacteria. The structure of the skin was discussed in Chapter 18. Bacteria, fungi, and even mites populate the surface of the body, but are unable to penetrate the protective layers of dead skin cells to cause infection. The skin contains glands that secrete acidic oils that inhibit bacterial growth, and sweat, tears, and saliva contain enzymes that break down the outer walls of many bacteria.

The mucus membranes that line the respiratory, digestive, urinary, and reproductive tracts secrete **mucus** that forms a barrier against infection. The respiratory system has an additional barrier in the form of cilia, which sweep out bacteria trapped in the mucus. These physical barriers form the first line of defense against infection with disease-causing agents.

### Nonspecific Responses Are Activated by the Inflammatory Reaction

If microorganisms penetrate the skin or epithelial layers lining the respiratory, digestive, or urinary systems through a break or injury, a reaction called **inflammation** results ( Figure 22.1). Damaged cells release chemical signals including **histamine.** These chemicals increase capillary blood flow in the affected area (the action of these chemicals is responsible for the heat and redness that develops around a cut or scrape). The increased heat creates an environment that is unfavorable for the growth of microorganisms, causes an increase in the mobility of white blood cells, and raises the metabolic rate in nearby cells, promoting healing. In addition, white blood cells such as macrophages and other phagocytes migrate to the area in response to the chemical signals and the increased capillary circulation, where they engulf and destroy the invading microorganisms.

The capillary beds in the area of an infection become leaky, allowing plasma to flow into the injured tissue, often causing the affected area to become swollen. Clotting factors in the plasma trigger a cascade of small blood clots to seal off the injured area, preventing the escape of invading organisms. Finally, the area becomes the target for another type of white blood cell, the **monocytes,** which clean up dead viruses, bacteria, or fungi and dispose of dead cells and debris.

This chain of events, beginning with the release of chemical signals and ending with the cleanup by monocytes, makes up the **inflammatory response.** This response is an active defense mechanism employed by the body to resist infection. This localized and limited reaction is usually enough to stop the spread of infectious agents. If, however, this system fails, other more powerful systems, the complement system and the immune response, are called into action.

## The Complement System Kills Microorganisms Directly

The **complement system** is a chemical defense system that kills microorganisms directly, supplements the inflammatory response, and works with the immune system (✦ Figure 22.2). The name of the system is derived from the fact that it complements the action of the immune system. Complement consists of a number of proteins that are synthesized in the liver and circulate in the bloodstream as inactive precursors. When activated, the first component (C1) activates the second (C2), and so forth, in a cascade of activation. The final five components (C5–C9) form a large, cylindrical multiprotein complex, called the **membrane-attack complex** (MAC). The MAC embeds itself in the plasma membrane of

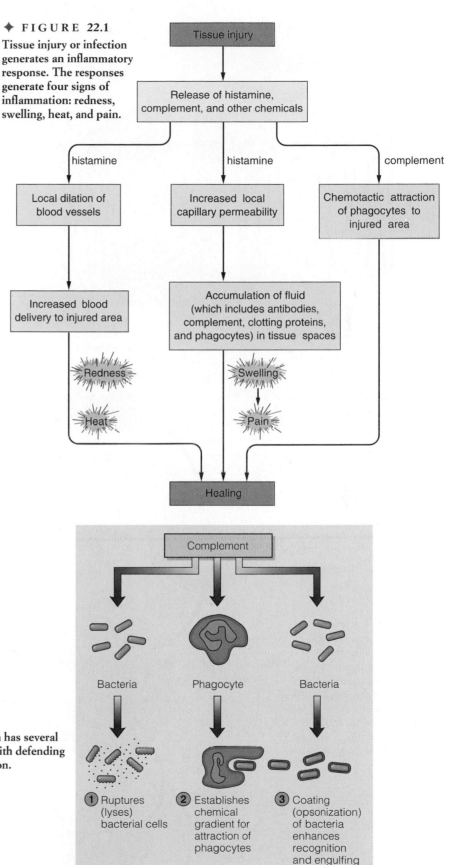

✦ **FIGURE 22.1**

**Tissue injury or infection generates an inflammatory response. The responses generate four signs of inflammation: redness, swelling, heat, and pain.**

✦ **FIGURE 22.2**

**The complement system has several actions, all associated with defending the body against infection.**

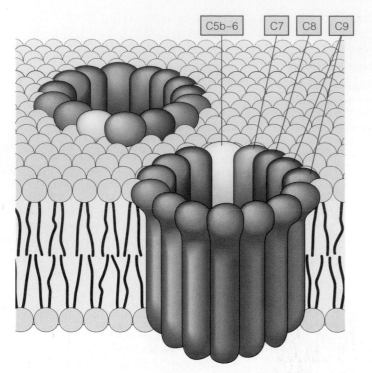

C5b–6    C7    C8    C9

**✦ FIGURE 22.3**

The membrane-attack complex (MAC). Complement proteins insert into the membrane of an invading microorganism. Flow of fluids into the cell causes it to swell and burst.

**✦ FIGURE 22.4**

Immature lymphocytes (lymphoblasts) produced in bone marrow that travel to the thymus for maturation become T cells. Those that remain and mature in bone marrow become B cells. Once mature, lymphocytes migrate through the body in the circulatory system as part of the immune system.

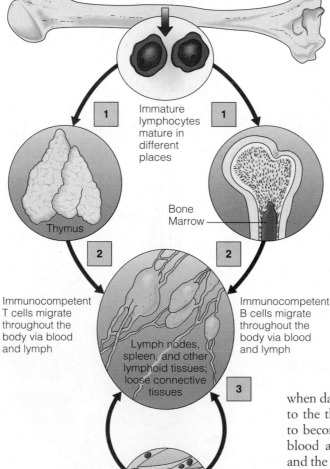

Bone marrow

1 — Immature lymphocytes mature in different places — 1

Thymus

Bone Marrow

2                2

Immunocompetent T cells migrate throughout the body via blood and lymph

Immunocompetent B cells migrate throughout the body via blood and lymph

Lymph nodes, spleen, and other lymphoid tissues; loose connective tissues

3

Immunocompetent and mature B and T cells recirculate in blood and lymph

an invading microorganism, creating a pore (✦ Figure 22.3). Fluid flows through the pore in response to an osmotic gradient, bursting the cell.

In addition to destroying microorganisms directly, some of the complement proteins supplement the inflammation response by guiding phagocytes to the site of microbial invasion. Other components aid the immune response by binding to the outer surface of microorganisms and marking them for phagocytosis.

## THE IMMUNE RESPONSE IS A SPECIFIC DEFENSE AGAINST INFECTION

The immune system generates a response designed to neutralize and/or destroy specific foreign substances including viruses, bacteria, fungi, and cancer cells. The immune system is more effective than the nonspecific defense system, and it has a memory component that remembers previous encounters with infectious agents. Immunological memory allows a rapid, massive response to a second exposure to a foreign substance.

### An Overview of the Immune Response

Immunity is a state of resistance brought about by the production of proteins called **antibodies.** Antibodies bind to foreign molecules and microorganisms and can inactivate them in several ways. Molecules carried by or produced by microorganisms that initiate antibody production are called **antigens** (*anti*body *gen*erators). Most antigens are themselves proteins or proteins combined with polysaccharides, but *any* molecule, regardless of its source, that causes antibody production is an antigen.

The immune response is mediated by white blood cells called **lymphocytes.** These cells arise in the bone marrow, by mitotic division of **stem cells** (✦ Figure 22.4). One type of lymphocyte is formed when daughter cells migrate from the bone marrow to the thymus gland, where they are programmed to become **T cells.** Mature T cells circulate in the blood and become associated with lymph nodes and the spleen. The **B cells** mature in the bone marrow and move directly to the circulatory system and the lymph system. B cells are genetically programmed to produce antibodies; each B cell produces only one type of antibody.

The immune system has two components, **antibody-mediated** (humoral) immunity, regulated by B cells and antibody production, and **cell-mediated** immunity, controlled by T cells (◆ Table 22.1). The primary task of antibody-mediated reactions is to defend the body against invading viruses and bacteria. Cell-mediated immunity is directed against cells of the body that have been infected by viruses and bacteria. T cells also protect against infection by parasites, fungi, and protozoans. One group of T cells can also kill cells of the body if they become cancerous.

## Antibody-Mediated Immunity Uses Molecular Weapons

The antibody-mediated immune response involves several stages: antigen detection, activation of helper T cells, and antibody production by B cells. Each of these stages is directed by a specific cell type of the immune system.

White blood cells called **macrophages** continuously wander through the circulatory system and the interstitial spaces between cells searching for foreign (nonself) antigenic molecules, viruses, or microorganisms. When such an antigen is encountered, it is engulfed, ingested by the macrophage (◆ Figure 22.5), and destroyed by enzymes. Small fragments of

◆ FIGURE 22.5

**Macrophages engulf invading microorganisms and process their antigens for presentation to the helper T cells of the immune system. The activated T cell, in turn, stimulates division of the B cell, producing antibodies against the antigen presented by the macrophage. Daughter B cells become differentiated as plasma cells for the production of large quantities of antibody. Macrophages can also directly stimulate antibody production by interaction with a B cell.**

| TABLE 22.1  COMPARISON OF HUMORAL AND CELL-MEDIATED IMMUNITY | |
| --- | --- |
| HUMORAL | CELL-MEDIATED |
| Principal cellular agent is the B cell. B cell responds to bacteria, bacterial toxins, and some viruses. | Principal cellular agent is the T cell. T cells respond to cancer cells, virus-infected cells, single-celled fungi, parasites, and foreign cells in an organ transplant. |
| When activated, B cells form memory cells and plasma cells, which produce antibodies to these antigens. | When activated, T cells differentiate into memory cells, cytotoxic cells, suppressor cells, and helper cells; cytotoxic T cells attack the antigen directly. |

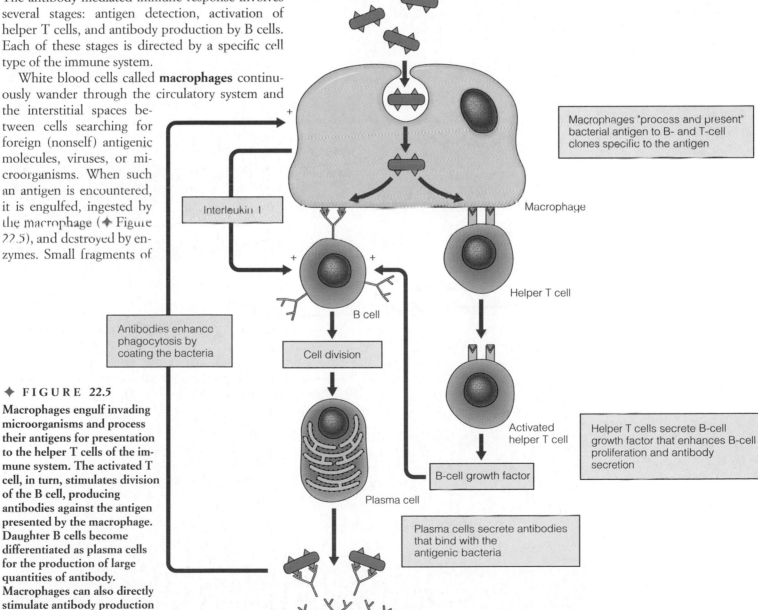

Invading bacteria

Macrophages "process and present" bacterial antigen to B- and T-cell clones specific to the antigen

Macrophage

Interleukin 1

Helper T cell

Antibodies enhance phagocytosis by coating the bacteria

B cell

Cell division

Activated helper T cell

Helper T cells secrete B-cell growth factor that enhances B-cell proliferation and antibody secretion

B-cell growth factor

Plasma cell

Plasma cells secrete antibodies that bind with the antigenic bacteria

Antibodies

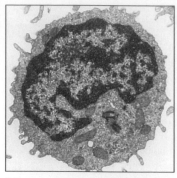

(a)

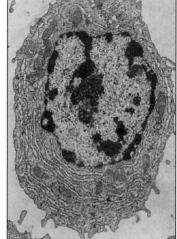

(b)

the antigen are displayed on the outer surface of the macrophage's plasma membrane, along with cell surface markers that identify the cell as a macrophage.

As the macrophage moves about, it may encounter a lymphocyte called a **helper T cell.** Surface receptors on the T cell make contact with the antigen, activating the T cell. The activated T cell, in turn, identifies and activates B cells. The activated B cells divide, forming two types of daughter cells, one of which is the **plasma cell,** which synthesizes and releases 2000 to 20,000 antibody molecules per

*second* into the bloodstream during its life span of four to five days (✦ Figure 22.6). A smaller number of **B memory cells** are also produced at this time. These cells have a life span of months and even years. They are part of the immune memory system and are described in a later section.

This same cascade of events results when a macrophage presents an antigen directly to a B cell.

## Antibodies Are Molecular Weapons against Antigens

Antibodies are protein molecules that bind to specific antigens in a lock-and-key fashion to form an antigen–antibody complex (✦ Figure 22.7). They are secreted by plasma cells and circulate in the blood and lymph, or are attached to the surface of B cells. Antibodies belong to a class of proteins

✦ **FIGURE 22.6**

**Electron micrographs of (a) a mature, unactivated B cell, and (b) a differentiated plasma cell (an activated B cell).**

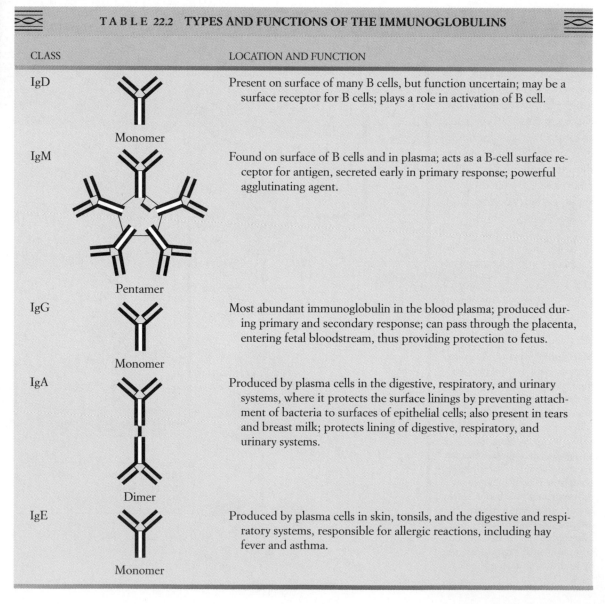

| CLASS | | LOCATION AND FUNCTION |
|---|---|---|
| **T A B L E** *22.2* | **TYPES AND FUNCTIONS OF THE IMMUNOGLOBULINS** | |
| IgD | Monomer | Present on surface of many B cells, but function uncertain; may be a surface receptor for B cells; plays a role in activation of B cell. |
| IgM | Pentamer | Found on surface of B cells and in plasma; acts as a B-cell surface receptor for antigen, secreted early in primary response; powerful agglutinating agent. |
| IgG | Monomer | Most abundant immunoglobulin in the blood plasma; produced during primary and secondary response; can pass through the placenta, entering fetal bloodstream, thus providing protection to fetus. |
| IgA | Dimer | Produced by plasma cells in the digestive, respiratory, and urinary systems, where it protects the surface linings by preventing attachment of bacteria to surfaces of epithelial cells; also present in tears and breast milk; protects lining of digestive, respiratory, and urinary systems. |
| IgE | Monomer | Produced by plasma cells in skin, tonsils, and the digestive and respiratory systems, responsible for allergic reactions, including hay fever and asthma. |

known as **immunoglobulins.** The five classes of immunoglobulins are IgG, IgA, IgM, IgD, and IgE. Each class has a unique structure, size, and function (◆ Table 22.2). In general, antibody molecules are Y-shaped structures composed of two identical long polypeptides (H chains), and two identical short polypeptides (L chains). The chains are held together by chemical bonds.

The structure of antibodies is related to their functions: (1) to recognize and bind to antigens and (2) to inactivate the antigen. At one end of the antibody, there is an antigen-combining site formed by the ends of the H and L chains. This unique antigen-combining site recognizes and binds to a site on the antigen, called the **antigenic determinant.** The formation of an antigen–antibody complex leads to the destruction of an antigen in several ways (◆ Figure 22.8).

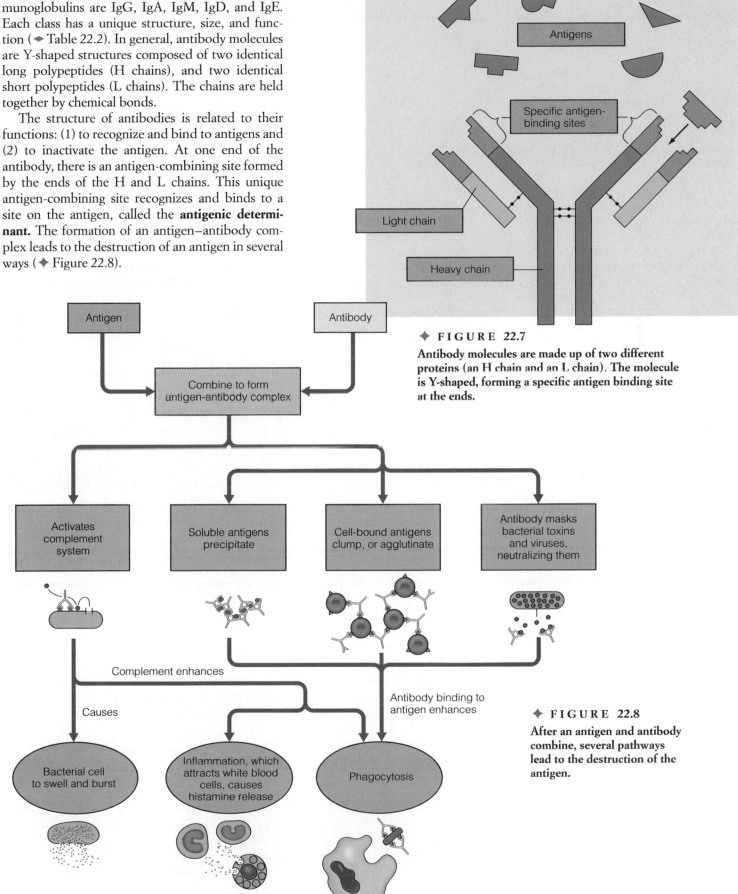

◆ **FIGURE 22.7**

Antibody molecules are made up of two different proteins (an H chain and an L chain). The molecule is Y-shaped, forming a specific antigen binding site at the ends.

◆ **FIGURE 22.8**

After an antigen and antibody combine, several pathways lead to the destruction of the antigen.

## Genomic Research

J. CRAIG VENTER

My high school days were devoted more to competitive swimming and the building of hydroplanes than to my studies. As a result, I barely graduated. After high school, I moved to Newport Beach, California, to concentrate on surfing. Because I was not in school, I was drafted in 1965, but I took the option to enlist in the Navy, where I was to serve on the swim team, until Lyndon Johnson further escalated the war in Vietnam.

Out of an interest in medicine, I became a medical corpsman. One and a half years of training later, I ended up in Da Nang, Vietnam, at the Navy hospital. The schedule was twelve hours a day, seven days a week. In addition, there was a nearby Vietnamese orphanage without any medical care, and I volunteered to spend one day a week there. I was learning at an incredible rate, making a difference, and I loved it. I turned 21 in Vietnam, and I left the Navy with a clear, new direction. I wanted to go into medicine.

I completed my undergraduate work in three years and went on to do the same in graduate school, receiving a Ph.D. in physiology and pharmacology in 1975. My route to medical practice was altered by my introduction to fundamental science. I was fascinated by how cells worked at the molecular level and by how little was actually known. I was very fortunate to be able to study under Nathan O. Kaplan, the discoverer of co-enzyme A and the lactate dehydrogenase isoenzymes. I learned from Kaplan not to be afraid to take on any new technique, an approach that has helped me throughout my career.

In 1984, my lab moved to the National Institutes of Health (NIH). Using the standard manual DNA sequencing technique, we sequenced the gene for one of the first human receptors from the brain.

The Human Genome Project was just being discussed. I found the prospect of identifying the entire human genome with its 50,000 to 100,000 genes exciting. Having the human genes characterized would alter medicine, change fundamental, basic science, and rewrite molecular evolution. I was hooked.

Because of our team's sequencing experience and capacity, in 1991 we were asked by the Secretary of Health and Human Services and the Centers for Disease Control to sequence the 186,000 bp genome of the small pox virus (variola) as a prelude to its final destruction. We have since embarked on a major program to sequence the genomes of a number of microbes. As of 1995, we have sequenced the first two genomes of free living organisms, *Haemophilus influenzae* (1,831,000 bp, *Science*, July 18, 1995) and *Mycoplasma genitalium* (581,000 bp, *Science*, October 1995).

Because of the conservation of genes throughout evolution, many human genes have counterparts in other organisms, including animals and plants. Thus, it is possible to assign probable function to newly discovered genes by their sequence similarity to genes that have been studied in other organisms. This occurred in December 1993 at our institute when in collaboration with Bert Vogelstein and his coworkers at Johns Hopkins University we identified the function of three new human genes in our database through their sequence similarity to mismatch repair genes in bacteria. The human genes are associated with non-polyposis colon cancer. This ability to rapidly compare gene sequences from different organisms comes from the emerging science of bioinformatics, which will serve as a powerful tool for all of the biological sciences, including evolutionary biology.

*J. Craig Venter is the founder, director, and president of The Institute for Genomic Research, a not-for-profit research center located in Rockville, Maryland. He earned a B.S. and Ph.D. from the University of California, San Diego. The Institute applies high-throughput sequencing methods and bioinformatics to identify and characterize genes and entire genomes in a variety of organisms, including plant, animal, and human, with the overall goal of describing evolution at the whole genome level.*

### T Cells Mediate the Cellular Immune Response

There are several types of T cells in the immune system (➤ Table 22.3). Helper T cells, described earlier, activate B cells to produce antibodies. **Suppressor T cells** slow down and stop the immune response of B cells and other T cells, and act as the off switch for the immune system. A third type of T cell is the **cytotoxic** or **killer T cells.** These cells target and destroy cells of the body that are infected with a virus or with bacteria (✦ Figure 22.9). If a cell is infected with a virus, it will display viral antigens on its surface. The foreign antigens are recognized by receptors on the surface of a killer T cell. The T cell attaches to the infected cell and secretes a protein that punches holes in the plasma membrane of the infected cell. The cytoplasmic contents of the infected cell leak out through these holes, and the infected cell dies and is removed by phagocytes. Cytotoxic T cells will also bind to and kill cells of transplanted organs if it recognizes them as foreign. The nonspecific and specific reactions of the immune system are summarized in ✦ Table 22.4.

### The Immune System Has a Memory Function

As described in the opening section of the chapter, ancient writers observed that exposure to certain diseases conferred immunity to second infections by the same disease. This resistance is called **secondary immunity** and results from the production of B and T memory cells during the first exposure to the antigen. A second exposure to the same anti-

## TABLE 22.3
## SUMMARY OF T CELLS

| CELL TYPE | ACTION |
|---|---|
| Cytotoxic T cells | Destroy body cells infected by viruses, and attack and kill bacteria, fungi, parasites, and cancer cells. |
| Helper T cells | Produce a growth factor that stimulates B-cell proliferation and differentiation and also stimulates antibody production by plasma cells; enhance activity of cytotoxic T cells. |
| Suppressor T cells | May inhibit immune reaction by decreasing B- and T-cell activity and B- and T-cell division. |
| Memory T cells | Remain in body awaiting reintroduction of antigen, at which time they proliferate and differentiate into cytotoxic T cells, helper T cells, suppressor T cells, and additional memory cells. |

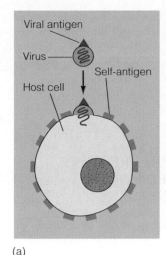
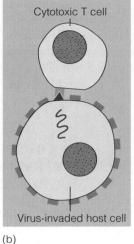
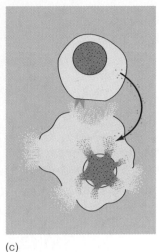

(a)  (b)  (c)

✦ **FIGURE 22.9**

**(a) Cytotoxic T cells can recognize virus-infected cells of the body by the presence of viral proteins on the surface of the infected cell. (b) The T cell can bind to the infected cell and release chemicals that lead to the destruction of the infected cell before the virus can begin to replicate (c).**

gen results in an immediate, large-scale production of antibodies and killer T cells. Because of the presence of the memory cells, the second reaction is faster, more massive, and lasts longer than the primary immune response.

The secondary immune response is the basis for **vaccination** against infectious diseases. A **vaccine** is designed to stimulate the production of memory cells against a disease-causing agent. To do this, the disease-causing antigen is administered orally or by injection, and provokes a primary immune response and the production of memory cells. Often, a second or booster dose is administered to elicit another secondary response that raises or boosts the number of memory cells (which is why such shots are called "booster" shots).

| TABLE 22.4 NONSPECIFIC AND SPECIFIC IMMUNE RESPONSES TO BACTERIAL INVASION | |
|---|---|
| NONSPECIFIC IMMUNE MECHANISMS | SPECIFIC IMMUNE MECHANISMS |
| **Inflammation** | Processing and presenting of bacterial antigen by macrophages to B cells specific to the antigen |
| Engulfment of invading bacteria by resident tissue macrophages | Proliferation and differentiation of activated B-cell clone into plasma cells and memory cells |
| Histamine-induced vascular responses to enhance delivery of increased blood flow to area, bringing in additional immune effector cells and plasma proteins | Secretion by plasma cells of customized antibodies, which specifically bind to invading bacteria |
| Walling off of invaded area by fibrin clot | Enhancement by interleukin 1 secreted by macrophages |
| Emigration of neutrophils and monocytes/macrophages to the area to engulf and destroy foreign invaders and to remove cellular debris | Enhancement by helper T cells, which have been activated by the same bacterial antigen processed and presented to them by macrophages |
| Secretion by phagocytic cells of chemical mediators, which enhance both nonspecific and specific immune responses and induce local and systemic symptoms associated with infection | Binding of antibodies to invading bacteria and enhancement of nonspecific mechanisms that lead to their destruction |
| **Nonspecific Activation of the Complement System** | Action as opsonins to enhance phagocytic activity |
| | Activation of lethal complement system |
| Formation of hole-punching membrane attack complex that lyses bacterial cells | Stimulation of killer cells, which directly lyse bacteria |
| Enhancement of many steps of inflammation | Persistence of memory cells capable of responding more rapidly and more forcefully should the same bacteria be encountered again |

Vaccines are made from killed pathogens or from weakened strains that are able to stimulate the immune system, but unable to produce life-threatening symptoms of the disease. Recombinant DNA techniques are now being used to prepare vaccines against a number of diseases that affect humans and farm animals.

## BLOOD TYPES ARE DETERMINED BY CELL SURFACE ANTIGENS

The presence or absence of certain antigens on the surface of blood cells is the basis of blood transfusions. Each of the 30 or more known antigens on blood cells constitutes a **blood group** or **blood type.** In transfusions, the blood types of the donor and recipient must be matched. If transfused red blood cells have a foreign antigen on their surface, the immune system of the recipient will already have or will produce antibodies to this antigen. These antibodies will cause the transfused cells to clump and block circulation in capillaries and other small blood vessels, with severe and often fatal results. Two blood groups are of major significance: the ABO system and the Rh blood group.

### ABO Blood Typing Allows Safe Blood Transfusions

ABO blood types are determined by a gene *I* (for isoagglutinin), which encodes a cell surface protein. This gene has three alleles, $I^A$, $I^B$, and $I^O$, often written as A, B, and O. The A and B alleles each produce a slightly different version of the gene product, and O produces no gene product. The proteins produced by the A and B alleles are antigenic and can stimulate the production of antibodies. Individuals with type A blood carry the A antigen on their red blood cells, so they will not make antibodies against this cell surface marker. But type A individuals have antibodies against the antigen encoded by the B allele (◆ Table 22.5). Those with type B blood have B antigens on their red cells, and have antibodies against the A antigen. If you have type AB blood, both antigens are present on the surface of red cells, and no antibodies against A or B are made. In those with type O blood, neither antigen is present, but antibodies against *both* the A and B antigens are present in the blood.

In transfusions, AB individuals have no serum antibodies against A or B and can receive blood of any type. Type O individuals have neither red cell antigen, and can donate blood to anyone, even

**Sidebar**

**CONTROLLING THE IMMUNE SYSTEM**

Franz Schubert, Elizabeth Barrett Browning, and John Keats all had two things in common. Each was extremely creative and each had lives cut short by diseases that today could have been cured or prevented. Immunological research is discovering innovative methods for treating cancer and conditions that occur due to abnormalities of the immune system.

Cancer cells wreak havoc by avoiding attacking lymphocytes, even when they carry distinctive antigens that are normally recognized by the immune system. The failure of the immune system to respond properly might be due to the fact that cancerous cells lack accessory chemical signals on their cell surfaces. Researchers are attempting to induce the immune system to combat tumor formation by inserting a highly potent trigger molecule (called B7) into cancer cells. When inserted into tumor cells, B7 allows the immune system to recognize and destroy them. The discovery of this helper molecule has renewed scientific interest in developing anticancer vaccines.

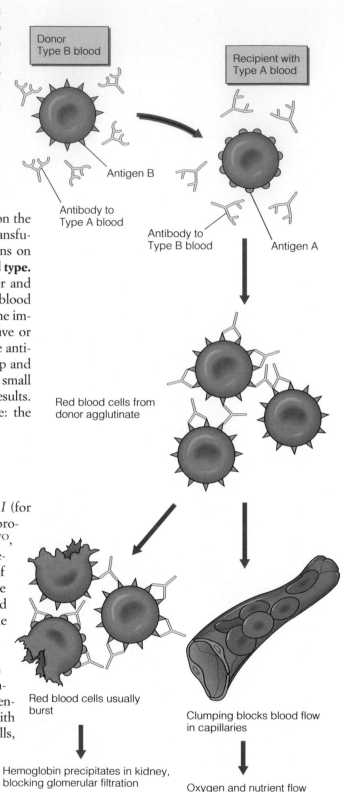

◆ FIGURE 22.10

A transfusion reaction that results when type B blood is transfused into a recipient with type A blood.

though their plasma contains antibodies against A and B. After transfusion, the concentration of these antibodies is too low to cause problems.

When transfusions are made between incompatible blood types, several problems arise. ✦ Figure 22.10 shows the cascade of reactions that results from transfusing someone who is type A with type B blood. Antibodies to the B antigen are in the blood of the recipient; these bind to the transfused red blood cells, causing them to clump together and burst. The clumped cells restrict blood flow in capillaries, reducing oxygen delivery. Lysis of red blood cells releases large amounts of hemoglobin into the blood. In the kidneys, the hemoglobin crystallizes, blocking the tubules of the kidney, often causing kidney failure.

## Rh Blood Types Can Cause Immune Reactions between Mother and Fetus

The Rh blood group (named for the rhesus monkey in which it was discovered) consists of those who can make the Rh antigen (*Rh positive,* Rh+), and those who cannot make the antigen (*Rh negative,* Rh−).

The Rh blood group is of concern when it leads to immunological incompatibility between mother and fetus, a condition known as **hemolytic disease of the newborn** or HDN. This occurs when the

**TABLE 22.5   SUMMARY OF BLOOD TYPES**

| ↑ BLOOD TYPE | ↑ ANTIGENS ON PLASMA MEMBRANES OF RBCs | ↑ ANTIBODIES IN BLOOD | SAFE TO TRANSFUSE | |
|---|---|---|---|---|
| | | | To | From |
| A | A | Anti-B | A, AB | A, O |
| B | B | Anti-A | B, AB | B, O |
| AB | A + B | none | AB | A, B, AB, O |
| O | — | Anti-A Anti-B | A, B, AB, O | |

*Lowercase indicate antibody to B antigen.

mother is Rh− and the fetus is Rh+. If the mother is Rh−, and Rh+ blood from the fetus enters the maternal circulation, antibodies against the Rh antigen will be made. This mixing most commonly occurs during the process of birth, so that the first Rh+ positive child is sometimes not affected. However, the maternal circulation now contains antibodies against the antigen, and a subsequent Rh+ fetus evokes a secondary response from the maternal immune system, producing massive amounts of antibodies that cross the placenta in late stages of the pregnancy and destroy the red blood cells of the fetus (✦ Figure 22.11).

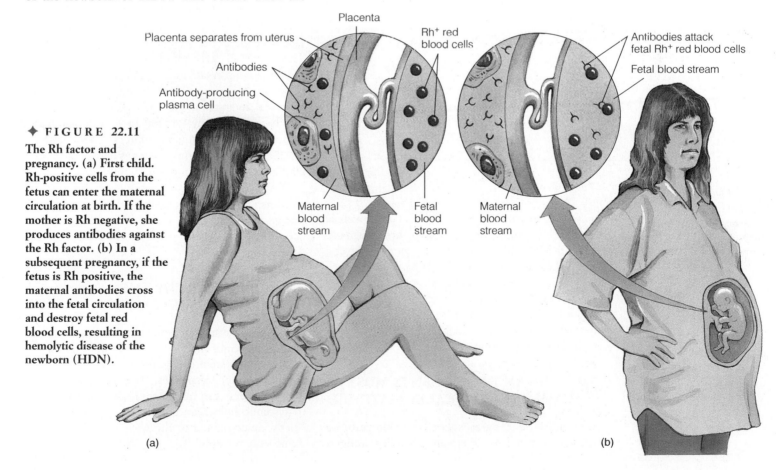

✦ **FIGURE 22.11**
**The Rh factor and pregnancy. (a) First child. Rh-positive cells from the fetus can enter the maternal circulation at birth. If the mother is Rh negative, she produces antibodies against the Rh factor. (b) In a subsequent pregnancy, if the fetus is Rh positive, the maternal antibodies cross into the fetal circulation and destroy fetal red blood cells, resulting in hemolytic disease of the newborn (HDN).**

Placenta
Placenta separates from uterus
Antibodies
Antibody-producing plasma cell
Rh⁺ red blood cells
Antibodies attack fetal Rh⁺ red blood cells
Fetal blood stream
Maternal blood stream
Fetal blood stream
Maternal blood stream

(a)

(b)

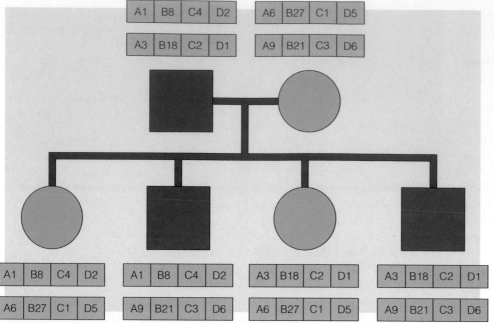

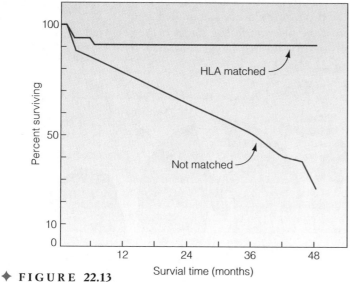

**✦ FIGURE 22.12**

The transmission of HLA haplotypes. Each haplotype contains four genes, each encoding a different antigen.

**✦ FIGURE 22.13**

The outcome of kidney transplants with (upper curve) and without (lower curve) HLA matching.

found on the surface of all cells in the body. In humans, antigens produced by a cluster of genes on chromosome 6 known as the human leucocyte antigen (HLA) complex play a critical role in the outcome of transplants. The HLA complex consists of four neighboring genes: HLA-A, HLA-B, HLA-C, and HLA-D. A large number of alleles have been identified for each of the HLA genes, making possible literally millions of allele combinations. The array of HLA alleles on a given copy of chromosome 6 is known as a **haplotype.** Because we each carry two copies of chromosome 6, we each have two HLA haplotypes (✦ Figure 22.12).

Because of the large number of alleles that are possible, it is rare that anyone's HLA genes will be genetically identical to anyone else's. In the example in Figure 22.12, a child receives one haplotype from each parent. The result is four new haplotype combinations represented in the children.

## Successful Transplants Depend on HLA Matching

Successful transplantation of organs and tissues depends to a large extent on matching HLA haplotypes between donor and recipient. Because there is such a large number of HLA alleles, the best chance for a match is usually between related individuals, with identical twins having a 100% match. The order of preference for organ and tissue donors among relatives is identical twin > sibling > parent > unrelated donor. Among unrelated donors and recipients, the chances for a successful match are only 1 in 100,000 to 1 in 200,000. Because HLA allele frequency differs widely between racial and ethnic groups, matches across racial and ethnic lines are often more difficult.

When HLA types are matched, the survival of transplanted organs is dramatically improved. ✦ Figure 22.13 shows survival rates for matched and unmatched kidney transplants over a four-year period.

## ✿ DISORDERS OF THE IMMUNE SYSTEM

The immune system is vital in protecting the body against infectious disease. Unfortunately, failures in the immune system can result in abnormal or even absent immune responses. The consequences of these disorders can range from mild inconvenience to systemic failure and death. In this section, we briefly catalog some of the ways in which the immune system can fail.

To prevent HDN, Rh− mothers are given an Rh-antibody preparation during the first pregnancy with an Rh+ child and all subsequent Rh+ children. These antibodies move through the maternal circulatory system and destroy any fetal cells that may have entered the mother's circulation. To be effective, this antibody must be administered before the maternal immune system can make its own antibodies against the Rh antigen.

## ✿ ORGAN TRANSPLANTS MUST BE IMMUNOLOGICALLY MATCHED

The success of organ transplants and skin grafts depends on matching **histocompatibility antigens**

## Overreaction in the Immune System Causes Allergies

Allergies are the result of immunological hypersensitivity to weak antigens that do not provoke an immune response in most people (✦ Figure 22.14). These weak antigens are known as **allergens,** and include a wide range of substances: house dust, pollen, cat dander, certain foods, and even medicines such as penicillin. It is estimated that up to 10% of the U.S. population suffers from at least one allergy. Typically, allergic reactions develop after exposure to an allergen. When exposed to an allergen, some people make antibodies belonging to the IgE class of immunoglobulins are produced by plasma cells, and memory cells are produced.

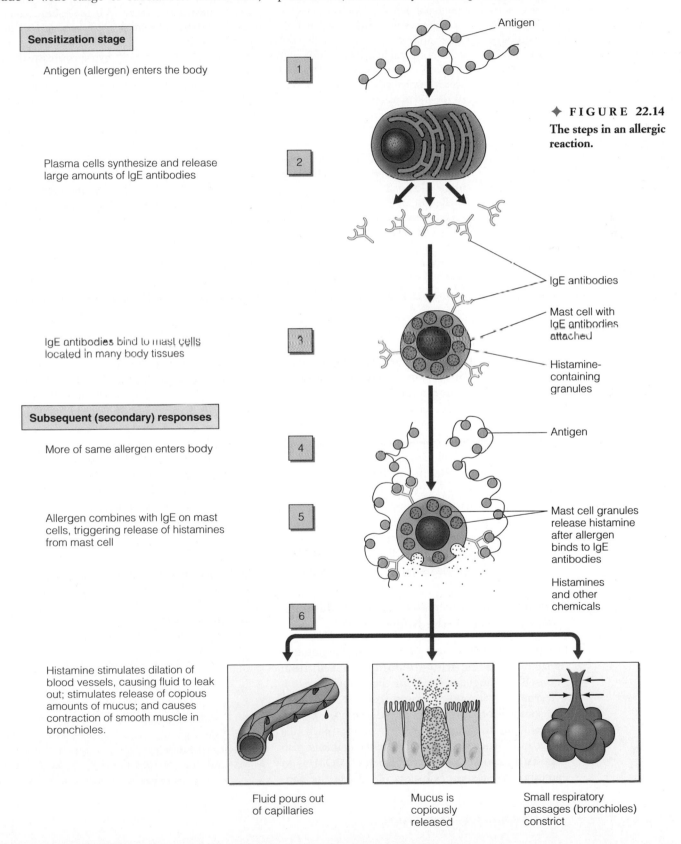

**Sensitization stage**

Antigen (allergen) enters the body

Plasma cells synthesize and release large amounts of IgE antibodies

IgE antibodies bind to mast cells located in many body tissues

**Subsequent (secondary) responses**

More of same allergen enters body

Allergen combines with IgE on mast cells, triggering release of histamines from mast cell

Histamine stimulates dilation of blood vessels, causing fluid to leak out; stimulates release of copious amounts of mucus; and causes contraction of smooth muscle in bronchioles.

Antigen

✦ FIGURE 22.14
The steps in an allergic reaction.

IgE antibodies

Mast cell with IgE antibodies attached

Histamine-containing granules

Antigen

Mast cell granules release histamine after allergen binds to IgE antibodies

Histamines and other chemicals

Fluid pours out of capillaries

Mucus is copiously released

Small respiratory passages (bronchioles) constrict

# WHY BEE STINGS CAN BE FATAL

In most cases, an allergy is an inconvenience, not a life-threatening condition. Typically, the response to a contact allergen involves localized itching, swelling, and reddening of the skin, often in the form of hives. Inhaled allergens usually result in a localized response with itching and watery eyes, runny nose, and constricted bronchial tubes. The most common form of allergy involving inhaled allergens is hay fever; in the U.S., upwards of 20 million people are affected.

In response to injected allergen, such as venom from a bee sting, or certain drugs such as penicillin, the reaction can be systemic, rather than localized, and result in anaphylactic shock. Anaphylaxis is usually a two stage process. In the first stage, such as a bee sting, the venom enters the body, and IgE antibodies are made in response, causing sensitization to future stings. It is not clear why some people become sensitized and others do not. If sensitized, a second exposure to bee venom can provoke a life-threatening allergic reaction. When a sensitized individual is stung by a bee, the allergen in the venom enters the body and binds to immunoglobulin E antibodies made in response to a previous exposure. Within 1–15 minutes after exposure, the IgE antibodies activate mast cells. The stimulated mast cells release large amounts of histamines and chemotactic factors which attract other white blood cells as part of the inflammatory response. In addition, the mast cells release prostaglandins and the slow reacting substance of anaphylaxis (SRS-A). SRS-A is more than 100 times more powerful than histamine and prostaglandins in eliciting an allergic reaction, and intensifies the response to the allergen.

Release of SRS-A, histamine and other factors into the bloodstream causes a systemic reaction. In response, the bronchial tubes constrict, closing the airways, and fluids pass from the tissues into the lungs, making breathing difficult. Blood vessels dilate, dropping blood pressure, and plasma escapes into the tissues, causing shock. Heart arrhythmias and cardiac shock can develop, and cause death within 1–2 minutes after the onset of symptoms. Immediate treatment with epinephrine, antihistamines, and steroids is effective in reversing the symptoms.

In the case of hypersensitivity to insect stings, desensitization can be used as a form of immunotherapy. This process consists of injecting small quantities of the allergen, with a gradual increase in the dose. It is thought that exposure over time to small doses of the allergen causes the production of IgG antibodies against the allergen, rather than IgE antibodies. There is some evidence that desensitization also reduces synthesis of IgG antibodies made against the allergen. When desensitization is complete, subsequent exposure to the allergen causes the more numerous IgG antibodies to bind to the mast cells, prevents the binding of IgE antibodies, and the resulting anaphylactic reaction.

---

Subsequent exposure to the same antigen causes a secondary immune response, which releases a massive amount of IgE antibodies. These antibodies bind to **mast cells,** which are found most often in connective tissues surrounding blood vessels. In response to antibody binding, the mast cells release histamine, a chemical signal that starts an inflammatory response, resulting in fluid accumulation, tissue swelling, and mucus secretion.

In some individuals, the secondary reaction is severe, and histamine is released into the circulatory system, causing a life-threatening reaction called **anaphylaxis** or **anaphylactic shock.** Prompt treatment of anaphylaxis can reverse the reaction.

## Autoimmune Reactions Cause the Immune System to Attack the Body

One of the most elegant properties of the immune system is its capacity to distinguish self from nonself. During development the immune system "learns" not to react against cells of the body. In some disorders, this immune tolerance breaks down, and the immune system attacks and kills cells and tissues in the body. Juvenile diabetes, also known as insulin-dependent diabetes (IDDM) is an autoimmune disease. Clusters of cells in the pancreas normally produce insulin, a hormone that lowers blood sugar levels. In IDDM, the immune system attacks and kills the insulin-producing cells, resulting in diabetes and the need for insulin injections to control blood sugar levels.

Other forms of autoimmunity, such as systemic lupus erythematosus (SLE), are directed against many of the major organ systems in the body, instead of just one cell type. Some autoimmune disorders are listed in  Table 22.6.

| TABLE 22.6 SOME AUTOIMMUNE DISEASES |
| --- |
| Addison disease |
| Autoimmune hemolytic anemia |
| Diabetes mellitus—insulin-dependent |
| Graves' disease |
| Membranous glomerulonephritis |
| Multiple sclerosis |
| Myasthenia gravis |
| Polymyositis |
| Rheumatoid arthritis |
| Scleroderma |
| Sjögren's syndrome |
| Systemic lupus erythematosus |

## Immunodeficiency Diseases Impair the Immune System

Immunodeficiency disorders cause one or more of the parts of the immune system to be missing or nonfunctional. Affected individuals are highly susceptible to infections that would not bother most people. Immunodeficiency can result from diseases such as Hodgkin's disease, from cancer chemotherapy, or radiation therapy. A number of genetic disorders also cause immunodeficiency.

A rare genetic disorder causes a complete absence of both the cell-mediated and antibody-mediated immune response. This condition is called **severe combined immunodeficiency** (SCID). Affected individuals are susceptible to recurring and severe bacterial, viral, and fungal infections, and usually die at an early age from seemingly minor infections. One of the longest known survivors of this condition was David, the "boy in the bubble" who died at 12 years of age, after being isolated in a sterile plastic bubble for all but the last 15 days of his life (✦ Figure 22.15).

One form of SCID is associated with a deficiency of an enzyme known as adenosine deaminase (ADA). A small group of children affected with ADA-deficient SCID are currently undergoing **gene therapy** to provide them with a normal copy of the gene. Genetically modified white blood cells are replaced into their circulatory systems by transfusion. Expression of the normal ADA gene stimulates the development of functional T and B cells, and at least partially restores a functional immune system. The recombinant DNA techniques used in gene therapy are reviewed in Chapter 9.

## AIDS Attacks the Immune System

The immunodeficiency disorder that is currently receiving the most attention is **acquired immunodeficiency syndrome** (AIDS). AIDS is a collection of disorders that develop as a result of infection with a retrovirus known as the human immunodeficiency virus (HIV) (✦ Figure 22.16). The HIV virus consists of a protein coat enclosing an RNA molecule that serves as the genetic material, and an enzyme, reverse transcriptase. The entire viral particle is enclosed in a coat derived from the plasma membrane of a T cell. The virus selectively infects

✦ **FIGURE 22.15**

David, the "boy in the bubble," the longest known survivor of severe combined immunodeficiency disease.

and kills the T4 helper cells of the immune system. Inside the cell, the RNA is transcribed into a DNA molecule by reverse transcriptase, and the viral DNA is inserted into a human chromosome, where it can remain for months or years.

Later, when the infected T cell is called on to participate in an immune response, the viral genes are activated. Viral RNA and proteins are made, and new viral particles are formed. These bud off the surface of the T cell, rupturing and killing the cell, and setting off a new round of T-cell infection

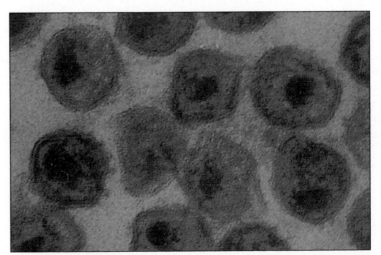

✦ **FIGURE 22.16**

Particles of the HIV virus.

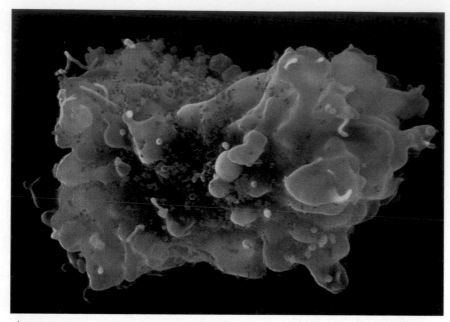

(✦ Figure 22.17). Gradually, over the course of HIV infection, there is a decrease in the number of helper T4 cells. T4 cells act as the master "on" switch for the immune system, and as the T4 cell population falls, there is a decrease in the ability to mount an immune response.

The result is greater susceptibility to infection, and increased risk of certain forms of cancer. Eventually, the outcome is premature death brought about by any of a number of diseases that overwhelm the body and its compromised immune system.

✦ **FIGURE 22.17**

**A colorized electron micrograph showing particles of HIV virus (purple) on the outside of its principal target, a helper T-lymphocyte.**

# SUMMARY

1. The immune system protects the body against infection by a graded series of responses that attack and inactivate foreign molecules and organisms. The lowest level involves a nonspecific, local, inflammatory response.
2. The immune system has two components, antibody-mediated immunity, regulated by B cells and antibody production, and cell-mediated immunity, controlled by T cells.
3. The primary task of antibody-mediated reactions is to defend the body against invading viruses and bacteria. Cell-mediated immunity is directed against cells of the body that have been infected by viruses and bacteria.
4. The presence or absence of certain antigens on the surface of blood cells is the basis of blood transfusions and blood types. Two blood groups are of major significance: the ABO system and the Rh blood group.
5. The matching of ABO blood types is important in blood transfusions. In some cases, mother–fetus incompatibility in the Rh system can cause maternal antigens to destroy red blood cells of the fetus, causing hemolytic disease of the newborn.
6. The success of organ transplants and skin grafts depends on matching histocompatibility antigens found on the surface of all cells in the body. In humans, the antigens produced by a group of genes on chromosome 6 known as the HLA complex play a critical role in the outcome of transplants.
7. Allergies are the result of immunological hypersensitivity to weak antigens that do not provoke an immune response in most people. These weak antigens are known as allergens, and include a wide range of substances: house dust, pollen, cat hair, certain foods, and even medicines such as penicillin.
8. Acquired immunodeficiency syndrome (AIDS) is a collection of disorders that develop as a result of infection with a retrovirus known as the human immunodeficiency virus (HIV). The virus selectively infects and kills the T4 helper cells of the immune system.

acquired immunodeficiency syndrome (AIDS)
allergens
anaphylactic shock
anaphylaxis
antibodies
antibody-mediated immunity
antigenic determinant
antigens
B memory cells
blood group
blood type
cell-mediated immunity

complement system
gene therapy
haplotype
helper T cells
hemolytic disease of the newborn
histamine
histocompatibility antigens
immunoglobulins
inflammation
inflammatory response
killer (cytotoxic) T cells
lymphocytes

macrophages
mast cells
membrane-attack complex
monocytes
mucus
plasma cell
secondary immunity
severe combined immunodeficiency
stem cells
suppressor T cells
vaccination

# QUESTIONS AND PROBLEMS

## MATCHING

1. antibody
2. antigen
3. B cells
4. cell-mediated immune response
5. humoral immune response
6. T cells
7. monocytes

____ molecule that elicits an immune response
____ immunity mediated by T cells
____ white blood cells that phagocytize bacteria and viruses in the body
____ molecule produced in response to the presence of antigens
____ cells that fight invaders directly by binding to them
____ cells that fight invaders by producing antibodies
____ immunity mediated by antibodies

## TRUE/FALSE

1. The first response by the immune system to a foreign substance is a nonspecific defense system. The second, characterized by attacks aimed at a particular intruder, is a specific response.
2. B cells and T cells have the same ancestral origin.
3. An adult with type A blood would normally have anti-A antibodies in the plasma.
4. The constant region of the polypeptide chains of an antibody molecule have binding sites, allowing the antibody to attach to a specific antigen.
5. The body's first line of defense is the skin and mucus membranes.

6. A nonidentical twin is a better organ or tissue donor than an identical twin.

## FILL IN THE BLANK

1. The release of chemical signals called _____ start the chain of events leading to the _____ response.
2. Stem cells that mitotically divide in the _____ _____ give rise to white blood cells called _____.
3. Antibodies belong to a class of proteins known as _____. The five classes of immunoglobulins are _____, _____, _____, _____, and _____.
4. The three types of T cells in the immune system are _____, _____, and _____.
5. A vaccine provokes a _____ immune response and stimulates the production of _____ cells.

## SHORT ANSWER

1. Describe the phenomenon of immunological incompatibility between mother and fetus. How is it prevented?
2. A young child is stung by a bee, swells up, and has great difficulty catching her breath. What has happened? What can be done to help her?
3. What is the difference between passive immunity and active immunity?
4. Why is immunity not developed to diseases such as colds or influenza?
5. How does HIV affect the immune system?
6. Why is an individual with type O blood called a universal donor?

## SCIENCE AND SOCIETY

1. The immune system of newborns is poorly developed and lacks the necessary immunoglobulins for fighting foreign substances. A newborn baby depends on passive immunity from its mother to protect it from the dangerous world in which viruses and bacteria abound. During fetal development, antibodies travel through the maternal bloodstream to the fetus. Antibodies also travel to the infant in breast milk. Breast milk contains IgA—an immunoglobulin that prevents bacteria ingested by the infant from adhering to the epithelium and gaining entrance. Breast milk also contains an enzyme (lysozyme) that destroys bacteria by eroding the cell wall.

Unfortunately, breast milk is extremely low in iron, which has led some medical professionals to question the benefits of breast milk. Researchers have also found that giving iron supplements to breast-fed infants increases the incidence of harmful bacterial infections. The irony of this situation is that low levels of iron in breast milk may reduce bacterial infections in newborns but may also cause developmental problems in these infants. What is your opinion on this issue? What additional information would you need to form an opinion? Can you think of additional reasons why breast feeding would be better than bottle feeding? What limitations does breast feeding have? Do you think the advantages outweigh the disadvantages? Who should make the decision to breast feed or bottle feed—the doctor, the mother, the insurance agency, the government, etc.?

2. Exercise physiology is the study of the functional changes that occur in response to exercise. These changes are the body's attempt to reduce the stress that has been placed on it due to exercise. Very little information is known about the effects of exercise on the immune system. People engaging in moderate levels of regular exercise appear to have fewer infections than their sedentary counterparts. One possible, yet not clearly demonstrated, reason for this is an apparent increase in the number of lymphocytes for short periods of time. Evidence also suggests that very strenuous exercise weakens the immune system and may increase susceptibility to respiratory infections. What additional information is needed on the individuals who were subjects in these experiments? Why is more information needed? Would it be reasonable to say that moderate regular exercise is a proven means of fighting colds? Should someone "catching a cold" start on an exercise program to combat the cold more effectively?

# DIGESTION AND NUTRITION

OPENING IMAGE

*Villi increase the surface area of cells*

*in the small intestine.*

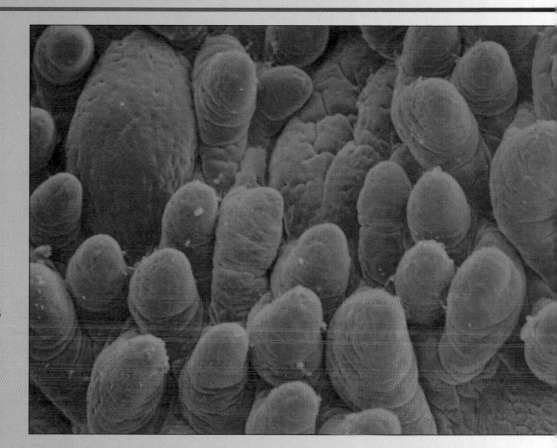

The American Fur Company's store on Mackinac Island, Michigan, was crowded with traders, trappers, and voyageurs on a day in early June 1822. A customer examining a shotgun accidentally fired the weapon in the packed store. Less than three feet away, a young voyageur named Alexis St. Martin received the full force of the blast in his chest and abdomen. Dr. William Beaumont, a surgeon at the nearby fort, was summoned. He dressed the gaping wound and moved St. Martin to the fort hospital. Examining the wound, Beaumont discovered that the shotgun blast created a hole in the abdominal wall and that part of the stomach had been blown away. To Beaumont's surprise, St. Martin survived, although the wound did not close completely. Over the next year, a fold of flesh grew over the opening in St. Martin's abdomen, but inside, the stomach remained open.

Beaumont found that he could raise the skin flap and look directly into the stomach and observe the process of digestion. Seizing this unique opportunity, the surgeon undertook a series of some 238 detailed observations about the physiology of digestion, using St. Martin's stomach as his laboratory. These experiments stretched over several years and were conducted at frontier forts and trading posts in the East and Upper Midwest, whenever Beaumont came in contact with St. Martin.

In his experiments, Beaumont tied pieces of food to a silk string, and placed them into the stomach. At regular intervals, he would pull on the string, recover the food from the stomach, and observe and record the rate at which different foods were digested. He also

recorded changes in the stomach's temperature during digestion, siphoned off gastric juices to study digestion outside the stomach, compared the digestion of chewed and unchewed food, and the rate at which animal and plant foods were digested. In 1833 Beaumont published a book called Experiments and Observations on the Gastric Juice and the Physiology of Digestion, *which outlined the basic principles of digestion, and helped establish the physiological basis of medicine.*

*In this chapter we examine how food is broken down to the molecular level, transferred into the body's internal environment, and used by cells in the processes of energy conversion and biosynthesis. The metabolism of food molecules within cells and the mechanisms involved in energy conversion were discussed in Chapter 4; here we emphasize the function of the digestive system by which food is converted into nutrients, water, and electrolytes. We will also outline nutritional requirements and the role of diet in maintaining homeostasis.*

## FEEDING AND DIGESTION PROCESS FOOD INTO NUTRIENTS

In many single-celled organisms, contact with the environment allows direct uptake of nutrient molecules. However, when nutrient molecules are absent or present in low concentrations, complex molecules or other organisms cannot be used as a food source without a mechanism of feeding and digestion. In multicellular animals, most cells are removed from direct contact with the external environment, and are often specialized for one or a small number of functions. These cells must be provided with nutrients for energy conversion, growth, and repair. To obtain these nutrients, animals depend on two processes: **feeding** and **digestion.**

### Animals Use a Variety of Feeding Mechanisms

Animals are heterotrophs, and must absorb organic nutrients or ingest food sources. Animals exhibit many adaptations in feeding, but several basic features can be distinguished. The majority of animals use a mouth for ingesting food. **Ingestive** eaters feed on plants, animals, or both. A small number of

animals, including tapeworms, live in the digestive tracts of other animals, and absorb nutrients directly through the body wall. In this case, **absorptive** feeding represents a specialized adaptation to a parasitic existence.

Many aquatic species are **filter feeders,** collecting particles or small organisms from the surrounding water. Oysters and mussels are filter feeders, using cilia to pump up to 10 liters of water per hour through their bodies. Some burrowing animals are **substrate feeders,** eating the dirt (earthworms) or wood (termites) through which they burrow. **Fluid feeders** such as aphids, ticks, leeches, and mosquitoes have specialized structures for piercing the body of a plant or animal. They obtain nutrients from the fluids of the host animal. Most animals use bills, claws, tentacles, jaws, and teeth to ingest food (✦ Figure 23.1).

## ANIMAL DIGESTIVE SYSTEMS HAVE TWO MAJOR FEATURES

A digestive system uses a combination of mechanical and chemical processes to transform food into nutrient molecules. Animals employ one of two plans for digestive systems: **sac-like** or **tube-within-a-tube.** In sac-like digestive systems, there is only one opening, used for both food intake and the discharge of wastes (✦ Figure 23.2). Vertebrates use

✦ **FIGURE 23.1**
**Most animals eat their food with teeth.**

## Beyond the Basics

# DINING OUT: THE ASSASSIN BUG

Insects belonging to the order *Hemiptera* live on a liquid diet. Many are vegetarians and that have mouthparts specialized to pierce the leaves and stems of plants, and suck the juices and sap from the plants. Others feed on animals, obtaining a blood meal from their victims. Charles Darwin expressed the sentiment of most of those who have been the victims of such a bug after his South American encounter with one called *Triatoma:* "It was quite disgusting to feel the soft, wingless insects about an inch long, crawling over one's body."

Among the Hemiptera are the assassin bugs, which are efficient predators of other insects. One such species is *Platymeris,* which lives in Africa and feeds on the rhinoceros beetle. The bug lies in wait, and then leaps onto and grips its prey with adhesive pads present on the lower surface of the legs. As the rhinoceros beetle struggles, the assassin bug moves the tip of its mouthparts over the body, searching for a joint in the beetle's hard cuticle. When it finds such a joint, it forces its needle-like mouthparts into the body, and administers an injection of salivary secretions that causes the beetle to become paralyzed in a few seconds.

The mouthparts of *Platymeris* are shaped to form two needle-like channels, one of which serves to inject the saliva. The saliva is drawn from a storage sac adjacent to the salivary glands by a pump-like structure in the head, and injected through one of the channels. This action is much like that of a bicycle pump or syringe, with one stroke drawing up the fluid, and the other injecting the fluid into the prey.

In addition to a paralyzing agent, the saliva of the bug contains powerful digestive enzymes that reduce the internal organs, muscles, and nerves into a semiliquid soup within an hour. After digestion, only the respiratory trachea and the external cuticle of the prey remain. When its food has been digested, the assassin bug feeds by reinserting its mouthparts into the beetle's body, and draws up the liquified food by means of the second channel, which connects to a storage organ, the crop. The empty shell of the rhinoceros beetle is discarded, and the food stored in the crop can be moved as needed to the intestine and hind gut of the digestive system, allowing the assassin bug to live for a week or more on one large meal.

---

the more efficient tube-within-a-tube plan, with two body openings, one for ingestion of food and the other for waste removal (✦ Figure 23.3).

A second major feature of digestive systems is the location of digestion, which can be **intracellular** or **extracellular.** For intracellular digestion, food is taken into cells by phagocytosis. Enzymes are secreted into the phagocytic vacuoles to digest and release nutrients from the food. Intracellular digestion is found in sponges, and in most Protozoa and coelenterates.

In extracellular digestion, found in annelids, crustaceans, and chordates (including vertebrates) digestion takes place within the lumen of the digestive system. The resulting nutrient molecules are

✦ **FIGURE 23.2**

In animals with a sac-like digestive system, like cnidarians, food and waste pass through the same opening. Inside the body, enzymes released by cells lining the sac digest the food.

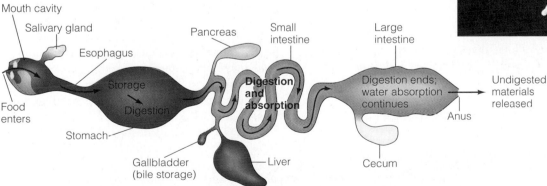

Mouth cavity
Salivary gland
Esophagus
Pancreas
Small intestine
Large intestine
Storage
Digestion and absorption
Digestion ends; water absorption continues
Undigested materials released
Food enters
Digestion
Stomach
Gallbladder (bile storage)
Liver
Cecum
Anus

✦ **FIGURE 23.3**

**In animals with a tube-within-a-tube digestive system, food enters the mouth and undigested residue passes through the anus at the other end of the tube. Specialized organs digest the food and absorb nutrients.**

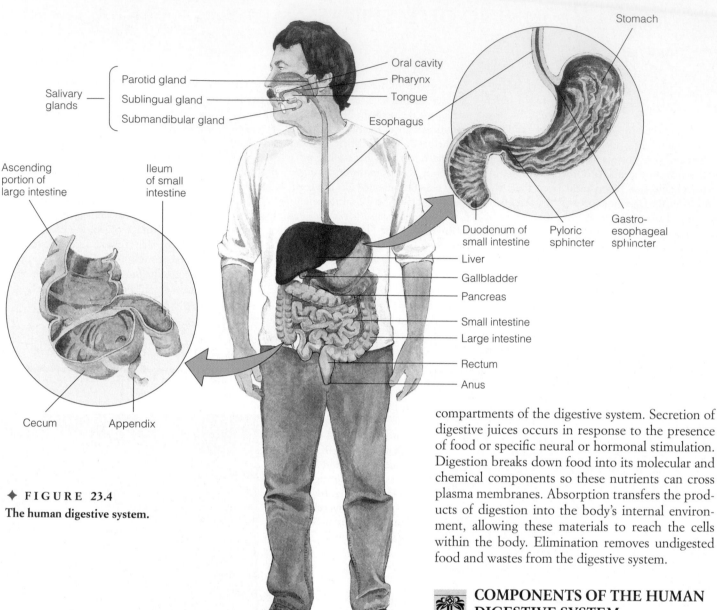

Salivary glands
  Parotid gland
  Sublingual gland
  Submandibular gland

Oral cavity
Pharynx
Tongue
Esophagus

Stomach

Duodenum of small intestine
Pyloric sphincter
Gastro-esophageal sphincter

Liver
Gallbladder
Pancreas
Small intestine
Large intestine
Rectum
Anus

Ascending portion of large intestine
Ileum of small intestine

Cecum
Appendix

✦ **FIGURE 23.4**
**The human digestive system.**

transferred into the blood or body fluid. The evolutionary trend in animals has been the adaptation of extracellular digestion, and it appears that this mechanism may have evolved independently several times.

### ❀ FOOD PROCESSING AND DIGESTION OCCUR IN STAGES

The digestive system performs five basic functions: **movement, secretion, digestion, absorption,** and **elimination.** Movement propels food through the digestive tract so that it can be acted on in stages. It also mixes secreted juices with the food in different compartments of the digestive system. Secretion of digestive juices occurs in response to the presence of food or specific neural or hormonal stimulation. Digestion breaks down food into its molecular and chemical components so these nutrients can cross plasma membranes. Absorption transfers the products of digestion into the body's internal environment, allowing these materials to reach the cells within the body. Elimination removes undigested food and wastes from the digestive system.

### ❀ COMPONENTS OF THE HUMAN DIGESTIVE SYSTEM

In humans, as in other vertebrates, the digestive tract is an elongated, coiled tubular system that extends from mouth to anus (✦ Figure 23.4). When stretched out, the human digestive tract is 6 to 9 meters (20 to 30 feet) in length, and consists of several specialized compartments. These are the **mouth, pharynx** (throat), **esophagus, stomach, small intestine** (composed of the duodenum, jejunum, and ileum), **large intestine** (composed of the cecum, appendix, colon, and rectum), and the **anus.** In addition, there are several accessory digestive organs located outside the digestive tract which are connected to it via ducts. These accessory organs include the **salivary glands,** portions of the **pancreas,** and the **biliary system,** which includes the liver and gallbladder. Together, the digestive tract and the accessory organs make up the digestive system.

## THE MOUTH AND PHARYNX: FEEDING AND SWALLOWING

The mouth or oral cavity is the entrance to the digestive system. In the mouth, jaws and teeth mechanically break down food into smaller particles. Most vertebrates (birds being the most notable exception) have teeth that are used for tearing, chewing, and grinding food (◆ Figure 23.5). The tongue, attached to the floor of the mouth, is used to manipulate food during chewing and swallowing. In mammals, the tongue contains chemical receptors clustered into taste buds. In the first stage of digestion, food is chewed, broken down into smaller pieces for swallowing, and mixed with saliva.

### Digestion Begins in the Mouth

Saliva is secreted by paired salivary glands connected to the mouth via short ducts. The process of digestion begins with the breakdown of complex sugars and starches through the action of the enzyme **salivary amylase. Mucus** moistens food and lubricates the esophagus, preventing abrasion and making it easier to swallow. Acids in food are neutralized by **bicarbonate ions** ($HCO_3^-$) present in saliva. The salivary glands normally produce up to 1 to 2 liters of saliva each day.

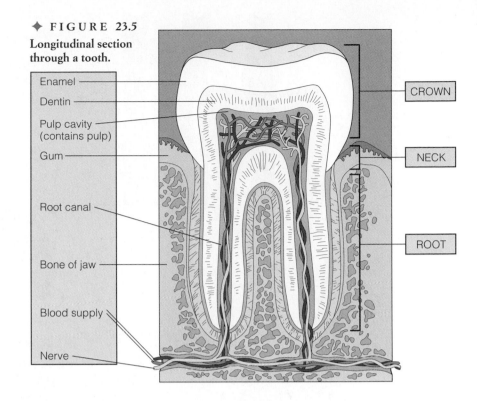

◆ FIGURE 23.5
Longitudinal section through a tooth.

Enamel
Dentin
Pulp cavity (contains pulp)
Gum
Root canal
Bone of jaw
Blood supply
Nerve

CROWN
NECK
ROOT

### The Digestive System and Respiratory System Open into the Pharynx

Swallowing is the movement of food from the mouth through the pharynx and esophagus into the stomach (◆ Figure 23.6). In the first stage of swallowing, a mass of chewed food (called a **bolus**), softened by saliva, is moved to the back of the mouth by the tongue as a voluntary action. As the bolus enters the pharynx, it triggers an involuntary

◆ FIGURE 23.6
**During swallowing, as a food mass enters the back of the throat, the trachea is lifted up against the epiglottis, preventing food from entering the lower respiratory system.**

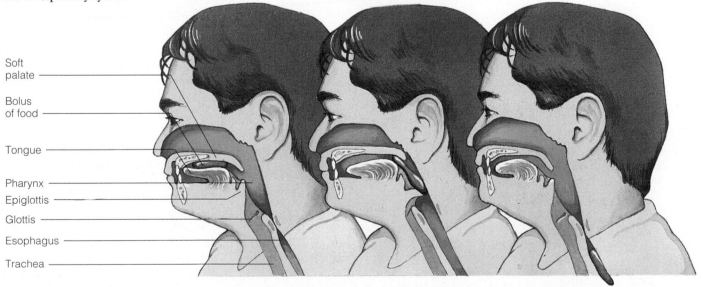

Soft palate
Bolus of food
Tongue
Pharynx
Epiglottis
Glottis
Esophagus
Trachea

Esophagus closed; glottis open; food in mouth.

Esophagus open; glottis closed; food in pharynx.

Esophagus closed; glottis open; food in esophagus.

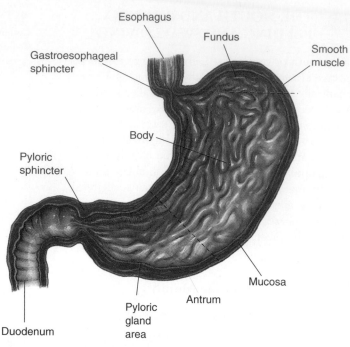

Esophagus

Gastroesophageal sphincter

Fundus

Smooth muscle

Body

Pyloric sphincter

Mucosa

Pyloric gland area

Antrum

Duodenum

✦ **FIGURE 23.8**

**The stomach is divided into three sections, the fundus, the body, and the antrum.**

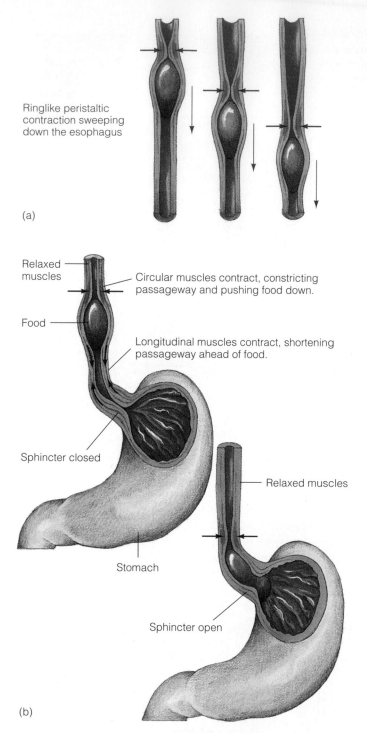

Ringlike peristaltic contraction sweeping down the esophagus

(a)

Relaxed muscles

Circular muscles contract, constricting passageway and pushing food down.

Food

Longitudinal muscles contract, shortening passageway ahead of food.

Sphincter closed

Relaxed muscles

Stomach

Sphincter open

(b)

✦ **FIGURE 23.7**

**(a) In the esophagus, peristalsis propels food downward into the stomach. (b) As food reaches the stomach, the gastroesophageal sphincter opens, allowing food to enter.**

swallowing reflex. This reflex prevents food from entering the lungs and directs the bolus into the esophagus.

The second stage of swallowing begins as food enters the esophagus, a muscular tube that extends from the pharynx to the stomach (✦ Figure 23.7). Muscles in the esophagus are arranged in an inner circular layer, and an outer longitudinal layer. As a bolus of food enters the esophagus, it initiates a wave of involuntary muscular contractions that propels food through the esophagus into the stomach. This wave of muscular contraction is known as **peristalsis** and involves the relaxation and contraction of circular smooth muscle fibers in the wall of the esophagus. The esophagus also secretes mucus, to moisten food and to protect the walls of the esophagus.

At the junction of the esophagus and stomach, the food mass passes through a ring-like muscular structure known as the **gastroesophageal sphincter.** Except when swallowing, this sphincter remains closed, preventing the contents of the stomach from entering the esophagus. When acidic gastric juices leak through the sphincter into the esophagus, the result is an irritation of the esophagus known as **heartburn.**

## THE STOMACH STORES FOOD AND BREAKS IT DOWN

The stomach is a J-shaped, expandable region of the digestive system (✦ Figure 23.8) between the esophagus and the small intestine. When empty, the lining of the stomach is highly folded, and has a capacity of 50 to 100 ml. During a meal, as the stomach fills, it gradually relaxes and expands to a capacity of about 1 liter. At the price of discomfort, the stomach can be distended beyond this capacity to hold 2 liters or more.

## The Stomach Stores, Mixes, and Digests Food

The stomach performs several important functions in digestion: It stores food, it mixes and further digests food, and it controls passage of food into the small intestine. The stomach is lined by a layer of epithelial cells, thrown into a series of folds and grooves known as **gastric pits** (✦ Figure 23.9). This lining secretes about 2 liters of gastric juices per day. These secretions contain components important in digestion, including hydrochloric acid, pepsinogen, and mucus.

Gastric secretions are controlled by both the nervous system and the endocrine system. Increased secretions can be stimulated by a variety of factors, including sensory stimuli such as seeing or smelling food, or even thinking of food (✦ Figure 23.10). Stimuli in the stomach such as distension caused by stomach filling, and certain food types such as protein and protein fragments, or foods containing caffeine can also stimulate secretion of gastric juices.

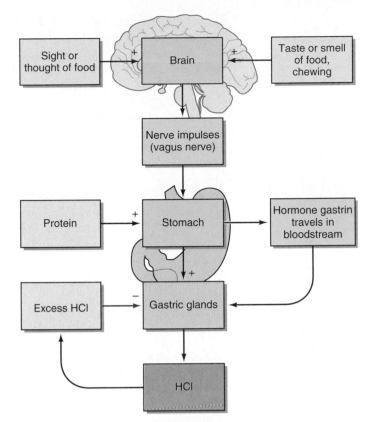

✦ **FIGURE 23.10**

Signals from several pathways can trigger release of stomach acid.

✦ **FIGURE 23.9**

The lining of the stomach and the gastric pits. The lining of the stomach (mucosa) is folded, forming gastric pits. Cells near the top of the fold secrete mucus, while those along the walls and at the base secrete acid and pepsinogen, a digestive enzyme.

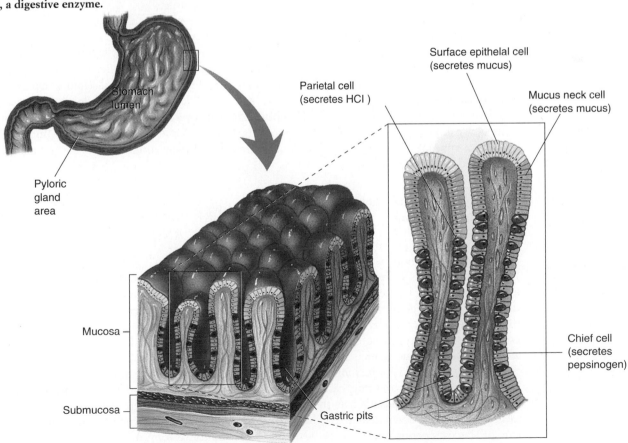

The secreted hydrochloric acid (HCl) dissociates into hydrogen ions ($H^+$) and chloride ions ($Cl^-$), making the gastric juices very acidic, with a pH of 1.5 to 2.5. The acid does not directly function in digestion, but it does promote the breakdown and separation of muscle fibers and connective tissues in the food. The acid also kills microorganisms that may have been ingested with the food, and activates **pepsinogen,** an enzyme that initiates protein digestion.

Pepsinogen is synthesized and stored in cells lining the gastric pits. It is secreted from these cells into the stomach lumen when food is present. Acid conditions in the stomach cleave off a portion of the pepsinogen molecule, producing the enzyme **pepsin.** Pepsin begins protein digestion by splitting off peptide fragments from protein molecules.

## Two Digestive Processes Take Place in the Stomach

Two digestive processes take place in the stomach. As food enters the stomach, carbohydrate digestion started in the mouth by salivary amylase continues inside the bolus. As food is propelled into the lower part of the stomach by peristalsis, it mixes with gastric juices and the bolus breaks down into a semiliquid mass, called **acid chyme.** As food masses become liquified, amylase action is inhibited, and protein digestion by pepsin begins.

Although significant amounts of digestion take place in the stomach, there is no absorption of nutrients. Two exceptions to this are the absorption of alcohol and certain compounds such as aspirin. Alcohol diffuses into the epithelial cells lining the stomach, and enters the bloodstream through capillaries located below these cells. Compounds that are weak acids, such as aspirin, also cross the plasma membrane of the epithelial cells and are transferred to the bloodstream.

## Gastric Juices Can Digest the Stomach Lining

The stomach lining is normally protected from acidity and the action of pepsin in several ways. First, epithelial cells secrete mucus, which serves as a protective barrier to prevent contact between gastric juices and the cells of the stomach lining. Furthermore, pepsin is inactivated when it comes in contact with this protective mucus, and bicarbonate ions in the mucus reduce acidity. In addition, the cells of the gastric epithelium are held together by tight junctions, preventing gastric juices from infiltrating between cells. When these protective mechanisms break down, damage to the epithelium can produce a **peptic ulcer.** If tissue damage is

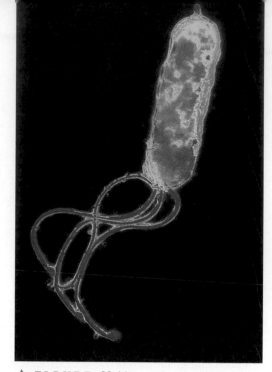

◆ **FIGURE 23.11**

A bacterium, *Helicobacter pylori,* **is responsible for most cases of peptic ulcers.**

extensive, it may cause bleeding into the stomach (a bleeding ulcer). In extreme circumstances acid may create a hole in the stomach wall (perforated ulcer), a life-threatening condition requiring immediate surgery.

What causes the protective mechanisms to break down? It is now known that infection by the bacterium *Helicobacter pylori* is the cause of at least 90% of all peptic ulcers (◆ Figure 23.11). Infection with this bacterial strain weakens the epithelial layer, leading to a breakdown of the gastric epithelial layer. Other factors including aspirin, stress, and infections elsewhere in the body can also bring about ulcer formation.

##  THE SMALL INTESTINE: THE MAJOR SITE OF DIGESTION AND ABSORPTION

Food is mixed in the lower part of the stomach by peristaltic waves of muscular contractions (◆ Figure 23.12). This motion pushes the acid-chyme mixture against the **pyloric sphincter,** a ring of muscles which regulates movement of food from the stomach into the small intestine. As peristalsis increases in intensity, small amounts of semiliquid food are pushed into the small intestine. Successive waves of peristalsis force food through the sphincter, gradually emptying the stomach over a 1- to 2-hour period. With a high-fat diet, this time can be significantly longer.

Sidebar

### ADAPTING TO STRESS

Human beings are not the only creatures who experience daily stress. Animals deal with predators, cold and hot spells, food shortages, and stressful seasonal conditions. Animals respond to the stress of seasonal variations by migration, hibernation, torpor, and aestivation.

Some animals, such as ducks and geese, migrate great distances, moving from colder to warmer climates to find ample food and suitable temperatures for survival. Mammals, such as chipmunks and bears, enter a period of reduced activity called hibernation during the winter months. Metabolic, heart, and breathing rates decrease during this period of suspended animation. Bats save energy by decreasing their metabolic rate during periods of inactivity. This adaptation, known as torpor, reduces the loss of body heat and saves enormous amounts of energy for the animal. Aestivation is much like hibernation, except it is a period of summer dormancy. This adaptation allows animals to cope with food shortages and temperature extremes. Aestivation, like hibernation, reduces the animal's metabolic rate, which permits the organism to survive extended periods of extreme temperatures.

Animals, like humans, develop adaptations for successful survival. The most effective adaptations are those that save energy and ensure the survival of the species.

## Chemical Digestion Is Completed in the Small Intestine

The small intestine is the major site of chemical digestion and absorption of nutrients. The small intestine averages about 6 meters in length, and is the longest segment of the digestive system (the small intestine is named for its diameter, about 2 to 3 cm, not its length). The upper part of the small intestine, the **duodenum,** is most active in digestion. Secretions from the liver and pancreas are used for digestion in the duodenum (✦ Figure 23.13). Muscle contractions in the walls of the duodenum mix the chyme with these secretions.

The intestinal epithelial cells secrete a watery mucus that protects the lining of the intestine and provides water to solubilize food particles for digestion. The pancreas secretes digestive enzymes into the intestine through the pancreatic duct (Figure 23.13). It also secretes an alkaline solution (bicarbonate) that helps neutralize the acid in the liquified food. The liver produces bile, which solubilizes fats for digestion. Bile is stored in the gallbladder, which empties into the intestine through the bile duct (Figure 23.13).

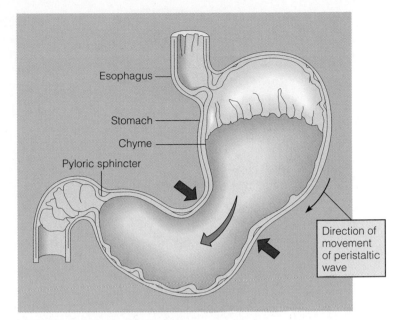

(a)

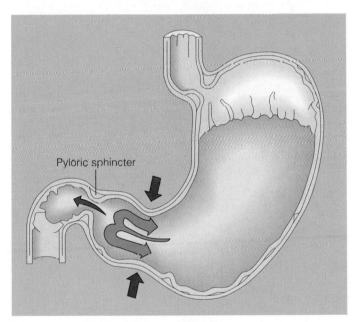

(b)

✦ **FIGURE 23.12**

(a) Waves of peristalsis travel from the upper to the lower part of the stomach, mixing the contents. (b) At the pyloric sphincter, only a small amount of food enters the duodenum. Most of the food is thrown back into the stomach, further mixing the contents.

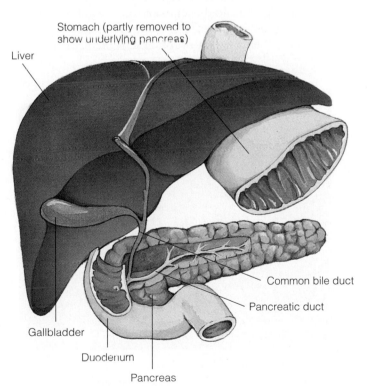

✦ **FIGURE 23.13**

The liver, pancreas, and gallbladder play important roles in digestion.

The digestion of carbohydrates, proteins, and fats continues in the small intestine. Digestion of carbohydrates (started in the mouth) is continued by the action of enzymes secreted by the pancreas. Starch and glycogen are broken down into maltose, a sugar composed of two linked glucose molecules (a disaccharide). The digestion of proteins, started in the stomach, continues in the small intestine. Enzymes (called proteases) secreted by the pancreas break down proteins into small peptide fragments and some amino acids.

Digestion of fats is enhanced by the action of bile (a mixture of salts, pigments, bilirubin, cholesterol, and phospholipids), which acts as an emulsifying agent to break down fats into smaller and smaller globules, so they can be acted on by **lipases** secreted by the pancreas. Fats are completely digested in the lumen of the small intestine, but the final stages of carbohydrate and protein digestion are completed *inside* the cells of the epithelial layer of the small intestine, which is specialized for absorption and digestion.

## The Small Intestine Is Well Adapted to Absorb Nutrients

The breakdown products of digestion are absorbed in the small intestine. Most absorption occurs in the duodenum and the **jejunum** (the second portion of the small intestine). The lining of the small intestine is specialized for absorption with an increased sur-

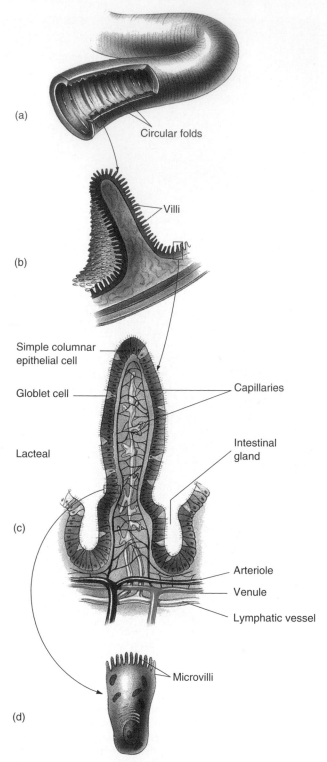

(a) Circular folds

(b) Villi

Simple columnar epithelial cell

Globlet cell

Lacteal

(c)

Capillaries

Intestinal gland

Arteriole

Venule

Lymphatic vessel

Microvilli

(d)

✦ **FIGURE 23.14**

**The small intestine is specialized for absorption and digestion of food. (a) Circular folds on the inner surface increase the absorptive area of the intestine. (b) The inner surface is thrown into finger-like projections called villi. (c) Each villus contains a capillary network from the circulatory system and a network from the lymphatic system (lacteal). (d) The surface of each epithelial cell has extensions of the plasma membrane called microvilli or the brush border.**

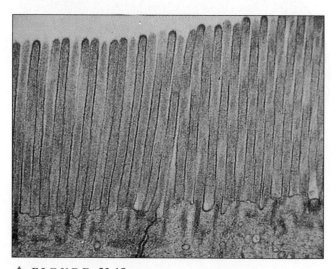

✦ **FIGURE 23.15**

**Electron micrograph of the brush border of an intestinal epithelial cell.**

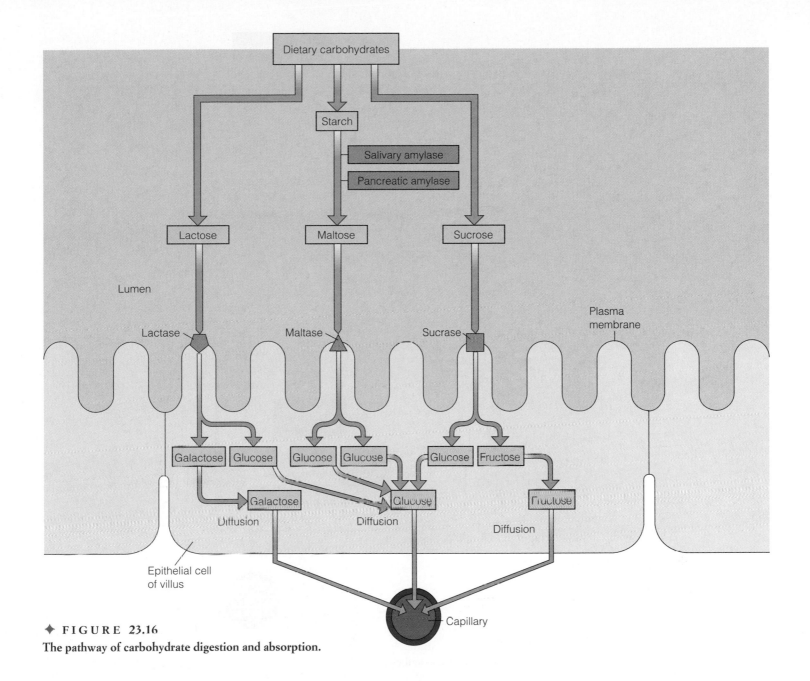

**✦ FIGURE 23.16**

**The pathway of carbohydrate digestion and absorption.**

face area and specialized transport mechanisms. The inner surface of the small intestine has circular folds that increase the surface area by a factor of 3 or more (✦ Figure 23.14). Finger-like projections called **villi,** covered with epithelial cells, increase the surface area by another factor of 10. The surface of the epithelial cells in the villi is covered with thousands of hair-like **microvilli,** further increasing the surface area of the small intestine. The microvilli form a microscopic border on the intestinal side of the epithelial cell known as a **brush border** (✦ Figure 23.15) Altogether, the 6 or so meters of the small intestine have a surface area of some 300 square meters, about a 600-fold increase in surface for the absorption of nutrients.

Each villus contains a capillary network supplied by a small arteriole. Substances to be absorbed into the blood pass from the epithelial cell, across the underlying connective tissue, and then through the wall of the capillary into the blood. Some of these steps involve passive diffusion, whereas others require the expenditure of energy.

The dietary carbohydrates present in the small intestine are mainly maltose, sucrose, and lactose. These sugars are absorbed by the microvilli of the epithelial cell (✦ Figure 23.16). Enzymatic action inside the cell converts them into simple sugars (monosaccharides). The monosaccharides leave the intestinal cell and enter the capillary network associated with the villi.

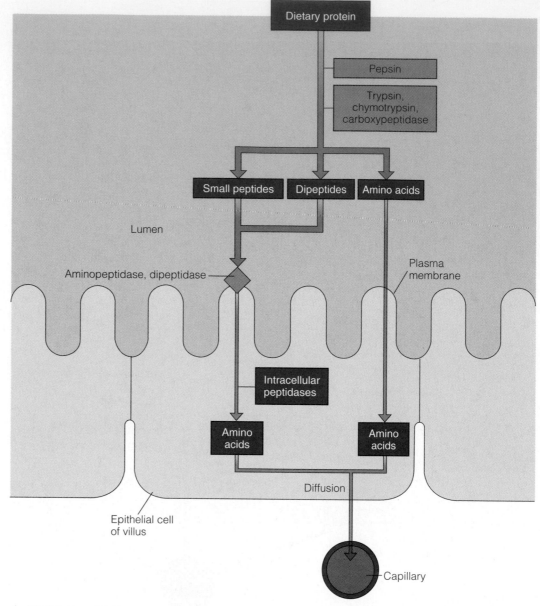

**◆ FIGURE 23.17**
**The pathway of protein digestion and absorption.**

The genetic trait called **lactose intolerance** is associated with a lack of the enzyme lactase in the epithelial cells of the small intestine. As a result, lactose, the main sugar in milk and other dairy products, remains in the small intestine. The lactose molecules create an osmotic gradient, causing water to be drawn into the intestine. As the undigested lactose passes into the large intestine, bacteria use the lactose as a food source, producing carbon dioxide and methane as by-products. The accumulation of gas along with the watery fluid received from the small intestine produces the cramps and diarrhea characteristic of lactose intolerance.

Peptide fragments and amino acids cross the plasma membrane of the epithelial cell by active transport (◆ Figure 23.17). Inside the cell, the peptides are broken down into amino acids. Like the monosaccharides, amino acids enter the capillary network in the villi and are carried away by the bloodstream.

Some people are unable to absorb gluten, a protein found in wheat and some other grains. This condition is called **gluten enteropathy.** In affected individuals, gluten damages the microvilli, causing them to collapse and become flattened. How gluten causes this in sensitive individuals is not understood, but as a result, absorption of all nutrients is

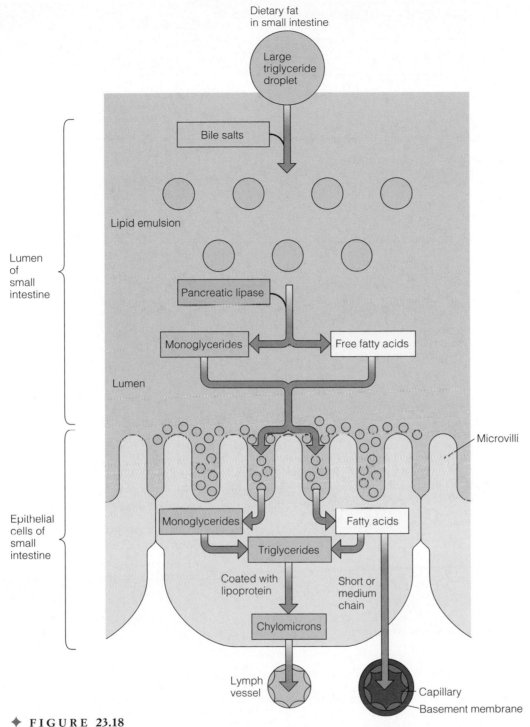

Dietary fat
in small intestine

Large
triglyceride
droplet

Bile salts

Lipid emulsion

Lumen
of
small
intestine

Pancreatic lipase

Monoglycerides        Free fatty acids

Lumen

Microvilli

Epithelial
cells of
small
intestine

Monoglycerides        Fatty acids

Triglycerides

Coated with
lipoprotein

Short or
medium
chain

Chylomicrons

Lymph
vessel

Capillary
Basement membrane

✦ FIGURE 23.18

**The pathway of lipid digestion and absorption.**

affected. The condition can be treated by eliminating all gluten from the diet.

Fats are absorbed differently than carbohydrates or proteins. Digested fats are not very soluble. Bile salts surround the fat molecules to form **micelles** (✦ Figure 23.18), which are taken into the epithelial cells. The bile salts are left on the surface of the

cells and return to the lumen to repeat the formation of micelles. By the time the contents of the intestine reach the **ileum** (the third section of the small intestine), fats have been absorbed. In the last part of the ileum, the bile salts are absorbed by active transport, allowing them to be recycled through the liver and into the gallbladder. Fats pass from the

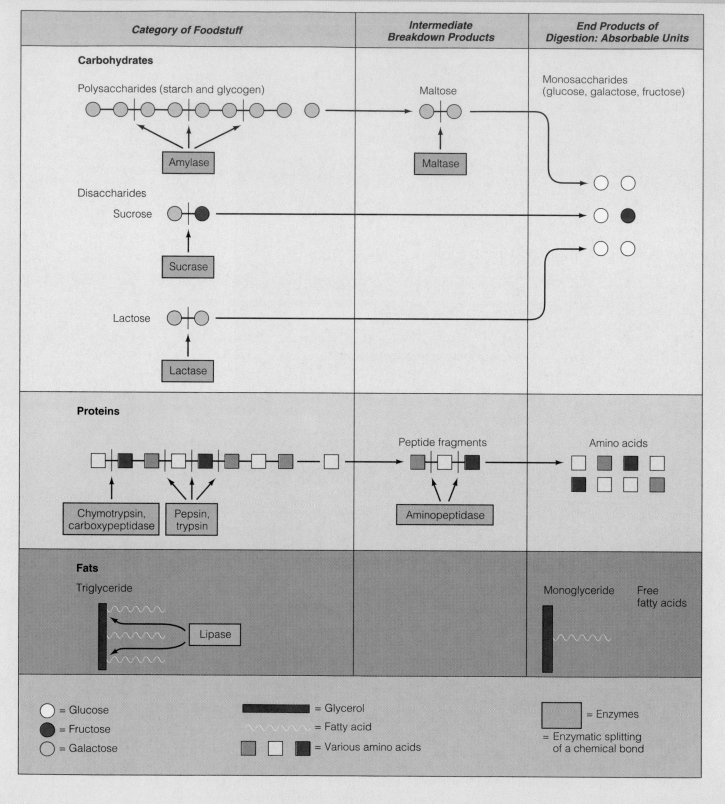

epithelial cells into the lymphatic system rather than into the capillaries. The process of digestion is summarized in ➤ Table 23.1.

## THE LARGE INTESTINE ABSORBS WATER AND STORES WASTE

The large intestine consists of the colon, cecum, appendix, and rectum (✦ Figure 23.19). Because digestion and absorption have taken place in the small intestine, the material received by the large intestine consists mainly of undigested residue and liquid. Movements within the large intestine consist of involuntary contractions that slowly shuffle the contents back and forth, and propulsive contractions that move materials through the large intestine. The secretions of the large intestine consist of an alkaline mucus that protects the epithelial lining and neutralizes acids produced by bacterial metabolism.

The inner surface of the colon is fairly smooth, and has less absorptive surface than the small intestine. Nevertheless, some absorption occurs in this region of the digestive tract, primarily water, salts, and vitamins. The remaining contents of the lumen are **feces.** The large intestine contains symbiotic bacteria, which metabolize materials that were not digested and absorbed in the small intestine. These bacterial populations synthesize a number of vitamins (including vitamin K) which are absorbed by the large intestine.

The fecal matter eliminated from the body through the anus consists mainly of cellulose, bacterial cells, bilirubin (a pigment present in bile), trace amounts of salts, and very little metabolic waste.

## REGULATION OF APPETITE, DIGESTION, AND NUTRIENTS

### Appetite Is Controlled by Internal and External Factors

The intake of food in response to hunger is controlled by two centers in a region of the brain known as the **hypothalamus.** One is an appetite center, and the other, a satiety center. The mecha-

✦ **FIGURE 23.19**

**The large intestine contains four parts: cecum, appendix, colon, and rectum.**

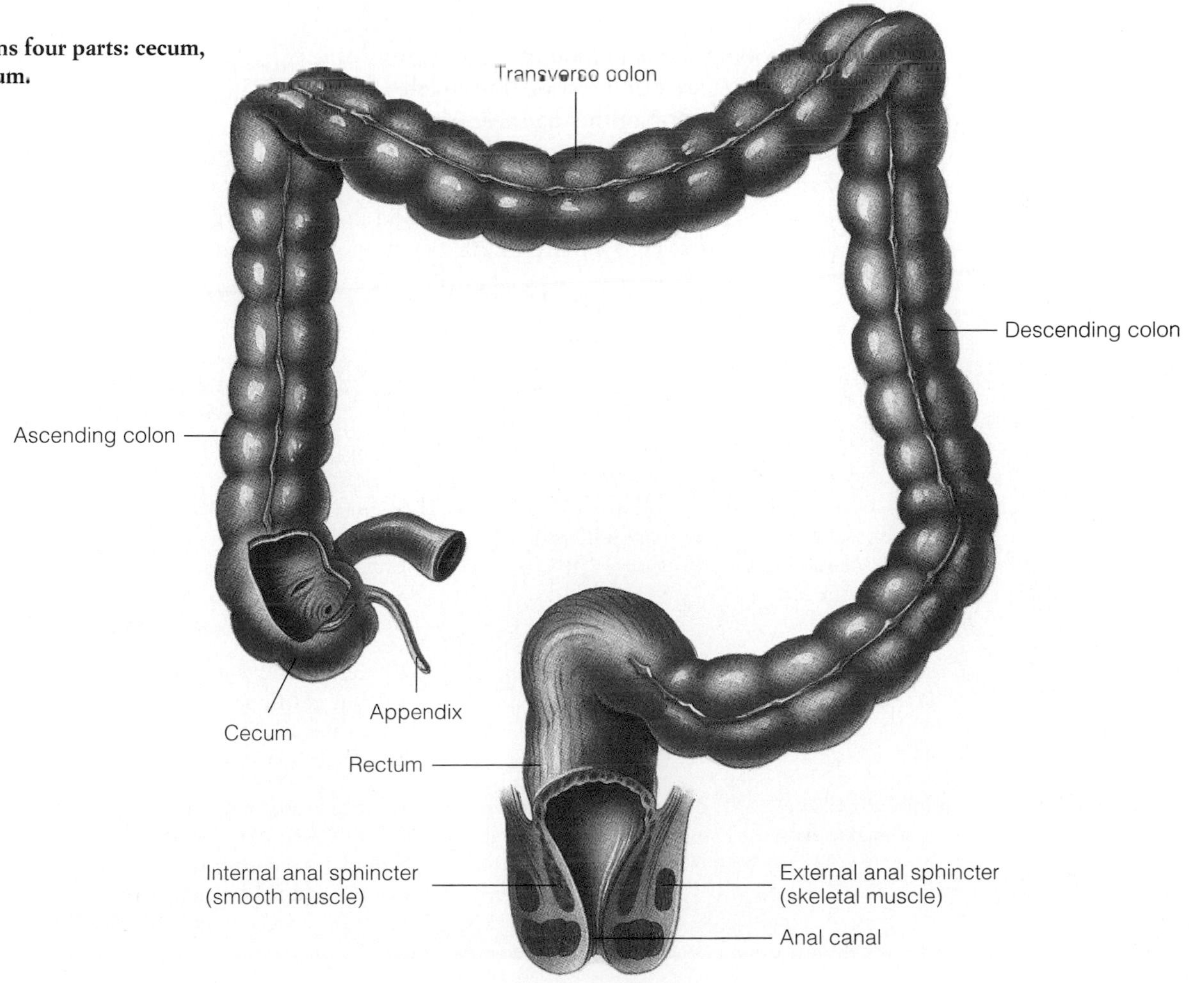

- Transverse colon
- Descending colon
- Ascending colon
- Cecum
- Appendix
- Rectum
- Internal anal sphincter (smooth muscle)
- External anal sphincter (skeletal muscle)
- Anal canal

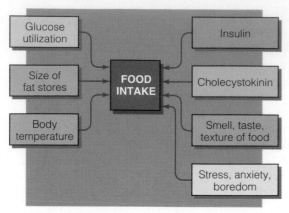

**✦ FIGURE 23.20**

**Some of the factors known to influence food intake. Increased glucose utilization after a meal decreases food intake. Low blood glucose levels cause hunger sensations. Large fat stores provide long-term reduction in food intake, while low levels of fat stores increase feeding over the long term. Increases in body temperature slow eating, while lowered body temperature increase feeding. Higher blood levels of hormones such as insulin, and cholecystokinin depress food intake. Psychosocial factors such as the smell, appearance, and texture of food, as well as stress, anxiety, and boredom can also positively or negatively influence food intake.**

nisms controlling the operation of these regions of the brain are not well understood, but involve factors such as distension of the digestive tract, blood glucose levels, ATP production, and also psychological and social influences, such as custom, stress, and boredom. Factors known to influence food intake are summarized in ✦ Figure 23.20.

## Digestion Is Hormonally Regulated

Many stages in digestion are regulated by gastrointestinal hormones, including **gastrin, secretin,** and **cholecystokinin.** The sources and actions of these hormones are summarized in ◆ Table 23.2. The presence of protein in the stomach stimulates the secretion of gastrin. Gastrin circulates through the blood to the epithelial cells of the gastric pits, prompting the release of more HCl and pepsinogen. Gastrin also stimulates motility of the stomach, the small intestine, and colon, ensuring that food moves through the digestive tract in a timely manner.

As the stomach empties into the duodenum, the acidic mixture causes the release of secretin, a hormone that performs several interrelated functions. Secretin stimulates the pancreatic duct to release alkaline secretions into the duodenum to neutralize the acid. It also inhibits the release of additional food from the stomach until the acid is neutralized. Cholecystokinin (CCK) is released into the blood from the duodenal epithelium in response to the presence of fats. CCK causes contraction of the gallbladder to release bile to emulsify fats in the lumen. CCK also stimulates the pancreas to release large amounts of the fat-digesting enzyme lipase.

## Storage and Release of Glucose for Metabolism Are Hormonally Controlled

Although we may eat only three meals a day, the amount of glucose in the blood (the primary source of energy for cellular metabolism) remains fairly constant. Glucose absorbed into the blood after a

| TABLE 23.2 | THE GASTROINTESTINAL HORMONES | | |
|---|---|---|---|
| HORMONE | SOURCE | PRIMARY STIMULUS FOR SECRETION | FUNCTIONS |
| **Gastrin** | Endocrine cells in pyloric gland area of stomach | Protein in the stomach | Stimulates secretion of acid by cells in gastric pits<br>Enhances gastric motility<br>Stimulates ileal motility<br>Induces colonic mass movements |
| **Secretin** | Endocrine cells in duodenal mucosa | Acid in duodenal lumen | Inhibits gastric emptying<br>Inhibits gastric secretion<br>Stimulates aqueous $NaHCO_3$ secretion by pancreatic duct cells<br>Stimulates secretion of $NaHCO_3$-rich bile by liver |
| **Cholecystokinin** | Endocrine cells in duodenal mucosa | Nutrients in duodenal lumen, especially fat products and to a lesser extent protein products | Inhibits gastric emptying<br>Inhibits gastric secretion<br>Stimulates digestive enzyme secretion by pancreatic cells<br>Causes gallbladder contraction<br>Contributes to satiety |

meal moves to the liver, where it is taken up and converted into glycogen (a large molecule made up of interconnected glucose molecules). After a meal, the glucose produced by digestion is absorbed in about 4 hours. Between meals, the levels of blood glucose are maintained by reversing this process. Glucose levels in the blood are regulated by several hormones, and by interaction between the pancreas and the liver. This relationship will be explored in detail in Chapter 25.

## NUTRITION AND THE HUMAN DIET

Digestion is the mechanical and chemical breakdown of food into its component organic molecules. Nutrition deals with the composition of food, its energy content, and especially with organic molecules that cannot be synthesized by the organism, or cannot be synthesized at a rate that meets demand. Organisms fall into three classes according to their energy requirements: **chemotrophs,** mostly bacteria that derive energy from inorganic reactions; **phototrophs,** such as cyanobacteria, algae, and plants that use sunlight to synthesize organic nutrients; and **heterotrophs,** animals (and some plants) that eat to obtain energy from the breakdown of organic molecules in food.

## THE WELL-BALANCED DIET

Foods required on a large scale each day are often called macronutrients (✦ Table 23.3). These include carbohydrates, proteins, and lipids. Water is sometimes included, because adequate water intake is vital to all chemical reactions that occur within an organism; water helps maintain body temperature and helps to control blood levels of nutrients.

### Carbohydrates Should Be the Primary Energy Source

The primary energy source for cellular metabolism is carbohydrates. Carbohydrates are present as starch in vegetables, grains, and grain products such as bread, as glycogen in meats, as sugars such as lactose in milk and nonfermented milk products, and the sugars in fruits. About 60% of the diet should consist of carbohydrates (✦ Figure 23.21).

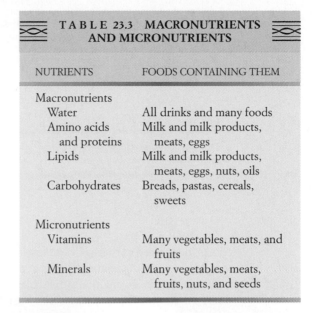

| TABLE 23.3 | MACRONUTRIENTS AND MICRONUTRIENTS |
| --- | --- |
| NUTRIENTS | FOODS CONTAINING THEM |
| **Macronutrients** | |
| Water | All drinks and many foods |
| Amino acids and proteins | Milk and milk products, meats, eggs |
| Lipids | Milk and milk products, meats, eggs, nuts, oils |
| Carbohydrates | Breads, pastas, cereals, sweets |
| **Micronutrients** | |
| Vitamins | Many vegetables, meats, and fruits |
| Minerals | Many vegetables, meats, fruits, nuts, and seeds |

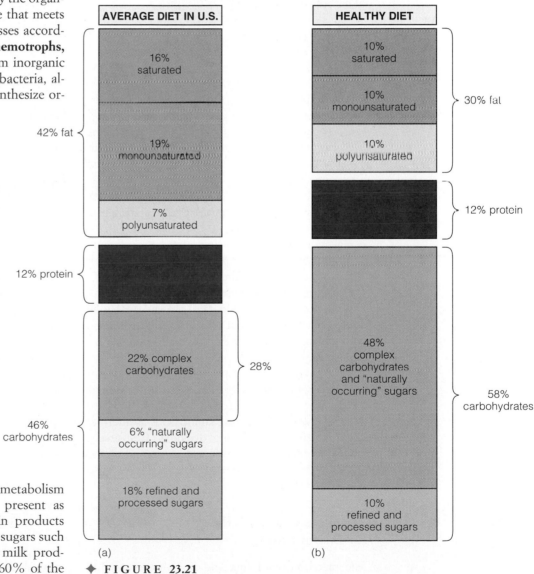

✦ **FIGURE 23.21**

**(a) The composition of a typical U.S. diet. (b) The composition of a healthy diet with lower fat intake and reduced levels of fat.**

| TABLE 23.4 ESSENTIAL AMINO ACIDS* | |
|---|---|
| Histidine† | Phenylalanine |
| Isoleucine | Threonine |
| Leucine | Tryptophan |
| Lysine | Valine |
| Methionine | |

*Arginine is also sometimes considered to be an essential amino acid; although it can be synthesized, the rate of synthesis is often too slow to meet the demands of growth.
†An essential amino acid only during periods of rapid growth.

As mentioned earlier, glucose levels in the blood are regulated within narrow limits at all times, even between meals. To maintain this balance, it is recommended that the diet contain at least 100 grams of carbohydrates per day.

## Proteins Provide Essential Amino Acids

The amino acids that result from protein digestion are used to synthesize new proteins that have a variety of functions, including structural components and enzymes. For example, when a female mosquito sucks up a blood meal from your arm, the blood proteins are broken down into amino acids and used to synthesize the yolk proteins stored in mosquito eggs.

As a dietary component, proteins are necessary for cellular growth and repair. Proteins are found in foods such as meat, poultry, fish, dairy products, cereal grains, and legumes, including beans. Twenty common amino acids are found in proteins. Of these, humans can synthesize eleven. The remaining nine are **essential amino acids** and must be supplied in the diet (◆ Table 23.4). To ensure that all essential amino acids are included, it is recommended that the diet utilize several different protein sources.

Under normal circumstances, proteins are not used as a source of energy. However, in conditions of starvation, muscle protein is broken down to provide energy by utilizing amino acids as intermediates in energy-producing reactions. Conversely, when proteins are present in excess, surplus amino acids can be used to generate energy and are also converted into fat.

## Lipids Serve as a Method for Energy Storage

The metabolism of fats and lipids generates the highest energy yields, so they are used by plants and animals as molecular energy reserves. Lipids and fats are found in a variety of foods, including oils, butter, many meats, and some plant foods such as the avocado. Just as there are certain essential amino acids, some fatty acids, primarily linoleic acid, are essential and must be included in the diet.

In addition to serving as cellular energy reserves, lipids serve as insulation, preventing loss of body heat, and are structural components of the cell, primarily in membranes. Lipids also serve as biochemical precursors to hormones and, when present in the intestine, they increase the absorption of certain vitamins such as vitamins A, D, E, and K.

## A WELL-BALANCED DIET INCLUDES VITAMINS AND MINERALS

### Vitamins Are Required to Carry Out Metabolic Reactions

**Vitamins** are organic molecules required for metabolic reactions (◆ Table 23.5) and, in general, cannot be synthesized in the body. Vitamins act as **enzyme cofactors** or **coenzymes** in all cells of the body. Because they can be reused many times over, they are only needed in trace amounts in the diet. Although vitamins participate in energy-releasing reactions, they are not in themselves a source of energy. As they are released from foods during digestion, vitamins are absorbed intact from the digestive tract. There are two broad categories of vitamins: water-soluble and fat-soluble vitamins. The water-soluble vitamins include vitamin C and the different forms of vitamin B. Because these vitamins are eliminated by the kidneys, they are not stored in the body, and are required on a daily basis. The fat-soluble vitamins can be stored in body fat. In excess, they can accumulate to toxic levels. Thirteen vitamins are essential components of a well-balanced human diet (Table 23.5).

TABLE 23.5   IMPORTANT INFORMATION ON VITAMINS

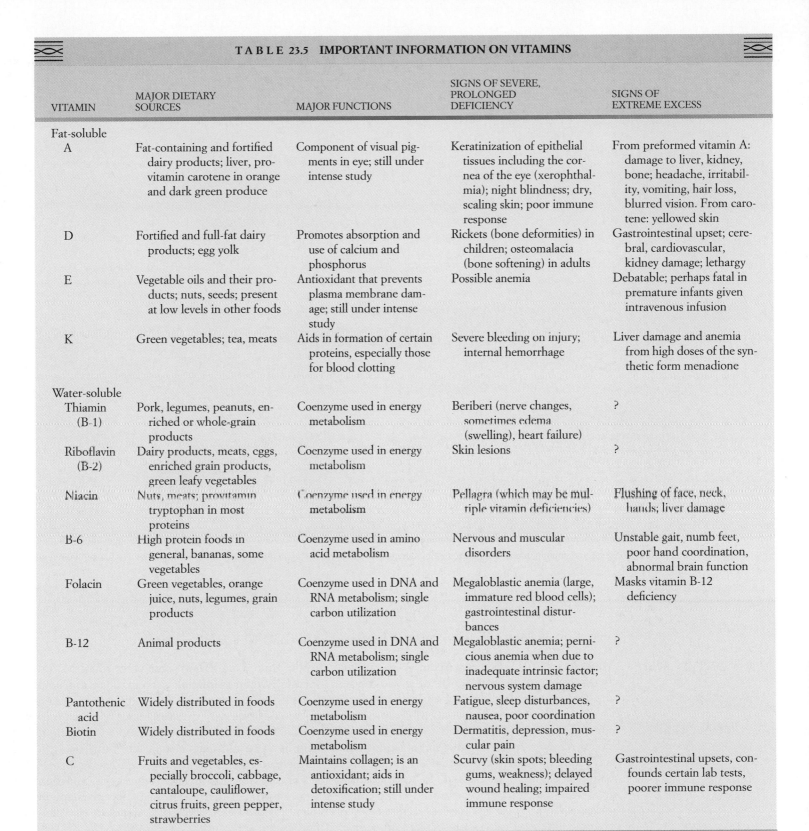

| VITAMIN | MAJOR DIETARY SOURCES | MAJOR FUNCTIONS | SIGNS OF SEVERE, PROLONGED DEFICIENCY | SIGNS OF EXTREME EXCESS |
|---|---|---|---|---|
| **Fat-soluble** | | | | |
| A | Fat-containing and fortified dairy products; liver, provitamin carotene in orange and dark green produce | Component of visual pigments in eye; still under intense study | Keratinization of epithelial tissues including the cornea of the eye (xerophthalmia); night blindness; dry, scaling skin; poor immune response | From preformed vitamin A: damage to liver, kidney, bone; headache, irritability, vomiting, hair loss, blurred vision. From carotene: yellowed skin |
| D | Fortified and full-fat dairy products; egg yolk | Promotes absorption and use of calcium and phosphorus | Rickets (bone deformities) in children; osteomalacia (bone softening) in adults | Gastrointestinal upset; cerebral, cardiovascular, kidney damage; lethargy |
| E | Vegetable oils and their products; nuts, seeds; present at low levels in other foods | Antioxidant that prevents plasma membrane damage; still under intense study | Possible anemia | Debatable; perhaps fatal in premature infants given intravenous infusion |
| K | Green vegetables; tea, meats | Aids in formation of certain proteins, especially those for blood clotting | Severe bleeding on injury; internal hemorrhage | Liver damage and anemia from high doses of the synthetic form menadione |
| **Water-soluble** | | | | |
| Thiamin (B-1) | Pork, legumes, peanuts, enriched or whole-grain products | Coenzyme used in energy metabolism | Beriberi (nerve changes, sometimes edema (swelling), heart failure) | ? |
| Riboflavin (B-2) | Dairy products, meats, eggs, enriched grain products, green leafy vegetables | Coenzyme used in energy metabolism | Skin lesions | ? |
| Niacin | Nuts, meats; provitamin tryptophan in most proteins | Coenzyme used in energy metabolism | Pellagra (which may be multiple vitamin deficiencies) | Flushing of face, neck, hands; liver damage |
| B-6 | High protein foods in general, bananas, some vegetables | Coenzyme used in amino acid metabolism | Nervous and muscular disorders | Unstable gait, numb feet, poor hand coordination, abnormal brain function |
| Folacin | Green vegetables, orange juice, nuts, legumes, grain products | Coenzyme used in DNA and RNA metabolism; single carbon utilization | Megaloblastic anemia (large, immature red blood cells); gastrointestinal disturbances | Masks vitamin B-12 deficiency |
| B-12 | Animal products | Coenzyme used in DNA and RNA metabolism; single carbon utilization | Megaloblastic anemia; pernicious anemia when due to inadequate intrinsic factor; nervous system damage | ? |
| Pantothenic acid | Widely distributed in foods | Coenzyme used in energy metabolism | Fatigue, sleep disturbances, nausea, poor coordination | ? |
| Biotin | Widely distributed in foods | Coenzyme used in energy metabolism | Dermatitis, depression, muscular pain | ? |
| C | Fruits and vegetables, especially broccoli, cabbage, cantaloupe, cauliflower, citrus fruits, green pepper, strawberries | Maintains collagen; is an antioxidant; aids in detoxification; still under intense study | Scurvy (skin spots; bleeding gums, weakness); delayed wound healing; impaired immune response | Gastrointestinal upsets, confounds certain lab tests, poorer immune response |

SOURCE: *Adapted from J. L. Christian and L. L. Greger,* Nutrition for Living, *2d ed., copyright © 1988 by the Benjamin/Cummings Publishing Company. Used with permission.*

| MINERAL | MAJOR DIETARY SOURCES | MAJOR FUNCTIONS | SIGNS OF SEVERE, PROLONGED DEFICIENCY | SIGNS OF EXTREME EXCESS |
|---|---|---|---|---|
| **Major minerals** | | | | |
| Calcium | Milk, cheese, dark green vegetables, legumes | Bone and tooth formation; blood clotting; nerve transmission | Stunted growth; maybe bone loss | Depressed absorption of some other minerals |
| Phosphorous | Milk, cheese, meat, poultry, whole grains | Bone and tooth formation; acid-base balance; component of coenzymes | Weakness; demineralization of bone | Depressed absorption of some minerals |
| Magnesium | Whole grains, green leafy vegetables | Component of enzymes | Neurological disturbances | Neurological disturbances |
| Sodium | Salt, soy sauce, cured meats, pickles, canned soups, processed cheese | Body water balance; nerve function | Muscle cramps; reduced appetite | High blood pressure in genetically predisposed individuals |
| Potassium | Meats, milk, many fruits and vegetables, whole grains | Body water balance; nerve function | Muscular weakness; paralysis | Muscular weakness; cardiac arrest |
| Chloride | Salt, many processed foods (as for sodium) | Plays a role in acid-base balance; formation of gastric juice | Muscle cramps; reduced appetite; poor growth | Vomiting |
| **Trace minerals** | | | | |
| Iron | Meats, eggs, legumes, whole grains, green leafy vegetables | Component of hemoglobin and enzymes | Iron-deficiency anemia, weakness, impaired immune function | Acute: shock, death; chronic: liver damage, cardiac failure |
| Iodine | Marine fish and shellfish; dairy products; iodized salt; some breads | Component of thyroid hormones | Goiter (enlarged thyroid) | Iodide goiter |
| Fluoride | Drinking water, tea, seafood | Maintenance of tooth (and maybe bone) structure | Higher frequency of tooth decay | Mottling of teeth; skeletal deformation |

SOURCE: *Adapted from J. L. Christian and L. L. Greger,* Nutrition for Living, *2d ed., copyright © 1988 by the Benjamin/Cummings Publishing Company. Used with permission.*

## Minerals Are Required for Many Functions

**Minerals** are trace elements required for normal metabolism, as components of cells and tissues, and in nerve conduction and muscle contraction (Table 23.6). Minerals can only be obtained from our diet. Some minerals have specific functions: Iron is required for the production of hemoglobin, and iodine is needed for synthesis of thyroxin, a hormone of the thyroid gland. In recent years, a great deal of attention has been paid to mineral requirements and their relationship to diseases. For example, intake of calcium is required to maintain bone and skeletal components. Low levels of calcium can result in osteoporosis, a degenerative bone disease. Sodium is required for nerve conduction and in maintaining the osmotic balance of cells and the extracellular fluid. Sodium is usually in excess in the human diet, resulting in widespread high blood pressure.

## Diet and Its Links to Disease

In the United States and some other countries, increasing attention is being given to the relationship between proper nutrition and health. To maintain homeostasis, there must be an intake of proper nutrients. In addition, there is a quantitative relationship between nutrients and health. Imbalances in nutrients either in the form of deficiencies or excesses can cause disease. Many studies on nutrition have concluded that diet is a significant factor in many diseases such as cardiovascular disease, hypertension, and cancer.

In recent decades it has become clear that controlling the intake of dietary cholesterol and fats can reduce the risk of cardiovascular disease, and dietary guidelines issued by federal agencies have suggested that fat intake be limited to no more than 30% of daily calories, and that cholesterol intake be limited to 300 mg/day. Without question these measures can be effective in lowering blood levels of lipids and cholesterol.

However, the situation may not be as simple as it appears, and other dietary factors may play important roles in controlling heart disease. For example, the mineral magnesium has been implicated in controlling lipid blood levels and the deposition of cholesterol on the inner surface of blood vessels as plaque. Rabbits fed a high-cholesterol, high-magnesium diet have 80 to 90% less cardiovascular disease than rabbits fed on a high-cholesterol, low-magnesium diet. In addition, certain vitamins with antioxidant properties such as vitamins E, C, and A appear to offer significant protection against the development of cardiovascular disease and plaque formation, in ways that may be independent of dietary levels of fats and cholesterol. In fact, antioxidant vitamins and drugs may be used to intervene as part of treatment to halt progress of the disease.

Instead of concentrating on avoiding heart disease or cancer via fad diets, attention should be centered on following a healthy, well-balanced diet containing the proper levels of nutrients as a way of maintaining good health and avoiding diseases with links to nutritional excesses or deficiencies.

## SUMMARY

1. Cells of multicellular organisms must be provided with nutrients for energy conversion, growth, and repair. To obtain nutrients, animals depend on two processes: feeding and digestion.
2. Animals exhibit many adaptations in feeding. Most animals use a mouth, and are ingestive feeders. A small number of animals, including tapeworms, live in the digestive tracts of other animals, and are absorptive feeders.
3. A digestive system uses a combination of mechanical and chemical processes to transform food into nutrient molecules. Animals employ one of two plans for digestive systems: sac-like or tube-within-a-tube.
4. The digestive tract in humans is an elongated, coiled tubular system that extends from mouth to anus. The digestive system performs five basic functions: movement, secretion, digestion, absorption, and elimination.
5. Digestion begins in the mouth with the action of salivary enzymes on carbohydrates. Chewed food is swallowed and passed to the stomach through the esophagus. The stomach mixes food with acid and digestive enzymes that begin protein digestion.
6. Digestion continues in the small intestine, where carbohydrates and proteins are further broken down. Complete digestion of fats takes place in the small intestine. Fragments of carbohydrates and proteins are digested in the cells lining the small intestine. The sugars, amino acids, and digested fats are absorbed into the bloodstream in the small intestine.
7. Undigested food is passed to the large intestine where water is absorbed and feces are produced and eliminated.
8. A well-balanced diet should provide sources of energy, nutrients that cannot be synthesized, and materials for synthesis of new molecules.
9. Vitamins are organic molecules required for many metabolic reactions. Minerals are trace elements that help maintain normal metabolism. Most vitamins and all minerals are obtained from a well-balanced diet.
10. Dietary choices are linked to disease. Fad diets should be avoided, and emphasis should be placed on a well-balanced diet with diverse sources of nutrients.

## KEY TERMS

absorption
absorptive feeders
acid chyme
anus
bicarbonate ions
biliary system
bolus
brush border
chemotrophs
cholecystokinin
coenzymes
digestion
duodenum
elimination
enzyme cofactors
esophagus

essential amino acids
extracellular
feces
feeding
filter feeders
fluid feeders
gastric pits
gastrin
gastroesophageal sphincter
gluten enteropathy
heartburn
heterotrophs
hypothalamus
ileum
ingestive feeders
intracellular

jejunum
lactose intolerance
large intestine
lipases
micelles
microvilli
minerals
mouth
movement
mucus
pancreas
pepsin
pepsinogen
peptic ulcer
peristalsis
pharynx

phototrophs
pyloric sphincter
salivary amylase
salivary glands

secretin
secretion
small intestine
stomach

substrate feeders
tube-within-a-tube system
villi
vitamins

 # QUESTIONS AND PROBLEMS

## TRUE/FALSE

1. Vertebrates use the sac-like digestive system, with two body openings, one for ingestion of food, and the other for waste removal.
2. Bile is produced in the gallbladder.
3. The point of exit from the stomach, where it joins with the small intestine, is guarded by the pyloric sphincter.
4. Most digestive activity and almost all absorption occur within the large intestine.
5. Heartburn is an irritation of the esophagus.
6. The hypothalamus is the region of the brain that controls the intake of food in response to hunger.
7. To maintain appropriate weight and overall general health, caloric intake must exceed energy output.
8. Minerals can only be obtained through our diet.

## MATCHING

Match the digestive agents appropriately.

1. gastrin
2. secretin
3. lipases
4. salivary amylase
5. HCl
6. bile
7. proteases
8. cholecystokinin

_____ solubilizes fats
_____ causes contraction of gallbladder for release of bile
_____ prompts release of HCl and pepsinogen
_____ breaks down proteins
_____ stimulates release of alkaline secretions into duodenum
_____ pancreatic enzyme that digests fat
_____ breaks down complex sugars and starches
_____ promotes breakdown of muscle fibers and connective tissue in food

## FILL IN THE BLANK

1. Food is taken into cells by phagocytosis in _____ digestion.
2. The small intestine is composed of the _____, _____, and _____.
3. The enzyme _____ begins protein digestion by splitting off peptide fragments from protein molecules.
4. Carbohydrates, proteins, and fats are all digested in the _____.
5. The moving wave of muscular contraction and relaxation that pushes food through the esophagus is called _____.
6. The _____ normally remains closed, but it opens during swallowing to allow food to pass into the esophagus.

## SHORT ANSWER

1. How is the stomach lining protected from acidity?
2. Why are fats, proteins, carbohydrates, minerals, and vitamins nutritionally important?
3. Explain the difference between filter feeders, substrate feeders, fluid feeders, and absorptive feeders. Give examples of each.
4. Name the components of the digestive system and their major functions.

## SCIENCE AND SOCIETY

1. Kidney stones are tiny crystals made of calcium and other ions that grow and can eventually obstruct the flow of urine, causing pain and considerable discomfort. Because stones are primarily made of calcium, physicians often suggest to patients who have already had a stone to reduce their intake of calcium. Can you think of any possible problems that might develop from a reduction in calcium intake? Would different effects be seen between men and women? Do you think there might be a genetic predisposition to stone forming? What, if any, additional information do you think a physician needs about stone formers before asking the patients to reduce their calcium intake? What other characteristics of kidney stones need to be evaluated before a dietary solution should be suggested?

2. Diseases of the heart and arteries are the leading causes of death in the United States. Cholesterol deposits impair the flow of blood in the heart and other organs, cutting off oxygen to vital tissues. Cholesterol is, however, essential to normal body function. It is part of the plasma membrane and is needed to synthesize hormones. A high cholesterol level or hypercholesterolemia tends to run in families. If a parent died of a heart attack, his or her offspring should take precautionary measures to reduce their chances of this same outcome. High cholesterol levels are also surprisingly common in children, leading health experts to believe that steps should be taken to prevent problems later in life. What measures should people take to reduce their blood cholesterol levels? Do you think that if everyone reduced their dietary cholesterol, the incidence of heart disease would decrease? Would a reduction in dietary cholesterol in one person result in a similar reduction in another person? If not, what could the difference in response be attributed to? What effects on children could a restrictive diet have on physical growth and development?

# EXCRETION AND TRANSPORT

In June 1972, a 43-foot schooner, the Lucette, was rammed by killer whales in the Pacific Ocean about 300 miles southwest of the Galapagos Islands. The boat sank in no more than 60 seconds, and the six crew, including the captain, Dougal Robertson, his wife, Lyn, their sons (one 18 years old, and 12-year-old twins), and a young crew member were left crammed together in a 9-foot dinghy. In midocean, far from shipping lanes, they had food and water for only three days, and few other supplies. With the prevailing winds and currents, their only hope of survival lay in sailing for Costa Rica, some 1000 miles to the northeast. Their ordeal ended some 37 days later, after sailing 750 miles toward their goal, when they were rescued by a Japanese fishing boat about 300 miles off the coast of Central America.

How did they survive? What did they do for water? Why not drink seawater? Knowing the answers to these questions played a critical role in the survival of the Lucette's crew. Drinking seawater causes a net loss of water from the body, contributing to dehydration and death. Seawater has a much higher salt concentration than human blood. Drinking seawater raises the level of salt in the blood. When blood is filtered in the kidney, the additional salt added by drinking seawater causes more water to be put into the urine and less water returned to the blood. Rather than a source of additional water, seawater actually causes water loss from the body. For every liter of seawater consumed, about 1.3 liters of urine must be produced to maintain salt balance in the blood.

So what did the crew of Lucette do for water during their ordeal? They obtained water

OPENING IMAGE

*A small schooner similar to the*

*Lucette.*

*from several sources. They caught fish and drank the body fluids squeezed from the meat, and sucked moisture from the eyes and the spine. They also caught sea turtles, and drank their blood. Sea turtles are acceptable sources of water because these marine animals have blood and body fluids with solute (especially salt) concentrations only one-third that of seawater, similar in solute concentration to human body fluids. Intake of fluids from these marine animals, unpleasant as it may sound, does not raise the salt concentration of blood. Keeping the solute concentration of the blood within a normal range allows the kidney to return a high proportion of water to the blood, conserving the body's supply of water. In addition, the crew supplemented their water supply by catching small amounts of freshwater from tropical rain showers.*

*The story of the* Lucette *and its crew's shipwreck and survival has been told in the book* Survive the Savage Sea *by Dougal Robertson (New York: Prager Press, 1973).*

*Both aquatic and terrestrial animals, including humans, use energy in controlling water and ionic concentrations in their body fluids. In this chapter, we examine how animals use excretory systems to balance the uptake and loss of water and solutes.*

## EXCRETORY SYSTEMS ELIMINATE EXCESS WATER, SALTS, AND WASTES

The role of excretion is best thought of in its relationship to metabolism. During energy conversion in the cell, a variety of chemicals, including sugars, fats, and proteins, are utilized. Before they can yield energy, these large, complex molecules are reduced to a smaller size and stripped of chemical groups such as nitrogen (in the form of ammonia), sulfur (in the form of sulfate) and phosphorus (as phosphates). Cells produce water and carbon dioxide as a by-product of metabolism. The continuous production of waste materials inside the cell sets up a steep concentration gradient across the plasma membrane, causing wastes to diffuse out of the cells into the extracellular fluid. In single-celled organisms and aquatic animals with a simple body plan, waste products diffuse into the surrounding water through the plasma membrane or across the body surface, maintaining a balance between the intracellular and extracellular environments.

In multicellular animals, most cells are not in direct contact with the environment, but are surrounded by an interstitial extracellular fluid such as blood or lymph. The composition of this extracellular fluid is critical in maintaining proper condition for cellular functions. Removal of cellular wastes from this fluid is necessary to support life.

### Excretion Regulates the Composition of Extracellular Fluids

Excretory systems regulate the chemical composition of body fluids by removing metabolic wastes and retaining the proper amounts of water, salts, and nutrients. In vertebrates, the components of this system include:

*Kidneys:* paired organs that regulate water and salts and dispose of nitrogen wastes
*Liver:* regulates the nutrient level in the blood and breaks down waste products from the blood for disposal by the kidneys
*Lungs:* respiratory organs that dispose of carbon dioxide and other gases produced during metabolism
*Skin:* aids in disposal of water and salts.

## EXCRETION AND HOMEOSTASIS

The wastes produced by cellular metabolism in humans and their method of excretion are summarized in ➤ Table 24.1. Not all animals use the same routes of excretion nor do they excrete wastes in the same form as humans. To a certain extent, the form of wastes and their method of excretion are adaptations to the environment in which the animal lives. The term **excretion** applies to metabolic waste products that cross a plasma membrane, and should not be confused with **elimination,** the removal of undigested food residues in the form of feces.

### Nitrogen Wastes Are Produced in Protein Metabolism

The production of nitrogen wastes is a by-product of protein metabolism. Proteins are first broken down into amino acids. Next, the amino group ($-NH_2$) is removed, and the remaining compound can be used for energy conversion, or transformed into fats or carbohydrate for storage and later utilization. The amino group combines with a proton ($H^+$) to form ammonia ($NH_3$) (✦ Figure 24.1).

## Animals Use Several Methods of Nitrogen Excretion

Ammonia is a highly toxic compound, and is usually excreted directly by aquatic animals. These animals are able to mix the ammonia with large quantities of water before excretion, and produce copious quantities of relatively dilute urine.

In a terrestrial environment, water is often limited, and so the transition from aquatic life-forms to land has been accompanied by metabolic adaptations that convert ammonia to other nitrogen compounds (such as urea), which can be tolerated at higher concentrations in body fluids or can be excreted using a minimum of water (uric acid and related compounds). These adaptations are associated with the expenditure of energy to convert ammonia into other forms, but this expenditure is compensated for by water conservation.

Ammonia is made at little metabolic cost, but must be diluted with large amounts of water for excretion. Urea is made from ammonia by a series of metabolic reactions in the liver, is less toxic than ammonia, and can be excreted with less water loss. Uric acid is made through a complex metabolic process requiring large energy expenditures, but can be excreted with very little water loss.

Among land animals, mammals and adult amphibians excrete urea. Urea is formed in the liver and, to a lesser extent, the brain and kidneys. In the liver, amino acids are removed from the blood and stripped of amino groups, which are converted to urea. The urea is put into the blood and removed by the kidneys. Although urea is the primary method of nitrogen excretion in mammals, small amounts of uric acid are produced as a by product of nucleotide metabolism.

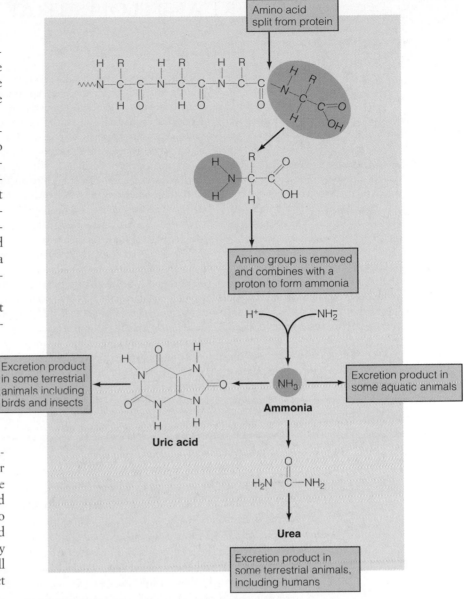

### ✦ FIGURE 24.1

Animals generate ammonia by the breakdown of amino acids. In some aquatic species, ammonia is excreted directly. In many terrestrial animals, ammonia is converted to urea before excretion. In other animals, ammonia is converted to uric acid before excretion.

**T A B L E 24.1   IMPORTANT METABOLIC WASTES AND SUBSTANCES EXCRETED FROM THE BODY**

| CHEMICAL | SOURCE | ORGAN OF EXCRETION |
|---|---|---|
| Ammonia | Deamination of amino acids in liver | Kidneys |
| Urea | Derived from ammonia | Kidneys, skin |
| Uric acid | Nucleotide catabolism | Kidneys |
| Bile pigments | Hemoglobin breakdown in liver | Liver (into small intestine) |
| Carbon dioxide | Breakdown of glucose in cells | Lungs |
| Water | Food and water; breakdown of glucose | Kidneys, skin, and lungs |
| Inorganic ions* | Food and water | Kidneys and sweat glands |

*Ions are not a metabolic waste product like the other substances shown in this table. Nonetheless, ions are excreted to help maintain constant levels in the body.

# DESERT ANTELOPES THAT DO NOT DRINK

Most animals drink water in order to survive. Water is absorbed in the intestine from solid and liquid food, and is also produced as a by-product of cellular metabolism. Some small animals such as the kangaroo rat are able to live in the desert without drinking. These rodents require little water to survive, and obtain water from the seeds they eat and from water produced by metabolism. In fact, the kangaroo rat obtains almost 90% of its water as a product of metabolism.

Kangaroo rats and other desert rodents have adaptations that minimize water loss; they are small (they weigh only a few grams) and escape the dehydrating heat of the day by hiding in underground burrows. The rats forage at night when it is cooler.

Two species of antelopes, the eland and the oryx, live in the African deserts and require no drinking water to survive. Unlike the kangaroo rat, these animals are not small (the oryx is about four feet at the shoulder; the eland is the largest African antelope and weighs about 500 kilograms), and they are unable to escape the desert heat. To regulate their body temperatures, these antelopes evaporate substantial quantities of water every day. Water is also needed for maintaining the proper concentration of solutes in blood and extracellular fluids. What strategies do they use to regulate their body temperature? Since these antelopes can survive without drinking, how do they meet their need for water?

These animals have a variety of adaptations to control their body temperatures. As the heat builds during the day, both animals increase their body temperatures to reduce water loss by evaporation. The eland's body temperature increases to just over 41°C (106°F), and the oryx's temperature can rise to 45°C (113°F). To reduce heat production, both animals have slower metabolisms during the day. In the oryx, lowered metabolism reduces evaporative water loss by 17%. The eland does not lower its metabolism as much as the oryx, but it seeks shade under acacia trees at the height of the sun.

Typically, the desert environment of these animals has a pattern of cooler nights (34°C or 93°F) and hot days (41 to 45°C or 106 to 113°F). To balance the water lost by evaporation and excretion in these conditions, the eland requires nearly 5.5 liters of water per 100 kilograms per day, and the oryx needs just under 3 liters per 100 kilograms per day. These antelopes are able to survive in the desert because the food they eat contains all the water they require. The eland prefers to eat the leaves of the acacia, which contain about 58% water. The oryx eats grasses and shrubs, which have low water content, but collect water from the desert air by condensation at night. Food containing an average of 30% moisture is sufficient to provide the water needs of the oryx.

These animals have developed several physiological and behavioral adaptations that enable them to survive in a desert environment. These mechanisms minimize the amount of water lost to the environment, lowering the amount of water they need to take in.

Animals whose ancestors evolved on land, but who now live in the sea (like whales), retain the nitrogen excretion pattern of their ancestors and excrete urea (✦ Figure 24.2a). Freshwater snails evolved from terrestrial snails and excrete uric acid, like their terrestrial ancestors (Figure 24.2b).

Reptiles, birds, and adult insects excrete uric acid in a concentrated, sometimes crystalline form. The formation of uric acid from ammonia requires more than a dozen metabolic reactions, and consumes large amounts of energy, but excretion of uric acid uses very little water. For animals that need to conserve water, this trade-off is effective. Some physiologists have argued that uric acid excretion was a prerequisite to the evolution of flight in insects and birds, since the animals do not have to carry extra weight in the form of water, which would serve only as a medium for excretion. The next time you are decorated by a passing bird, you may want to contemplate the adaptive value of this form of excretion.

✦ **FIGURE 24.2**
(a) Although whales live in the sea, their ancestors evolved on land and excreted nitrogen as urea. Whales retain this link with their terrestrial ancestors. (b) Freshwater snails evolved from terrestrial ancestors, and although they live in water, they excrete uric acid, as did their ancestors.

(a)

(b)

Although habitat plays an important role in determining the most efficient means of nitrogen excretion, methods of reproduction are also a factor in the mode of nitrogen removal. Reptiles and birds lay eggs that contain all the water needed for embryonic development. The embryo develops inside a fluid-filled sac, the amnion, which is connected to two other sacs (✦ Figure 24.3). One of these sacs provides food, and the other holds metabolic waste products. Uric acid is produced and stored in the egg, allowing the limited amount of water present in the egg to be used for other metabolic purposes.

## Water and Salt Balance Is an Important Function of Excretion

The form of nitrogen excreted by an animal is closely related to the regulation of water and solute balance in the extracellular fluids. The excretory system is responsible for maintaining the water and solute content of the blood and other body fluids. Animals in an aquatic habitat are surrounded by water, and must maintain water and salt balance in such an environment, a process called **osmoregulation.**

Some marine invertebrates such as jellyfish, scallops, and crabs have body fluids that are similar in composition to saltwater (✦ Figure 24.4a). As a result, there is no net gain or loss of water between the body fluids and seawater, and these body fluids are **isotonic** with saltwater. Such organisms have no system of osmoregulation, and their body fluids change composition if the surrounding water changes composition. These animals are said to be **osmoconformers.**

Marine vertebrates, on the other hand, have body fluids with only about one-third the solute concentration of seawater, and are said to be **os-**

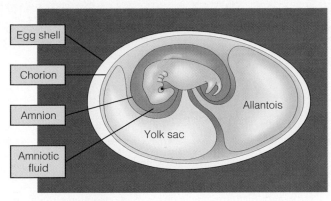

✦ **FIGURE 24.3**

In birds and reptiles, the embryo is enclosed in a sac formed from the amnion, which contains amniotic fluid. This sac is connected to the yolk sac, which contains food reserves used during development, and the allantois, which gradually fills with metabolic wastes.

**moregulators.** These organisms have two problems: They must prevent water loss to the environment via osmosis, and must prevent diffusion of salts from the environment into the body fluids. Several different adaptations have evolved to solve these problems. In most marine fish (especially the bony fish), water is lost through the gills by osmosis. Although these fish produce only small amounts of urine, it represents another source of water loss. To replace lost water, marine bony fish drink large amounts of seawater, which is absorbed as it passes through the intestine. Salts absorbed along with the water are excreted through the gills by active transport.

Cartilaginous fish such as sharks and rays are unique in that the solute concentration in their blood is higher than seawater, most of which is due to high levels of urea (Figure 24.4b). Because of

✦ **FIGURE 24.4**

(a) The body fluids of crabs are similar in composition to saltwater. As osmoconformers, their body fluids change composition if the osmotic concentration of the surrounding water changes. (b) Cartilaginous fish such as sharks have blood solute concentrations higher than seawater. As seawater moves into the body by osmosis, it is used for excretion.

(a)

(b)

# GUEST ESSAY: RESEARCH AND APPLICATIONS

## Sea Squirts and Kidney Stones

MARY BETH SAFFO

In college, my scientific interests focused on marine organisms, as well as subjects such as history, music, and Russian. And here I am now, a biologist.

My circuitous route into science has turned into a more efficient path than one might initially think. As my career has progressed, I have been increasingly impressed at how many skills go into "doing" biology. Biologists do not just run machines, make calculations, and know facts. They give talks, they teach, they chair committees, they travel to other counties. They interact with colleagues, with students, and with the public. They write papers and books and grant proposals. They may pursue fieldwork requiring physical skills in addition to intellectual and technical ones. For such activities, mastery of laboratory techniques is not enough. Even when doing "pure" research, I find myself drawing on the entire breadth of my education and experience as I ask new questions, learn new techniques, or seek connections between my work and other fields. For me, a broad perspective is the fuel for creative growth.

I have been studying some filter-feeding invertebrate animals called sea squirts—especially one particular family, the Molgulidae, and the microbes that live inside them. These organisms interest me because they don't obey textbook rules about the ways they're supposed to organize their lives: Compared to *E. coli*, or a rat, they seem to do things "wrong." But Mol-

gulidae are abundant animals at the bottoms of oceans and estuaries all over the world (one molgulid species even lives in New York harbor!). To be such successes, these biological "outlaws" must be doing something *right*—they also suggest that textbook rules can be wrong. I am using molgulids as provocateurs to challenge our views of two biological phenomena: *nitrogen excretion* and *symbiosis* (the "living together" of two or more species). As I have probed these questions, molgulids have led me to other questions as well.

Molgulids supposedly have a kidney, the so-called "renal sac." Its "waste" products—solid deposits of urate—would be unsurprising for a terrestrial animal, but are quite surprising for a marine invertebrate. The renal sac has no openings, leading to the paradoxical notion that the renal sac stores its wastes permanently. These odd features led me to ask: "Is the renal sac really a kidney and urate a waste product, or do these serve other functions? Is urate really a permanent deposit, or is it metabolized further?"

In probing these questions, I found more surprises. First, I found cells of a eukaryotic fungus-like protist, *Nephromyces*, floating in the renal sac fluid. The life cycle and structural features of *Nephromyces* are so odd that the taxonomic relationships of this protist are still unclear.

*Nephromyces* is not only present in all molgulids, but it has never been found anywhere

except molgulids; this suggests that molgulids and *Nephromyces* are *mutualistically* associated—that is, that they both benefit from this symbiosis. How and why did these organisms team up together, and what ecological and evolutionary consequences has this association had, especially for molgulids? Could the presence of *Nephromyces* even be the *key* to renal sac function and to the evolution of this organ as well?

Molgulids also remind us how important it is for biologists to study the full range of organismal diversity, rather than concentrating on just a few laboratory-tamed model species. All those *other* species out in nature are a rich source of fresh questions, and of fresh answers to old questions.

*Mary Beth Saffo is an associate professor in the Department of Life Sciences at Arizona State University West. She received her B.A. in biology in 1969 from Cowell College, University of California–Santa Cruz, and her Ph.D. in biological sciences in 1977 from Stanford University. In addition to other honors, she is a Fellow of the American Association for the Advancement of Science. Her research interests include marine biology, physiological ecology, biology of invertebrates and protists, and the biology of symbiosis.*

---

this, water moves into their body fluids from seawater, and this water is used for excretion.

Freshwater vertebrates have blood that is higher in solute concentration than the surrounding water, and must prevent water gain from the environment, and prevent loss of salts from their bodies by diffusion. Freshwater fish do not drink water, and the skin is covered with a thin film of mucus to make the skin impermeable. Water enters the body through the gills, and the excretory system of the fish removes this water, producing large amounts of dilute urine. Salts are lost to the environment through the gills and the kidney, and this salt loss must be replaced. Salt is obtained from food, and by absorption from water by secretory cells in the gills.

Animals that live on land are faced with the problem of water loss to the environment and the

need to maintain a salt balance. Some animals such as amphibians and annelids prevent water loss by living in moist habitats. Others such as insects, reptiles, and birds have impermeable body coverings that conserve water. Humans, like other mammals, are able to live in a wide range of environments, including moist jungles, and deserts, and over a wide range of temperatures from the Arctic to the tropics because of a highly efficient strategy for water and salt balance.

This system is so effective that some mammals such as whales and dolphins are able to live in saltwater. In these animals, water and salt balance are carefully regulated. As an example, consider a typical pattern of water output in humans ( ➤ Table 24.2). Actual water loss depends on several factors such as the temperature, humidity, amount of exercise performed, and the amount of water in the body. Water loss can range from less than 1 liter to more than 9 liters per day. For example, a man walking in the desert at 110°F will lose more than 1 liter of water per hour.

## EXCRETORY SYSTEMS PERFORM THREE MAIN FUNCTIONS

Life independent of the sea is possible to a large extent because of organ systems that maintain the equilibrium and composition of the extracellular fluids. These organ systems adjust the amount of water and solutes within narrow optimal ranges. In addition, these organ systems are responsible for removing toxic metabolic wastes and foreign compounds that have entered the body. Excretory organs perform several specific functions:

1. They collect and filter body fluids.
2. They remove and concentrate waste products of metabolism from the body fluids, and return other substances to the body fluids as necessary to maintain homeostasis.
3. They eliminate excretory products from the body.

With few exceptions, animals use tubular-shaped organs to control the composition of body fluids.

**TABLE 24.2   DAILY WATER BALANCE**

| WATER INPUT | | WATER OUTPUT | |
|---|---|---|---|
| Avenue | Quantity (ml/day) | Avenue | Quantity (ml/day) |
| Fluid intake | 1250 | Insensible loss (from lungs | |
| H₂O in food intake | 1000 | and non-sweating skin) | 900 |
| Metabolically produced | | Sweat | 100 |
| H₂O | 350 | Feces | 100 |
| | | Urine | 1500 |
| Total input | 2600 | Total output | 2600 |

### Invertebrates Have Tubular Excretory Organs

In flatworms and many other invertebrates, the excretory organ is a called a **nephridium** (✦ Figure 24.5). These are blind-ended tubules that open to the environment from the body by means of an excretory pore. The blind end of the nephridium has a specialized cell, the **flame cell.** Cilia on one side of the cell pump body fluids into the tubule. As fluid passes down the tubule, solutes are reabsorbed through the walls and returned to the body fluids. Wastes are expelled through the excretory pore.

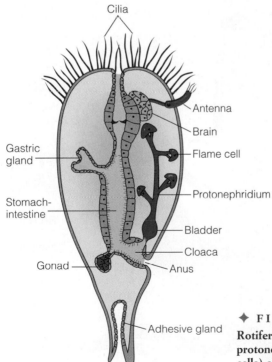

✦ **FIGURE 24.5**
**Rotifers have a tubular excretory structure (the protonephridium), which filters body fluids (in the flame cells) and excretes wastes.**

In insects, the excretory organ is a set of long tubules, the **Malpighian tubules,** which open into the gut (✦ Figure 24.6). Body fluids are drawn into the tubules by osmosis in response to a high concentration of potassium inside the tubule. As the fluid moves down the tubule, solutes and water pass through the wall of the tubule and return to the extracellular fluid, and nitrogenous waste (in this case, uric acid) empties into the gut, where any residual water is absorbed before the waste is eliminated.

## Vertebrates Have Paired Kidneys for Excretion

All vertebrates have paired kidneys. It is worth pointing out that excretion of waste is not the primary function of kidneys, even though kidneys are referred to as excretory organs. As mentioned earlier, kidneys are one type of tubular organs that regulate the composition of body fluids in animals. We will discuss the human kidney and its associated organs as an example of a vertebrate excretory system.

# THE HUMAN EXCRETORY SYSTEM HAS FOUR MAIN COMPONENTS

## Overview of the System

In humans, the urinary system is composed of kidneys, ureters, bladder, and urethra (✦ Figure 24.7). Kidneys are oval-shaped organs, indented on one side. They are surrounded by a protective layer of fat, and located at the back of the abdominal cavity just below the diaphragm. The functional unit of the kidney is the **nephron,** a tubular structure that is an evolutionary adaptation of the nephridium. Waste filtered from the blood forms **urine,** which leaves the kidney through the **ureter** (✦ Figure 24.8). Contractions in the smooth muscles lining the ureters move urine to the **bladder.** The walls of the bladder contain a thick layer of smooth muscle that allows the bladder to distend as it fills with urine. Urine is expelled from the bladder through the **urethra.**

✦ **FIGURE 24.6**

**Excretory systems in insects. (a) Insects have Malpighian tubules that connect to the digestive system. (b) In the tubules, water, ions, and other metabolites are resorbed.**

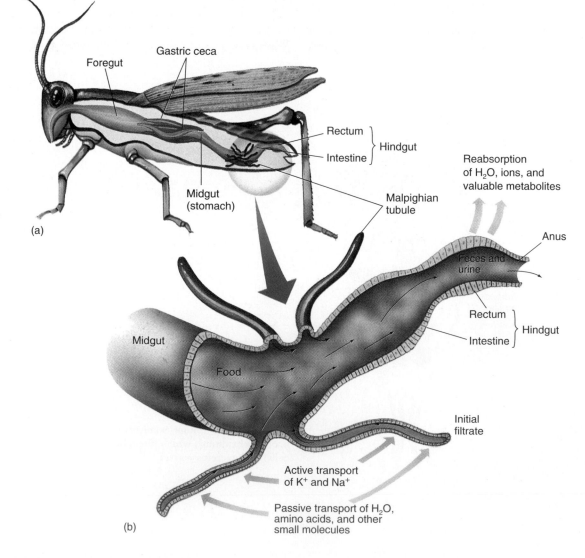

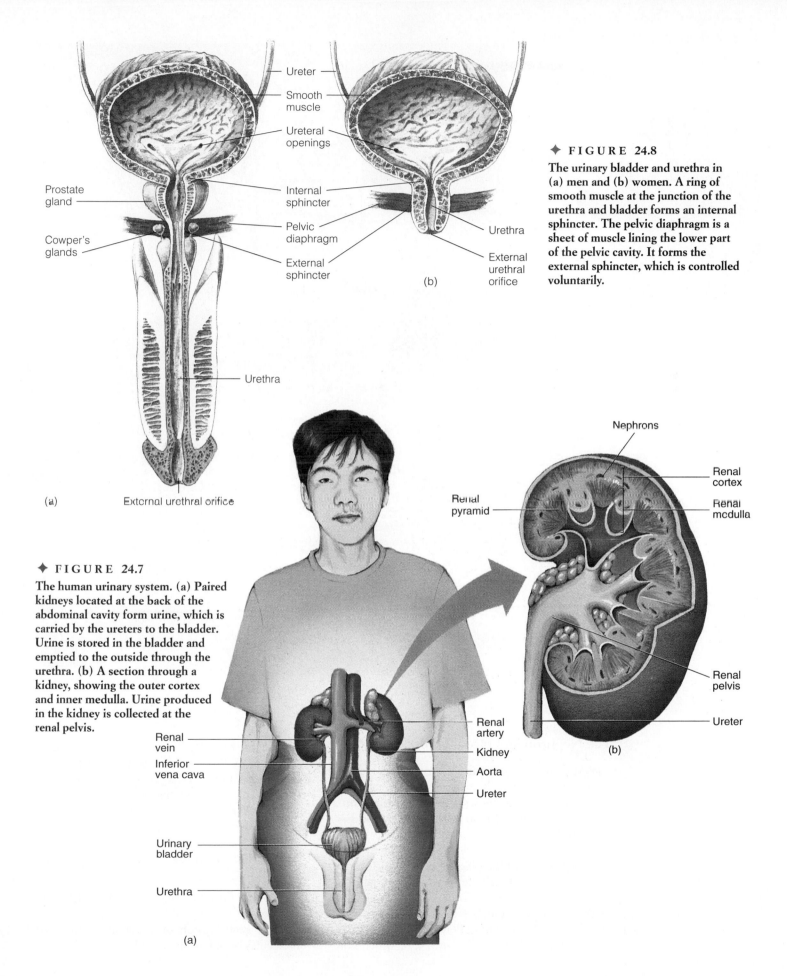

Ureter

Smooth muscle

Ureteral openings

Prostate gland

Internal sphincter

Cowper's glands

Pelvic diaphragm

Urethra

External sphincter

External urethral orifice

(b)

Urethra

External urethral orifice

(a)

Nephrons

Renal cortex

Renal pyramid

Renal medulla

Renal pelvis

Renal artery

Kidney

Renal vein

Aorta

Inferior vena cava

Ureter

Urinary bladder

Urethra

Ureter

(b)

**◆ FIGURE 24.8**

The urinary bladder and urethra in (a) men and (b) women. A ring of smooth muscle at the junction of the urethra and bladder forms an internal sphincter. The pelvic diaphragm is a sheet of muscle lining the lower part of the pelvic cavity. It forms the external sphincter, which is controlled voluntarily.

**◆ FIGURE 24.7**

The human urinary system. (a) Paired kidneys located at the back of the abdominal cavity form urine, which is carried by the ureters to the bladder. Urine is stored in the bladder and emptied to the outside through the urethra. (b) A section through a kidney, showing the outer cortex and inner medulla. Urine produced in the kidney is collected at the renal pelvis.

(a)

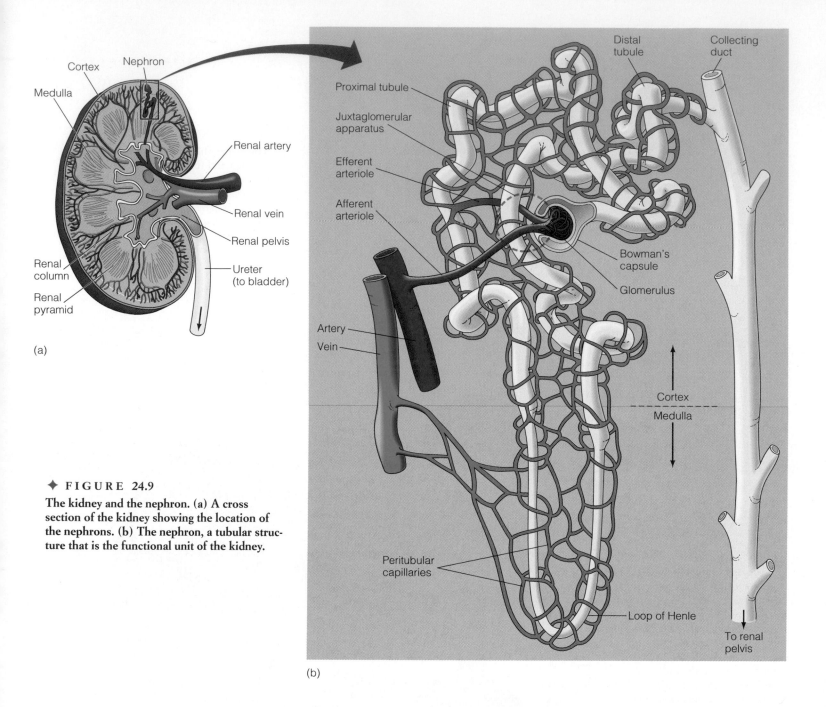

Cortex
Nephron
Medulla

Renal artery

Renal vein

Renal pelvis

Ureter
(to bladder)

Renal column

Renal pyramid

(a)

Proximal tubule

Juxtaglomerular apparatus

Efferent arteriole

Afferent arteriole

Artery

Vein

Peritubular capillaries

Distal tubule

Collecting duct

Bowman's capsule

Glomerulus

Cortex

Medulla

Loop of Henle

To renal pelvis

(b)

◆ **FIGURE 24.9**

**The kidney and the nephron. (a) A cross section of the kidney showing the location of the nephrons. (b) The nephron, a tubular structure that is the functional unit of the kidney.**

## The Nephron and How It Works

Each nephron is a long tubular structure composed of two parts: a cup-shaped capsule, containing a tangle of capillaries known as a **glomerulus,** and a long **renal tubule.** Blood flows into the kidney through the renal artery, which enters the kidney at the indented central region. The renal artery branches into smaller vessels, eventually forming the capillaries of the glomerulus. The arterial pressure of blood flowing through the glomerulus causes water and solutes from the blood to filter into the capsule. The fluid flows through the winding **proximal tubule** (◆ Figure 24.9), which con-

nects to a thin U-shaped **loop** (the loop of Henle). From the loop, fluid drains into the **distal tubule,** which empties into a collecting duct. The arteriole leaving the capsule branches again to form a network of capillaries surrounding the nephron tubule. These capillaries receive fluid and solutes from the tubule as they are returned to the blood.

The nephron performs three specialized functions:

1. Glomerular filtration of water and solutes from the blood
2. Tubular reabsorption of water and conserved molecules back into the blood

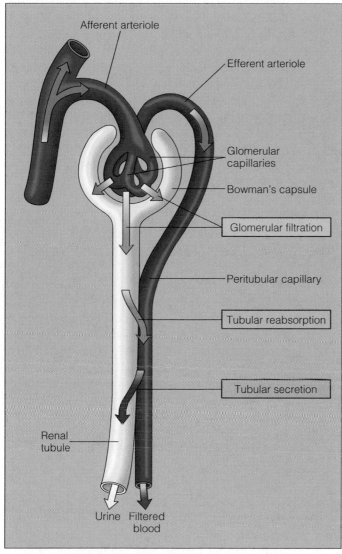

**✦ FIGURE 24.10**

**A diagrammatic representation of the nephron showing where the processes of filtration, reabsorption, and secretion occur.**

Labels in figure: Afferent arteriole; Efferent arteriole; Glomerular capillaries; Bowman's capsule; Glomerular filtration; Peritubular capillary; Tubular reabsorption; Tubular secretion; Renal tubule; Urine; Filtered blood

nephron capsule. The walls in the glomerular capillaries are very permeable (think of them as leaky), and under pressure from heart contractions, water and solutes (such as glucose, urea, amino acids, ammonia, and sodium) are forced from the capillaries into the nephron capsule. Left behind in the capillary are blood cells, antibodies and other protein molecules, and a variety of other solutes.

The filtrate leaves the capsule and flows the proximal tubule (Figure 24.9). The remaining blood components leave the glomerulus and flow through the capillary network that surrounds the nephron tubules. In the proximal tubule, **reabsorption** begins. Most of the water, sodium, amino acids, and sugar are returned to the blood. In this second step in urine formation, sodium ions are actively transported from the tubule into the capillaries, an energy-requiring process. As these ions are pumped across the wall of the tubule into the capillary, the resulting osmotic gradient causes water to flow from the tubule into the capillary by diffusion.

The reabsorption step restores normal levels of water and other solutes to the blood and conserves the body's supply of these materials. The process of reabsorption takes place along the entire length of the tubule, but most takes place in the proximal tubule. Large quantities of filtrate are reabsorbed in the tubules.

The fluid remaining in the tubule flows through the U-shaped loop between the proximal and distal tubules (Figure 24.9) and undergoes the third step in urine formation, **tubular secretion.** Wastes such as excess potassium ions and hydrogen ions that were not filtered from the blood in the capsule are transported from the capillaries into the distal tubule of the nephron. Other materials such as uric acid and the by-products of hemoglobin breakdown are removed from the blood at this stage. Tubular secretion is the main mechanism by which the acid–base balance of the blood and other body fluids is regulated. Recall that hydrogen ions (protons) are produced as a result of cellular metabolism. The processes that occur in the compartments of the nephron are summarized in ✦ Figure 24.10 and ➛ Table 24.3.

3. Tubular secretion of ions and other waste products from the surrounding capillaries into the distal tubule.

The nephrons filter about 125 ml of body fluid every minute, so that the total extracellular fluid of the body is filtered through the kidneys about 16 times each day. In a 24-hour period, the nephrons produce about 180 liters of filtrate, of which about 178.5 liters is resorbed, with the rest being the roughly 1.5 liters of urine produced by your body each day.

The first step in urine production is **filtration,** a process that takes place at the glomerulus and

*Sidebar*

**WILSON'S DISEASE**

There are several genetic conditions that cause improper excretion and transport of metabolic wastes. One such disorder involves the accumulation of copper in bodily tissues. This autosomal recessive condition is known as Wilson's disease and is estimated to affect about 1 in 30,000 individuals in the United States.

The diagnosis of this disease is frequently missed unless the managing physician looks for it specifically in patients experiencing neurological problems, psychiatric problems, or liver disease. Once diagnosed, treatment with agents to complex with and remove copper from the body can reverse or prevent the clinical signs of the disease. Without treatment, Wilson's disease progresses to acute liver failure, degeneration of the central nervous system, and death.

The failure of the body to properly metabolize copper results in cirrhosis of the liver, degeneration of the brain and greenish-brown rings in the cornea of the eye. Individuals with Wilson's disease lack the copper-binding protein serum ceruloplasmin, and bodily tissues become overloaded with loosely bound copper. The primary cause for this defect is unknown.

The gene has been mapped to the long arm of chromosome 13 (13q14). Prenatal diagnosis for a fetus at risk for this disorder is possible using DNA linkage analysis. Genetic counseling can help families understand this disease and educate them about the natural history of Wilson's disease.

**TABLE 24.3 COMPONENTS OF THE NEPHRON AND THEIR FUNCTION**

| COMPONENT | FUNCTION |
|---|---|
| Glomerulus | Mechanically filters the blood |
| Bowman's capsule | Mechanically filters the blood |
| Proximal convoluted tubule | Reabsorbs 75% of the water, salts, glucose, and amino acids |
| Loop of Henle | Participates in counter-current exchange, which maintains the concentration gradient |
| Distal convoluted tubule | Site of tubular secretion of $H^+$, potassium, and certain drugs |

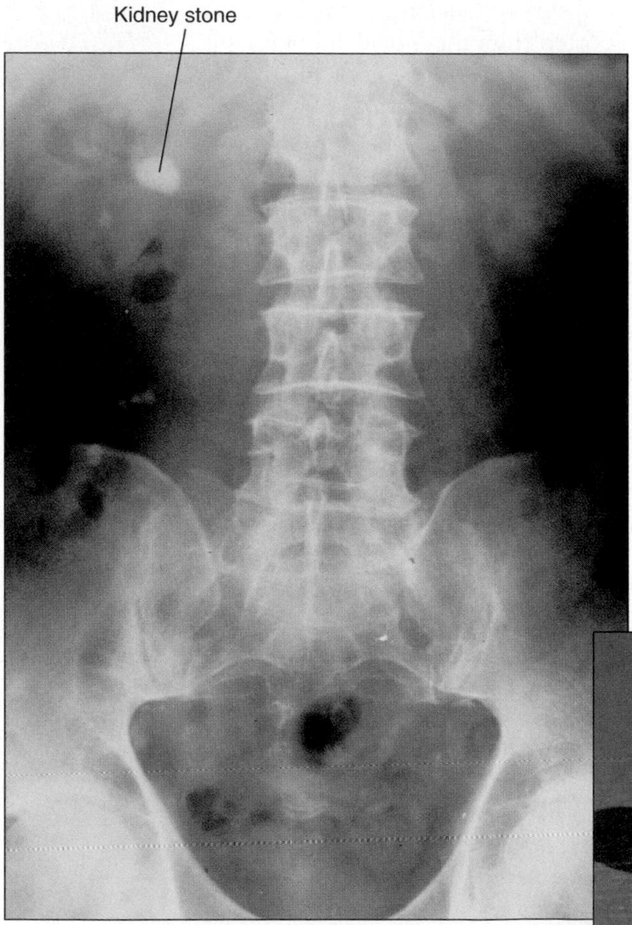

Kidney stone

(a)

✦ **FIGURE 24.11**

**Kidney stones. (a) An X-ray of a kidney stone. (b) Kidney stones removed by surgery. The nail provides a perspective for size.**

(b)

In some cases, as water is reabsorbed and the urine becomes more concentrated, excess wastes such as uric acid, calcium, and magnesium do not remain in solution, and crystallize out in the kidney to form **kidney stones,** which grow by the deposition of materials on their outer surface (✦ Figure 24.11). Some of these stones pass through the ureter into the bladder, and out through the urethra. If the stones are large enough, they may cause intense pain as they move from the kidney to the outside. The largest stones may become lodged in the kidney or its ducts, blocking the flow of urine, and requiring surgery or destruction of the stones by ultrasound.

## KIDNEY FUNCTION REGULATES INTERNAL FLUID COMPOSITION

The kidneys perform a number of functions that maintain homeostasis:

1. They maintain the volume of extracellular fluid. This is one of the ways in which blood pressure is regulated.
2. They maintain ionic balance in the extracellular fluid. Ionic levels are crucial in a number of ways. For example, if potassium levels fall too low, cardiac failure can result.
3. They maintain the acid–base balance and the solute concentration (osmotic concentration) of body fluids.
4. They excrete toxic by-products of metabolism such as urea, ammonia, and uric acid.

The kidney does not act alone in performing these homeostatic functions. Hormones produced by the brain, adrenal glands, and the heart, all work to regulate kidney function.

## Water and Salt Levels Are Controlled by Hormones

Water reabsorption in the nephron tubules is regulated by **antidiuretic hormone** (ADH) in a negative feedback system. ADH is produced by cells in the brain, and released by an endocrine gland known as the pituitary gland (✦ Figure 24.12). Release of ADH is controlled by two sets of receptors. One set, in a region of the brain known as the hypothalamus, monitors the osmotic concentration of the blood. Other receptors in the heart monitor changes in blood volume and can cause the release of ADH. On a hot day when fluids are lost by sweating, blood volume decreases by water loss. As water leaves the blood, the osmotic concentration rises because proteins and glucose are left in the blood. These signals cause the release of ADH. In

the kidney, ADH acts to increase water absorption by making the nephron tubule more permeable. This increases water absorption, putting more water back into the blood and producing a more concentrated urine.

In the opposite situation, when a large amount of water is consumed, the blood volume is increased and the osmotic concentration of the blood is reduced. These changes reduce ADH secretion, making the tubules more impermeable to water. Less water is reabsorbed from the tubule, lowering water concentration in the blood and producing large quantities of a dilute urine.

Regulation of water volume by ADH is an important mechanism in controlling the concentration of solutes such as ions in the body. Another hormone, **aldosterone,** secreted by the adrenal glands (located on top of the kidneys), regulates the trans-

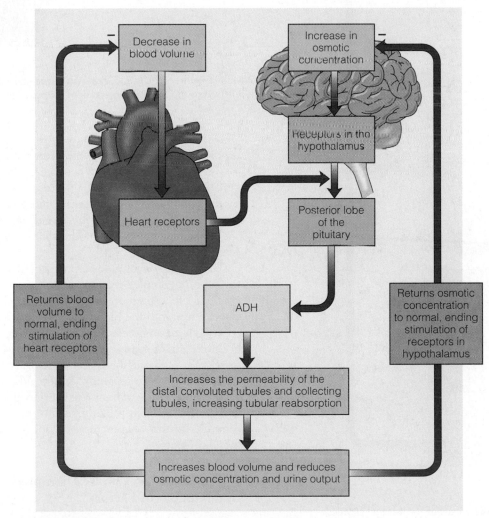

✦ **FIGURE 24.12**

**ADH secretion and action. Receptors in the heart and brain monitor the volume and osmotic concentration of the blood. When blood volume falls, or osmotic concentration rises, ADH is released from the posterior lobe of the pituitary. ADH action returns blood volume and osmotic concentration to normal.**

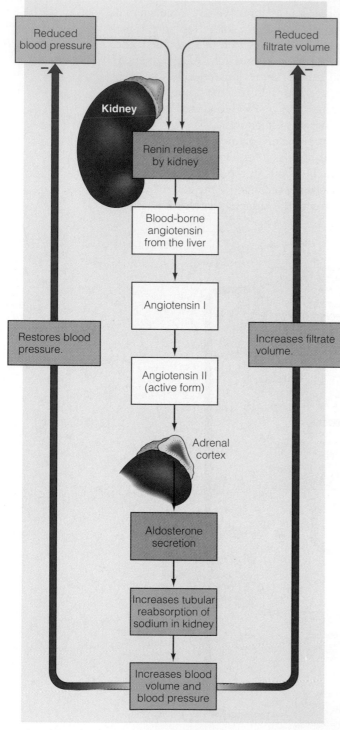

**FIGURE 24.13**

Aldosterone is released by the cortex of the adrenal glands; it controls reabsorption of sodium in the nephron tubules.

The diagram contains the following labeled boxes and text:

- Reduced blood pressure
- Reduced filtrate volume
- Kidney
- Renin release by kidney
- Blood-borne angiotensin from the liver
- Angiotensin I
- Restores blood pressure.
- Increases filtrate volume.
- Angiotensin II (active form)
- Adrenal cortex
- Aldosterone secretion
- Increases tubular reabsorption of sodium in kidney
- Increases blood volume and blood pressure

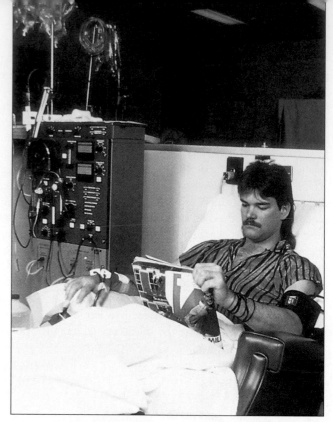

**FIGURE 24.14**

Kidney dialysis filters the blood, removing urea and other metabolic wastes, to replace the action of damaged kidneys.

fer of sodium ions from the nephron tubule of the blood (✦ Figure 24.13). When the concentration of sodium in the blood falls, aldosterone is released by the adrenal cortex and causes more sodium to pass from the tubule into the blood. By osmosis, water follows, increasing blood volume and pressure.

The regulation of aldosterone secretion is controlled in a complex manner. When there is a decrease in blood pressure, the kidneys secrete an enzyme, **renin,** into the blood. Renin acts on a blood protein called **angiotensinogen,** converting it to the hormone **angiotensin II.** This hormone raises blood pressure by constricting blood vessels and by stimulating aldosterone secretion. When aldosterone is present, increased sodium resorption causes more water to move from the tubules into the blood by osmosis. In turn, this increases blood volume and blood pressure.

## Kidney Function Can Be Disrupted by a Number of Factors

If kidney function is disrupted by infection, environmental toxins (such as mercury), or by a genetic disease, the result can be devastating. Slow loss of kidney function is difficult to detect because up to 75% of the kidney can be inactivated or even de-

## TABLE 24.4 COMMON URINARY DISORDERS

| DISEASE | SYMPTOMS | CAUSE |
|---|---|---|
| Bladder infections | Especially prevalent in women; pain in lower abdomen; frequent urge to urinate; blood in urine; strong smell to urine | Nearly always bacteria |
| Kidney stones | Large stones lodged in the kidney often create no symptoms at all; pain occurs if stones are being passed to the bladder; pains come in waves a few minutes apart | Deposition of calcium, phosphate, magnesium, and uric acid crystals in the kidney, possibly resulting from inadequate water intake |
| Kidney failure | Symptoms often occur gradually: more frequent urination, lethargy, and fatigue; should the kidney fail completely, patient may develop nausea, headaches, vomiting, diarrhea, water buildup, especially in the lungs and skin, and pain in the chest and bones | Immune reaction to some drugs, especially antibiotics; toxic chemicals; kidney infections; sudden decreases in blood flow to the kidney—for example, resulting from trauma |
| Pyelonephritis | Infection of the kidney's nephrons; sudden, intense pain in the lower back immediately above the waist, high temperature, and chills | Bacterial infection |

stroyed before any symptoms appear. Kidney failure and other urinary disorders have a variety of causes, some of which are summarized in ➡ Table 24.4. Many of these symptoms can be treated by kidney dialysis. A kidney dialysis machine acts as a kidney to remove wastes and maintain solute concentrations. In dialysis, blood is pumped through permeable tubing immersed in a large volume of fluid with a composition that simulates plasma (✦ Figure 24.14). Wastes such as urea move through the wall of the tubing into the fluid, and solutes above their normal concentration also move into the fluid. The blood that is returned to the body has low levels of waste and normal solute concentrations. An alternative to dialysis is the transplantation of a kidney from a donor. Kidneys are the organ most commonly transplanted in the United States.

 SUMMARY

1. Cells constantly alter their immediate environment by removing materials for metabolism and by producing metabolic by-products that are moved from the cells into the environment. Animals have developed several adaptations to maintain an optimum environment for cells. These systems, collectively called excretory systems, to a large extent make life possible in a terrestrial environment.

2. The excretory system functions to remove nitrogen wastes generated by metabolism and to maintain water and salt balance in marine, freshwater, and terrestrial environments. The form of nitrogen excretion is tied closely to the environment and to the mechanisms of water balance.

3. Both invertebrates and vertebrates use a system of one or more tubules to control the volume and composition of body fluids and to excrete wastes. In invertebrates, excretion is accomplished by nephridia and Malpighian tubules.

4. In vertebrates, the kidney is responsible for maintaining the composition of body fluids. The functional unit of the kidney is the nephron, a differentiated tubule that is closely associated with a network of capillaries.

5. In the nephron, water and solutes are filtered from the blood into the tubule of the nephron. As the filtrate passes through the tubule, some materials are returned to the surrounding capillaries, while waste products are retained for removal from the body. Other substances are secreted from the capillaries into the tubule for disposal. The filtrate retained by the tubules is collected and excreted as urine.

6. The functions of the kidney are regulated by a series of hormones that controls the composition and volume of the urine in order to maintain the composition of body fluids within a homeostatic range.

7. Kidney failures caused by infection, environmental toxins, or genetic disease have widespread and devastating effects on many organ systems. Kidney failure can be treated by dialysis or by transplantation.

aldosterone
angiotensin II
angiotensinogen
antidiuretic hormone
bladder
distal tubule
excretion
elimination
filtration
flame cell

glomerulus
isotonic
kidney stones
loop
Malpighian tubules
nephridium
nephron
osmoconformers
osmoregulation
osmoregulators

proximal tubule
reabsorption
renal tubule
renin
tubular secretion
ureter
urethra
urine

# QUESTIONS AND PROBLEMS

## SHORT ANSWER

1. Define the following terms: kidneys, Malphigian tubules, nephron, glomerulus, aldosterone.

2. What is the difference between excretion and elimination of wastes?

3. The hormone ADH controls water and salt levels. What happens to urine when there is no antidiuretic hormone?

4. Describe the differences between urea, uric acid, and ammonia.

5. The urinary system consists of the kidney, ureters, bladder, and urethra. Describe the functions of each of these components.

6. What three specialized functions does the nephron perform?

## FILL IN THE BLANK

1. The production of _____ is a by-product of protein metabolism.

2. _____ may cause urine retention, pain, and infection due to blockage of the ureters.

3. Small amounts of _____ are produced as a by-product of nucleotide metabolism in land animals.

4. _____ is the process of maintaining a balance between water and salt that is used by aquatic animals.

5. In flatworms, the excretory organ is called a _____. Wastes are expelled through the _____.

6. In humans, blood enters the kidney from the _____ through the _____, which branches to form the capillary system.

7. In the production of urine, resorption begins in the _____ and most of the water, sodium, amino acids, and sugars are returned to the _____.

8. The third step in urine formation is _____. This is the main mechanism by which _____ balance of the blood and other body fluids is regulated.

9. Antidiuretic hormone (ADH) is produced by a group of cells in the _____. These cells monitor the _____ of blood.

10. When blood pressure decreases, _____ is secreted by the kidneys into the blood.

## TRUE/FALSE

1. The kidney is surrounded by a protective layer of fat and is located at the front of the abdominal cavity below the diaphragm.

2. The first step in urine production is filtration which takes place at the glomerulus and nephron capsule.

3. The reabsorption step in urine production restores normal levels of water and other solutes to the blood, and conserves the body's supply of these nutrients.

4. Kidney stones form due to a build up of by-products of hemoglobin.

5. By maintaining the volume of the extracellular fluid, the kidneys regulate blood osmolarity.

## SCIENCE AND SOCIETY

1. Kidney failure is life-threatening. When the kidneys stop functioning, water and toxic substances begin to accumulate in the body. If untreated, a person can die within three days. In this chapter, we discussed kidney dialysis and how this procedure rids the body of wastes. A more recent method for the treatment of renal failure is called continuous ambulatory peritoneal dialysis or CAPD. In this procedure, two liters of dialysis fluid are injected into a patient's abdomen through a permanently implanted tube. Waste products diffuse out of the blood vessels into the abdominal cavity across a thin membrane that lines the organs and walls of the peritoneum (abdominal cavity). The fluid which contains waste products is drained from the abdominal cavity a few times each day. What are the advantages of CAPD versus the advantages of conventional kidney dialysis? Which method seems more restrictive? Which method presents a higher potential for contamination or infection? What type of patient would be a better candidate for CAPD than for the conventional dialysis method? If you had to choose one of these methods of treatment for yourself, which would you select? Explain your reasoning.

2. The kidneys rid the body of wastes and play a key role in regulating the chemical composition of blood. Substances removed from the blood by the kidneys show up in the urine. Many employers require a urine drug test when hiring an individual and at various intervals during their employment. An employee whose urine contains a chemical substance, such as cocaine or marijuana, can lose their job. What is your opinion on drug testing? Professional athletes are tested for drugs frequently. Is drug testing for an athlete more warranted than drug testing for an accountant? Why or why not? Is passing a drug test valid criteria for employment? Why do you think there is opposition to drug testing? Would you consent to be tested for drugs to get a job?

# 25

# NERVOUS SYSTEM AND SENSORY SYSTEM

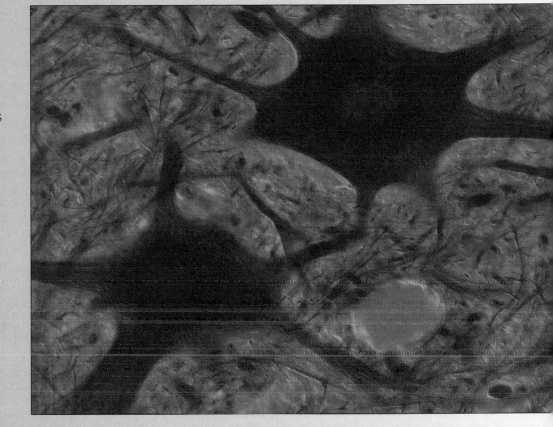

OPENING IMAGE
*Nerve cells with multiple cellular
projections.*

S ometime in the late 1850s, an 8-year-old boy and his father, a physician, drove
 along a road through the woods on eastern Long Island. They met two women
 walking along the road, and this chance encounter had a profound effect on the
young boy. Years later, he recalled that meeting:

> I recall it as vividly as though it had occurred but yesterday. It made a most enduring impression
> upon my boyish mind, an impression every detail of which I recall today. . . . Driving with my
> father through a wooded road leading from East Hampton to Amagansett, we suddenly came
> upon two women, mother and daughter, both tall, thin, almost cadaverous, both bowing, twist-
> ing, and grimacing. I stared in wonderment, almost in fear. What could it mean? My father
> paused to speak to them, and we passed on. . . . From this point on, my interest in the disease
> has never wholly ceased.

The young boy, George Huntington, went on to study medicine at Columbia University.
In 1872, a year after graduation, he published an account of this disorder, which became
known as Huntington's chorea, or as it is called today, Huntington disease.
 Huntington accurately described this disorder of the nervous system and its characteristics: (1) the hereditary nature of the disease, (2) the tendency toward insanity and suicide,
and (3) its appearance in adult life. Huntington disease is inherited as an autosomal
dominant condition, usually first expressed in mid-adult life as involuntary muscular

*movements and jerky motions of the arms, legs, and torso. As the condition progresses, brain structures degenerate, the personality changes, behavior becomes agitated, and dementia sets in. Most affected individuals die within 15 years after onset of symptoms.*

*This disorder affects basic functions of the nervous system, and its fatal outcome is a graphic illustration of the importance of the nervous system in homeostasis. In this chapter, we consider the organization and functions of the nervous system. Three functions of the nervous system will be considered: sensory input, integration of input signals, and motor responses. The following chapter will cover the endocrine system, which works closely with the nervous system in sensing and responding to environmental stimuli. A later chapter will survey behavior and the integrated action of the nervous and endocrine systems.*

## THE NERVOUS SYSTEM: SENSATION, INTEGRATION, AND COORDINATION

To survive and reproduce, multicellular animals must maintain a constant internal environment for activities such as respiration, circulation, digestion, and excretion. In addition, they must respond to stimuli from the external environment for activities such as the pursuit of prey, mates, and avoidance of predators. In many animals, homeostasis and response to environmental stimuli are integrated and coordinated by two systems: the nervous system and the endocrine system. These systems are interconnected and work together to maintain the internal environment and to coordinate responses to changes in the environment.

### Nervous Systems Perform Three Basic Functions

Nervous systems have three basic functions: receive sensory input, integrate the input, and respond to stimuli. These stimuli are received from the internal and external environments. The components and overall operation of this system are shown in ✦ Figure 25.1. Receptors are parts of the nervous system that sense changes in the internal or external environment. This **sensory input** can be in many forms, including pressure, taste, sound, light, blood pH, or hormone levels. These stimuli are converted or transduced into a signal, which is transmitted to processing centers in the brain or spinal cord. Here, the sum of incoming information is **integrated,** and a response is generated. The response, a **motor output,** is a signal transmitted to organs that can convert this signal into some form of action, such as movement, changes in heart rate, excretion, or constriction of blood vessels.

In some animals, a second control system, the endocrine system, is present. The nervous system coordinates rapid responses to external stimuli, such as movement of the body. The endocrine system controls slower, longer lasting responses to internal stimuli, such as solute concentration in the extracellular fluid. The activity of the endocrine system is integrated with the nervous system. This chapter discusses the structure and function of the nervous system, and the endocrine system is discussed in the next chapter. Throughout this and the next chapter, interactions between the two systems will be highlighted to emphasize their coordinated control of the body.

### Nervous Systems Have Two Divisions

The nervous system regulates, monitors, and controls almost every organ system. This system works through a series of both positive and negative feedback loops to establish and maintain homeostasis. In many animals, two divisions of the nervous system work in a reciprocal fashion to adjust the levels of organ function to meet changing needs. One division is called the **central nervous system** (CNS). In vertebrates, the CNS includes the brain and the spinal cord. The second system is called the **peripheral nervous system.** The peripheral nervous system connects the CNS to other parts of the body. It

✦ **FIGURE 25.1**
**The nervous system has three components, each with a separate function.**

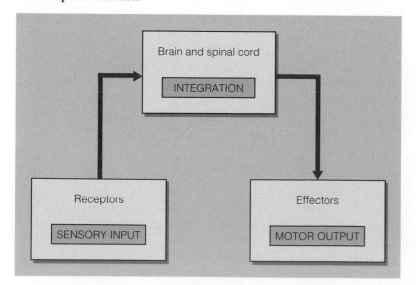

is composed of **nerves** that carry signals to and from the CNS. Nerves are bundles of cellular processes enclosed by a connective tissue covering. The processes are cytoplasmic extensions of **neurons,** the functional unit of the nervous system. The number of neurons present in an organism can vary widely. In some nematodes, the nervous system is composed of several hundred cells, while in humans, the nervous system contains about 100 billion cells.

## NERVOUS SYSTEMS HAVE SEVERAL LEVELS OF ORGANIZATION

Not all animals have highly organized nervous systems; those with a simple nervous system tend to be small and highly mobile, or large and immobile (sessile). Animals that are both large and very mobile have complex and highly organized nervous systems, indicating that the nervous system is an important adaptation in the evolution of body size and mobility. In general, the range of functions of a nervous system is a function of the size and complexity of the system.

### Nerve Nets Are Simple Nervous Systems

In marine invertebrates such as coelenterates (jellyfish), cnidarians (hydra), and echinoderms (starfish), neurons are organized into a **nerve net** (✦ Figure 25.2). These organisms are radially symmetrical and have no head region or brain. The interconnected meshwork of neurons sends signals in all directions, and there is no division of the nervous into central and peripheral regions. In spite of their relatively simple organization, nerve nets are capable of supporting a complex set of responses and behaviors. Although composed of few cells, these primitive nervous systems exhibit many of the integrative properties found in more complex systems, indicating that organization as well as cell number is an important property of the nervous system.

### Bilateral Symmetry Is Associated with More Complex Nervous Systems

Animals with bilateral symmetry have a body plan with a defined head and tail region. The development of bilateral symmetry is accompanied by **cephalization,** the accumulation of sensory and integrative functions at the anterior end of the animal. In the flatworm, neurons in the head are organized into clusters called **ganglia,** which form a small **brain.** Two nerve cords (bundles of neurons) run the length of the body, and with the brain, form the central nervous system (✦ Figure 25.3). Nerves that form cross connections along with other, smaller

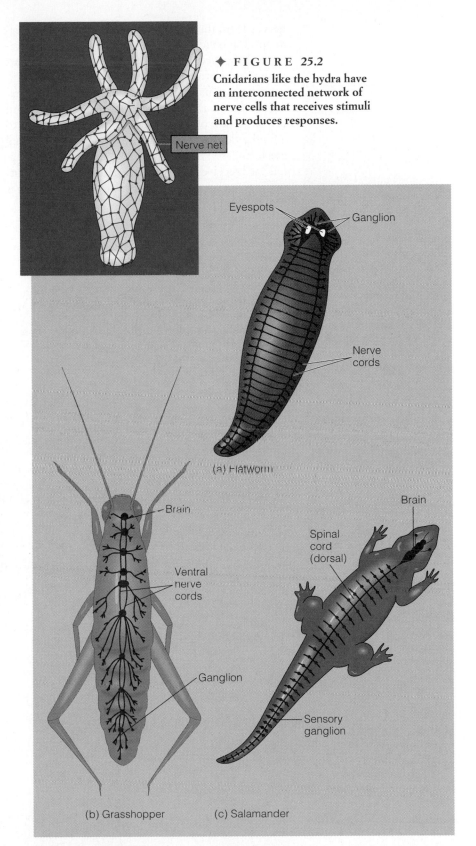

✦ **FIGURE 25.2**
Cnidarians like the hydra have an interconnected network of nerve cells that receives stimuli and produces responses.

Nerve net

Eyespots
Ganglion
Nerve cords
(a) Flatworm

Brain
Ventral nerve cords
Ganglion
(b) Grasshopper

Brain
Spinal cord (dorsal)
Sensory ganglion
(c) Salamander

✦ **FIGURE 25.3**
(a) In flatworms the nerve net is modified to form two nerve cords that run the length of the body. Nerve fibers from the cords innervate muscles. The head contains an aggregation of nerve cells called ganglia. (b) In arthropods such as the grasshopper a well-developed brain and ventral nerve cords are evident. (c) In vertebrates there is a well-organized brain and spinal cord, with nerves branching to organs.

nerves form the peripheral nervous system. In segmented annelids (like the earthworm), ganglia receive and process signals that control a limited part of the body, often a single segment.

Other animal groups, including vertebrates, exhibit a trend toward increased cephalization and centralization of the nervous system. In addition to a highly developed brain, vertebrates have a **spinal cord.** Encased in a bony vertebral column, the spinal cord receives sensory information and sends output motor signals. The vertebrate nervous system is described in later sections.

## NEURONS ARE THE FUNCTIONAL UNITS OF THE NERVOUS SYSTEM

The functional nervous system has a complex organization that includes several tissue types. Nervous tissue itself is composed of two main cell types, **neurons** and **glial cells.** Glial cells are in direct contact with neurons and often surround them. They serve as support cells in the nervous system, and help protect neurons. Nerve cells are further protected by layers of connective tissue. Some parts of the nervous system including the brain and spinal cord are surrounded by a layer of extracellular fluid known as **cerebrospinal fluid.**

### Neurons Have Distinctive Structural Features

There are three functional types of neurons, corresponding to the three functions performed by the nervous system. **Sensory neurons** carry signals from receptors and transmit information about the envi-

**Sidebar**

**HORMONAL TREATMENT FOR NEUROMUSCULAR DISORDERS**

Multiple sclerosis (MS) is a severe demyelinating disorder which affects about 1 in 1,000 individuals in the United States; females are affected twice as often as males. This condition results from the destruction of myelin which surrounds and protects nerve fibers. This electrical insulator keeps nerve impulses from going astray. In MS, loss of myelin results in erratic nerve signals that cause weakness and movement disturbances. Current research suggests that new therapeutic approaches for demyelinating diseases, such as MS, may be on the horizon.

Evidence from studies conducted on mice shows that one of the female sex hormones, progesterone, is locally produced in the peripheral nervous system, where it promotes the formation of the myelin sheath that covers nerve fibers. The French researchers involved in this study state that mice showed that the source of the progesterone that affected peripheral nerves was Schwann cells near the nerves and not hormones, produced by sex glands, that normally circulate in the body. This finding indicates that a form of hormonal therapy might be developed for neuromuscular disorders involving demyelination.

◆ **FIGURE 25.4**

**Anatomy of a nerve cell (neuron). (a) Most neurons have a series of dendrites that carry impulses to the body of the nerve cell, and one or more axons that carry signals away from the cell body. (b) A scanning electron micrograph of neurons from the central nervous system.**

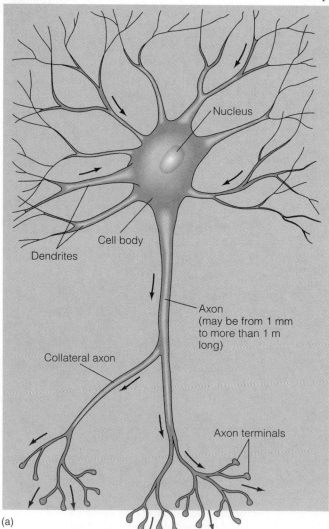

(a)

Arrows indicate direction in which nerve signals are conveyed.

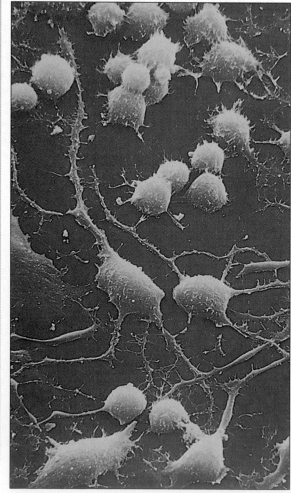

(b)

ronment to processing centers in the brain and spinal cord. **Interneurons** process signals from one or more sensory neurons and relay signals to **motor neurons.** Motor neurons transfer signals to effector cells that produce a response.

All three types of neurons share some basic structural features (✦ Figure 25.4). The **cell body** contains the nucleus and most of the cytoplasm and the organelles. All neurons have two types of cellular extensions: **dendrites** and **axons.** Dendrites carry signals toward the cell body and are often numerous, short, and highly branched. Axons carry signals away from the cell body and are often long and generally few in number. Some axons are wrapped in several layers of a **myelin sheath** formed by the plasma membranes of specialized glial cells, called **Schwann cells** (✦ Figure 25.5). The gap between each Schwann cell along the axon

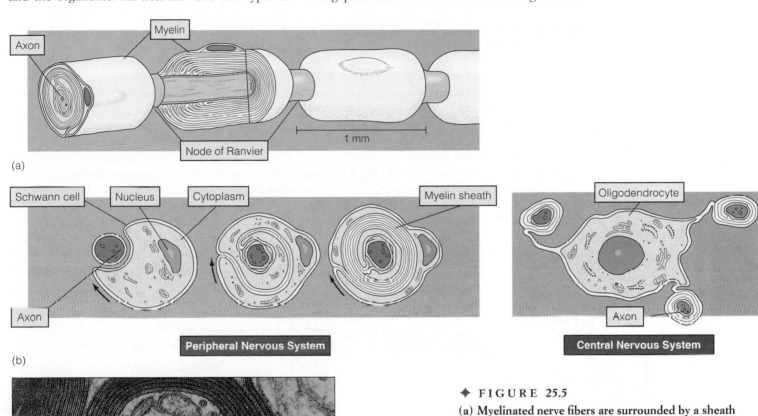

(a)

(b)

**Peripheral Nervous System**

**Central Nervous System**

(c)

✦ **FIGURE 25.5**

(a) Myelinated nerve fibers are surrounded by a sheath at regular intervals. The nonmyelinated intervals are known as the nodes of Ranvier. (b) In the peripheral nervous system, the sheath is formed by a Schwann cell wrapping around a single nerve fiber. In the central nervous system, oligodendrocytes can wrap around regions of several nerve fibers. (c) A transmission electron micrograph of a myelinated nerve fiber viewed in cross section. (d) In a myelinated fiber, a membrane potential exists only at the nodes, where the plasma membrane of the nerve is in contact with the extracellular fluid.

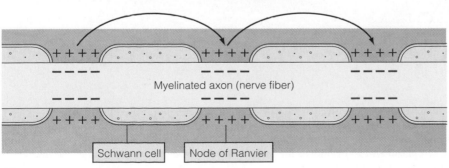

(d)

is known as the **node of Ranvier.** The nodes serve as points along the neuron for generating a signal. Signals that jump from node to node travel hundreds of times faster than those traveling the entire surface of the axon. This adaptation allows the brain to communicate with the extremities (such as your toes) within a few thousandths of a second (a *millisecond* is 1/1000 of a second), and is important in coordinated movement.

## Neurons Maintain an Electrical Potential across Their Membranes

The plasma membrane of neurons, like all other cells, has an unequal distribution of ions and electrical charges between the outside and inside surfaces. The outside of the membrane is positively charged, and the inside is negatively charged. This difference in charge between the inside and outside of the membrane is known as a **resting potential,** and is measured in *millivolts* (a millivolt is 1/1000 of a volt). In a resting nerve cell, the resting potential is −60 to 70 millivolts (about 4% of the energy in a size AA battery).

The resting potential is the result of differences in the concentration of sodium ions ($Na^+$), potas-

sium ions ($K^+$), and negatively charged molecules in the cytoplasm (✦ Figure 25.6). This imbalance is maintained through the expenditure of energy to pump sodium ions out of the cell and to pump potassium ions into the cell. As a result, the concentration of potassium inside the cell is about 30 times higher than in the extracellular fluid, and the concentration of sodium is about 10 times higher in the extracellular fluid than it is inside the neuron.

## Nerve Signals Cause a Temporary Change in Resting Potential

When a nerve cell is stimulated, there is a reversal of the electrical potential in the membrane for a few milliseconds. This reversal is called an **action potential.** It is generated by rapid changes in membrane permeability to sodium and potassium. These ions do not cross membranes by passing through the lipid bilayer; instead they move through transport proteins that span the membrane. These proteins, called **channels,** act as if they have "gates" that can be opened or closed to control ion movement.

When an action potential is triggered, sodium channels open, and sodium ions flow inward (✦ Figure 25.7). In less than a millisecond, the polarity

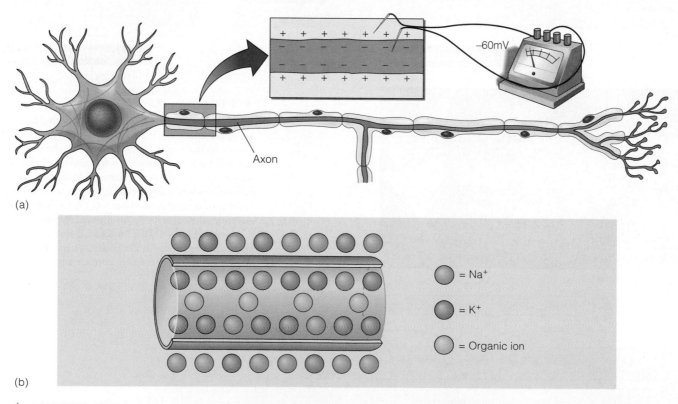

## ✦ FIGURE 25.6

(a) Electrodes placed on either side of the plasma membrane of a neuron can record the potential difference (called the resting potential) across the membrane, which is typically about −60 millivolts (mV). (b) The resting potential results from an imbalance of ions on either side of the membrane. Sodium is actively transported out of the cell, and potassium is selectively transported into the cell. Other organic ions in the cytoplasm also contribute to the ionic imbalance.

of the membrane is reversed so it becomes more positive on the inside than on the outside. At the peak of the reversal, the sodium channels close, and the potassium channels open, causing potassium ions to rush out of the cell in response to a concentration gradient, carrying positive charges outside the cell.

When enough potassium ions have moved out of the cell, the original resting potential is restored. Then the cell begins to pump sodium ions out of the cell and potassium ions back into the cell to restore the original gradients of sodium and potassium.

## Action Potentials Move along the Neuron

Originally, an action potential involves only a small patch of the cell membrane at the site of stimulation. Once started, depolarization of the membrane and ion flow spreads to adjacent membrane regions, moving the impulse along the membrane away from the original site of stimulation. During recovery from the action potential, the membrane has a brief period when it cannot be stimulated (a refractory period). This prevents the action potential from moving backwards. In summary, an action potential follows a series of steps:

1. At rest, the outside of a membrane is more positive than the inside.
2. Stimulation results in an action potential that moves sodium into the cell, making the inside of the membrane more positive than the outside.
3. Potassium ions flow out of the cell, restoring a net positive charge to the outside of the cell membrane.
4. Sodium ions are pumped out of the cell, and potassium ions are pumped into the cell, restoring the original distribution of ions.

Action potentials travel along dendrites toward the cell body, and along axons away from the cell

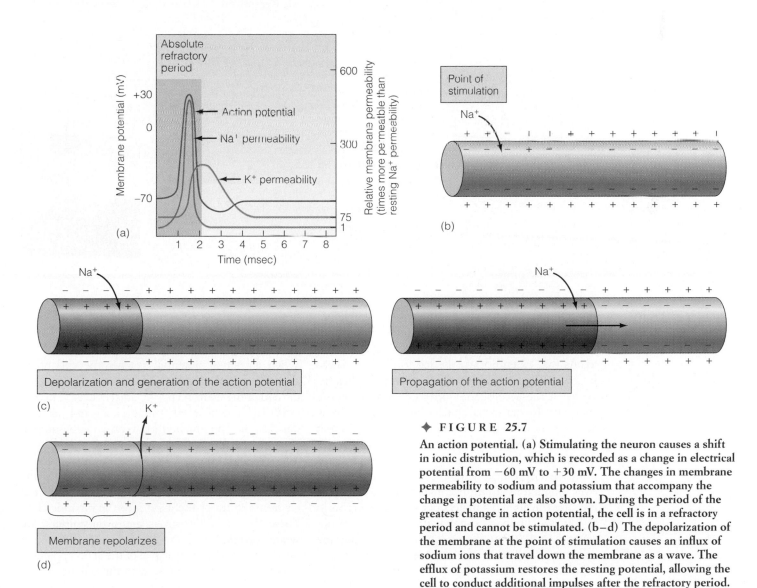

(a)

(b)

(c) Depolarization and generation of the action potential

Propagation of the action potential

(d) Membrane repolarizes

### ◆ FIGURE 25.7

An action potential. (a) Stimulating the neuron causes a shift in ionic distribution, which is recorded as a change in electrical potential from −60 mV to +30 mV. The changes in membrane permeability to sodium and potassium that accompany the change in potential are also shown. During the period of the greatest change in action potential, the cell is in a refractory period and cannot be stimulated. (b–d) The depolarization of the membrane at the point of stimulation causes an influx of sodium ions that travel down the membrane as a wave. The efflux of potassium restores the resting potential, allowing the cell to conduct additional impulses after the refractory period.

## Turning Marine Toxins into Human Medicines

WILLIAM R. KEM

The fabulous variety of living organisms has long fascinated me. As a child I started collecting rocks, fossils, snail shells, and eventually wild plants for a garden, and kept a variety of native fishes, amphibians, and reptiles at my home.

During my freshman year at Swarthmore College, my zoology professor asked me if I would be interested in spending two months of the summer working on a research project with him, first at Swarthmore and then at the University of New Hampshire, where he would be teaching a course on marine invertebrates. What I saw and learned that summer about marine animals pushed me, without knowing, further toward a career in neuropharmacology.

First, I was assigned to do a seemingly simple project investigating the actions of certain chemicals on the pulsating blood vessels of a freshwater nemertine worm. As soon as I observed the contracting blood vessels of this minute, semitransparent worm through the microscope, I wondered if the pulsations of this most primitive animal heart were directly initiated within the muscle cells as in the vertebrate heart, or if they resulted from signals in nerve cells innervating the muscle cells. To locate the pacemaker, I decided to investigate the influence of two putative neurotransmitters (acetylcholine and norepinephrine) on the blood vessel pulsations. I used agonist (stimulant) and antagonist drugs and enzyme inhibitors that modified neurotransmitter availability.

The project greatly stimulated my interest in pharmacology, a science that undergraduate students other than pharmacy majors rarely study. Later, as a physiology graduate student at the University of Illinois in Urbana, I found two short articles, published 30 years earlier by a Belgian pharmacologist, reporting the discovery of nicotine-like toxins in certain nemertines. I subsequently isolated and identified a nemertine toxin with these same properties for my doctoral dissertation. This was an important experience for me, because I learned that it was useful to combine marine biological, chemical, and pharmacological approaches to investigate marine toxins. I have subsequently used this same interdisciplinary approach when investigation other animal and plant toxins.

At the time the nemertine toxin anabaseine did not seem to possess any unique application, so I started investigating other more promising marine toxins while doing postdoctoral research in pharmacology at Duke University. However, in 1985, a possible pharmaceutical application for anabaseine appeared—the brains of persons who had died from Alzheimer's disease were found to possess abnormally low levels of nicotine receptors. Since I had shown that anabaseine stimulates muscle nicotine receptors, a similar compound might be able to selectively stimulate the remaining nicotine receptors in the brain, and thereby at least partially compensate for the lack of cholinergic stimulation. I eventually succeeded in preparing a derivative of anabaseine that selectively stimulates a brain nicotinic receptor, which appears to be particularly important for learning and memory, called "alpha 7." This compound, now undergoing clinical tests in humans, may find use in treating the cognitive dysfunction occurring in Alzheimer's patients.

So an undergraduate curiosity about marine invertebrates led to a compound that now serves as a molecular model for designing novel drugs acting on specific cholinergic receptors in the brain. Sheer curiosity still leads to discoveries in biological science. Although the significance of most discoveries is not immediately obvious, applications for even some of the most esoteric discoveries will eventually be found!

*William R. Kem is a professor in the Department of Pharmacology and Therapeutics at the University of Florida, Gainesville. He earned his B.A. degree at Swarthmore College, his Ph.D. at the University of Illinois in Urbana, and did postdoctoral research at Duke University before joining the University of Florida faculty in 1971.*

---

body. Axons can vary in length from a few millimeters to several meters. Nerves in the toes may have their cell bodies in the base of the spinal cord, and axons that extend down the leg to the toe. These long axons are generally wrapped in myelin sheaths. Between these cells, the nodes of Ranvier expose the axon to the extracellular fluid, and it is only at these nodes (spaced about 1 mm apart) that changes in membrane polarity can occur. The result is that the action potential (nerve impulse) jumps from node to node, speeding transmission over long axons. In unmyelinated fibers, the nerve impulse travels at 0.5 meters per second, whereas in myelinated fibers, it can travel at 200 meters per second, or 400 times faster. The difference is in the amount of cell membrane that is depolarized and repolarized in conducting the impulse and the distance between sites of depolarization.

### Signals Are Transferred between Neurons at Synapses

When an action potential reaches the end of an axon, the signal must travel across the gap between the end of the neuron and the adjacent cell. This adjacent cell may be another neuron, a gland cell, or a muscle cell. The junction between an axon and an adjacent neuron is called a **synapse** (✦ Figure 25.8). The space separating the two cells is known as a **synaptic cleft.** As a nerve impulse reaches the synapse, it causes the release of a chemical transmitter from the tip of the axon into the synaptic cleft. These chemicals, known as **neurotransmitters,** are synthesized and stored in small vesicles clustered at the tip of the axon. The arrival of the action potential causes these vesicles to fuse with the cell membrane and empty their contents into the synaptic cleft.

| TABLE 25.1 SOME COMMON NEUROTRANSMITTERS |
| --- |
| Acetylcholine |
| Dopamine |
| Norepinephrine |
| Epinephrine |
| Serotonin |
| Histamine |
| Glycine |
| Glutamate |
| Gamma-aminobutyric acid (GABA) |

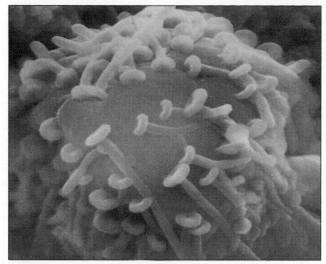

(a)

The released neurotransmitters diffuse across the cleft and make contact with receptors on the surface of the receptor cell's plasma membrane. The binding of the neurotransmitter to the receptor causes specific ion channels in the membrane of the adjacent cell to open. The altered ionic permeability generates a response in the receiving cell. Some neurotransmitters produce an action potential in the cell on the far side of the synapse, while others are inhibitory. Neurotransmitters that open sodium channels generate an action potential in the receiving cell. Other neurotransmitters open other ion channels (like Cl ), and decrease the ability of the receiving cell to generate an action potential.

After release, the time required for a neurotransmitter to act is about 0.5 to 1 millisecond. To prevent continued stimulation of the receiving cell, neurotransmitter molecules are destroyed by specific enzymes present in the cleft, diffuse out of the cleft, or are reabsorbed by the axon. More than 30 different organic molecules are known or thought to be neurotransmitters. (Some of these are listed in ◆ Table 25.1). Most are small molecules, some of which also act as hormones. Others are amino acids such as glutamate and glycine. Even gases such as nitric oxide (NO) and carbon monoxide (CO) can act as neurotransmitters.

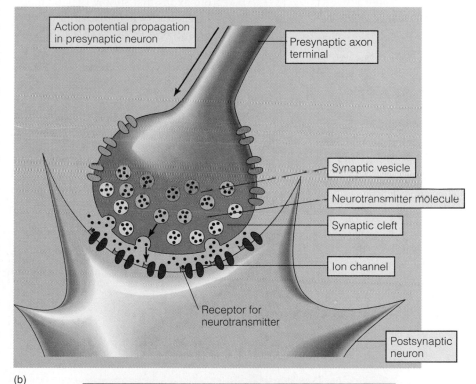

(b)

✦ **FIGURE 25.8**

**The synapse and synaptic transmission. (a) A scanning electron micrograph of the endings (terminal boutons) of an axon on another cell. (b) An impulse arriving at the end of the axon triggers the release of neurotransmitters from synaptic vesicles. The neurotransmitters travel across the synaptic cleft to the membrane of the postsynaptic cell, where they trigger another action potential. (c) A transmission electron micrograph of a synapse.**

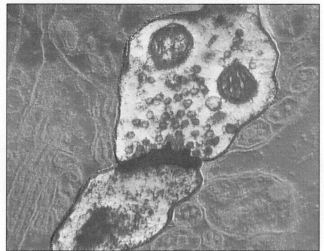

(c)

## Chemicals and Diseases Affect the Nervous System at Synapses

Diseases and infections that affect the function of signal transmission across the synapse can have serious consequences. In Parkinson's disease, there is a deficiency of the neurotransmitter dopamine, caused by the death of certain brain cells. The result is a progressive loss of movement, rigidity, tremor, and unstable posture. Treatment with **L-dopa,** a chemical related to dopamine, eases symptoms, but cannot reverse the progression of symptoms. In the brain, L-dopa is taken up by axons, stored in vesicles, and released as a substitute neurotransmitter.

The toxin produced by the bacterium *Clostridium tetani* prevents release of gamma-aminobutyric acid (GABA) from axons that supply skeletal muscles. GABA is an inhibitory neurotransmitter that prevents muscle contraction. Without GABA release, control over muscle stimulation is lost, causing muscles to lock into contracted spasms. When the muscles responsible for breathing are affected, the condition is fatal.

A toxin produced by a related bacterium, *Clostridium botulinum,* has an opposite, but equally fatal effect. This toxin, found in improperly canned foods, binds to the ends of axons that innervate muscles, preventing the release of neurotransmit-ters. Muscles become progressively relaxed, and when the pharynx and diaphragm become paralyzed, respiratory failure can cause death.

A wide range of chemicals and drugs act in the synaptic cleft to alter transmission of nerve impulses, including insecticides, anesthetics, caffeine, and psychoactive drugs such as cocaine. Psychoactive drugs such as LSD exert their effects by binding to neurotransmitter receptors on the surface of brain cells.

## AN OVERVIEW OF THE VERTEBRATE NERVOUS SYSTEM

As noted in Chapter 14, the evolutionary appearance of chordates is associated with a dorsal, rather than a ventral, nervous system. Several evolutionary trends are apparent in the nervous systems of chordates. These include a spinal cord, a continuation of cephalization in the form of larger and more complex brains, and the development of a more elaborate nervous system.

In vertebrates, the nervous system is subdivided into a number of components (✦ Figure 25.9). The primary subdivisions are the central nervous system (brain and spinal cord) and the peripheral nervous system (sensory and motor pathways). The motor neuron pathways are of two types: the **somatic system,** which innervates skeletal muscle, and the **au-**

✦ **FIGURE 25.9**
**The nervous system is divided into the central nervous system (CNS) and the peripheral nervous system (PNS). The PNS consists of the somatic and autonomic divisions.**

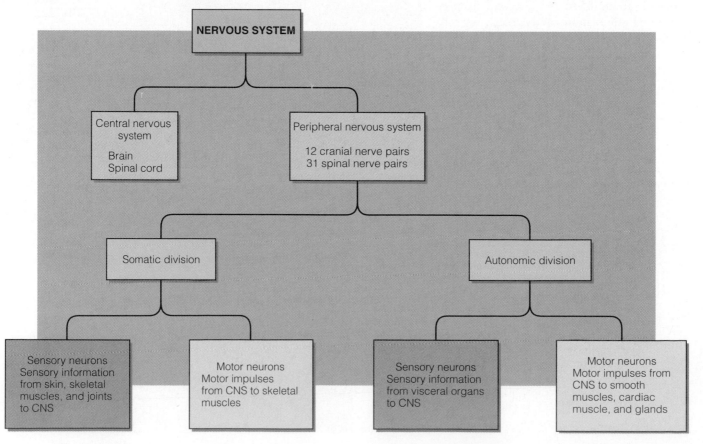

**tonomic system,** which stimulates smooth muscle, cardiac muscle, and glands. Finally, the autonomic system is divided into the **sympathetic system** and the **parasympathetic system.**

## VERTEBRATE BRAINS HAVE A HIGH CAPACITY FOR SIGNAL INTEGRATION

During embryonic development, the central nervous system of vertebrates first forms a tube; the anterior end of this tube enlarges into three hollow swellings that form the brain, and the posterior portion of the tube develops into the spinal cord. In vertebrate evolution, some parts of the brain have changed very little, while other parts have undergone substantial remodeling, increases in size, and alterations in function.

Three trends characterize evolutionary changes in the vertebrate brain. The first trend is an increase in brain size relative to body size in the progression from fish through amphibians, reptiles, and birds to mammals (✦ Figure 25.10). The second is the subdivision and increasing specialization of the forebrain, midbrain, and hindbrain. The third trend

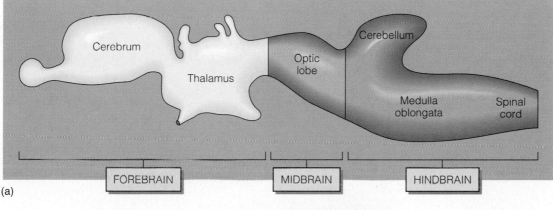

(a)

✦ **FIGURE 25.10**

**One of the main trends in vertebrate evolution is an increase in the size of the forebrain (cerebrum). (a) The regions and subdivisions of the vertebrate brain. (b-f) Changes in the brain during vertebrate evolution.**

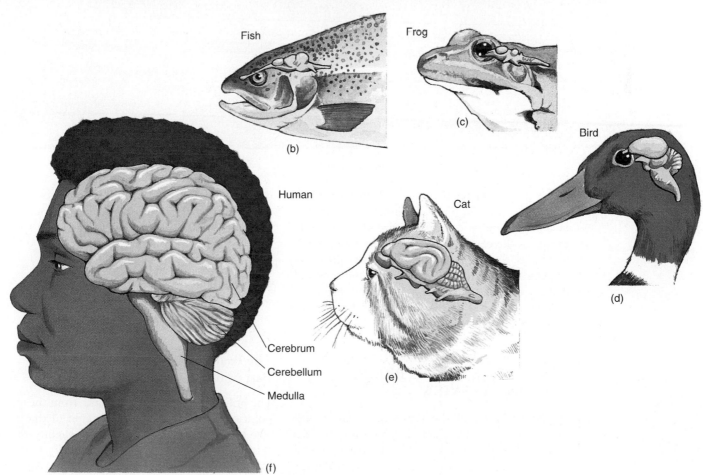

(b) Fish
(c) Frog
(d) Bird
(e) Cat
(f) Human

TABLE 25.2   THE BRAIN

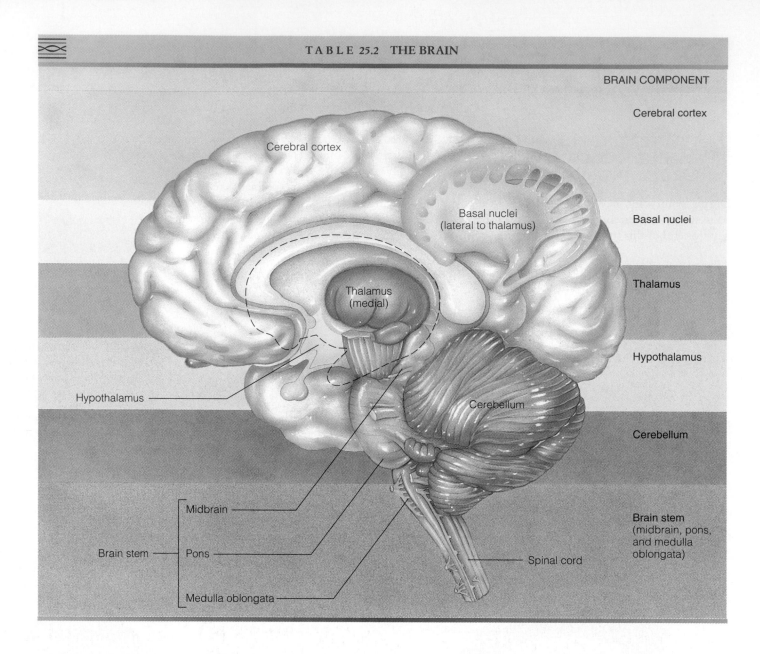

BRAIN COMPONENT

Cerebral cortex

Basal nuclei

Thalamus

Hypothalamus

Cerebellum

Brain stem
(midbrain, pons,
and medulla
oblongata)

is the growth in relative size and function of the forebrain. There is a parallel between an increase in complex behavior and the size and integrative power of the **cerebrum,** a part of the forebrain. In mammals, the cerebrum is highly folded, greatly increasing its surface area. In humans, the cerebrum makes up more than 80% of the brain's mass. The following sections discuss the human brain as the most complex vertebrate brain.

## STRUCTURE AND FUNCTION OF THE HUMAN BRAIN: AN OVERVIEW

In the evolution of the vertebrate brain, newer, more specialized areas are placed on top of older, more primitive regions. In this discussion, regions of the human brain are discussed in an order that

combines anatomical location (from bottom to top) with evolution and level of function (from oldest and least specialized to newest and more specialized) ( ◆ Table 25.2):

1. Brain stem
2. Cerebellum
3. Forebrain
   A. Diencephalon
   B. Cerebrum

## The Brain Stem Is the Most Primitive Part of the Brain

The **brain stem** is the smallest and, from the standpoint of evolution, the oldest region of the brain. The brain stem has changed only slightly during vertebrate evolution, and is functionally similar in organisms as diverse as fish and humans. Anatomi-

## MAJOR FUNCTIONS

1. Sensory perception
2. Voluntary control of movement
3. Language
4. Personality traits
5. Sophisticated mental events, such as thinking, memory, decision making, creativity, and self-consciousness

---

1. Inhibition of muscle tone
2. Coordination of slow, sustained movements
3. Suppression of useless patterns of movement

---

1. Relay station for all synaptic input
2. Crude awareness of sensation
3. Some degree of consciousness
4. Role in motor control

---

1. Regulation of many homeostatic functions such as temperature control; thirst, urine output, and food intake
2. Important link between nervous and endocrine systems
3. Extensive involvement with emotion and basic behavioral patterns

---

1. Maintenance of balance
2. Enhancement of muscle tone
3. Coordination and planning of skilled voluntary muscle activity

---

1. Origin of majority of peripheral cranial nerves
2. Cardiovascular, respiratory, and digestive control centers
3. Regulation of muscle reflexes involved with equilibrium and posture
4. Reception and integration of all synaptic input from spinal cord; arousal and activation of cerebral cortex
5. Sleep centers

---

✦ **FIGURE 25.11**

**The brain stem consists of the midbrain and the pons, and the medulla oblongata of the hindbrain.**

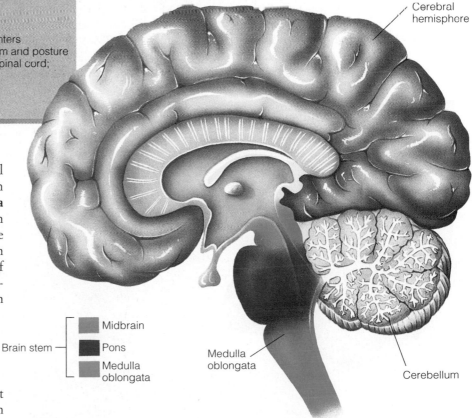

cally, the brain stem is continuous with the spinal cord, and is made up of parts of the hindbrain and the midbrain (✦ Figure 25.11). The **medulla oblongata** and the **pons** of the hindbrain contain cells that relay sensory and motor signals from the spinal cord to other parts of the brain. This region controls heart rate, constriction and dilation of blood vessels, respiration (depth and rate of breathing), coughing, sneezing, swallowing, and digestion (Table 25.2).

The **midbrain** contains a network of neurons that connects with the forebrain and relays sensory signals (including sound) to other integrating centers. In some vertebrates, this region also coordinates sensory input from the eyes. In mammals, sight is coordinated in the forebrain, and the midbrain controls only eye reflexes (Table 25.2).

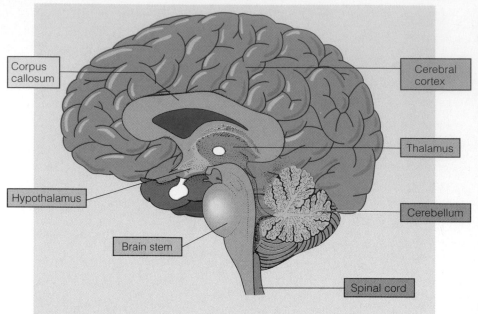

**FIGURE 25.12**

A section through a human brain showing the location of the cerebellum and hypothalamus.

Labels in figure: Corpus callosum, Hypothalamus, Brain stem, Cerebral cortex, Thalamus, Cerebellum, Spinal cord

## The Cerebellum Coordinates Body Movement

The **cerebellum** is a third part of the hindbrain, but is not part of the brain stem. It is attached to the rear portion of the brain stem (✦ Figure 25.12) and is concerned with fine motor coordination and body movement, posture, and balance. The cerebellum receives nerves from the eyes, balance organs in the ear, and the tendons, muscles, and joints. Not surprisingly, this region of the brain is enlarged in birds, and coordinates the fine muscle control necessary for flight.

## The Forebrain Is a Sophisticated Processing Center

The **forebrain** consists of two regions, the **diencephalon** and the **cerebrum.** The diencephalon itself has two components: the **thalamus** and the **hypothalamus** (Figure 25.12). The thalamus serves as a switching center for sensory signals that pass from the brain stem to other regions of the brain, and also processes some sensory signals.

Located below the thalamus, the hypothalamus is the center for homeostatic control (Table 25.2). It integrates information about the internal environment and is an important link between the nervous system and the endocrine system. In addition to receiving and processing neural signals, the hypothalamus synthesizes and releases hormones that serve as signals to the endocrine system. The hypothalamus also plays a role in emotion and behavior, and is the region of the brain that converts emotional stimuli (such as fright) into action (such as flight).

## The Cerebrum is the Largest Portion of the Human Brain

The cerebrum, which in humans is the largest portion of the brain, is divided into the right and left hemispheres. They are connected to each other by millions of tightly bundled nerve fibers called the **corpus callosum** (✦ Figure 25.13). Each hemisphere is composed of a thin, outer layer of gray matter called the **cerebral cortex.** This outermost layer is the most recently evolved region of the vertebrate brain. In the lower vertebrates such as fish, there is no cortex, and amphibians and reptiles

**FIGURE 25.13**

A section through the brain showing the right and left hemispheres and the gray and white matter of the cortex.

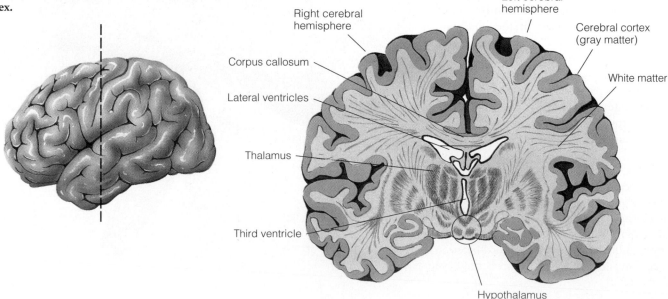

Labels in figure: Right cerebral hemisphere, Corpus callosum, Lateral ventricles, Thalamus, Third ventricle, Left cerebral hemisphere, Cerebral cortex (gray matter), White matter, Hypothalamus

## Beyond the Basics

# THE FOUR-EYED FISH

Many species of vertebrates are able to see well in both air and water. The cormorant, for example, is a diving bird that sees well enough under water to capture fish without apparent problems. There are a more limited number of species that can see simultaneously in air and water. One of these is the so-called "four-eyed" fish, *Anableps*. Species of *Anableps* live along the northeast coast of South America, from Guiana to Brazil. They live in the brackish water of estuaries, but also venture upriver into freshwater.

*Anableps* spends much of its time swimming at the water surface, with the upper parts of its bulging eyes in the air, and the lower half in the water. *Anableps* feeds on small crustaceans captured under the water, and on floating insects.

To carry out these tasks, the eye of the fish is specialized for work in two visual worlds. Finger-like projections of the iris separate the eye into two pupillary openings. The cornea of each eye is divided at the water line by a horizontal stripe of pigment into an upper and a lower half (thus the name, four-eyed fish).

The visual abilities of this fish raise several questions, including how the fish forms images in water and air in a single eye, and how images are projected and interpreted in the brain. To compensate for the refractive differences between air and water, the lens of the eye is asymmetrically placed. It is closer to the part of the retina used for underwater vision, and farther from the retinal area used for vision in air. This allows images to be in focus whether they are viewed in air or water.

Results from mapping the brain regions that process visual stimuli indicate that the eye of *Anableps* is unique. The brain contains two optical fields, one for aerial vision and one for aquatic vision. In this way, images are formed separately and processed separately. Having this information integrated separately allows the fish to interpret whether prey is located above or below the water and to move accordingly. Like bifocals, the brain area for vision in air has a band of greatly magnified vision just above the waterline, allowing the fish a close-up look at nearby objects.

This highly specialized visual system allows the fish to occupy a niche in the shoreline aquatic environment and to feed on prey in or on the water with equal success.

---

have only a rudiment of this region. The cortex is larger and more complex in primates. Humans have a large, highly developed cortical region. The gray matter in the cortex consists mainly of neuronal cell bodies and dendrites. This cortex overlays a large region of white matter containing myelinated axons that transfer signals from one region of the cortex to another, or between the cortex and other regions of the central nervous system.

## THE CEREBRAL CORTEX IS A COMPLEX MOSAIC OF FUNCTIONAL REGIONS

In each hemisphere, the cortex is a sheet of cells, ranging from 1 to 4 mm thick. It contains a series of folds that divide the hemisphere into four lobes: the **occipital, temporal, parietal,** and **frontal** lobes (✦ Figure 25.14). These regions have been intensively studied, and discrete functions have been assigned to each. Before beginning a brief discussion of these functions, remember that no region of the brain exists or functions alone. A function may be mapped to a specific region, but this region interacts and communicates with other regions of the cortex and other parts of the central nervous system, and acts as part of a complete nervous system.

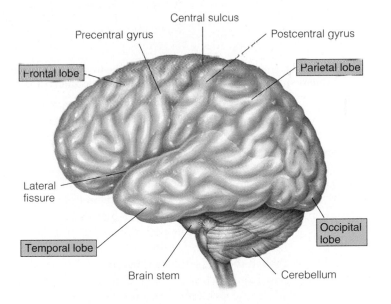

✦ **FIGURE 25.14**
**The lobes of the cerebral cortex.**

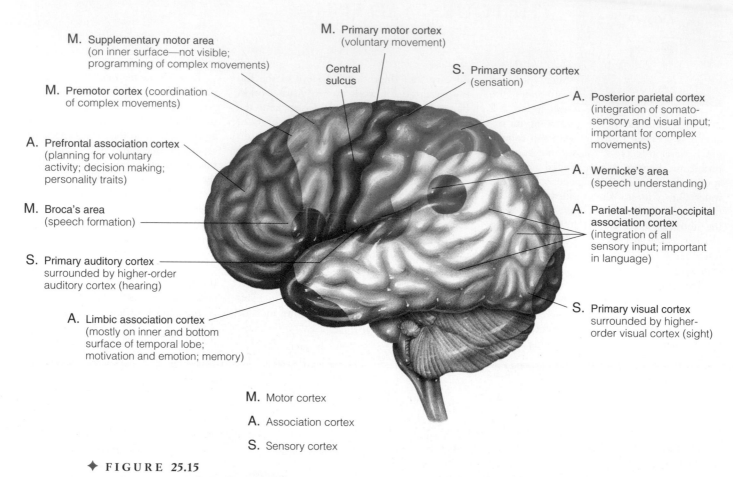

M. Supplementary motor area
(on inner surface—not visible;
programming of complex movements)

M. Primary motor cortex
(voluntary movement)

Central
sulcus

S. Primary sensory cortex
(sensation)

M. Premotor cortex (coordination
of complex movements)

A. Posterior parietal cortex
(integration of somato-
sensory and visual input;
important for complex
movements)

A. Prefrontal association cortex
(planning for voluntary
activity; decision making;
personality traits)

A. Wernicke's area
(speech understanding)

M. Broca's area
(speech formation)

A. Parietal-temporal-occipital
association cortex
(integration of all
sensory input; important
in language)

S. Primary auditory cortex
surrounded by higher-order
auditory cortex (hearing)

A. Limbic association cortex
(mostly on inner and bottom
surface of temporal lobe;
motivation and emotion; memory)

S. Primary visual cortex
surrounded by higher-
order visual cortex (sight)

M. Motor cortex

A. Association cortex

S. Sensory cortex

✦ FIGURE 25.15

Functional regions of the cerebral cortex. Three functions of the cortex include receiving sensory
information, integrating the information in the association cortex, and generating motor responses.

## The Lobes of the Cerebral Cortex Are Specialized for Different Functions

The **occipital lobe,** located at the back of the head, is responsible for receiving and processing visual information (✦ Figure 25.15). Signals from the eye are separated in this pathway and then integrated by the visual cortex in the occipital lobe. Even though there is significant overlap in the visual signals from each eye, slight differences in the perspective provide the basis for depth perception or for viewing objects in a three-dimensional spatial arrangement.

The **temporal lobe** receives auditory signals from the ears. Different regions of this cortex respond to different sound frequencies (Figure 25.15). Part of the temporal lobe is responsible for processing language input and the meaning of words.

The **parietal lobe** is associated with the **sensory cortex.** It processes information about touch, taste, pressure, pain, and heat and cold. Each sector of the parietal lobe receives signals from a specific body region (✦ Figure 25.16). Note that there is

differential representation of different body parts; for example, the relative sizes of the mouth, face, hands, and genitals are related to the high level of sensory perception of these regions. The parietal cortex in each hemisphere receives signals from the opposite side of the body because the sensory neurons ascending the spinal cord cross over to the opposite side of the brain. Damage to the left side of the brain in a stroke or aneurysm causes loss of sensation on the right side of the body.

The **frontal lobe** is associated with three main functions: (1) motor activity and the integration of muscle movements, (2) speech, and (3) thought processes. The area at the rear of the frontal lobe, adjacent to the sensory cortex of the parietal lobe, is the primary motor cortex, which sends signals to skeletal muscles.

In the vast majority of people studied, speech and language map to a region of the left hemisphere. Language comprehension is found in Wernicke's area (✦ Figure 25.17), and speaking ability is found in Broca's area. Damage to Broca's area causes speech difficulties, but does not impair com-

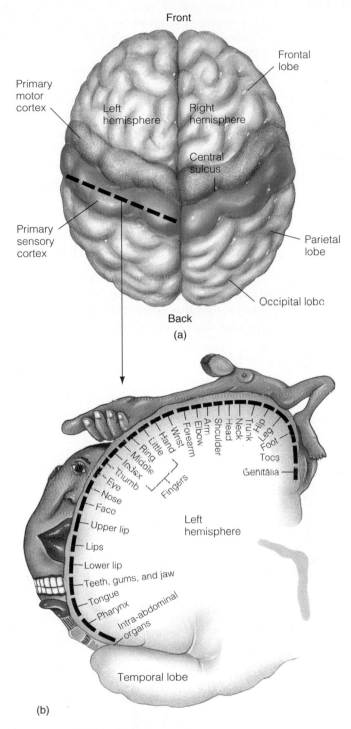

Front

Primary motor cortex

Left hemisphere

Right hemisphere

Frontal lobe

Central sulcus

Primary sensory cortex

Parietal lobe

Occipital lobe

Back

(a)

Hip
Leg
Foot
Toes
Trunk
Neck
Head
Shoulder
Arm
Elbow
Forearm
Wrist
Hand
Little
Ring
Middle
Index
Thumb
Fingers
Eye
Nose
Face
Upper lip
Lips
Lower lip
Teeth, gums, and jaw
Tongue
Pharynx
Intra-abdominal organs
Genitalia

Left hemisphere

Temporal lobe

(b)

✦ **FIGURE 25.16**

(a) **A top view of the brain showing the location of the sensory cortex.**
(b) **A map of the input received in the sensory cortex from various regions of the body.**

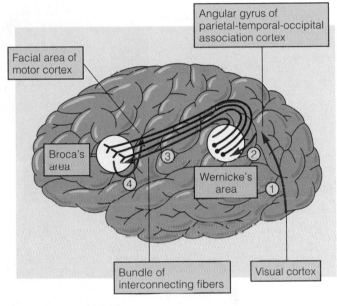

Facial area of motor cortex

Angular gyrus of parietal-temporal-occipital association cortex

Broca's area

Wernicke's area

Bundle of interconnecting fibers

Visual cortex

✦ **FIGURE 25.17**

**Steps in the process of naming a visual object or speaking a written word. In step 1, information is transferred from the visual cortex to an integrating area (angular gyrus). In step 2, information is transferred to Wernicke's area, where words to describe the visual information are selected. The language command is then sent to Broca's area (step 3). Here the signal is converted into a sound pattern and sent to the appropriate regions (step 4) to activate the facial muscles necessary to form the words.**

prehension, while lesions in Wernicke's area impair comprehension of both verbal and written words, but do not affect speech.

What is roughly the remaining half of the cortex is known as the association areas, and these are involved in higher thought processes, planning, memory, personality traits, and other human activities. In primates, a high proportion of the cortex is given over to these functions, and covers a very large area in humans.

## THE SPINAL CORD LINKS THE BRAIN AND THE BODY

The spinal cord extends from the brain stem along the dorsal side of the body and provides a link between the brain and the rest of the body. It is enclosed by the bones of the vertebral column (✦ Figure 25.18). The butterfly-shaped region of gray matter seen in the cross section of ✦ Figure 25.19a contains cell bodies and dendrites. The surrounding white matter consists of bundles of interneu-

ronal axons, called tracts. Some tracts are ascending, carrying signals from the body to the brain; descending tracts carry impulses from the brain to a particular part of the body.

At intervals, paired spinal nerves connect to each side of the spinal cord. At the junction with the spinal cord, each spinal nerve splits into a dorsal root, carrying nerve fibers into the spinal cord, and a ventral root, carrying impulses from the spinal cord.

In addition to serving as a link between the brain and the rest of the body, the spinal cord also integrates sensory and motor activity without involving the brain, in a reflex arc (Figure 25.19b).

## THE PERIPHERAL NERVOUS SYSTEM: AN OVERVIEW

The peripheral nervous system is the second main branch of the nervous system in vertebrates. It contains neurons that connect the central nervous system with the organs of the body. The peripheral

✦ FIGURE 25.18

The spinal cord is part of the central nervous system, and extends from the base of the brain to the upper lumbar region.

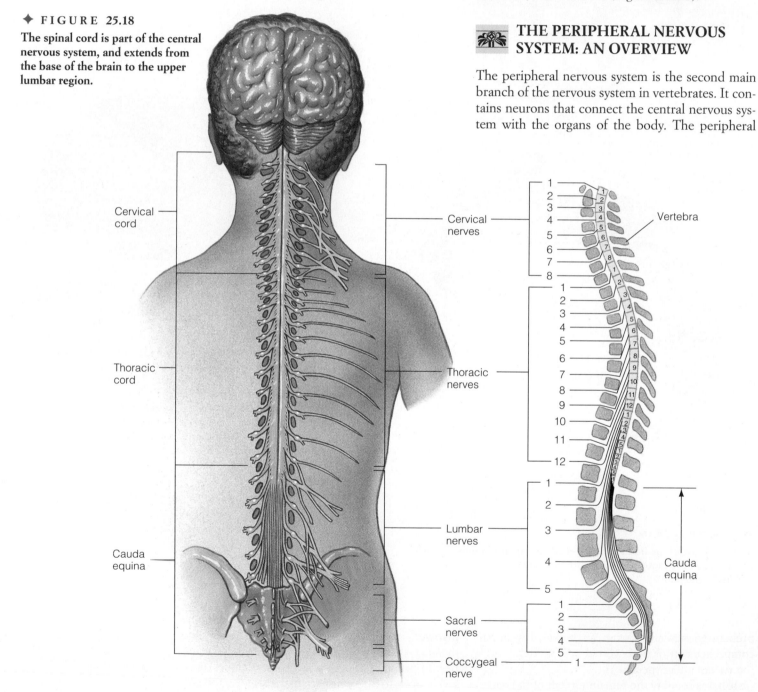

Cervical cord

Thoracic cord

Cauda equina

Cervical nerves

Thoracic nerves

Lumbar nerves

Sacral nerves

Coccygeal nerve

Vertebra

Cauda equina

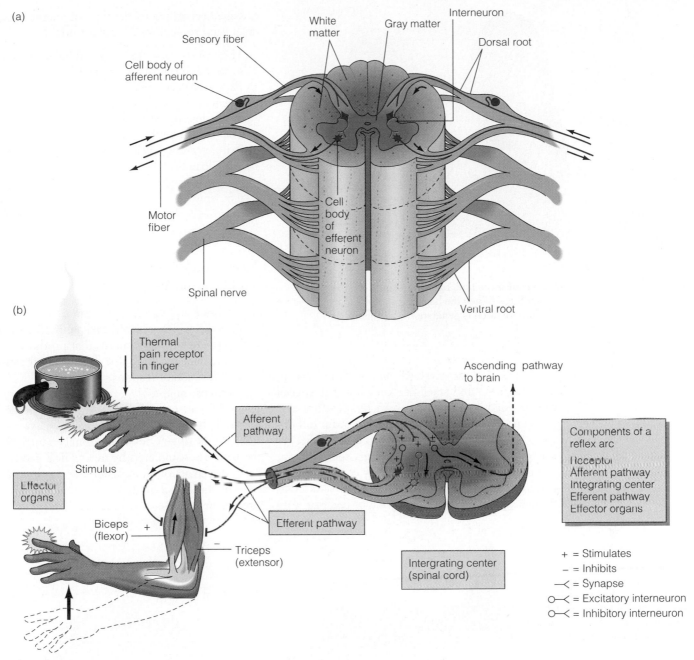

(a)

Cell body of
afferent neuron

Sensory fiber

White
matter

Gray matter

Interneuron

Dorsal root

Motor
fiber

Cell
body
of
efferent
neuron

Spinal nerve

Ventral root

(b)

Thermal
pain receptor
in finger

Ascending pathway
to brain

Afferent
pathway

Stimulus

Effector
organs

Efferent pathway

Biceps
(flexor)

Triceps
(extensor)

Intergrating center
(spinal cord)

Components of a
reflex arc

Receptor
Afferent pathway
Integrating center
Efferent pathway
Effector organs

+ = Stimulates
– = Inhibits
—< = Synapse
O—< = Excitatory interneuron
O—< = Inhibitory interneuron

✦ **FIGURE 25.19**

**(a) A section of the spinal cord and its nerves. The dorsal roots carry sensory information into the spinal cord, and the ventral roots carry motor signals from the spinal cord. Spinal nerves can carry both sensory and motor nerves. (b) In a reflex arc, an action such as withdrawing your hand from a hot stove occurs before sensory information is relayed to the brain. In the reflex illustrated, sensory fibers send impulses to the spinal cord resulting in motor impulses to contract the biceps (+) and prevent contraction of the triceps (–). Secondary impulses send sensory information to the brain.**

nervous system has two main components, the **sensory (afferent) pathways,** which provide input from the body to the central nervous system, and the **motor (efferent) pathways,** which carry signals to muscles and glands. The afferent system is the link that provides the brain with information about the internal and external environment, and the efferent system carries signals to tissues and organs of the body to establish and maintain homeostasis.

## The Peripheral Nervous System Transmits Sensory Input

Much of the sensory information carried by the peripheral nervous system never reaches the level of conscious awareness. Input that does reach the level of consciousness contributes to **perception,** our interpretation of the external environment. This input comes from special sensors, and is interpreted

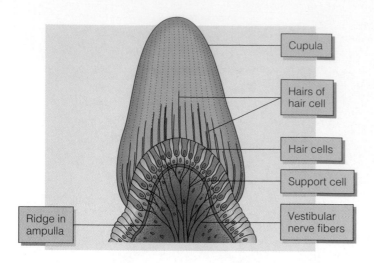

Labels on figure: Cupula, Hairs of hair cell, Hair cells, Support cell, Ridge in ampulla, Vestibular nerve fibers

**✦ FIGURE 25.20**

**At one end, hair cells have one or more fine, hair-like projections. At the other end are neuronal endings. When the hairs are stimulated by mechanical pressure, an action potential is generated in the dendrites at the base of the cell.**

classified according to the type of energy they can detect and respond to. These include:

*Mechanoreceptors:* hearing and balance, as well as stretch receptors in skeletal muscles

*Photoreceptors:* sensitive to light

*Chemoreceptors:* smell and taste, as well as those in the digestive system that detect the composition of gastric juices, and those in blood vessels that detect the levels of oxygen and carbon dioxide in blood

*Thermoreceptors:* sensitive to changes in temperature

*Electroreceptors:* detect electric currents in the surrounding environment.

## Mechanoreceptors Are Involved with Hearing and Balance

Invertebrate and vertebrate organisms have a wide range of mechanoreceptors. Mechanoreceptors vary widely in the kind of stimulus that produces an action potential, and they also differ in their sensitivity and duration of response to stimuli.

The most adaptable vertebrate mechanoreceptor is the hair cell (✦ Figure 25.20). At one end, hair cells have one or more fine, hair-like projections. At the other end are neuronal endings. When the hairs are stimulated by mechanical pressure, it generates an action potential in the dendrites at the base of the cell. This type of mechanoreceptor is present in the lateral line organ of fish and some

as pain, vision, taste, smell, and hearing. Vision, hearing, taste, and smell are known as the **special senses.** All others, including pain, temperature, and pressure are known as the **somatic senses.** In both cases, sensory input begins with receptors that react to stimuli in the forms of energy such as light, heat, chemicals, and sound. This energy is converted into an action potential and is transmitted to the central nervous system. Sensory receptors are

**✦ FIGURE 25.21**

**Hair cells are present in the lateral line organ of fish. Pressure waves in the water stimulate the hair cells, generating an action potential that makes the fish aware of the presence of other organisms or obstacles.**

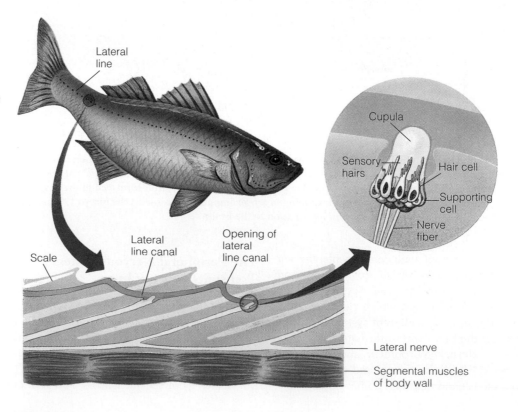

Labels on figure: Lateral line, Scale, Lateral line canal, Opening of lateral line canal, Cupula, Sensory hairs, Hair cell, Supporting cell, Nerve fiber, Lateral nerve, Segmental muscles of body wall

immature amphibians (♦ Figure 25.21). In humans and other mammals, hair cells are involved in the detection of sound, and gravity, and in providing sensory information about orientation and balance.

Hearing involves the action of several structures, including the external ear, the eardrum, ossicles, and cochlea (♦ Figure 25.22). The components of the ear and their functions are summarized in

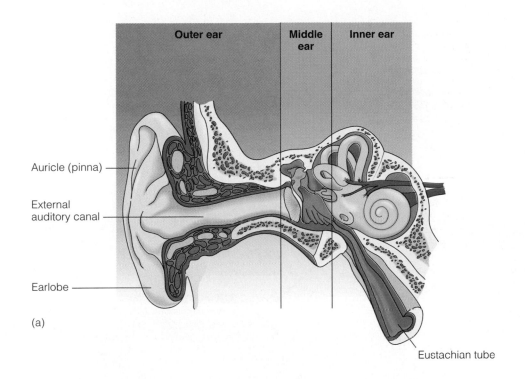

Outer ear | Middle ear | Inner ear

Auricle (pinna)

External auditory canal

Earlobe

(a)

Eustachian tube

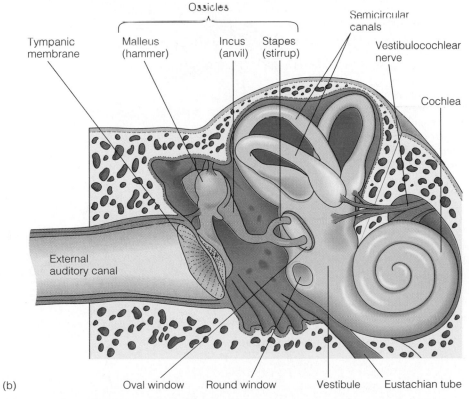

Ossicles

Tympanic membrane

Malleus (hammer)

Incus (anvil)

Stapes (stirrup)

Semicircular canals

Vestibulocochlear nerve

Cochlea

External auditory canal

(b)

Oval window | Round window | Vestibule | Eustachian tube

♦ FIGURE 25.22

(a) A cross section of the ear, showing the structures of the outer, middle, and inner ear.
(b) The structures of the middle and inner ear are involved in hearing and balance.

Table 25.3. In hearing, sound waves (mechanical vibrations of the air) are converted into vibrations of a liquid, then into movement of hair cells in the cochlea, and finally into action potentials in a sensory dendrite connected to the auditory nerve (Figure 25.23). Sensory neurons that synapse with hair cells are part of a pathway that leads from the ear to the auditory cortex in the temporal lobe, where the brain interprets these action potentials as sound.

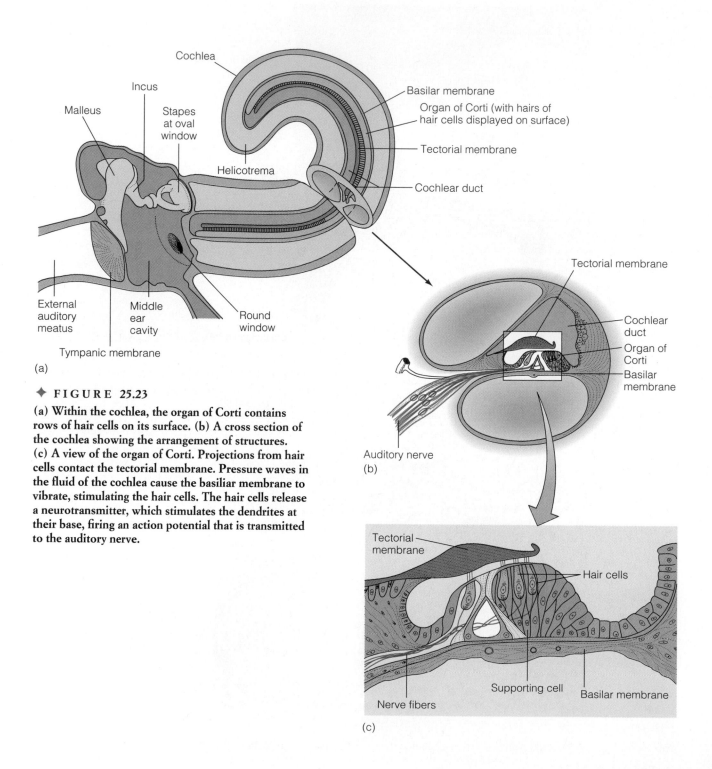

### ✦ FIGURE 25.23

**(a) Within the cochlea, the organ of Corti contains rows of hair cells on its surface. (b) A cross section of the cochlea showing the arrangement of structures. (c) A view of the organ of Corti. Projections from hair cells contact the tectorial membrane. Pressure waves in the fluid of the cochlea cause the basiliar membrane to vibrate, stimulating the hair cells. The hair cells release a neurotransmitter, which stimulates the dendrites at their base, firing an action potential that is transmitted to the auditory nerve.**

Very loud sounds (high intensity) can cause violent vibrations in the membrane underlying the hair cells, causing them to shear off or become permanently distorted (✦ Figure 25.24). This results in permanent hearing loss.

Orientation and gravity are detected in the semicircular canals and the structure at their base (Figure 25.22). In the semicircular canals, hair cells activated by movement of the liquid inside the canals respond by triggering increases or decreases in action potentials in synapsed dendrites. The canals are oriented in three different planes, providing a three-dimensional sense of equilibrium. Hair cells in the base of the semicircular canals are stimulated

**TABLE 25.3  STRUCTURES AND FUNCTIONS OF THE EAR**

| PART | STRUCTURE | FUNCTION |
|------|-----------|----------|
| Outer ear | Auricle | Funnels sound waves into external auditory canal |
| | Ear lobe | |
| | External auditory canal | Directs sound waves to the eardrum |
| Middle ear | Tympanic membrane, or eardrum | Vibrates when struck by sound waves |
| | Ossicles | Transmit sound to the cochlea in the inner ear |
| Inner ear | Cochlea | Converts fluid waves to nerve impulses |
| | Semicircular canals | Detect head movement |
| | Saccule and utricle | Detect head movement and linear acceleration |

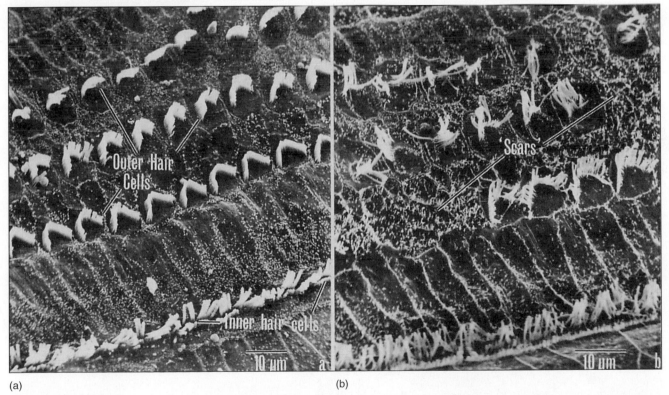

(a)    (b)

✦ **FIGURE 25.24**

**Scanning electron micrograph showing portions of the organ of Corti in guinea pigs. (a) A normal guinea pig. (b) A guinea pig exposed to a noise of 120 decibels, a level commonly achieved at music concerts and in clubs.**

by movement of small crystals of calcium carbonate that shift in response to gravity (✦ Figure 25.25). These shifts cause the stimulated hair cells to trigger action potentials that provide sensory information about gravity and acceleration.

## Photoreceptors Monitor Vision and Light Sensitivity

The ability to respond to light is almost a universal property of animals. The human eye is sensitive to only a small portion of this energy, and can detect light at wavelengths from about 400 to 700 nanometers (✦ Figure 25.26). Light at shorter wavelengths is ultraviolet light and at longer wavelengths is infrared. Many organisms, including insects, have the capacity to detect light over a wider portion of the spectrum than humans.

The structures of the human eye and their functions are summarized in ✦ Table 25.4. In the eye, two types of photoreceptor cells are grouped in a layer that forms the **retina** (✦ Figure 25.27). These

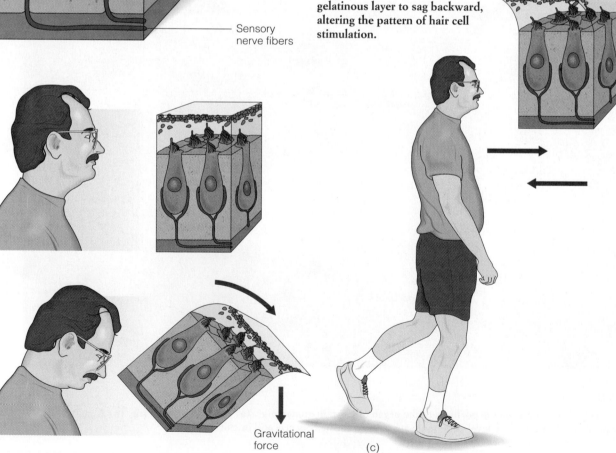

Kinocilium    Stereocilia

Otoliths

Gelatinous layer

Hair cells

Supporting cells

Sensory nerve fibers

(a)

(b)

(c)

Gravitational force

### ✦ FIGURE 25.25

(a) In the utricle and saccule, clusters of hair cells are surrounded by support cells. Over the hair cells is a gel containing otoliths, small crystals of calcium carbonate. Standing upright, the hair cells are vertical, and have projecting cilia. (b) When the head is bent forward, gravity causes the gelatinous cap to sag forward, stimulating the hair cells to fire impulses, signaling the brain about the position of the head.
(c) Moving forward causes the gelatinous layer to sag backward, altering the pattern of hair cell stimulation.

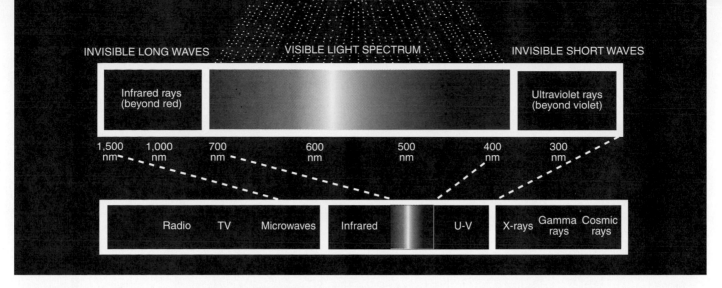

✦ FIGURE 25.26

Visible light makes up only a small portion of the electromagnetic spectrum. The numbers are wavelength measured in nanometers.

**TABLE 25.4  STRUCTURES AND FUNCTIONS OF THE EYE**

| STRUCTURE | | FUNCTION |
|---|---|---|
| **Wall** | | |
| Outer layer | Sclera | Anchors for extrinsic eye muscles |
| | Cornea | Allows light to enter; bends incoming light |
| Middle layer | Choroid | Absorbs stray light; provides nutrients to eye structures |
| | Ciliary body | Regulates lens, allowing it to focus images |
| | Iris | Regulates amount of light entering the eye |
| Inner layer | Retina | Responds to light, converting light to nerve impulses |
| **Accessory structures and components** | | |
| | Lens | Focuses images on the retina |
| | Vitreous humor | Holds retina and lens in place |
| | Aqueous humor | Supplies nutrients to structures in contact with the anterior cavity of the eye |
| | Optic nerve | Transmits impulses from the retina to the brain |

✦ FIGURE 25.27

Anatomy of the eye.

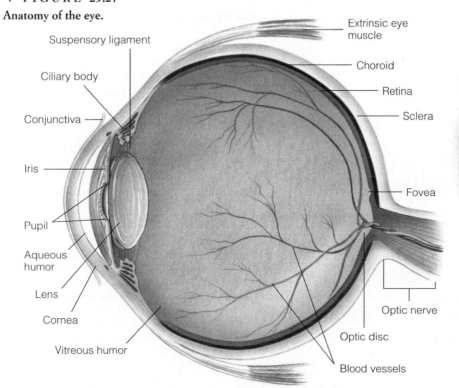

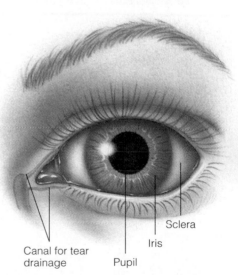

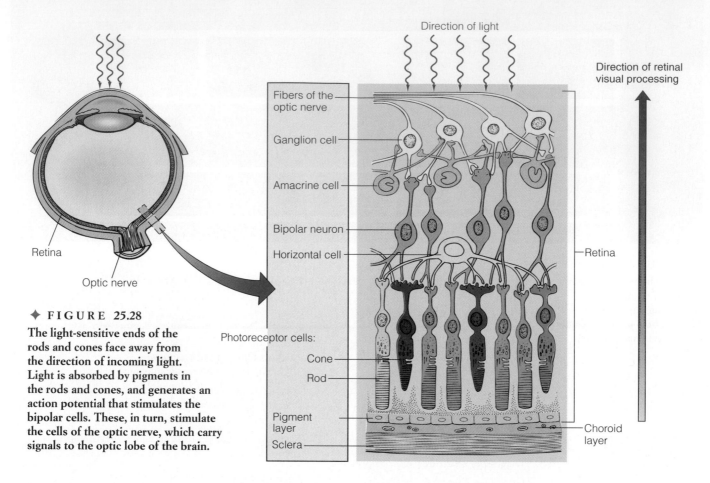

**Direction of light**

Fibers of the optic nerve

Ganglion cell

Amacrine cell

Bipolar neuron

Horizontal cell

Photoreceptor cells:

Cone

Rod

Pigment layer

Sclera

Retina

Choroid layer

Direction of retinal visual processing

✦ **FIGURE 25.28**

**The light-sensitive ends of the rods and cones face away from the direction of incoming light. Light is absorbed by pigments in the rods and cones, and generates an action potential that stimulates the bipolar cells. These, in turn, stimulate the cells of the optic nerve, which carry signals to the optic lobe of the brain.**

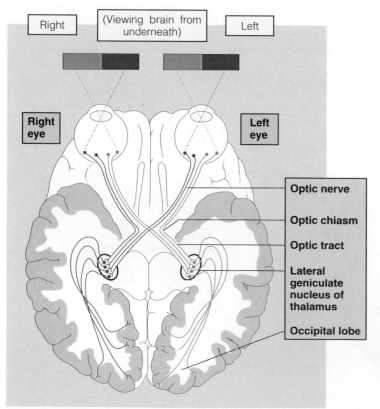

Right

(Viewing brain from underneath)

Left

Right eye

Left eye

Optic nerve

Optic chiasm

Optic tract

Lateral geniculate nucleus of thalamus

Occipital lobe

cells, the **rods** and the **cones,** have apparently evolved from hair cells (✦ Figure 25.28). Rods are sensitive to differences in light intensity, and the cones are sensitive to differences in color. Rods and cones are differentially distributed in the eye; cones predominate near the center of the retina (fovea centralis), whereas the rods are most concentrated in a circular zone near the edge of the eye.

When light enters the eye and strikes the photoreceptor cells, it initiates a chemical breakdown of a visual pigment called **rhodopsin.** This, in turn, changes the membrane polarity of the receptor cell and generates an action potential. The action potential is transmitted to cells synapsed to the receptor cell, and to ganglia whose axons form the optic nerve. The optic nerve connects the eye to the occipital lobe of the brain (✦ Figure 25.29).

✦ **FIGURE 25.29**

**The visual pathway from the eyes to the brain. The left half of the visual cortex receives information from the right half of *each* eye. The right half of the visual cortex receives information from the left half of each eye. Since each eye views an object from a slightly different perspective, the integration of the information from each eye provides depth perception.**

## Humans Have Color Vision

Rods are more sensitive to dim light than cones, but do not produce images of high resolution. In bright light, cones provide finely focused images and color vision. In humans, color vision is possible because there are three different types of cones: one for red light, one for green, and one for blue light. Molecules called **opsins** present in cone cells bind to pigments, making the opsin/pigment complex sensitive to light of a given wavelength. In humans, there are three different forms of opsin, encoded by three different genes on the X chromosome. Defective or absent forms of one of these opsins causes color blindness. If the red-opsin gene product is missing or nonfunctional, red-detecting cones are impaired, and the result is red color blindness. The most common form of color blindness, red/green blindness, affects about 8% of the males in the United States (✦ Figure 25.30).

## ![] THE PERIPHERAL NERVOUS SYSTEM: MOTOR OUTPUT

Sensory input from the peripheral nervous system is processed by the brain and spinal cord. Responses to sensory input are transmitted from the central nervous system by motor neurons to the organs of the body. There are two subdivisions of the motor pathways in the peripheral nervous system: the somatic and the autonomic ( ◆ Table 25.5).

The somatic system controls the skeletal muscles, and responses such as limb movement, walking, running, or throwing. The motor neurons of the somatic system are distinct from those of the autonomic system. The cell bodies of the somatic system are in the central nervous system, with long axons extending directly to the innervated muscles. The motor neurons of the somatic system act to stimulate the organs they innervate. They cannot send inhibitory signals.

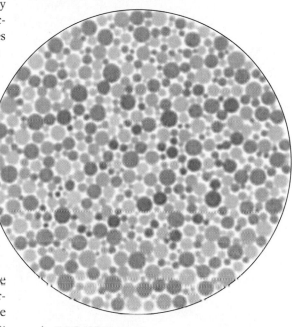

✦ **FIGURE 25.30**
**Individuals with red/green color blindness cannot distinguish the number 29 in the figure.**

| TABLE 25.5 COMPARISON OF THE AUTONOMIC NERVOUS SYSTEM AND THE SOMATIC NERVOUS SYSTEM | | |
|---|---|---|
| FEATURE | AUTONOMIC NERVOUS SYSTEM | SOMATIC NERVOUS SYSTEM |
| Site of origin | Brain or spinal cord | Spinal cord |
| Number of neurons from origin in CNS to effector organ | Two-neuron chain (preganglionic and post-ganglionic) | Single neuron (motor neuron) |
| Organs innervated | Cardiac muscle, smooth muscle, and glands | Skeletal muscle |
| Type of innervation | Most effector organs dually innervated by the two antagonistic branches of this system (sympathetic and parasympathetic) | Effector organs innervated only by motor neurons |
| Neurotransmitter at effector organs | May be acetylcholine (parasympathetic terminals) or norepinephrine (sympathetic terminals) | Only acetylcholine |
| Effects on effector organs | Either stimulation or inhibition (antagonistic actions of two branches) | Stimulation only (inhibition possible only centrally through IPSPs on cell body of motor neuron) |
| Types of control | Under involuntary control; may be voluntarily controlled with biofeedback techniques and training | Subject to voluntary control; much activity subconsciously coordinated |
| Higher centers involved in control | Spinal cord, medulla, hypothalamus, prefrontal association cortex | Spinal cord, motor cortex, basal nuclei, cerebellum, brain stem |

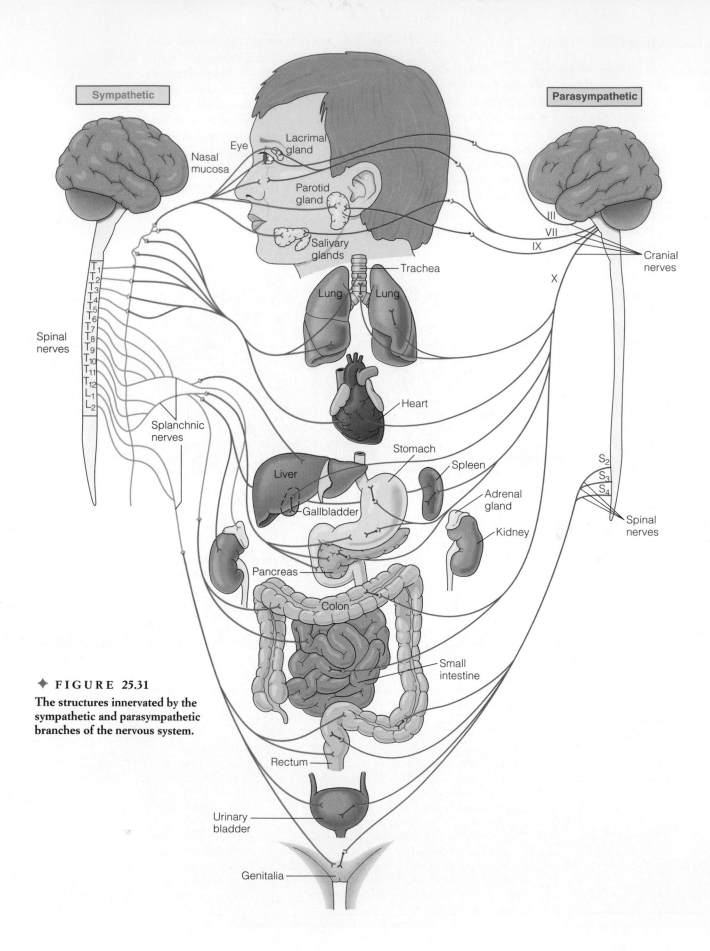

Eye
Lacrimal gland
Nasal mucosa
Parotid gland
Salivary glands
Trachea
Lung
Lung

III
VII
IX
X
Cranial nerves

Spinal nerves

T₁
T₂
T₃
T₄
T₅
T₆
T₇
T₈
T₉
T₁₀
T₁₁
T₁₂
L₁
L₂

Splanchnic nerves

Heart

Stomach

Liver
Spleen
Gallbladder
Adrenal gland
Kidney

S₂
S₃
S₄

Spinal nerves

Pancreas
Colon
Small intestine

Rectum

Urinary bladder

Genitalia

◆ **FIGURE 25.31**

**The structures innervated by the sympathetic and parasympathetic branches of the nervous system.**

The autonomic system controls muscles in the heart, and smooth muscle in internal organs including the intestine, bladder, and uterus. In the autonomic system, motor nerves originate in the central nervous system, but axons from these nerves do not reach their target organs. Motor neurons in these nerves synapse with a second motor neuron, which innervates the target organ. In addition, motor neurons of the autonomic system can either stimulate or inhibit the target organ.

The autonomic system has two structurally and functionally distinct components: the sympathetic and parasympathetic divisions (◆ Figure 25.31). Major organs of the body are innervated by motor neurons from both divisions. Functionally, the sympathetic and parasympathetic divisions act in opposite directions to maintain homeostasis. When you are afraid, the sympathetic system causes the heart to beat faster and stronger, the respiration rate to increase, and the digestive system to halt peristalsis. Motor neurons of the parasympathetic system can reverse these effects. These neurons slow the heart and respiratory rate, increase peristalsis, and counteract the effects of sympathetic stimulation. The net effect is that these two systems interact to maintain the function of body systems within a homeostatic range.

# SUMMARY

1. To survive and reproduce, multicellular animals must maintain a constant internal environment and respond to stimuli from the external environment. Homeostasis and response to environmental stimuli are integrated and coordinated by the nervous system.

2. Nervous systems have three basic functions: to receive sensory input, integrate the input, and respond to stimuli.

3. The nervous system works through a series of positive and negative feedback loops to establish and maintain homeostasis. In many animals, there are two divisions of the nervous system: the central nervous system (CNS) consists of the brain, and in vertebrates, the spinal cord. The second system is called the peripheral nervous system, and connects the CNS to other parts of the body.

4. Nervous tissue itself is composed of two main cell types, neurons and glial cells.

There are three functional types of neurons, corresponding to the three functions performed by the nervous system. These are sensory, interneurons, and motor neurons.

5. The outside of the plasma membrane of neurons is positively charged, and the inside is negatively charged. When a nerve cell is stimulated, there is a reversal of the electrical potential in the membrane called an action potential.

6. Action potentials pass from one side of the neuron to the other. At the end of an axon, chemical signals are released across a synapse to communicate the nerve impulse to another cell.

7. The nervous system is subdivided into the central nervous system (brain and spinal cord) and the peripheral nervous system (sensory and motor pathways). The motor pathways are of two types: the somatic system and the autonomic system. The

autonomic system is divided into the sympathetic system and the parasympathetic system.

8. The cerebral cortex is the most specialized portion of the human brain, and is associated with higher functions including language and abstract thought.

9. The peripheral nervous system has two main components, the sensory (afferent) pathways, which provide input from the body to the central nervous system, and the motor (efferent) pathways, which carry signals to muscles and glands.

10. Sensory receptors are classified according to the type of energy they can detect and respond to and include the following types: mechanoreceptors (hearing and balance), photoreceptors (light), chemoreceptors (smell and taste), thermoreceptors (changes in temperature), and electroreceptors (electric currents in the surrounding environment).

# KEY TERMS

action potential
autonomic system
axons
brain
brain stem
cell body
central nervous system
cephalization
cerebellum
cerebral cortex
cerebrospinal fluid
cerebrum
channels

cones
corpus callosum
dendrites
diencephalon
forebrain
frontal lobe
ganglia
glial cells
hypothalamus
integration
interneurons
L-dopa
medulla oblongata

midbrain
motor (efferent) pathways
motor neurons
motor output
myelin sheath
nerve net
nerves
neurons
neurotransmitters
node of Ranvier
occipital lobe
opsins
parasympathetic system

parietal lobe
perception
peripheral nervous system
pons
resting potential
retina
rhodopsin
rods

Schwann cells
sensory (afferent) pathways
sensory cortex
sensory input
sensory neurons
somatic senses
somatic system

special senses
spinal cord
sympathetic system
synapse
synaptic cleft
temporal lobe
thalamus

# QUESTIONS AND PROBLEMS

## MATCHING

Match the following nervous system components appropriately.

1. dendrite
2. axon
3. Schwann cell
4. central nervous system
5. nervous tissue
6. node of Ranvier
7. synapse
8. peripheral nervous system
9. forebrain
10. Wernicke's area
11. cones

_____ sensory and motor pathways
_____ sensitive to color differences
_____ composed of neurons and glial cells
_____ junction between axon and adjacent cell
_____ brain and spinal cord
_____ diencephalon and cerebrum
_____ receptive portion of neuron
_____ language comprehension
_____ conducting process of neuron
_____ gap between Schwann cell along axon
_____ produces myelin sheath that are associated with some axons

## TRUE/FALSE

1. There are two systems that serve primarily as means of internal communications—the nervous system and the endocrine system.
2. Both the somatic and the autonomic divisions are under the control of centers located in the peripheral nervous system.
3. The sympathetic and parasympathetic subdivision of the autonomic nervous system work in a synergistic manner to maintain homeostasis.
4. The development of bilateral symmetry is accomplished by cephalization, the accumulation of sensory and integrative functions at the dorsal end of the animal.

5. Glial cell fluid surrounds and protects some parts of the nervous system including the brain and spinal cord.
6. Motor neurons transfer signals from receptors, and transmit information about the environment.

## FILL IN THE BLANK

The nervous system is divided into two main parts. Complete the outline for the organization of the nervous system.

I. _____
  A. brain
  B. _____
II. _____
  A. sensory (afferent) system
  B. _____
    1. _____
    2. autonomic nervous system
      a. _____
      b. _____

## SHORT ANSWER

1. What are the three basic functions of the nervous system?
2. Describe the movement of sodium and potassium ions across the plasma membrane of neurons during the generation of an action potential.
3. Compare the effects of _Clostridium botulinum_ and _Clostridium tetani_ on the nervous system.
4. Describe the functions of the four hemispheres of the cerebral cortex.

## SCIENCE AND SOCIETY

1. Multiple sclerosis is a progressive illness in which the myelin sheath around the axons is gradually destroyed. The cause of demyelination in this disease is not known. Although the neurons are not directly affected, the loss of the myelin associated with them impairs the conduction of nerve impulses and results in a faulty system for the input and output of information. Knowing this information, what would you predict are the manifestations of this disease? Is damage that occurs to the neurons of the central nervous system permanent? Do you think the destruction caused by this condition is localized or widespread? Would the speed of the nerve signal be slower or faster in people with multiple sclerosis?

2. Research in rodents has shown that addictive drugs, such as alcohol and cocaine, stimulate the brain's pleasure center. Stimulation of this region of the brain, researchers believe, may be the cause of all drug addictions. Experiments in rats showed that they can be trained to press a lever to stimulate their pleasure center. Some rats pressed the lever for hours, ignoring water, food, and sex. These animals continued this self-stimulation until they fell over with exhaustion. Upon waking, they immediately restarted this addictive behavior. Experiments in rats demonstrated that cocaine stimulates the release of large amounts of the neurotransmitter dopamine. Cocaine, ethanol, and nicotine cause increased dopamine concentrations in the brain. These levels, however, are much higher in the pleasure center. Is there a problem using rodents as a model for learning about disorders of the nervous system? If so, identify the specific problems. Is it possible to state that neurotransmitter abnormalities are responsible for all drug addictions? Why or why not? Could genetics be a factor in one's susceptibility to alcoholism or other addictive behaviors? How much influence do social, economic, and physical characteristics have on a person's addictive behavior? Is it reasonable to state that addictions are the result of habits rather than of a disease? Explain your reasoning. Do you think addictions have a psychological or biological basis?

# 26

# THE ENDOCRINE SYSTEM

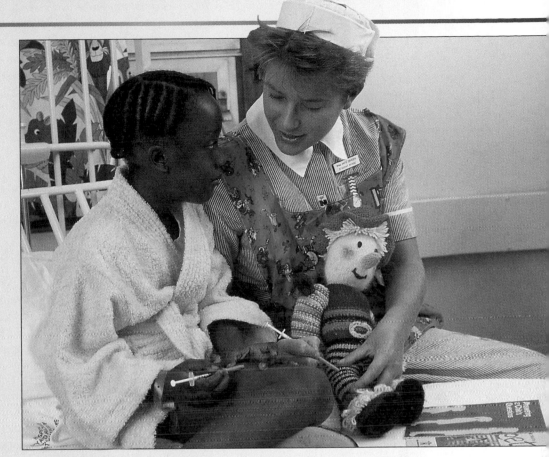

OPENING IMAGE

*A child with juvenile diabetes*

*receives instruction on this disease.*

The symptoms of diabetes were accurately described by the ancient Romans. However, the relationship between diabetes and the pancreas was not established until the late nineteenth century, by means of experimental surgery. When the pancreas was removed from dogs, they developed symptoms similar to those of diabetes in humans, including high levels of sugar in the urine. This work established that diabetes is associated with a defect in carbohydrate metabolism, and that the pancreas regulates glucose metabolism.

To treat diabetes in experimental animals, pancreatic extracts were prepared and administered, but were never successful in regulating glucose levels. Although clusters of cells, called islets, embedded in the pancreas were thought to be involved in regulating sugar levels, they could not be isolated from the rest of the organ.

In the early 1920s, F. G. Banting and his colleagues undertook a series of experiments demonstrating that secretions of the pancreatic islet cells were responsible for regulating carbohydrate metabolism. They tied off the pancreatic duct in dogs, which caused most of the gland to degenerate, but left the islet cells intact. Extracts were prepared from these glands and used to treat diabetic animals. These treatments were successful in lowering glucose levels in blood and urine to within the normal range.

Banting and his mentor, J. J. Macleod, were awarded the Nobel Prize in 1923 for their work on diabetes. Five years after Banting's work, the active component of their pancreas

*extracts, insulin, was identified, isolated, and crystallized. Since then, insulin has played a critical role in several scientific advances. In 1954, Fred Sanger and his colleagues reported the amino acid sequence and molecular structure of insulin, the first protein to have its amino acid sequence decoded. Sanger was awarded the Nobel Prize in 1958 for his work on protein structure. In 1982, human insulin sold under the name Humulin became the first drug produced by recombinant DNA technology.*

*In this chapter, we consider how insulin and other chemicals called hormones act to maintain homeostasis in the internal environment. We discuss how hormone-producing glands are organized to form the endocrine system, and how this system interacts with the nervous system.*

## THE ENDOCRINE SYSTEM COORDINATES LONG-TERM RESPONSES TO STIMULI

Multicellular animals are constantly bombarded with sensory information about the external and internal environment. These sensory stimuli require both short-term and long-term responses. The nervous system coordinates rapid and precise responses to stimuli using action potentials. The speed and duration of these responses are measured in seconds or milliseconds. The endocrine system maintains homeostasis and control over long-term events and uses chemical signals. The duration of these responses is measured in hours, days, months, or even years. The endocrine system works in parallel with the nervous system to maintain homeostasis and to regulate long-term processes such as growth and maturation (◆ Table 26.1).

The **endocrine system** (*endo-* is a prefix meaning "within") is a collection of glands that secrete chemical messages called **hormones** (◆ Figure 26.1). These signals are transmitted through the blood to target organs. Exocrine glands (*exo-* is a prefix meaning "outside") produce secretions that travel through ducts and are released outside the body. Sweat glands, salivary glands, and digestive glands (remember the digestive system is a tube open at both ends traveling through the body, but is not strictly speaking, inside the body) are exocrine glands.

## ENDOCRINE SYSTEMS HAVE EVOLVED SEVERAL TIMES

Most animal species with well-developed circulatory systems and nerve cells that generate and conduct action potentials have an endocrine system, or the components of an endocrine system. Endocrine systems are found in crustaceans, arthropods, and vertebrates. Although there are similarities in the organization and function of endocrine systems among these groups, most of these similarities are examples of convergent evolution. Endocrine systems offer distinct adaptive advantages and have evolved several times. In vertebrates, the endocrine system is composed of more than a dozen glands such as the pituitary, thyroid, and adrenal glands, as well as diffuse groups of cells scattered throughout epithelial tissues, such as the lining of the digestive

**TABLE 26.1 COMPARISON OF THE NERVOUS SYSTEM AND THE ENDOCRINE SYSTEM**

| PROPERTY | NERVOUS SYSTEM | ENDOCRINE SYSTEM |
|---|---|---|
| Anatomical arrangement | A "wired" system; specific structural arrangement between neurons and their target cells; structural continuity in the system | A "wireless" system; endocrine organs widely dispersed and not structurally related to one another or to their target cells |
| Type of chemical messenger | Neurotransmitters released into synaptic cleft | Hormones released into blood |
| Distance of action of chemical messenger | Very short distance (diffuses across synaptic cleft) | Long distance (carried by blood) |
| Means of specificity of action on target cells | Dependent on close anatomical relationship between nerve cells and their target cells | Dependent on specificity of target cell binding and responsiveness to a particular hormone |
| Speed of response | Rapid (milliseconds) | Slow (minutes to hours) |
| Duration of action | Brief (milliseconds) | Long (minutes to days or longer) |
| Major functions | Coordinates rapid, precise responses | Controls activities that require long duration rather than speed |
| Influence on other major control system? | Yes | Yes |

system. These glands secrete more than 50 different hormones.

Endocrine glands arise during development from each of the primary layers: ectoderm, mesoderm and endoderm. Some glands are formed from one layer, while others have a dual origin. The chemical nature of the hormone secreted by the gland is related to the embryological origin of the gland. Glands of ectodermal and endodermal origin secrete hormones made from amino acids (peptide and amine hormones), whereas glands of mesodermal origin secrete hormones based on lipids, including steroid hormones.

## ENDOCRINE SYSTEMS USE FEEDBACK SYSTEMS AND CYCLES FOR REGULATION

The nervous system and the endocrine system are the primary homeostatic mechanisms in the vertebrate body. The endocrine system uses two principles to maintain homeostasis and to regulate physiological functions: **negative feedback** and **cycles.**

Negative feedback regulates the secretion of almost every hormone in the endocrine system (see Chapter 18 for a discussion of negative feedback).

To review briefly this principle, consider the secretion of thyroid-stimulating hormone. Thyroid-stimulating hormone (TSH) is secreted by the anterior pituitary gland. TSH causes the thyroid gland to release thyroid hormone. Increased levels of thyroid hormone in the blood slow and eventually inhibit further release of TSH. In effect, thyroid hormone is inhibiting further secretion of more thyroid hormone (✦ Figure 26.2).

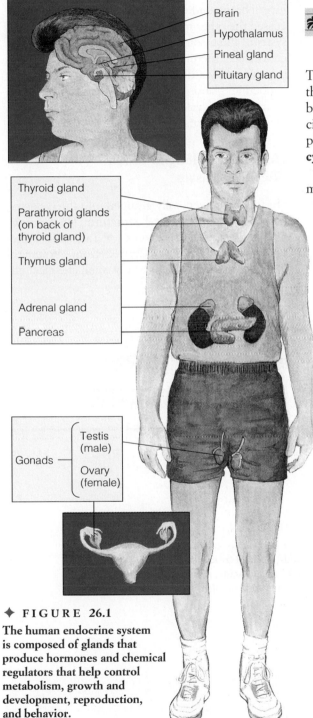

**✦ FIGURE 26.1**

**The human endocrine system is composed of glands that produce hormones and chemical regulators that help control metabolism, growth and development, reproduction, and behavior.**

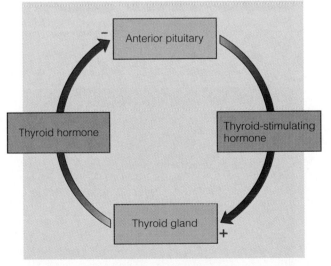

**✦ FIGURE 26.2**

**Negative control governs secretion of thyroid-stimulating hormone.**

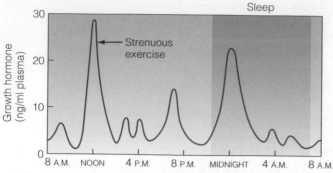

ng/ml = nanograms per mililiter

✦ FIGURE 26.3

**Growth hormone is released at night in a circadian rhythm. During the day, release is affected by exercise.**

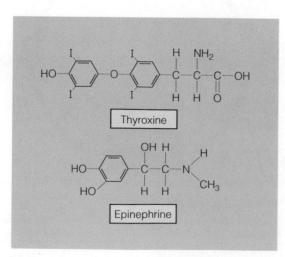

(a)

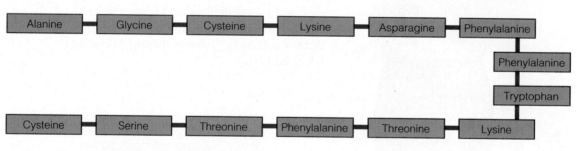

Somatostatin, a growth hormone

(b)

(c)

The endocrine system maintains homeostasis and controls physiological processes through cycles of secretion (✦ Figure 26.3). These cycles can have durations of hours, days, weeks, or months. Long-range control of migration, hibernation, mating, and reproduction are achieved by cyclic secretion of hormones, often in response to environmental cues.

In this chapter, we examine the endocrine system in its role as one of the major control systems in invertebrates and vertebrates. (We have already discussed the other major control system, the nervous system.) First, some of the primary mechanisms of endocrine structure and function are outlined. This is followed by a discussion of some of the primary endocrine glands, the disorders associated with them, and the cycles of control that are characteristic of the endocrine system.

## Hormones Belong to Three Chemical Classes

Hormones are generally grouped into three classes, based on their chemical structure: (1) **steroids,** (2) **peptides,** and (3) **amines** (➥ Table 26.2). Steroids are lipids and are derived from cholesterol (✦ Figure 26.4). There are only small chemical dif-

✦ FIGURE 26.4

**Three chemical classes of hormones include (a) steroids, derived from cholesterol, (b) peptides, made up of amino acids, and (c) amines, which are derivatives of amino acids.**

ferences between some steroid hormones. Even so, these small differences are responsible for important and distinctive effects. Testosterone is the hormone responsible for many male sex characteristics. A similar steroid, estradiol, is responsible for many female sex characteristics. Steroid hormones are secreted by the gonads, part of the adrenal gland, and the placenta.

Peptides and amines are both derived from amino acids. Peptides are short chains of amino acids, and most hormones are peptide hormones (Table 26.2), including those secreted by the pituitary, parathyroid, heart, stomach and small intestine, pancreas, liver, and kidneys.

Amines are derived from the amino acid tyrosine and are secreted by the thyroid and the portion of the adrenal gland known as the adrenal medulla.

The manner in which a hormone is synthesized, stored, secreted, and transported and the mechanism by which it exerts it effects are all related to its chemical structure. The chemical property that is most important in defining the hormone is its solubility inside and outside the cell. Steroid hormones and amines of the thyroid are lipid soluble, but highly insoluble in water. Peptides and the amines of the adrenal medulla are water soluble, but insoluble in lipids.

## How Hormones Are Synthesized, Stored, and Secreted

Steroid hormones are derived from cholesterol by a series of biochemical reactions. Defects in the synthesis of steroid hormones often lead to hormone imbalances that have serious consequences.

Once synthesized, steroid hormones move through the lipid bilayer of the cell membrane and enter the bloodstream. Steroid hormones are not

| | | AMINES | | |
|---|---|---|---|---|
| PROPERTIES | PEPTIDES | Catecholamines | Thyroid Hormone | STEROIDS |
| Structure | Chains of specific amino acids, for example: (vasopressin) | Tyrosine derivative, for example: (epinephrine) | Iodinated tyrosine derivative, for example: (thyroxine, $T_4$) | Cholesterol derivative, for example: (cortisol) |
| Synthesis | In rough endoplasmic reticulum; packaged in Golgi complex | In cytosol | In colloid, an inland extracellular site | Stepwise modification of cholesterol molecule in various intracellular compartments |
| Storage | Large amounts in secretory granules | In chromaffin granules | In colloid | Not stored; cholesterol precursor stored in lipid droplets |
| Secretion | Exocytosis of granules | Exocytosis of granules | Endocytosis of colloid | Simple diffusion |
| Transport in blood | As free hormone | Half bound to plasma proteins | Mostly bound to plasma proteins | Mostly bound to plasma proteins |
| Receptor site | Surface of target cell | Surface of target cell | Inside target cell | Inside target cell |
| Mechanism of action | Channel changes or activation of second-messenger system to alter activity of preexisting proteins that produce the effect | Activation of second-messenger system to alter activity of preexisting proteins that produce the effect | Activation of specific genes to produce new proteins that produce the effect | Activation of specific genes to produce new proteins that produce the effect |
| Hormones of this type | All hormones from the hypothalamus, anterior pituitary, posterior pituitary, pancreas, parathyroid gland, gastrointestinal tract, kidneys, liver, thyroid C cells, heart | Only hormones from the adrenal medulla | Only hormones from the thyroid follicular cells | Hormones from the adrenal cortex and gonads plus most placental hormones (vitamin D is steroidlike) |

**TABLE 26.2 CHEMICAL CLASSIFICATION OF HORMONES**

stored in the cells that produce them, and the blood levels of steroid hormones are controlled directly by their rate of synthesis.

Peptide hormones are synthesized as precursor molecules (called prohormones) from mRNA. The precursors pass into the interior of the endoplasmic reticulum, and are transported to the Golgi complex. In the Golgi, the precursors are packaged and stored in secretory granules. When needed, the hormones are released into the bloodstream by the fusion of the secretory granule membrane with the plasma membrane, a process that is often controlled by other hormones.

Prohormones are converted to active hormones by enzymatic cleavage. In some cases, different hormones can be released from the same precursor by different processing steps.

The major amine hormone of the adrenal gland is epinephrine (originally called adrenaline). It is synthesized from tyrosine and stored within the cell in granules. When the cells are neurally stimulated, the resulting depolarization of the plasma membrane causes the granules to fuse with the cell membrane, releasing the contents into the extracellular fluid.

## Hormones Act on Target Cells by Two Mechanisms

As noted earlier, the endocrine system acts through the release of hormones, which in turn trigger responses in specific target cells (✦ Figure 26.5). The response of target cells depends on receptors located on the cell surface, embedded in the plasma membrane and in the cytoplasm. Receptors bind only to one type of hormone. The presence of specific receptors defines the ability of a cell to respond to a given hormone.

More than 50 human hormones have been identified. All act by binding to receptor molecules. Binding of the hormone to the receptor causes a change in the shape of the receptor molecule. Receptors with altered shapes initiate a chain of events that generate a response to the hormone. These events may include changes in membrane channel

**Endocrine gland**

**Hormone**

**Binding with receptor**

**(Target cell)**

Intracellular events:

1. Alters channel permeability by acting on preexisting channel-forming proteins
   or
2. Acts through second-messenger system to alter activity of preexisting proteins
   or
3. Activates specific genes to cause formation of new proteins

**Physiological response**

✦ **FIGURE 26.5**

**Hormones work by binding to receptors. Responses to hormones can be generated in three ways.**

✦ **FIGURE 26.6**

**Hormone action by a second messenger. Peptide hormones bind to a receptor on the cell membrane, activating adenylate cyclase, which converts ATP into cyclic AMP. Cyclic AMP activates enzymes called protein kinases, which add phosphate groups to intracellular enzymes. Adding phosphate groups activates some enzymes and inactivates others, bringing about a response to the presence of the hormone. The response is turned off by converting cyclic AMP into AMP.**

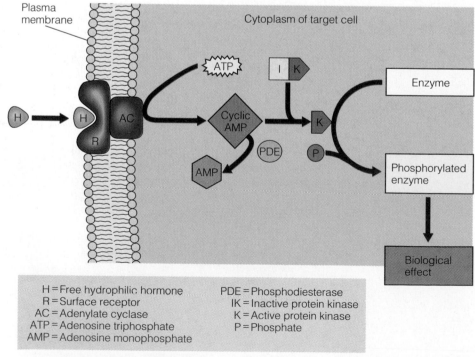

H = Free hydrophilic hormone
R = Surface receptor
AC = Adenylate cyclase
ATP = Adenosine triphosphate
AMP = Adenosine monophosphate

PDE = Phosphodiesterase
IK = Inactive protein kinase
K = Active protein kinase
P = Phosphate

proteins, start-up or shutdown of enzyme action, and/or changes in gene expression. Over time, these changes can have significant effects on homeostasis, growth, development, and behavior. There are two general mechanisms by which all hormones act on target cells. The first involves the action of nonsteroid hormones.

Nonsteroid hormones do not enter the cell. They bind to plasma membrane receptors and exert their effect by generating a chemical signal inside the target cell. The signal is known as a **second messenger** (✦ Figure 26.6). Five different molecules that serve as second messengers have been identified: cyclic AMP, cyclic GMP, inositol triphosphate, diacrylglycerol, and calcium. Second messengers activate other intracellular molecules to produce a response in the target cell (✦ Figure 26.7). Because there are many hormones that bind to plasma membrane receptors, the action of several different receptors can generate the same second messenger.

The second general mechanism of hormone action involves steroid hormones. Steroid hormones pass through the plasma membrane and act in a two-step process (✦ Figure 26.8). Inside the cell, steroid hormones move to the nucleus where they bind to receptor molecules. The binding of a steroid hormone to its receptor changes the shape of the receptor, producing an activated hormone-receptor complex. The hormone-receptor complex binds to DNA and activates specific genes. The re-

sult is an increase in transcription and mRNA production, and the subsequent production of new protein gene products.

Problems in the endocrine system can be grouped into three general categories: (1) overproduction of a hormone, (2) underproduction of a hormone, and (3) nonfunctional receptors that cause target cells to become insensitive to hormones. We discuss examples of each of these conditions in the following section.

✦ **FIGURE 26.7**

**The chemical structure of (a) cyclic AMP compared to (b) ATP.**

✦ **FIGURE 26.8**

**Hormone action by steroid hormones. Steroid hormones enter the cell by passing through the plasma membrane and binding to receptors in the cytoplasm. The receptor-hormone complex moves to the nucleus and binds to acceptor sites on the DNA. This activates adjacent genes, causing transcription and translation of new gene products that generate a response to the presence of the hormone.**

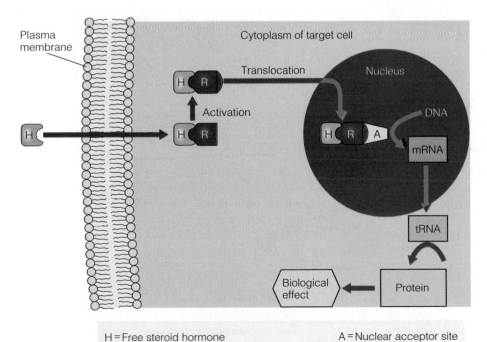

H = Free steroid hormone
R = Cytoplasmic receptor
A = Nuclear acceptor site
mRNA = Messenger RNA

## THE HYPOTHALAMUS-PITUITARY AXIS CONNECTS THE NERVOUS AND ENDOCRINE SYSTEMS

The **pituitary gland,** located in a small bony cavity at the base of the brain, is attached to the brain by a thin stalk (◆ Figure 26.9). The stalk contains nerves and blood vessels that link the pituitary to the hypothalamus. The pituitary gland has two lobes: an anterior and a posterior lobe. The anterior lobe is composed of glandular tissue, and is connected to the hypothalamus by a common set of blood vessels. The posterior lobe is made up of nervous tissue. It is connected to the hypothalamus by neural connections. The release of hormones by the anterior and posterior lobes of the pituitary is controlled by the hypothalamus. The nervous system, acting through the hypothalamus, regulates the pituitary, often called the master gland of the endocrine system. The pituitary gland secretes hormones that regulate organ function and the internal environment.

| TABLE 26.3 MAJOR HORMONES OF THE HYPOTHALAMUS | |
|---|---|
| HORMONE | EFFECT ON ANTERIOR PITUITARY |
| Thyrotropin-releasing hormone (TRH) | Stimulates release of TSH (thyrotropin) and prolactin |
| Corticotropin-releasing hormone (CRH) | Stimulates release of ACTH (corticotropin) |
| Gonadotropin-releasing hormone (GnRH) | Stimulates release of FSH and LH (gonadotropins) |
| Growth hormone-releasing hormone (GHRH) | Stimulates release of growth hormone |
| Growth hormone-inhibiting hormone (GHIH) | Inhibits release of growth hormone and TSH |
| Prolactin-releasing hormone (PRH) | Stimulates release of prolactin |
| Prolactin-inhibiting hormone (PIH) | Inhibits release of prolactin |

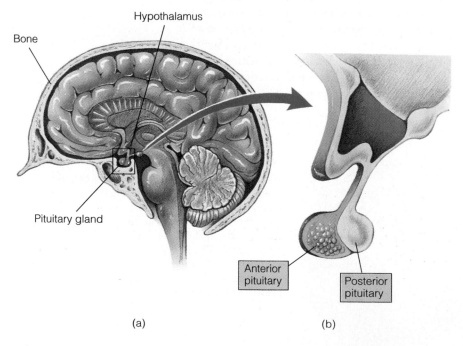

(a)

(b)

◆ **FIGURE 26.9**

**(a) A section through the brain, showing the location of the pituitary gland.**
**(b) An enlargement, showing the two lobes of the pituitary and their connection to the brain.**

### Anterior Pituitary Hormone Release Is Controlled by Neurosecretion

The hypothalamus contains neurons that synthesize, store, and release hormones that act on the anterior pituitary (◆ Figure 26.10). At least seven different hypothalamic hormones that affect the anterior pituitary have been identified (◆ Table 26.3). These hormones are released into a capillary system in the hypothalamus that connects to veins in the pituitary stalk. These veins open into a second capillary network, located in the anterior lobe (Figure 26.10). This arrangement, in which capillaries drain into a vein that opens into another capillary network, is called a **portal system.** The other main portal system in the body is in the liver.

Hormones from the hypothalamus are delivered directly to target cells in the anterior pituitary gland. In response, the anterior pituitary gland secretes a number of hormones (◆ Table 26.4).

### Growth Hormone Stimulates Bone Growth

**Growth hormone (GH)** is a peptide hormone from the anterior pituitary that is essential for growth. Two hypothalamic hormones regulate the serum concentration of GH. GH-releasing hormone stimulates release of growth hormone, and GH-inhibit-

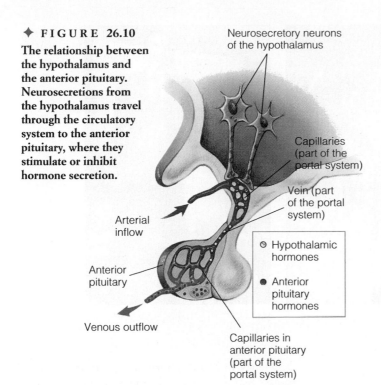

**✦ FIGURE 26.10**

The relationship between the hypothalamus and the anterior pituitary. Neurosecretions from the hypothalamus travel through the circulatory system to the anterior pituitary, where they stimulate or inhibit hormone secretion.

Neurosecretory neurons of the hypothalamus

Capillaries (part of the portal system)

Vein (part of the portal system)

Arterial inflow

Anterior pituitary

Venous outflow

⊙ Hypothalamic hormones

● Anterior pituitary hormones

Capillaries in anterior pituitary (part of the portal system)

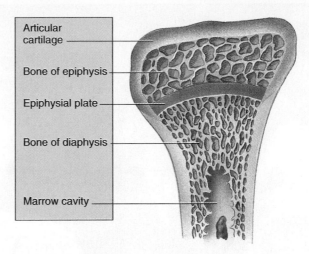

Articular cartilage

Bone of epiphysis

Epiphysial plate

Bone of diaphysis

Marrow cavity

**✦ FIGURE 26.11**

Growth in long bones takes place at the epiphyseal plate. Action of the growth hormone in late adolescence stops cartilage formation in these plates, and bone formation closes the plate, stopping growth.

ing hormone suppresses the release of GH. The hypothalamus contains receptors that maintain homeostatic levels of GH. Interestingly, GH levels show a daily cycle (discussed later).

Growth of tissues under the influence of GH involves cellular enlargement (hypertrophy) through increases in the rate of protein synthesis, and increases in cell number (hyperplasia) through stimulation of mitosis. GH causes an increase in bone length and bone thickness. Recall (from Chapter 19) that bone growth occurs in a band of cartilage near each end of a long bone. GH stimulates the formation of this cartilage, and stimulates osteoblasts, the cells that lay down the mineral portion of the bone. During childhood, new cartilage forms continuously in these plates, causing bone elongation and body growth. In late adolescence, sex hormones slow and then stop cartilage formation in the plates. Bone formation continues from each side, gradually eliminating the cartilage, and halting further bone growth, even though GH production continues (✦ Figure 26.11).

| | |
|---|---|
| ✄ | **TABLE 26.4   HORMONES SECRETED BY THE PITUITARY GLAND** ✄ |

| HORMONE | FUNCTION |
|---|---|
| **Anterior pituitary** | |
| Growth hormone (GH) | Stimulates cell growth. Primary targets are muscle and bone, where GH stimulates amino acid uptake and protein synthesis. It also stimulates fat breakdown in the body. |
| Thyroid-stimulating hormone (TSH) | Stimulates release of thyroxine and triiodothyronine. |
| Adrenocorticotropic hormone (ACTH) | Stimulates secretion of hormones by the adrenal cortex, especially glucocorticoids. |
| Gonadotropins (FSH and LH) | Stimulate gamete production and hormone production by the gonads. |
| Prolactin | Stimulates milk production. |
| Melanocyte-stimulating hormone (MSH) | Function in humans is unknown. |
| **Posterior pituitary** | |
| Antidiuretic hormone (ADH) | Stimulates water reabsorption by nephrons of the kidney. |
| Oxytocin | Stimulates ejection of milk from breasts and uterine contractions during birth. |

(a)

(b)

✦ FIGURE 26.12
Pituitary disorders include (a) dwarfism and (b) gigantism.

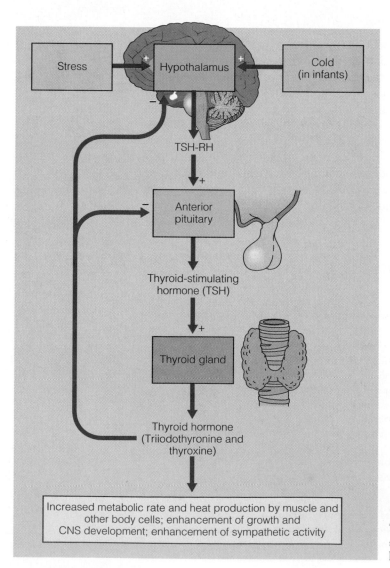

Deficiency in growth hormone production can be caused by malfunctions in the hypothalamus or a defect in the anterior pituitary. The result is a form of dwarfism associated with deficient skeletal growth (✦ Figure 26.12a). Excesses of growth hormone secretion before the closure of the epiphyseal plates causes gigantism, and affected individuals may grow to more than eight feet tall (Figure 26.12b).

## Thyroid-Stimulating Hormone Regulates the Thyroid

The hypothalamus contains receptors that monitor serum levels of thyroid hormones, and when levels are low, TSH-releasing hormone is secreted into the portal system. In the anterior pituitary, this stimulates the release of **thyroid-stimulating hormone (TSH).** TSH travels through the circulatory system to the thyroid, and triggers the release of thyroid hormones. Thyroid hormones regulate metabolic rates and body temperature (✦ Figure 26.13).

## The Anterior Pituitary Secretes Hormones Involved in Reproduction

Other hormones of the anterior pituitary include **gonadotropins** and **prolactin.** Gonadotropins, including **follicle-stimulating hormone (FSH)** and

✦ FIGURE 26.13
The sequence of events in the secretion and regulation of thyroid hormone production.

**luteinizing hormone (LH),** affect the testis and ovary. These hormones stimulate gamete formation and steroid hormone production by the gonads. In mammals, prolactin is secreted by the anterior pituitary at the end of pregnancy and activates milk production by the mammary glands. The anterior pituitary also secretes **adrenocorticotropic hormone (ACTH).** ACTH affects the adrenal glands, in particular, the adrenal cortex. Its actions are discussed later.

## The Posterior Pituitary Is an Extension of the Brain

The posterior pituitary is an extension of the brain. It is composed of neurosecretory cells that store and release hormones into the blood (✦ Figure 26.14). The two hormones of the posterior pituitary, **antidiuretic hormone (ADH)** and **oxytocin,** are actually produced in the hypothalamus, where they are packaged into secretory vesicles and transported to axon endings in the posterior pituitary (Figure 26.14).

ADH, also known as **vasopressin,** controls water balance in the body. ADH acts on cells in the distal tubule of the kidney to control water balance by increasing water absorption in the convoluted distal tubule (see Chapter 24). ADH also plays a role in controlling blood pressure by causing constriction of smooth muscle in arterioles.

Oxytocin is a small peptide hormone (nine amino acids long) secreted by the posterior pituitary. It causes contractions of the smooth muscles in the uterus during childbirth. The response of the uterus to oxytocin in pregnant women is controlled by the number of oxytocin receptors present in uterine cells. When the number of uterine receptors reaches a critical threshold, it allows a response to normal levels of oxytocin. The number of uterine receptors, therefore, plays a role in stimulating labor contractions.

## OTHER ENDOCRINE ORGANS HAVE SPECIALIZED FUNCTIONS

### The Adrenal Glands Regulate Responses to Stress

There are two adrenal glands, one located above each kidney, embedded in fat. Each adrenal gland has two components, an inner **medulla,** and an outer **cortex** (✦ Figure 26.15). The medulla secretes amine hormones, and the cortex synthesizes and secretes a number of steroid hormones.

*Sidebar*

**PREMENSTRUAL SYNDROME**
Premenstrual syndrome or PMS plagues many women world-wide. Researchers have recently identified a potential cause for many women's severe premenstrual discomfort. An allergic reaction to the female hormone, progesterone, may be the monthly culprit.

Researchers studied women who experience symptoms such as hives, breathing difficulties, vomiting, diarrhea, bloating, faintness, low blood pressure, shock and flushing each month before the onset of menstruation. They reproduced these monthly symptoms in women by giving them doses of luteinizing hormone-releasing hormone (LHRH), which stimulates progesterone production. When women were given an agent that prevents LHRH release, progesterone secretion was blocked and no symptoms appeared.

Allergists frequently observe an increase in severity in allergic symptoms at premenstrual times, but until this discovery there was no explanation for why this might occur.

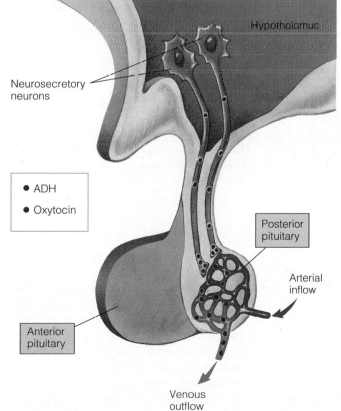

Hypothalamus

Neurosecretory neurons

● ADH
● Oxytocin

Posterior pituitary

Arterial inflow

Anterior pituitary

Venous outflow

✦ **FIGURE 26.14**

The posterior pituitary is in direct contact with neurons from the hypothalamus that synthesize oxytocin and ADH. Hormones from these cells are released into the bloodstream in the posterior pituitary.

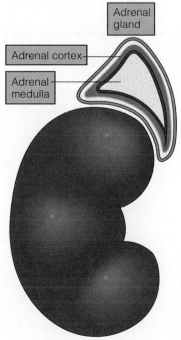

Adrenal gland

Adrenal cortex

Adrenal medulla

✦ **FIGURE 26.15**

The adrenal glands are embedded in fat on top of the kidneys. Each gland consists of an outer layer of cells, the cortex, and an inner medulla. The cortex produces steroid hormones, and the medulla produces adrenaline and noradrenaline.

The adrenal medulla is made up of modified neurons, and is regarded as an extension of the sympathetic nervous system. Unlike axons of the sympathetic system, which reach a target organ, secretory axons of the adrenal medulla release their contents into the blood. The adrenal medulla synthesizes two hormones, **epinephrine** (adrenaline) and **norepinephrine.** The hormones are stored in secretory granules and released into the bloodstream when the adrenal medulla cells are stimulated by the sympathetic system. They produce a rapid, short-term response to stress, including the "fight or flight" response (◆ Table 26.5). The circulating hormones reach cells and tissues not innervated by the sympathetic system, and stimulate the breakdown of carbohydrates and fats for use in metabolic reactions during periods of stress-related physical activity.

The adrenal cortex makes up about 80% of the adrenal gland and produces several steroid hormones. Three functional classes of adrenal steroids are produced: **mineralocorticoids, glucocorticoids,** and **sex hormones.** The major role of mineralocorticoids is in maintainence of electrolyte balance, through action on the distal tubules of the kidneys.

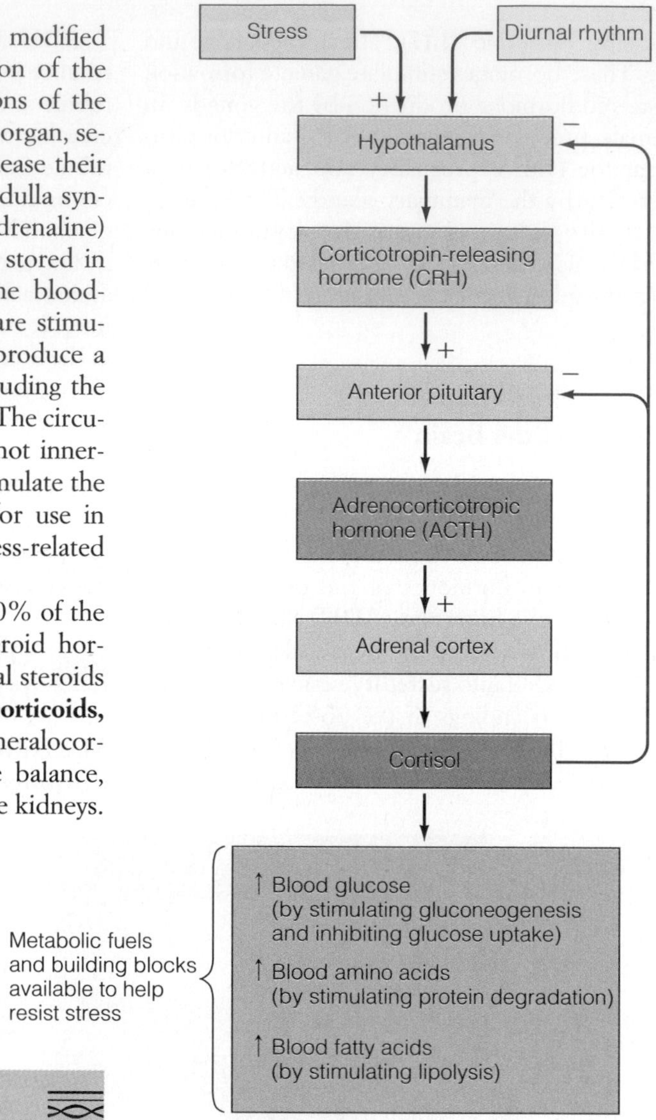

◆ **FIGURE 26.16**

**The sequence of events in the production and control of cortisol secretion.**

The glucocorticoids produce a slow, long term response to stress (◆ Figure 26.16). They are released in response to hormonal stimulation. The hypothalamus secretes a hormone that causes the anterior pituitary to release the hormone ACTH. The adrenal cortex responds to the presence of ACTH by releasing several hormones including **cortisol,** the primary glucocorticoid.

Glucocorticoids are important in regulating the metabolism of carbohydrates, fats, and proteins. The net effect of their action is to increase blood levels of glucose through the breakdown of fats and amino acids. The released glucose is used as an energy source during periods of stress. Glucocorticoids are used as drugs to inhibit most of the steps

| TABLE 26.5 MAJOR HORMONAL CHANGES DURING THE STRESS RESPONSE | | |
|---|---|---|
| HORMONE | CHANGE | PURPOSE SERVED |
| Epinephrine | ↑ | Reinforces sympathetic nervous system to prepare the body for "fight or flight" |
| | | Mobilizes carbohydrate and fat energy stores; increases blood glucose and blood fatty acids |
| CRH-ACTH-cortisol | ↑ | Mobilizes energy stores and metabolic building blocks for use as needed; increases blood glucose, blood amino acids, and blood fatty acids |
| | | ACTH facilitates learning and behavior |
| Renin-angiotensin-aldosterone | ↑ | Conserve salt and $H_2O$ to expand plasma volume; help sustain blood pressure when acute loss of plasma volume occurs |
| Vasopressin | ↑ | Angiotensin II and vasopressin cause arteriolar vasoconstriction to increase blood pressure |
| | | Vasopressin facilitates learning |

in the inflammatory reaction (described in Chapter 22). This makes them valuable in the treatment of conditions such as arthritis, where the inflammatory response becomes destructive. Glucocorticoids also suppress the immune response (Chapter 22), and are used in treating some allergic reactions and in suppressing the rejection of organ transplants.

The adrenal cortex also produces steroid sex hormones, but the amounts produced are insignificant compared to those produced by the ovary or the testis.

## The Thyroid Gland Helps Regulate Metabolism and Growth

The thyroid gland lies over the trachea just below the larynx, and is composed of a collection of spherical structures called **follicles** (✦ Figure 26.17). These follicles contain a gel-like colloid, surrounded by a single layer of cells. Thyroglobulin, a storage form of thyroid hormone, is secreted by the follicle cells into the colloid, where it is stored until needed.

Thyroid-stimulating hormone (TSH) produced in the anterior pituitary causes the conversion of thyroglobulin into the thyroid hormones $T_4$ and $T_3$ (✦ Figure 26.18). These amine hormones diffuse through the cell membrane into the blood.

Almost all tissues in the body are targets of thyroid hormone, and respond to hormone stimulation in several ways. First, thyroid hormone increases the overall metabolic rate, which causes increased heat production, raising the body's temperature. Perhaps the most important long-range effect of thyroid hormone is on growth and development. In children, thyroid hormone is essential for growth of bone and muscle, and for development of the central nervous system. Normal levels of thyroid hormone are also necessary for the onset of sexual maturation.

Reduced levels of thyroid hormone in children cause retardation of physical growth and mental development. This condition is reversible by treatment with thyroid hormone. In adults, low levels of thyroid hormone slow down metabolism, producing fatigue, chills, lowered heart rate, and slowed neural reflexes.

The thyroid also secretes **calcitonin** from large cells interspersed in the follicular layer. This hormone, as its name implies, plays a role in regulating levels of calcium. As serum levels of calcium rise, calcitonin secretion increases. Calcitonin increases calcium excretion by the kidneys. As calcium levels in the serum fall, the release of calcitonin is inhibited, keeping calcium levels within a homeostatic range.

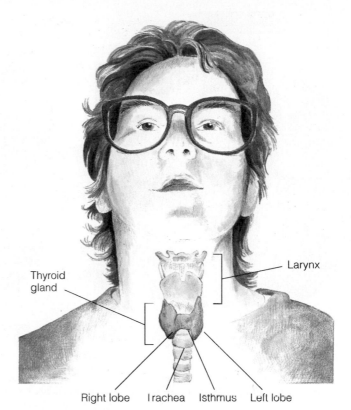

✦ **FIGURE 26.17**
**The thyroid gland is located in the neck. The gland is composed of follicles that produce thyroxine and triiodothyronine.**

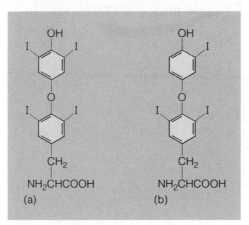

✦ **FIGURE 26.18**
**Chemical structures of the thyroid hormones.**
**(a) Thyroxine ($T_4$). (b) Triiodothyronine ($T_3$).**

## The Pancreas Regulates Glucose Metabolism

The pancreas is composed of exocrine cells that secrete digestive enzymes into a duct system connected to the small intestine, and clusters of en-

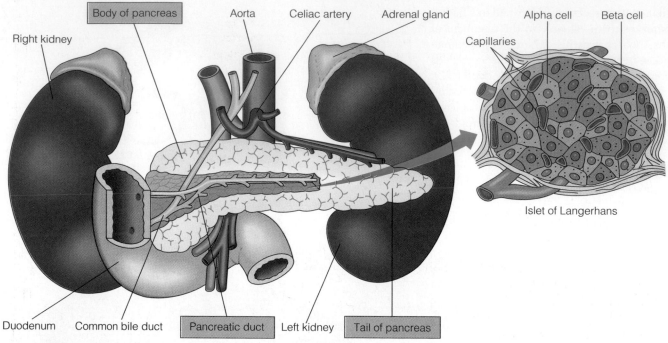

**Body of pancreas** · Aorta · Celiac artery · Adrenal gland · Alpha cell · Beta cell

Right kidney · Capillaries

Islet of Langerhans

Duodenum · Common bile duct · **Pancreatic duct** · Left kidney · **Tail of pancreas**

✦ **FIGURE 26.19**

**The pancreas produces digestive enzymes, which are released into the small intestine via the pancreatic duct. Embedded in the pancreas are clusters of cells, the islets of Langerhans, that produce insulin and glucagon. These hormones are released into the blood.**

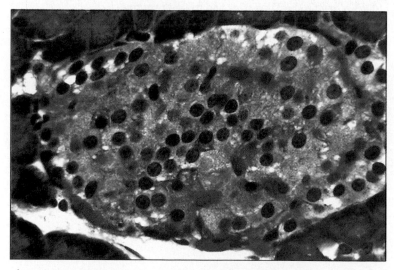

✦ **FIGURE 26.20**
**An islet of Langerhans.**

docrine cells, called **pancreatic islets** ( ✦ Figure 26.19). The adult pancreas has about one to two million islets that secrete hormones including **insulin** and **glucagon** ( ✦ Figure 26.20). These hormones interact to regulate blood glucose levels.

Glucose levels in the blood remain fairly constant between meals. After eating, glucose levels rise, and excess glucose must be removed from the blood and stored. Stored glucose is released between meals in a controlled fashion to keep blood glucose levels within a homeostatic range. This metabolic task is complicated, and involves many aspects of protein and fat metabolism as well.

After a meal, rising blood glucose levels trigger the release of insulin. Insulin enhances glucose uptake by cells, helping to lower blood glucose levels. Insulin also stimulates the formation of glycogen, the storage form of glucose in cells of the liver and skeletal muscle ( ✦ Figure 26.21).

Insulin secretion is regulated by the concentration of glucose in the bloodstream. High levels of glucose cause the synthesis and release of insulin by cells of the pancreatic islets. As glucose levels are lowered by the action of insulin, the synthesis and secretion of insulin is inhibited.

Glucagon is a peptide hormone secreted by islet cells that also regulates blood glucose levels. The combined action of insulin and glucagon maintains blood glucose levels within a homeostatic range. The synthesis and release of glucagon is directly linked to blood glucose levels in a pattern opposite to that of insulin. Glucagon synthesis and secretion is stimulated when blood glucose levels fall, leading to an increase in glucose production. When blood glucose levels rise, glucagon secretion is inhibited.

**Diabetes** is a disorder resulting from inadequate insulin levels or action. It is the most common endocrine disorder. **Diabetes mellitus** is associated with defects in insulin action, and results in elevated levels of glucose in the blood and urine. There are two forms of diabetes mellitus. **Type I** (insulin-dependent or juvenile-onset) diabetes is characterized by inadequate levels of insulin secretion. This form of diabetes accounts for 10 to 20% of all cases. In many cases, there is a genetic predisposition for Type I diabetes.

**Type II diabetes** (noninsulin-dependent or adult-onset) diabetes usually develops in adults, and both genetic and environmental factors play a major role in this disorder. In Type II diabetes, insulin secretion can be normal or even elevated. The defect in this form of diabetes involves an impaired insulin secretion in response to elevated levels of blood glucose, or loss of response to insulin by target cells. Major symptoms include elevated glucose levels in blood and urine (◆ Table 26.6).

Diabetes mellitus causes disturbances in carbohydrate, fat, and protein metabolism. Over time, these metabolic disturbances cause serious, life-threatening problems in several organ systems, including the circulatory system, the nervous system, the respiratory system, the eyes, and the kidneys. As a result, diabetics have a shorter life expectancy. Complications include degenerative changes in the circulatory and nervous systems, leading to loss of feeling in the extremities, and gangrene, caused by impaired circulation. Amputation of toes, feet, and limbs is often necessary. Other degenerative changes in the nervous system involve the brain, spinal cord, and peripheral nerves. Kidney failure is a long-term compliction, and retinal changes make blindness from diabetes the second leading cause of blindness in the United States.

Treatment for diabetes involves daily injections of insulin (for Type I), a controlled diet, and exercise to help regulate blood glucose levels. The symptoms in Type II diabetes are usually milder and have a slower onset. Control in Type II diabetes can often be accomplished by diet and weight reduction, with meals at regular intervals to maintain blood glucose levels within the normal range. In some cases, this form of diabetes can be controlled by drugs that increase insulin secretion, and lower blood glucose levels.

## OTHER CHEMICAL MESSENGERS SHARE CHARACTERISTICS OF HORMONES

Other types of chemical messengers are also produced in mammals, but are not the product of endocrine glands. Since these messengers share several characteristics of hormones, they are discussed briefly in the section below.

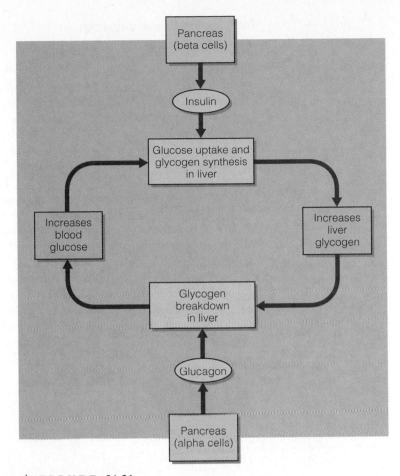

◆ **FIGURE 26.21**
The liver and the pancreas regulate blood glucose levels.

| TABLE 26.6 COMPARISON OF TYPE I AND TYPE II DIABETES MELLITUS | | |
|---|---|---|
| CHARACTERISTIC | TYPE I DIABETES | TYPE II DIABETES |
| Level of insulin secretion | None or almost none | May be normal or exceed normal |
| Typical age of onset | Childhood | Adulthood |
| Percentage of diabetics | 10–20% | 80–90% |
| Basic defect | Destruction of $\beta$ cells | Reduced sensitivity of insulin's target cells |
| Associated with obesity? | No | Usually |
| Genetic and environmental factors important in precipitating overt disease? | Yes | Yes |
| Speed of development of symptoms | Rapid | Slow |
| Development of ketosis | Common if untreated | Rare |
| Treatment | Insulin injections; dietary management | Dietary control and weight reduction; occasionally oral hypoglycemic drugs |

## Interferons Protect against Viral Infections

**Interferons** are proteins released by cells in response to viral infection. These proteins move through the extracellular space and circulatory system to neighboring cells and distant target cells to signal the presence of a viral infection. Interferons activate the synthesis and secretion of antiviral proteins.

The antiviral proteins secreted by cells in response to interferon stimulation are enzymes that remain dormant until a cell is infected by a virus. Once infected, the cells respond by activating the enzymes to destroy the viruses (✦ Figure 26.22). Interferons also inhibit the growth of certain types of cancer cells, and they enhance the activity of killer cells of the immune system that attack cancer cells. Several types of interferon are now commercially produced by genetic engineering, and are being used in clinical trials as anticancer drugs.

## Prostaglandins Are Local-Acting Signals

**Prostaglandins** belong to a class of fatty acids that have many of the properties of hormones. They are considered to be locally active hormones, even though they are not produced by endocrine glands. Prostaglandins are synthesized and secreted by most tissues of the body, and act on nearby cells. The actions of prostaglandins are diverse and sometimes overlapping (✦ Table 26.7). Prostaglandins are used in clinical settings to treat asthma and ulcers and in the induction of labor.

## Pheromones Act as Signals between Organisms

**Pheromones** are chemical signals that travel between organisms rather than between cells within an organism. Pheromones are present in a wide range of organisms, from yeasts to insects and mammals. Pheromones in yeasts and insects are dispersed into the environment to signal the presence of potential mates. In social insects such as bees and ants, pheromones are used as a means of communication and as a way of maintaining the social structure of the colony. In mammals, pheromones dispersed in urine or feces or exuded from special glands are used to mark territory (✦ Figure 26.23). The common sight of a male dog urinating on poles, trees, and bushes is an example of marking a territory by depositing a pheromone contained in urine. Whether there are human sex attractant pheromones is an unresolved question.

## THE ENDOCRINE SYSTEM REGULATES MANY BIOLOGICAL CYCLES

Animals ranging from invertebrates through mammals exhibit biological cycles. These cycles can extend over minutes or years. Cycles involve functions as diverse as mating and reproduction, hibernation, migration, changes in body temperature, blood glucose content, hormone secretion, and many other physiological processes. These cycles are regulated by the endocrine system through the hypothalamus.

## Circadian Rhythms Are Daily Cycles

Although hormone levels are maintained within a homeostatic range, there is often a regular cycle of fluctuation that is a function of time (✦ Figure 26.24). Hormone release can show cyclic changes every few hours, or on a daily basis. Rhythms or cycles that show fluctuations about every 24 hours are called **circadian rhythms.** Many hormones

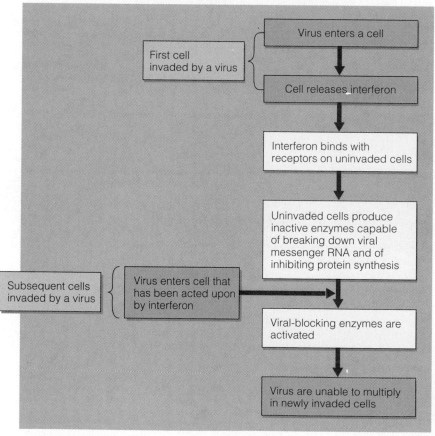

✦ **F I G U R E  26.22**

**Interferon is released from virally infected cells and binds to uninfected cells. This binding triggers the release of intracellular enzymes that block viral replication. If the cell is invaded by a virus, the enzymes are activated, and the invading virus is unable to replicate.**

including the ACTH-cortisol system, thyroid-stimulating hormone, and growth hormone show circadian rhythms.

In most cases, when the animal is in its normal environment, circadian rhythms have a cycle that averages about 24 hours. If the animal is kept in constant darkness at constant temperature, the cycle is maintained, but the length may vary a few hours in either direction.

## Many Biological Processes Are Regulated by Rhythms

The circadian rhythm is not the only endocrine cycle observed in vertebrates. Other cycles are much shorter than a day, and others are longer. The monthly human menstrual cycle is controlled by a number of hormones secreted in a cyclical fashion. In humans, thyroid hormone secretion is on an annual cycle, with secretion higher in winter than in summer. Childbirth, a hormonally mediated process, is highest between 2 A.M. and 7 A.M., and lowest between 2 P.M. and 8 P.M. Interestingly, the death rate is highest between 2 A.M. and 7 A.M.

Other animals exhibit cycles tied to the 29.5-day lunar month (the time between new moons). Egg production in the Mediterranean sea urchin peaks at the time of the full moon in the summer. Some

✦ FIGURE 26.23
White-tail deer marking its territory with chemicals from a scent gland.

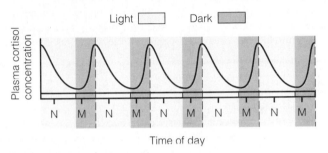

N = Noon
M = Midnight

✦ FIGURE 26.24
Cortisol is a hormone secreted on a diurnal rhythm, with peaks of release in the morning.

| TABLE 26.7 | KNOWN OR SUSPECTED ACTIONS OF PROSTAGLANDINS |
|---|---|
| BODY SYSTEM ACTIVITY | ACTIONS OF PROSTAGLANDINS |
| Reproductive system | Promote sperm transport by action on smooth muscle in male and female reproductive tracts |
| | Important in menstruation |
| | Play role in ovulation |
| | Contribute to preparation of maternal portion of placenta |
| | Contribute to birth process |
| Respiratory system | Some promote bronchodilation, others bronchoconstriction |
| Urinary system | Increase renal blood flow |
| | Increase excretion of water and salt |
| Digestive system | Inhibit HCl secretion by stomach |
| | Stimulate intestinal motility |
| Nervous system | Influence neurotransmitter release and action |
| | Act at hypothalamic "thermostat" to increase body temperature |
| Endocrine system | Enhance cortisol secretion |
| | Influence tissue responsiveness to hormones in many instances |
| Circulatory system | Influence platelet aggregation |
| Fat metabolism | Inhibit fat breakdown |
| Defense system | Promote many aspects of inflammation, including fever and development of pain |

# HIBERNATING BEARS AND HUMAN DISEASE

In the fall, bears retreat to their dens to begin a period of winter hibernation. During the four to five months they are in a deep sleep, bears do not eat or drink, nor do they urinate or defecate. After hibernation, they emerge with no signs of urea in their blood, a waste product of protein metabolism.

In humans, months of inactivity lead to dangerous bone thinning and loss of muscle mass. Six months of bed rest can result in a loss of about 25% of bone mass. Yet bears emerge from hibernation with their bones and muscles intact. In fact, bears actually gain lean muscle mass while in hibernation, whereas humans must use exercise and nutrition to increase muscle mass. Understanding how hibernating bears maintain bone structure may help in the development of treatments for osteoporosis, a bone thinning disorder of humans.

Bone is a dynamic material; it is constantly being formed by osteoblasts and resorbed by osteoclasts. During most of adult life, there is a balance between bone formation and breakdown. This balance is affected by exertion and inactivity. Regular exercise increases bone density, and inactivity results in bone loss. By age 60, bone resorption exceeds bone formation, resulting in a gradual reduction of bone mass known as osteoporosis.

Hibernating bears do not lose bone mass, and osteoblasts produce new bone matrix all through the winter, as if the bears were physically active. Bone growth and remodeling is hormonally controlled, and involves growth hormone and hormones regulating calcium levels in the blood. The work on bone growth suggests that hibernating bears produce a factor (perhaps a hormone) that promotes new bone growth during periods of inactivity. Isolation and purification of such a factor may provide a new method for treating osteoporosis, which causes more than 1 million bone fractures each year, mostly in elderly individuals. Because bone loss occurs in the weightlessness of space, a hormone that regulates bone growth could be used to treat astronauts, who lose 2% of their bone mass for every month in space.

---

mammals (bears, hamsters) have an annual cycle of hibernation that is tied to hormonal fluctuations. Events such as bird migrations are closely tied to reproduction and hormonal changes that precede mating behavior.

## Hypothalamic Mechanisms Control Cycles and Rhythms

Internal cycles of endocrine secretion are controlled by the hypothalamus, and in particular a region called the **suprachiasmic nucleus (SCN).** According to one model, external stimuli such as light reach the retina and generate signals to the SCN. Integration of sensory information in the SCN sets up the cycle and its timing. Nerve impulses from the SCN are sent to the **pineal gland,** located between the cerebral hemispheres. This gland secretes melatonin, which stimulates the hypothalamus to regulate hormone secretion to the pituitary gland. Secretions from the pituitary regulate the activities associated with the cycle.

---

# SUMMARY

1. The endocrine system maintains homeostasis and control over long-term events, and uses chemical signals known as hormones. The endocrine system works in parallel with the nervous system to maintain homeostasis and to regulate long-term processes such as growth and maturation.

2. Endocrine systems offer distinct adaptive advantages and have evolved several times. The vertebrate endocrine system is composed of glands including the pituitary, thyroid, and adrenal glands, as well as diffuse groups of cells scattered throughout epithelial tissues.

3. The endocrine system uses two principles to maintain homeostasis and to regulate physiological functions: negative feedback and cycles. Negative feedback regulates the secretion of almost every hormone in the endocrine system. The endocrine system controls physiological processes through cycles of secretion that can have durations of hours, days, weeks, or months.

4. The nervous system, acting through the hypothalamus, regulates the pituitary, often called the master gland of the endocrine system. The pituitary gland secretes hormones that regulate organ function and the internal environment.

5. Nonsteroid hormones do not enter their target cells. They act by binding to receptors at the cell surface and initiating the formation of a second messenger inside the cell. Steroid hormones enter the cell and interact with a receptor in the nucleus. The receptor/hormone complex binds to DNA, and initiates a program of gene action.

6. Animals ranging from invertebrates through mammals exhibit biological cycles. These cycles can extend over minutes or years. Cycles involve functions as diverse as mating and reproduction, hibernation, migration, changes in body temperature, blood glucose content, hormone secretion, and many other physiological processes. These cycles are regulated by the endocrine system through the hypothalamus.

adrenocorticotropic hormone (ACTH)
amines
antidiuretic hormone (ADH)
calcitonin
circadian rhythms
cortex
cortisol
cycles
diabetes mellitus, Type I and Type II
endocrine system
epinephrine
follicle-stimulating hormone
follicles
glucagon

glucocorticoids
gonadotropins
growth hormone
hormones
insulin
interferons
luteinizing hormone
medulla
mineralocorticoids
negative feedback
norepinephrine
oxytocin
pancreatic islets
peptides

pheromones
pineal gland
pituitary gland
portal system
prolactin
prostaglandins
second messenger
sex hormones
steroids
suprachiasmic nucleus (SCN)
thyroid-stimulating hormone
vasopressin

 QUESTIONS AND PROBLEMS

## MULTIPLE CHOICE

Circle the correct terms.

1. The thyroid gland secretes the hormone calcitonin which in turn (*stimulates or inhibits*) the activity of (*osteoblasts, osteoclasts, or hormones*), which are cells that degrade bone tissue. Therefore, calcitonin acts to (*lower or raise*) blood calcium levels.

2. ADH also known as (*vasopressin or thyroxine*) controls water balance in the body.

3. Glands of ectodermal and endodermal origin secrete hormones made from (*carbohydrates, proteins, or amino acids*), while glands of mesodermal origin secrete hormones based on (*lipids, proteins, or sugars*).

4. The (*hypothalamus or pineal gland*) located above the pituitary gland controls the release of hormones from the (*posterior or anterior*) pituitary.

5. The anterior lobe of the pituitary is composed of (*glial or glandular*) tissue and the posterior lobe is made up of (*connective or nervous*) tissue.

6. The two adrenal glands are each located above each of the (*lungs, kidneys, or liver*). The inner (*medulla or cortex*) portion secretes amine hormones and the outer (*cortex or medulla*) secretes (*growth or steroid*) hormones.

7. The rate of secretion of the thyroid hormones is regulated by another hormone, thyroid-stimulating hormone which is secreted by the (*hypothalamus, anterior pituitary, or adrenal cortex*).

8. Insulin enhances (*glucose or glucagon*) uptake by cells, helping to lower blood glucose levels. Insulin also (*suppresses or stimulates*) the formation of glycogen, the storage form of glucose in skeletal muscle and liver cells.

9. Chemical messengers that are secreted from special glands and used to mark one's territory are known as (*pheromones, interferons, or prostaglandins*).

10. Most hormones show (*lunar, circadian, or seasonal*) rhythms.

## TRUE/FALSE

1. The endocrine system is regulated by positive feedback.

2. Testosterone is a masculinizing hormone and estradiol is a feminizing hormone.

3. The chemical property that is most important in defining a hormone is its target site.

4. The major amine hormone of the adrenal gland is adrenaline.

5. The hormones responsible for the "fight or flight" response are adrenaline and norepinephrine.

6. TSH acts on the thyroid gland to increase secretion of the thyroid hormones.

7. Diabetes mellitus is a disorder characterized by low blood glucose levels.

## SHORT ANSWER

1. Describe the differences between the four basic types of chemical messengers used by the endocrine system.

2. What causes the variation in height and body build in humans?

3. Compare how non-steroid and steroid hormones act on target cells.

4. Name the pituitary gland hormones and describe their function.

5. Describe the effects of having hyposecretion of growth hormones and hypersecretion of growth hormones.

6. Name the three functional classes of the adrenal steroids.

## SCIENCE AND SOCIETY

1. Low levels of growth hormone (GH) in infancy or early life can result in the failure to grow. This low rate of secretion of GH causes pituitary dwarfism. Using GH made by recombinant DNA technology, victims of pituitary dwarfism can be treated, permitting them to grow more or less normally. The side effects of recombinant GH are not yet known. Some parents and physicians are treating healthy children who might be normally small with GH. What is your opinion on the use of GH for children? Explain your reasoning. Can you think of any long-term side effects from the use of GH? Should parents or physicians be able to use GH on clinically normal children? Why or why not? Why would parents want to use GH? Is this a form of eugenics? Why or why not?

2. Our understanding of the diverse roles that prostaglandins play in the local control of cellular function is incomplete. Prostaglandins are important in the development of inflammation, pain, fever, and blood clotting. Prostaglandins contribute

to the formation of blood clots by acting on platelets. The most common painkiller, aspirin, prevents the formation of prostaglandin by blocking the action of an enzyme involved in prostaglandin formation. What effects would aspirin have on prostaglandins? Should an individual with a fever take prostaglandins or aspirin? Why is one better than the other? How does aspirin help people who have suffered from strokes or heart attacks? There are many different forms of prostaglandins. Do you think each form would have different actions or potencies? Can you think of other areas of the body that may be influenced by prostaglandins?

# VI

# REPRODUCTION, DEVELOPMENT, AND BEHAVIOR

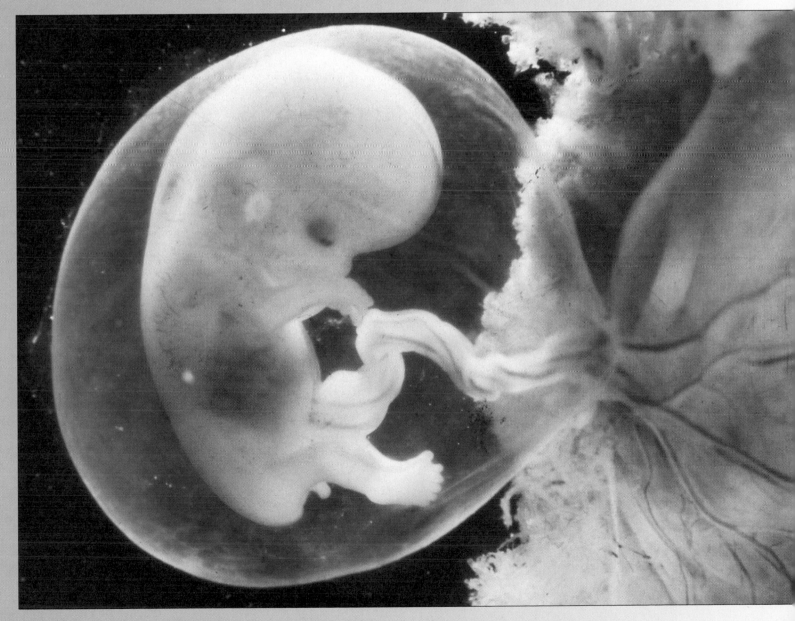

A HUMAN EMBRYO NEAR THE END OF THE FIRST TRIMESTER OF DEVELOPMENT.

# 27

# PLANT REPRODUCTION AND DEVELOPMENT

OPENING IMAGE

*Insects help pollinate flowering plants.*

Murals from the early dynasties of ancient Egypt show women wearing cones on top of their heads. These cones, made from resins and flower extracts, served as perfumes. As body heat warmed the cones, they slowly melted and perfumed the body. Although they may have been the height of fashion in 2400 B.C., perfumed resin cones that melt and run down the body would probably not be popular today. From then to now, even though the form is different, humans have used floral extracts for a wide range of social and cultural purposes.

In ancient Rome, the water in the public baths was scented with extracts of Mediterranean lavender. Its original use is preserved in its name. Lavender comes from the Latin word lavare, *meaning "to wash." In the Middle Ages, flowers were strewn in the street at public gatherings because they were thought to prevent disease.*

Cologne was originally a specific scent produced in the city of Cologne, Germany, in the early eighteenth century. It was made from a mixture of citrus oils, such as lemon and oranges, combined with lavender. Later, as the formula became widely known and copied, the term cologne *became used for many different scents and is now a generic term.*

Today, perfumes derived from flowers are a multimillion dollar industry. In addition to personal use, industrial perfumes are used to cover undesirable odors from paints and cleaning solutions and to sell products. Fresh bread odors are added to the wrappers of loaves of commercially produced bread.

*In nature, flower scents attract pollinators. Flowers have coevolved with their pollinators, and often have distinctive characteristics. Flowers that are pollinated by beetles are often dull in color, but have a rich, spicy odor, or are similar to the odors of fermentation. Beetles have a highly developed sense of smell, but have a less well-developed sense of vision, so color is less important than odor in attracting such pollinators.*

*This chapter reviews plant reproduction, with emphasis on the flowering plants. The evolutionary origins of flowers are not clear, but recent fossil studies indicate they originated some 200 million years ago or earlier. The world we see today, with flowering plants as a major life-form, was partly shaped by flowers.*

## REPRODUCTION IN PLANTS: A PREVIEW

Angiosperms became the dominant plant group on land about 80 to 90 million years ago. One of the main reasons for the success of angiosperms is the evolution of flowers and fruits as reproductive structures. The life cycle of angiosperms involves an alternation between a sporophyte generation and a gametophyte generation (✦ Figure 27.1). Flowers form on the sporophyte generation. Haploid spores form in the flower and develop into the gametophyte generation. Eggs form in the female gametophyte, which is enclosed within the flower, and are fertilized by pollen, the male gametophyte, to form a seed. The seed germinates and forms the sporophyte generation.

Flowers promote the spread of pollen from one plant to another, and they help angiosperms achieve and maintain genetic diversity. Fruits surround seeds and aid in seed dispersal in a variety of ways. Seeds protect plant embryos from harsh conditions and from predators, and most seeds contain a food supply to nourish the developing seedling. These adaptations have allowed the angiosperms to spread across many different environments to become the dominant form of terrestrial plants.

Although angiosperms reproduce sexually, they can also reproduce asexually (vegetatively) in a variety of ways. Some plants such as the strawberry or the violet have long runners that grow along the surface of the soil and produce new plants. Other plants like onions and many weeds use underground stems that produce new plants at intervals. The houseplant *Kalanchoe* has reproductive leaves (✦ Figure 27.2) that form new plants in the notches of the leaf margins. Other plants can be propagated

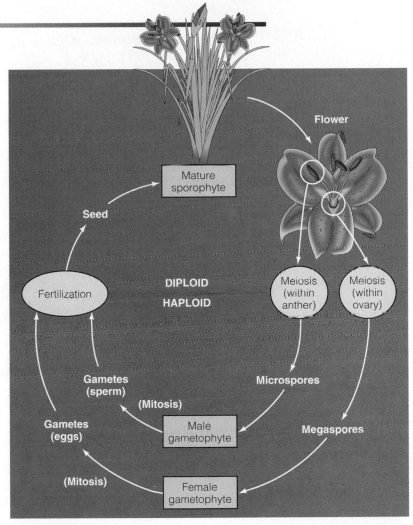

✦ **FIGURE 27.1**

**The life cycle of angiosperms includes an alternation of generations and the formation of seeds.**

✦ **FIGURE 27.2**

**Some angiosperms can reproduce vegetatively. New *Kalanchoe* plants form in the leaf notches of a mature plant.**

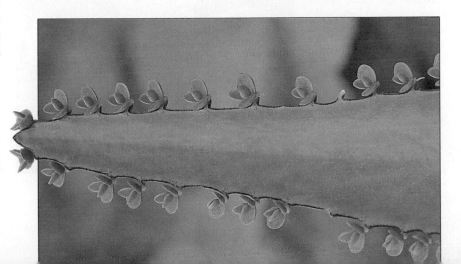

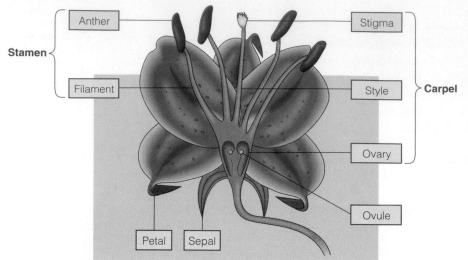

**✦ FIGURE 27.3**

Cross section of a flower. The sepals and petals are nonreproductive structures. The carpel is a female reproductive structure, and the stamen is a male reproductive structure.

from cuttings. Many varieties of fruits such as the McIntosh apple and varieties of pink grapefruit were discovered as **bud sports,** buds producing fruit that is different from the rest of the fruit on the tree. Bud sports are vegetatively propagated by grafting cuttings onto another plant.

Because angiosperms are a familiar form of plants, the emphasis in this chapter is on the reproduction and development of flowering plants.

## ✿ GAMETE FORMATION TAKES PLACE IN FLOWERS

For successful sexual reproduction in plants, as in animals, three processes must occur: (1) gamete production, (2) fusion of gametes in fertilization,

and (3) the development of a new individual. In flowering plants, the first two stages and part of the third stage occur in the flower.

### Flowers: Structure and Function

What is a flower? To humans, it is something to enjoy; for bees and other nectar-gathering animals, it is a source of food. For the plant, a flower is an organ of reproduction in which male and female gametes are formed. The flower is a complex organ that is a modified shoot, on which the flower parts are thought to be modified leaves (✦ Figure 27.3). Flowers are attached to stems by the base or **receptacle.** The outermost parts of the flower are the **sepals** and **petals.** Sepals are modified leaves that protect the inner petals and reproductive structures. The petals are usually brightly colored and may produce fragrant oils; both features attract pollinators. Insects and birds are the most frequent pollinators. Insects are attracted by odor and to a lesser extent by colors visible to humans. Many insects are also attracted to flowers by ultraviolet light, which is a highly visible color to insects (✦ Figure 27.4). Most birds do not have a well-developed sense of smell, so they are attracted to flowers by color. To birds, red is a highly visible and attractive color. Flowers pollinated by wind dispersal of gametes do not need to attract pollinators. These flowers usually lack conspicuous colors and in some cases have no petals at all.

Inside the petals are the reproductive structures of the flower. These include the **stamens** and the **carpels** (Figure 27.3). The stamens are the male reproductive structures, and usually consist of a slender, thread-like stalk, called the **filament,** which is topped by an **anther.** Internally, the anther is divided into a number of pollen sacs. The pollen sacs are chambers in which the **pollen grains** (the male gametophyte) form (✦ Figure 27.5).

**✦ FIGURE 27.4**
Flower photographed in (a) normal light and (b) ultraviolet light. Many insects can see ultraviolet light, and some flowers appear brighter when seen in this kind of light.

(a)

(b)

# GUEST ESSAY: RESEARCH AND APPLICATIONS

## *Improving Plants by Genetic Engineering*

ROBERT J. GRIESBACH

I always wanted to be a geneticist. My father was a university professor and conducted research on ornamental plant breeding. While in elementary school, I spent many hours with my brother caring for our dad's plants. We worked nearly every weekend during the spring planting bulbs. In summer, the chores switched to weeding and in fall to digging for winter storage. The work did not stop in the winter. Winter was the time when the bulbs were cleaned and prepared for spring planting. Rain or shine we worked. I can still remember one Thanksgiving when we had to shovel the snow off the ground before digging!

In high school, I became more involved in the laboratory aspects of plant breeding. My father's research centered around the use of polyploidy to increase the fertility of sterile seedlings. When two widely divergent species are used to create a hybrid, the chromosomes from the two parents usually do not pair during meiosis, causing sterility. This sterility can be overcome through chromosome doubling. I learned to double the chromosome number of sterile lily hybrids through colchicine treatment. In addition, he taught me how to prepare root tip and pollen cells for chromosome analysis.

In 1973 during my last year in high school, *National Geographic* magazine published an article on genetic engineering. Dr. Peter Carlson's research on cell fusion was described. After reading that article, I decided to be a plant breeder. I wanted to utilize genetic engineering to create improved plants—just like my father had done back in the 1950s.

It was obvious that strong backgrounds in chemistry and biology were essential. I double majored in chemistry and biology at DePaul University in Chicago, the same university at which my dad taught. It was quite unusual taking classes from my father. He was tough—especially on me! On a genetics test he took off one point because my name was not clearly written.

It was obvious to me which graduate school I wanted to attend—Michigan State University. Dr. Peter Carlson had just moved there from Brookhaven National Laboratory in New York. My research in graduate school focused on developing a technique to isolate individual chromosomes. The individual chromosome could be introduced into cells and the cells grown back into whole plants. It was hoped that such a procedure would break most hybridization barriers.

After completing graduate school, I spent a year with Dr. Ken Sink at Michigan State University learning the biology of the petunia. This plant is an ideal model system to test new genetic engineering technology. In 1980, I moved to the U.S. Department of Agriculture (USDA) in Beltsville, Maryland, to develop new technology to improve ornamental plants.

At the USDA, I developed a procedure to inject isolated chromosomes into petunia cells and to produce whole plants from those injected cells. The recovered plants expressed several traits from the injected chromosomes. Since then, I have developed several other new technologies to expand the range of germplasm that can be used by plant breeders. It was hard to believe that I found a job doing exactly what I had wanted to since elementary school!

*Robert J. Griesbach earned his Ph.D. degree in genetics from Michigan State University. He is employed by the Agricultural Research Service of the U.S. Department of Agriculture as a research geneticist in the Floral and Nursery Plant Research Unit of the U.S. National Arboretum in Beltsville, Maryland. His research interests include the use of genetic engineering in ornamental plant improvement, orchid breeding, and the biochemistry of flower color.*

## ◆ FIGURE 27.5

**Pollen grains encase the male gametophyte. The pollen tube will grow from one of the pores in the grain. (a) Grass pollen. (b) Poppy pollen. (c) Mustard pollen.**

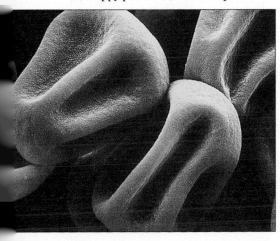

(a)

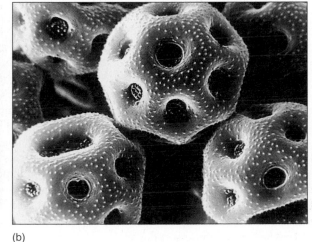

(b)

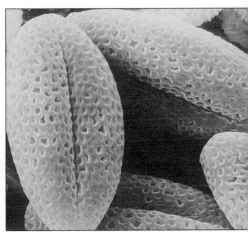

(c)

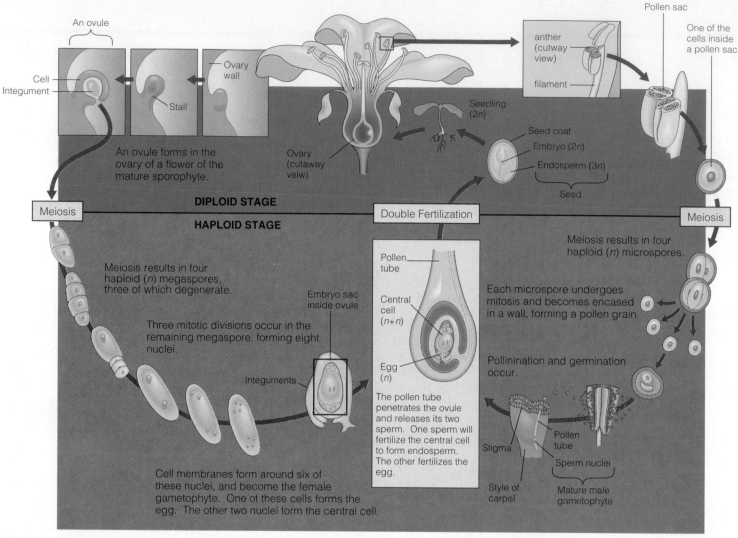

An ovule

Cell
Integument

Ovary wall

Stall

Ovary (cutaway veiw)

Seedling (2n)

Seed coat
Embryo (2n)
Endosperm (3n)

Seed

An ovule forms in the ovary of a flower of the mature sporophyte.

Pollen sac

anther (cutway view)

filament

One of the cells inside a pollen sac

**DIPLOID STAGE**

**HAPLOID STAGE**

Meiosis

Double Fertilization

Meiosis

Meiosis results in four haploid (n) megaspores, three of which degenerate.

Embryo sac inside ovule

Three mitotic divisions occur in the remaining megaspore, forming eight nuclei.

Integuments

Pollen tube

Central cell (n+n)

Egg (n)

The pollen tube penetrates the ovule and releases its two sperm. One sperm will fertilize the central cell to form endosperm. The other fertilizes the egg.

Cell membranes form around six of these nuclei, and become the female gametophyte. One of these cells forms the egg. The other two nuclei form the central cell.

Meiosis results in four haploid (n) microspores.

Each microspore undergoes mitosis and becomes encased in a wall, forming a pollen grain.

Pollinination and germination occur.

Stigma

Pollen tube

Sperm nuclei

Style of carpel

Mature male gametophyte

✦ **F I G U R E  27.6**

**Gametophyte formation, gamete production, and fertilization in an angiosperm. Development of the male gametophyte and pollen are shown on the right side, and development of the female gametophyte and egg are shown on the left. After pollination, sperm nuclei fuse with the egg and the central cell. This double fertilization event gives rise to a diploid embryo (2n) and a triploid endosperm (3n).**

At the center of the flower are the female reproductive structures, consisting of one or more carpels. Each carpel contains three parts: (1) the **ovary,** (2) the **style,** a tube formed from the ovary wall, and (3) the terminal **stigma** (Figure 27.3). The ovary contains clusters of cells called **ovules,** each of which contains the cell that will form the female gametophyte. The tip of the style carries the stigma, a sticky surface to which pollen grains attach. After fertilization, the ovary forms the fruit, and the ovules form the seeds.

Many plants produce flowers with both male and female reproductive structures (stamens and carpels). These are known as perfect flowers. Other plants produce flowers carrying only male or female parts; these are called imperfect flowers.

## Meiosis Produces Gametophytes, Not Gametes

Meiosis takes place within the anthers and ovules. The haploid products of meiosis are *not* gametes, as in animals; instead, they grow to form a haploid gametophyte generation. When mature, the gametophyte forms gametes, which will fuse to form a diploid zygote that grows into a sporophyte.

The male gametophyte forms in the chambers (pollen sacs) of the anther. A typical anther contains four pollen sacs (✦ Figure 27.6). Within the pollen sacs, each cell that undergoes meiosis forms four haploid cells called **microspores.** Each microspore undergoes at least one mitotic division and becomes encased in a thick protective wall, forming a pollen grain. In some cases, one of the haploid nuclei divides again, forming three haploid cells. Some pollen grains contain only two haploid cells. Such pollen grains are immature gametophytes; one of the cells will divide after the pollen grain has landed on the stigma, forming three haploid cells.

The female gametophyte forms in the ovules. Within the ovule, a large, centrally located cell, protected by several surrounding, smaller cells undergoes meiosis, producing four haploid cells, known as **megaspores** (Figure 27.6). Usually, three of these cells degenerate, and the remaining cell becomes the gamete. This haploid cell divides three times, forming eight nuclei. Cell membranes form around six of these cells, and make up the female gametophyte. One of these cells becomes the haploid egg. The remaining two nuclei are incorporated into a single cell, called the central cell. After fertilization, this cell forms **endosperm,** a tissue that provides nutrients to the developing embryo.

## POLLINATION IS FOLLOWED BY FERTILIZATION

Pollen grains are structures adapted for dispersal by wind or animals. Pollination is the transfer of pollen grains to the stigma. After transfer, water moves from the stigma into the pollen grain, and other chemical signals are exchanged. In immature pollen grains, this triggers division of one haploid nuclei. These events are followed by the development of the pollen tube. Germinating pollen grains contain three haploid nuclei, which make up the mature male gametophyte generation. One nucleus is associated with the pollen tube, and the other two are the sperm nuclei.

### Flowering Plants Employ Double Fertilization

The pollen tube grows through the tissues of the stigma and the style, pushing between cells, to reach the female gametophyte contained in the ovule (✦ Figure 27.7). One of the characteristics of angiosperms is **double fertilization.** One of the two sperm nuclei carried by the pollen tube fertilizes the central cell, forming a triploid ($3n$) cell that gives rise to the endosperm. The other nucleus fertilizes the

✦ **FIGURE 27.7**

**Pollination and fertilization. (a) In pollination, a pollen grain is deposited on the sticky surface of the stigma. (b) Chemical cues trigger the growth of the pollen tube, carrying two sperm nuclei. (c) The pollen tube grows between the cells of the style and reaches the ovule, where double fertilization takes place.**

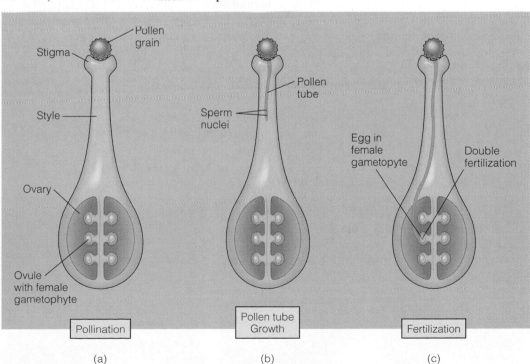

(a) Pollination

(b) Pollen tube Growth

(c) Fertilization

*Sidebar*

**UNISEXUALITY IN PLANTS**

Sex determination in humans is a complex process involving an interplay between genetic determinants and hormones. This process is as equally intricate in flowering plants. Most flowering plants bear flowers containing both female and male sex organs within each flower, However, patterns of unisexuality have arisen in the plant kingdom. Monoecious plants are those with unisexual flowers on the same plant and dioecious plants are those with unisexual flowers on separate plants.

Some species of plants, such as spinach, accomplish unisexuality by bypassing the formation of the inappropriate sex organ. Other species, such as asparagus and maize, stop the development of sex organs at various stages of maturation. The genetic determinants of unisexuality show great diversity in plants. This developmental and genetic variability suggests that the cellular controls of sex determination in plants may vary considerably.

Unisexuality has evolved in different species of plants, allowing scientists to study genetic, molecular, and cellular processes. Many amazing similarities exist between plants and animals in the process of sex determination. Understanding sex determination in plants is valuable for agricultural purposes. Unisexuality in maize, for example, facilitates large-scale production of hybrid offspring and is responsible for tremendous increases in crop yields.

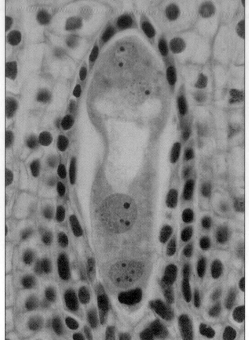

✦ FIGURE 27.8
Fertilization in lily. The fertilized egg at the bottom contains the egg nucleus and one sperm nucleus. These will fuse and then undergo mitosis, beginning the formation of the embryo. The triploid fertilized central cell, just above the egg, will divide to form endosperm.

haploid egg, forming a diploid (2*n*) embryo (✦ Figure 27.8).

After fertilization, each ovule differentiates into a seed. The ovary of the flower differentiates into the fruit, containing the seeds.

## Plants and Animals Use Different Reproductive Strategies

One of the differences between plants and animals is that animals contain a group of cells called germ cells, set aside early in development, that undergo meiosis and form gametes in the adult. In plants, the gametes come from a population of cells that were previously active in forming the body of the plant. In addition, embryo development in animals results in a nearly complete, but smaller version of the adult.

In plants, embryonic development establishes the basic body plan, but the plant embryo contains only a fraction of the components of the mature plant body, and no reproductive structures and no germ line are present. The plant embryo grows from the seed by cell division and cell enlargement to form a seedling. The seedling then initiates organ formation from meristems.

## SEEDS AND FRUITS AID IN DISPERSAL AND PROTECT THE PLANT EMBRYO

In the seed plants, a new individual (the embryo) is formed inside the tissues that are still attached to one parent. The embryo is protected and nourished by the parent, enclosed in a structure that becomes a seed. Seeds can be distributed over long distances. Dispersal is an adaptation that offers several advantages (✦ Figure 27.9). It lowers competition between parents and offspring for space and other resources, and it allows for exploitation of distant environments.

### The Developmental and Evolutionary Advantages of Seeds

After fertilization, the ovule and its internal structures begin to develop into seeds (✦ Figure 27.10). In its general organization, the seed contains the embryo, which is the developing sporophyte generation. The embryo does not develop continuously, but enters a state of dormancy, during which it is usually resistant to many environmental stresses. The embryo is always accompanied by nutrient reserves, formed by the endosperm. These structures are enclosed in a protective seed coat.

The development of seeds represents a major evolutionary advance for plants. Because seeds can

✦ FIGURE 27.9
Adaptations have evolved to disperse seeds by (a) air, by (b) passing animals, and by (c) water.

(a)

(b)

(c)

withstand extreme conditions, a species can survive by having an annual cycle that involves photosynthesis and growth during spring and summer, followed by survival as a seed during winter, when conditions are too harsh. Seeds have another advantage besides stress tolerance—they provide a means of species dispersal. If conditions become too extreme, or a new environment becomes available, animals can quickly migrate. Plants, on the other hand, are not capable of long-distance migration. The dispersal of seeds by animals, wind, and water allows a preformed plant (the embryo) to move to a new location, where if conditions are favorable, a new plant can be established.

## Fruits Protect Seeds and Promote Seed Dispersal

The growth of the seed in the ovule occurs at the same time that the ovary becomes transformed into the fruit. The ovarian tissue expands and becomes modified during fruit formation (✦ Figure 27.11). The purpose of the fruit is to (1) protect the embryo and the seed and (2) help disperse the seed.

There is no such thing as a typical fruit; they may be fleshy (like apples) or dry (wheat). The fleshy fruits may be soft or hard, and the dry fruits may be papery or hard (✦ Table 27.1).

Fruits that develop only from the ovary are called true fruits, whereas fruits that incorporate other structures (the sepals, receptacle, etc.) are called false fruits. In the strawberry, the fruits are the small, dark, hard structures (called achenes) embedded in the fleshy, red part, which is derived from the receptacle. The strawberry is therefore a false fruit, because the fleshy part is not formed from the ovary.

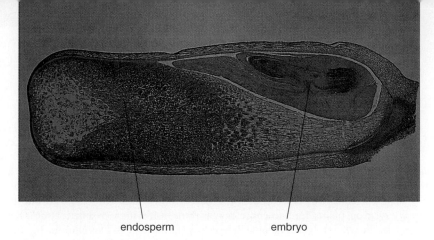

✦ FIGURE 27.10

Cross section of a seed. The embryo develops in the seed and forms the shoot tip, the cotyledons, and the root tip. The endosperm is at the left.

| TABLE 27.1 SOME FRUIT TYPES | |
| --- | --- |
| TYPE | EXAMPLE |
| **SIMPLE** Usually formed from single carpel. Can be dry or fleshy. | Dry: Sunflower, milkweed pea, soybean Fleshy: Grape, bananas, tomatoes |
| **AGGREGATE** Formed from multiple carpels of a single flower. | Berries (raspberry, blackberry) |
| **MULTIPLE** Formed from the ovaries of two or more flowers. | Pineapples, osage oranges |
| **ACCESSORY** Incorporates other plant tissue in addition to the ovary. | Apples, pears, strawberries |

✦ FIGURE 27.11

Stages in the formation of fruit. (a) Mature flowers. (b) After fertilization, the flower petals drop away, and the formation of fruits and seeds begins. (c) Fruit forms as the ovary expands.

(a)

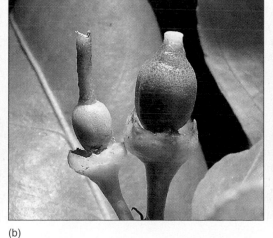

(b)

(c)

*Beyond the Basics*

# REDISCOVERING LOST CROP PLANTS

There are at least 75,000 species of edible plants in the world. Historically, about 2000 to 3000 of these have been used as food crops. Currently, only about 30 plants account for the majority of crop plants for the entire human population. In 1989, the U.S. National Research Council issued a report that evaluated the use of crops grown by the Incas of Peru before the arrival of Europeans. In the sixteenth century, the Incas cultivated some 70 species of plants as food crops. Aside from potatoes and lima beans, most of these crops were replaced by species of European plants and then forgotten.

There are several advantages associated with reintroducing these lost crops, including nutrition, dietary variety, and a reduction of the dangers of crop failure. The Inca crop plants were originally grown in mountainous regions, and they can be transferred to cool, temperate climates almost anywhere in the world. In the years since the report was issued, some of these crops have been planted as specialty crops around the world. These include the cherimoya, a fruit with a creamy texture when ripe and a flavor that is a mixture of pineapple, banana, and papaya; the pepino, a small, yellow-purple fruit whose taste resembles a honeydew melon; and quinona, a grain that is one of the best vegetable sources of protein.

Other crops may be developed into more widely available food, rather than being found only in health food stores or specialty shops. One such plant, the oca, a tuber that looks like a wrinkled carrot, has already become widely planted in New Zealand, where it is marketed as a New Zealand yam. Cultivation of this species may soon result in a commercial crop in North America, Japan, and Europe. Another species, the nuna or popping bean, can be prepared by toasting, using far less fuel than boiling. This crop may be useful in parts of the Third World where deforestation has made wood for cooking a scarce commodity, and where other fuels like kerosene are expensive.

The diet of the Incas from five centuries ago may make another contribution to feeding the world's population. The potato, cultivated by the Incas, was introduced to Europe in the sixteenth century. Since then, it has spread around the world, and is now the fourth leading crop plant, with over 285 million metric tons produced every year. The use of additional species may help feed the growing world population as we enter the twenty-first century.

## PATTERNS OF GROWTH IN EMBRYOS AND SEEDLINGS

The developing embryo grows into a seedling, which forms a sporophyte. The mature sporophyte forms the flowers, seeds, and fruits of the next generation. The dormant embryo within the seed begins to develop when water is absorbed by the seed, but other factors including temperature, day length, and light intensity also affect development.

Germination begins when the embryo breaks through the seed coat and resumes growth. Usually the first structure that emerges from the seed is the embryonic root. This serves to anchor the developing plant, and it also absorbs water. At the opposite end of the embryo, the shoot emerges, bearing the **cotyledons,** sometimes referred to as seed leaves. As the names imply, the monocots have one cotyledon, and the dicots have two. The cotyledons contain nutrients that are released during germination, and they often wither and drop off the plant after their nutrient supply is exhausted. By this time, the shoot has produced its first foliage leaves and is actively engaged in photosynthesis. The basic patterns of growth and development in dicots are summarized in ◆ Figure 27.12.

As the plant develops, there is an orderly production of leaves, nodes, and internodes by the apical meristem of the shoot. At the same time, the root system develops, expanding the primary root and the lateral root system. This growth is called vegetative growth. As the plant matures, one or more of the apical meristems develops into a flower. The mature sporophyte is then ready to form a new generation, in the form of the gametophyte, and initiate reproduction.

## PLANT GROWTH REGULATORS CONTROL DEVELOPMENT

Development and other functions in plants are controlled by chemical messengers that are similar to hormones in animals. There are five broad categories of plant growth regulators: auxins, gibberellins, cytokinins, abscisic acid, and ethylene. Each is discussed briefly in the following section.

### The Five Known Types of Plant Hormones

**Auxins** are a group of naturally occurring and synthetic plant growth regulating substances involved in controlling cell growth. All plant cells have the ability to produce auxins, but only a few (including young leaves and developing embryos) actually produce them. Auxins move from the young leaves downward through the rest of the plant, causing the shoot and the root to elongate (◆ Figure 27.13). Auxin is also produced in developing embryos and moves into the fruit, causing expansion and growth. Auxin flow from cell to cell can be influenced by light and gravity.

✦ FIGURE 27.12

Stages in the development of a dicot. The root grows first. This anchors the plant and provides water and nutrients for development. The shoot then emerges above the soil and the cotyledons begin photosynthesis. After foliage leaves form, the cotyledons shrink and fall off.

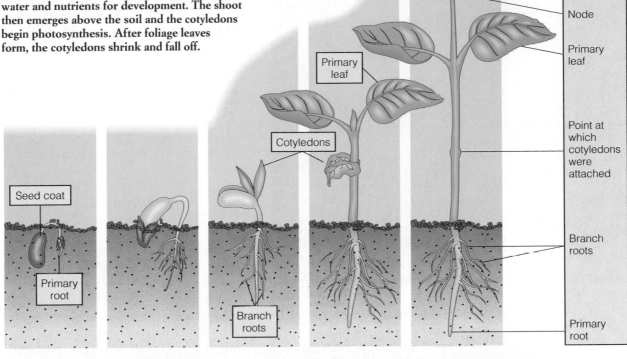

✦ FIGURE 27.13

The effect of auxins on plant growth. (a) Untreated plant. (b), (c) Plants treated with auxins.

Auxins can also inhibit the growth of lateral buds. When you pinch off the top of a plant in your flowerbox, small lateral buds begin to elongate, and new branches form. This is caused by a phenomenon called apical dominance. Auxins are produced at the tip of the apical meristem and block the growth of lateral buds. If the top of the plant (including the apical meristem) is removed, the auxin source is removed, allowing the growth of lateral buds.

**Gibberellins** are required for normal elongation of the shoot. They are thought to be produced in young leaves and developing embryos, and they stimulate cell division as well as cell elongation. When some plants are sprayed with gibberellins, fruit size increases. Gibberellins are used commer-

✦ FIGURE 27.14

**Gibberellins are used commercially to produce large clusters of Thompson grapes.**

cially to produce large fruit size in Thompson seedless grapes (✦ Figure 27.14).

**Cytokinins** are produced in roots and transported to shoots by transpiration; they are also synthesized by developing embryos. Cytokinins promote cell division and inhibit aging of green tissues, especially leaves. Cytokinins are used in the floral industry to maintain green tissue in cut flowers.

**Abscisic acid** is produced in leaves. It promotes dormancy in perennial plants and causes rapid closure of leaf stomata when a leaf begins to wilt.

**Ethylene** gas is also considered to be a plant hormone. The site of ethylene production is still uncertain. It stimulates fruit ripening and the dropping of leaves. Bananas, honeydew melons, and tomatoes are often harvested before ripening, and then kept in airtight storage containers. Before shipping, ethylene is injected into the container, so the fruit will ripen in transport.

## Other Substances Act as Plant Hormones

These five hormones are not the only molecules that can influence plant growth and development. An increasing number of compounds (including polyamines, phytochrome, and calcium) are reported to produce effects on plant tissues. A hormone named florigen, as yet unidentified, may trigger the flowering process. Evidence suggests that other hormones are produced by roots and leaves.

## VEGETATIVE REPRODUCTION AND CLONING

In addition to sexual reproduction, plants have many ways of vegetative reproduction. They can reproduce by means of runners, or stolons, which are long stems that run along the surface of the ground. In the strawberry, leaves and roots are produced at every other node on the runner. This location gives rise to a new shoot, which, in turn, can produce new runners (✦ Figure 27.15).

Rhizomes are underground stems that can produce new plants at each node. Grasses and many weed species use rhizomes to spread from the parent plant. Irises and other flowers are propagated from rhizomes.

Roots can also produce new plants by asexual reproduction. The roots of raspberry plants and apple trees produce "suckers" that sprout and form new plants. The commercial variety of bananas does not produce seeds; instead, commercial bananas are maintained by planting suckers that develop on underground stems.

✦ FIGURE 27.15

**Strawberry plants reproduce vegetatively by runners that periodically root and start new plants.**

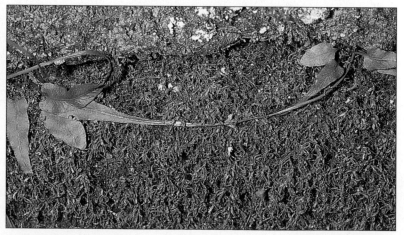

✦ FIGURE 27.16

**The walking fern reproduces vegetatively wherever a leaf tip touches the ground.**

As mentioned earlier, the leaves of some species like *Kalanchoe* can vegetatively reproduce. In the walking fern, new plants form where the leaf tips touch the ground (✦ Figure 27.16).

Vegetative reproduction is an adaptation that allows the rapid production of genetically identical individuals that are well suited to a particular habitat. Organisms produced by some forms of vegetative reproduction are candidates as the largest organisms in the world.

In April 1992, scientists announced the discovery of an underground fungus in Michigan that might be the largest organism on the planet. The fungus is estimated to weigh 100 tons, cover 30 acres, and may be up to 10,000 years old. It started from a single spore, and grew on the surface as familiar mushrooms, and underground by a network of root-like mycelia. DNA testing has shown that the fungus is genetically identical from one end to the other.

Even this organism may be dwarfed by other vegetatively grown plants. In November 1992, another report demonstrated that a 6000-acre stand of aspens located in the Wasatch Mountains south of Salt Lake City is genetically identical. This cluster of more than 47,000 aspen trees was formed by sprouts forming from roots (✦ Figure 27.17).

## Plants Can Be Grown from Single Cells

One of the distinctive features of plants is that new adult plants can be generated from single cells. This phenomenon was first discovered when single phloem cells from carrots were grown in the laboratory. Each cell divides a number of times and forms a mass called a callus. When the callus is transferred to a different growth medium, the calluses form mature plants (✦ Figure 27.18). The growth of genetically identical individuals from a common ancestor is called cloning, and the organisms are said to be clones.

By conventional genetic crosses, a loblolly pine tree was developed that had rapid growth characteristics, was disease resistant, and had a high wood content. Cells from this tree were grown individually until they formed calluses, and then each was converted into thousands of copies of the original tree. The result is a forest of genetically identical trees that will mature at the same time, allowing the forest to be harvested on a predictable schedule.

Another application of single cells is the production of synthetic seeds. Single cells from some plants can be grown in the laboratory to form embryos. The embryos are then encapsulated in a covering that allows them to be handled and planted in the

✦ **FIGURE 27.17**

**Aspens reproduce vegetatively by means of underground stems. This grove of genetically identical trees represents the descendants of a single tree.**

✦ **FIGURE 27.18**

**Tissue masses called calluses are derived from the mitotic division of single cells. These calluses will grow into mature plants, producing clones of the parent plant.**

*Sidebar*

**DEVELOPMENTAL SWITCHES IN PLANT DEVELOPMENT**

In flowering plants, the meristem at the tip of the growing stem contains undifferentiated cells that form leaves, flower clusters (called inflorescences), and flowers. Normally, developing plants undergo vegetative growth and leaf formation, and then reproductive development and the formation of inflorescences and flowers. Organs form from the meristem in a predetermined order: leaves, inflorescences, and flowers. In the switch from vegetative to reproductive growth, leaf formation stops, and inflorescences and flowers develop. Several genes are active in the early stages of flower development, including AP1 and LFY.

Recent work has demonstrated that alterations in the expression of either of these two genes cause several changes in the normal developmental program. Genetic manipulation that results in constant expression of AP1 or LFY causes the transition from vegetative growth to reproductive growth to occur much earlier than normal. As a result, the developing plant forms fewer leaves, single flowers develop instead of flower clusters, and the shoot meristem forms a flower at its tip, closing off vegetative growth.

The early expression of these genes causes undifferentiated cells that would normally form leaves to alter their developmental fate, bypass the formation of flower clusters, and form flowers instead. This action suggests that AP1 and LFY are master genes that control important switch points in developmental pathways, activating a cascade of gene action. Manipulation of flowering time in crop plants may be useful for increasing the productivity of agricultural land, thus boosting food production.

same way as seeds (✦ Figure 27.19). Cloned crops can be raised from these synthetic seeds, producing uniformly sized plants with desirable characteristics such as disease resistance and high yield.

✦ FIGURE 27.19
Cloned plant embryos grown in tissue culture can be encapsulated in a protective coat and handled just like seeds. (a) Synthetic seed capsules. (b) Synthetic seed germination.

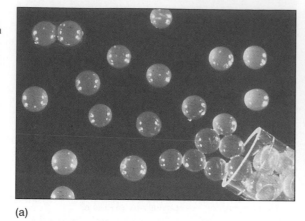
(a)

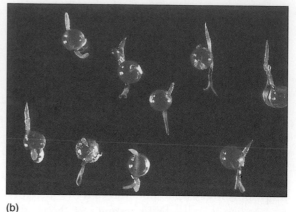
(b)

# SUMMARY

1. Angiosperms have a life cycle that involves alternating between a sporophyte generation and a gametophyte generation. The reproductive structures of the sporophyte generation are flowers. Haploid spores form in the flower and develop into the next gametophyte generation.

2. The outermost parts of the flower are the sepals and petals. These nonreproductive parts protect the inner, reproductive parts and attract pollinators. Inside the petals are the reproductive structures of the flower, the stamens and the carpels. The stamens are the male reproductive structures, and contain anthers, in which the pollen grains (the male gametophyte) form.

3. At the center of the flower are the female reproductive structures, consisting of one or more carpels. Each carpel contains the ovary, where egg development occurs in the ovules; a slender style, formed from the ovary wall; and the terminal stigma, to which pollen grains stick. The ovary contains clusters of cells called ovules, each of which contains the cell that will form the female gametophyte.

4. Pollen grains are transferred to the stigma, and the pollen tube grows down between cells of the style to the ovule. One sperm nucleus fertilizes the egg nucleus to form a diploid zygote. The other fuses with the two nuclei in the central cell to form a triploid cell that becomes endosperm, a food storage tissue.

5. After fertilization, the ovule forms the seed, and the ovary becomes the fruit. Seeds germinate to form a new sporophyte generation. When mature, they form flowers, which then undergo meiosis to form a new gametophyte generation and gametes.

6. Development in plants is controlled by hormones. Five types of hormones have been identified. Auxins promote stem elongation and inhibit growth of lateral buds. Gibberellins regulate shoot length and fruit size. Cytokinins stimulate mitosis and delay leaf aging. Abscisic acid regulates water loss by closing stomata and can induce dormancy. Ethylene promotes the ripening of fruit.

7. In addition to sexual reproduction, plants can reproduce vegetatively. This form of reproduction is an evolutionary adaptation that allows the rapid production of genetically identical individuals well suited to a particular habitat.

8. Plants can also be grown from single mature cells in a process called cloning. In the laboratory, single cells divide to form a cell mass called a callus. The callus grows to form a new adult plant. Identical copies of a single plant (clones) can be made by this method.

# KEY TERMS

abscisic acid
anther
auxins
bud sports
carpel
cotyledon
cytokinins
double fertilization

endosperm
ethylene
gibberellins
megaspore
microspore
ovary
ovule
petal

pollen grains
receptacle
sepal
stamen
stigma
style

## TRUE/FALSE

1. Reproduction in angiosperms can occur sexually and asexually.
2. Cloned organisms vary in their genetic material.
3. The strawberry is a true fruit.
4. Cytokinins are transported to the shoots of the plant by osmosis.
5. The reproductive parts of the flower are located inside the petals.

## MATCHING

Match the term with the appropriate description.

1. receptacle
2. sepals
3. stamens
4. stigma
5. carpels
6. endosperm
7. abscisic acid
8. stolons
9. anthers
10. flowers

_____ long stems that run along the ground
_____ female reproductive structures
_____ mature seeds
_____ location for pollen grain formation
_____ male reproductive structures
_____ modified leaves
_____ base of flower
_____ located at tip of style
_____ hormone that promotes dormancy
_____ tissue providing nutrients

## SHORT ANSWER

1. How are the haploid products of meiosis in plants different from those produced in animals?
2. How does mitosis differ between microspores and megaspores?
3. What are the two basic differences in the reproductive designs of plants and animals?
4. What are the advantages of seeds?
5. Describe the various functions of the five known types of plant hormones.

## SCIENCE AND SOCIETY

1. Plant breeders try to develop strains of plants with genetic features that make plants more useful to consumers. Researchers have given some experimental crop plants genes for weed, disease, and insect resistance. More recent work focuses on the legume family, which houses nitrogen-fixing bacteria in its roots. These bacteria convert nitrogen from the atmosphere into a form the plant can use. Researchers are hoping to transplant the gene responsible for nitrogen fixation into other plants. What are the advantages and disadvantages of plant bioengineering? How can transplanting nitrogen-fixing genes into plants help the ecosystem? What are the problems of conducting plant studies in a laboratory?

2. In the tundra, winds gust up to 200 miles per hour, temperatures are below freezing, the landscape is exposed to high doses of ultraviolet light, and the subsoil remains permanently frozen. This harsh biome has a growing season of only one and a half months, and most of the precipitation is in the form of snow, which is quickly blown away by the strong winds. Despite these adverse conditions, wild flowers grow in extraordinary diversity. What adaptations do you think evolved in order for plant life to be so successful in the tundra? Many plants in the tundra have small leaves and short stems. What advantages would these characteristics give plants? Plant life in the tundra is very similar to plant life in the desert. Describe how the plants in these two very different biomes are similar and how they are different.

# 28

# ANIMAL REPRODUCTION AND DEVELOPMENT

OPENING IMAGE

*Sperm surrounding a mammalian oocyte*

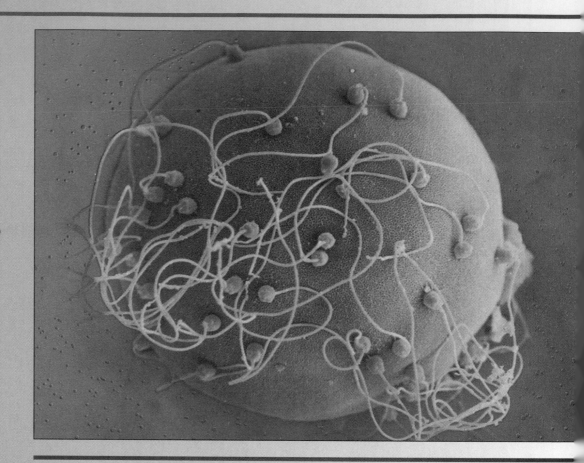

In August 1848, A. Berthold began a series of experiments on using organ transplants to establish that substances carried in the blood can affect both development and behavior. His experiments used six young roosters. He removed both testes from the first pair of roosters. Each rooster in the second pair had only one testis removed. In the third pair, both testes were removed, but each rooster received a testis transplant from the other rooster.

Berthold observed that the animals with both testes removed underwent a series of developmental and behavioral changes. These roosters had poorly developed secondary sex characteristics (the red wattle and comb on the head), and did not develop the characteristic call of normal roosters. In addition, they were not aggressive in battling with other roosters, and showed no interest in mating with hens. The other four roosters crowed normally, they were combative toward other roosters, were attracted to hens, and had well-developed wattles and combs.

During the next few months, the roosters were killed and examined. Only a small scar remained at the position normally occupied by the testes in the roosters with both testes removed. In the roosters with one testis removed, the remaining testis had a normal appearance. In the roosters that received a transplant, the testis had been placed on the surface of the intestine, and Berthold found that the transplant had developed connections with the circulatory system. The testes were normal in size and shape, and contained mature spermatozoa.

*Because the roosters that received transplanted testes looked and behaved like normal roosters, Berthold concluded that the transplanted testes produced a substance other than the spermatozoa. He postulated that this substance traveled through the blood, controlled the development of the comb, wattle, and crowing, and affected the behavior and sex drive of the recipients.*

*It took almost a hundred years to identify and isolate the diffusible substance postulated by Berthold. The steroid hormone testosterone was isolated and crystallized in 1935. The existence of other hormones involved in reproduction was inferred by similar experiments, and follicle-stimulating hormone and lutenizing hormone were isolated as crystallized proteins. The hormonal control of reproduction, especially human reproduction, continues to be a major area of research today.*

*This chapter considers animal reproduction, with emphasis on human reproduction and development. Reproduction is one of the most critical adaptations for the survival of a species. The chapter begins with an analysis of reproductive strategies and their costs and benefits, and then considers some basic mechanisms of development. Next, the human reproductive system and its interactions with the endocrine system are considered, and the chapter concludes with a survey of human prenatal and postnatal development.*

##  PRINCIPLES OF ANIMAL DEVELOPMENT

One of the basic properties of living systems is the ability to reproduce. Two major systems of reproduction have evolved: asexual and sexual reproduction.

### Asexual Reproduction Produces Offspring from a Single Parent

**Asexual reproduction** produces genetically identical offspring from a single parent. Asexual reproduction occurs by many mechanisms, including fission, budding, fragmentation, and the formation of rhizomes and stolons. Many invertebrates, including the hydra, reproduce by budding (✦ Figure 28.1). In this process, small outgrowths develop

✦ **FIGURE 28.1**
Hydras can reproduce asexually by budding.

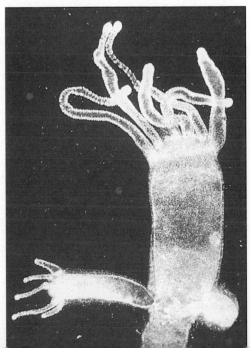

into fully formed organisms that separate from the parent. Some plants reproduce by budding, including lilies and ferns. Like other forms of asexual reproduction, budding produces genetically identical organisms.

Another form of asexual reproduction is fragmentation, in which the body of an adult is broken into several pieces, each of which grows to form an intact organism. This process is associated with the ability to regenerate missing parts. In sea stars (✦ Figure 28.2), a single arm with part of the central disk attached can regenerate into an adult, and a sea star missing an arm can regenerate to form an intact adult over a period of several weeks. Fragmentation is also common in mosses and liverworts.

Asexual reproduction is an adaptation that allows a single animal to produce many offspring quickly, without time and resources devoted to the formation of specialized sexual structures and processes, finding mates, and engaging in courtship behaviors. All reproductive time and energy can be used directly to produce offspring. Because all offspring are genetically identical to the parent, asexual reproduction forms populations that may be well adapted to stable environments, and such populations can rapidly exploit available resources.

Producing genetically uniform populations of individuals is one of the advantages of asexual reproduction, but it is also one of its main disadvantages. If conditions change, and the environment becomes less favorable, there is little genetic variation in the population, so that all members may be detrimentally affected.

✦ **FIGURE 28.2**
Small fragments of sea stars can regenerate to form a complete adult.

## Sexual Reproduction Increases Genetic Variability among Offspring

In **sexual reproduction,** new individuals are produced by the fusion of two haploid sex cells or **gametes** to form a diploid **zygote.** The zygote contains a new combination of parental genes carried by the male gamete (**sperm**), and the female gamete (**ovum**). As discussed in Chapter 5, meiosis and the random combination of gametes at fertilization generate genetic diversity among the offspring. This variation produces an increased ability to adapt to changing conditions. If the environment becomes less favorable, there is a greater likelihood that some members of the population will be less affected and therefore more able to survive and reproduce.

Some organisms, such as the rotifer (✦ Figure 28.3) are able to reproduce both asexually and sexually, offering benefits that depend on environmental conditions. When food is abundant and temperatures are moderate, female rotifers reproduce by laying eggs produced by mitosis. These eggs hatch without fertilization into genetically identical female offspring. When conditions are unfavorable, rotifers reproduce sexually, producing zygotes encased in a shell resistant to desiccation and cold temperatures. When conditions are favorable, these hatch into diploid adults.

Sexual reproduction offers the benefit of generating genetic variation among offspring, which enhances survival of the population under changing conditions, but it is not without some costs. These costs include the need for two mating types, mechanisms for gamete production, and often, elaborate courtship rituals (described in Chapter 29). Sexual reproduction also depends on a number of basic mechanisms to convert the fertilized egg into an adult. These mechanisms are reviewed in the following sections.

✦ **FIGURE 28.3**
**Rotifers can reproduce sexually and asexually.**

## FERTILIZATION AND CLEAVAGE ARE INITIAL STAGES OF DEVELOPMENT

Fertilization is the fusion of two gametes to produce a zygote, which develops into a new individual with a genetic heritage derived from two parents. Fertilization has three functions: (1) the transmission of genes from both parents to the offspring, (2) the restoration of the diploid number of chromosomes that was reduced in meiosis, and (3) the initiation of development in the offspring. Fertilization involves a number of steps, including contact between sperm and egg, entry of the sperm into the egg, fusion of sperm and egg nuclei, and activation of development.

Once formed, the zygote initiates steps that result in the production of a multicellular organism, which can (like humans) contain trillions of cells. In all multicellular animals, the first step is cleavage, the conversion of the zygote into a multicellular embryo, via mitosis (✦ Figure 28.4). Usually, there is no net increase in embryonic size during cleavage; the zygote cytoplasm is divided among smaller and smaller cells. At the first division, the embryo is divided into two cells, then into four, then into eight, and so on. In frogs, about 37,000 cells are produced in just over 40 hours.

Cleavage produces a ball of cells, called the **blastula,** surrounding a fluid-filled cavity (the **blastocoel**). During cleavage, the surface-to-volume ratio of each cell is increased, facilitating oxygen exchange with the environment. Unfertilized eggs of many species contain informational molecules in the form of mRNA and proteins that determine the developmental fate of the embryo. These molecules are distributed to different embryo regions during cleavage. This molecular differentiation sets the stage for laying out the body plan in the next phases of development.

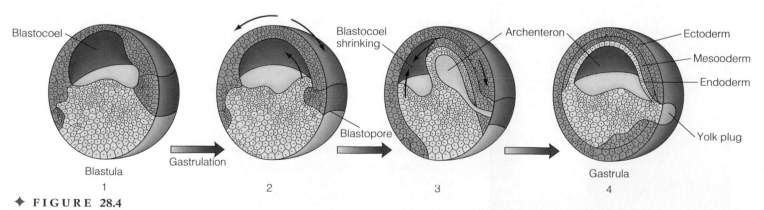

✦ **FIGURE 28.4**
**Stages in the formation of the three primary layers of the embryo. Cell migration forms an inner endoderm, an intermediate mesoderm, and an outer ectoderm. All the major organs of the body will form from these layers.**

## GASTRULATION PRODUCES AN ORGANIZED, THREE-LAYERED EMBRYO

Gastrulation is the next major step in development, and it involves a series of cell migrations. Cells move to new positions in the embryo, and form three primary cell layers: the **ectoderm, mesoderm, and endoderm.** The ectoderm forms the outer layer and the endoderm forms the inner layer of the gastrula. The mesoderm is the middle layer. The organs and tissues formed from these layers are listed in ➤ Table 28.1. For the most part, ectoderm forms structures associated with the outer layers of the body such as skin. Ectoderm also gives rise to the brain and the nervous system. The mesoderm forms structures associated with support, movement, transport, and reproduction. The endoderm forms organs and tissues associated with digestion and respiration, including the liver, pancreas, and lungs. In frogs, the processes of cleavage and gastrulation are completed about 20 hours after fertilization.

Immediately after gastrulation, the body axis of the embryo begins to form. In chordates, the ectoderm region that will form the nervous system folds to form the **neural tube.** At the same time, the mesoderm beneath this region forms the **notochord** (➤ Figure 28.5). The formation of the neural tube and the notochord provides a preview of the general plan of the body. The dorsal and ventral axes and the anterior and posterior ends of the embryo are now visible. The neural tube will form the spinal cord (the bundle of nerves carrying impulses to and from the brain), and the notochord will become the vertebrae of the spinal column.

Mesodermal structures called **somites** also form at this time. These blocks of mesodermal cells will form segmented body parts including the muscles of the body wall.

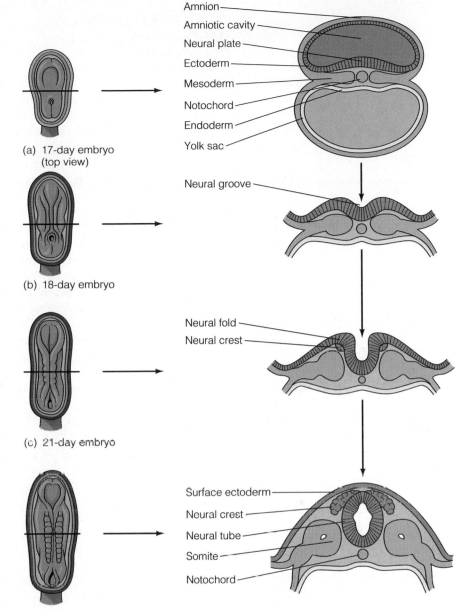

(a) 17-day embryo (top view)

(b) 18-day embryo

(c) 21-day embryo

(d) 23-day embryo

✦ **FIGURE 28.5**

**(a) A top view and side view of a 17-day human embryo showing the relationship among the ectoderm, mesoderm, and endoderm. (b) Ectoderm begins to fold into the neural groove, which progressively deepens (c). Finally, (d) the neural tube forms.**

---

**TABLE 28.1   ORGANS AND TISSUES FORMED FROM GERM LAYERS**

| ECTODERM | MESODERM | ENDODERM |
|---|---|---|
| Epidermis | Dermis | Lining of the digestive system |
| Hair, nails, sweat glands | All muscles of the body | Lining of the respiratory system |
| Brain and spinal cord | Cartilage | Urethra and urinary bladder |
| Cranial and spinal nerves | Bone | Gallbladder |
| Retina, lens, and cornea of eye | Blood | Liver and pancreas |
| Inner ear | All other connective tissue | Thyroid gland |
| Epithelium of nose, mouth, and anus | Blood vessels | Parathyroid gland |
| Enamel of teeth | Reproductive organs | Thymus |
| | Kidneys | |

## PATTERN FORMATION AND INDUCTION RESULT IN ORGAN FORMATION

With the main axis of the body established by the events of blastulation and gastrulation, organ formation is the next major stage of embryonic development. During organ formation, cell division is accompanied by migration and aggregation.

### Pattern Formation Lays Out the Body Plan

Pattern formation is the result of positional information, the ability of cells to "sense" their position in the embryo relative to other cells and to form structures that are appropriate for that position. How cells acquire positional information and how a program of differentiation begins is the subject of intense investigation. One model suggests the existence of gradients of informational molecules within the embryo. Pattern genes, called **homeobox genes,** are similar in organisms ranging from fruit flies to sea urchins, frogs, mice, and humans. The coordinated action of these genes in response to gradients of regulatory molecules establishes the body plan and the development of organs.

### Induction Starts Organ Formation

**Induction** is a process in which one cell or tissue type affects the developmental fate of another cell or tissue. As a cell begins to form specific structures in the embryo, certain genes are turned on and become active, while others are inactivated by being switched off. Induction affects the pattern of gene expression by physical contact or by the secretion of chemical signals. In most cases, the link between induction and the activation or inactivation of specific genes is not yet understood.

One of the best known examples of induction occurs during the formation of the eye in vertebrate (including human) embryos (✦ Figure 28.6). Eye formation begins when an outgrowth of the brain called the optic vesicle comes in contact with ectoderm on the embryo surface. Cells of the optic vesicle thicken and roll inward to form a cup. The cells in the cup will form the retina of the eye. Cells of the head ectoderm elongate and roll inward to form the lens of the eye. Finally, the lens induces the overlying ectoderm to form the cornea of the eye.

Induction plays an important role in the early formation of organs from the ectoderm, mesoderm, and endoderm. Understanding how genes are turned on and off during this process is one of the goals of developmental biology.

## HUMAN REPRODUCTION AND DEVELOPMENT

The basic principles of development as just outlined apply to a large number of animal species, including humans. In this section, we consider how humans reproduce and develop.

Human reproduction employs internal fertilization, a method found in vertebrates including some fish and amphibians, reptiles, birds and mammals. Human reproduction depends on the integrated action of the hypothalamus, the endocrine system, especially the anterior pituitary, and the reproductive organs. Males and females each possess sex organs, a pair of **gonads,** with associated accessory glands and ducts. The gonads of males are the **testes,** which produce spermatozoa and male sex hormones. The **ovaries** of females produce eggs (ova) and female sex hormones.

Within the gonads, cells produced by meiosis mature into gametes, and by fertilization, gametes from two parents unite to form a zygote, from which a new individual develops.

✦ **FIGURE 28.6**

**Induction plays an important role in organ formation. In this example, eye formation begins when an outgrowth of the brain called the optic vesicle comes in contact with ectoderm. Cells of the optic vesicle thicken and roll inward to form the optic cup. The cells in the cup form the retina of the eye. Cells of the head ectoderm elongate and roll inward to form the lens of the eye. In the last step, the lens induces the overlying ectoderm to form the cornea of the eye.**

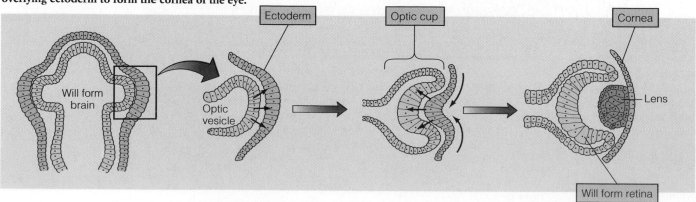

## Anatomy of the Male Reproductive System

The testes form in the abdominal cavity during male embryonic development and descend into the **scrotum,** a pouch of skin located outside the body cavity. The temperature in the scrotum is slightly lower than internal body temperature, and allows proper sperm development.

The interior of the testis is divided into a series of lobes, each of which contains tightly coiled lengths of **seminiferous tubules,** where sperm are produced (✦ Figure 28.7). Altogether, about 250 meters (850 feet) of tubules are packed into each testis. Cells called spermatocytes in the tubules divide by meiosis to produce haploid spermatids, which, in turn, differentiate into mature sperm

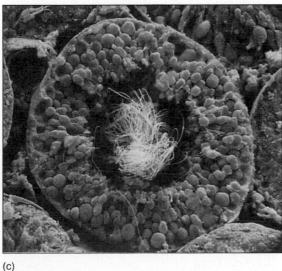

(a)

Nerve
Blood vessels
Ductus deferens
Seminiferous tubules
Epididymis
Testis

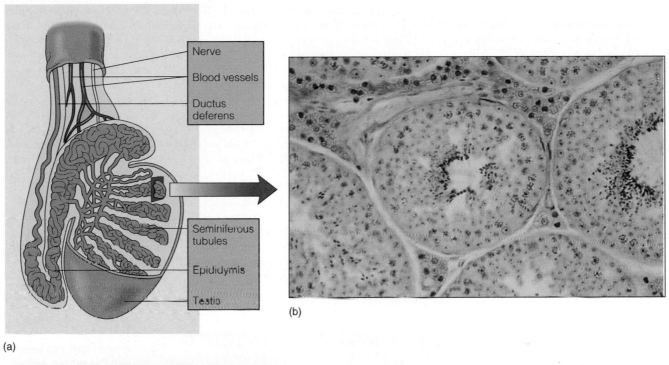

(b)

(c)

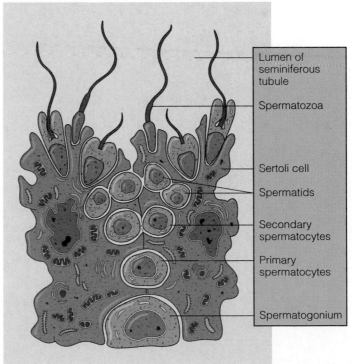

Lumen of seminiferous tubule
Spermatozoa
Sertoli cell
Spermatids
Secondary spermatocytes
Primary spermatocytes
Spermatogonium

(d)

✦ **FIGURE 28.7**

**(a) Section through the testis showing the location and arrangement of the seminiferous tubules and the epididymis. (b) Light micrograph of a section through a seminiferous tubule. The undifferentiated spermatogonia are at the periphery, the mature spermatozoa in the lumen, with various stages of development between. (c) A scanning electron micrograph of a section through a seminiferous tubule. (d) The location and relationship between Sertoli cells and the developing sperm cells.**

*Beyond the Basics*

# SEX TESTING IN INTERNATIONAL ATHLETICS—IS IT NECESSARY?

Success in amateur athletics such as the Olympics is often a prelude to financial rewards as a professional athlete. Several methods are used to guard against cheating in competition. Olympic competitors must submit urine samples (collected while someone watches) for drug testing, and other organizations depend on random drug testing to prevent the use of steroids and other drugs that enhance performance.

In the 1960s, concerns about males attempting to compete as females led the International Olympic Committee (IOC) to institute sex testing beginning with the 1968 Olympics. Sex testing may seem reasonable, given that men are usually larger, stronger, and faster than females, and males competing as females might have an unfair advantage. The IOC test involves analysis of epithelial cells recovered by scraping the inside of the mouth. In genetic females (XX) one chromosome is inactive, and when the cells are stained, shows up as a coiled bit of chromatin at the edge of the nucleus. Genetic males (XY) show no stained chromatin in their nuclei. No physical examination is required, the procedure is non-invasive, and females are not required to undress and submit to a physical examination of their genitals. If sexual identity is called into question as a result of the test, a full chromosome analysis is done, and if necessary, a gynecological examination follows. In practice, the IOC test has been noticeably unsuccessful. The test itself is somewhat unreliable, and leads to both false positives and false negatives. It fails to take into account XY females (who have a genetically determined insensitivity to male hormones and so develop as females) and XXY males (with Klinefelter syndrome). In addition, the test does not take into account the psychological, sociological, and cultural factors that enter into one's identity as a male or female. The test has not identified any males attempting to compete as females, but has barred several women from competition at all the Olympics since 1968. An analysis of testing on over 6000 female athletes has found that 1 in 500 females has had to withdraw from competition because of the sex tests.

In response to criticism, the IOC and the International Amateur Athletic Federation (IAAF) have developed different responses. Beginning in 1991, all IAAF athletes are required to undergo a physical examination and have their sexual status certified. The IOC has instituted a new test, based on recombinant DNA technology, to detect the presence or absence of a male-determining gene (SRY) carried on the Y chromosome. A positive test makes the athlete ineligible to compete as a female. Even this new technology remains controversial, and fails to recognize several chromosomal combinations that result in a female phenotype.

The basic question remains unanswered: Why are such sex tests needed in the first place? The IAAF and the IOC continue to debate this question and review their policies, but at this point, are not willing to give up sex testing.

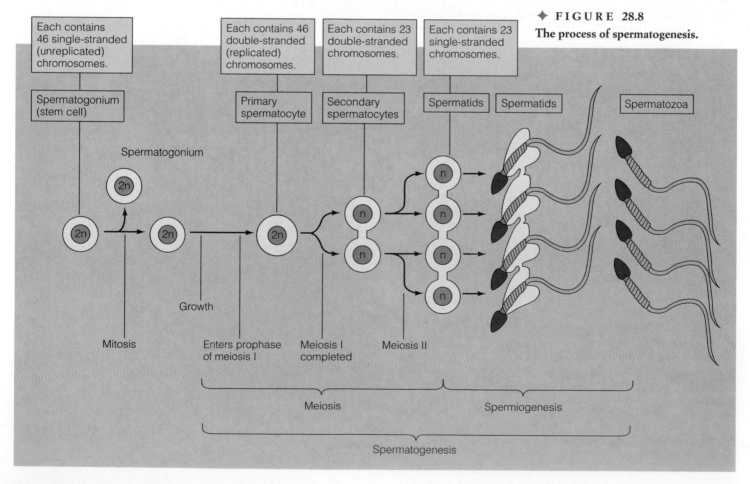

◆ **FIGURE 28.8**

**The process of spermatogenesis.**

Each contains 46 single-stranded (unreplicated) chromosomes.

Each contains 46 double-stranded (replicated) chromosomes.

Each contains 23 double-stranded chromosomes.

Each contains 23 single-stranded chromosomes.

Spermatogonium (stem cell)

Primary spermatocyte

Secondary spermatocytes

Spermatids

Spermatids

Spermatozoa

Spermatogonium

Growth

Mitosis

Enters prophase of meiosis I

Meiosis I completed

Meiosis II

Meiosis

Spermiogenesis

Spermatogenesis

(✦ Figure 28.8). Sperm production begins at puberty and continues throughout life; each day, several hundred million sperm are in various stages of maturation. Once formed, sperm move from the seminiferous tubules to the **epididymis,** where they mature and are stored.

Sperm production is controlled by hormones from the anterior pituitary and the testes (✦ Figure 28.9). The anterior pituitary secretes **follicle-stimulating hormone (FSH)** and **luteinizing hormone (LH).** LH secretion is controlled by a hormone from the hypothalamus known as **gonadotropin-releasing hormone (GnRH).** LH stimulates cells in the seminiferous tubules to secrete testosterone, which plays an important role in sperm production. Testosterone has other effects as well, including the development of secondary sexual characteristics such as muscle growth and the male pattern of hair distribution. FSH acts on cells in the seminiferous tubules to enhance maturation of sperm. The regulation of sperm production is controlled by a negative feedback system (see Chapter 25). Negative feedback by testosterone levels in the blood controls the secretion of GnRH.

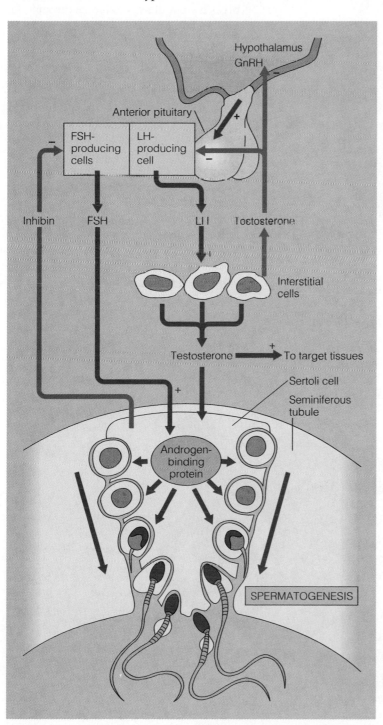

## ✦ FIGURE 28.9

**Cells in the hypothalamus release gonadotropin-releasing hormone (GnRH), which stimulates the anterior pituitary to release luteinizing hormone (LH) and follicle-stimulating hormone (FSH). LH acts to stimulate interstitial cells in the testis to release testosterone. FSH stimulates the Sertoli cells to make androgen-binding protein, which concentrates testosterone in the Sertoli cells. The testosterone, in turn, stimulates spermatogenesis. Elevated testosterone levels feed back to the hypothalamus and inhibit GnRH release, and inhibin, produced by Sertoli cells, inhibits FSH release.**

In addition to the testes, the male reproductive system also includes (1) a duct system that transports sperm out of the body, (2) three sets of glands that secrete fluids to maintain sperm viability and motility and (3) the penis (✦ Figure 28.10).

✦ **FIGURE 28.10**
**The anatomy of the male reproductive organs.**

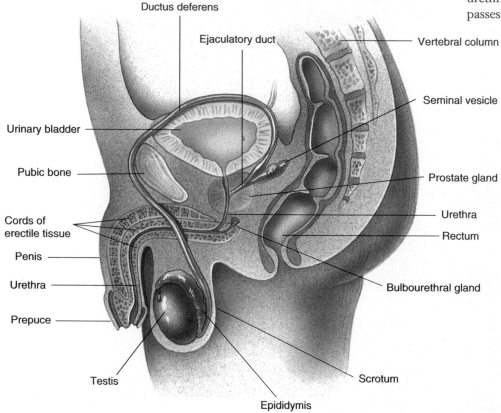

Sperm move through the male reproductive system in stages. When a male is sexually aroused, sperm move from the epididymis into the **vas deferens,** a duct connected to the epididymis. The walls of the vas deferens are lined with muscles, which contract rhythmically to move sperm forward. The vas deferens from each testis join to form a short **ejaculatory duct** that connects to the **urethra.** The urethra (which also functions in urine transport) passes through the penis and opens to the outside.

In the second stage, sperm are propelled by muscular contractions accompanying orgasm from the vas deferens through the urethra and expelled from the body.

As sperm are transported through the duct system in the first stage, secretions are added from three sets of glands. The **seminal vesicles** contribute fructose, a sugar that serves as an energy source for the sperm, and **prostaglandins,** locally acting chemical messengers that stimulate contraction of the female reproductive system to assist in sperm movement. The **prostate gland** secretes a milky, alkaline fluid that neutralizes acidic vaginal secretions and enhances sperm viability. The **bulbourethral glands** secrete a mucus-like substance that provides lubrication for intercourse. Together, the sperm and these various glandular secretions make up **semen,** a mixture that is about 95% secretions and about 5% spermatozoa. The components and functions of the male reproductive system are summarized in ➥ Table 28.2.

✦ **FIGURE 28.11**
**The ovary contains follicles in various stages of development.**

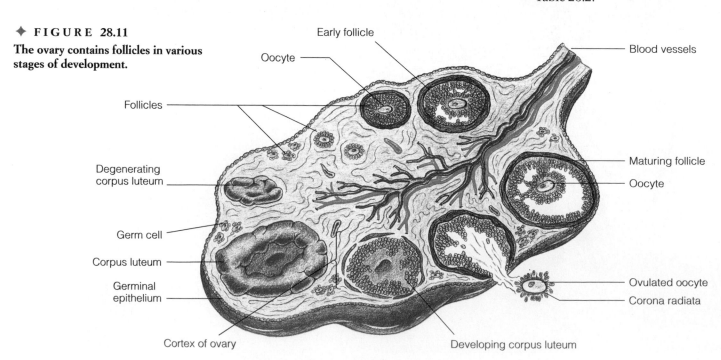

## TABLE 28.2 THE MALE REPRODUCTIVE SYSTEM

| COMPONENT | FUNCTION |
|---|---|
| Testes | Produce sperm and male sex steroids |
| Epididymis | Stores sperm |
| Vas deferens | Conducts sperm to urethra |
| Sex accessory glands | Produce seminal fluid that nourishes sperm |
| Urethra | Conducts sperm to outside |
| Penis | Organ of copulation |
| Scrotum | Provides proper temperature for testes |

## Anatomy of the Female Reproductive System

The female gonads are a pair of oval-shaped **ovaries** about 3 cm long, located in the abdominal cavity (✦ Figure 28.11). The ovary contains many **follicles,** consisting of a developing egg surrounded by a outer layer of follicle cells (✦ Figure 28.12). At the beginning of oogenesis, the developing egg is a primary oocyte. It begins meiosis in the third

✦ **FIGURE 28.12**

(a) Development of follicles from primary stages through the mature follicle. (b) A scanning electron micrograph of a developing oocyte. (c) Drawing of an ovary (actual size) showing a mature follicle.

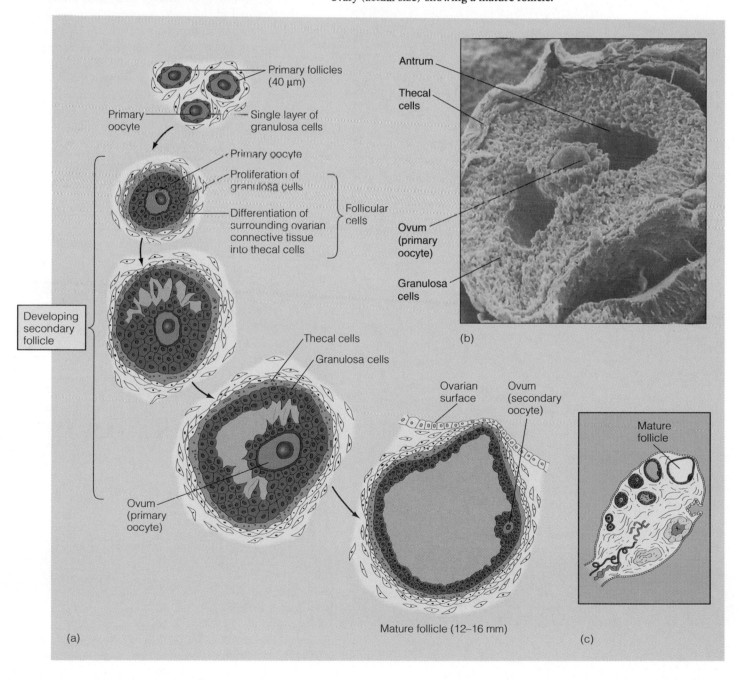

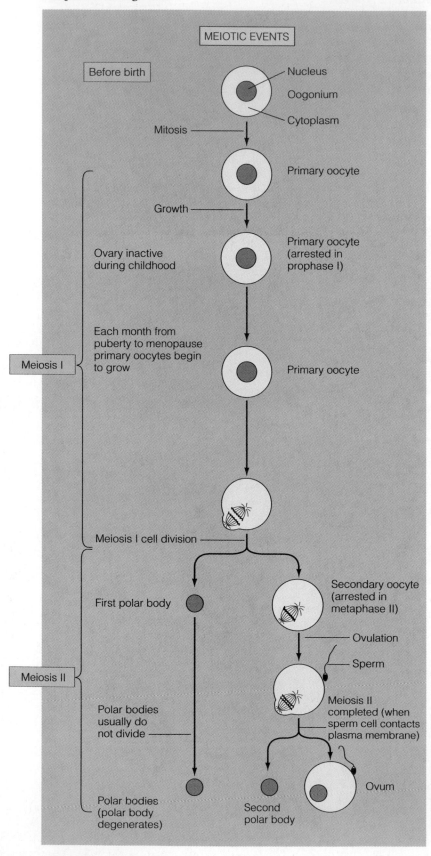

month of female prenatal development. At birth, the female carries a lifetime supply of developing oocytes, each of which is in the prophase of the first meiotic division (✦ Figure 28.13). A developing egg, called a secondary oocyte, is released by ovulation each month beginning at puberty, and over a woman's reproductive lifetime, about 400–500 of these gametes will be produced.

After puberty, the ovary cycles between two phases: the **follicular phase,** associated with maturing follicles, and the **luteal phase,** characterized by the presence of a corpus luteum. These cyclic phases are interrupted only by pregnancy, and continue until menopause, when reproductive capacity ceases.

The ovarian cycle usually lasts about 28 days. In the first half of the cycle, an oocyte matures within the follicle, and completes meiosis I. At midpoint in the average cycle (about 14 days), the maturing follicle bulges out, and ruptures to release the oocyte onto the surface of the ovary, a process called **ovulation** (✦ Figure 28.14). Following ovulation, the follicle forms the **corpus luteum,** which synthesizes

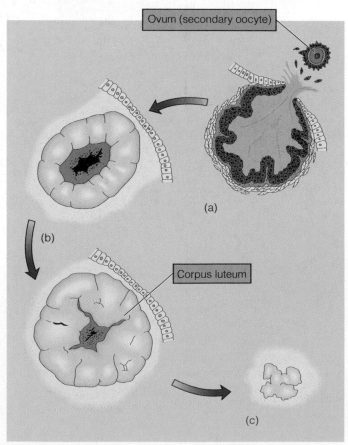

✦ **FIGURE 28.14**

**(a) Ovulation and release of a secondary oocyte. (b) Formation of a corpus luteum from a mature follicle following ovulation. (c) The corpus luteum degenerates if the oocyte is not fertilized.**

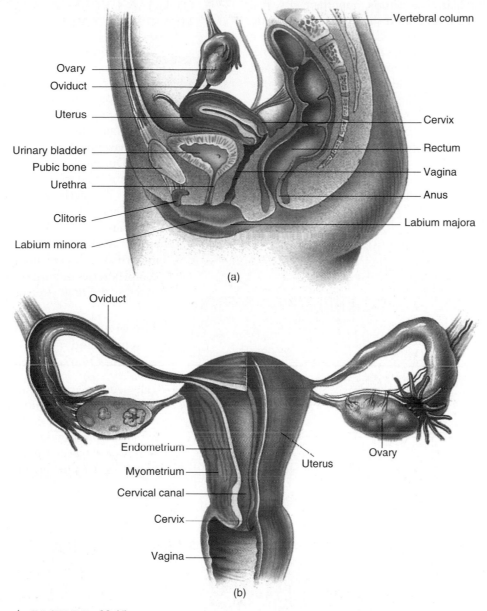

Ovary

Oviduct

Uterus

Urinary bladder

Pubic bone

Urethra

Clitoris

Labium minora

Vertebral column

Cervix

Rectum

Vagina

Anus

Labium majora

(a)

Oviduct

Endometrium

Myometrium

Cervical canal

Cervix

Vagina

Uterus

Ovary

(b)

✦ **FIGURE 28.15**

**The anatomy of the female reproductive system. (a) External anatomy. (b) Internal components of the reproductive system.**

and secretes hormones that prepare the uterus for pregnancy in case the egg is fertilized.

The ovulated cell, called a secondary oocyte, is swept by ciliary action into the **oviduct** (also called the fallopian tube or uterine tube) (✦ Figure 28.15a). At one end, the oviduct has finger-like projections that partially surround the ovary.

The oviduct is connected to the **uterus,** a hollow, pear-shaped muscular organ about 7.5 centimeters (3 inches) long and 5 centimeters (2 inches) wide (Figure 28.15b). The uterus has a thick, muscular outer layer and an inner layer, the **endometrium.**

The inner layer of the endometrium is shed at menstruation if fertilization has not occurred. The lower neck of the uterus, the **cervix,** opens into the **vagina.** The vagina receives the penis during intercourse and also serves as the birth canal. The vagina opens to the outside of the body behind the urethra (Figure 28.15).

The external genitals are known collectively as the **vulva.** The **labia minora** are thin, membranous folds of skin just outside the vaginal opening (Figure 28.15a). A pair of fleshy outer folds known as the **labia majora** cover and protect the genital region. The surface of the labia majora are covered with pubic hair. The **clitoris,** important in sexual arousal, consists of a short shaft with a sensitive tip covered by a fold of skin. Glands near the opening of the vagina secrete fluids that lubricate the vaginal canal during sexual arousal. The components and functions of the female reproductive system are summarized in ➡ Table 28.3.

# HORMONES REGULATE THE OVARIAN CYCLE AND THE MENSTRUAL CYCLE

The ovarian cycle is hormonally regulated in two phases. Before ovulation, events are regulated by an estrogen-secreting follicle. After ovulation, events are controlled by the corpus luteum, which secretes estrogen and progesterone. The ovarian cycle itself is regulated by hormones from the hypothalamus and the anterior pituitary. During the ovarian cycle, a series of related events occurs in the uterus. The cyclic secretion of hormones and the associated uterine tissue changes are known as the **menstrual cycle.** The ovarian and menstrual cycles are interrelated by a complex series of hormonal feedback systems (✦ Figure 28.16).

## The Menstrual Cycle Is Synchronized with the Ovarian Cycle

The menstrual cycle is highly variable in length from one individual to another, and from one cycle to another, but the typical cycle is 28 days. By convention, the first day of the cycle (day 0) is the first day of blood flow

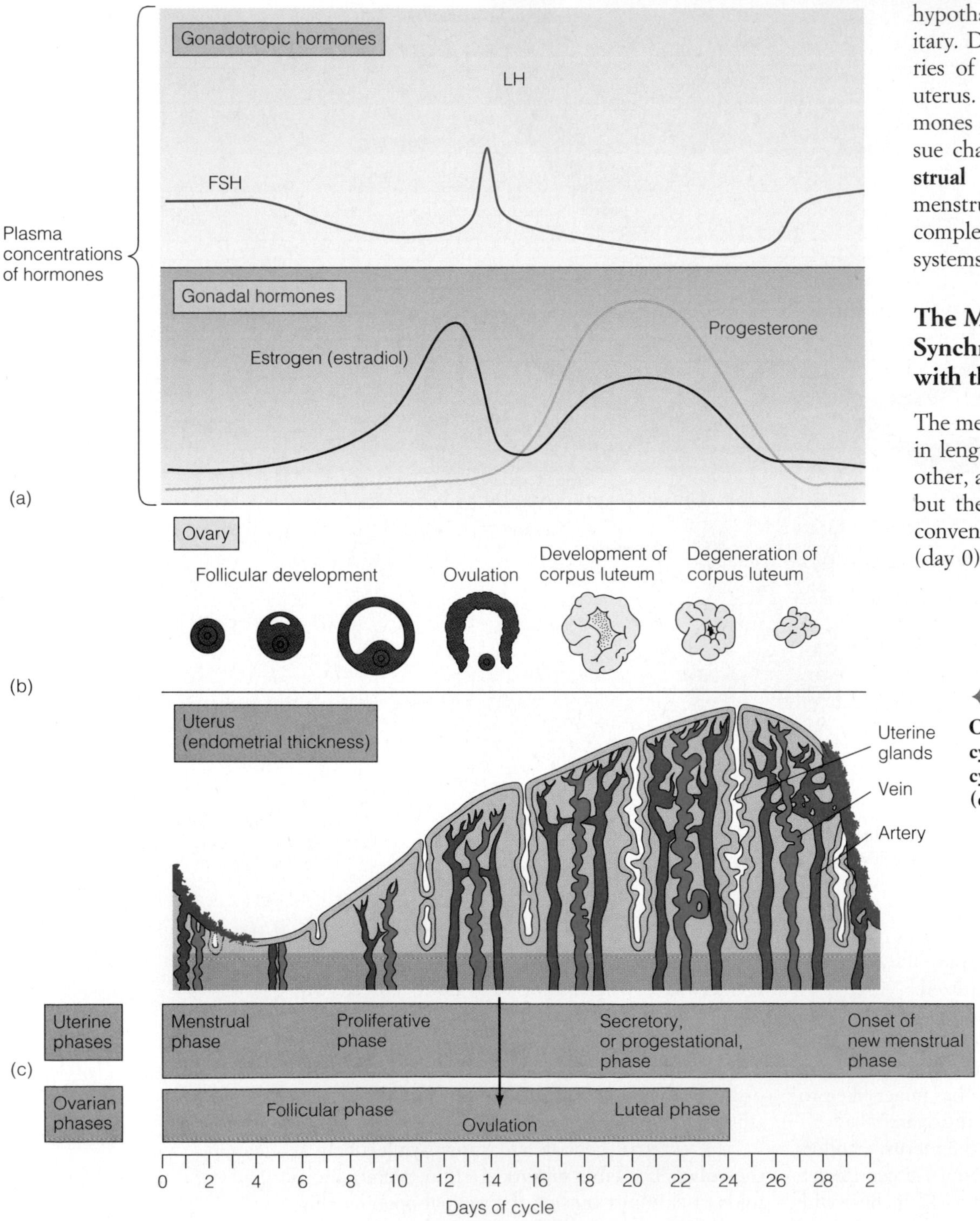

(a) Plasma concentrations of hormones — Gonadotropic hormones (LH, FSH); Gonadal hormones (Estrogen (estradiol), Progesterone)

(b) Ovary — Follicular development, Ovulation, Development of corpus luteum, Degeneration of corpus luteum

(c) Uterus (endometrial thickness) — Uterine glands, Vein, Artery

Uterine phases: Menstrual phase, Proliferative phase, Secretory, or progestational, phase, Onset of new menstrual phase

Ovarian phases: Follicular phase, Ovulation, Luteal phase

Days of cycle: 0 2 4 6 8 10 12 14 16 18 20 22 24 26 28 2

✦ **FIGURE 28.16**
**Overview of the menstrual cycle. (a) The hormonal cycle. (b) The ovarian cycle. (c) The uterine cycle.**

(**menstruation**) in the menstrual period. Menstruation is associated with the breakdown and shedding of cells in the uterine endometrium. These cells, along with blood and mucus, pass out of the body through the vagina as menstrual flow. The first day of menstruation is also the beginning of the ovarian cycle (Figure 28.16b). Follicle-stimulating hormone (FSH) and luteinizing hormone (LH) are secreted by the anterior pituitary in the first half of the cycle. FSH stimulates a single follicle in the ovary to mature, and both FSH and LH stimulate the follicle to secrete the hormone **estrogen.** As the blood level of estrogen rises, it triggers the secretion of LH by the pituitary. This burst of LH secretion stimulates follicle maturation and ovulation, which takes place about midcycle.

Under the influence of LH, the remaining cells in the follicle form the corpus luteum, which synthesizes progesterone and estrogen. In the second half of the ovarian cycle, these hormones secreted by the corpus luteum act on the uterus. This causes the endometrium to develop a new inner layer of cells with a rich supply of blood vessels for implantation of a fertilized egg. Rising levels of estrogen and progesterone from the corpus luteum inhibit the secretion of FSH and LH by the pituitary. If pregnancy does not occur, the drop in FSH and LH levels causes the corpus luteum to degenerate. As it does so, hormonal output from the corpus luteum falls, causing the inner layer of the endometrium and its associated blood vessels to die. Mild, rhythmic contractions of the uterus expel the endometrial debris and blood as a new period of menstrual flow. The uterine contractions are caused by uterine prostaglandins. Oversecretion of these local messengers can cause the menstrual cramps experienced by many women.

Without the continued presence of inhibiting hormones from the corpus luteum, the pituitary begins again to secrete FSH and LH, and a new ovarian cycle is initiated. This sequence of cycles is interrupted only when fertilization and a resulting pregnancy occur.

##  HUMAN SEXUAL RESPONSES

In most mammals, females are fertile only during a restricted time period, once or a few times a year. During this time, the female is receptive to the male. In some higher primates, including humans, females are sexually receptive at all times. The physical events of mating in humans occurs in four stages: arousal, plateau, orgasm, and resolution.

During sexual arousal in males, blood flows into the three cylinders of spongy tissue within the penis, causing it to elongate and become erect (◆ Table 28.4). Sexual arousal in the female causes

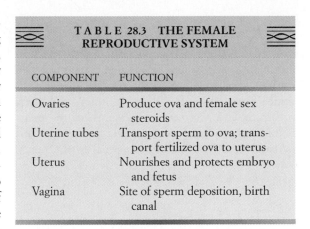

**TABLE 28.3   THE FEMALE REPRODUCTIVE SYSTEM**

| COMPONENT | FUNCTION |
|---|---|
| Ovaries | Produce ova and female sex steroids |
| Uterine tubes | Transport sperm to ova; transport fertilized ova to uterus |
| Uterus | Nourishes and protects embryo and fetus |
| Vagina | Site of sperm deposition, birth canal |

**TABLE 28.4   MALE RESPONSES DURING THE SEXUAL RESPONSE CYCLE**

**Arousal Phase**
1. Vasocongestion (accumulation of blood) erects the penis.
2. Scrotal skin tightens.
3. Testes start to increase in size.
4. Testes and scrotum become elevated.
5. Nipples may become erect.
6. Some increase occurs in muscle tension, heart rate, and blood pressure.

**Plateau Phase**
1. There is a slight increase in the area of the glans penis.
2. Vasocongestion purples the glans.
3. Testes continue to elevate up into the scrotal sac until they are positioned close against the body.
4. Testes increase in size as much as 50%.
5. Cowper's glands secrete a few drops of fluid.
6. Possible sex flush and muscle tension are present, breathing is rapid, and heart rate increases (100 to 160 beats per minute).

**Orgasm**
1. Contractions of the vas deferens, seminal vesicles, ejaculatory duct, and prostate gland cause semen to collect in the base of the urethra.
2. Collection of semen in the base of the urethra produces feelings of ejaculatory inevitability.
3. The internal sphincter in the prostate contracts, preventing passage of urine.
4. The external sphincter in the prostate relaxes, allowing passage of semen.
5. Contractions in the urethra propel the semen out of the penis. The contractions occur four or five times at intervals of eight-tenths of a second.
6. Muscles go into spasm throughout the body, respiration increases, and blood pressure and heart rate reach a peak (about 180 beats per minute).

**Resolution Phase**
1. Body returns gradually to its prearoused state.
2. Male gradually loses his erection.
3. Testes and scrotum return to normal size.
5. Male enters a refractory period (unresponsive to further sexual stimulation).
6. Blood pressure, heart rate, and respiration become normal.
7. About one-third of males find their palms and soles or entire body covered with perspiration.

the clitoris and nipples to become erect, and the areas around the vaginal opening to become swollen ( Table 28.5). In addition, lubricating fluids are secreted into the vagina. In both sexes, during the plateau phase, arousal responses continue, muscles in the limbs contract, and both breathing and heart rate increase. After the penis is inserted into the vagina, pelvic thrusts by both partners stimulate sensory receptors in the penis, vaginal walls, and clitoris. This stimulation of the penis causes sperm to move out of the epididymis and secretions from the seminal vesicles and prostate to enter the urethra. Continued stimulation may result in **orgasm** for both partners, involving rhythmic muscular contractions of the genitals and waves of intense pleasurable sensations. Orgasm in males results in the ejaculation of sperm. Resolution reverses the responses of the previous phases. Muscles relax, breathing and heart rates return to normal, and blood flows out of the penis, which returns to normal size.

## T A B L E 28.5   FEMALE RESPONSES DURING THE SEXUAL RESPONSE CYCLE

### Arousal Phase

1. Vaginal lubrication begins.
2. Vasocongestion swells the external genitalia.
3. Labia majora flatten and retract from the vaginal opening.
4. Inner two-thirds of the vagina expands.
5. Vaginal walls thicken.
6. Breasts enlarge, and blood vessels near the surface become more prominent.
7. Nipples become erect.
8. Muscle tension, heart rate, and blood pressure increase somewhat.

### Plateau Phase

1. Vasocongestion produces a narrow vaginal pathway in the outer third of the vagina.
2. Inner part of the vagina expands fully.
3. Uterus becomes elevated.
4. Clitoris withdraws beneath the clitoral hood.
5. Rosy appearance may occur on the stomach, thighs, and back.
6. Nipples become more erect.
7. Muscles tense, breathing is rapid, and heart rate increases (100 to 160 beats per minute).

### Orgasm

1. Swelling of the tissues of the outer part of the vagina constricts the vaginal opening.
2. Contractions begin in the outer third of the vagina. First contractions may last 2 to 4 seconds, and later ones may last 3 to 15 seconds. They occur at intervals of 0.8 of a second.
3. Inner two-thirds of the vagina expands slightly.
4. Uterus contracts.
5. Muscles go into spasm throughout the body, respiration increases, and blood pressure and heart rate reach a peak (about 180 beats per minute).

### Resolution Phase

1. Blood is released from engorged areas.
2. Rosy appearance disappears.
3. Clitoris descends to normal position.
4. Vagina, uterus, and labia gradually shrink to normal size.
5. Blood pressure, heart rate, and respiration become normal.
6. About one-third of females find their palms and soles or entire body covered with perspiration.

## Sexually Transmitted Diseases

Sexually transmitted diseases (STDs) are common in the United States and other parts of the world. On any given day, more than 50 million people in the United States have an STD, and each year more than $7 billion is spent on treatment of STDs, not counting the funds spent on research and treatment of AIDS. Sexual partners are not the only ones at risk. STDs affect the fetus and newborn infants, and can cause blindness, mental retardation, severe neurological defects and death of the mother and/or child.

STDs are grouped into three classes. The first includes those that produce inflammation of the urethra, epididymis, cervix, or oviducts. The most common of these are **gonorrhea** and **chlamydia.** Gonorrhea is caused by a bacterium that inflames and damages epithelial cells. Chlamydia is caused by a parasitic bacterium that lives inside cells of the reproductive tract. This disease is particularly insidious, because infected individuals may not have any noticeable symptoms and unknowingly spread the disease by sexual contact. In addition, females with this infection can develop pelvic inflammatory disease (PID), which can cause infertility by blocking and scarring the oviducts. Once diagnosed, both these diseases can be treated and cured with antibiotics.

The second class of STDs includes those that produce sores on the external genitals. **Genital herpes** is the most common disease in this class, with more than 25 million people in the United States affected. Genital herpes is caused by direct contact with the herpes I or herpes II virus. About 15% of all cases of genital herpes are caused by herpes I and 85% by herpes II. Symptoms are recurring and involve small, often painful sores on mucus membranes of the mouth or genital system. About 85% of all herpes infections in the mouth are caused by herpes I, and 25% by herpes II. Antiviral drugs can be used to treat symptoms of herpes I and II infections, but cannot cure the infections. **Syphilis** is another disease in this class. It is caused by a bacterial

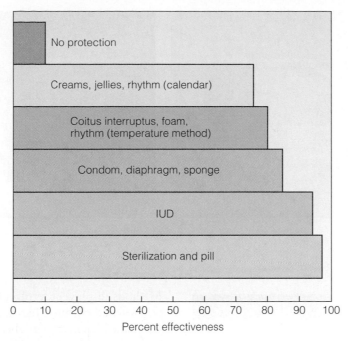

Percent effectiveness

✦ **FIGURE** **28.17**

**Effectiveness of various methods of birth control. The percent effectiveness is a measure of how many women in a group of 100 will not become pregnant in a year when using a given method of birth control.**

infection, and produces an ulcer on the genitals, although in women, the sore may be on the cervix and not visible. This disease can be treated and cured with antibiotics, but if left untreated, can affect the heart, eyes, kidneys, brain, and liver, and cause serious symptoms and death.

The third class of STDs includes viral diseases that affect organ systems other than those of the reproductive system. These include **AIDS** and **hepatitis B.** In addition to sexual contact, these diseases are spread by contact with blood. In the early stages, infected individuals have no symptoms and can unknowingly spread the infection. AIDS was discussed in detail in Chapter 22. Hepatitis B affects the liver and can cause a progressive fatal disease. There is no cure for either of these diseases.

## REPRODUCTION AND TECHNOLOGY

Advances in genetics, physiology, and molecular biology have led to the development of techniques that enhance or reduce the chances of conception. Reproductive technologies can correct defective functions and manipulate the physiology of reproduction. This technology has developed more rapidly than social conventions and laws governing its use, leading to controversy about the moral, ethical, and legal grounds for the use of these techniques.

## Contraception Uncouples Sexual Intercourse from Pregnancy

The uncoupling of sexual intercourse from fertilization and pregnancy employs methods that block one of three stages in reproduction: release and transport of gametes, fertilization, or implantation. None of these methods is completely successful in preventing pregnancy or STDs. The only completely effective method is abstinence from sexual intercourse (✦ Figure 28.17).

Aside from abstinence, methods that physically prevent the release and transport of gametes (such as vasectomy and tubal ligation) are the most effective. In **tubal ligation,** the oviduct is cut and the ends tied off to prevent ovulated eggs from reaching the uterus. In **vasectomy,** the vas deferens is withdrawn through a small incision in the scrotum and cut. The cut ends are sealed to prevent transport of sperm.

Birth control pills are an effective method for preventing the development and release of eggs. The most common birth control pills contain a combination of hormones that prevent the release of LH and FSH by the pituitary and hypothalamus. As a result, ovarian follicle development is inhibited, and no oocytes are released. Time-release capsules (sold as Norplant) that can be implanted under the skin release hormones slowly and offer long-term suppression of ovulation.

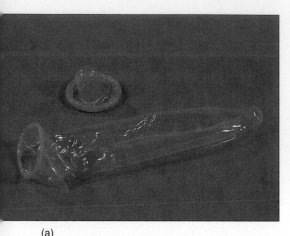
(a)

(b)

(c)

✦ **FIGURE 28.18**

**Methods of contraception. (a) Condom. (b) Diaphragm. (c) Spermicidal gel.**

| NEW WAYS TO MAKE BABIES | |
|---|---|
| **Artificial Insemination and Embryo Transfer** | **In Vitro Fertilization** |
| 1. Father is infertile. Mother is inseminated by donor and carries child.<br><br>● + ● = ☖ | 1. Mother is fertile but unable to conceive. Ovum from mother and sperm from father are combined in laboratory. Embryo is placed in mother's uterus.<br><br>● + ● = ☖ |
| 2. Mother is infertile but able to carry child. Donor of ovum is inseminated by father; then embryo is transferred and mother carries child.<br><br>○ + ● = ☖ | 2. Mother is infertile but able to carry child. Ovum from donor is combined with sperm from father.<br><br>○ + ● = ☖ |
| 3. Mother is infertile and unable to carry child. Donor of ovum is inseminated by father and carries child.<br><br>○ + ● = ☖ | 3. Father is infertile and mother is fertile but unable to conceive. Ovum from mother is combined with sperm from donor.<br><br>● + ● = ☖ |
| 4. Both parents are infertile, but mother is able to carry child. Donor of ovum is inseminated by sperm donor; then embryo is transferred and mother carries child.<br><br>○ + ● = ☖ | 4. Both parents are infertile, but mother is able to carry child. Ovum and sperm from donors are combined in laboratory (also see number 4, column at left).<br><br>○ + ● = ☖ |
|  | 5. Mother in infertile and unable to carry child. Ovum of donor is combined with sperm from father. Embryo is transferred to donor (also see number 2, column at left).<br><br>○ + ● = ☖ |
|  | 6. Both parents are fertile, but mother is unable to carry child. Ovum from mother and sperm from father are combined. Embryo is transferred to donor.<br><br>● + ● = ☖ |
|  | 7. Father is infertile; mother is fertile but unable to carry child.<br><br>● + ● = ☖ |

| LEGEND: | |
|---|---|
| Sperm from father | ● |
| Ovum from mother | ● |
| Baby born of mother | ☖ |
| Sperm from donor | ● |
| Ovum from donor | ○ |
| Baby born of donor (Surrogate) | ☖ |

Methods to prevent fertilization include the use of physical and chemical barriers (✦ Figure 28.18). Condoms are latex or gut sheaths worn over the penis during intercourse. Only latex condoms prevent sexually transmitted diseases, including AIDS. Diaphragms are caps that fit over the cervix and are inserted before intercourse. Chemical barriers include the use of spermicidal jellies or foams that kill sperm on contact. They are placed into the vagina just before intercourse. Contraceptive sponges, filled with spermicides, combine physical and chemical barriers. Condoms treated with a chemical spermicide are more successful at preventing pregnancy than either condoms or chemical barriers alone.

A European drug, RU-486, is being tested in the United States for use as a contraceptive. This drug, chemically related to reproductive hormones, interferes with events following fertilization and may inhibit implantation.

## Reproductive Technology Can Enhance Fertility

About one in six couples is infertile, that is, unable to have children after a year of trying to conceive. Physical and physiological conditions prevent the production of gametes, fertilization, or implantation. Technologies to reduce or overcome these problems have been developed in the last two decades.

Blocked oviducts, often the result of untreated STDs, are the leading cause of infertility in females.

✦ **FIGURE 28.19**

**Some of the ways in which gametes can be combined to produce babies using *in vitro* fertilization.**

In males, a low sperm count, low motility, and blocked ducts are causes of infertility. Hormone therapy can often be used to increase egg production, and surgical procedures can sometimes be used to open closed ducts in the reproductive tracts. Overall, about 40% of infertility is related to problems in the male reproductive tract, 40% is associated with the female tract, and in 20% of the cases, the infertility is of unknown origin.

One method to recover and fertilize gametes outside the body, known as *in vitro* fertilization, is a widely used method of reproductive technology. Many variations of this technology are available (✦ Figure 28.19).

Methods to enhance fertility have enabled a growing number of couples to become parents, but has also generated some controversies that break new legal and ethical ground.

Reproductive technology has also altered accepted patterns of reproduction. It has been discovered that the age of ova, not the reproductive system, is responsible for infertility as women age. Women are now becoming mothers in their late fifties and early sixties by receiving implanted zygotes, produced by fertilization of donated eggs from younger women. Fertilized eggs can now be collected and frozen for later use. This allows women to collect eggs for fertilization while risks for chromosome abnormalities in the offspring are low, and to use them over a period of years, including menopause, to become mothers.

## HUMAN DEVELOPMENT FROM FERTILIZATION TO BIRTH

**Fertilization,** the fusion of male and female gametes, usually occurs in the upper third of the oviduct. Sperm deposited into the vagina travel through the cervix, up the uterus, and into the oviduct. About 30 minutes after ejaculation, sperm are present in the oviduct (✦ Figure 28.20). Sperm travel this distance by swimming via whip-like contractions of their tails. Muscular contractions of the uterus aid in movement and help disperse the sperm throughout the uterine cavity. Although several hundred million sperm are released in a single ejaculation, only a few thousand reach the upper area of each oviduct.

Only one sperm will fertilize the egg, but many other sperm assist in this process (✦ Figure 28.21), perhaps by triggering chemical changes in the egg. One sperm contacts receptors on the surface of the secondary oocyte, and fuses with its outer membrane. This attachment triggers a series of chemical

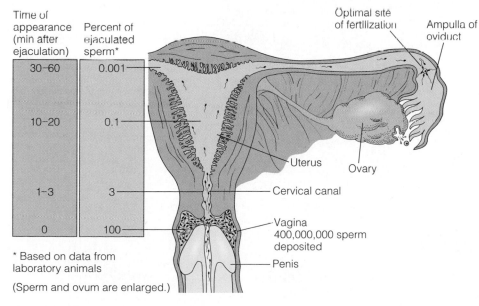

✦ FIGURE 28.20

The time and relative amounts of sperm transport to the oviduct.

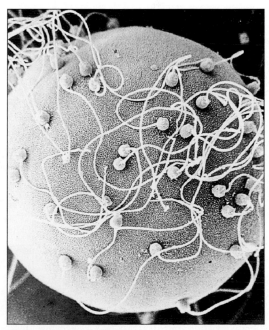

✦ FIGURE 28.21

Scanning electron micrograph of an oocyte surrounded by sperm. Only one sperm will enter the egg.

changes in the outer membrane and prevents any other sperm from entering the oocyte (✦ Figure 28.22). Movement of the sperm into the oocyte cytoplasm initiates the second meiotic division of the oocyte. Fusion of the haploid sperm nucleus with the resulting haploid oocyte nucleus forms a diploid zygote.

The zygote travels down the oviduct to the uterus over the next three to four days. While in the oviduct, the zygote undergoes a series of rapid cell divisions to form a solid ball of cells called the morula. Once the zygote begins to divide, it becomes an embryo. The morula descends into the uterus and floats unattached in the uterine interior for several days, drawing nutrients from the uterine fluids, and continuing to divide to form a **blastocyst**

(✦ Figure 28.23). If the morula does not move out of the oviduct, it can continue to develop in the oviduct, causing a **tubal pregnancy.** This condition causes sharp abdominal pain as the developing embryo stretches the oviduct. A potentially lethal hemorrhage can result, and these pregnancies are usually terminated by surgical intervention.

The blastocyst is made up of the inner cell mass, an internal cavity and an outer layer of cells (the **trophoblast**). During the week or so that the zygote is dividing, the endometrium of the uterus continues to enlarge and differentiate. During **implantation,** the trophoblast cells adhere to the endometrium and secrete enzymes that allow fingers of trophoblast cells to penetrate the endometrium, locking the embryo into place (✦ Figure 28.24).

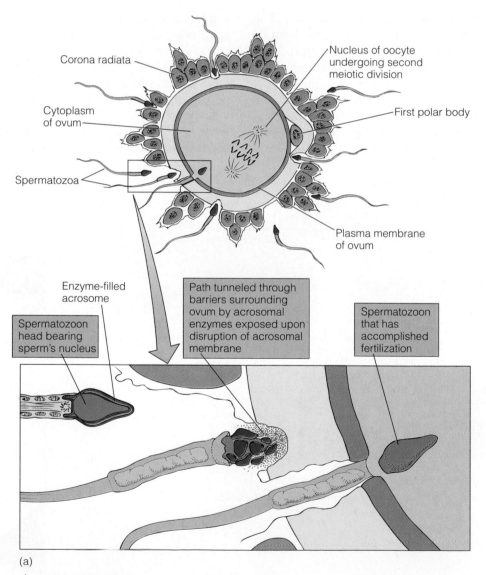

Corona radiata

Cytoplasm of ovum

Spermatozoa

Nucleus of oocyte undergoing second meiotic division

First polar body

Plasma membrane of ovum

Enzyme-filled acrosome

Spermatozoon head bearing sperm's nucleus

Path tunneled through barriers surrounding ovum by acrosomal enzymes exposed upon disruption of acrosomal membrane

Spermatozoon that has accomplished fertilization

(a)

✦ **FIGURE 28.22**

**The process of fertilization. The tip of the sperm head, known as the acrosome, contains enzymes that dissolve the outer barriers surrounding the oocyte. Only the head of the sperm enters the egg.**

At 12 days after fertilization, the embryo is embedded in the endometrium, and the trophoblast has formed a two-layered structure, the **chorion.** One of the first events that follows implantation is the secretion of a peptide hormone, **human chori-** **onic gonadotropin (hCG),** by the chorion. This hormone prolongs the life of the corpus luteum, and prevents breakdown of the uterine lining. The corpus luteum continues to secrete greater amounts of estrogen and progesterone for 10 weeks, until

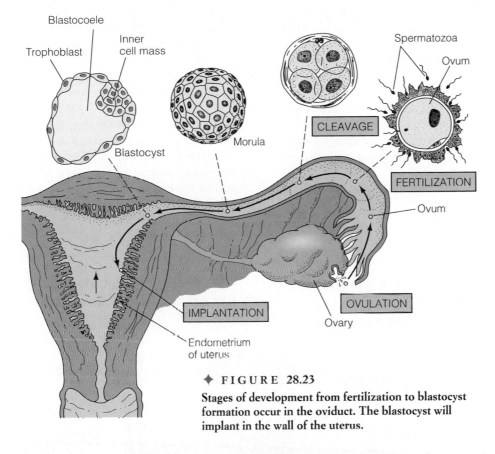

✦ **FIGURE 28.23**

**Stages of development from fertilization to blastocyst formation occur in the oviduct. The blastocyst will implant in the wall of the uterus.**

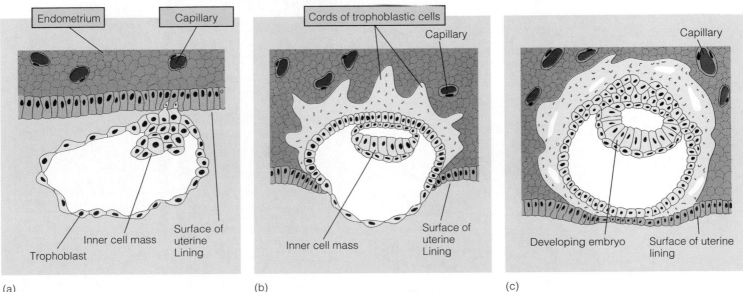

(a)  (b)  (c)

✦ **FIGURE 28.24**

**The process of implantation. (a) A blastocyst attaches to the endometrial lining of the uterus. (b) As the blastocyst implants, cords of chorionic cells form. (c) When implantation is complete, the blastocyst is buried in the endometrium.**

the placenta begins secreting these hormones. Excess hCG is eliminated in the urine. Home pregnancy tests work by detecting elevated hCG levels, as early as two weeks after the first missed menstrual period.

As the chorion grows and expands, it forms a series of finger-like projections that extend into cavities filled with maternal blood. Embryonic capillaries extend into these projections, or **villi.** The embryonic circulation and the maternal pools of blood are separated by only a thin layer of cells, allowing exchange of nutrients between the embryonic and maternal circulation. Further development of this interlocking structure forms the **placenta.** The membranes connecting the embryo to the placenta develop to form the **umbilical cord,** containing two umbilical arteries and a single umbilical vein ( ✦ Figure 28.25).

## Human Development Is Divided into Three Stages

The period from conception to birth is divided into three trimesters, each about three months (12 weeks) in length. During this period of about 38 weeks, the single-celled zygote undergoes about 40 to 44 rounds of mitosis, producing an infant containing trillions of cells organized into highly specialized tissues and organs.

## Rapid and Complex Events Occur in the First Trimester

The first 12 weeks of development are a period of radical changes in the size, shape, and complexity of the embryo. In the week after implantation, the ectoderm, mesoderm, and endoderm are formed, and by the end of the third week, cellular differentiation begins to form organ systems.

At the end of the first month, the embryo is about 5 mm in length, and much of the body is composed of paired somite segments ( ✦ Figure 28.26). Also at this stage, the embryo has a shape similar to that of other vertebrate embryos, and even includes a tail, which is later resorbed.

✦ **FIGURE 28.25**

**The maternal and embryonic structures interact to form the placenta.**

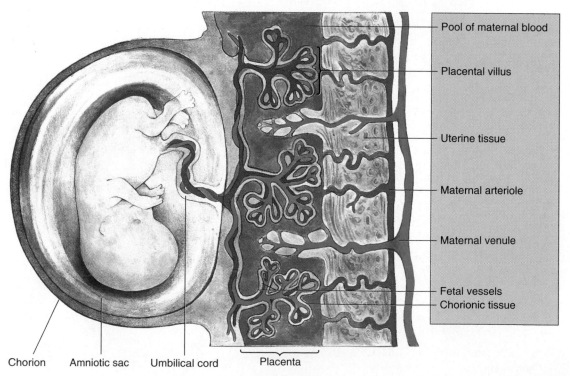

Pool of maternal blood

Placental villus

Uterine tissue

Maternal arteriole

Maternal venule

Fetal vessels
Chorionic tissue

Chorion  Amniotic sac  Umbilical cord  Placenta

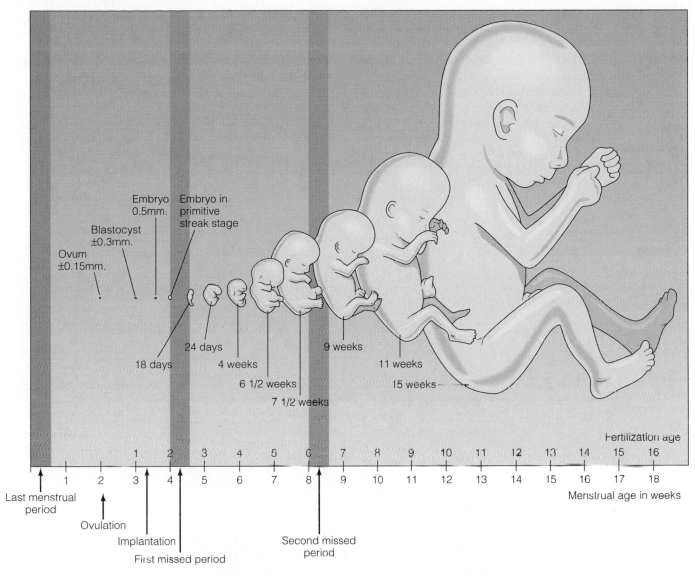

Ovum
±0.15mm.

Blastocyst
±0.3mm.

Embryo
0.5mm.

Embryo in
primitive
streak stage

24 days

18 days

4 weeks

6 1/2 weeks

7 1/2 weeks

9 weeks

11 weeks

15 weeks

Fertilization age

1　2　3　4　5　6　7　8　9　10　11　12　13　14　15　16

1　2　3　4　5　6　7　8　9　10　11　12　13　14　15　16　17　18

Last menstrual
period

Ovulation

Implantation

First missed period

Second missed
period

Menstrual age in weeks

✦ **FIGURE 28.26**
**Growth of the embryo and fetus during the first 16 weeks of development. Sizes in the figure represent actual sizes.**

In the second month, the embryo grows dramatically to a length of about 3 cm, and undergoes a 500-fold increase in mass. Most of the major organ systems, including the four chambers of the heart, are formed. The limb buds develop into arms and legs, complete with fingers and toes. The head is very large in relation to the rest of the body, because of the rapid development of the nervous system.

By about seven weeks, the embryo is now called a fetus. Although chromosomal sex (XX females and XY males) is determined at the time of fertilization, the fetus is sexually neutral at the end of seven weeks. Beginning in the eighth week, devel-

opmental pathways activate different gene sets and initiate sexual development.

In XY fetuses, genes present on the Y chromosome cause the undifferentiated gonad to begin development as a testis. Once formed, testes synthesize **testosterone** and **Mullerian inhibiting hormone (MIH),** hormones that control further male sexual development. Testosterone promotes development of the male duct system, and MIH causes the degeneration of the female duct system.

In XX embryos, absence of a Y chromosome and the presence of a second X chromosome initiates formation of ovaries from the undifferentiated gonads. In the absence of testosterone, the male duct system degenerates, and in the absence of MIH, the female duct system survives and develops.

After male or female gonads are established, the external genitalia develop (✦ Figure 28.27). In males, a hormone, dihydroxytestosterone (DHT), causes the indifferent genitalia to develop into the penis and scrotum. In the female, absence of DHT causes the indifferent genitalia to form the clitoris and labia.

By the end of the third month, the fetus is about 9 cm (about 3.5 inches) long, and weighs about 12 to 15 g (about half an ounce), and all the major organ systems have formed.

## The Second Trimester Is a Period of Organ Maturation

In the second trimester, major changes involve an increase in size and the further development of organ systems. Bony parts of the skeleton begin to form, and the heartbeat can be heard with a stethoscope. Fetal movements begin in the third month, and by the fourth month, movements of the arms and legs can be felt by the mother. At the end of the second trimester, the fetus weighs about 700 g (27 ounces), and is 30 to 40 cm (about 13 inches) long. It has a well-formed face, toes and fingers with nails, and its eyes can open.

✦ **FIGURE 28.27**

**Human sexual differentiation. (a) Undifferentiated genitalia at seven weeks of development. (b) Pathway of male sexual development. (c) Pathway of female sexual development.**

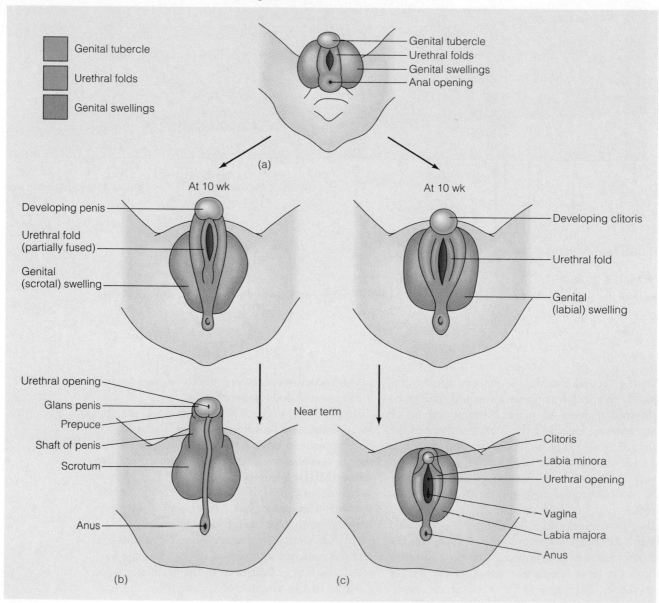

## The Last Trimester Is a Period of Rapid Growth

In the third trimester the fetus grows rapidly, and the circulatory system and the respiratory systems mature in preparation for air breathing. During this period of rapid growth, maternal nutrition is important, since a large fraction of the protein the mother eats will be used for growth and development of the fetal brain and nervous system. Similarly, much of the calcium in the mother's diet will be used for development of the fetal skeletal system.

The fetus doubles in size during the last two months, and chances for survival outside the uterus increase rapidly during this time. In the last month, antibodies from the maternal circulation pass into the fetal circulation, conferring temporary immunity on the fetus. In the first months after birth, the baby's immune system will mature and begin to make its own antibodies, and the maternal antibodies disappear. At the end of the third trimester, the fetus is about 50 cm (19 inches) in length, and weighs from 2.5 to 4.8 kg (5.5 to 10.5 pounds).

## Birth Is Hormonally Induced and Occurs in Stages

Birth is a hormonally induced process involving a positive feedback system. During birth, the cervical opening dilates to allow passage of the fetus, and uterine contractions expel the fetus. The cervix softens in the last trimester, and the fetus shifts downward, usually with its head pressed against the cervix. In birth, the head usually emerges first; if any other body part enters the birth canal first, the result is a breech birth.

Mild uterine contractions begin in the third trimester, but at the start of the birth process, contractions become more frequent and intense. Birth involves three stages (✦ Figure 28.28). The hormonal control of birth involves the release of oxytocin from the posterior pituitary under control of the hypothalamus. Oxytocin and uterine prostaglandins stimulate uterine contractions. In a positive feedback loop, increased uterine contractions stimulate release of more oxytocin and prostaglandins, in turn, producing more contractions.

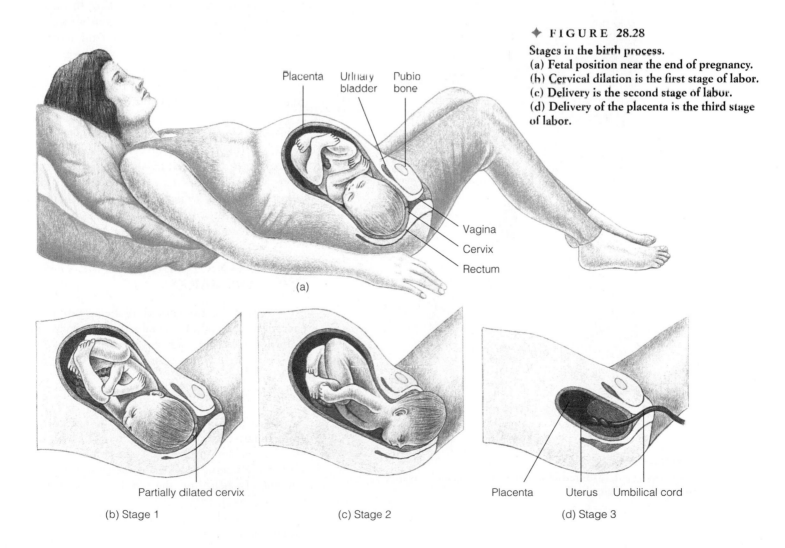

✦ **FIGURE 28.28**
Stages in the birth process.
(a) Fetal position near the end of pregnancy.
(b) Cervical dilation is the first stage of labor.
(c) Delivery is the second stage of labor.
(d) Delivery of the placenta is the third stage of labor.

Placenta    Urinary bladder    Pubic bone

Vagina
Cervix
Rectum

(a)

Partially dilated cervix

(b) Stage 1

(c) Stage 2

Placenta    Uterus    Umbilical cord

(d) Stage 3

The first stage of birth lasts from the beginning of contractions until the cervix is fully dilated (to about 10 centimeters). This is the longest stage of birth, and can last from 10 to 16 hours or even much longer. During the first stage, membranes of the amniotic sac rupture, and the fluid escaping from the cervix helps lubricate the vagina.

The second stage, from full cervical dilation until delivery, involves strong uterine contractions lasting a minute or more. These contractions occur at intervals of two to three minutes. As the infant moves down through the cervix and enters the vagina, stretch receptors in the vagina trigger a series of contractions of the abdominal muscles in synchrony with uterine contractions. This stage lasts from 30 to 90 minutes and ends with the delivery of the baby. The newborn is still attached to the placenta by the umbilical cord, which is clamped and cut by those attending the birth.

A short time after delivery, a second round of uterine contractions begins the third stage of birth, delivery of the placenta. These contractions separate the placenta from the lining of the uterus, and the placenta is expelled through the vagina.

## Milk Production Begins Shortly after Birth

Levels of progesterone and estrogen in the mother decrease after birth, and the uterus gradually shrinks and returns to its normal size. In nursing mothers, this process takes about four weeks, but requires about six weeks in non-nursing mothers.

During pregnancy, high levels of progesterone and estrogen promote the development of secretory lobules and ducts in the breasts (✦ Figure 28.29). The breasts are capable of milk production at about the midpoint of pregnancy, but secretion does not begin until after delivery. Following birth, as levels of progesterone and estrogen fall, the hormone **prolactin**, released by the anterior pituitary, promotes milk production. Suckling by the infant releases **oxytocin**, which causes contractions in breast tissue, forcing the milk into the ducts that empty at the nipple.

## POSTNATAL DEVELOPMENTAL LANDMARKS

The stages of life are marked by the acquisition of characteristic physical, mental and social skills. The major landmarks in each stage are briefly summarized in the following sections.

### Infancy and Childhood

Infancy lasts from birth through the age of two years. Infants are born with several reflexes, including the suckling reflex and the rooting reflex. Within a few minutes after birth, infants can suckle a nipple. To help find the nipple, infants have a rooting reflex that turns their head toward a tactile stimulus. If you touch an infant's face, it will turn toward the stimulus.

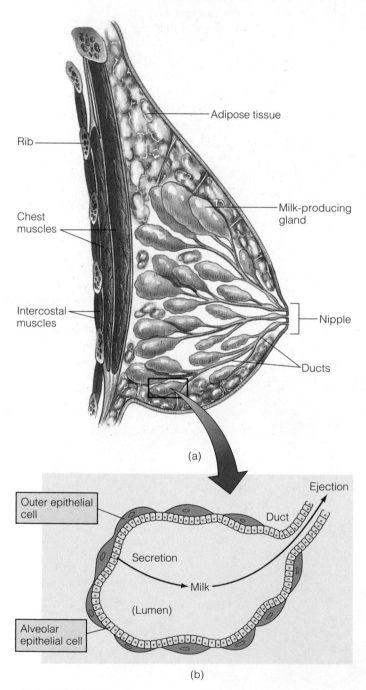

✦ **FIGURE 28.29**

**The mammary gland. (a) Internal anatomy of the mammary gland. (b) Structure of a single gland.**

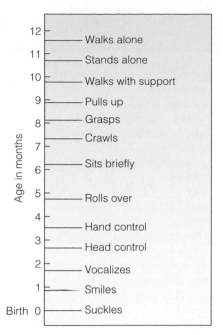

| Age in months | |
|---|---|
| 12 | Walks alone |
| 11 | Stands alone |
| 10 | Walks with support |
| 9 | Pulls up |
| 8 | Grasps |
| 7 | Crawls |
| 6 | Sits briefly |
| 5 | Rolls over |
| 4 | Hand control |
| 3 | Head control |
| 2 | Vocalizes |
| 1 | Smiles |
| Birth 0 | Suckles |

 **FIGURE 28.30**

**Major landmarks of motor development in the first year of life.**

During the first year, infants pass through a series of developmental stages ( Figure 28.30). The order of the stages and the timing for each child may show some variability. The transition between infancy and childhood is marked by the first signs of independence (a period known as the "terrible twos"), and the development of language skills.

Childhood begins at three years of age and lasts until puberty. It is a time of physical growth, with increased development of motor and language skills. Children at this stage develop the ability to represent reality and to engage in make-believe play. They also show increases in memory and attention span.

## Adolescence and Adulthood

During childhood, growth is fairly constant, but during puberty, there is a remarkable acceleration in the development of the skeletal system. Height increases in boys by three to five inches per year, and in girls, increases of two to four inches per year are common. Growth in adolescence involves other systems; the heart doubles in size during the teen years, and the circulatory and respiratory systems reach peak efficiency. Hormonal changes initiate sexual maturity, and adolescents begin to perceive themselves as adults.

Adulthood begins at the end of adolescence. Physical growth ends, and social and emotional development continues. Aging begins in the late twenties, and is associated with a decline in cell numbers and a reduction in the efficiency of cellular function. The control of cell divisions and its relation to aging were considered in Chapter 5. The role of environmental factors in aging is illustrated by the dramatic increases in life expectancy in this century. In 1900, life expectancy was 47 years for males and 50 years for females. Today, it is 74 years for males and 81 years for females. Although most of this increase is due to a reduction in infant mortality, medical advances including the use of antibiotics, and changes in diet and exercise, are also responsible for lengthening the life span.

*Sidebar*

**AGING**

Aging is the result of normal cellular development and metabolic processes. Longevity varies between and within species. The existence of species-specific limits to life-span and its partial heritability indicates the existence of genetic factors that influence the aging process. It has been proposed that longevity determinant genes (LDGs) may govern the rate of aging. Experimental data indicates that the byproducts of ATP production (oxyradicals) play a causative role in aging. Models of cellular aging have been tested to determine whether genes that control tested oxyradicals represent LDGs in mammals. Researchers found that the amount of oxidative damage to DNA correlates inversely with life span of mammals. These results suggest a role of oxyradicals in causing aging and that the oxyradical status of an individual could give clues to determine frequency of age-dependent diseases.

## SUMMARY

1. Asexual reproduction produces genetically identical offspring from a single parent. In sexual reproduction, new individuals are produced by the fusion of two haploid sex cells or gametes to form a diploid zygote. Sexual reproduction offers the benefit of generating genetic variation among offspring, which enhances the survival of the population under changing conditions.

2. Fertilization involves a number of steps, including contact between sperm and egg, entry of the sperm into the egg, fusion of sperm and egg nuclei, and activation of development.

3. In all multicellular animals, the first step after fertilization is cleavage, a series of rapid cell divisions that forms a hollow ball of cells, the blastula. In gastrulation, cell migrations form three layers: ectoderm, mesoderm, and endoderm.

4. Through pattern formation and induction, the body axis is determined, and organ formation begins.

5. The human reproductive system consists of gonads (testes in males and ovaries in females), ducts to carry gametes, and genital structures for copulation and fertilization.

6. Human development begins with fertilization, cleavage, and blastocyst formation. The embryo implants in the uterine wall, and a placenta develops to nourish the developing embryo.

7. Human development is divided into three stages, or trimesters of about 12 weeks each. The first trimester is the period of the most rapid changes in development. Organ formation occurs during this period, and the embryo grows to become a fetus. Growth and maturation of organ systems occurs in the second trimester. The third trimester is a period of rapid growth.

AIDS
asexual reproduction
blastocoel
blastocyst
blastula
bulbourethral glands
cervix
chlamydia
chorion
clitoris
corpus luteum
ectoderm
ejaculatory duct
endoderm
endometrium
epididymis
estrogen
fertilization
follicle-stimulating hormone
follicles
follicular phase
gametes
genital herpes
gonadotropin-releasing hormone
gonads

gonorrhea
hepatitis B
homeobox genes
human chorionic gonadotropin (hCG)
implantation
induction
labia majora
labia minora
luteal phase
luteinizing hormone
menstrual cycle
menstruation
mesoderm
Mullerian inhibiting hormone
neural tube
notochord
oogenesis
orgasm
ovaries
oviduct
ovulation
ovum
oxytocin
placenta
prolactin

prostaglandins
prostate gland
scrotum
semen
seminal vesicles
seminiferous tubules
sexual reproduction
somites
sperm
syphilis
testes
testosterone
trophoblast
tubal ligation
tubal pregnancy
umbilical cord
urethra
uterus
vagina
vas deferens
vasectomy
villi
vulva
zygote

## SHORT ANSWER

1. What are the advantages and disadvantages of asexual reproduction?
2. Gastrulation results in the formation of three primary layers of cells. Name these layers and describe structures associated with each.
3. Briefly describe the two regulatory phases of the ovarian cycle.
4. The female reproductive system is composed of ovaries, uterine tubes, the uterus, and the vagina. Briefly describe the function of each component.
5. Describe the three stages of human development.

## TRUE/FALSE

1. Some organisms are able to reproduce both asexually and sexually.
2. The first step in fertilization is gastrulation.
3. It is possible for one cell or tissue type to affect the developmental fate of another cell or tissue.
4. Oversecretion of estrogen can cause menstrual cramps in women.
5. Menstruation is associated with the destruction of abnormal ova.
6. Most STDs have noticeable symptoms and do not harm the fetus or newborn infants.
7. All forms of contraception prevent sexually transmitted diseases, like AIDS.
8. The immune system is the first organ system to be formed during the first weeks of development.
9. When Mullerian inhibiting hormone is present in an XX embryo, the female duct system survives and develops.
10. Infancy lasts from birth to the age of two years.

## MULTIPLE CHOICE

1. During cleavage the surface to volume ratio is _____, facilitating oxygen exchange with the environment.
   a. reduced
   b. increased
   c. constant
   d. fluctuating
2. The gonads of males are _____ and of females are _____.
   a. sperms; eggs
   b. scrotum; vagina
   c. testes; ovaries
   d. seminiferous tubules; follicles
3. Sperm are stored in the _____ and delivered to the urethra during ejaculation via the _____ , two muscular ducts.
   a. epididymis; vas deferens
   b. penis; ejaculatory duct
   c. scrotum; seminiferous tubules
   d. prostate gland; fallopian tubes
4. The _____ are the flaps of skin covered with hair that are part of the external genitalia in women.
   a. labia minora
   b. vagina
   c. endometrium
   d. labia majora
5. The four stages of mating in humans occur in what order?
   a. plateau, orgasm, resolution, arousal
   b. arousal, plateau, orgasm, resolution
   c. plateau, resolution, arousal, orgasm
   d. resolution, orgasm, arousal, plateau
6. Fertilization usually occurs in the _____.
   a. uterus

b. fallopian tube
c. upper third of the oviduct
d. cervix

7. While in the oviduct, the zygote undergoes rapid cell division to form a ball of cells called the _____.
   a. morula
   b. trophoblast
   c. blastocyst
   d. oocyte

8. Birth is a hormonally induced process involving a(n) _____ system.
   a. inhibitory
   b. positive feedback
   c. stimulatory
   d. negative feedback

## SCIENCE AND SOCIETY

1. What do you think are the legal and ethical issues surrounding the development of *in vitro* fertilization? How could these issues be resolved? Should the child, once he or she is old enough, be told how they were conceived? Why or why not? What should be done with the extra gametes that are removed from the woman's body but never implanted in her uterus? Should this technique be restricted to women of a certain age, race, socioeconomic class, or health background? Why or why not?

2. The chance of having a child affected with Down syndrome increases as a woman ages. Researchers attribute about 95% of Down syndrome babies to maternal nondisjunction (failure of chromosomes to separate during meiosis) of chromosome number 21. Therefore, the cause of Down syndrome has been attributed to the age of the oocyte in women. A recent international study has shown that older women have a higher incidence of Down syndrome because they are more likely to carry a Down syndrome fetus to term than younger women. These investigators also found that the extra chromosome 21 is most often introduced during the first meiotic division. They also revealed that the frequency of nondisjunction was almost identical in older and younger women.

If there is no difference in frequency of nondisjunction between older and younger women, why do you think older women have a higher incidence of Down syndrome babies? Does the uterine environment contribute to the continuation of a pregnancy? If it does, what is the process involved? If a younger woman had an older woman's ova implanted in her uterus, would the younger woman's chance of having a Down syndrome child change? Explain your reasoning. Can you think of ways to test whether the results of the international study are valid? Would comparing the ages of women who spontaneously miscarry a Down syndrome fetus help in testing this hypothesis? Why or why not? Should a woman's genetic, family, and medical history be taken into account in understanding the reasoning why older women have a higher percentage of Down syndrome babies than younger women? Why or why not?

# ANIMAL BEHAVIOR

OPENING IMAGE

*Meerkats turn toward the sun*

*to warm themselves.*

H*ere's a simple test. Sit in a room, and have someone show you pictures of
32 unfamiliar faces, one at a time, for 4 seconds each. After 15 minutes,
you are asked to view the photos again, but they are mixed in with some
new photos you have not seen before. You are asked to pick out faces you saw in the first
session.*

*Memorization is associated with activity in the hippocampus, a region of the brain that
collects and organizes information from the cerebral cortex and other regions of the brain.
The hippocampus acts as an encoding center, making memorized items easier to recall.*

*Scientists at the National Institutes of Health (NIH) recently conducted this memory
test on two groups of subjects, one group ranging in age from 23 to 27 years of age, the
other ranging in age from 64 to 76 years. While the subjects were viewing the photos, the
researchers were monitoring brain activity in the hippocampus and adjacent areas of the
brain. This monitoring was done using positron emission tomography (PET) scans to accurately measure the rate of blood flow, an indication of brain activity.*

*In the brains of the younger subjects, the hippocampus and nearby areas showed ma-
jor increases in blood flow when the photos were being viewed. In the older subjects,
there was little or no increase in blood flow to these brain regions when memorizing or*

*reviewing the photos. The older subjects had less ability to memorize the faces and recognize them when viewed later.*

*These results suggest that altered brain physiology is associated with the diminished memory in the older subjects. This example illustrates the close link between the nervous system and behavior. This chapter examines several aspects of behavior, including instinct and learning, and the development and components of social behavior. Behavior also has adaptive value, and survival of the individual and the species depends on behavior. The chapter closes with a discussion of the links between genetics and behavior. Underlying these topics is the role of the nervous system in receiving and processing stimuli and in generating a response through the muscular system.*

 ## TWO COMPONENTS OF BEHAVIOR: INSTINCT AND LEARNING

Behavior can be defined as a reaction to a stimulus. By this definition, almost everything an animal does can be regarded as a form of behavior. Animals flee or fight in the presence of a predator, engage in elaborate courtship rituals, and develop complex social structures associated with certain forms of behavior. Beginning in the nineteenth century, scientists studying animals developed two general ideas about the origins of their behavior.

Because some behaviors are recognized as innate or instinctive, one idea is that some behavior is a result of the organization and development of the nervous system. Instinctive behavior is thought to be controlled ultimately by genes acting within and on the nervous system. The second idea arose from the observation that some behaviors can be modified by experience, and therefore learning is a source of behavior.

For many years, these two approaches generated a great deal of debate, known as the nature versus nurture controversy. It is now recognized that both instinct and learning are important in behavior, and that for individual behaviors, one factor or another may be more important.

### Reflexes and Taxes Are Simple Forms of Behavior

A **reflex** is a response that occurs without conscious effort, and is one of the simplest forms of behavior. In reflex behavior, a **stimulus** is a physical or chemical change in the environment that leads to a response controlled by the nervous system. When someone touches a hot stove or receives any painful stimulus, a reflex pulls the injured body part away.

The knee jerk is a classic example of a reflex (✦ Figure 29.1). Tapping the tendon just below the kneecap causes the upper leg muscle known as the quadriceps to stretch. This activates receptors in the muscle, generating a nerve impulse that trav-

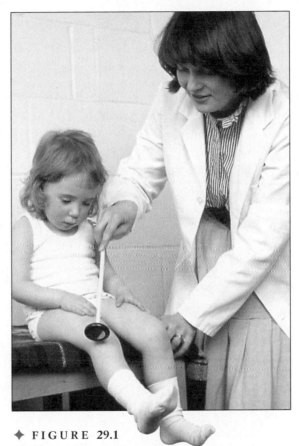

✦ **FIGURE 29.1**
The knee jerk is a reflex arc that uses only two neurons.

els up a sensory neuron to the spinal cord. In the spinal cord, the sensory impulse activates a motor neuron, causing the muscle to contract. This contraction extends the knee, resulting in the knee-jerk reflex.

When an animal turns and begins to move toward or away from an external stimulus, this behavior is called a **taxis** (pl., **taxes**). Stimuli that evoke taxes can include physical stimuli such as light or gravity, and chemical stimuli, such as odor. The fruit fly *Drosophila* responds to light by moving toward it, a response known as a positive phototaxis

(a)

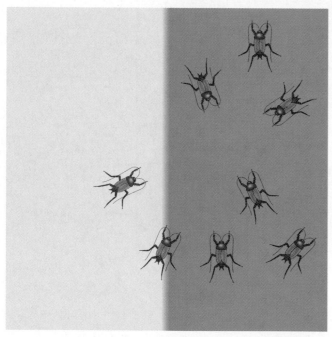

(b)

♦ FIGURE 29.2

**Positive and negative phototaxis. (a) The fruit fly *Drosophila* is attracted to light and is positively phototactic. (b) The cockroach is negatively phototactic so it moves away from light.**

(♦ Figure 29.2a). Cockroaches, on the other hand, move away from light and exhibit a negative phototaxis (Figure 29.2b). In a taxis, the information provided by the environmental stimulus is received by sensory organs, processed by the nervous system, and a response is then generated through the muscular system.

## Fixed Action Patterns Are Complex, Unlearned Behaviors

Reflexes and taxes are innate responses to external stimuli mediated by the nervous system. This finding led to the idea that more complex forms of behavior were also innate responses, built up as a series of reflexes or taxes strung together. When this idea proved largely unworkable, the concept of a fixed action pattern was developed. A **fixed action pattern** (FAP) is a sequence of complex, unlearned behaviors found within a species.

In a freshwater fish, the three-spined stickleback, each mature male establishes a territory and builds a nest. Whenever a female enters a male's territory, the male engages in an elaborate courtship dance. The male's dance acts as a signal to the female, who in turn engages in a fixed action pattern of courtship. If successful, this series of behaviors results in mating and the production of offspring.

Males engage in their courtship dance in response to the sight of a female's swollen, egg-filled belly. In fact, they will respond with a courtship dance to a model "fish" lacking eyes, fins, and tail, if it has a swollen abdomen. During the breeding season, male sticklebacks have a red abdomen. Males defend their territory with displays of aggression to drive away other breeding males. This behavior is triggered in response to a red abdomen on the intruder, since males will respond aggressively to a model "fish" having a red abdomen, even though it has no fins or tail.

Although fixed action patterns appear to be unlearned and apparently inherited, components of experience (that is, learning) are part of these behaviors. Experience is a source of information that can be used to change or modify behavioral responses to stimuli.

In the laughing gull, parents present their beaks pointed downward to their chicks. The chicks respond to this stimulus with a pecking motion to grasp the bill and pull it toward them. The parent then regurgitates food, which is eaten by the chick. Chicks raised in the light increase the accuracy of the grasping motion within a few days after hatching. If chicks are raised in the dark, however, their accuracy increases, but never reaches the level of those chicks raised in the light. The conclusion from these experiments is that visual learning is an important component in the development of this fixed action pattern.

## Learning Alters Behavior through Experience

The ability of an animal to change its responses to stimuli as a result of experience is learned behavior.

Ethology is the study of animal behavior in natural settings. One of the earliest forms of learning to be defined is **associative learning.** Ivan Pavlov was a Russian physiologist who studied digestion. For his work, Pavlov needed to collect dog saliva. He did this by placing meat in the dog's mouth. He noticed that the dog began salivating in anticipation of feeding as soon as the researcher entered the room. Pavlov then began ringing a bell while feeding the dogs. Soon, they salivated at the sound of the bell even without being fed. Thus, the dogs had learned to respond to the bell by associating it with feeding. This type of associative learning is called a **conditioned response.**

The identification and study of distinct forms of behavior in different animals indicate that both innate and external factors influence behavior. The nervous system collects and processes information about the external environment. Responses are generated through several systems, including the endocrine system and the muscular system. As illustrated earlier, behavior is also affected by experience and learning. The nature versus nurture debate has largely faded away with the realization that behavior is not a rigid, predetermined set of genetically programmed responses, nor is it solely derived from interaction with the environment. The current view is that animals are born with a genetically predisposed set of behavioral patterns that is modified and refined by experience. However, the extent of each influence on the innate aspects of behavior and the modification of behavior by learning are still under debate.

### Learning Has Genetic Components

For a study of the genetics of learning, the fruit fly *Drosophila* has several advantages, including the fact that well-developed methods of genetic analysis are available for this organism. But can *Drosophila* learn anything? Perhaps surprisingly, the answer is yes. If flies are presented with two odors, and one of those odors is associated with an electrical shock, they quickly learn to avoid the odor associated with the shock. This behavior is a form of learning because the response is reversible; flies can be trained to select an odor they previously avoided. In addition, flies show a short-term memory for this learned response.

This test for learning has been used to collect mutant flies deficient in both learning and memory. Some of these mutated genes include *dunce, rutabaga,* and *turnip.* Each represents a defect in a specific aspect of learning and memory. Analysis of mutant flies carrying these genes has shown them to be defective in signaling systems within nerve cells (review second messengers in Chapter 25). Many of the mutants have defective second messenger systems, indicating that the intracellular second messenger, cyclic AMP, may play a role in learning and memory. This same pathway may be involved in human learning and memory, so the use of *Drosophila* as a model system may provide insight into how humans learn and remember.

## SOCIAL BEHAVIOR INVOLVES INTERACTION WITH OTHERS

Behavior often takes place in a social context, and results from the interaction between and among individuals. This behavior, known as **social behavior,** has received a great deal of attention from scientists who study natural selection and evolution. Animals exhibit a wide range of social behaviors. Some organisms are largely solitary, some live in populations with other, unrelated members of the same species, while honeybees, termites, and mole rats live in highly organized social groups of related individuals.

Levels of social behavior all have adaptive value, and contribute to relative evolutionary **fitness,** which can be defined as the chance that an individual will leave more offspring in the next generation than other individuals. Although there are different levels of social organization and different kinds of social behavior, they jointly contribute to the fitness of individuals and, ultimately, to the success or failure of a species.

### Imprinting and the Stages of Social Behavior

The development of social bonds can be studied by observing how parents and offspring form attachments. Within a few hours after hatching, some young birds will follow their mother. Konrad Lorenz demonstrated that newly hatched geese do not recognize their mother, but will follow the first moving object they see after hatching (◆ Figure 29.3). **Imprinting** in geese is visually controlled, and is limited to a critical time period after hatching. In geese, the first 43 hours after hatching is the sensitive phase, during which imprinting occurs. The first third of this phase, roughly the first 13 to 16 hours is a critical period, during which imprinting is most intense.

Imprinting and the formation of bonds is not limited to birds, but has been found in other animals, including salmon, which hatch in freshwater streams and migrate to the ocean to feed and mature. Olfactory imprinting guides the salmon back

◆ **FIGURE 29.3**
These goslings imprinted on Konrad Lorenz as their mother.

**FIGURE 29.4**

Animals at a water hole are an aggregate. They are gathered here in response to thirst as a stimulus.

to their home stream to spawn. Human infants need a certain amount of social contact to develop normally. Premature infants that are touched and massaged gain weight faster than those who are not. The formation of social attachments is a step in normal behavioral development in many animals, including humans.

## There Are Three Classes of Social Organization

The pattern of social relationships among individuals can be difficult to classify, but it is agreed that there are at least three broad classes of social organization: **aggregates, groups,** and **societies.** The difficulty of classifying social organizations arises in part from their dynamic nature; they can change in response to both internal and external conditions.

Aggregates are fairly random associations of animals with little or no internal organization (✦ Figure 29.4). Aggregates are formed in response to a simple stimulus, such as a grouping of moths and other insects that congregate around a light source. Once the external stimulus is removed, the aggregate disperses.

Groups, on the other hand, have some internal organization and may have a division of labor (✦ Figure 29.5). A school of fish, a flock of geese, and a herd of wild animals are examples of groups. The social behavior of groups can include reproductive self-interest, cooperation against predators, or self-sacrifice that helps other individuals survive and reproduce.

The most highly organized of social groups, and the one receiving the most attention, is the animal society. There are several different types of societies, but almost all consist of individuals that show varying degrees of cooperation and communication with one another, and often have a rigid division of labor.

One of the most familiar examples of an animal society is a hive of honeybees. This society may consist of 25,000 to 60,000 individuals, all of whom share the same mother, the queen bee. Within the hive, worker bees are infertile females who forage for food and guard the colony against predators. The workers also build and maintain the hive, and feed the queen and the larvae. At certain times of the year, male drones are produced. These stingless males do no work in the colony, but mate once with a new queen, providing her with enough sperm for her reproductive life.

Within the society, communication among the workers is essential to carry out the functions of the colony. The methods of communication include chemical, visual, auditory, and tactile (touch) signals.

## Communication Is a Means of Modifying Behavior

Communication is a means of modifying behavior through signals passed from one animal to another. Chemical signals in the form of **pheromones** are an evolutionarily ancient form of communication. Pheromones are found in a wide range of animal species, including mammals. They can convey specific signals to initiate or suppress aggressive behav-

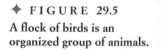

**FIGURE 29.5**

A flock of birds is an organized group of animals.

ior, spread an alarm, attract a mate, or act in the recognition of territorial boundaries. There are several classes of pheromones. Signaling pheromones produce an immediate response on the part of the recipient. A second class of pheromones, called priming pheromones, acts over a longer time period and results in physiological changes in the recipient.

Because pheromones are so widely distributed in the animal kingdom, do humans have pheromones? None have been identified with certainty, but the synchronization of menstrual periods between females living together, such as mothers and daughters or female roommates, may be caused by a pheromone.

In addition to pheromones, animals use other forms of communication, including visual signals, auditory signals, and tactile signals. Individually, these signals convey distinct, species-specific infor-

mation. More complex forms of communication may include combinations of signals. One of the best known systems of animal communication is that of honeybees.

Observers all the way back to Aristotle have commented on the ability of honeybees to communicate information about food sources to their fellow workers. In this century, Karl von Frisch studied the behavior of bees and discovered meaning in bee movements (he called them "dances"). More recent work using robotic bees (✦ Figure 29.6) placed into hives has revealed that in addition to providing visual signals, the dance is associated with auditory signals, and tasting samples of the food provided by the foraging bee.

A forager bee performs the dance on the vertical surface of a honeycomb, tracing out a figure eight (✦ Figure 29.7). She pauses in each loop of the figure eight to waggle her body from side to side.

✦ **FIGURE 29.7**

**The bee dance. (a) The dance is performed on a vertical surface. (b) The bee moves in a figure-eight. (c) Movements during the straight run (wavy line) communicate information.**

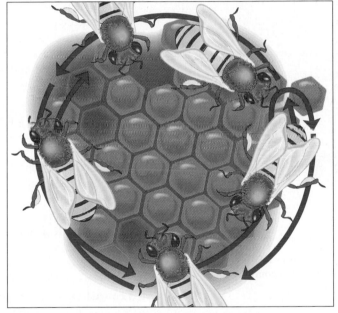

(a)

✦ **FIGURE 29.6**

**A robotic bee, controlled by a series of small motors, can mimic the dance of the honeybee. A plastic tube with a syringe delivers sugar water, mimicking the food offered by a foraging bee. Dances by robotic bees are successful in directing foragers to a specific food source.**

(b)

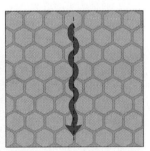

When bee moves straight up comb, recruits fly toward the sun

When bee moves straight down comb, recruits fly directly away from the sun.

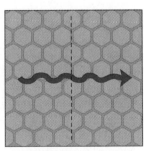

When bee moves to right horizontally, recruits fly to the right, at 90° angle to sun.

(c)

✦ **FIGURE 29.8**
Even animals in groups keep a minimum distance from each other to avoid aggression.

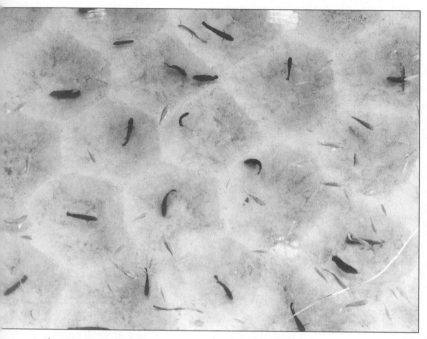

✦ **FIGURE 29.9**
Male mouth-breeding fish have established saucer-shaped territories in the sand. They defend these territories against intruding males, and court females that enter the territory.

The speed of the dance is related to the distance of the food source, with closer food sources generating faster dances. The direction faced during the waggle provides information about the direction of the food source from the hive. If the dancer faces straight up during the waggle (12 o'clock on a clock face), then the food is found in the direction of the sun. If she waggles 60 degrees to the left of 12 o'clock (10 o'clock on the clock face), then the food is found 60 degrees to the left of the sun.

As noted earlier, this communication sequence is not entirely visual. While performing, the dancer emits sounds that help observers determine how the dancer is moving, reinforcing the visual information about the direction and distance to the food source (Figure 29.7). The observer bees also emit sounds that cause the forager to stop the dance and provide samples of the food to the observers. This gives the dance observers information about the taste and quality of the food source.

The workers observing the dance then leave the hive to search for the food source. If they are successful, they return to the hive to dance, recruiting more workers to the food. In this way, the best food sources are discovered and exploited by the members of the hive.

## Some Animals Have Elaborate Courtship Behaviors

Even animals that live in groups tend to maintain a certain distance from others in the group (✦ Figure 29.8). Intrusion into an individual's space usually results in the invader being attacked, often after warning signals are given. As you might imagine, this tendency to attack can make it difficult for animals to get close enough for mating. To overcome this problem, animals use behavioral sequences called **courtship behaviors** that precede mating. In some species, males establish a territory (✦ Figure 29.9), and then set about attracting a receptive female. Often, males compete for territory, resulting

✦ **FIGURE 29.10**
Males often engage in ritual combat to defend territory or to establish mating rights.

✦ FIGURE 29.11

Sea gulls have a species-specific ritual of courting.

in fights (✦ Figure 29.10). These ritualized battles are usually not fights to the death, but allow the victorious males to occupy sites that are associated with high reproductive success.

Courtship behavior has been extensively studied in many animals, including gulls. Male and female gulls are similar in appearance, so behavioral cues are important in recognition, courtship, and mating (✦ Figure 29.11). Males establish a nesting site, and as other gulls approach, they are greeted with a characteristic posture and call. Other males turn away and avoid males using this call, but females may land nearby. When this occurs, the birds stand parallel to each other (but not facing each other) and lean forward. After a few seconds, one and then the other gull will make an aggressive gesture by jerking into an upright position, immediately followed by a head turn away from the other.

The female usually flies off after a short time to visit other males. After sampling the available males, the female revisits one of the males, repeating the ritual. Each time the ritual is repeated, the aggressive displays by each bird decrease, and appeasement gestures increase, so that the female remains at the side of the male for a longer time period. Finally, the displays cease, the male regurgitates food to feed the female, and the pair mate.

When the birds first meet, they are subject to three impulses: to attack, to escape, and to remain near each other. Repetition of the courtship behavior helps resolve this conflict by suppressing the first two behaviors, and allows them to remain close enough for mating.

## BEHAVIOR HAS AN ADAPTIVE VALUE

Behavior is related to evolution through the adaptive nature of individual and social behavior. Behavior can be a trait used to distinguish two closely related species or two species that are anatomically similar. The study of behavior is usually interpreted in terms of fitness. Relative fitness is the chance that an individual will leave more offspring as compared to another individual. Any behavior that increases reproductive success will generate more offspring. These offspring carry alleles associated with that successful behavior, increasing the fitness of that behavior. Consequently, successful behavior will eventually predominate in the population. Unfortunately, the reproductive consequences of a behavior are often difficult to measure.

### Selection and Behavior Are Related

Charles Darwin suggested that ornamental structures such as the male peacock's tail (✦ Figure 29.12) might be the result of preferences expressed by females in

✦ FIGURE 29.12

The tail of the peacock is an ornament used to attract females.

**◆ FIGURE 29.13**

The male widowbird. Males with artificially elongated tails are more successful in mating, indicating a preference by females for this ornamentation.

*Sidebar*

**HATCHING BEHAVIOR THAT MAY INCREASE SURVIVAL**

Red-eyed frogs of Costa Rica have the ability to hatch and escape into the water if a threatening snake is nearby. When egg clutches were placed in cages with their predator, the cat-eyed snake, eggs that were near the hatching stage hatched in moments. These hatchlings dropped from leaves into the water below and developed into tadpoles. These escapees demonstrated a 74% escape rate compared to a 21% rate for a closely related species that is unable to hatch on demand. The triggering factor for early hatching seems to involve pressure on the eggs, as the snake begins eating them.

Typically, changes in the environment of vertebrate embryos, such as warming or increased light, can speed up hatching. However hatching behavior has never before been attributed to adaptation that might improve survival.

selecting mates. For such a selective system to work, two conditions must exist. First, females must demonstrate a behavioral preference for males with extravagant ornaments and, second, the species must have a polygamous mating system, with males competing with each other, resulting in a few successful males inseminating a larger number of females than the less successful males.

In the widowbird (◆ Figure 29.13), a polygamous species, males with experimentally elongated tails had more mating success than males whose tails had been shortened. From this and similar studies, it seems that the behavior of females in selecting mates can affect the evolution of male sexual ornaments. Studies in monogamous species show that female choice can also be a selective factor in male ornamentation. In this case, selection results in earlier breeding and increases the quantity of offspring in a season.

## Social Behavior Contributes to Evolutionary Fitness

In the preceding examples, males with more elaborate ornamentation had greater fitness, since they left more offspring than males with less elaborate ornamentation. Social behavior also makes a contribution to fitness, just as the examples of individual behavior discussed earlier. Social behavior contributes to fitness in several ways. One is through predator avoidance. A group of animals that cooperate in defense against a predator reduces the overall risk to any individual. Musk oxen, which live in herds in the Arctic region, are preyed on by packs of roving wolves. When threatened, the musk oxen gather into a circle, with the adults facing outward and the young inside (◆ Figure 29.14). Wolves, on the other hand, engage in cooperative behavior in hunting the musk ox, and will help each other bring down and kill prey. In these cases, the benefits to an individual (increased chance of reproductive success) outweigh any costs. In other cases, members of social groups engage in behavior patterns that help other individuals survive and reproduce at personal cost, or at what appears to be personal cost. This behavior, called altruism, is described in a later section.

## The Evolution of Behavior Is Difficult to Study

Behavior evolved because of its adaptive value, but the stages of its development are difficult to study because there is no fossil record, and ancestral species are most often extinct. However, by studying behavior patterns in related species, ethologists have been able to discover clues about how behavior patterns have evolved.

In the honeybee dance, both visual and auditory information are communicated. The dance and its variations have been studied in four species, including the common honeybee, *Apis mellifera*. All four species use visual cues, but only three use sounds. The three species that use sounds often dance in the interior of dark hives, making auditory signals an important means of communication. The species that dances without sound, the dwarf bee (*Apis florea*), dances in the open, only during daylight.

**◆ FIGURE 29.14**

Animals such as musk oxen engage in cooperative behavior in defending themselves against predators.

## Exploring the Oceans with DNA Sequences

BRIAN W. BOWEN

My fascination with the marine realm began as a youth on the shores of Cape Cod Bay, where I spent summer afternoons with mask and snorkel, exploring the submerged rocks and seaweed patches of the New England coast. This interest lead to aquarium keeping, scuba diving, and eventually to a career in marine biology. While I was earning a M.S. degree at Virginia Institute of Marine Science in the early 1980s, a revolution was beginning in molecular biology.

To learn the new science of molecular genetics, I entered a Ph.D. program at the University of Georgia. There, Dr. Avise had harnessed the information in mitochondrial (mt) DNA to learn about the oceanic movements of eels and other fishes. To test a theory that green sea turtles return to nest on their natal beach, we compared mtDNA samples from several Atlantic nesting colonies. Pivotal to our study was the nesting colony on Ascension Island, a solitary volcanic speck on the mid-Atlantic ridge. Some of the green turtles that feed along the coast of Brazil are known to nest on Ascension Island, while others are known to nest in Surinam, on the coast of South America. We collected turtle eggs on the isolated shores of Ascension Island and compared their mtDNA to the nesting population in Surinam. The results were startling—all Surinam nesting turtles had a DNA sequence change that was not carried by the Ascension turtles. Despite the fact that these two populations feed together along the coast of Brazil, they are genetically distinct. These data demonstrated that exchange of females between nesting populations is extremely low, which is strong evidence in favor of the natal homing theory.

These findings about the homing behavior of green turtles were very exciting, but they were only the beginning of my marine turtle odyssey. I became a researcher at the Biotechnology Center at the University of Florida, where I began to use genetic markers to identify sea turtles during oceanic migrations. For example, juvenile loggerhead sea turtles are known to feed off the coast of Baja California, but the presence of these turtles is a prominent mystery because loggerhead turtles do not nest anywhere in the eastern Pacific. The closest nesting colonies are in Japan and Australia, over 10,000 kilometers from Baja California. Where are these juvenile turtles coming from? Genetic markers indicated that loggerhead turtles cross the North Pacific, approximately one-third of this planet, in the course of their juvenile migrations. This voyage is made more remarkable by the fact that loggerhead turtles begin the journey as 3-inch-long hatchlings.

Like the demonstration of natal homing behavior, these genetic results carried a strong conservation message. Thousands of sea turtles are drowned by driftnet fishing operations in the Pacific Ocean. Through the driftnet observer program operated by the National Marine Fisheries Service, we obtained tissue samples from 34 drowned loggerhead turtles. Over 95% of these samples matched the Japanese nesting population, with perhaps a small fraction also coming from Australia. Genetic markers demonstrate which nesting populations are impacted by commercial fisheries on the high seas. This information allows conservationists to evaluate the threat that high seas fisheries pose to the welfare of sea turtles.

As the biodiversity crisis continues to deepen, genetic technology will be a valuable weapon in the defense of our natural heritage.

*Brian W. Bowen is a marine biologist and conservation geneticist in the Biotechnology Center at the University of Florida. He earned a B.S. in biology from Providence College, and then hiked the Appalachian Trail 2140 miles from Georgia to Maine. This experience provided the impetus for additional studies in biology, including a M.A. from the Virginia Institute of Marine Science and a Ph.D. in genetics from University of Georgia.*

This species is probably the most primitive of the four, indicating that the development of auditory signals was an adaptation acquired when enclosed hives arose in a common ancestor of the other three species.

The tie between behavior and adaptation is illustrated by the removal of eggshells by gulls. The outer eggshells of gulls are colored to camouflage them (✦ Figure 29.15). When breeding, gulls are careful to remove broken eggshells from the nest and dispose of them at a distance. When they are not nesting, the gulls make no effort to remove broken shells. To investigate this behavior, Nicholas Tinbergen placed gull eggs near broken eggshells, and placed other gull eggs away from shell fragments. About two-thirds of the eggs near broken shells were discovered by predators, but only one-

✦ **FIGURE 29.15**

**The coloration of gull eggs protects them from predators. Broken shells are removed by the parents to protect the hatchlings against predation.**

# TESTOSTERONE AND AGGRESSION

In popular culture, testosterone and aggressive or combative behavior by males are often thought to be closely linked. In the moments before a football game that is critical to a team's ranking, sportscasters often remark about how the testosterone is really flowing. In fact, the relationship between aggression and hormones is more complicated, and testosterone itself may play only a minor role. This conclusion comes from studying the endocrinology of the spotted hyena, a species in which the female is aggressively dominant.

In humans and in most other mammals, the developing testes secrete testosterone and related hormones to direct sexual development. In the absence of testosterone, female sex organs develop. The system is so sensitive that the small amount of testosterone in the mater-nal circulation is inactivated at the placenta and does not affect sexual development. In the hyena, however, the hormone androstenedi-one, a precursor to both estrogen and testosterone, is converted at the placenta into testosterone, and both male and female hyenas are exposed to high levels of testosterone and other male hormones.

The result is that female hyenas are born with masculinized genitals, an enlarged clitoris and vaginal lips that look like testes. The cubs are usually born in pairs. At birth, they have their eyes open and their teeth are fully developed. Newborn cubs fight with each other, and frequently kill their sibling, especially if it is of the same sex.

As the females age, blood levels of testosterone fall far below that of males, but the females retain their aggressive behavior and dominate the males. Hyenas live in clans, with a related group of females forming the core of the group, and only a limited number of males allowed as members. Females defend the clan's territory and drive off unwanted males. In feeding, males wait until all females in the clan, even the youngest cubs, have eaten, before they begin feeding.

Since testosterone levels in adult females are low, researchers believe that other hormones are involved in maintaining the aggressive behavior of female spotted hyenas. One possibility is that androstenedione can itself influence behavior and that its action sustains female aggressive behavior. Because human females also have significant levels of androstenedione, research into the sex hormones of the spotted hyena may shed light on how the balance of sex hormones in humans affects behavior.

---

fifth of the other eggs were discovered. Tinbergen concluded that eggshell removal is an adaptive behavior that helps ensure reproductive success.

##  SOCIOBIOLOGY: GENETICS AND BEHAVIOR

In 1975, a book entitled *Sociobiology: The New Synthesis,* written by Edward O. Wilson, focused attention on the evolution of social behavior. Drawing on his work with ant societies, Wilson reviewed and updated the idea that social behavior has a genetic component and can be influenced by natural selection. Most of the book deals with the social behavior of nonhuman animals, but at the end of the book, Wilson considers the evolutionary history of some forms of human social behavior.

The application of sociobiology to human behavior touched off a debate about the role of genetics and natural selection in social behavior. Critics, especially in the fields of anthropology and social sciences, have pointed out that a narrow definition of sociobiology, in which all social behavior is assigned a genetic basis, ends up offering a genetic explanation for social problems such as crime. Sociobiologists, in turn, replied that critics often assume that genes play little or no role in the observed diversity of human behavior, and place too much emphasis on environmental causes of behavior.

Leaving aside the debate about human applications of sociobiology, the study of the biological basis of social behavior has made a contribution to our understanding of behavior in nonhuman species. Sociobiology has been especially useful in explaining the relationship of the individual to a social structure, and in explaining the existence of altruistic behavior.

### Social Structure Offers Advantages to the Individual

A cluster of animals does not necessarily constitute a social group. Only when there is communication and cooperative interaction among individuals is there social behavior. Social behavior can offer several advantages to an individual, increasing its chances of survival and reproduction. As mentioned earlier, wolves hunt in packs, and by cooperation are able to kill large animals such as musk ox, elk, or moose that a single wolf would be unable to kill. This makes more food available to individual wolves, increasing the probability of survival for each individual.

In a social group, aggressive behavior often establishes a social structure known as a **dominance hierarchy.** Flocks of hens, wolf packs, and baboon troops all show dominance hierarchies (✦ Figure 29.16). In wolves, there is a dominance hierarchy among the females. The dominant female is the only one to breed under conditions where food is scarce. When food is abundant, she shares the males with the other females, increasing the size of the pack.

What then are the advantages to an individual who occupies a subordinate, nonbreeding position in a dominance hierarchy? One is a conservation of energy. Instead of constant battles, members of a hierarchy can spend time and effort on activities such as locating food and guarding against predators. Remember that the social organization may increase the chances of individual survival, and subordinates may be able to reproduce when food supplies are adequate. In addition, hierarchies are dynamic, and subordinate individuals can move up when dominant members die or age.

## How Did Altruism Originate?

If selection has been a major force in the evolution of behavior in a given species, then individuals would be expected to behave in ways that would promote the propagation of their genes. Against this expectation, social animals have evolved behavior that is cooperative and even self-sacrificing. The origins and persistence of **altruism** have been explained in several ways.

William Hamilton has proposed the idea of indirect selection, in which selection favors behavior for the benefit of close relatives. In general, siblings share about half their genes with each other, and an aunt or uncle will share about one-fourth of their genes with a niece or nephew. If indirect selection operates, then altruistic behavior toward relatives helps an individual make an indirect genetic contribution to the next generation. This contribution can be measured by the degree of genetic relatedness. The geneticist J. B. S. Haldane was once asked in a pub if he would ever sacrifice his life to help another on evolutionary grounds. Grabbing a coaster and a pencil, he made some quick calculations and concluded that he would lay down his life if it would save more than two brothers, four half-brothers, or eight cousins.

Aside from pub conversations, altruistic behavior and indirect selection can be observed directly in animal societies. In a bee hive, worker bees are all sisters and have more than half their genes in common. Worker bees are also infertile; their genes can be passed on to future generations only through their mother (the queen bee) or through sisters who may become queens. Any altruistic behavior that protects the queen or the hive will help pass on the worker's genes to a future generation. Workers who act as guard bees give up their lives by stinging invaders, but this self-sacrificing act protects the hive and their resident sisters and mother, the queen.

According to Robert Trivers, altruistic behavior toward nonrelatives can also have selective value if the behavior has an eventual benefit. In this case, aid is offered to others, including nonrelatives, if the risk is not too great. The expectation is that similar aid will be offered in the future. If the entire population is exposed to risks, and rescue by others is widespread, then this altruistic behavior will be selected for over the behavior of those who face risks alone. There is now evidence from field studies that support the notion of reciprocal altruism.

## GENETICS AND HUMAN BEHAVIOR

A number of human genetic disorders have associated and characteristic behaviors. However, observations and pedigree analysis indicate that not all behaviors are inherited as simple Mendelian traits. In addition to situations where several genes contribute to behavior, the degree of gene expression often depends on interactions with the environment.

✦ **FIGURE 29.16**
**Many animals living in groups establish a dominance hierarchy, reinforced by submissive behavior.**

In discussing the inheritance of human behavior traits, it is useful to review some essential concepts. Recall from Chapter 6 that the genotype represents the set of genes present in an individual that affects any particular character; it is fixed at the moment of conception and, barring mutations, is unchanging. The phenotype is the sum of the observable characteristics of an organism and is derived only in part from the genotype. The phenotype is variable and undergoes continuous change throughout the lifetime of the individual. The environment of a gene is composed of genetic factors, including all other genes in the genotype, with their effects and interactions, and all nongenetic factors, whether physical or social, that have the capacity to interact with the genotype.

This review of concepts is necessary because in discussing human behavior, the term "inherited traits" is often used as if *phenotypic traits* are passed from parent to child. In fact, it is genes that are passed from generation to generation, not phenotypes. In fact, the phenotype in human behavior genetics can be difficult to define. For some mental illnesses, clinical definitions of phenotypes are established by guidelines such as those of the American Psychiatric Association. For other behaviors, the phenotypes are poorly defined and may have little relationship to the underlying biochemical and molecular basis of the behavior.

## Single Genes and Human Behavior

Often, single gene disorders disrupt the development, structure, or function of cells. If the affected cells are part of the nervous system, then alterations in behavior may be part of the phenotype. Other cases have a more complex interaction between the genotype and environmental factors. Often, the number and action of genes is less well known, and effects on the nervous system may be less obvious.

An autosomal dominant disorder that affects behavior in humans is **Huntington disease.** It is first expressed in mid-adult life as involuntary muscular twitches and jerky motions of the arms and legs. As the condition progresses, personality changes, agi-

tated behavior, and dementia occur. Most affected individuals die within 10 to 15 years after onset of symptoms.

Brain autopsies of affected individuals show that cells in several regions are altered in shape or destroyed. Brains of affected individuals also accumulate excess amounts of a metabolic by-product, quinolonic acid, that is toxic to nerve cells. It appears that the defective gene leads to abnormal brain metabolism that produces a toxic substance, which in turn destroys brain cells that control behavior.

## Genetics and Social Behavior

Many forms of human behavior have complex phenotypes and are not inherited as single genes. These behaviors, many of which have social components, are being studied to identify their genetic components (if any) and their associated environmental factors.

The question of whether these behaviors are genetically influenced is controversial, and has renewed the nature versus nurture debate described at the beginning of the chapter. In this case, the answer sought is not either or, but to what degree these traits are influenced by genes and the environment (including culture and experience).

Studies of sexual orientation have been investigated by twin studies. In a study involving twins, 52% of identical twins were found to be concordant for homosexuality. That is, if one twin was homosexual, in 52% of the cases, the other identical twin was homosexual. In the case of nonidentical twins, 22% were found to be concordant for homosexuality. Identical twins share a common genome, whereas nonidentical twins share, on average, about half their genes. Because of the different distributions of this behavior, these results have been interpreted to indicate that there is a genetic component to homosexual behavior. Although these studies are intriguing, further studies, especially on twins reared apart, will be needed to determine whether these results are valid.

# SUMMARY

1. Behavior can be defined as a reaction to a stimulus. Some behavior is a result of the organization and development of the nervous system (instinctive), while other behavior is modified by experience, and therefore learning is a source of behavior.
2. A reflex is a response that occurs without conscious effort, and is one of the simplest forms of behavior. A fixed action pattern is a sequence of complex, unlearned behaviors found within a species.
3. Learned behavior is the ability of an animal to change its responses to stimuli as a result of experience.

4. Social behavior has adaptive value, and contributes to fitness. There are three broad classes of social organizations: aggregates, groups, and societies. Communication is a means of modifying social behavior.
5. Social behavior makes a contribution to fitness in several ways. One is through cooperation to avoid predators; another is altruism, which occurs when members of social groups engage in behavior that help other individuals survive and reproduce at personal cost.
6. One form of altruism, indirect selection, favors behavior for the benefit of close

relatives. Altruistic behavior toward relatives helps an individual make an indirect genetic contribution to the next generation, measured by the degree of genetic relatedness. Altruistic behavior toward nonrelatives can also have selective value if the behavior has an eventual benefit.
7. A number of human genetic disorders have associated and characteristic behaviors. For many behaviors, the phenotypes are poorly defined and may have little relationship to the underlying biochemical and molecular basis of the behavior.

# KEY TERMS

aggregates
altruism
associative learning
conditioned response
courtship behaviors
dominance hierarchy

ethology
fitness
fixed action pattern
groups
Huntington disease
imprinting

pheromones
reflex
social behavior
societies
stimulus
taxis

# QUESTIONS AND PROBLEMS

## MATCHING

Match the list of types of behavior patterns that correspond to the following situations.

a. altruism
b. social behavior
c. adaptive behavior
d. dominance
e. territoriality
f. cooperation
g. courtship behavior
h. fixed action pattern
i. associative learned behavior
j. reflex

____ A cricket will leap forward in a standard escape reaction when receptors on its abdomen are stimulated.
____ A male pigeon engages in a specific dance that acts as a signal to the female pigeon, who in turn engages in a fixed pattern of behaviors.
____ A cat who is fed canned cat food comes running into the kitchen when he hears the electric can opener.

____ Dogs urinating on trees in their neighborhood.
____ The head chicken occasionally pecks other chickens.
____ In a herd of foraging gazelles, a few animals are always alert to potential predators.
____ Many species of shore birds remove empty eggshells from the nest soon after the young are hatched.
____ If a predator gets too close to the nest of a killdeer, the female will flutter around, calling and dragging its wing as if it were broken.
____ Human babies grasp a finger tightly when it is placed in their hand.
____ The interactions seen in a hive of honeybees.

## SHORT ANSWER

1. Describe the three broad classes of social organization. Give an example of each.
2. What three impulses do birds experience when they first meet?

3. Why is the evolution of behavior difficult to study?
4. Explain how guard bees behave in an altruistic manner.
5. Discuss the late-onset disorder in humans that affects behavior.
6. Why are identical twins useful for studying human behavior?

## TRUE/FALSE

1. Behaviors that must be produced perfectly the first time they are performed are usually innate.
2. Many behavior patterns become programmed into the skeletal-muscular system.
3. Pheromones are chemical signals that are a form of communication between animal species.
4. Behavior changes as a result of experience.
5. Honeybees communicate information about food sources to their fellow workers solely by visual signals.

6. A behavior that decreases reproductive success will ensure the successful continuation of the population.

7. Learning is more effective for species in environments that are more unpredictable, whereas instinct is more efficient where environments are more constant.

8. Sociobiology is the general biology of social systems that takes into account the influences of social behavior, genetics, and environment.

## SCIENCE AND SOCIETY

1. Alzheimer's disease is the most common form of chronic, progressive irreversible dementia. It affects 2–3% of the population and approximately 50–60% of adult dementia cases in the U.S. are due to Alzheimer's disease. The disease may have several genetic etiologies. The first noticeable symptoms are generally loss of memory, shortened attention span, language disturbances, and difficulty comprehending oral and written speech. Personality disturbances tend to accompany increasing loss of intellectual function. The age of onset is variable, usually from 50 to 75 years. The disease tends to be more severe in younger patients. The course of the disease can run from 2 to 20 years. Imagine that one of your parents suffered from Alzheimer's disease. Would you have a higher risk of falling victim to this condition than someone who does not have a family history of Alzheimer's disease? Why or why not? If there was a test available to diagnose Alzheimer's disease before symptoms appear, would you want to be tested? Explain your reasoning. Suppose a definitive gene responsible for one form of Alzheimer's disease was discovered and you tested positive for the mutant allele. In what ways, if any, would this information change your life? Do the symptoms of Alzheimer's disease overlap with other conditions that you have learned about? If so, do you think this overlapping of symptoms could make it difficult to provide an accurate diagnosis? What would you do if you were diagnosed with Alzheimer's disease?

2. Cities of the world are noisy, polluted, and overcrowded, and can be the centers of poverty, crime, and social instability. Sociobiologists believe that inner city life can have adverse effects on a variety of behaviors. The stress of overcrowding has been implicated in divorce, social problems, mental illness, and drug and alcohol abuse. Rising crime rates and infant mortality may also be attributable to overcrowding. Do you think overcrowding influences the way we act toward one another? How do you feel when you are in a crowded elevator? Is this a feeling that you could live with every day of your life? Can you think of ways to test whether overcrowding contributes to behavioral disorders? Can you think of other factors that may contribute to socially deviant behavior? Do you foresee any legal problems with attributing crime to overcrowding?

# VII

# ECOLOGY AND
# THE ENVIRONMENT

ECOSYSTEMS ARE COMPOSED OF COMMUNITIES
AND THEIR PHSICAL ENVIRONMENT.

# 30 POPULATIONS

**OPENING IMAGE**

*A population of fish near a reef in the Bahamas.*

Sugar cane is a major crop in parts of Australia, and in one state, Queensland, it is the main agricultural export. One of the insect pests that damages the cane plants is the gray-back beetle. To combat the beetle using a biological weapon, Australian farmers imported the cane toad, Bufo marinus. The first shipment of toads arrived in 1934 from Hawaii. Unfortunately, the experiment in biological control was not a success. The beetles live on the tops of the cane plants, and do not descend to the ground, where the toads live.

More unfortunate is the fact that the toads thrive in Australia, where they have few natural predators. Cane toads are large and poisonous, and animals such as birds and snakes that would be expected to be predators are killed by a poisonous venom sprayed by the toad from glands located on its back.

The toad has spread over about 40% of Queensland, a state in northeast Australia, and is spreading west and south into the Northern Territory and New South Wales, advancing about 17 miles each year. As the toad spreads into more populated areas, it is becoming a pest, killing dogs and cats it comes in contact with.

Proposals for controlling and hopefully eliminating the toad are being considered, but no selective poisons or organisms have been found. Some, like the xenodon snake, from South America, are predators, but have been ruled out because they would also kill native Australian animals. Officials hope to avoid creating another biological problem in efforts to control the toads.

*The quandary faced by Australia in dealing with an imported species is not unique. Here in the United States, plants ranging from the dandelion to the kudzu vine and animals including the sparrow and the beetle carrying Dutch elm disease were imported, either by accident or design.*

*To understand how newly introduced species interact with native organisms, we must first examine the interactions between individuals and between populations. This chapter begins our consideration of how organisms interact with their physical environment and with other organisms in that environment. The chapter focuses on the population, the lowest level of biological organization above the individual, and reviews the factors that control the life cycle of populations and the role of human intervention in the growth and decline of populations.*

 ## ECOLOGY AND THE HIERARCHY OF LIFE

In previous chapters, we focused on the biology of individual organisms, their components, and their evolution. In this and the remaining two chapters, we will take a broader view of life on planet Earth and how humans interact with and affect it.

The first observation in our broader view is that living systems have a hierarchical organization. From the most basic level, organisms are composed of atoms, molecules, cells, tissues, organs, and organ systems. Individual organisms are grouped into **populations.** In turn, populations form **communities,** which compose **ecosystems.** Ecosystems make up the **biosphere,** which includes all life on Earth (✦ Figure 30.1). This hierarchical organization demonstrates that living systems at all levels are interconnected, and that each level provides the foundation for the level above. Furthermore, the hierarchy is not static: Each level is dynamic, with many interactions occurring among the units. For instance, organisms that compose populations interact with each other and individuals of other species in many ways.

**Ecology** is the study of how organisms interact with each other and their physical environment. Because interactions occur at many levels, ecology is a very complex subject. In addition, human activity often disturbs living systems and affects interactions among organisms and their environment. As a result, ecology often yields generalities and predictions that are less precise than we would like. Nevertheless, the science of ecology has made significant advances since its origin more than 100 years ago, and provides a general framework for understanding interactions in the hierarchy of life.

✦ **FIGURE 30.1**

**Living systems are organized into hierarchies, from the subatomic level to the biosphere.**

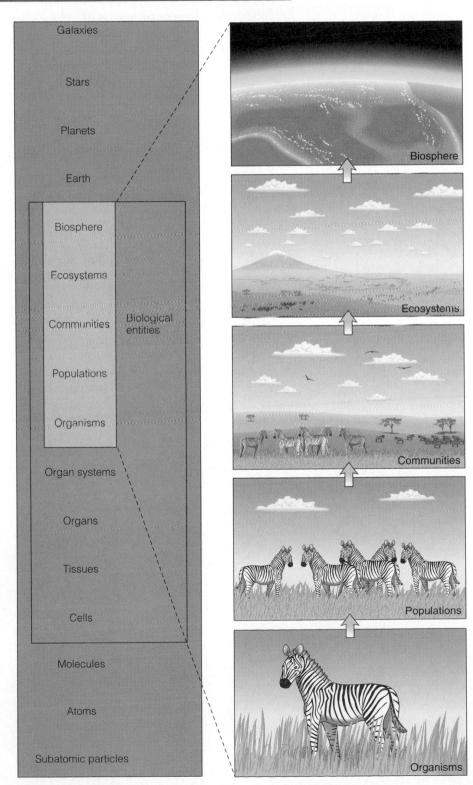

Populations have three distinct phases: growth, stability, and decline. Often, between growth and stability, is a period where the population overshoots its stabilized level.

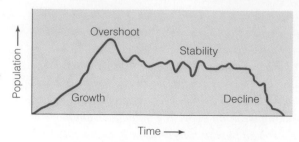

This and the next two chapters will focus on only one or two levels of the hierarchy of life. This will allow us to dissect some of the complex interactions among living systems. We will begin with populations as the basic entity of the hierarchy and work up through communities and ecosystems to the biosphere.

## POPULATIONS HAVE A LIFE CYCLE WITH THREE PHASES

A **population** is a group of individuals of the same species living in the same geographic area. A population can include all the bass in a lake, or all the black bears in Alaska. Whatever the species or area, all populations undergo three distinct phases as part of their life cycle: **growth, stability,** and **decline** (♦ Figure 30.2)

Population growth occurs when available resources exceed the number of individuals able to exploit them. This often occurs when a population occupies a new geographic area. In this situation, individuals tend to reproduce rapidly and death rates are low because resources are relatively abundant. Eventually, growth rates level off, and the population size becomes stable. In many cases, however, stability is preceded by a population "crash" because rapid population growth can abruptly overshoot the available resources (Figure 30.2). Even stable populations are dynamic, and undergo fluc-

tuation in numbers. Stability is usually the longest phase in the life cycle of a population. Decline is decrease in the number of individuals in a population, which, in the long term, leads to localized population extinction.

Many factors affect the growth, stability, and decline of populations. The study of these factors and their interactions is called **population dynamics.** In the rest of this chapter, we discuss each of the phases that characterize the life cycle of populations and their dynamics.

## POPULATION GROWTH IS CONTROLLED BY SEVERAL FACTORS

Nearly all populations have one trait in common: If left unchecked, they will grow exponentially as long as resources are available (♦ Figure 30.3). This is because reproduction is generally a multiplicative process. If each individual produced only a single offspring before death, the population size would remain constant. However, individuals usually produce many more than one offspring in their lifetime. Furthermore, each of those offspring can, in turn, have many more progeny. The result is that most populations have the potential to increase very rapidly, often at an **exponential rate.** The intrinsic rate of increase varies among populations, depending on many factors, some of which are genetically determined. Two of the most basic factors are **birth rate** and **death rate:**

Intrinsic rate of increase = Birth rate − Death rate

This relationship shows that the net rate of increase in a population is governed by the rate at which the population can reproduce *and* by the rate at which individuals die. In bacterial populations, millions of new individuals are produced each day, whereas elephant populations have a birth rate of only one

♦ **FIGURE** 30.3

Elephant seal populations have undergone exponential growth since they were hunted into near extinction in 1890.

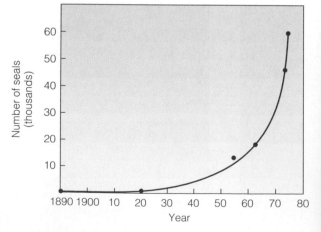

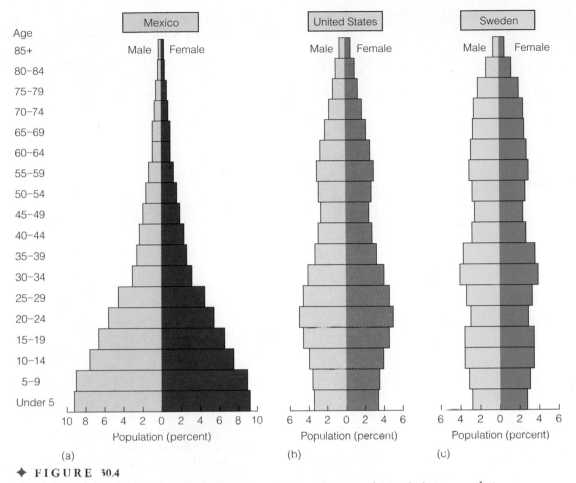

**♦ FIGURE 30.4**

Age structures for (a) a growing population, (b) a stable population, and (c) a declining population.

or two individuals per couple of decades. But bacteria also die by the millions per day so the intrinsic rate of increase in a bacterial population is dramatically lower than if determined by birth rates alone.

## Population Growth Potential Is Related to Life History

Reproductive traits such as age of reproduction and number of offspring are major determinants of the birth rate and thus affect the intrinsic rate of increase. A population that reproduces at an early age and has many offspring will produce more offspring over a given time than a population that delays reproduction and has fewer offspring. **Life history** is a term that refers to the age of sexual maturation, age of death, and other events in an individual's lifetime that influence reproductive traits.

Ecologists have traditionally used two extremes to describe life histories and population growth. At one extreme are organisms that grow fast, reproduce quickly, and have many offspring in each reproductive cycle. At the other extreme are organisms that grow slowly, reproduce at a late age, and have few offspring per cycle.

These two extremes are clearly a simplification. Most organisms have intermediate life history traits. Indeed, the relationship between life history and population growth is more complex than these extremes imply. Nevertheless, these examples point out that early reproduction and more offspring per reproductive cycle will raise a population's intrinsic rate of increase.

## Age Structure Can Predict Growth Potential

The rate of population growth is strongly influenced by the proportion of individuals of reproductive age. **Age structure** refers to the relative proportion of individuals in each age group in a population. An age structure with individuals of reproductive and prereproductive age has a much greater potential for population growth than one that has more older individuals. Thus, a pyramid-shaped age structure has more individuals that have yet to reproduce, while one that is tapered at the bottom has more individuals that have already reproduced (♦ Figure 30.4). Stable populations that maintain roughly constant size tend to have similar

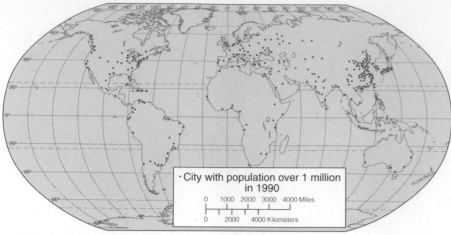

**✦ FIGURE 30.5**

A map showing human population density distributions on earth.

## Human Populations Are in a Growth Phase

The fossil and genetic evidence reviewed in Chapter 15 indicates that our species arose about 200,000 years ago. Since then, humans have spread over a wide range of geographic areas that includes most of the land surfaces of the earth (✦ Figure 30.5). The growth of the human population up to about 1650 was slow. About then, world population growth entered a phase of rapid, exponential growth (✦ Figure 30.6). Several factors have contributed to this growth pattern. Soon after their appearance, human populations were organized into hunter-gatherer groups with an effective population size of about 500 individuals. During the late Pleistocene (an epoch that ended about 10,000 years ago), several cultural innovations resulted in population growth. These included the development of new weapons (spears, harpoon, and possibly the bow and arrow) and the domestication of plants and animals. Still, population growth was slow, so that by the beginning of the Christian era, (1 A.D.), the world population was estimated to be 200 million (the present U.S. population is about 260 million). It took 1800 years to reach one billion, but only another 130 years to reach two billion, and another 45 years to reach four billion.

This growth was made possible by the development of technology (the Industrial Revolution), which made more resources available to a rapidly growing population. In the last 50 to 100 years, viability and survival have improved through the de-

proportions of reproductive and prereproductive individuals.

Age structure can be related to life history: Populations of individuals that mature and reproduce early have a short prereproductive age. Therefore, populations with the highest potential for intrinsic rate of increase are those having an early time of reproduction combined with many prereproductive individuals (pyramid-shaped age structure). Such populations are truly explosive, and are capable of growing at rates that are exponentially faster than most other life history and age structure combinations.

**✦ FIGURE 30.6**

Growth of the human population during the last 2000 years.

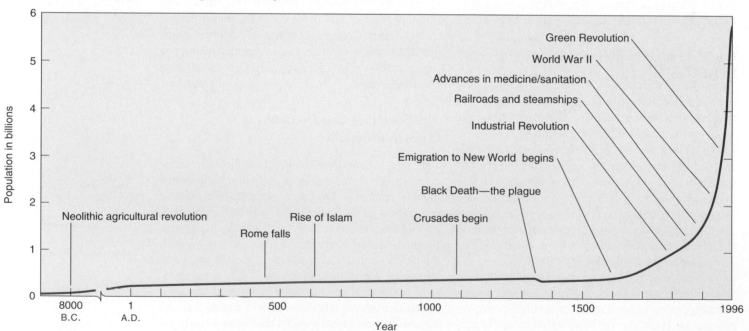

velopment and application of medical and public health measures. This has extended the life span, allowing individuals to leave more offspring. In July 1987, the human population was estimated at five billion, and before the year 2000, six billion humans will be living on planet Earth. Although technology has reduced the impact of physical and biological controls on the growth of the human population, our species, like all others, is ultimately subject to the limitation of physical and biological resources. The following sections explore the factors that limit population growth.

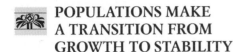

## POPULATIONS MAKE A TRANSITION FROM GROWTH TO STABILITY

Common sense tells you that nothing grows forever and this is certainly true of populations as well. In most environments, insects are one of the most common animal groups. However, even insects have natural limits on their population size. What are these limits? It is not simply space and food; rather it is a complex interaction of many factors, including the presence of other species.

A simple way to examine population size limits is to consider a species of bacteria living in a laboratory flask. If a few adult individuals are placed into the container, exponential growth will occur, for the reasons already discussed. At some point, the limited space and food in the flask will cause the exponential growth phase to stop, and the population will enter a slower growth phase and eventually stabilize. By "stabilize" we mean that population size remains roughly constant, fluctuating around some average density. This population growth curve fits a **logistic growth model,** where exponential growth is slowed and eventually limited by some factor (✦ Figure 30.7). This point where the population size levels off is called the **carrying capacity.** The carrying capacity is the maximum population size that can be regularly sustained by an environment.

However, the population in the flask will not stabilize indefinitely. Unless waste is withdrawn and new food is added, the population will undergo a decline and eventual extinction.

### Several Basic Controls Govern Population Size

What causes population growth to level off and stabilize? Clearly, the ultimate cause is to be found in the environment: Factors external to the population must be limiting the rate at which individuals can reproduce and survive. Ecologists typically divide these environmental factors into two basic categories, the (l) physical and (2) biological envi-

✦ **FIGURE 30.7**

**Carrying capacity in two different populations. (a) After introduction in the early 1800s, the sheep population of Tasmania stabilized at about 1.7 million, after overshooting the carrying capacity of the environment. As the population stabilized, fluctuations in the population diminished. (b) In a bacterial culture, the growth curve shows a lag phase (1), an exponential growth phase (2), and (3) a stable phase.**

ronment. Within the biological environment, we can identify three subcategories of biological interactions that can limit population growth: **competition, predation,** and **symbiosis.** These factors along with the physical environment work independently and in combinations to control the growth of populations.

### Resources in the Physical Environment Can Limit Growth

Limitations in the physical environment include constraints such as food, shelter, water supply, space availability, or in the case of plants, soil and light. Any single aspect of the physical environment

*Sidebar*

## THE WAR BETWEEN THE GECKOS

Native Hawaiians have grown accustomed to sharing their urban homes with the three inch long mourning gecko. These creatures are all females and virtually impossible to keep out of one's home. They reproduce asexually by laying eggs without male assistance.

Another species of gecko, known as the house gecko, has displaced the mourning gecko in many areas. The house geckos are native to the Philippines and reached the Hawaiian islands by hitchhiking on boats and planes. This species reproduces sexually. The introduction of house geckos has pushed the mourning geckos out of urban areas and into the forests and rural areas.

Researchers set up artificial environments to study how the two species interact. Some environments had both species; others only one. Some areas were well lit to attract insects; others were kept dark. They observed fighting between male house geckos and among the mourning geckos. Larger house geckos did not bother the smaller mourning geckos. Instead, the house geckos chose to congregate around the lights to gather insects. Scientists found the smaller mourning geckos were wasting away because they had nothing to eat. In darkened areas, both species coexisted as they do in the forests. Well-lit homes attracting insects give house geckos a competitive edge over mourning geckos, pushing them out of this habitat.

can limit population growth. Even if light, soil, temperature, and many other resources are abundant, and can support many more plants, population growth may be constrained far below that limit if only one resource, such as water, is scarce. This observation was first made by the German botanist Justus Liebig in 1840 and has come to be called the **law of the minimum:** Population growth is limited by the resource in the shortest supply.

## Competition Is a Biological Factor Controlling Growth

In the biological environment, each species has a distinctive habitat, food preferences, behaviors, etc. The biological role played by a species is called its **niche.** Competitors are organisms that require similar resources, and thus compete with members of other populations for these limited resources. **Niche overlap** specifies the extent to which similar resources are required and therefore the strength of competition between two species. **Competitive exclusion** occurs when competition is so intense that one species completely eliminates the second species from an area. This happens when both species require many of the same resources, especially food. In such cases, ecologists say that there is a great deal of niche overlap between the two species.

Although competitive exclusion has received much attention, many ecologists think it is relatively rare in nature. In contrast, there are many cases where species compete for only one or a few resources and overlap is minimal. Foxes and owls in a forest may compete for mice but each will also eat other prey that the other does not usually pursue. Such cases of minor niche overlap do not result in the competitive exclusion of one species. Instead, each competitor merely limits the abundance of the other. When the ciliated protozoan, *Paramecium aurelia,* is raised alone, its population size is nearly twice as large as when it must share resources with a second species (✦ Figure 30.8).

An important consequence of competitive exclusion is that removal of one of two competing species will cause the remaining species to increase in number. Thus, the number of *P. aurelia* would nearly double if a competitor were removed (Figure 30.8). This effect is called **competitive release** because the remaining species is "released" from one of the factors that limits its population size. Competitive release is a relatively common effect of human activity as we will see later.

## Predators Help Stabilize Populations

Predators are organisms that kill and consume other living organisms. There are two basic categories of predators: (1) carnivores, which prey on animals, and (2) herbivores, which consume plants. As with competition, there is much variation in the extent to which predation limits the abundance of prey. The most extreme cases occur when a predator drives the prey species to extinction. Some of these cases have resulted from human activity, such as the introduction of a predator into a new area (especially an island). More commonly, predators simply limit the size of a population and do not drive the prey to extinction.

But why? Why shouldn't predators reproduce until they have eaten all the prey? There are at least three major reasons. First, prey species often evolve protective traits, such as camouflage, poisons, spines, large size, and so on. Second, prey species often have **refuges,** such as burrows or treetops, where predators cannot reach them. The importance of refuges has been shown in experiments where an insect or microbe predator population in a container will drive the prey population into extinction unless the prey can escape to a refuge. The third reason for nonextinction of prey is **prey switching** by predators. When one prey species becomes rare, predators will tend to switch to another, easily available prey. This allows the first prey species to rebound in numbers.

✦ **FIGURE 30.8**

**The ciliated protozoan,** *Paramecium aurelia,* **achieves a population size twice that as when it must share resources with a second species. This difference is an example of competitive release.**

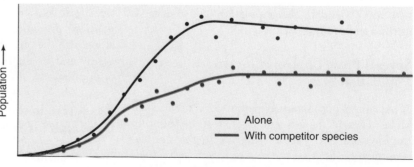

| TABLE 30.1 TYPES OF SYMBIOSES | | |
|---|---|---|
| TYPE OF RELATIONSHIP | SPECIES A | SPECIES B |
| Mutualism | + | + |
| Parasitism | + | − |
| Commensalism | + | 0 |
| Amensalism | − | 0 |

0 *No significant relationship.*
+ *Positive relationship to population indicated.*
− *Negative or inhibitory relationship to population indicated.*

## The Real World: Complex Interaction of Population Controls

A brief description of the four main controls on population growth (the physical environment, competition, predation, and symbiosis) has oversimplified what happens to populations in the "real world" by discussing each control separately. Natural populations are not governed by any single control. Instead, many controls simultaneously play a role in determining the size of most natural populations. For example, when two populations of beetles interact in a laboratory environment, competition leads to a large population size for one species, and a small population size for the other species. If a third species (a parasite) is introduced, the formerly abun-

Ecological release is also seen in the relationship between predators and prey. If predator populations are greatly reduced by disease, or other causes, or if prey populations migrate to an area without major predators (such as islands), the prey species may show **predatory release.** This occurs in the same way as competitive release.

## Symbiotic Interactions Can Affect Population Size and Stability

Symbiosis is a general term that has become a "catch-all" for almost all other interactions beside predation and competition. These interactions are classified according to whether they benefit or inhibit one or both populations involved in the interaction (● Table 30.1). **Mutualism** is a form of symbiosis that benefits both species. Algae coexist within the tissues of the animals that build coral reefs. The algae are provided with a protected living space, and the coral animals are supplied with nutrients from the algae's photosynthesis. Mutualism regulates the population size of algae because their numbers are influenced by the abundance of the coral-building host.

**Parasitism** is a form of symbiosis that is similar to predation because it benefits the population of one species while harming the other species. Parasitism differs from predation in that parasites act more slowly than predators and do not always kill the prey (host). **Commensalism** is a symbiotic relationship in which one species benefits and the other is not affected. "Spanish moss" (a lichen) hangs from trees for support but causes the trees no great harm or benefit. Lichens themselves are mutualistic combinations of algae and fungi. Barnacles attach to crab shells in a similar way, and derive food from the surrounding water, but do not affect the crab. **Amensalism** occurs when members of one population inhibit growth of another while being unaffected themselves (◆ Figure 30.9).

(a)

(b)

◆ **FIGURE 30.9**

**Commensalism and amensalism. (a) Spanish moss grows in a commensal relationship with trees. (b) Sage brush produces terpenes, a chemical that inhibits the growth of other plants.**

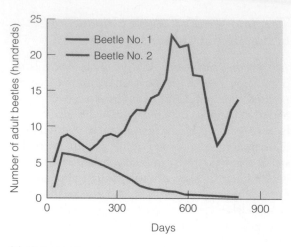 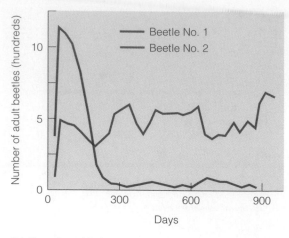

(a) Parasite absent

(b) Parasite added

✦ **FIGURE 30.10**

**Competition between two species of flour beetles. (a) In the absence of parasites,** *Tribolium castaneum* **(Beetle No. 1) will eliminate a competing species,** *T. confusum* **(Beetle No. 2). (b) In the presence of a parasite, the outcome is reversed, because** *T. castaneum* **is more susceptible to the parasite.**

dant species becomes rare because it is more susceptible to the parasite (✦ Figure 30.10).

In nature, communities are often composed of hundreds to many thousands of species. Any single population may have many competitors, predators, and symbionts. Changes in any of the controlling factors can have a major influence on the size of that population. At the same time, changes in physical conditions can also influence any of the populations in the community, with each responding differently to a single physical change. This complex interaction of controls on population size in natural ecosystems means that small changes in one control, such as the abundance of a prey species, can have a cascading "domino effect" throughout the ecosystem, causing larger changes in the population structure of other species.

## Physical versus Biological Control Systems in Population Growth Control

For more than 60 years ecologists have debated whether physical or biological factors, if either, are more important in controlling population growth. Some argue that physical factors such as climate play the major role. This is sometimes called density-independent regulation because physical processes usually operate independently of current population size. A severe drought or storm can drastically reduce the size of a population without regard to how many individuals exist when it strikes. Density-independent factors can cause population numbers to fluctuate widely. Mosquito populations peak in the summer, but frost and freezing temperatures kill all adults, leaving only eggs over the winter. In

the spring, when the temperature is warmer, and water is available, the population undergoes a period of exponential growth.

At the other extreme, others argue that biological interactions, especially competition and predation, are the major controls. This is called density-dependent regulation because current population size plays a role in determining population change. The effect of a predator in reducing the size of a prey species population depends on how many prey there are. As noted earlier, if the population of a prey species becomes too low, the predator may switch to another prey.

Considering the complexity of control interactions, it is easy to see why the debate has gone on so long. In reality, population size is regulated by both physical and biological constraints. Therefore, most ecologists today recognize that neither is more important in all cases. Physical or biological controls may dominate in certain environments with a continuum in between (✦ Figure 30.11). At one end of the continuum are environments that are often subjected to physical disturbance and stress. A beach with wave action from storms and hurricanes can often undergo great population shifts. At the other extreme are environments such as the offshore tropical waters around a coral reef, where physical disturbances are relatively rare. In these cases, rapid changes in physical parameters such as current flow, wave action, and temperature are less common and less severe. This leaves biological interactions as the major determinant controlling the size of the many reef species. But even in these extreme situations, several factors interact to control population size. Along the beach, even though physical processes may play a dominant role in regulating population size, competition, predation, and other biological interactions still occur, and affect population size.

## POPULATIONS CAN DECLINE AND BECOME EXTINCT

Even though stable populations undergo fluctuations in size, they can eventually decline to zero size and become extinct. **Extinction** is the elimination of all individuals in a group. Local extinction refers to the loss of all individuals in a population. It is termed local because a new population can be re-established from other populations of that species living elsewhere. Species extinction occurs when all populations of the species become extinct. Species extinction is obviously a much greater loss because once gone, species cannot be replaced. Even local extinction can result in the permanent loss of some alleles in the species' gene pool because most local populations have a few alleles that differ from other populations of the same species.

The widespread occurrence of population decline is seen in the fact that more than 99% of all species that have ever existed are now extinct. The ultimate cause of decline and extinction is environ-mental change: One or more of the controlling factors in the physical or biological environment becomes altered. This leads to a decline in population size. Physical changes such as environmental cooling are well documented as factors in many extinctions (Chapter 11). Similarly, there are cases recorded from fossils where competition from new groups apparently caused population decline. For example, the invasion of South America by placental mammals led to the decline and extinction of most marsupial species in that region.

Dramatic declines in human populations have periodically occurred as infectious diseases have gained access to susceptible populations. Waves of bubonic plague introduced from Asia killed between one-third to one-half of Europe's population in the years between 1346 and 1350. Later epidemics of plague lasting to around 1700 killed a quarter of the European population. Smallpox and other infectious diseases transferred from Europe to North and South America reduced aboriginal populations in the early sixteenth century. The role of disease in human history has been documented in a number of books including *Plagues and Peoples* by William McNeil. More recently, potential effects of changing the biological environment have been discussed in *The Coming Plague,* a book by by Laurie Garrett.

✦ FIGURE 30.11

Two extremes in the regulation of population abundance. At the left, density-independent physical factors control the population size. At the right, physical factors are less important than biological factors, and population size is density dependent.

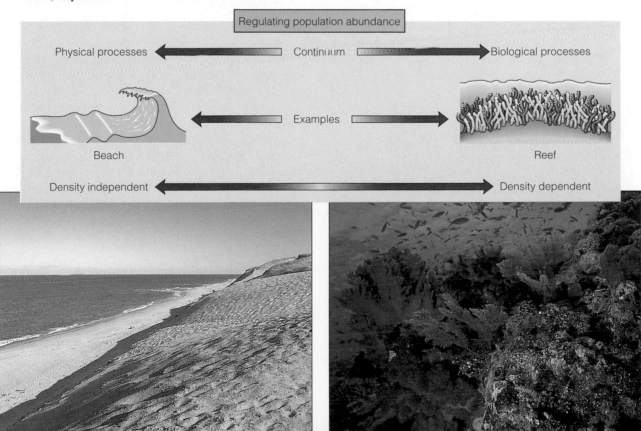

## HUMAN IMPACT ON POPULATION GROWTH, STABILITY, AND DECLINE

An increasing human population size and the use of technology have caused progressively greater amounts of disruption to natural populations. Pollution, agriculture, and other human alterations of the environment have destabilized some populations by affecting the various control systems discussed earlier. This destabilization leads to two possible outcomes, depending on the population: (1) population growth as previous limitations are removed, or (2) population decline as new limitations are imposed.

Population growth occurs because there are excess resources relative to the number of individuals available to exploit them. This condition can be created in four ways (◆ Table 30.2). Each causes exponential population growth until a new carrying capacity is reached. Each of these factors involves removing controls that limit population size.

### Altering Resources Can Selectively Influence Population Growth

An increase in resources available to a population can be either planned or unplanned. Agriculture and animal domestication are examples of how the population size of favored organisms can be greatly increased. This is accomplished by providing more food and other resources than these organisms would have available in their natural state. The influence of human intervention on population size can be demonstrated by considering cat populations. In England alone, domestic cats are so abundant that more than 300,000 cats must be destroyed per year. Yet before humans began to domesticate

them a few thousand years ago, the small, wild ancestors of domesticated cats were relatively rare. They probably occupied a relatively small area of the Middle East and Europe compared to their present range. Similar points could be made about the smaller ancestral ranges of corn, potatoes, and many other domesticated species (◆ Figure 30.12).

Unplanned increases in available resources that are by-products of human activity can also have major effects on animal and plant populations. Pollutants generally represent unplanned release of substances into the water or air. These substances are often nutrients that are in short supply. Recall from Liebig's law of the minimum that resources in shortest supply are the ones that limit growth. Phosphorus and nitrogen are often limiting nutrients for aquatic and terrestrial plants. Runoff from agricultural fertilizers containing these minerals is carried into rivers, lakes and ponds, chemically altering the physical environment. The result is often "runaway" plant growth, called **eutrophication.** Eutrophication is a classic example of "too much of a good thing." As the plant population increases, the death rate also increases, and bacterial decomposition of dead plants uses up increasing amounts of oxygen, causing fish and other organisms to suffocate.

Agricultural runoff can also cause large-scale changes in populations and ecosystems. In the northern Everglades, where runoff from agricultural fertilizers is high, cattails have replaced the native sawgrass, altering an entire ecosystem.

### Pesticides Can Alter Competition

Competitive release occurs when populations of one species that compete for resources with another species are removed. Removal allows the remaining species to undergo population growth. Competitive release is common in situations where humans attempt to eliminate a population of one species. Often, the result is an increase in populations of its competitors. An economically important example occurs when farmers use pesticides to eradicate insect pests that destroy crops (◆ Figure 30.13). Because poison tolerance varies among species, some pests will not be killed by the poisons and will actually increase in numbers when their competitors are gone. This is called a secondary pest outbreak.

A dramatic example of this phenomenon occurred in the 1950s when cotton crops in Central America were sprayed with pesticides to kill boll weevils. This was successful until 1955, when cotton aphids and cotton boll worms underwent explosive population growth. When a new pesticide was used to remove them, five further secondary

| TABLE 30.2 FOUR WAYS THAT HUMANS CAUSE POPULATION GROWTH | |
|---|---|
| | EXAMPLES |
| Increase available resources | Agriculture, nutrient pollution in lakes |
| Competitive release | Poisoning of insect pests |
| Predator release | Overhunting of large carnivores |
| Introduction to new areas | Game releases |

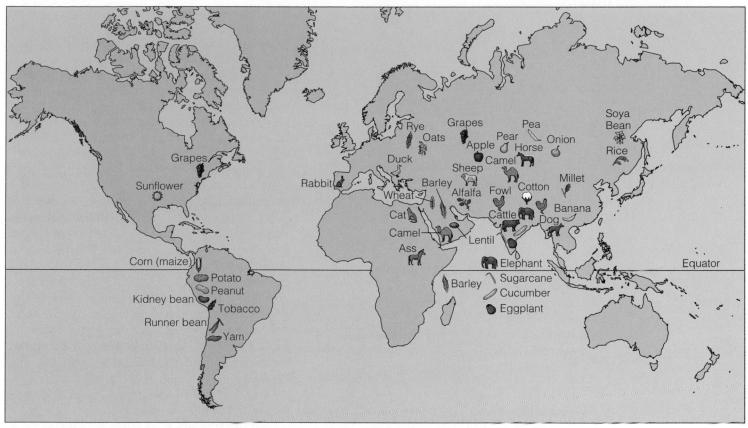

**✦ FIGURE 30.12**

Areas of the world where various species of plants and animals were originally distributed and first domesticated.

pests emerged. Such experiences have led to the development of other methods for pest control, which will be discussed in Chapter 31.

## Removing Predators Can Have Unforeseen Consequences

Predator release is common where humans hunt, trap, or otherwise reduce populations of predators, allowing the prey species population to increase. Large predators such as wolves and panthers have long been the target of ranchers and farmers because they prey on domesticated animals. The result has been a rapid population increase in the predator's natural prey. In areas where these predators are depleted, deer populations have shown an especially spectacular increase. Most experts estimate that there are now more deer in the United States than were present before Europeans arrived. The unfortunate lesson from the depletion of predators is that excess populations of deer often cause overgrazing of plants and subsequent death by starvation.

For this reason, selective game hunting can be ecologically useful, but is controversial among some environmental activists. By prudently "culling" deer and other prey, humans are essentially carrying out the role of predators that no longer exist. Reintroduction of predators is a long-term solution to such problems. Programs to reintroduce wolves to areas of Montana and surrounding states are now under way in an attempt to balance predator/prey interactions.

**✦ FIGURE 30.13**

**Pesticide use can cause outbreaks of a secondary pest.**

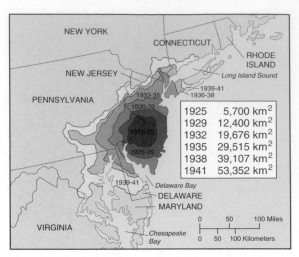

**✦ FIGURE 30.14**

Spread of the Japanese beetle in the Eastern United States after its accidental introduction.

| | |
|---|---|
| 1925 | 5,700 km² |
| 1929 | 12,400 km² |
| 1932 | 19,676 km² |
| 1935 | 29,515 km² |
| 1938 | 39,107 km² |
| 1941 | 53,352 km² |

## Introduction of New Species Can Alter Population Structure

Deliberate or accidental introduction of non-native (also called "exotic" or "alien") species into new areas has been perhaps the single greatest alteration of natural populations. Few people realize the enormous scale on which humans have, either accidentally or purposely, transferred organisms from one area to another. More than 1500 non-native insect species and more than 24 families of non-native fish have been introduced into North America. Virtually almost any given lake or bay contains nonnative species of plants or animals (Figure 30.12). In a California study, 50 of the 133 fish species captured at one location represented non-native species. Even more pervasive has been the introduction of new plant species, both for agricultural and ornamental purposes. More than 3000 species of plants have been introduced into North America, including the Norway maple, the eucalyptus tree, clover, and ryegrass.

The extent of this environmental alteration is more impressive when the survival rate of introduced populations is considered. Of 424 documented attempts to introduce game bird species into new environments, 360 (or 85%) failed even though dozens of individuals were released in several waves, and were released into environments considered to be habitable to those species. This may mean that the vast majority of accidental introductions probably fail, since they generally involve fewer individuals than planned introductions, and environmental suitability is not always taken into account.

Once introduced species are established, their growth can be truly explosive. The Japanese beetle arrived in New Jersey in 1916 on a shipment of plants. By 1941, they had spread over an area of 53,000 square kilometers (✦ Figure 30.14).

## Altering Population Growth Can Be Destructive or Beneficial

Is human-induced population growth of organisms bad? Natural populations are regulated by four basic physical and biological factors that humans can remove or alter. From a practical point of view, it is clear that such imbalances can be both destructive and beneficial.

On the negative side, human-induced population growth can be destructive to both humans and to natural ecosystems. Secondary outbreaks of crop-eating pests can occur when competitors are removed by pesticide spraying. About 17% of the 1500 insect species introduced into North America are pests requiring the use of pesticides for control. "Killer bees," a strain of bees resulting from the hybridization of African bees with native South American bees, are undergoing a population growth and expansion. From their point of origin in South America they are now moving into the United States (✦ Figure 30.15).

Although killer bees have received much attention, the destruction of native honeybee populations by these hybrid bees has occurred. The damage caused by "killer bees" is on a much smaller scale than the effects generated by the accidental

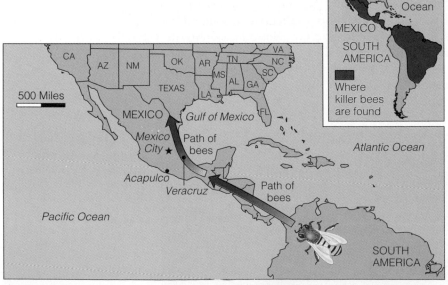

**✦ FIGURE 30.15**

Killer bees originated as a hybrid between African bees and South American bees in Brazil in the early 1960s. Since then, Africanized strains have moved northward, entering the United States from its southern border.

## Beyond the Basics

# FISHING IN EMPTY WATERS

In 1981, in an attempt to save its fishing industry, the United States declared a 200 mile (322 kilometers) boundary around its shores. This effectively banned most foreign fishing fleets that had previously fished in these waters. This drastic action was taken in an attempt to save the declining populations of fish from extinction, and to provide an economic boost to New England.

This measure might have worked, except for the fact that American fishermen began overfishing the protected area. A federal loan program allowed the construction of more boats and the purchase of new electronic technology to find fish. The fishing industry lobbied to have quotas on catches removed. Over the short term, commercial fishing boomed. The catch from fishing trawlers peaked in 1983, and has declined sharply since then. Stocks of some fish such as the flounder and haddock are at or near record low levels. The Atlantic bluefin tuna is now on the World Wildlife Federation's list of the ten most endangered species. Other fish populations such as the swordfish and cod are seriously depleted, and the cod may be soon placed on the endangered species list.

The fishing industry has always had booms and busts, but this time, scientists say things are different. There are so many boats and so much fish-finding technology that the fish have no place to hide. In the record years of 1983, the catch reached 410 million pounds (128 million kg), up 66% from 1976. In 1990, the catch had declined to 282 million pounds (105 million kg). This figure is somewhat deceptive, since it included huge numbers of juvenile fish that had just reached the minimum size for harvesting, and before they had a chance to reproduce. Without action, both the fish and the fishing industry may become extinct.

This problem is not unique to the United States. In 1995, Canada used armed naval vessels to drive European trawlers from its fishing grounds, and the resulting feud has been the subject of diplomatic negotiations. The oceans are a resource used by many individuals and nations. To preserve and harvest the resources in a rational way, many nations must agree on laws that govern access to resources such as fish, and devise ways of enforcing these laws to prevent overfishing. In the United States, the size of the fleet may be reduced by buying out vessels with money from a tax on marine diesel fuel, by restricting catches, and occasional closure of overfished waters within our territorial boundaries.

---

introduction of zebra mussels into Lake Erie in the mid-1980s (✦ Figure 30.16). These fast-growing and fast-breeding mussels from Asia have spread to the other Great Lakes. They attach to water intake pipes, boat hulls, and other underwater structures. They compete with native species of animals for oxygen and plankton and are causing billions of dollars in damage as they spread through the Great Lakes and into the vast Mississippi River drainage system.

On the positive side, human-induced population growth of organisms can be beneficial by providing needed resources for human populations. Agriculture produces more food per acre of land than would be produced by plant populations under natural conditions. Even natural populations can be effectively managed by human activity to increase yields with minimum damage to the populations. Forestry and fishery managers have long applied a concept called **maximum sustainable yield (MSY)**, which seeks to permit the maximum harvest of a food or game population without harming the population's ability to grow back.

MSY is based on the premise that the maximum growth rate of the "S-shaped" or sigmoidal population curve occurs at exactly one-half the carrying capacity (✦ Figure 30.17). The maximum sustainable harvest of a fish population would therefore be obtained if enough fish were harvested each year to bring the population back to one-half the carrying capacity (Figure 30.17). Unfortunately, there have been many problems in applying this attractive concept. It is difficult to measure the carrying capacity of wild populations. In addition, the age structure of the population to be managed must be

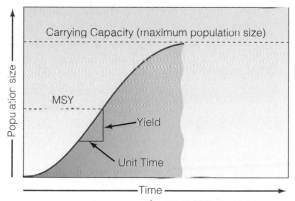

✦ **FIGURE 30.17**

**Growth in a fish population. The population size at one-half the maximum population density is called the point of maximum sustainable yield (MSY).**

✦ **FIGURE 30.16**

**Zebra mussels were accidently introduced into the Great Lakes in the 1980s, and are rapidly displacing native species.**

**✦ FIGURE 30.18**

**The brown tree snake has undergone a population explosion on Guam, and has been responsible for the extinction of many of the native species of birds.**

known. Nevertheless, MSY continues to be widely used and many ecologists agree that if prudently applied, it can be a useful component of an effective ecosystem management strategy.

## Intervention Can Cause Population Decline

Although extinctions are part of the life cycle of populations, human population density and the use of technology are causing the decline and extinction of animal and plant species at thousands of times the natural rate. This impact is occurring in nearly all parts of the biosphere. We will examine the overall, global ramifications of this in Chapter 32. To understand the causes of this large-scale loss of diversity, we need to examine how human activity causes the local population declines that lead to extinction. Recall that all extinction ultimately results from environmental change that reduces pop-

| **TABLE 30.3 FOUR WAYS THAT HUMANS CAUSE POPULATION DECLINE AND EXTINCTION\*** | |
|---|---|
| | EXAMPLES |
| Change physical environment: | |
| Habitat disruption | Drain swamp, toxic pollution |
| Change biological environment: | |
| Species introduction | New predator |
| Overhunting | Big-game hunting |
| Secondary extinctions | Loss of food species |

*\*Listed in approximate order of importance in causing extinctions. In other words, habitat destruction causes most extinctions today; secondary extinctions cause the least.*

ulations to zero. Extinction is therefore caused by alteration of a population's environment in a deleterious way.

Human activity can alter the environment to cause extinction in four basic ways: habitat disruption, species introduction, overhunting, and secondary extinctions (✦ Table 30.3). These are directly related to the factors that control population size discussed in earlier sections of this chapter.

**Habitat disruption** is a disturbance of the physical environment in which a population lives. This disruption can range from small, gradual disturbances such as chemical changes generated by air pollution, to major events such as total destruction of a forest by bulldozers or fire. Small-scale disturbances tend to affect only populations of the most susceptible species. For example, oxygen loss from eutrophication by sewage (which is rich in minerals and nutrients) in lakes first kills populations of susceptible gamefish, such as trout and bass, because they need more oxygen than carp or garfish.

Habitat disruption is currently the major cause of extinction. This is not only because of the scale of the disturbance generated by human population growth, but also because populations of many species can decline simultaneously when the environment is disturbed.

Changes in the biological environment as a cause of extinctions occurs in three ways (Table 30.3). **Species introduction** has been common in many parts of the world. This has often occurred through accidental introductions, such as the discharge of zebra mussels into Lake Erie from a tanker that took on water ballast in Europe. In other cases, plants and animals have been purposely transported to new environments. These organisms were often introduced for agricultural or recreational reasons. In some cases, there has been little impact on native populations, leading to a net increase in diversity. In other cases, newly introduced species reduce the size of native species populations, sometimes to the point of extinction, reducing biological diversity. This occurs because introduced species become added competitors, predators, or symbionts (including diseases and parasites) in the native biological system.

A striking example of predatory decline is the introduction of the brown tree snake to the Pacific island of Guam (✦ Figure 30.18). This species of snake is native to the Solomon Islands, New Guinea, and the northern coast of Australia. It was apparently brought to Guam by U.S. military transports from New Guinea. From its introduction in the early 1960s, the predator has undergone a rapid population growth, with as many as 30,000 snakes per square mile in some areas of the island. The snakes feed on birds and other small animals, and because the birds have not developed adaptations to prevent predation, the snakes have caused the extinction of 11 of the 18 native bird species on the island.

Recently, brown tree snakes have been found in Hawaii, apparently carried in the wheel wells or cargo bays of planes from Guam. There are no native snakes in Hawaii, and the brown tree snake could rapidly colonize one or more of the islands and eliminate many of Hawaii's unique bird species.

## Overkills and Secondary Extinctions Can Alter the Biological Environment

**Overkills** and **secondary extinctions** are external causes of extinction generated by altering the biological environment (Table 30.3). Overkill is the shooting, trapping, or poisoning of certain populations, usually for sport or economic reasons. It is difficult to cause the extinction of some "pest" species, such as roaches or mice, by overkill because such species have large populations and reproduce rapidly. However, overkill has eliminated some populations of large animals because of their relatively small populations and slow rates of reproduction.

Economic motives for overkills include protecting domesticated animals from predators (such as wolves), but also because the hunted animals have economic value. Elephant ivory, leopard skins, and tropical bird feathers are but a few such animal products generated by killing for economic reasons. The effect of this is illustrated by the decline in rhinoceros populations. These animals have been hunted to near extinction because their horns are thought to have sexual or curative powers (✦ Figure 30.19). It seems unlikely that populations of the Javan or Sumatran rhinos will ever rebound from overkills (✦ Figure 30.20).

✦ **FIGURE 30.19**

The rhino has been hunted to near extinction for its horn, which is reputed to be an aphrodisiac.

✦ **FIGURE 30.20**

Estimated population density for various species of rhinoceros. From 30 species found in the fossil record, there are only 5 species today, all of which are threatened or endangered.

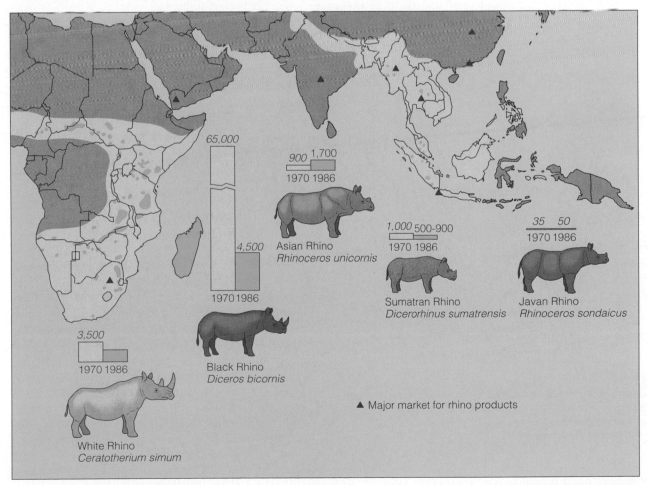

**◆ FIGURE 30.21**

Pandas are close to extinction because their habitat, bamboo forests, are being destroyed.

Secondary extinctions occur when the death of one population causes the extinction of another. Often, this involves the loss of a food species. For example, the giant panda of China subsists largely on a diet of bamboo (◆ Figure 30.21). Because bamboo forests are being destroyed by expansion of human populations, the panda may become extinct from that cause alone. Other examples are more subtle, reflecting the complex and unpredictable interactions among populations that make the effects of environmental disturbances by human activity so difficult to predict. The extinction of the dodo bird in the eighteenth century has caused the Calveria tree to become unable to reproduce (◆ Figure 30.22). The dodo ate the seeds of the tree and the dodo's digestive system removed the outer seed covering, which allowed the seed to germinate.

## Populations Have a Minimum Viable Size

It is not necessary for environmental changes to reduce a population to zero in order to cause extinction. Even if many individuals survive environmen-

tal disturbances, the population may never recover if it becomes too small. There are two basic reasons for this. One is that small populations may have breeding problems. There may not be enough females left, or individuals may not be able to locate each other for mating if the population is too dispersed. Even if there are enough mates and they can find each other, genetic inbreeding can be a major problem in small populations, causing a reduction in fertility or viability.

Secondly, small populations are susceptible to random environmental fluctuations, such as an abnormally harsh winter, which would not significantly affect larger populations. The chance of becoming extinct increases exponentially with decreasing population size.

The **minimum viable population (MVP)** refers to the smallest population size that can avoid extinction by these two sets of problems. If a population falls below this minimum size, it is said to be no longer viable, and long-term breeding problems and environmental fluctuations will eventually drive the population to extinction. The concept of MVP is in many ways only a rough estimate, because there is no certain number below which a population will definitely die out. If no severe environmental fluctuations occur for a long period, a small population may have time to recover; but no one can predict if this will happen.

Organisms vary widely in their ability to rebound from low numbers. Organisms that breed and grow rapidly will tend to be able to recover from lower population sizes than slow-growing organisms. Thus, an MVP for some species may consist of about 50 individuals, while for other species, it may consist of 500 individuals. In spite of its limitations, the MVP is valuable as a concept that conveys the perils of small population size as a factor in extinction.

## RANGE AND DENSITY ARE PROPERTIES OF POPULATIONS

Major changes in populations can occur through time because of variations in factors that regulate population size. However, the size of a population can vary through time and geographic range. Populations tend to have a maximum density near the center of their **geographic range,** which is the total area occupied by the population (◆ Figure 30.23). This central maximum is where physical and biological factors permit the greatest density. As one moves away from this central optimum, density generally begins to decline into the **zone of physiological stress,** where members of the population are rare. The geographic decline of density is usually

**◆ FIGURE 30.22**

The dodo bird, now extinct, was crucial in the reproduction of the Calvaria tree. The dodo ate the seeds of the tree, and a protective coating was removed by the bird's digestive system, allowing the seed to germinate.

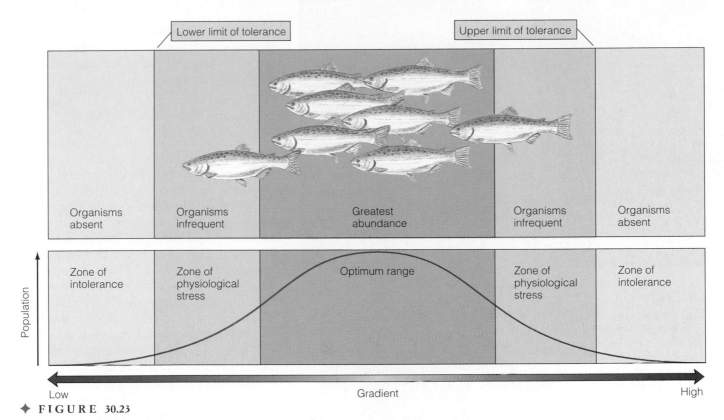

Upper limit of tolerance

| Organisms absent | Organisms infrequent | Greatest abundance | Organisms infrequent | Organisms absent |

| Zone of intolerance | Zone of physiological stress | Optimum range | Zone of physiological stress | Zone of intolerance |

Population

Low | Gradient | High

✦ **FIGURE 30.23**

**The population density of a species varies over its geographic range.**

gradual because both physical and biological limiting factors tend to follow a gradient.

Factors such as soil acidity, porosity, moisture, and other traits tend to change gradually over distance. Similarly, biological limits such as the presence of competitors and predators tend to follow a gradient. Eventually, the zone of physiological stress grades into the **zone of intolerance,** where the population is absent. The zone of intolerance occurs because some limiting factor has become so great that the species can no longer survive in that area. For example, soil moisture may become so low that individuals of a plant population can no longer tolerate it.

✦ **FIGURE 30.24**

**Patterns of population distribution. (a) Uniform. (b) Random. (c) Clumped.**

## Populations Have Several Dispersion Patterns

Populations are almost never uniformly distributed throughout their geographic range. This is because the environment is rarely uniform enough to follow perfect gradients. Instead, there are many irregularities. The way populations are distributed over their geographic range is called the dispersion pattern. In populations that inhabit a very patchy environment, individuals tend to be clumped together, with gaps in between. The clumped pattern is the most common one found in natural populations. Other patterns are uniform and random (✦ Figure 30.24).

The geographic range of populations varies greatly among species. Some species may consist of just one population that inhabits only a small

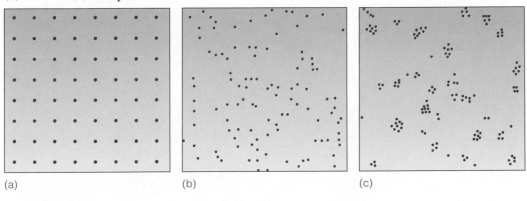

(a)  (b)  (c)

(a)

## ✦ FIGURE 30.25

**Limited population ranges. (a) Populations of Hawaiian** *Drosophila* **live in groves of trees separated by old lava flows, and are isolated from one another. (b) Hawaiian** *Drosophila* **(left) are much larger than** *Drosophila melanogaster* **(right).**

(b)

## ✦ FIGURE 30.26

**Original distribution of the brown trout and areas where it has been introduced, greatly altering its range of distribution.**

area. This is especially common among tropical organisms, such as highly specialized plant and insect species that are found in areas covering only a few acres.

## Population Ranges Are Dynamic

The geographic ranges of populations can be expanded or contracted as a result of natural forces and by human activity. Range decrease by habitat destruction has often led to declining populations and extinction. In Hawaii, some species of *Drosophila,* the fruit fly, live in isolated groves of trees surrounded by barren lava (✦ Figure 30.25). The population range (and the species range) can be restricted to only one or two groves. These groves are occasionally destroyed by new lava flows, causing extinction of the population. Clear cutting of forests is another form of habitat destruction that can contract population ranges. In such cases, the population's geographic range can be reduced to zero, and it becomes extinct. This is a major reason why tropical extinctions are particularly destructive; with the high number of endemic populations, it does not take much habitat destruction to reduce the geographic range to zero.

Geographic range can be expanded by domesticating and introducing wild species into new areas. The brown trout is one of many species that has had its range greatly expanded, far beyond its native habitat, by human intervention (✦ Figure 30.26).

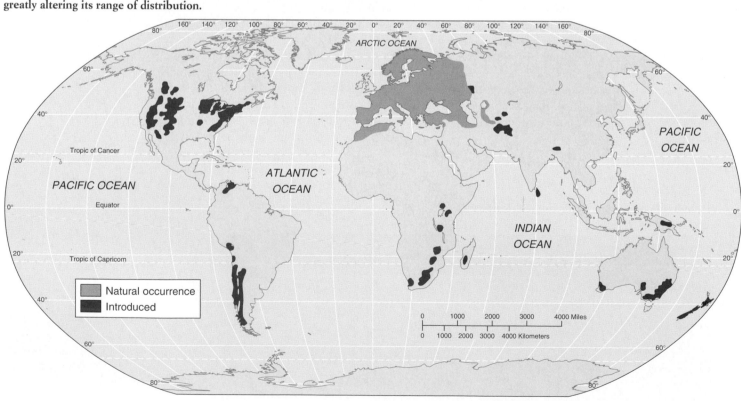

Natural occurrence

Introduced

# SUMMARY

1. Living systems have a hierarchical organization. Organisms are grouped into populations, which in turn form communities, which compose ecosystems. Ecosystems form the biosphere, which includes all life on earth. The hierarchy is not static: Each level has many interactions among the units that compose each entity.

2. Populations are groups of individuals of the same species living in the same geographic area. All populations undergo three distinct phases as part of their life cycle: growth, stability, and decline. These natural processes can be altered by human activity, and these alterations can directly affect higher levels of organization in the hierarchy of the biosphere.

3. Left unchecked, population growth tends to be exponential, but few populations sustain this growth pattern for long. Several factors influence the growth of populations, including birth rate, death rate, life history, and age structure.

4. Population growth eventually levels off and the population size becomes stable. Causes for this transition include the physical environment, competition, predation, and symbiosis.

5. All populations eventually decline to zero size and become extinct. Extinction is the elimination of all individuals in a group. Local extinction refers to the loss of all individuals in a population. It is termed local because a new population can be reestablished from other populations of that species living elsewhere. Species extinction occurs when all populations of the species become extinct. Species extinction is obviously a much greater loss because once gone, species cannot be replaced. Even local extinction can result in the permanent loss of some genes because most local populations have a few genes that differ from other populations of the same species in other areas.

6. Increasing human population size and the use of technology are causing increasing amounts of disruption to natural populations. Pollution, agriculture, population growth, and other human alterations of the environment have destabilized populations by affecting control systems that maintain population stability. Destabilization leads to two possible outcomes, depending on the population: (1) population growth as previous limitations are removed or (2) population decline and possible extinction as new limitations are imposed.

7. Populations are almost never uniformly distributed throughout their geographic range, and may have uniform, clumped, or random dispersion patterns. Dispersion patterns and population range are dynamic properties of populations that can be altered by natural changes and by human activity.

# KEY TERMS

age structure
amensalism
biosphere
birth rate
carrying capacity
commensalism
communities
competition
competitive exclusion
competitive release
death rate
decline
ecology
ecosystems

eutrophication
exponential rate
extinction
geographic range
growth
habitat disruption
law of the minimum
life history
logistic growth model
maximum sustainable yield (MSY)
minimum viable population (MVP)
mutualism
niche
niche overlap

overkills
parasitism
population dynamics
populations
predation
predatory release
prey switching
refuges
secondary extinctions
species introduction
stability
symbiosis
zone of physiological stress
zone of intolerance

# QUESTIONS AND PROBLEMS

## TRUE/FALSE

1. Individual organisms are grouped into populations. Populations are grouped into communities, which compose ecosystems.

2. A population is a group of individuals of different species living in the same geographic area.

3. Population dynamics is the study of how organisms interact with each other and their physical environment.

4. Eutrophication is a good thing for the ecosystem.

5. The most common form of dispersion pattern in populations is the clumped pattern.

## SHORT ANSWER

1. Describe the three basic phases that populations undergo as part of their life cycle.

2. What biological interactions cause population growth to level off and stabilize?

3. Briefly describe the predator-prey relationship.

4. What are the four basic ways in which human activity can alter the environment to cause extinction?
5. How does the "zone of intolerance" occur in populations?

## MULTIPLE CHOICE

1. If populations are left unchecked, they will grow _____.
   a. slowly
   b. at a constant rate
   c. exponentially
   d. none of the above
2. The intrinsic rate of increase in a population is _____.
   a. (birth rate + immigration) − (death rate + emigration)
   b. migration − birth rate
   c. carrying capacity − death rate
   d. prey − predators
3. The law of the minimum has the same meaning as _____.
   a. "you only need the bare essentials to survive"
   b. "only the strong survive"
   c. "waste not, want not"
   d. "you are only as strong as your weakest link"
4. The effect of a predator in reducing the size of a prey species population depends on how many prey there are. This is an example of _____.
   a. destabilization
   b. density-dependent regulation
   c. species extinction
   d. density-independent regulation
5. The smallest population size that can avoid extinction is called the _____.
   a. minimum viable population
   b. maximum sustainable yield
   c. zone of physiological stress
   d. species load

## MATCHING

Match the symbiotic relationship with the appropriate examples.

1. A shrub called the bull-horn acacia produces a sugary secretion that feeds the ants living on it, and the ants protect the acacia from predators.
2. Along California's coast, species of sage produce a toxin that is washed into the soil and prevents germination and growth of other plants.
3. In the New England forests, different species of wood warblers forage for insects in different parts of a coniferous tree, thus reducing the competition for food.
4. A volcanic explosion on the Caribbean island of Martinique exterminated the Martinique rice rat.
5. Driving an off-road vehicle along the sandy beaches of Florida.

   a. extinction
   b. parasitism
   c. competitive exclusion
   d. mutualism
   e. amensalism

## SCIENCE AND SOCIETY

1. Parasites depend on other organisms for resources, defense, movement, and many other necessities of life while they concentrate on reproducing. Parasites have evolved strategies that keep their host alive while they feed on it.

   The human immunodeficiency virus (HIV) that causes AIDS acts in a parasitic manner. The HIV virus attacks helper T cells, severely damaging the immune system. AIDS patients grow progressively weaker and fall victim to infectious agents. Do you think this parasite has evolved in a manner that is beneficial to its survival? Why? Do you think HIV will become extinct by its own means? Why or why not? Can you think of other poorly adapted parasites?
2. Vegetarians refuse to eat meat and dairy products to demonstrate their unwillingness to kill other organisms. As you have learned, plants are just as alive, as active, and as productive as animals are in the ecosystem. What does this lifestyle choice mean to you? Is being a plant predator better than being an animal predator? If vegetarians are going to be true to their beliefs, should they not buy or use products made from animals, such as leather shoes? Why or why not?

# COMMUNITIES
# AND ECOSYSTEMS

OPENING IMAGE

*Animals and plants make up*

*communities.*

Located in the Sonoran Desert outside Tucson, Arizona, Biosphere 2 is a self-contained, closed ecosystem constructed of glass and steel. This complex, rising to 26 meters (85 ft) above the desert floor, and described as looking like "an octopus with lumps," covers 1.28 hectares (3.15 acres). It contains desert, marsh, ocean and rainforest regions. This facility was designed to simulate Biosphere 1, our Earth, and to create a working model of our planet. In September, 1991, a team of four men and four women entered the biosphere to live and work for two years. In that time, they were to maintain the environment, and conduct research on the complex interactions among the living systems and the physical environment.

The original plan called for large scale recycling, with the biomes removing carbon dioxide and pollutants and contributing oxygen to the internal atmosphere. In addition, waste water was to be recycled as well. The team inside the sphere was to grow a wide variety of crops as food. Although designed to be a sealed, self-contained system, Biosphere 2 proved to be more complex than anticipated.

The first problem encountered was a rise in carbon dioxide levels, and a fall in oxygen levels. This forced one of the team members to use an oxygen mask almost nightly. The problem was later traced to oxidizing bacteria in the compost-rich soil used in the facility. To solve the problem, a carbon dioxide scrubber was installed. This was somewhat controversial, since Biosphere 2 was supposed to be a closed system.

A second problem was a decline in the productivity of the agricultural crops. This caused the team to go on a strict diet, and all eight team members lost weight. On average, *the male members of the team lost in excess of 11 kg (25 lbs) over a period of six months. The decline in food productivity was largely attributed to more cloud cover than predicted, and to pest-related crop damage. The team also had to spend more time than anticipated in maintaining the biomes. Problems arose with overgrowth of algae on the ocean and with weedy species crowding out other plant species on land. Plant pollinator species began to die off, forcing the team members to hand pollinate plants. As a result, there was little time left for research.*

*In spite of these problems, over the two year period, 100% of the waste and water were recycled, the marsh, coral reef and ocean biomes were healthy and intact, and the team produced 80% of their food supply. In addition, the Biosphere lost only 9% of its atmosphere. After some changes, including the use of new, shade tolerant food crops, and new pest control species (toads and geckos), a second team entered the Biosphere in March, 1994.*

*This large scale experiment using a closed system to simulate our larger biosphere, illustrates the complexity of the relationships between living systems and the physical environment.*

*This chapter considers how populations are organized into communities, and how these communities interact with the physical environment to form ecosystems and biomes. The work in Biosphere 2 provides an insight into the importance of understanding this organization and the flow of energy among the components of these systems.*

## WHAT ARE COMMUNITIES AND ECOSYSTEMS?

**◆ FIGURE 31.1**

**Ecosystems are composed of organisms, their environment, and all of the interactions that occur among these components.**

In the previous chapter, we discussed how populations change through time and are distributed in space. In this chapter, we shift to a larger scale, and consider the many populations that comprise communities and ecosystems (see ◆ Figure 31.1).

A **community** is defined as the set of all populations that inhabit a certain area. The size of this area can be very small, such as a puddle of water, or encompass many hundreds of square miles. An **ecosystem** is a community plus its physical environment (Figure 31.1).

Because ecosystems include both organisms and their physical environment, the study of ecosystems tends to focus on the movement of physical components through the "system," especially the flow of energy and the cycling of matter. In contrast, because communities consist only of organisms, the focus is on describing how these organisms are distributed through time and space. This community focus on population distribution is often called the "structural" view, whereas the ecosystem focus on processes such as energy flow and matter cycling is called the "functional" view. We turn first to the community, or structural, view.

## COMMUNITY STRUCTURE DEPENDS ON POPULATION DISPERSION PATTERNS

Many populations are not randomly distributed in communities. Indeed, discovering the various patterns by which species are spatially distributed is one of the major accomplishments of ecology during the last century. Often, these patterns are surprisingly simple, given the apparent chaos we sometimes attribute to the "wilderness." Two of the most important patterns are the open structure of communities and the relative rarity of most species in communities.

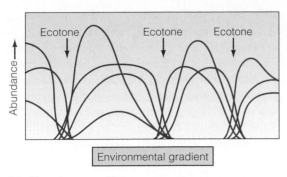

(a) Closed communities

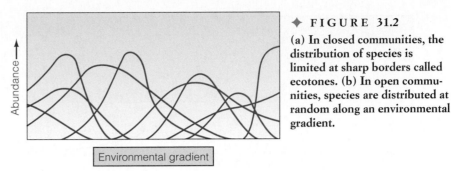

(a) In closed communities, the distribution of species is limited at sharp borders called ecotones. (b) In open communities, species are distributed at random along an environmental gradient.

(b) Open communities

## Most Communities Have an Open Structure

We saw in Chapter 30 that populations can range over very wide or very local areas. We also saw that population density peaks near the center and tapers off toward the edges of a population's geographic range. A key question about community structure is whether the populations that comprise a community have similar geographic range and density peaks. Ecologists debated this question for many years. On one side it was argued that in most communities populations have similar range boundaries and density peaks. This type of organization is called a **closed community.** A closed community is thus a discrete unit, with sharp boundaries (✦ Figure 31.2), called **ecotones.** In contrast, others argued that most communities have populations that differ in their density peaks and range boundaries, and form **open communities.** Populations in open communities are distributed more or less randomly (Figure 31.2).

Decades of study have finally resolved this debate, showing that most communities are indeed open. For example, the populations of the many plants comprising a typical forest community show a pattern of random distribution (✦ Figure 31.3). In the case of forests, moisture tolerance is a major determinant of structure: Some species do best in wetter areas and others thrive in drier ones. As a result, population ranges often overlap to varying degrees, depending on similarities and differences in moisture tolerances.

The major exception to the open structure of communities occurs in cases where the physical environment makes an abrupt transition. This causes sharp boundaries to develop between communities because population tolerances are abruptly exceeded. A beach, where land and sea come together is an example, with an ecotone separating land and water. Transitions in the physical environment are usually more gradual (for example, note

the moisture gradient in Figure 31.3), and population densities between communities also change gradually.

Because communities are not closed, it is more difficult to characterize, describe, and even name a community as it changes gradually in nearly every geographic direction. Ecologists often use advanced statistical methods, such as "gradient analysis," to describe quantitatively the spatial changes in species composition. The name of the community at any point along the gradient is usually assigned on the basis of the most common species. One might thus refer to an oak–hickory forest community, or barnacle–blue mussel tidal zone community.

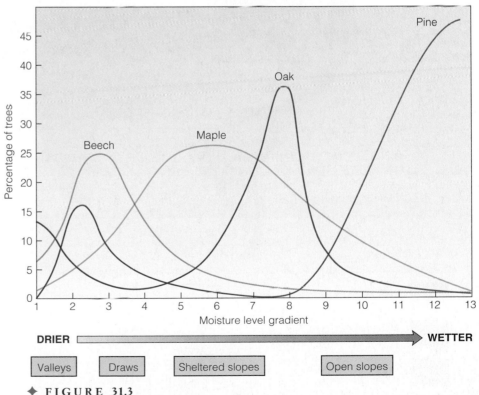

✦ FIGURE 31.3

**Trees in a forest are distributed along a gradient of moisture.**

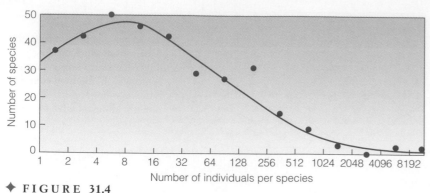

✦ FIGURE 31.4
Distribution of moth species versus number of individuals per species captured in light traps.

✦ FIGURE 31.5
The terrestrial biomes and their subdivisions.

## Open Communities Are Not Tightly Knit Groups of Organisms

Another consequence of an open structure in communities is that they are not tightly integrated assemblages of organisms that can be destroyed in all-or-nothing fashion. Some ecologists once argued that communities were closed, highly integrated units that formed a "superorganism." They argued that if we destroyed just one or a few populations, the whole "superorganism" would die, just as an organism dies if one key organ is removed. However, the fact that species come and go in communities means that communities are generally not as integrated as the concept of a superorganism requires. In reality, communities are generally less fragile and more flexible than the superorganismic concept would imply, as we discuss later.

## Most Species Are Rare in Communities

In addition to an open structure, a second basic characteristic of communities is the relatively low density of most populations in communities. We have already said that populations tend to have their peak abundance near the center of their range, where optimum conditions prevail. In the example of the forest community mentioned earlier, optimum moisture level seems to be a primary

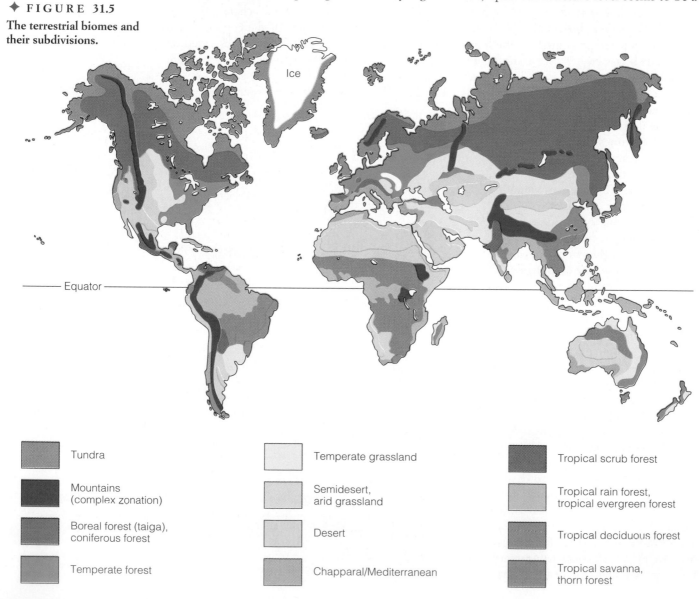

Tundra

Mountains
(complex zonation)

Boreal forest (taiga),
coniferous forest

Temperate forest

Temperate grassland

Semidesert,
arid grassland

Desert

Chapparal/Mediterranean

Tropical scrub forest

Tropical rain forest,
tropical evergreen forest

Tropical deciduous forest

Tropical savanna,
thorn forest

determinant of density (Figure 31.3). At any given point on the gradient of moisture, only a few tree species compose a large percentage of individuals. Even where they are at their maximum density, most populations are much less abundant than the few dominant populations such as beech, maple, oak, and pine.

High population density of only a few species is found in nearly all natural communities across many kinds of organisms. From deep ocean to mountain communities, we almost always find that populations of just a few species are dominant, with many species being represented by just a few individuals (✦ Figure 31.4).

What causes this pattern of abundance by a few species? The reasons are a subject of current debate among ecologists. Most agree that the general cause is related to **resource partitioning.** In any environment, available resources are limited. Because of their evolutionary history, only a small number of species are able to exploit a large part of the available resources. Usually these species were the first to evolve the ability to obtain and eat an abundant food in the community. Other species are left to partition the remaining resources, and thus are present in fewer numbers. However, a species that is rare in one community may be abundant in another, nearby community. This contrasting situation arises because of resource distributions in the two communities. In one community, a certain species is better adapted for exploiting the available resources than in a neighboring community.

## COMMUNITIES ARE GROUPED INTO CLASSIFICATIONS

There are many thousands of communities on Earth. Rather than describe each one in detail, we will examine the basic categories into which communities are grouped.

The most basic categories distinguish between terrestrial (land) and aquatic (water) communities. These two basic divisions contain eight biomes:

1. *Terrestrial:* tundra, grassland, desert, taiga, temperate forest, tropical forest
2. *Aquatic:* marine, freshwater

A **biome** (pronounced "by-ohm") is a large-scale category that includes many communities of a similar nature (✦ Figure 31.5).

Both terrestrial and aquatic biomes (and thus the communities within them) are largely determined by climate, especially temperature. Climate affects many aspects of the physical environment: rainfall, air and water temperature, soil conditions, and so on. Many secondary factors are also important, including local availability of nutrients. Biomes il-

lustrate the point that species will adapt to a set of physical conditions in similar ways, no matter what their evolutionary heritage. A desert biome in the western United States looks very similar to a desert biome in northern Africa. In outward appearance, the plants look very similar ("cactus-like," for example) but may have very different evolutionary histories.

### There Are Six Terrestrial Biomes

✦ Table 31.1 and Figure 31.5 describe and illustrate the six basic types of land biomes. The tundra and desert biomes represent adaptations to the most

**TABLE 31.1 SIX MAJOR LAND BIOMES***

1. **Tropical Forest:** The most complex and diverse biome, the tropical forest, contains more than 50% of the world's species while occupying only 7% of the land area. This high diversity is largely due to the relatively constant temperatures at all times: Daily and seasonal changes are usually less than a total of 5°C. Rainfall is very heavy, more than 200 cm (80 inches) annually. Major plants include deciduous trees that form a multilayered canopy, including understory trees. Herbs and shrubs that tolerate intense shade form the ground flora. Insects are extremely abundant; perhaps 96% of insect species are found here.

2. **Grasslands:** Rainfall is scarce, about 25 to 75 cm (10 to 30 inches) per year, causing grasses to be the most prominent plants. Fires are common. Major animals include grazers, such as the bison in North America. Economically, this is the most important biome, providing grazing land for sheep, cattle, and other food animals, as well as providing the richest cropland in the world. This is because of the rich soils formed by the lack of rain and held in place by the grass roots.

3. **Deserts:** Rainfall is very scarce, less than 25 cm (10 inches) per year. Temperature can be very hot or very cold. Desert plants are widely spaced to allow maximum moisture per plant. Plant adaptations to arid conditions include (1) storage of water as in cacti, (2) deciduous shrubs that shed leaves in dry periods, and (3) rapid growth and reproduction during rare rainy periods. Animal adaptations are similar to plants, being able to store large amounts of water (such as camels) and rapid growth and reproduction after rains (such as desert toads).

4. **Temperate Forests.** Rainfall is abundant (75 to 150 cm or 30 to 60 inches per year), with distinct seasonal change. Deciduous trees dominate, such as oak, hickory (western U.S.), beech, and maple (north-central U.S.). Ground cover of shrubs and herbs. Lacks spectacular diversity of tropical forests, but still more diverse than coniferous forests. Animals include deer, foxes, squirrels, raccoons, and many other familiar forms.

5. **Taiga.** Also called coniferous forests, these occur in broad belts in northern North America and Asia. Diversity is relatively low. Plants dominated by conifers (evergreens), which are tolerant of dry, cold conditions. Prominent types of evergreens include spruce, firs, and pines, whose needles conserve water and withstand freezing better than leaves. Animals include moose, snowshoe hare, wolves, and grouse. Acidic and thin soils mean that cleared taiga makes poor cropland.

6. **Tundra.** An extensive treeless plain, tundra's topsoil is frozen all year except for about six weeks in summer. Below this is permafrost soil, frozen all year long, and a hazard for building. During the brief summer thaw, life grows and reproduces rapidly. Lichens (algae and fungi symbionts), grasses, and small shrubs are dominant. Prominent animals include caribou (reindeer), arctic hare, arctic fox, and snowy owl. Many migratory birds arrive for the rich summer growing season, characterized by billions of insects, bright flowers, and marshy conditions.

*See Figure 31.5 for illustrations. The biomes are listed in approximate order going from the equator to the poles.*

extreme conditions of very low temperature and low water, respectively. Not surprisingly, communities in these biomes tend to have the least number of species, because organisms have a difficult time adapting to the extreme physical conditions.

**Tropical rain forests** (✦ Figure 31.6) tend to be the richest in the number of species, in part because the tropics have the most moderate conditions. About half of all terrestrial species, and perhaps more, live in tropical biomes. The rainfall in tropical rain forests occurs in all seasons, and totals 200 to 400 centimeters (about 80 to 160 inches) per year. Typically, trees in rain forests have shallow roots and wide bases that support their weight. The tops of the trees form a dense canopy that blocks light from reaching the floor of the forest. As a result, the treetops are where most life is found.

Somewhat paradoxically, soils in tropical regions are deficient in minerals, including phosphorus, potassium, and calcium. Topsoil layers are thin, and most of the available nutrients are in the plants of the rain forest themselves. When plant parts (such as leaves) or plants die, they are rapidly decomposed by bacteria, fungi, and insects. The nutrients released are quickly recycled into other plants.

Tropical rain forests are found near the equator in South America and in Africa. About 17 million hectares (1 hectare = about 2.5 acres) of rain forest are destroyed each year (an area equal to the state of Washington) for lumber and agriculture. At this rate, they will be gone in less than 100 years (✦ Figure 31.7). Destruction of tropical rain forests will be accompanied by the extinction of a significant proportion of all species on earth, and may significantly reduce rainfall in the temperate zones, with a substantial impact on agriculture.

✦ **FIGURE 31.6**
The tropical rain forest is a species-rich biome.

The **temperate forest** biome is spread across regions of the Northern Hemisphere with abundant rainfall and long growing seasons (✦ Figure 31.8). This biome covers the eastern United States and Canada, and a large portion of Europe and western Asia. Rainfall in the temperate forest ranges from 75 to 200 centimeters (about 30 to 80 inches) per year. The dominant plants of this biome are deciduous, broad-leaved trees (oak, maple, beech) that shed their leaves each year.

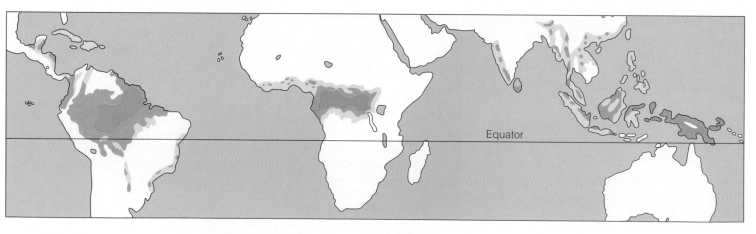

Tropical rain forests

Original extent  Present extent  Year 2000 at current deforestation rate

✦ **FIGURE 31.7**
**The declining rain forests, and their projected extent in the year 2000.**

**FIGURE 31.8**
The temperate deciduous forest is dominated by broadleaf trees.

**FIGURE 31.9**
The grasslands do not receive enough rainfall to support the growth of trees.

**FIGURE 31.10**
Animals of the grasslands include large herbivores.

**Grasslands** develop in temperate and tropical regions with reduced rainfall or prolonged dry seasons (✦ Figure 31.9). The grassland biome stretches across North and South America, Africa, Asia, and Australia, and receives between 23 to 75 centimeters (about 10 to 30 inches) of rainfall each year. The soil in these regions is deep and rich from thousands of years of growth and decay.

The open landscapes of the grasslands are almost entirely devoid of trees, because of the lack of water. This biome can support large herds of grazing animals (✦ Figure 31.10). In North America, bison, elk, deer, and antelope once grazed in these grasslands. Farming has altered this biome in much of the United States (✦ Figure 31.11).

The **desert** biome is characterized by dry conditions and a wide range of temperatures (✦ Figure 31.12). This biome usually receives less than 25 centimeters (about 10 inches) of rain annually. The temperature in desert regions fluctuates widely in large part because the air is dry, reducing heat retention, and clear skies allow heat radiation at night. From near freezing at night, temperatures during the day can reach more than 50°C (122°F).

**FIGURE 31.11**
Farming has greatly altered the landscape of the grasslands.

**FIGURE 31.12**
The desert biome is characterized by dry, hot conditions.

## How Snails Shape Shorelines

MARK BERTNESS

Growing up on the shores of Puget Sound, I spent a lot of my time scouring beaches for shells and poking around tide pools. As an undergraduate, however, I was not initially drawn to biology or ecology. Instead, I tried a little bit of everything. Chemistry, psychology, and anthropology all interested me. It wasn't until my senior year in college when I took a coral reef biology course and a natural history course that I realized that my childhood passion for seashores was what really captured my imagination.

After graduating from college, I enrolled in a master's degree program at Western Washington University, to try out marine ecology. I next traveled to Maryland to study with Gary Vermeij, noted snail guru and marine evolutionary biologist. After half a year, I ended up at the Smithsonian Tropical Research Institute in Panama eager to study the ecology of hermit crabs. Hermit crabs rely entirely on snails to produce shells to live in, and they have ritualized fights for shells. But best of all, empty shells are easy to add or remove from populations. I spent the next two years in Panama experimentally examining the relationship between hermit crabs and their shells. By adding empty shells to hermit crab populations in the field, I was able to show that shell supply strongly regulates hermit crab population growth and reproduction. I also found that the particular shell occupied by a hermit crab de-termined how vulnerable it was to being eaten or to overheating and drying out during low tide exposure to terrestrial conditions. I had the time of my life living in an interesting foreign country with a stimulating group of scientists. I was often broke and slept a lot on my lab floor, but was totally engaged in my work.

I lucked out and got a job right out of graduate school—a good job as an assistant professor at Brown University. In graduate school, I had become a tropical biologist enthralled with the complexity of tropical reefs and shorelines. New England shorelines were drab compared to the tropical beaches of Panama. They had been nailed so frequently in the past by Ice Age glaciers that what remained was pretty simple. It took me a few years to realize how advantageous simple systems could be to an experimental ecologist. In simple systems, the interactions among organisms are strong and not diluted by a myriad of other species. Simple systems are also typically much easier to manipulate than more complex ones. The first aspect of southern New England shorelines that attracted my attention was the number of dominant species that were relatively recent invaders. In the early 1980s, my students and I found that the European periwinkle—a simple herbivorous snail that had been introduced to North America in the mid-nineteenth century maintained cobble beaches by cleaning rock surfaces of algae and sediment and eating marsh plant roots. Without these simple snails, many rocky beaches in southern New England would become muddy beaches or salt marshes.

Twenty years after deciding to study marine ecology, I still find my work driven by a basic intellectual curiosity about the forces that generate patterns in nature—essentially the same curiosity that I had growing up while roaming beaches on Puget Sound. Over the years, however, I have come to appreciate how important it is to understand how natural populations and communities work. If we don't understand how natural communities work, how will we make prudent decisions concerning our constantly shrinking natural resources?

*Mark Bertness is professor of marine ecology in the Department of Ecology and Evolutionary Biology at Brown University. He received his undergraduate training at the University of Puget Sound and his Ph.D. from the University of Maryland, College Park. His research focuses on the organization of natural communities.*

Plants in the desert have developed several characteristic adaptations to low moisture levels and wide temperature ranges. Shrubs often have deep root systems that reach water deep below the surface. Water is stored in succulent tissues that provide a supply of moisture during extended dry periods. Cacti have spines that reduce water loss and offer some shade against the sun.

Desert animals also have adaptations that allow them to survive the low moisture levels and temperature extremes of the desert. These include thick scales to reduce water loss and coloring that reflects heat. Many vertebrate species live in deep burrows and are active at night, when temperatures are cooler (✦ Figure 31.13).

The **taiga** is a biome of coniferous forest extending across much of the northern regions of North America, Europe, and Asia (✦ Figure 31.14a). The taiga receives between 35 to 100 centimeters (about 14 to 40 inches) of rain annually and is characterized by long, cold winters, and short, cool summers.

Conifers are well adapted to the taiga, and have narrow, pointed leaves called needles that are retained throughout the year. The needles are coated with a waxy layer that prevents water loss. Deciduous trees such as alders, birch, and willow occur in dense clusters.

Many large mammals such as wolves, grizzly bears, moose, and caribou are common inhabitants of the taiga (Figure 31.14b), but species diversity is low compared to biomes in the deciduous forests.

The **tundra** biome stretches across the northern regions of North America, Europe, and Asia be-

(a)

(b)

✦ FIGURE 31.13

Animals of the desert have thick scales for protection against heat and water loss like the gila monster (a), or live in burrows and forage at night like the kangaroo rat (b).

(a)

✦ FIGURE 31.14
(a) The taiga is characterized by short growing seasons and long cold winters.
(b) Timberwolf.

(b)

tween the taiga to the south and the regions of permanent ice to the north (✦ Figure 31.15). This region receives less than 25 centimeters (about 10 inches) of precipitation annually, and the available water is frozen most of the time. Because of the long winters and short summers, much of the soil remains permanently frozen, and is called permafrost. In the brief summer, water on the surface melts, forming a boggy landscape.

The permafrost prevents growth of deep-rooted plants such as trees. The tundra is covered with patches of grasses, shrubs, and vegetation including mosses, lichens, and sedges. Large grazing mammals including musk ox, reindeer, and caribou exist along with wolves, lynx, and rodents.

## Climate and Altitude Affect Terrestrial Biomes

The climatic variables of temperature and moisture exert primary control over land biomes. This control occurs in two ways: the **altitudinal gradient** and the **latitudinal gradient.** With increases in either altitude or latitude, a gradient of cooler and

✦ FIGURE 31.15
In the summer, the tundra becomes moist and boggy.

drier conditions occurs (✦ Figure 31.16). Cooler conditions cause **aridity** because cooler air can hold less water vapor than warmer air. Deserts are an exception to this trend; they tend to occur in relatively warm climates and form because of local and global influences that block rainfall. Large mountain ranges cause a "rain shadow" by blocking out moisture-laden clouds (✦ Figure 31.17).

The general trend of the latitudinal gradient can be seen in the distribution of biomes on the present-day Earth.

Tropical forests are found mainly in warm moist equatorial areas. North and south of the equator, the forests grade into deserts or grasslands, depending on local conditions (such as mountain ranges and prevailing wind patterns). Further north and south, the deserts and grasslands are gradually replaced by temperate forests, which, near the poles, give way to taiga, tundra, and then ice and snow.

## There Are Two Types of Aquatic Biomes

In general, conditions in water are much less harsh than those on land. The aquatic environment has fewer temperature fluctuations and more buoyancy as support against gravity than biomes on land. In addition, living in water eliminates the danger of desiccation, as opposed to living on land,

✦ **FIGURE 31.16**

**The distribution of biomes is related to both latitude and altitude.**

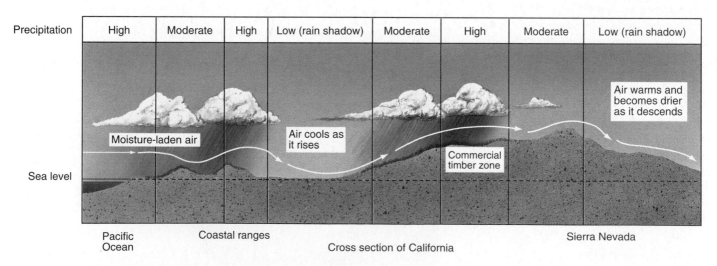

✦ **FIGURE 31.17**

**Deserts often form on the downwind side of mountain ranges.**

where water is often scarce. It should be no surprise then that life originated in the oceans and took many millions of years to adapt to land. Yet even though water covers about 71% of the Earth's surface, most of it contains relatively little life. This is because most of the open ocean is a vast aquatic desert, with few nutrients.

It is more difficult to subdivide aquatic communities into the distinctive biomes found on land. This is because the liquid state of water makes non-climatic conditions more important in determining what species can live in a particular environment. Water is a powerful solvent so it readily carries many chemicals (from nutrients to toxins) in solution. The distribution of these dissolved materials influences life at local levels. Unlike air, water also readily transports heat so that currents, such as the Gulf Stream, can warm large areas even near the poles. This prevents a simple latitudinal gradient from forming. Therefore, ecologists generally designate only two aquatic biomes: the marine biome and freshwater biome.

The **marine biome** differs from the freshwater biome by containing more dissolved minerals ("salts") of various kinds. On average, marine waters contain 3.5% salt, mainly sodium chloride. The marine biome is the largest biome by far, covering more than 70% of the earth's surface, but it is easily subdivided (✦ Figure 31.18). The two most basic categories of this biome are **benthic** (bottom-dwelling) and **pelagic** (water column) zones. Benthic communities are further subdivided by depth: the shore/continental shelf community and the deep-sea community. Benthic organisms range from fish, to burrowers (such as worms), to crawlers (such as snails), to stationary filter-feeders (such as barnacles). Pelagic organisms include (1) **planktonic** organisms ("floaters") and (2) **nektonic** organisms ("swimmers"). The **euphotic zone** is the upper part of the marine biome where light penetrates; this is usually about 200 meters below the water surface. The euphotic zone is crucial to the community because it is the main zone of photosynthesis. Life in the euphotic zone is easily contrasted with land communities. In this zone, plants that form the base of the food pyramid are mainly tiny planktonic organisms, such as diatoms and dinoflagellates.

*Sidebar*

**SURVIVING THE ODDS**

Agricultural development and encroaching forest have led to the disappearance of the prairie ecosystem. In Oregon less than one half of one percent of the Willamette Valley's native prairie remains, according to professor Mark Wilson at Oregon State University.

Small populations of a rare species of butterfly that was believed to be extinct was recently rediscovered in a remote area of Oregon. The Fender's blue butterfly (*Icaricia icarioides fenderi*) dates back to the Ice Age and was last seen in 1937. The male is a brilliant iridescent blue and the female is a dull brown. These delicate insects were found when researchers were searching for the last remains of native prairie wildlife. Scientists plan to petition the United States Fish and Wildlife Service to have Fender's butterfly declared a threatened species.

The near extinction of this butterfly species was linked to the decline of a blue wild flower that is the only food source for the caterpillars. This flower is found almost exclusively in the Willamette Valley. The prairie and butterfly relationship is a classic example of the demise of an ecosystem causing the disappearance of animal and plant species that rely on it.

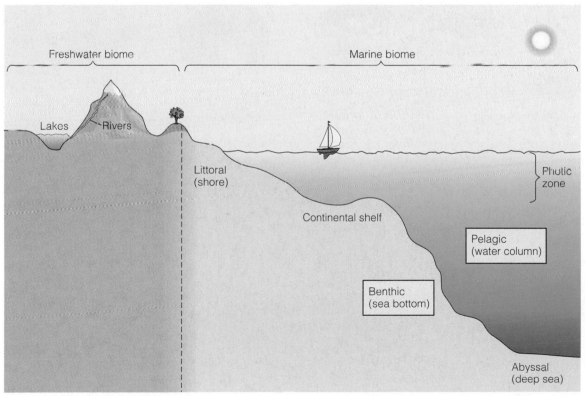

✦ **FIGURE 31.18**
**The marine life zones and their characteristics.**

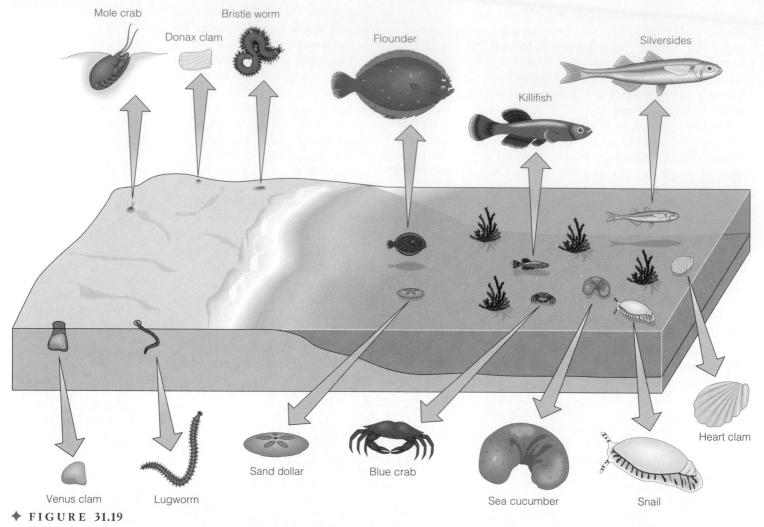

Mole crab

Donax clam

Bristle worm

Flounder

Silversides

Killifish

Venus clam

Lugworm

Sand dollar

Blue crab

Sea cucumber

Snail

Heart clam

✦ FIGURE 31.19

Organisms in a beach community along the Northeast coast of the United States.

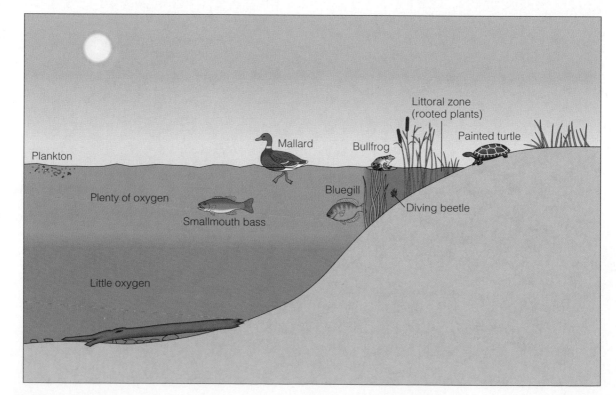

✦ FIGURE 31.20

Oxygen distribution in a freshwater lake ecosystem.

Plankton

Mallard

Littoral zone (rooted plants)

Bullfrog

Painted turtle

Plenty of oxygen

Bluegill

Smallmouth bass

Diving beetle

Little oxygen

♦ Figure 31.19 illustrates some of the benthic, nektonic, and planktonic animals common in a shore community. Most of these live either by directly filtering plankton from the water (such as clams) or by eating the organic detritus from dead plankton on the sediment (such as worms and sand dollars).

The **freshwater biome** is subdivided into two zones: running waters (rivers, streams) and standing waters (lakes, ponds). The distinction between rivers and streams or lakes and ponds is not sharply defined. In general, rivers and lakes are larger and more permanent than streams and ponds. The mo-

tion of running waters tends to keep them more oxygenated and more difficult to pollute. The slower motion of water in lakes and ponds leads to **stratification** of the water: Oxygen levels decrease with depth (♦ Figure 31.20). The uppermost layer of water has an abundant supply of oxygen, but lower levels are oxygen-poor. The uppermost layer is warmer during the summer and cooler during the winter. The temperature at lower depths changes very slowly. Mixing between the uppermost and deeper layers occurs during seasonal changes called spring and fall overturn (♦ Figure 31.21).

♦ **FIGURE 31.21**

**Deep lakes are stratified into several layers in the summertime (top), with different temperatures and oxygen content identified in parts per million (ppm). In the fall (right), the water cools, and the stratification breaks down. In the winter (bottom), ice forms at the top, but the temperature remains almost uniform. In the spring (left), the ice melts, and the water mixes. The longer days cause the lake to gain heat and become stratified.**

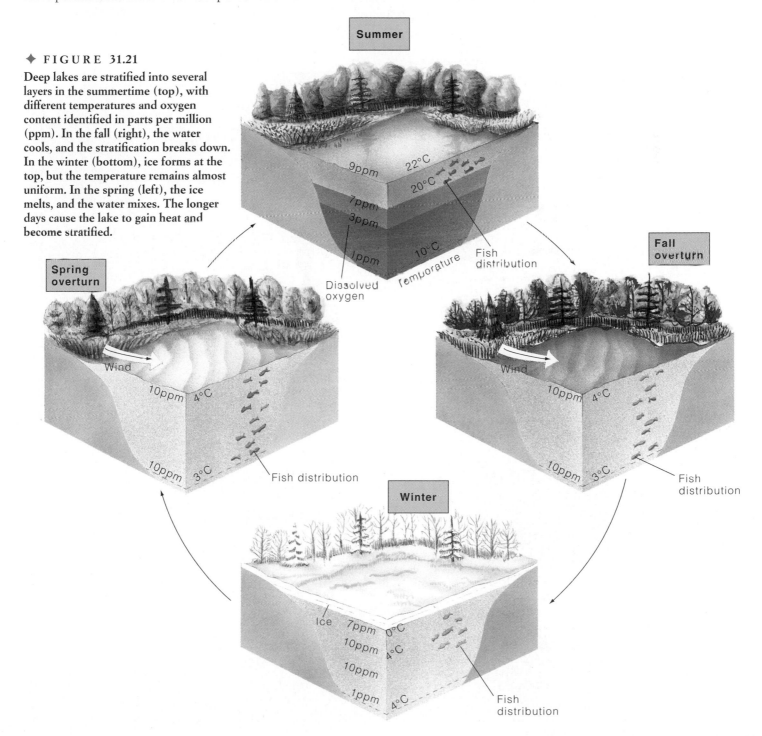

The freshwater biome is similar to marine waters in having benthic (bottom-dwelling) and pelagic (swimming and planktonic) organisms and communities. Most of these have relatives in the marine realm (clams, snails, and fish, for example), where the groups originated hundreds of millions of years ago.

## COMMUNITIES HAVE TWO PROPERTIES: DIVERSITY AND STABILITY

The communities we have discussed have two important properties: diversity and stability. These characteristics help define the community and are important factors in environmental discussions.

## Community Diversity Is Measured by Species Richness

**Diversity** refers to the different types of organisms that occur in a community. Diversity is often expressed in terms of **species richness,** the number of species present in a community. Many factors influence diversity and the importance of any single factor varies with the particular community, but two important trends help define diversity. The first is the **latitudinal diversity gradient:** species richness

◆ **FIGURE 31.22**

**Latitudinal patterns of species distribution for (a) lizards and (b) birds. Variations in species distribution in marine environments for (c) snails and (d) polychaete worms.**

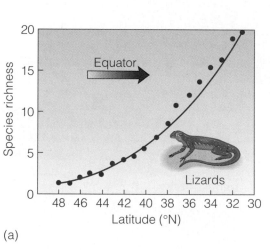
(a)

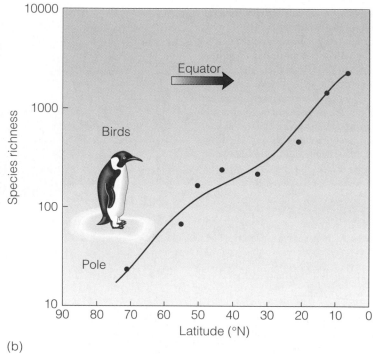
(b)

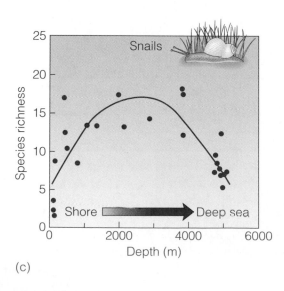
(c)

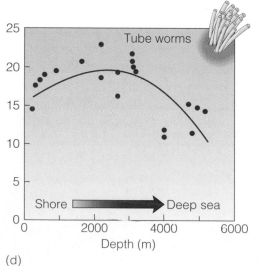
(d)

decreases steadily going away from the equator (✦ Figures 31.22a and b). This leads to richer communities in tropical areas. A hectare (about 2.5 acres) of tropical forest typically contains from 40 to 100 tree species; a typical temperate zone forest has about 10 to 30 tree species, and a taiga forest in northern Canada has only 1 to 5 species per hectare. In addition, the number of insect and other animal species living on those trees increases with the kinds of trees (resources) available. As a result, tropical forests have more kinds of insects and other animals per hectare. This means that habitat destruction in tropical countries generally leads to many more extinctions per hectare (Chapter 32).

Ecologists generally agree that three interrelated factors help cause this gradient: **environmental stability, community age,** and **length of growing season.** There is greater environmental stability in equatorial areas. This means that tropical communities are exposed to less environmental change on a daily, seasonal, and even hundred-year basis. This stability allows more kinds of species to thrive in this region because high levels of disturbance or stress tend to reduce diversity. Second, equatorial communities are older because they have been less distributed by advancing ice sheets and other climatic changes over the long span of geologic time. As a result, there has been more time in these ancient communities for new species to evolve. Third, the longer growing season in equatorial areas leads to more photosynthesis and plant growth, which forms the food base for all life. As we will see later, higher plant productivity supports a greater diversity of organisms that depend on the plants.

The second important trend that defines diversity is the **depth diversity gradient** found in aquatic communities. This gradient consists of increasing species richness with increasing water depth. The gradient exists from the surface to a point about 2000 m deep, where species richness begins to decline (Figures 31.22c and d). This gradient is established by two factors: **environmental stability** and **nutrients.** Greater environmental stability in the water occurs as one moves away from the higher energy levels of the beach and shoreline. As we have seen in tropical settings, stability allows more species to thrive. Similarly, as one moves offshore, the amount of nutrients from land runoff begins to diminish. Thus, even though deep water is very stable, there are not enough nutrients to permit the high productivity seen in shallower waters. Marine life is especially dependent on land runoff to supply limited nutrients such as phosphorus and nitrogen.

In summary, we have cited four major factors that increase diversity in any community: increasing environmental stability, age, growing season, and nutrients. Stability provides an environment for diversity to proliferate. Age provides the time for diversity to develop, and the last two factors provide the energy and nutrients to supply many types of organisms. In general, the more of these factors that a community has, the more species-rich it will be.

## Community Stability Is Related to Structure

**Stability** refers to the ability of a community to persist unchanged. In the 1950s and 1960s ecologists generally believed that community stability increased as community diversity increased. Simple observation showed, for instance, that agricultural ecosystems that contained only a few species were easily disrupted by pests and physical disturbances. In contrast, species-rich natural ecosystems, such as a forest community, were much more stable in the face of changes because of "redundancy": the reduction of one species, such as a plant, was often compensated by an increase in another plant species.

Beginning in the 1970s the correlation of stability with diversity began to be questioned. Mathematical modeling showed that increasing diversity can actually decrease stability because as diversity develops, it can produce an increasing interdependence among species. This can lead to "cascade collapses" of the whole ecosystem when "keystone species" are removed. This modeling conclusion has been supported by evidence from communities in nature: the more diverse tropical and reef ecosystems are often more easily disrupted than simpler natural ecosystems in temperate areas. However, this is not always true and the issue is far from resolved. Suffice it to say that the relation between diversity and stability is much more complex than previously thought.

## COMMUNITIES CHANGE THROUGH TIME

Biological communities change through time. This is not surprising since the physical environment that ultimately supports life is always changing. Whether we perceive change as being "fast" or "slow" depends almost entirely on what timescale we are using. In examining change in communities, two timescales are of particular interest: ecological time and geological time. **Ecological time** focuses on community events that occur on the order of tens to hundreds of years. These events are most relevant to our own human timescale. **Geological time** focuses on events on the order of thousands of years or more.

## Community Succession Is a Form of Species Replacement

**Community succession** is the sequential replacement of species in a community by immigration of new species and by the local extinction of old ones. Community succession is initiated by a disturbance that creates unoccupied habitats for colonizing species. These colonizers are usually adapted for widespread dispersal, rapid growth, and have a "hardy" nature, making them the first species to appear and thrive. This initial community of colonizing species is called the **pioneer community.** Eventually, other species migrate into the community. Species in the second wave of colonization are usually not efficient at dispersal, and are slower growing. Usually they are more efficient specialists and better competitors, and begin to replace the colonizers. This process continues, as still newer species migrate in, until the **climax community** is reached. Some ecolo-

gists once thought that there was no further change in community composition after the climax community was reached (unless the community was disturbed again). However, most ecologists now agree that change continues in the climax community, although it is at a much slower pace.

Succession has been most fully documented in forest communities (✦ Figure 31.23). Pioneering plant species include grasses, which give way to shrubs, small trees, and finally large trees in the climax community. Various animals, such as the bird and mammal species shown, also appear in sequence. The animal sequence is determined largely by the appearance of the plants on which they rely.

Succession is characterized by a number of trends including a decrease in productivity. Pioneering plants tend to be smaller in size and grow rapidly, which maximizes productivity. As more specialized, slower growing species migrate into the

✦ **FIGURE 31.23**

**The stages in community succession, beginning with barren ground through the climax forest.**

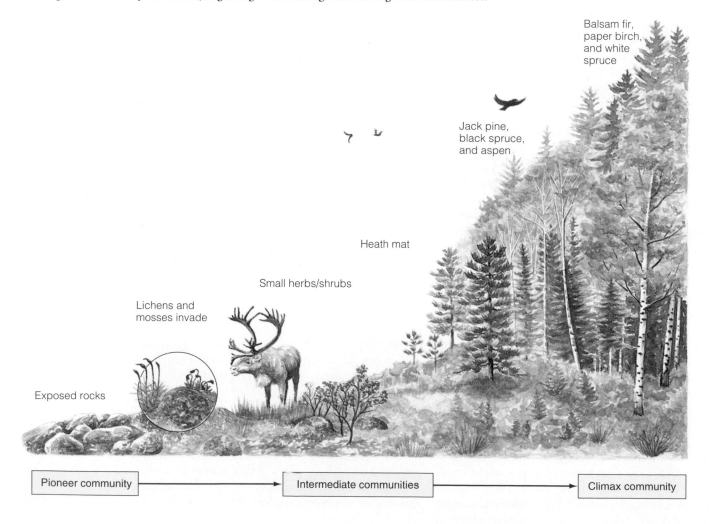

Exposed rocks

Lichens and mosses invade

Small herbs/shrubs

Heath mat

Jack pine, black spruce, and aspen

Balsam fir, paper birch, and white spruce

Pioneer community → Intermediate communities → Climax community

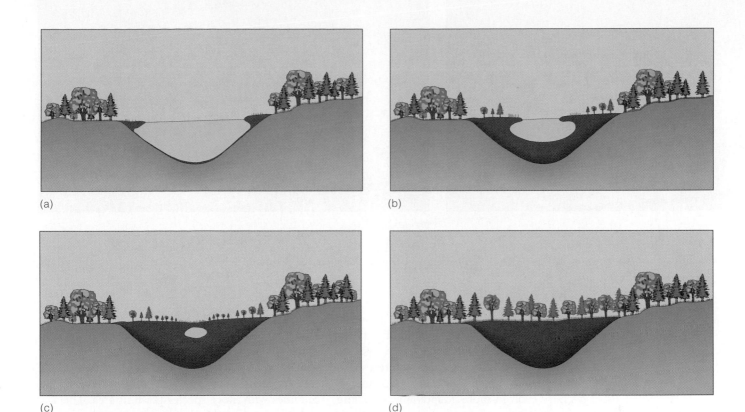

(a)

(b)

(c)

(d)

✦ **FIGURE 31.24**

**Succession (a-d) and growth of a climax forest as a pond becomes filled.**

community, productivity declines. A second trend is increased diversity. As later species immigrate into the community, diversity increases as the more specialized species finely subdivide the resources. Still other trends are the larger size and longer life cycles of species migrating in later stages.

These trends lead to an increase in biomass in later stages. **Biomass** is the total weight of living tissue in a community. Finally, later stages tend to have populations controlled mainly by biological or density-dependent factors such as competition and predation. In contrast, early stages mainly show physical or density-independent controls, such as physical disturbance (Chapter 30). Some communities, such as a beach, are in a generally constant state of physical disturbance so that early successional populations are permanent dwellers.

Succession occurs because each community stage, from pioneer to climax, prepares the way for the stage that follows. Soil conditions, nutrient availability, temperature, and many other environmental traits are altered by each preceding community. In the pioneering stage of a forest (Figure 31.23), the soil in a bare patch of land stabilizes, begins to accumulate nutrients, retains water, and

provides ground shade. These events make the environment more livable for later stages. This preparation is even more dramatic where succession in an aquatic (lake) community permits the eventual development of a climax terrestrial forest community (✦ Figure 31.24). The process of "preparation" is ironic in the sense that species in each community often bring about their own demise. Nearly all natural landscapes (including sea-bottom areas) consist of a mosaic of undisturbed patches intermixed with patches that are disturbed to varying degrees. This "patchiness" of the natural environment is crucial for maintaining diversity because it allows species from different stages of succession to exist simultaneously.

## Community Evolution Occurs over Long Time Spans

We saw in Chapter 11 how species evolve over time spans often involving millions of years. Since groups of species make up communities, communities must also evolve. To see this, compare a typical community from the Paleozoic Era, about 500 million years ago, with a typical community from the Cenozoic

(a)

(b)

✦ **FIGURE 31.25**

(a) A reconstruction of an Ordovician marine community. (b) A Cenozoic (present-day) marine community.

Era, living today (✦ Figure 31.25). Not only does the species composition of the communities differ, but most paleobiologists agree that there is an important qualitative difference: communities today generally contain more species. This is because organisms have evolved more adaptations over time. In the Paleozoic Era, there were fewer burrowers on the sea floor. In the present era, the ocean floor contains organisms, with enhanced digging muscles (such as clams), capable of burrowing to great depths in the sediment. This phenomenon is called **species packing.** Over time, as species become more specialized, natural selection produces communities with more species per unit area. This is a main reason why the global diversity of life has increased through time (Chapters 13 and 14).

✦ **FIGURE 31.26**

**Distribution of diatom species and individuals per species in unpolluted and polluted waters.**

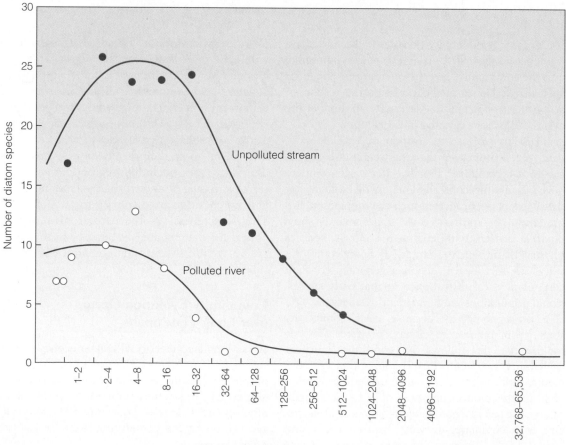

## Beyond the Basics

# WHERE HAVE ALL THE SONGBIRDS GONE?

In the tradition of nineteenth century naturalists, many amateur birdwatchers keep lists of species they have observed, along with information on how many individuals of a given species they have seen, and where they were sighted. Sometime in the 1980s, birdwatchers noticed a decline in the abundance of many songbird species across North America. This decline has been confirmed by statistical studies comparing bird sightings in an area over two different decades. The table shows results of one such study comparing bird sightings in the 1940s with those of the 1980s in the Washington, D.C. area.

In 1994, the Worldwatch Institute reported that two-thirds of the songbird species in the world are declining in abundance. Why are such birds disappearing, and what concerns should this raise? Songbirds can add to the quality of life, and many people enjoy their beauty and singing. Based on this consideration alone, a decline in songbirds would not seem to be all that serious, given many of the other problems facing our society. From another point of view, the decline of songbirds may be an alarm signal, giving us an early warning of danger. The decline of bird populations in the 1950s was traced to the buildup of pesticides, especially DDT, in the bodies of vertebrates, including humans. A ban on DDT led to a gradual recovery of bird populations, and DDT levels in many animals, including birds and fish, are falling.

### SIGHTINGS OF SONGBIRD SPECIES IN ROCK CREEK PARK, DISTRICT OF COLUMBIA

| Species | Mean Number of Pairs Sighted | | Percentage Change |
| --- | --- | --- | --- |
| | In 1940s | In 1980s | |
| **Migrants** | | | |
| Red-eyed vireo | 41.5 | 5.8 | −86.0 |
| Ovenbird | 38.8 | 3.3 | −91.5 |
| Acadian flycatcher | 21.5 | 0.1 | −99.5 |
| Wood thrush | 16.3 | 3.9 | −76.1 |
| Yellow-throated vireo | 6.0 | 0.0 | −100.0 |
| Hooded warbler | 5.0 | 0.0 | −100.0 |
| Scarlet tanager | 7.3 | 3.5 | −52.1 |
| Black-and-white warbler | 3.0 | 0.0 | −100.0 |
| **Nonmigrants** | | | |
| Carolina chickadee | 5.0 | 4.3 | −14.0 |
| Tufted titmouse | 5.0 | 4.5 | −10.0 |
| Downy woodpecker | 3.5 | 3.0 | −14.3 |
| White-breasted nuthatch | 3.5 | 3.1 | −11.4 |

*Source: National Audubon Society.*

In some cases, songbird populations are declining because of the introduction of predators, and the loss of food species. But the greatest danger appears to be loss of habitats and the fragmentation of habitats. North American songbird species that migrate to the tropics are declining faster than those that do not. Tropical deforestation is destroying their winter nesting sites, and they are also losing habitats in their summer nesting grounds, hitting them with a 'one-two' punch.

Fragmentation of habitats is affecting the abundance of many songbirds that do not migrate in the winter. As forests become fragmented, the edges grow larger relative to the rest of the forest. This provides opportunity for predators such as raccoons, opossums and housecats to have more access to nests along the edge of the forest. Nests located deep in the forest experience less predation. Parasitic species of birds such as the cowbird also prefer nests on the edge of the forest. Cowbirds lay their eggs in the nests of songbirds, and the cowbird chicks often push the young songbirds out of the nest.

The net result is a decline in biodiversity, and if unchecked, may lead to the extinction of many bird species in the not so distant future.

## DISTURBANCE CAN CAUSE COMMUNITIES TO LOSE THEIR DIVERSITY

Community structure is determined by species distributions. Whenever species distributions are changed by human intervention, the structure of the community is altered. The basic effect of nearly all human activity on communities is **community simplification:** the reduction of overall species diversity (number of species).

In many cases, this intervention is purposeful. The farmer's agricultural and the suburbanite's horticultural communities of plants, insects, and other animals are common examples. In such cases, we seek to grow only certain species, creating a much lower diversity than is normally found in that area. In the extreme case, this is called **monoculture,** growth of only one particular species. A cornfield is an example of monoculture. Monocultures and other forms of extreme community simplification are susceptible to diseases and other forms of destruction, such as the Irish potato famine of the nineteenth century. It is interesting to note that most of the plant species cultivated for food and pleasure are species from pioneering communities. For example, corn, wheat, and many other plants are grasses that were originally adapted to colonizing disturbed areas. Humans favor them as food because they are fast-growing, rapidly reproducing organisms.

In other cases, humans inadvertently simplify communities. Construction, road building, and pollution act as disturbances that simplify communities. An important aspect of such **stressed communities** is that they are simplified not only because they have fewer species, but because some species become superabundant. Some organisms thrive in highly polluted waters, even using the pollutants as food (◆ Figure 31.26). Even inadvertent disturbances often favor early successional species. In building roads, farms, cities, or lawns, one of the first actions is to bulldoze or otherwise remove the climax community. Because early successional species are adapted to colonizing disturbed environments, they tend to thrive as human settlement has

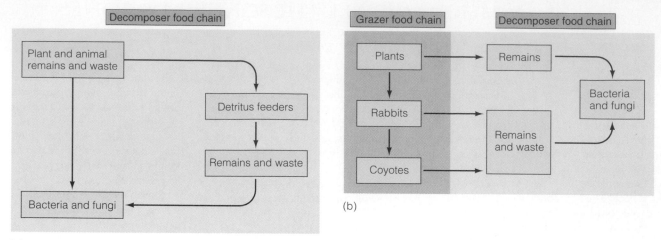

**✦ FIGURE 31.27**

Two different food chains. (a) A decomposer food chain. (b) A grazer and decomposer food chain, showing the connections between the two.

**✦ FIGURE 31.28**

Food webs show the complex interactions that take place in an ecosystem.

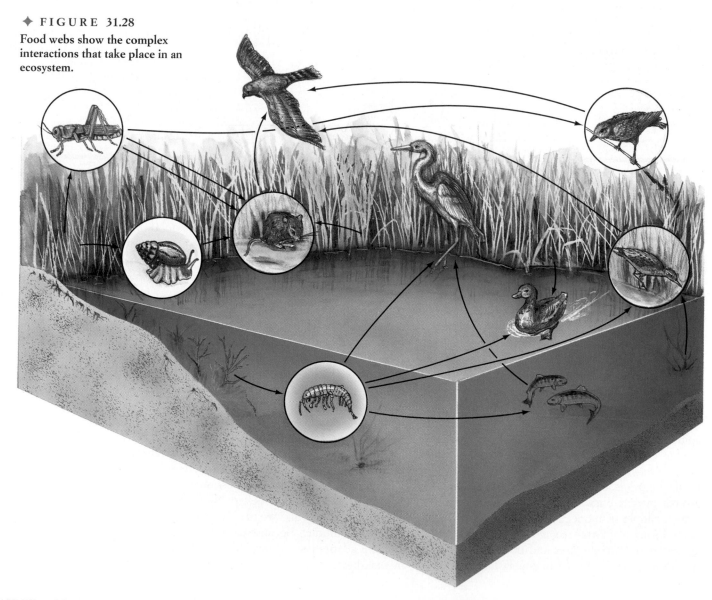

expanded. The term "weed" is virtually synonymous with early successional species. This also includes "weedy" animals, such as many insects, mice, and some birds.

## ECOSYSTEMS AND COMMUNITY FUNCTION

Although communities vary in structure, there are basic processes that unite them. The most basic processes are energy flow and matter cycling. All organisms must take in energy and matter to stay alive, causing energy and matter to move through the community. All energy and matter ultimately come from, and return to, the physical environment. We must therefore observe the ecosystem (community plus physical environment) to understand the complete process.

### Energy Flows through Ecosystems

**Energy flows** are routed through feeding relationships. The **ecological niche** occupied by an organism refers to how it functions in the ecosystem, and is closely associated with feeding. There are three primary ways of representing energy flow through any ecosystem. These are:

1. Food chain
2. Food web
3. Food pyramid.

A **food chain** is a series of organisms, each feeding on the one preceding it. There are two types of food chains: decomposer and grazer (✦ Figure 31.27). Grazer chains begin with algae and plants, which are eaten by grazers (herbivores). The grazers, in turn, are consumed by carnivores. Waste and decomposing organisms from the grazer chain enter the decomposer chain (Figure 31.27b). Bacteria and fungi break down this material to generate nutrients that are reincorporated into the plants at the base of the grazer chain. Although the food chain concept is widely used, it is an oversimplification, because most animals do not eat just one kind of organism.

The **food web** is a more accurate depiction of energy flow in natural ecosystems. Food webs are networks of feeding interrelations among species in most ecosystems (✦ Figure 31.28).

The **food pyramid** provides a detailed understanding of energy flow in an ecosystem (✦ Figure 31.29). The first level in the pyramid consists of

✦ FIGURE 31.29

Two energy pyramids in food chains. (a) Food chain representing the typical U.S. diet. (b) Food chain in other countries where grains are a principal part of the diet. This second chain allows more energy transfer to the top consumers.

Feeds one human 2000 kilocalories

Carnivore

Herbivore

Steer

Producer

Grains

20,000 Kilocalories

(a)

Feeds 10 humans
2000 kilocalories each

2000 Kilocalories

Producer

Grains

20,000 Kilocalories

(b)

**producers,** which generate the food used by all other organisms in the system. Usually, producers are plants, making food by photosynthesis. All higher levels in the pyramid contain **consumers.**

Primary consumers directly consume the producers, deriving energy from the chemical energy stored in the producer's body. In the marine environment, primary consumers include crustaceans and other organisms that eat phytoplankton. In a forest, primary consumers include deer and other plant-eaters (herbivores). Above the primary consumers are the secondary consumers, which feed on the primary consumers. In a marine ecosystem, secondary consumers consist of fish, lobsters, and other species. In a forest ecosystem, secondary consumers include wolves, panthers, and other meat-eaters (carnivores) that eat the deer and other primary consumers. Third, fourth, and even higher order consumers can occur in some ecosystems.

In human food chains, grains serve as the producers. In those chains where the diet is primarily grains, more energy is available, and more people can be fed. In those chains where herbivores such as cattle or sheep consume the primary producers, and people eat the herbivores, less energy is available, and fewer people can be fed. This illustrates the principle that the shorter the food chain, the more food is available to the top-level consumers.

Feeding relationships form a pyramid because there is progressively less food available at each higher level. This occurs because much of the food eaten by a herbivore is either lost as undigested waste or "burned up" by the animal's metabolism. Some herbivores excrete about 25% of ingested calories in an unused form. Of the 75% that is digested, most is lost as metabolic waste products (such as urine) and body heat generated from movement. Of all the calories eaten, fewer than 2% are converted into body tissue that can be eaten by top-level consumers. Because much of the human population is at the top of the energy pyramid, it has a limited amount of food energy available to consume.

The result is a "leakage" of energy between each feeding level (✦ Figure 31.30). This is why feeding relationships form a pyramid. In general, about 80

✦ **FIGURE 31.30**
**The flow of energy and biomass through a food chain. Energy is lost at each level, making less energy available to higher levels in the chain.**

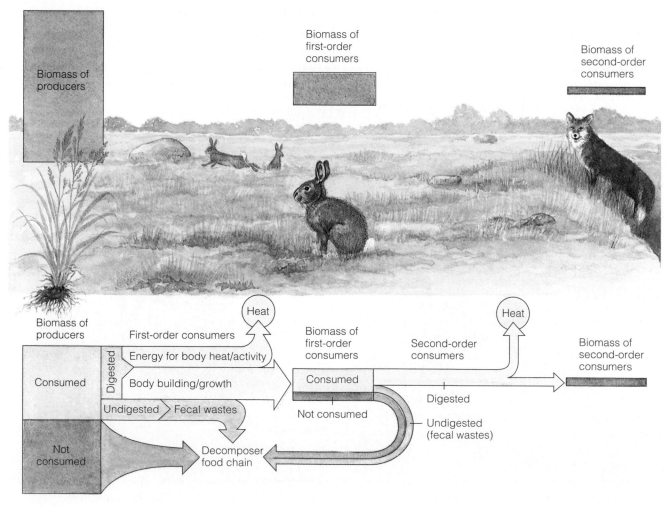

Biomass of producers

Biomass of first-order consumers

Biomass of second-order consumers

| Ecosystem type | Area (million square kilometers) | NET PRIMARY PRODUCTIVITY PER UNIT AREA | | World net primary production (billions of tons) |
| --- | --- | --- | --- | --- |
| | | Normal range | Mean | |
| Tropical rainforest | 17.0 | 1000–3500 | 2200 | 37.4 |
| Tropical seasonal forest | 7.5 | 1000–2500 | 1600 | 12.0 |
| Temperate deciduous forest | 7.0 | 600–2500 | 1200 | 8.4 |
| Savanna | 15.0 | 200–2000 | 900 | 13.5 |
| Temperate grassland | 9.0 | 200–1500 | 600 | 5.4 |
| Tundra and alpine | 8.0 | 10–400 | 140 | 1.1 |
| Desert and semidesert shrub | 18.0 | 10–250 | 90 | 1.6 |
| Cultivated land | 14.0 | 100–3500 | 650 | 9.1 |
| Swamp and marsh | 2.0 | 800–3500 | 2000 | 4.0 |
| Lake and stream | 2.0 | 100–1500 | 250 | 0.5 |
| Total continental | 149 | | 773 | 115 |
| Open ocean | 332.0 | 2–400 | 125 | 41.5 |
| Upwelling zones | 0.4 | 400–1000 | 500 | 0.2 |
| Continental shelf | 26.6 | 200–600 | 360 | 9.6 |
| Reefs | 0.6 | 500–4000 | 2500 | 1.6 |
| Estuaries | 1.4 | 200–3500 | 1500 | 2.1 |
| Total marine | 361 | | 152 | 55.0 |
| Total: marine + continental | 510 | | 333 | 170 |

to 95% of the energy is lost in the transfer between each level, depending on the organisms involved. As a result, very few ecosystems have food pyramids with more than five levels, because there is so little energy left at the top. This is also why large carnivores are rare: They are the organisms at the top of the pyramid.

## Ecosystem Productivity Begins at the Base of the Food Pyramid

The amount of food generated by producers at the base of the food pyramid varies greatly among ecosystems. Productivity is the rate at which biomass is produced in a community. **Net primary productivity (NPP)** is the rate at which producer (usually plants) biomass is created. The most productive terrestrial ecosystems are tropical forests and swamps (➥ Table 31.2). These produce plant biomass (NPP) at many times the rate of deserts. Temperate communities such as grasslands and temperate forests have intermediate productivity levels. Most of this pattern is explained because productivity on land increases with length of growing season. There is a general trend toward increasing productivity nearer the equator where winters are milder and shorter.

Deserts are the exception to this trend because lack of water limits growth, even if the growing season is long.

The most productive aquatic ecosystems are estuaries and reefs. These are nearly ten times more productive than freshwater ecosystems and other marine ecosystems (Table 31.2). The open ocean is the least productive ecosystem. This is a crucial point because the open ocean constitutes about 90% of the ocean. Therefore, 90% of the ocean, which means over half of the earth's surface, is essentially a "marine desert" in terms of productivity.

This is why marine ecosystems near shore need to be protected from damage. Reefs and estuaries (nutrient-rich zones near river mouths) are where most productivity occurs. Marine productivity (unlike terrestrial) is not as strongly determined by length of growing season. Instead, nutrient availability tends to be the limiting factor on productivity in marine ecosystems. The open ocean is relatively "starved" for some nutrients, especially phosphorus, because the source of the nutrients is runoff from land.

Zones of upwelling along the continental shelf are highly productive areas because the upwelling currents often carry nutrients from the ocean bottoms (✦ Figure 31.31).

**Net secondary productivity (NSP)** is the rate at which consumer and decomposer biomass is produced. Net secondary productivity includes all biomass except plants. Primary and secondary net productivity are correlated: Communities that have high primary productivity almost always have high secondary productivity. If the base of the food pyramid is producing large amounts of biomass, those organisms that consume and decompose plants will usually produce more biomass too.

## Human Activity Can Disturb Energy Flow and Productivity

The extent of ecosystem energy flow alteration by human activity is dramatic: humans redirect about 40% of the net primary productivity of all land plants on earth. Humans directly or indirectly use about 40% of the land food pyramid base on the planet. This means that nearly half of the energy converted by plants from sunlight is not available to species in natural ecosystems. This energy loss to nature is particularly striking considering that up to 90% of all species on Earth live on land.

## Matter Cycles through Ecosystems

**Matter cycling** is an important process in ecosystems. Unlike energy, matter is not converted into less useful forms when used. More than 30 chemical elements are cycled through ecosystems in **biogeochemical cycles,** which carry the elements through living tissue and the physical environment such as water, air, and rocks. An example is the carbon cycle (✦ Figure 31.32).

Most of the chemical elements that cycle through ecosystems are trace elements, used in small amounts by organisms. Carbon, hydrogen, oxygen, nitrogen, sulfur, and phosphorus are the six elements used in large amounts by living things. Because organisms both metabolize and store these elements, ecosystems exert great control over how fast elements cycle. Some elements cycle in a matter of days, while some may be buried for millions of years. Carbon may spend millions of years underground, being stored in fossil fuels such as coal and oil, or as limestone.

Ecosystems are generally efficient in cycling matter over and over within the ecosystem itself (✦ Figure 31.33). The carbon atoms removed from the atmosphere by a plant via photosynthesis will be in-

✦ **FIGURE 31.31**

Ocean waters rise from the depths along the continental shelf, carrying nutrients that support microscopic plants (phytoplankton) and animals (zooplankton). The plankton, in turn, are a rich source of food for many fish. Upwellings are biologically rich zones in the ocean.

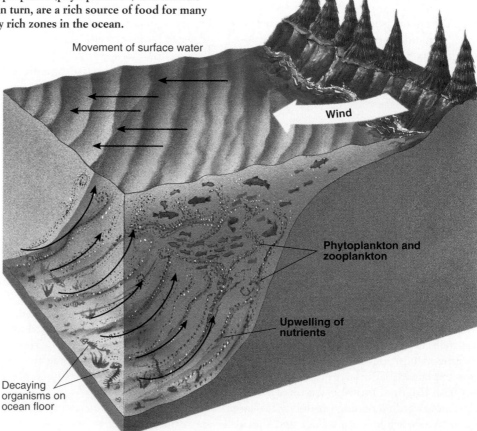

Movement of surface water

Wind

Phytoplankton and zooplankton

Upwelling of nutrients

Decaying organisms on ocean floor

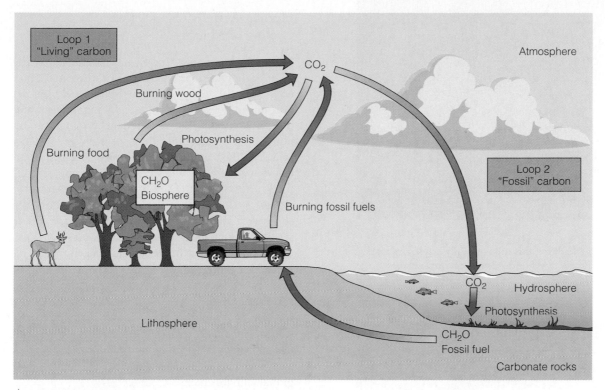

The carbon cycle illustrates the flow of this element through an ecosystem.

corporated into a catepillar. These, in turn, will be incorporated into the tissue of a bird that eats the catepillar. When the bird dies, decomposers (bacteria and fungi) will incorporate the same carbon atoms. When the decomposers die, their remains enter the soil as nutrients that are taken up by growing producers. All of this takes place within the ecosystem. Nevertheless, a small amount of matter in the ecosystem will be lost through time. Runoff and leaching from rainfall will carry off carbon in the form of decaying organic matter, leaves, and so on. In undisturbed ecosystems, this output loss is roughly balanced by an equal input gain of the same matter. Carbon can originate via weathering of rocks and is carried into the ecosystem by rainwater. In undisturbed natural ecosystems, both the input and output is small relative to the amount of matter "locked up" and recycled within the biomass of the ecosystem itself.

Both the rate and efficiency of matter cycling vary between ecosystems. The cycle of matter is generally faster and more efficient in tropical ecosystems, such as tropical rain forests and coral reefs. It is faster because biochemical reaction rates tend to increase with temperature. Although matter cycling is efficient in most natural ecosystems, it is especially efficient in tropical ecosystems. High rainfall in the tropics will leach elements from the

soil unless rain forest plants incorporate them quickly and efficiently into their tissues. Similarly, coral reefs thrive mainly in nutrient-poor tropical waters so that the elements in the nutrients must be utilized and recycled efficiently in the tissues of the organisms in the reef community.

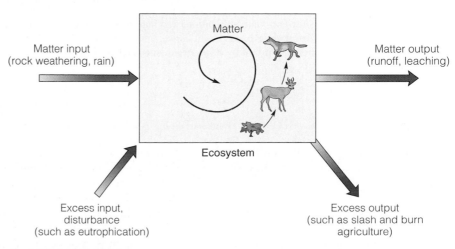

✦ FIGURE 31.33

**Most undisturbed biosystems are in a state of equilibrium, balancing input and output. Excess input or excess output are disturbances that can have far reaching consequences.**

**✦ FIGURE 31.34**

As rain forests are clear-cut for agriculture, the soil is rapidly eroded by rains, leaving nutrient-poor soil that may require hundreds of years to recover.

**✦ FIGURE 31.35**

Excess input of nutrients (eutrophication) can lead to overgrowth of some organisms, such as bacteria and algae, which consume oxygen and nutrients required by other organisms such as fish.

## Human Activity Can Disturb Cycling of Matter

Matter cycling is disturbed when humans alter the balance between input and output of matter in ecosystems. This occurs in two basic ways: excess output or excess input (Figure 31.33). **Excess output** occurs when the large quantity of matter retained in the biomass of the ecosystem is suddenly released. **Slash and burn agriculture** consists of cutting down and burning trees. The burning releases the nutrients into the soil for agriculture. Unfortunately, the nutrients are quickly leached out of the soil where rainfall is heavy (as in the tropics where slash and burn techniques are common). This output of matter from the ecosystem is not fully replaced by input for hundreds, perhaps thousands, of years (✦ Figure 31.34). Meanwhile, the area can sustain only a relatively deficient ecosystem with a fraction of its former richness and diversity. Many tons of carbon are released into the atmosphere annually from the burning of tropical forests and oil and gas. This is contributing to the likelihood of global climate change (Chapter 32).

Disturbance by **excess input** commonly occurs from agricultural activity, which carries large amounts of fertilizer, organic waste, and other nutrients into natural ecosystems. This also reduces species diversity because excess nutrients cause eutrophication, leading to unrestrained growth of some organisms, such as algae in a lake. When the algae die, the decay of their now-abundant bodies by bacteria uses up so much oxygen that fish and many other organisms perish (✦ Figure 31.35).

Rates of both matter cycling and energy flow in an ecosystem have a great effect on the ecosystem's ability to recover from human disturbance. Generally speaking, the faster that matter cycles and energy flows through an ecosystem, the faster it can recover from human disturbance.

---

 SUMMARY

1. Populations comprise communities and ecosystems. A community is defined as the set of all populations that inhabit a certain area. The size of this area can be very small, such as a puddle of water, or encompass large regional areas containing many hundreds of square miles. An ecosystem is defined as the community plus its physical environment. Ecosystems can also be defined at many levels.

2. Most communities have populations that differ in their density peaks and range

boundaries, forming open communities. Populations in open communities are distributed more or less randomly. The major exception to the open structure of communities occurs in cases where the physical environment makes an abrupt transition, such as a beach. Transitions in the physical environment are usually gradual and population densities between communities also change gradually.

3. Communities are divided into aquatic and terrestrial. These two basic divisions are

grouped into biomes. The aquatic biomes are marine and freshwater. The terrestrial biome contains tundra, grassland, desert, taiga, temperate forest, and tropical forest. A biome is a large-scale category that includes many communities of a similar nature.

4. Communities have two basic properties: diversity and stability. Diversity is measured by species richness, the number of different species in a community. Four factors increase diversity in a community:

environmental stability, age, growing season, and nutrients. The relationship between diversity and stability is much more complex than previously thought.

5. Community succession is the gradual replacement of one species with another. This process occurs through competition and other biological factors. Community evolution occurs over longer time spans, and can involve thousands or millions of years.

6. Communities are interconnected by energy flow among them. Energy flow is represented by food chains, food webs, and food pyramids. Food pyramids are the most accurate depiction of energy flow, and they consist of producers and consumers.

7. Matter is cycled through ecosystems by biogeochemical cycles. Six major cycles are important to communities and ecosystems: carbon, hydrogen, oxygen, nitrogen, sulfur, and phosphorus. These elements are used in large amounts by living things, and are recycled over time.

# KEY TERMS

altitudinal gradient
aridity
benthic zone
biogeochemical cycle
biomass
biome
climax community
closed community
community
community age
community simplification
community succession
consumers
depth diversity gradient
desert biome
diversity
ecological niche
ecological time
ecosystem
ecotones

energy flow
environmental stability
euphotic zone
excess input
excess output
food chain
food pyramid
food web
freshwater biome
geological time
grasslands biome
growing season
latitudinal diversity gradient
latitudinal gradient
length of growing season
marine biome
matter cycling
monoculture
nektonic organisms
net primary productivity (NPP)

net secondary productivity (NSP)
nutrients
open community
pelagic zone
photic zone
pioneer community
planktonic organisms
producers
resource partitioning
slash and burn agriculture
species packing
species richness
stability
stratification
stressed community
taiga biome
temperate forest biome
tropical rain forest biome
tundra biome

 # QUESTIONS AND PROBLEMS

## SHORT ANSWER

1. Groups of different organisms may be considered at three progressively more inclusive levels of organization. These levels are communities, ecosystems, and biomes. Define each.

2. Describe the moisture-holding capacity of air in warm and cold climates.

3. Why is there greater environmental stability in equatorial areas?

4. Define a food chain, food web, and food pyramid.

5. What are the six elements used in large amounts by living things?

## MULTIPLE CHOICE

1. The community focus on population distribution is called the _____ view.
   a. chemical
   b. functional
   c. physical
   d. structural

2. The two most important patterns seen in populations in communities are
   a. closed structure and relative rarity of most species
   b. open structure and abundance of species
   c. open structure and relative rarity of most species
   d. closed structure and species diversity

3. In the marine biome, the benthic community is divided into
   a. shore/continental shelf and deep-sea community
   b. bottom-dwelling and water column
   c. floaters and swimmers
   d. running waters and standing waters

4. Colonizers that are adapted for widespread dispersal, rapid growth, and have a "hardy" nature are referred to as the
   a. pioneer community
   b. climax community
   c. stressed community
   d. producing community

5. The cycle of matter is generally faster and more efficient in
   a. tundra ecosystems
   b. tropical ecosystems
   c. deciduous ecosystems
   d. disturbed ecosystems

## TRUE/FALSE

1. Most communities have populations that differ in their density peaks and range boundaries.

2. Terrestrial and aquatic biomes are largely determined by their distance from the equator.

3. More than three-fourths of the earth's surface is covered by oceans.
4. Community succession is characterized by an increase in productivity and an increase in diversity.
5. Community simplification is the reduction of overall species diversity.

## MATCHING

Match the appropriate description with its major land biomes.

1. Abundant rainfall with distinct seasonal changes; deciduous trees dominate.
2. Economically, this is the most important biome.
3. Also referred to as coniferous forests; low diversity; acidic and thin soils.
4. The most complex and diverse biome.
5. Extensive treeless plain whose topsoil is frozen.
6. Plant and animal adaptations for large amounts of water storage; rapid growth and reproduction after it rains.

a. tropical forest
b. grasslands
c. deserts
d. temperate forests
e. taiga
f. tundra

## SCIENCE AND SOCIETY

1. In southern California two major plant communities dominate the coastline. The coastal sage community is composed of low shrubs that often whither in the summer. Higher up, the chaparral community is dominated by tall evergreens. Which community do you suppose receives the most rainfall? Does the consistency of the soil play a role in which plant community occupies which area? If so, which community grows on the better soil? Which community is more productive on a yearly basis? Explain your reasoning. Where would you expect to find the most annual plants? Why?
2. Rapid population growth in both developing and developed countries is our most serious environmental problem. Although this belief is commonly regarded as true by many influential people, U.S. political pressures have caused the denial of funds to any organization that informs women that abortion is one of their legal options. What kinds of environmental problems occur when a population's growth exceeds the amount of available resources? Do you feel overpopulation would cause greater environmental damage in developed countries or in developing countries? How do you feel about the United States' policy of not informing women about abortion? Do you think by informing women about this option that the rate of abortion in the United States would dramatically increase? Why or why not? What effect does the United States' policy have on economically poor women? How could the U.S. government's actions affect population growth? How could this action affect the environment?

# 32

# BIOSPHERE AND HUMAN IMPACT

O P E N I N G   I M A G E

*Human population growth is*

*endangering wildlife, including*

*the rhinoceros.*

V*isitors to Disneyland can stroll past figures of several cartoon characters sculpted from shrubbery, including Mickey Mouse and Pluto. These figures, known as topiary, are carefully shaped from the Australian eugenia tree. This plant, with its lush leaves, is highly prized by gardeners and is widely planted throughout California as an ornamental tree, shrub and hedge. At Disneyland, over 800 meters (a distance equal to about a half-mile) of hedges as well as the topiary figures are shaped from eugenia.*

*In the late 1980s, a less welcome immigrant from Australia arrived in Southern California—the eugenia psyllid. Females of this rapidly developing aphid-like insect lay eggs in the edges of new leaves on eugenia stems. The newly hatched nymphs feed on the leaves, causing them to become blistered and discolored. Infestation distorts the shape of new stems and leaves, and is often followed by the growth of a black, sooty mold that further spoils the appearance of the plant. Although chemical pesticides were intensively used, the eugenia psyllid had spread throughout the state by 1990.*

*Scientists at Disney Imagineering Research and Development along with members of the Laboratory of Biological Control at the University of California, Berkeley decided to explore the use of biological predators to control the psyllids. The eugenia tree grows in sandy soils and sand dunes of New South Wales, Australia. Eugenia psyllids in all stages of the life cycle were collected from these regions in Australia and shipped to California where they were placed under quarantine. Careful study showed that some of the psyllid*

*nymphs had been parasitized by a small wasp. Female wasps lay an egg under a psyllid nymph as it feeds on leaf tissue. The egg hatches into a wasp larvae which kills and feeds on a psyllid nymph, crawling into the nymph's carcass to complete development. To hatch, the adult wasp drills a hole in the carcass, and emerges, ready to mate and begin a new generation.*

*After a year of testing, permission was received from the U.S. Department of Agriculture to release adult wasps at Disneyland in an attempt to control the psyllids. In July, 1992, 103 adult wasps were released, and in each year since, the psyllid population has been reduced, without affecting native plants or animals. The eugenia trees at Disneyland have rebounded, and no pesticides have been used on the hedges since early in 1993. The wasps are now being used at the San Diego Zoo to control psyllid infection. It is too early to tell whether this is a success story, since the psyllid population may rebound, or the wasp may discover other hosts.*

*This chapter deals with the impact of human activity on the biosphere, which is the highest level of biological organization. The eugenia tree was imported from Australia as a decorative plant, and seems to have had little impact on native species. The arrival of the eugenia psyllid also took place by human activity, although how the insect was transported from Australia is not yet known. The use of the parasitic wasp as a predator was a deliberate step taken to control psyllid pest. This example serves to illustrate how human activity alters the biosphere.*

## THE BIOSPHERE AND ITS PHYSICAL ENVIRONMENT

In this chapter, we consider the highest level in the hierarchy of life. The **biosphere** is the sum of all living matter on Earth. The boundaries of the biosphere range from the depths of the oceans beyond the peaks of the highest mountains. However, only highly specialized life-forms have adapted to such extremes and the majority of life exists within a few dozen feet of the Earth's surface.

The biosphere is interconnected with three other "spheres" that comprise the physical environment: the lithosphere, hydrosphere, and atmosphere. These represent the three states of matter: solid, liquid, and gas, respectively (✦ Figure 32.1).

### The Lithosphere Is the Outer Layer of the Earth

The **lithosphere** is the solid outer layer of the Earth, including rocks and their erosion products such as sand and soil. This includes the land area that covers about 30% of the Earth, and the solid crust beneath the oceans, lakes, and other parts of the hydrosphere that covers the remaining 70% of the Earth. Unlike most other planets, the Earth's lithosphere is dynamic. The tectonic cycle describes the creation of new crust in parts of the lithosphere and

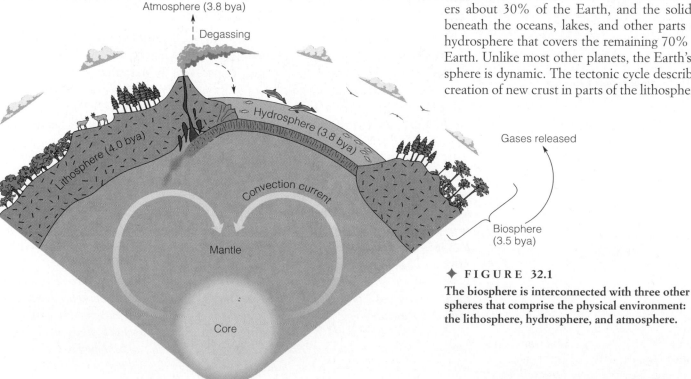

✦ **FIGURE 32.1**

The biosphere is interconnected with three other spheres that comprise the physical environment: the lithosphere, hydrosphere, and atmosphere.

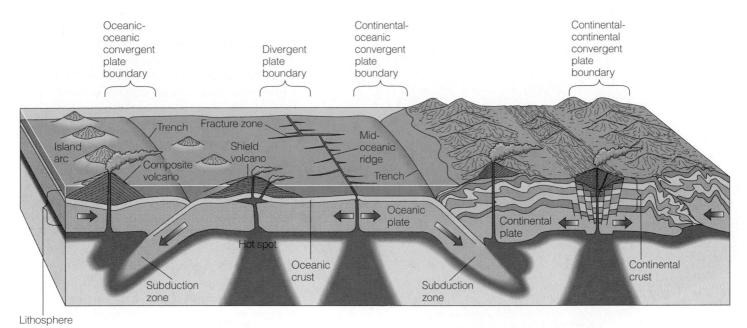

◆ FIGURE 32.2

**Movement of plates creates several types of boundaries, including convergent and divergent boundaries.**

the destruction of old crustal rocks in other parts. This cycle begins when new crust is created by up-welling and cooling of molten magma from the deeper, hotter parts of the Earth (◆ Figure 32.2). Magma is molten rock generated deep within the Earth. Upwelling of magma breaks apart the hard rocky layer so that the Earth's lithosphere is not a continuous layer. It actually consists of about 10 or so large plates and a number of small ones, so that geologists often call this cycle "plate tectonics." Because the plates are pushed apart by the upwelling, this is called a **divergent plate boundary.** As this new material is added, the older crust at the other end of the plate is pushed downward, underneath another plate. This is called a **convergent plate boundary** because plates are being pushed together. As the older part of the plate is pushed progressively deeper, it encounters hotter temperatures and begins to melt, producing magma, which once again rises to the surface and completes the cycle.

Because the plates essentially "float" on the molten material beneath them, the landmasses move through time, leading to continental drift. While

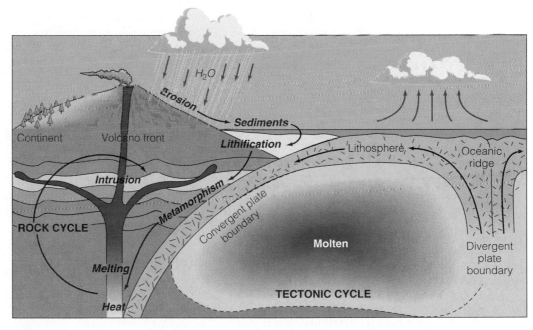

◆ FIGURE 32.3

**The tectonic cycle releases gases and molten material to provide a continuous source of chemicals for the hydrosphere and atmosphere.**

the rate of motion is slow, averaging a few inches per decade, this movement has produced major climatic and geographic changes over the billions of years of geological time. The release of gases and molten material in the tectonic cycle provides a continuous source of chemicals for the hydrosphere and atmosphere (◆ Figure 32.3). By driving

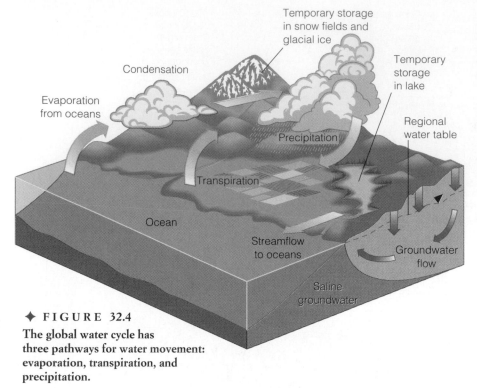

**◆ FIGURE 32.4**

**The global water cycle has three pathways for water movement: evaporation, transpiration, and precipitation.**

a changing environment in all these other spheres, the tectonic cycle has also been a major force in the evolution of life. This cycle has moved landmasses, generated new mountain ranges and changed weather patterns, caused extinctions, and provided conditions for the evolution of new species.

### The Hydrosphere Is the Water At or Near the Earth's Surface

The **hydrosphere** includes all the liquid and solid water at or near the Earth's surface. Water composes 60 to 95% of all plants and animals (depending on the type of organism). Humans are about 65% water. Because of its abundance and chemical properties, water is probably the single most important substance on Earth. Water is an excellent solvent and can dissolve a wide range of substances. This makes water very chemically active in a range of activities from biochemical reactions to erosion.

The hydrologic cycle is the continuous recycling of water from the oceans, through the atmosphere, to the continents, and back into the oceans. This cycle operates within the hydrosphere and includes two components: **evaporation** of liquid water into the atmosphere as water vapor (including **transpiration,** which is evaporation of water used by plants), and **precipitation** of liquid water from the atmosphere as rain or snow (◆ Figure 32.4). Precipitation usually falls directly into the oceans, which contain more than 97% of all the world's water. However, it may fall on land where it eventually returns, by streams or subsurface transportation

("groundwater"), to a collecting basin such as an ocean or a lake. More than 100 trillion gallons of water per year move through the Earth's hydrologic cycle. Even so, this represents only 5% of the Earth's total water; the other 95% is bound up in the rocks and magma of the lithosphere.

### The Atmosphere Envelops the Earth

The **atmosphere** is the envelope of gases that surrounds the Earth. Several gases compose the atmosphere, but most are present only in very small amounts. Together, nitrogen (78%) and oxygen (21%) compose about 99% of the atmosphere. The atmosphere is densest near the Earth's surface where the pull of gravity is greatest and holds on to the most number of gas molecules. At higher altitudes, the atmosphere becomes progressively "thinner." At an altitude of several hundred miles, the atmosphere grades into the vacuum of space, where gas molecules are extremely rare.

### A LIVING PLANET: INTEGRATION OF THE SPHERES

Volcanic action transfers water from the lithosphere to the atmosphere, where precipitation moves it into the hydrosphere. Water vapor and virtually all matter on Earth cycles among the spheres. The movement of chemical elements through the atmosphere, lithosphere, hydrosphere, and biosphere are called biogeochemical cycles (Chapter 31). For our survey of the biosphere, the most important biogeochemical cycles are those that transport the six elements most important to life: carbon, hydrogen, oxygen, nitrogen, phosphorus, and sulfur. Of the more than 90 elements that occur naturally on Earth, these six comprise the vast majority of atoms

| TABLE 32.1 RELATIVE ABUNDANCE OF SOME CHEMICAL ELEMENTS IN THE EARTH'S CRUST | |
|---|---|
| ELEMENT (CHEMICAL SYMBOL) | RELATIVE ABUNDANCE (%) |
| Oxygen (O) | 62.5 |
| Silicon (Si) | 21.2 |
| Aluminum (Al) | 6.47 |
| Sodium (Na) | 2.64 |
| Calcium (Ca) | 1.94 |
| Iron (Fe) | 1.92 |
| Magnesium (Mg) | 1.84 |
| Phosphorus (P) | 1.42 |
| Carbon (C) | 0.08 |
| Nitrogen (N) | 0.0001 |

| TABLE 32.2 ATOMIC COMPOSITION OF THREE REPRESENTATIVE ORGANISMS | | | |
|---|---|---|---|
| ELEMENT | MAN (%) | ALFALFA (%) | BACTERIUM (%) |
| Oxygen | 62.81 | 77.90 | 73.68 |
| Carbon | 19.37 | 11.34 | 12.14 |
| Hydrogen | 9.31 | 8.72 | 9.94 |
| Nitrogen | 5.14 | 0.83 | 3.04 |
| Phosphorus | 0.63 | 0.71 | 0.60 |
| Sulfur | 0.64 | 0.10 | 0.32 |
| Total | 97.90 | 99.60 | 99.72 |

in the cells of all living things. Oxygen makes up more than 62% of all the atoms in the human body, and more than 77% in the alfalfa plant (◆ Table 32.1). If carbon is added, more than 80% of the atoms in the human body are accounted for. The relative abundance of these elements in the Earth's crust is very different (◆ Table 32.2). Although oxygen is the most common (about 62%, the same as in the human body), the second most common element in the human body, carbon, is hundreds of times rarer in the Earth's crust. Instead, silicon, which is virtually absent from the human body, is extremely abundant in the crust. These discrepancies demonstrate the contrast between life and its environment. Without biogeochemical cycles to transport and store concentrations of matter for food and other uses, life could not survive.

## Biogeochemical Cycles Connect the Biosphere with the Physical Spheres

Biogeochemical cycles integrate the biosphere with the physical spheres. This integration of the Earth with living systems has led to the speculation that the entire system of all four spheres forms a single "superorganism" called **Gaia.** This analogy has been criticized on many counts, including the fact that the spheres are not as highly integrated as the cells in a body.

 ## THE BIOSPHERE IS BEING DISRUPTED AT MANY LEVELS

In an earlier chapter, the relationship between human activity and direct alteration of the biosphere was outlined. Human population growth and technology can also disturb the biosphere indirectly, by modifying the other spheres. Much has been written about the possibility of global climate change, and how it may affect life. In the following sections we discuss how human activity is indirectly disturbing the biosphere by altering the atmosphere

and hydrosphere. Being gaseous and liquid, respectively, these two physical spheres readily transmit alterations to other spheres.

## Global Changes in the Atmosphere

Although nitrogen and oxygen comprise about 99% of the atmosphere, the trace gases and particles in the remaining 1% play an important role in global climate and in shielding life from solar radiation. Gases added to the atmosphere through agriculture and industry may affect the global climate and the radiation-shielding capacity of the atmosphere.

The most important trace gas added to the atmosphere by technology is carbon dioxide ($CO_2$). Carbon dioxide is a common trace gas and a major part of the carbon cycle (Chapter 31). Carbon dioxide has many natural sources, including respiration by animals and combustion of organic matter. There are two main **sinks** that absorb atmospheric carbon dioxide: plants and the oceans. Plants absorb carbon dioxide during photosynthesis and incorporate carbon into storage products and plant tissue. The oceans are a larger sink, absorbing carbon dioxide and over long time periods depositing it as limestone (calcium carbonate $CaCO_3$). Under natural conditions, the amount of carbon dioxide released by sources and absorbed by the two sinks is about the same so that atmospheric carbon dioxide levels are relatively stable over short spans of time.

However, with the onset of the Industrial Revolution and modern agriculture, ever increasing amounts of carbon dioxide are being released into the atmosphere. With the growth of industrialization, the amount has increased dramatically, to about six billions tons per year (◆ Figure 32.5).

◆ **FIGURE 32.5**
**Global emissions from burning fossil fuels.**

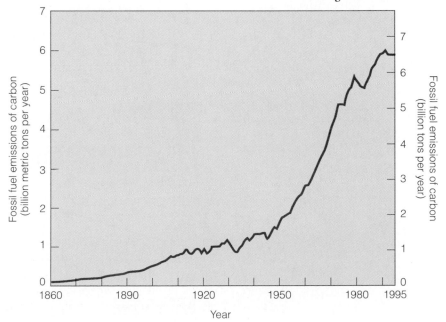

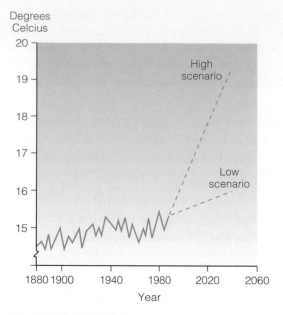

**◆ FIGURE 32.6**

**Global average temperatures, with projections to 2040.**

## Carbon Dioxide Emissions May Contribute to Global Warming

Is there any harm related to increasing carbon dioxide levels in the atmosphere? Carbon dioxide and a few other gases have properties that make them transparent to sunlight but not transparent to heat. Carbon dioxide thus acts like glass in a greenhouse by permitting sunlight to penetrate the atmosphere but absorbing and trapping the resulting heat generated when the additional sunlight warms the Earth. Appropriately enough, this is called the **greenhouse effect.** At the present time, carbon dioxide is relatively rare in the atmosphere, and the Earth is warmed to generally hospitable temperatures. As more carbon dioxide is added, more heat is trapped. No one is sure how much carbon dioxide must be added to significantly heat the Earth's climate. The average global surface air temperature has risen along with carbon dioxide content, indicating a direct relationship between carbon dioxide levels and temperature (◆ Figure 32.6). The unresolved question is whether the rise in temperature is caused by the increase in carbon dioxide, or whether the two trends are occurring by sheer coincidence. The average global temperature has risen about 0.5°C during the last 100 years, but we cannot say how much (if any) of this increase was caused by carbon dioxide emission. Although this seems like a minor change, it has been calculated that an increase of 2°C would be enough to cause melting of the polar ice caps.

Fossil fuels (coal, oil, natural gas), which accounted for about 80% of world energy use in 1990, are the major source, releasing carbon that was taken out of the atmosphere many millions of years ago. Accumulation has been enhanced by the destruction of vast tracts of tropical forests, reducing the number of plants available to absorb carbon dioxide.

**◆ FIGURE 32.7**

**Levels of greenhouse gases. (a) Carbon dioxide. (b) Methane. (c) Nitrogen oxides. (d) CFC-11.**

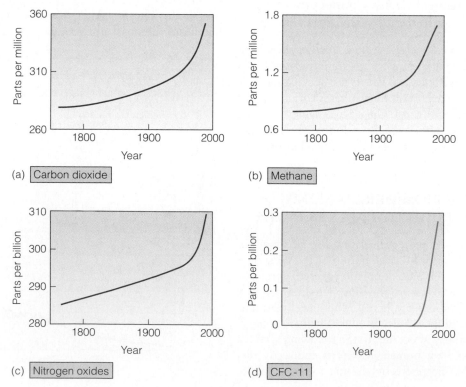

Other factors also influence global climate, including the release of other "greenhouse gases." Carbon dioxide causes an estimated 56% of greenhouse heating, but other gases, including methane, contribute significantly. Methane is expelled in great quantities by cows. As cattle production has increased, methane release has increased dramatically (✦ Figure 32.7b). Methane output from all sources is increasing by about 1% each year. In addition to greenhouse gases, solar output varies through time, changing the amount of radiation that reaches Earth.

It will probably be years before the heating effects of carbon dioxide and other greenhouse gases are conclusively established. Computer models show an inevitable rise in temperature as carbon dioxide is added (Figure 32.6).

A global warming trend will affect some areas more than others, shifting wind and rainfall patterns. Most computer models show that temperate areas, such as the United States and Canada, will become drier and warmer, and increased rainfall will occur in equatorial and polar areas (✦ Figure 32.8). The effect on the biosphere would be enormous. The rich grain-growing regions of the Midwest and Central Asia would probably no longer be the "breadbaskets of the world." Many natural communities would have to migrate, as they did to keep pace with temperature changes during the Ice Age glaciations. Global warming may change temperatures some ten times faster than during the Ice Age. Some plants and other organisms that disperse slowly may not be able to migrate fast enough. Pine trees adapted to moderate temperatures may not be able to disperse seeds fast enough and far enough to avoid becoming extinct.

✦ **FIGURE 32.8**

**Predicted rainfall patterns that would result from global warming.**

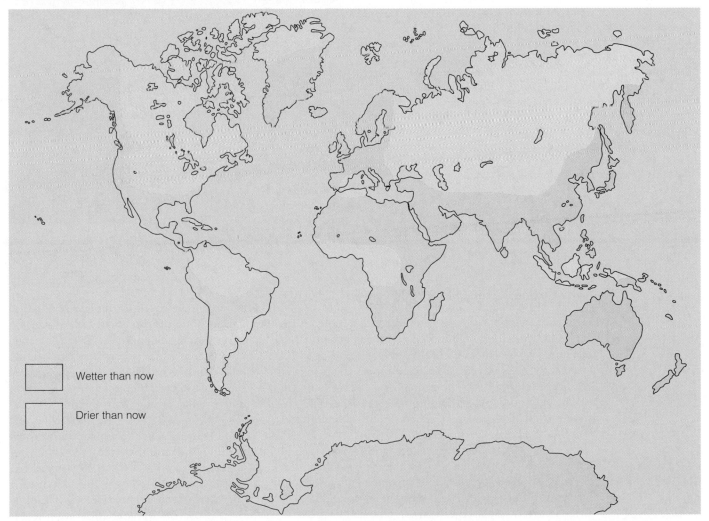

Wetter than now

Drier than now

## Guest Essay: Molecular Biology and Endangered Species

MARY V. ASHLEY

After receiving my bachelor's degree at Kenyon College, a small liberal arts college in Ohio, I started graduate school at the University of California at San Diego (UCSD), with no clear research direction, other than a continuing interest in evolution, especially evolutionary genetics. In a department that was dominated by molecular biologists, it was natural for me to develop an interest in applying techniques of molecular biology to the study of evolution. I had to piece together my graduate education, learning DNA techniques from one source, and evolutionary theory and population genetics from other sources. It was difficult, but in the end I believe the independence that I was forced to develop has served me very well.

One of my main research interests involves applying molecular evolution to the conservation of endangered species. Often it is difficult to identify the "units" to manage for conservation, whether it is a population from a particular geographic location, a subspecies endemic to a region, or the species as a whole. I addressed this question for the endangered black rhinoceros, again using genetic information gleaned from the mitochondrial genome. Black rhinos once roamed throughout much of sub-Saharan Africa, but have disappeared faster than any large mammal during this century. From numbers in the hundreds of thousands at the turn of the century, there are less than 3000 individuals alive today. The overwhelming cause of their

demise is slaughter for the sole purpose of supplying rhino horn to markets in the Far East. Many Asian cultures believe that rhino horn has remarkable medicinal powers, and it can be sold at values up to $15,000 per kilogram.

The few remaining black rhinos in Africa generally exist in small, isolated populations. Early in the century, when rhinos were still abundant, taxonomists had divided the species, *Diceros bicornis,* into several subspecies based on slight morphological differences such as horn shape and body size. Should each of these named subspecies be managed separately? Or, alternatively, should all remaining black rhinos be considered as a single population for breeding purposes? We reasoned that if the rhino subspecies actually did have separate evolutionary histories, this separation should be reflected in the divergence of the rapidly evolving mitochondrial genome. The results of the study were quite remarkable; black rhinos from east, central, and southern Africa differed very little if at all in their mitochondrial DNA sequences. We concluded that if management would proceed more efficiently by translocating remaining animals from different regions into secure "rhino sanctuaries," that there was no reason from an evolutionary or genetic standpoint not to do so.

My interests have led me to work on a variety of other exotic species, including tapirs, owl monkeys, and armadillos. People always ask

whether I collect my own samples, and I always wish that I had exciting stories to tell of facing off with an African rhino to collect a blood sample, or dangling from a tree in Brazil to dart a monkey. The truth, however, is that I rarely do field work. My exotic specimens, generally extracted from zoo or museum collections, are reduced to tiny tubes of DNA in my freezer. I'm not interested in the sequence of "gene *x*" from "species *y*" (usually a laboratory organism such as a rat or fruit fly), but how that gene sequence varies from species to species, what that variation tells us about the evolution history of these species, and what it tells us about how evolutionary forces have shaped that gene.

*Mary V. Ashley has been an assistant professor at the University of Illinois at Chicago in the Department of Biological Sciences since 1992. She teaches evolution to undergraduates, advises graduate students, and conducts research in the areas of molecular evolution and conservation genetics. She received her B.A. from Kenyon College in Ohio in 1981 and a Ph.D. from the University of San Diego in 1986.*

### The Ozone Layer Is Being Destroyed

Destruction of the ozone layer is another potential atmospheric change caused by human activity. **Ozone** ($O_3$) is generated in the stratosphere when sunlight strikes oxygen atoms and causes them to combine temporarily. The ozone helps filter much of the high-energy radiation from the sun. As ozone is removed from the atmosphere, more radiation reaches the Earth's surface. Each 1% drop in ozone levels increases human skin cancer rates by an estimated 4 to 6%. Within the last decade, scientists have discovered a great "hole" in the ozone layer above the Earth over Antarctica (◆ Figure 32.9). Much current research is focusing on whether this "hole" will spread to other areas, or if it is only a temporary phenomenon.

Ozone is being destroyed by the release of gases, including **chlorofluorocarbons** (**CFCs**), which are used in refrigerators, air conditioners, and solvents. One CFC molecule in the atmosphere can destroy about 100,000 ozone molecules. Growing awareness of these destructive abilities led to a reduction in the rate of CFC production beginning in the early 1970s (◆ Figure 32.10). The 1987 Montreal Protocol and later agreements among nations will phase out CFC production by the year 2000. But since it takes CFC molecules 20 to 30 years to rise into the upper atmosphere where they survive for about 100 years, ozone destruction will continue well after CFC use is stopped. CFCs are a major greenhouse gas and contribute to global warming effects.

## Acid Deposition and Smog Are Local Atmospheric Changes

Climatic warming and ozone depletion represent global effects of human activity on the biosphere. Humans also alter the atmosphere on a local level. (➥ Table 32.3 summarizes three of the most important types of local air pollution and their sources.

Acid deposition is most commonly called **acid rain.** The burning of coal and other fossil fuels re-

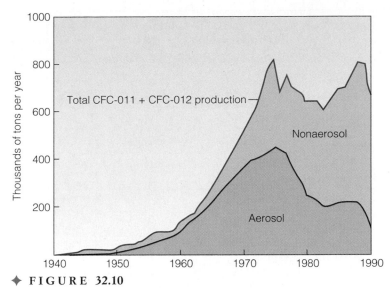

**✦ FIGURE 32.10**

World production of the two most widely used CFCs. Production of CFCs in aerosols declined when CFC-containing aerosol sprays were banned in the United States in 1978.

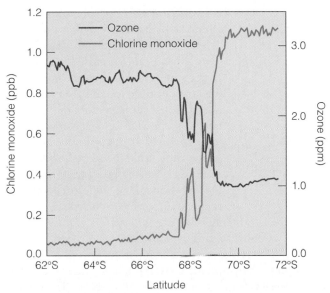

**✦ FIGURE 32.9**

(a) The increase in reactive chlorine (chlorine monoxide) across latitude is accompanied by a decrease in ozone levels. (b) Ozone levels in Antarctica have been falling over a 25-year span. (Measurements taken at Halley Bay, Antarctica.)

**TABLE 32.3    MAJOR AIR POLLUTANTS, HUMAN-CAUSED SOURCES, AND ENVIRONMENTAL EFFECTS**

| AIR POLUTANT | MAJOR HUMAN-CAUSED SOURCES | GREENHOUSE WARMING | STRATOSPHERIC OZONE DEPLETION | ACID DEPOSITION | SMOG | DAMAGE TO VEGETATION |
|---|---|---|---|---|---|---|
| Carbon dioxide ($CO_2$) | Fossil fuels, deforestation | + | +/− | | | |
| Methane ($CH_4$) | Rice fields, cattle, landfills, fossil fuels | + | +/− | | | |
| Nitric oxide (NO), nitrogen oxide ($NO_2$) | Fossil fuels, biomass burning | | +/− | + | + | + |
| Nitrous oxide ($N_2O$) | Nitrogenous fertilizers, deforestation, biomass burning | + | +/− | | | |
| Sulfur dioxide ($SO_2$) | Fossil fuels, ore smelting | − | | + | | + |
| Chlorofluorocarbons (CFCs) | Aerosol sprays, refrigerants, solvents, foams | + | + | | | |
| Ozone ($O_3$) | Fossil fuels | + | | | + | + |

Source: *Thomas E. Graedel and Paul J. Crutzen, "The Changing Atmosphere,"* Scientific American, *September 1989, p. 62; and James J. MacKenzie and Mohamed T. El-Ashry,* Ill Winds: Airborne Pollution's Toll on Trees and Crops *(Washington, DC: World Resources Institute, 1988).*

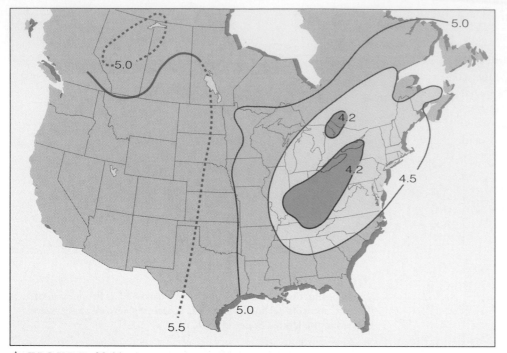

**✦ FIGURE 32.11**

Acid rain patterns in the United States in 1988. The numbers represent the measured pH values of rain. A pH below 7 is acidic.

**✦ FIGURE 32.12**

Smog is a local alteration in the atmosphere caused by human activity.

leases nitrogen and sulfur. Sulfur forms sulfuric and other types of acids when combined with water vapor in the atmosphere. This acid may be deposited many hundreds of miles downwind from its source (✦ Figure 32.11). Many lakes in the northeastern United States are acidified from coal-burning in the Midwest and cannot support game fish. Forests also suffer as acid-sensitive trees die.

**Smog** is mainly an urban problem caused by fuel combustion (✦ Figure 32.12). Pollutants such as sulfur oxides, nitrogen oxides, and hydrocarbons react in the presence of sunlight to form more than a hundred secondary pollutants that cause respiratory problems in humans, but also harm plant tissues.

Aside from CFCs, which are being phased out, nearly all of the problems, both global and local, are linked to **fossil fuels** (Table 32.3). The first step suggested by most experts is to reduce the use of fossil fuel through conservation. The United States derives nearly 90% of its energy from fossil fuels; about half of that total is wasted. Much of this energy could be saved by using more efficient cars, factories, and other improved technologies. The second, longer term step would be to switch to nonfossil fuels, which release virtually none of the pollutants in Table 32.3. Examples include wind, hydro, solar, geothermal, and nuclear power. Switching to other forms of power is inevitable because world petroleum supplies will be exhausted within 50 to 100 years, and coal reserves will be used up within the next few hundred years.

## Water Pollution Is Altering the Hydrosphere

Alteration of the atmosphere indirectly alters the hydrosphere. Sulfur pollution of the air causes acid rain that affects lakes, rivers, and ponds. On a much larger scale, climate changes have a major impact on the oceans. During periods of global cooling, more rain precipitates as snowfall and accumulates as ice, especially at the poles in the form of ice caps. This ice serves as a water reservoir, causing sea levels to drop. During the Ice Age, widespread glaciation caused sea levels to drop, exposing the continental shelves.

Global warming would have the reverse effect. An increase in global temperature would partially melt the polar ice caps and many glaciers, causing sea levels to rise. In the past 50 years, sea levels have risen 10 to 12 centimeters (4 to 6 inches), and some coastal areas have already been affected (✦ Figure 32.13). At the present rate of warming, models predict that sea levels will rise another 50 to 100 centimeters (2 to 3 feet) by 2050. The major cities of most countries are located on low-lying coastal areas or on large rivers. In the United States, about half the population lives within 50 miles of the ocean. Some cities like Miami are only a few feet above sea level. Rising sea levels could cause widespread damage and population displacement in low-lying coastal areas. In countries such as Bangladesh, about 17% of the country would be under water if sea levels rose by 3 feet.

The hydrosphere can be directly altered through water pollution. Although water pollution can occur from natural causes, most water pollution occurs from discharge of wastes generated by humans (✦ Figure 32.14). There are three basic sources of water pollutants: (1) Municipal sewage discharge is a main source of nutrients that can cause eutrophication, and is a source of disease-causing organisms. (2) Agricultural, mining, and logging activities discharge sediments that reduce light penetration, cover aquatic organisms, and eventually fill and block waterways. (3) Industries are the source of the most concentrated, toxic pollutants in the form of inorganic compounds that contain heavy metals

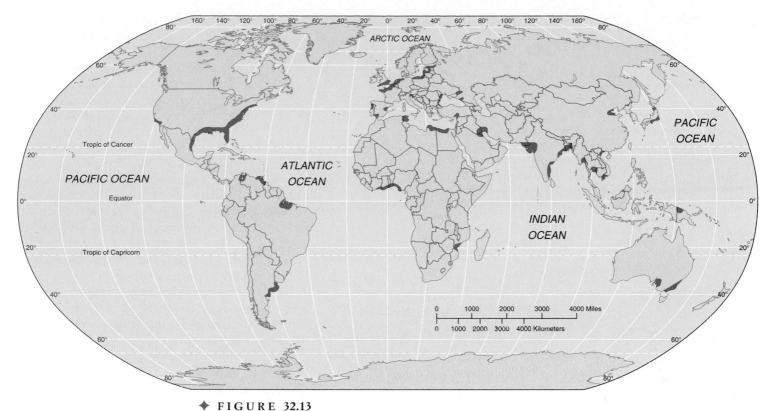

✦ FIGURE 32.13

Areas of the world coastline that have been sinking in recent decades.

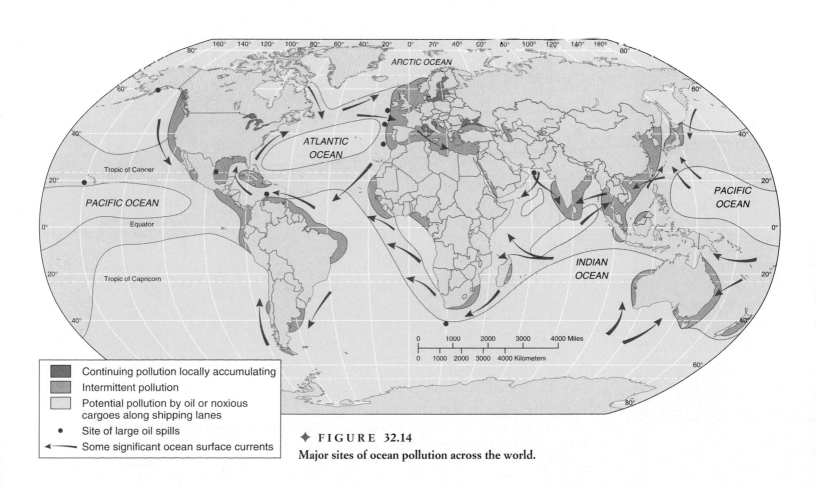

Continuing pollution locally accumulating

Intermittent pollution

Potential pollution by oil or noxious cargoes along shipping lanes

• Site of large oil spills

← Some significant ocean surface currents

✦ FIGURE 32.14

Major sites of ocean pollution across the world.

### TABLE 32.4  PROBABLE CAUSES OF PAST MASS EXTINCTIONS

| EXTINCTION EVENT | PROBABLE CAUSE |
| --- | --- |
| Major mass extinctions: | |
| End Mesozoic | Meteorite impact |
| Early Mesozoic | Global cooling, low sea level |
| End Paleozoic | Global cooling, low sea level due to Pangea forming |
| Middle Paleozoic | Global cooling, low sea level |
| Early Paleozoic | Global cooling, glaciation |

Source: *Stephen K. Donovan*, Mass Extinctions *(New York: Columbia University Press, 1989).*

*Sidebar*

**READING THE SKY**

In April of 1995 the National Weather Service began issuing predictions of UV intensity for 58 cities in the United States. These levels are calculated based on altitude, latitude, cloud cover, time of year, and the most critical factor, the amount of ozone in the atmosphere. A computer crunches these values to determine the UV-A and UV-B rays for 1:00 pm the next day. Next, it calculates a number called the UV Index, which gauges how badly the predicted level of ultraviolet light will damage human skin. Zero to two indicates minimal exposure and ten and up indicates very high exposure.

The index is formulated for skins that tan but will also burn. Researchers recommend that when the UV Index tops six, light-skinned people who plan to go outdoors should cover up and wear sunscreen. Darker-skinned individuals have more leeway, but should still keep an eye on the sun.

Check your local weather reports on the television and radio or in the newspaper for the UV Index before you plan your next outing at the beach.

such as mercury and zinc and many hazardous organic chemicals.

## ARE WE ON THE VERGE OF A SIXTH MASS EXTINCTION?

We saw in Chapter 30 that extinction occurs when the environment changes too fast for a group of organisms to adapt. Changes in the physical environment, such as climate change, or in the biological environment, such as the introduction of alien species, can trigger extinction. One or two species can go extinct if there is a local disturbance, whereas major regional or global changes will cause many species to go extinct.

Environmental changes of many kinds have caused extinctions in the past. More than 99% of all species that have existed were extinct long before humans arrived. Paleontologists estimate that 5 to 50 billion species have existed since life began 3.5 billion years ago, and that less than 100 million species exist today. Most of these extinctions were caused by local or regional environmental changes, such as minor fluctuations in sea level or volcanic activity. Recall from Chapter 11 that there have been five environmental changes of global proportions resulting in **mass extinctions,** meaning that more than 50% of all species living then became extinct. Recovery from these five sharp decreases in the diversity of life required millions of years. Most of these mass extinctions were caused by global cooling, which led to loss of the highly diverse tropical life (◆ Table 32.4). The one spectacular exception is the last mass extinction, about 65 million years ago, which included loss of the dinosaurs. Evidence is accumulating that this extinction was associated with the impact of a large meteorite that struck the earth near the Yucatán peninsula of Mexico (◆ Figure 32.15).

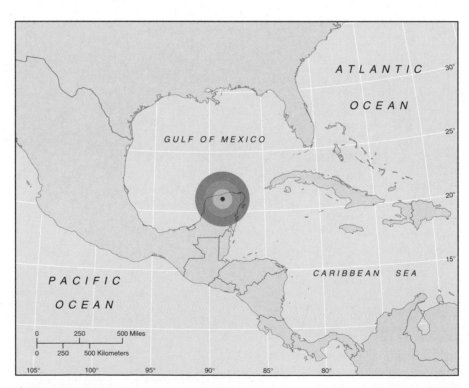

**◆ FIGURE 32.15**

The last mass extinction may have been caused by a meteorite that struck the Earth off the coast of present-day Mexico.

*Beyond the*
*Basics*

# WHAT HAPPENED IN MILWAUKEE?

The Clean Water Act was passed in 1972. Since then, more than $575 billion dollars has been spent in the United States on the construction and upgrading of urban water and sewage treatment plants, and on industrial treatment plants designed to treat single sources of industrial waste. Controlling wastes from these single point discharges has improved water quality across the nation. But as recent events in Milwaukee, Wisconsin demonstrate, single point discharges are not the only source of problems in our water supply.

Non-point sources of water pollution are those with multiple points of origin. Runoff of fertilizer from farms and suburban lawns, animal wastes, and agricultural herbicides and pesticides are some examples of non-point sources. These materials not only run off into streams and lakes, but can contaminate groundwater sources that are used as drinking water.

In 1993, people in Wilwaukee, Wisconsin became ill from contaminated drinking water.

More than half of the 800,000 residents served by the water system became ill. Symptoms varied in severity, and the illness generally lasted about two weeks. Individuals with compromised immune systems, including those with AIDS or cancer, were more susceptible to infection. Nine deaths were linked to the illness.

Within several days of the outbreak, the source of the contamination was identified as an intestinal parasite known as cryptosporidium. These small protistan organisms parasitize vertebrates, and inhabit the spaces between microvilli on the intestinal epithelial cells of the small intestine. The city issued an order for citizens to boil their water before using it for drinking or preparing food. Within eight days, steps were taken to control the parasite in the drinking water, but unresolved questions remained. Where did the infection originate? How did this parasite contaminate the drinking water of an entire city?

Investigation revealed that the cryptosporidium most likely came from infected cattle on farms upstream, whose waste runoff flowed into streams and then into the Milwaukee River, and from it, into Lake Michigan, where the city obtains its water. Because the city's sewage discharges into Lake Michigan a few miles from the water intake, this could have contributed to the infection.

Whatever the source, the city water treatment system should have eliminated the parasite before it entered the water system. At the time, however, one of the water treatment plants was using polyaluminum chloride to reduce the corrosiveness of the water. This inadvertently reduced the effectiveness of the sand filters that normally would have removed the cryptosporidium.

The incident in Milwaukee was not unique. Between 1984 and 1992 there were three other cases of cryptosporidian contamination of water supplies. These incidents took place in Braun Station, Texas, Carrolton, Georgia and Medford, Oregon. These cases point out the need to identify and control non-point sources of water pollution, and to carefully evaluate changes in water treatment methods.

## Development is Accelerating in Tropical Regions

Environmental changes caused by human activity are now causing extinctions in many parts of the globe, at an accelerating pace (✦ Figure 32.16). No one knows exactly how many species are becoming extinct, but most estimates range from 10 to 100 species per day. If this accelerating pace continues, many biologists predict that more than half of all species (the "sixth mass extinction") could be extinct by the middle of the twenty-first century.

A major contributor to the rise in extinction rates is the species diversity present in tropical ecosystems. When industrialization and its accompanying environmental disturbances occurred mainly in nontropical areas (North America, Europe, Northern Asia), the number of species lost was relatively low because these areas are relatively species-poor. Development is accelerating fastest in tropical regions, which are estimated to contain 80% of the world's species.

In the Amazon forests of South America, as many as 300 species of trees can occur in a typical hectare (about 2.5 acres). In Southeast Asia about 200 tree species occur per hectare, and in Central America, the number is about 120 species. In contrast, the average hectare of forest in North America (and Europe) ranges from only 1 tree species to a maximum of about 12 species. Other organisms show the same pattern: One insect specialist recently counted 43 species of ants on one tree in Peru, equal to about the total number of ant species in the entire British Isles.

Although rain forests cover only about 7% of the Earth's land surface, they contain well over one-half of the Earth's terrestrial species. A similar pattern is found in ocean waters; as a result, marine pollution is much more damaging in tropical waters. For example, biologically diverse coral reefs, which occur only in warm water, are especially sensitive to organic and sediment pollution.

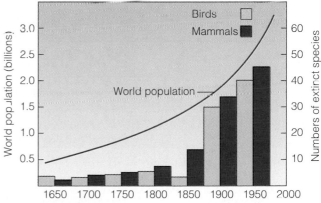

✦ **FIGURE 32.16**

**The rate of extinction of birds and mammals can be correlated with the growth of the human population.**

As of 1990, it was estimated that at least 30% of the Earth's rain forests were already gone. Using satellite photos and other methods, it has been estimated that about 15 hectares per minute are being destroyed in Brazil alone. Globally, an area of rain forest totaling about the size of the state of Washington is being destroyed each year, often by slash and burn agriculture. At these rates, all remaining rain forest is expected to be eliminated by sometime in the next 100 years, perhaps as early as the year 2010 if the rates continue to increase as human populations soar.

## Some Species Are More Susceptible to Extinction

Species are not equally likely to become extinct. For many reasons, some species are more able than others to survive environmental change. Cockroaches represent a particularly resilient group of species; they have existed almost unchanged for more than 300 million years and will probably be alive for millions more.  ◆ Table 32.5 lists characteristics that make species susceptible to extinction. Island species are sensitive to the introduction of new species because they have been isolated for a long time. Indeed, most of the recent bird and mammal extinctions have involved island-dwelling

species (Figure 32.16). Species with limited habitats become extinct easily simply because they have so little habitat to destroy or disturb that human activity can quickly eliminate it.

Many species possess traits that place them at greater risk of becoming extinct. Food pyramids tends to concentrate pollutants at the top, so that predators, in addition to being relatively rare, often die from pollution. However, there are many more species of small organisms: Insects make up more species than all other species combined (see later discussion). This vast diversity of small organisms means that the large majority of the 10 to 100 species per day becoming extinct consists mostly of localized insect and small plant species in the tropics.

Many of the seven characteristics in Table 32.5 can be tested by analyzing past extinctions. Paleontologists have found that species in tropical habitats with local distribution and large body size, or in marine habitats (especially those in shallow water), were often disproportionately affected in past extinctions.

## Why Try to Save Species from Extinction?

Why worry about extinctions? On one side, some people argue that humans have little need of wildlife—elephants, exotic tropical insects or plants,

| **TABLE 32.5** | **CHARACTERISTICS OF EXTINCTION SUSCEPTIBILITY** | |
|---|---|---|
| CHARACTERISTICS THAT CAUSE SOME SPECIES TO BE SUSCEPTIBLE TO EXTINCTION | REASON CHARACTERISTICS TEND TO CAUSE EXTINCTION | EXAMPLES |
| Island species | Unable to compete with invasion from continental species | More than half of the 2000 plant species in Hawaii |
| Species with limited habitats | Some species are found in only a few ecosystems | Woodland caribou, Everglades crocodile, millions of species in the tropical rainforest |
| Species that require large territories to survive | Large-scale habitat destruction in the modern world | California condor, blue whale, Bengal tiger |
| Species with low reproductive rates | Many species evolved low reproductive rates because predation was low, but in modern times, people have become effective predators against some of these species | Blue whale, California condor, polar bear, rhinoceros |
| Species that are economically valuable or hunted for sport | Hunting pressures by humans | Snow leopard, blue whale, elephant, rhinoceros |
| Predators | Often killed to reduce predation of domestic stock | Grizzly bear, timber wolf, Bengal tiger |
| Species that are susceptible to pollution | Some species are more susceptible than others to industrial pollution | Bald eagle (susceptible to certain pesticides) |

and the like are best viewed as curiosities with no practical value. Others take the view at the other extreme and want to preserve all nature for its own sake. They see humans as intruders and argue that extinction rates must be slowed at any cost. Between these two extremes lie many practical realities, including the view that humans as a species should seek a balance with other organisms on Earth.

Several arguments are often put forth as reasons to save species, based on aesthetics, ethics, economics, and ecology. In the limited space available, we will discuss only the last two of these.

Economic reasons to save species are the most immediately persuasive to most of the world's population. For discussion, the economic reasons can be divided into two groups: food and nonfood uses of biological diversity.

## Crop Plants Have Lost Their Genetic Diversity

Historically, humans have used about 7000 plant species as food. At the present time, only about 30 plant species provide 95% of the world's nutrition (✦ Figure 32.17). Just four of these—wheat, maize, rice, and potatoes—provide most of the world's food and these four have been subjected to centuries of inbreeding. The effect of this reliance on a few species can have many undesirable consequences. One is a matter of taste. Do we want our diet to be so bland and monotonous forever? Botanists estimate that at least 75,000 edible plant species exist, many superior in flavor and nutrition to the ones we use.

Other consequences are more serious. Low genetic diversity makes organisms more vulnerable to changing conditions. The inbreeding of most crop species makes them notoriously susceptible to diseases and insects. The infamous Irish potato famine, which killed millions of people in the nineteenth century and led to massive displacement of the Irish population, was caused by a fungus that wiped out a widely used but highly inbred strain of potato.

By the late 1960s, the corn crop in the United States was almost entirely derived from hybrid strains carrying a male sterility gene. This gene for male sterility was originally derived from a single line, greatly reducing the genetic diversity in the corn seed stock. Unfortunately, the line used as the source of the male sterility gene was highly susceptible to a leaf fungus. An outbreak of this fungus in 1970 wiped out 15% of the nation's corn crop in a single year.

Wild, related species can often be interbred with cultivated forms to improve resistance to disease

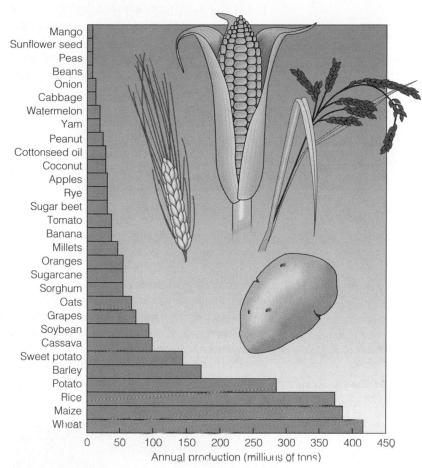

✦ FIGURE 32.17

Only about 30 crop plants make up 95% of the human diet, and four of these (rice, potato, wheat, and maize) predominate.

and general hardiness. This method was used to rescue the U.S. corn crops, using a rare wild strain found only in Mexico. An estimated $300 million per year was saved as a result. Improvement using wild strains is not limited to species that can be interbred. Genetic engineering allows genes to be transferred from one species to another, even when they are unrelated (Chapter 9). This increases the value of wild strains since useful traits can be combined among many species without being limited to those related strains that can naturally interbreed.

It is economically advantageous to maximize genetic diversity not only within crops that we grow, but to have as many different crop species at our disposal as possible. If current rates of extinction continue, some 25,000 plant species are expected to die out by the year 2000, before we have a chance to study them. This has led to the proposal that **gene banks** be established to save seeds, spores, sperm, and other genetic materials for those species that cannot be saved in the wild. Currently, seed banks contain only a small fraction of most plant varieties and many wild species have yet to be collected and described (✦ Figure 32.18).

Similar arguments can also be made for more diversity in domesticated animal species. Breeding of cattle with buffalo ("beefalo") to strengthen the stamina of cattle breeds and their resistance to predation is one of many examples where wild genes have been useful (✦ Figure 32.19). Similarly, there are many wild animal species that are themselves potentially tasty and healthy food sources, even without being interbred with existing food species.

## Managing Ecosystems Can Have Economic Benefits

Nonfood uses also provide economic reasons to preserve species. Wild species directly provide many nonfood products, ranging from medicines to rubber. A large number of modern medicines, such as aspirin, began as plants used by premodern societies. Forty-seven major drugs have been produced from flowering plants gathered in tropical forests. With so many plant species left to be identified, it is certain that many potentially useful medicines will be lost as tropical forests are destroyed. A recent

study estimated that at least 320 drugs, with a value of $147 billion, remain to be discovered. Finding useful drugs derived from plants is not a thing of the past. Taxol, discovered in the last few years, is a compound extracted from the rare Pacific yew that can be used in treating ovarian cancer.

Many plant products imported by industrial societies would be lost if native ecosystems were destroyed, because the plants do not grow well in domesticated conditions. Rubber trees are a highly valuable resource in the tropics, and rubber can be economically extracted from living trees year after year. Economists call such products **forest-sustainable resources** because they contrast with products that call for the forest to be cut down. Other forest-sustainable resources are exotic fruits, oils, and fiber. A recent report has shown that in many cases, such products bring in more money than the wood, so it is more economical to leave some forests intact.

## Ecosystems Are Environmental Support Systems

In many ways, ecology is the ultimate reason for conserving species because species compose the ecosystems that provide us with many of life's essentials. Ecosystems are environmental support systems that provide us with things that we now take for granted: oxygen to breathe (from plants), drinkable water (purified by microbial activity), and many other natural chemical cycles that occur via ecosystem functions (Chapter 31). If we remove too many species from an ecosystem, it will cease to function. If we do this to too many ecosystems, functions of the entire biosphere will be impaired.

 ## TAKING ACTION TO REDUCE SPECIES EXTINCTION AND BIOSPHERE DESTRUCTION

Concluding that biological diversity is worth preserving, how can the rate of extinction be slowed? In 1992, representatives from 178 countries gathered in Rio de Janeiro, Brazil, to discuss how to manage the resources of the world. At the end of the conference, 166 nations signed a binding treaty pledging them to cooperate in finding new and effective ways of using the finite resources of our planet. In 1994, a follow-up conference in Cairo, Egypt, on population and development focused attention on the need to control the growth of the human population as an important step in stabilizing the biosphere.

The following section contains a description of some of the initial steps that are being taken to preserve diversity and reverse damage to the biosphere.

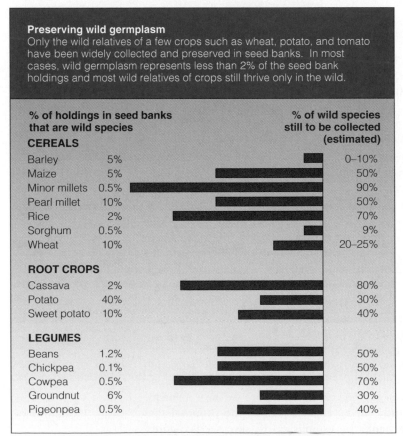

**Preserving wild germplasm**
Only the wild relatives of a few crops such as wheat, potato, and tomato have been widely collected and preserved in seed banks. In most cases, wild germplasm represents less than 2% of the seed bank holdings and most wild relatives of crops still thrive only in the wild.

| % of holdings in seed banks that are wild species | | % of wild species still to be collected (estimated) |
|---|---|---|
| **CEREALS** | | |
| Barley | 5% | 0–10% |
| Maize | 5% | 50% |
| Minor millets | 0.5% | 90% |
| Pearl millet | 10% | 50% |
| Rice | 2% | 70% |
| Sorghum | 0.5% | 9% |
| Wheat | 10% | 20–25% |
| **ROOT CROPS** | | |
| Cassava | 2% | 80% |
| Potato | 40% | 30% |
| Sweet potato | 10% | 40% |
| **LEGUMES** | | |
| Beans | 1.2% | 50% |
| Chickpea | 0.1% | 50% |
| Cowpea | 0.5% | 70% |
| Groundnut | 6% | 30% |
| Pigeonpea | 0.5% | 40% |

✦ **FIGURE 32.18**
**Wild plants make up only a small percentage of species preserved in seed banks.**

## Research and Description of Species

One of the first steps toward preserving biological diversity, or **biodiversity,** is to measure how much now exists. Because life is hierarchical, biodiversity can be measured at a number of levels: **Genetic diversity** is the number of genes and their alleles in the biosphere, **species diversity** is the number of species, and **ecosystem diversity** is the number of local ecosystems (or communities). Genes and ecosystems are often more difficult to measure than species number, so species diversity is most often studied. The sheer numbers of species on Earth makes this measurement of biodiversity difficult. Only about 1.7 million species have been described. Some biologists estimate that up to 100 million species now exist.

Most undescribed species are tropical, mainly plants and small animals, especially insects (✦ Figure 32.20). One expert, E. O. Wilson, has estimated that it would take the entire lifetimes of 25,000 specialists to describe them all. Currently, there are about 1500 specialists with the required knowledge in tropical biology.

## Establishment of Preserves

Even before completing the assessment of species diversity, it is essential to preserve the habitats of

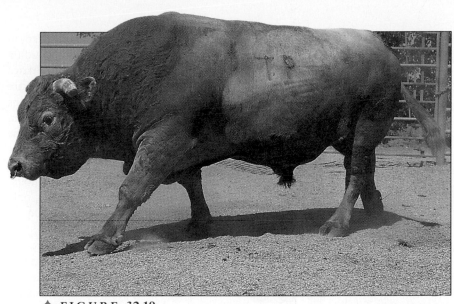

✦ **FIGURE 32.19**
**The beefalo is a hybrid between buffalo and cattle.**

endangered species. It is usually very difficult to reestablish a species after its habitat has been restored. A more straightforward approach is to simply preserve areas before they are disturbed or developed. Preserves are essential in saving communities: No community or ecosystem has ever been completely restored to its former state after being damaged or destroyed.

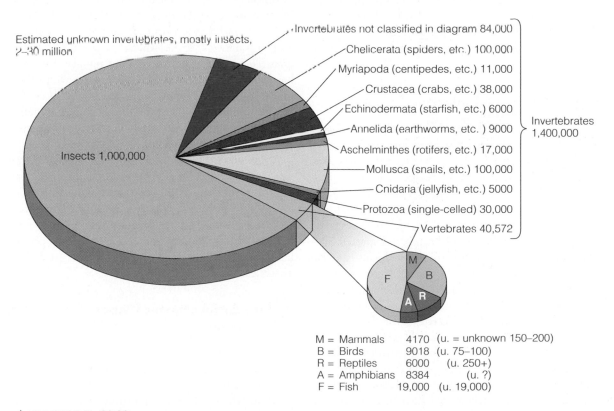

Estimated unknown invertebrates, mostly insects, 2–30 million

Invertebrates not classified in diagram 84,000
Chelicerata (spiders, etc.) 100,000
Myriapoda (centipedes, etc.) 11,000
Crustacea (crabs, etc.) 38,000
Echinodermata (starfish, etc.) 6000
Annelida (earthworms, etc. ) 9000
Aschelminthes (rotifers, etc.) 17,000
Mollusca (snails, etc.) 100,000
Cnidaria (jellyfish, etc.) 5000
Protozoa (single-celled) 30,000
Vertebrates 40,572

Invertebrates 1,400,000

Insects 1,000,000

M = Mammals     4170   (u. = unknown 150–200)
B = Birds       9018   (u. 75–100)
R = Reptiles    6000   (u. 250+)
A = Amphibians  8384        (u. ?)
F = Fish        19,000 (u. 19,000)

✦ **FIGURE 32.20**
**The estimated distribution of unknown animal species.**

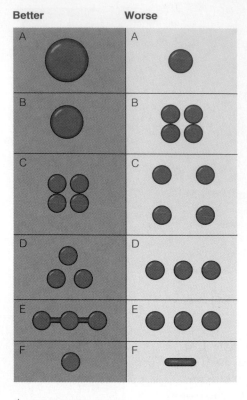

**Better**      **Worse**

✦ **FIGURE 32.21**

**The design of animal preserves is an important factor in determining the size of the breeding populations that can be established.**

Preserves are well established in many parts of the world, including the developed countries and Africa and Asia. However, preserves are badly lacking in the tropical regions of Central and South America. Even where they are established in these regions, poachers present a major problem. Many countries in tropical regions cannot afford adequate policing of the preserves. Currently, only about 1% of the Earth's land surface has been set aside as perserves.

Preserves involve more than just setting aside a section of land or water. A preserve must be designed to include prime nesting areas, migratory pathways, and so on. Species preservation is maximized by the size, shape, and location of the areas preserved. Larger areas, clustering of areas, connecting corridors for migration, and a small circumference to minimize "edge effects" tend to be the most effective (✦ Figure 32.21). Edge effects mean that organisms from the surrounding area can diffuse into the preserve too easily.

Preserving enough of the natural ecosystem is usually the biggest problem. Larger animals usually have more territorial needs because they require more food per individual and more space to forage. It is often necessary to set aside immense tracts of land to save large animals.

## Laws Protecting Endangered Species

It is difficult to catch and effectively punish poachers, especially in large preserves located in less developed countries. A more effective tactic has been to make it illegal to trade, transport, and sell products made from endangered species. In 1975, 81 countries signed the Convention on International Trade in Endangered Species, which outlaws trade in products derived from endangered species. This agreement has been effective in reducing the impact of poaching. One of the success stories in this approach resulted from a 1989 world ban on selling or trading ivory from elephant tusks. Before the ban, elephant ivory sold in Africa for about $100 a pound and elephant populations were rapidly decreasing. Following the ban, the price of ivory dropped to about $5 a pound on the black market, and the decline in elephant populations has been greatly slowed. Until more effective bans arc implemented for other organisms, poaching will continue. The sale of outlawed products including leopard skin coats and crocodile shoes generates an estimated $2 to $5 billion per year worldwide.

Protection of species in the United States is also a growing problem. Some biologists predict that up to 4000 species in the United States are in danger of extinction by the year 2000. The main legal apparatus to protect species in the United States is the **Endangered Species Act** of 1973, which directs the U.S. Fish and Wildlife Service to maintain a list of species that are endangered or threatened. Endangered species are in immediate danger of extinction. Threatened species are likely to be endangered soon. The Endangered Species Act has generated controversy over the cost of saving an ever-increasing number of species. Critics argue that instead of making costly efforts to save every population and subspecies, the act should focus on saving entire ecosystems instead of following a "one species at a time" approach. This would be more cost effective and take social and economic considerations into greater account. Furthermore, they argue that the current approach is often futile anyhow: Most species on the list are closer to extinction now than when originally listed.

## Breeding in Captivity

One of the last resorts in preventing extinction is to breed endangered species in captivity. This is less desirable than other means, but there is often no other choice if the genetic diversity is to be preserved. Conditions in captivity are often unnatural and populations do not thrive or reproduce. It is possible to artificially inseminate females but this has been used successfully on only a few species. Gene banks for endangered species can be established to freeze sperm, eggs, and cultured cells. This preserves the genomes of endangered species and makes them available for genetic engineering.

Even when species reproduce in captivity, the offspring often cannot survive in the wild because they were not raised under natural conditions. Finally, captive breeding is expensive and requires large areas of land. It is doubtful whether any modern zoo could maintain healthy breeding populations of more than 900 species, and probably much less.

## Reduce Socioeconomic Causes of Extinction

Ultimately, to preserve biodiversity, basic social and economic issues must be addressed. Rapid human population growth in developing countries is currently the single most important cause of extinction. The present world population is about 5.4 billion (✦ Figure 32.22a). It is estimated that this will

| TABLE 32.6 THE FINANCIAL SIZE OF DIFFERENT INDUSTRIES IN RAIN FORESTS | | |
|---|---|---|
| INDUSTRY | ESTIMATED WORLD MARKET (US$) BILLIONS | GROWTH RATE |
| Forestry | | |
|   Total | 85 | 1–2 |
|   Tropical | 7 | |
| Pharmaceuticals | ? | ? |
| Tourism[a] | 2000 (1987) | 4–5 |
| Entertainment[b] | 150 (1988) | 10–15 |

[a]Tourism is estimated to employ 6.3% of the global workforce.
[b]In 1983, American film studios earned $800 million from video cassettes, against $2.6 billion at the box office; the comparable figures for 1988 were $4.5 and $2.9 billion.
Source: "How to Pay for Tropical Rain Forests," Trends in Ecology and Evolution, 1991, p. 348.

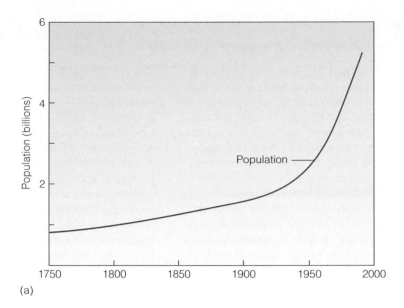
(a)

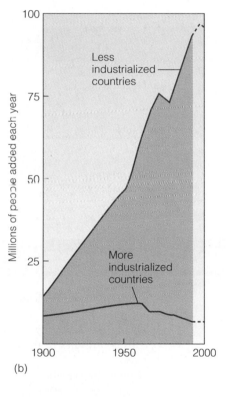
(b)

✦ FIGURE 32.22
(a) World population is estimated to be about 5.4 billion. (b) The number of people added to the world population each year will mainly be in less industrialized countries.

double by the middle of the twenty-first century, mainly from population growth in developing countries (Figure 32.22b). As these countries industrialize, pollution and resource depletion will exacerbate the problem. To minimize destruction of the biosphere in the future, population growth must be reduced. It will be equally important to promote industrialization that uses renewable resources that produce little pollution.

Slowing population growth will be an enormous task and will likely take many decades. In the meantime, actions are being taken to address the immediate effects of population growth on the biosphere. It is possible to remove economic incentives that cause many species extinctions. Laws are in place to accomplish this, but the goal can also be accomplished in other ways. In the U.S., public reluctance to buy tuna caught with drift nets that kill many dolphins resulted in a major reduction of dolphin deaths. On a larger scale, a rapidly growing branch of economics, called ecological economics, has shown ways to make ecosystems more profitable if left intact rather than be destroyed. Forest-sustainable products derived from tropical rain forests and other uses of the rain forests are an example (✦ Table 32.6). All told, these forest-sustainable uses seem to be much more valuable than destroying the rain forest by harvesting trees.

In Costa Rica, a pharmaceutical company has invested more than $2 million in searching for new drugs in the rain forests. This provides an incentive not to destroy the forests. In other cases, pharmaceutical companies have negotiated arrangements to pay a percentage of profits from drugs made from tropical plants that have not yet been identified or described. Because many such plants will probably be found, it pays for countries to set aside undisturbed areas for this purpose.

The pressure of rapid population growth makes the need for a subsistence diet so pressing that long-term economic incentives may not have time to work. An increasingly common practice is the **debt-for-nature swap.** In this arrangement, a country agrees to set aside a certain area of pristine land as a preserve in exchange for reduction in the financial debt owed to another country. Thus, richer countries lose money on loans made to the poorer countries but natural preserves are gained for the world's benefit.

1. The biosphere is interconnected with three other spheres that comprise the physical environment: the lithosphere, hydrosphere, and atmosphere, representing the three states of matter (solid, liquid, and gas, respectively). Without biogeochemical cycles to move and store concentrations of matter for food and other uses, life on earth could not survive.

2. Human activity can directly affect the biosphere by altering populations and communities. Human population growth and technology can also disturb the biosphere indirectly, by modifying the other spheres, particularly the atmosphere and hydrosphere.

3. The atmosphere is being changed by carbon dioxide pollution from burning fossil fuels. This may be causing the earth's temperature to warm. Ozone depletion by CFCs and other pollutants may trigger major climate changes in the next century.

4. A reduction in biological diversity is occurring as species become extinct under pressure from human population growth, increasing development, and the use of technology. Habitat destruction is the leading cause of extinction; about 30% of the tropical rain forests have already been destroyed.

5. Economic incentives must be created to assist the developing nations of the world in managing biological diversity, especially in tropical regions.

6. A schedule for reducing species extinction and preserving biodiversity is being developed by almost 200 of the world's nations. This program involves identifying and describing new species, establishing preserves, passing laws to protect endangered species, starting captive breeding programs, and working to slow human population growth.

acid rain
atmosphere
biodiversity
biosphere
chlorofluorocarbons (CFCs)
convergent plate boundary
debt-for-nature swap
divergent plate boundary
ecosystem diversity

Endangered Species Act
evaporation
forest-sustainable resources
fossil fuels
Gaia
gene banks
genetic diversity
greenhouse effect
hydrosphere

lithosphere
mass extinctions
ozone
precipitation
sinks
smog
species diversity
transpiration

## FILL IN THE BLANK

1. The movements of nutrient elements through the biosphere, or through any particular ecosystem, by physical or biological processes, are called _____ _____ .

2. _____ _____ is rain that has been made highly acidic by pollutants in the air.

3. The two main sinks that absorb atmospheric carbon dioxide are _____ and the _____ .

4. _____ species are in immediate danger of extinction. _____ species are likely to be endangered soon.

5. The _____ _____ _____ of 1973 is the legal apparatus that protects species in the United States.

## TRUE/FALSE

1. The greenhouse effect is the heating of the earth caused by gases in the atmosphere that trap infrared radiation from the earth and prevent it from escaping into space.

2. The greater diversity of tropical ecosystems contributes to the rise in extinction rates.

3. Species preservation is maximized by larger areas, clustering of areas, connecting corridors for migration, and minimum circumference to minimize "edge effects."

4. Captive breeding provides an economical means for preserving the genomes of endangered species.

5. Economic incentives would help developing countries manage biological diversity.

## SHORT ANSWER

1. What two steps would reduce the atmospheric alterations caused by fossil fuels?

2. What are the three basic sources of water pollutants?

3. Why do species with limited habitats become extinct easily?

4. Explain the concept of "forest-sustainable" resources.

5. Briefly cite the initial steps that are being taken to preserve diversity and reverse damage to the biosphere.

## MATCHING

Match each term appropriately.

1. lithosphere
2. hydrosphere
3. atmosphere
4. ozone
5. genetic diversity
6. species diversity
7. ecosystem diversity

____ number of genes in the biosphere
____ gases that surround the earth
____ filters high-energy radiation from sun
____ solid outer layer of the earth
____ number of local ecosystems

_____ represents liquid form of matter

_____ number of species in biosphere

## SCIENCE AND SOCIETY

1. There is heated debate among atmospheric scientists about the link between global warming and carbon dioxide. Most environmental scientists believe that there is substantial evidence that atmospheric constituents trap heat escaping from the earth's surface, causing the greenhouse effect. Government-sponsored assessments have agreed that there is better than a 50% chance that current trends in population growth, fossil fuel use, and destruction of land will cause dramatic climatic changes in the next century. Opponents to this line of thinking feel that more observations, better theories, and more extensive calculations are needed before the finger can be pointed at greenhouse gases as the cause for climate changes. This faction feels that there is little risk in delaying policy responses to greenhouse warming because there is reasonable expectation that scientific understanding will vastly improve within the next few years. What is your opinion about each side of this debate? Can you side with one of the factions based on the information presented here? What information is needed that would help with your decision? Would early preventive measures to combat global warming have any disadvantages? What would be the advantages of doing something now to slow global warming? What are the disadvantages and advantages of waiting to complete the ongoing and recently expanded research on global warming? Can you think of ways to test whether there is a link between carbon dioxide and global warming?

2. During the 1950s and 1960s a revolution called the Green Revolution depended on the development of new, improved strains of wheat and rice at international centers with the assistance of many governments. As a result of these centers, the production of wheat in Mexico, for example, increased almost 10-fold over a 20-year span. Food production in India was able to outpace a population growth of about 2% annually, and China became self-sufficient in food during that same time period. What do you think are the disadvantages and advantages of the Green Revolution? In what ways are the economy of nature and the human economy working against each other? How can this antagonistic relationship be reduced or eliminated? Do you think the Green Revolution could cause nutritional problems in these countries? In some regions the introduction of the Green Revolution methods have actually worsened poverty for many people and reduced their access to food and fuel. Explain how this could happen.

# APPENDIX A
# CLASSIFICATION OF ORGANISMS

The classification scheme presented below follows the five kingdom system, which recognizes a separate kingdom (Monera) for the bacteria. The eukaryotes are classed into four kingdoms, the Protista, Fungi, Plantae and Animalia.

The organization is not inclusive, and emphasizes groups of organisms discussed in the text. This system may differ from others you have seen in other texts or laboratory manuals. Taxonomists often debate whether to call groups superphyla or subkingdoms, or over the name of a phylum or class. This reflects the dynamic nature of the field, and the increasing use of evolutionary relationships as the basis of classification.

## KINGDOM MONERA

Bacteria. Prokaryotic organisms that are autotrophs or heterotrophs.

Subkingdom Archaebacteria. Methanogens (methane-producing), halophiles (live in salty environments), and thermophiles (bacteria in hot springs and deep sea vents). These organisms differ from other bacteria in the cell wall, some membranes, and ribosomes.

Subkingdom Eubacteria. All other bacteria, including photosynthetic autotrophs, chemosynthetic autotrophs, and hetcrotrophs.

## KINGDOM PROTISTA

A diverse group of single-celled and multicellular autotrophs and heterotrophs. These are eukaryotic organisms with a membrane-bound nucleus, cytoplasmic membranous organelles and complex flagella.

Phylum Dinoflagellata: Dinoflagellates
Phylum Rhizopoda: Amoebas
Phylum Sporozoa: Sporozoans
Phylum Euglenophyta: Euglenoids
Phylum Phaeophyta: Brown algae
Phylum Chlorophyta: Green algae
Phylum Rhodophyta: Red algae
Phylum Ciliphora: ciliates
Phylum Oomycota: oomycetes, water molds, downy mildews
Phylum Zoomastigina: diverse group of flagellated organisms
Phylum Acrasiomycota: Cellular slime molds
Phylum Myxomycota: Plasmodial slime molds

## KINGDOM FUNGI

Heterotrophic eukaryotes, mostly multicellular organisms, with cell walls that contain chitin. Some fungi are saprophytes, feeding on dead organisms or decaying organic material. Other fungi are parasitic. In this kingdom, divisions are the equivalent of phyla.

Division Zygomyceta: Zygote-forming fungi (black bread mold)
Division Ascomyceta: Sac fungi, yeast, molds, cup fungi
Division Basidiomyceta: Club fungi, mushrooms, bracket fungi
Division Fungi Imperfecta: fungi without a sexual stage
Lichens: Associations between a fungi (usually an ascomycete or a basidiomycete) and either green algae or cyanobacteria.

## KINGDOM PLANTAE

Photosynthetic, multicellular, eukaryotic organisms, primarily terrestrial. Cell walls contain cellulose. Use chlorophyll $a$ and $b$ for photosynthesis. Life cycle involves an alternation of generations, between a gametophyte and a sporophyte.

Division Bryophyta: Mosses, liverworts, hornworts
Division Psilophyta: Whisk ferns, a primitive vascular plant
Division Lycophyta: Club mosses
Division Sphenophyta: Horsetails
Division Pterophyta: Ferns
Division Coniferophyta: Conifers, belong to the gymnosperms
Division Cycadophyta: Cycads
Division Ginkophyta: Ginkos
Division Gnetophyta: a diverse group of gymnosperms
Division Anthophyta: Angiosperms, the flowering plants
    Class Monocotyledons: (monocots) one seedling leaf
    Class Dicotyledons: (dicots) two seedling leaves

## KINGDOM ANIMALIA

Heterotrophic, multicellular organisms that ingest their food.

Phylum Porifera: Sponges
Phylum Cnidaria: Corals, jellyfish, hydras. Radial symmetry
    Class Hydrozoa: Hydra
    Class Scyphozoa: Jellyfish
    Class Anthozoa: Corals, anemones
Phylum Ctenophera: Comb jellies
Phylum Platyhelminthes: Flatworms. Bilateral symmetry
    Class Turbellaria: Free-living flatworms
    Class Trematoda: Flukes, intestinal parasites
    Class Cestoda: Tapeworms
Phylum Nematoda: Roundworms
Phylum Rotifera: Rotifers
Phylum Bryozoa: Bryozoans

Phylum Brachiopoda: Brachiopods (clam-like organisms)

Phylum Mollusca: Bilaterally symmetrical, protostome coelomates

 Class Polyplacophora: Chitons

 Class Gastropoda: Snails

 Class Bivalvia: Clams, oysters, scallops, mussels

 Class Cephalopoda: Octopus, squid, nautilus

Phylum Annelida: Segmented bilaterally symmetrical protostomes

 Class Polychaeta: Polychetes, some other marine worms

 Class Oligochaeta: Earthworms

 Class Hirudinea: Leeches

Phylum Arthropoda: Largest phylum of animals

 Subphylum Cheliocrata

 Class Arachnida: Spiders, mites, scorpions, ticks

 Class Merostomata: Horseshoe crabs

 Class Pycnogonida: Sea spiders

 Subphylum Crustacea

 Class Crustacea: Shrimp, lobsters, crayfish, crabs

 Subphylum Unirama

 Class Chilopoda: Centipedes

 Class Millipeda: Millipedes

 Class Insecta: Insects. The largest group of organisms

Phylum Echinodermata: Deuterostome coelomates, most radially symmetrical

 Class Crinoidea: Sea lilies, crinoids

 Class Asteroidea: Sea stars

 Class Ophiuroidea: Brittle stars

 Class Echinoidea: Sea urchins, sand dollars

 Class Holothuroidea: Sea cucumbers

Phylum Chordata: Animals that have a notochord at some stage of their life

 Subphylum Urochordata: Tunicates, sea squirts

 Subphylum Cephalochordata: Lancelets

 Subphylum Vertebrata: Notochord replaced with cartilage or bone

 Class Agnatha: Hagfish and lampreys

 Class Chondrichthyes: Cartilaginous fish (sharks, rays, skates)

 Class Osteichthyes: Bony fish (trout, perch, pike)

 Class Amphibia: Salamanders, frogs, toads

 Class Reptilia: lizards, snakes, crocodiles, turtles

 Class Aves: Birds

 Class Mammalia: Mammals

  Subclass Prototheria: Egg laying mammals

  Subclass Metatheria: Marsupials

  Subclass Eutheria: Placental mammals

   Order Primates: Lemurs, monkeys, apes, humans

    Superfamily Hominoidea

     Family Hylobatidae: Gibbons, orangutans, chimpanzees

     Family Pongidae: Gorillas

     Family Hominidae: Humans

# APPENDIX B
# PERIODIC TABLE OF ELEMENTS

| | Group IA | Group IIA | Group IIIA | Group IVA | Group VA | Group VIA | Group VIIA | Group VIII | | | Group IB | Group IIB | Group IIIB | Group IVB | Group VB | Group VIB | Group VIIB | Group 0 |
|---|---|---|---|---|---|---|---|---|---|---|---|---|---|---|---|---|---|---|
| Period 1 | 1 H 1.008 | | | | | | | | | | | | | | | | | 2 He 4.003 |
| Period 2 | 3 Li 6.941 | 4 Be 9.012 | | | | | | | | | | | 5 B 10.81 | 6 C 12.01 | 7 N 14.01 | 8 O 16.00 | 9 F 19.00 | 10 Ne 20.18 |
| Period 3 | 11 Na 22.99 | 12 Mg 24.31 | | | | | | | | | | | 13 Al 26.98 | 14 Si 28.09 | 15 P 30.97 | 16 S 32.06 | 17 Cl 35.45 | 18 Ar 39.95 |
| Period 4 | 19 K 39.10 | 20 Ca 40.08 | 21 Sc 44.96 | 22 Ti 47.90 | 23 V 50.94 | 24 Cr 52.00 | 25 Mn 54.94 | 26 Fe 55.85 | 27 Co 58.93 | 28 Ni 58.70 | 29 Cu 63.55 | 30 Zn 65.38 | 31 Ga 69.72 | 32 Ge 72.59 | 33 As 74.92 | 34 Se 78.96 | 35 Br 79.90 | 36 Kr 83.80 |
| Period 5 | 37 Rb 85.47 | 38 Sr 87.62 | 39 Y 88.91 | 40 Zr 91.22 | 41 Nb 92.91 | 42 Mo 95.94 | 43 Tc (98) | 44 Ru 101.1 | 45 Rh 102.9 | 46 Pd 106.4 | 47 Ag 107.9 | 48 Cd 112.4 | 49 In 114.8 | 50 Sn 118.7 | 51 Sb 121.8 | 52 Te 127.6 | 53 I 126.9 | 54 Xe 131.3 |
| Period 6 | 55 Cs 132.9 | 56 Ba 137.3 | 57 La 138.9 | 72 Hf 178.5 | 73 Ta 180.9 | 74 W 183.9 | 75 Re 186.2 | 76 Os 190.2 | 77 Ir 192.2 | 78 Pt 195.1 | 79 Au 197.0 | 80 Hg 200.6 | 81 Tl 204.4 | 82 Pb 207.2 | 83 Bi 209.0 | 84 Po (209) | 85 At (210) | 86 Rn (222) |
| Period 7 | 87 Fr (223) | 88 Ra (226.0) | 89 Ac (227) | 104 Unq | 105 Unp | 106 Unh | 107 Uns | | 109 Une | | | | | | | | | |

### Lanthanides (rare earth metals)

| 58 Ce 140.1 | 59 Pr 140.9 | 60 Nd 144.2 | 61 Pm (145) | 62 Sm 150.4 | 63 Eu 152.0 | 64 Gd 157.3 | 65 Tb 158.9 | 66 Dy 162.5 | 67 Ho 164.9 | 68 Er 167.3 | 69 Tm 168.9 | 70 Yb 173.0 | 71 Lu 175.0 |
|---|---|---|---|---|---|---|---|---|---|---|---|---|---|

### Actinides

| 90 Th 232.0 | 91 Pa (231) | 92 U 238.0 | 93 Np (244) | 94 Pu (242) | 95 Am (243) | 96 Cm (247) | 97 Bk (247) | 98 Cf (251) | 99 Es (252) | 100 Fm (257) | 101 Md (258) | 102 No (259) | 103 Lr (260) |
|---|---|---|---|---|---|---|---|---|---|---|---|---|---|

KEY

16 —— Atomic number
S —— Symbol of element
32.06 —— Atomic mass

Metals

Nonmetals

Metalloids

Noble Gases

# TABLE OF ATOMIC WEIGHTS AND NUMBERS

| NAME | SYMBOL | ATOMIC NUMBER | ATOMIC WEIGHT | NAME | SYMBOL | ATOMIC NUMBER | ATOMIC WEIGHT |
|------|--------|---------------|---------------|------|--------|---------------|---------------|
| Actinium | Ac | 89 | (227) | Neon | Ne | 10 | 20.18 |
| Aluminum | Al | 13 | 26.98 | Neptunium | Np | 93 | 237.0 |
| Americium | Am | 95 | (243) | Nickel | Ni | 28 | 58.71 |
| Antimony | Sb | 51 | 121.8 | Niobium | Nb | 41 | 92.91 |
| Argon | Ar | 18 | 39.95 | Nitrogen | N | 7 | 14.01 |
| Arsenic | As | 33 | 74.92 | Nobelium | No | 102 | (255) |
| Astatine | At | 85 | (210) | Osmium | Os | 76 | 190.2 |
| Barium | Ba | 56 | 137.3 | Oxygen | O | 8 | 16.00 |
| Berkelium | Bk | 97 | (247) | Palladium | Pd | 46 | 106.4 |
| Beryllium | Be | 4 | 9.012 | Phosphorus | P | 15 | 30.97 |
| Bismuth | Bi | 83 | 209.0 | Platinum | Pt | 78 | 195.1 |
| Boron | B | 5 | 10.81 | Plutonium | Pu | 94 | (244) |
| Bromine | Br | 35 | 79.90 | Polonium | Po | 84 | (210) |
| Cadmium | Cd | 48 | 112.4 | Potassium | K | 19 | 39.10 |
| Calcium | Ca | 20 | 40.08 | Praseodymium | Pr | 59 | 140.9 |
| Californium | Cf | 98 | (251) | Promethium | Pm | 61 | (147) |
| Carbon | C | 6 | 12.01 | Protactinium | Pa | 91 | 231.0 |
| Cerium | Ce | 58 | 140.1 | Radium | Ra | 88 | 226.0 |
| Cesium | Cs | 55 | 132.9 | Radon | Rn | 86 | (222) |
| Chlorine | Cl | 17 | 35.45 | Rhenium | Re | 75 | 186.2 |
| Chromium | Cr | 24 | 52.00 | Rhodium | Rh | 45 | 102.9 |
| Cobalt | Co | 27 | 58.93 | Rubidium | Rb | 37 | 85.47 |
| Copper | Cu | 29 | 63.55 | Ruthenium | Ru | 44 | 101.1 |
| Curium | Cm | 96 | (247) | Samarium | Sm | 62 | 150.4 |
| Dysprosium | Dy | 66 | 162.5 | Scandium | Sc | 21 | 44.96 |
| Einsteinium | Es | 99 | (254) | Selenium | Se | 34 | 78.96 |
| Erbium | Er | 68 | 167.3 | Silicon | Si | 14 | 28.07 |
| Europium | Eu | 63 | 152.0 | Silver | Ag | 47 | 107.9 |
| Fermium | Fm | 100 | (257) | Sodium | Na | 11 | 22.99 |
| Fluorine | F | 9 | 19.00 | Strontium | Sr | 38 | 87.62 |
| Francium | Fr | 87 | (223) | Sulfur | S | 16 | 32.06 |
| Gadolinium | Gd | 64 | 157.3 | Tantalum | Ta | 73 | 180.9 |
| Gallium | Ga | 31 | 69.72 | Technetium | Tc | 43 | 98.91 |
| Germanium | Ge | 32 | 72.59 | Tellurium | Te | 52 | 127.6 |
| Gold | Au | 79 | 197.0 | Terbium | Tb | 65 | 158.9 |
| Hafnium | Hf | 72 | 178.5 | Thallium | Tl | 81 | 204.4 |
| Helium | He | 2 | 4.003 | Thorium | Th | 90 | 232.0 |
| Holmium | Ho | 67 | 164.9 | Thulium | Tm | 69 | 168.9 |
| Hydrogen | H | 1 | 1.008 | Tin | Sn | 50 | 118.7 |
| Indium | In | 49 | 114.8 | Titanium | Ti | 22 | 47.90 |
| Iodine | I | 53 | 126.9 | Tungsten | W | 74 | 183.9 |
| Iridium | Ir | 77 | 192.2 | Unnilennium | Une | 109 | (266) |
| Iron | Fe | 26 | 55.85 | Unnilhexium | Unh | 106 | (263) |
| Krypton | Kr | 36 | 83.80 | Unniloctium | Uno | 108 | (265) |
| Lanthanum | La | 57 | 138.9 | Unnilpentium | Unp | 105 | (262) |
| Lawrencium | Lr | 103 | (256) | Unnilquadium | Unq | 104 | (261) |
| Lead | Pb | 82 | 207.2 | Unnilseptium | Uns | 107 | (262) |
| Lithium | Li | 3 | 6.941 | Uranium | U | 92 | 238.0 |
| Lutetium | Lu | 71 | 175.0 | Vanadium | V | 23 | 50.94 |
| Magnesium | Mg | 12 | 24.31 | Xenon | Xe | 54 | 131.3 |
| Manganese | Mn | 25 | 54.94 | Ytterbium | Yb | 70 | 173.0 |
| Mendelevium | Md | 101 | (258) | Yttrium | Y | 39 | 88.91 |
| Mercury | Hg | 80 | 200.6 | Zinc | Zn | 30 | 65.37 |
| Molybdenum | Mo | 42 | 95.94 | Zirconium | Zr | 40 | 91.22 |
| Neodymium | Nd | 60 | 144.2 | | | | |

*A value in parentheses is the mass number of the isotope of longest half-life.*

# THE METRIC SYSTEM

In the United States, the metric system is a system of measurement used principally by scientists. In our day-to-day lives, though, most Americans use the English system of measurement—miles, inches, feet, pounds, tons, and so on.

Many countries, such as New Zealand, use the metric system for weights and other measures. If you travel abroad, road signs will list distance in kilometers rather than miles. Other linear measurements will be given in meters and centimeters instead of yards, feet, and inches. The weight of objects will be expressed in kilograms or grams instead of pounds and ounces. Liquids will be measured in liters rather than quarts or gallons.

It can be confusing if you don't know how to convert from one system of weights and measures to another. The lists below show some of the most common units you will encounter in biology and other sciences and compare them to their English equivalents.

### ⋊⋉ MOST COMMON ENGLISH UNITS AND THE CORRESPONDING METRIC UNITS ⋊⋉

|  | ENGLISH UNIT | METRIC UNIT |
| --- | --- | --- |
| Weight | tons | metric tons |
|  | pounds | kilograms |
|  | ounces | grams |
| Length | miles | kilometers |
|  | yards | meters |
|  | inches | centimeters |
| Square Measure | acres | hectares |
|  | square miles | square kilometers |
| Volume | quarts and gallons | liters |
|  | fluid ounces | milliliters |

### ⋊⋉ CONVERTING ENGLISH UNITS TO METRIC UNITS ⋊⋉

|  | ENGLISH UNIT | METRIC UNIT |
| --- | --- | --- |
| Weight | 1 ton (2000 pounds) | = 0.9 metric tons |
|  | 1 pound | = 0.454 kilograms |
|  | 1 ounce | = 28.35 grams |
| Length | *1 mile | = 1.6 kilometers |
|  | 1 yard | = 0.9 meters |
|  | *1 inch | = 2.54 centimeters |
| Square Measure | 1 acre | = 0.4 hectares |
|  | 1 square mile | = 2.59 square kilometers |
| Volume | 1 quart | = 0.95 liters |
|  | *1 gallon | = 3.78 liters |
|  | 1 fluid ounce | = 29.58 milliliters |

*Most useful conversions to know.*

—See the next page for more Metric System tables

## CONVERTING METRIC UNITS TO ENGLISH UNITS

|  | METRIC UNIT | ENGLISH UNIT |
|---|---|---|
| Weight | *1 metric ton | = 2204 pounds |
|  | 1 metric ton | = 1.1 tons |
|  | *1 kilogram | = 2.2 pounds |
|  | 1 gram | = 28.35 ounces |
| Length | *1 kilometer | = .6 miles |
|  | 1 meter | = 1.1 yards |
|  | 1 centimeter | = .39 inches |
| Square Measure | *1 hectare | = 2.47 acres |
|  | 1 square kilometer | = 0.386 square miles |
| Volume | 1 liter | = 1.057 quarts |
|  | 1 liter | = .26 gallons |
|  | 1 milliliter | = 0.0338 fluid ounces |

*Most useful conversions to know.*

# APPENDIX D
# ANSWERS TO QUESTIONS AND PROBLEMS

## CHAPTER 2

### Multiple Choice

1. (b) elements
2. (a) atom
3. (d) protons . . . neutrons
4. (c) two electrons from carbon atom are shared with the oxygen atom and two electrons from oxygen atom are shared with carbon.
5. (d) acids . . . bases

### True/False

1. True.
2. True.
3. False, new bonds and molecules are formed during anabolic reactions.
4. False, minimum amount of energy needed to disrupt chemical bonds in reactions.
5. False, several factors can change the chemical equilibrium in a reaction.
6. False, all enzymes are proteins, but not all proteins are enzymes.

### Matching

1. ATP — energy storage molecule
2. buffers — pH regulators
3. fatty acids — simplest lipid
4. DNA — genetic information storage molecule
5. monosaccharide — simplest carbohydrate

## CHAPTER 3

### True/False

1. True.
2. False, Pasteur and others provided evidence that replaced ideas of abiogenesis and provided the foundation for modern biology.
3. False, two subdivisions of life: prokaryotic and eukaryotic.
4. True.
5. True.
6. False, the number of chromosomes present in an organism is characteristic for each species.
7. True.

### Matching

1. nucleolus — ribosomal subunit assembly
2. ribosome — protein synthesis
3. endoplasmic reticulum — protein modification and lipid synthesis
4. lysosome — intracellular digestion
5. mitochondria — "powerhouse" of cell
6. cilia — movement
7. nucleus — organization of DNA

### Short Answer

1. Animal cells do not contain chloroplasts but plant cells do contain mitochondria. Chloroplasts are the site of photosynthesis in plants.

This process converts solar energy in the form of ATP into food such as glucose for the plant. Animal cells undergo cellular respiration within mitochondria in order to convert energy into ATP molecules, carbon dioxide, and water.

2. (b) fluid-mosaic model
3. (1) All living things are composed of at least one cell and that cell is the fundamental unit of function in all organisms. (2) All cells are fundamentally alike in their chemical composition. (3) All cells arise from preexisting cells through a process of cell division.
4. (d) small cells are able to provide for an adequate rate of exchange with the environment.
5. *Prokaryotes*
   smaller, less complex
   lack membrane-bound nucleus
   DNA in direct contact with cytoplasm
   lack organelles

   *Eukaryotes*
   most highly complex
   membrane-bound nucleus
   DNA packaged in nucleus
   organelles
6. (d) Proteins are synthesized in the endoplasmic reticulum (ER). The rough ER is the site of synthesis of lysosomal enzymes and proteins for extracellular use. The smooth ER is involved in production of phospholipids and various other functions in different cells. Protein chains travel by vesicles to the Golgi complex where they are sorted, chemically modified, and packaged. Vesicles again carry the modified packaged proteins to the plasma membrane where the membrane of the vesicle fuses with the plasma membrane and the fused membranes dissolve away, releasing the contents of the secretory vesicle into the extracellular space.

## CHAPTER 4

### Short Answer

1. The first law of thermodynamics states that work and heat are both forms of energy and that energy can be changed from one form to another, but never created or destroyed. The second law states that all objects in the universe tend to become progressively more disordered.
2. Inside.
3. The conversion of light into chemical energy (light reaction) is accompanied by harvesting energy from sunlight and transferring this energy into chemical bonds of ATP. This reaction can only take place in the light. The conversion of chemical energy into food (dark reaction) occurs by a series of chemical reactions that use ATP from the light reaction to form organic molecules from $CO_2$. This reaction can take place in the dark providing ATP is present.
4. **proteins**———amino acids
   carbohydrates———**glycolosis**———energy + acetyl CoA———
   **Krebs cycle**
   Fats———glycerol + **fatty acids**———Energy
5. Enzymes work by lowering the energy of activation, allowing a reaction to start with a lower energy input. Reactions with lower activation energies are more likely to proceed; catalysis results in more rapid reaction rates.
6. Endergonic.
7. Ecosystems in the depths of the sea rely on vents on the seafloor, where the water temperature can reach 350°C and is rich in miner-

als for the generation of energy. These vents release large amounts of hydrogen sulfide and water. Bacteria surrounding these vents use the sulfur in hydrogen sulfide and combine it with oxygen to generate energy.

8. (1) Glycolysis, (2) the Krebs cycle (or tricarboxylic acid cycle), and (3) electron transport and ATP formation.

9. Water.

10. The proteins of the mitochondria electron transfer chain pump protons from the inner compartment into the outer compartment. The accumulation of protons in outer mitochondrial compartment results in their outflow into the inner compartment through inner membrane channels. As they do, they pass enzymes that use energy from proton movement to add phosphate to ADP, creating ATP. Photosynthesis works in a similar way. As the concentration of protons increases in the thylakoid, they begin to flow outward into the chloroplasts and, as they do so, this proton movement is used to convert ADP into ATP.

## True/False

1. True.
2. True.
3. True.
4. False, as disorder increases, entropy increases.
5. False: 

| *TCA Cycle* | *Glycolosis* |
|---|---|
| ATP—2 molecules | ATP—2 molecules |
| NADH—6 molecules | NADH—2 molecules |
| FADH$_2$—2 molecules | pyruvate—2 molecules |

6. False, prokaryotes lack mitochondria. Electron transport systems are located in plasma membrane of cell.

## Matching

1. (a) respiration
   (b) glycolysis
   (c) TCA cycle
   (d) photosynthesis

2. glycolysis — cytoplasm
   TCA cycle — mitochondria
   electron transport and ATP formation — mitochondria
   fat digestion — intestines
   lactate fermentation — skeletal muscle
   enzyme — substrate

## CHAPTER 5

### True/False

1. True.
2. False, a cell that contains a single basic complement is haploid (*n*).
3. True, this type of life cycle is known as alteration of generations.
4. False, most organisms pass through the cell cycle continuously and divide on a regular basis. However, some cell types become permanently arrested at G1 causing them to never undergo division. Other cells can become arrested in G1 or G2 but can divide under particular cases.
5. True.

### Multiple Choice

1. (d) homologous chromosomes
2. (c) binary fission
3. (d) identical to parent cell; same
4. (a) 46
5. (c) meiosis

## Short Answer

1. Asexual reproduction results in progeny genetically identical to the parent, whereas sexual reproduction produces offspring that are genetically different from the parent. Sexual reproduction generates species diversity. When environmental conditions change, this genetic diversity helps ensure that some members of the population will have combinations of genetic traits that enable their survival. Asexual reproduction is useful when environmental conditions are constant, which allows a species to produce offspring rapidly that are extremely well adapted to that environment.

2. Chromatin is the strand of DNA and associated histone protein. Chromosome is the structure consisting of one or two chromatin fibers. Chromatid refers to one of the chromatin fibers of a replicated chromosome. Centromere is the region of each chromatid to which a sister chromatid attaches.

3. The cell cycle is divided into two distinct phases: cell division and interphase. Interphase is a time of intense metabolic activity and is divided into three parts: G1, S, and G2. G1 (gap 1) begins immediately after mitosis. During this stage RNA, protein, and other molecules are synthesized. The S (synthesis) stage is where DNA replication occurs. During G2 (gap 2) mitochondria divide and the precursors of spindle fibers are synthesized. The mitotic phase of the cell cycle is divided into five distinct steps that result in distinct morphological changes in the nucleus. Prophase is characterized by chromosome condensation, nuclear envelope disintegration, centriole division and migration to opposite poles of the dividing cell, and spindle fiber formation and attachment to chromosomes. The chromosomes line up on the equatorial plate during metaphase. Chromosome separation signals the anaphase stage. Telophase is signified by chromosome migration to opposite poles, the newly formed nuclear envelope, and the uncoiling of chromosomes. Cytokinesis or cytoplasmic division is a separate process that occurs at the end of mitosis. The cleavage furrow forms and deepens, resulting in cytoplasmic division.

4. In plants, although spindle fibers are formed from microtubules, no centrioles are present. Cytokinesis is different in plant cells because they are surrounded by cell walls outside their plasma membrane. The division of cytoplasm begins with the synthesis of vesicles in the Golgi complex that migrate to the cell equator. These aligned vesicles fuse to produce a membrane-bound space called the cell plate. This plate grows outward fusing with the plasma membrane of the cell, dividing the cell in half. A new middle lamella forms in the space between the two cells, and the new cell wall material is laid down by each daughter cell.

5. chiasma

6. prophase I

## CHAPTER 6

### True/False

1. True.
2. False, genotype is the genetic makeup of an organism that determines its outward appearance or phenotype.
3. False, alleles are alternate forms of a gene.
4. False, Mendel was the first to describe independent assortment and segregation.
5. True.

### Matching

1. Darwin — gemmules
2. P$_1$ — parental generation

3. Mendel      pea plant
4. ss      homozygous state
5. $F_1$      first filial generation
6. AaBb × AaBb      dihybrid cross
7. Ss      heterozygous state
8. AA × aa      monohybrid cross

## Short Answer

1. (1) An experimental organism should have a number of different traits that can be studied; (2) the plant should be self-fertilizing and have a flower structure that minimizes accidental contamination with foreign pollen; and (3) the offspring of self-fertilized plants should be fully fertile so that further crosses can be made.

2. The law of allele segregation: the factors specifying a pair of alternative attributes (alleles) are separate, and only one may be carried in a particular gamete; gametes combine randomly in forming offspring. Chromosomes had not been observed in Mendel's time and meiosis had not been described. Modern form of this law is that alleles segregate as chromosomes do.

3.

|   | F | f |
|---|---|---|
| f | Ff | ff |

approximately 50% of all children *will* have freckles
approximately 50% of all children *will not* have freckles

4.

|   (father)   |   | A | a | (mother) |
|---|---|---|---|---|
|   | A | AA | Aa |   |
|   | a | Aa | aa (albino offspring) |   |

5.

|   | FA | Fa | fA | fa |
|---|---|---|---|---|
| FA | FFAA* | FFAa* | FfAA* | FfAa* |
| Fa | FFAa* | FFaa$^x$ | FfAa* | Ffaa$^x$ |
| fA | FfAA* | FfAa* | ffAA$^\#$ | ffAa$^\#$ |
| fa | FfAa* | Ffaa$^x$ | ffAa$^\#$ | ffaa$^-$ |

F = freckles    A = attached

9   * = freckled and attached
3   x = freckled and unattached
3   # = no freckles and attached
1   − = no freckles and unattached

6. a. mother—FfRr     F = freckles
    father—ffrr     R = roll tongue
    first child—ffrr

   b.

|   | FR | Fr | fR | fr |
|---|---|---|---|---|
| fr | FfRr | Ffrr | ffRr | ffrr |

Ffrr = have freckles and cannot roll tongue ($\frac{1}{4}$ or 25% chance)

## CHAPTER 7

### Matching

1. dominant     allele always expressed in heterozygote
2. incomplete dominance     two alleles, neither one dominant over other
3. codominance     each allele in heterozygote is separately expressed
4. discontinuous variation     phenotypic variation in a genetic trait that fits into distinct categories
5. cytology     study of cell structure
6. zygote     fertilized egg
7. linkage     situation of genes located on same chromosome stays together during meiosis and end up together in same gamete
8. autosomes     chromosomes 1 through 22
9. continuous variation     small degree of phenotypic variation that occurs over a range

10. nondisjunction     failure of chromosomes to separate during division

1. Klinefelter syndrome     47, XXY
2. intelligence     polygenic inheritance
3. Down's syndrome     aneuploidy
4. Leber's optic atrophy     mitochondrial disorder
5. Turner syndrome     XO
6. Hemophilia A     X-linked
7. Huntington's disease     autosomal dominant

### True/False

1. False, aneuploidy is a change in chromosome number that results in less than the complete set of chromosomes.
2. True.
3. False, as the number of classes increases, there is less and less phenotypic difference between classes.
4. True.
5. False, the male phenotype requires the presence of a Y chromosome but there can be more than one X chromosome.

### Short Answer

1. (1) Traits are usually quantified by measurement rather than by counting. (2) Two or more gene pairs contribute to the phenotype. (3) The phenotypic expression of polygenic traits varies within a wide range. This variation can be produced in a number of ways and is best studied in populations rather than individuals.

2. The high level of variation in phenotypic expression cannot be explained by lack of stability, but by interaction between different genes and between genes and the environment.

3. There must be present on the X chromosome genes that control other necessary life functions besides the ones controlling sex determination. This genetic information is vital for survival regardless of one's sex. This must not be the case for the Y chromosome due to the fact that females survive in its absence. The Y chromosome's major role via the testis-determining factor (TDF) gene is to initiate the development of male phenotype during embryonic development from what would have been a female.

4. The gene for hemophilia A is located on the X and not the Y chromosome. A male has only one X chromosome. Therefore, it would be impossible for a male to be a carrier; the allele for hemophilia A would always be expressed.

## CHAPTER 8

### Fill in the Blank

1. enzyme
2. one gene, one enzyme
3. polynucleotide strands; hydrogen; purine
4. covalent; phosphate; deoxyribose
5. messenger RNA, ribosomal RNA, transfer RNA; protein
6. uracil; thymine
7. transcription; translation
8. anticodon; codon
9. RNA polymerase; promoter
10. histones; nucleosomes
11. Semiconservative replication
12. mutations, gene product

## Short Answer

1. In DNA replication, the two strands unwind with the aid of a special enzyme. After unwinding, the strands serve as templates for the production of complementary DNA strands, a process aided by the enzyme DNA polymerase. After chromosomal replication, each chromosome consists of two chromatids joined by a common centromere. Each chromatid contains a DNA molecule composed of one old strand and one newly synthesized strand of DNA. Before a cell can divide, it must first make an exact copy of all of its DNA. This process occurs during the S phase of mitosis and ensures that a cell about to divide has two sets of genetic material, one for each daughter cell.

2. Messenger RNA (mRNA) molecules carry genetic information that is transferred to it from the DNA molecule to the cytoplasm. Transfer RNA (tRNA) molecules deliver amino acid molecules to the mRNA and insert them in the growing polypeptide chain. Ribosomal RNA (rRNA) is produced in the nucleolus located in the nucleus. It combines with proteins to form the ribosome. Ribosomes are required to produce proteins on the mRNA template.

3. tyr-arg-thr-ser-gly-stop-asp

4. *DNA*
   Double-stranded
   Contains sugar deoxyribose
   Contains adenine, guanine, cytosine, and thymine
   Functions primarily in nucleus

   *RNA*
   Single-stranded
   Contains sugar ribose
   Contains adenine, guanine, cytosine, and uracil
   Functions primarily in cytoplasm

5. AUGCUGGACUAC

6. Six

7. Cytoplasm

8. $20^{15}$, $(20^n)$ where "$n$" is the number of amino acids in the protein and 20 is the number of types of amino acids.


## CHAPTER 9

### Short Answer

1. Recombinant DNA technology (genetic engineering) is a means by which scientists remove segments of DNA from one organism and insert them into the DNA of another organism. The advantages include (1) gene therapy for human genetic disorders, (2) the transfer of genes into plants and animals to increase disease resistance, stimulate growth, or induce other desirable traits, and (3) the mass production of commercially important chemicals, such as insulin. Potential disadvantages include (1) transgenic plants and animals may experience adverse side effects, (2) genetically altered bacteria or viruses may be unleashed, and (3) future health effects on humans.

2. Genetic engineering and selective breeding frequently achieve the same end points, however, the two techniques are quite different. Selective breeding alters gene frequencies. Genetic engineering often seeks to introduce new genes into species, producing new alleles (new genetic combinations).

3. (1) Restriction enzymes cut DNA at specific base sequences. (2) DNA fragments are linked to vectors that are self-replicating to create recombinant DNA molecules. (3) The recombinant DNA molecule is replicated in host organism to produce large amounts of clones on the DNA inserted fragment. (4) Cloned DNA segments are retrieved in large quantities. (5) Gene product encoded by cloned DNA segment is produced and purified.

4. DNA polymerase is used to make copies of a specific region of DNA.

The DNA to be amplified is selected through the use of short DNA fragments (primers) that bind to either side of the region to be amplified. These primers guide the DNA polymerase to copy only the region between the primers. The first round of copies is then used to make more copies, so that the number of copies doubles with each round of amplification.

5. (1) The ability of restriction enzymes to cut DNA at specific base sequences and (2) the existence of large, multigenerational families in which a genetic disorder is expressed.

6. (1) The development of high-resolution genetic maps of each of the 23 human chromosomes. (2) The construction of physical maps of each chromosome that show the actual location and distance between markers. (3) The DNA from each chromosome will be cloned. (4) Nucleotide sequences of each chromosome will be determined and used to construct a catalog of the more than three billion base pairs in the human genome.

7. (1) Individuals afflicted with hemophilia A are unable to manufacture a clotting factor. Pooled blood samples can be used to extract this necessary blood product for hemophiliacs. However, many of these individuals have been infected with HIV because they received blood products prior to HIV testing in blood components. Recombinant DNA techniques can be implemented to manufacture the missing clotting factor in hemophiliacs and at the same time ensure that it is free from contamination.

   (2) The late-onset progressive genetic disorder called Huntington's disease (HD) is also aided by recombinant DNA techniques. This condition can be diagnosed in individuals prior to the onset of symptoms providing there is a family history and other family members are willing to participate. Presymptomatic testing for this condition raises several social and ethical issues.

   (3) In the autosomal recessive condition of sickle cell anemia (SCA), recombinant DNA techniques can be used to determine the genotypes of normal homozygous, heterozygous, and affected homozygous individuals. The mutation responsible for SCA destroys a restriction enzyme cutting site, resulting in a change in the DNA fragment length generated by the enzyme. By analyzing the restriction fragments in the DNA, scientists can determine an individual's genotype. However, because this condition can be quite variable, the phenotype for a particular individual is impossible to make.


### Fill in the Blank

1. plasmids
2. codominant; genetic markers
3. gene transfer
4. retrovirus; DNA
5. severe combined immune deficiency (SCID)
6. nuclei; genes


### Matching

| | |
|---|---|
| 1. restriction enzymes | DNA cutting enzymes |
| 2. cloning vehicles | vectors |
| 3. Huntington's disease | gene location on chromosome #4 |
| 4. amplification of DNA fragments | PCR |
| 5. BST | synthetic growth hormone |
| 6. heritable differences in DNA fragments | RFLP |
| 7. missing/defective protein in cystic fibrosis | CFTCR |

## CHAPTER 10

### Matching

| | | |
|---|---|---|
| 1. | mutation | heritable change in DNA |
| 2. | genetic drift | random fluctuations in allele frequency due to chance |
| 3. | gene flow | change in allele frequency due to immigration and/or emigration |
| 4. | natural selection | change or stabilization of allele frequency due to differential production of variant members of a population |
| 5. | recombination | exchange of genetic material |
| 6. | directional selection | shifts in allele frequency in response to the environment |
| 7. | Carolus Linnaeus | taxonomy |
| 8. | Comte de Buffon | advocated that each species was individually created by a Supreme Being |
| 9. | James Hutton | uniformitarianism |
| 10. | Alfred Russel Wallace | first to publish the role of natural selection in species formation |
| 11. | Greek philosophers | Enlightenment |
| 12. | Erasmus Darwin | speculated about evolution |
| 13. | Jean-Baptiste Lamarck | simple organisms give rise to complex organisms |
| 14. | Charles Darwin | *The Origin of Species* |
| 15. | Charles Lyell | linked geological time with uniformitarianism |

### Fill in the Blank

1. stabilizing
2. DDT; directional
3. disruptive
4. punctuated
5. Mendel; meiosis

### True/False

1. False, it is believed to be approximately 4.6 billion years old.
2. False, Wallace and Darwin proposed that the formation of a new species requires populations of organisms and long periods of time. Natural selection provided an explanation of how evolution occurs, but it did not explain the origin of variations in populations or how these variations were maintained. Once Mendelian genetics was applied to populations, the evolution of species diversity involving the genetic structure of populations was understood.
3. True.
4. True.
5. False, the basis of selection is the differential survival and reproduction of the most adapted individuals, not on survival of the fittest. This is true because these better adapted individuals survive and leave more offspring.

## CHAPTER 11

### Short Answer

1. Microevolution is a small-scale evolutionary event characterized by the formation of a species from a preexisting one or the divergence of separated populations into new species. Macroevolution is described when the origin, diversification, extinction, and interactions of organisms result in evolutionary changes.
2. During Darwin's time there was no means for establishing the age of fossils or the rocks in which they were encased. It was accepted that fossils in the lower rock layers were older, while those in higher levels were younger.

    (2) Geologists and paleontologists can date a fossil with a high level of accuracy by using radioactive decay. By measuring the amount of unstable element and its daughter element present in a rock or fossil and knowing the element's half-life, scientists can determine the fossil's age.
3. Carbon is the basic element of all living things and all cellular activities are driven by carbon-based compounds.

    (2) Carbon-14 has a relatively short half-life (5730 years plus or minus 30 years) allowing scientists to use this technique to date once living material. Determining the ratio of $C_{14}$ to $C_{12}$ provides archaeologists with the means to date more recent fossils.
4. DNA from one species can be heated so the two strands dissociate into single strands. These strands are combined with dissociated strands from another species and allowed to cool and form double-stranded hybrid DNA molecules. Closely related species have more complementary sequences in common, which would be represented by more stable double-stranded DNA molecules. The relatedness is calculated by measuring the amount of newly formed double-stranded molecules and their stability.
5. Skeletal support, protection from desiccation, and increased gravity.

### Matching

| | | |
|---|---|---|
| 1. | homologous structures | limbs of gophers and limbs of elephants |
| 2. | analogous structures | wings of birds and fins of penguins |
| 3. | morphological convergence | bird wings and insect wings |
| 4. | vestigial structures | appendix and dewclaw |
| 5. | adaptive radiation | Galapagos Island finches |
| 6. | convergent evolution | brachiopods and clams |
| 7. | Archean | no ozone layer, evolution of photosynthesis |
| 8. | Proterozoic | accumulation of oxygen, appearance of eukaryotic cells |
| 9. | Paleozoic | shifting continental masses |
| 10. | Cambrian | development of major animal groups |
| 11. | Mesozoic | age of the dinosaur |
| 12. | Cenozoic | encompasses the present |
| 13. | Devonian | rapid speciation of fish |

### Fill in the Blank

1. adapting; changing
2. fewer; more
3. chimpanzees; gorillas
4. biogeography
5. extinction
6. plate tectonics
7. ectotherms; endotherms

## CHAPTER 12

### Fill in the Blank

1. virus
2. Prions; viroids
3. methanogens, halophiles, and thermoacidophiles
4. saprophyte
5. locomotion
6. Phytoplankton
7. protists; plants and animals

## True/False

1. True.
2. False, according to the symbiotic theory, eukaryotes evolved in several steps.
3. False, reproduction is usually by binary fission after mitotic division of the nucleus, but many protozoans can also reproduce sexually following meiosis.
4. False, many ciliates have barbed, thread-like organelles called trichocysts that are used for defense and capturing prey.
5. True.

## Matching

1. viruses — intracellular parasites
2. autotrophs — synthesize organic molecules
3. heterotrophs — require organic molecules
4. methanogens — involved in recycling carbon
5. halophiles — found in salty environments
6. thermoacidophiles — found in hot, acidic environments
7. cyanobacteria — chlorophyll *a*
8. symbiosis — interactive relationship
9. slime molds — resemble protists and fungi

## Short Answer

1. Bacteria reproduce by a process called binary fission. Under the appropriate conditions, bacterial cells enlarge, replicate their genetic information, and split into two identical cells. If conditions are not suitable, some bacteria may form spores and remain encapsulated until conditions improve.
2. Bacteria are classified according to the sequence of nucleotides in their DNA. This method attempts to base the classification on evolutionary relationships.
3. First stage involves the formation of a symbiotic relationship between two or more prokaryotes. After the formation of a mutualistic relationship, the development of a progressive interdependence over many generations occurred until the symbionts could not survive without their partner. This transition may have involved the loss of duplicate genes in one partner and/or the movement of genes between partners. Such symbiotic events may have arisen more than once, resulting in mitochondria and chloroplasts as independent events. The end product is a eukaryotic cell.

## CHAPTER 13

### True/False

1. True.
2. True.
3. False, sac fungi are important in decomposing and recycling organic molecules from dead and decaying organisms. They break down these materials and return the substances locked in those molecules to circulation in the ecosystem.
4. False, the plant kingdom is highly diverse, however, all plant cells contain organelles called plastids and are surrounded by cell walls containing cellulose.
5. False, the three divisions of algae are based on the idea that they arose independently at several different times from protist ancestors.
6. False, mosses are the most common and familiar bryophyte.
7. True.

### Matching

1. fungi — absorptive heterotrophs
2. mycorrhiza — "fungus root"
3. brown algae — fucoxanthin
4. red algae — phycoerythrin
5. green algae — chlorophyll *a* and *b*
6. bryophytes — rhizoids
7. gymnosperm — seed plants
8. angiosperm — flowering plants
9. nectaries — secreting organs that serve as insect feeding stations
10. xylem — vascular transport system moving nutrients from roots to leaves
11. phloem — vascular transport system moving food through plant
12. stomata — pores on leaves that prevent water loss
13. holdfast — anchoring system for *Fucus*
14. basidiospores — reproductive structures

### Short Answer

1. The hyphae of zygomycetes are multinucleate, with septa only where gametogonia or sporangia are cut off. The ascomycetes or sac fungi are composed of densely interwoven hyphae. Basidiomycetes or club fungi have an inconspicuous mycelial mass that forms a large reproductive structure.
2. Plant evolution occurred in distinct stages, with each stage representing an adaptation that better suited plants for survival on land. These include protection from desiccation, leaf development for increased photosynthesis efficiency, root development for nutrient retrieval from the soil, stems to elevate photosynthetic parts above other plants competing for sunlight, a unique vascular transport system to move nutrients from soil through the plant, and reproductive phases of alteration of generations.
3. Variability in body form, photosynthetic pigments, and type of food storage molecules are used to classify red, brown, and green algae.
4. The development of the root-leaf-vascular system axis, reduction in the size of the gametophyte generation, and seeds were the three evolutionary advances in vascular plants.

## CHAPTER 14

### Fill in the Blank

1. heterotrophs; autotrophs
2. asymmetrical; radially; bilaterally
3. epidermal; collar
4. polyp; medusa
5. annelids
6. deuterostomes
7. cartilage
8. fluid-filled
9. endothermy
10. placenta

### Short Answer

1. The tube-within-a-tube body plan is a "complete" digestive system with regional specializations and an opening at each end. One opening is for food and the other for digestive waste elimination. This body plan consists of a specialized digestive system and is about 10% more efficient at recovering nutrients from food than in a sac-like body plan.
2. Acoelomate organisms lack a fluid-filled cavity between the gut and body wall. Pseudocoelomates have a body cavity, but it does not arise by splitting of the mesoderm. Coelomate animals have a body cavity lined with mesoderm.

3. (1) Hard exoskeleton
   (2) Presence of jointed appendages
   (3) Complex nervous system
   (4) Unique respiratory system
   (5) Complex and adaptable life cycle.
4. During the life cycle of chordates, three characteristics will be present at some point. A notochord, a hollow nerve cord, and gill slits or pouches that may be present only in early development or retained in the adult.
5. The evolution of jaws is an example of existing structures being modified for new functions. Jaws evolved from the first three gills of jawless fish. These arches became modified into supports for jaws. The combination of bones and muscles in the lobe fin of fish is another structure modified for a new function, this time as legs in the amphibians.
6. The success of the reptiles was due to their advantage in reproduction, development of more efficient jaws, teeth for seizing and biting prey, and dry, rough skin that resists desiccation.
7. The presence of hair and mammary glands for nourishing young distinguishes mammals from other vertebrates.
8. Primates are somewhat less specialized than other mammals but are considered the most advanced mammals based on the development of the opposable thumb and the enlargement of the cerebrum of the brain, which gives humans the ability to reason.

## Matching

1. sponges                filter feeders
2. arthropods             joint-footed
3. echinoderms            deuterostomes
4. Agnatha                jawless fish
5. crossopterygians       ancestral to amphibians
6. crocodiles             related to birds
7. monotremes             mammals that lay eggs

## CHAPTER 15

### Short Answer

1. genus: homo
   species: sapiens
   order: primate
   family: Hominide
   class: mammalia
2. Highly developed forelimbs and hands, stereoscopic color vision, versatile teeth, upright posture, large brains, and eventually language and culture are the adaptive measures that make primates the most advanced mammals.
3. The jaw became shorter from front to back, and deeper from top to bottom, which altered the face from an elongated snout to a flatter appearance. Changes in tooth shape and size occurred simultaneously with jaw changes in response to changing diets. Early fossil primate teeth show the diet consisted of mainly fruit, and later fossil records show a dentition pattern that is better designed for grinding food.
4. The group of primates known as hominoids underwent a series of rapid evolutionary changes that were associated with a series of climatic changes, which gave rise to an interconnected array of species. This diversity as well as large gaps in the fossil record make it difficult to draw conclusions about evolutionary relationships among early hominoids and to identify the ancestral lines leading to modern hominoids.
5. The physical similarities between *Homo erectus* and modern humans include height and walking patterns. The differences include a lower skull that is more pointed in the back, not dome-shaped as in *Homo sapiens*. *Homo erectus* had a receding chin, a prominent brow ridge, and differences in teeth.
6. The Noah's ark or out of Africa hypothesis argues that modern human populations are all derived from a single speciation event that took place in a restricted region in Africa. It argues that after *H. erectus* moved out of Africa, populations remained and continued to evolve, giving rise to *H. sapiens*. The second idea called the multiregional model postulates that the transition from *H. erectus* to *H. sapiens* occurred in multiple sites outside of Africa and that *H. erectus* gradually transformed into *H. sapiens*. It argues that *H. erectus* spread from Africa over most of Europe and Asia and modern humans arose at multiple sites as part of an interbreeding network of lineages descended from the original colonizing populations of *H. erectus*.
7. Neandertals were large-brained, with low forehead, prominent brow ridges, receding chin, a face that projects from the skull as if pulled forward, and a prominent nose.
8. The genetic model, based on several assumptions, approximates how the genetic variation, population density, and geographic distribution of Native Americans could have been generated. The model assumes several waves of migration crossed from Siberia, and these migrants were in bands of no greater than 500 individuals. If the bands reproduced and split off into new bands, doubling in size every 25 years, and migrated an average of five miles each year, then the Americas could have been occupied in a few thousand years. Once established, neighboring populations underwent numerous fusions and splits, creating a flow of genes and spreading of new alleles across geographic regions.

### Multiple Choice

1. b
2. d
3. a
4. d
5. a

### Fill in the Blank

1. prehensile movement; Opposable
2. upright; semiupright
3. environment
4. *Australopithecus afarensis*
5. Mosaic evolution
6. increased; flatter; prominent
7. Neandertals
8. *Homo sapiens neandertalensis; Homo neandertalensis*
9. cultural; physical
10. Berengia

## CHAPTER 16

### True/False

1. False, flowering plants are divided into the monocots and the dicots.
2. True.
3. True.
4. False, herbaceous plants are mostly composed of primary tissue.
5. True.
6. False, the stomata are openings in the epidermis of a leaf that are surrounded by guard cells, which control the movement of gases into and out of the leaves.

7. False, a fibrous root system is more efficient at absorption than a tap-root system. A taproot system is better for anchorage.
8. True.
9. False, root hairs grow in the region of maturation.

**Matching**

1. c
2. i
3. f
4. h
5. b
6. e
7. g
8. j
9. d
10. a

**Compare and Contrast**

1. Dicot
2. Monocot
3. Monocot
4. Both
5. Dicot
6. Monocot

## CHAPTER 17

**Matching**

1. e
2. h
3. b
4. i
5. g
6. d
7. j
8. f
9. a
10. c

**True/False**

1. False, the soil is the primary source of nutrients for plants.
2. False, the transportation of nutrients in the plant is dependent upon concentration gradients, some of which require energy.
3. True.
4. True.
5. True.

**Short Answer**

1. (a) as structural components
   (b) for organic molecules important in metabolism
   (c) for enzyme activators
   (d) to help maintain osmotic balance.
2. A waxy, impermeable layer called the Casparian strip prevents the passage of water and solutes into the xylem.
3. The sieve elements, companion cells, and the vascular parenchyma compose the phloem.
4. The rate of transpiration would decrease in high humidity.
5. In short- and long-day plants, flower induction is regulated by the duration of the night, which is detected by the pigment phyto-chrome.

## CHAPTER 18

**True/False**

1. False, homeostasis is maintaining the internal environment of an organism between narrow limits compatible with life.
2. True.
3. False, elevated carbon dioxide blood levels result in an increase in acidity in the intracellular and extracellular fluids, which is a decrease in pH.
4. True.
5. True.

**Fill in the Blank**

1. vitamin D; dehydrocholesterol; cholecalciferol
2. Meissner's capsules; Pacinian corpuscles
3. negative feedback control
4. eccrine glands; Apocrine glands
5. ectoderm; mesoderm

**Matching**

| | | |
|---|---|---|
| 1. | circulatory | transports gases, nutrients, and wastes |
| 2. | digestive | receives food |
| 3. | endocrine | controls activity of internal environment |
| 4. | excretory | regulates internal fluid volumes |
| 5. | integument | outermost protection |
| 6. | immune | defends internal environment against intruders |
| 7. | muscular | movement |
| 8. | nervous | coordinates actions of body |
| 9. | reproductive | species survival |
| 10. | respiratory | moves oxygen and removes carbon dioxide |
| 11. | skeletal | body support |

**Short Answer**

1. The skin is composed of an outer layer called the epidermis that arises from embryonic ectoderm. The keratinocyte cell is the epithelial cell type that makes up the epidermis. Division of the innermost cell of the epidermis (basal cells) produce keratinocytes. Melanocytes, which produce the pigment melanin that gives rise to skin color and protection, are also components of the epidermis. Desmosomes link the keratinocyte cells to each other. As keratinocytes mature and move toward the skin surface, they become filled with keratin. The outer layer of keratinocytes becomes elongated and flattened to form a protective layer as it reaches the skin surface. Cell death eventually occurs and a thin outer layer of dead cells is constantly being sloughed away. The dermal layer of cells lies under the epidermis and is composed of elastin, collagen fibers, capillary networks, and nerve endings. Nutrient and waste exchange occur by diffusion between the dermis and epidermis.
2. This form of communication in cats is similar to the development of "goose bumps" or "goose pimples" in humans. Hair follicles have a muscle, called the arrector pili, running from the follicle to the dermis. As this muscle contracts, it forces the hair follicles upright, causing the hair or fur to stand on end or perpendicular to the skin surface.
3. The skin functions in homeostasis include protection from microorganisms, body temperature regulation, sensory reception, water balance, vitamin synthesis, hormone synthesis, UV protection, and absorption of materials.
4. Heat and cold receptors in the skin send messages to the hypothalamus in the brain that stimulate the following responses. When body temperature rises, the hypothalamus signals sweat glands in the skin, via the nervous system, to release sweat, which cools the body by evaporation. Hypothalamic-induced dilatation of blood vessels in

the dermis allows more blood to be carried to the skin, and the heat from the blood is conducted to the body surface and radiates into the atmosphere. Decreases in body temperature cause the hypothalamus to shut down sweat glands and constrict dermal blood vessels. If the body temperature continues to fall even after blood flow through the skin has been restricted, the hypothalamus can generate heat by raising the body's metabolic rate (thermiogenesis) and by shivering.

## CHAPTER 19

### Short Answer

1. Skeletal muscle—voluntary control
   Smooth muscle—involuntary control
   Cardiac muscle—involuntary control
2. Skeletal muscles attach to the bones of the skeleton. Contractions are under the control of the nervous system and are voluntary. Skeletal muscles are composed of alternating light and dark bands that give them a striated appearance. Smooth muscle cells lack striations and are under involuntary control in most situations. Regulation of smooth muscle is controlled by intrinsic factors of the muscle, hormones, and autonomic nervous system. Cardiac muscles are composed of striated cells and are under involuntary control. Like smooth muscle, cardiac muscle is regulated by factors intrinsic to the muscle itself, hormones, and the autonomic nervous system.
3. Immovable joints are found only between flat bones of the skull (sutures). In this type of joint the bone edges tightly interlock and are held together by fibers, or bony processes that cross the gap between bones. Partly movable joints have cartilage between bones that allows for more flexibility and movement. Joints between vertebrae and the ribs are of this type. Synovial joints are the most movable and can be found throughout most of the body including the shoulder, knee, hip, and elbow.
4. Coelenterates and annelids use a hydrostatic skeletal system that is composed of internal fluid-filled chambers that provide support and maintain body shape. Spiders use a combination of a hydrostatic and an exoskeletal system. Exoskeletons, or external skeletons, encase the body of the organism in a hard, jointed covering, and all muscles and organs are encased by the skeleton. Most arthropods and all insects have exoskeletons. Animals with exoskeletons are restricted in movement and as a result the exoskeleton needs to be shed periodically. These animals are usually smaller than other groups of animals without exoskeletons. Exoskeletons in mollusks are composed of calcium carbonate. Vertebrates have an internal supporting endoskeleton, with muscles on the outside. Vital organs, such as ribs and brain, are protected by this skeletal system. It is composed of cartilage or a combination of bone and cartilage.
5. Skeletal muscle contractions are controlled by the nervous system. A motor neuron attaches at a specialized point along the plasma membrane of the skeletal muscle fiber, forming the neuromuscular junction. When the muscle becomes excited, nerve impulses reach the neuromuscular junction, and acetylcholine is released and binds to receptors on muscle fiber membrane. This binding changes the permeability of muscle fiber plasma membrane at the junction, causing a wave of electrical activity. This electrical impulse causes calcium ions to be released from the sarcoplasmic reticulum of the muscle cell. Calcium ions bind to actin, which leads to the interaction between the thick filaments (myosin) and the thin filaments (actin) and to muscle contraction. During contraction, actin filaments at each side of the sarcomere slide past myosin filaments toward the center of the sarcomere, until they meet in the middle. This causes the sarcomere

to shorten, putting the Z lines closer together. The energy for muscle contraction is provided by ATP. ATP is converted to ADP, releasing energy when it binds to the cross bridges between myosin heads and actin binding sites. Calcium released into the sarcomere exposes the actin binding sites, allowing myosin to form cross bridges to actin. When muscle is no longer stimulated to contract, calcium ions are pumped from the sarcomere back into storage. Myosin heads swivel toward the center of the sarcomere. This moves actin inward toward the center of the sarcomere. Myosin heads detach and reattach to the next available actin binding site, and swivel again, moving actin filaments closer to the center of sarcomere.

6. The skeletal system in conjunction with the endocrine, nervous, urinary, and digestive systems maintains calcium and phosphorous levels in the extracellular fluid. Mediated by the interactions of the osteoblasts and osteoclasts, the skeletal system stimulates the release of calcium from bone tissue into the extracellular fluid when levels are low and triggers calcium storage in bones when levels are high.
7. Osteoarthritis is the "wear-and-tear" type of arthritis involving the degeneration of cartilage. It is also referred to as degenerative joint disease. Rheumatoid arthritis is caused by an autoimmune reaction that can be severely crippling. It can lead to the build-up of inflamed joint tissue and eventually destroy the joints. It seems to affect women more often than men and has an underlying genetic component.

### Multiple Choice

1. b
2. c
3. a
4. c
5. e

### Fill in the Blanks

1. 33; 12
2. wrist; tarsals
3. Haversian canal
4. red marrow; yellow marrow
5. muscle; ligaments; bone

## CHAPTER 20

### Short Answer

1. Cellular respiration involves the breakdown of nutrients (fats, carbohydrates, and proteins) to release energy for cell functioning. Water is formed from the release of carbon dioxide and the consumption of oxygen. External respiration involves the exchange of gases at the cellular level. This process provides oxygen for cellular energy conversion and the elimination of carbon dioxide that is the byproduct of metabolism.
2. d, f, g, e, h, b, c, a
3. Fish have an adaptation that maximizes the transfer of oxygen from water to blood. Countercurrent flow moves water off the gills in one direction while blood flows through capillaries in the gill in the opposite direction.
4. *Inhalation:* Nerve impulses from breathing center stimulate muscles of inhalation— intercostal and diaphragm.
   Thoracic cavity volume increases due to the contraction of diaphragm and elevation of ribs.
   Lung volume increases.
   Intrapulmonary pressure falls below atmospheric pressure.
   Air flows into lungs through nose and mouth.

*Exhalation:* Nerve impulses feed back on breathing center, shutting off stimuli to inspiration muscles.

Intercostal muscles relax, decreasing volume of thoracic cavity, and the rib cage falls.

Diaphragm relaxes and rises.

Lowered lung volume increases intrapulmonary pressure above atmospheric pressure, causing air flow out of the lungs.

5. The alveoli are microscopic air sacs located at the end of the branching air tubes in the lungs. They serve as the site for gas exchange between air and blood. Decreasing the number of small alveoli by forming fewer, larger alveoli also decreases the surface area available for gas exchange. This causes decreased diffusion rates and therefore areas of the body are not receiving enough oxygen to function properly.

6. Holding your breath increases blood pH or makes the blood more basic. Holding your breath prevents the release of carbon dioxide resulting in its increase in the blood and an increased pH level. Increased blood carbon dioxide levels causes decreased oxygen levels.

7. Oxygen diffusion from blood into cells
Movement of oxygen-containing medium so that contact is made with a moist membrane overlying blood vessels
Oxygen diffusion from medium into blood
Oxygen transport to tissues and cells of body

8. The nasal passages contain hairs that screen for large airborne materials. Ciliated epithelium and mucus-producing cells line the nasal cavity and remove dirt and microorganisms. Immunological defense is provided by the tonsils and adenoids against pathogens. The respiratory airways are lined with cilia, which, through a wave-like motion, carry inhaled particulates away.

9. Elevated altitudes have decreased atmospheric pressure. As a result, the oxygen pressure is lower than at sea level. This lowered oxygen pressure causes oxygen loading to hemoglobin to be lower, resulting in lowered oxygen levels in blood. This is compensated by increasing the breathing rate to move more gas through the lungs.

## True/False

1. True.
2. False, the lungs are situated in the thoracic cavity and the outer surface of the lungs is covered with pleura. The space between the two sheets of pleura is the pleural cavity. The diaphragm is located at the bottom of the thoracic cavity and separates it from the abdomen.
3. False, forced expiration reduces the volume of the thoracic cavity beyond normal expiration. However, even when you try to empty all the air from your lungs, a residual amount remains to allow a continuous flow of gas exchange.
4. True.
5. True.

## Matching

1. apnea     periods of breathing cessation during sleep
2. cystic fibrosis     thick, viscous mucus clogging airways
3. asthma     narrowing of smaller lung airways
4. pneumonia     infection of lung tissue
5. SIDS     newborn respiratory abnormality
6. bronchitis     inflammatory response from irritants

## CHAPTER 21

### Fill in the Blank

1. pulmonary; systemic
2. Amphibians

3. atrioventricular valve
4. arteries; veins
5. capillaries; venules
6. atrium; sinoatrial
7. semilunar; ventricles
8. Atherosclerosis
9. Plasma; 90%
10. liver; spleen

## Matching

1. red blood cells     erythrocytes
2. white blood cells     leukocytes
3. phagocytes     cell eaters
4. platelets     clot formers
5. lymph nodes     small masses of spongy tissue
6. angina pectoralis     chest pain
7. hypertension     blood pressure consistently above 140/90
8. elephantiasis     swelling of tissues
9. sickle cell anemia     hemoglobin mutation
10. hemophilia     clotting disorder

## Short Answer

1. Arteries are vessels that carry blood away from the heart. The arteries have thick flexible walls, made up of three layers. The elastic connective tissue of arteries allow the vessel to expand and contract as blood pressure changes. The elastic arteries branch to form smaller and smaller vessels. Veins carry blood toward the heart. Unlike the arteries, the veins start off small and converge with other veins, forming larger and larger vessels. Blood pressure in veins is low, and veins have relatively thin walls and fewer smooth muscle cells than arteries.

2. The lymphatic system is a network of vessels that drains interstitial fluid from bodily tissues and transports it to the blood. It consists of several components, such as lymph nodes, the spleen, the thymus, and the tonsils, which primarily provide immune protection. Lymph nodes located along the lymphatic vessels filter the lymph. Lymph is removed from tissues at a rate proportional to its production, keeping tissues from swelling.

3. Injured cells in the wall of blood vessels release the chemical thromboplastin (a lipoprotein), which converts inactive plasma enzyme, prothrombin, found in plasma, into thrombin. Thrombin stimulates conversion of plasma protein fibrinogen into fibrin. The fibrin network captures platelets and red blood cells. Platelets in the blood clot release platelet thromboplastin, which converts additional plasma prothrombin into thrombin. Thrombin, in turn, stimulates the production of additional fibrin.

4. The circulatory system is one of the body's chief homeostatic systems. It helps transport oxygen to body cells, distributes body heat, maintains a constant level of nutrients and wastes, protects against microorganisms, and, through clotting, protects against blood loss.

5. In an open circulatory system, the heart propels blood through vessels, which empty their contents into body cavities to bathe organs. In a closed system, blood remains in blood vessels. The exchange of nutrients and wastes occurs through capillaries.

## CHAPTER 22

### Matching

1. antibody     molecule produced in response to the presence of antigens
2. antigen     molecule that elicits an immune response
3. B cells     cells that fight invaders by producing antibodies

| 4. cell-mediated immune response | immunity mediated by T cells |
| 5. humoral immune response | immunity mediated by antibodies |
| 6. T cells | cells that fight invaders directly by binding to them |
| 7. monocytes | white blood cells that phagocytize bacteria and viruses in the body |

### True/False

1. True.
2. True.
3. False, adults with type A blood carry the A antigen on their red blood cells, so they will not make antibodies against this cell surface marker.
4. False, the variable region of the antibody possesses different amino acid sequences for each kind of antibody. The conformation of the variable region enables an antibody to bind to a specific antigen in a lock-and-key fashion.
5. True.
6. False, identical twins theoretically share 100% of their genetic material making them better organ donors than nonidentical twins, which share only 50% of their genetic material.

### Fill in the Blank

1. histamine; inflammatory
2. bone marrow; lymphocytes
3. immunoglobulins; IgG, IgA, IgM, IgD, IgE
4. helper T cells, suppressor T cells, killer T cells
5. primary; memory

### Short Answer

1. Hemolytic disease of the newborn (HDN) occurs when the mother is Rh− and the fetus is Rh+. If blood from the Rh+ fetus enters the maternal circulation of an Rh− mother, antibodies against the Rh+ antigen will be made. This mixing usually occurs during delivery of the fetus, so that the first Rh+ child is not affected. The maternal circulation now contains antibodies against this antigen and a subsequent Rh+ fetus will evoke a secondary response from the mother's immune system. This response results in the production of antibodies that cross the placenta in the third trimester of pregnancy and attack the red blood cells of the fetus. To prevent HDN, Rh− mothers are given an injection of an Rh− antibody preparation prior to invasive genetic testing (amniocentesis or CVS) and/or after birth of the first Rh+ child and all subsequent Rh+ children. These antibodies destroy fetal cells that have entered the maternal circulation before the maternal immune system can make its own antibodies against the Rh antigen.
2. The child is experiencing anaphylactic shock. The bee sting caused the release of massive amounts of IgE antibodies and these antibodies bound to mast cells, causing the release of histamine. The release of histamine into the circulatory system resulted in fluid accumulation, tissue swelling, drop in blood pressure, and constriction of the air passages in the lungs. Injections of epinephrine prevent the release of substances from mast cells and antagonize the actions of histamines.
3. A solution containing dead or weakened virus, bacterium, or bacterial toxin that is injected into people to create active immunity is called a vaccine. Active immunity lasts at most only a few months. Passive immunity is achieved by injecting antibodies into a patient or by the transfer of antibodies from a mother to her baby through the bloodstream or breast milk.
4. First, a number of different viruses produce the symptom of a "cold." Second, the viruses causing colds or influenza are constantly changing or undergoing mutations. These mutations alter the viral genetic material, which may alter the part of the virus that acts as an antigen. These alterations may be so extreme that the immune system no longer recognizes it as a foreign substance. For example, a cold caught this year will probably not produce memory cells capable of recognizing a cold virus next year.
5. The human immunodeficiency virus (HIV) that causes AIDS is transmitted in blood, semen, and vaginal secretions. The virus selectively infects and kills the T4 helper cells of the immune system. Inside the helper cells, the RNA is transcribed into a DNA molecule by reverse transcriptase, and the viral DNA is inserted into a human chromosome. The virus can remain dormant at this stage for many months or even years later. Later, when the infected T cell is called on to respond to an antigen, the viral genes become activated. Viral RNA proteins are made, and new viral particles are formed. These bud off the surface of the T cell, replicating and killing the cell, and causing a new round of T-cell infection. Gradually, the number of helper T4 cells decreases and the immune system is unable to respond effectively to infections.
6. Red blood cells from individuals with type O blood have neither A nor B antigens. Therefore, type O blood can be transferred into individuals with all four types: A, B, AB, and O.

## CHAPTER 23

### True/False

1. False, vertebrates use the more efficient tube-within-a-tube digestive system, with two body openings, one for ingestion of food, and the other for waste removal.
2. False, bile is produced in the liver.
3. True.
4. False, the greatest amount of digestive activity and most of absorption occur in the small intestine.
5. True.
6. True.
7. False, caloric intake must balance energy output to maintain overall health and weight
8. True.

### Matching

| 1. gastrin | prompts release of HCl and pepsinogen |
| 2. secretin | stimulates release of alkaline secretions into duodenum |
| 3. lipases | pancreatic enzyme that digests fat |
| 4. salivary amylase | breaks down complex sugars and starches |
| 5. HCl | promotes breakdown of muscle fibers and connective tissue in food |
| 6. bile | solubilizes fats |
| 7. proteases | breaks down proteins |
| 8. cholycystokinin | causes contraction of gallbladder for release of bile |

### Fill in the Blank

1. intracellular
2. duodenum; jejunum; ileum
3. pepsin
4. small intestine
5. peristalsis
6. epiglottis

## Short Answer

1. The stomach lining is composed of epithelial cells that secrete mucus that serves as a protective barrier, preventing contact between gastric juices and the cells of the stomach lining. Pepsin is inactivated when it comes in contact with this mucus, and bicarbonate ions in the mucus induce acidity. The cells of the gastric epithelium are held together by tight junctions, preventing gastric juices from infiltrating between cells.

2. Vitamins are organic molecules that participate in energy-releasing metabolic reactions. Minerals are inorganic substances required for normal metabolism, as components of cells and tissues, and in nerve conduction and muscle contraction. The metabolism of fats generates the highest energy yields and fat is used as molecular energy reserves. Lipids also serve as insulation, preventing loss of body heat, and are structural components of the cell. Proteins are necessary for growth and repair. In extreme situations, proteins can be used as a source of energy. Carbohydrates are the primary source of energy for cellular metabolism.

3. Absorptive feeders such as tapeworms live in the digestive tracts of other animals and absorb nutrients directly through the body wall. Filter feeders such as oysters and mussels collect particles or small organisms from the surrounding water. Substrate feeders like worms and termites eat the dirt or wood that they burrow through. Fluid feeders such as ticks, leeches, and aphids have specialized structures for piercing the body of a plant or animal.

4.
| | |
|---|---|
| mouth | mechanically break down food, mix it with saliva |
| salivary glands | moisten food, start complex sugar and starch breakdown; neutralize acidic food |
| stomach | store, mix, dissolve food; start protein digestion |
| small intestine | digest and absorb most nutrients |
| pancreas | enzymatically break down food molecules, buffer HCl from stomach |
| liver | secrete bile for fat absorption; secrete bicarbonate in order to neutralize HCl from stomach |
| gallbladder | store and concentrate bile from liver |
| large intestine | store, concentrate undigested matter by absorbing water and salts |
| rectum | control over elimination of undigested and unabsorbed materials |

## CHAPTER 24

### Short Answer

1. Kidneys: paired organs that regulate water and salts and dispose of nitrogen wastes.
Malphigian tubules: excretory organ in insect that is composed of long tubules that open into the gut.
Nephron: the functional part of the kidney.
Glomerulus: a knot of coiled capillaries in the kidney.
Aldosterone: a hormone produced by the adrenal glands that is important in sodium retention and reabsorption.

2. Excretion applies to metabolic waste products that cross a plasma membrane. Elimination is removal of undigested food residues in the form of feces.

3. With no ADH secretion, tubules become impermeable to water, lowering the water concentration in the blood and large quantities of dilute urine are produced.

4. Urea is made from ammonia by a series of metabolic reactions in the liver, is less toxic than ammonia, and can be excreted with less water loss than ammonia. Ammonia is made at little metabolic cost, but must be diluted with large amounts of water for excretion. Uric acid is made through a complex metabolic process requiring large energy expenditures, but can be excreted with little water loss.

5. Kidneys: eliminate wastes from the blood; help regulate body water concentration; help regulate blood pressure; help maintain constant blood pH.
Ureters: transport urine to urinary bladder.
Urinary bladder: stores urine; contracts to eliminate stored urine.
Urethra: transports urine to outside of body.

6. (1) Filtration of water and solutes from the blood.
(2) Reabsorption of water and conserved molecules back into the blood.
(3) Receives ions and other wastes secreted into the distal tubule by the surrounding capillaries.

### Fill in the Blank

1. nitrogen wastes
2. kidney stones
3. uric acid
4. osmoregulation
5. nephridium: excretory pore
6. aorta; renal artery
7. proximal tubule; blood
8. tubular secretion; acid–base
9. brain; osmotic concentration
10. renin

### True/False

1. False, the kidneys are surrounded by a protective fat layer but are located at the back of the abdominal cavity below the diaphragm.
2. True.
3. True.
4. False, kidney stones form when excess wastes such as uric acid, calcium, and magnesium do not remain in solution and crystallize out in the kidney. Kidney stones grow by the deposition of more materials on their outer surface.
5. False, the kidneys maintain the volume of extracellular fluid. This is one way in which blood pressure is regulated.

## CHAPTER 25

### Matching

| | |
|---|---|
| 1. dendrite | receptive portion of neuron |
| 2. axon | conducting process of neuron |
| 3. Schwann cell | produces myelin sheaths that are associated with some axons |
| 4. central nervous system | brain and spinal cord |
| 5. nervous tissue | composed of neurons and glial cells |
| 6. node of Ranvier | gap between Schwann cell along axon |
| 7. synapse | junction between axon and adjacent cell |
| 8. peripheral nervous system | sensory and motor pathways |
| 9. forebrain | diencephalon and cerebrum |
| 10. Wernicke's area | language comprehension |
| 11. cones | sensitive to color differences |

### True/False

1. True.
2. True.
3. False, functionally, the sympathetic and parasympathetic divisions of the autonomic system act in opposite directions to maintain homeostasis.

4. False, the development of bilateral symmetry is accompanied by cephalization, the accumulation of sensory and integrative functions at the anterior end of the animal.
5. False, some parts of the nervous system including the brain and spinal cord are protected by a layer of extracellular fluid known as cerebrospinal fluid.
6. False, motor neurons transfer signals to effector cells that produce a response. Sensory neurons carry signals from receptors and transmit information about the environment.

## Fill in the Blanks

Central nervous system
  A. brain
  B. spinal cord
Peripheral nervous system
  A. sensory (afferent) system
  B. motor (efferent) system
    1. somatic
    2. autonomic
      a. sympathetic
      b. parasympathetic

## Short Answer

1. The three basic functions of the nervous system are to (1) receive sensory input, (2) integrate the input, and (3) respond to stimuli.
2. (1) Depolarization of a neuron by graded potential reaches threshold. (2) Activation gate of sodium (Na+) channel opens quickly; potassium (K+) channel slowly begins to open; inactivation gate of Na+ channel begins to slowly close. (3) Na+ rapidly enters the cell as positive feedback cycle operates, and interior of cell becomes less negative. (4) Inactivation gate of Na+ channel completely closes, and membrane permeability to Na+ decreases. (5) Cell returns to resting membrane potential, and Na+-K+ pump moves Na+ back out of and K+ back into the neuron.
3. *Clostridium tetani* prevents the release of the inhibitory neurotransmitter GABA from axons that supply skeletal muscles. Without the release of GABA, muscles lock into contracted spasms. Death can occur when muscles responsible for breathing are affected.
   *Clostridium botulinum* binds to the ends of axons that innervate muscles, preventing the release of neurotransmitters. Muscles become progressively relaxed and the effect can eventually lead to death.
4. The occipital lobe, located at the back of the head, is responsible for receiving and processing visual information. The temporal lobe receives auditory signals from the ears. The parietal lobe processes information about touch, taste, pressure, pain, heat, and cold. The frontal lobe is associated with thought processes, motor activity, and the integration of muscle movements, and speech.

## CHAPTER 26

### Fill in the Blank

1. inhibits; osteoclasts; lower
2. vasopressin
3. amino acids; lipids
4. hypothalamus; anterior
5. glandular; nervous
6. kidney; medulla; steroid
7. anterior pituitary
8. glucose; stimulates
9. pheromones
10. circadian

### True/False

1. False, the endocrine system is regulated by negative feedback and cycles.
2. True.
3. False, the chemical property that is most important in defining a hormone is its solubility inside and outside the cell.
4. True.
5. True.
6. True.
7. False, diabetes mellitus is associated with elevated levels of blood glucose.

### Short Answer

1. Hormones are chemical messages secreted into the blood by endocrine glands. They exert their effect on target sites far from their point of release. Paracrines are chemical messages that affect cells in the immediate vicinity. Neurotransmitters are short-range chemical messengers that are synthesized and released by neurons in response to action potentials. Neurohormones are long-range chemical signals that are synthesized by a neuron and released into the blood.
2. The differences in height and body build between people is largely attributable to growth hormone (GH), a protein hormone produced by the anterior pituitary. GH affects all body cells, but it acts primarily on bone and muscles. As a rule, the more growth hormone one produces during the growth phase of an individual's life cycle, the taller and more muscular he or she will be. Growth hormone production is determined by one's genes.
3. Nonsteroid hormones bind to the plasma membrane receptors and exert their effect by generating a chemical messenger known as a second messenger. Secondary messengers activate other intracellular molecules to produce a response in the target cell. Steroid hormones pass through the plasma membrane and bind to receptor molecules located in the cell nucleus. Binding of the steroid hormone to its receptor alters the receptor shape, producing an activated hormone–receptor complex. This complex binds to DNA and activates gene transcription and mRNA production, and the production of protein gene products.
4. *Anterior Pituitary*
   Growth hormone (GH): Stimulates cell growth. Primary targets are muscle and bone, where GH stimulates amino acid uptake and protein synthesis. GH also stimulates fat breakdown.
   Thyroid-stimulating hormone (TSH): Stimulates release of thyroxine and triiodothyronine.
   Adrenocorticotropic hormone (ACTH): Stimulates secretion of hormones by adrenal cortex.
   Gonadotropins (FSH and LH): Stimulate gamete production and hormone production by gonads.
   Prolactin: Stimulates milk production by the breast.
   *Posterior Pituitary*
   Antidiuretic hormone (ADH): Stimulates water reabsorption by nephrons of the kidney.
   Oxytocin: Stimulates ejection of milk from breasts and uterine contractions during childbirth.
5. Hyposecretion of growth hormone (GH) is a deficiency in growth hormone production either due to malfunctions in the hypothalamus or a defect in the anterior pituitary. A form of dwarfism associated with deficient skeletal growth is the result of a deficient amount of GH.

Hypersecretion of GH results in the overproduction of this peptide hormone. Affected individuals can be over eight feet tall (gigantism).
6. Mineralcorticoids, glucocorticoids, and sex hormones.

## CHAPTER 27

### True/False

1. True.
2. False, cloned organisms are genetically identical individuals.
3. False, in the strawberry, the fruits are the hard structures embedded in the soft, fleshy red part, which is derived from the receptacle. The soft fleshy part is not derived from the ovary, therefore, the strawberry is considered a false fruit.
4. False, cytokinins are produced in the roots and reach the shoots by transpiration.
5. True.

### Matching

| | | |
|---|---|---|
| 1. | receptacle | base of flower |
| 2. | sepals | modified leaves |
| 3. | stamens | male reproductive structures |
| 4. | stigma | located at tip of style |
| 5. | carpels | female reproductive structures |
| 6. | endosperm | tissue providing nutrients |
| 7. | abscisic acid | hormone that promotes dormancy |
| 8. | stolons | long stems that run along the ground |
| 9. | anthers | location for pollen grain formation |
| 10. | flowers | mature seeds |

### Short Answer

1. The haploid products of meiosis are not gametes in plants as they are in animals. These products in plants grow to form a haploid gametophyte generation, which when mature forms gametes. These gametes will fuse to form a diploid zygote that grows into a sporophyte.
2. Each microspore (four haploid cells) undergoes at least one mitotic division and becomes encased in a protective wall, forming a pollen grain. In some cases, one of the haploid nuclei divides again, forming three haploid cells. Such pollen grains are immature gametophytes; one of the cells will divide after the pollen grain has landed on the stigma, forming three haploid cells. The four haploid cells in the female gametophyte known as megaspores form in the ovules. Usually, three of these cells degenerate, and the remaining cell becomes the gamete. This haploid cell divides three times, forming eight nuclei. Cell membranes form around six of these cells, and make up the female gametophyte. One of these cells becomes the haploid egg.
3. Animals contain germ cells that are set aside early in development and undergo meiosis and form gametes in the adult. In plants, the gametes come from a population of cells previously active in forming the plant. The development of the embryo in animals results in a nearly complete, but much smaller, version of the adult. In plants, embryonic development establishes the rudimentary body plan, but the plant embryo contains only a fraction of the elements of the mature plant.
4. Seeds can withstand extreme conditions and provide a means of species dispersal.
5. Auxins are growth-regulating substances involved in controlling cell growth. Gibberellins are required for normal elongation of the shoot, and they stimulate cell division as well as cell elongation. Cytokinins promote cell division and inhibit aging of green tissues. Abscisic acid promotes dormancy in perennial plants and causes rapid closure of leaf stomata when a leaf begins to wilt. Ethylene gas stimulates fruit ripening and the dropping of leaves.

## CHAPTER 28

### Short Answer

1. Asexual reproduction allows a single animal, living in isolation or attached to a surface, to reproduce quickly, without having to waste energy and time finding mates. Offspring are well adapted to stable environments and can rapidly use surrounding resources. The major disadvantage to asexual reproduction is that if environmental conditions change, there is no genetic variation among the population and the population may die off.
2. Ectoderm: (1) outer layer
   (2) hair, skin, feathers, brain and nervous system
   Mesoderm: (1) middle layer, between ectoderm and endoderm
   (2) structures involved in movement, support, transport, and reproduction
   Endoderm: (1) innermost layer
   (2) organs and tissues associated with respiration and digestion, including the liver, pancreas, and lungs
3. Before ovulation, the ovarian cycle is regulated by an estrogen-secreting follicle. After ovulation, events are controlled by the corpus luteum, which secretes estrogen and progesterone.
4. The ovaries produce ova and female sex steroids. The uterine tubes transport sperm to ova and transport fertilized ova to the uterus. The uterus nourishes and protects the embryo and fetus. The vagina is the site of sperm deposition and the birth canal.
5. First trimester: cleavage, implantation begins; nervous system, gut, and blood vessels start to develop; testes differentiate in males; arms and legs move; primary oocyte enters first meiotic division in females
   Second trimester: all major organ systems formed and growing; heart can be heard with a stethoscope
   Third trimester: temperature regulation; central nervous system and lungs completely developed; birth

### True/False

1. True.
2. False, the first step in fertilization is cleavage. Gastrulation is the second major developmental step.
3. True.
4. False, oversecretion of uterine prostaglandins may cause menstrual cramps in women.
5. False, menstruation is associated with the breakdown of cells in the uterine endometrium.
6. False, most STDs do not have noticeable symptoms and can severely affect the fetus and newborn infants.
7. False, only latex condoms and abstinence prevent STDs.
8. False, the central nervous system is the first organ system to be formed in embryonic development.
9. False, MIH controls male development. MIH in an XX embryo causes degeneration of the female duct system.
10. True.

### Multiple Choice

1. a
2. c
3. a
4. d
5. b
6. c

7. a
8. b

## CHAPTER 29

### Matching

1. h
2. g
3. i
4. e
5. d

6. f
7. c
8. a
9. j
10. b

### Short Answer

1. Aggregates are animal groups that lack internal organization. They are formed in response to a simple stimuli and disperse once that stimuli is removed. Moths around a light are an example of this form of social organization. Groups have some internal organization and may exhibit some division of labor. A herd of buffalo, a school of fish, and a flock of ducks are examples of groups. Societies are the most highly organized social group in the animal kingdom. Societies vary among species, but almost all consist of individuals that display cooperation and communication with one another, and have a division of labor. The most well-known example of a society is the hive of bees.
2. When birds first meet, they want to attack, escape, and then remain near each other.
3. Behavior evolved due to its adaptive value. The steps of its development are difficult to study because there is no fossil record, and ancestral species are extinct.
4. Altruistic behavior favors the reproductive success of another member of the same species. Any altruistic behavior that protects the queen or the hive will help pass on the workers' genes to future generations. Worker bees who act as guards give up their life by stinging intruders, but this act of self-sacrifice protects the hive.
5. Huntington's disease is an autosomal dominant condition that affects human behavior. Onset is usually in the third or fourth decade of life and most affected individuals die within 10 to 15 years after onset of symptoms. It is first expressed as involuntary muscular twitches (chorea) and jerky motions of the arms and legs. As the disease advances, personality changes, agitated behavior, forgetfulness, and dementia appear. The defective gene in this condition is believed to produce a toxic substance that destroys brain cells that control behavior.
6. Identical twins share 100% of their genes. Therefore, any differences in behavior between identical twins are believed to be due to environmental influences rather than genetic influences.

### True/False

1. True.
2. False, many behavior patterns become programmed into the nervous system.
3. True.
4. True.
5. False, the communication sequence of honeybees is based on visual and auditory signals and tasting samples of the food provided by the foraging bee.
6. False, a behavior that increases reproductive success will generate more offspring. These offspring carry genes associated with that successful behavior, thereby increasing the fitness of that behavior. Consequently, this successful behavior will gradually spread through the population.

7. True.
8. True.

## CHAPTER 30

### True/False

1. True.
2. False, a population is a group of the same species living in the same geographical area.
3. False, population dynamics is the study of how various factors affect population growth, stability, and decline. Ecology is the study of how organisms interact with each other and their physical environment.
4. False, eutrophication is "too much of a good thing." It causes plant population increases, death rate increases, and bacterial decomposition if dead plants use up increasing amounts of oxygen, causing fish and other organisms to suffocate.
5. True.

### Short Answer

1. Population growth occurs when available resources exceed the number of individuals able to exploit them. Individuals in this situation tend to reproduce rapidly and death rates are low because resources are abundant. Population stability is frequently preceded by a rapid decline in population because growth abruptly overshoots the available resources. Stability is usually the longest phase of the life cycle. Population decline is the inevitable decrease in the number of individuals in a population. Over a long period of time, population decline leads to extinction.
2. The biological interactions that can limit population growth are competitors, predators, and symbiosis.
3. Through evolutionary mechanisms prey have developed protective traits, such as poisons, spines, and camouflage. Prey have protective niches, and burrows where predators cannot reach them. When one prey species becomes reduced in number, the predator switches to another more easily available prey.
4. Habitat disruption, species introduction, overkills, and secondary extinctions are external causes of extinction generated by altering the biological environment.
5. The "zone of intolerance" is where the population is absent. It occurs because some limiting factor has become so great that the species can no longer survive.

### Multiple Choice

1. c
2. a
3. d
4. b
5. a

### Matching

1. d
2. b
3. c
4. a
5. e

## CHAPTER 31

### Short Answer

1. A community is the interacting group of different kinds of organisms that occur together in a particular area. An ecosystem is that set of

organisms along with the nonliving factors with which it interacts. A biome is an assemblage of organisms that has a characteristic appearance and that occurs over a wide geographical area on land.

2. The moisture-holding capacity of air increases when it is warmed and decreases when it is cooled. Cooler conditions cause aridity because cooler air can hold less water vapor than warmer air. Deserts are an exception to this trend; they tend to occur in warm climates and form because of local and global influences that block rainfall.

3. Tropical communities are exposed to less environmental change on a daily and/or yearly basis. This stability allows more kinds of species to survive and thrive in this region because high levels of disturbance or stress tend to reduce diversity. Equatorial communities are older because they have been less distributed by advancing ice sheets and other climatic changes over geologic time. This gives evolution more time in these communities to create new species. Longer growing seasons in equatorial areas lead to more photosynthesis and plant growth. Higher plant growth supports greater diversity of organisms that depend on plants.

4. A food chain is the simplest representation of energy flow through any ecosystem. It is a series of organisms, each one feeding on the organism preceding it. A food web is a more accurate depiction of energy flow in ecosystems because most animals eat more than one kind of organism. A food web is a complex of feeding interactions among species in an ecosystem. A food pyramid provides a more detailed understanding of energy flow through an ecosystem. It is composed of producers and consumers.

5. Carbon, hydrogen, oxygen, nitrogen, sulfur, and phosphorus.

## Multiple Choice

1. d
2. c
3. a
4. a
5. b

## True/False

1. True.
2. False, terrestrial and aquatic biomes are largely determined by climate, especially temperature.
3. True.
4. False, community succession is characterized by a decrease in productivity and an increase in diversity.
5. True.

## Matching

1. d
2. b
3. e

4. a
5. f
6. c

## CHAPTER 32

### Fill in the Blank

1. biogeochemical cycles
2. Acid rain
3. plants; oceans
4. Endangered; Threatened
5. Endangered Species Act

### True/False

1. True.
2. True.
3. True.
4. False, captive breeding is expensive and requires large amounts of land.
5. True.

### Short Answer

1. The first step is to reduce the use of fossil fuel through conservation. Secondly, switching to nonfossil fuels, such as wind, solar, and nuclear power, would be a long-term step.
2. Municipal sewage discharge, agricultural sources, and industrial toxic pollutants are the three wastes generated by humans that cause water pollution.
3. Species with limited habitats become extinct easily because they have so little habitat to destroy or disturb that human activity can quickly eliminate it.
4. "Forest-sustainable" resources are those products that are extracted from living trees year after year. This concept contrasts with products that call for the forest to be cut down.
5. (1) Research and description of species
   (2) Establishment of preserves
   (3) Laws protecting endangered species
   (4) Breeding in captivity
   (5) Reduce socioeconomic causes of extinction

### Matching

| | |
|---|---|
| 1. lithosphere | solid layer of the earth |
| 2. hydrosphere | represents liquid form of matter |
| 3. atmosphere | gases that surround the ecosystem |
| 4. ozone | filters high-energy radiation from sun |
| 5. genetic diversity | number of genes in the biosphere |
| 6. species diversity | number of species in biosphere |
| 7. ecosystem diversity | number of local ecosystems |

# GLOSSARY

**abiogenesis** Early theory that held that some organisms originated from nonliving material.

**abscisic acid** A plant hormone that promotes dormancy in perennial plants and causes rapid closure of leaf stomata when a leaf begins to wilt.

**absorption** The process by which the products of digestion are transferred into the body's internal environment, enabling them to reach the cells.

**absorptive feeders** Animals such as tapeworms that ingest food through the body wall.

**acetylcholine** A chemical released at neuromuscular junctions that binds to receptors on the surface of the plasma membrane of muscle cells, causing an electrical impulse to be transmitted. The impulse ultimately leads to muscle contraction.

**acetyl CoA** An intermediate compound formed during the breakdown of glucose by adding a two-carbon fragment to a carrier molecule (Coenzyme A or CoA).

**acid** A substance that increases the number of hydrogen ions in a solution.

**acid rain** The precipitation of sulfuric acid and other acids as rain. The acids form when sulfur dioxide and nitrogen oxides released during the combustion of fossil fuels combine with water and oxygen in the atmosphere.

**acoelomates** Animals that do not have a coelom or body cavity; e.g., sponges and flatworms.

**acquired immunodeficiency syndrome (AIDS)** A collection of disorders that develop as a result of infection by the human immunodeficiency virus (HIV), which attacks helper T cells, crippling the immune system and greatly reducing the body's ability to fight infection; results in premature death brought about by various diseases that overwhelm the compromised immune system.

**actin** The protein from which microfilaments are composed; forms the contractile filaments of sarcomeres in muscle cells.

**action potential** A reversal of the electrical potential in the plasma membrane of a neuron that occurs when a nerve cell is stimulated; caused by rapid changes in membrane permeability to sodium and potassium.

**active transport** Transport of molecules against a concentration gradient (from regions of low concentration to regions of high concentration) with the aid of proteins in the cell membrane and energy from ATP.

**adaptive radiation** The development of a variety of species from a single ancestral form; occurs when a new habitat becomes available to a population.

**adenosine triphosphate (ATP)** A common form in which energy is stored in living systems; consists of a nucleotide with three phosphate groups.

**adhesion** The ability of molecules of one substance to adhere to a different substance.

**adrenocorticotropic hormone (ACTH)** A hormone produced by the anterior pituitary that stimulates the adrenal cortex to release several hormones including cortisol.

**adventitious roots** Roots that develop from the stem following the death of the primary root. Branches from the adventitious roots form a fibrous root system in which all roots are about the same size; occur in monocots.

**age structure** The relative proportion of individuals in each age group in a population.

**aggregates** Fairly random associations of animals with little or no internal organization; form in response to a single stimulus and disperse when the stimulus is removed; one of the three broad classes of social organization.

**aldosterone** A hormone secreted by the adrenal glands that controls the reabsorption of sodium in the renal tubule of the nephron.

**alleles** Alternate forms of a gene.

**allergens** Antigens that provoke an allergic reaction.

**alternation of generations** A life cycle in which a multicellular diploid stage is followed by a haploid stage and so on; found in land plants and many algae and fungi.

**altitudinal gradient** As altitude increases, a gradient of cooler, drier conditions occurs.

**altruism** Self-sacrificing behavior by animals in social groups that helps other individuals in the group survive and reproduce.

**alveoli** Tiny, thin-walled, inflatable sacs in the lungs where oxygen and carbon dioxide are exchanged.

**amensalism** A symbiotic relationship in which members of one population inhibit the growth of another population without being affected.

**amino acids** The subunits from which proteins are assembled.

**amniocentesis** A method of prenatal testing in which amniotic fluid is withdrawn from the uterus through a needle. The fluid and the fetal cells it contains are analyzed to detect biochemical or chromosomal disorders.

**amniote egg** An egg with compartmentalized sacs (a liquid-filled sac in which the embryo develops, a food sac, and a waste sac) that allowed vertebrates to reproduce on land.

**amoebocytes** Amoeboid cells in sponges that occur in the matrix between the epidermal and collar cells. They transport nutrients.

**anabolic reactions** Reactions in cells in which new chemical bonds are formed and new molecules are made; generally require energy, involve reduction, and lead to an increase in atomic order.

**anaerobic** Refers to organisms that are not dependent on oxygen for respiration.

**analogous structures** Body parts that serve the same function in different organisms, but differ in structure and embryological development; e. g., the wings of insects and birds.

**anaphase** Phase of mitosis in which the chromosomes begin to separate.

**anaphylactic shock** *See* anaphylaxis.

**anaphylaxis** A severe allergic reaction in which histamine is released into the circulatory system; occurs upon subsequent exposure to a particular antigen; also called anaphylactic shock.

**aneuploidy** Variation in chromosome number involving one or a small number of chromosomes; commonly involves the gain or loss of a single chromosome.

**angina pectoralis** Chest pain, especially during physical exertion or emotional stress, that is caused by gradual blockage of the coronary arteries.

**angiosperms** Flowering plants.

**angiotensin II** A hormone that is the active form of angiotensinogen; raises blood pressure by constricting blood vessels and stimulating aldosterone secretion.

**angiotensinogen** A blood protein that is converted to angiotensin II by renin.

**antagonistic muscles** A pair of muscles that work to produce opposite effects—one contracts as the other relaxes.

**anther** The top of a stamens's filament; divided into pollen sacs in which the pollen grains form.

**antibiotics** Substances produced by microorganisms and some plants and vertebrates that kill or inhibit the growth of bacteria.

**antibodies** Proteins produced by immune system cells that bind to foreign molecules and microorganisms and inactivate them.

**antibody-mediated immunity** Immune reaction that protects primarily against invading viruses and bacteria through antibodies produced by plasma cells; also known as humoral immunity.

**anticodon** A sequence of three nucleotides on the transfer RNA molecule that recognizes and pairs with a specific codon on a messenger RNA molecule;

helps control the sequence of amino acids in a growing polypeptide chain.

**antidiuretic hormone (ADH)** A hormone produced by the hypothalamus and released by the pituitary gland that increases the permeability of the renal tubule of the nephron and thereby increases water reabsorption; also known as vasopressin.

**antigenic determinant** The site on an antigen to which an antibody binds, forming an antigen-antibody complex.

**antigens** Molecules carried or produced by microorganisms that initiate antibody production; mostly proteins or proteins combined with polysaccharides.

**antinutrients** Chemicals produced by plants as a defense mechanism; inhibit the action of digestive enzymes in insects that attack and attempt to eat the plants.

**anus** The posterior opening of the digestive tract.

**aorta** The artery that carries blood from the left ventricle for distribution throughout the tissues of the body.

**apical meristem** A meristem (embryonic tissue) at the tip of a shoot or root that is responsible for increasing the plant's length.

**apnea** A disorder in which breathing stops for periods longer than 10 seconds during sleep; can be caused by failure of the automatic respiratory center to respond to elevated blood levels of carbon dioxide.

**apocrine glands** Sweat glands that are located primarily in the armpits and groin area; larger than the more widely distributed eccrine glands.

**apoptosis** The programmed death of cells that is brought about by a gene, which initiates apoptosis rather than allow a cell with damaged DNA to divide and become cancerous.

**appendicular skeleton** The bones of the appendages (wings, legs, and arms or fins) and of the pelvic and pectoral girdles that join the appendages to the rest of the skeleton; one of the two components of the skeleton of vertebrates.

**Archean/Proterozoic Era** The period of time beginning 4.6 billion years ago with the formation of the Earth and ending 570 million years ago.

**aridity** The condition of receiving sparse rainfall; associated with cooler climates because cool air can hold less water vapor than warm air. Many deserts occur in relatively warm climates, however, because of local or global influences that block rainfall.

**arrector pili** A muscle running from a hair follicle to the dermis. Contraction of the muscle causes the hair to rise perpendicular to the skin surface, forming "goose pimples."

**arteries** Thick-walled vessels that carry blood away from the heart.

**arterioles** The smallest arteries; usually branch into a capillary bed.

**artificial selection** The process in which breeders choose the variants to be used to produce succeeding generations.

**asexual reproduction** A system of reproduction in which genetically identical offspring are produced from a single parent; occurs by many mechanisms, including fission, budding, and fragmentation.

**associative learning** The ability of an animal to change its response to a stimulus as a result of associating the stimulus with a certain effect.

**assortment** A way in which meiosis produces new combinations of genetic information. Paternal and maternal chromosomes line up randomly during synapsis, so each daughter cell is likely to receive an assortment of maternal and paternal chromosomes rather than a complete set of one or the other.

**asthma** A respiratory disorder caused by allergies that constrict the bronchioles by inducing spasms in the muscles surrounding the lungs, by causing the bronchioles to swell, or by clogging the bronchioles with mucus.

**asymmetrical** In animals, refers to organisms that lack a general body plan or axis of symmetry that divides the body into mirror-image halves.

**atmosphere** The envelope of gases that surrounds the Earth; consists largely of nitrogen (78%) and oxygen (21%).

**atom** The smallest indivisible particle of matter that can have an independent existence.

**atomic number** The number of protons in the nucleus of an atom.

**atomic weight** Approximately equal to the sum of the weights of an atom's protons and neutrons.

**atrioventricular (AV) node** Tissue in the right ventricle of the heart that receives the impulse from the atria and transmits it through the ventricles by way of the bundles of His and the Purkinje fibers.

**atrioventricular (AV) valve** The valve between each auricle and ventricle of the heart.

**auricle** The chamber of the heart that receives blood from the body via the veins.

**autonomic system** The portion of the peripheral nervous system that stimulates smooth muscle, cardiac muscle, and glands; consists of the parasympathetic and sympathetic systems.

**autosomes** The chromosomes other than the sex chromosomes.

**autotrophic** Refers to organisms that synthesize their nutrients from inorganic raw materials.

**autotrophs** Organisms that synthesize their own nutrients; include some bacteria that are able to synthesize organic molecules from simpler inorganic compounds.

**auxins** A group of hormones involved in controlling plant growth; responsible for phototropism by causing the cells on the shaded side of a plant to elongate, thereby causing the plant to bend toward the light.

**axial skeleton** The skull, vertebral column, and rib cage; one of the two components of the skeleton in vertebrates.

**axons** Long fibers that carry signals away from the cell body of a neuron.

**bacteriophages** Viruses that attack and kill bacterial cells; composed only of DNA and protein.

**bark** The outer layer of the stems of woody plants; composed of an outer layer of dead cells (cork) and an inner layer of phloem.

**basal body** A structure at the base of a cilium or flagellum; consists of nine triplet microtubules arranged in a circle with no central microtubule.

**base** A substance that lowers the hydrogen ion concentration in a solution.

**basidia** Specialized club-shaped structures on the underside of club fungi within which spores form (sing.: basidium).

**basidiospores** The spores formed on the basidia of club fungi (basidiomycetes).

**B cells** Type of lymphocyte responsible for antibody-mediated immunity; mature in the bone marrow and circulate in the circulatory and lymph systems where they transform into antibody-producing plasma cells when exposed to antigens.

**benthic zone** One of the two basic subdivisions of the marine biome; includes the sea floor and bottom-dwelling organisms.

**bicarbonate ions** A weak base present in saliva that helps to neutralize acids in food.

**big bang theory** A model for the evolution of the universe that holds that all matter and energy in the universe were concentrated in one point, which suddenly exploded. Subsequently, matter condensed to form atoms, elements, and eventually galaxies and stars.

**bilateral symmetry** In animals, refers to those that have a single axis of symmetry.

**biliary system** The bile-producing system consisting of the liver, gallbladder, and associated ducts.

**binary fission** The method by which bacteria reproduce. The circular DNA molecule is replicated; then the cell splits into two identical cells.

**binomial system of nomenclature** A system of taxonomy developed by Linnaeus in the early eighteenth century. Each species of plant and animal receives a two-term name; the first term is the genus, and the second is the species.

**biochemical cycle** The flow of an element through the living tissue and physical environment of an ecosystem; e. g., the carbon, hydrogen, oxygen, nitrogen, sulfur, and phosphorus cycles.

**biodiversity** Biological diversity; can be measured in terms of genetic, species, or ecosystem diversity.

**biogeography** The study of the distribution of plants and animals across the Earth.

**bioluminescent** Refers to organisms that emit light under certain conditions.

**biomass** The total weight of living tissue in a community.

**biome** A large-scale grouping that includes many communities of a similar nature.

**biosphere** All ecosystems on Earth as well as the Earth's crust, waters, and atmosphere on and in which organisms exist; also, the sum of all living matter on Earth.

**birth rate**   The ratio between births and individuals in a specified population at a particular time.

**bladder**   A hollow, distensible organ with muscular walls that stores urine and expels it through the urethra.

**blastocoel**   The fluid-filled cavity at the center of a blastula.

**blastocyst**   The developmental stage of the fertilized ovum by the time it is ready to implant; formed from the morula and consists of an inner cell mass, an internal cavity, and an outer layer of cells (the trophoblast).

**blastula**   A ball of cells surrounding a fluid-filled cavity (the blastocoel) that is produced by the repeated cleavage of a zygote.

**blood group or type**   One of the classes into which blood can be separated on the basis of the presence or absence of certain antigens; notably, the ABO types and the Rh blood group.

**B memory cells**   Long-lived B cells that are produced after an initial exposure to an antigen and play an important role in secondary immunity. They remain in the body and respond rapidly if the antigen reappears.

**body fossil**   The actual remains of an organism; includes bones, shells, and teeth.

**bottlenecks**   Drastic short-term reductions in population size caused by natural disasters, disease, or predators; can lead to random changes in the population's gene pool.

**brain**   The most anterior, most highly developed portion of the central nervous system.

**brain stem**   The portion of the brain that is continuous with the spinal cord and consists of the medulla oblongata and pons of the hindbrain and the midbrain.

**bronchi**   Tubes that carry air from the trachea to the lungs (sing.: bronchus).

**bronchioles**   Small tubes in the lungs that are formed by the branching of the bronchi; terminate in the alveoli.

**bronchitis**   A respiratory disorder characterized by excess mucus production and swelling of the bronchioles; caused by long-term exposure to irritants such as cigarette smoke and air pollutants.

**brush border**   The collection of microvilli forming a border on the intestinal side of the epithelial cells of the small intestine.

**bud sports**   Buds that produce fruit that is different from the rest of the fruit on the tree; vegetatively propagated by grafting cuttings onto another plant.

**buffers**   Chemicals that maintain pH values within narrow limits by absorbing or releasing hydrogen ions.

**bulbourethral glands**   Glands that secrete a mucus-like substance that is added to sperm and provides lubrication during intercourse.

**bursae**   Small sacs lined with synovial membrane and filled with synovial fluid; act as cushions to reduce friction between tendons and bones.

**calcitonin**   A hormone produced by the thyroid that plays a role in regulating calcium levels.

**campodactyly**   A dominant trait in which a muscle is improperly attached to bones in the little finger, causing the finger to be permanently bent.

**capillaries**   Small, thin-walled blood vessels that allow oxygen to diffuse from the blood into the cells and carbon dioxide to diffuse from the cells into the blood.

**capillary bed**   A branching network of capillaries supplied by arterioles and drained by venules.

**carbohydrates**   Organic molecules composed of carbon, hydrogen, and oxygen that serve as energy sources and structural materials for cells of all organisms.

**cardiac cycle**   One heartbeat; consists of atrial contraction and relaxation, ventricular contraction and relaxation, and a short pause.

**cardiac muscle**   The type of muscle that is found in the walls of the heart.

**cardiovascular system**   The human circulatory system consisting of the heart and the vessels that transport blood to and from the heart.

**carpals**   The bones that make up the wrist joint.

**carpels**   The female reproductive structures of a flower; consist of the ovary, style, and stigma.

**carrying capacity**   The maximum population size that can be regularly sustained by an environment; the point where the population size levels off in the logistic growth model.

**Casparian strip**   In plants, an impermeable waxy layer between the cells of the endodermis that stops water and solutes from entering the xylem, except by passing through the cytoplasm of adjacent cells.

**catabolic reactions**   Reactions in cells in which existing chemical bonds are broken and molecules are broken down; generally produce energy, involve oxidation, and lead to a decrease in atomic order.

**cell body**   In a neuron, the part that contains the nucleus and most of the cytoplasm and the organelles.

**cell cycle**   The sequence of events from one division of a cell to the next; consists of mitosis, or division, and interphase.

**cell-mediated immunity**   Immune reaction directed against body cells that have been infected by viruses and bacteria; controlled by T cells.

**cell plate**   In plants, a membrane-bound space produced during cytokinesis by the vesicles of the Golgi apparatus. The cell plate fuses with the plasma membrane, dividing the cell into two compartments.

**cells**   The smallest structural units of living matter capable of functioning independently.

**cell theory**   Holds that all living things are composed of at least one cell and that the cell is the fundamental unit of function in all organisms. Corollaries: the chemical composition of all cells is fundamentally alike; all cells arise from preexisting cells through cell division.

**cellular respiration**   The transfer of energy from various molecules to produce ATP; occurs in the mitochondria. In the process, oxygen is consumed and carbon dioxide is generated.

**cellulose**   A polysaccharide that is composed of unbranched chains of glucose; the major structural carbohydrate of plants, insoluble in water, and indigestible in the human intestine.

**Cenozoic Era**   The period beginning at the end of the Mesozoic Era 65 million years ago and encompassing the present.

**central nervous system (CNS)**   The division of the nervous system that includes the brain and spinal cord.

**centromere**   A specialized region on each chromatid to which kinetochores and sister chromatids attach.

**cephalization**   The concentration of sensory tissues in the anterior part of the body.

**cerebellum**   Structure of the brain that is concerned with fine motor coordination and body movement, posture, and balance; is part of the hindbrain and is attached to the rear portion of the brain stem.

**cerebral cortex**   The outer layer of gray matter in the cerebrum; consists mainly of neuronal cell bodies and dendrites in humans; associated with higher functions, including language and abstract thought.

**cerebrospinal fluid**   A layer of extracellular fluid that surrounds the brain and spinal cord.

**cerebrum**   The part of the forebrain that includes the cerebral cortex; the largest part of the human brain.

**cervix**   The lower neck of the uterus that opens into the vagina.

**channels**   Transport proteins that act as gates to control the movement of sodium and potassium ions across the plasma membrane of a nerve cell.

**chemical equilibrium**   The condition when the forward and reverse reaction rates are equal and the concentrations of the products remain constant.

**chemiosmosis**   The process by which ATP is produced in the inner membrane of a mitochondrion. The electron transport system transfers protons from the inner compartment to the outer; as the protons flow back to the inner compartment, the energy of their movement is used to add phosphate to ADP, forming ATP.

**chemotrophs**   Organisms (usually bacteria) that derive energy from inorganic reactions.

**chiasma**   The site where the exchange of chromosome segments between homologous chromosomes takes place (crossing over) (pl.: chiasmata).

**chitin**   A polysaccharide contained in fungi; also forms part of the hard outer covering of insects.

**chlamydia**   A sexually transmitted disease caused by a parasitic bacterium that lives inside cells of the reproductive tract.

**chlorofluorocarbons (CFCs)**   Chemical substances used in refrigerators, air conditioners, and solvents that drift to the upper stratosphere and dissociate. Chlorine released by CFCs reacts with ozone, eroding the ozone layer.

**chlorophyll** The pigment in green plants that absorbs solar energy.

**chloroplasts** Disk-like organelles with a double membrane found in eukaryotic plant cells; contain thylakoids and are the site of photosynthesis.

**cholecystokinin** A hormone secreted in the duodenum that causes the gallbladder to release bile and the pancreas to secrete lipase.

**chorion** The two-layered structure formed from the trophoblast after implantation; secretes human chorionic gonadotropin.

**chorionic villi sampling (CVS)** A method of prenatal testing in which fetal cells from the fetal side of the placenta (chorionic villi) are extracted and analyzed for chromsomal and biochemical defects.

**chromatid** Generally refers to a strand of a replicated chromosome; consists of DNA and protein.

**chromatin** A complex of DNA and proteins in eukaryotic cells that is dispersed throughout the nucleus during interphase and condensed into chromosomes during meiosis and mitosis.

**chromosomes** Structures in the nucleus of a eukaryotic cell that consist of DNA molecules that contain the genes.

**chromosome theory of inheritance** Holds that chromosomes are the cellular components that physically contain genes; proposed in 1903 by Walter Sutton and Theodore Boveri.

**cilia** Hair-like organelles extending from the membrane of many eukaryotic cells; often function in locomotion (sing.: cilium).

**circadian rhythms** Biorhythms that occur on a daily cycle.

**circulatory system** One of eleven major body systems in animals; transports oxygen, carbon dioxide, nutrients, and waste products between cells and the respiratory system and carries chemical signals from the endocrine system; consists of the blood, heart, and blood vessels.

**circulatory system, closed** A system that uses a continuous series of vessels of different sizes to deliver blood to body cells and return it to the heart; found in echinoderms and vertebrates.

**circulatory system, open** A system in which the circulating fluid is not enclosed in vessels at all times; found in insects, crayfish, some mollusks, and other invertebrates.

**classes** Taxonomic subcategories of phyla.

**clavicle** The collar bone.

**cleavage furrow** A constriction of the cell membrane at the equator of the cell that marks the beginning of cytokinesis in animal cells. The cell divides as the furrow deepens.

**climax community** The stage in community succession where the community has become relatively stable through successful adjustment to its environment.

**clitoris** A short shaft with a sensitive tip located where the labia minora meet; consists of erectile tissue and is important in female sexual arousal.

**clone** An exact copy of a DNA segment; produced by recombinant DNA technology.

**closed community** A community in which populations have similar range boundaries and density peaks; forms a discrete unit with sharp boundaries.

**codominance** A type of inheritance in which heterozygotes fully express both alleles.

**codon** A sequence of three nucleotides in messenger RNA that codes for a single amino acid.

**coelom** In animals, a body cavity between the body wall and the digestive system that forms during preadult development.

**coelomates** Animals that have a coelom or body cavity lined with mesoderm.

**coenzymes** Chemicals required by a number of enzymes for proper functioning; also known as enzyme cofactors.

**cohesion** The force that holds molecules of the same substance together.

**cohesion-adhesion theory** Describes the properties of water that help move it through a plant. Cohesion is the ability of water molecules to stick together (held by hydrogen bonds), forming a column of water extending from the roots to the leaves; adhesion is the ability of water molecules to stick to the cellulose in plant cell walls, counteracting the force of gravity and helping to lift the column of water.

**collenchyma** One of the three major cell types in plants; are elongated with thicker walls than parenchyma cells and are usually arranged in strands; provide support and are generally in a region that is growing.

**commensalism** A symbiotic relationship in which one species benefits and the other is not affected.

**community** All species or populations living in the same area.

**community age** One of the factors that helps cause the latitudinal diversity gradient. Tropical communities have had more time to evolve because they have been less disrupted by advancing ice sheets and other climatic changes.

**community simplification** The reduction of overall species diversity in a community; generally caused by human activity.

**community succession** The sequential replacement of species in a community by immigration of new species and by local extinction of old ones.

**compact bone** The outer dense layer that forms the shaft of the long bones; made up of concentric layers of mineral deposits surrounding a central opening.

**companion cells** Specialized cells in the phloem that load sugars into the sieve elements and help maintain a functional plasma membrane in the sieve elements.

**competition** One of the biological interactions that can limit population growth; occurs when two species vie with each other for the same resource.

**competitive exclusion** Competition between species that is so intense that one species completely eliminates the second species from the area.

**competitive release** Occurs when one of two competing species is removed from an area, thereby releasing the remaining species from one of the factors that limited its population size.

**complement system** A chemical defense system that kills microorganisms directly, supplements the inflammatory response, and works with, or complements, the immune system.

**complete dominance** The type of inheritance in which both heterozygotes and dominant homozygotes have the same phenotype.

**compound** A substance formed by two or more elements combined in a fixed ratio.

**conditioned response** The response to a stimulus that occurs when an animal has learned to associate the stimulus with a certain positive or negative effect.

**cones** Light receptors in primates' eyes that operate in bright light; provide color vision and visual acuity.

**consumers** The higher levels in a food pyramid; consist of primary consumers, which feed on the producers, and secondary consumers, which feed on the primary consumers.

**continuous variation** Occurs when the phenotypes of traits controlled by a single gene cannot be sorted into two distinct phenotypic classes, but rather fall into a series of overlapping classes.

**contrast** In relation to microscopes, the ability to distinguish different densities of structures.

**convergent evolution** The development of similar structures in distantly related organisms as a result of adapting to similar environments and/or strategies of life.

**convergent plate boundary** The boundary between two plates that are moving toward one another.

**coprolites** Fossilized feces.

**cork** The outer layer of the bark in woody plants; composed of dead cells.

**cork cambium** A layer of lateral meristematic tissue between the cork and the phloem in the bark of woody plants.

**coronary arteries** Arteries that supply the heart's muscle fibers with nutrients and oxygen.

**corpus callosum** Tightly bundled nerve fibers that connect the right and left hemispheres of the cerebrum.

**corpus luteum** A structure formed from the ovulated follicle in the ovary; secretes progesterone and estrogen.

**cortex** The outer part of an organ, e.g., the adrenal cortex, which produces several steroid hormones.

**cortisol** The primary glucocorticoid hormone; released by the adrenal cortex.

**cotyledon** A leaf-like structure that is present in the seeds of flowering plants; appears during seed germination and sometimes is referred to as a seed leaf.

**countercurrent flow** An arrangement by which fish obtain oxygen from the water that flows through their gills. Water flows across the respiratory surface of the gill in one direction while blood flows in the other direction through the blood vessels on the other side of the surface.

**courtship behavior**  Behavioral sequences that precede mating.

**covalent bond**  A bond created by the sharing of electrons between atoms.

**cranium**  The braincase; composed of several bones fitted together at immovable joints.

**cristae**  Structures formed by the folding of the inner membrane of a mitochondrion (sing.: crista).

**crossing over**  During the first meiotic prophase, the process in which part of a chromatid is physically exchanged with another chromatid to form chromosomes with new allele combinations.

**crossopterygians**  A type of lobe-finned fish with lungs that were ancestral to amphibians.

**crustaceans**  A large class of arthropods that includes lobsters, shrimps, and crabs.

**cuticle**  A film composed of wax and cutin that occurs on the external surface of plant stems and leaves and helps to prevent water loss.

**cyanobacteria**  Blue-green bacteria; unicellular or filamentous chains of cells that carry out photosynthesis.

**cycle**  A recurring sequence of events; e. g., the secretion of certain hormones at regular intervals.

**cyclin**  A protein found in the dividing cells of many organisms that acts as a control during cell division.

**cystic fibrosis**  A genetic disorder that causes the production of mucus that clogs the airways of the lungs and the ducts of the pancreas and other secretory glands.

**cytokinesis**  The division of the cytoplasm during cell division.

**cytokinins**  A group of hormones that promote cell division and inhibit aging of green tissues in plants.

**cytology**  The branch of biology dealing with cell structure.

**cytoplasm**  The viscous semiliquid inside the plasma membrane of a cell; contains various macromolecules and organelles in solution and suspension.

**cytoskeleton**  A three-dimensional network of microtubules and filaments that provides internal support for the cells, anchors internal cell structures, and functions in cell movement and division.

**cytoxic T cells**  T cells that destroy body cells infected by viruses or bacteria; also attack bacteria, fungi, parasites, and cancer cells and will kill cells of transplanted organs if they are recognized as foreign; also known as killer T cells.

**dark reaction**  The photosynthetic process in which food (sugar) molecules are formed from carbon dioxide from the atmosphere with the use of ATP; can occur in the dark as long as ATP is present.

**death rate**  The ratio between deaths and individuals in a specified population at a particular time.

**debt-for-nature swap**  An arrangement whereby a country agrees to set aside pristine land as a preserve in exchange for a reduction in the debt it owes to another country.

**decline**  One of the phases of a population's life cycle. The number of individuals decreases, leading in the long run to localized population extinction.

**deletion**  The loss of a chromosome segment without altering the number of chromosomes.

**dendrites**  Short, highly branched fibers that carry signals toward the cell body of a neuron.

**dendrochronology**  The process of determining the age of a tree or wood used in structures by counting the number of annual growth rings.

**deoxyribonucleic acid (DNA)**  A nucleic acid composed of two polynucleotide strands wound around a central axis to form a double helix; the repository of genetic information.

**depth diversity gradient**  The increase in species richness with increasing water depth until about 2000 meters below the surface, where species richness begins to decline.

**dermatoglyphics**  The science of the study of fingerprints and other skin patterns.

**dermis**  One of the two layers of skin; a connective tissue layer under the epidermis containing elastic and collagen fibers, capillary networks, and nerve endings.

**desert biome**  Characterized by dry conditions and plants and animals that have adapted to those conditions; found in areas where local or global influences block rainfall.

**desmosome**  A circular region of membrane cemented to an adjacent membrane by a molecular glue made of polysaccharides; found in tissues that undergo stretching

**deutcrostomes**  Animals in which the first opening that appears in the embryo becomes the anus while the mouth appears at the other end of the digestive system.

**diabetes mellitus, Types I and II**  A disorder associated with defects in insulin action. Type I diabetes is characterized by inadequate insulin secretion; Type II diabetes is characterized by impaired insulin secretion in response to elevated blood glucose levels or by loss of sensitivity to insulin by target cells.

**diaphragm**  A dome-shaped muscle that separates the thoracic and abdominal cavities.

**diastole**  The filling of the ventricle of the heart with blood.

**diatomaceous earth**  Fossilized deposits of diatoms; used for abrasives, polishes and as a filtering agent.

**dicots**  One of the two main types of flowering plants; characterized by having two cotyledons, floral organs arranged in cycles of four or five, and leaves with reticulate veins; include trees (except conifers) and most ornamental and crop plants.

**dictyosomes**  Organelles in plant cells composed of a series of flattened membrane sacs that sort, chemically modify, and package proteins produced on the rough endoplasmic reticulum.

**diencephalon**  Part of the forebrain; consists of the thalamus and hypothalamus.

**diffusion**  The spontaneous movement of particles from an area of higher concentration to an area of lower concentration.

**digestion**  The process of breaking down food into its molecular and chemical components so that these nutrient molecules can cross plasma membranes.

**digestive system**  One of eleven major body systems in animals; converts food from the external environment into nutrient molecules that can be used and stored by the body and eliminates solid wastes; involves five functions: movement, secretion, digestion, absorption, and elimination.

**dihybrid cross**  In genetics, a cross that involves two sets of characteristics.

**diploid**  Cells that contain homologous chromosomes. The number of chromosomes in the cells is the diploid number and is equal to $2n$ ($n$ is the number of homologous pairs).

**directional selection**  A process of natural selection that tends to favor phenotypes at one extreme of the phenotypic range.

**disaccharides**  Sugars made up of two monosaccharides held together by a covalent bond; e.g., sucrose and lactose.

**discontinuous variation**  Occurs when the phenotypes of traits controlled by a single gene can be sorted into two distinct phenotypic classes.

**disruptive selection**  A process of natural selection that favors individuals at both extremes of a phenotypic range.

**distal tubule**  The section of the renal tubule where tubular secretion occurs.

**divergent evolution**  The evolution of a single interbreeding population or species into two or more descendant species.

**divergent plate boundary**  The boundary between two tectonic plates that are moving apart.

**diversity**  The different types of organisms that occur in a community.

**DNA hybridization**  The formation of hybrid DNA molecules that contain a strand of DNA from two different species. The number of complementary sequences in common in the two strands is an indication of the degree of relatedness of the species.

**DNA ligase**  In recombinant DNA technology, an enzyme that seals together two DNA fragments from different sources to form a recombinant DNA molecule.

**DNA polymerase**  In DNA replication, the enzyme that links the complementary nucleotides together to form the newly synthesized strand

**dominance**  The property of one of a pair of alleles that suppresses the expression of the other member of the pair in heterozygotes.

**dominance hierarchy**  A social structure among a group of animals in which one is dominant and the others have subordinate nonbreeding positions.

**dominant**  Refers to an allele of a gene that is always expressed in heterozygotes.

**double fertilization**  A characteristic of angiosperms in which a pollen tube carries two sperm nuclei to the female gametophyte in the ovule. One nucleus gives rise to a diploid embryo, and the other to a triploid cell that forms the endosperm.

**double helix**   The structure of the DNA molecule in which two polynucleotide chains are coiled around a central axis.

**duodenum**   The upper part of the small intestine.

**duplication**   An extra copy of a chromosome segment without altering the number of chromosomes.

**eccrine glands**   Sweat glands that are linked to the sympathetic nervous system and are widely distributed over the body surface.

**ecological niche**   The role an organism occupies and the function it performs in an ecosystem; closely associated with feeding.

**ecological time**   A timescale that focuses on community events that occur on the order of tens to hundreds of years.

**ecology**   The study of how organisms interact with each other and their physical environment.

**ecosystem**   The community living in an area and its physical environment.

**ecosystem diversity**   The number of local ecosystems (or communities).

**ecotones**   Well-defined boundaries typical of closed communities.

**ecotype**   A subdivision of a species; a stage in the formation of a species such that reproductive isolation has occurred.

**ectoderm**   The outer layer of cells in embryonic development; gives rise to the skin, brain, and nervous system. Also, the outermost tissue layer in flatworms.

**ectotherms**   Animals with a variable body temperature. Their body temperature is determined by the environment.

**effector**   In a closed system, the element that initiates an action in response to a signal from a sensor.

**ejaculatory duct**   In males, a short duct that connects the vas deferens from each testis to the urethra.

**electron**   A subatomic particle with a negative charge. Electrons circle the atom's nucleus.

**electron acceptor**   A molecule that forms part of the electron transport system that transfers electrons ejected by chlorophyll during photosynthesis. Part of the energy carried by the electrons is transferred to ATP, part is transferred to NADPH, and part is lost in the transfer system.

**electrostatic attraction**   The attraction between atoms of opposite charge that holds the atoms together in ionic bonds.

**element**   A substance composed of atoms with the same atomic number; cannot be broken down in ordinary chemical reactions.

**elephantiasis**   The swelling of tissue due to a buildup of lymph caused by blocked lymph nodes; caused by a parasitic infection.

**elimination**   The removal of undigested food and water from the digestive system in the form of feces.

**elongation**   During protein synthesis, the growth of the polypeptide chain through the addition of amino acids; the second step in translation.

**Endangered Species Act**   A statute enacted in 1973 by the U.S. Congress that directs the U.S. Fish and Wildlife Service to maintain a list of endangered and threatened species; the main legal apparatus for protecting species in the United States.

**endochondral ossification**   The process by which human bones form from cartilage.

**endocrine system**   One of eleven major body systems in animals; a system of glands that works with the nervous system in controlling the activity of internal organs, especially the kidneys, and in coordinating the long-range response to external stimuli.

**endocytosis**   The incorporation of materials from outside the cell by the formation of vesicles in the plasma membrane. The vesicles surround the material so the cell can engulf it.

**endoderm**   The inner layer of cells in embryonic development that gives rise to organs and tissues associated with digestion and respiration. Also, the inner tissue layer in flatworms.

**endodermis**   A layer of cells surrounding the vascular cylinder of plants.

**endometrium**   The inner lining of the uterus.

**endoplasmic reticulum (ER)**   A network of membranous tubules in the cytoplasm of a cell; involved in the production of phospholipids and other functions. Rough ER is studded with ribosomes; smooth ER is not.

**endoskeleton**   An internal supporting skeleton with muscles on the outside; in vertebrates, consists of the skull, spinal column, ribs, and appendages.

**endosperm**   A food storage tissue that provides nutrients to the developing embryo in angiosperms; formed from the triploid cell produced when a sperm nucleus fertilizes the central cell.

**endothermic**   A reaction that gives off energy. The product is in a lower energy state than the reactants.

**endotherms**   Animals that have the ability to maintain a constant body temperature over a wide range of environmental conditions.

**endothermy**   The internal control of body temperature; the ability to generate and maintain internal body heat.

**energy**   The ability to bring about changes or to do work.

**energy flow**   The movement of energy through a community via feeding relationships.

**energy of activation**   The minimum amount of energy required for a given reaction to occur; varies from reaction to reaction.

**entropy**   The degree of disorder in a system. As energy is transferred from one form to another, some is lost as heat; as the energy decreases, the disorder in the system—and thus the entropy—increases.

**environmental stability**   One of the factors that helps cause the latitudinal and depth diversity gradients. Tropical communities are exposed to less envi-

ronmental change on a daily, seasonal, and hundred-year basis than other communities, and the lack of disturbances encourages diversity; similarly, deeper water is more stable than the shallower water along the shoreline, which has higher energy levels.

**enzyme cofactors**   *See* coenzymes.

**enzymes**   Protein molecules that act as catalysts in biochemical reactions.

**epidermis**   The outermost layer of skin consisting of several layers of epithelial cells—notably, keratinocytes—and, in the inner layer of the epidermis, basal cells and melanocytes.

**epididymis**   A long, convoluted duct on the testis where sperm are stored.

**epiglottis**   A flap of tissue that closes off the trachea during swallowing.

**epinephrine**   A hormone produced by the adrenal medulla and secreted under stress; contributes to the "fight or flight" response.

**erythrocytes**   Red blood cells; doubly concave, enucleated cells that transport oxygen in the blood.

**esophagus**   The muscular tube that extends from the pharynx to the stomach.

**essential amino acids**   Nine amino acids that cannot be synthesized by the body and must be supplied in the diet.

**essential elements**   Elements that plants require to survive and acquire from the atmosphere during photosynthesis: carbon, hydrogen, and oxygen.

**estrogen**   A female sex hormone that performs many important functions in reproduction.

**ethology**   The study of animal behavior in natural settings.

**ethylene**   A gaseous plant hormone that stimulates fruit ripening and the dropping of leaves.

**eukaryote**   A type of cell found in many organisms including single-celled protists and multicellular fungi, plants, and animals; characterized by a membrane-bounded nucleus and other membraneous organelles; an organism composed of such cells.

**euphotic zone**   The upper part of the marine biome where light penetrates and photosynthesis occurs; usually extends to about 200 meters below the water surface.

**eutrophication**   "Runaway" growth of aquatic plants that occurs when agricultural fertilizers containing phosphorus and nitrogen run off into lakes and ponds; also ultimately increases the plant death rate with the result that the bacterial decomposition of the dead plants uses up oxygen, causing fish and other organisms to suffocate.

**evaporation**   The part of the hydrologic cycle in which liquid water is converted to vapor and enters the atmosphere.

**evolutionary tree**   A diagram showing the evolutionary history of organisms bases on differences in amino acid sequences. Organisms with fewer differences are placed closer together while those with more differences are further apart.

**excess input** A disturbance of matter cycling that occurs when humans add a large quantity of matter (e.g., fertilizer and organic waste) to an ecosystem.

**excess output** A disturbance of matter cycling that occurs when humans suddenly release a large quantity of the matter contained in the biomass of an ecosystem.

**excretion** The process of removing the waste products of cellular metabolism from the body.

**excretory system** One of eleven major body systems in animals; regulates the volume and molecular and ionic constitution of internal body fluids and eliminates metabolic waste products from the internal environment.

**exocytosis** The process in which a membrane-enclosed vesicle first fuses with the plasma membrane and then opens and releases its contents to the outside.

**exoskeleton** A hard, jointed, external covering that encloses the muscles and organs of an organism; typical of many arthropods including insects.

**exothermic** A reaction where the product is at a higher energy level than the reactants.

**exponential rate** An extremely rapid increase, e.g., in the rate of population growth.

**expression** In relation to genes, the phenotypic manifestation of a trait. Expression may be age-dependent (e.g., Huntington disease) or affected by environmental factors (e.g., dark fur on Siamese cats).

**external respiration** In large animals with complex body plans, the processes of breathing, gas transport, and exchange of gases at the cellular level.

**extinction** The elimination of all individuals in a group.

**extracellular digestion** A form of digestion found in annelids, crustaceans, and chordates including vertebrates; takes place within the lumen of the digestive system, and the resulting nutrient molecules are transferred into the blood or body fluid.

**eyespot** A pigmented photoreceptor in euglenoids. The eyespot senses light and orients the cell for maximum rates of photosynthesis.

**families** Subcategories within orders.

**fats** Triglycerides that are solid at room temperature.

**feces** Semisolid material containing undigested foods, bacteria, bilirubin, and water that is produced in the large intestine and eliminated from the body.

**feeding** The ingestion of food.

**femur** The upper leg bone.

**fermentation** The synthesis of ATP in the absence of oxygen through glycolysis.

**fertilization** The fusion of two gametes (sperm and ovum) to produce a zygote that develops into a new individual with a genetic heritage derived from both parents.

**fibrous root** A root system found in monocots in which branches develop from the adventitious roots, forming a system in which all roots are about the same size and length.

**filaments** Slender, thread-like stalks that make up the stamens of a flower; topped by the anthers.

**filter feeders** Organisms such as sponges that feed by removing food from water that filters through their body.

**filtration** The removal of water and solutes from the blood; occurs in the glomerulus of the nephron.

**fitness** A measure of an individual's ability to survive and reproduce; the chance that an individual will leave more offspring in the next generation than other individuals.

**fixed action pattern (FAP)** A sequence of complex, unlearned behaviors found within a species.

**flame cell** A specialized cell at the blind end of a nephridium that filters body fluids.

**flowers** The reproductive structures in angiosperms where gametophytes are generated.

**fluid feeders** Animals such as aphids, ticks, and mosquitoes that pierce the body of a host plant or animal and obtain food from ingesting its fluids.

**fluid-mosaic** Widely accepted model of the plasma membrane in which proteins (the mosaic) are embedded in lipids (the fluid).

**follicles (ovary)** Structures in the ovary consisting of a developing egg surrounded by a layer of follicle cells.

**follicles (thyroid)** Spherical structures that make up the thyroid gland; contain a gel-like colloid surrounded by a single layer of cells, which secrete thyroglobulin into the colloid.

**follicle-stimulating hormone (FSH)** A hormone secreted by the anterior pituitary that promotes gamete formation in both males and females.

**fontanels** Membranous areas in the human cranial bones that do not form bony structures until the child is 14 to 18 months old; know as "soft spots."

**food chain** The simplest representation of energy flow in a community. At the base is energy stored in plants, which are eaten by small organisms, which in turn are eaten by progressively larger organisms; the food chain is an oversimplification in that most animals do not eat only one type of organism.

**food pyramid** A way of depicting energy flow in an ecosystem; shows producers (mostly plants) on the first level and consumers on the higher levels.

**food web** A complex network of feeding interrelations among species in a natural ecosystem; more accurate and more complex depiction of energy flow than a food chain.

**foraminifera** Single-celled protists with a shell. Accumulations of the shells of dead foraminifera form chalk deposits.

**forebrain** The part of the brain that consists of the diencephalon and cerebrum.

**forest-sustainable resources** Products that can be economically and repeatedly harvested from living trees over a long period of time, e.g., rubber, oils, and fruits, in contrast to products that require the tree to be cut down.

**fossil fuels** Fuels that are formed in the Earth from plant or animal remains; e.g., coal, petroleum, and natural gas.

**founder effect** The difference in gene pools between an original population and a new population founded by one or a few individuals randomly separated from the original population, as when an island population is founded by one or a few individuals; often accentuates genetic drift.

**fovea** The area of the eye in which the cones are concentrated.

**freshwater biome** The aquatic biome consisting of water containing fewer salts than the waters in the marine biome; divided into two zones: running waters (rivers, streams) and standing waters (lakes, ponds).

**frontal lobe** The lobe of the cerebral cortex that is responsible for motor activity, speech, and thought processes.

**Gaia** A hypothetical superorganism composed of the Earth's four spheres: the biosphere, hydrosphere, lithosphere, and atmosphere.

**gametes** Reproductive cells (ovum and sperm).

**gametophyte** The haploid stage of a plant exhibiting alternation of generations.

**ganglia** Clusters of neurons that receive and process signals; found in flatworms and earthworms.

**gap junctions** Junctions between the plasma membranes of animal cells that allow communication between the cytoplasm of adjacent cells.

**gastric pits** The folds and grooves into which the stomach lining is arranged.

**gastrin** A hormone produced by the pyloric gland area of the stomach that stimulates the secretion of gastric acids.

**gastroesophageal sphincter** A ring of muscle at the junction of the esophagus and the stomach that remains closed except during swallowing to prevent the stomach contents from entering the esophagus.

**gene banks** Storage centers where seeds, spores, sperm, and other genetic materials of wild species can be stored as a source of genetic diversity.

**gene pool** The sum of all the genetic information carried by members of a population.

**genera** Taxonomic subcategories within families (sing.: genus).

**genes** Specific segments of DNA that control cell structure and function; the functional units of inheritance.

**gene therapy** The insertion of normal or genetically altered genes into cells through the use of recombinant DNA technology; usually done to replace defective genes as part of the treatment of genetic disorders.

**genetic code** The linear series of nucleotides, read as triplets, that specifies the sequence of amino acids in proteins. Each triplet specifies an amino acid, and the same codons are used for the same amino acids in almost all life-forms, an indication of the universal nature of the code.

**genetic counseling**  Counseling to help individuals understand the ramifications of testing for genetic disorders and cope with the results if they choose to undergo such testing.

**genetic divergence**  The separation of a population's gene pool from the gene pools of other populations due to mutation, genetic drift, and selection. Continued divergence can lead to speciation.

**genetic diversity**  The number of genes and their alleles in the biosphere.

**genetic drift**  Random changes in the frequency of alleles from generation to generation; especially in small populations, can lead to the elimination of a particular allele by chance alone.

**genetic maps**  Diagrams showing the order of and distance between genes; constructed using crossover information.

**genetics**  The study of the structure and function of genes and the transmission of genes from parents to offspring.

**genital herpes**  A sexually transmitted disease caused by the herpes virus; results in sores on the mucus membranes of the mouth or genitals.

**genome**  The set of genes carried by an individual.

**genotype**  The genetic makeup of an organism with regard to an observed trait.

**geographic range**  The total area occupied by a population.

**geological time**  The span of time that has passed since the formation of the Earth and its physical structures; also, a timescale that focuses on events on the order of thousands of years or more.

**geotropism**  Plants' response to gravity: roots grow downward, showing positive geotropism, while shoots grow upward in a negative response.

**germ cells**  Cells in the reproductive organs of multicellular organisms that divide by meiosis to produce gametes.

**gibberellins**  A group of hormones that stimulate cell division and elongation in plants.

**gill slits**  Opening or clefts between the gill arches in fish. Water taken in by the mouth passes through the gill slits and bathes the gills. Also, rudimentary grooves in the neck region of embryos of air-breathing vertebrates such as humans; a characteristic of chordates.

**glial cells**  Nonconducting cells that serve as support cells in the nervous system and help to protect neurons.

**glomerulus**  A tangle of capillaries that makes up part of the nephron; the site of filtration.

**glucagon**  A hormone released by the pancreas that stimulates the breakdown of glycogen and the release of glucose, thereby increasing blood levels of glucose.

**glucocorticoids**  A group of steroid hormones produced by the adrenal cortex that are important in regulating the metabolism of carbohydrates, fats, and proteins.

**glycolipids**  Polysaccharides formed of sugars linked to lipids.

**glycoproteins**  Polysaccharides formed of sugars linked to proteins. On the outer surface of a membrane, they act as receptors for molecular signals originating outside the cell.

**glyoxysomes**  In cells of green plants, membrane-bound vesicles containing oxidative enzymes.

**Golgi complex**  Organelles in animal cells composed of a series of flattened sacs that sort, chemically modify, and package proteins produced on the rough endoplasmic reticulum.

**gonadotropin-releasing hormone (GnRH)**  A hormone produced by the hypothalamus that controls the secretion of luteinizing hormone.

**gonadotropins**  Hormones produced by the anterior pituitary that affect the testis and ovary; include follicle-stimulating hormone and luteinizing hormone.

**gonads**  The male and female sex organs.

**gonorrhea**  A sexually transmitted disease that is caused by a bacterium that inflames and damages epithelial cells of the reproductive system.

**grana**  A series of stacked disks containing chlorophyll; found in the inner membrane of chloroplasts.

**grasslands biome**  Occurs in temperate and tropical regions with reduced rainfall or prolonged dry seasons; characterized by deep, rich soil, an absence of trees, and large herds of grazing animals.

**greenhouse effect**  The heating that occurs when gases such as carbon dioxide trap heat escaping from the Earth and radiate it back to the surface; so-called because the gases are transparent to sunlight but not to heat and thus act like the glass in a greenhouse.

**ground system**  Tissue system, composed mainly of parenchyma cells with some collenchyma and sclerenchyma cells, that occupies the space between the epidermis and the vascular system; is involved in photosynthesis, water and food storage, and support; one of the four main tissue systems in plants.

**groups**  Associations of animals with some internal organization and possibly some division of labor; one of the three broad classes of social organization.

**growing season**  The period of the year during which plants are actively growing; is longer in the tropics and thus allows more plant growth in those areas.

**growth hormone (GH)**  A peptide hormone produced by the anterior pituitary that is essential for growth.

**gymnosperms**  Flowerless, seed-bearing land plants; the first seed plants.

**habitat disruption**  A disturbance of the physical environment in which a population lives.

**hair bulb**  The base of a hair; contains cells that divide mitotically to produce columns of hair cells.

**hair root**  The portion of a hair that extends from the skin's surface to the hair bulb.

**hair shaft**  The portion of a hair that extends above the skin's surface.

**half-life**  The time required for one-half of an original unstable radioactive element to be converted to a more stable daughter element.

**haploid**  Cells that contain only one member of each homologous pair of chromosomes (haploid number = $n$). At fertilization, two haploid gametes fuse to form a single cell with a diploid number of chromosomes.

**haplotype**  Closely linked genes or markers on a chromosome; in the immune system, the array of human leucocyte antigen (HLA) alleles on chromosome 6. These alleles must be matched between donor and recipient if organ transplants are to be successful. Each individual has two copies of chromosome 6 and thus two HLA haplotypes.

**Haversian canal**  The central opening of compact bone; contains nerves and blood vessels.

**heart**  The multicellular, chambered, muscular structure that pumps blood through the circulatory system by alternately contracting and relaxing.

**heartburn**  An irritation of the esophagus caused by acidic gastric juices leaking through the gastroesophageal sphincter.

**heartwood**  Inner rings of xylem that have become clogged with metabolic by-products and no longer transport water; visible as the inner darker areas in the cross section of a tree trunk.

**helper T cells**  A type of lymphocyte that stimulates the production of antibodies by activating B cells when an antigen is present.

**hemizygous**  Having one or more genes that have no allele counterparts.

**hemoglobin**  A red pigment in red blood cells that can bind with oxygen and is largely responsible for the blood's oxygen-carrying capacity.

**hemolytic disease of the newborn (HDN)**  A condition of immunological incompatibility between mother and fetus that occurs when the mother is Rh$^-$ and the fetus is Rh$^+$.

**hemophilia**  A genetic disorder that results in the absence of certain blood-clotting factors.

**hepatitis B**  A potentially serious viral disease that affects the liver; can be transmitted through sexual contact or through contact with infected blood.

**heterocyst**  A specialized cell that is present in some filamentous species of cyanobacteria and can fix nitrogen directly from the atmosphere.

**heterotrophic**  Refers to organisms such as animals that depend on preformed organic molecules from the environment as a source of nutrients.

**heterotrophs**  Organisms that obtain their nutrition by breaking down organic molecules in foods; include animals and some plants that eat organisms.

**heterozygous**  Having two different alleles (one dominant, one recessive) of a gene pair.

**histamine**  A chemical released during the inflammatory response that increases capillary blood flow in the affected area, causing heat and redness.

**histocompatibility antigens**  Antigens found on the surface of all cells in the body that must be matched in organ transplants and skin grafts to avoid the rejection of the transplant.

**homeobox genes**  Patterns genes that establish the body plan and position of organs in response to gradients of regulatory molecules.

**homeostasis**  The ability to maintain a relatively constant internal environment.

**homologous structures**  Body parts in different organisms that have similar bones and similar arrangements of muscles, blood vessels, and nerves and undergo similar embryological development, but do not necessarily serve the same function; e.g., the flipper of a whole and the forelimb of a horse.

**homologues**  A pair of chromosomes in which one member of the pair derived from the organism's maternal parent and the other from the paternal parent; found in diploid cells.

**homozygous**  Having identical alleles for a given gene.

**hormones**  Chemical substances that are produced in the endocrine glands and travel in the blood to target organs where they elicit a response.

**human chorionic gonadotropin (hCG)**  A peptide hormone secreted by the chorion that prolongs the life of the corpus luteum and prevents the breakdown of the uterine lining.

**Huntington disease**  A progressive and fatal disorder of the nervous system that develops between the ages of 30 and 50 years; caused by an expansion of a trinucleotide repeat and inherited as a dominant trait.

**hydrogen bond**  A weak bond between two atoms (one of which is hydrogen) with partial but opposite electrical charges.

**hydrophilic**  Water-loving.

**hydrophobic**  Water-fearing.

**hydrophytic leaves**  The leaves of plants that grow in water or under conditions of abundant moisture.

**hydrosphere**  The part of the physical environment that consists of all the liquid and solid water at or near the Earth's surface.

**hydrostatic skeleton**  Fluid-filled closed chambers that give support and shape to the body in organisms such as jellyfish and earthworms.

**hypertension**  High blood pressure; blood pressure consistently above 140/90.

**hypertonic**  A solution having a high concentration of solute.

**hyphae**  The multinucleate or multicellular filaments that make up the mycelium of a fungus (sing.: hypha).

**hypothesis**  An idea that can be experimentally tested.

**hypothalamus**  A region in the brain beneath the thalamus; consists of many aggregations of nerve cells and controls a variety of autonomic functions aimed at maintaining homeostasis.

**hypotonic**  A solution having a low concentration of solute.

**ileum**  The third and last section of the small intestine.

**immovable joint**  A joint in which the bones interlock and are held together by fibers or bony processes that prevent the joint from moving; e.g., the bones of the cranium.

**immune system**  One of the eleven major body systems in vertebrates; defends the internal environment against invading microorganisms and viruses and provides defense against the growth of cancer cells.

**immunoglobulins**  The five classes of protein to which antibodies belong (IgD, IgM, IgG, IgA, IgE).

**implantation**  The process in which the blastocyst embeds in the endometrium.

**imprinting**  A rapid learning process that takes place early in the life of a social animal and establishes a behavior pattern, such as recognition of and attraction to its own kind of animal or a substitute.

**incomplete dominance**  A type of inheritance in which the heterozygote has a phenotype intermediate to those of the homozygous parents.

**induction**  The process by which one cell or tissue type affects the developmental fate of another cell or tissue; occurs by activating certain genes and inactivating others.

**inflammation**  A reaction to the invasion of microorganisms through the skin or through the epithelial layers of the respiratory, digestive, or urinary system; characterized by four signs: redness, swelling, heat, and pain.

**inflammatory response**  The body's reaction to invading infectious microorganisms; includes an increase in blood flow to the affected area, the release of chemicals that draw white blood cells, an increased flow of plasma, and the arrival of monocytes to clean up the debris.

**ingestive feeders**  Animals that ingest food through a mouth.

**initiation**  The first step in translation; occurs when a messenger RNA molecule, a ribosomal subunit, and a transfer RNA molecule carrying the first amino acid bind together to form a complex; begins at the start codon on mRNA.

**insulin**  A hormone secreted by the pancreas that stimulates the uptake of glucose by body cells.

**integration**  The process of combining incoming information; one of the functions of the nervous system.

**integument**  Something that covers or encloses, e.g., the skin.

**integumentary system**  The skin and its derivatives (hair, nails, feathers, horns, antlers, and glands), which in multicellular animals protect against invading foreign microorganisms and prevent the loss or exchange of internal fluids.

**interferons**  Proteins released by cells in response to viral infection; activate the synthesis and secretion of antiviral proteins.

**intermediate filaments**  Components of the cytoskeleton that help to anchor organelles; composed of protein subunits.

**internal environment**  In multicellular organisms, the aqueous environment that is outside the cells but inside the body.

**interneurons**  Neurons that process signals from one or more sensory neurons and relay signals to motor neurons.

**internodes**  The stem regions between nodes in plants.

**interphase**  The period between cell divisions when growth and replacement occur in preparation for the next division; consists of gap 1 (G1), synthesis (S), and gap 2 (G2).

**interstitial**  Being situated within a particular organ or tissue.

**interstitial fluid**  Fluid surrounding the cells in body tissues; provides a path through which nutrients, gases, and wastes can travel between the capillaries and the cells.

**intracellular digestion**  A form of digestion in which food is taken into cells by phagocytosis; found in sponges and most protozoa and coelenterates.

**intracellular parasites**  Viruses that enter a host cell and take over the host's cellular machinery to produce new viral particles.

**inversion**  A reversal in the order of genes on a chromosome segment.

**ion**  An atom that has lost or gained electrons from its outer shell and therefore has a positive or negative charge, respectively; symbolized by a superscript plus or minus sign, e.g., $Na^+$, $Cl^-$.

**ionic bond**  A chemical bond in which atoms of opposite charge are held together by electrostatic attraction.

**isotonic**  Solutions with equal solute concentrations.

**isotopes**  Atoms with the same atomic number but different numbers of neutrons; indicated by adding the mass number to the element's name, e.g., carbon 12 or $^{12}C$.

**jejunum**  The second portion of the small intestine.

**karyotype**  The chromosomal characteristics of a cell; also, a representation of the chromosomes aligned in pairs.

**keratin**  A fibrous protein that fills mature keratinocytes near the skin's surface.

**keratinocytes**  The basic cell type of the epidermis; produced by basal cells in the inner layer of the epidermis.

**kidney stones**  Crystallized deposits of excess wastes such as uric acid, calcium, and magnesium that may form in the kidney.

**killer T cells**  See cytoxic T cells.

**kilocalorie**  The energy needed to heat 1000 grams of water from 14.5 to 15.5°C.

**kinetochores**  Structures at the centromeres of the chromosomes to which the fibers of the mitotic spindle connect.

**kingdoms**  Five broad categories (Monera, Proista, Plantae, Fungi, Animalia) into which organisms are grouped, based on common characteristics.

**Klinefelter syndrome**  In human, a genetically determined condition in which the individual has two X and one Y chromosome. Affected individuals are male and typically tall and infertile.

**labia majora** The outer folds of skin that cover and protect the genital region in women.

**labia minora** Thin membranous folds of skin outside the vaginal opening.

**lactose intolerance** A genetic trait characterized by the absence of the enzyme lactase, which breaks down lactose, the main sugar in milk and other dairy products.

**Langerhans' cells** Epidermal cells that participate in the inflammatory response by engulfing microorganisms and releasing chemicals that mobilize immune system cells.

**large intestine** Consists of the cecum, appendix, colon, and rectum; absorbs some nutrients, but mainly prepares feces for elimination.

**larva** A stage in the development of many insects and other organisms including sea urchins and sponges. In sponges, sexual reproduction results in the production of motile ciliated larvae.

**larynx** A hollow structure at the beginning of the trachea. The vocal cords extend across the opening of the larynx.

**latitudinal diversity gradient** The decrease in species richness that occurs as one moves away from the equator.

**latitudinal gradient** As latitude increases, a gradient of cooler, drier conditions occurs.

**law of the minimum** Holds that population growth is limited by the resource in shortest supply.

**L-dopa** A chemical related to dopamine that is used in the treatment of Parkinson's disease.

**leaves** The site of photosynthesis; one of the three major organs in plants.

**leukocytes** White blood cells; primarily engaged in fighting infection.

**lichens** Autotrophic organisms composed of a fungus (sac or club fungus) and a photosynthetic unicellular organism (e.g., a cyanobacterium or alga) in a symbiotic relationship; are resistant to extremes of cold and drought and can grow in marginal areas such as Arctic tundra.

**life history** The age at sexual maturity, age at death, and age at other events in an individual's lifetime that influence reproductive traits.

**ligaments** Dense parallel bundles of connective tissue that strengthen joints and hold the bones in place.

**light reactions** The photosynthetic process in which solar energy is harvested and transferred into the chemical bonds of ATP; can occur only in light.

**lignin** A polymer in the secondary cell wall of woody plant cells that helps to strengthen and stiffen the wall.

**linkage** The condition in which the inheritance of a specific chromosome is coupled with that of a given gene. The genes stay together during meiosis and end up in the same gamete.

**lipases** Enzymes secreted by the pancreas that are active in the digestion of fats.

**lithosphere** The solid outer layer of the Earth; includes both the land area and the land beneath the oceans and other water bodies.

**lobe-finned** Fish with muscular fins containing large jointed bones that attach to the body; one of the two main types of bony fish.

**logistic growth model** A model of population growth in which the population initially grows at an exponential rate until it is limited by some factor; then, the population enters a slower growth phase and eventually stabilizes.

**long-day plants** Plants that flower in the summer when nights are short and days are long; e.g., spinach and wheat.

**loop of Henle** A U-shaped loop between the proximal and distal tubules in the kidney.

**lungfish** A type of lobe-finned fish that breathe by a modified swim bladder as well as by gills.

**lungs** Sac-like structures of varying complexity where blood and air exchange oxygen and carbon dioxide; connected to the outside by a series of tubes and a small opening. In humans, the lungs are situated in the thoracic cavity and consist of the internal airways, the alveoli, the pulmonary circulatory vessels, and elastic connective tissues.

**lunula** The crescent-shaped area at the base of the nail where dividing cells are linked together and keratinized to form the nail.

**luteal phase** The second half of the ovarian cycle when the corpus luteum is formed; occurs after ovulation.

**luteinizing hormone (LH)** A hormone secreted by the anterior pituitary that stimulates the secretion of testosterone in men and estrogen in women.

**lymph** Interstitial fluid in the lymphatic system.

**lymphatic circulation** A secondary circulatory system that collects fluids from between the cells and returns it to the main circulatory system; the circulation of the lymphatic system, which is part of the immune system.

**lymphatic system** A network of glands and vessels that drain interstitial fluid from body tissues and return it to the circulatory system.

**lymph hearts** Contractile enlargements of vessels that pump lymph back into the veins; found in fish, amphibians, and reptiles.

**lymphocytes** White blood cells that arise in the bone marrow and mediate the immune response; include T cells and B cells.

**lysosomes** Membrane-enclosed organelles containing digestive enzymes. The lysosomes fuse with food vacuoles and digest their contents.

**macroevolution** The combination of events associated with the origin, diversification, extinction, and interactions of organisms which produced the species that currently inhabit the Earth.

**macromolecules** Large molecules made up of many small organic molecules; e.g., carbohydrates, lipids, proteins, and nucleic acids.

**macrophages** A type of white blood cell derived from monocytes that engulf invading antigenic molecules, viruses, and microorganisms and then display fragments of the antigen to activate helper T cells; ultimately stimulate the production of antibodies against the antigen.

**Malpighian tubules** The excretory organs of insects; a set of long tubules that open into the gut.

**mantle** In mollusks, a membranous or muscular structure that surrounds the visceral mass and secretes a shell if one is present.

**marine biome** The aquatic biome consisting of waters containing 3.5% salt on average; includes the oceans and covers more than 70% of the Earth's surface; divided into benthic and pelagic zones.

**marsupials** Pouched mammals. The young develop internally, but are born while in an embryonic state and remain in a pouch on the mother's abdomen until development is complete; e.g., kangaroos and opossums.

**mass extinction** A time during which extinction rates are generally accelerated so that more than 50% of all species then living become extinct; results in a marked decrease in the diversity of organisms.

**mast cells** Cells that synthesize and release histamine, as during an allergic response; found most often in connective tissue surrounding blood vessels.

**matter** Anything that occupies space and has mass.

**matter cycling** The flow of matter through various organisms and the physical environment of an ecosystem.

**maximum sustainable yield (MSY)** The maximum number of a food or game population that can be harvested without harming the population's ability to grow back.

**medulla** A term referring to the central portion of certain organs; e.g., the medulla oblongata of the brain and the adrenal medulla, which synthesizes epinephrine and norepinephrine.

**medulla oblongata** The region of the brain that, with the pons, makes up the hindbrain; controls heart rate, constriction and dilation of blood vessels, respiration, and digestion.

**medusa** The motile bell-shaped form of body plan in cnidarians; e.g., jellyfish.

**megakarocytes** Cells found in the bone marrow that produce platelets.

**megaspores** Four haploid cells produced by meiosis in the ovule of a flower. Usually, three degenerate, and the remaining cell becomes a gamete.

**meiosis** Cell division in which the chromosomes replicate, followed by two nuclear divisions. Each of the resulting gametes receives a haploid set of chromosomes.

**Meissner's corpuscles** Sensory receptors concentrated in the epidermis of the fingers and lips that make these areas very sensitive to touch.

**melanin** A pigment that gives the skin color and protects the underlying layers against damage by ultraviolet light; produced by melanocytes in the inner layer of the epidermis.

**melanocytes**   The cells in the inner layer of the epidermis that produce melanin.

**membrane-attack complex (MAC)**   A large cylindrical multiprotein complex formed by the complement system; kills invading microorganisms by embedding in their plasma membrane, creating a pore through which fluid flows, ultimately causing the cell to burst.

**menstrual cycle**   The recurring secretion of hormones and associated uterine tissue changes; typically 28 days in length.

**menstruation**   The process in which the uterine endometrium breaks down and sheds cells, resulting in bleeding; occurs approximately once a month. The first day marks the beginning of the menstrual and ovarian cycles.

**meristematic tissue**   Embryonic tissue located at the tips of stems and roots and occasionally along their entire length; can divide to produce new cells; one of the four main tissue systems in plants.

**mesoderm**   The middle layer of cells in embryonic development; gives rise to muscles, bones, and structures associated with reproduction.

**mesoglea**   A gel-like matrix that occurs between the outer and inner epithelial layers in cnidarians.

**mesophytic leaves**   The leaves of plants that grow under moderately humid conditions with abundant soil and water.

**Mesozoic Era**   The period beginning 245 million years ago and ending 65 million years ago; the age of the dinosaurs; falls between the Paleozoic and Cenozoic Eras and includes the Triassic, Jurassic, and Cretaceous Periods.

**metabolic pathway**   A series of individual chemical reactions in a living system that together serve one or more important functions. The product of one reaction in a pathway serves as the substrate for the following reaction.

**metabolism**   The sum of all chemical reactions (energy exchanges) in cells.

**metamorphosis**   The process of changing from one form to another; e.g., in insects, from the larval stage to the pupal stage to the reproductive adult stage.

**metaphase**   The stage of mitosis in which the chromosomes line up at the equator of the cell.

**metastasis**   The process in which cancer cells break away from the original tumor mass and establish new tumor sites elsewhere in the body.

**micelles**   Structures formed when bile salts surround digested fats in order to enable the water-insoluble fats to be absorbed by the epithelial cells.

**microevolution**   A small-scale evolutionary event such as the formation of a species from a preexisting one or the divergence of separated populations into new species.

**microfilaments**   Rods composed of actin that are found in the cytoskeleton and are involved in cell division and movement.

**micronutrients**   Elements that are required by plants in very small quantities, but are toxic in large quantities: iron, manganese, molybdenum, copper, boron, zinc, and chloride.

**microspores**   Four haploid cells produced by the meiotic division in the pollen sacs of flowers; undergo mitotic division and become encased in a thick protective wall to form pollen grains.

**microtubules**   Filaments about 25 nanometers in diameter found in cilia, flagella, and the cytoskeleton.

**microvilli**   Hair-like projections on the surface of the epithelial cells of the villi in the small intestine; increase the surface area of the intestine.

**midbrain**   A network of neurons that connects with the forebrain and relays sensory signals to other integrating centers.

**middle lamella**   A layer composed of pectin that cements two adjoining plant cells together.

**mineralocorticoids**   A group of steroid hormones produced by the adrenal cortex that are important in maintaining electrolyte balance.

**minerals**   Trace elements required for normal metabolism, as components of cells and tissues, and in nerve conduction and muscle contraction.

**minimum viable population (MVP)**   The smallest population size that can avoid extinction due to breeding problems or random environmental fluctuations.

**mitochondria**   Self-replicating membrane-bound cytoplasmic organelles in plant and animal cells that complete the breakdown of glucose, producing NADH and ATP (sing.: mitochondrion).

**mitosis**   The division of the nucleus and nuclear material of a cell; consists of four stages: prophase, metaphase, anaphase, and telophase.

**mitotic spindle**   A network of microtubules formed during prophase. Some microtubules attach to the centromeres of the chromosomes and help draw the chromosomes apart during anaphase.

**molecules**   Units of two or more atoms held together by chemical bonds.

**monocots**   One of the two major types of flowering plants; characterized by having a single cotyledon, floral organs arranged in cycles of three, and parallel-veined leaves; include grass, cattails, lilies, and palm trees.

**monoculture**   The growth of only one species in a given area; e.g., a cornfield.

**monocytes**   White blood cells that clean up dead viruses, bacteria, and fungi and dispose of dead cells and debris at the end of the inflammatory response.

**monohybrid cross**   In genetics, a cross that involves only one characteristic.

**monosaccharides**   Simple carbohydrates, usually with a five- or six-carbon skeleton; e.g., glucose and fructose.

**monotremes**   Egg-laying mammals; e.g., the spiny anteater and the duck-billed platypus.

**morph**   A distinct phenotypic variant within a population.

**morphological convergence**   The evolution of basically dissimilar structures to serve a common function.

**mosaic evolution**   A pattern of evolution where all features of an organism do not evolve at the same rate. Some characteristics are retained from the ancestral condition while others are more recently evolved.

**motor neurons**   Neurons that receive signals from interneurons and transfer the signals to effector cells that produce a response.

**motor output**   A response to the stimuli received by the nervous system. A signal is transmitted to organs that can convert the signals into action, such as movement or a change in heart rate.

**motor (efferent) pathways**   The portion of the peripheral nervous system that carries signals from the central nervous system to the muscles and glands.

**motor units**   Consist of a motor neuron with a group of muscle fibers; form the units into which skeletal muscles are organized; enable muscles to contract on a graded basis.

**mouth**   The oral cavity; the entrance to the digestive system where food is broken into pieces by the teeth and saliva begins the digestion process.

**mucus**   A thick, lubricating fluid produced by the mucous membranes that line the respiratory, digestive, urinary, and reproductive tracts; serves as a barrier against infection and, in the digestive tract, moistens food, making it easier to swallow.

**Mullerian inhibiting hormone (MIH)**   A hormone that promotes male sexual development in the fetus by causing the degeneration of the female duct system.

**muscle fibers**   Long, multinucleated cells found in skeletal muscles; made up of myofibrils.

**muscular system**   One of eleven major body systems in animals; allows movement and locomotion, powers the circulatory, digestive, and respiratory system, and plays a role in regulating temperature.

**mutation**   Any heritable change in the nucleotide sequence of DNA; can involve substitutions, insertions, or deletions of one or more nucleotides.

**mutualism**   A form of symbiosis in which both species benefit.

**mycelium**   The mass of interwoven filaments of hyphae in a fungus.

**mycorrhiza**   Occurs when a fungus (basidiomycete or zygomycete) weaves around or into a plant's roots and forms a symbiotic relationship. Fungal hyphae absorb minerals from the soil and pass them on to the plant roots while the fungus obtains carbohydrates from the plant (pl.: mycorrhizae).

**myelin sheath**   Layers of specialized glial cells, called Schwann cells, that coat the axons of many neurons.

**myofibrils**   Striated contractile microfilaments in skeletal muscle cells.

**myosin**   Thick protein filaments in the center sections of sarcomeres.

**nares**   Nostrils; the openings in the nose through which air enters.

**nastic movement**   A plant's response to a stimulus in which the direction of the response is independent of the direction of the stimulus.

**natural selection**   The process of differential survival and reproduction of fitter genotypes; can be stabilizing, directional, or disruptive. Better adapted individuals are more likely to survive to reproductive age and thus leave more offspring and make a larger contribution to the gene pool than do less fit individuals.

**nectaries**  Nectar-secreting organs in flowering plants that serve as insect feeding stations and thus attract insects, which then assist in the transfer of pollen.

**negative feedback control**  Occurs when information produced by the feedback reverses the direction of the response; regulates the secretion of most hormones.

**nektonic organisms**  "Swimmers"; one of the two main types of organisms in the pelagic zone of the marine biome.

**nephridium**  The excretory organ in flatworms and other invertebrates; a blind-ended tubule that expels waste through an excretory pore.

**nephron**  A tubular structure that is the filtering unit of the kidney; consists of a glomerulus and renal tubule.

**nerve cord**  A dorsal tubular cord of nervous tissue above the notochord of a chordate.

**nerve net**  An interconnected mesh of neurons that sends signals in all directions; found in radially symmetrical marine invertebrates, such as jellyfish and starfish, that have no head region or brain.

**nerves**  Bundles of neuronal processes enclosed in connective tissue that carry signals to and from the central nervous system.

**nervous system**  One of eleven major body systems in animals; coordinates and controls actions of internal organs and body systems, receives and processes sensory information from the external environment, and coordinates short-term reactions to these stimuli.

**net primary productivity (NPP)**  The rate at which producer (usually plants) biomass is created in a community.

**net secondary productivity (NSP)**  The rate at which consumer and decomposer biomass is produced in a community.

**neural tube**  A tube of ectoderm in the embryo that will form the spinal cord.

**neuromuscular junction**  The point where a motor neuron attaches to a muscle cell.

**neurons**  Highly specialized cells that generate and transmit bioelectric impulses from one part of the body to another; the functional unit of the nervous system.

**neurotransmitters**  Chemicals released from the tip of an axon into the synaptic cleft when a nerve impulse arrives; may stimulate or inhibit the next neuron.

**neutron**  An uncharged subatomic particle in the nucleus of an atom.

**niche**  The biological role played by a species.

**niche overlap**  The extent to which two species require similar resources; specifies the strength of the competition between the two species.

**nicotine adenine dinucleotide phosphate (NADP+)**  A substance to which electrons are transferred from photosystem I during photosynthesis; then combines with hydrogen to form NADPH, which is a storage form of energy.

**node**  The stem region of a plant where one or more leaves attach.

**node of Ranvier**  A gap between two of the Schwann cells that make up an axon's myelin sheath; serves as a point for generating a nerve impulse.

**nondisjunction**  The failure of chromosomes to separate properly during cell division.

**norepinephrine**  A hormone produced in the adrenal medulla and secreted under stress; contributes to the "fight or flight" response.

**notochord**  In chordates, a cellular rod that runs the length of the body and provides dorsal support. Also, a structure of mesoderm in the embryo that will become the vertebrae of the spinal column.

**nuclear area**  In prokaryotic cells, a region containing the cell's genetic information. Unlike the nucleus in eukaryotic cells, it is not surrounded by a membrane.

**nuclear pores**  Openings in the membrane of a cell's nuclear envelope that allow the exchange of materials between the nucleus and the cytoplasm.

**nucleic acids**  Polymers composed of nucleotides; e.g., DNA and RNA.

**nucleolus**  A round or oval body in the nucleus of a eukaryotic cell; consists of DNA and RNA and produces ribosomal RNA (pl.: nucleoli).

**nucleosomes**  Spherical bodies formed by coils of chromatin. The nucleosomes in turn are coiled to form the fibers that make up the chromosomes.

**nucleotides**  The subunits of nucleic acids; composed of a phosphate, a sugar, and a base.

**nucleus (atom)**  An atom's core; contains protons and one or more neutrons (except hydrogen, which has no neutrons).

**nucleus (cell)**  The largest, most prominent organelle in eukaryotic cells; a round or oval body that is surrounded by the nuclear envelope and contains the genetic information necessary for control of cell structure and function.

**occipital lobe**  The lobe of the cerebral cortex located at the rear of the head; is responsible for receiving and processing visual information.

**oils**  Triglycerides that are liquid at room temperature.

**oncogenes**  Genes that can activate cell division in cells that normally do not divide or do so only slowly.

**"one gene, one enzyme hypothesis"**  Holds that a single gene controls the production, specificity, and activity of each enzyme in a metabolic pathway. Thus, mutation of such a gene changes the ability of the cell to carry out a particular reaction and disrupts the entire pathway.

**"one gene, one polypeptide hypothesis"**  A revision of the one gene, one enzyme hypothesis. Some proteins are composed of different polypeptide chains encoded by separate genes, so the hypothesis now holds that mutation in a gene encoding a specific polypeptide can alter the ability of the encoded protein to function and thus produce an altered phenotype.

**oogenesis**  The production of ova.

**open community**  A community in which the populations have different density peaks and range boundaries and are distributed more or less randomly.

**opposable**  The capability of being placed against the remaining digits of a hand or foot; e.g., the ability of the thumb to touch the tips of the fingers on that hand.

**opsins**  Molecules in cone cells that bind to pigments, creating a complex that is sensitive to light of a given wavelength.

**orders**  Subcategories of classes.

**organelles**  Cell components that carry out individual functions; e.g., the cell nucleus and the endoplasmic reticulum.

**organism**  An individual composed of organ systems. Multiple organisms make up a population.

**organs**  Differentiated structures consisting of tissues and performing some specific function in an organism.

**organ systems**  Groups of organs that perform related functions.

**orgasm**  Rhythmic muscular contractions of the genitals combined with waves of intense pleasurable sensations; in males, results in the ejaculation of semen.

**osmoconformers**  Marine organisms that have no system of osmoregulation and must change the composition of their body fluids as the composition of the water changes; include invertebrates such as jellyfish, scallops, and crabs.

**osmoregulation**  The regulation of the movement of water by osmosis into and out of cells; the maintenance of water balance within the body.

**osmoregulators**  Marine vertebrates whose body fluids have about one-third the solute concentration of seawater; must therefore undergo osmoregulation.

**osmosis**  Diffusion of water molecules across a membrane in response to differences in solute concentration. Water moves from areas of high-water/low-solute concentration to areas of low-water/high-solute concentration.

**osteoarthritis**  A degenerative condition associated with the wearing away of the protective cap of cartilage at the ends of bones. Bone growths or spurs develop, restricting movement and causing pain.

**osteoblasts**  Bone-forming cells.

**osteoclasts**  Cells that remove material to form the central cavity in a long bone.

**osteocytes**  Bone cells that lay down new bone; found in the concentric layers of compact bone.

**osteoporosis**  A disorder in which the mineral portion of bone is lost, making the bone weak and brittle; occurs most commonly in postmenopausal women.

**out of Africa hypothesis**  Holds that modern human populations (*Homo sapiens*) are all derived from a single speciation event that took place in a restricted region in Africa.

**ovaries** In animals, the female gonads, which produce eggs (ova) and female sex hormones. In flowers, part of the female reproductive structure in the carpel; contain the ovules, where egg development occurs.

**overkill** The shooting, trapping, or poisoning of certain populations, usually for sport or economic reasons.

**oviducts** Tubes that connect the ovaries and the uterus; transport sperm to the ova, transport the fertilized ova to the uterus, and serve as the site of fertilization; also called the fallopian tubes or uterine tubes.

**ovulation** The release of the oocyte onto the surface of the ovary; occurs at the midpoint of the ovarian cycle.

**ovule** In seed plants, a protective structure in which the female gametophyte develops, fertilization occurs, and seeds develop; contained within the ovary.

**ovum** The female gamete.

**oxidation** The loss of electrons from the outer shell of an atom; often accompanied by the transfer of a proton and thus involves the loss of a hydrogen atom.

**oxytocin** A peptide hormone secreted by the posterior pituitary that stimulates the contraction of the uterus during childbirth.

**ozone** A triatomic form of oxygen that is formed in the stratosphere when sunlight strikes oxygen atoms. This atmospheric ozone helps filter radiation from the sun.

**pacemaker.** *See* sinoatrial node.

**Pacinian corpuscles** Sensory receptors located deep in the epidermis that detect pressure and vibration.

**Paleozoic Era** The period of time beginning 570 million years ago ending 245 million years ago; falls between the Proterozoic and Mesozoic Eras and is divided into the Cambrian, Ordovician, Silurian, Devonian, Carboniferous, and Permian Periods.

**palindrome** A sequence that reads the same in either direction; in genetics, refers to an enzyme recognition sequence that reads the same on both strands of DNA.

**pancreas** A gland in the abdominal cavity that secretes digestive enzymes into the small intestine and also secretes the hormones insulin and glucagon into the blood, where they regulate blood glucose levels.

**pancreatic islets** Clusters of endocrine cells in the pancreas that secrete insulin and glucagon; also known as islets of Langerhans.

**parallel evolution** The development of similar characteristics in closely related organisms as a result of adapting to similar environments and/or strategies of life.

**parasites** Organisms that live in, with, or on another organism. The parasites benefit from the association without contributing to the host.

**parasitism** A form of symbiosis in which the population of one species benefits at the expense of the population of another species; similar to predation,

but differs in that parasites act more slowly than predators and do not always kill the host.

**parasympathetic system** The subdivision of the autonomic nervous system that reverses the effects of the sympathetic nervous system.

**parenchyma** One of the three major cell types in plants; have thin usually multisided walls, are unspecialized but carry on photosynthesis and respiration and can store food; form the bulk of the plant body; found in the fleshy tissue of fruits and seeds, photosynthetic cells of leaves, and the vascular system.

**parietal lobe** The lobe of the cerebral cortex that lies at the top of the brain; processes information about touch, taste, pressure, pain, and heat and cold.

**passive transport** Diffusion across a plasma membrane in which the cell expends no energy.

**pectin** A substance in the middle lamella that cements adjoining plant cells together.

**pectoral girdle** In humans, the bony arch by which the arms are attached to the rest of the skeleton; composed of the clavicle and scapula.

**pedigree analysis** A type of genetic analysis in which a trait is traced through several generations of a family to determine how the trait is inherited. The information is displayed in a pedigree chart using standard symbols.

**pelagic zone** One of the two basic subdivisions of the marine biome; consists of the water above the sea floor and its organisms.

**pelvic girdle** In humans, the bony arch by which the legs are attached to the rest of the skeleton; composed of the two hipbones.

**pelvis** The hollow cavity formed by the two hip bones.

**pepsin** An enzyme produced from pepsinogen that initiates protein digestion by breaking down protein into large peptide fragments.

**pepsinogen** An inactive form of pepsin; synthesized and stored in cells lining the gastric pits of the stomach.

**peptic ulcer** Damage to the epithelial layer of the stomach lining; generally caused by bacterial infection.

**peptide bond** A covalent bond that links two amino acids together to form a polypeptide chain.

**peptides** Short chains of amino acids; include many hormones.

**perichondrium** A layer of connective tissue that forms around the cartilage during bone formation. Cells in the perichondrium lay down a peripheral layer that develops into compact bone.

**periosteum** A fibrous membrane that covers bones and serves as the site of attachment for skeletal muscles; contains nerves, blood vessels, and lymphatic vessels.

**peripheral nervous system** The division of the nervous system that connects the central nervous system to other parts of the body.

**peristalsis** Involuntary contractions of the smooth muscles in the walls of the esophagus, stomach, and intestines that propel food along the digestive tract.

**peroxisomes** Membrane-bound vesicles that contain oxidative enzymes.

**petals** Usually brightly colored elements of a flower that may produce fragrant oils; nonreproductive structures that attract pollinators.

**PGA (phosphoglycerate)** A three-carbon molecule formed when carbon dioxide is added to ribulose biphosphate during the dark reaction of photosynthesis. PGA is converted to PGAL, using ATP and NADPH.

**PGAL (phosphoglyceraldehyde)** A substance formed from PGA during the dark reaction of photosynthesis. Some PGAL is converted to glucose while other PGAL is used to form ribulose biphosphate to continue the dark reaction.

**phagocytes** Cells that can engulf and destroy microorganisms including viruses and bacteria; include neutrophils and monocytes.

**phagocytosis** A form of endocytosis in which blood cells surround and engulf invading bacteria or viruses.

**pharynx** The passageway between the mouth and the esophagus and trachea. Food passes from the pharynx to the esophagus, and air passes from the pharynx to the trachea.

**phenotype** The observed properties or outward appearance of a trait.

**pheromones** Chemical signals that travel between organisms rather than between cells within an organism; serve as a form of communication between animals.

**phloem** Tissue in the vascular system of plants that moves dissolved sugars and other products of photosynthesis from the leaves to other regions of the plant; consists of cells called sieve tubes.

**phospholipids** Asymmetrical lipid molecules with a hydrophilic head and a hydrophobic tail.

**photic zone** The layer of the ocean that is penetrated by sunlight; extends to a depth of about 200 meters.

**photoperiodism** The ability of certain plants to sense the relative amounts of light and dark in a 24-hour period; controls the onset of flowering in many plants.

**photosynthesis** The process by which plant cells use solar energy to produce ATP.

**photosystems** Clusters of several hundred molecules of chlorophyll in a thylakoid in which photosynthesis takes place. Eukaryotes have two types of photosystems: I and II.

**phototrophs** Organisms that use sunlight to synthesize organic nutrients; e.g., cyanobacteria, algae, and plants.

**phototropism** The reaction of plants to light in which the plants bend toward the light.

**phylum** The broadest taxonomic category within kingdoms (pl.: phyla).

**phytoalexins** Antibiotics produced by plants in response to leaf damage caused by an invading fungus; attack and kill the fungal hyphae.

**phytochrome** A pigment in plant leaves that detects day length and generates a response; partly responsible for photoperiodism.

**phytoplankton** A floating layer of photosynthetic organisms including algae that are an important source of atmospheric oxygen and form the base of the aquatic food chain.

**pineal gland** A small gland located between the cerebral hemispheres of the brain that secretes melatonin.

**pioneer community** The initial community of colonizing species.

**pituitary gland** A small gland located at the base of the brain; consists of an anterior and a posterior lobe and produces numerous hormones.

**placenta** An organ produced from interlocking maternal and embryonic tissue in placental mammals; supplies nutrients to the embryo and fetus and removes wastes.

**planaria** Small free-living flatworms with bilateral symmetry and cephalization. The freshwater type is often used as an experimental organism.

**planktonic organisms** "Floaters"; one of the two main types of organisms in the pelagic zone of the marine biome.

**plasma** The liquid portion of the blood. Along with the extracellular fluid, it makes up the internal environment of multicellular organisms.

**plasma cells** Cells produced from B cells that synthesize and release antibodies.

**plasmids** Self-replicating, circular DNA molecules found in bacterial cells; often used as vectors in recombinant DNA technology.

**plasmodesmata** Junctions in plants that penetrate cell walls and plasma membranes, allowing direct communication between the cytoplasm of adjacent cells (sing.: plasmodesma).

**plastids** Membrane-bound organelles in plant cells that function in storage or food production.

**platelets** In vertebrates, cell fragments that bud off from the megakaryocytes in the bone marrow; carry chemicals needed for blood clotting.

**plate tectonics** The movement of the plates that make up the surface of the Earth.

**pleura** A thin sheet of epithelium that covers the inside of the thoracic cavity and the outer surface of the lungs.

**pleural cavity** The space between the sheets of pleura (one covering the inside of the thoracic cavity, the other covering the outside of the lungs).

**polar covalent bond** A bond in which atoms share electrons in an unequal fashion. The resulting molecule has regions with positive and negative charges.

**pollen grains** The male gametophytes of flowers.

**polygenic inheritance** Occurs when a trait is controlled by several gene pairs; usually results in continuous variation.

**polymerase chain reaction (PCR)** A method of amplifying or copying DNA fragments that is faster than cloning. The fragments are combined with DNA poly-

merase, nucleotides, and other components to form a mixture in which the DNA is cyclically amplified.

**polynucleotides** Long chains of nucleotides formed by chemical links between the sugar and phosphate groups.

**polyp** The sessile form of body plan in cnidarians; e.g., the freshwater hydra.

**polyploidy** Abnormal variation in the number of chromosome sets.

**polysaccharides** Long chains of monosaccharide units; e.g., glycogen, starch, and cellulose.

**pons** The region that, with the medulla oblongata, makes up the hindbrain, which controls heart rate, constriction and dilation of blood vessels, respiration, and digestion.

**population** A group of individuals of the same species living in the same area at the same time and sharing a common gene pool.

**population dynamics** The study of the factors that affect the growth, stability, and decline of populations, as well as the interactions of those factors.

**portal system** An arrangement in which capillaries drain into a vein that opens into another capillary network.

**positive feedback control** Occurs when information produced by the feedback increases and accelerates the response.

**precipitation** The part of the hydrologic cycle in which the water vapor in the atmosphere falls to Earth as rain or snow.

**predation** One of the biological interactions that can limit population growth; occurs when organisms kill and consume other living organisms.

**predatory release** Occurs when a predator species is removed from a prey species such as by great reduction in the predator's population size or by the migration of the prey species to an area without major predators. The removal of the predator releases the prey from one of the factors limiting its population size.

**prehensile movement** The ability to seize or grasp.

**prenatal testing** Testing to detect the presence of a genetic disorder in an embryo or fetus; commonly done by amniocentesis or chorionic villi sampling.

**presymptomatic screening** Testing to detect genetic disorders that only become apparent later in life. The tests are done before the condition actually appears.

**prey switching** The tendency of predators to switch to a more readily available prey when one prey species becomes rare; allows the first prey population to rebound and helps prevent its extinction.

**primary cell wall** The cell wall outside the plasma membrane that surrounds plant cells; composed of cellulose.

**primary macronutrients** Elements that plants require in relatively large quantities: nitrogen, phosphorus, and potassium.

**primary root** The first root formed by a plant.

**primary structure** The sequence of amino acids in a protein.

**principle of independent assortment** Mendel's second law; holds that during gamete formation, alleles in one gene pair segregate into gametes independently of the alleles of other gene pairs. As a result, gametes contain all combinations of alleles.

**principle of segregation** Mendel's first law; holds that each pair of factors of heredity separate during gamete formation so that each gamete receives one member of a pair.

**prions** Infectious agents composed only of one or more protein molecules without any accompanying genetic information.

**producers** The first level in a food pyramid; consist of organisms that generate the food used by all other organisms in the ecosystem; usually consist of plants making food by photosynthesis.

**prokaryote** Type of cell that lacks a membrane-bound nucleus and has no membrane organelles; a bacterium. Prokaryotes are more primitive than eukaryotes.

**prolactin** A hormone produced by the anterior pituitary; secreted at the end of pregnancy when it activates milk production by the mammary glands.

**promoter** The specific nucleotide sequence in DNA that marks the beginning of a gene.

**prophase** The first stage of mitosis during which chromosomes condense, the nuclear envelope disappears, and the centrioles divide and migrate to opposite ends of the cell.

**prostaglandins** A class of fatty acids that has many of the properties of hormones; synthesized and secreted by many body tissues and have a variety of effects on nearby cells.

**prostate gland** A gland that is located near and empties into the urethra; produces a secretion that enhances sperm viability.

**proteinoids** Polymers of amino acids formed spontaneously from organic molecules; have enzyme-like properties and can catalyze chemical reactions.

**proteins** Polymers made up of amino acids that perform a wide variety of cellular functions.

**prothallus** In ferns, a small heart-shaped gametophyte.

**protists** Single-celled organisms; a type of eukaryote.

**proton** A subatomic particle in the nucleus of an atom that carries a positive charge.

**protostomes** Animals in which the first opening that appears in the embryo becomes the mouth; e.g., mollusks, annelids, and arthropods.

**proximal tubule** The winding section of the renal tubule where most reabsorption of water, sodium, amino acids, and sugar takes place.

**pseudocoelom** In nematodes, a closed fluid-containing cavity that acts as a hydrostatic skeleton to maintain body shape, circulate nutrients, and hold the major body organs.

**pseudocoelomates** Animals that have a body cavity that is in direct contact with the outer muscular layer of the body and does not arise by splitting of the mesoderm; e.g., roundworms.

**pseudopodia** Temporary cytoplasmic extensions from a cell that enables it to move (sing.: pseudopodium).

**pulmonary artery** The artery that carries blood from the right ventricle to the lungs.

**pulmonary circuit** The loop of the circulatory system that carries blood to and from the lungs.

**pulmonary vein** The vein that carries oxygenated blood from the lungs to the left atrium of the heart.

**punctuated equilibrium** A model that holds that the evolutionary process is characterized by long periods with little or no change interspersed with short periods of rapid speciation.

**pyloric sphincter** The ring of muscle at the junction of the stomach and small intestine that regulates the movement of food into the small intestine.

**quantum models of speciation** Models of evolution that hold that speciation sometimes occurs rapidly as well as over long periods, as the classical theory proposed.

**quaternary structure** In some proteins, a fourth structural level created by interactions with other proteins.

**race** A subdivision of a species that is capable of interbreeding with other members of the species.

**radially symmetrical** In animals, refers to organisms with their body parts arranged around a central axis. Such animals tend to be circular or cylindrical in shape.

**radiation** Energy emitted from the unstable nuclei of radioactive isotopes.

**ray-finned** Fish, such as trout, tuna, salmon, and bass, that have thin, bony supports holding the fins away from the body and an internal swim bladder that changes the buoyancy of the body; one of the two main types of bony fishes.

**reabsorption** The return to the blood of most of the water, sodium, amino acids, and sugar that were removed during filtration; occurs mainly in the proximal tubule of the nephron.

**receptacle** The base that attaches a flower to the stem.

**recessive** Refers to an allele of a gene that is expressed when the dominant allele is not present.

**recombinant DNA molecules** New combinations of DNA fragments formed by cutting DNA segments from two sources with restriction enzyme and then joining the fragments together with DNA ligase.

**recombinant DNA technology** A series of techniques in which DNA fragments are linked to self-replicating forms of DNA to create recombinant DNA molecules. These molecules in turn are replicated in a host cell to create clones of the inserted segments.

**recombination** A way in which meiosis produces new combinations of genetic information. During synapsis, chromatids may exchange parts with other chromatids, leading to a physical exchange of chromosome parts; thus, genes from both parents may be combined on the same chromosome, creating a new combination.

**reduction** The gain of an electron or a hydrogen atom.

**reductional division** The first division in meiosis; results in each daughter cell receiving one member of each pair of chromosomes.

**reflex** A response to a stimulus that occurs without conscious effort; one of the simplest forms of behavior.

**region of division** The area of cell division in the tip of a plant root.

**region of elongation** The area in the tip of a plant root where cells grow by elongating, thereby increasing the length of the root.

**region of maturation** The area where primary tissues and root hairs develop in the tip of a plant root.

**renal tubule** The portion of the nephron where urine is produced.

**renin** An enzyme secreted by the kidneys that converts angiotensinogen into angiotensin II.

**reproductive isolating mechanism** Biological or behavioral characteristics that reduce or prevent interbreeding with other populations; e.g., the production of sterile hybrids.

**reproductive system** One of eleven major body systems in animals; is responsible for reproduction and thus the survival of the species.

**resolution** In relation to microscopes, the ability to view adjacent objects as distinct structures.

**resource partitioning** The division of resources such that a few dominant species exploit most of the available resources while other species divide the remainder; helps explain why a few species are abundant in a community while others are represented by only a few individuals.

**respiratory surface** A thin, moist, epithelial surface that oxygen can cross to move into the body and carbon dioxide can cross to move out of the body.

**respiratory system** One of eleven major body systems in animals; moves oxygen from the external environment into the internal environment and removes carbon dioxide from the body.

**resting potential** The difference in electrical charge across the plasma membrane of a neuron.

**restriction enzymes** A series of enzymes that attach to DNA molecules at specific nucleotide sequences and cut both strands of DNA at those sites.

**restriction fragment length polymorphism (RFLP)** A heritable difference in DNA fragment length and fragment number; passed from generation to generation in a codominant way.

**retina** The inner, light-sensitive layer of the eye; includes the rods and cones.

**retroviruses** Viruses that contain a single strand of RNA as their genetic material and reproduce by copying the RNA into a complementary DNA strand using reverse transcriptase. The single-stranded DNA is then copied, and the resulting double-stranded DNA is inserted into a chromosome of the host cell.

**reverse transcriptase** An enzyme used in the replication of retroviruses; aids in copying the retrovirus's RNA into a complementary strand of DNA in the host cell.

**rheumatoid arthritis** A crippling form of arthritis that begins with inflammation and thickening of the synovial membrane, followed by bone degeneration and disfigurement.

**rhizoids** Filamentous structures in bryophytes that attach to a substrate and absorb moisture.

**rhizome** In ferns, a horizontal stem with upright leaves containing vascular tissue.

**rhodopsin** A visual pigment contained in the rods.

**ribosomes** Small organelles made of RNA and protein in the cytoplasm of prokaryotic and eukaryotic cells; aid in the production of proteins on the rough endoplasmic reticulum and ribosome complexes.

**RNA polymerase** During transcription, an enzyme that attaches to the promoter region of the DNA template, joins nucleotides to form the synthesized strand of RNA and detaches from the template when it reaches the terminator region.

**rods** Light receptors in primates' eyes that provide vision in dim light.

**root-leaf-vascular system axis** Refers to the arrangement in vascular plants in which the roots anchor the plant and absorb water and nutrients, the leaves carry out photosynthesis, and the vascular system connects the roots and leaves, carrying water and nutrients to the leaves and carrying sugars and other products of photosynthesis from the leaves to other regions of the plant.

**roots** Organs, usually occurring underground, that absorb nutrients and water and anchor the plant; one of the three major plant organs.

**salivary amylase** An enzyme secreted by the salivary glands that begins the breakdown of complex sugars and starches.

**salivary glands** Glands that secrete salvia into the mouth.

**saprophytes** Organisms that obtain their nutrients from decaying plants and animals. Saprophytes are important in recycling organic material.

**sapwood** Layers of xylem that are still functional in older woody plants; visible as the outer lighter areas in the cross section of a tree trunk.

**sarcomeres** The functional units of skeletal muscle; consist of filaments of myosin and actin.

**Schwann cells** Specialized glial cells that form the myelin sheath that coats many axons.

**sclerenchyma** One of the three major cell types in plants; have thickened, rigid, secondary walls that are hardened with lignin; provide support for the plant. Sclerenchyma cells include fiber and sclerids.

**scrotum** In mammals, a pouch of skin located outside the body cavity into which the testes descend; provides proper temperature for the testes.

**secondary cell wall** In woody plants, a second wall inside the primary cell wall; contains alternating layers of cellulose and lignin.

**secondary compounds** Plant products that are not important in metabolism but serve other purposes, such as attracting animals for pollination or killing parasites.

**secondary extinction** The death of one population due to the extinction of another, often a food species.

**secondary immunity** Resistance to an antigen the second time it appears. Because of the presence of B and T memory cells produced during the first exposure to the antigen, the second response is faster and more massive and lasts longer than the primary immune response.

**secondary macronutrients** Elements that plants require in relatively small quantities: calcium, magnesium, and sulfur.

**secondary structure** The structure of a protein created by the formation of hydrogen bonds between different amino acids; can be a pleated sheet, alpha helix, or random coil.

**second messenger** The mechanism by which nonsteroid hormones work on target cells. A hormone binds to receptors on the cell's plasma membrane activating a molecule—the second messenger—that activates other intracellular molecules that elicit a response. The second messenger can be cyclic AMP, cyclic GMP, inositol triphosphate, diacrylglycerol, or calcium.

**secretin** A hormone produced in the duodenum that stimulates alkaline secretions by the pancreas and inhibits gastric emptying.

**secretion** The release of a substance in response to the presence of food or specific neural or hormonal stimulation.

**sedimentary rock** Any rock composed of sediment, i.e., solid particles and dissolved minerals. Examples include rocks that form from sand or mud in riverbeds or on the sea bottom.

**segments** Repeating units in the body parts of some animals.

**selective breeding** The selection of individuals with desirable traits for use in breeding. Over many generations, the practice leads to the development of strains with the desired characteristics.

**semen** A mixture of sperm and various glandular secretions.

**semiconservative replication** Process of DNA replication in which the DNA helix is unwound and each strand serves as a template for the synthesis of a new complementary strand, which is linked to the old strand. Thus, one old strand is retained in each new molecule.

**semilunar valve** A valve between each ventricle of the heart and the artery connected to that ventricle.

**seminal vesicles** Glands that contribute fructose to sperm. The fructose serves as an energy source.

**seminiferous tubules** Tubules on the interior of the testes where sperm are produced.

**sensor** In a closed system, the element that detects change and signals the effector to initiate a response.

**sensory cortex** A region of the brain associated with the parietal lobe.

**sensory input** Stimuli that the nervous system receives from the external or internal environment; includes pressure, taste, sound, light, and blood pH.

**sensory neurons** Neurons that carry signals from receptors and transmit information about the environment to processing centers in the brain and spinal cord.

**sensory (afferent) pathways** The portion of the peripheral nervous system that carries information from the organs and tissues of the body to the central nervous system.

**sepals** Modified leaves that protect a flower's inner petals and reproductive structures.

**severe combined immunodeficiency (SCID)** A genetic disorder in which afflicted individuals have no functional immune system and are prone to infections. Both the cell-mediated immune response and the antibody-mediated response are absent.

**sex chromosomes** The chromosomes that determine the sex of an organism. In humans, females have two X chromosomes, and males have one X chromosome and one Y chromosome.

**sex hormones** A group of steroid hormones produced by the adrenal cortex.

**sex linkage** The condition in which the inheritance of a sex chromosome is coupled with that of a given gene; e.g., red-green color blindness and hemophilia in humans.

**sexual reproduction** A system of reproduction in which two haploid sex cells (gametes) fuse to produce a diploid zygote.

**shoot** The plant stem; provides support for the leaves and flowers; one of the three major plant organs.

**short-day plants** Plants that flower during early spring or fall when nights are relatively long and days are short; e.g., poinsettia and dandelions.

**sieve elements** Tubular, thin-walled cells that form a system of tubes extending from the roots to the leaves in the phloem of plants; lose their nuclei and organelles at maturity, but retain a functional plasma membrane.

**sieve plates** Pores in the end walls of sieve elements that connect the sieve elements together.

**sink** A body or process that acts as a storage device or disposal mechanism; e.g., plants and the oceans act as sinks absorbing atmospheric carbon dioxide. Also, a location in a plant where sugar is being consumed, either in metabolism or by conversion to starch.

**sinoatrial (SA) node** A region of modified muscle cells in the right atrium that sends timed impulses to the heart's other muscle cells, causing them to contract; the heart's pacemaker.

**sister chromatids** Chromatids joined by a common centromere and carrying identical genetic information.

**sleep movement** In legumes, the movement of the leaves in response to daily rhythms of dark and light. The leaves are horizontal in daylight and folded vertically at night.

**skeletal muscle** Muscle that is generally attached to the skeleton and causes body parts to move; consists of muscle fibers.

**skeletal system** One of eleven major body systems in animals; supports the body, protects internal organs, and, with the muscular system, allows movement and locomotion.

**skin** One of eleven major body systems in animals; the outermost layer protecting multicellular animals from the loss or exchange of internal fluids and from invasion by foreign microorganisms; composed of two layers: the epidermis and dermis.

**slash and burn agriculture** The cutting down and burning of trees to clear land for agricultural use; an example of excess output from the ecosystem.

**sliding filament model** Model of muscular contraction in which the actin filaments in the sarcomere slide past the myosin filaments, shortening the sarcomere and therefore the muscle.

**small intestine** A coiled tube in the abdominal cavity that is the major site of chemical digestion and absorption of nutrients; composed of the duodenum, jejunum, and ileum.

**smog** A local alteration in the atmosphere caused by human activity; mainly an urban problem that is often due to pollutants produced by fuel combustion.

**smooth muscle** Muscle that lacks striations; found around circulatory system vessels and in the walls of such organs as the stomach, intestines, and bladder.

**social behavior** Behavior that takes place in a social context and results from the interaction between and among individuals.

**societies** The most highly organized type of social organization; consist of individuals that show varying degrees of cooperation and communication with one another; often have a rigid division of labor.

**somatic** Relating to the non-gonadal tissues and organs of an organism's body.

**somatic senses** All senses except vision, hearing, taste, and smell; include pain, temperature, and pressure.

**somatic system** The portion of the peripheral nervous system consisting of the motor neuron pathways that innervate skeletal muscles.

**somites** Mesodermal structures formed during embryonic development that give rise to segmented body parts such as the muscles of the body wall.

**special senses** Vision, hearing, taste, and smell.

**species** One or more populations of interbreeding or potentially interbreeding organisms that are reproductively isolated in nature from all other organisms.

**species diversity** The number of living species on Earth.

**species packing** The phenomenon in which present-day communities generally contain more species than earlier communities because organisms have evolved more adaptations over time.

**species richness** The number of species present in a community.

**sperm** The male gamete.

**spicules** Needle-shaped skeletal elements in sponges that occur in the matrix between the epidermal and collar cells.

**spinal cord** A cylinder of nerve tissue extending from the brain stem; receives sensory information and sends output motor signals; with the brain, forms the central nervous system.

**spongy bone** The inner layer of bone; found at the ends of long bones and is less dense than compact bone. Some spongy bone contains red marrow.

**sporangia** The structures in which spores are produced (sing.: sporangium).

**spores** Impervious structures formed by some cells that encapsulate the cells and protect them from the environment; haploid cells that can survive unfavorable conditions and germinate into new haploid individuals or act as gametes in fertilization.

**sporophyte** The diploid stage of a plant exhibiting alteration of generations.

**stability** One of the phases of a population's life cycle. The population's size remains roughly constant, fluctuating around some average density. Also, the ability of a community to persist unchanged.

**stabilizing selection** A process of natural selection that tends to favor genotypic combinations that produce an intermediate phenotype.

**stalk** A leaf's petiole; the slender stem that supports the blade of a leaf and attaches it to a larger stem of the plant.

**stamens** The male reproductive structures of a flower; usually consist of slender, thread-like filaments topped by anthers.

**start codon** The codon on a messenger RNA molecule where protein synthesis begins.

**stem cells** Cells in bone marrow that produce lymphocytes by mitotic division.

**sternum** The breastbone.

**steroids** Compounds with a skeleton of four rings of carbon to which various side groups are attached; one of the three main classes of hormones.

**stigma** Part of the female reproductive structure of the carpel of a flower; the sticky surface at the tip of the style to which pollen grains attach.

**stimulus** A physical or chemical change in the environment that leads to a response controlled by the nervous system.

**stomach** The muscular organ between the esophagus and small intestine that stores, mixes, and digests food and controls the passage of food into the small intestine.

**stomata** Pores on the underside of leaves that can be opened or closed to control gas exchange and water loss.

**stop codon** The codon on a messenger RNA molecule where protein synthesis begins.

**stratification** The division of water in lakes and ponds into layers with different temperatures and oxygen content. Oxygen content declines with depth, while the uppermost layer is warmest in summer and coolest in winter.

**stressed community** A community that is disturbed by human activity, such as road building or pollution, and is inadvertently simplified. Some species become superabundant while others disappear.

**stroma** The matrix surrounding the grana in the inner membrane of chloroplasts.

**style** Part of the female reproductive structure in the carpel of a flower; formed from the ovary wall. The tip of the style carries the stigma to which pollen grains attach.

**subatomic particles** The three kinds of particles that make up atoms: protons, neutrons, and electrons.

**subspecies** A subdivision of a species; a population of a particular region genetically distinguishable from other such populations and capable of interbreeding with them.

**substrate feeders** Animals such as earthworms or termites that eat the soil or wood through which they burrow.

**sudden infant death syndrome (SIDS)** A disorder resulting in the unexpected death during sleep of infants, usually between the ages of two weeks and one year. The causes are not fully understood, but are believed to involve failure of automatic respiratory control.

**suppressor T cells** T cells that slow down and stop the immune response of B cells and other T cells.

**suprachiasmic nucleus (SCN)** A region of the hypothalamus that controls internal cycles of endocrine secretion.

**symbiosis** An interactive association between two or more species living together; may be parasitic, commensal, or mutualistic.

**sympathetic system** The subdivision of the autonomic nervous system that dominates in stressful or emergency situations and prepares the body for strenuous physical activity, e.g., causing the heart to beat faster.

**synapse** The junction between an axon and an adjacent neuron.

**synapsis** The alignment of chromosomes during meiosis I so that each chromosome is beside its homologue.

**synaptic cleft** The space between the end of a neuron and an adjacent cell.

**synovial joint** The most movable type of joint. The bones are covered by connective tissue, the interior of which is filled with synovial fluid, and the ends of the bones are covered with cartilage.

**syphilis** A sexually transmitted disease caused by a bacterial infection that produces an ulcer on the genitals and can have potentially serious effects if untreated.

**systematics** The classification of organisms based on information from observations and experiments; includes the reconstruction of evolutionary relatedness among living organisms. Currently, a system that divides organisms into five kingdoms (Monera, Protista, Plantae, Fungi, Animalia) is widely used.

**systemic circuit** The loop of the circulatory system that carries blood through the body and back to the heart.

**systole** The contraction of the ventricles that opens the semilunar valve and forces blood into the arteries.

**taiga biome** The region of coniferous forest extending across much of northern Europe, Asia, and North America; characterized by long, cold winters and short, cool summers and by acidic, thin soils.

**tap root** A primary root that grows vertically downward and gives off small lateral roots; occurs in dicots.

**tarsals** The bones that make up the ankle joint.

**taxis** The behavior when an animal turns and moves toward or away from an external stimulus (pl.: taxes).

**taxonomy** A systematic method of classifying plants and animals.

**T cells** The type of lymphocyte responsible for cell-mediated immunity; also protects against infection by parasites, fungi, and protozoans and can kill cancerous cells; circulate in the blood and become associated with lymph nodes and the spleen.

**telophase** The final stage of mitosis in which the chromosomes migrate to opposite poles, a new nuclear envelope forms, and the chromosomes uncoil.

**temperate forest biome** Extends across regions of the Northern Hemisphere with abundant rainfall and long growing seasons. Deciduous, broad-leaved trees are the dominant plants.

**temporal lobe** The lobe of the cerebral cortex that is responsible for processing auditory signals.

**tendons** Bundles of connective tissue that link muscle to bone.

**termination** The end of translation; occurs when the ribosome reaches the stop codon on the messenger RNA molecule and the polypeptide, the messenger RNA, and the transfer RNA molecule are released from the ribosome.

**tertiary structure** The folding of a protein's secondary structure into a functional three-dimensional configuration.

**testes** The male gonad; produce spermatozoa and male sex hormones.

**testosterone** Male sex hormone that stimulates sperm formation, promotes the development of the male duct system in the fetus, and is responsible for secondary sex characteristics such as facial hair growth.

**tetrad** The four chromatids in each cluster during synapsis; formed by the two sister chromatids in each of the two chromosomes.

**thalamus** The brain region that serves as a switching center for sensory signals passing from the brain stem to other brain regions; part of the diencephalon.

**theory** A hypothesis that has withstood extensive testing by a variety of methods.

**thermiogenesis** The generation of heat by raising the body's metabolic rate; controlled by the hypothalamus.

**thermoregulation** The regulation of body temperature.

**thigmotropism** Plants' response to contact with a solid object; e.g., tendrils' twining around a pole.

**thoracic cavity** The chest cavity in which the heart and lungs are located.

**thorax** In many arthropods, one of three regions formed by the fusion of the segments (others are the head and abdomen).

**thoroughfare channels** Shortcuts within the capillary network that allow blood to bypass a capillary bed.

**thylakoids** The specialized membrane structures in which photosynthesis takes place.

**thyroid-stimulating hormone.** A hormone produced by the anterior pituitary that stimulates the production and release of thyroid hormones.

**tight junctions** Junctions between the plasma membranes of adjacent cells in animals that form a barrier, preventing materials from passing between the cells.

**tissues** Groups of similar cells organized to carry out one or more specific functions.

**trace fossil** Any indication of prehistoric organic activity, such as tracks, trails, burrows, or nests.

**trachea** In insects and spiders, a series of tubes that carry air directly to cells for gas exchange; in humans, the air-conducting duct that leads from the pharynx to the lungs.

**tracheids** Long, tapered cells with pitted walls that form a system of tubes in the xylem and carry water and solutes from the roots to the rest of the plant.

**transcription** The synthesis of RNA on a DNA template.

**transfer RNAs (tRNAs)** Small, single-stranded molecules that bind to amino acids and deliver them to the proper codon on messenger RNA.

**transformation** In Griffith's experiments with strains of pneumonia bacterium, the process by which hereditary information passed from dead cells of one strain into cells of another strain, causing them to take on the characteristic virulence of the first strain.

**transforming factor** Griffith's name for the unknown material leading to transformation; later found to be DNA.

**translation** The synthesis of protein on a template of messenger RNA; consists of three steps: initiation, elongation, and termination.

**translocation** The movement of a segment from one chromosome to another without altering the number of chromosomes. Also, the movement of fluids through the phloem from one part of a plant to another, with the direction of movement depending on the pressure gradients between source and sink regions.

**transpiration** The loss of water molecules from the leaves of a plant; creates an osmotic gradient; producing tension that pulls water upward from the roots.

**trichocysts** Barbed, thread-like organelles of ciliated protozoans that can be discharged for defense or to capture prey.

**trophoblast** The outer layer of cells of a blastocyst that adhere to the endometrium during implantation.

**tropical rain forest biome** The most complex and diverse biome; found near the equator in South America and Africa; characterized by thin soils, heavy rainfall, and little fluctuation in temperature.

**tropism** The movement of plant parts toward or away from a stimulus in the plant's environment.

**true-breeding** Occurs when self-fertilization gives rise to the same traits in all offspring, generation after generation.

**tubal ligation** A contraceptive procedure in women in which the oviducts are cut, preventing the ova from reaching the uterus.

**tubal pregnancy** Occurs when the morula remains in the oviduct and does not descend into the uterus.

**tube-within-a-tube system** A type of body plan in animals. The organism has two openings—one for food and one for the elimination of waste—and a specialized digestive system.

**tubular secretion** The process in which ions and other waste products are transported into the distal tubules of the nephron.

**tubulins** The protein subunits from which microtubules are assembled.

**tumor suppressor genes** Genes that normally keep cell division under control, preventing the cell from responding to internal and external commands to divide.

**tundra biome** Extensive treeless plain across northern Europe, Asia, and North American between the taiga to the south and the permanent ice to the north. Much of the soil remains frozen in permafrost, and grasses and other vegetation support herds of large grazing mammals.

**Turner syndrome** In humans, a genetically determined condition in which an individual has only one sex chromosome (an X). Affected individuals are always female and are typically short and infertile.

**umbilical cord** The structure that connects the placenta and the embryo; contains the umbilical arteries and the umbilical vein.

**unicellular** Single celled.

**uniformitarianism** The idea that geological processes have remained uniform over time and that slight changes over long periods can have large-scale consequences; proposed by James Hutton in 1795 and refined by Charles Lyell.

**ureter** A muscular tube that transports urine by peristaltic contractions from the kidney to the bladder.

**urethra** A narrow tube that transports urine from the bladder to the outside of the body. In males, it also conducts sperm and semen to the outside.

**urine** Fluid containing various wastes that is produced in the kidney and excreted from the bladder.

**uterus** The organ that houses and nourishes the developing embryo and fetus.

**vaccination** The process of protecting against infectious disease by introducing into the body a vaccine that stimulates a primary immune response and the production of memory cells against the disease-causing agent.

**vaccine** A preparation containing dead or weakened pathogens that elicit an immune response when injected into the body.

**vacuoles** Membrane-bound fluid-filled spaces in plant and animal cells that remove waste products and store ingested food.

**vagina** The tubular organ that is the site of sperm deposition and also serves as the birth canal.

**vascular cambium** A layer of lateral meristematic tissue between the xylem and phloem in the stems of woody plants.

**vascular cylinder** A central column formed by the vascular tissue of a plant root; surrounded by parenchymal ground tissue.

**vascular parenchyma** Specialized cells in the phloem of plants.

**vascular system** Specialized tissues for transporting fluids and nutrients in plants; also plays a role in supporting the plant; one of the four main tissue systems in plants.

**vas deferens** The duct that carries sperm from the epididymis to the ejaculatory duct and urethra.

**vasectomy** A contraceptive procedure in men in which the vas deferens is cut and the cut ends are sealed to prevent the transportation of sperm.

**vasopressin** See antidiuretic hormone.

**vault** An elongated octagonal membrane-enclosed vesicle that may transport materials between the nucleus and the cytoplasm.

**vectors** Self-replicating DNA molecules that can be joined with DNA fragments to form recombinant DNA molecules.

**veins** Thin-walled vessels that carry blood to the heart.

**ventilation** The mechanics of breathing in and out through the use of the diaphragm and muscles in the wall of the thoracic cavity.

**ventricle** The chamber of the heart that pumps the blood into the blood vessels that carry it away from the heart.

**venules** The smallest veins. Blood flows into them from the capillary beds.

**vernalization** Artificial exposure of seeds or seedlings to cold to enable the plant to flower.

**vertebrae** The segments of the spinal column; separated by disks made of connective tissue (sing.: vertebra).

**vertebrate** Any animal having a segmented vertebral column; members of the subphylum Vertebrata; include reptiles, fishes, mammals, and birds.

**vesicles** Small membrane-bound spaces in most plant and animal cells that transport macromolecules into and out of the cell and carry materials between organelles in the cell.

**vessel elements** Short, wide cells arranged end to

end, forming a system of tubes in the xylem that moves water and solutes from the roots to the rest of the plant.

**vestigial structures** Nonfunctional remains of organs that were functional in ancestral species and may still be functional in related species; e.g., the dewclaws of dogs.

**villi** Finger-like projections of the lining of the small intestine that increase the surface area available for absorption. Also, projections of the chorion that extend into cavities filled with maternal blood and allow the exchange of nutrients between the maternal and embryonic circulations.

**viroids** Infective forms of nucleic acid without a protective coat of protein; unencapsulated single-stranded RNA molecules.

**vitamins** A diverse group of organic molecules that are required for metabolic reactions and generally cannot be synthesized in the body.

**vulva** A collective term for the external genitals in women.

**wood** The inner layer of the stems of woody plants; composed of xylem.

**xerophytic leaves** The leaves of plants that grow under arid conditions with low levels of soil and water.

**xylem** Tissue in the vascular system of plants that moves water and dissolved nutrients from the roots to the leaves; composed of various cell types including tracheids and vessel elements.

**zebroid** A hybrid animal that results from breeding zebras and horses.

**Z lines** Dense areas in myofibrils that mark the beginning of the sarcomeres. The actin filaments of the sarcomeres are anchored in the Z lines.

**zone of intolerance** The area outside the geographic range where a population is absent; grades into the zone of physiological stress.

**zone of physiological stress** The area in a population's geographic range where members of population are rare due to physical and biological limiting factors.

**zygospore** In fungi, a structure that forms from the diploid zygote created by the fusion of haploid hyphae of different mating types. After a period of dormancy, the zygospore forms sporangia, where meiosis occurs and spores form.

**zygote** A fertilized egg.

# INDEX

## A

Abiogenesis, 43
Abnormalities, chromosome, 132–34, 136
ABO blood typing, 416–17
Abscisic acid, 524
Absorption, 428
Absorptive feeding, 426
Acetylcholine, 366, 471
Acetyl CoA, 75
Achenes, 521
Acid chyme, 432
Acid deposition, 629–30
Acid rain, 629–30
Acids, 25–26
Acoelomate organisms, 261
Acquired immunodeficiency syndrome (AIDS),
    421–22
  first appearance of, 408
  replication of HIV virus in, 222, 223
  sexual transmission of, 542, 543, 544
Actin, 60, 88, 364
Action potential, 468–70
Active transport, 49
Acute bronchitis, 382
Adams, Thomas, 331
Adam's apple, 379
Adaptation, impact of, on organisms, 3
Adaptive radiation and rapid speciation, 203–4
Adaptive value of behavior, 563–66
Adenosine deaminase (ADA), 421
Adenosine triphosphate (ATP), 25, 365
  bacteria use of phosphorylation in generating, 77
  calculating energy yield in converting glucose to,
      77–78
  chemiosmosis in forming, 77
  in converting food into energy, 76–77
  and energy transfer within cells, 66–67
  formation of, 59
  as product of glycolysis, 73
Adhesion, 25
Adolescence, 553
Adrenal glands in regulating stress response, 503–5
Adrenocorticotropic hormone (ACTH), 503
Adrenoleukodystrophy, 130
Adulthood, 553
Adventitious roots, 311
Aerial roots, 312
Aerobic exercises, 393
Aeschylus, 107
African sleeping sickness, 230
African swallowtail butterfly, 186
*Agaricus campestris,* 242
Age structure in predicting growth potential,
    575–76
Aggregates in social organization, 560
Aggression, testosterone in, 566

Aging, 553
Agriculture, 630
  slash and burn, 618
  transition to, as cultural adaptation, 296–97
Agriculture runoff, and changes in population, 582
Air pollution, 380
Albinism, 116, 118
Alcoholic fermentation, 80
Aldosterone, 459–60
Algae
  brown, 233–34
  evolution of, 232–33
  green, 234–35, 303
  red, 233
Alginate, 234
Alkapton, 139
Alkaptonuria, 139, 140
Alleles, 112
  codominant, 122
  in genes, 122
  mutation in generation of new, 182
Allergens, 419
Allergies, 419–20
Alligators, 274
Alternation of generations, 101, 244
Altitude, effect of, on terrestrial biomes, 601–2
Altitudinal gradient, 601–2
Altruism, 564
  origins and persistence of, 567
Alveoli, 380, 382
Amensalism, 579
American Indians, groups of, 298
Amerinds, 298
Amine, 496, 497, 498
*Aminita phalloides,* 242
Amino acids, 442
  in forming proteins, 33–34
Ammonia, excretion of, 449
Amniocentesis, 163
Amniote egg, 211
Amoebocytes, 262
Amoeboid protozoa, 229–30
Amphibians, 274
  evolution of, 210–11
  heart in, 391
*Anableps,* 477
Anabolic reactions, 23
Anabolic steroids. *See* Steroids
Anaerobic organisms, 206
Analogous structures, 198
Analytical thinking, and healthy skepticism, 13
Anaphase, 92–93
Anaphylactic shock, 420
Anaphylaxis, 420
Anatomy, comparative, use of, to trace evolutionary
    change, 197–99
Anemia, sickle cell, 150–52, 163–64, 403

Aneuploidy, 132–33
Angina pectoralis, 398
Angiosperms, 213, 251–53, 254
  asexual reproductions in, 515–16
  dividing, 303–4
Angiotensin II, 383
Animal development
  asexual reproduction in, 529
  gastrulation in, 531
  induction, 532
  pattern formation in, 532
  principles of, 529–30
  sexual reproduction in, 530
Animalia, 4, 215, 219
Animals
  courtship behaviors in, 562–63
  desert, 600
  evolutionary origins and classification of,
      258–59
  reproductive strategy in, 520
  sexual reproduction in, 530
  trends in evolution in, 259
Annelids, 266, 353
  respiration in, 373
Annuals, 314
Antagonistic muscles, 362
Anterior pituitary
  neurosecretion in controlling, 500
  and reproduction, 502–3
Anther, 516
Antibiotics, 222, 227
Antibodies, 410, 412–13
Antibody-mediated immunity, 411–12
Anticancer agents, 320
Anticodon, 147
Antidiuretic hormone (ADH), 459, 503
Antigenic determinant, 413
Antigens, 410
Antlers, 345
Anus, 428
Aorta, 396
Apical dominance, 523
Apical meristem, 313–14
Apnea, 385
Apocrine glands, 345
Appendicular skeleton, 355–57
Appetite, control of, 439–40
Aquatic biomes
  definition of, 597
  types of, 602–5
Archaea, 224
Archaebacteria, 219, 224
Archean/Proterozoic era, 196, 206–7
*Archeopteryx,* 199, 212, 276
Aridity, 602
Aristotle, 17, 561
Arrector pili, 345

Sugar cane, 254
Sulfur
    in pollution, 630, 632
    as secondary macronutrients, 320
Sunburn, 348
Superorganism, 596
Suppressor T cells, 414, 415
Suprachiasmic nucleus (SCN), 510
Survival of the fittest, 174
Sutton, Walter, 120–21, 125
Swallowing, 429–30
Sweat glands, 345
Symbiosis, 207, 226, 228–29, 577
    lichens in, 243–44
    and mineral uptake, 321–22
    and population size and stability, 579
Symmetry, 351–52
Sympathetic system, 473
Synapses
    chemicals and disease affecting nervous system
        at, 472
    transfer of signals between neurons as,
        470–71
Synapsis, 96
Synaptic cleft, 470
Synovial joint, 360
Synthesis phase, 88
Synthetic seeds, 525
Syphilis, 542–43
Systematics, 215
Systemic circuit, 392
Systemic lupus erythematosus (SLE), 420
Systole, 396

# T

T4, 149
Tactile signals, in communication, 561–62
Taiga, 597
Taiga biome, 600, 602
Tanzania, 282
Tapeworms, 264
Tap root, 311
Tarsals, 357
Tatum, Edward, 140
Taxis in behavior, 557–58
Taxol, 320, 636
Taxonomy, 175
Tay-Sachs disease, 56, 118
T cells, 410
    cytotoxic, 414, 415
    helper, 415
    memory, 415
    suppressor, 414, 415
Tea, 254
Technology, and reproduction, 543–45
Tectonic cycle, 622–23, 623–24
Teeth in primates, 286, 288
Telophase, 93
Temin, Howard, 222
Temperate forest, 597, 598, 602
Temperature regulation, control of, 346
Temporal lobe, 478–79
Tendons, 360
Termination, 148
Terminator region, 146
Termite, 230
Terrapins, 275

Terrestrial biomes, 597–601
    definition of, 597
    impact of climate and altitude on, 601–2
Tertiary structure, 34
Test anxiety, Mendel and, 110
Testes, 532
Testis-determining factor (TDF), 130
Testosterone, 529, 549
    and aggression, 566
    negative feedback by, 535
    in skin production, 535
Tetrad, 96
Tetrahymena, 150, 335
Thalamus, 476
Thalassemia, 118
Theory, 6–7
Thermiogenesis, 346
Thermoacidophiles, 224
Thermodynamics, laws of, 65
Thermoreceptors, 482
Thermoregulation, 346
Thiamin, 443
Thigmotropism, 328
Thoracic cavity, 380
Thoroughfare channels, 395
Threonine, 442
Thucydides, 407
Thylakoids, 67, 225
    pigments in, 68
Thyroid gland
    in regulating metabolism and growth, 505
    secretion of, 338
Thyroid-stimulating hormone (TSH), 495
    in homeostasis, 339
    in regulating thyroid, 502
Thyroxin, 338
Tight junctions, 60
Time
    ecological, 607
    geological, 607
Tinbergen, Nicholas, 565–66
Tissue, 5
    meristematic, 306
    organization of cells into, 61
Tissue systems in plants, 306
Toads, 274
Tonguestones, 193–94
Tonsils, 402
Tortoises, 275
Toxic pollutants, as source of water pollution, 630,
        632
Toxic shock syndrome, 407
Toxoplasmosis, 231
Trace fossils, 194
Tracheae, 374, 375, 376
Tracheal systems, role of, in respiration, 374, 376
Tracheids, 323–24
Traits, genetic control of, 122–24
Transcription, 146
Transfer RNAs (tRNAs), 146, 147
Transformation, 140
Transforming factor, 140
Transgenic pigs, 169
Translation, 146–48
Translocation, 134–35, 326
Transmission electron microscope (TEM), 50
Transpiration, 324, 326, 624
    guard cells in regulation of, 324–26

Transport
    active, 49
    passive, 48–49
Tree rings, 315
Trichocysts, 231–32
Trichonymphs, 230
Triglycerides, 32
Trilobites, 209
Trisomy 21, 133
Trivers, Robert, 567
Trophoblast, 546
Tropical rain forest, 597, 598, 602
Tropical regions, acceleration of development in,
        633–34
Tropism, 327
True-breeding, 108
True fruits, 521
Trypanosomes, parasitic, 230
Tryptophan, 442
Tubal ligation, 543
Tubal pregnancy, 546
Tube within a tube, 259
Tube-within-a-tube digestive system, 427
Tubular reabsorption, 456
Tubular secretion, 457
Tubulins, 60
Tumor suppressor genes, 96
Tundra, 597, 600–601, 602
    lichens in, 243
Turner syndrome, 128, 132
Turtles, 274, 275
    evolution of, 275
Type I diabetes, 507
Type II diabetes, 507

# U

Ulcer, peptic, 432
Ultraviolet (UV) radiation, 348
Umbilical cord, 548
Unicellular organisms, 86
Uniformitarianism, 177
Unisexuality in plants, 519
Unsaturated fatty acids, 32
Urea, 449
Ureter, 454
Urethra, 454, 536
Urey, Harold, 29
Uric acid, 449
    excretion of, 450
Urine, 454
Uterine contractions, causes of, 541
Uterus, 539
UV Index, 632

# V

Vaccine, 407–8, 415–16
Vacuoles, function of, 57
Vagina, 539
Valine, 442
Value, adaptive in behavior, 563–66
Valve
    atrioventricular, 392
    semilunar, 392
Variation
    continuous, 122
    discontinuous, 122
    and natural selection, 184–87

# PHOTO CREDITS

## CHAPTER 1

**Opener** ©Rod Planck/Photo Researchers, Inc.
**1.1** ©Kirtley-Perkins/Visuals Unlimited
**1.2** Photo courtesy of the Minnesota Timberwolves
**1.3** ©R. Calentine/Visuals Unlimited
**1.4** ©Gary Retherford/Photo Researchers, Inc.
**1.5** ©T. Kuwabara D. Fawcett/Visuals Unlimited
**1.6** ©David Newman/Visuals Unlimited
**1.7** ©Dr. Gopal Murti/Science Photo Library/Photo Researchers, Inc.
**1.8a** ©Wm. S. Ormerod, Jr./Visuals Unlimited
**1.8b** ©John D. Cunningham/Visuals Unlimited
**1.8c** ©Kjell B. Sandved/Visuals Unlimited
**1.9** ©Michael Gabridge/Visuals Unlimited
**1.10** ©T.E. Adams/Visuals Unlimited
**1.11a** ©Nada Pecnik/Visuals Unlimited
**1.11b** ©Fletcher & Baylis/Photo Researchers, Inc.
**1.12** ©Mack Henley/Visuals Unlimited
**1.13** ©Glenn Oliver/Visuals Unlimited
**1.14** ©Craig K. Lorenz/Photo Researchers, Inc.
**1.15a** *Chemical Industry Upheld by Pure Science.* 1937 painting by Leon H. Soderston (1894–1955).
**1.15b** *Frankentomato* by Terry Miura, National Museum of American History
**1.16a** Bettmann Archive
**1.16b** Bettmann Archive
**1.17** ©SIU/Visuals Unlimited

## PART 1

**Opener** ©Dr. Don Fawcett/Science Source/Photo Researchers, Inc.

## CHAPTER 2

**Opener** ©J. Alcock/Visuals Unlimited
**2.1** ©Dr. Jean Lorre/SPL/Science Source/Photo Researchers, Inc.
**2.2** Photo courtesy Herb Orth, Life Magazine ©Time Warner Inc.; painting by Chesley Bonestell, ©The Estate of Chesley Bonestell
**2.12** ©E.R. Degginger/Photo Researchers, Inc.
**Sidebar** ©NIH Science Source/Photo Researchers, Inc.
**2.18** ©Simon Fraser/Science Photo Library/Photo Researchers, Inc.
**2.23a** ©Carolina Biological/Visuals Unlimited
**2.23b** ©William Ober/Visuals Unlimited
**2.24** ©Kjell B. Sandved/Visuals Unlimited
**2.32a–b** Photo courtesy of Sidney W. Fox, Distinguished Research Professor, Southern Illinois at Carbondale.
**2.33a–b** Photo courtesy of Dr. J. William Schopf, UCLA Center for the Study of Evolution and the Origin of Life

## CHAPTER 3

**Opener** ©NIBSC/Science Photo Library/Photo Researchers, Inc.
**3.1b** Armed Forces Institute of Pathology, Neg. No. 66-1836-1
**3.3a–d** ©Dr. Lloyd M. Beidler/Science Photo Library/Photo Researchers, Inc.
**3.5a** ©Science Source/Photo Researchers, Inc.
**3.6** ©David M. Phillips/Visuals Unlimited
**3.8** ©Don W. Fawcett/Visuals Unlimited

**Beyond the Basics** (p. 50, left) ©W. Ormerod/Visuals Unlimited
**Beyond the Basics** (p. 50, right) ©M. Abbey/Visuals Unlimited
**Beyond the Basics** (p. 51, upper left) ©T.E. Adams/Visuals Unlimited
**Beyond the Basics** (p. 51, upper right) ©Sinclair Stammers/Science Photo Library/Photo Researchers, Inc.
**Beyond the Basics** (p. 51, lower left) ©M. Schliwa/Visuals Unlimited
**Beyond the Basics** (p. 51, lower right) ©Stanley Flegler/Visuals Unlimited
**3.13** ©Prof. Marcel Bessis, Science Source/Photo Researchers, Inc.
**3.14** ©S. Ito & D.W. Fawcett/Visuals Unlimited
**3.15a** ©Biophoto Associates/Photo Researchers, Inc.
**3.16** ©C.G. Van Dyke/Visuals Unlimited
**3.17** ©Biophoto Associates/Science Source/Photo Researchers, Inc.
**3.18** ©Don W. Fawcett/Visuals Unlimited
**3.19** ©Don W. Fawcett/Visuals Unlimited
**3.20b** ©K.G. Murti/Visuals Unlimited
**3.21b** ©David M. Phillips/Visuals Unlimited
**3.23b** Photo courtesy of Dr. Leonard H. Rome, UCLA School of Medicine
**3.24** ©Cabisco/Visuals Unlimited
**3.25a** ©Bill Longcore/Photo Researchers, Inc.
**3.26a** ©George B. Chapmen & Priscilla Devadcoss/Visuals Unlimited
**3.27b** ©I. Gibbons & D. Fawcett/Visuals Unlimited
**3.27c** ©David M. Phillips/Visuals Unlimited
**3.28a** ©M. Schliwa/Visuals Unlimited

## CHAPTER 4

**Opener** ©Bullaty/Lomeo/The Image Bank
**4.1** ©A. Gurmankin/Visuals Unlimited
**4.4a** ©John Sohlden/Visuals Unlimited
**4.4b** ©Walt Anderson/Visuals Unlimited
**4.4c** ©Hank Levine/Visuals Unlimited
**4.5a** ©M. Powell/Visuals Unlimited
**4.11a** ©WH01/D.Foster/Visuals Unlimited
**4.11b** ©WH01/J. Edmond/Visuals Unlimited

## CHAPTER 5

**Opener** ©Dr. Elena A. Smirnova, University of Oregon
**5.1** ©G. Shih-R. Kessel/Visuals Unlimited
**5.2** ©David M. Phillips/Visuals Unlimited
**5.5** Photo courtesy of Cytogenetics Laboratory, Loyola University Medical Center, Maywood, Illinois
**5.6b–k** ©Dr. Andrew S. Bajer, University of Oregon
**5.7b** ©Science VU/Visuals Unlimited
**5.9a–b** ©David M. Phillips/Visuals Unlimited
**5.11** ©David M. Phillips/Visuals Unlimited
**5.12** ©Ken Greer/Visuals Unlimited
**5.19b** ©Cabisco/Visuals Unlimited
**5.20a–b** ©Cabisco/Visuals Unlimited

## PART II

**Opener** ©Dr. Andrew S. Bajer, University of Oregon

## CHAPTER 6

**Opener** ©Dr. Jeremy Burgess/Science Photo Library/Photo Researchers, Inc.
**6.1a** ©James W. Richardson/Visuals Unlimited

**6.1b** ©David Sieren/Visuals Unlimited
**6.1c** ©R. Calentine/Visuals Unlimited
**6.2** Photo by K. Libal. From Moravske Museum, Bruno

## CHAPTER 7

**Opener** ©Biophoto Assoc./Science Source/Photo Researchers, Inc.
**7.1a–b** ©E.R. Degginger/Photo Researchers, Inc
**7.1c** ©Leslie Holzer/Photo Researchers, Inc.
**7.4b** ©Joe McDonald/Visuals Unlimited
**7.4c** ©Tim Hauf/Visuals Unlimited
**7.4d** ©M. Long/Visuals Unlimited
**7.4e** ©Marek Litman/Visuals Unlimited
**7.4f** ©Mark D. Cunningham/Visuals Unlimited
**7.7** ©Renee Lynn/Photo Researchers, Inc.
**7.11a–b** Photo courtesy of Dr. Irene Uchida, Genetic Services, Oshawa General Hospital, Hamilton, Ontario, Canada
**7.12** ©R. Calentine/Visuals Unlimited
**7.19a** ©Bernd Wittich/Visuals Unlimited
**7.19b** Photo courtesy of Dr. Irene Uchida, Genetic Services, Oshawa General Hospital, Hamilton, Ontario, Canada
**7.20a** ©G. Büttner/Naturbild/OKAPIA/Photo Researchers, Inc.
**7.20b** ©Holt Studios International (Nigel Cattlin)/Photo Researchers, Inc.
**7.20c** ©Alan L. Detrick/Photo Researchers, Inc.
**7.21a** ©Seth Joel/Science Photo Library/Photo Researchers, Inc.
**7.21b** ©Adrienne Hart-Davis/Science Photo Library/Photo Researchers, Inc.
**7.21c** ©G. Büttner/Naturbild/OKAPIA/Photo Researchers, Inc.
**7.22** Photo courtesy of Dr. Irene Uchida, Genetic Services, Oshawa General Hospital, Hamilton, Ontario, Canada

## CHAPTER 8

**Opener** ©Will and Deni McIntyre/Photo Researchers, Inc.
**8.2** ©John D. Cunningham/Visuals Unlimited
**8.4a** ©Lee D. Simon/Science Source/Photo Researchers, Inc.
**8.4b** ©M. Wurtz/Biozentrum, University of Basel/Science Photo Library/Photo Researchers, Inc.
**8.9** ©J.R. Paulson, U.K. Laemmli, D.W. Fawcett/Visuals Unlimited
**8.17** ©Stanley Flegler/Visuals Unlimited

## CHAPTER 9

**Opener** ©K.G. Murti/Visuals Unlimited
**9.1a** ©John Sohlden/Visuals Unlimited
**9.1b** ©Renee Lynn/Photo Researchers, Inc.
**9.1c** ©Alexander Lowry/Photo Researchers, Inc.
**9.1d** ©Jacana/Photo Researchers, Inc.
**9.13** ©Philippe Plailly/Eurelios/Science Photo Library/Photo Researchers, Inc.
**9.15a** Photo courtesy of Bio Rad
**9.16** Photo courtesy of Calgene
**9.17** Photo courtesy of Agricultural Research Service, USDA

## PART III

**Opener** ©Joseph van Os/The Image Bank

## CHAPTER 10

**Opener** ©Christian Grzimek/Okapia/Photo Researchers, Inc.
**10.1a** ©Bruce Gaylord/Visuals Unlimited
**10.1b** ©Leonard Lee Rue III/Visuals Unlimited
**10.1c** ©Joe McDonald/Visuals Unlimited
**10.1d** ©Joe McDonald/Visuals Unlimited
**10.1e** ©Milton H. Tierney, Jr./Visuals Unlimited
**10.2a–c** Photo courtesy of History & Special Collections Division, Louise M. Darling Biomedical Library, UCLA
**10.4** ©Frank Hanna/Visuals Unlimited
**10.5** ©Science VU/Visuals Unlimited
**10.7a** ©William J. Weber/Visuals Unlimited
**10.7b** ©Science VU/Visuals Unlimited
**10.9** ©Stan Elems/Visuals Unlimited
**10.10** ©Fred McConnaughey/Photo Researchers, Inc.
**10.11** ©Kirtley-Perkins/Visuals Unlimited
**10.13a** ©Tim Davis/Photo Researchers, Inc.
**10.13b** ©Frans Lanting/Photo Researchers, Inc.
**10.13c** ©Tim Davis/Photo Researchers, Inc
**10.16a–b** ©Michael Tweedie/Photo Researchers, Inc.
**10.18** ©Rod Planck/Photo Researchers, Inc.
**10.19a** ©Thomas Gula/Visuals Unlimited
**10.19b** ©W. Mike Howell/Visuals Unlimited
**10.20** ©Mark Boulrow/Visuals Unlimited

## CHAPTER 11

**Opener** ©John Agnew, courtesy of the Academy of Natural Sciences
**11.1a** ©Don W. Fawcett/Visuals Unlimited
**11.1b** Photo courtesy of Dr. Raul J. Cano, California Polytechnic State University
**11.2** Photo courtesy of Wayne Moore.
**11.12a** ©James R. McCullagh/Visuals Unlimited
**11.12b** ©Glenn M. Oliver/Visuals Unlimited
**11.17a–b** ©A. Kerstitch/Visuals Unlimited
**11.18** Photo from NHK Special *Planet of Life,* Tokyo, Japan
**11.19** Photo courtesy of the Field Museum of Natural History, Chicago, #Geo80820c
**11.20** Photo courtesy of Sue Monroe
**11.21** Negative number K17092. Photo by M. Ellison. Courtesy of the Department of Library Services, American Museum of Natural History.
**11.24b** Photo courtesy of D. J. Nichols, U.S. Geological Survey

## CHAPTER 12

**Opener** ©Science Source/Photo Researchers, Inc.
**12.1b** ©K.G. Murti/Visuals Unlimited
**12.1c** ©Robert Caughhey/Visuals Unlimited
**12.1d** ©K.G. Murti/Visuals Unlimited
**12.1e** ©B. Heggeler/Visuals Unlimited
**12.3a–c** ©Hans Gelderblom/Visuals Unlimited
**Sidebar** ©Science VU-Wayside/Visuals Unlimited
**12.5** ©Ralph Slepecky/Visuals Unlimited
**12.6a–b** ©David M. Phillips/Visuals Unlimited
**12.6c** ©Stanley Flegler/Visuals Unlimited
**12.7b** ©E. White/Visuals Unlimited
**12.8a** ©R. Calentine/Visuals Unlimited
**12.8b** ©Ron Dengler/Visuals Unlimited
**12.9** ©T.J. Beveridge & S.S. Schultze/Visuals Unlimited
**12.10** ©Philip Sze/Visuals Unlimited
**12.11** ©A. Gurmankin/Visuals Unlimited
**12.13** ©George Chapman & Peter Chen/Visuals Unlimited
**12.14** ©A.M. Siegelman/Visuals Unlimited
**12.15** ©M. Abbey/Visuals Unlimited
**12.16a** ©Science VU/Visuals Unlimited
**12.16b** ©A.M. Siegelman/Visuals Unlimited
**12.16c** ©G.R. Roberts/Photo Researchers, Inc.
**12.17** ©Fred Marsik/Visuals Unlimited
**12.18a** ©M. Abbey/Visuals Unlimited
**12.18b** ©George Loun/Visuals Unlimited
**12.19** ©Daniel Snyder/Visuals Unlimited
**12.20** ©M. Abbey/Visuals Unlimited
**12.22a** ©R. Calentine/Visuals Unlimited
**12.22b** ©Philip Sze/Visuals Unlimited
**12.23a** ©A.M. Siegelman/Visuals Unlimited
**12.23b** ©Sanford Berry/Visuals Unlimited

12.24   ©D. Gotshall/Visuals Unlimited
12.25   ©John D. Cunningham/Visuals Unlimited
12.26a  ©R. Calentine/Visuals Unlimited
12.26b  ©L.L. Sims/Visuals Unlimited
12.26c  ©John D. Cunningham/Visuals Unlimited
12.27   ©R. Kessel & G. Shih/Visuals Unlimited
12.28   ©Cabisco/Visuals Unlimited
12.29   ©Cabisco/Visuals Unlimited


## CHAPTER 13

Opener  ©Charles A. Mauzy/Tony Stone Images
13.1a   ©Dana Richter/Visuals Unlimited
13.2a   ©Phil Dotson/Visuals Unlimited
13.2b   ©John D. Cunningham/Visuals Unlimited
13.4    ©D. Newman/Visuals Unlimited
13.5    ©David M. Phillips/Visuals Unlimited
13.6a   ©George Loun/Visuals Unlimited
13.6b   ©John D. Cunningham/Visuals Unlimited
13.6c   ©James W. Richardson/Visuals Unlimited
13.7b   ©S. Flegler/Visuals Unlimited
13.8    ©Sylvia Duran Sharnoff/Visuals Unlimited
13.9    ©Bud Lehnhausen/Photo Researchers, Inc.
13.10   ©Dana Richter/Visuals Unlimited
13.12   ©C. Gerald Van Dyke/Visuals Unlimited
13.13a  ©Doug Sokell/Visuals Unlimited
13.13b  ©Gregory K. Scott/Photo Researchers, Inc.
13.13c  ©Pat Armstrong/Visuals Unlimited
13.15   ©W. Ormerod/Visuals Unlimited
13.16   ©Doug Sokell/Visuals Unlimited
13.17a  ©Allen H. Benton/Visuals Unlimited
13.17b  ©Stan Elmes/Visuals Unlimited
13.19   ©John D. Cunningham/Visuals Unlimited
13.20   ©John Gerlach/Visuals Unlimited
13.22a  ©R. Calentine/Visuals Unlimited
13.22b  ©George J. Wilder/Visuals Unlimited
13.23   ©Doug Sokell/Visuals Unlimited


## CHAPTER 14

Opener  ©Rod Planck/Photo Researchers, Inc.
14.1    ©Nada Pecnik/Visuals Unlimited
14.4a   ©R. Wallace/Visuals Unlimited
14.4b   ©James R. McCullash/Visuals Unlimited
14.4c   ©A. Kerstitch/Visuals Unlimited
14.10   ©Hal Beral/Visuals Unlimited
14.11a  ©R. Calentine/Visuals Unlimited
14.12   ©Cabisco/Visuals Unlimited
14.13   ©James King-Holmes/Science Photo Library/Photo
        Researchers, Inc.
14.15   ©Gregory Ochocki/Photo Researchers, Inc.
14.16a  ©Cabisco/Visuals Unlimited
14.16b  ©Glenn M. Oliver/Visuals Unlimited
14.18a  ©Science VU/Visuals Unlimited
14.18b  ©John D. Cunningham/Visuals Unlimited
14.19a–c ©John D. Cunningham/Visuals Unlimited
14.19d  ©A. Kerstitch/Visuals Unlimited
14.20a  ©Richard Walters/Visuals Unlimited
14.20b  ©Richard Thom/Visuals Unlimited
14.21a  ©Kjell B. Sandved/Visuals Unlimited
14.21b  ©D. Cavagnaro/Visuals Unlimited
14.21c  ©Gary Retherford/Photo Researchers, Inc.
14.23   ©Marty Snyderman/Visuals Unlimited
14.24   ©R. Calentine/Visuals Unlimited
14.25a  ©A. Kerstitch/Visuals Unlimited
14.25b  ©C.P. Hickman/Visuals Unlimited
14.27   ©Peter Scones/Planet Earth Pictures
14.28a  ©John D. Cunningham/Visuals Unlimited

14.28b  ©Richard L. Carlton/Visuals Unlimited
14.29   ©Patrice Ceisel/Visuals Unlimited
14.32a  ©John Gerlach/Visuals Unlimited
14.32b  ©Joe McDonald/Visuals Unlimited
14.32c  ©Stephen Krasemann/Photo Researchers, Inc.
14.33   ©John D. Cunningham/Visuals Unlimited
14.34   ©George Holton/Photo Researchers, Inc.
14.35a  ©Ron Spomer/Visuals Unlimited
14.35b  ©F.S. Westmorland/Photo Researchers, Inc.
14.35c  ©Malcolm Boulton/Photo Researchers, Inc.
14.36a–b ©Tom McHugh/Photo Researchers, Inc.
14.38a  ©M. Loup/Jacana/Photo Researchers, Inc.
14.38b  ©M. Philip Kahl/Photo Researchers, Inc.


## CHAPTER 15

Opener  ©Bios (D. Escartin)/Peter Arnold, Inc.
15.1    ©Milton H. Tierney, Jr./Visuals Unlimited
15.2a   ©M. Loup/Jacama, Photo Researchers, Inc.
15.2b   ©Tim Davis/Photo Researchers, Inc.
15.3a   Courtesy John Oates
15.3b   Lynn Kilgore
15.6b   ©Ron Austing/Photo Researchers, Inc.
15.8    ©Jerry Wachter/Photo Researchers, Inc.
15.11b  Photo courtesy R. Jurmain
15.13   Institute of Human Origins
15.14   Photo courtesy of the Cleveland Museum of Natural History
15.16   ©National Museum of Kenya/Visuals Unlimited
15.17a  ©John D. Cunningham/Visuals Unlimited
15.17b  Photo courtesy of David Frayer
15.18   Photo courtesy of the Field Museum of Natural History, Chicago;
        Neg. A76851c
15.19a (Spy)                    Courtesy of Milford Wolpoff
15.19b (Shanidar)              Courtesy of H. Nelson
15.19c–f (La Ferrassie,
La Chapelle, St. Cesaire, Amud)   Courtesy of Fred Smith


## PART IV

Opener  ©P. Dayanandan/Photo Researchers, Inc.


## CHAPTER 16

Opener (bottom)  Bettmann Archive
Opener (top)     ©Will & Deni McIntyre/Tony Stone Images
16.4    ©E. Webber/Visuals Unlimited
16.5a–b ©John Kaprielian/Photo Researchers, Inc.
16.6a   ©John Kaprielian/Photo Researchers, Inc.
16.6b   ©Doug Sokell/Visuals Unlimited
16.6c   ©Scott Camazine/Photo Researchers, Inc.
16.7    ©Dean Morris,Oxford Scientific Films/Animals Animals/Earth
        Scenes
16.8a   ©David M. Phillips/Visuals Unlimited
16.8b   ©C. Gerald Van Dyke/Visuals Unlimited
16.11a  ©Wm. Ormerod/Visuals Unlimited
16.11b  ©Roger Cole/Visuals Unlimited
16.12b  ©Cabisco/Visuals Unlimited
16.12c  ©George J. Wilder/Visuals Unlimited
16.13b  ©Cabisco/Visuals Unlimited
16.13c  ©Jack M. Bostrack/Visuals Unlimited
16.14a  ©John D. Cunningham/Visuals Unlimited
16.14b–c ©R. Calentine/Visuals Unlimited
16.15a–b ©R. Calentine/Visuals Unlimited
16.16   ©John D. Cunningham/Visuals Unlimited
16.17   ©E.R. Degginger/Photo Researchers, Inc.
16.18   ©Hank Andrews/Visuals Unlimited
16.21a  ©Jack M. Bostrack/Visuals Unlimited
16.22a  ©John D. Cunningham/Visuals Unlimited

## CHAPTER 27

**Opener**  ©Holt Studios International (Nigel Cattlin)/Photo Researchers, Inc.
**27.2**  ©David Sieren/Visuals Unlimited
**27.4a–b**  ©N.A.S./M.W.F. Tweedie/Photo Researchers, Inc.
**27.5a–b**  ©David M. Phillips/Visuals Unlimited
**27.5c**  ©Fred Hossler/Visuals Unlimited
**27.8**  ©R. Calentine/Visuals Unlimited
**27.9a**  ©Ken Wagner/Visuals Unlimited
**27.9b**  ©John D. Cunningham/Visuals Unlimited
**27.9c**  ©John Kaprielian/Photo Researchers, Inc.
**27.10**  ©John D. Cunningham/Visuals Unlimited
**27.11a**  ©William J.Weber/Visuals Unlimited
**27.11b**  ©Stan Elems/Visuals Unlimited
**27.11c**  ©Walt Anderson/Visuals Unlimited
**27.13**  ©Joe Eakes/Visuals Unlimited
**27.14**  ©Holt International Studios (Inga Spence)/Photo Researchers, Inc.
**27.15**  ©Alan L. Detrick/Photo Researchers, Inc.
**27.16**  ©R.F. Ashley/Visuals Unlimited
**27.17**  ©Rod Planck/Photo Researchers, Inc.
**27.18**  ©Pfizer, Inc./Phototake
**27.19a–b**  ©Plantek/Photo Researchers, Inc.

## CHAPTER 28

**Opener**  ©David Phillips/Photo Researchers, Inc.
**28.1**  ©John Gerlach/Visuals Unlimited
**28.2**  ©R. Myers/Visuals Unlimited
**28.3**  ©T.E. Adams/Visuals Unlimited
**28.7b**  Elizabeth R. Walker, Associate Professor, and Dennis O. Overman, Associate Professor, Department of Anatomy, School of Medicine, West Virginia University
**28.7c**  From *Tissues and Organs: A Text-Atlas of Scanning Electron Microscopy* by Richard G. Kessel and Randy H. Kardon. Copyright ©1979 by W.H. Freeman and Company. Reprinted with permission.
**28.12b**  Photo courtesy of Dr. P. Bagavandoss, Developmental and Reproductive Biology, University of Michigan Medical School
**28.18a**  ©Stanley Flegler/Visuals Unlimited
**28.18b**  ©SIU/Visuals Unlimited
**28.18c**  ©M. Long/Visuals Unlimited
**28.21**  ©David M. Phillips/Visuals Unlimited

## CHAPTER 29

**Opener**  ©Bios (Klien-Hubert)/Peter Arnold, Inc.
**29.1**  ©Jim Selby/Science Photo Library/Photo Researchers, Inc.
**29.3**  Thomas McAvoy, Life Magazine ©Time Warner, Inc.
**29.4**  ©John S. Flannery/Visuals Unlimited
**29.5**  ©William J. Weber/Visuals Unlimited
**29.6**  ©Mark Moffett/Minden Pictures
**29.8**  ©D. Newman/Visuals Unlimited
**29.9**  Photo courtesy of Dr. G.W. Barlow, Department of Integrative Biology, University of California, Berkeley
**29.10**  ©John Gerlach/Visuals Unlimited
**29.11**  ©Wm. Grenfell/Visuals Unlimited
**29.12**  ©Pat & Tom Leeson/Photo Researchers, Inc.
**29.13**  ©Nigel J. Dennis/Photo Researchers, Inc.
**29.14**  ©Tom J. Ulrich/Visuals Unlimited
**29.15**  ©Alan Desbonnet/Visuals Unlimited
**29.16**  ©Joe McDonald/Visuals Unlimited

## PART VII

**Opener**  ©Lewis Kemper/Tony Stone Images

## CHAPTER 30

**Opener**  ©Lane Photo/The Image Bank
**30.3b**  ©Robert W. Hernandez/Photo Researchers, Inc.
**30.7b**  ©Kees van der Berg/Photo Researchers, Inc.
**30.9a**  ©Richard Walters/Visuals Unlimited
**30.9b**  ©Frank T. Awbrey/Visuals Unlimited
**30.11b**  ©David Weintraub/Photo Researchers, Inc.
**30.11c**  ©Nancy Sefton/Photo Researchers, Inc.
**30.13**  ©Holt Studios International/Photo Researchers, Inc.
**30.16**  ©Scott Camazine/Photo Researchers, Inc.
**30.18**  ©John Mitchell/Photo Researchers, Inc.
**30.19**  ©Daniel J. Cox/Tony Stone Images
**30.21**  ©Kjell B. Sandved/Visuals Unlimited
**30.22**  The Granger Collection, New York
**30.25a**  ©Steven Montgomery, Conservation Biologist
**30.25b**  Photo courtesy of William P. Mull

## CHAPTER 31

**Opener**  ©Art Wolfe/Tony Stone Images
**31.1**  ©Carr Clifton/Minden Pictures
**31.6**  ©Nada Pecnik/Visuals Unlimited
**31.8**  ©Bill Beatty/Visuals Unlimited
**31.9**  ©Ron Spomer/Visuals Unlimited
**31.10**  ©Jeffrey H. Black/Visuals Unlimited
**31.11**  ©Patricia K. Armstrong/Visuals Unlimited
**31.12**  ©Art Stein/Photo Researchers, Inc.
**31.13a**  ©A. Kerstitch/Visuals Unlimited
**31.13b**  ©Tom McHugh/Photo Researchers, Inc.
**31.14a**  ©Rod Kieft/Visuals Unlimited
**31.14b**  ©Tom & Pat Leeson/Photo Researchers, Inc.
**31.15**  ©Michael Giannechini/Photo Researchers, Inc.
**31.25a**  Photo courtesy of the Field Museum of Natural History, Chicago, #Geo80821c
**31.25b**  ©Hal Beral/Visuals Unlimited
**31.34**  ©Tim Hauf/Visuals Unlimited
**31.35**  ©David S. Addison/Visuals Unlimited

## CHAPTER 32

**Opener**  ©Joseph van Os/The Image Bank
**32.12**  ©John Mead/Science Photo Library/Photo Researchers, Inc.
**32.19**  ©Tom McHugh/Photo Researchers, Inc.

3 5282 00406 9202